Radio Shack®
DICTIONARY
of
ELECTRONICS

Rudolf F. Graf has been in the electronics industry for more than 30 years, in capacities ranging from instructor, sales engineer, and magazine editor to director of engineering and consultant. He is a graduate in communications engineering from Polytechnic Institute of Brooklyn and received his M.B.A. at New York University. He is a senior member of the IEEE and holds a first-class radiotelephone operator's license. Mr. Graf has written numerous books and articles of interest to amateur radio engineers. He is the coauthor of *Automotive Electronics, Solid-State Ignition Systems, Electronics Quizbook, Build-It Book of Car Electronics, Build-It Book of Safety Electronics, Build-It Book of Home Electronics,* all published by Howard W. Sams & Co., Inc.

Radio Shack® DICTIONARY of ELECTRONICS

Rudolf F. Graf

Radio Shack®
A TANDY CORPORATION COMPANY

FIFTH EDITION
SECOND PRINTING—1978

International Standard Book Number: 0-672-21314-1
Library of Congress Catalog Card Number: 77-71678

Printed in the United States of America.

Preface

In recent decades, technological advances that profoundly affect our daily lives have taken place at a feverish pace in electronics and closely related fields. Invariably, those who work in these fields find that they need new vocabulary terms to effectively communicate thoughts and ideas about their fields of specializations. The originators of these new words give them their initial meaning, but *exact* definitions change with technological advances and through actual use by others. The content of a dictionary is thus an analysis of words and their meanings as determined by common usage.

Therefore, it should come as no surprise that this fifth edition of the *Modern Dictionary of Electronics*—probably the most up-to-date electronics dictionary in the world—contains definitions of approximately *20,000 terms* unique to electronics and closely related fields. This includes 3000 more entries than were found in the fourth edition published in 1972, and nearly *twice* as many terms as were explained in the first edition of this Dictionary published 15 years ago! All earlier definitions were reviewed and modified or expanded, where necessary, to further enhance the intelligibility of each entry and to ensure meaningful, concise definitions requiring no further interpretation. The illustrations have been updated and modified as needed to help give greater clarity to the definitions.

While this volume is as up-to-date as possible at the time of writing, the field of electronics is expanding so rapidly that new terms are constantly being developed and old terms are taking on broader or more-specialized meanings. It is the intention of the publishers to periodically issue revised editions of this dictionary; thus, suggestions for new terms and definitions will always be welcomed.

Acknowledgement and thanks are due several technical and engineering societies—notably the IEEE and ASA—who generously aided in defining many terms during the initial preparation of this work. In particular I want to express my appreciation to my good friend George J. Whalen for his invaluable comments and constructive suggestions.

RUDOLF F. GRAF

To Bettina—a many faceted gem
whose matchless brilliance becomes more abundant
with every passing year

How to Use This Dictionary

This *Modern Dictionary of Electronics* follows the standards accepted by prominent lexicographers. All terms of more than one word are treated as one word. For example, "bridged-T network" appears between "bridge circuit" and "bridge duplex system." Abbreviations are also treated alphabetically; the initials "ARRL" follow the term "arrester" rather than appearing at the beginning of the A's.

For ease in quickly locating a specific term, catchwords for the first and last entries on each page are shown at the top of the page.

Where more than one definition exists for a term, the different meanings are arranged numerically. This method, however, does not necessarily imply a preferred order of meaning.

Illustrations have been positioned with the terms they depict and are clearly captioned so they can be immediately associated with the proper definition. When a term has more than one meaning, the number of the corresponding definition is included in the caption.

Moderate cross-referencing has been used as an aid in locating terms that you might look for in more than one place. For example, when looking up "Esaki diode" you'll be referred to "tunnel diode." However, occasionally you may look for a term and not find it. In such instances, always think of the term in its most logical form; e.g., you will find "acoustic resonator" in the A's and not in the R's. In other words, when looking up the definition for a specific device, such as a "dipole antenna," refer to the modifier "dipole" rather than to "antenna."

Because the policy of the United States is to increase the use of the metric system, the International System of Units has been listed on pages 828 and 829.

Schematic symbols are shown on pages 830 and 831 and a table on page 832 lists the letters of the Greek alphabet, along with technical terms for which these letters are used as symbols.

Since it follows the most authoritative standards of the industry, this dictionary will serve as an excellent guide on spelling, hyphenation, abbreviation, capitalization, etc.

It is hoped you will find the *Modern Dictionary of Electronics* helpful, informative, and satisfactory in every way. Should you care to pass along any comments or suggestions that come to mind as a result of its use, we will be most happy to hear from you.

A

A—1. Abbreviation for angstrom unit, used in expressing wavelength of light. Its length is 10^{-8} centimeter. 2. Chemical symbol for argon, an inert gas used in some electron tubes. 3. Symbol for area of a plane surface. 4. Symbol for ampere.

a—Abbreviation for atto (10^{-18}).

A− (A-minus or A-negative)—Sometimes called F−. Negative terminal of an A-battery or negative polarity of other sources of filament voltage. Denotes the terminal to which the negative side of the filament-voltage source should be connected.

A+ (A-plus or A-positive)—Sometimes called F+. Positive terminal of an A-battery or positive polarity of other sources of filament voltage. The terminal to which the positive side of the filament voltage source should be connected.

ab—The prefix attached to names of practical electric units to indicate the corresponding unit in the cgs (centimeter-gram-second) electromagnetic systems—e.g., abampere, abvolt, abcoulomb.

abac—*See* Alignment Chart.

abampere—Centimeter-gram-second electromagnetic unit of current. The current which, when flowing through a wire one centimeter long bent into an arc with a radius of one centimeter, produces a magnetic field intensity of one oersted. One abampere is equal to 10 amperes.

A-battery—Source of energy which heats the filaments of vacuum tubes in battery-operated equipment.

abbreviated dialing—A system using special-grade circuits that require fewer than the usual number of dial pulses to connect two or more subscribers.

abc—Abbreviation for automatic bass compensation, a circuit used in some equipment to increase the amplitude of the bass notes to make them appear more natural at low volume settings.

abcoulomb—Centimeter-gram-second electromagnetic unit of electrical quantity. The quantity of electricity passing any point in an electrical circuit in one second when the current is one abampere. One abcoulomb is equal to 10 coulombs.

aberration—In lenses a defect that produces inexact focusing. Aberration may also occur in electron optical systems, causing a halo around the light spot.

abfarad—Centimeter-gram-second electromagnetic unit of capacitance. The capacitance of a capacitor when a charge of one abcoulomb produces a difference of potential of one abvolt between its plates. One abfarad is equal to 10^9 farads.

abhenry—Centimeter-gram-second electromagnetic unit of inductance. The inductance in a circuit in which an electromotive force of one abvolt is induced by a current changing at the rate of one abampere per second. One abhenry is equal to 10^9 henrys.

abmho—Centimeter-gram-second electromagnetic unit of conductance. A conductor or circuit has a conductance of one abmho when a difference of potential of one abvolt between its terminals will cause a current of one abampere to flow through the conductor. One abmho is equal to 10^9 mho.

abnormal glow—In a glow tube, a current discharge of such magnitude that the cathode area is entirely surrounded by a glow. A further increase in current results in a rise in its density and a drop in voltage.

abnormal propagation—The phenomenon of unstable or changing atmospheric and/or ionspheric conditions acting upon transmitted radio waves. Such waves are prevented from following their normal path through space, causing difficulties and disruptions of communications.

abnormal reflections—*See* Sporadic Reflections.

abohm—Centimeter-gram-second electromagnetic unit of resistance. The resistance of a conductor when, with an unvarying current of one abampere flowing through it, the potential difference between the ends of the conductor is one abvolt. One abohm is equal to 10^{-9} ohm.

abort—To cut short or break off (an action, operation, or procedure) with an aircraft, guided missile, or the like—especially because of equipment failure. An abort may occur at any point from start of countdown or takeoff to the destination. An abort can be caused by human technical or meteorological errors, miscalculation, or malfunctions.

AB power pack—Assembly in a single unit of the A- and B-batteries of a battery-operated circuit. Also, a unit that supplies the necessary A and B voltages from an ac source of power.

abrasion machine—A laboratory device for determining the abrasive resistance of wire or cable. The two standard types of machines are the squirrel cage with square steel bars and the abrasive grit types.

abrasion resistance—A measure of the ability of a wire or wire covering to resist damage due to mechanical causes. Usually expressed as inches of abrasive tape travel.

abscissa—Horizontal, or X-, axis on a chart or graph.

absence-of-ground searching selector—In dial telephone systems, an automatic switch that rotates, or rises vertically and rotates, in search of an ungrounded contact.

absolute accuracy—The tolerance of the full scale set point referred to as the absolute voltage standard.

absolute address—1. An address used to specify the location in storage of a word in a computer program, not its position in the program. 2. A binary number assigned permanently as the address of a storage location in a computer.

absolute altimeter—Electronic instrument which furnishes altitude data with regard to the surface of the earth or any other surface immediately below the instrument —as distinguished from an aneroid altimeter, the readings of which depend on air pressure.

absolute altitude—Altitude with respect to the earth's surface, as differentiated from the altitude with respect to sea level.

absolute code—A code using absolute addresses and absolute operation codes—i.e., a code that indicates the exact location where the reference operand is to be found or stored.

absolute coding—Coding written in machine language. It can be understood by the computer without processing.

absolute delay—The time interval between the transmission of two synchronized radio, loran, or radar signals from the same or different stations.

absolute digital position transducer—A digital position transducer, the output signal of which is indicative of absolute position. Also called encoder.

absolute efficiency—Ratio of the actual output of a transducer to that of a corresponding ideal transducer under similar conditions.

absolute error—1. The amount of error expressed in the same units as the quantity containing the error. 2. Loosely, the absolute value of the error—i.e., the magnitude of the error without regard to its algebraic sign.

absolute gain of an antenna—The gain in a given direction when the reference antenna is an isotropic antenna isolated in space.

absolute humidity—Amount of water vapor present in a unit volume of atmosphere.

absolute maximum rating—Limiting values of operating and environmental conditions, applicable to any electron device of a specified type as defined by its published data, and not to be exceeded under the worst probable conditions. Those ratings beyond which the life and reliability of a device can be expected to decline.

absolute maximum supply voltage—The maximum supply voltage that may be applied without the danger of causing a permanent change in the characteristics of a circuit.

absolute minimum resistance—The resistance between the wiper and the termination of a potentiometer, when the wiper is adjusted to minimize that resistance.

absolute Peltier coefficient—The product of the absolute temperature and the absolute Seebeck coefficient of a material.

absolute power—Power level expressed in absolute units (e.g., watts or dBm).

absolute pressure—Pressure of a liquid or gas measured relative to a vacuum (zero pressure).

absolute pressure transducer—1. A pressure transducer that accepts simultaneously two independent pressure sources, and the output of which is proportional to the pressure difference between the sources. 2. A transducer that senses a range of pressures which are referenced to a fixed pressure. The fixed pressure is normally total vacuum.

absolute scale—*See* Kelvin Scale.

absolute Seebeck coefficient—The integral from absolute zero to the given temperature of the quotient of the Thomson coefficient of a material divided by its absolute temperature.

absolute spectral response—Output or response of a device, in terms of absolute power levels, as a function of wavelength.

absolute system of units—Also called coherent system of units. A system of units in which a small number of units is chosen as fundamental—e.g., units of mass, length, time, and charge. Such units are termed absolute units. All other units are derived from them by taking a definite proportional factor in each of those laws chosen as the basic laws for expressing the relationships between the physical quantities. The proportional factor is generally taken as unity.

absolute temperature—Temperature measured from absolute zero, a theoretical level defined as −273.2°C or −459.7°F or 0 K.

absolute tolerance (accuracy)—The maximum deviation from the nominal resistance (or capacitance) value, usually given as a percentage of the nominal value.

absolute units—A system of units based on physical principles, in which a small number of units are chosen as fundamental and all other units derived from them—i.e. abohm, abcoulomb, abhenry, etc.

absolute value—The numerical value of a number or symbol without reference to its algebraic sign. Thus, $|3|$ is the absolute value of +3 or −3. An absolute value is signified by placing vertical lines on both sides of the number or symbol.

absolute value device—A computing element that produces an output equal to the magnitude of the input signal, but always of one polarity.

absolute zero—Lowest possible point on the scale of absolute temperature; the point at which all molecular activity ceases. Absolute zero is defined as −273.2°C, −459.7°F, or 0 K.

absorber—1. In a nuclear reactor, a substance that absorbs neutrons without reproducing them. Such a substance may be useful in control of a reactor or, if unavoidably present, may impair the neutron economy. 2. Any material or device which absorbs and dissipates radiated energy. 3. In microwave terminology, a material or device that takes up and dissipates radiated energy. It may be used for shielding, to prevent reflection, or to transmit selectively one or more radiation components.

absorption—1. Dissipation of the energy of a radio or sound wave into other forms as a result of its interaction with matter. 2. The process by which the number of particles or photons entering a body of matter is reduced by interaction of the particle or radiation with the matter. Similarly, the reduction of the energy of a particle while traversing a body of matter. This term is sometimes erroneously used for capture. 3. Penetration of a substance into the body of another.

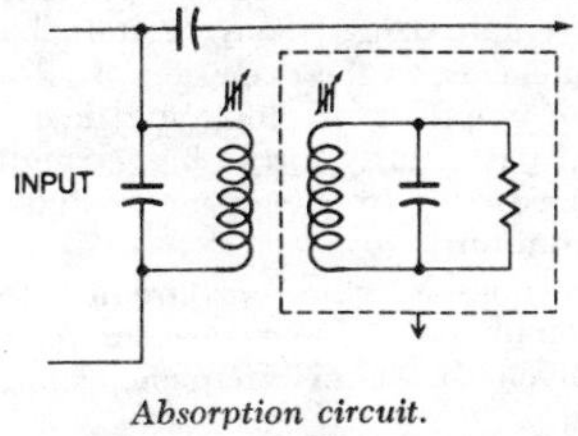

Absorption circuit.

absorption circuit—A tuned circuit that dissipates energy taken from another circuit.

absorption coefficient — 1. Measure of sound-absorbing characteristics of a unit area of a given material, compared with the sound-absorbing characteristics of an open space (total absorption) having the same area. 2. Ratio of loss of intensity caused by absorption, to the total original intensity of radiation.

absorption control—Control of a nuclear reactor by use of a neutron absorber. Adjustment is made by varying the effective amount of absorber in or near the core. The most common arrangement is to incorporate the absorber in rods which can be moved in or out to produce the desired effect.

absorption current—The current flowing into a capacitor following its initial charge, due to a gradual penetration of the electric stress into the dielectric. Also, the current which flows out of a capacitor following its initial discharge.

absorption dynamometer—An instrument for measuring power, in which the energy of a revolving wheel or shaft is absorbed by the friction of a brake.

absorption fading—A slow type of fading, primarily caused by variations in the absorption rate along the radio path.

absorption frequency meter—*See* Absorption Wavemeter.

absorption loss—That part of transmission loss due to dissipation or conversion of electrical energy into other forms (e.g., heat), either within the medium or attendant upon a reflection.

absorption marker—A sharp dip on a frequency-response curve due to the absorption of energy by a circuit sharply tuned to the frequency at which the dip occurs.

absorption modulation—Also called loss modulation. A system for amplitude-modulating the output of a radio transmitter by means of a variable-impedance device (such as a microphone or vacuum-tube circuit) inserted into or coupled to the output circuit.

absorption trap—A parallel-tuned circuit coupled either magnetically or capacitively to absorb and attenuate interfering signals.

absorption wavemeter—Also called absorption frequency meter. An instrument for measuring frequency. Its operation depends on the use of a tuned electrical circuit or cavity loosely coupled to the source. Maximum energy will be absorbed at the resonant frequency, as indicated by a meter or other device. Frequency can then be determined by reference to a calibrated dial or chart.

absorptivity—A measure of the portion of incident radiation or sound energy absorbed by a material.

A-B test—1. Direct comparison of two sounds by playing first one and then the other. May be done with two tape recorders playing identical tapes (or the same tape), two speakers playing alternately from the same tape recorder, or two amplifiers playing alternately through one speaker, etc. 2. An audio comparison test for evaluating the relative performance of two or more components or systems by quickly changing from one to the other. The left- and right-hand channels or the record and replay sound signals are often designated A and B. A and B test facilities are installed at most high fidelity dealers.

abvolt — Centimeter-gram-second electromagnetic unit of potential difference. The potential difference between two points when one erg of work is required to transfer one abcoulomb of positive electricity from a lower to a higher potential. An abvolt is equal to 10^{-8} volt.

ac—Abbreviation for alternating current.

ac bias—The alternating current, usually of a frequency several times higher than the highest signal frequency, that is fed to a record head in addition to the signal current. The ac bias serves to linearize the recording process.

accelerated life test—Test conditions used to bring about, in a short time, the deteriorating effect obtained under normal service conditions.

accelerated service test—A service or bench test in which some service condition is exaggerated to obtain a result in a shorter time than that which elapses in normal service.

accelerating conductor or relay—One which causes the operation of a succeeding device to begin in the starting sequence after the proper conditions have been established.

accelerating electrode—An electrode in a cathode-ray or other electronic tube to which a positive potential is applied to increase the velocity of electrons or ions toward the anode. A klystron tube does not have an anode but does have accelerating electrodes.

accelerating time—The time required for a motor to reach full speed from a standstill (zero speed) position.

acceleration—The rate of change in velocity. Often expressed as a multiple of the acceleration of gravity ($g = 32.2$ ft/s^2).

acceleration at stall—The value of servomotor angular acceleration calculated from the stall torque of the motor and the moment of inertia of the rotor. Also called torque-to-inertia ratio.

acceleration switch—A type of limit switch which senses acceleration forces above a preset value and in a particular direction. Commonly used in aerospace equipment, it triggers sensors or arming networks when a specific gravitational force has been reached.

acceleration time—In a computer, the elapsed time between the interpretation of instructions to read or write on tape and the possibility of information transfer from the tape to the internal storage, or vice versa.

acceleration torque—Numerical difference between motor torque produced and load torque demanded at any given speed during the acceleration period. It is this net torque which is available to change the speed of the driven load.

acceleration voltage—Potential between a cathode and anode or other accelerating element in a vacuum tube. Its value determines the average velocity of the electrons.

accelerator—A device for imparting a very high velocity to charged particles such as electrons or protons. Fast-moving particles of this type are used in research or in studying the structure of the atom itself.

accelerator dynamic test—A test performed on an accelerometer by means of which information is gathered pertaining to the overall behavior frequency response and/or natural frequency of the device.

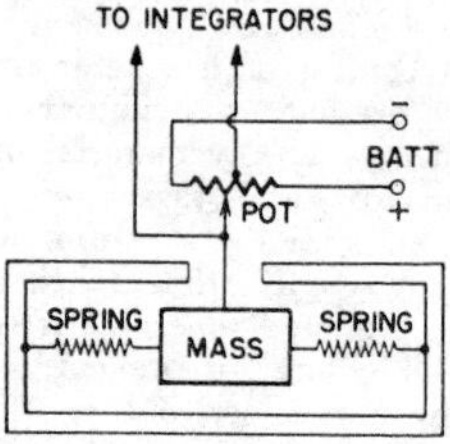

Accelerometer, 1.

accelerometer—1. An instrument or device, often mounted in an aircraft, guided missile, or the like, used to sense accelerative forces and convert them into corresponding electrical quantities, usually for measuring, indicating, or recording purposes. It does not measure velocity or distance, only changes in velocity. 2. A transducer which measures acceleration and/or gravitational forces capable of imparting acceleration.

accentuation—Also called pre-emphasis. The emphasizing of any certain band of frequencies, to the exclusion of all others, in an amplifier or electronic device. Applied particularly to the higher audio frequencies in frequency-modulated (fm) transmitters.

accentuator—Network or circuit used for pre-emphasis increase in amplitude of a given band of frequencies, usually audio.

acceptable-environmental-range test — A test to determine the range of environmental conditions for which an equipment maintains at least the minimum required reliability.

acceptable quality level—The maximum percentage of defective components considered to be acceptable as an average for a process or the lowest quality a supplier is permitted to present continually for acceptance. (Abbreviated AQL.)

acceptance sampling plan—A plan for the inspection of a sample as a basis for acceptance or rejection of a lot or taking another sample.

acceptance test—A test to demonstrate the degree of compliance of a purchaser's equipment with his requirements and specifications.

acceptor—Also called acceptor impurity. An impurity lacking sufficient valence electrons to complete the bonding arrangement in the crystal structure. When added to a semiconductor crystal, it accepts an electron from a neighboring atom and thus creates a hole in the lattice structure of the crystal.

acceptor circuit—1. A circuit which offers minimum opposition to a given signal. 2. A circuit tuned to respond to a single frequency.

acceptor impurity—*See* Acceptor.

access arm—In a computer storage unit, a mechanical device which positions the reading and writing mechanism.

access method—A data-management technique available for use in transferring data between the main storage and an input/output device.

access mode—A technique used in COBOL to obtain a specific logic record from, or to place it into, a file assigned to a mass storage device.

access time—1. Also called waiting time. The time interval (called read time) between the instant of calling for data from a storage device and the instant of completion of delivery. 2. The time interval (called write time) between the instant of requesting storage of data and the instant of completion of storage. 3. In a memory system, the time delay, at specified thresholds, from the presentation of an enable or address input pulse until the arrival of the memory data output.

accidental jamming—Jamming caused by transmission from friendly equipment.

ac circuit breaker—A device which is used to close and interrupt an ac power circuit under normal conditions, or to interrupt this circuit under faulty or emergency conditions.

accompaniment manual—In an organ, the keyboard used for playing the accompaniment to the melody. Also called the lower manual, or great manual.

accompanying audio (sound) channel—Also known as co-channel sound frequency. The rf carrier frequency which supplies the sound to accompany a television picture.

accordion—A type of contact used in some printed-circuit connectors. The contact spring is given a *z* shape to permit high deflection without excessive stress.

ac-coupled flip-flop—A flip-flop that changes state when triggered by the rise or fall of a clock pulse. There is a maximum allowable rise or fall time for proper triggering.

ac coupling—Coupling of one circuit to another circuit through a capacitor or other device which passes the varying portion but not the static (dc) characteristics of an electrical signal.

accumulation key—In a calculator, it automatically accumulates products and totals of successive calculations.

accumulator—1. In an electronic computer, a device which stores a number and which, on receipt of another number, adds the two and stores the sum. An accumulator may have properties such as shifting, sensing signals, clearing, complementing, etc. 2. A chemical cell able to store electrical energy (British). Also called secondary cell. 3. The "scratch pad" section of the computer, in which arithmetic operations are carried out.

accuracy—1. The maximum error in the measurement of a physical quantity in terms of the output of an instrument when referred to the individual instrument calibration. Usually given as a percentage of full scale. 2. The quality of freedom from mistake or error in an electronic computer—that is, of conformity to truth or to a rule. 3. The closeness with which a measured quantity approaches the true value of that quantity. (*See* True Value.) 4. The degree to which a measured or calculated value conforms to the accepted standard or rule. 5. The measure of a meter's ability to indicate a value corresponding to the absolute value of electrical energy applied. Accuracy is expressed as a percentage of the meter's rated full-scale value.

accuracy rating of an instrument—The limit, usually expressed as a percentage of full-scale value, not exceeded by errors when the instrument is used under reference conditions.

ac/dc—Electronic equipment capable of operation from either an ac or dc primary power source.

ac/dc receiver—A radio receiver designed to operate directly from either an ac or a dc source.

ac/dc ringing—A method of telephone ringing in which alternating current is used to operate a ringing device, and direct current is used to aid the action of a relay that stops the ringing when the called party answers.

ac directional overcurrent relay—A device which functions on a desired value of ac overcurrent flowing in a predetermined direction.

ac dump—The intentional, accidental, or conditional removal of all alternating-current power from a system or component. An ac dump usually results in the removal of all power, since direct current is usually supplied through a rectifier or converter.

ac erasing head—In magnetic recording, a device using alternating current to produce the magnetic field necessary for removal of previously recorded information.

acetate—A basic chemical compound in the mixture used to coat recording discs.

acetate base—The transparent plastic film which forms the tough backing for acetate magnetic recording tape.

acetate disc—A mechanical recording disc, either solid or laminated, made mostly from cellulose nitrate lacquer plus a lubricant.

acetate tape—A sound-recording tape with a smooth, transparent acetate backing. One side is coated with an oxide capable of being magnetized.

ac generator—1. A rotating electrical machine that converts mechanical power into alternating current. Also known as an alternator. 2. A device, usually an oscillator, designed for the purpose of producing alternating current.

A-channel—One of two stereo channels, usually the left.

achieved reliability—Reliability determined on the basis of actual performance of nominally identical items under equivalent environmental conditions. Also called operational reliability.

achromatic—1. In color television, a term meaning a shade of gray from black to white, or the absence of color (without color). 2. Black-and-white television, as distinguished from color television. 3. A term applied to lenses, signifying their more or less complete correction for chromatic aberration.

achromatic lens—A lens which has been corrected for chromatic aberration. Such a lens is capable of bringing all colors of light rays to approximately the same point of focus. This it does by combining a concave lens of flint glass with a convex lens of crown glass.

achromatic locus—Also called achromatic region. On a chromaticity diagram an area that contains all points representing acceptable reference white standards.

achromatic region—*See* Achromatic Locus.

acicular—Needle-shaped, descriptive of the shape of the magnetizable particles comprising the coating of a recording tape. Modern tapes are premagnetized during the coating process to line the "needles" up with the direction of the tape, thus providing maximum sensitivity from the oxide.

acid—A chemical compound which dissociates and forms hydrogen ions when in aqueous solution.

acid depolarizer—An acid, such as nitric acid, sometimes introduced into a primary cell to prevent polarization.

aclinic line—Also called isoclinic line. On a magnetic map, an imaginary line which connects points of equal magnetic inclination or dip.

ac magnetic biasing—In magnetic recording, the method used to remove random noise and/or previously recorded material from the wire or tape. This is done by introducing an alternating magnetic field at a substantially higher frequency than the highest frequency to be recorded.

ac noise—Noise which displays a rate of change which is fast relative to the response capability of the device.

ac noise immunity—A measure of a logic circuit's ability to maintain the prescribed logic state in the presence of such noise. It is defined in terms of the amplitude and pulse width of an input noise signal to which the element will not respond.

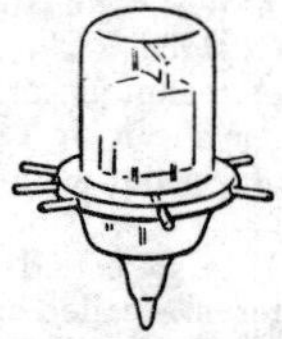

Acorn tube.

acorn tube—A button- or acorn-shaped vacuum tube with no base, for uhf applications. Electrodes are brought out through the glass envelope on the side, top, and bottom.

acoustic—Also acoustical. Pertaining to sound or the science of sound.

acoustic absorption loss—The energy lost by conversion into heat or other forms when sound passes through or is reflected by a medium.

acoustic absorptivity—The ratio of sound energy absorbed by a surface to the sound energy arriving at the surface. Equal to 1 minus the reflectivity of the surface.

acoustical attenuation constant—The real part of the acoustical propagation constant. The commonly used unit is the neper per section or per unit distance.

acoustical-electrical transducer—A device designed to transform sound energy into electrical energy and vice versa.

acoustical mode—A mode of crystal-lattice vibration that does not produce an oscillating dipole.

acoustical ohm—A measure of acoustic resistance, reactance, or impedance. One acoustical ohm is equal to a volume velocity of 1 cubic centimeter per second when produced by a sound pressure of 1 microbar.

acoustical phase constant—The imaginary part of the acoustical propagation constant. The commonly used unit is the radian per section or per unit distance.

acoustical reflectivity—*See* Sound-Reflection Coefficient.

acoustical transmittivity—*See* Sound-Transmission Coefficient.

acoustic burglar alarm—Also called acoustic intrusion detector. A burglar alarm that is responsive to sounds produced by an intruder. Concealed microphones connected to an audio amplifier trip an alarm when sounds exceed a predetermined normal level.

acoustic capacitance—In a sound medium, a measure of volume displacement per dyne per square centimeter. The unit is centimeter to the fifth power per dyne.

acoustic clarifier—A system of cones loosely attached to the baffle of a speaker and designed to vibrate and absorb energy during sudden loud sounds, thereby suppressing them.

acoustic compliance—1. The measure of volume displacement of a sound medium when subjected to sound waves. 2. That type of acoustic reactance which corresponds to capacitive reactance in an electrical circuit.

acoustic coupler—A device utilizing a speaker and/or microphone into which a telephone handset is placed to transfer audio-range signals to or from a telephone line, without an electrical connection to the telephone line.

acoustic coupling—Coupling resonator elements by mechanical means through the use of wires, rods, or nonelectroded sections of quartz or ceramic. The terms acoustic and mechanical can be used interchangeably.

acoustic delay line—A device which retards one or more signal vibrations by causing them to pass through a solid or liquid.

acoustic depth finder—*See* Fathometer.

acoustic dispersion—The change of the speed of sound with frequency.

acoustic elasticity—1. The compressibility of the air in a speaker enclosure as the cone moves backward. 2. The compressibility of any material through which sound is passed.

acoustic feedback—Also called acoustic regeneration. The mechanical coupling of a portion of the sound waves from the output of an audio-amplifying system to a preceding part or input circuit (such as the microphone) of the system. When excessive, acoustic feedback will produce a howling sound in the speaker.

acoustic filter—A sound-absorbing device that selectively suppresses certain audio frequencies while allowing others to pass.

acoustic frequency response—The voltage-attenuation frequency measured into a resistive load, producing a bandwidth approaching sufficiently close to the maximum.

acoustic generator—A transducer such as a speaker, headphones, or a bell, which converts electrical, mechanical, or other forms of energy into sound.

acoustic homing system—A missile guidance system which responds to noise radiated by the target.

acoustic horn—Also called horn. A tube of varying cross section having different terminal areas which change the acoustic impedance to control the directivity of the sound pattern.

acoustic impedance—Total opposition of a medium to sound waves. Equal to the force per unit area on the surface of the medium, divided by the flux (volume velocity or linear velocity multiplied by area) through that surface. Expressed in ohms and equal to the mechanical impedance divided by the square of the surface area. One unit of acoustic impedance is equal to a volume velocity of one cubic centimeter per second produced by a pressure of 1 microbar. Acoustic impedance contains both acoustic resistance and acoustic reactance.

acoustic inertance—A type of acoustic reactance which corresponds to inductive reactance in an electrical circuit. (The resistance to movement or reactance offered by the sound medium because of the inertia of the effective mass of the medium.) Measured in acoustic ohms.

acoustic intensity—The limit approached by the quotient of acoustical power being transmitted at a given time through a given area divided by the area as the area approaches zero.

acoustic interferometer—An instrument for measuring the velocity or frequency of sound waves in a liquid or gas. This is done by observing the variations of sound pressure in a standing wave, established in the medium between a sound source and a reflector, as the reflector is moved or the frequency is varied.

acoustic intrusion detector—*See* Acoustic Burglar Alarm.

acoustic labyrinth—A loudspeaker enclosure in which the rear of the loudspeaker is coupled to a tube which, at the resonant frequency of the loudspeaker, is one quarter of a wavelength long. The tube, folded upon itself in order to save space, gives the appearance of a labyrinth.

acoustic lens—1. An array of obstacles that refract sound waves in the same way that an optical lens refracts light waves. The dimensions of these obstacles are small compared to the wavelengths of the sounds being focused. 2. A device that produces convergence or divergence of moving sound waves. When used with a loudspeaker, the acoustic lens widens the beam of the higher-frequency sound waves.

acoustic line—Mechanical equivalent of an electrical transmission line. Baffles, labyrinths, or resonators are placed at the rear of a speaker to help reproduce the very low audio frequencies.

acoustic memory—A computer memory using an acoustic delay line. The line employs a train of pulses in a medium such as mercury or quartz.

acoustic mine—Also called sonic mine. An underwater mine that is detonated by sound waves, such as those from a ship's propeller or engines.

acoustic mirage—The distortion of a sound wavefront by a large temperature gradient in air or water. This creates the illusion of two sound sources.

acoustic ohm—The unit of acoustic resistance, reactance, or impedance. One acoustic ohm is present when a sound pressure of 1 dyne per square centimeter produces a volume velocity of 1 cubic centimeter per second.

acoustic phase constant—The imaginary part of the acoustic propagation constant. The commonly used unit is the radian per section or per unit distance.

acoustic pickup—In nonelectrical phonographs, the method of reproducing the material on a record by linking the needle directly to a flexible diaphragm.

acoustic radiator—In an electroacoustic transducer, the part that initiates the radiation of sound vibration. A speaker cone or headphone diaphragm is an example.

acoustic radiometer—An instrument for measuring sound intensity by determining the unidirectional steady-state pressure caused by the reflection or absorption of a sound wave at a boundary.

acoustic reactance—That part of acoustic impedance due to the effective mass of the medium—that is, to the inertia and elasticity of the medium through which the sound travels. The imaginary component of acoustic impedance and expressed in acoustic ohms.

acoustic reflectivity—The ratio of the rate of flow of sound energy reflected from the surface on the side of incidence, to the incident rate of flow.

acoustic refraction—A bending of sound waves when passing obliquely from one medium to another in which the velocity of sound is different.

acoustic regeneration—*See* Acoustic Feedback.

acoustic resistance—That component of acoustic impedance responsible for the dissipation of energy due to friction between molecules of the air or other medium through which sound travels. Measured in acoustic ohms and analogous to electrical resistance.

acoustic resonance—An increase in sound intensity as reflected waves and direct waves which are in phase combine. May also be due to the natural vibration of air columns or solid bodies at a particular sound frequency.

acoustic resonator—An enclosure which intensifies those audio frequencies at which the enclosed air is set into natural vibration.

acoustics—1. Science of production, transmission, reception, and effects of sound. 2. In a room or other location, those characteristics which control reflections of sound waves and thus the sound reception in it.

acoustic scattering—The irregular reflection, refraction, or diffraction of a sound wave in many directions.

acoustic shock—Physical pain, dizziness, and sometimes nausea brought on by hearing a loud, sudden sound.

acoustic surface-wave component—A passive electroacoustic device that has metallized interdigital transducer elements on the surface of a piezoelectric substrate. The device allows acoustic energy to be generated, manipulated, and detected on the substrate surface. Most of the acoustic energy is confined to a region within one wavelength of the surface of the substrate.

acoustic suspension—1. A loudspeaker system in which the moving cone is held by an overcompliant suspension, the stiffness required for proper operation being supplied by air that is trapped behind the cone in a sealed enclosure. While relatively inefficient, such a system permits good bass reproduction in a unit of moderate size. 2. A speaker enclosure design in which the speaker cone is "suspended" in an airtight box. This enables the acoustic pressure of the air enclosed therein to provide the principal restoring force for the diaphragm of the speaker. It needs somewhat more power from the amplifier than a "free" speaker but has better low frequency performance.

acoustic system—Arrangement of components in devices designed to reproduce audio frequencies in a specified manner.

acoustic transmission system—An assembly of elements adapted for the transmission of sound.

acoustic treatment—Use of certain sound-absorbing materials to control the amount of reverberation in a room, hall, or other enclosure.

acoustic wave—A traveling vibration by which sound energy is transmitted in air, in water, or in the earth. The characteristics of these waves may be described in terms of change of pressure, of particle displacement, or of density.

acoustic wave filter—A device designed to separate sound waves of different frequencies. (Through electroacoustic transducers, such a filter may be associated with electric circuits.)

acoustoelectric effect—Generation of an electric current in a crystal by a traveling longitudinal sound wave.

ac plate resistance—Also called dynamic plate resistance. Internal resistance of a vacuum tube to the flow of alternating current. Expressed in ohms, the ratio of a small change in plate voltage to the resultant change in plate current, other voltages being held constant.

ac power supply—A power supply that provides one or more ac output voltages—e.g., ac generator, dynamotor, inverter, or transformer.

acquisition—1. The process of pointing an antenna or telescope so that it is properly oriented to allow gathering of tracking or telemetry data from a satellite or space probe. 2. In radar, the process between the initial location of a target and the final alignment of the tracking equipment on the target.

acquisition and tracking radar—A radar set which locks onto a strong signal and tracks the object emitting or reflecting the signal. May be airborne or on the ground. Tracking radars use a dish-type antenna reflector to produce a searchlight-type beam.

acquisition radar—A radar set that detects an approaching target and feeds approximate position data to a fire-control or missile-guidance radar, which then takes over the function of tracking the target.

ac receiver—A radio receiver designed to operate from an ac source only.

ac reclosing relay—A device which controls the automatic reclosing and locking out of an ac circuit interrupter.

ac relay—A relay designed to operate from an alternating-current source.

ac resistance—Total resistance of a device in an ac circuit. (*Also see* High-Frequency Resistance.)

across-the-line starting—Connection of a motor directly to the supply line for starting. (Also called full-voltage starting.)

ac time overcurrent relay—A device which has either a definite or an inverse time characteristic and functions when the current in an ac circuit exceeds a predetermined value.

actinic—In radiation, the property of producing a chemical change, such as the photographic action of light.

actinium—A radioactive element discovered in pitchblende by the French chemist Debierne in 1889. Its atomic number is 89; its atomic weight, 227; its symbol, Ac.

actinodielectric—A photoconductive dielectric.

actinoelectric effect—The property of some special materials whereby when an electric current is impressed on them, their resistance changes with light.

actinoelectricity—Electricity produced by the action of radiant energy on crystals.

actinometer—An instrument that measures the intensity of radiation by determining the amount of fluorescence produced by that radiation.

action area—In the rectifying junction of a metallic rectifier, that portion which carries the forward current.

action current—A brief and very small electric current which flows in a nerve during a nervous impulse.

action potential—1. The instantaneous value of the voltage between excited and resting portions of an excitable living structure. 2. The voltage variations in a nerve or muscle cell when it is excited or "fired" by an appropriate stimulus. After a short time, the cell recovers its normal resting potential, typically about 80 millivolts. The interior of the cell is negative relative to the outside.

activation—1. Making a substance artificially radioactive by placing it in an accelerator such as a cyclotron, or by bombarding it with neutrons. 2. To treat the cathode or target of an electron tube in order to create or increase its emission. 3. The process of adding electrolyte to a cell to make it ready for operation.

activation time—In a cell or battery, the time interval from the moment activation is initiated to the moment the desired operating voltage is obtained.

activator—An additive that improves the action of an accelerator.

active—1. Controlling power from a separate supply. 2. Requiring a power supply separate from the controls.

active area—The portion of the rectifying junction of a metallic rectifier that carries forward current.

active balance—In operation of a telephone repeater, the summation of all return currents at a terminal network balanced against the local circuit or drop impedance.

active circuit—A circuit that contains active elements such as transistors, diodes, or ICs.

active communications satellite—A communications satellite in which on-board receivers and transmitters receive signals beamed at them from a ground terminal, amplify them greatly, and retransmit them to another ground terminal. Less sensitive receivers and less powerful transmitters can be used on the ground than are needed for passive satellites.

active component—1. Those components in a circuit which have gain, or direct current flow, such as SCRs transistors, thyristors, or tunnel diodes. They change the basic character of an applied electrical signal by rectification, amplification, switching, etc. (Passive elements have no gain characteristics. Examples: inductors, capacitors, resistors.) 2. A device, the output of which is dependent on a source of power other than the main input signal.

active computer—The one of two or more computers in an installation that is on-line and processing data.

active current—In an alternating current, a component in phase with the voltage. The working component as distinguished from the idle or wattless component.

active decoder—A device that is associated with a ground station and automatically indicates the radar beacon reply code that is received in terms of its number or letter designation.

active device—*See* Active Component.

active ECM—*See* Jamming.

active electric network—An electric network containing one or more sources of energy.

active element—*See* Active Component.

active filter—A device employing passive network elements and amplifiers. It is used for transmitting or rejecting signals in certain frequency ranges, or for controlling the relative output of signals as a function of frequency.

active guidance—*See* Active Homing.

active homing—Also called active guidance. A missile system using a radar system in the missile itself to provide target information and to guide itself to the target.

active infrared detection—An infrared detection system in which a beam of infrared rays is transmitted toward one or more possible targets, and the rays reflected from the target are detected.

active jamming—Intentional radiation or reradiation of electromagnetic waves to impair the use of a specific portion of the electromagnetic-wave spectrum.

active junction—In a semiconductor, a change in "n" type to "p" type doping or conversely, by a diffusion step. On discrete transistors there are two active junctions, the collector-base junction and the emitter-base junction.

active leg—Within a transducer, an electrical element which changes its electrical characteristics as a function of the applied stimulus.

active line—A horizontal line which produces the tv picture, as opposed to the lines occurring during blanking (horizontal and vertical retrace).

active maintenance downtime—The time during which work is actually being done on an item from the recognition of an occurrence of failure to the time of restoration to normal operation. This includes both preventive and corrective maintenance.

active material—1. In the plates of a storage battery, lead oxide or some other active substance which reacts chemically to produce electrical energy. 2. The fluorescent material, such as calcium tungstate, used on the screen of a cathode-ray tube.

active mixer and modulator—A device requiring a source of electrical power and using nonlinear network elements to heterodyne or combine two or more electrical signals.

active network—A network containing passive and active (gain) elements.

active pressure—In an ac circuit, the pressure which produces a current, as distinguished from the voltage impressed upon the circuit.

active pull-up—An arrangement in which a transistor is used to replace the pull-up resistor in an integrated circuit in order to provide low output impedance without high power consumption.

active RC network—A network formed by resistors, capacitors, and active elements.

active repair time—That portion of corrective maintenance downtime during which repair work is being done on the item, including preparation, fault-location, part-replacement, adjustment and recalibration, and final test time. It may also include part procurement time under shipboard or field conditions.

active satellite—A satellite which receives, regenerates, and retransmits signals between stations. *See also* Communications Satellite.

active sonar—*See* Sonar.

active substrate—A substrate in which active elements are formed to provide discrete or integrated devices. Examples of active substrates are single crystals of semiconductor materials within which are transistors, resistors, and diodes, or combinations of these elements. Another example is ferrite substrates within which electromagnetic fields are used to perform logical, gating, or memory functions.

active swept-frequency interferometer radar—A dual radar system for air surveillance. It provides angle and range information of high precision for pinpointing target locations by trigonometric techniques.

active systems—In radio and radar, systems which require transmitting equipment, such as a beacon or transponder, to be carried in the vehicle.

active tracking system—Usually, a system which requires the addition of a transponder or responder on board the vehicle to repeat or retransmit information to the tracking equipment—e.g., dovap, secor, azusa.

active transducer—1. A type of transducer in which its output waves depend on one or more sources of power, apart from the actuating waves. 2. A transducer that requires energy from local sources in addition to that which is received.

active wire—The wire of an armature winding that produces useful voltage. That portion of the winding in which induction takes place.

activity—1. In a piezoelectric crystal, the magnitude of oscillation relative to the exciting voltage. 2. The intensity of a radioactive source. 3. Operations that result in the use or modification of the information in a computer file.

activity curve—A graph showing how the activity of a radioactive source varies with time.

activity ratio—The ratio of the number of records in a computer file which have activity to the total number of records in the file.

ac transducer—A transducer which, for proper operation, must be excited with alternating currents only. Also a device, the output of which appears in the form of an alternating current.

actual height—The highest altitude at which refraction of radio waves actually occurs.

actual power—The average of values of instantaneous power taken over 1 cycle.

actuating device—A mechanical or electrical device, either manual or automatic, that operates electrical contacts to bring about signal transmission.

actuating system—1. In a device or vehicle, a system that supplies and transmits energy for the operation of a mechanism or other device. 2. A manually or automatically operated mechanical or electrical device which operates electrical contacts to effect signal transmission.

actuating time—The time at which a specified contact functions.

actuator—1. In a servo system, the device which moves the load. 2. The part of a relay that converts electrical energy into mechanical motion.

ACU—Abbreviation for automatic calling unit.

ac voltage—*See* Alternating Voltage.

acyclic machine—A direct-current machine in which the voltage generated in the active conductors maintains the same direction with respect to those conductors at all times.

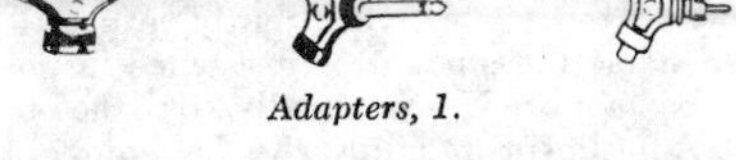

Adapters, 1.

adapter—1. A fitting designed to change the terminal arrangement of a jack, plug, socket, or other receptacle, so that other than the original electrical connections are possible. 2. An intermediate device that permits attachment of special accessories or provides special means for mounting.

adaptive communication — A method in which automatic changes in the communications system allow for changing inputs or changing characteristics of the device or process being controlled. Also called self-adjusting communication, or self-optimizing communication.

adaptive control system—A device the parameters of which are automatically adjusted to compensate for changes in the dynamics of the process to be controlled. An afc circuit utilizing temperature-compensating capacitors to correct for temperature changes is an example.

adaptive telemetry—Telemetry having the ability to select certain vital information or any change in a given signal.

a/d—Abbreviation for analog-to-digital.

adc—Abbreviation for analog-to-digital converter.

Adcock antenna—A pair of vertical antennas separated by one-half wavelength or less and connected in phase opposition to produce a figure-8 directional pattern.

Adcock direction finder—A radio direction finder using one or more pairs of Adcock antennas for directional reception of vertically polarized radio waves.

Adcock radio range—A type of radio range utilizing four vertical antennas (Adcock antennas) placed at the corners of a square, with a fifth antenna in the center.

add-and-subtract relay—A stepping relay capable of being operated so as to rotate the movable contact arm in either direction.

addend—A quantity which, when added to another quantity (called the augend), produces a result called the sum.

adder—1. A device which forms the sum of two or more numbers, or quantities, impressed on it. 2. In a color tv receiver, a circuit which amplifies the receiver primary signal coming from the matrix. Usually there is one adder circuit for each receiver primary channel. 3. An arrangement of logic gates that adds two binary digits and produces sum and carry outputs.

addition record—A new record created during the processing of a file in a computer.

additive—Sometimes referred to as the key. A number, series of numbers, or alphabetical intervals added to a code to put it in a cipher.

additive color—A system which combines two colored lights to form a third.

additive primaries—Primary colors which can be mixed to form other colors, but which cannot themselves be produced by mixing other primaries. Red, green, and blue are the primaries in television because, when added in various proportions, they produce a wide range of other colors.

additive process—A printed-circuit manufacturing process in which a conductive pattern is formed on an insulating base by electrolytic chemical deposition.

additron — An electrostatically focused, beam-switching tube used as a binary adder in high-speed digital computers.

add mode—Allows entry of numbers in a calculator, to two decimal places without the need to enter the decimal point.

address—1. An expression, usually numerical, which designates a specific location in a storage or memory device or other source or destination of information in a computer. 2. An identification, as represented by a name, label, or number, for a register, location in storage, or any other data source or destination such as the location of a station in a communications network. 3. Loosely, any part of an instruction that specifies the location of an operand for the instruction. 4. To select the location of a stored information set for access. 5. *See also* Instruction Code.

address bus—*See* Bus System.

address characters—Blocks of alphanumeric characters that identify users or stations uniquely.

address computation — The process by which the address part of an instruction in a digital computer is produced or modified.

address constant—*See* Base Address.

addressed memory—In a computer, memory sections containing each individual register.

address field—The portion of an instruction that specifies the location of a particular piece of information in a computer memory.

address modification—In a computer, a change in the address portion of an instruction or command such that, if the routine which contains that instruction or command is repeated, the computer will go to a new address or location for data or instructions.

address part—In an electronic-computer instruction, a portion of an expression designating location. (*See also* Instruction Code.)

add-subtract time—The time required by a digital computer to perform addition or subtraction. It does not include the time required to obtain the quantities from storage and put the result back into storage.

add time—The time required in a digital computer to perform addition. It does not include the time required to obtain the quantities from storage and put the result back into storage.

a/d encoder—Analog-to-digital encoder; a device that changes an analog quantity into equivalent digital representation.

adf—*See* Automatic Direction Finder.

adiabatic damping—A reduction in the size of an accelerator beam as the energy of the beam is increased.

adiabatic demagnetization—A technique used to obtain temperatures within thousandths of a degree of absolute zero. It consists of applying a magnetic field to a substance at a low temperature and in good thermal contact with its surroundings, insulating the substance thermally, and then removing the magnetic field.

A-display—Also called A-scan. A radar scope presentation in which time (distance or range) is one coordinate (usually horizontal) and the target appears displaced perpendicular to the time base.

adjacency—In character recognition, a condition in which the character-spacing reference lines of two characters printed consecutively on the same line are less than a specified distance apart.

adjacent- and alternate-channel selectivity—A measure of the ability of a receiver to differentiate between a desired signal and signals which differ in frequency from the desired signal by the width of one channel or two channels, respectively.

adjacent audio (sound) channel—The rf carrier frequency which contains the sound modulation associated with the next lower-frequency television channel.

adjacent channel—That frequency band immediately above or below the one being considered.

adjacent-channel attenuation—*See* Selectance, 2.

adjacent-channel interference—Undesired signals received on one communication channel from a transmitter operating on a channel immediately above or below.

adjacent-channel selectivity—The ability of a receiver to reject signals on channels adjacent to the channel of the desired station.

adjacent video carrier—The rf carrier that carries the picture modulation for the television channel immediately above the channel to which the receiver is tuned.

adjustable motor tuning—An arrangement by which the motor tuning of a receiver may be confined to a portion of the total frequency range.

adjustable resistor—A resistor which has the resistance wire partly exposed to enable the amount of resistance in use to be adjusted occasionally by the user. Adjustment requires the loosening of a screw, the subsequent moving of the lug, and retightening of the screw.

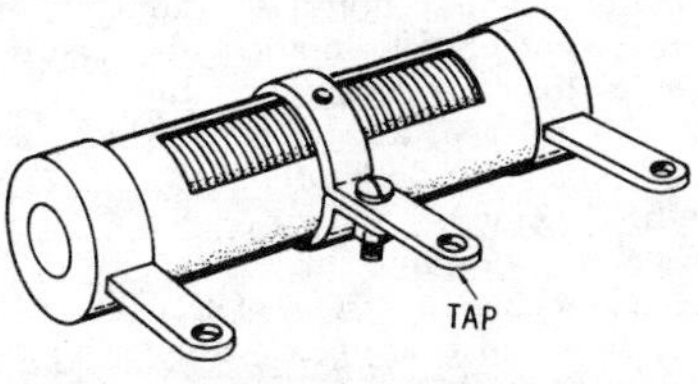

Adjustable resistor.

adjustable voltage divider—A wirewound resistor with one or more movable terminals that can be slid along the length of the exposed resistance wire until the desired voltage values are obtained.

adjusted circuit—Also called bolted-fault level. In a circuit the current measured under short-circuit conditions with the leads that are normally connected to the circuit breaker bolted together.

adjusted decibels—An expression of the ratio of the noise level to a reference noise at any point in a transmission system, when the noise meter has been adjusted to allow for the interfering effect under specified conditions.

admittance—The ease with which an alternating current flows in a circuit. The reciprocal of impedance and usually expressed in mhos. Symbol is Y or y.

ADP—Abbreviation for automatic data processing.

adsorption—The deposition of a thin layer of gas or vapor particles or gas onto the surface of a solid. The process is known as chemisorption if the deposited material is bound to the surface by a simple chemical bond.

ADU—Abbreviation for automatic dialing unit.

advance ball—In mechanical recording, a rounded support (often sapphire) which is attached to a cutter and rides on the surface of the recording medium. Its purpose is to maintain a uniform mean depth of cut and to correct for small irregularities on the surface of the disc.

advanced license—A license issued by the FCC to amateur radio operators who are capable of sending and receiving Morse code at the rate of 13 words per minute, and are familiar with general and intermediate radio theory and practice. Its privileges include exclusive use of certain frequencies.

advance wire—An alloy of copper and nickel, used in the manufacture of electric heating units and some wirewound resistors.

aeolight—A glow lamp which employs a cold cathode and a mixture of inert gases and in which the intensity of illumination varies with the applied signal voltage. This lamp is used to produce a modulated light for motion-picture sound recordings.

aerial—*See* Antenna.

aerial cable—A cable installed on a pole line or similar overhead structure.

aerodiscone antenna—An aircraft antenna that is aerodynamically shaped and is physically small compared to other antennas having similar electrical characteristics. Its radiation pattern is omnidirectional and linearly polarized.

aerodrome control radio station—A radio station providing communications between an aerodrome control tower and aircraft or mobile aeronautical radio stations.

aerodynamics—The science of the motion of air and other gases. Also, the forces acting on bodies when they move through such gases, or when such gases move against or around the bodies.

aeromagnetic—Pertaining to the magnetic field of the earth as surveyed from the air.

aeronautical advisory station—A station used for civil defense and advisory communications with private aircraft stations.

aeronautical broadcasting service — The broadcasting service intended for the transmission of information related to air navigation.

aeronautical broadcast station—A radio station which broadcasts meteorological information and notices to airmen.

aeronautical fixed service—A fixed service intended for the transmission of information relating to air navigation and preparation for and safety of flight.

aeronautical fixed station—A station operating in the aeronautical fixed service.

aeronautical ground station—A radio station operated for the purpose of providing air-to-ground communications in connection with the operation of aircraft.

aeronautical marker-beacon station — A land station operating in the aeronautical radionavigation service and providing a signal to designate a small area above the station.

aeronautical mobile service—A radio service between aircraft and land stations or between aircraft stations.

aeronautical radio beacon station—A radionavigation land station in the aeronautical radionavigation service, the emission of which enables an aircraft or other mobile service to determine its bearing or its position in relation to the aeronautical radio beacon station.

aeronautical radionavigation service—A radionavigation service intended for use in the operation of aircraft.

aeronautical radio service—1. Service carried on between aircraft stations and/or land stations. 2. Special radio for air navigation.

aeronautical station—A land station (or in

certain instances a shipboard station) in the aeronautical mobile service that carries on communications with aircraft stations.

aeronautical telecommunication agency—An agency to which is assigned the responsibility for operating a station or stations in the aeronautical telecommunication service.

aeronautical telecommunication log—A record of the activities of an aeronautical telecommunication station.

aeronautical telecommunications—Any telegraph or telephone communications of sign signals, writing, images, and sounds of any nature, by wire, radio or other system or process of signaling, used in the aeronautical service.

aeronautical telecommunication service—Telecommunication service provided for aeronautical purposes.

aeronautical telecommunication station—A station in the aeronautical telecommunication service.

aeronautical utility land station—A land station located at an airport control tower and used for communications connected with the control of ground vehicles and aircraft on the ground.

aeronautical utility mobile station—A mobile station used at an airport for communications with aeronautical utility land stations, ground vehicles, and aircraft on the ground.

aerophare—*See* Radio Beacon.

AES—Abbreviation for Audio Engineering Society. A professional group; the official association of technical personnel, scientists, engineers, and executives in the audio field.

af—*See* Audio Frequency.

afc—*See* Automatic Frequency Control.

afterglow—Also called phosphorescence. The light that remains in a gas-discharge tube after the voltage has been removed, or on the phosphorescent screen of a cathode-ray tube after the exciting electron beam has been removed.

afterheat—The heat produced by the continuing decay of radioactive atoms in a reactor after fission has stopped. Most of the afterheat is due to the radioactive decay of fission products.

afterpulse—In a photomultiplier, a spurious pulse induced by a preceding pulse.

agc—*See* Automatic Gain Control.

age—To maintain an electrical component in a specified environment as with respect to pressure, temperature, applied voltage, etc. until its characteristics stabilize.

aging—1. Storing a permanent magnet, capacitor, rectifier, meter, or other device, sometimes with voltage applied, until its desired characteristics become essentially constant. 2. The change of a component or a material with time under defined environmental conditions, leading to improvement or deterioration of properties.

agonic line—An imaginary line on the earth's surface, all points of which have zero magnetic declination.

AGREE—Advisory Group on Reliability of Electronics Equipment.

aided tracking—A system of tracking a target signal in bearing, elevation, or range (or any combination of these variables) in which manual correction of the tracking error automatically corrects the rate at which the tracking mechanism moves.

AIEE—Abbreviation for American Institute of Electrical Engineers. Now merged with IRE to form IEEE.

airborne intercept radar—Short-range airborne radar employed by fighter and interceptor planes to track down their targets.

airborne long-range input—Airborne equipment designed to extend air-surveillance coverage seaward so that long-range interceptors may be used.

airborne moving target indicator—A type of airborne-radar display that does not present essentially stationary objects.

airborne noise—Undesired sound in the form of fluctuations of air pressure about the atmospheric pressure as a mean.

airborne radar platform—Airborne surveillance and height-finding radar for early warning and control.

air capacitor—A capacitor in which air is the only dielectric material between its plates.

aircarrier aircraft station—A radio station aboard an aircraft that is engaged in or essential to the transportation of passengers or cargo for hire.

air cell—A cell in which depolarization at the positive electrode is accomplished chemically by reduction of the oxygen in the air.

air column—The air space within a horn or acoustic chamber.

air condenser—*See* Air Capacitor.

air-cooled tube—An electron tube in which the generated heat is dissipated to the surrounding air directly, through metal heat-radiating fins, or with the aid of channels or chimneys that increase the air flow.

air-core coil—A number of turns of spiral wire in which no metal is used in the center.

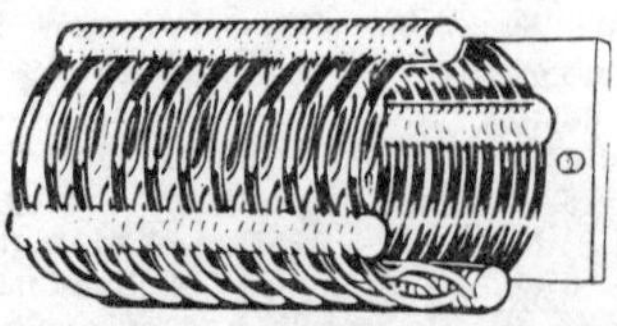

Air-core coil.

air-core transformer—A transformer (usually rf) having two or more coils wound around a nonmetallic core. Transformers wound around a solid insulating substance or on an insulating coil form are included in this category.

aircraft bonding—Electrically connecting together all of the metal structure of the aircraft, including the engine and metal covering of the wiring.

aircraft flutter—Flickering in a tv picture as the signal is reflected from flying aircraft. The reflected signal arrives in or out of phase with the normal signal and thus strengthens or weakens the latter.

aircraft station—A radio station installed on aircraft and continuously subject to human control.

airdrome control station—A station used for communication between an airport control tower and aircraft.

air environment—In communications electronics, all airborne equipment that is part of the communications-electronics system, as distinguished from the equipment on the ground, which belongs to the ground environment.

air gap—1. A nonmagnetic discontinuity in a ferromagnetic circuit. For example, the space between the poles of a magnet—although filled with brass, wood or any other nonmagnetic material—is nevertheless called an air gap. This gap reduces the tendency toward saturation. 2. The air space between two magnetically or electrically related objects.

air/ground control radio station—An aeronautical telecommunication station with the primary responsibility of handling communications related to the operation and control of aircraft in a given area.

air lock—A small chamber, located at the entrance to an area, the doors of which are so interlocked that only one can be opened at a time. This acts as an air seal to maintain the condition of the air within the area.

air navigation radio aids—Aeronautical ground stations, radio beacons, direction finders, and similar facilities.

airport beacon—A beacon (light or radio) the purpose of which is to indicate the location of an airport.

airport control station—A station that furnishes communications between an airport control tower and aircraft in the immediate vicinity; messages are limited to those related to actual aviation needs.

airport radar control—The surveillance-radar portion of radar approach control.

airport runway beacon—A radio-range beacon which defines one or more approaches to an airport.

airport surface detection equipment—(Abbreviated ASDE.) Radar which shows the movement of aircraft and other vehicles on the ground at an airport. Valuable tool at night and during low visibility.

airport surveillance radar — Abbreviated asr. A short-range radar system that maintains constant surveillance over aircraft at the lower levels of flight, normally within a thirty-mile radius of an airport. Distinct from air route surveillance radar (arsr) which is long-range radar—150-mile radius—to control traffic between terminals.

air-position indicator—Airborne computing system which presents a continuous indication of aircraft position on the basis of aircraft heading, air speed, and elapsed time.

air-spaced coax—A coaxial cable in which air is basically the dielectric material. The conductor may be centered by means of a spirally wound synthetic filament, by beads, or by braided filaments. This construction is also referred to as an air dielectric.

air-to-ground communication — Transmission of radio signals from an aircraft to stations or other locations on the earth's surface, as differentiated from ground-to-air, air-to-air, or ground-to-ground.

air-to-ground radio frequency — The frequency or band of frequencies agreed upon for transmission from an aircraft to an aeronautical ground station.

air-to-surface missile—A missile designed to be dropped from an aircraft. An internal homing device or the aircraft's radio guides it to a surface target.

airwaves—Slang expression for radio waves used in radio and television broadcasting.

alacritized switch—A mercury switch treated to yield a low adhesional force between the rolling surface and mercury pool, resulting in a decreased differential angle.

alarm—A device that signals the existence of an abnormal condition by means of an audible or a visible discrete change, or both, intended to attract attention.

alarm hold—A means of holding an alarm once sensed. The typical magnetic trap does not hold or "latch" and thus the reclosing of a trapped door resets the typical magnetic trap. A hold circuit applied to such a device indicates the door has been opened and continues to so indicate until reset.

alarm relay—A relay, other than an annunciator, used to operate, or to operate in connection with, a visual or audible alarm.

albedo—The reflecting ability of an object. It is the ratio of the amount of light reflected compared to the amount received.

alc—Abbreviation for automatic level (volume) control. A special compressor circuit included in some tape recorders, for automatically maintaining the recording volume within the required limits regardless of changes in the volume of the sound.

Alexanderson alternator—An early mechanical generator used as a source of low-frequency power for transmission or induction heating. It is capable of generating frequencies as high as 200,000 hertz.

Alexanderson antenna—A vlf antenna consisting of a horizontal wire connected to ground at equally spaced points by vertical wires with base-loading coils; the transmitter is coupled to an end coil.

Alford antenna—A square loop antenna comprising four linear sides with their ends bent inward so that capacitive loading is provided to equalize the current around the loop.

algebraic adder—A computer circuit which can form an algebraic sum.

algebraic logic—A calculator mode permits all calculations to be done in the order in which they are written.

ALGOL—An international problem language designed for the concise, efficient expression of arithmetic and logical processes and the control (iterative, etc.) of these processes. From ALGOrithmic Language.

algorithm—1. A set of rules or processes for solving a problem in a finite number of steps (for example, a full statement of an arithmetic procedure for finding the value of sin x with a stated precision). *See also* Procedure. 2. A series of equations, some of which may state inequalities which cause decisions to be made and the computational process to be altered based on these decisions.

algorithmic language—An arithmetic language by which a numerical procedure may be presented to a computer precisely and in a standard form.

alias—An alternate label. For instance, a label and one or more aliases may be used to identify the same data element or point in a computer program.

aliasing—The introduction of error into the Fourier analysis of a discrete sampling of continuous data when components with frequencies too great to be analyzed with the sampling interval being used contribute to the amplitudes of lower-frequency components.

align—To adjust the tuned circuits of a receiver or transmitter for maximum signal response.

aligned-grid tube—A multigrid vacuum tube in which at least two of the grids are aligned one behind the other to give such effects as beam formation and noise suppression.

alignment—1. The process of adjusting components of a system for proper interrelationship. The term is applied especially to (1) the adjustment of tuned circuits in a receiver to obtain the desired frequency response, and (2) the synchronization of components is a system. 2. In a tape recorder, the physical positioning of a tape head relative to the tape itself. Alignment in all respects must conform to rigid requirements in order for a recorder to function properly. 3. The accuracy of proper relative position of an image on a photomask with respect to an existing image on a substrate, as in a photoresist coating, or etched in the oxide of an oxidized silicon wafer.

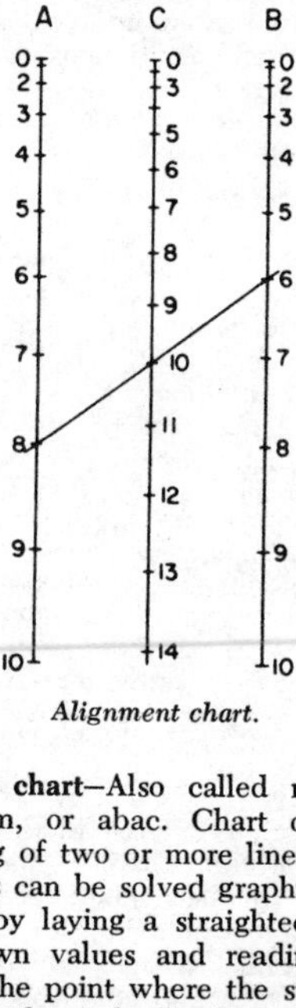

Alignment chart.

alignment chart—Also called nomograph, nomogram, or abac. Chart or diagram consisting of two or more lines on which equations can be solved graphically. This is done by laying a straightedge on the two known values and reading the answer at the point where the straightedge intersects the scale for the value sought.

alignment pin—1. A pin in the center of the base of a tube. A projecting rib on the pin assures that the tube is correctly inserted into its socket. 2. Any pin or device that will ensure the correct mating of two components designed to be connected.

alignment protractor—An instrument that indicates error in a pickup's lateral alignment. It fits on the center spindle of the turntable and the pickup stylus fits into a small hole on the device. The correct indication is shown when the angle of lateral movement of the pickup head is at 90° to the tangent of the groove at any point, although minimal tracking error is expected with most pickup arms.

alignment tool—A special screwdriver or socket wrench used for adjusting trimmer or padder capacitors or cores in tuning inductances. It is usually constructed partly or entirely of nonmagnetic material. *See also* Neutralizing Tool.

alive—Electrically connected to a source of potential difference, or electrically charged to have a potential different from that of the earth.

alive circuit—One which is energized.

alkali—A compound which forms hydroxyl ions when in aqueous solution. Also called a base.

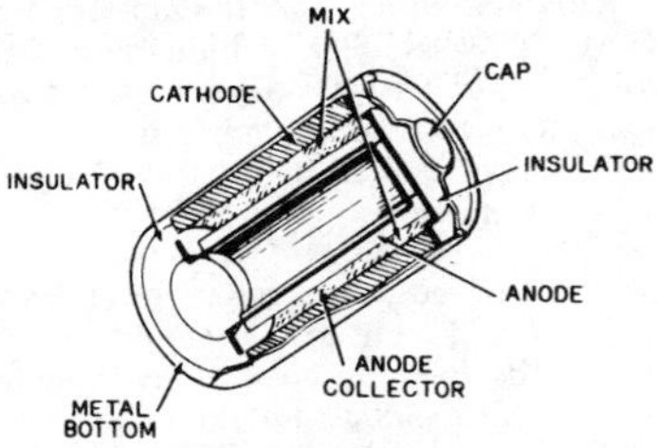

Alkaline cell.

alkaline cell—Also called alkaline-manganese cell. A primary cell similar to the zinc-carbon cell except that a potassium hydroxide (KOH) electrolyte is used. This cell has about 50 to 100 percent more capacity than the zinc-carbon cell. Nominal cell voltage is 1.5 volts.

all-diffused monolithic integrated circuit—Also called compatible monolithic integrated circuit. A microcircuit consisting of a silicon substrate into which all of the circuit parts (both active and passive elements) are fabricated by diffusion and related processes.

Allen screw—A screw having a hexagonal hole or socket in its head. Often used as a setscrew.

Allen wrench—A straight or bent hexagonal rod used to turn an Allen screw.

Alligator clip.

alligator clip—A spring-loaded metal clip with long, narrow meshing jaws, used for making temporary electrical connections.

allocate—In a computer, to assign storage locations to main routines and subroutines, thus fixing the absolute values of symbolic addresses.

allocated channel—A channel assigned to a specific user.

allocated frequency band—A segment of the radio-frequency spectrum established by competent authority designating the use which may be made of the frequencies contained therein.

allocated-use circuit—1. A circuit in which one or more channels have been allocated for the exclusive use of one or more services by a proprietary service; may be a unilateral or joint circuit. 2. Communication link specifically assigned to user(s) warranting such facilities.

allochromatic—Exhibiting photoelectric effects due to the inclusion of microscopic impurities, or as a result of exposure to various types of radiation.

allotter—In a telephone system, a distributor, associated with the finder control group relay assembly, that allots an idle linefinder in preparation for an additional call.

allotter relay—In a telephone system, a relay of the linefinder circuit, the function of which is to preallot an idle linefinder to the next incoming call from the line, and to guard relays.

alloy—1. A composition of two or more elements, of which at least one is a metal. It may be a solid solution, a heterogeneous mixture, or a combination of both. 2. Method of making pn junctions by melting a metallic dopant so that it dissolves some of the semiconductor material, and then hardens to produce a doped "alloy."

alloy deposition—The process of depositing an alloy on a substrate.

alloy-diffused transistor—A transistor with a diffused base and alloyed emitter.

alloyed contact—An ohmic contact formed by an alloy process.

alloy junction—Also called fused junction. A junction produced by alloying one or more impurity metals to a semiconductor. A small button of impurity metal is placed at each desired location on the semiconductor wafer, heated to its melting point, and cooled rapidly. The impurity metal alloys with the semiconductor material to form a p or n region, depending on the impurity used.

alloy-junction photocell—A photodiode in which an alloy junction is produced by alloying (mixing) an indium disc with a thin wafer of n-type germanium.

alloy-junction transistor—Also called fused-junction transistor. A semiconductor wafer of p- or n-type material with two dots containing p- or n-type impurities fused, or alloyed, into opposite sides of the wafer to provide emitter and base junctions. The base region comprises the original semiconductor wafer.

alloy process—A fabrication technique in which a small part of the semiconductor material is melted together with the desired metal and allowed to recrystallize. The alloy developed is usually intended to form a pn junction or an ohmic contact.

alloy transistor—A transistor in which the emitter and collector junctions are both alloy junctions.

all-pass filter—A network designed to produce a delay (phase shift) and an attenuation that is the same at all frequencies; a lumped-parameter delay line. Also called all-pass network.

all-pass network—A network designed to introduce phase shift or delay but not appreciable attenuation at any frequency.

all-relay central office—An automatic central-office dial switchboard in which relay circuits are used to make the line interconnections.

all-wave antenna—A receiving antenna suitable for use over a wide range of frequencies.

all-wave receiver—A receiver capable of receiving stations on all the commonly used wavelengths in short-wave bands as well as in the broadcast band.

alnico—An alloy consisting mainly of ALuminum, NIckel, and CObalt plus iron. Capable of very high flux density and magnetic retentivity. Used in permanent magnets for speakers, magnetrons, etc.

alpha—1. Emitter-to-collector current gain of a transistor connected as a common-base amplifier. For a junction transistor, alpha is less than unity, or 1. 2. Brain wave signals whose frequency is approximately 8 to 12 Hz. The associated mental state is relaxation, heightened awareness, elation, and in some cases, dreamlike.

alphabet—1. An ordered set of all the letters and associated marks used in a language. 2. An ordered set of the letters used in a language, for example, the Morse code alphabet, the 128 characters of the USASCII alphabet.

alphabetic coding—A system of abbreviation used in preparing information for input into a computer. Information may then be reported in the form of letters and words as well as in numbers.

alphabetic-numeric—Having to do with the alphabetic letters, numerical digits, and special characters used in electronic data processing work.

alphabetic string—A character string containing only letters and special characters.

alpha cutoff frequency—The frequency at which the current gain of a common-base transistor stage has decreased to 0.707 of its low-frequency value. Gives a rough indication of the useful frequency range of the device.

alphameric (alphanumeric)—Generic term for alphabetic letters, numerical digits, and special characters which are machine-processable.

alphameric characters—Used to include numeric digits, alphabetic characters, and special characters.

alphanumeric—Pertaining to a character set that contains both letters and numerals and usually other characters.

alphanumeric code—A code used to express numerically the letters of the alphabet.

alphanumeric keys—Keys resembling those on a standard keyboard used on a data entry device. Usually they are used to manually input or edit text for the display system, although they can also be used in a function key mode.

alpha particle — A small, electrically charged particle thrown off at a very high velocity by many radioactive materials including uranium and radium. Identical to the nucleus of a helium atom, it is made up of two neutrons and two protons. Its electrical charge is positive and is equal in magnitude to twice that of an electron.

alpha ray—A stream of fast-moving alpha particles which produce intense ionization in gases through which they pass, are easily absorbed by matter, and produce a glow on a fluorescent screen. The lowest-frequency radioactive emissions.

alpha system—A signaling system in which the signaling code to be used is designated by alphabetic characters.

alpha-wave detector—A device that detects and displays alpha-wave segments of brain wave output. Used in biofeedback. Also called alpha-wave meter.

alpha-wave meter—*See* Alpha-Wave Detector.

alteration switch—A manual switch on a computer console or program-simulated switch which can be set on or off to control coded machine instructions.

alternate channel—A channel located two channels above or below the reference channel.

alternate-channel interference — Interference caused in one communication channel by a transmitter operating in the channel after an adjacent channel. (*See also* Second-Channel Interference.)

alternate frequency—The frequency assigned for use at a certain time, or for a certain purpose, to replace or supplement the frequency normally used.

alternate mode—A means of displaying on an oscilloscope the output signals of two or more channels by switching the channels, in sequence, after each sweep.

alternate routing—A secondary or backup communications path to be used if the normal (primary) routing is not possible.

alternating-charge characteristic — The function relating, under steady-state conditions, the instantaneous values of the alternating component of transferred charge to the corresponding instantaneous values of a specified periodic voltage applied to a nonlinear capacitor.

alternating current—Abbreviated ac. A flow of electricity which reaches maximum in one direction, decreases to zero, then reverses itself and reaches maximum

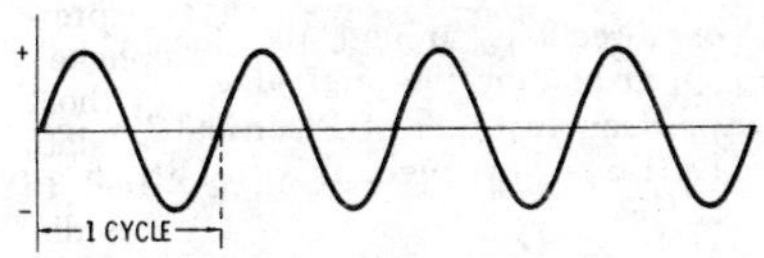

Alternating current.

in the opposite direction. The cycle is repeated continuously. The number of such cycles per second is the frequency. The average value of voltage during any cycle is zero.

alternating current/direct current—A term applied to electronic equipment indicating it is capable of operation from either an alternating-current or direct-current primary power source.

alternating-current erasing head—An erasing head, used in magnetic recording in which alternating current produces the magnetic field necessary for erasing. Alternating-current erasing is achieved by subjecting the medium to a number of cycles of a magnetic field of a decreasing magnitude. The medium is, therefore, essentially magnetically neutralized.

alternating-current pulse—An alternating-current wave of brief duration.

alternating-current transmission—In television, that form of transmission in which a fixed setting of the controls makes any instantaneous value of signal correspond to the same value of brightness only for a short time.

alternating quantity—A periodic quantity which has alternately positive and negative values, the average value of which is zero over a complete cycle.

alternating voltage—Also called ac voltage. Voltage that is continually varying in value and reversing its direction at regular intervals, such as that generated by an alternator or developed across a resistance or impedance through which alternating current is flowing.

alternation—One-half of a cycle—either when an alternating current goes positive and returns to zero, or when it goes negative and returns to zero. Two alternations make one cycle.

alternator—A device for converting mechanical energy into electrical energy in the form of an alternating current.

alternator transmitter—A radio transmitter that generates power by means of a radio-frequency alternator.

altimeter—An instrument that indicates the altitude of an aircraft above a specific reference level, usually sea level or the ground below the aircraft. It may be similar to an aneroid barometer that utilizes the change of atmospheric pressure with altitude, or it may be electronic.

altimeter station—An airborne transmitter, the emissions from which are used to determine the altitude of an aircraft above the surface of the earth.

altitude—The vertical distance of an aircraft or other object above a given reference plane such as the ground or sea level.

altitude delay—The synchronization delay introduced between the time of transmission of the radar pulse and the start of the trace on the indicator. This is done to eliminate the altitude circle on the plan-position-indicator display.

alto-troposphere—A portion of the atmosphere about 40 to 60 miles above the surface of the earth.

alu—Abbreviation for arithmetic and logic unit. A device that performs the basic mathematical operations such as addition, subtraction, multiplication, and division of numbers (usually binary) presented to its inputs, and provides an output that is an appropriate function of the inputs.

alumina—A ceramic used for insulators in electron tubes or substrates in thin-film circuits. It can withstand continuously high temperatures and has a low dielectric loss over a wide frequency range.

aluminized-screen picture tube—A cathode-ray picture tube which has a thin layer of aluminum deposited on the back of its fluorescent surface to improve the brilliance of the image and also prevent ion-spot formation.

aluminum-electrolytic capacitor—A capacitor with two aluminum electrodes (the anode has the oxide film) separated by layers of absorbent paper saturated with the operating electrolyte. The aluminum-oxide film or dielectric is repairable in the presence of an operating electrolyte.

am—*See* Amplitude Modulation.

amateur—Also called a ham. A person licensed to operate radio transmitters as a hobby. Any amateur radio operator.

amateur bands—Certain radio frequencies assigned exclusively to radio amateurs. In the United States of America, the Federal Communications Commission (FCC) makes these assignments.

amateur extra license—A license issued by the FCC to amateur radio operators who are able to send and receive Morse code at the rate of 20 words per minute, and who are familiar with general, intermediate, and advanced radio theory and practice. Its privileges include all authorized amateur rights and the exclusive rights to operate on certain frequencies.

amateur station—A radio transmitting station operated by one or more licensed amateur operators.

amateur-station call letters—A group of numbers and letters assigned exclusively to a licensed amateur operator to identify his station.

ambience—Reverberant or reflected sound that reaches a listener's ear from all directions as sound waves "bounce" successively off the various surfaces of a listening area—the walls, ceiling, etc. The term is usually reserved for large areas such as auditoriums and concert halls, though home listening-rooms have their own ambience effects.

ambient—Surrounding. (*See also* Ambient Noise *and* Ambient Temperature.)

ambient level—The level of interference emanating from sources other than the test sample, such as inherent noise of the measuring device and extraneous radiated fields.

ambient light—Normal room light.

ambient-light filter—A filter used in front of a television picture-tube screen to reduce the amount of ambient light reaching the screen, and to minimize the reflections of light from the glass face of the tube.

ambient noise—1. Acoustic noise in a room or other location. Usually measured with a sound-level meter. The term "room noise" commonly designates ambient noise at a telephone station. 2. Unwanted background noise picked up by a microphone—i.e., any extraneous clatter in a room. Also any acoustic coloration that influences sounds, brought about by the acoustic properties of a room in which a recording is being made or replayed.

ambient pressure—The general surrounding atmospheric pressure.

ambient temperature—Temperature of air or liquid surrounding any electrical part or device. Usually refers to the effect of such temperature in aiding or retarding removal of heat by radiation and convection from the part or device in question.

ambient temperature range—The range of environmental temperatures in the vicinity of a component or device, over which it may be operated safely and within specifications. For forced-air cooled operation, the ambient temperature is measured at the air intake.

ambiguity—1. An undesirable tendency of a synchro or servo system to seek a false null position, in addition to the proper null position. 2. Inherent error resulting from multiple-bit changes in a polystrophic code. (Proper logic design prevents such errors.)

ambiguous count—A count on an electronic scaler that is obviously impossible.

ambisonic reproduction—The re-creation of the ambience of an original recording situation with associated directionality. Sound from every direction is picked by a tetrahedral microphone array which is then encoded onto two channels which, upon decoding, produce sound through several speakers in a continuous range of directions around the listener, thus approximating the original.

American Institute of Electrical Engineers (AIEE)—Now merged with IRE to form IEEE.

American Morse code—A system of dot-and-dash signals originated by Samuel F. B. Morse and still used to a limited extent for wire telegraphy in North America. It differs from the International Morse code used in radiotelegraph transmission.

American National Standards Institute, Inc.—Abbreviated ANSI. An independent, industry-wide association that establishes standards for the purpose of promoting consistency and interchangeability among the products of different manufacturers. Formerly United States of America Standards Institute (USASI) and American Standards Association (ASA).

American Radio Relay League (ARRL)—An organization of amateur radio operators.

American Standards Association—Abbreviated ASA. *See* American National Standards Institute.

American Wire Gage (AWG)—The system of notation generally adopted in the United States for measuring the size of solid wires.

am/fm receiver—A device capable of converting either amplitude- or frequency-modulated signals into audio frequencies.

am/fm tuner—A device capable of converting either amplitude- or frequency-modulated signals into low-level audio frequencies.

ammeter—An instrument for measuring either direct or alternating electric current (depending on its construction). Its scale is usually graduated in amperes, milliamperes, microamperes, or kiloamperes.

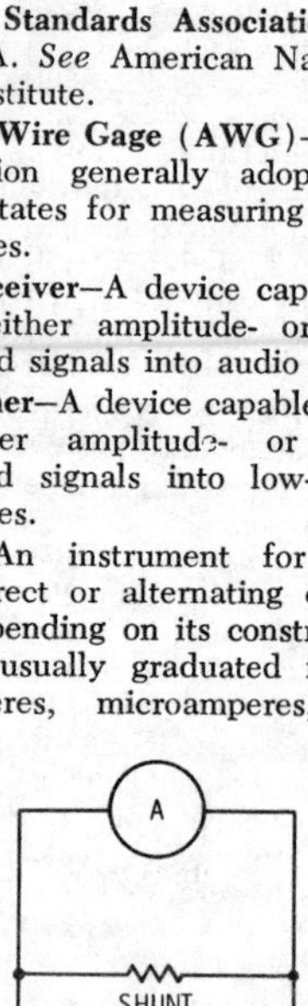

Ammeter shunt.

ammeter shunt—A low-resistance conductor placed in parallel with the meter movement so most of the current flows through this conductor and only a small part passes through the movement itself. This extends the usable range of the meter.

amorphous—A characteristic, particularly

of a crystal, determining that it has no regular structure.

amortisseur winding—*See* Damper Winding.

amp—Abbreviation for ampere.

ampacity—Current-carrying capacity expressed in amperes.

amperage—The number of amperes flowing in an electrical conductor or circuit.

ampere—1. A unit of electrical current or rate of flow of electrons. One volt across 1 ohm of resistance causes a current flow of 1 ampere. A flow of 1 coulomb per second equals 1 ampere. An unvarying current is passed through a solution of silver nitrate of standard concentration at a fixed temperature. A current that deposits silver at the rate of .001118 gram per second is equal to 1 ampere, or 6.25×10^{18} electrons per second passing a given point in a circuit. 2. The constant current which, if maintained in two straight parallel conductors of infinite length, of negligible circular sections, and placed 1 meter apart in a vacuum will produce between these conductors a force equal to 2×10^{-7} newton per meter of length.

ampere-hour—A current of one ampere flowing for one hour. Multiplying the current in amperes by the time of flow in hours gives the total number of ampere-hours. Used mostly to indicate the amount of energy a storage battery can deliver before it needs recharging, or the energy a primary battery can deliver before it needs replacing. One ampere-hour equals 3,600 coulombs.

ampere-hour capacity—The amount of current a battery can deliver in a specified length of time under specified conditions.

ampere-hour meter—An electrical meter which measures the amount of current (amperes) per unit of time (hours) which has been consumed in a circuit.

Ampere's rule—Current in a certain direction is equivalent to the motion of positive charges in that direction. The magnetic flux generated by a current in a wire encircles the current in the counterclockwise direction when the current is approaching the observer.

ampere-turn—A measure of magnetomotive force, especially as developed by an electric current, defined as the magnetomotive force developed by a coil of one turn through which a current of one ampere flows; that is, 1.26 Gilberts.

amp-hr—Abbreviation for ampere-hour or ampere-hours.

amplidyne—A special direct-current generator used extensively in servo systems as a power amplifier. The response of its output voltage to changes in field excitation is very rapid, and its amplification factor is high.

amplification—1. Increase in size of a medium in its transmission from one point to another. May be expressed as a ratio or, by extension of the term, in decibels. 2. An increase in the magnitude of a signal brought about by passing through an amplifier.

amplification factor (μ)—1. In a vacuum tube, the ratio of a small change in plate voltage to a small change in grid voltage required to produce the same change in plate current (all other electrode voltages and currents being held constant). 2. In any device, the ratio of output magnitude to input magnitude.

amplified agc — An automatic gain-control circuit in which the control voltage is amplified before being applied to the tube or transistor, the gain of which is to be controlled in accordance with the strength of the incoming signal.

amplified back bias—Degenerative voltage developed across a fast time-constant circuit within a stage of an amplifier and fed back into a preceding stage.

amplifier—1. A device which draws power from a source other than the input signal and which produces as an output an enlarged reproduction of the essential features of its input. The amplifying element may be an electron tube, transistor, magnetic circuit, or any of various devices. 2. A device for increasing the magnitude of a signal by means of a varying control voltage, maintaining the signal's characteristic form as closely as possible to the original. 3. An electronic device for magnifying (and usually controlling) electrical signals. High fidelity amplifiers consist of a preamplifier equalizer section, plus a power or basic amplifier section. In an integrated amplifier, both sections are built on one chassis and made available as a single unit. Alternately, the two sections are available as separate units. 4. Device for increasing power associated with a signal (voltage or current). Basic types include dc, ac, audio, linear, radio, video, differential, pulse, logarithmic.

amplifier noise—All spurious or unwanted signals, random or otherwise, that can be observed in a completely isolated amplifier in the absence of a genuine input signal.

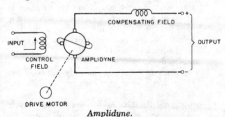

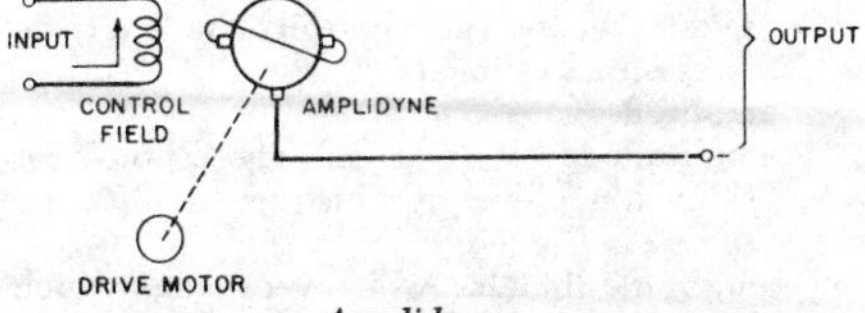

Amplidyne.

amplifier nonlinearity—The inability of an amplifier to produce an output at all times proportionate to its input.

amplify—To increase in magnitude or strength, usually said of a current or voltage.

amplifying delay line—A delay line used in pulse-compression systems to amplify delayed superhigh-frequency signals.

amplistat—A self-saturating type of magnetic amplifier.

Amplitron—A broad-band crossed-field amplifier with a re-entrant electron stream. The electron stream interacts with the backward wave of a nonre-entrant rf structure (Raytheon.)

amplitude—1. The magnitude of variation in a changing quantity from its zero value. The word must be modified with an adjective such as peak, rms, maximum, etc., which designates the specific amplitude in question. 2. The level of an audio or other signal in voltage or current terms.

amplitude-controlled rectifier—A rectifier circuit in which a thyratron is the rectifying element.

amplitude density distribution—A function that gives the fraction of time that a voltage is within a narrow range.

amplitude distortion—Distortion that is present in an amplifier when the amplitude of the output signal fails to follow exactly any increase or decrease in the amplitude of the input signal. It results from nonlinearity of the transfer function and gives rise to harmonic and intermodulation distortion. No amplifier is completely free from the effect because its transfer function is slightly curved. The nature of the curvature determines the order of the distortion produced, but negative feedback and other circuit configurations help minimize the curvature within the dynamic range and hence keep the distortion at a very low level.

amplitude distribution function—A function that gives the fraction of time that a time-varying voltage is below a given level.

amplitude fading—Fading in which the amplitudes of all frequency components of a modulated carrier wave are uniformly attenuated.

amplitude-frequency response—The variation of gain, loss, amplification, or attenuation of a device or system as a function of frequency. Usually measured in the region where the transfer characteristic is essentially linear.

amplitude gate—*See* Slicer.

amplitude-level selection—The choice of the voltage level at which an oscilloscope sweep is triggered.

amplitude limiter—A circuit or stage which automatically reduces the amplification to prevent signal peaks from exceeding a predetermined level.

amplitude-modulated transmitter—A transmitter in which the amplitude of its radio-frequency wave is varied at a low frequency rate—usually in the audio or video range. This low frequency is the intelligence (information) to be conveyed.

amplitude-modulated wave—A constant-frequency waveform in which the amplitude varies in step with the frequency of an impressed signal.

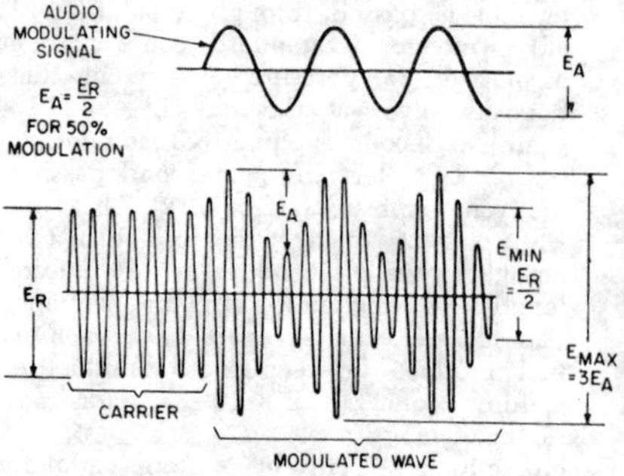

Amplitude modulation.

amplitude modulation – Modulation in which the amplitude of a wave is the characteristic subject to variation. Those systems of modulation in which each component frequency (f) of the transmitted intelligence produces a pair of sideband frequencies at carrier frequency plus (f) and carrier frequency minus (f). In special cases, the carrier may be suppressed; either the lower or upper sets of sideband frequencies may be suppressed; the lower set of sideband frequencies may be produced by one or more channels of information. The carrier may be transmitted without intelligence-carrying sideband frequencies. The resulting emission bandwidth is proportional to the highest frequency component of the intelligence transmitted.

amplitude-modulation noise level—Undesired amplitude variations of a constant radio-frequency signal, especially in the absence of any intended modulation.

amplitude noise—The effect on radar accuracy of the fluctuations in amplitude of the signal returned by the target. These fluctuations are caused by any change in aspect if the target is not a point source.

amplitude of noise—When impulse-type noise is of random occurrence and so closely spaced that the individual waveshapes are not separated by the receiving equipment, then the noise has the waveshape and characteristics of random noise. Random-noise amplitude is proportional to the square root of the bandwidth. If the impulses are separated, the noise no longer has the waveshape of random noise and its amplitude is directly proportional to the bandwidth of the transmission system.

amplitude permeability—The relative permeability at a stated value of field strength and under stated conditions, the field strength varying periodically with time and no direct magnetic-field component being present.

amplitude range—The ratio, usually expressed in decibels, between the upper and lower limits of program amplitudes which contain all significant energy contributions.

amplitude resonance—The condition that exists when any change in the period or frequency of the periodic agency (but not its amplitude) decreases the amplitude of the oscillation or vibration of the system.

amplitude response—The maximum output amplitude that can be obtained at various points over the frequency range of an instrument operated under rated conditions.

amplitude selection—The process of selecting that portion of a waveform which lies above or below a given value, or between two given values.

amplitude separator—A television-receiver circuit that separates the control impulses from the video signal.

amplitude shift keying—Abbreviated ask. The modulation of digital information on a carrier by changing the amplitude of the carrier.

amplitude-suppression ratio—In frequency modulation, the ratio of the magnitude of the undesired output to the magnitude of the desired output of an fm receiver when the applied signal is simultaneously amplitude- and frequency-modulated. Generally measured with an applied signal that is amplitude-modulated 30% at a 400-hertz rate and is frequency-modulated 30% of the maximum system deviation at a 1000-hertz rate.

amplitude versus frequency distortion — Distortion caused by the nonuniform attenuation or gain of the system, with respect to frequency under specified terminal conditions.

am rejection—The ratio of the recovered audio output produced by a desired fm signal with specified modulation, amplitude, and frequency to that produced by an am signal, on the same carrier, with specified modulation index.

am tuner—A device capable of converting amplitude-modulated signals into low-level audio frequencies.

amu—Abbreviation for atomic mass unit.

anacoustic—Without sound, designating in particular the regions on the fringes of the earth's atmosphere and beyond, where sound propagation is negligible, or zero.

anacoustic zone—Zone of silence in space where distances between air molecules are so great that sound waves are not propagated.

analog—1. In electronic computers a physical system in which the performance of measurements yields information concerning a class of mathematical problems. 2. Of or pertaining to the general class of devices or circuits in which the output varies as a continuous function of the input. 3. The representation of numerical quantities by means of physical variables, e.g. translation, rotation, voltage, resistance, contrasted with "digital."

analog channel—A computer channel in which the transmitted information can have any value between the defined limits of the channel.

analog communications—A system of telecommunications employing a nominally continuous electrical signal that varies in frequency, amplitude, etc., in some direct correlation to nonelectrical information (sound, light, etc.) impressed on a transducer.

analog computer—1. A computer operating on the principle of creating a physical (often electrical) analogy of the mathematical problem to be solved. Variables such as temperature or flow are represented by the magnitude of a physical phenomenon such as voltage or current. The computer manipulates these variables in accordance with the mathematical formulas "analogued" on it. 2. A computer system both the input and output of which are continuously varying signals. 3. A computing machine that works on the principle of measuring, as distinguished from counting. 4. A computer that solves problems by setting up equivalent electric circuits and making measurements as the variables are changed in accordance with the corresponding physical phenomena. An analog computer gives approximate solutions, whereas a digital computer gives exact solutions. 5. A nondigital computer that manipulates linear (continuous) data to measure the effect of a change in one variable on all other variables in a particular problem.

analog data—A physical representation of information such that the representation bears an exact relationship to the original information. The electrical signals on a telephone channel are an analog data representation of the original voice. 2. Data represented in a continuous form, as contrasted with digital data represented in a discrete (discontinuous) form. Analog data are usually represented by physical variables, such as voltage, resistance, rotation, etc.

analog multiplexer—Circuit used for time sharing of analog-to-digital converters between a number of different analog information channels. Consists of a group of analog switches arranged with inputs connected to the individual analog channels and outputs connected in common. 2. Two or more analog switches with separate inputs and a common output, with each gate separately controllable. Multiplexing is performed by sequentially turning on each switch one at a time, switching each individual input to a common output.

analog network—A circuit or circuits that represent physical variables in such a manner as to permit the expression and solution of mathematical relationships between the variables, or to permit the solution directly by electric or electronic means.

analog output—As distinguished from digital output. Here the amplitude is continuously proportionate to the stimulus, the proportionality being limited by the resolution of the device.

analog recording—A method of recording in which some characteristic of the record current, such as amplitude or frequency, is continuously varied in a manner analogous to the time variations of the original signal.

analog representation — A representation that does not have discrete values, but is continuously variable.

analog signal—An electrical signal that varies continuously, as obtained from temperature or pressure, or speed transducers. A voltage level that changes in proportion to the change in a physical variable.

analog switch—1. A device that either transmits an analog signal without distortion, or completely blocks it. 2. Any solid-state device, with or without a driver, capable of bilaterally switching voltages or current. It has an input terminal, output terminal, and, ideally, no offset voltage, low *on* resistance, and extreme isolation between the signal being gated and control signals.

analog-to-digital conversion—The process of converting a continuously variable (analog) signal to a digital signal (binary code) that is a close approximation of the original signal.

analog-to-digital converter—1. A circuit that changes a continuously varying voltage or current into a digital output. The input may be ac or dc, and the output may be serial or parallel, binary or decimal. 2. Device that translates analog signals (voltages, pressures, etc.) into numerical digital form (binary, decimal, etc.).

analogue computing—Computing system where continuous signals represent mechanical (or other) parameters.

analytical engine—An early form of general-purpose digital computer invented in 1833 by Charles Babbage.

analyzer—1. An instrument or other device designed to examine the functions of components, circuits, or systems, and their relations to each other, as contrasted with an instrument designed to measure some specific parameter of such a system or circuit. 2. Of computers, a routine the purpose of which is to analyze a program written for the same or a different computer. This analysis may consist of summarizing instruction references to storage and tracing sequences of jumps.

anastigmat—A lens system designed so as to be free from the aberration called astigmatism.

anchor—An object, such as a metal rod, set into the ground to hold the end of a guy wire.

ancillary equipment—Equipment not directly employed in the operation of a system, but necessary for logistic support, preparation for flight, or assessment of target damage—e.g., test equipment, vehicle transport.

AND circuit—Synonym for AND gate.

AND device—A device that has its output in the logic 1 state if and only if all the control signals are in the logic 1 state.

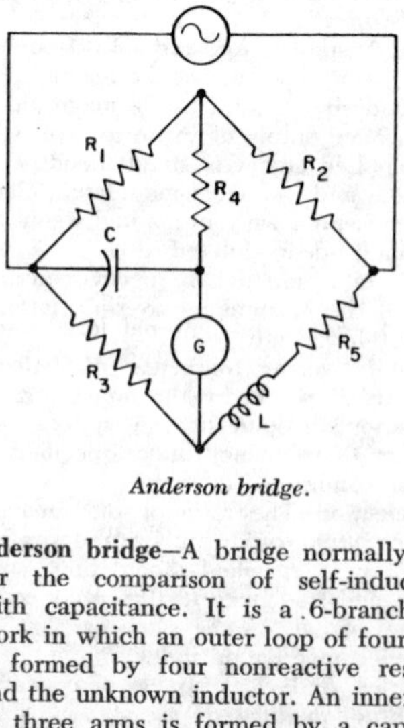

Anderson bridge.

Anderson bridge—A bridge normally used for the comparison of self-inductance with capacitance. It is a 6-branch network in which an outer loop of four arms is formed by four nonreactive resistors and the unknown inductor. An inner loop of three arms is formed by a capacitor and a fifth resistor in series with each other and in parallel with the arm opposite the unknown inductor. The detector is connected between the junction of the capacitor and the fifth resistor and at that end of the unknown inductor separated from a terminal of the capacitor by only one resistor. The source is connected to the other end of the unknown inductor and to the junction of the capacitor with two resistors of the outer loop. The balance is independent of frequency.

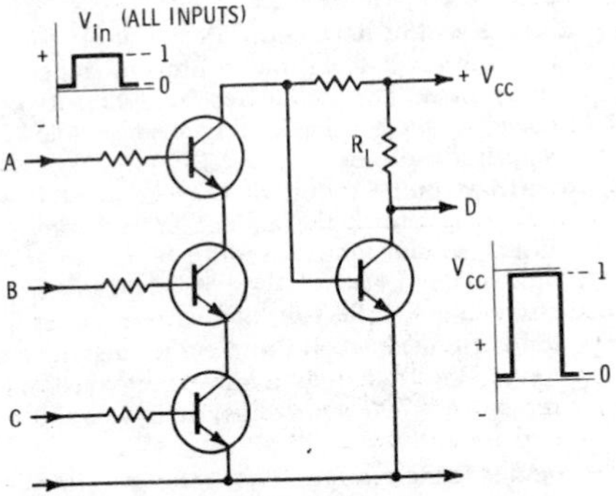

Schematic.

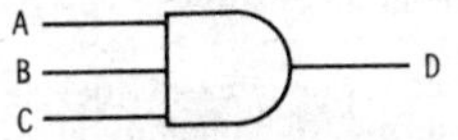

Logic diagram.

AND gate with three inputs.

AND gate—1. In an electronic computer, a gate circuit with more than one control (input) terminal. No output signal will be produced unless a pulse is applied to all inputs simultaneously. 2. A binary circuit, with two or more inputs and a single output, in which the output is logic 1 only when all inputs are logic 1, and the output is logic 0 if any one of the inputs is logic 0.

AND/NOR gate – A single logic element that performs the operation of two AND gates with outputs feeding a NOR gate. No access to the internal logic elements is provided (i.e., no connection is available at the outputs of the AND gates).

AND/OR circuit—A gating circuit that produces a prescribed output condition when several possible combinations of input signals are applied. It exhibits the characteristics of the AND gate and the OR gate.

anechoic – Nonreflective, producing no echoes.

anechoic enclosure—A special echo-free enclosure used for testing audio transducers, where all wall surfaces have been covered with acoustically absorbent materials so that reflections of the sound waves are eliminated. Also known as a dead room or an anechoic room.

anechoic room—A room whose walls have been treated so as to make them absorb a particular kind of radiation almost completely, used for testing components of sound systems, radar systems, etc. in an environment free of reflections.

anelectrotonus—The reduced sensitivity produced in a nerve or muscle in the region of contact with the anode when an electric current is passed through it.

anelectrotonic—*See* anelectrotonus.

anemometer—An instrument used for measuring the force or speed of wind.

angels—Short-duration radar reflections in the lower atmosphere. Most often caused by birds, insects, organic particles, tropospheric layers, or water vapor.

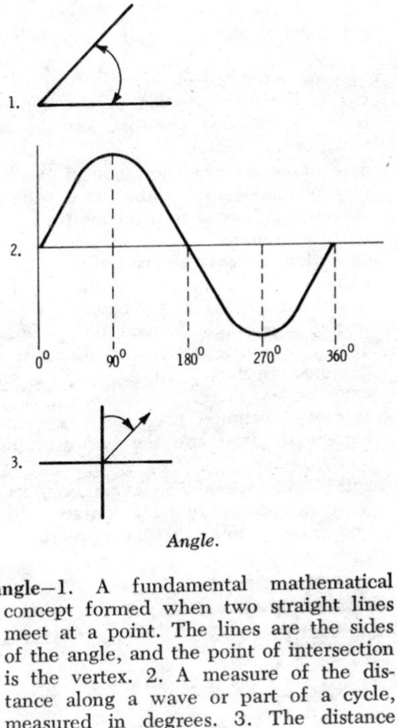

Angle.

angle—1. A fundamental mathematical concept formed when two straight lines meet at a point. The lines are the sides of the angle, and the point of intersection is the vertex. 2. A measure of the distance along a wave or part of a cycle, measured in degrees. 3. The distance through which a rotating vector has progressed.

angle jamming—An ECM technique in which azimuth and elevation information present in the modulation components of the returning echo pulse of a scanning fire-control radar is jammed by transmitting a pulse similar to the radar pulse but with angle information of erroneous phase.

angle modulation—Modulation in which the angle of a sine-wave carrier is the characteristic varied from its normal value. Phase and frequency modulation are particular forms of angle modulation.

angle noise—Tracking error introduced into radar by variations in the apparent angle of arrival of the echo from a target due to finite target size. (This effect is caused by variations in the phase front of the radiation from a multiple-point target as the target changes its aspect with respect to the observer.)

angle of arrival—Angle made between the line of propagation of a radio wave and the earth's surface at the receiving antenna.

angle of azimuth—The angle measured clockwise in a horizontal plane, usually from the north. The north used may be true north, Y-north, or magnetic north.

angle of beam—The angle which encloses most of the transmitted energy from a directional-antenna system.

angle of convergence—Angle formed by the lines of sight of both eyes when focusing on an object.

angle of deflection—The angle formed between the new position of the electron beam in a cathode-ray tube and the normal position before deflection.

angle of departure—The angle of the line of propagation of a radio wave with respect to a horizonal plane at the transmitting antenna.

angle of divergence—In cathode-ray tubes, a measure of its spread as the electron beam travels from the cathode to the screen. The angle formed by an imaginary center line and the border line of the electron beam. In good tubes, this angle is less than 2°.

angle of elevation—The angle between the horizontal plane and the line ascending to the object.

angle of incidence—The angle between a wave or beam striking a surface and a line perpendicular to that surface.

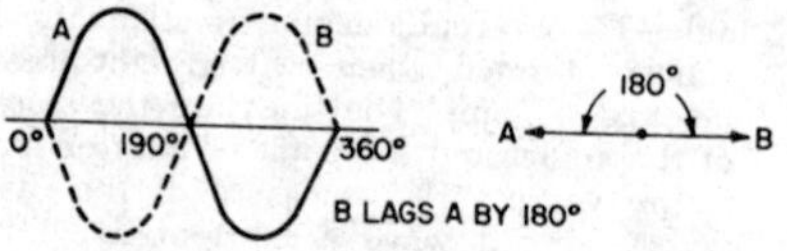

Angle of lag.

angle of lag—The angular phase difference between one sinusoidal function and a second having the same frequency. Expressed in degrees, the amount the second function must be retarded to coincide with the first.

angle of lead—1. The time or angle by which one alternating electrical quantity leads another of the same cyclic period. 2. The angle through which the commutator brushes of a generator or motor must be moved from the normal position to prevent sparking.

angle of radiation—The angle between the surface of the earth and the center of the beam of energy radiated upward into the sky from a transmitting antenna.

angle of reflection—The angle between a wave or beam leaving a surface and a line perpendicular to that surface.

angle of refraction—The angle between a wave or beam as it passes through a medium and a line perpendicular to the surface of that medium.

angle tracking noise—Any deviation of the tracking axis from the center of reflectivity of a target. The resultant of servo noise, receiver noise, angle noise, and amplitude noise.

angstrom unit—A unit of measurement of wavelength of light and other radiation. Equal to one ten-thousandth of a micron or one hundred-millionth of a centimeter (10^{-8} cm). The visible spectrum extends from about 4000 to 8000 angstrom units. Blue light has a wavelength in the region of 4700 angstroms; yellow, 5800; and red, 6500.

angular acceleration—The rate at which angular velocity changes with respect to time, generally expressed in radians per second.

angular accelerometer—A device capable of measuring the magnitude of, and/or variations in, angular acceleration.

angular aperture—The largest angular extent of wave surface which an objective can transmit.

angular deviation loss—The ratio of the response of a microphone or speaker on its principal axis to the response at a specified angle from the principal axis (expressed in decibels).

angular distance—The angle subtended by two bodies at the point of observation. It is equal to the distance in wavelengths multiplied by 2π radians or by 360°.

angular frequency—Frequency expressed in radians per second. It is equal to the number of hertz (cycles per second) multiplied by 2π.

angular length—Length expressed in radians or equivalent angular measure equal to 2π radians, or 360°, multiplied by the length in wavelengths.

angular momentum—The momentum that a

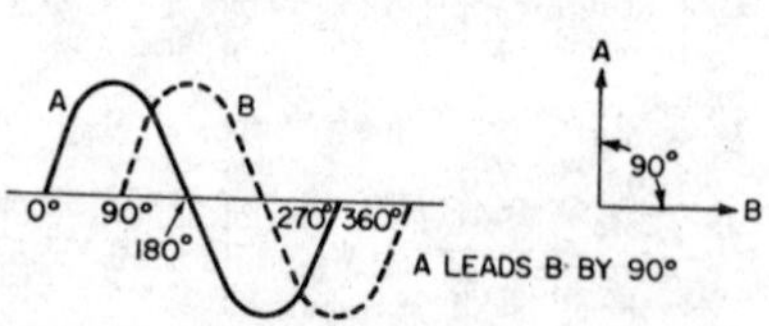

Angle of lead, 1.

body has by virtue of its rotational movement.

angular phase difference—Phase difference between two sinusoidal functions expressed as an angle.

angular rate—The rate of change of bearing.

angular resolution—The ability of a radar to distinguish between two targets solely on the basis of angular separation.

angular velocity—The rate at which an angle changes. Expressed in radians per second, the angular velocity of a periodic quantity is the frequency multiplied by 2π. If the periodic quantity results from uniform rotation of a vector, the angular velocity is the number of radians per second passed over by the rotating vector. Generally designated by the Greek letter omega (ω).

anharmonic oscillator—An oscillating system in which the restoring force is a nonlinear function of the displacement from equilibrium.

anhysteresis—The process whereby a material is magnetized by applying a unidirectional field upon which is superimposed an alternating field of gradually decreasing amplitude.

anion—1. A negatively charged ion which, during electrolysis, is attracted toward the anode. A corresponding positive ion is called a cation. 2. A negative ion that moves toward the anode in a discharge tube, electrolytic cell, or similar device.

anisotropic—1. Describing a substance that exhibits different properties when tested along axes in different directions. 2. A material that has characteristics such as wave propagation constant, magnetic permeability, conductivity, etc. that vary with direction—i.e., not isotropic.

anisotropic body—A body in which the value of any given property depends on the direction of measurement, as opposed to a body that is isotropic.

anisotropic magnet—A magnetic material having a better magnetic characteristic along the preferred axis than along any other.

anisotropic material—A material having preferred orientation so that the magnetic characteristics are superior along a particular axis.

anisotropy — Directional dependence of magnetic properties, leading to the existence of easy or preferred directions of magnetization. Anisotropy of a particle may be related to its shape, to its crystalline structure, or to the existence of strains within it.

annealed wire—A wire which has been softened by heating (soft-drawn wire).

annealing—The process of heating any solid material such as glass or metal, followed by slow cooling. This generally lowers the tensile strength and thereby improves the ductility.

annular—Ringed; ring-shaped.

annular conductor—A conductor consisting of a number of wires stranded in three reversed concentric layers surrounding a saturated hemp core.

annular transistor—A mesa transistor in which the semiconductor regions are arranged in concentric circles about the emitter.

annulling network—An arrangement of impedance elements connected in parallel with filters to annul or cancel capacitive or inductive impedance at the extremes of the passband of a filter.

annunciation relay—1. An electromagnetically operated signaling apparatus which indicates whether a current is flowing or has flowed in one or more circuits. 2. A nonautomatic reset device which gives a number of separate visual indications upon the functioning of protective devices, and which may also be arranged to perform a lockout function.

annunciator—A visual device consisting of a number of pilot lights or drops. Each light or drop indicates the condition which exists or has existed in an associated circuit and is labeled accordingly.

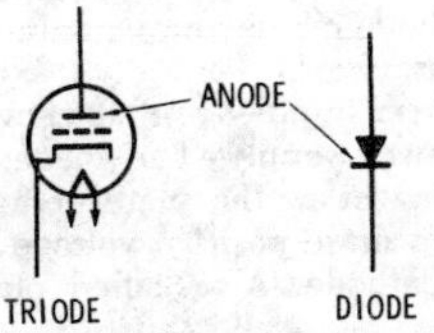

Anode.

anode—1. The positive electrode such as the plate of a vacuum tube; the element to which the principal stream of electrons flows. 2. In a cathode-ray tube, the electrodes connected to a source of positive potential. These anodes are used to concentrate and accelerate the electron beam for focusing. 3. The less noble and/or higher-potential electrode of an electrolytic cell at which corrosion occurs. This may be an area on the surface of a metal or alloy, the more active metal in a cell composed of two dissimilar metals, or the positive electrode of an impressed-current system.

anode-balancing coil—A set of mutually coupled windings used to maintain approximately equal currents in anodes operating in parallel from the same transformer terminal.

anode breakdown voltage—The potential required to cause conduction across the main gap of a gas tube when the starter gap is not conducting and all other tube elements are held at cathode potential.

anode-bypass capacitor—Also called plate-bypass capacitor. A capacitor connected between the anode and ground in an electron tube circuit. Its purpose is to bypass high-frequency currents and keep them out of the load.

anode characteristic curve—A graph that shows how the anode current of an electron tube is affected by changes in the anode voltage.

anode circuit breaker—A device used in the anode circuits of a power rectifier for the primary purpose of interrupting the rectifier circuit if an arc-back should occur.

anode current—The electron flow in the element designated as the anode. Usually signifies plate current.

anode dark space—In a gas tube, a narrow, dark zone next to the surface of the anode.

anode dissipation—The power dissipated as heat in the anode of an electron tube because of the bombardment by electrons and ions.

anode efficiency—*See* Plate Efficiency.

anode-load impedance—*See* Plate-Load Impedance.

anode modulation—*See* Plate Modulation.

anode neutralization—Also called plate neutralization. A method of neutralization in which a portion of the anode-cathode ac voltage is shifted 180° and applied to the grid-cathode circuit through a neutralizing capacitor.

anode power input—*See* Plate Power Input.

anode power supply—The means for supplying power to the plate of an electron tube at a more positive voltage than that of the cathode. Also called plate power supply.

anode pulse modulation—*See* Plate Pulse Modulation.

anode rays—Positive ions coming from the anode of an electron tube; these ions are generally due to impurities in the metal of the anode.

anode saturation—*See* Plate Saturation.

anode sheath—A layer of electrons surrounding the anode in mercury-pool arc tubes.

anode strap—A metallic connector between selected anode segments of a multicavity magnetron, used principally for mode separation.

anode supply—Also called plate supply. The direct-voltage source used in an electon-tube circuit to place the anode at a high positive potential with respect to the cathode.

anode terminal — The semiconductor-diode terminal that is positive with respect to the other terminal when the diode is biased in the forward direction.

anode voltage—The potential difference existing between the anode and cathode.

anode voltage drop (of a glow-discharge, cold-cathode tube)—Difference in potential between cathode and anode during conduction, caused by the electron flow through the tube resistance (IR drop).

anodize—To deposit a protective coating of oxide on a metal by means of an electrolytic process in which it is used as the anode.

anodizing — An electrochemical oxidation process used to improve the corrosion resistance or to enhance the appearance of a metal surface. Aluminum and magnesium parts are frequently anodized.

anomalous propagation — 1. Propagation that is unusual or abnormal. 2. The conduction of uhf signals through atmospheric ducts or layers in a manner similar to that of a waveguide. These atmospheric ducts carry the signals with less than normal attenuation over distances far beyond the optical path taken by uhf signals. Also called super-refraction. 3. In sonar, pronounced and rapid variations in the strength of the echo due to large, rapid focal fluctuations in propagation conditions.

A-N radio range—A navigational aid which provides four equisignal zones for aircraft guidance. Deviation from the assigned course is indicated aurally by the Morse code letters A (• —) or N (— •). On-course position is indicated by an audible merging of the A and N code signals into a continuous tone.

ANSI standards—A series of standards recommended by the American National Standards Institute.

A-N signal—A radio-range, quadrant-designation signal which indicates to the pilot whether he is on course or to the right or left.

answerback—The response of a terminal to remote-control signals. *See also* Handshaking.

answering cord—The cord nearest the face of a telephone switchboard. It is used for answering subscriber's calls and calls on incoming trunks.

answer lamp—In a telephone switchboard, a lamp that lights when an answer cord is plugged into a line jack; it extinguishes when the telephone answers and lights when the call is complete.

antenna—1. Also called aerial. That portion, usually wires or rods, of a radio transmitter or receiver station used for radiating waves into or receiving them from space. 2. A section of wire or a metallic device designed to intercept radio waves in the air and convert them to an electric signal for feeding to a receiver. Under relatively difficult reception conditions, such as created by location, terrain, obstructions, etc., an antenna becomes fairly critical and should be one especially designed for its intended purpose.

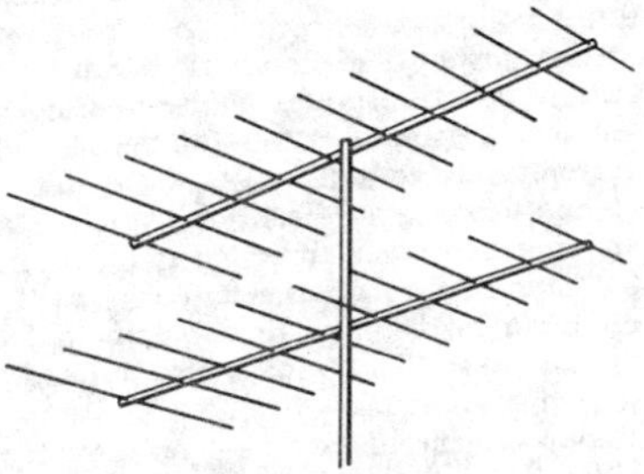

Antenna array.

antenna array—A combination of antennas assembled to obtain a desired pickup or rejection pattern.

antenna bandwidth—The range of frequencies over which the impedance characteristics of the antenna are sufficiently uniform that the quality of the radiated signal is not significantly impaired.

antenna beam width—The angle, in degrees, between two opposite half-power points of an antenna beam.

antenna coil—In a radio receiver or transmitter, the inductance through which antenna current flows.

antenna coincidence—That instance when two rotating, highly directional antennas are pointed toward each other.

antenna-conducted interference—Any signal that is generated within a transmitter or receiver, and appears as an undesired signal at the antenna terminals of the device—e.g., harmonics of a transmitter signal, or the local-oscillator signal of the receiver.

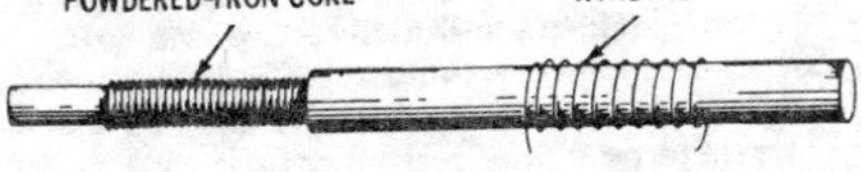

Antenna core.

antenna cores — Ferrite cores of various cross-sections for use in radio antennas.

antenna coupler — 1. A radio-frequency transformer used to connect an antenna to a transmission line or to connect a transmission line to a radio receiver. 2. A radio-frequency transformer, link circuit, or tuned line used to transfer radio-frequency energy from the final plate-tank circuit of a transmitter to the transmission line feeding the antenna.

antenna cross talk—A measure of undesired power transfer through space from one antenna to another. Usually expressed in decibels, the ratio of power received by one antenna to the power transmitted by the other.

antenna current—The radio-frequency current that flows in an antenna.

antenna detector—A device consisting of an antenna and electronic equipment to warn aircraft crew members of their being observed by radar sets. (Usually located in the nose or tail of the aircraft and illuminates a light on one or more panels when radar signals are detected.)

antenna-directivity diagram—A curve representing, in polar or Cartesian coordinates, a quantity proportional to the gain of an antenna in the various directions in a particular plane or cone.

antenna disconnect switch—A safety switch or interlock plug used to remove driving power from the antenna to prevent rotation while work is being performed.

antenna duplexer — A circuit that permits two transmitters to transmit simultaneously from the same antenna without interaction between them.

antenna effect—1. Cause of error in a loop antenna due to the capacitance to ground. 2. In a navigational system, any undesirable output signal that results when a directional antenna acts as a nondirectional antenna.

antenna effective area—In any specified direction, the square of the wavelength multiplied by the power gain (or directive gain) in that direction, and divided by 4π. (When power gain is used, the effective area is that for power reception; when directive gain is used, the effective area is that for directivity.)

antenna efficiency—The relative ability of an antenna to convert rf energy from a transmitter into electromagnetic waves. If the gain rating of a directional antenna is 10 dB, for example, it is often assumed that the effective radiated power will be 10 times greater than the rf power fed to it. However, if the antenna efficiency is say 50%, a loss of 3 dB, the true gain will be only 7 dB ($10 - 3 = 7$ dB).

antenna factor—The value in dB that must be added to a two-terminal voltmeter reading to obtain the actual induced antenna open-circuit voltage or the electric-field strength.

antenna field—1. The region defined by a group of antennas. 2. A group of antennas placed in a geometric configuration which is specific for a particular trajectory measuring system. 3. The effective free-space energy distribution produced by an antenna or group of antennas.

antennafier—An integrated low-profile antenna and amplifier for use with compact, portable communications systems.

antenna gain—1. The effectiveness of a directional antenna in a particular direction, compared against a standard (usually an isotropic antenna). The ratio of standard antenna power to the directional antenna power that will produce the same field strength in the desired direction. 2. The increase in signal level at the antenna terminals with reference to the level at the terminals of a half-wave dipole antenna, expressed in dB.

antenna ground system—That portion of an antenna closely associated with the earth and including an extensive conducting surface which may be the earth itself.

antenna height — The average antenna height above the terrain from two to ten miles from the antenna. In general, the antenna height will be different in each direction from the antenna. The average of these various heights is considered the antenna height above average terrain.

antenna induced microvolts — The voltage that exists across the open-circuited antenna terminals, as calculated from a measurement.

antenna lens — An arrangement of metal vanes or dielectric material used to focus a microwave beam in a manner similar to an optical lens.

antenna lobe—*See* Lobe.

antenna matching—Selection of components to make the impedance of an antenna equal to the characteristic impedance of its transmission line.

antennamitter — An integrated low-profile antenna and oscillator for use with compact, portable communications systems.

antenna pair—Two antennas located on a base line of accurately surveyed length. The signals received by these antennas are used to determine quantities related to a target position.

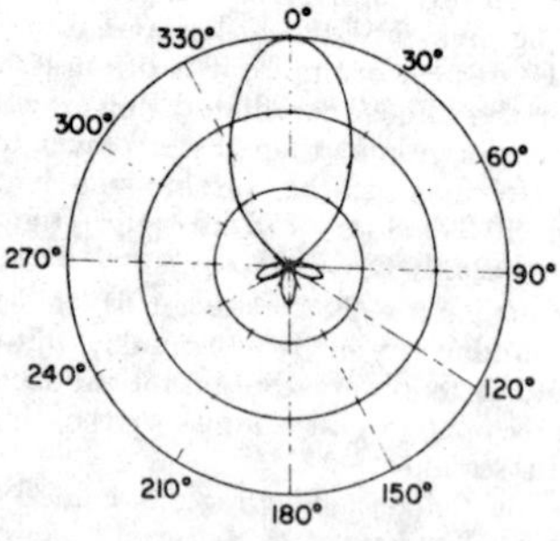

Antenna pattern.

antenna pattern—Also called antenna polar diagram. A plot of angle *versus* free-space field intensity at a fixed distance in the horizontal plane passing through the center of the antenna.

antenna-pattern measuring equipment—Devices used to measure the relative field strength or intensity existing at any point or points in the space immediately surrounding an antenna.

antenna pedestal—A structure which supports an antenna assembly (motors, gears, synchros, rotating joints, etc.).

antenna polar diagram—*See* Antenna Pattern.

antenna power—The square of the antenna current of a transmitter, multiplied by the antenna resistance at the point where the current is measured.

antenna power gain—The power gain of an antenna in a given direction is four times the ratio of the radiation intensity in that direction to the total power delivered to the antenna. (The term is also applied to receiving antennas.)

antenna preamplifier—A low-noise rf amplifier, usually mast-mounted near the terminals of the receiving antennas, used to compensate for transmission-line loss and thereby improve the overall noise figure.

antenna reflector—In a directional-antenna array, an element which modifies the field pattern in order to reduce the field intensity behind the array and increase it in front. In a receiving antenna, the reflector reduces interference from stations behind the antenna.

antenna relay—A relay used in radio stations to automatically switch the antenna to the receiver or transmitter and thus protect the receiver circuits from the rf power of the transmitter.

antenna resistance—The total resistance of a transmitting antenna system at the operating frequency. The power supplied to the entire antenna circuit, divided by the square of the effective antenna current referred to the feed point. Antenna resistance is made up of such components as radiation resistance, ground resistance, radio-frequency resistance of conductors in the antenna circuit, and equivalent resistance due to corona, eddy currents, insulator leakage, and dielectric power loss.

antenna resonant frequency—The frequency (or frequencies) at which an antenna appears to be a pure resistance.

antenna stabilization—A system for holding a radar beam steady despite the roll and pitch of a ship or airplane.

antenna switch—Switch used for connecting an antenna to or disconnecting it from a circuit.

antenna system—An assembly consisting of the antenna and the necessary electrical and mechanical devices for insulating, supporting, and/or rotating it.

antenna terminals — On an antenna, the points to which the lead-in (transmission line) is attached.

antenna tilt error—The angular difference between the antenna tilt angle shown on the mechanical indicator, and the electrical center of the radar beam.

antennaverter — A receiving antenna and converter combined in a single unit that feeds directly into the receiver i-f amplifier.

antenna wire—A wire, usually of high tensile strength such as copperweld, bronze, etc., with or without insulation, used as an antenna for radio and electronic equipment.

antiaircraft missile — A guided missile launched from the surface against an airborne target.

anticapacitance switch — A switch with widely separated legs, designed to keep capacitance at a minimum in the circuits being switched.

anticathode—Also called target. The target of an X-ray tube on which the stream of electrons from the cathode is focused and from which the X rays are radiated.

anticlutter circuit—In a radar receiver, an auxiliary circuit which reduces undesired reflection, to permit the detection of targets which otherwise would be obscured by such reflections.

anticlutter gain control — A device which automatically and gradually increases the gain of a radar receiver from low to maximum within a specified period after each transmitter pulse. In this way, short-range echoes producing clutter are amplified less than long-range echoes.

anticoincidence—A nonsimultaneous occurrence of two or more events (usually, ionizing events).

anticoincidence circuit—1. A counter circuit that produces an output pulse when either of two input circuits receives a pulse, but not when the two inputs receive pulses simultaneously. 2. A circuit that provides an output only when all inputs are absent; a NAND circuit.

anticollision radar—A radar system used in an aircraft or ship to warn of possible collision.

antiferroelectricity—The property of a class of crystals which also undergo phase transitions from a higher to a lower symmetry. They differ from the ferroelectrics in having no electric dipole moment.

antiferroelectric materials—Those materials in which spontaneous electric polarization occurs in lines of ions; adjacent lines are polarized in an antiparallel arrangement.

antiferromagnetic materials — Those materials in which spontaneous magnetic polarization occurs in equivalent sublattices; the polarization in one sublattice is aligned antiparallel to the other.

antiferromagnetic resonance — The absorption of energy from an oscillating electromagnetic field by a system of processing spins located on two sublattices, with the spins on one sublattice going in one direction and the spins on the other sublattice in the opposite direction.

antihunt—A stabilizing signal or equalizing circuit used in a closed-loop feedback system of a servomechanism to prevent the system from hunting, or oscillating. Special types of antihunt circuits are the anticipator, derivative, velocity feedback, and damper.

antihunt circuit—A circuit used to prevent excessive correction in a control system.

antihunt device—A device used in positioning systems to prevent hunting or oscillation of the load around an ordered position. The device may be mechanical or electrical. It usually involves some form of feedback.

antijamming—1. Art of minimizing the effect of enemy electronic countermeasures to permit echoes from targets detected by radar to be visible on the indicator. 2. Controls or circuit features incorporated to minimize jamming.

antijamming radar data processing—Use of data from one or more radar sources to determine target range in the presence of jamming.

antilogarithm—The number from which a given logarithm is derived. For example, the logarithm of 4261 is 3.6295. Therefore the antilogarithm of 3.6295 is 4261.

antimagnetic—Made of alloys that will not remain in a magnetized state.

antimicrophonic — Specifically designed to prevent microphonics. Possessing the characteristic of not introducing undesirable noise or howling into a system.

antimissile missile — A missile which is launched to intercept and destroy another missile in flight.

antinodes—Also called loops. The points of maximum displacement in a series of standing waves. Two similar and equal wave trains traveling at the same velocity in opposite directions along a straight line result in alternate antinodes and nodes along the line. Antinodes are separated from their adjacent nodes by half the wavelength of the wave motion.

antinoise carrier-operated device—A device commonly used to mute the audio output of a receiver during standby or no carrier periods. Usually the automatic volume control voltage is used to control a squelch tube which, in turn, controls the bias applied to the first audio tube so that it is permitted to operate only when a carrier is present at the receiver input. Thus, the receiver output is heard when a signal is received, and is muted when no signal is present.

antinoise microphone—A microphone which discriminates against acoustic noise. A lip or throat microphone is an example.

antiphase—Two identical signals disposed in 180° phase opposition. When superimposed, they tend to cancel each other because their waveform patterns are of equal magnitude but opposite polarity.

antiproton—An elementary atomic particle

which has the same mass as a proton but is negatively charged.

antirad—A material that inhibits damage caused by radiation.

antiresonance — A type of resonance in which a system offers maximum impedance at its resonant frequency.

antiresonant circuit—A parallel-resonant circuit offering maximum impedance to the series passage of the resonant frequency.

antiresonant frequency—1. The frequency at which the impedance of a system is very high. 2. Of a crystal unit, the frequency for a particular mode of vibration at which, neglecting dissipation, the effective impedance of the crystal unit is infinite.

antisidetone—In a telephone circuit, special circuits and equipment which are so arranged that only a negligible amount of the power generated in the transmitter reaches the associated receiver.

antisidetone circuit—A telephone circuit which prevents sound, introduced in the local transmitter, from being reproduced in the local receiver. (Reduces sidetones.)

antisidetone induction coil—An induction coil designed for use in an antisidetone telephone set.

antisidetone telephone set—A telephone set with an antisidetone circuit.

antiskating bias—A bias force applied to a pivoted pickup arm to counteract the inward force (toward the center of the record) resulting from the drag of the stylus in the groove, and the offset angle of the head.

antiskating device—A mechanism found on modern pickups which provides a small outward force on a pickup arm. This counteracts the arm's tendency to move toward the turntable center (inward) due to offset geometry, and reduces stylus/groove friction.

antistatic agents—Methods employed to minimize static electricity in plastic materials. Such agents are of two basic types: Metallic devices which come into contact with the plastics and conduct the static to earth give complete neutralization initially, but because it is not modified, the surface of the material can become prone to further static accumulation during subsequent handling. Chemical additives, which are mixed with the compound during processing, give a reasonable degree of protection to the finished products.

antistatic cleaner—Substance used on phonograph records which helps to prevent the buildup of a static charge which attracts dust.

antistickoff voltage—A small voltage, usually applied to the rotor winding of the coarse synchro control transformer in a two-speed system. The antistickoff voltage acts to eliminate the possibility of ambiguous behavior in the system.

antitransmit-receive box—A second transmit-receive switch used in a radar antenna system to minimize absorption of the echo signal in the transmitter circuit during the interval between transmitted pulses.

antitransmit-receive switch — Abbreviated atr switch. An automatic device employed in a radar system to prevent received energy from being absorbed in the transmitter.

antitransmit-receive tube—A gas-filled tube used as a switching element in a duplexer.

antivoice-operated transmission—A method of radiocommunication in which a voice-activated circuit prevents the operation of the transmitter during reception of messages on an associated receiver.

A-1 or **A.1**—The atomic time scale maintained by the U.S. Naval Observatory; presently it is based on weighted averages of frequencies from cesium-beam devices operated at a number of laboratories.

aperiodic—Having no fixed resonant frequency or repetitive characteristics or no tendency to vibrate. A circuit that will not resonate within its tuning range is often called aperiodic.

aperiodic antenna—An antenna designed to have a constant impedance over a wide frequency range (for example, a terminated rhombic antenna).

aperiodic damping—Also called overdamping. The condition of a system when the amount of damping is so large that, when the system is subjected to a single disturbance, either constant or instantaneous, the system comes to a position of rest without passing through that position. While an aperiodically damped system is not strictly an oscillating system, it has such properties that it would become an oscillating system if the damping were sufficiently reduced.

aperiodic function—A function having no repetitive characteristics.

aperture—1. In a unidirectional antenna, that portion of the plane surface which is perpendicular to the direction of maximum radiation and through which the major part of the radiation passes. 2. In an opaque disc, the hole or window placed on either side of a lens to control the amount of light passing through. 3. Also called aperture time. The amount of uncertainty about the exact time when the encoder input was at the value represented by a given output code. In general, the aperture is equal to the conversion time; it may be reduced by the use of sample-and-hold circuits.

aperture antenna—A type of antenna the beam width of which is determined by the dimensions of a horn, lens, or reflector.

aperture compensation—Reduction of aperture distortion by boosting the high-frequency response of a television-camera video amplifier.

aperture distortion—In a television signal, the distortion due to the finite dimension of the camera-tube scanning beam. The beam covers several mosaic globules simultaneously, resulting in a loss of picture detail.

aperture illumination—The field distribution in amplitude and phase through the aperture.

vibrations are in the range of 0.1 to 20 Hz.

APL—Abbreviation for average picture level. The average luminance level of the part of a television line between blanking pulses.

A power supply—A power supply used as a source of heating current for the cathode or filament of a vacuum tube.

apparent bearing — The direction from which the signal arrives with respect to some reference direction.

apparent power—The product of voltage and current of a single-phase circuit in which the two reach their peaks at different times.

apparent power loss—For voltage-measuring instruments, the product of nominal end-scale voltage and the resulting cur-

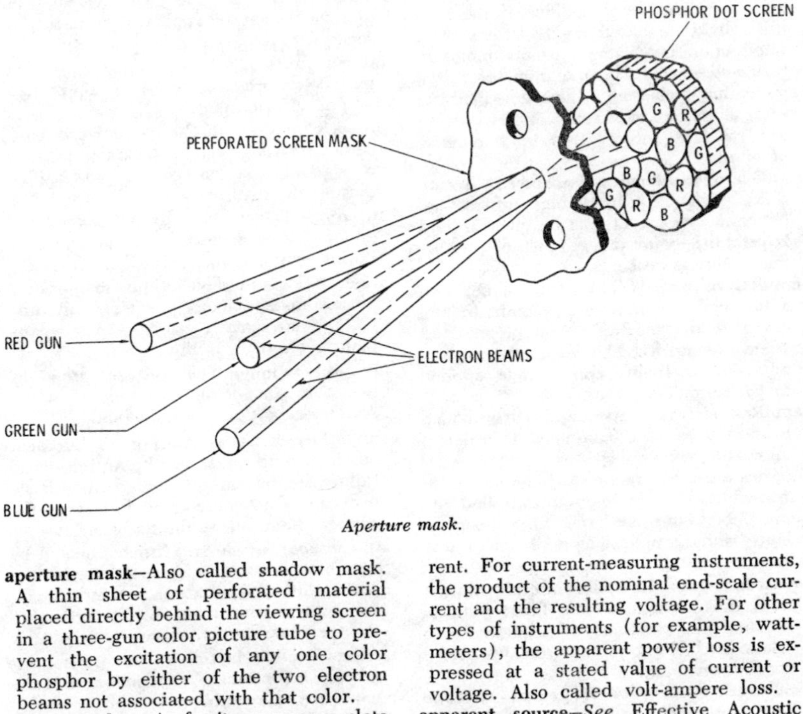

Aperture mask.

aperture mask—Also called shadow mask. A thin sheet of perforated material placed directly behind the viewing screen in a three-gun color picture tube to prevent the excitation of any one color phosphor by either of the two electron beams not associated with that color.

aperture plate—A ferrite memory plate containing a large number of uniformly spaced holes arranged in parallel rows and interconnected by plated conductors to provide a magnetic memory plane.

aperture time—*See* Aperture, 3.

apexcardiography—The recording and interpretation of movement of the chest directly over the apex of the heart. The recording, called an apexcardiogram, provides diagnostically important information about the functioning of the chambers and valves in the heart. The recorded

rent. For current-measuring instruments, the product of the nominal end-scale current and the resulting voltage. For other types of instruments (for example, wattmeters), the apparent power loss is expressed at a stated value of current or voltage. Also called volt-ampere loss.

apparent source—*See* Effective Acoustic Center.

Applegate diagram—A graphical representation of electron bunching in a velocity-modulated tube, showing their positions along the drift space. This bunching is plotted on the vertical coordinate, against time along the horizontal axis.

Appleton layer—In the ionosphere, a region of highly ionized air capable of reflecting or refracting radio waves back to earth. It is made up of the F_1 and F_2 layers.

apple tube—A color-television picture tube

in which the three colors of phosphors are laid in fine vertical strips along the screen. The intensity of the electron beam is modulated as its sweeps over them so that each color is produced with appropriate brightness.

appliance—Any electrical equipment used in the home and capable of being operated by a nontechnical person. Included are units that perform some task that could be accomplished by other, more difficult means, but usually not those used for entertainment (radios, tv's, hi-fi sets, etc.).

application—System or problem to which a computer is applied. An application may be of the computational type, in which arithmetic computations predominate, or of the data-processing type, in which data-handling operations predominate.

application factor—A modifier of the failure rate. It is based on deviations from rated operating stress (usually temperature and one electrical parameter).

application schematic diagram—Pictorial representation using symbols and lines to illustrate the interrelation of a number of circuits.

applicator (applicator electrodes)—Appropriately shaped conducting surfaces between which an alternating electric field is established for the purpose of producing dielectric heating.

applied voltage—1. The potential between a terminal and a reference point in any circuit or device. 2. The voltage obtained when measuring between two given point in a circuit with voltage applied to the complete circuit.

applique circuit—A special circuit provided to modify existing equipment in order to allow for some special usage.

approach-control radar—Any radar set or system used in a ground-controlled approach system—e.g., an airport-surveillance radar, precision approach radar, etc.

approach path—In radio aircraft navigation, that portion of the flight path in the immediate vicinity of a landing area where such flight path terminates at the touchdown point.

approved circuit—*See* Protected Wireline Distribution System.

AQL—Abbreviation for acceptable quality level.

aquadag—Trademark of Acheson Industries, Inc. A conductive graphite coating on the inner side walls of some cathode-ray tubes. It serves as an electrostatic shield or as a postdeflection and an accelerating anode. Also applied to outer walls and grounded; here it serves, with the inner coating, as a capacitor to filter the applied high voltage.

arbitrary function fitter—A circuit having an output voltage or current that is a presettable, adjustable, usually nonlinear function of the input voltage(s) or current(s) fed to it.

arc—1. A luminous discharge of electricity through a gas. Characterized by a change in space potential in the immediate vicinity of the cathode; this change is approximately equal to the ionization potential of the gas. 2. A prolonged electrical discharge, or series of prolonged discharges, between two electrodes. (Both produce a bright-colored flame, as contrasted to a dim corona-glow discharge).

arcback—Also called backfire. Failure of the rectifying action in a tube, resulting from the flow of a principal electron stream in the reverse direction due to the formation of a cathode spot on the anode. This action limits the peak inverse voltage which may be applied to a particular rectifier tube.

arc converter—A form of oscillator utilizing an electric arc to generate an alternating or pulsating current.

arc discharge—A discharge between electrodes in gas or vapor. Characterized by a relatively low voltage drop and a high current density.

arc drop—The voltage drop between the anode and cathode of a gas rectifier tube during conduction.

arc-drop loss—In a gas tube, the product of the instantaneous values of arc-drop voltage and current averaged over a complete cycle of operation.

arc-drop voltage—The voltage drop between the anode and cathode of a gas rectifier tube during conduction.

arc failure—1. A flashover in the air near an insulation surface. 2. An electrical failure in the surface heated by a flashover arc. 3. An electrical failure in the surface damaged by the flashover arc.

arc furnace—An electric furnace heated by arcs between two or more electrodes.

architecture—Organizational structure of a computing system, mainly referring to the CPU or microprocessor.

arcing—The production of an arc—for example, at the brushes of a motor or at the contact of a switch.

arcing contacts—Special contacts on which the arc is drawn after the main contacts

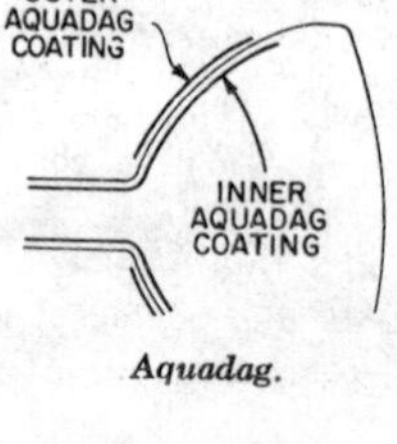

Aquadag.

of a switch or circuit breaker have opened.

arcing time—1. The interval between the parting, in a switch or circuit breaker, of the arcing contacts and the extension of the arc. 2. The time elapsing, in a fuse, from the severance of the fuse link to the final interruption of the circuit under the specified condition.

arc lamp—Source of brilliant artificial light obtained by an electric arc passing between two carbon rods. The arc is struck by bringing the two rods together and then rapidly separating them. As the arc burns the carbon rods are vaporized away. A mechanism is employed to keep the space between the two rods constant. This type of lamp is used extensively in motion picture projectors and spotlights.

arc oscillator—A negative-resistance oscillator comprising a sustained dc arc and a resonant circuit.

arcover voltage—Under specified conditions, the minimum voltage required to create an arc between electrodes separated by a gas or liquid insulation.

arc percussive welding—A type of welding in which the materials to be welded are separated by a gap, across which an arc is struck; the arc melts the surfaces of the materials, and the materials are simultaneously brought together. *See also* Pulse Arc Welding.

arc resistance—The length of time that a material can resist the formation of a conductive path by an arc adjacent to the surface of the material. Also called tracking resistance.

arc suppressor—*See* Spark Suppressor.

arc-through—In a gas tube, a loss of control with the result that a principal electron stream flows in the normal direction during what should be a nonconducting period.

area code—A three-digit number identifying one of the geographic areas of the USA, Canada, and Mexico to permit direct distance dialing on the telephone system. *See also* Direct Distance Dialing and Numbering Plan.

area redistribution—A method of measuring the duration of irregularly shaped pulses. A rectangle is drawn having the same peak amplitude and the same area as the original pulse under consideration. Because the same time units are used in measuring the original and the new pulse, the width of the rectangle is considered the duration of the pulse.

A register—The accumulator for all arithmetical operations in a computer. Also called A accumulator.

argon—An inert gas used in discharge tubes and some electric lamps. It gives off a purple glow when ionized; its symbol, Ar.

argon glow lamp—A glow lamp containing argon gas which produces a pale blue violet light.

argument—A variable upon which the value of a function depends. The arguments of a function are listed in parentheses after the function name. The computations specified by the function definition are made with the variables specified as arguments.

arithmetic check—A check of a computation making use of the arithmetical properties of the computation.

arithmetic element—Synonym for arithmetic unit.

arithmetic mean—Usually, the same as average. It is obtained by first adding quantities together and then dividing by the number of quantities involved. It also means a figure midway between two extremes and is found by adding the minimum and maximum together and dividing by two.

arithmetic operation—In an electronic computer, the operations in which numerical quantities form the elements of the calculation, including the fundamental operations of arithmetic (addition, subtraction, multiplication, comparison, and division).

arithmetic organ—*See* Arithmetic Unit.

arithmetic shift—In a digital computer, the multiplication or division of a quantity by a power of the base used in the notation.

arithmetic symmetry—Filter response showing mirror-image symmetry about the center frequency when frequency is displayed on an arithmetic scale. Constant envelope delay in bandpass filters usually is accompanied by arithmetic symmetry in the phase and amplitude responses, and generally requires a computer design. *See also* Geometric Symmetry.

arithmetic unit—Also called arithmetic element or arithmetic organ. In an automatic digital computer, that portion in which arithmetical and logical operations are performed on elements of information.

armature—The moving element in an electromechanical device such as the rotating part of a generator or motor, the movable part of a relay, or the spring-mounted, iron portion of a bell or buzzer.

armature contacts — 1. Contacts mounted directly on the armature. 2. Sometimes used for movable contacts.

armature control of speed—The varying of voltage applied to the armature of a shunt-wound motor to control the motor's speed over the basic speed range.

armature core—An assembly of laminations forming the magnetic circuit of an armature.

armature gap—The space between the armature and pole face.

armature hesitation—A delay or momentary reversal of the motion of the armature.

armature-hesitation contact chatter—Chatter caused by delay or momentary reversal in direction of the armature motion of a relay during either the operate or the release stroke.

armature-impact contact chatter – Chatter caused by impact of the armature of a relay on the pole piece in operation, or on the backstop in release.

armature overtravel—That portion of the available stroke occurring after the contacts of a relay have touched.

armature reaction—In an armature, the reaction of the magnetic field produced by the current on the magnetic lines of force produced by the field coil of an electric motor or generator.

armature rebound—Return motion of a relay armature after striking the backstop.

armature-rebound contact chatter—Chatter caused by the partial return of the armature of a relay to its operated position as a result of rebound from the backstop in release.

armature relay—A relay operated by an electromagnet which, when energized, causes an armature to be attracted to a fixed pole or poles.

armature slot—In the core of an armature, a slot or groove into which the coils or windings are placed.

armature stud—In a relay, an insulating member that transmits the motion of the armature to an adjacent contact member.

armature travel—The distance traveled during operation by a specified point on the armature of a relay.

armature voltage control—A means of controlling the speed of a motor by changing the voltage applied to its armature windings.

armature wire – Stranded annealed copper wire, straight lay, soft loose white cotton braid. It is used for low-voltage, high-current rotor winding motors and generators. Straight lay permits forming in armature slots, and compressibility.

armchair copy – Amateur term for clear, static-free signals.

armed sweep—*See* Single Sweep.

arming the oscilloscope sweep—Closing a switch which enables the oscilloscope to trigger on the next pulse.

armor—A metallic cover placed over the insulation of wire or cable to protect it from abrasion or crushing.

armor clamp—A fitting for gripping the armor of a cable at the point where the armor terminates or where the cable enters a junction box.

armored cable – Two or more insulated wires collectively provided with a metallic covering, primarily to protect the insulated wires from damage.

Armstrong frequency-modulation system – A phase-shift modulation system originally proposed by E. H. Armstrong.

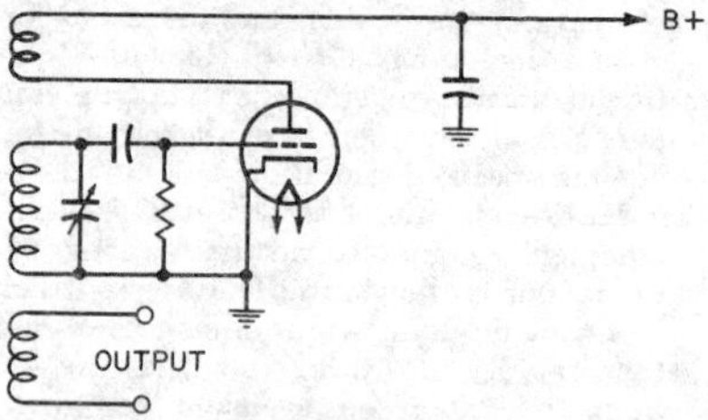

Armstrong oscillator.

Armstrong oscillator – An inductive feedback oscillator that consists of a tuned grid circuit and an untuned tickler coil in the plate circuit. Feedback is controlled by varying the coupling between the tickler and the grid circuit.

array—1. In an antenna, a group of elements arranged to provide the desired directional characteristics. These elements may be antennas, reflectors, directors, etc. 2. A series of items, not necessarily arranged in a meaningful pattern. 3. *See* Random-Access Memory.

array antenna – An antenna comprising a number of radiating elements, generally similar, arranged and excited to obtain directional effects.

array device—A group of many similar, basic, complex, or integrated devices without separate enclosures. Each has at least one of its electrodes connected to a common conductor, or all are connected in series.

arrester—1. Also called a lightning arrester. A protective device used to provide a bypass path directly to ground for lightning discharges that strike an antenna or other conductor. 2. A power-line device capable of reducing the voltage of a surge applied to its terminals, interrupting current if present, and restoring itself to original operating conditions.

ARRL – Abbreviation for American Radio Relay League.

arrowhead—A linearly polarized, frequency-independent, log-periodic antenna.

ARSR—Abbreviation for Air Route Surveillance Radar.

articulation—(Sometimes called intelligibility.) In a communications system, the percentage of speech units understood by a listener. The word "articulation" is customarily used when the contextual relationships among the units of speech material are thought to play an unimportant role; the word "intelligibility," when the context is thought to play an important role in determining the listener's perception.

articulation equivalent—The articulation of speech reproduced over a complete telephone connection, expressed numerically in terms of the trunk loss of a working reference system which is adjusted to give equal articulation.

artificial antenna—Also called dummy antenna. A device which simulates a real antenna in its essential impedance characteristics and has the necessary power-handling capabilities, but which does not radiate or receive radio waves. Used mainly for testing and adjusting transmitters.

artificial ear—A microphone-equipped device for measuring the sound pressures developed by an earphone. To the earphone it presents an acoustic impedance equivalent to the impedance presented by the human ear.

artificial echo—1. Received reflections of a transmitted pulse from an artificial target, such as an echo box, corner reflector, or other metallic reflecting surface. 2. A delayed signal from a pulsed radio-frequency signal generator.

artificial horizon—A gyroscopically operated instrument that shows, within limited degrees, the pitching and banking of an aircraft with respect to the horizon. Lines or marks on the face of the instrument represent the aircraft and the horizon. The relative positions of the two are then easily discernible.

artificial intelligence – 1. The design of computer and other data-processing machinery to perform increasingly higher-level cybernetic functions. 2. The capability of a device to perform functions that are normally associated with human intelligence, such as reasoning, learning, and self-improvement. Related to machine learning.

artificial ionization—Introduction of an artificial reflecting or scattering layer into the atmosphere to permit beyond-the-horizon communications.

artificial language—In computer terminology, a language designed for ease of communication in a particular area of activity, but one that is not yet natural to that area (as contrasted with a natural language evolved through long usage).

artificial line—A lumped-constant network designed to simulate some or all the characteristics of a transmission line over a desired frequency range.

artificial line duct—A balancing network simulating the impedance of the real line and distant terminal apparatus. It is employed in a duplex circuit to make the receiving device unresponsive to outgoing signal currents.

artificial load—Also called dummy load. A dissipative but essentially nonradiating device having the impedance characteristics of an antenna, transmission line, or other practical utilization circuit.

artificial radioactivity – Radioactivity induced in stable elements under controlled conditions by bombarding them with neutrons or high-energy, charged particles. Artificially radioactive elements emit beta and/or gamma rays.

artificial voice—A small speaker mounted in a specially shaped baffle that is proportioned to simulate the acoustical constants of the human head. It is used for calibrating and testing close-talking microphones.

artos stripper – A machine which, when properly adjusted, will automatically measure to a predetermined length, cut, strip, count, and tie wire in bundles.

artwork—1. A topological pattern of an integrated circuit, made with accurate dimensions so that it can be used in mask making. Generally, it is a large multiple of the final mask size, and final reduction is accomplished through the use of a step-and-repeat camera. 2. Detailed, original drawing (often developed with the aid of a computer) showing layout of an integrated circuit. 3. The images formed by drawing, scribing, or by cutting and stripping on a film or glass support which are reduced, contact-printed, or stepped and repeated to make a photomask or intermediate.

ASA—Abbreviation for American Standards Association. *See* American National Standards Institute.

ASA code—A code that was recommended by the American Standards Association for industry-wide use in the transmission of information. Now ANSI code.

asbestos—A nonflammable material generally used for heat insulation, such as in a line-cord resistor.

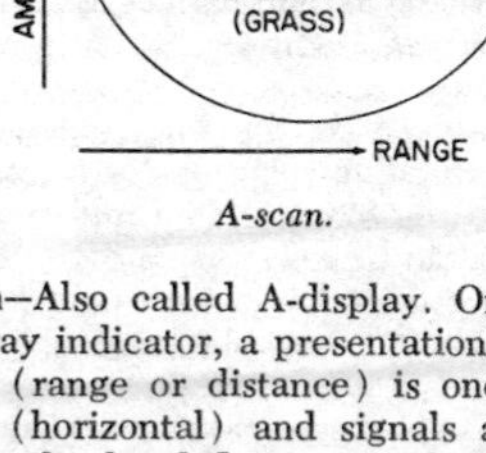

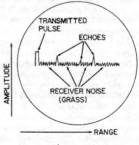

A-scan.

A-scan—Also called A-display. On a cathode-ray indicator, a presentation in which time (range or distance) is one co-ordinate (horizontal) and signals appear as perpendicular deflections to the time scale (vertical).

ASCII – Abbreviation for American Standard Code for Information Interchange (pronounced "ask-ee"). A standard code used extensively in data transmission, in which 128 numerals, letters, symbols, and special control codes are each represented by a 7-bit binary number. For example numeral 5 is represented by 011 0101, letter K by 100 1011, percent symbols (%) by 010 0101 and start of text (STX) control code by 000 0010.

A-scope–An oscilloscope that uses an A-scan to present the range of a target as the distance along a horizontal line from the transmitted pulse pip to the target, or echo pip. Signals appear as vertical excursions of the horizontal line, or trace.

ASI–An abbreviation for Standards published by the United States of America Standards Institute. Now American National Standards Institute.

ask–*See* Amplitude Shift Keying.

aspect ratio–1. Ratio of frame width to frame height. In the United States the television standard is 4/3. 2. The ratio of an object's height to its width. In graphics this ratio usually pertains to the face of a rectangular crt or to the characters or symbols drawn by the character generator.

asperities–Local microscopic points on an electrode surface at which there is considerable field enhancement. They lead to a dependence of electric strength on electrode area (area effect).

aspheric–Not spherical; an optical element having one or more surfaces that are not spherical.

asr–Abbreviation for automatic send and receive. A terminal equipped with recording devices, usually a paper tape reader and punch, which is capable of answering a call, recording a message, or sending data loaded in its tape reader without the need for an operator in attendance at the time of the call. Also used to specify terminals that have paper-type equipment used by the operator.

ASRA–Abbreviation for automatic stereophonic recording amplifier. An instrument developed by Columbia Broadcasting System for stereo recording. Compression of the vertical component of the stereo recording signal is automatically decreased or increased as required by the recording conditions.

assemble–1. To collect, interpret, and coordinate the data required for a computer program, translate the data into computer language, and project it into the master routine for the computer to follow. 2. To translate from a symbolic program to a binary program by substituting binary operation codes for symbolic operation codes and replacing symbolic addresses with absolute or relocatable addresses.

assembler–1. A program that prepares a program in machine language from a program in symbolic language by substituting absolute operation codes for symbolic operation codes and absolute or relocatable addresses for symbolic addresses. 2. A unit that converts the assembly language of a computer program into the machine language of the computer.

assembler program–Software usually supplied by the computer manufacturer to convert an assembly-language application program into machine language.

assembly – 1. A complete operating unit, such as a radio receiver, made up of subassemblies such as an amplifier and various components. 2. Process in which instructions written in symbolic form by the programmer are changed to machine language by the computer.

assembly language – 1. A computer language that has one-to-one correspondence with an assembly program. The assembly program directs a computer to operate on a program in symbolic language to produce a program in machine language. *See also* Higher Order Language; Machine Language, 3; and Source Language. 2. Grouped alphabet characters, called mnemonics, that replace the numeric instructions of machine language. These mnemonics are easier to remember than machine instructions and hence easier to develop into a working program.

assembly-language programming–*See* Symbolic-Language Programming.

assembly program–A program that enables a computer to assemble mnemonic language into machine language; for example, a FORTRAN assembly program. Also called assembly routine.

assembly routine–*See* Assembly Program.

assignable cause – A definitely identified factor contributing to a quality variation.

assigned frequency–The center of the frequency band assigned to a station.

assigned frequency band – The frequency band, the center of which coincides with the frequency assigned to the station, and the width of which equals the necessary bandwidth plus twice the absolute value of the frequency tolerance.

associative memory–A memory where the storage locations are identified by their contents rather than by their addresses. Enables faster interrogation to retrieve a particular data element.

associative storage – Computer storage in which locations may be identified by specification of part or all of their contents. Also called parallel-search storage or content-addressed storage.

astable–1. Pertaining to a device that has two temporary states; the device alternates between these states with a period and duty cycle determined by circuit time constants. *See also* Bistable. 2. Refers to a device which has two temporary states. The device oscillates between the two states with a period and duty cycle predetermined by time constants.

astable circuit – A circuit which continuously alternates between its two unstable states at a frequency determined by the circuit constants. It can be readily synchronized by applying a repetitive input signal of slightly higher frequency. A blocking oscillator is an example of an astable circuit.

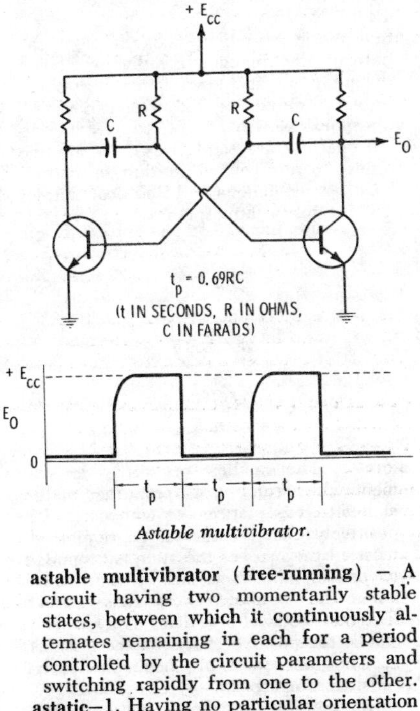

Astable multivibrator.

astable multivibrator (free-running) – A circuit having two momentarily stable states, between which it continuously alternates remaining in each for a period controlled by the circuit parameters and switching rapidly from one to the other.

astatic–1. Having no particular orientation or directional characteristics; such as a vertical antenna. 2. Being in neutral equilibrium; having no tendency toward any change of position.

astatic galvanometer – A sensitive galvanometer used for detecting small currents. Consists of two small magnetized needles of equal size and strength arranged in parallel and with their north and south poles adjacent, suspended inside the galvanometer coil. Since the resultant magnetic moment is zero, the earth's magnetic field does not affect the system.

A station–One of a pair of transmitting stations in a loran system. The A-station signal always occurs less than half a repetition period after the immediately preceding signal of the other station of the pair and more than half a repetition period before the next succeeding signal of the other station.

astigmatism–A type of spherical aberration in which the rays from a single point of an object do not converge on the image, thereby causing a blurred image. Astigmatism in an electron-beam tube is a focus defect in which electrons in different axial planes come to focus at different points.

astrionics–Electronics as involved with astronautics.

astrocompass–An instrument for determining direction relative to the stars. It is unaffected by the errors to which magnetic or gyrocompasses are subject.

astrodome–A rigid hemispherical structure used to cover large tracking instruments to protect them from the elements. It is usually constructed so that the dome rotates with the instrument.

astronautics–The science and art of operating space vehicles.

astrotracker–A device for tracking stars.

A-supply – The A-battery, transformer filament winding, or other voltage source that supplies power for heating the filaments of vacuum tubes.

asymmetrical cell–A cell, such as a photoelectric cell, in which the impedance to the flow of current is greater in one direction than in the other direction.

asymmetrical distortion – Distortion affecting a two-condition or binary modulation or restitution, in which all the significant intervals corresponding to one of the two significant conditions have longer or shorter durations than the corresponding theoretical durations of the excitation. If this particular requirement is not met, distortion is present.

asymmetric-sideband transmission – *See* Vestigial-Sideband Transmission.

asymmetry control–In pH meters, an adjustment sometimes provided to compensate for differences in the electrodes.

asymptote–A line which comes nearer and nearer a given curve but never touches it.

asymptotic breakdown voltage – A voltage that will break down insulation if applied over a long period of time.

asynchronous–Lacking a regular time relationship; hence, as applied to computer program execution, unexpected or unpredictable with respect to the instruction sequence.

asynchronous computer–An automatic digital computer in which an operation is started by a signal denoting that the previous operation has been completed.

asynchronous device–A device in which the speed of operation is not related to any frequency in the system to which it is connected.

asynchronous input/output–The ability to

accept input data while simultaneously delivering output data.

asynchronous inputs – The terminals that affect the output state of a flip-flop independently of the clock terminals. Called set, preset, reset, or clear; sometimes referred to as dc inputs.

asynchronous logic—Logic networks the operational speed of which depends only on the signal propagation through the network, rather than on clock pulses as in synchronous logic.

asynchronous machine – Any machine in which its speed of operation is not proportionate to the frequency of the system to which the machine is connected.

asynchronous operation – 1. Generally, an operation that is started by a signal at the completion of a previous operation. It proceeds at the maximum speed of the circuits until it is finished and then generates its own completion signal. 2. A mode in which entry of data into a flip-flop does not require a gating or clock pulse. 3. Operation of a switching network by a free-running signal that triggers successive instructions. The completion of one instruction triggers the next.

asynchronous transmission – Transmission in which each character of the information is synchronized individually (usually by the use of start and stop elements).

AT-cut crystal—A quartz-crystal slab cut at a 35° angle wtih respect to the optical, or Z-axis, of the crystal. It has practically a zero temperature coefficient and is used at frequencies of about 0.5 to 10 megahertz.

atmosphere—1. The body of air surrounding the earth. 2. A unit of pressure defined as the pressure of 760 mm of mercury at 0° C. Approximately 14.7 pounds per square inch.

atmospheric absorption—The energy lost in the transmission of radio waves due to dissipation in the atmosphere.

atmospheric absorption noise – The dominant noise factor, at frequencies above 1000 MHz, caused by the absorption of energy from radio waves by oxygen and water vapor in the atmosphere.

atmospheric duct—Within the troposphere, a condition in which the variation of refractive index is such that the propagation of an abnormally large proportion of any radiation of sufficiently high frequency is confined within the limits of a stratum. This effect is most noticeable above 3000 MHz.

atmospheric electricity – Static electricity between clouds, or between clouds and the earth.

atmospheric noise—Also called atmospherics. The noise heard during radio reception because of atmospheric interference.

atmospheric pressure—The barometric pressure of air at a particular location on the earth's surface. The nominal, or standard, value of atmospheric pressure is 760 mm of mercury (14.7 pounds per square inch) at sea level. Atmospheric pressure decreases at higher altitudes.

atmospheric radio wave—A radio wave that is propagated by reflections in the atmosphere. May include the ionospheric wave, the tropospheric wave, or both.

atmospheric radio window—That portion of the frequency spectrum that will allow radio-frequency waves to pass through the earth's atmosphere (approximately 10 to 10,000 MHz).

atmospheric refraction—The bending of the path of electromagnetic radiation from a distant point as the radiation passes obliquely through varying air densities.

atmospherics—Also referred to as static, atmospheric noise, and strays. In a radio tuner or receiver, noise due to natural weather phenomena and electrical charges existing in the atmosphere.

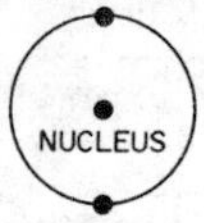

Atom.

atom—1. The smallest portion of an element which exhibits all properties of the element. It is pictured as composed of a positively charged nucleus containing almost all the mass of the atom, surrounded by one or more electrons. In the neutral atom, the number of electrons is such that their total charge (negative) exactly equals the positive charge in the nucleus. 2. The basic unit of chemical element consisting of a positively charged nucleus surrounded by a number of electrons sufficient to counterbalance the charge of the nucleus. The identity of an element, in a chemical sense, depends on the number of positive charges in the nucleus of its atom. The nucleus also contains particles that contribute mass but no charge. The stability of a nucleus depends on its ratio of charge to mass.

atomic battery—*See* Nuclear Battery.

atomic charge—The electronic charge of an ion, equal to the number of ionization multiplied by the charge on one electron.

atomic energy—*See* Nuclear Energy.

atomic fission—*See* Fission.

atomic frequency – The natural vibration frequency of an atom.

atomic fuel – A fissionable material—i.e., one in which the atomic nucleus may be split to release energy.

atomic fusion—*See* Fusion.

atomic mass unit—Abbreviated amu. Used to express the relative masses of isotopes. It is so proportioned that the mass of a neutral atom of the naturally most abundant isotope of oxygen (O16) is 16.00000 atomic mass units.

atomic migration—The progressive transfer of a valence electron from one atom to another within the same molecule.

atomic number — The number of protons (positively charged particles) in the nucleus of an atom. All elements have different atomic numbers, which determine their positions in the periodic table. For example, the atomic number of hydrogen is 1; that of oxygen is 8; iron, 26; lead, 82; uranium, 92.

atomic power—*See* Nuclear Energy.

atomic ratio—The ratio of quantities of different substances to the number of atoms of each.

atomic reactor—*See* Nuclear Reactor.

atomic theory—A generally accepted theory concerning the structure and composition of substances and compounds. It states that everything is composed of various combinations of ultimate particles called atoms.

atomic time—Time scales based on molecular or atomic resonance effects, which are apparently constant and equivalent (or nearly equivalent) to ephemeris time.

atomic weight—The approximate weight of the number of protons and neutrons in the nucleus of an atom. The atomic weight of oxygen, for example, is approximately 16 (actually it is 16.0044)—it contains 8 neutrons and 8 protons. Aluminum is 27 and contains 14 neutrons and 13 protons. If expressed in grams, these weights are called gram atomic weights.

atr tube — Abbreviation for antitransmit-receive tube. A gas-filled, radio-frequency switching tube used to isolate the transmitter while a pulse is being received.

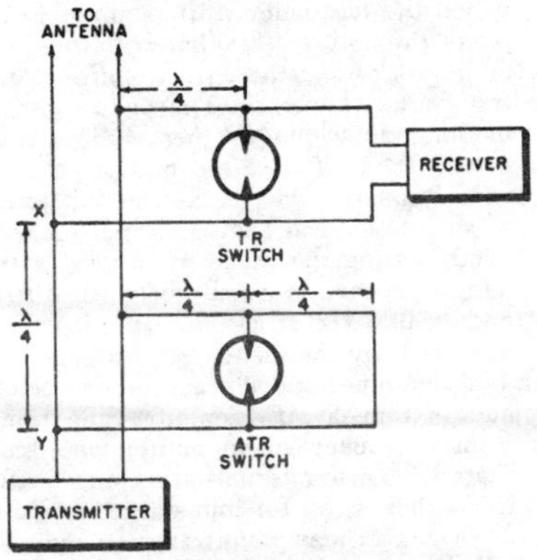

ATR (switch) tube.

attached foreign material — In a semiconductor, a foreign substance that cannot be removed when subjected to a nominal gas flow. Lint, silicon dust, etc., are not considered attached since they can be removed after die mounts.

attachment cap—*See* Attachment Plug.

attachment cord—*See* Patch Cord.

attachment plug — An assembly consisting of two or more blades projecting from a small insulating base, with provision for connecting the plug to a cord. Also called attachment cap.

attack—1. The length of time it takes for a tone in an organ to reach full intensity after a key is depressed. On most organs this effect is adjustable by either a switch or potentiometer. 2. The action of a control system in response to a sudden error condition. 3. The "responsiveness" of an amplifier to signals with a fast rise-time, such as produced by percussive sounds of a transient nature.

attack time—The interval required for an input signal, after suddenly increasing in amplitude, to attain a specified percentage (usually 63%) of the ultimate change in amplification or attenuation due to this increase.

attendant's switchboard—A switchboard, of one or more positions, that permits an operator in the central office to receive, transmit, or cut in on a call to or from one of the lines serviced by the office.

attention display — A computer-generated, tabular or vector message, placed on the display tubes of a control facility to draw attention to a particular situation.

attenuate—To obtain a fractional part or reduce in amplitude an action or signal.

attenuating—Decreasing electrical current, voltage, or power in a communicating channel. Refers to audio, radio, or carrier frequencies.

attenuation—1. The decrease in amplitude of a signal during its transmission from one point to another. It may be expressed as a ratio or, by extension of the term, in decibels. 2. *See* Insertion Loss. 3. The decrease in amplitude of a signal at a specified frequency due to its transmission through a filter. This is expressed as a function of the amplitude ratio V_1/V_2, which is the reciprocal of the magnitude of the transmission function.

attenuation constant—1. The real component of the propagation constant. 2. For a traveling plane wave at a given frequency, the rate at which the amplitude of a field component (or the voltage or current) decreases exponentially in the direction of propagation, in nepers or decibels per unit length.

attenuation distortion—1. In a circuit or system, its departure from uniform amplification or attenuation over the frequency range required for transmission. 2. Distortion that causes a decrease in the amplitude of a field component (voltage or current) in the direction of propagation.

attenuation equalizer—A corrective network designed to make the absolute value of the transfer impedance of two chosen pairs of terminals substantially constant for all frequencies within a desired range.

attenuation-frequency distortion—A form of wave distortion in which the relative magnitudes of the different frequency components of the wave are changed.

attenuation network—1. A network providing relatively slight phase shift and substantially constant attenuation over a range of frequencies. 2. The arrangement of circuit elements, usually impedance elements, inserted in circuitry to introduce a known loss or to reduce the impedance level without reflections.

attenuation ratio—The magnitude of the propagation ratio which indicates the relative decrease in energy.

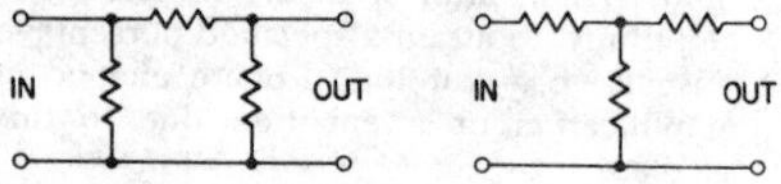

Attenuators, 1.

attenuator—1. A resistive network that provides reduction of the amplitude of an electrical signal without introducing appreciable phase or frequency distortion. 2. A distributed network that absorbs part of a signal and transmits the remainder with a minimum of distortion or delay. 3. Network for reducing signal level. Sometimes necessary in the input circuit of a tuner to avoid overloading by strong, local signals.

attenuator tube — A gas-filled, radio-frequency switching tube in which a gas discharge, initiated and regulated independently from the radio-frequency power, is used to control this rf power by reflection or absorption.

atto—Prefix meaning 10^{-18}. Abbreviated a.

audibility—1. The ability to be heard, usually construed as being heard by the human ear. 2. The ratio of the strength of a specific sound to the strength of a sound that can barely be heard. Usually expressed in decibels.

audible—Capable of being heard, in most contexts by the average human ear.

audible ringing tone—That tone received by the calling telephone indicating that the called telephone is being rung.

audible tones — Sounds composed of frequencies which the average human can detect.

audio — Pertaining to frequencies corresponding to a normally audible sound wave. These frequencies range roughly from 15 to 20,000 hertz.

audio amplifier—*See* Audio-Frequency Amplifier.

audio band—The range of audio frequencies passed by an amplifier, receiver, transmitter, etc. (*See also* Audio Frequency.)

audio-channel wire—A small-diameter wire, shielded and jacketed, used primarily in radio and television for wiring consoles, panels, etc.

audio component — That portion of any wave or signal which contains frequencies in the audible range (between 15 and 20,000 hertz).

audio frequency—Abbreviated af. Any frequency corresponding to a normally audible sound wave. Audio frequencies range roughly from 15 to 20,000 hertz.

audio-frequency amplification—An increase in voltage, current, or power of a signal at an audio frequency.

audio-frequency amplifier—Also called audio amplifier. A device which contains one or more electron tubes or transistors (or both) and which is designed to amplify signals within a frequency range of about 15 to 20,000 hertz.

audio-frequency choke—An inductance used to impede the flow of audio-frequency currents.

audio-frequency noise — In the audio-frequency range, any electrical disturbance introduced from a source extraneous to the signal.

audio-frequency oscillator — An oscillator circuit using an electron tube, transistor, or other nonrotating device capable of producing audio signals.

audio-frequency peak limiter — A circuit generally used in the audio system of a radio transmitter to prevent overmodulation. It keeps the signal amplitude from exceeding a predetermined value.

audio-frequency shift modulator—A system of facsimile transmission over radio, in which the frequency shift required is applied through an 800-hertz shift of an audio signal, rather than shifting the transmitter frequency. The radio signal is modulated by the shifting audio signal, usually at 1500 to 2300 hertz.

audio frequency spectrum—The full range of sounds we can hear. In a person with good hearing the range of sounds in the audio spectrum is usually between 30 Hz and 16,000 Hz. In older people it is usually 50 Hz to 10,000 Hz.

audio-frequency transformer — Also called audio transformer. An iron-core transformer for use with audio-frequency currents to transfer signals from one circuit to another. Used for impedance matching or to permit maximum transfer of power.

audiogram — Also called threshold audiogram. A graph showing hearing loss, percent of hearing loss, or percent of hearing as a function of frequency.

audio-level meter—An instrument that measures audio-frequency power with reference to a predetermined level. Usually calibrated in decibels.

audiometer – An electronic instrument for measuring hearing acuity. In simple units, the listener is provided (usually through earphones) with an audio signal (commonly a pure tone) of known intensity and frequency. More complex instruments provide a variety of signals (pure tones, white noise, and speech) through a variety of output transducers (such as earphones, bone vibrators, or loudspeakers).

audion—A three-electrode vacuum tube introduced by Dr. Lee de Forest.

audio oscillator—*See* Audio-Frequency Oscillator.

audio output—The output signal from any audio equipment. It is generally measured in volts or watts, rms.

audio patch bay—Specific patch panels provided for termination of all audio circuits and equipment used in the channel and technical-control facility. This equipment can also be found in transmitting and receiving stations.

audio peak limiter – *See* Audio-Frequency Peak Limiter.

audiophile—A person who is interested in good musical reproduction for his own personal listening and who uses the latest audio equipment and techniques.

audio signal—An electrical signal the frequency of which is within the audio range.

audio spectrum—The continuous range of audio frequencies extending from the lowest to the highest (from about 15 to 20,000 hertz).

audio taper—Semilogarithmic change of resistance. Used on tone controls in audio amplifiers to compensate for the lower sensitivity of the human ear when listening to low-volume sounds.

audio transformer – *See* Audio-Frequency Transformer.

audiovisual – Involving both sight and sound. (For example, audiovisual education uses films, slides, phonograph records, and the like, to supplement instruction.)

audiovisual system—A system of communications which simultaneously transmits pictorial and audio signals.

augend – In an arithmetic addition, the number increased by having another number (called the addend) added to it.

augmented operation code—In a computer, an operation code that is further defined by information contained in another portion of an instruction.

aural—Pertaining to the ear or to the sense of hearing. (*See also* Audio.)

aural radio range – A radio range the courses of which are normally followed by interpretation of an aural signal.

aural signal—The signal corresponding to the sound portion of a television program. In general, the audible component of a signal.

aural transmitter—The equipment used to transmit the aural (sound) signals from a television broadcast station.

aurora—Sheets, streamers, or streaks of pale light often seen in the skies of the northern and southern hemispheres. The aurora borealis and aurora australis.

auroral absorption – Absorption of radio waves due to auroral activity. (*See also* Aurora.)

auroral absorption index—A factor that relates the average auroral absorption with the geographic location of the points of reflection from the ionosphere.

authorized carrier frequency – A specific carrier frequency authorized for use, from which the actual carrier frequency is permitted to deviate, solely because of frequency instability, by an amount not to exceed the frequency tolerance.

autoalarm—Also called automatic alarm receiver. A device which is tuned to the international distress frequency of 500 kHz and which automatically actuates an alarm if any signal is received.

auto call—An alerting device which sounds a preset code of signals in a building, to page those persons whose code is being sounded.

autocondensation—A method of introducing high-frequency alternating current into living tissue for therapeutic purposes. The patient is connected as one plate of a capacitor to which the current is applied.

autoconduction – A method of introducing high-frequency alternating currents into living tissues for therapeutic purposes. The patient is placed inside a coil and acts essentially as the secondary of a transformer.

autocorrelation – 1. The correlation of a waveform with itself. It gives the Fourier transform of the power spectrum of the waveform (the power-density spectrum in the case of random signals). 2. A mathematical technique to measure the degree of rhythmic activity in physical phenomena that vary in a complex manner as a function of time.

autocorrelation function—A measure of the similarity between time-delayed and undelayed versions of the same signal, expressed as a function of delay.

autodyne circuit – A vacuum-tube circuit which serves simultaneously as an oscillator and as a heterodyne detector.

autodyne reception—A type of radio reception employed in regenerative receivers for the reception of cw code signals. In this system the incoming signal beats with the signal from an oscillating detector to produce an audible beat frequency.

auto iris control—An accessory unit which measures the video level of a tv camera and opens and closes the iris of the lens to compensate for light changes.

auto-man—A locking switch which controls the method of operation—i.e., automatic or manual.

automata—A plural form of automaton.

automated communications – Combination of techniques and facilities by which intelligence is conveyed from one point to another without human effort.

automatic—1. Self-regulating or self-acting; capable of producing a desired response to certain predetermined conditions. 2. Self-acting and self-regulating; operating without human intervention; often implying the presence of a feedback control system. 3. Pertaining to a process or device that, under specific conditions performs its functions without intervention by a human operator.

automatic-alarm receiver—Complete receiving, selecting, and warning device capable of being actuated automatically by intercepted radio-frequency signals forming the international automatic-alarm signal.

automatic-alarm-signal keying device – A device that automatically keys the radiotelegraph transmitter on board a vessel to transmit the international automatic-alarm signal.

automatic back bias—A radar-receiver technique which consists of one or more automatic gain-control loops to prevent large signals from overloading a receiver, whether by jamming or by actual echoes.

automatic bass compensation – A circuit used in a receiver or audio amplifier to make the bass notes sound more natural at low volume settings. The circuit, which usually consists of resistors and capacitors connected to taps on the volume control, automatically compensates for the poor response of the human ear to weak sounds.

automatic bias—*See* Self-Bias, 2.

automatic brightness control – A circuit used in television receivers to keep the average brightness of the reproduced image essentially constant. Its action is similar to that of an automatic volume-control circuit.

automatic calling unit – A dialing device, supplied by the communication common carrier, that permits a business machine to dial calls automatically over the communication networks. Abbreviated ACU.

automatic check – An operation performed by equipment built into an electronic computer to automatically verify proper operation.

automatic chrominance control – A color-television circuit which automatically controls the gain of the chrominance bandpass amplifier by varying the bias.

automatic circuit breaker – A device that automatically opens a circuit, usually by electromagnetic means, when the current exceeds a safe value. Unlike a fuse, which must be replaced once it blows, the circuit breaker can be reset manually when the current is again within safe limits.

automatic coding—A technique by which a digital computer is programmed to perform a significant portion of the coding of a problem.

automatic color purifier – *See* Automatic Degausser.

automatic computer—A computer capable of processing a specified volume of work without a need for human intervention other than program changes.

automatic connections – Connections between users made by electronic switching equipment without human intervention.

automatic constant—In a calculator, a provision that allows the user to multiply or divide a series of numbers by the same divider or multiplier without reentering each time.

automatic contrast control—A television circuit which automatically changes the gain of the video if and rf stages to maintain proper contrast in the television picture.

automatic control engineering—That branch of science and technology which deals with the design and use of automatic control devices and systems.

automatic controller – A device or instrument for measuring and regulating that operates by receiving a signal from a sensing device, comparing this signal with a desired value, and issuing signals for corrective action.

automatic crossover—1. A type of current-limiting circuit on a power supply provided with an adjustment for setting the short-circuit current to an adjustable maximum value. 2. A term applied to bimodal power supplies (constant voltage/constant current) that describes the transferal from one operating mode to the other at a predetermined value of load resistance. Usually the crossover point is preset by means of front panel controls.

automatic current limiting – An overload-protection mechanism designed to limit the maximum output current of a power supply to a preset value. Usually it automatically restores the output when the overload is removed.

automatic cutout – A device operated by electromagnetism or centrifugal force, to automatically disconnect some parts of an equipment after a predetermined operating limit has been reached.

automatic data processing—The processing of digital information by automatic computers and other machines. Abbreviated ADP. Also called integrated data processing.

automatic data-processing system — A system that includes electronic data-processing equipment together with auxiliary and connecting communications equipment.

automatic degausser — Also called automatic color purifier and degausser. An arrangement of degaussing coils mounted around a color-television picture tube. These coils are energized only for a short while after the set is turned on. They serve to demagnetize any parts of the picture tube that may have been affected by the earth's magnetic field or the magnetic field of any nearby home appliance.

automatic dialing unit — A device capable of generating dialing digits automatically. Abbreviated ADU.

automatic direction finder — Abbreviated ADF. Also called an automatic radio compass. An electronic device, usually for marine or aviation application, which provides a radio bearing to any transmitter whose frequency is known but whose direction and location are not.

automatic electronic data-switching center —Communications center designed specifically for the automatic, electronic transmission, reception, relay, and switching of digitalized data.

automatic error correction — A technique, usually requiring the use of special codes and/or automatic retransmission, which detects and corrects errors occurring in transmission. The degree of correction depends upon coding and equipment configuration.

automatic exchange—A telephone exchange in which connections are made between subscribers by means of devices set in operation by the originating subscriber's instrument and without the intervention of an operator.

automatic focusing — A method of electrostatically focusing a television picture tube; the focusing anode is internally connected through a resistor to the cathode and thus requires no external focusing voltage.

automatic frequency control — 1. Abbreviated afc. A system that produces an error voltage in proportion to the amount by which an oscillator drifts away from its correct frequency, the error voltage acting to reverse the drift. 2. A control circuit in a receiver or tuner which compensates for small variations in the carrier signal frequency to provide a stable audio output. A circuit function in a tuner or receiver which keeps the unit accurately tuned to the desired station, eliminating any tendency to drift.

automatic frequency correction—*See* Automatic Frequency Control.

automatic function key correction — When the wrong function key is depressed in a calculator, pressing correct function key automatically replaces it.

automatic gain control—1. Abbreviated agc. A type of circuit used to maintain the output volume of a receiver constant, regardless of variations in the signal strength applied to the receiver. 2. A self-acting compensation device which maintains the output of a transmission system constant within narrow limits in the face of wide variations in the attenuation of the system. 3. A radar circuit which prevents saturation of the radar receiver by long blocks of received signals, or by a carrier modulated at low frequency.

automatic gain stabilization — A circuit, used in certain identification friend-or-foe equipment and radar beacon systems, which serves to maintain optimum sensitivity in a superregenerative stage by keeping the noise-pulse load constant. The system prevents random noises from triggering the automatic transmitter associated with the receiver.

automatic grid bias—Grid-bias voltage provided by the difference in potential across a resistance (or resistances) in the grid or cathode circuit due to grid or cathode current or both.

automatic intercept — Automatic recording of messages a caller may wish to leave when the called party is away from his telephone.

automatic level compensation—A system which automatically compensates for variations in the circuit. *See also* Automatic Volume Control.

automatic light control — The process by which the illumination incident upon the face of a television pickup device is automatically adjusted as a function of scene brightness.

automatic message-switching center — A center in which messages are routed automatically according to information they contain.

automatic modulation control—A transmitter circuit that reduces the gain for excessively strong audio input signals without affecting the strength of normal signals, thus permitting higher average modulation without overmodulation; this is equivalent to an increase in the carrier-frequency power output.

automatic noise limiter—A circuit that automatically clips off all noise peaks above the highest peak of the desired signal being received. This circuit prevents strong atmospheric or man-made interference from being troublesome.

automatic numbering equipment — Equipment used in association with tape transmitters to transmit a channel number.

automatic pedestal control—The process by which the pedestal height of a television signal is automatically adjusted as a function of the input or other specified parameter.

automatic phase control—A circuit used in color television receivers to synchronize the burst signal with the 3.58-MHz color oscillator.

automatic pilot—*See* Autopilot.

automatic programming — Any technique designed to simplify the writing and execution of programs in a computer. Examples are assembly programs which translate from the programmer's symbolic language to the machine language, those which assign absolute addresses to instruction and data words, and those which integrate subroutines into the main routine.

automatic quality control—A technique in which the quality of a product being processed is evaluated in terms of a predetermined standard, and proper corrective action is taken automatically if the quality falls below the standard.

automatic radio compass—A radio direction finder which automatically rotates the loop antenna to the correct position. A bearing can then be secured from the indicator dials without mechanical adjustments or calculation. (*See also* Automatic Direction Finder.)

automatic record changer — An electrically operated mechanism which automatically feeds, plays, and rejects a number of records in a preset sequence. It consists of a motor, turntable, pickup arm, and changer. Modern changers are designed to play automatically 16⅔-, 33⅓-, 45-, and 78-rpm records.

automatic relay — A means of selective switching which causes automatic equipment to record and retransmit communications.

automatic reset relay — Also called automatic reset. 1. A stepping relay that returns to its home position either when it reaches a predetermined contact position or when a pulsing circuit fails to energize the driving coil within a given time. It may either pulse forward or be spring-reset to the home position. 2. An overload relay that restores the circuit as soon as an overcurrent situation is corrected.

automatic reverse—Ability of some four-track stereo tape recorders to play the second pair of stereo tracks automatically, in the reverse direction, without need to interchange the empty and full reels after the first pair of stereo tracks has been played.

automatic scanning — A variable-speed sweep of the entire frequency range of an rfi meter. It may also include the scanning of a portion of this frequency range over a predetermined sector.

automatic scanning receiver — A receiver which can automatically and continuously sweep across a preselected frequency, either to stop when a signal is found or to plot signal occupancy within the frequency spectrum being swept.

automatic secure voice communications — A network that provides cryptographically secure voice communications through the use of a combination of wideband and narrow-band voice-digitizing techniques.

automatic selective control (or transfer) relay—A device which operates to select automatically between certain sources or conditions in an equipment, or performs a transfer operation automatically.

automatic send/receive — A teletypewriter unit that includes a keyboard, printer, paper tape, reader/transmitter, and paper-tape punch. This combination of facilities may be used on line or off line, and, in some cases, on line and off line simultaneously.

automatic sensitivity control—1. A circuit used for automatically maintaining the receiver sensitivity at a predetermined level. It is similar to an automatic gain control, but it affects the receiver constantly rather than during the brief interval selected by the range gate. 2. The self-acting mechanism which varies the system sensitivity as a function of the specified control parameters. This may include automatic target control, automatic light control, or any combination thereof.

automatic sequencing — The ability of a computer to perform successive operations without additional instructions from a human being.

automatic short-circuiter—A device used in some forms of single-phase commutator motors to short-circuit the commutator bars automatically.

automatic short-circuit protection—An automatic current-limiting system which enables a power supply to continue operating at a limited current, and without damage, into any output overload, including a short circuit. The output voltage is restored to normal when the overload is removed, as distinguished from a fuse or circuit-breaker system which opens with overload and must be replaced or reclosed manually to restore power.

automatic shutoff — In a tape recorder, a switching arrangement which automatically shuts the recorder off when the tape breaks or runs out. Also, a switching arrangement which stops the record changer after the last record.

automatic starter—1. A device which, after being given the initial impulse by means of a push button or similar device, starts a system or motor automatically in the proper sequence. 2. A self-acting starter that is completely controlled by master or pilot switches or some other sensing device.

automatic switchboard—Telephone switchboard in which the connections are made by the operation of remotely controlled switches.

automatic switch center—A switch center in which messages originating at any subscriber terminal are relayed automatically through one or more switching centers to their destinations.

automatic target control — The self-acting mechanism which controls the vidicon target potential as a function of the scene brightness.

automatic telegraph transmission—A form of telegraphy in which signals are transmitted mechanically from a perforated tape.

automatic telegraphy—A form of telegraphy in which signals are transmitted and/or received automatically.

automatic threshold variation—A constant false-alarm rate scheme which is an open-loop type of automatic gain control in which the decision threshold is varied continuously in proportion to the incoming intermediate frequency and video noise level.

automatic time switch—A combination of a switch with an electric or spring-wound clock arranged to turn an apparatus on and off at predetermined times.

automatic tracking — In radar, the process whereby a mechanism, actuated by the echo, automatically keeps the radar beam locked on the target, and may also determine the range simultaneously.

automatic transfer equipment—Equipment which automatically transfers a load so that a source of power may be selected from one of several incoming lines.

automatic tuning—An electrical, mechanical, or electrical/mechanical system that automatically tunes a circuit to a predetermined frequency when a button or other control is operated.

automatic video-noise leveling—A constant false-alarm rate scheme in which the video-noise level at the output of the receiver is sampled at the end of each range sweep and the receiver gain is readjusted accordingly to maintain a constant video-noise level at the output.

automatic voltage regulator — A device or circuit which maintains a constant voltage, regardless of any variation in input voltage or load.

automatic volume compression — *See* Volume Compression.

automatic volume control (avc)—1. A self-acting compensation device which maintains the output of a transmission system constant within narrow limits in the face of wide variations in attenuation in that system. 2. A self-acting device which maintains the output of a radio receiver or amplifier substantially constant within relatively narrow limits while the input voltage varies over a wide range. 3. *See* Automatic Level Compensation.

automatic volume expansion — Also called volume expansion. An audio-frequency circuit that automatically increases the volume range by making loud portions louder and weak ones weaker. This is done to make radio reception sound more like the actual program, because the volume range of programs is generally compressed at the point of broadcast.

automatic zero and full-scale calibration correction—A system of zero and sensitivity stabilization in which electronic servos are used that compare demodulated "zero" and "full-scale" signals with reference voltages.

automation—1. The method or act of making a manufacturing or processing system partially or fully automatic. 2. The entire field of investigation, design, development, application, and methods of rendering or making processes or machines self-acting or self-moving; rendering automatic. 3. Automatically controlled operation of an apparatus process, or system by mechanical or electronic devices that take the place of observation, effort, and decision by a human operator.

automaton—1. A device that automatically follows predetermined operations or responds to encoded instructions. 2. Any communication-linked set of elements. 3. A machine which exhibits living properties.

automonitor—1. To instruct an automatic digital computer to produce a record of its information-handling operations. 2. A program or routine for this purpose.

automotive electronics—The branch of engineering science that deals with the generation, control, conversion, and application of electricity in self-propelled vehicles.

autopilot—Also called automatic pilot, gyropilot, or robot pilot. A device containing amplifiers, gyroscopes, and servomotors which automatically control and guide the flight of an aircraft or guided missile. The autopilot detects any deviation from the planned flight and automatically applies the necessary corrections to keep the aircraft or missile on course.

autopolarity—A feature of a digital voltmeter or digital multimeter wherein the correct polarity (either negative or positive) for a measured quantity is automatically indicated on the display.

auto radio—A radio receiver designed to be

installed in an automobile and powered by the storage battery of the automobile.

autoradiography — Self-portraits of radioactive sources made by placing the radioactive material next to photographic film. The radiations fog the film and thus leave an image of the source.

autoregulation induction heater—An induction heater in which a desired control is effected by the change in characteristics of a magnetic charge as it is heated at or near its Curie point.

auto single play—A turntable in which records are played one at a time and the arm performs a playing cycle, lifting itself from the record at the end (and in some cases shutting off the motor).

autostarter—*See* Autotransformer Starter.

Autosyn—A trade name of the Bendix Corp. for a remote-indicating instrument or system based on the synchronous-motor principle in which the angular position of the rotor of a motor at the measuring source is duplicated by the rotor of the indicator motor.

auto tracking—Also called automatic tracking operation. A master/slave connection of two or more power supplies, each of which has one of its output terminals in common with one of the output terminals of all of the other supplies, such connections being characterized by one-knob control and proportional output voltage from all supplies. Useful where simultaneous turn-up, turn-down, or proportional control of all power supplies in a system is required.

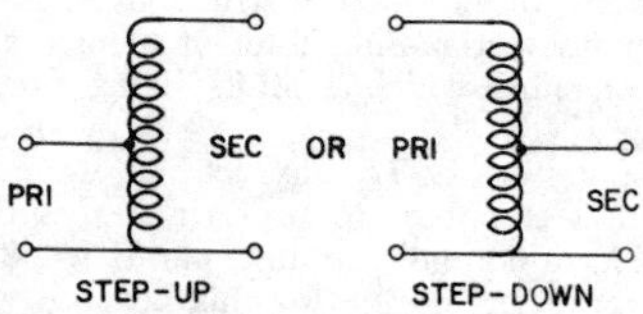

Autotransformers, 1.

autotransformer—1. A transformer with a single winding (electrically) in which the whole winding acts as the primary winding, and only part of the winding acts as the secondary (step-down); or part of the winding acts as the primary and the whole winding acts as the secondary (step-up). 2. A voltage, current, or impedance transforming device in which parts of one winding are common to both the primary and secondary circuits.

autotransformer starter—Also called a compensator or autostarter. A motor starter having an autotransformer to furnish a reduced voltage for starting. Includes the necessary switching mechanism.

aux—Abbreviation for auxiliary. Often applied to amplifier inputs and refers to an extra input facility as distinct from "mic," "tuner," "pickup," etc.

auxiliary actuator — A mechanism which may be attached to a switch to modify its characteristics.

auxiliary bass radiator—A parasitic (non-electrically driven) unit resembling a bass speaker unit located in a loudspeaker enclosure, as if it were an ordinary unit, to increase the movement of air and hence enhance the bass performance at a given enclosure volume.

auxiliary circuit — Any circuit other than the main circuit.

auxiliary contacts—In a switching device, contacts, in addition to the main circuit contacts, which function with the movement of the latter.

auxiliary electrodes — Metallic electrodes pushed or driven into the earth to provide electrical contact for the purpose of performing measurements on grounding electrodes of ground grid systems.

auxiliary equipment — Equipment not directly controlled by the central processing unit of a computer.

auxiliary function—In automatic control of machine tools, a machine function other than the control of the motion of a workpiece or cutter, e.g., control of machine lubrication and cooling.

auxiliary memory—*See* Auxiliary Storage.

auxiliary operation — An operation performed by equipment not under continuous control of the computer central processing unit.

auxiliary relay—1. A relay which responds to the opening or closing of its operating circuit, to assist another relay or device in the performance of a function. 2. A relay actuated by another relay and used to control secondary circuit functions such as signals, lights, or other devices. Also called slave relay.

auxiliary-station line filter—A line filter for use at repeater points to separate frequencies of different carrier systems, using the same line pair. For example, such a filter might be used at a high-frequency carrier-system repeater to bypass the low-frequency carrier system and voice frequencies around the repeater.

auxiliary storage — Storage capacity, such as magnetic tape, disc, or drum, in addition to the main memory of a computer. Also called auxiliary memory.

auxiliary switch — A switch actuated by some device such as a circuit breaker, for signaling, interlocking, or other purpose.

auxiliary transmitter—A transmitter held in readiness in case the main transmitter of a broadcasting station fails.

availability—1. The ratio, expressed as a percent, of the time during a given period that an equipment is correctly operating to the total time in that period. Also called operating ratio. 2. The probability that a system is operating satisfactorily at any point in time when used under stated conditions, where the total time considered includes operating time, active repair time, administrative time, and logistic time.

available conversion gain—Ratio of available output-frequency power from the output terminals of a transducer to the available input-frequency power from the driving generator, with terminating conditions specified for all frequencies which may affect the result. Applies to outputs of such magnitude that the conversion transducer is operating in a substantially linear condition.

available gain—The ratio of the available power at the output terminals of the network to the available power at the input terminals of the network.

available line—In a facsimile system, that portion of a scanning line which can be used for picture signals. Expressed as a percentage of the length of the scanning line.

available machine time—Time after the application of power during which a computer is operating correctly.

available power—1. The mean square of the open-circuit terminal voltage of a linear source, divided by four times the resistive component of the source impedance. 2. Of a network, the power that would be delivered to a conjugately matched load. It is the maximum power that a network can deliver. Available power, though defined in terms of an output load impedance, is independent of that impedance.

available power gain — Sometimes called completely matched power gain. Ratio of the available power from the output terminals of a linear transducer, under specified input-termination conditions, to the available power from the driving generator. The available power gain of an electrical transducer is maximum when the input-termination admittance is the conjugate of the driving-point admittance at the input terminals of the transducer.

available signal-to-noise ratio—Ratio of the available signal power at a point in a circuit, to the available random-noise power.

avalanche—Rapid generation of a current flow with reverse-bias conditions as electrons sweep across a junction with enough energy to ionize other bonds and create electron-hole pairs, making the action regenerative.

avalanche breakdown—In a semiconductor diode, a nondestructive breakdown caused by the cumulative multiplication of carriers through field-induced impact ionization.

avalanche conduction—A form of conduction in a semiconductor in which charged-particle collisions create additional hole-electron pairs.

avalanche diode — Also called breakdown diode. A silicon diode that has a high ratio of reverse-to-forward resistance until avalanche breakdown occurs. After breakdown the voltage drop across the diode is essentially constant and is independent of the current. Used for voltage regulating and voltage limiting. Originally called zener diode, before it was found that the Zener effect had no significant role in the operation of diodes of this type.

avalanche impedance—*See* Breakdown Impedance.

avalanche noise — 1. A phenomenon in a semiconductor junction in which carriers in a high-voltage gradient develop sufficient energy to dislodge additional carriers through physical impact. 2. Noise produced when a p-n junction diode is operated at the onset of avalanche breakdown.

avalanche transistor — A transistor that, when operated at a high reverse bias voltage, supplies a chain generation of hole-electron pairs.

avc—*See* Automatic Volume Control.

average—*See* Arithmetic Mean.

average absolute pulse amplitude — The average of the absolute value of instantaneous amplitude taken over the pulse duration. Absolute value means the arithmetic value regardless of algebraic sign.

average brightness—The average illumination in a television picture.

average calculating operation — A typical computer calculating operation longer than an addition and shorter than a multiplication, often taken as the mean of nine additions and one multiplication.

average electrode current—The value obtained by integrating the instantaneous electrode current over an averaging time and dividing by the average time.

average life—*See* Mean Life, 1.

average noise factor — *See* Average Noise Figure.

average noise figure—Also called average noise factor. In a transducer, the ratio of total output noise power to the portion attributable to thermal noise in the input termination, with the total noise being summed over frequencies from zero to infinity and the noise temperature of the input termination being standard (290K).

average power output of an amplitude-modulated transmitter — The radio-frequency power delivered to the transmitter output terminals, averaged over a modulation cycle.

average pulse amplitude—The average of the instantaneous amplitudes taken over the pulse duration.

average rate of transmission – Effective speed of transmission.

average value—The value obtained by dividing the sum of a number of quantities by the number of quantities. The average value of a sine wave is 0.637 times the peak value.

average voltage—The sum of the instantaneous voltages in a half-cycle waveshape, divided by the number of instantaneous voltages. In a sine wave, the average voltage is equal to 0.637 times the peak voltage.

aviation channels—A band of frequencies, below and above the standard broadcast band, assigned exclusively for aircraft and aviation applications.

aviation services—The aeronautical mobile and radionavigation services.

avionics – 1. An acronym designating the field of AVIation electrONICS. 2. The branch of electronics which is concerned with aviation applications.

Avogadro's number—The actual number of molecules on one gram-molecule, or of atoms in one gram-atom of an element or any pure substance (6.023×10^{23}).

AWG—American Wire Gage. A means of specifying wire diameter. The higher the number, the smaller the diameter.

axial leads—Leads coming out the ends and along the axes of a resistor, capacitor, or other axial part, rather than out the side.

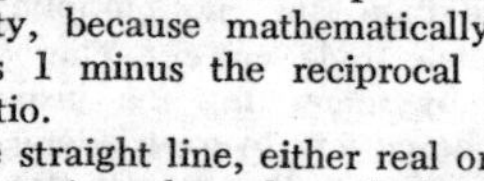

Axial leads.

axial ratio—Ratio of the major axis to the minor axis of the polarization ellipse of a waveguide. This term is preferred over ellipticity, because mathematically ellipticity is 1 minus the reciprocal of the axial ratio.

axis—The straight line, either real or imaginary, passing through a body around which the body revolves or around which parts of a body are symmetrically arranged.

Ayrton-Perry winding – Two conductors connected in parallel so that the current flows in opposite directions in each conductor and thus neutralizes the inductance between the two.

Ayrton shunt—Also called universal shunt. A high-resistance parallel connection used to increase the range of a galvanometer without changing the damping.

azel display—A modified type of plan-position indicator presentation showing two separate radar displays on one cathode-ray screen. One display presents bearing information, and the other shows elevation.

azimuth—1. The angular measurement in a horizontal plane and in a clockwise direction. 2. In a tape recorder, the angle which recording and playback head gaps make with the line along which the tape moves. The head is oriented until this angle is 90°. 3. The vertical setting (alignment) of the head in a tape recorder.

azimuth alignment—Alignment of the recording and reproducing gaps so that their center lines lie parallel with one another. Misalignment of the gaps causes a loss in output at short wavelengths.

azimuth blanking – Blanking of the crt screen in a radar receiver as the antenna scans a selected azimuth region.

azimuth gain reduction—A technique which allows control of the radar receiver system throughout any two azimuth sectors.

azimuth rate—The rate of change of true bearing.

azimuth resolution—The angle or distance by which two targets must be separated in azimuth to be distinguished by a radar set, when the targets are at the same range.

azimuth stabilization—The presentation of indications on a radar display so that north, or any specific reference line of direction, is always at the top of the screen.

azimuth-stabilized, plan-position indicator —A ppi scope on which the reference bearing (usually true or magnetic north) remains fixed with respect to the indicator, regardless of the vehicle orientation.

azimuth versus amplitude – An electronic counter-countermeasures receiver with plan-position indicator-type display attached to the main antenna, used to display strobes due to jamming aircraft. It is useful in making passive fixes when two or more radar sites can operate together.

azusa – A short base-line continuous-wave phase-comparison electronic tracking system operating on the C-band, wherein a single station provides two direction cosines and slant range.

B

B—1. Symbol for the base of a transistor. 2. Symbol for magnetic flux. 3. Abbreviation for photometric brightness. 4. B or b. Abbreviation for susceptance.

B– (**B-minus or B-negative**)—Negative terminal of a B battery or the negative polarity of other sources of anode voltage. Denotes the terminal to which the negative side of the anode-voltage source should be connected.

B+ (B-plus or B-positive)—Positive terminal of a B battery or the positive polarity of other sources of anode voltage. The terminal to which the positive side of the anode voltage source should be connected.

babble—1. The aggregate cross talk from a large number of disturbing channels. 2. In a carrier, or other multiple-channel system, the unwanted disturbing sounds which result from the aggregate cross talk or mutual interference from other channels.

babble signal—A type of electronic deception signal used to confuse enemy receivers. Generally it has characteristics of energy transmission signals. It can be composed by superimposing incoming signals on previously recorded intercepted signals. This composite signal can then be radiated as a jamming signal.

BABS (Blind Approach Beacon System)—A pulse-type ground-based navigation beacon used for runway approach. The BABS ground beacon is installed beyond the far end of the runway on the extended center line. When interrogated by an aircraft, it retransmits two diverging beams, one of short and the other of long duration pulses. The beams are transmitted alternately, but because of the fast switching, the aircraft receives what appears to be a continuous transmission of both beams. The cathode-ray tube in the aircraft displays both long and short pulses superimposed on each other. When the aircraft is properly aligned with the runway, the pulses will be of equal amplitude.

back bias—1. A degenerative or regenerative voltage which is fed back to circuits before its originating point. Usually applied to a control anode of a tube. 2. A voltage applied to a grid of a tube (or tubes) to restore a condition which has been upset by some external cause. 3. *See also* Reverse Bias.

backbone — A high-voltage, high-capacity transmission line or group of lines having a limited number of large-capacity connections between loads and points of generation.

back contact—Relay, key, jack, or other contact designed to close a circuit and permit current to flow when, in the case of a relay, the armature has released or fallen back, or in other cases, when the equipment is inoperative.

back current—Also called reverse current. The current which flows when reverse bias is applied to a semiconductor junction.

back diode—A tunnel diode that is usually chosen for its reverse-conduction characteristics.

back echo—An echo due to the back lobe of an antenna.

back emission—Also called reverse emission. Emission from an electrode occurring only when the electrode has the opposite polarity from that required for normal conduction. A form of primary emission common to rectifiers during the inverse portion of their cycles.

backfire—*See* Arcback.

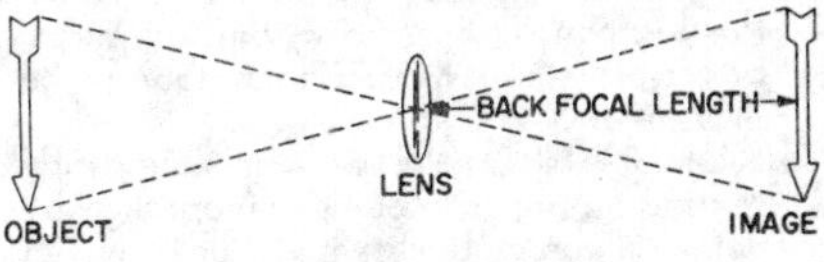

Back focal length.

back focal length—Distance from the center of a lens to its principal focus on the side of the lens away from the object.

back focus—The distance from the rear vertex of a lens to the focal plane, with the subject at infinite distance.

background—1. The picture white of the copy being scanned when the picture is black and white only. Also undesired printing in the recorded copy of the picture being transmitted, resulting in shading of the background area. 2. Noise heard during a radio program; this noise is caused by atmospheric interference or operation of the receiver at such high gain that inherent tube and circuit noises become noticeable.

background control—In color television, a potentiometer used as a means of controlling the dc level of a color signal at one input of a tricolor picture tube. The setting of this control determines the average (or background) illumination produced by the associated color phosphor.

background count—Count caused by radiation from sources other than the one being measured.

background noise — 1. The total system noise, independent from the presence or absence of a signal. The signal is not included as part of the noise. 2. In a receiver, the noise in the absence of signal modulation in the carrier. 3. Any unwanted sound that intrudes upon program material, such as sounds produced as a

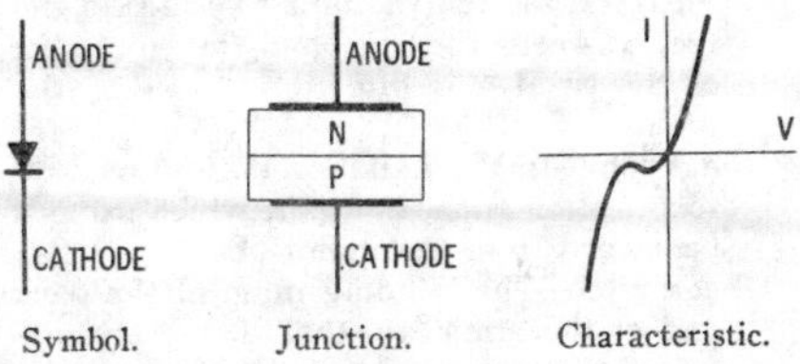

Back diode.

result of surface imperfections of a disc record.

background processing—The automatic execution of lower-priority programs in a computer when the system resources are not being used for higher-priority programs.

background radiation—Radiation due to the presence of radioactive material in the vicinity of the measuring instrument.

background response—In radiation detectors, response caused by ionizing radiation from sources other than that to be measured.

back-haul—Use of excess circuit mileage by routing through switching centers that are not in a direct facility path between an originating office and a terminating office.

backing – Flexible material (usually cellulose acetate or polyester) on which is deposited the magnetic-oxide coat that "records" the signal on magnetic recording tape. Also known as base.

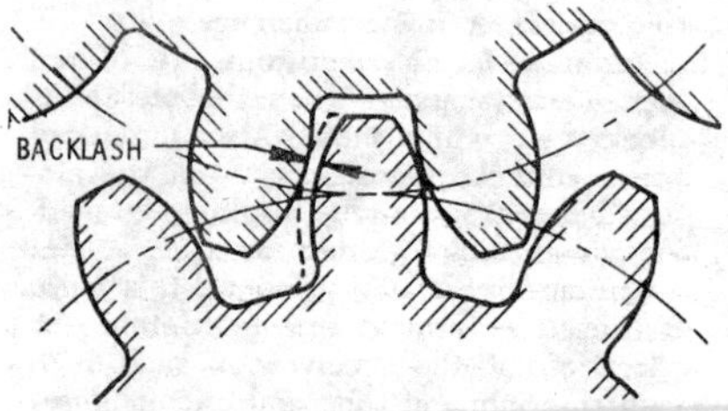

Backlash, 2.

backlash—1. In a potentiometer, the maximum difference that occurs in shaft position when the shaft is moved to the same actual output-ratio point from opposite directions. Resolution and contact-width effects must be excluded from this measure. 2. The shortest distance between nondriving tooth surfaces of adjacent teeth in mating gears or the amount of clearance between the teeth of mating gears. Usually measured at the common pitch circle.

backloaded horn—A speaker enclosure arranged so the sound from the front of the cone feeds directly into the room, while the sound from the rear feeds into the room via a folded horn.

back loading—A form of horn loading particularly applicable to low-frequency speakers; the rear radiating surface of the speaker feeds the horn and the front part of the speaker is directly exposed to the room.

back lobe—In the radiation pattern of a directional antenna, that part which extends backward from the main lobe.

back pitch—The winding pitch of the back end of the armature—that is, the end opposite the commutator.

backplane—Area of a computer or other equipment where various logic and control elements are interconnected. Often takes the form of a "rat's nest" of wires interconnecting printed-circuit cards in the back of computer racks or cabinets.

backplate—In a camera tube, the electrode to which the stored-charge image is capacitively coupled.

back porch—In a composite picture signal, that portion which lies between the trailing edge of a horizontal-sync pulse and the trailing edge of the corresponding blanking pulse. A color burst, if present, is not considered part of the back porch.

back-porch effect—The continuation of collector current in a transistor for a short time after the input signal has dropped to zero. The effect is due to storage of minority carriers in the base region. It also occurs in junction diodes.

back-porch tilt – The slope of the back porch from its normal horizontal position. Positive and negative refer, respectively, to upward and downward tilt to the right.

back scattering—1. Radiation of unwanted energy to the rear of an antenna. 2. The reflected radiation of energy from a target toward the illuminating radar.

back-shunt keying—A method of keying a transmitter, in which the radio-frequency energy is fed to the antenna when the telegraph key is closed and to an artificial load when the key is open.

backside illumination – A charge coupled device fabrication technique employing "thinned" silicon where the image is impressed on the side opposite the MOS electrodes.

backstop—That part of the relay which limits the movement of the armature away from the pole face or core. In some relays, a normally closed contact may serve as a backstop.

backswing—The amplitude of the first maximum excursion in the negative direction after the trailing edge of a pulse expressed as a percentage of the 100% amplitude.

backtalk—Transfer of information to the active computer from a standby computer.

back-to-back circuit—Two tubes or semiconductor devices connected in parallel but in opposite directions so that they can be used to control current without introducing rectification. Also called inverse-parallel connection.

backup—1. An item kept available to replace an item which fails to perform satisfactorily. 2. An item under development intended to perform the same general function performed by another item also under development.

backup facility – A communications-electronics facility which is established for the purpose of replacing or supplementing another facility or facilities, under real or simulated emergency conditions.

backup item—An additional item to perform the general functions of another item. It may be secondary to an identified primary item or a parallel development to improve the probability of success in performing the general function.

backwall—The plate in a pot core which connects the center post to the sleeve.

backwall photovoltaic cell—A cell in which light must pass through the front electrode and a semiconductor layer before reaching the barrier layer.

backward-acting regulator—A transmission regulator in which the adjustment made by the regulator affects the quantity which caused the adjustment.

backward diode—A highly doped, alloyed germanium junction that operates on the principle of quantum-mechanical tunneling. The diode is "backward" because its easy-current direction is in the negative-voltage rather than the positive-voltage region of the i-v curve. The backward diode has a negative-resistance region, but the resultant "valley" of its i-v curve is much less pronounced than in tunnel diodes.

backward wave—In a traveling-wave tube, a wave having a group velocity opposite the direction of electron-stream motion.

backward-wave oscillator — An oscillator employing a special vacuum tube in which oscillatory currents are produced by using an oscillatory electromagnetic field to bunch the electrons as they flow from cathode to anode.

backward-wave tube — A traveling-wave tube in which the electrons travel in a direction opposite to that in which the wave is propagated.

back wave—*See* Spacing Wave.

baffle—1. In acoustics, a shielding structure or partition used to increase the effective length of the external transmission path between two points (for example, between the front and back of an electroacoustic transducer). A baffle is often used in a speaker to increase the acoustic loading of the diaphragm. (Although this term sometimes is used to designate the entire cabinet, or enclosure, that houses a loudspeaker it is—strictly speaking—the panel on which the speaker is mounted, usually the front panel of such an enclosure. The term derives from its original use in preventing or "baffling" the speaker's rear sound waves from interfering with its front waves.) 2. In a gas tube, an auxiliary member placed in the arc path and having no separate external connection. 3. A device for deflecting oil or gas in a circuit breaker. 4. A single shielding device designed to reduce the effect of ambient light on the operation of an optical transmission link.

baffle plate—A metal plate inserted into a waveguide to reduce the cross-sectional area for wave-conversion purposes.

bail—A loop of wire used to prevent permanent separation of two or more parts assembled together—e.g., the bail holding dust-caps on round connectors.

Bakelite — A trademark of the Bakelite Corp. for its line of plastic and resins. Formerly, the term applied only to its phenolic compound used as an insulating material in the construction of radio parts.

balance—1. The effect of blending the volume of various sounds coming over different microphones in order to present them in correct proportion. 2. The maintenance of equal average volume from both speaker systems of a stereo installation. 3. Relative volume, as between different voices or instruments bass and treble, or left and right stereo channels. 4. Either a condition of symmetry in an electrical circuit, such as a Wheatstone Bridge, or the condition of zero output from a device when properly energized. In the latter sense, depending upon the nature of the excitation, two general categories of balance may be encountered: for dc excitation, resistive balance; for ac excitation, resistive and/or reactive balance.

balance control—1. On a stereo amplifier, a differential gain control used to vary the volume of one speaker system relative to the other without affecting the overall volume level. As the volume of one speaker increases and the other decreases, the sound appears to shift from left to center to right, or vice versa. 2. A variable resistor used to compensate for any slight loss of signal in the right or left channel of a stereo amplifier. To some extent, this control can compensate for unbalanced speakers and be used for adjustment when the listener is not in an equidistant position between the two loudspeakers.

balanced — 1. Electrically alike and symmetrical with respect to ground. 2. Arranged to provide balance between certain sets of terminals.

balanced amplifier — An amplifier circuit with two identical signal branches, connected to operate in phase opposition and with their input and output connections each balanced to ground; for example, a push-pull amplifier.

balanced armature—An armature which is approximately in equilibrium with respect to both static and dynamic forces.

balanced-armature unit—The driving unit used in magnetic speakers, consisting of an iron armature pivoted between the poles of a permanent magnet and sur-

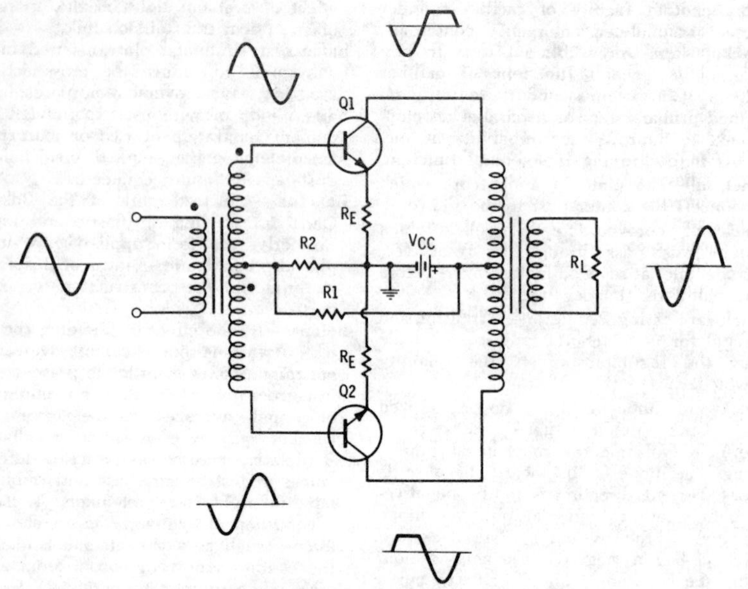

Balanced (push-pull) amplifier.

rounded by coils carrying the audio-frequency current. Variations in the audio-frequency current cause corresponding changes in the armature magnetism and corresponding movements of the armature with respect to the poles of the permanent magnet.

balanced bridge—A bridge circuit with its components adjusted so that it has an output voltage of zero.

balanced circuit—1. A circuit with two sides electrically alike and symmetrical to a common reference point, usually ground. 2. A circuit terminated by a network that has infinite impedance losses.

balanced converter—*See* Balun.

balanced currents — Also called push-pull currents. In the two conductors of a balanced line, currents which are equal in value and opposite in direction at every point along the line.

balanced detector—A demodulator for frequency-modulation systems. In one form, the output consists of the rectified difference of the two voltages produced across two resonant circuits, one circuit being tuned slightly above the carrier frequency and the other slightly below.

balanced line—A line or circuit utilizing two identical conductors. Each conductor is operated so that the voltages on them at any transverse plane are equal in magnitude and opposite in polarity with respect to ground. Thus, the currents on the line are equal in magnitude and opposite in direction. A balanced line is preferred where minimum noise and cross talk are desired.

balanced-line system—A system consisting of a generator, balanced line, and load adjusted so that the voltages of the two conductors at all transverse planes are equal in magnitude and opposite in polarity with respect to ground.

balanced low-pass filter—A low-pass filter designed to be used with a balanced line.

balanced method—A method of measurement in which the reading is taken at zero. It may be a visual or audible reading, and in the latter case the null is the no-sound setting.

balanced modulator—An amplitude modulator in which the control grids of two tubes are connected for parallel operation, and the screen grids and plates for push-pull operation. After modulation, the output contains the two sidebands without the carrier.

balanced network—A hybrid network in which the impedances of the opposite branches are equal.

balanced oscillator—Any oscillator in which (1) the impedance centers of the tank circuits are at ground potential and (2) the voltages between either end and the centers are equal in magnitude and opposite in phase.

balanced output—A three-conductor output (as from a microphone) in which the signal voltage alternates above and below a third neutral circuit. This symmetrical arrangement tends to cancel any hum picked up by long lengths of interconnecting cable.

balanced telephone line—A telephone line which is floated with respect to ground so that the impedance measured from either side of the line to ground is equal to that of the other side to ground.

balanced termination—For a system or network having two output terminals, a load presenting the same impedance to ground for each output terminal.

balanced transmission line—A transmission line having equal conductor resistances per unit length and equal impedances from each conductor to earth and to other electrical circuits.

balanced voltages — Also called push-pull voltages. On the two conductors of a balanced line, voltages (relative to ground) which are equal in magnitude and opposite in polarity at every point along the line.

balanced-wire circuit—A circuit with two sides electrically alike and symmetrical to ground and other conductors. Commonly refers to a circuit the two sides of which differ only by chance.

balancer—In a direction finder, that portion used for improving the sharpness of the direction indication. It balances out the capacitance effect between the loop and ground.

balance stripe — A magnetic sound stripe placed on the edge of a motion-picture film opposite the main stripe; it provides mechanical balance for the film.

balance-to-unbalance transformer — A device for matching a pair of lines, balanced with respect to earth, to a pair of lines not balanced with respect to earth. *See also* Balun.

balancing network—An electrical network designed for use in a circuit in such a way that two branches of the circuit are made substantially conjugate (i.e., such that an electromotive force inserted into one branch produces no current in the other).

balancing unit — 1. An antenna-matching device used to permit efficient coupling of a transmitter or receiver having an unbalanced output circuit to an antenna having a balanced transmission line. 2. A device for converting balanced to unbalanced transmission lines, and vice versa, by placing suitable discontinuities at the junction between the lines instead of using lumped components.

ball—In face bonding, a method of providing chips with contact.

ballast lamp—A lamp which maintains a nearly constant current by increasing its resistance as the current increases.

ballast resistor—A special type of resistor used to compensate for fluctuations in alternating-current power-line voltage. It is usually connected in series with the power supply to a receiver or amplifier. The resistance of a ballast resistor increases rapidly with increases in current through it, thereby tending to maintain an essentially constant current despite variations in the line voltage.

ballast tube — A current-controlling resistance device designed to maintain a substantially constant current over a specified range of variations in the applied voltage to a series circuit.

ball bond — A type of thermocompression bond in which a gold wire is flame-cut to produce a ball-shaped end which is then bonded to a metal pad by pressure and heat.

ball bonding — A bonding technique that uses a capillary tube to feed the bonding wire. The end of the wire is heated and melts, thus forming a large ball. The capillary and ball are then positioned on the

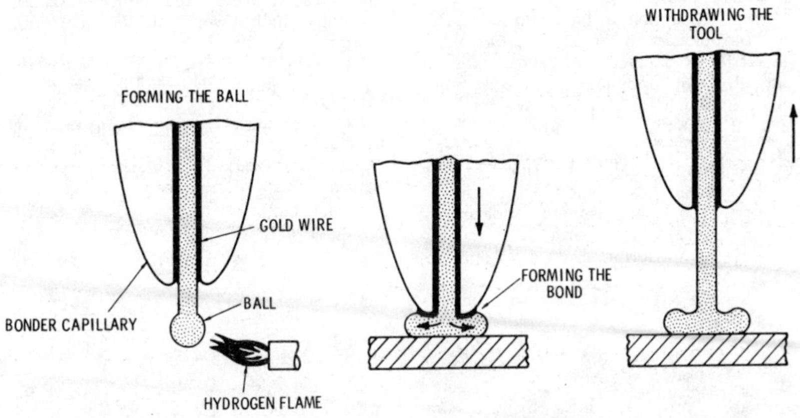

Ball bonding.

contact area and the capillary is lowered. This forms a large bond. The capillary is then removed and a flame is applied severing the wire and forming a new ball.

ballistic galvanometer—An instrument that indicates the effect of a sudden rush of electrical energy, such as the discharge current of a capacitor.

ballistic-missile early-warning system—An electronic system for providing detection and early warning of attack by enemy intercontinental ballistic missiles.

ballistics—A general term used to describe the dynamic characteristics of a meter movement—most notably, response time, damping, and overshoot.

ballistic trajectory—In the trajectory of a missile, the curve traced after the propulsive force is cut off and the body of the missile is acted upon only by gravity, aerodynamic drag, and wind.

ballistocardiogram—A waveform of the impulse imparted to or incurred by the body as a result of the displacement of blood upon each heartbeat. The period of the cycle of this waveform is the time interval between heart beats. Typically taken in one of two ways—by measuring the seismic disturbance imparted to the table upon which the patient is lying supine, or by measuring the deflection of a heavy metal bar placed across the patient's ankles while he is lying supine.

ballistocardiograph—An instrument used to record the movements imparted to the body by the beating of the heart.

ballistocardiography – The recording and interpretation of the movements imparted to the body by the beating of the heart and the movement of the blood.

balop—Contraction of balopticon, an apparatus for the projection of opaque images in conjunction with a television camera.

balopticon—*See* Balop.

balun—1. Also called balanced converter or "bazooka." An acronym from BALanced to UNbalanced. A device used for matching an unbalanced coaxial transmission line to a balanced two-wire system. 2. Usually a transformer designed to accept 75-ohm unbalanced input (coaxial cable) and deliver the signal at 300-ohm balanced (twin lead). Usable in the converse sense, and sometimes necessary for matching a tuner with 300-ohm balanced antenna terminals to a 75-ohm coaxial line.

banana jack—A jack that accepts a banana plug. Generally designed for panel mounting.

banana plug—A plug with a banana-shaped spring-metal tip and with elongated springs to provide a low-resistance compression fit.

band—1. Any range of frequencies which lies between two defined limits. 2. A group of radio channels assigned by the FCC to a particular type of radio service.

Very low freq.	(vlf)	10-30 kHz.
Low freq.	(lf)	30-300 kHz.
Medium freq.	(mf)	300-3,000 kHz.
High freq.	(hf)	3-30 MHz.
Very high freq.	(vhf)	30-300 MHz.
Ultrahigh freq.	(uhf)	300-3,000 MHz.
Superhigh freq.	(shf)	3,000-30,000 MHz.
Extremely high freq.	(ehf)	30-300 GHz.

3. A group of tracks or channels on a magnetic drum in an electronic computer. (*See also* Track, 2.) 4. In instrumentation, a range of values that represents the scope of operation of an instrument.

bandage – Rubber ribbon, about 4 inches wide, used as temporary moisture protection for a splice in telephone or coaxial cable.

band center—The geometric mean between the limits of a band of frequencies.

banded cable—Two or more cables banded together by stainless-steel strapping.

band-elimination filter—Also called bandstop filter. A wave filter with a single attenuation band, neither of the cutoff frequencies being zero or infinite. The filter passes frequencies on either side of this band.

band gap—The energy difference between the conduction band and the valence band in a material.

band-gap energy—The difference in energy between the conduction band and the valence band.

band I messages—Messages prepared in native machine symbols on a magnetic tape of standard length and converted to punch common language characters.

bandpass—The number of hertz expressing the difference between the limiting frequencies at which the desired fraction (usually half-power) of the maximum output is obtained.

bandpass-amplifier circuit – A stage designed to uniformly amplify signals of certain frequencies only.

bandpass filter—A wave filter with a single transmission band, neither of the cutoff frequencies being zero or infinite. The filter attenuates frequencies on either side of this band.

bandpass flatness—The variations in gain in the bandpass of a filter or tuned circuit.

bandpass response—Also called flat-top response. The response characteristic in which a definite band of frequencies is transmitted uniformly.

band-reject filter—A filter that does not pass a band of frequencies but passes both higher and lower frequencies. Sometimes called a notch filter.

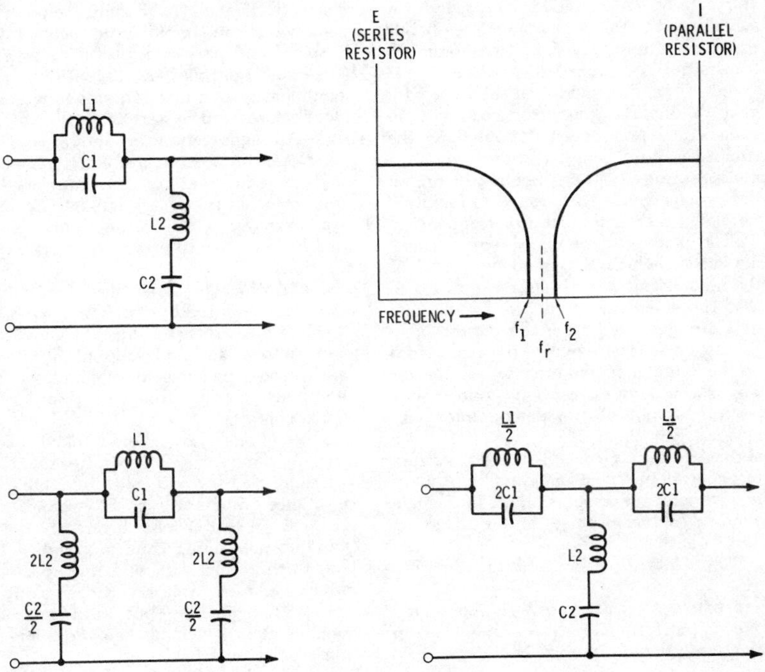

Band-elimination filters.

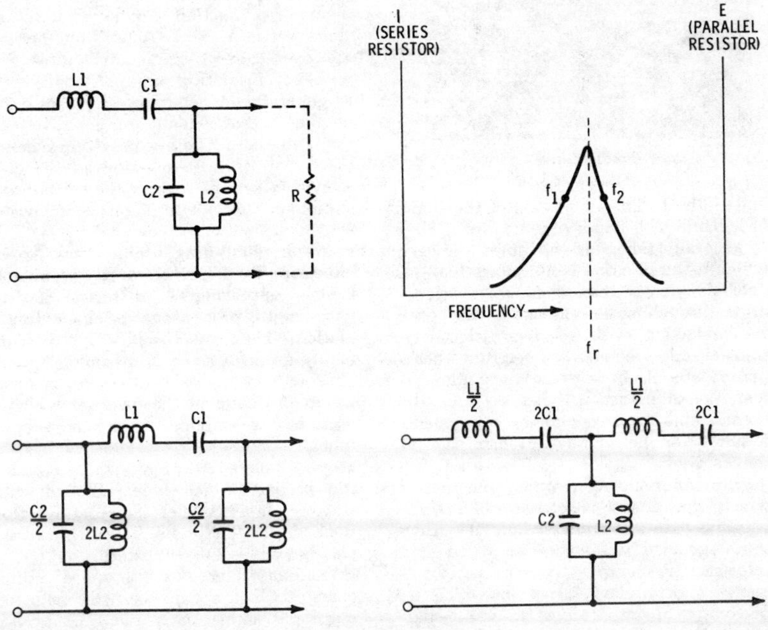

Bandpass filters.

band selector—Also called bandswitch. A switch used to select any one of the bands in which a receiver, signal generator, or transmitter is designed to operate.

B and S gage—Brown and Sharpe wire gage, where the conductor sizes rise in geometrical progression. Adopted as the American Wire Gage standard.

bandspreading—1. The spreading of tuning indications over a wide range to facilitate tuning in a crowded band of frequencies. 2. The method of double-sideband transmission in which the frequency band of the modulating wave is shifted upward so that the sidebands produced by modulation are separated from the carrier by a frequency at least equal to the bandwidth of the original modulating wave. In this way, second-order distortion products may be filtered out of the demodulator output.

bandspread tuning control—A separate tuning control provided on some short-wave receivers to spread the stations in a single band of frequencies over an entire tuning dial.

band-stop filter—*See* Band-Elimination Filter.

bandswitch—A switch used to select any one of the frequency bands in which an electronic apparatus may operate.

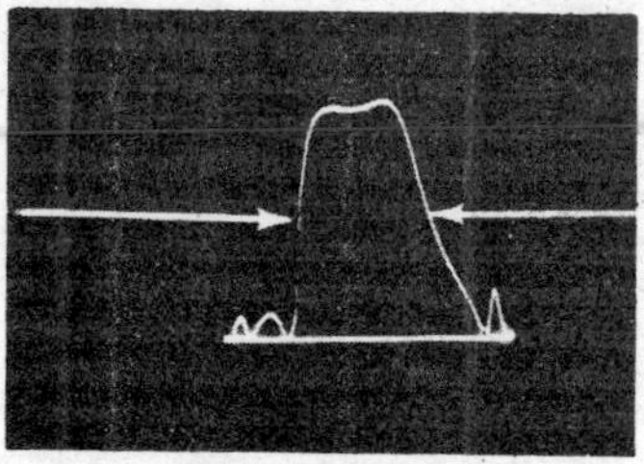

Bandwidth, 1.

bandwidth—1. The range within the limits of a band. The width of a bandpass filter is generally taken as the limits between which its attenuation is not more than 3.0 decibels greater than its average attenuation throughout its passband. Also used in connection with receiver selectivity, transmitted frequency spectrum occupancy, etc. 2. In a given facsimile system, the difference in hertz between the highest and lowest frequency components required for the adequate transmission of the facsimile signals. 3. The least frequency interval of a wave, outside of which the power spectrum of a time-varying quantity is everywhere less than some specified fraction of its value at a reference frequency. 4. The range of frequencies of a device, within which its performance, with respect to some characteristic, conforms to a specified standard. 5. The range of audio frequencies over which an amplifier or receiver will respond and provide a useful output.

bang-bang controller—A discontinuity-type nonlinear system that contains time delay, dead space, and hysteresis.

bank—An aggregation of similar devices (e.g. transformers, lamps, etc.) connected and used together. In automatic switching, a bank is an assemblage of fixed contacts over which one or more wipers or brushes move to establish electric connections.

bank-and-wiper switch—A switch in which the electromagnetic ratchets or other mechanisms are used, first, to move the wipers to a desired group of terminals, and second, to move the wipers over the terminals of this group to the desired bank contacts.

bank winding—Also called banked winding. A compact multilayer form of coil winding used for reducing distributed capacitance. Single turns are wound successively in two or more layers, the entire winding proceeding from one end of the coil to the other without being returned.

bantam tube—A compact tube having a standard octal base but a considerably smaller glass envelope than the standard glass octal tube has.

bar—1. *See* Microbar. 2. A subdivision of a crystal slab. 3. A vertical or horizontal line on a television screen, used for testing. 4. A symbol, placed over a letter, used to indicate the inverse, or complement, of a function. For example, inversion of A is $\bar{A}$, read "A bar" or "not A."

bare conductor—A conductor not covered with any insulating material.

bar generator—A generator of pulses or repeating waves which are equally separated in time. These pulses are synchronized by the synchronizing pulses of a television system so that they produce a stationary bar pattern on a television screen.

bar-graph monitoring oscilloscope—An oscilloscope for observation of commutated signals appearing as a series of bars with lengths proportional to channel modulation. The same oscilloscope is commonly used for setup and troubleshooting observations.

barium—An element the oxide of which is used in the cathode coating of vacuum tubes.

barium titanate—A ceramic that has electric properties and is capable of withstanding much higher temperatures than Rochelle salt crystals. Used in crystal pickups and sonar transducers.

Barkhausen effect—A succession of abrupt changes which occur when the magnetizing force acting on a piece of iron or other magnetic material is varied.

Barkhausen interference – Interference caused by Barkhausen oscillations.

Barkhausen-Kurz oscillator – Circuit for generating ultrahigh frequencies. Its operation depends on the variation in the electrical field around the positive grid and less positive plate of a triode; the variation is caused by oscillatory electrons in the interelectrode spaces.

Barkhausen oscillation–A form of parasitic oscillation in the horizontal-output tube of a television receiver; it results in one or more narrow, dark, ragged vertical lines near the left side of the picture or raster.

Barkhausen oscillator – *See* Barkhausen-Kurz Oscillator.

Barkhausen tube–*See* Positive-Grid Oscillator Tube.

bar magnet–A bar of metal that has been so strongly magnetized that it holds its magnetism and thereby serves as a permanent magnet.

barn–A unit of measure of nuclear cross sections. Equal to 10^{-24} square centimeter.

Barnett effect–The magnetization resulting from the rotation of a magnetic specimen. The rotation of a ferromagnet produces the same effect as placing the ferromagnet in a magnetic field directed along the axis of rotation. On the macroscopic model, the domains of a ferromagnet can be considered a group of electron systems, each acting as an independent gyroscope or gyrostat.

barometer–An instrument for measuring atmospheric pressure. There are two types of barometers commonly used in meteorology–the mercury barometer and the aneroid barometer.

barometric pressure–The weight of the atmosphere per unit of surface. The standard barometer reading at sea level and 59°F is 29.92 inches of mercury absolute.

bar pattern–A pattern of repeating lines or bars on a television screen. When such a pattern is produced by pulses which are equally separated in time, the spacing between the bars on the television screen can be used to measure the linearity of the horizontal or vertical scanning systems.

bar quad–*See* B-quad.

barrage jamming – Simultaneous jamming of a number of adjacent channels or frequencies.

barrel–The cylindrical portion of a solderless terminal, splice, or contact in which the conductor is accommodated.

barrel effect–The boomy or hollow voice quality obtained when the voice is transmitted from a reverberant environment; usually accompanied by a loss of intelligibility and a sense of loss of privacy for far end users.

bar relay–A relay in which a bar actuates several contacts simultaneously.

barrel distortion – In camera or image tubes, the distortion which results in a monotonic decrease in radial magnification in the reproduced image away from the axis of symmetry of the electron optical system. In tv receivers, barrel distortion makes all four sides of the raster curve out like a barrel.

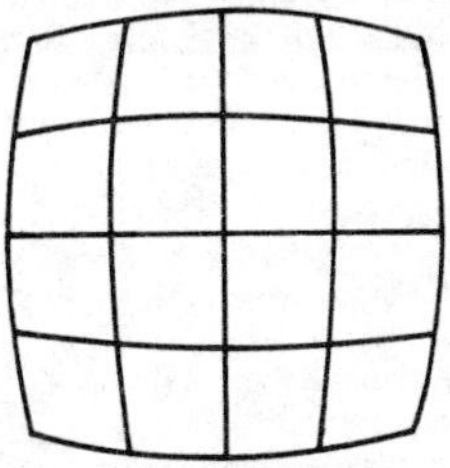

Barrel distortion.

barretter–1. A voltage-regulator tube consisting of an iron-wire filament in a hydrogen-filled envelope. The filament is connected in series with the circuit to be regulated and maintains a constant current over a given voltage variation. 2. A positive coefficient resistor whose resistance increases as temperature increases.

barretter mount – A waveguide mount in which a barretter can be inserted to measure electromagnetic power.

barricade shield–A type of movable shield for protection from radiation.

barrier–1. A partition for the insulation or isolation of electric circuits or electric arcs. 2. In a semiconductor, the electric field between the acceptor ions and the donor ions at a junction. *See* Depletion Layer.

barrier capacitance – *See* Depletion-Layer Capacitance.

barrier-film rectifier–A rectifier in which a film having unilateral (single-direction) conductivity is in contact with metal or other normally conducting plates.

barrier grid–A grid close to, or in contact with, a storage surface of a charge storage tube. It establishes an equilibrium voltage for secondary-emission charging and serves to minimize redistribution.

barrier height–In a semiconductor, the difference in potential from one side of a barrier to the other.

barrier layer–*See* Depletion Layer.

barrier-layer cell–A type of photovoltaic cell in which light acting on the surface of the contact between layers of copper and cuprous oxide causes an electromotive force to be produced.

barrier-layer rectification – *See* Depletion-Layer Rectification.

barrier plate – A layer of slow-diffusing

metal (usually palladium or nickel) placed between two fast-diffusing materials to slow or prevent their interdiffusion.

barrier region–*See* Depletion Region.

barrier shield–A wall or enclosure shielding the operator from an area where radioactive material is being used or processed by remote-control equipment.

barrier voltage – The voltage necessary to cause electrical conduction in a junction of two dissimilar materials, such as a pn junction diode.

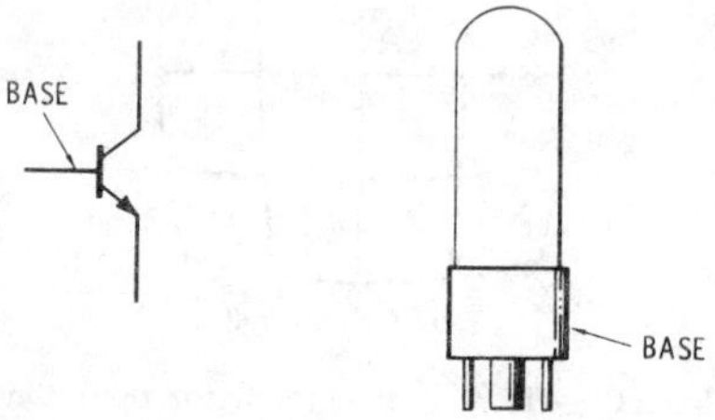

Base.

base–1. The region between the emitter and collector of a transistor which receives minority carriers injected from the emitter. It is the element which corresponds to the control grid of an electron tube. 2. In a vacuum tube, the insulated portion through which the electrodes are connected to the pins. 3. On a printed-circuit board, the portion that supports the printed pattern. 4. *See also* Positional Notation *and* Radix. 5. *See* Alkali. *See* Backing.

base address–A given address from which an absolute address is obtained by combination with a relative address. Also called address constant.

baseband–1. The frequency band occupied by the aggregate of the transmitted signals used to modulate a carrier. 2. In CD-4 records, the left- or right-channel's band containing musical information for the front and back channels, recorded at listening frequencies in the standard sound spectrum (from 30 to 15,000 hertz).

baseband frequency response – Response characteristics over the frequency band occupied by all of the signals which modulate a transmitted carrier.

base electrode–An ohmic or majority-carrier contact to the base region of a transistor.

base film–The plastic substrate that supports the coating of magnetic recording tape. The base film of most instrumentation and computer tapes is made of polyester. For less critical uses, cellulose acetate and polyvinyl chloride are employed.

basegroup–Designation for a number of carrier channels in a particular frequency range that forms a basic unit (channel bank) for further modulation to a final frequency band.

base insulator–Heavy-duty insulator used to support the weight of an antenna mast and to insulate the mast from the ground or some other surface.

base line–1. In radar displays, the visual line representing the track of the radar scanning beam. 2. In graphical presentations, the horizontal scale, often representing time, bias, or some other variable.

base-line break – In radar, a technique which uses the characteristic break in the base line on an A-scope display due to a pulse signal of significant strength in noise jamming.

base load–In a dc converter, the current which must be taken from the base to maintain a saturated state.

base-loaded antenna – A vertical antenna the electrical height of which is increased by adding inductance in series at the base.

base material – An insulating material (usually a copper-clad laminate) used to support a conductive pattern.

base number–The radix of a number system (10 is the base number, or radix, for the decimal system; 2 for the binary system).

base-One peak voltage–The peak voltage measured across a resistor in series with base-One when a unijunction transistor is operated as a relaxation oscillator in a specified circuit.

base pin–*See* Pin.

base point–*See* Radix Point.

base region–In a transistor, the interelectrode region into which minority carriers are injected.

base resistance–Resistance in series with the base lead in the common-T equivalent circuit of a transistor.

base ring–Ohmic contact to the base region of power transistors; so called because it is ring-shaped.

base spreading resistance–In a transistor, the resistance of the base region caused by the resistance of the bulk material of the base region.

base station–A land station, in the land mobile service, carrying on a service with land mobile stations. (A base station may secondarily communicate with other base stations incident to communications with land mobile stations.) Sometimes defined as a station in a land mobile system which remains in a fixed location and communicates with mobile stations.

base-timing sequencing–Sharing of a transponder on a time basis between several ground transmitters through the use of coded timing signals.

base voltage – The voltage between the

base terminals of a transistor and the reference point.

BASIC – A simplified computer language intended for use in engineering applications.

basic access method–A method of computer access in which each input/output statement results in a corresponding machine input/output operation.

basic frequency – In any wave, the frequency which is considered the most important. In a driven system, it would in general be the driving frequency, while in most periodic waves it would correspond to the fundamental frequency.

basic linkage – In a computer, a linkage that is used repeatedly in one routine, program, or system and that follows the same set of rules each time it is used.

basic processing unit–The principal section for control and data processing within a communications system.

basic protection – Fundamental lightning protection measures and/or devices, such as the use of gas tubes or carbon-block protectors, which are applied directly to transmission media at apparatus locations to provide initial voltage limitation.

basic *Q*–*See* Nonloaded *Q*.

basic rectifier–A metallic rectifier in which each rectifying element consists of a single metallic rectifier cell.

basic speed range–The range over which a motor and control are capable of delivering full load torque without overheating or cogging and obtained by armature voltage control.

basket winding–A coil winding in which adjacent turns are separated except at the points of crossing.

bass–Sounds in the low audio-frequency range. On the standard piano keyboard, all notes below middle C (261.63 hertz).

bass boost–A deliberate adjustment of the amplitude-frequency response of a system or component to accentuate the lower audio frequencies.

bass-boosting circuit–A circuit that attenuates the higher audio frequencies in order that low or bass frequencies will be emphasized by comparison.

bass compensation–Emphasizing the low-frequency response of an audio amplifier at low volume levels to compensate for the lowered sensitivity of the human ear to weak low frequencies.

bass control–A manual tone control that has the effect of changing the level of bass frequencies reproduced by an audio amplifier.

bass-reflex enclosure–A type of speaker enclosure in which the rear wave from the speaker emerges through an auxiliary opening or port of critical dimensions to reinforce the bass tones. *See also* Vented Baffle.

bass response–1. The extent to which a speaker or audio-frequency amplifier handles low audio frequencies. 2. The ability of any device to pick up or reproduce low audio frequencies.

bassy–Term applied to sound reproduction in which the low-frequency notes are overemphasized.

batch–A group of documents to be processed; an arbitrary subdivision of a job by the supervisor into smaller, more manageable parts. Batch is the smallest group of such documents accessible by name (number) for data entry, data verify, peripheral device transfers, etc.

batch control sample – A representative batch extracted either at random or at specific intervals from a process or product for quality-control purposes. Results of equivalent tests of the batches are averaged to interpolate quality of the total process.

batch process – A method of fabricating monolithic resistors, capacitors, and diodes with the same process at the same time.

batch processing – 1. In a computer, a method of processing in which a number of similar input items are grouped for processing during the same machine run. 2. Pertaining to the technique of executing a set of computer programs such that each is completed before the next program of the set is started. Loosely, the execution of computer programs serially.

bat handle–Standard form of a toggle-switch lever having a shape similar to that of a baseball bat.

bathtub capacitor–A type of capacitor enclosed in a metal housing having broadly rounded corners like those on a bathtub.

bathyconductorgraph–A device used from a moving ship to measure the electrical conductivity of sea water at various depths.

bathythermograph–A device that automatically plots a graph showing temperature as a function of depth when lowered into the sea.

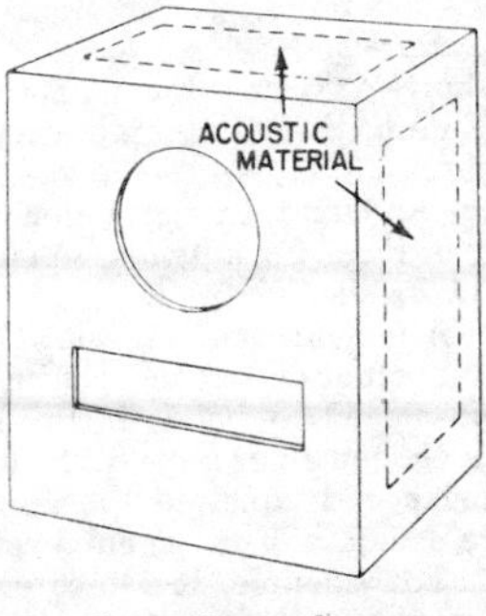

Bass-reflex enclosure.

Batten system—A method developed by W. E. Batten for coordinating single words in a computer to identify a document. Sometimes called peek-a-boo system.

battery—1. A dc voltage source consisting of two or more cells which converts chemical, nuclear, solar, or thermal energy into electrical energy. 2. In communications, a source (not necessarily a storage device) of direct current or the current itself. 3. Two or more cells coupled together in series or parallel. In the former the arrangement gives a greater voltage (two cells give twice the voltage, three cells give three times the voltage, and n cells give n times the voltage), and the latter arrangement gives the same voltage as the individual cell but a greater current.

battery acid—A solution that serves as the electrolyte in a storage battery. In the common lead-acid storage battery, the electrolyte is diluted sulfuric acid.

battery cable – A single conductor cable, either insulated or uninsulated, used for carrying current from batteries to the point where power is needed. May also be used for grounding.

battery capacity—The amount of energy obtainable from a storage battery, usually expressed in ampere-hours.

battery charger—Device used to convert alternating current into a pulsating direct current which can be used for charging a storage battery.

battery clip—A metal clip with a terminal to which a connecting wire can be attached, and with spring jaws that can be quickly snapped onto a battery terminal or other point to which a temporary connection is desired.

battery life—The number of times that a battery can be charged and discharged. One complete charge and one complete discharge is called a cycle. The number of complete cycles a battery will give depends on construction of the battery, charging procedure, maintenance, and operation.

battery receiver—A radio receiver that obtains its operating power from one or more batteries.

battle short—A switch for short-circuiting safety interlocks and lighting a red warning light.

bat wing—An element on an fm or tv transmitting or receiving antenna, so called because of its shape.

baud—1. A unit of signaling speed derived from the duration of the shortest code element. Speed in bauds is the number of code elements per second. 2. A unit of signaling speed equal to the number of discrete conditions or signal events per second. For example, in Morse code one baud equals one-half dot cycle per second; in a train of binary signals, one baud is one bit per second; and in a train of signals each of which can assume one of eight different states, one baud is one three-bit value per second. 3. A measurement of communication channel capacity as a function of time. For example, a 110-baud line is divided into 110 equal parts. Within each of these parts a certain amount of data can be placed, typically, one bit. This means that a speed of 110 baud is 110 bits per second.

Baudot code—A data-transmission code in which one character is represented by five equal-length bits. This code is used in most dc teletypewriter machines in which 1 start element and 1.42 stop elements are added.

bay—1. A portion of an antenna array. 2. A vertical compartment in which a radio transmitter or other equipment is housed.

bayonet base—A base having two projecting pins on opposite sides of a smooth cylindrical base; the pins engage corresponding slots in a bayonet socket and hold the base firmly in the socket.

bayonet coupling—A quick-coupling device. Connection is accomplished by rotating two parts under pressure. Pins on the side of the male connector engage slots on the side of the female connector.

bayonet socket—A socket for bayonet-base tubes or lamps; it has slots on opposite sides and one or more contact buttons at the bottom.

bazooka—*See* Balun.

B battery—The battery that furnishes the required dc voltages to the plate and screen-grid electrodes of the vacuum tubes in a battery-operated circuit.

bcd—Abbreviation for binary-coded decimal. A system of representing numerical, alphabetic, and special characters in which individual decimal digits are represented by some binary code. For example, in an 8-4-2-1 bcd notation, 16 might be represented as 0001 (for 1) and 0110 (for 6). In pure binary notation, 16 is 10000.

bcd counter—A counter in which each section consists of four flip-flops or stages, each section of which counts to nine (binary 1001) and then resets to zero (binary 0000). The outputs are in bcd form.

B channel – One of two stereo channels, usually the right, together with the microphone, speakers, or other equipment associated with this channel.

bci – Abbreviation for broadcast interference, a term denoting interference by transmitters with reception of broadcast signals on standard broadcast receivers.

B-display—On a radarscope, a type of presentation in which the target appears as a bright spot. Its bearing is indicated by the

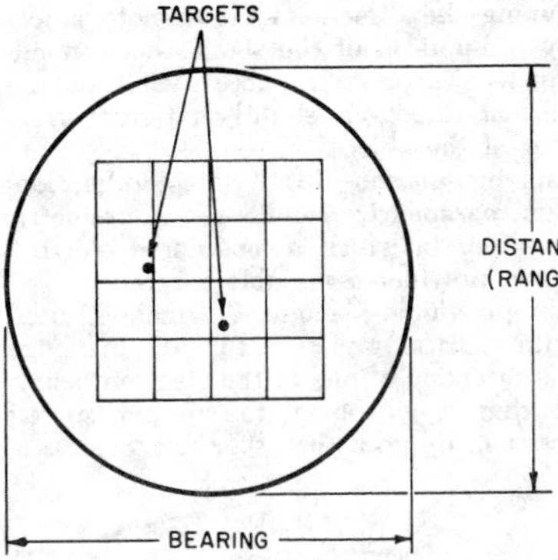

B-display.

horizontal coordinate and its range by the vertical coordinate.

beacon—A device which emits a signal for use as a guidance or warning aid. Radar beacons aid the radar set in locating and identifying special targets which may otherwise be difficult or impossible to sense.

beacon delay—The amount of inherent delay within the beacon—i.e., the time between the arrival of a signal and the response of the beacon. In a pulse beacon, delay ordinarily is measured between the leading 3-dB points of the triggering pulse and the reply pulse.

beacon receiver—A radio receiver for converting into perceptible signals the waves emanating from a radio beacon.

beacon skipping—A term used to describe a condition where beacon return pulses are missing at the interrogating radar. Beacon skipping can be caused by interference, overinterrogation of the beacon, antenna nulls, or pattern minima.

beacon stealing—The loss of beacon tracking by a (desired) radar due to (interfering) interrogation signals from another radar.

beacon time sharing — A technique by which two or more radars may interrogate and track a long-recovery type of beacon without exceeding the duty cycle of the beacon. This technique is accomplished by the proper sequencing of the various radar interrogations. It is necessary to ensure that the total of all interrogations does not exceed the beacon duty cycle and that enough time is allowed for the modulator section of the beacon to recover before it receives the next interrogation.

beaded coax—A coaxial cable in which the dielectric consists of beads made of various insulating materials.

beaded support—Ceramic and plastic beads used to support the inner conductor in coaxial transmission lines.

beaded transmission line — A line using beads to support the inner conductor in coaxial transmission lines.

bead thermistor — A thermistor consisting of a small bead of semiconducting material such as germanium placed between two wire leads. Used for microwave power measurement, temperature measurement, and as a protective device. The resistance decreases as the temperature increases.

beam—1. A flow of electromagnetic radiation concentrated in a parallel, converging, or diverging pattern. 2. The unidirectional or approximately unidirectional flow of radiated energy or particles.

beam alignment — The adjustment of the electron beam in a camera tube (on tubes employing low-velocity scanning) to cause the beam to be perpendicular to the target at the target surface.

beam angle—The angle between the directions, on either side of the axis, at which the intensity of the radio-frequency field drops to one-half the value it has on the axis.

beam antenna — An antenna that concentrates its radiation into a narrow beam in a definite direction.

beam bender—*See* Ion Trap.

beam bending—Deflection of the scanning beam of a camera tube by the electrostatic field of the charges stored on the target.

beam blanking—Interruption of the electron beam in a cathode-ray tube by the application of a pulse to the control grid or cathode.

beam candlepower—The candlepower of a bare source which, if located at the same distance as the beam, would produce the same illumination as the beam.

beam convergence—The converging of the three electron beams of a three-gun color picture tube at a shadow mask opening.

beam-coupling coefficient—In a microwave tube, the ratio of the amplitude, expressed in volts, of the velocity modulation produced by a gap to the radio-frequency gap voltage.

beam crossover—The point of overlap of a beam from an antenna that is nutated or rotated about the center line of the antenna radiation direction. The crossover point is normally at the half-power point. The received energy, when commutated into four quadrants, provides the necessary information for the servoamplifier error signal used to align the antenna to a target.

beam current—The current carried by the electron stream that forms the beam in a cathode-ray tube.

beam cutoff—In a television picture tube or cathode-ray tube, the condition in which the control-grid potential is so negative with respect to the cathode that electrons cannot flow and thereby form the beam.

beam-deflection tube – An electron-beam tube in which current to an output electrode is controlled by the transverse movement of an electron beam.

beam droop—A form of distortion of the normal rectilinear fan-shaped radiation pattern of a detection radar in which a portion of the fan is at a lower elevation than the rest of the fan.

beam-forming electrode—Electron-beam focusing elements in power tetrodes and cathode-ray tubes.

beam hole—An opening through a reactor shield and, generally, through the reactor reflector which permits a beam of radioactive particles or radiation to be used for experiments outside the reactor.

beam-index color tube – A color picture tube in which the signal generated by an electron beam after deflection is fed back to a control device or element in such a way that an image in color is provided.

beam lead—A metal beam deposited directly onto the surface of the die as part of the wafer processing cycle in the fabrication of an integrated circuit. Upon separation of the individual die (normally by chemical etching instead of the conventional scribe-and-break technique), the cantilevered beam is left protruding from the edge of the chip and can be bonded directly to interconnecting pads on the circuit substrate without the need for individual wire interconnections.

beam-lead bonding – 1. A free-bonding technique in which thick, gold extensions of the thin-film terminals of the semiconductor devices and circuits are electroformed so they extend beyond the edges of the chips. 2. A method of interconnecting ICs in a circuit by bonding beam leads located on the IC chip's back surface to the circuit's conducting paths.

beam-lead isolation—The method in which electrical isolation between IC elements is produced by interconnecting the elements with thick gold leads and selectively etching the silicon from between elements without affecting the gold leads. This process leaves the elements as separate units supported by the gold leads.

beam leads—1. A generic term describing a system in which flat metallic leads extend from the edges of a chip component much as wooden beams extend from a roof overhang. These are then used to interconnect the component to film circuitry. 2. Techniques for attachment of lead frames to silicon chips, including vacuum and chemical deposition, diffusion thermal-compression techniques, welding, etc.

beam-lobe switching—A method of determining the direction of a remote object by comparison of the signals corresponding to two or more successive beam angles at directions slightly different from that of the object.

beam modulation—*See* Z-Axis Modulation.

beam parametric amplifier—A parametric amplifier in which a modulated electron beam provides a variable reactance.

beam-positioning magnet—A magnet used with a tricolor picture tube to influence the direction of one of the electron beams so that it will have the proper spatial relationship with the other two beams.

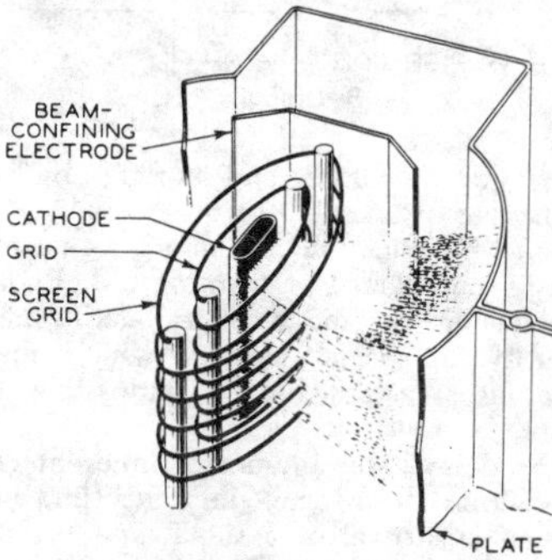

Beam-power tube.

beam-power tube—An electron-beam tube in which directed electron beams are used to contribute substantially to its power-handling capability, and in which the control and screen grids essentially are aligned.—*See* illustration below.

beam relaxor—A type of sawtooth scanning-oscillator circuit which generates but does not amplify the current wave required for magnetic deflection in a single beam-power pentode.

beam-rider control system – A system whereby the control station sends a beam to the target, and the missile follows this beam until it collides with the target.

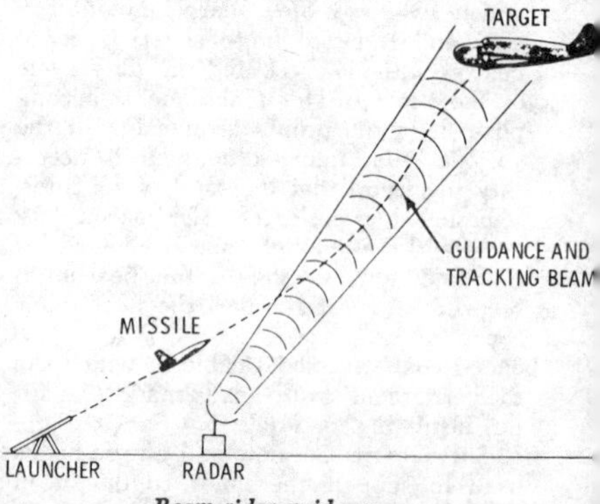

Beam-rider guidance.

beam-rider guidance—A form of missile guidance wherein a missile, through a self-contained mechanism, automatically guides itself along a beam transmitted by a radar.

beam splitting – A process for increasing the accuracy in locating targets by radar. By noting the azimuths at which one radar scan first discloses a target and at which radar data from it ceases, beam splitting calculates the mean azimuth for the target.

beam spreader – An optical element the purpose of which is to impart a small angular divergence to a colliminated incident beam.

beam switching – A method of obtaining more accurately the bearing and/or elevation of an object by comparing the signals received when the beam is in a direction differing slightly in bearing and/or elevation. When these signals are equal, the object lies midway between the beam axes.

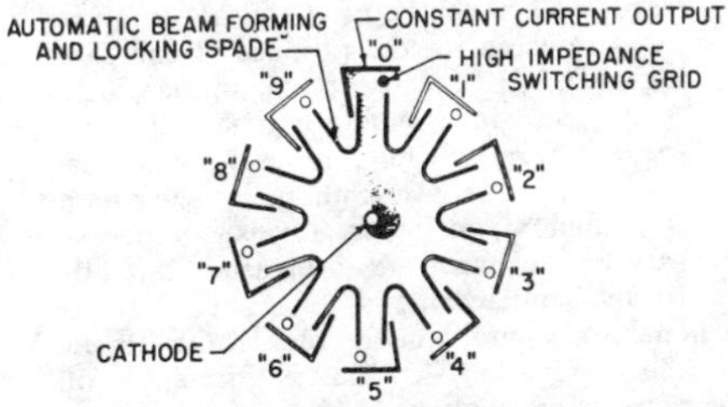

Beam-switching tube.

beam-switching tube – A multiposition, high-vacuum, constant-current distributor. The beam-switching tube consists of many identical "arrays" around a central cathode. Each array comprises a spade which automatically forms and locks the electron beam, a target-output electrode which gives the beam current its constant characteristics, and a high-impedance switching grid which switches the beam from target to target. A small cylindrical magnet, permanently attached to the glass envelope, provides a magnetic field. This field, in conjunction with an applied electric field, comprises the crossed fields necessary for operation of this tube. It is used in electronic switching and in distributing such as counting, timing, sampling, frequency dividing, coding, matrixing, telemetering, and controlling.

beam width—1. The angular width of a radio, radar, or other beam measured between two reference lines. 2. The width of a radar beam measured between lines of half-power intensity.

bearing—1. The horizontal direction of an object or point, usually measured clockwise from a reference line or direction through 360°. 2. Support for a rotating shaft. 3. The horizontal angle at a given point, measured from a specific reference datum, to a second point relative to another as measured from a specific reference datum.

bearing cursor—A mechanical bearing line of a plan-position indicator type of display for reading the target bearing.

bearing loss—The loss of power through friction in the bearings of an electric motor (brushes removed and no current in the windings).

bearing resolution—The minimum angular separation in a horizontal plane between two targets at the same range that will allow an operator to obtain data on either individual target.

beat—Periodic variations that result from the superimposition of waves having different frequencies. The term is applied both to the linear addition of two waves, resulting in a periodic variation of amplitude, and to the nonlinear addition of two waves, resulting in new frequencies, of which the most important usually are the sum and difference of the original frequencies.

beat frequency—One of the two additional frequencies produced when two different frequencies are combined. One beat frequency is the sum of the two original frequencies; the other is the difference between them.

beat-frequency oscillator—Abbreviated bfo. An oscillator that produces a signal which mixes with another signal to provide frequencies equal to the sum and difference of the combined frequencies.

beating—The combining of two or more frequencies to produce sum and difference frequencies called beats.

beating-in—Interconnecting two transmitter oscillators and adjusting one until no beat frequency is heard in a connected receiver. The oscillators are then at the same frequency.

beating oscillator—*See* Local Oscillator.

beat note—The difference frequency produced when two sinusoidal waves of different frequencies are applied to a nonlinear device.

beat reception—*See* Heterodyne Reception.

beats—1. Beat notes which are generally at a sufficiently low audio frequency that they can be counted. 2. The signal formed when two signals of different frequencies are simultaneously present in a nonlinear device. The frequency of the beat is equal to the difference in frequency of the two primary signals. For example, beats are produced in superheterodyne receivers, where the beat is between the incoming signal and the local oscillator in the receiver.

beat tone—Musical tone due to beats, produced by the heterodyning of two high-frequency wave trains.

beaver tail – A fan-shaped radar beam, wide in the horizontal plane and narrow in the vertical plane. The beaver tail is swept up and down for height finding.

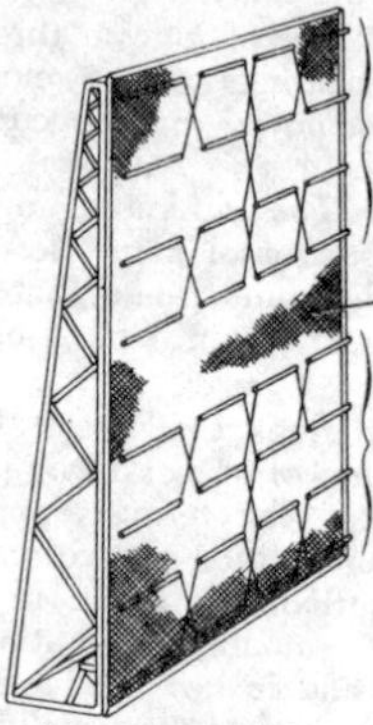

Bedspring antenna.

bedspring—A broadside antenna array with a flat reflector.

before start—The interval before the starting circuit to a timer has been operated. The timer is fully reset and all contacts are in the precycle position.

bel—The fundamental unit in a logarithmic scale for expressing the ratio of two amounts of power. The number of bels is equal to the $\log_{10} P_1/P_2$, where P_1 is the power level being considered and P_2 is an arbitrary reference level. The decibel, equal to 1/10 bel, is a more commonly used unit.

B eliminator—A power pack that changes the ac power-line voltage to the dc source required by the vacuum tubes. In this way, batteries can be eliminated.

bell—An electrical device consisting of a hammer vibrated by an electromagnet. The hammer strikes the sides of the bell and emits a ringing noise. The electromagnet attracts an armature or piece of soft iron forming part of the hammer lever. A contact breaker then opens the circuit and cuts off the attraction. A spring draws the hammer back to its original position, closing the circuit and repeating the action.

Bellini-Tosi antenna – A direction-finding antenna comprising two vertical orthogonal triangular loops installed with their bases over ground and used with a goniometer.

Bellini-Tosi direction finder—An early radio direction-finder system consisting of two loop antennas at right angles to each other and connected to a goniometer.

bellows – A pressure-sensing element consisting of a ridged metal cylinder closed at one end. A pressure difference between its outside and inside will cause the cylinder to expand or contract along its axis.

bellows contact – A contact in which a multileaf spring is folded. This type provides a more uniform spring rate over the full tolerance range of the mating unit.

bell transformer—A small iron-core transformer; its primary coil is connected to an ac primary line, and its secondary coil delivers 10 to 20 volts for operation of a doorbell, buzzer, or chimes.

bell wire—Cotton-covered copper wire, usually No. 18, used for doorbell and thermostat connections in homes and for similar low-voltage work.

belt drive—A drive system used to rotate a turntable in which the motor pulley drives the platter with a belt.

benchmark—In connection with microprocessors, a frequently used routine or program selected for the purpose of comparing different makes of microprocessors. A flowchart in assembly language is written out for each microprocessor and the execution of the benchmark by each unit is evaluated on paper. (It is not necessary to use hardware to measure capability by benchmark.)

benchmark problem—A problem used in the evaluation of the performance of computers relative to each other.

benchmark program – A sample program used to evaluate and compare computers. In general, two computers will not use the same number of instructions, memory words, or cycles to solve the same problems.

bench test—A test in which service conditions are approximated, but the equipment is conventional laboratory equipment and not necessarily identical with that in which the product will be employed in normal service.

bend—A change in the direction of the longitudinal axis of a waveguide.

bend waveguide—A section of waveguide in which the direction of the longitudinal axis is changed.

Benito—A cw navigational system in which the distance to an aircraft is determined on the ground by measuring the phase difference of an audio signal transmitted from the ground and retransmitted by the aircraft. Bearing information is obtained by ground direction finding of the aircraft signals.

Bessel function—A mathematical function used in the design of a filter for maximally constant time delay with little consideration for amplitude response. This function is very close to a Gaussian function.

beta—Symbolized by the Greek letter *beta* (β). Also called current-transfer ratio. 1. The current gain of a transistor connected as a grounded-emitter amplifier; it is the ratio of a small change in collector current to the corresponding change in base current, with the collector voltage constant. 2. A symbol used to denote B quartz. 3. Brainwave signals whose frequency is approximately 13 to 28 Hz. The associated mental state is irritation, anger, jitter, frustration, worry, tension, etc.

beta circuit—In a feedback amplifier, the circuit which transmits a portion of the amplifier output back to the input.

beta cutoff frequency — The frequency at which the beta of a transistor is 3 decibels below the low-frequency value.

beta particle—A small electrically charged particle thrown off by many radioactive materials. It is identical to the electron and possesses the smallest negative electrical charge found in nature. Beta particles emerge from radioactive material at high speeds, sometimes close to the speed of light.

beta ray—1. A stream of beta particles. 2. Electrons or positrons given off by a radioactive nucleus in the process of decay.

betatron—A large doughnut-shaped accelerator which produces artificial beta radiation. Electrons (beta particles) are whirled through a changing magnetic field. They gain speed with each trip and emerge with high energies (on the order of 100 million electron volts in some instances).

bev—A billion electron volts. An electron possessing this much energy travels at a speed close to that of light—186,000 miles a second.

bevatron—A very large circular accelerator in which protons are whirled between the poles of a huge magnet to produce energies in excess of 1 billion electron volts.

Beverage antenna—*See* Wave Antenna.

beyond-the-horizon propagation—*See* Scatter Propagation.

beyond-the-horizon transmission—*See* Scatter Propagation.

bezel—1. A holder designed to receive and position the edges of a lens, meter, window, or dial glass. 2. The flange or cover used for holding an external graticule or crt cover in front of the crt in an oscilloscope. May also be used for mounting a trace-recording camera or other accessory item.

bfo—Abbreviation for beat-frequency oscillator.

B-H curve—Curve plotted on a graph to show successive states during magnetization of a ferromagnetic material. A normal magnetization curve is a portion of a symmetrical hysteresis loop. A virgin magnetization curve shows what happens the first time the material is magnetized.

B-H meter—A device for measuring the intrinsic hysteresis loop of a sample of magnetic material.

bias—1. The electrical, mechanical, or magnetic force applied to a relay, semiconductor, vacuum tube, or other device for the purpose of establishing an electrical or mechanical reference level for the operation of the device. 2. Direct-current potential applied to the control grid of a vacuum tube. 3. Bias derived from a direct current, used on signaling or telegraph relays or electromagnets to secure the desired time spacing of transitions from marking to spacing. 4. A method of restraining a relay armature, by means of spring tension, to secure a desired time spacing of transitions from marking to spacing. 5. The average direct-current voltage between the control grid and cathode of a vacuum tube. 6. The effect on teletypewriter signals produced by the electrical characteristics of the line and the equipment. 7. Energy applied to a relay to hold it in a given position. 8. A high-frequency signal applied to the audio signal at the tape recording head to make the audio signal magnetize the tape over the linear part of head's magnetic characteristic. Although sometimes dc (fixed magnetic polarity) is used, the bias signal is usually above 40 kHz to avoid audible intermodulation distortion. 9. The sidways thrust of a pickup arm.

bias cell—A dry cell used in the grid circuit of a vacuum tube to provide the necessary C-bias voltage.

bias compensator—A device which counteracts the inward bias of a pickup arm as it tracks the record. The compensator exerts an outward force on the arm and generally can be adjusted to have a definite relationship to the playing weight of the pickup.

bias current—The current through the base-emitter junction of a transistor. It is adjusted to set the operating point of the transistor.

bias distortion—1. Distortion resulting from operation on a nonlinear portion of the

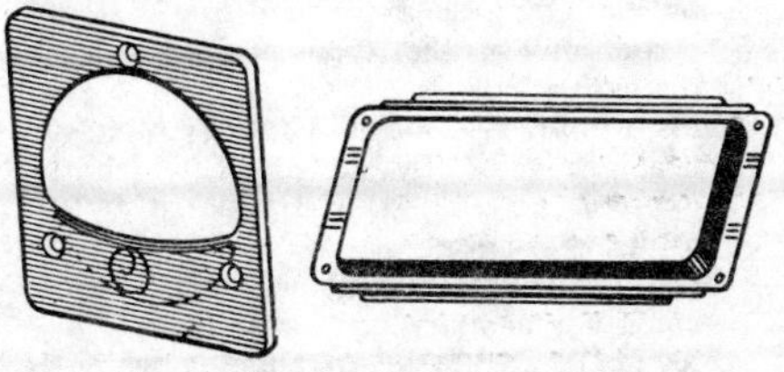

Bezels, 1.

characteristic curve of a vacuum tube, semiconductor, or other device, due to improper biasing. 2. In teletype circuits, the uniform shifting of mark pulses from their proper position in relationship to the start pulses.

biased induction—Symbolized by B_b. The biased induction at a point in a magnetic material which is subjected simultaneously to a periodically varying magnetizing force and a biasing magnetizing force and is the algebraic mean of the maximum and minimum values of the magnetic induction at the point.

bias-induced noise—The difference between bulk-erased and zero-modulation noise.

biasing magnetizing force—Symbolized by H_b. A biasing magnetizing force at a point in a magnetic material which is subjected simultaneously to a periodically varying magnetizing force and a constant magnetizing force and is the algebraic mean of the maximum and minimum values of the combined magnetizing forces.

bias meter—A meter used in teletypewriter work for determining signal bias directly in percent. A positive reading indicates a marking signal bias; a negative reading indicates a spacing signal bias.

bias oscillator—An oscillator used in magnetic recorders to generate an ac signal in the range of 40 to 80 kHz for the purpose of magnetic biasing to obtain a linear recording characteristic. Usually the bias oscillator also serves as the erase oscillator.

bias port—In a fluidic device, the port at which a biasing signal is applied.

bias resistor—A resistance connected into a self-biasing vacuum-tube or semiconductor circuit to produce the voltage drop necessary to provide a desired biasing voltage.

bias-set frequency—In direct magnetic tape recording, a specified recording frequency employed during the adjustment of bias level for optimum record performance (not the frequency of the bias).

bias telegraph distortion — Distortion in which all mark pulses are lengthened (positive bias) or shortened (negative bias). It can be measured with a steady stream of "unbiased reversals" (square waves having equal-length mark and space pulses). The average lengthening or shortening does not give true bias distortion unless other types of distortion are negligible.

bias windings—Control windings of a saturable reactor, by means of which the operating condition is translated by an arbitrary amount.

biax—Two-hole, orthogonal cubical ferrite computer memory elements.

biconical antenna — An antenna which is formed by two conical conductors, having a common axis and vertex, and excited at the vertex. When the vertex angle of one of the cones is 180°, the antenna is called a discone.

bidirectional—Responsive in opposite directions. An ordinary loop antenna is bidirectional because it has maximum response from the opposite directions in the plane of the loop.

bidirectional antenna—An antenna having two directions of maximum response.

bidirectional current—A current which is both positive and negative.

bidirectional diode-thyristor—A two-terminal thyristor having substantially the same switching behavior in the first and third quadrants of the principal quadrants of the principal voltage-current characteristic.

bidirectional microphone—1. A microphone in which the response predominates for sound incidences of 0° and 180°. 2. A microphone which is equally sensitive to sounds arriving at it from in front or in back, but discriminates against sounds arriving at it from the sides.

bidirectional pulses—Pulses, some of which rise in one direction and the remainder in the other direction.

bidirectional pulse train — A pulse train, some pulses of which rise in one direction and the remainder in the other direction.

bidirectional transducer — *See* Bilateral Transducer.

bidirectional transistor—A transistor which is specified with parameter limits in both the normal and inverted configuration and has substantially the same electrical characteristics when the terminals normally designed as emitter and collector are interchanged. (Bi-directional transistors are sometimes called symmetrical transistors. This term, however, is deprecated as it might give the incorrect impression of an ideally symmetrical transistor.)

bidirectional triode thyristor—A three-terminal thyristor having substantially the same switching behavior in the first and third quadrants of the principal voltage-current characteristic.

bifilar resistor — A resistor wound with a wire doubled back on itself to reduce the inductance.

bifilar suspension—A type of galvanometer movement that is highly resistant to overloads in which a D'Arsonval moving coil is supported at each end by two taut

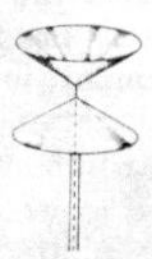

Biconical antenna.

wires. The elimination of the pivot, with its attendant friction, results in superior sensitivity and precision.

bifilar transformer—A transformer in which the turns of the primary and secondary windings are wound together side-by-side and in the same direction. This type of winding results in near unity coupling, so that there is a very efficient transfer of energy from primary to secondary.

bifilar winding—A method of winding non-inductive resistors in which the wire is folded back on itself and then wound double, with the winding starting from the point at which the wire is folded.

bifurcate—Having to do with the lengthwise slotting of a flat spring contact in a printed-circuit connector, to provide additional independently operating points of contact.

bifurcated—Usually fork-shaped. Refers to physical construction of a contact whereby two mating portions make physical contact. Yet, if one tip section of the contact fails, the remaining section maintains the physical and electrical connection.

bifurcated contact — A movable contact which is forked (divided) to provide two contact-mating surfaces in parallel for a more reliable contact.

bilateral—Having a voltage-current characteristic curve that is symmetrical with respect to the origin, that is, being such that if a positive voltage produces a positive current magnitude, an equal negative voltage produces a negative current of the same magnitude.

bilateral amplifier—An amplifier capable of receiving as well as transmitting signals; it is used primarily in transceivers.

bilateral antenna—An antenna, such as a loop, having maximum response in exactly opposite directions (180° apart).

bilateral bearing — A bearing which indicates two possible directions of wave arrival. One of these is the true bearing, and the other is a bearing displaced 180° from the true bearing.

bilateral circuit—A circuit wherein equipment at opposite ends is managed, operated, and maintained by different services.

bilateral element—A two-terminal element, the voltage-current characteristic of which has odd symmetry around the origin.

bilateral network—A network in which a given current flow in either direction results in the same voltage drop.

bilateral transducer—1. Also called bidirectional transducer. A transducer capable of transmission simultaneously in both directions between at least two terminations. 2. A device capable of measuring stimuli in both a positive and a negative direction from a reference zero or rest position.

billboard antenna—An antenna array consisting of several bays of stacked dipoles spaced ¼ to ¾ wavelength apart, with a large reflector placed behind the entire assembly. The required spacing of the dipoles tends to make the array inconveniently large at frequencies below the vhf range.

bimag—*See* Tape-Wound Core.

bimetal cold junction compensation—Automatic mechanical correction for ambient temperature change at the cold junction of a thermocouple which would normally cause erroneous readings.

bimetallic strip—A strip formed of two dissimilar metals welded together. Because the metals have different temperature coefficients of expansion, the strip bends or curls when the temperature changes.

bimetallic thermometer—A device containing a bimetallic strip which expands or contracts as the temperature changes. A calibrated scale indicates the amount of change in temperature.

bimetallic thermostat — A temperature-responsive device in which the sensing element is a strip formed of two dissimilar metal pieces. The two pieces are welded together, and, because of the unequal thermal coefficients of expansion, they deform or curl when the temperature is changed.

bimetal mask—A mask formed by chemically etching openings in a metal film or plate where it is not protected by photoresist or other chemically resistant material.

bimorph cell—Two crystal elements (usually Rochelle salt) in rigid combination, arranged to act as a mechanical transforcer in headphones, microphones, pickups, and speakers.

binary—1. A numbering system using a base number, or radix, of 2. There are two digits (1 and 0) in the binary system. 2. Pertaining to a characteristic or property involving a selection, choice, or condition in which there are two possibilities. 3. A bistable multivibrator.

binary card—A card that contains data in column binary or row binary form.

binary cell—In an electronic computer, an elementary unit of storage which can be placed in either of two stable states.

binary chain—A series of binary circuits, each of which can exist in either one of two states, arranged so each circuit can affect or modify the condition of the next circuit.

binary channel—A transmission facility limited to the use of two symbols.

binary code — A method of representing numbers in a scale of two (on or off, high level, or low level, one or zero, presence or absence of a signal) rather than the more familiar scale of ten used in normal arithmetic. Electronic circuits designed to work in two defined states are much simpler and more reliable than those working in ten such states.

binary-coded—Expressed by a series of binary digits (0's and 1's).

binary-coded character — A decimal digit, alphabetic letter, punctuation mark, etc., represented by a fixed number of consecutive binary digits.

binary coded decimal — Abbreviated bcd. A coding system in which each decimal digit from 0 to 9 is represented by four binary digits:

decimal digit	*binary code*
0	0000
1	0001
2	0010
3	0011
4	0100
5	0101
6	0110
7	0111
8	1000
9	1001

binary-coded digit—One element of a notation system for representing a decimal digit by a fixed number of binary positions.

binary-coded octal system—An octal numbering system in which each octal digit is represented by a three-place binary number.

binary counter—*See* Binary Scaler.

binary digit—1. A character that represents one of the two digits in the number system that has a radix of two. Also called bit. 2. Either of the digits, 0 or 1, that may be used to represent the binary conditions on or off. 3. A whole number in the binary scale of notation; this digit may be only 0 (zero) or 1 (one). It may be equivalent to an "on" or "off" condition, "yes" or "no," etc.

binary incremental representation—Incremental representation in which the value of an increment is rounded to plus or minus one quantum and is represented by one binary digit.

binary magnetic core—A ring-shaped magnetic material which can be made to take either of two stable states of magnetic polarization.

binary notation—*See* Binary Number System.

binary number system—A number system using two symbols (usually denoted 0 and 1) and having 2 as its base, just as the decimal system uses 10 symbols (0, 1, . . . , 9) and has a base of 10. Also called binary notation.

binary numeral—The binary representation of a number—e.g., "101" is the binary numeral and "V" is the Roman numeral of the number of fingers on one hand.

binary point—The point which marks the place between integral powers of two and fractional powers of two in a binary number.

binary pulse-code modulation—A form of pulse-code modulation in which the code for each element of information consists of one of two distinct kinds, e.g., pulses and spaces.

binary scaler—Also called binary counter. 1. A counter which produces one output pulse for every two input pulses. 2. A counting circuit, each stage of which has two distinguishable states. 3. A flip-flop having a single input (called a T flip-flop). Each time a pulse appears at the input, the flip-flop changes state.

binary search — Also called dichotomizing search. A search in which a set of items is divided into two parts; one part is rejected, and the process is repeated on the accepted part until those items with the desired property are found.

binary signal—A voltage or current that carries information in the form of changes between two possible values.

binary signaling—A communications mode in which information is passed by the presence and absence, or plus and minus variations, of one parameter of the signaling medium.

binary system—A system of mathematical computation based on powers of 2.

binary-to-decimal conversion—The process of converting a number written in binary notation to the equivalent number written in the ordinary decimal notation.

binary word—A related grouping of ones and zeroes that has a meaning assigned by definition, or that has a weighted numerical value in the natural binary number system.

binaural — Two-channel sound in which each channel recorded is heard only through one ear. Microphones are spaced in recording, to approximate the distance between a person's own ears. To hear the recording binaurally, the listener must use headphones. (Compare with stereo.)

binaural disc—A stereo record with two separate signals recorded in its grooves. Stereophonic sound is obtained by feeding each signal into its own speaker or headphone.

binaural effect—The effect which makes it possible for a person to distinguish the difference in arrival time or intensity of sound at his ears and thereby determine the direction from which a sound is arriving.

binaural recorder—A tape recorder which employs two separate recording channels, or systems, each with its own microphone, amplifier, recording and playback heads, and earphones. Recordings using both channels are made simultaneously on a single magnetic tape having two parallel tracks. During playback, the original sound is reproduced with depth and realism. For a true binaural effect, headphones are necessary.

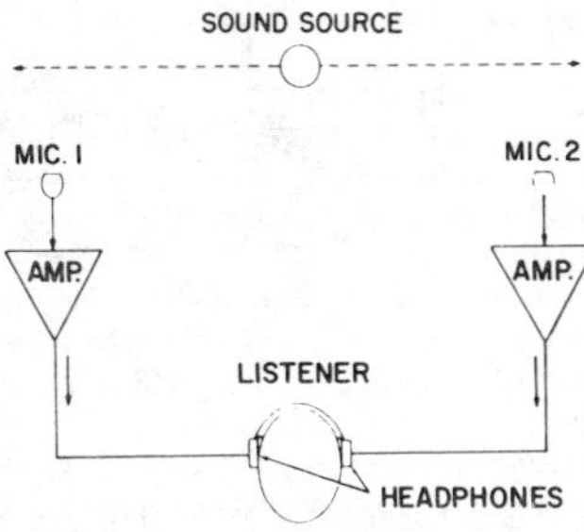

Binaural sound reproducing system.

binaural sound—Sound recorded or transmitted by pairs of equipment so as to give the listener the effect of having heard the original sound.

binder—A substance, like cement, used to hold particles together and thus provide mechanical strength in, for example, carbon resistors and phonograph records.

binding energy—The minimum energy required to dissociate a nucleus into its component neutrons and protons. Neutron or proton binding energies are those energies required to remove a neutron or a proton, respectively, from a nucleus. Electron binding energy is that energy required to remove an electron from an atom or a molecule.

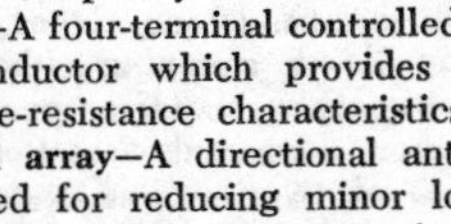

Binding post.

binding post—A bolt-and-nut terminal for making temporary electrical connections.

binistor—A four-terminal controlled rectifier semiconductor which provides bistable, negative-resistance characteristics.

binomial array—A directional antenna array used for reducing minor lobes and providing maximum response in opposite directions.

biochemical fuel cell—An electrochemical generator of electrical power in which bi-organic matter is used as the fuel source. In the usual electrochemical reaction, air serves as the oxidant at the cathode and microorganisms are used to catalyze the oxidation of the bi-organic matter at the anode.

bioelectricity—Electric currents and potential differences which occur in living tissues. Muscle and nerve tissue, for example, are generators of bioelectricity, although the potential registered may be less than one millivolt in some cases.

bioelectric potential—*See* Bioelectricity.

bioelectrogenesis—The practical application of electricity drawn directly from the bodies of animals, including humans, to power electronic devices and appliances.

bioelectronics—1. The application of electronic theories and techniques to the problems of biology. 2. The integrated, long-term electronic control of various, impaired, physiologic systems by means of small, low-power, electrical, and electromechanical devices. (The pacemaker is therefore a bioelectronic instrument.)

bioengineering—*See* Bionics.

biogalvanic battery—A device that makes use of reactions between metals and the oxygen and fluids in the body to generate electricity.

biological shield—A mass of absorbing material placed around a reactor or radioactive source to reduce the radiation to a level that is safe for human beings.

biologic energy—Energy that is produced by bodily processes and that can be used to supply electrical energy for implanted devices such as electronic cardiac pacemakers, bladder stimulators, etc. The biologic energy can result from muscle movement (such as that of the diaphragm), temperature differences, pressure differences, expansion of the aorta, oxidation of materials within the gastrointestinal tract, and other processes.

biomechanics—The application of mechanical laws to living animals. In common usage, this term is applied only to the locomotor system. Analysis of the motion of limbs permits the design of electronic devices that allow paralyzed individuals to perform useful tasks. Actual motions are accomplished either by stimulation of muscles or by programmed movement of braces.

biomedical oscilloscope — An oscilloscope designed or modified to be used in medical applications. Such oscilloscopes have slow sweep rates and long-persistence screens because of the low frequencies of many biological signals.

biometrics—The science of statistics as applied to biological observations.

bionics—1. The study of living systems so that their characteristics and functions can be related to the development of mechanical and electronic hardware. 2. The reduction of various life processes to mathematical terms to make possible duplication or simulation with systems hardware. Also called bioengineering. 3. The art which treats electronic simulation of biological phenomena.

biotelemetry—The process of remote measurement or recording of such biological variables as pulse rate, temperature, etc.

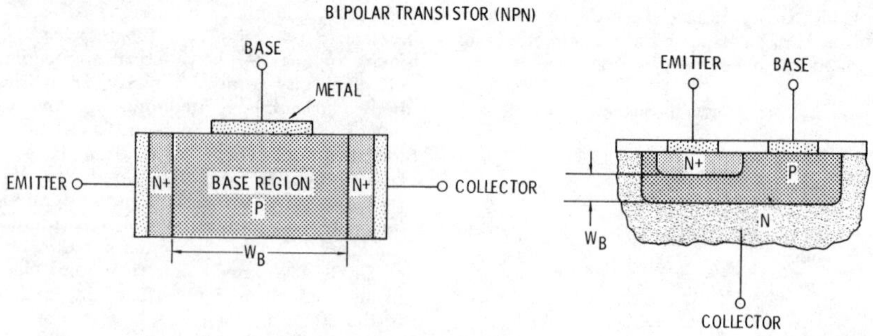

Bipolar transistor.

Typically, the information is transmitted between the patient and the receiving equipment by a radio link.

biotelescanner—A device that can analyze and radio data on life forms during space exploration.

bipolar—1. Having two poles. 2. Having to do with a device in which both majority and minority carriers are present. In connection with ICs, the term describes a specific type of construction; bipolar and MOS are the two most common types of IC construction. 3. The semiconductor technology employing two-junction transistors. 4. A transistor structure whose electrical properties are determined within the silicon material. Memories using this technology are characteristically high-speed devices. 5. General name for npn and pnp transistors since working current passes through semiconductor material of both polarities (p and n). Also applied to integrated circuits that use bipolar transistors.

bipolar device—A semiconductor device in which there are both majority and minority carriers. (This is the case in all npn and pnp transistors.)

bipolar electrode — An electrode without metallic connection with the current supply, one face of which acts as an anode surface and the opposite face as a cathode surface when an electric current is passed through a cell.

bipolar magnetic driving unit — A headphone or speaker unit having two magnetic poles acting directly on a flexible iron diaphragm.

bipolar memory cell—In a computer, a system comprising a storage latch, a pair of control gates, and an output gate. The control and output gates need not be part of the storage cell, but they usually are included because each latch requires both control and output gating.

bipolar pulse—A pulse that has appreciable amplitude in both directions from the reference axis.

bipolar transistor — A transistor that uses both negative and positive charge carriers.

biquinary code—A mixed-radix notation in which each decimal digit to be represented is considered to be the sum of two digits, the first of which is zero or one with significance five and the second of which is 0, 1, 2, 3, or 4 with significance one.

biradial—Having an elliptical cross section. A term used with reference to phonograph styli.

biradial stylus—Also called elliptical stylus. A stylus tip that has a small radius where it touches the walls of the record grooves as distinguished from a conventional stylus which has a hemispherical tip used with lightweight pickup arms to reduce tracking distortion.

bird-dogging—*See* Hunting, 1.

birdnesting—Clumping together of chaff dipoles after they have been dropped from an aircraft.

biscuit—*See* Preform.

bistable—1. A circuit element with two stable operating states—e.g., a flip-flop in which one transistor is saturated while the other is turned off. It changes state for each input pulse or trigger. 2. A device capable of assuming either one of two stable states. 3. Of or pertaining to the general class of devices that operate in either of two possible states in the presence or absence of the setting input.

bistable contacts—A contact combination in which the movable contact remains in its last operated position until the magnetic polarity of the coil is reversed.

bistable multivibrator (flip-flop)—A circuit having two stable states; it will stay in either one indefinitely until appropriately triggered, after which it immediately switches to the other state.

bistable relay—A relay that requires two pulses to complete one cycle composed of two conditions of operation. Also called locked, interlocked, and latching relay.

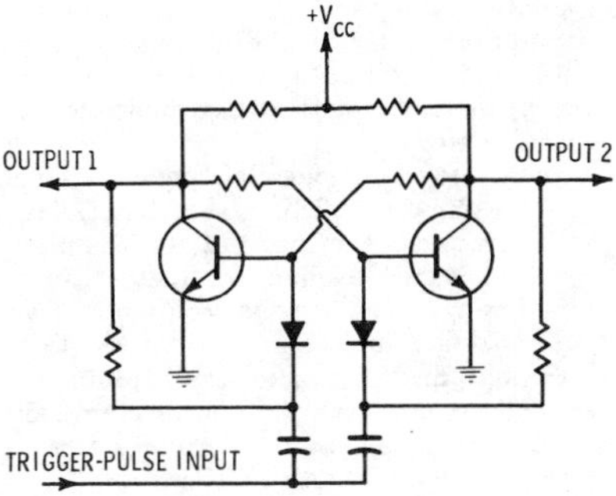

Bistable multivibrator.

bistatic radar—Radar system in which the receiver and transmitter have separate antennas and are some distance apart.

biswitch—A two-terminal integrated device which basically performs the function of two pnpn switches interconnected so as to provide bilateral switching.

bit—1. Abbreviation for binary digit. A unit of information equal to one binary decision, or the designation of one of two possible and equally likely values or states (such as 1 or 0) of anything used to store or convey information. It may also mean "yes" or "no." 2. The smallest part of information in a binary notation system. A bit is either a ONE ("1") or a ZERO (0). In a bcd system, four bits represent one decimal digit.

bit density—The number of bits of information contained in a given area, such as the number of bits written along an inch of magnetic tape.

bit interleave—A technique in time-division multiplexing in which bits of data are transmitted in one frame.

bit rate—The number of binary bits transmitted per unit time; for example, a bit rate of 80 means that 80 binary bits are transmitted per second.

bit slice—As in a 4-bit slice and a 2-bit slice. A multichip microprocessor in which the control section is contained on one chip, and one or more identical arithmetic and logic unit alu sections and register sections are contained on separate chips called slices. For example, three 4-bit slices connected in parallel with the control section produce a 12-bit word microprocessor.

bit stream—A binary signal without regard to grouping according to character.

bit string—A string of binary digits in which each bit position is considered an independent unit.

bit time—In a serial binary computer, the time during which each bit appears.

black—A signal produced at any point in a facsimile system by the scanning of a selected area of subject copy having maximum density.

black and white—*See* Monochrome.

black and white transmission—*See* Monochrome Transmission.

black area—An area with encrypted signal only present.

blackbody – 1. An idealized emitter for which total radiated energy and the spectral distribution of the energy are accurately known functions of temperature. These relationships are given by Planck's law. 2. A solid that radiates or absorbs energy with no internal reflection of the energy at any wavelength. Physically it may be a hollow sphere coated on the inside with lampblack and with an opening through which energy may enter or leave. 3. An ideal body that absorbs all incident light and therefore appears perfectly black at all wavelengths. The radiation emitted from such a body when it is hot is called blackbody radiation. The spectral energy density of blackbody radiation is the theoretical maximum for a body in thermal equilibrium.

blackbody luminous efficiency—As a function of temperature, the efficiency of an incandescent blackbody in terms of visible light.

blackbody radiation—*See* Blackbody.

black box—1. A term used loosely to refer to any subcomponent that is equipped with connects and disconnects so that it can be readily inserted into or removed from a specified place in a larger system (e.g., the complete missile or some major subdivision) without benefit of knowledge of its detailed internal structure. 2. A term pertaining to either the functional transformation that acts upon a specified input to give a particular output or to the apparatus for accomplishing this transformation (without regard to the detailed circuitry used). 3. A useful mathematical approach to an electronic circuit which concerns itself only with the input and output, and ignores the interior elements, discrete or integrated.

black compression—Also called black saturation. The reduction in the gain of a television picture signal at those levels corresponding to dark areas in the picture with respect to the gain at that level corresponding to the midrange light value in the picture. The overall effect of black compression is to reduce contrast in the low lights of the picture.

blacker-than-black—The amplitude region of the composite video signal below the reference black level in the direction of the synchronizing pulses.

blacker-than-black level – A voltage value used in an electronic television system for control impulses. It is greater than the value representing the black portions of the image.

black level—That level of the picture signal corresponding to the maximum limit of black peaks.

black light—Invisible light radiation. May be either ultraviolet or infrared radiation, both of which are invisible.

black-light emitter—A source of electromagnetic radiation in the ultraviolet or infrared region, just outside the visible spectrum.

blackout—1. Interruption of radiocommunication due to excess absorption caused by solar flares. During severe blackouts, all frequencies above approximately 1500 kHz are absorbed excessively in the daylight zone. 2. Passive defense that consists of interrupting all forms of communication or identification.

black peak—A peak excursion of the picture signal in the black direction.

black saturation—*See* Black Compression.

black scope—Cathode-ray tube being operated at the threshold of luminescence with no video signals applied.

black signal—Also called picture black. A signal produced at any point in a facsimile system by the scanning of a maximum-density area of the subject copy.

black transmission — 1. In an amplitude-modulation facsimile system a form of transmission in which the maximum transmitted power corresponds to the maximum density of the copy. 2. In a frequency-modulation system, a form of transmission, in which the lowest transmitted frequency corresponds to the maximum density of the copy.

blank—1. The result of the final operation on a crystal. 2. To cut off the electron beam of a cathode-ray tube. 3. A code character to denote the presence of no information rather than the absence of information. In the Baudot code, it is composed of all spacing pulses. In paper tape, it is represented by a feedhole without intelligence holes.

blank coil—Tape for perforation in which only the feed holes have been punched.

blank deleter—A device used to eliminate the receiving of blanks in perforated paper tape.

blanked picture signal—The signal resulting from adding blanking to a picture signal. Adding the sync signal to the blanked picture signal forms the composite picture signal.

blanketing—The overriding of a signal by a more powerful one or by interference, so that a receiver is unable to receive the desired signal.

blank groove—*See* Unmodulated Groove.

blanking—The process of making a channel or device noneffective for the desired interval. In television, blanking is the substitution for the picture signal, during prescribed intervals, of a signal the instantaneous amplitude of which is such as to make the return trace invisible. *See also* Gating.

blanking level—Also called pedestal level. In a composite picture signal, the level that separates the range of the composite picture signal containing picture information from the range containing synchronizing information.

blanking pulse—A square wave (positive or negative) used to switch off electronically a part of a television or radar set for a predetermined length of time.

blanking signal—A wave made up of recurrent pulses related in time to the scanning process and used to effect blanking. In television, this signal is composed of pulses at line and field frequencies, which usually originate in a central sync generator and are combined with the picture signal at the pickup equipment in order to form the blanked picture signal. The addition of a sync signal completes the composite picture signal.

blanking time—The length of time the electron beam of a cathode-ray tube is cut off.

blanking zone—*See* Blanking Pulse.

blank instruction — *See* No-Operation Instruction.

blank tape — Tape on which nothing has been recorded. Also called raw tape or virgin tape.

blasting—Overloading of an amplifier or speaker, resulting in severe distortion of loud sounds.

bleeder current—The current drawn continuously from a power supply by a resistor. Used to improve the voltage regulation of the power supply.

bleeder resistor—1. A resistor used to draw a fixed current. Also used, as a safety measure, to discharge a filter capacitor after the circuit is de-energized. 2. A resistor placed in the power supply of a radio receiver or other electronic device to stabilize the voltage supply.

bleeding — 1. Migration of plasticizers, waxes, or similar materials to the surface to form a film or bead. 2. Poor edge acutance at the junction of two images on a photographic plate.

bleeding whites—An overloading condition in which white areas in a television picture appear to flow into the black areas.

bleedout—The tendency of absorbed electrolytes, impurities, base materials, and preplates to diffuse to the surface of gold plating.

blemish — On the storage surface of a charge-storage tube, an imperfection which produces a spurious output.

blended data—Q-point that results from the combination of scanning data and tracking data to form a vector.

blind approach – An aircraft landing approach when visibility is poor, usually made with the aid of instruments and radiocommunication.

blind zone – An area from which echoes cannot be received; generally, an area shielded from the transmitter by some natural obstruction and therefore from which there can be no return.

blinking–1. An ECM technique by which two aircraft separated a short distance and within the same azimuth resolution appear as one target to a tracking radar. The two aircraft alternately spot jam, causing the radar system to oscillate from one target to the other and making it impossible to obtain an accurate solution of a fire-control problem. 2. In pulse systems, a method of providing information in which the signal is modified at its source so that the presentation on the display scope alternately appears and disappears. In loran, this indicates that a station is malfunctioning.

blip–1. Sometimes referred to as pip. On a cathode-ray display, a spot of light or a base-line irregularity representing the radar reflection from an object. 2. A discontinuity in the insulation of a wire.

block–1. A group of computer words considered as a unit because they are in successive storage locations. 2. The set of locations or tape positions in which a block of words is stored. 3. A circuit assemblage that functions as a unit, such as a circuit building block of standard design or the logic block in a sequential circuit.

block address – A method of identifying words through use of an address that specifies the format and meaning of the words in a block of information.

block cancel character – A character used to signify that the preceding portion of the block is to be disregarded. Also called block ignore character.

block code – A special code or character used to separate blocks of data. A block code is used typically on paper tape and generally occurs at both the beginning and end of a block. Thus, the information on a paper tape containing a number of blocks would be started by a block code, there would be a block code between adjacent blocks, and the data would be ended by a block code.

block diagram–1. A diagram in which the essential units of any system are drawn

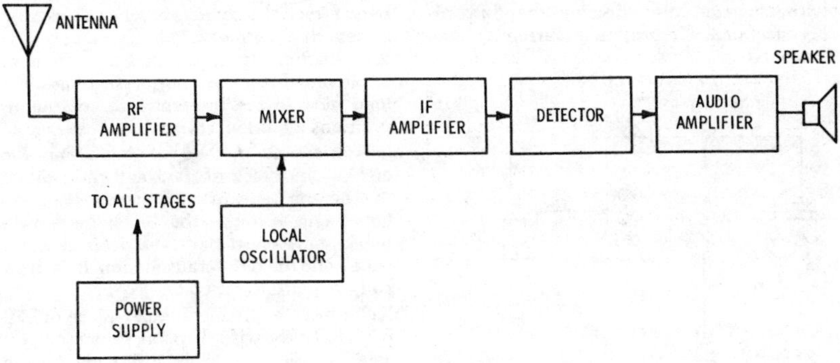

Block diagram, 1.

in the form of blocks, and their relationship to each other is indicated by appropriately connected lines. The path of the signal or energy may be indicated by lines and/or arrows. 2. In computer programming, a graphical representation of the data-processing procedures within the system. It is used by programmers as an aid to program development.

blip-frame ratio–The ratio of the number of computer frames during which radar data were obtained to the total number of computer frames.

blip-scan ratio–The ratio between a single recognizable blip on a radar scope and the number of scans necessary to produce it. The blip-scan ratio of any given radar set varies with the range, antenna tilt, level of operator and set performance, target aspect, wind, etc.

blister–The enclosure housing an airborne radar antenna.

blivet–An excess of coating material, such as a lump around a dust particle on a wire or a surface. *See also* Land, 2.

Bloch wall–The transition layer separating adjacent ferromagnetic domains.

blocked impedance–The input impedance of a transducer when its output is connected to a load of infinite impedance.

blocked resistance–Resistance of an audio-frequency transducer when its moving elements are restrained so they cannot move; it represents the resistance due only to electrical loss.

blockette – In digital computer programming, a subgroup, or subdivision, of a group of consecutive machine words transferred as a unit.

block gap—1. An area used to indicate the end of a block or record on a data medium. 2. An absence of data along a specified length of magnetic tape between adjacent blocks of data.

block-grid keying—A method of keying a continuous-wave transmitter by operating the amplifier stage as an electronic switch. During the spacing interval when the key is open, the bias on the control grid becomes highly negative and prevents the flow of plate current so that the tube has no output. During the marking interval when the key is closed, this bias is removed and full plate current flows.

block ignore character – *See* Block Cancel Character.

blocking—1. Application of an extremely high bias voltage to a transistor, vacuum tube, or metallic rectifier to prevent current from flowing in the forward direction. 2. Combining two or more records into one block.

blocking capacitor—A capacitor which introduces a comparatively high series impedance for limiting the flow of low-frequency alternating or direct current without materially affecting the flow of high-frequency alternating current.

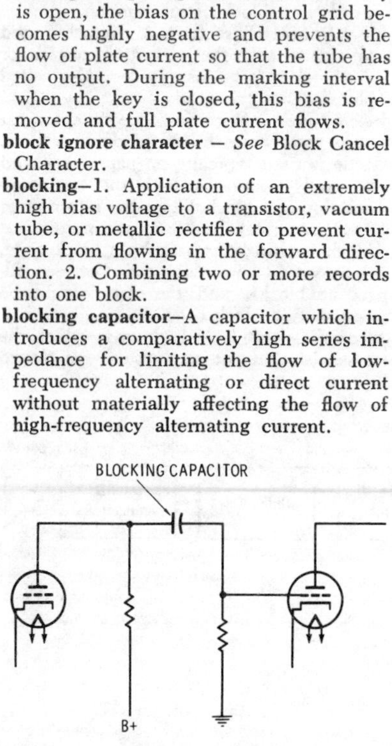

Blocking capacitor.

blocking layer—*See* Depletion Layer.

blocking oscillator—1. Also called squegging oscillator. An electron-tube oscillator that operates intermittently as its grid bias increases during oscillation to a point where the oscillations stop, and then decreases until oscillation resumes. 2. A relaxation oscillator consisting of an amplifier (usually a single stage) the output of which is coupled back to the input by means that include capacitance, resistance, and mutual inductance. 3. A relaxation-type oscillator that conducts for a short period of time and is cut off for a relatively long period of time.

blocking-oscillator driver—A circuit which develops a square pulse used to drive the modulator tubes, and which usually contains a line-controlled blocking oscillator that shapes the pulse into the square wave.

blocking relay—A device which initiates a pilot signal to block tripping on external faults in a transmission line or in other apparatus under predetermined conditions, or which cooperates with other devices to block tripping or to block reclosing on an out-of-step condition or on power swings.

block length—In a computer, the total number of records, words, or characters contained in one block.

block loading—In a computer, a form of fetch in which the control sections of a load module are brought into continuous positions of main storage.

block mark—A method of indicating the end of one block of data and the start of another on tape or in data transmission. On magnetic tape, the block mark is a block gap; on paper tape it is a block code; and in data transmission it is typically a pause or a code.

block protector—A rectangular piece of carbon, *Bakelite* with a metal insert, or porcelain with a carbon insert which, in combination with each other, make one

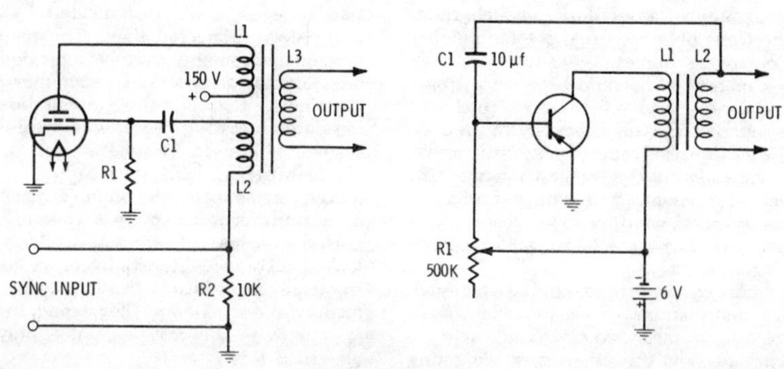

Blocking oscillator.

element of a protector. They form a gap which will break down and provide a path to ground for excessive voltages.

block sort—A computer-sorting technique in which the file is first divided according to the most significant character of the key, and the separate portions are then sorted one at a time. It is used particularly for large files.

block transfer—In a computer, the process of transmitting one or more blocks of data.

blooming—An increase in the size of the scanning spot on a cathode-ray tube, caused by defocusing when the brightness control is set too high. The result is expansion and consequent distortion of the image. May also be caused by insufficient high voltage.

blooper—A radio receiver that is oscillating and radiating an undesired signal.

blow—The opening of a circuit because of excessive current, particularly when the current is heavy and a melting or breakdown point is reached.

blower – An electric fan used to supply moving air for cooling purposes.

blown jacket—The common term given to an outer covering of insulation of a cable that was applied by the controlled inflation of the cured jacket tube and the pulling of the cable through it.

blowout coil – An electromagnetic device used to establish a magnetic field in the space where an electrical circuit is broken and thus displace and extinguish the arc.

blowout magnet—A strong permanent magnet or electromagnet used for reducing or deflecting the arc between electrodes or contacts.

blue-beam magnet – A small permanent magnet used to adjust the static convergence of the electron beam for blue phosphor dots in a three-gun color picture tube.

blue glow—The glow normally seen in vacuum tubes containing mercury vapor; it is due to ionization of the molecules of mercury vapor.

blue gun—In a three-gun, color picture tube, the electron gun whose beam strikes the phosphor dots emitting the blue primary color.

blue restorer—The dc restorer in the blue channel of a three-gun color-television picture-tube circuit.

blue video voltage—The signal voltage that controls the grid of the blue gun in a three-gun picture tube. This signal is a reproduction of the blue output signal of the color-tv camera.

bobbin – A small insulated spool which serves as a support for a coil or wire-wound resistor.

bobbin core—*See* Tape-Wound Core.

body capacitance—Capacitance introduced into an electrical circuit by the proximity of the human body.

body effect—Characteristic shift in threshold voltage resulting from bias applied to a semiconductor device's substrate.

body electrodes—Electrodes placed on or in the body to couple electrical impulses from the body to an external measuring or recording device.

boffle—A speaker enclosure, developed by H. A. Hartley, containing a group of stretched, resilient and sound-absorbing screens.

bogey—1. The average, or published, value for a tube characteristic. A bogey tube would be one having all characteristics of a bogey value. 2. An average, published, or nominal value for some characteristic of a device.

bogey electron device—An electron device the characteristics of which have the published nominal values for the type.

boilerplate—A full-size model that simulates the weight, size, and shape, but not all of the functional features, of the actual item.

boiling point—The temperature at which a liquid vaporizes when heated. The exact point depends on the absolute pressure at the liquid-vapor surface.

bolometer—A radiation detector that converts incident radiation into heat which, in turn, causes a temperature change in the material used in the detector. This change is then measured to give an indication of the amount of incident radiant energy.

bolted fault level—*See* Adjusted Circuit.

bombardment – 1. The directing of high-speed electrons at an electrode, causing secondary emission of electrons, fluorescence, disintegration, or the production of X-rays. 2. The process of directing high-speed particles at atoms to cause ionization or transmutation.

bond – 1. Electrical interconnection made with a low-resistance material between a chassis, metal shield cans, or cable shielding braid, in order to eliminate undesirable interaction and interference resulting from high-impedance paths between them. 2. *See* Valence Bond.

bonded assembly—An assembly, the supporting frame and metallic noncircuit elements of which are connected so as to be electrically shorted together.

bonded-barrier transistor – A transistor made by alloying the base with the alloying material on the end of a wire.

bonded nr diode—An n junction semiconductor device in which the negative resistance arises from a combination of avalanche breakdown and conductivity modulation due to the current through the junction.

bonded pickup—*See* Bonded Transducer.

bonded strain gage—A pressure transducer that uses a pressure-sensing system consisting of strain-gage elements firmly bonded to a pressure-responsive member. Thermal stability and insensitivity to shock and vibration are improved by means of this bonded construction.

bonded transducer — Also called bonded pickup. A transducer which employs the bonded strain-gage principle of transduction.

bonding—1. Soldering or welding together various elements, shields, or housings of a device to prevent potential differences and possible interference. 2. A method used to produce good electrical contact between metallic parts of any device. Used extensively in automobiles and aircraft to prevent static buildup. Also refers to the connectors and straps used to bond equipment. 3. The means employed to obtain an electromagnetically homogenous mass having an equipotential surface. 4. The attachment of wire to a circuit. (*See* Ball Bonding, Die Bonding, Stitch Bond, Thermal Compression Bond, Wedge Bonding, Wire Bond, and Wobble Bond.)

bonding conductor — A conductor which serves to connect exposed metal surfaces together.

bonding island—*See* Bonding Pad.

bonding pad—Also called bonding island. A relatively large metallic area at the edge of an integrated-circuit chip; this area is connected through a thin metallic strip to some specific circuit point to which an external connection is to be made. *See also* Beam Lead.

bond strength—A measure of the amount of stress required to separate a layer of material from the base to which it is bonded. Peel strength, measured in pounds per inch of width, is obtained by peeling the layer; pull strength, measured in pounds per square inch is obtained by a perpendicular pull applied to a surface of the layer.

bone conduction — The process by which sound is conducted to the inner ear through the cranial bones.

book capacitor—A two-plate trimmer capacitor that has its plates hinged together like the pages of a book. The capacitance is varied by changing the angle between the plates.

Boolean algebra—1. A system of mathematical logic dealing with classes, propositions, on-off circuit elements, etc., associated by operators as AND, OR, NOT, EXCEPT, IF . . . THEN, etc., thereby permitting computations and demonstration as in any mathematical system. Named after George Boole, famous English mathematician and logician, who introduced it in 1847. 2. Algebraic rules for manipulating logic equations.

Boolean calculus—Boolean algebra modified to include time.

Boolean function—A mathematical function in Boolean algebra.

boom—A mechanical support for a microphone, used in a television studio to suspend the microphone within range of the actors' voices, but out of camera range.

boost capacitor — A capacitor used in the damper circuit of a television receiver to supply a boosted B voltage.

boost charge—The partial charge of a storage battery, usually at a high current rate for a short period.

boosted B voltage—In television receivers the voltage resulting from the combination of the B-plus voltage from the power supply and the average value of voltage pulses coming through the damper tube from the horizontal deflection-coil circuit. The pulses are partially or wholly smoothed by filtering. This boosted voltage may be several hundred volts higher than the B-plus voltage.

booster—1. A carrier-frequency amplifier, usually a self-contained unit, connected between the antenna or transmission line and a television or radio receiver. 2. An intermediate radio or tv station which retransmits signals from one fixed station to another. 3. A small, self-contained transformer designed to be connected to a cathode-ray tube socket to increase the filament voltage and thereby extend the life of the tube.

booster amplifier—A circuit used to increase the output current or the voltage capabilities of an operational amplifier circuit without loss of accuracy (ideally) or inversion of polarity. Usually applied inside the loop for accuracy.

booster voltage — The additional voltage supplied by the damper tube of a television receiver to the horizontal-output, horizontal-oscillator, and vertical-output tubes, resulting in a greater sawtooth sweep output.

boot—1. A form placed around the wire termination of a multiple-contact connector for the purpose of containing the liquid potting compound until it hardens. 2. A protective housing, usually made from a resilient material, used to protect connector or other terminals from moisture.

boot loader—A program in a minicomputer that usually works on a simple data format called core image. The data format, in this context, is the organization of the data as it appears on the input device from which the program is being loaded. Core image data is binary, bit-for-bit identical to what will appear in memory after loadings. The boot loader is used to bring simple programs into memory and run them immediately. Also called bootstrap.

bootstrap—1. A technique or device that brings itself into a desired state through its own action; for example, a routine the first few instructions of which are sufficient to cause the rest of the routine to be brought into the computer from an input device. 2. *See* Boot Loader.

bootstrap circuit—A single-stage amplifier in which the output load is connected between the negative end of the plate supply and the cathode, the signal voltage being applied between the grid and cathode. The name "bootstrap" arises from the fact that the change in grid voltage also changes the potential of the input source (with respect to ground) by an amount equal to the output signal.

bootstrap driver—An electronic circuit used to generate a square pulse to drive a modulator tube.

bootstrap loader—Device for loading first instructions (usually only a few words) of a routine into memory; then using these instructions to bring in the rest of the routine.

bootstrapping—A feedback technique that tends to improve the linearity of circuits that operate over a wide input-signal range.

boresight error—The angular deviation of the electrical boresight of an antenna from its reference boresight.

boresighting—The initial alignment of a directional microwave or radar antenna system through use of an optical procedure or a fixed, known target.

boss—*See* Land, 2.

BOT—Abbreviation for beginning of tape. *See* Load Point.

bottom—To reach a point on an operating or characteristic curve where a negative change in the independent variable, as, for example, in input, no longer produces a constant change in the dependent variable, as, for example, output.

bottoming — A condition where a stylus reaches the bottom of a record groove because its tip radius is smaller than optimum for the groove. Also the opposite of the pinch effect.

bottom metalization — The metalization which may be provided over the back portion of an uncased IC chip facilitating its face-up attachment.

bounce—An unnatural, sudden variation in the brightness of a television picture.

boundary—An interface between p and n material at which donor and acceptor concentrations are equal.

boundary defect—In a crystal, the boundary area between two adjacent perfect crystal regions that are tilted slightly with respect to each other.

boundary marker — A transmitting device installed near the approach end of an airport runway and approximately on the localizer course line.

bound charge—On a conductor, the charge which, owing to the inductive action of a neighboring charge, will not escape to ground; residual charge.

bound circuit—A circuit designed to limit the excursion of a signal. The limit value it establishes may be nominal when used for protection, or highly precise when used operationally.

bound electron—An electron bound to the nucleus of an atom by electrostatic attraction.

Bow-tie antenna.

bow-tie antenna — An antenna generally used for uhf reception. It consists of two triangles in the same plane, usually with a reflector behind them. The transmission line is connected to the points which form a gap.

boxcar—One of a series of pulses having long duration in comparison to the spaces between them.

boxcar circuit—A circuit used in radar to sample voltage waveforms and store the latest value sampled.

boxcar detector—A signal recovery instrument that is used either to retrieve the waveform of a repetitive signal from noise or to measure the amplitude of a repetitive pulse buried in noise. The detector has two modes of operation, scan and single-point. The former is used for waveform retrieval and the later for pulse measurement.

boxcar integrator—A signal processor which uses a narrow filter to reduce the noise with little or no effect on the signal bandwidth. A simple integrator consists of a gated switch and a low-pass filter. During the time when the gate is closed, the repetitive input signal is applied to the low-pass filter which acts as an integrator.

boxcar lengthener—A circuit that lengthens a series of pulses without changing their heights.

B power supply — A power supply which provides the plate and screen voltages applied to a vacuum tube.

bps—Abbreviation for bits per second. In serial data transmission, the instantaneous bit speed within one character as transmitted.

B-quad—A quad arrangement similar to the

S-quad except for a short between the junction of the two sets of series elements. Also called bridge quad or bar quad.

Bragg's law—An expression of the conditions under which a system of parallel atomic layers in a crystal will reflect an X-ray beam with maximum intensity.

braid—1. A weave of organic or inorganic fiber used as a covering for a conductor or group of conductors. 2. A woven metal tube used as shielding around a conductor or group of conductors. When flattened, it is used as a grounding strap.

braided wire—1. A flexible wire made up of small strands woven together. 2. Woven bare or tinned copper wire used as shielding for wires and cables and as ground wire for batteries or heavy industrial equipment. There are many different types of construction.

brain waves—The patterns of lines produced on the moving chart of an electroencephalograph as the result of electrical potentials produced by the brain, picked up by electrodes, and amplified in the machine.

braking magnet—*See* Retarding Magnet.

branch—1. In an electronic network, a section between two adjacent branch points. 2. A portion of a network consisting of one or more two-terminal elements in series. 3. An instruction to a computer to follow one of several courses of action, depending on the nature of control events that occur later.

branch circuit—1. That portion of the wiring system between the final overcurrent device protecting the circuit and the outlet. 2. (As applied to appliances) A circuit designed for the sole purpose of supplying an appliance or appliances; nothing else can be connected to this circuit, including lighting. 3. (General purpose) A circuit that supplies lighting and appliances. 4. (Individual) A circuit that supplies just one piece of equipment such as a motor, an air conditioner, or a furnace. 5. (Multiwire) A circuit consisting of two or more underground conductors with a potential difference between them, and a grounded conductor with an equal potential between it and any one ungrounded conductor. (This may be a 120/240-volt three-wire circuit or a wye-connected circuit with two or more phase wires and a neutral.) It is not a circuit using two or more wires, connected to the same phase, and the neutral.

branch current—The current in the branch of a network.

branch impedance—In a passive branch, the impedance obtained by assuming a driving force across and a corresponding response in the branch, no other branch being electrically connected to the one under consideration.

branching—In a computer, a method of selecting, on the basis of the computer results, the next operation to execute while the program is in progress.

branch order—An instruction used to link subroutines into the main program of a computer.

branch point—1. In an electric network, the junction of more than two conductors. (*See also* Node, 1.) 2. In a computer, a point in the routine where one of two or more choices is selected under control of a routine.

branch voltage — The voltage across a branch of a network.

brass pounder—Amateur term for a Morse code operator, especially one who spends long hours handling "traffic."

braze bonding—The joining of similar or dissimilar metals by introducing a braze filler metal at the joint and establishing a conventional brazed joint, and then diffusing the braze filler into the base metals by subsequent heat treating. Melting occurs in the braze filler independent of the base metals. Base-metal fusion is not required.

breadboard construction—An arrangement in which electronic components are fastened temporarily to a board for experimental work.

breadboard model — 1. An assembly in rough form to prove the feasibility of a circuit, device, system, or principle. 2. An experimental model of a circuit in which the components are fastened temporarily to a chassis or board and electrically tested.

break—1. In a communication circuit, the taking control of the circuit by a receiving operator or listening operator. The term is used in connection with half-duplex telegraph circuits and two-way telephone circuits equipped with voice-operated devices. 2. In a circuit-opening device, the minimum distance between the stationary and movable contacts when these contacts are open.

break-before-make—The action of opening a switching circuit before closing another associated circuit.

break-before-make contacts — Contacts which interrupt one circuit before establishing another.

break contact—In a switching device, the contact which opens a circuit upon operation of the device (normally closed contact).

breakdown — 1. An electric discharge through an insulator, insulation on wire, or other circuit separator, or between electrodes in a vacuum or gas-filled tube. 2. The phenomenon occurring in a reverse-biased semiconductor diode. The start of the phenomenon is observed as a transition from a region of high dynamic resistance to one of substantially lower dynamic resistance. This is done to boost the reverse current.

breakdown diode—*See* Avalanche Diode.

breakdown impedance — Also called avalanche impedance. The small-signal impedance at a specified direct current in the breakdown region of a semiconductor diode.

breakdown region—The entire region of the volt-ampere characteristic beyond the initiation of breakdown due to reverse current in a semiconductor-diode characteristic curve.

breakdown strength — *See* Dielectric Strength.

breakdown torque—The maximum torque a motor will develop, without an abrupt drop in speed, as the rated voltage is applied at the rated frequency.

breakdown voltage — 1. That voltage at which an insulator or dielectric ruptures, or at which ionization and conduction take place in a gas or a vapor. 2. The voltage measured at a specified current in the breakdown region of a semiconductor diode at which there is a substantial change in characteristics from those at lower voltages. Also called zener voltage. 3. The voltage required to jump an air gap. 4. The reverse bias voltage applied to a pn junction for which large currents are drawn for relatively small increases in voltage.

break elongation—The relative elongation of a specimen of recording tape or base film at the instant it breaks after having been stretched at a given rate.

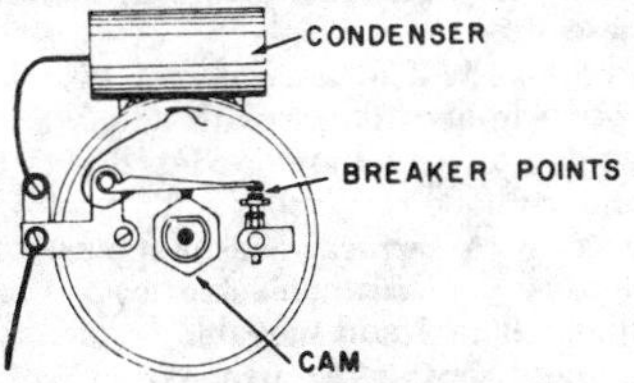

Breaker points.

breaker points—The low-voltage contacts that interrupt the current in the primary circuit of the ignition system of a gasoline engine.

break frequency—In a plot of log gain (attenuation) versus log frequency, the frequency at which the asymptotes of two adjacent linear slope segments meet.

break-in keying — In the operation of a radiotelegraph communication system, a method by which the receiver is capable of receiving signals during transmission spacing intervals.

break-in operation — A method of radiocommunication involving break-in keying which allows the receiving operator to interrupt the transmission.

break-in relay—A relay used for break-in operation.

break-make contact — Also called transfer contact. A contact form in which one contact opens its connection to another contact and then closes its connection to a third contact.

breakout—The exit point of a conductor or number of conductors along various points of a main cable of which the conductors are a part. This point is usually harnessed or sealed with some synthetic-rubber compound.

breakover—In a silicon controlled rectifier or related device, a transition into forward conduction caused by the application of an excessively high anode voltage. In some cases this is destructive to the device.

breakover voltage—The value of positive anode voltage at which an SCR with the gate circuit open switches into the conductive state.

break period — The time interval during which the circuit contacts of a telephone dial are open.

breakpoint—A place in a routine specified by an instruction, instruction digit, or other condition, where the routine may be interrupted by external intervention or by a monitor routine.

breakpoint instruction — In the programming of a digital computer an instruction which, together with a manual control, causes the computer to stop.

breakpoint switch — A manually operated switch which controls conditioned operation at breakpoints; it is used primarily in debugging.

breakup—*See* Color Breakup.

breathing—Amplitude variations similar to "bounce," but at a slow, regular rate.

breezeway—In the NTSC color system, that portion of the back porch between the trailing edge of the sync pulse and the start of the color burst.

B-register—A computer register that stores a word which will change an instruction before the computer carries out that instruction.

bremsstrahlung—Electromagnetic radiation emitted by a fast-moving charged particle (usually an electron) when it is slowed down (or accelerated) and deflected by the electric field surrounding a positively charged atomic nucleus. X rays produced in ordinary X-ray machines are bremsstrahlung. (In German, the term means "braking radiation.")

brevity code—A code which has as its sole purpose the shortening of messages rather than the concealment of their content.

Brewster angle—The angle of incidence at

which the reflection of parallel-polarized electromagnetic radiation at the interface between two dielectric media equals zero.

bridge—1. In a measuring system, an instrument in which part or all of a bridge circuit is used to measure one or more electrical quantities. 2. In a fully electronic stringed instrument, the part that converts the mechanical vibrations produced by the strings into electrical signals.

bridge circuit—A network arranged so that, when an electromotive force is present in one branch, the response of a suitable detecting device in another branch may be zeroed by suitable adjustment of the electrical constants of still other branches.

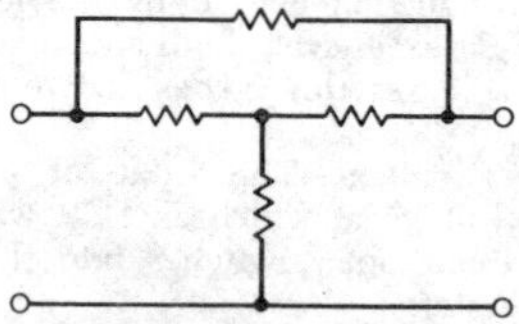

Bridged-T network.

bridged-T network—A T-network in which the two series impedances of the T are bridged by a fourth impedance.

bridge duplex system — A duplex system based on the Wheatstone-bridge principle in which a substantial neutrality of the receiving apparatus to the transmitted currents is obtained by an impedance balance. Received currents pass through the receiving relay, which is bridged between the points that are equipotential for the transmitted currents.

bridge hybrid—*See* Hybrid Junction, 2.

bridge quad—*See* B-quad.

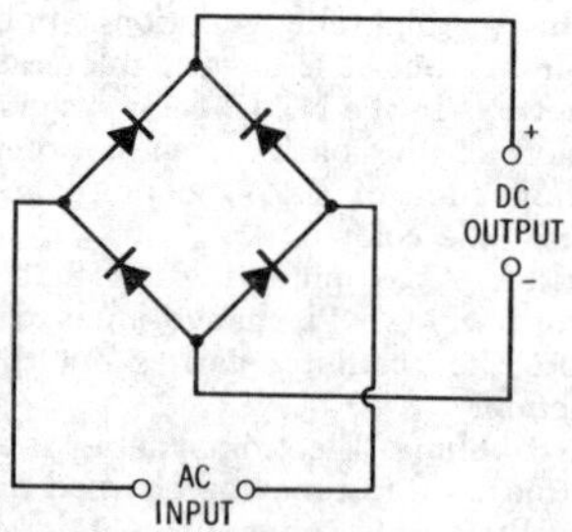

Bridge rectifier.

bridge rectifier—A full-wave rectifier with four elements connected in the form of a bridge circuit so that dc voltage is obtained from one pair of opposite junctions when an alternating voltage is applied to the other pair of junctions.

bridge tap—An unterminated length of line attached (bridged) at some point between the extremities of a communication line; bridge taps are undesirable.

bridge transformer—*See* Bridging Transformer.

bridging—The shunting of one electrical circuit by another.

bridging amplifier—An amplifier with an input impedance sufficiently high that its input may be bridged across a circuit without substantially affecting the signal level of the circuit.

bridging connection—A parallel connection by means of which some of the signal energy in a circuit may be withdrawn with imperceptible effect on the normal operation of the circuit.

bridging contacts — A set of contacts in which the moving contact touches two stationary contacts simultaneously during transfer.

bridging gain—Ratio between the power a transducer delivers to a specified load impedance under specified operating conditions, and the power dissipated in the reference impedance across which the transducer input is bridged. Usually expressed in decibels.

bridging loss — Ratio between the power dissipated in the reference impedance across which the input of a transducer is bridged, and the power the transducer delivers to a specified load impedance under specified operating conditions. Usually expressed in decibels.

bridging transformer—Also called bridge transformer and hybrid coil. A transformer designed to couple two circuits having at least nominal ohmic isolation and operating at different impedance levels, without introducing significant frequency or phase distortion, or significant phase shift.

brightness—1. The attribute of visual perception in accordance with which an area appears to emit more or less light. Used with cathode-ray tubes. 2. *See* Luminance. 3. A surface measurement of light intensity per unit projected area. Usually expressed in foot-lamberts.

brightness control—In a television receiver, the control which varies the average brightness of the reproduced image.

brightness signal—*See* Luminance Signal.

brilliance—1. The degree of brightness and clarity in a reproduced cathode-ray tube. 2. The degree to which the higher audio frequencies sound like the original when reproduced by a receiver or public-address amplifier.

brilliance control—A potentiometer used in a three-way speaker system to adjust the output level of the tweeter for proper relative volume between the treble and the lower audio frequencies produced by the complete speaker system.

British thermal unit—The energy required

to raise the temperature of one pound of water one degree Fahrenheit.

broad band—1. As applied to data transmission, the term denotes transmission facilities capable of handling frequencies greater than those required for high-grade voice communications (higher than 3 to 4 kilohertz). 2. Having an essentially uniform response over a wide range of frequencies. To design or adjust (an amplifier) for bandwidth.

broad-band amplifier—An amplifier which has an essentially flat response over a wide frequency range.

broad-band antenna—An antenna which is capable of receiving a wide range of frequencies.

broad-band electrical noise — Also called random noise. A signal that contains a wide range of frequencies and has a randomly varying instantaneous amplitude.

broad-band interference—Interference occupying a frequnecy range that is much greater than the bandwidth of the equipment being used to measure it.

broad-band klystron—A klystron in which three or more externally loaded, stagger-tuned resonant cavities are used to broaden the bandwidth.

broad-band random vibration—Single-frequency component vibrations that are random in both phase and amplitude at minute increments of frequency throughout a specified bandwidth. Typically it occurs when intense, high-power-level noise impinges on structures.

broad-band tube (tr and pre-tr tubes)—A gas-filled, fixed-tuned tube incorporating a bandpass filter suitable for radio-frequency switching.

broadcast—Radio or televisioin transmission intended for public reception, for which receiving stations make no receipt.

broadcast band—The band of frequencies extending from 535 to 1605 kHz.

broadcasting—The transmitting of speech, music, or visual programs for commercial or public-service motives to a relatively large audience (as opposed to two-way radio, for example, which is utilitarian and is directed toward a limited audience).

broadcasting service — A radiocommunication service in which the transmissions are intended for direct reception by the general public. This service may include sound transmissions, television transmissions, or other types of transmissions.

broadcasting station — A station in the broadcasting service.

broadside—A direction perpendicular to an axis or plane.

broadside array—An antenna array whose direction of maximum radiation is perpendicular to the line or plane of the array (depending on whether the elements lie on a line or a plane). A uniform broadside array is a linear array whose elements contribute fields of equal amplitude and phase.

broad tuning—A tuned circuit or circuits which respond to frequencies within one band or channel, as well as to a considerable range of frequencies on each side.

Bruce antenna—Original name of the rhombic end-fire antenna, which consists of a diamond-shaped arrangement of four wires with the feed line at one end and a resistive termination at the opposite end of the longer diagonal. Bruce's name also is given to a series-fed array of vertically polarized resonant rectangular loops (or half-loops over ground) with one-quarter-wave width and one-half-wave spacing.

brush—A piece of conductive material, usually carbon or graphite, which rides on the commutator of a motor and forms the electrical connection between it and the power source.

brush discharge—An intermittent discharge of electricity which starts from a conductor when its potential exceeds a certain value but is too low for the formation of an actual spark. It is generally accompanied by a whistling or crackling noise.

brush-discharge resistance—*See* Corona Resistance.

brush rocker—A movable rocker, or "yoke," on which the brush holders of a dynamo or motor are fixed so that the position of the brushes on the commutator can be adjusted.

brush station—In a computer, a position where the holes of a punched card are sensed, particularly when this is done by a row of brushes sweeping electrical contacts.

brute force—The use of seemingly inefficient design in order to achieve a desired result. Sometimes this is done in order to avoid involved design procedures, critical adjustments, or the like, but often it is the only possible approach. For example, the miniaturization of low-frequency loudspeakers requires "brute force" in the form of greatly increased amplifier power.

brute-force filter—A type of power-pack filter which depends on large values of capacitance and inductance to smooth out pulsations, rather than on the resonant effects of a tuned filter.

brute supply—One that is completely unregulated. It employs no circuitry to maintain the output voltage constant with a changing input line or load variations. The output voltage varies in the same percentage and in the same direction as the input line voltage. If the load current changes, the output voltage changes inversely—an increase in load current drops the output voltage.

B & S—Abbreviation for Brown and Sharpe

gage. A wire-diameter standard that is the same as AWG.

B-scope—A type of radarscope that presents the range of an object by a vertical deflection of the signal on the screen, and the bearing by a horizontal deflection.

"B" service – FAA service pertaining to transmission and reception by teletype or radio of messages containing requests for and approval to conduct an aircraft flight, flight plans, in-flight progress reports, and aircraft arrival reports.

B-supply—A source for supplying a positive voltage to the anodes and other positive electrodes of electron tubes. Sometimes called B+ supply.

bubble—*See* Dot, 2.

bubble logic—A (developmental) form of magnetic logic.

bubble memory – A storage medium in which the information is magnetically stored and moved from cell to cell by the application of an external magnetic field.

buck—To oppose, as one voltage bucking another, or the magnetic fields of two coils bucking each other.

bucket—A general term for a specific reference in storage in a computer, such as a section of storage, the location of a word, a storage cell, etc.

bucket brigade – 1. A shift register that transfers information from stage to stage in response to timing signals. 2. A technique of building a photosensing array using field effect transistors. Two FET amplifiers are at each sensing area and are interconnected as an element of a shift regsiter. The charges sensed by the FETs are successively clocked from element to element. The electrical charges representing bits of stored information are transferred from one element to the next by means of clock pulses that raise the level of each element in the correct sequence, in the same way as a firefighting brigade passes a bucket of water down the line.

bucking coil—A coil connected and positioned in such a way that its magnetic field opposes the magnetic field of another coil. The hum-bucking coil of an electrodynamic speaker is an example.

bucking voltage—A voltage which is opposite in polarity to another voltage in the circuit and hence bucks, or opposes, the latter voltage.

buckling—The warping of the plates of a battery due to an excessively high rate of charge or discharge.

Bucky diaphragm—A grid composed of narrow strips of lead arranged with X-ray–transparent spaces between adjacent strips and placed between the specimen and the X-ray film. Used to reduce scattered radiation.

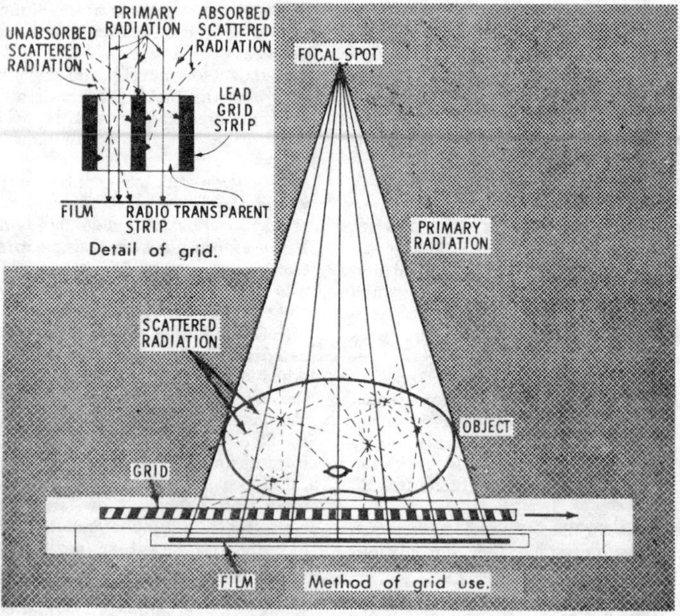

Bucky diaphragm.

buffer—1. A circuit or component which isolates one electrical circuit from another. Usually refers to electron-tube amplifiers used for this purpose. 2. A vacuum-tube stage used chiefly to prevent undesirable interaction between two other stages. In a transmitter, it generally follows the master-oscillator stage. 3. An isolating circuit used in an electronic computer to avoid reaction of a driven circuit on the corresponding driver circuit. 4. A storage device used to compensate for a difference in the rate of flow of information or the time or occurrence of events when transmitting information from one device to another. 5. A computer circuit having an output and a multiplicity of inputs and designed so that the output is energized whenever one or more inputs are energized. Thus, a buffer performs the circuit function equivalent to the logical OR. 6. A noninverting digital circuit element that may be used to handle a large fanout or to invert input and output levels. Normally, a buffer is an emitter follower.

buffer amplifier—An amplifier designed to isolate a preceding circuit from the effects of a following circuit.

buffer capacitor — A capacitor connected across the secondary of a vibrator transformer, or between the anode and cathode of a cold-cathode rectifier tube, to suppress voltage surges that might otherwise damage other parts in the circuit.

buffer circuitry — Circuitry necessary to adapt signals between two systems, e.g., between a test system and a board under test.

buffer computer—A computing system provided with a storage device so that input and output data may be stored temporarily to match the slower speed of input-output devices with the higher speed of the computer.

buffered terminal—A terminal which contains storage equipment so that the rate at which it sends or receives data over its line does not need to agree exactly with the rate at which the data is entered or printed.

buffer element—A low-impedance inverting driver circuit. Because of its very low source impedance the element can supply substantially more output current than the basic circuit. As a consequence, the buffer element is valuable in driving heavily loaded circuits or minimizing rise-time deterioration due to capacitive loading.

buffer gate—A logic gate with a high output-drive capability, or fan-out; a buffer gate is used when it is necessary to drive a large number of gate inputs from one gate function.

buffer storage—1. A synchronizing element used between two different forms of storage (usually internal and external). 2. An input device in which information from external or secondary storage is assembled and stored ready for transfer to internal storage. 3. An output device into which information from internal storage is copied and held for transfer to secondary or external storage. Computation continues during transfers from buffer storage to secondary or internal storage or vice versa. 4. A device for the purpose of storing information temporarily during data transfers.

buffer storage unit—A computer unit used for temporary storage of data before transmission to another destination. The unit often is able to accept and give back data at widely varying rates so that data transfer takes place efficiently for each device connected to it; that is, a high-speed device is not slowed by the presence of a slow-speed device in the same system unless the data transfer occurs directly between the two devices.

bug—1. A semiautomatic telegraph sending key consisting of a lever which is moved to one side to produce a series of correctly spaced dots, and to the other side to produce a dash. 2. A circuit fault due to improper design or construction. 3. An electronic listening device, generally small and concealed, used for commercial or military espionage. 4. To equip with a concealed listening device.

build—The increase of diameter due to insulation.

building-out—Addition to an electric structure of an element or elements electrically similar to an element or elements of the structure, to bring a certain property or characteristic to a desired value.

building-out circuit — A short section of transmission line, or a network which is shunted across a transmission line, for the purpose of impedance matching.

building-out network—A network designed to be connected to a basic network so that the combination will simulate the sending-end impedance, neglecting dissipation, of a line having a termination other than that for which the basic network was designed.

building-out section — A short section of transmission line, either open or short-circuited at the far end, shunted across another transmission line for use on an impedance-matching transformer.

bulb — The glass envelope which encloses an incandescent lamp or an electronic tube.

bulb-temperature pickup — A temperature transducer in which the sensing element is enclosed in a metal tube or sheath to protect it against corrosive liquids or other contaminants.

bulge — The difference between the actual characteristic and a linear characteristic of the attenuation-frequency characteristic of a transmission line.

bulk degausser—*See* Bulk Eraser.

bulk-erased noise—The noise arising when a bulk-erased tape is reproduced with the erase and record heads completely deenergized. Ideally, this noise is governed by the number of magnetic particles that pass by the head in unit time.

bulk eraser—Also called bulk degausser or degausser. Equipment for erasing a roll of tape. The roll is usually rotated while a 60-hertz ac erasing field is decreased, either by withdrawing the roll from an electromagnet or by reducing the ac supply to an electromagnet.

bulk noise—*See* Excess Noise.

bulk resistivity — Resistance measured between opposite faces of a cube of homogeneous material.

bulk resistor—A resistor made by providing ohmic contacts between two points of a homogeneous, uniformly doped crystal of silicon material.

bump contacts—Small amounts of material formed on the chip substrate, to register with terminal pads, as when the chip is employed in "flip-chip circuits."

bunched pair—A group of pairs tied together or otherwise associated for identification.

buncher—1. The input resonant cavity in a conventional klystron oscillator. 2. In a velocity-modulated tube, the electrode which concentrates the electrons in the constant-current electron beam into bunches.

buncher gap—*See* Input Gap.

buncher resonator—The input cavity resonator in a velocity-modulated tube. It serves to modify the velocity of the electrons in the beam.

bunching—1. Grouping pairs together for identification and testing. 2. In a velocity-modulated electron stream the action that produces an alternating convection-current component as a direct result of the differences of electron transit time produced by the velocity modulation.

bunching parameter—One-half of the product of the bunching angle in the absence of velocity modulation and the depth of velocity modulation.

bunching time—The time in the armature motion of a relay during which all three contacts of a bridging-contact combination are electrically connected.

bunching voltage — The radio-frequency voltage between the grids of the buncher resonator in a velocity-modulated tube such as a klystron. Generally, the term implies the peak value of this oscillating voltage.

bunch stranding — A method in which a number of wires are twisted together in a common direction and with a uniform pitch to form a finished, stranded wire.

bundled cable—Individual insulated wires laced together to form a bundle to facilitate handling.

buried cable — A cable installed underground and not removable except by disturbing the soil.

buried channel—Because charge trapping can occur at the surface of the $Si\text{-}SiO_2$ interface, a thin doped layer can be introduced in the silicon just below the oxide (typically by ion implantation) to prevent trapping of charges. (MIS technology term.)

buried layer—A layer of very low resistivity, usually of n+ material, between the high-resistivity n-type collector region and the p-type substrate of an integrated-circuit transistor. The buried layer tends to reduce the series collector resistance of the transistor without having an adverse effect on the breakdown voltage.

burned-in image—An image which remains in a fixed position in the output signal of a camera tube after the camera has been turned to a different scene.

burn-in—Operation of a device to stabilize its failure rate.

burn-in period—*See* Early Failure Period.

burst—1. A sudden increase in the strength of a signal. 2. The cosmic-ray effect upon matter, causing a sudden intense ionization that gives rise to great numbers of ion pairs at once. 3. *See* Color Burst.

burst pedestal—A rectangular pulselike television signal which may be part of the color burst. The amplitude of the color-burst pedestal is measured from the alternating-current axis of the sine-wave portion of the horizontal pedestal.

burst pressure—The maximum pressure to which a device can be subjected without rupturing.

burst separator output—The amplitude of the chroma reference burst at the output of the gated burst amplifier.

burst sequence—An arrangement of color burst signals in which the polarity of the burst signal is the same at the start of each field so that the stability of color synchronization is improved.

burst transmission — Radio transmission in which messages are stored and then released at 10 to 100 or more times the normal speed. The received signals are recorded and returned to the normal rate for the user.

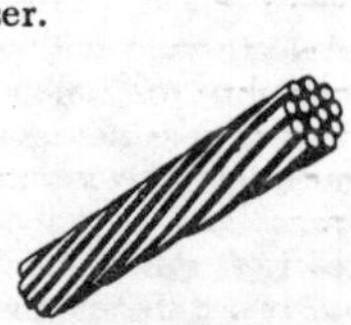

Bunch stranding.

bus—1. In a computer, one or more conductors used as a path over which information is transmitted from any of several sources to any of several destinations. 2. The term used to specify an uninsulated conductor (a bar or wire); may be solid, hollow, square, or round. 3. Sometimes used to specify a bus bar. 4. The communications path between two switching points.

bus architecture — A data-communications structure which consists of a common connection among a number of printer/plotter modules.

bus bar—A heavy copper strap or bar used to carry heavy currents or to make a common connection between several circuits.

bus driver—An integrated circuit which is added to the data bus system in a computer to facilitate proper drive to the CPU when several memories are tied to the data bus line. These are necessary because of capacitive loading which slows down the data rate and prevents proper time sequencing of microprocessor operation.

business data processing — 1. Automatic data processing used in accounting or management. 2. Data processing for business purposes, such as recording and summarizing the financial transactions of a business.

business machine — Customer-provided equipment that is connected to the communications services of a common carrier for the purpose of data movement.

busing—The joining of two or more circuits.

bus organization — The manner in which many circuits are connected to common input and output lines (buses).

bus reactor—A current-limiting reactor connected between two different buses or two sections of the same bus for the purpose of limiting and localizing the disturbance due to a fault on either bus.

bussback—The connection, by a common carrier, of the output portion of a circuit back to the input portion of a circuit. (*See also* Loopback Test.)

bus system — A network of paths inside a microprocessor which facilitate data flow. The important buses in a microprocessor are identified as data bus, address bus and control bus.

busy test—In telephony, a test to find out whether certain facilities which may be desired, such as a subscriber line or a trunk, are available for use.

busy tone—Interrupted low tone returned to the calling party to indicate that the called line is busy.

Butler antenna—An array antenna in which hybrid junctions are incorporated into the feed system to obtain a plurality of independent beams.

Butler oscillator — A two-tube (or transistor) crystal-controlled oscillator in which the crystal forms the positive feedback path when excited in its series-resonant mode.

butt contact — A hemispherically shaped contact designed to mate against a similarly shaped contact. When properly aligned the two convex surfaces form a reasonably good surface-to-surface contact.

butterfly circuit — Frequency-determining element having no sliding contacts and providing simultaneous change of both inductance and capacitance. It is used to replace conventional tuning capacitors and coils in ultrahigh-frequency oscillator circuits. The rotor of the device resembles the opened wings of a butterfly.

butterfly resonator — Also called butterfly capacitor. A tuning device that combines both inductance and capacitance in such a manner that it exhibits resonant properties at very high and ultrahigh frequencies (characterized by a high tuning ratio and Q). So called because the shape of the rotor resembles the opened wings of a butterfly.

Butterworth filter—A filter network that exhibits the flattest possible response in the passband. The response is monotonic, rolling off smoothly into the stop band, where it approaches a constant slope of 6 dB/octave.

Butterworth function — A mathematical function used in designing a filter for maximally constant amplitude response with little consideration for time delay or phase response.

butt joint—1. A splice or other connection formed by placing the ends of two conductors together and joining them by welding, brazing, or soldering. 2. A connection between two waveguides which maintains electrical continuity by providing physical contact between the ends.

button—1. The metal container in which the carbon granules of a carbon microphone are held. 2. Also called dot. A piece of metal used for alloying onto the base wafer in making alloy transistors.

button silver-mica capacitor — A stack of silvered mica sheets encased in a silver-plated brass housing. The high-potential terminal is connected through the center of the stack. The other capacitor terminal is formed by the metal shell, which connects at all points around the outer edge of the electrodes. This design permits the current to fan out in a 360° pattern from the center terminal, providing the shortest possible electrical path between the center terminal and chassis. The internal series inductance is thus kept small.

button stem—In a tube, the glass base onto which the mount structure is assembled. The pins may be sealed into the glass; if so, no base is needed. In some large tubes, the stiff wires are passed directly into the base pins to give added strength. (*See also* Pressed Stem.)

button up—To close or completely seal any operating device.

butt splice—A device for joining two conductors placed end to end with their axes in line (that is, conductors not overlapping).

buzzer—A signaling device in which an armature vibrates to produce a raucous, nonresonant sound.

BV_{EBO}—The reverse-breakdown voltage of the emitter-to-base junction of a transistor with the collector open-circuited.

BX cable.

BX cable—Insulated wires enclosed in flexible metal tubing, used in electrical wiring.

bypass—1. A shunt (parallel) path around one or more elements of a circuit. 2. A secondary channel that permits routing of data in a computer sample around the data compressor, regardless of the value of the sample, at intervals determined by the operator.

bypass capacitor—A capacitor used for providing a comparatively low-impedance ac path around a circuit element.

bypass filter—A filter providing a low-attenuation path around some other circuit or equipment.

B — Y signal—One of the three color-difference signals in color television. The B − Y signal forms a blue primary signal for the picture tube when combined—either inside or outside the picture tube—with a luminance, or Y, signal.

byte—1. A single group of bits processed together (in parallel). It can consist of a variable number of bits. 2. A sequence of adjacent binary digits, usually shorter than a word, operated on as a unit. 3. The number of bits that a computer processes as a unit. This may be equal to or less than the number of bits in a word. For example, both an 8-bit and a 16-bit length computer may process data in 8-bit bytes. 4. The smallest addressable unit of main storage in a computer system. The byte consists of eight data bits and one parity bit.

C

C—Symbol for capacitor, capacitance, carbon, coulomb, centigrade or Celsius, transistor collector, candle, velocity of light.

C− (C minus)—The negative terminal of a C battery, or the negative polarity of other sources of grid-bias voltage. Used to denote the terminal to which the negative side of the grid-bias voltage source should be connected.

C+ (C plus)—Positive terminal of a C battery, or the positive polarity of other sources of grid-bias voltage. The terminal to which the positive side of the grid-bias voltage source should be connected.

cabinet—A protective housing for electrical or electronic equipment.

cable—1. An assembly of one or more conductors, usually within a protective sheath, and so arranged that the conductors can be used separately or in groups. 2. A stranded conductor (single-conductor cable) or a combination of conductors insulated from one another (multiple-conductor cable).

cable armor—In cable construction, a layer of steel wire or tape, or other extra-strength material, used to reinforce the lead wall.

cable assembly—A cable with plugs or connectors on each end for a specific purpose. It may be formed in various configurations.

cable clamp—A device used to give mechanical support to the wire bundle or cable at the rear of a plug or receptacle.

cable complement—A group of cable pairs that have some common distinguishing characteristic.

cable core—That portion of an insulated cable lying under the protective covering or jacket.

cable coupler—A device used to join lengths of similar or dissimilar cable having the same electrical characteristics.

cable fill—The ratio of the number in use to the total number of pairs in a cable.

cable filler—Material used in multiple-conductor cables to occupy the spaces between the insulated conductors.

cable messenger—A stranded cable supported at intervals by poles or other structures and employed to furnish frequent points of support for conductors or cables.

cable Morse code—A three-element code used mainly in submarine-cable telegraphy. Dots and dashes are represented by positive and negative current impulses of equal length, and a space by the absence of current.

cable run—The path occupied by a cable on cable racks or other support from one termination to another.

cable sheath—A protective covering of rubber, neoprene, resin, or lead over a wire or cable core.

cable splice—A connection between two or more separate lengths of cable. The conductors in one length are individually connected to conductors in the other length, and the protecting sheaths are so connected that protection is extended over the joint.

cable terminal—A means of electrically connecting a predetermined number of cable conductors in such a way that they can be individually selected and extended by conductors outside the cable.

cable vault—A vault in which the outside plant cables are spliced to the tipping cables.

cabling—The assembly of wire bundles extended from one physical structure to another to interconnect the circuits within structures. Cabling differs from wire jumpers in that it is understood to be external to the physical structures and may include tubing sheaths, zipper tubing, or rubber jackets.

cache memory—1. A high-speed, low-capacity computer similar to a scratch-pad memory except that it has a larger capacity. 2. The fastest portion of the overall memory which stores only the data that the computer may need in the immediate future.

cactus needle—A phonograph needle made from the thorn of a cactus plant.

cad—Abbreviation for computer-aided design. 1. Use of a computer to aid in the design of complex MSI or LSI circuit arrays. Cad is especially useful for custom IC fabrication. 2. Man/computer interaction for the design and testing of customer MSI/LSI arrays and other complex engineering designs in a reasonable time frame.

cadmium—A metallic element widely used for plating steel hardware or chassis to improve its appearance and solderability and to prevent corrosion. It is also used in the manufacture of photocells.

cadmium cell—A standard cell used as a voltage reference; at 20°C its voltage is 1.0186 volts.

cadmium selenide photoconductive cell—A photoconductive cell that uses cadmium selenide as the semiconductor material. It has a fast response time and high sensitivity to longer light wavelengths, such as those emitted by incandescent lamps and some infrared light sources.

cadmium sulfide photoconductive cell—A photoconductive cell in which a small wafer of cadmium sulfide is used to provide an extremely high dark-light resistance ratio. Some of the cells can be used directly as a light-controlled switch operated directly from the 120-volt ac power line.

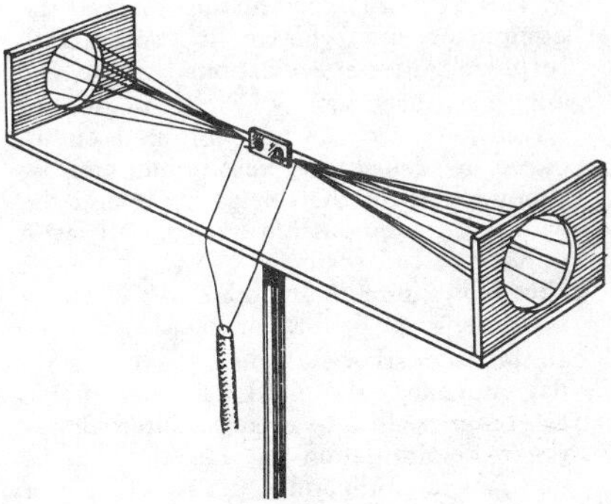

Cage antenna.

cage antenna—An antenna comprising a number of wires connected in parallel and arranged in the form of a cage. This is done to reduce the copper losses and increase the effective capacity.

calculating—Computing a result by multiplication, division, addition, or subtraction or by a combination of these operations. A data-processing function.

calculating punch—A punched-card machine that reads data from a group of cards and punches new data in the same or other cards.

calculator—1. A device capable of performing arithmetic. 2. A calculator as in 1 which requires frequent manual intervention. 3. Generally and historically, a device for carrying out logical and arithmetical digital operations of any kind.

calendar age—Age of an item or object measured in terms of time elapsed since it was manufactured.

calendar life — That period of time expressed in days, months, or years during which an item may remain installed and in operation, but at the end of which the item should be removed and returned for repair, overhaul, or other maintenance.

calibrate—To ascertain, by measurement or by comparison with a standard, any variations in the readings of another instrument, or to correct the readings.

calibrated interference measurements — Measurements in which the meter is calibrated against a known sine wave or impulse-generator output and interference readings are taken directly from the output indicating meter on the receiver.

calibrated triggered sweep—In a cathode-ray oscilloscope, a sweep that occurs only when initiated by a pulse and moves horizontally at a known rate.

calibration—1. The process of comparing an instrument or device with a standard to determine its accuracy or to devise a corrected scale. 2. Taking measurements of various parts of electronic equipment to determine the performance level of the equipment and whether it conforms to technical order specifications.

calibration accuracy — Finite degree to which a device can be calibrated (influenced by sensitivity, resolution, and reproducibility of the device itself and the calibrating equipment). Expressed as a percent of full scale.

calibration curve—A smooth curve connecting a series of calibration points.

calibration marker—On the screen of a radar indicator, the markings that divide the range scale into accurate intervals for range determination or checking against mechanical indicating dials, scales, or counters.

call—1. A transmission made for the purpose of identifying the transmitting station and the station for which the transmission is intended. 2. To transfer control to a specified closed subroutine. 3. In communication the action performed by the calling party, or the operations necessary in making a call, or the effective use made of a connection between two stations.

call announcer—Device for accepting pulses from an automatic telephone office and reproducing the corresponding number with speechlike sounds.

call circuit—A communication circuit between switching points used by traffic forces for transmitting switching instructions.

call in—To transfer control of a digital computer temporarily from a main routine to a subroutine, which is inserted in the sequence of calculating operations to fulfill a subsidiary purpose.

call indicator—Device for accepting pulses from an automatic switching system and displaying the corresponding called number before an operator.

calling device — Apparatus that generates the pulses used to control the establishment of connections in an automatic telephone switching system.

calling sequence — In a computer, a sequence of instructions required to enter a subroutine. It may contain information required by the subroutine.

call letters — A series of government-assigned letters, or letters and numbers, which identify a transmitting station.

call number—In computer operations, a set of characters that identifies a subroutine and contains information with respect to parameters to be inserted in the subroutine or information related to the operands.

call sign—Any combination of characters or pronounceable words which identifies a communication facility, a command, an authority, an activity, or a unit; used primarily for establishing and maintaining communications.

call word—A call number which is exactly the size of one machine word.

calomel electrode—An electrode consisting of mercury in contact with a solution of potassium chloride saturated with mercurous chloride (calomel). *See also* Glass Electrode.

calorimeter — An apparatus for measuring quantities of heat. Used to measure microwave power in terms of heat generated.

calorimeter system—A precision rf wattmeter as well as an efficient dummy load able to absorb energy at any frequency band. It can absorb and measure any level of microwave energy, and functions by circulating a known amount of liquid through a suitably designed low vswr load. The load is located in a waveguide or coaxial section that mates to the termination of the rf energy source being measured or absorbed.

CAM — Abbreviation for content addressable memory. A computer memory in which information is retrieved by addressing the content (the data actually stored in the memory) rather than by selecting a physical location. *See also* Associative Storage.

cam actuator—An electromechanical device in which a switch is closed when the high spot of a rotating cam, or eccentric, is in a certain position.

camera—*See* Television Camera.

camera cable—A cable or group of wires that carry the picture signal from the television camera to the control room.

camera chain—A television camera, associated control units, power supplies, monitor, and connecting cables necessary to deliver a picture for broadcasting.

camera signal—The video-output signal of a television camera.

camera tube — An electron-beam tube in which an electron current or charge-density image is formed from an optical image and scanned in a predetermined sequence to provide an electric signal.

camp-on—A method of holding a call for a line that is already in use and of signaling when the lien becomes free. Also called clamp-on.

can—1. A metal shield placed around a tube, coil, or transformer to prevent electromagnetic or electrostatic interaction.

2. A metal package for enclosing a device, as opposed to a plastic or ceramic package.

Canadian Standards Association—In Canada, a body that issues standards and specifications prepared by various voluntary committees of government and industry. Abbreviated CSA.

canal ray—Also called positive ray. Streams of positive ions that flow from the anode to the cathode in an evacuated tube.

candela (cd)—Formerly candle. The unit of luminous intensity. The luminous intensity of one/sixtieth of one square centimeter of projected area of a blackbody radiator operating at the temperature of solidification of platinum (2046K). Values for standards having other spectral distributions are derived by the use of accepted spectral luminous efficiency data for photopic vision.

candela/cm² — Luminance unit called "stilb."

candle — The unit of luminous intensity. One candle is defined as the luminous intensity of 1/60th square centimeter of a blackbody radiator operating at the solidification temperature of platinum.

candlepower — 1. Luminous intensity expressed in terms of standard candles. 2. A measure of the intensity of light produced by a source. This standard of measurement is used in France, Britain and the U.S. One candlepower corresponds approximately to the light produced in the horizontal direction by an ordinary sperm candle weighing six to the pound and burning at the rate of 120 gr/hr.

candoluminescence — A phenomenon that produces white light without need for very high temperatures.

canned cycle—The use of preparatory functions on a punched tape to initiate a complete machining sequence; the need for much repetitive information in the program is thereby eliminated.

cannibalization—A method of maintenance or modification in which the required parts are removed from one system or assembly for installation on a similar system or assembly.

capacitance — Abbreviated C. Also called capacity. In a capacitor or a system of conductors and dielectrics, that property which permits the storage of electrically separated charges when potential differences exist between the conductors. The capacitance of a capacitor is defined as the ratio between the electric charge that has been transferred from one electrode to the other and the resultant difference in potential between the electrodes. The value of this ratio is dependent on the magnitude of the transferred charge.

$$C(\text{farad}) = \frac{Q(\text{coulomb})}{V(\text{volt})}$$

capacitance between two conductors—The ratio between the charge transferred from one conductor to the other and the resultant difference in the potentials of the two conductors when insulated from each other and from all other conductors.

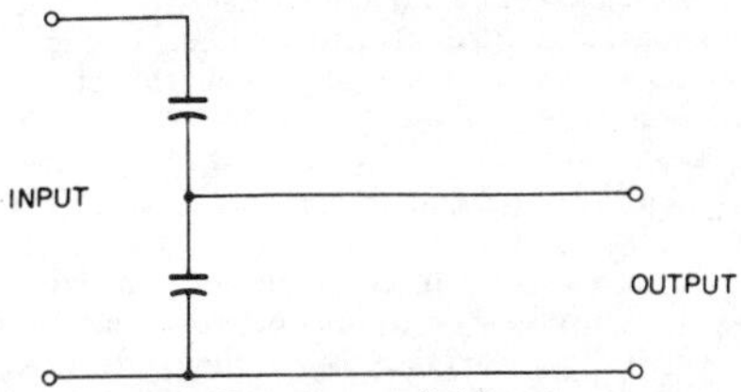

Capacitance divider.

capacitance divider—A circuit made up of capacitors and used for measuring the value of a high-voltage pulse by making available only a small, known fraction of the total pulse voltage for measurement.

capacitance meter—An instrument for measuring capacitance. If the scale is graduated in microfarads, the instrument is usually designated a microfaradmeter.

capacitance-operated intrusion detector — A boundary alarm system in which the approach of an intruder to an antenna wire encircling the protected area (a few feet above ground) changes the antenna-ground capacitance and thereby sets off the alarm.

capacitance ratio — The ratio of maximum to minimum capacitance, as determined from a capacitance characteristic, over a specified voltage range.

capacitance relay—An electronic circuit incorporating a relay which responds to a small change in capacitance, such as that created by bringing the hand or body near a pickup wire or plate.

capacitance tolerance—The maximum percentage deviation from the specified nominal value (at standard or stated environmental conditions) specified by the manufacturer.

capacitive coupling — Also called electrostatic coupling. The assocation of two or more circuits with one another by means of mutual capacitance between them. For example, between stages of an amplifier, that type of interconnection which employs a capacitor in the circuit, between the plate of one tube and the grid of the succeeding tube.

capacitive diaphragm—A resonant window placed in a waveguide to provide the effect of capacitive reactance at the frequency being transmitted.

capacitive divider—Two or more capacitors

placed in series across a source, making available a portion of the source voltage across each capacitor. The voltage across each capacitor will be inversely proportional to its capacitance.

capacitive feedback—The process of returning part of the energy in the plate or output circuit of a vacuum tube to the grid, or input, circuit by means of a capacitance common to both circuits.

capacitive load—A predominantly capacitive load—i.e., one in which the current leads the voltage.

capacitive post—A metal post or screw extending at right angles to the E field in a waveguide. It provides capacitive susceptance in parallel with the waveguide for purposes of tuning or matching.

capacitive reactance — Symbolized by X_c. The impedance a capacitor offers to ac or pulsating dc. Measured in ohms and equal to $\frac{1}{2\pi fC}$, where f is in hertz and C is in farads.

capacitive speaker — *See* Electrostatic Speaker.

capacitive storage welding — A particular type of resistance welding whereby the energy is stored in banks of capacitors, which are then discharged through the primary of the welding transformer. The secondary current generates enough heat to produce the weld.

capacitive transduction—Conversion of the measurand into a change in capacitance.

capacitive tuning—Tuning by means of a variable capacitor.

capacitive window — A conductive diaphragm extended into a waveguide from one or both sidewalls to introduce the effect of capacitive susceptance in parallel with the waveguide.

capacitivity—*See* Dielectric Constant.

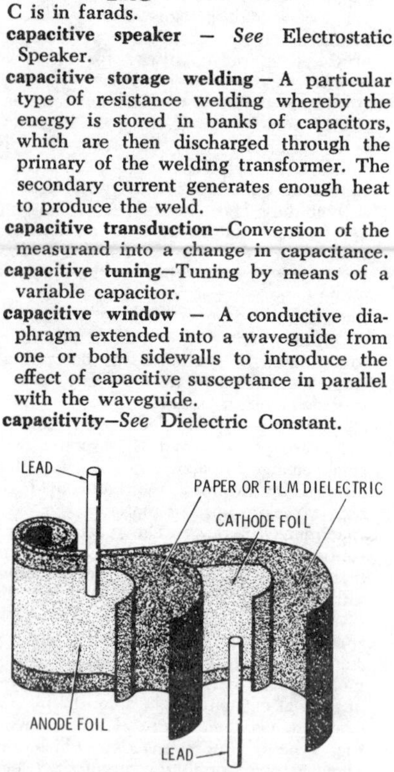

Capacitor (internal construction).

capacitor — A device consisting essentially of two conducting surfaces separated by an insulating material or dielectric such as air, paper, mica, glass, plastic film, or oil. A capacitor stores electrical energy, blocks the flow of direct current, and permits the flow of alternating current to a degree dependent essentially upon the capacitance and the frequency.

capacitor antenna — Also called condenser antenna. An antenna which consists of two conductors or systems of conductors and the essential characteristic of which is its capacitance.

capacitor bank — A number of capacitors connected together in series, parallel, or in series-parallel.

capacitor braking—A means of stopping an induction motor. The capacitor or capacitors can be applied to the winding after shut off. The capacitor acts as a short on a generator.

capacitor color code—Color dots or bands placed on capacitors to indicate one or more of the following: capacitance, capacitance tolerance, voltage rating, temperature coefficient, and the outside foil (on paper or film capacitors).

capacitor filtering—A method for improving the form factor of a direct current by means of a parallel capacitor. Also, a means for increasing the magnitude of a rectified voltage.

capacitor-input filter—A power-supply filter in which a capacitor is connected directly across, or in parallel with, the rectifier output.

capacitor microphone — *See* Electrostatic Microphone.

capacitor motor—A single-phase induction motor with the mean winding arranged for direct connection to the power source, together with an auxiliary winding connected in series with a capacitor.

capacitor pickup — A phonograph pickup which depends for its operation on the variation of its electrical capcitance.

capacitor series resistance—An equivalent resistance in series with a pure capacitance which gives the same resultant losses as the actual capacitor. This equivalent circuit does not represent the variation in capacitor losses with frequency.

capacitor speaker — *See* Electrostatic Speaker.

capacitor-start motor—An ac split-phase induction motor in which a capacitor is connected in series with an auxiliary winding to provide a means of starting. The auxiliary circuit opens when the motor reaches a predetermined speed.

capacitor voltage—The voltage across the terminals of a capacitor.

capacity—1. The current-output capability of a cell or battery over a period of time. Usually expressed in ampere-hours (amp-hr). 2. Capacitance. 3. The limits, both upper and lower, of the items or numbers which may be processed in a computer register, in the accumulator. When quantities exceed the capacity, a computer interrupt develops and requires special handling. 4. The total quantity of data that a part of a computer can hold or handle, usually expressed as words per unit of time. 5. The capability of a specific system to store data, accept transactions, process data, and generate reports. 6. In a calculator, the maximum number of digits that can be entered as one factor or obtained in a result. In most machines, the capacity is equivalent to the number of digits in the display. In a few machines, it is larger than the number of digits in the display and the flip-flop key is used to show the full result.

capstan—The driven spindle or shaft in a tape recorder—sometimes the motor shaft itself—which rotates against the tape, pulling it through the machine at a constant speed during recording and playback modes of operation. The rotational speed and diameter of the capstan thus determine the tape speed.

capstan idler—*See* Pressure Roller.

captive screw—Screw-type fastener that is retained when unscrewed and cannot easily be separated from the part it secured.

capture area—The area of the antenna elements that intercept radio signals.

capture bandwidth — The frequency range over which an unlocked, free-running oscillator can be brought into lock by either phase or injection-locking techniques.

capture effect — 1. The selection of the stronger of two frequency-modulated signals of the same frequency, with the complete rejection of the weaker signal. If both signals are of equal strength, both may be accepted and no intelligible signal will result. 2. An effect occurring in a transducer (usually a demodulator) whereby the input wave having the largest magnitude controls the output.

capture ratio—The ability of a tuner to reject unwanted fm stations and interference on the same frequency as a desired one, measured in dB. The lower the figure, the better the performance of the tuner. 2. The power ratio of two signals in the same channel required to keep the signal/interference ratio to a value of 30 dB referred to 100-percent modulation and 1-mV input signal level. The ratio of the powers of the two input signals is expressed in decibels and the smaller the dB number the better the capture ratio. Top-flight tuners have a value as low as 1 dB but 4.5 dB is usually sufficient.

carbon—One of the elements, consisting of a nonmetallic conductive material occurring as graphite, lampblack, diamond, etc. Its resistance is fifty to several hundred times that of copper and decreases as the temperature increases.

carbon brush — A current-carrying brush made of carbon, carbon and graphite, or carbon and copper.

carbon-contact pickup—A phonograph pickup which depends for its operation on the variation in resistance of carbon contacts.

carbon-film resistor—A resistor formed by vacuum-depositing a thin carbon film on a ceramic form.

carbonize—To coat with carbon.

carbonized filament — A thoriated-tungsten filament treated with carbon. A layer of tungsten carbide formed on the surface slows down the evaporation of the active emitting thorium and thus permits higher operating temperatures and much greater electron emission.

carbonized plate — An electron-tube anode that has been blackened with carbon to increase its heat dissipation.

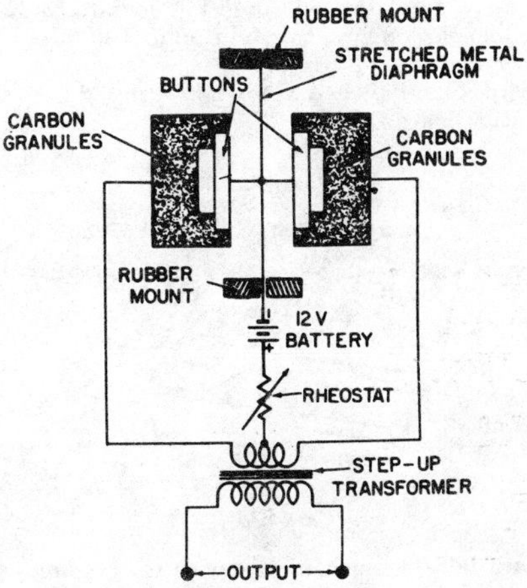

Carbon microphone.

carbon microphone — A microphone which depends for its operation on the variation in resistance of carbon contacts.

carbon-pile regulator — An arrangement of carbon discs the series resistance of which decreases as more pressure or compression is applied.

carbon resistor—Also called composition resistor. A resistor consisting of carbon particles which are mixed with a binder molded into a cylindrical shape, and then baked. Terminal leads are attached to opposite ends. The resistance of a carbon resistor decreases as the temperature increases.

carbon transfer recording—A type of facsimile recording in which carbon particles are deposited on the record sheet in response to the received signal.

Carborundum—A compound of carbon and silicon used in crystals to rectify or detect radio waves.

carcinotron — A voltage-tuned, backward-

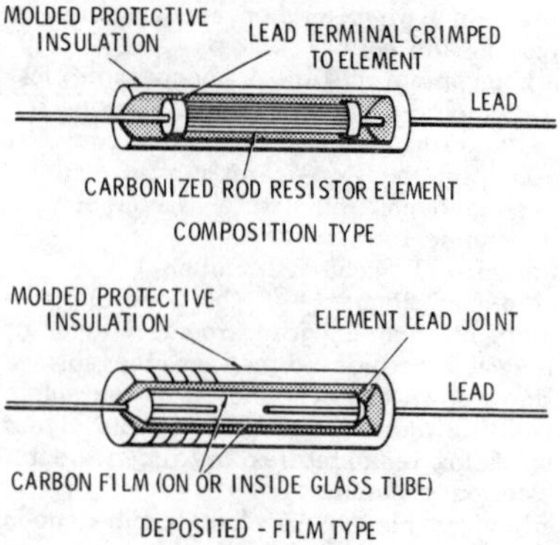

Carbon resistor.

wave oscillator tube used to geenrate frequencies ranging from uhf up to 100 GHz or more.

card—*See* Punched Card *and* Printed-Circuit Board.

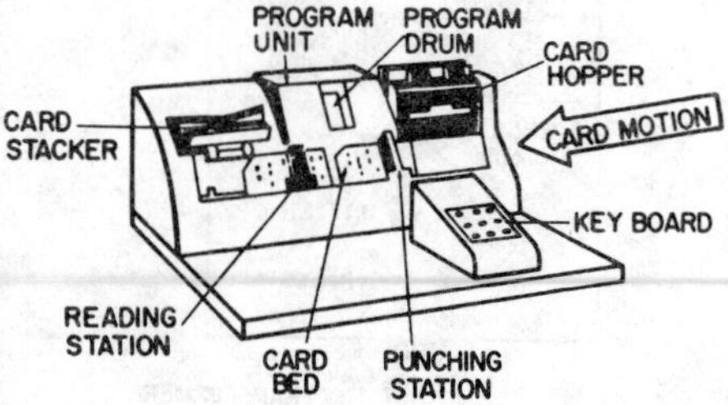

Card bed.

card bed—A mechanical device for holding punch cards to be transported past the punching and reading stations.

card code – An arbitrary code in which holes punched in a card are assigned numeric or alphabetic values.

card column—One of 20 to 90 single-digit columns in a tabulating card. When punched, a column contains only one digit, one letter, or one special code.

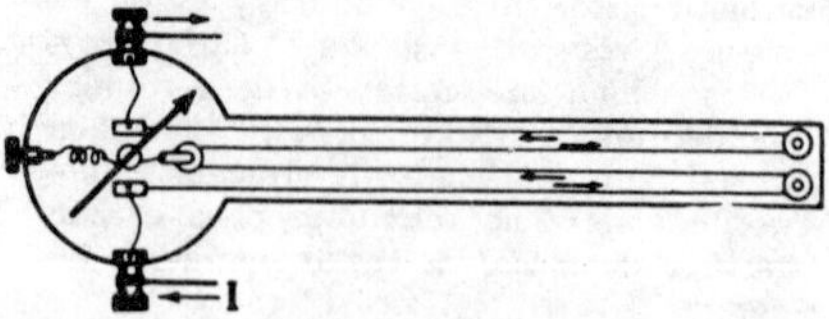

Cardew voltmeter.

Cardew voltmeter – The earliest type of hot-wire instrument. It consisted of a small-diameter platinum-silver wire sufficiently long to give a resistance high enough to be connected directly across the circuit being measured. The wire was looped over pulleys and it expanded as current flowed, causing the pointer to rotate.

card face—The printed side of a punched card if only one side is printed.

card feed—A mechanism that moves punch cards, one at a time, into a machine.

card field—On a punch card, the fixed columns in which the same type of information is routinely entered.

card hopper—*See* Card Stacker.

cardiac monitor—An instrument that usually has an oscilloscope display of the heart wave, and combines the features of several cardiac instruments such as an electrocardiograph, cardiotachometer, etc. May also allow upper and lower limits to be set and trigger audible and/or visual alarms when these limits are exceeded.

cardiac pacemaker—1. A device that controls the frequency of cardiac contractions. 2. A device that stimulates the heart and controls its rhythm by means of electrodes placed on the chest wall or implanted under the skin.

card image—1. A representation in storage of the holes punched in a card, in such a manner that the holes are represented by one binary digit and the unpunched spaces are represented by the other binary digit. 2. In machine language, a duplication of the data contained in a punch **card.**

cardiograph—An instrument (or recording of instruments) for measuring the form or force of heart motion.

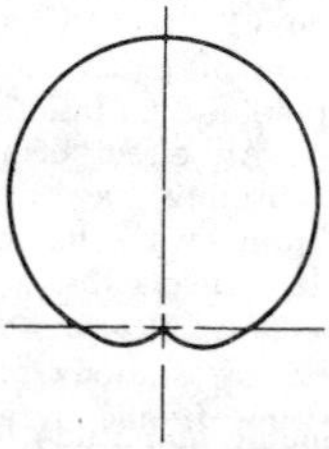

Cardioid diagram.

cardioid diagram—Heart-shaped polar diagram showing the response or radiation characteristic of certain directional antennas, or the response characteristic of certain types of microphones.

cardioid microphone – A microphone that has a heart-shaped response pattern that gives nearly uniform response for a range of about 180° in one direction and a minimum response in the opposite direction.

cardiostimulator—A device used to stimulate the heart and/or regulate its beat. *See also* Pacemaker *and* Defibrillator.

cardiotachometer—A measuring instrument that provides a meter reading proportional to the rate at which the heart beats.

card jam—A pile-up of cards in a machine.

card machine—A machine used to transfer information from or to punched cards.

card programmed—The capability of being programmed by punched cards.

card punch—1. A device to record information in cards by punching holes in the cards to represent letters, digits, and special characters. 2. Device used in data handling systems to enter data on cards according to a desired code.

card reader—1. A device designed to read punched cards and convert each hole into an electrical impulse for use in a computer system. 2. Device used in data handling systems to sense data on punched cards via mechanical or photoelectric technique.

card row—On a punched card, one of the horizontal lines of punching positions.

card sensing — The process of sensing or reading the information in punched cards and converting this information, usually into electrical pulses.

card stacker — A mechanism that stacks cards in a pocket or bin after they have passed through a computer. Also called card hopper.

card-to-tape—Having to do with equipment that transfers information directly from punched cards to punched tape or magnetic tape.

carillon—A bell tower designed to play from a keyboard. In an organ, this may be achieved by tube synthesis of bell-like tones struck with felt hammers, or completely electronically.

Carnot theorem—A thermodynamic principle which states that a cycle continuously operating between a low temperature and a high temperature can be no more efficient than a reversible cycle operating between the same temperatures.

carriage tape—*See* Control Tape.

carrier—1. A wave of constant amplitude, frequency, and phase which can be modulated by changing amplitude, frequency, or pulse. 2. An entity capable of carrying an electric charge through a solid (e.g., holes and conduction electrons in semiconductors). 3. A wave that has at least one characteristic that can be varied from a known reference value by a modulation process. 4. That part of the modulated wave that corresponds to the unmodulated wave in a specified way.

carrier-amplitude regulation — The change in amplitude of the carrier wave in an amplitude-modulated transmitter when symmetrical modulation is applied.

carrier band—In CD-4 discs, the left- or right-channel's band that contains musical information recorded at very high frequencies (in the 20-45 kilohertz range). In playback, the demodulator recovers those frequencies, which have been frequency modulated.

carrier beat—In facsimile transmission, an undesirable heterodyne of signals, each synchronous with a different stable oscillator, causing a pattern in received copy. Where one or more of the oscillators is fork-controlled, this is called fork beat.

carrier chrominance signal—In color television, sidebands of a modulated chrominance subcarrier, plus any unsuppressed subcarrier, added to the monochrome signal to convey color information.

carrier color signal — In color television, sidebands of a modulated chrominance subcarrier, plus the chrominance subcarrier, if not suppressed, which are added to the monochrome signal to convey color information.

carrier concentration—The number of carriers in a cubic centimeter of semiconductor material.

carrier control—A control by the presence or absence of an rf carrier.

carrier current—1. The current associated with a carrier wave. 2. High-frequency alternating current superimposed on ordinary telephone, telegraph, or power-line frequencies. The carrier may be modulated with voice signals for telephone communications between points in a power system, or it may be tone modulated to operate switching relays or transmit data.

carrier-current communication—The superimposing of a high-frequency alternating current on ordinary telephone, telegraph, and power-line frequencies for telephone communication and control.

carrier-current control — Remote control in which the receiver and transmitter are coupled together through power lines.

carrier frequency — 1. The frequency (hertz) of the wave modulated by the intelligence wave; usually a radio frequency (rf). 2. The reciprocal of the period of a periodic carrier. 3. The frequency of the unmodulated fundamental output from a radio transmitter.

carrier-frequency interconnection — In the formation of carrier networks, the transfer of groups of channels between terminals of wire-line cable or radio carrier systems at carrier frequencies.

carrier-frequency peak-pulse power — The power averaged over that carrier-frequency cycle which occurs at the maximum pulse of power (usually half the maximum instantaneous power).

carrier-frequency pulse — A carrier that is amplitude-modulated by a pulse. The amplitude of the modulated carrier is zero before and after the pulse.

carrier-frequency range — The continuous range of frequencies within which a transmitter may normally operate. A transmitter may have more than one carrier-frequency range.

carrier-frequency stereo disc—A stereo disc with two laterally cut channels. One channel is cut in the usual manner. The second channel is employed to frequency-modulate a supersonic carrier frequency. The playback cartridge delivers the signal for one channel plus the carrier frequency containing the other channel. The latter must then be demodulated to obtain the second channel.

carrier-isolating choke coil — An inductor inserted in series with a line on which carrier energy is applied to impede the flow of carrier energy beyond that point.

carrier leak—The carrier-frequency signal remaining after suppression in a suppressed carrier system.

carrier level — The strength, expressed in decibels, of an unmodulated carrier signal at a particular point in a system.

carrier lifetime—The time required for excess carriers doped into a semiconductor to recombine with other carriers of the opposite sign.

carrier line — A transmission line used for multiple-channel carrier communications.

carrier loading—The insertion of additional lump inductance in a cable section of a transmission line utilized for carrier transmission up to about 35 kHz. Loading serves to minimize impedance mismatch between cable and open wire, and to reduce the cable attenuation.

carrier mobility—The average drift velocity of carriers per unit electric field in a homogenous semiconductor. The mobility of electrons is usually different from that of holes.

carrier noise—Undesired variation of a radio-frequency carrier signal in the absence of intended modulation. Also called residual modulation.

carrier noise level — Also called residual modulation. The noise level produced by undesired variations of a radio-frequency signal in the absence of any intended modulation.

carrier on microwave—A means of transmitting many voice messages on one microwave radio channel. Transmission is point to point by microwave antennas mounted on towers or tall buildings.

carrier on wire—A means widely used by the telephone companies to transmit many voice messages on a single pair of wires. Circuits involving one or more carrier links never evidence dc continuity.

carrier-operated antinoise device — A device the purpose of which is to mute the audio output of a receiver during standby or intervals of no carrier.

carrier power output rating — The power available at the output terminals of a transmitter when the output terminals are connected to the normal load circuit or to a circuit equivalent thereto.

carrier repeater—An assembly, including an amplifier (or amplifiers), filters, equalizers, level controls, etc., used to raise the carrier signal level to a value suitable for traversing a succeeding line section while maintaining an adequate signal-to-noise ratio.

carrier shift—1. The transmission of radio teletypewriter messages by shifting the carrier frequency in one direction for a marking signal and in the opposite direction for a spacing signal. 2. The condition resulting from imperfect modulation, whereby the positive and negative excursions of the envelope pattern are unequal, thus effecting a change in the power associated with the carrier.

carrier signaling—In a telephone system, the method by which ringing, busy signals, or dial-signaling relays are operated by the transmission of a carrier-frequency tone.

carrier storage time (of a switching transistor)—The time interval between the beginning of the fall of the pulse applied to the input terminals and the beginning of the fall of the pulse generated by charge carriers at the output terminals. (The time is generally measured between the 90% values of the two pulse amplitudes.)

carrier suppression—The method of operation in which the carrier wave is not transmitted.

carrier swing—The total deviation of a frequency- or phase-modulated wave from the lowest to the highest instantaneous frequency.

carrier system — A means of obtaining a number of channels over a single path by modulating each channel upon a different carrier frequency and demodulating at the receiving point to restore the signals to their original form.

carrier tap choke coil—A carrier-isolating choke coil inserted in series with a line tap.

carrier tap transmission choke coil—An inductor inserted in series with a line tap to control the amount of carrier energy flowing into the tap.

carrier telegraphy—That form of telegraphy in which the transmitted signal is formed by modulating the alternating current, under control of the transmitting apparatus, before supplying it to the line.

carrier telephony—Ordinarily applied only to wire telephony. That form of telephony in which carrier transmission is used, the modulating wave being at an audio frequency.

carrier terminal—Apparatus at one end of a carrier transmission system, whereby the processes of modulation, demodulation, filtering, amplification, and associated functions are effected.

carrier-to-noise ratio—Ratio of the magnitude of the carrier to the magnitude of the noise after selection and before any nonlinear process such as amplitude limiting and detection. This ratio is expressed in many different ways—for example, in terms of peak values in the case of impulse noise, and in terms of root-mean-square values in the case of random noise.

carrier-transfer filter—A group of filters arranged to form a carrier-frequency crossover or bridge between two transmission circuits.

carrier transmission—That form of electrical transmission in which a single-frequency wave is modulated by another wave containing the information.

carrier type dc amplifier—An amplifier system which converts a dc input to modulated ac, amplifies, and synchronously detects to provide amplifier dc output.

carrier wave—1. The single-frequency wave which is transmitted and which is modulated by another wave containing the information. 2. The basic frequency or pulse repetition rate of a signal, bearing no intrinsic intelligence until it is modulated by another signal which does bear intelligence. A carrier may be amplitude, phase, or frequency modulated.

carry—1. A signal or expression produced in an electronic computer by an arithmetic operation on a one-digit place of two or more numbers expressed in positional notation and transferred to the next higher place for processing there. 2. A signal or expression—as defined in (1)—which arises when the sum of two digits in the same digit place equals or exceeds the base of the number system in use. If a carry into a digit place will result in a carry out of the same digit place and the normal adding circuit is bypassed when this new carry is generated, the result is called a high-speed or "stand-on-nines" carry. If the normal adding circuit is used, the result is called a cascade carry. If a carry resulting from the addition of carries is not allowed to propagate, the process is called a partial carry; if it is, a complete carry. A carry generated in the most significant digit place and sent directly to the least significant place is called an end-around carry. 3. In direct subtraction, a signal or expression—as defined in (1) above—which arises when the difference between the digits is less than zero. Such a carry is frequently called a borrow. 4. The action of forwarding a carry. 5. The command directing a carry to be forwarded.

carry time—The time required for a computer to transfer a carry digit to the next higher column and add it there.

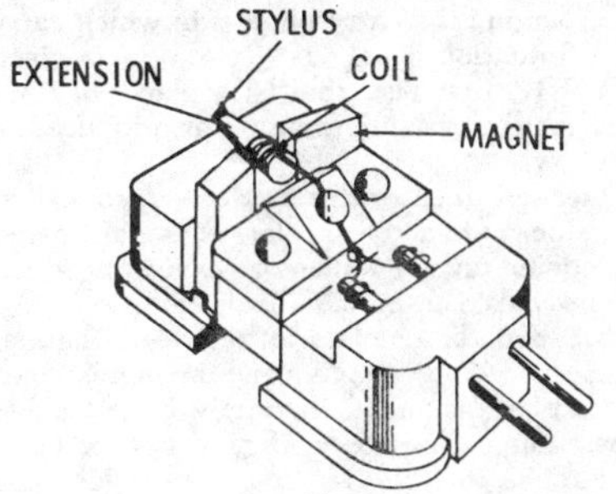

Cartridge.

cartridge—1. The electromechanical transducer of a pickup head that converts stylus vibrations to an electrical signal. It is generally detachable and fits into the head shell of a pickup. 2. Generally, any enclosed "package" containing a length of magentic tape and its basic winding receptacles, designed to eliminate the need for handling or threading the tape. Specifically, the word cartridge is used to describe that variety of package that contains a continuous (endless) loop of tape on a single reel. 3. A film or tape magazine containing only one spool.

cartridge fuse—A tubular fuse the end caps of which are enclosed in a glass or composition insulating tube to confine the arc or vapor when the fuse blows.

cascadable counter—A logic counting block that has available the necessary connections to permit more than one counter to be operated in series, thus increasing the modulus of the counter subsystem.

cascade—An arrangement of two or more similar circuits or amplifying stages in which the output of one circuit provides the input of the next. Also called tandem.

cascade amplifier—A multiple-stage amplifier in which the output of each stage is connected to the input of the next stage.

cascade-amplifier klystron—A klystron that has three resonant cavities to provide increased power amplification and output. The extra resonator is located between the input and output resonators and is excited by the bunched beam energizing from the first resonator gap, thereby producing further bunching of the beam.

cascade connection—Two or more similar component devices arranged in tandem, with the ouput of one connected to the input of the next.

cascade control—Also called piggyback control. An automatic control system in which the control units, linked in chain fashion, feed into one another in succession. Each unit thus regulates the operation of the next in line.

cascaded carry—In a computer, a system of

executing the carry process in which carry information cannot be passed on to place (N + 1) unless the Nth place has received carry information or produced a carry.

cascaded feedback canceler – Also called velocity-shaped canceler. A sophisticated moving-target indicator canceler which provides clutter and chaff rejection.

cascaded thermoelectric device—A thermoelectric device having two or more stages that are arranged thermally in series.

cascading—The connecting of two or more circuits in series so that the output from one provides the input to the next.

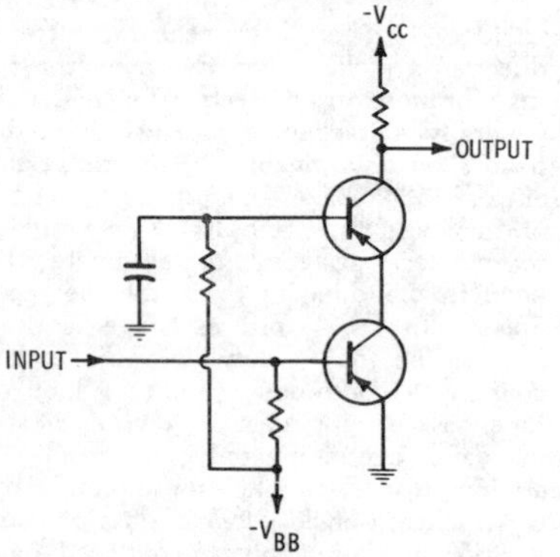

Cascode amplifier.

cascode amplifier – An amplifier using a neutralized grounded-cathode input stage followed by a grounded-grid output stage. The circuit has high gain, high input impedance, and low noise.

case—In a computer, a set of data for use in a particular program.

case pressure—The total differential pressure in the internal cavity of a transducer and the ambient pressure. The term is commonly used to summarize the limiting combined differential and/or line-pressure capabilities of differential transducers.

case temperature—The temperature on the surface of the case at a designated point.

Cassegrain antenna—An antenna the feed of which is positioned near the vertex of the reflector, with a small subreflector placed near the focal point. The feed illuminates the subreflector, and the subreflector redirects the waves toward the main reflector, which then forms the radiated beam (called the secondary beam or pattern). The shapes of the subreflector and main reflector are so chosen that the secondary rays will emerge parallel to the main-reflector axis.

Cassegrain feed—A method of feeding a reflector antenna in which a waveguide located in the center of the main reflector feeds energy to a small reflector which reflects it in turn to the main reflector.

cassette—1. A thin, flat, rectangular enclosure that contains a length of narrow magnetic recording tape permanently affixed to two flangeless, floating reels which, respectively, wind and unwind the tape while it passes an external recording and/or playback head. 2. A flat enclosure that contains two flangeless reels that link a narrow magnetic tape. 3. A sealed "package" instant-load cartridge containing a length of tape and separate supply and takeup reels or hubs. Cassettes, unlike continuous-loop cartridges, can be rewound as well as fast-forwarded. 4. A film or tape magazine containing two spools.

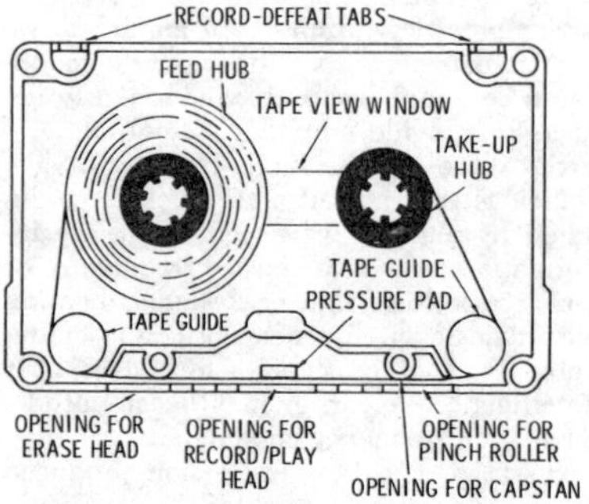

Cassette.

casting-out-nines check—A partial verification of an arithmetical operation on two or more numbers. It involves casting out nines from the numbers and from the results.

catalog—An ordered compilation of descriptions of items, including sufficient information to afford access to the items.

catalyst – Also called hardener and promoter. A substance which markedly speeds up the cure of a casting or molding compound when added in a minor quantity as compared to the amounts of primary reactants.

catastrophic failure—A sudden failure without warning, as opposed to degradation failure. Or, a failure the occurrence of which can prevent the satisfactory performance of an entire assembly or system.

catcher—In a velocity-modulated vacuum tube, an electrode on which the spaced electron groups induce a signal. The output of the tube is taken from this element.

catching diode—A diode connected to act as a short circuit when its anode becomes more positive than its cathode; the diode then tends to prevent the voltage of a circuit terminal from rising above the voltage at the cathode.

categorization—The process by which multiple addressed messages are separated to form individual messages for single addresses.

catelectrotonus—The increased sensitivity produced in a nerve or muscle in the region of contact with the cathode when an electric current is passed through it.

catena—A chain or connected series.

catenate—*See* Concatenate.

catenate—*See* Concatenate.

cathamplifier – A push-pull vacuum-tube amplifier in which the push-pull transformer is in the cathode circuit.

cathode—1. In an electron tube the electrode through which a primary source of electrons enters the interelectrode space. 2. General name for any negative electrode. 3. The lower-potential electrode of a corrosion cell where the action of the corrosion current may reduce or eliminate corrosion, or the negatively charged metallic parts of an impressed-current system. 4. When a semiconductor diode is biased in the forward direction, that terminal of the diode which is negative with respect to the other terminal.

cathode activity—Measure of the efficiency of an emitter. The mathematical relationship between two values of emission current measured under two conditions of cathode temperature.

cathode bias—A method of biasing a vacuum tube by placing the biasing resistor in the common cathode-return circuit, thereby making the cathode more positive —rather than the grid more negative— with respect to ground.

cathode-coupled amplifier—A cascade amplifier in which the coupling between two stages is accomplished by a common-cathode resistor.

cathode coupling—The use of an input or output element in the cathode circuit for coupling energy to another stage.

cathode current—*See* Electrode Current.

cathode-current density—The current per square centimeter of cathode area, expressed as amperes or milliamperes per centimeter squared.

cathode dark space—Also called Crookes' dark space. The relatively nonluminous region between the cathode and negative glow in a glow-discharge—cold-cathode tube.

cathode emission—The process whereby electrons are emitted from the cathode structure.

cathode follower—Also called grounded-plate amplifier. A vacuum-tube circuit in which the input signal is applied to the control grid, and the output is taken from the cathode. Electrically, such a circuit possesses a high input impedance and a low output impedance characteristic and a gain of less than unity. The equivalent

circuit using a transistor is called an emitter follower.

cathode glow—The apparent luminosity or "glow" that immediately envelops the cathode in a gas-discharge tube when operating at low pressures. The glow increases as the pressure decreases.

cathode guide—The element of a glow tube used in switching the neon glow from one indicated number to the next.

cathode heating time—The time required for the cathode to attain a specified condition—for example, a specified value of emission or a specified rate of change in emission.

cathode interface—A resistive and capacitive layer formed between the nickel sleeve and oxide coating of an indirectly heated cathode. Raising the cathode temperature will largely nullify the layer.

cathode keying—A method of keying a radiotelegraph transmitter by opening the plate return lead to the cathode or filament center tap.

cathode luminous sensitivity (**of a multiplier phototube**)—The photocathode current divided by the incident luminous flux.

cathode modulation—A form of amplitude modulation in which the modulating voltage is applied to the cathode circuit.

cathode pulse modulation—Modulation produced in an amplifier or oscillator by applying externally generated pulses to the cathode circuit.

cathode pulse modulation—Modulation produced in an amplifier or oscillator by applying externally generated pulses to the cathode circuit.

cathode radiant sensitivity — The current leaving the photocathode divided by the incident radiant power of a given wavelength.

cathode ray—A stream of electrons emitted, under the influence of an electric field, from the cathode of an evacuated tube or from the ionized region nearby.

$$Z_0 = \frac{R_0}{1 + g_m R_0} \approx \frac{1}{g_m}$$

$$R_0 = \frac{Z_0}{1 - g_m Z_0}$$

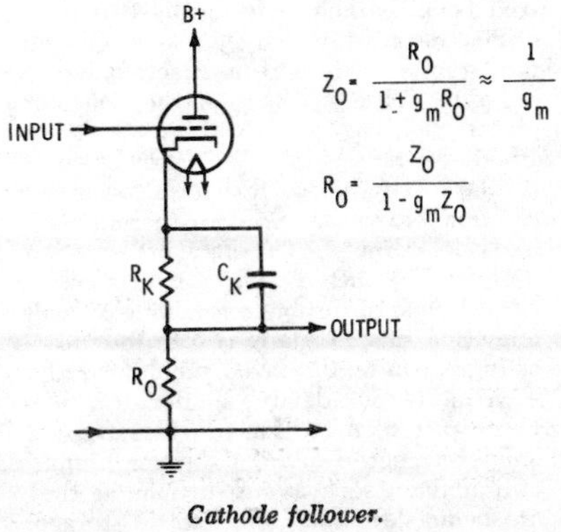

Cathode follower.

cathode-ray charge-storage tube—A charge-storage tube in which the desired information is written by means of a cathode-ray beam.

cathode-ray instrument—*See* Electron-Beam Instrument.

cathode-ray oscillograph — An apparatus capable of producing, from a cathode-ray tube, a permanent record of the value of an electrical quantity as a function of time.

cathode-ray oscilloscope—A test instrument which, when properly adjusted, makes possible the visual inspection of alternating-current signals. It consists of an amplifier, time-base generating circuits, and a cathode-ray tube for transformation of electrical energy into light energy.

cathode-ray storage—An electrostatic data-storage device in which a cathode-ray beam provides access to the data.

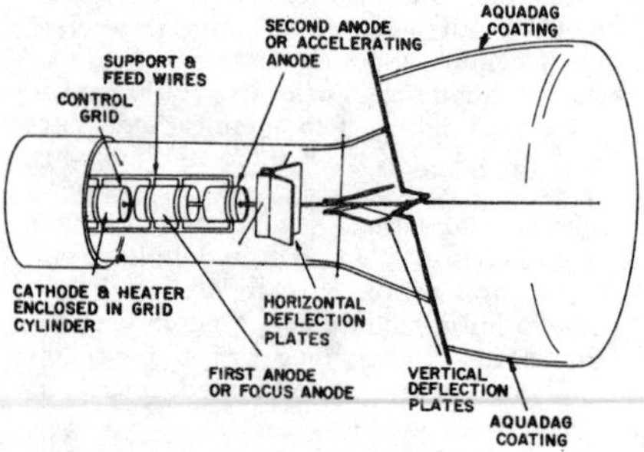

Cathode-ray tube.

cathode-ray tube—Abbreviated crt. 1. A tube in which its electron beam can be focused to a small cross section on a luminescent screen and can be varied in position and intensity to produce a visible pattern. 2. A vacuum tube with an electron "gun" at one end and a fluorescent screen at the other. Electrons emitted from a heated filament are accelerated by a series of annular anodes at progressively higher positive voltages. The electron beam is then deflected by two pairs of electrostatically charged plates between the "gun" and the screen. Electromagnets are often used in place of the deflector plates.

cathode-ray-tube display — Abbreviated crt display. 1. A device in which controlled electron beams are used to present data in visual form. 2. The data presentation produced by such a device. 3. In a calculator, a type of display resembling a small television tube. Usually two to four rows of digits can be displayed simultaneously. 4. A high-speed device, similar to a television picture tube, that provides a visual nonpermanent display of system input/output data, such as instructions as they are being developed and data in storage.

cathode-ray tuning indicator — Commonly called magic eye. A small-diameter cathode-ray tube that visually indicates whether an apparatus such as a radio receiver is tuned precisely to a station.

cathode resistor—A resistance connected in the cathode circuit of a tube so that the voltage drop across it will supply the proper cathode-biasing voltage.

cathode spot—On the cathode of an arc, the area from which electrons are emitted at a current density of thousands of amperes per square centimeter and where the temperature of the electrode is too low to account for such currents by thermionic emission.

cathode sputtering—*See* Sputtering, 1.

cathode protection — The control of the electrolytic corrosion of an underground or underwater metallic structure by the application of an electric current through a sacrificial anode in such a way that the structure is made to act as a cathode of an electrolytic cell.

cathodoluminescence — Luminescence produced by the bombardment with high-velocity electrons of a metal in a vacuum. Small amounts of the metal are vaporized in an excited state by the bombardment and emit radiation characteristic of the metal.

cation—A positive ion that moves toward the cathode in a discharge tube, electrolytic cell, or similar equipment. The corresponding negative ion is called an "anion."

catv—Abbreviation for community antenna television.

catwhisker—A small, pointed wire used to make contact with a sensitive area on the surface of a crystal or semiconductor.

cavitation — The production of gas-filled cavities in a liquid when the pressure is reduced below a certain critical value with no change in the temperature. Ordinarily this is a destructive effect as the high pressures produced when these cavities collapse often damage mechanical components of hydraulic systems, but the effect is turned to advantages in ultrasonic cleaning.

cavitation noise—The noise produced in a liquid by the collapse of the bubbles created by cavitation.

cavity—A metallic enclosure inside which resonant fields may be excited at a microwave frequency.

cavity filter—A selective tuned device having the proper coupling means for insertion into a transmission line to produce attenuation of unwanted off-frequency signals.

cavity impedance—The impedance that appears across the gap of the cavity of a microwave tube.

cavity magnetron—A magnetron having a

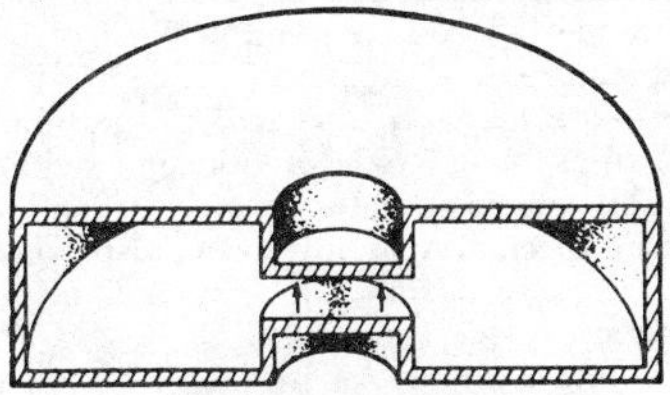

X-band reflex-klystron resonant cavity.

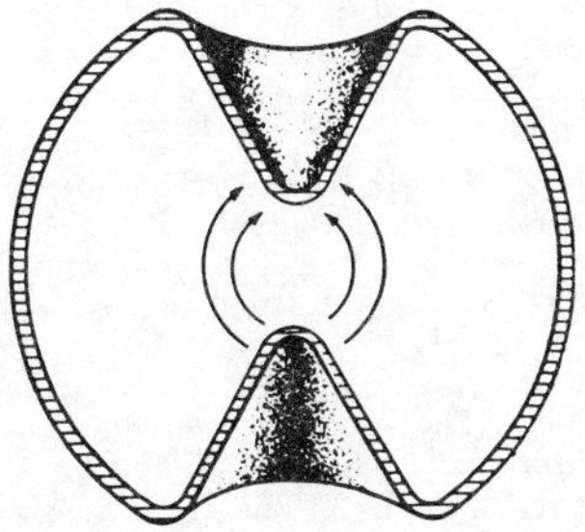

TR cavity.

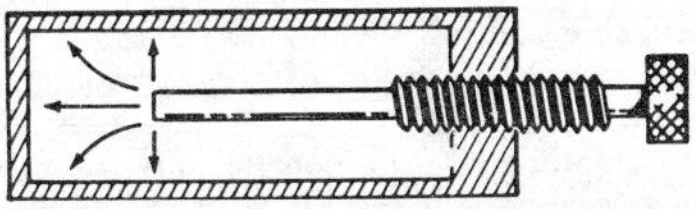

Re-entrant wavemeter cavity.

Cavities.

number of resonant cavities forming the anode; used as a microwave-transmitting oscillator.

cavity oscillator – Abbreviated CO. An oscillator in which the primary frequency-determining element is either a waveguide or coaxial cavity. Oscillator frequency can be mechanically tuned and voltage tuned (via a tuning varactor) over a relatively narrow band.

ccd—Abbreviation for charge coupled device. 1. A semiconductor storage device in which an electrical charge is moved across the surface of a semiconductor by electrical control signals. Zeroes or ones are represented by the absence or presence of a charge. 2. Semiconductor storage device that employs a charge transfer system in which charges created by either an input diode or by an impinging photon is contained in MOS or MIS capacitors fabricated on a single crystal wafer. By varying electrode voltages successively, charge packets are moved from capacitor to capacitor to a single output amplifier. Information can be stored temporarily in the device.

cavity radiation – The radiation (heat) emerging from a small hole leading to a constant-temperature enclosure. Such radiation is identical with blackbody radiation at the same temperature, no matter what the nature of the inner surface of the enclosure.

cavity resonator—1. A space which is normally bounded by an electrically conducting surface and in which oscillating electromagnetic energy is stored; the resonant frequency is determined by the geometry of the enclosure. 2. A section of coaxial line or waveguide completely enclosed by conducting walls; it is often made variable for use as a wavemeter.

cavity-resonator frequency meter—A cavity resonator used for determining the frequency of an electromagnetic wave.

cavity-tuned, absorption-type frequency meter—A device used for measuring frequency. Its operation depends on the use of an enclosure with a conductive inner wall; the resonant frequency of the wall is determined by its internal dimensions.

cavity tuned, heterodyne-type frequency meter—A device for measuring frequency. Its operation depends on the use of an enclosure, the resonant frequency of which is determined by its internal dimensions.

cavity-tuned, transmission-type frequency meter—A device for measuring frequency. Its operation depends on the use of an enclosure with a conductive inner wall; the resonant frequency of the wall is determined by its internal dimensions.

C-band—A radio-frequency band of 3.9 to 6.2 GHz, with wavelengths of 7.69 to 4.84 cm. It includes the top two sidebands of the S-band, and the bottom three sidebands of the X-band.

C-battery – Also called grid battery. The energy source which supplies the voltage for biasing the grid of a vacuum tube.

C-bias—*See* Grid Bias.

CCIF—Abbreviation for International Telephone Consultative Committee.

CCIR—Abbreviation for International Radio Consultative Committee.

CCIT—Abbreviation for International Telegraph Consultative Committee.

ccs—Abbreviation for continuous commercial service. Refers to the power rating of transformers, tubes, resistors, etc. Used for rating components in broadcasting stations and some industrial applications.

cctv camera—That part of a closed circuit tv system which captures and transmits the picture.

cctv monitor—That part of a closed circuit tv system which receives the picture from the cctv camera and displays it on the picture tube.

ccw—Abbreviation for counterclockwise.

CD-4—1. A record-playback system for discrete discs. Invented by the Victor Co. of Japan (JVC) and developed by JVC and RCA Records, the system needs a demodulator and special cartridge with a special stylus for discrete four-channel playback. The system is not compatible with matrix, quad discs and is not used for fm broadcasting. Also called quadradisc. 2. A recording and playback system similar in some respect to fm multiplex stereo broadcast and reception. Each wall of the record groove carries a single channel of information—left front plus left rear on the inner wall and right front plus right rear on the outer wall of the groove. In addition, each groove wall carries a 30-kHz fm subcarrier that is modulated by the front-minus-back difference signals that are needed to decode or demodulate the quadraphonic signal into four discrete channels.

CD-4 capability—The ability of a cartridge to reproduce the ultrasonic signals necessary for discrete four-channel disc reproduction using a CD-4 demodulator.

cdi—Abbreviation for collector-diffusion isolation.

C-display—A type of radar display in which the signal is a bright spot, with the bearing as the horizontal and the elevation angle as the vertical co-ordinate.

ceiling—The maximum voltage that may be attained by an exciter under specified conditions.

celestial guidance—A system of guidance in which star sightings that are automatically taken during the flight of a missile provide position information used by the guidance equipment.

cell—1. A single unit that produces a direct voltage by converting chemical energy into electrical energy. 2. A single unit that produces a direct voltage by converting radiant energy into electrical energy for example, a solar or photovoltaic cell. 3. A single unit that produces a varying voltage drop because its resistance varies with illumination. 4. Elementary unit of storage. 5. In corrosion processes, a source of electric potential that is responsible for corrosion. It consists of an anode and a cathode immersed in an electolyte and electrically bonded together. The anode and cathode may be separate metals or dissimilar areas on the same metal. The different metals will develop a difference in potential that is accompanied by corrosion of the anode. When this cell involves an electrolyte as it does in corrosion processes, it is referred to as an electrolytic cell.

cell counter—An electronic instrument used to count white or red blood cells or other very small particles.

cell-type enclosure—A prefabricated basic shielded enclosure of double-walled copper-mesh construction. The original screen-room design.

cell-type tube (tr, atr, and pre-tr tubes)—A gas-filled, radio-frequency switching tube which operates in an external resonant circuit. A tuning mechanism may be incorporated into the external resonant circuit or the tube.

cellulose acetate—An inexpensive transparent plastic film used as the backing material for many recording tapes.

cellulose-nitrate disc—*See* Lacquer Disc.

Celsius temperature scale—Also called centigrade temperature scale. A temperature scale based on the freezing point of water defined as 0°C and the boiling point defined as 100°C both under conditions of normal atmospheric pressure (760 mm of mercury).

cent—A measure of frequency, defined as equal to 100th of a semitone.

center frequency—1. Also called resting frequency. The average frequency of the emitted wave when modulated by a symmetrical signal. 2. The frequency at the center of a spectrum display (for linear frequency scanning). It is usually tunable.

centering control—One of two controls used to shift the position of the entire image on the screen of a cathode-ray tube. The horizontal-centering control moves the image to the right or left, and the vertical-centering control moves it up or down. *See also* Framing Control.

centering diode—A clamping circuit used in some types of plan-position indicators.

centering magnet—A magnet which centers the televised picture on the face of the tube. Also called framing magnet.

center of gravity—A point inside or outside a body and around which all parts of the body balance each other.

center of mass—On a line between two bodies, the point around which the two bodies would revolve freely as a system.

center poise—Scale of viscosity for insulating varnishes.

center ring—The part that supports the stator in an induction-motor housing. The motor end shields are attached to the ends of the center ring.

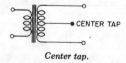

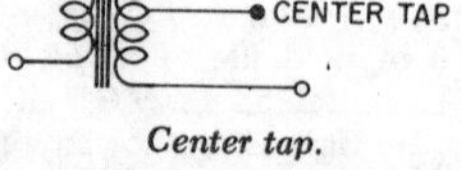

Center tap.

center tap—A connection at the electrical center of a winding, or midway between the electrical ends of a resistor or other portion of a circuit.

center-tapped winding—An internal tap at the center of a winding (three external leads).

center wire—A fine loop of wire used in proportional counters as an anode. A high voltage is applied to it to set the conditions for radiation measurement.

centi—One hundredth (10^{-2}) of a specific quantity or dimension.

centigrade temperature scale — The older name for a Celsius temperature scale in the English-speaking countries. Officially abandoned by international agreement in 1948, but still in common usage.

centimeter waves—Microwave frequencies between 3 and 30 GHz, corresponding to wavelengths of 1 to 10 centimeters.

central battery exchange — Manual telephone exchange in which a battery situated at the exchange is the source of current for operating supervisory signals, current for subscribers' calling signals, and the current required to enable a subscriber to speak over his line.

central office—The facility at which a communications common carrier terminates customer lines and locates the equipment for interconnecting those lines.

central-office line—*See* Subscriber Line.

central processing unit—Also called central processor. Part of a computer system which contains the main storage, arithmetic unit, and special register groups. Performs arithmetic operations, controls instruction processing, and provides timing signals and other housekeeping operations.

centrifugal force—The force which acts on a rotating body and which tends to throw the body farther from the axis of its rotation.

centripetal force—The force which compels a rotating body to move inward toward the center of rotation.

Ceracircuits—A trademark of the Sprague Electric Company for hybrid thick-thin film integrated circuits that consist of discrete passive and semiconductor active elements attached to precision resistor substrates to form functional electronic modules.

ceramic—1. A claylike material, consisting primarily of magnesium and aluminum oxides, which after molding and firing is used as an insulating material. It withstands high temperatures and is less fragile than glass. When glazed, it is called porcelain. 2. Pertaining to or made of clay or other silicates. 3. Piezoelectric part of a pickup, speaker or microphone that acts as a transducer. It has characteristics that are similar to a crystal transducer but it is more robust.

ceramic amplifier—An amplifier that makes use of the piezoelectric properties of ceramics such as barium titanate and the piezoresistive properties of semiconductors such as silicon. An ac signal applied through electrodes to a barium titanate bar produces deformation of the bar and the attached silicon strip, thereby producing a corresponding variation in resistance. This resistance change causes the load current to vary. The device is essentially a current amplifier with extremely high input impedance and low output impedance.

ceramic-based microcircuit—A microminiature circuit printed on a ceramic substrate. Usually consists of combinations of resistive, capacitive, or conductive elements fired on a waferlike piece of ceramic.

ceramic capacitor—A capacitor the dielectric of which is a ceramic material such as steatite or barium titanate, the composition of which can be varied to give a wide range of temperature coefficients. The electrodes are usually silver coatings, fired on opposite sides of the ceramic disc or slab, or fired on the inside and outside of a ceramic tube. After connecting leads are soldered to the electrodes, the unit it usually given a protective insulating coating.

ceramic filter — Electrically coupled, two-terminal piezoelectric ceramic resonators in ladder and lattice configurations. Monolithic filters with ceramic substrates are also called ceramic filters.

ceramic microphone—A microphone with a ceramic cartridge.

ceramic permanent magnet—A permanent, nonmetallic magnet made from pressed and sintered mixtures of metallic-oxide powders, usually oxides of barium and iron.

ceramic pickup—1. A phonograph pickup with a ceramic cartridge. 2. A pickup whose generator system is based on piezoelectricity produced by the stressing of natural and man-polarized crystals.

ceramic transducer — *See* Piezoelectric Transducer.

Cerenkov radiation — Light emitted when charged particles pass through a transparent material at a velocity greater than that of light in that material. In can be seen, for example, as a blue glow in the water around the fuel elements of pool reactors. P.A. Cerenkov was the Russian scientist who first explained the origin of this light.

Cerenkov rebatron radiator—A device in which a tightly bunched velocity-modulated electron beam is passed through a hole in a dielectric. The reaction between the higher velocity of the electrons passing through the hole and the slower velocity of the electromagnetic energy passing through the dielectric results in radiation at some frequency higher than the frequency of modulation of the electron beam.

cermet—A metal-dielectric mixture used in making thin-film resistive elements. The first half of the term is derived from *cer*amic and the second half from *met*al.

certified magnetic tape—Magnetic tape that has been tested and is certified to be free from error over its entire recording surface.

cesium—A chemical element having a low work function. Used as a getter in vacuum tubes and in cesium-oxygen-silver photocell cathodes.

cesium-vapor lamp — A low-voltage arc lamp for producing infrared radiation.

cev — Abbreviation for corona extinction voltage.

C_{gk}—Symbol for grid-cathode capacitance in a vacuum tube.

C_{gp}—Symbol for grid-plate capacitance in a vacuum tube.

cgs—Abbreviation for centimeter gram second. These quantities of space, mass, and time are the basis of absolute units.

cgs electromagnetic system of units—A coherent system of units for expressing the magnitude of electrical and magnetic quantities. The most common fundamental units of these quantities are the centimeter, gram, and second. Their unit of current (abampere) is of such a magnitude that if maintained constant in two straight parallel conductors having an infinite length and negligible circular sections and placed one centimeter apart in a vacuum, a force equal to 2 dynes per centimeter of length will be produced.

cgs electrostatic system of units—A coherent system of units for expressing the magnitude of electrical and magnetic quantities. The most common fundamental units of these quantities are the centimeter, gram, and second. Their unit of electrical charge (stacoulomb) is of such a magnitude that two equal unit point charges one centimeter apart in a vacuum will repel each other with a force of one dyne.

chad—The piece of material removed when a hole or notch is formed in a storage medium such as punched tape or punched cards.

chadless—Pertaining to tape in which the data holes are deliberately not punched through and a flap of material remains attached to the tape.

chadless tape—A type of punched paper tape in which each chad is left fastened by about a quarter of the circumference of the hole. Chadless punched paper tape must be sensed by mechanical fingers, because chad interferes with reliable electrical or photoelectrical reading.

chad tape—Tape, used in printing telegraph or teletypewriter operation, in which the perforations are severed from the tape to form holes that represent characters. Normally, the characters are not printed on chad tape.

chafe—Undesirable rubbing with friction.

chaff—A general name applied to radar-confusion reflectors which consist of thin, narrow, metallic strips of various length and frequency responses used to reflect radar echoes.

chain calculations—In a calculator, series of continued operations in a single mode. Example: $118 \times 94 \times 116 \times 395$.

chained list—A list in which the items may be dispersed but in which each item contains an identifier for locating the next item to be considered.

chaining—In a computer, a system of storing records such that each record belongs to a list or group of records and has a linking field for tracing the chain.

chaining search — A search technique in which each item contains an identifier for locating the next item to be considered.

chain printer—In a computer, a high-speed printer having type slugs carried on the links of a revolving chain.

chain radar beacon—A radar beacon with a very fast recovery time, so that simultaneous interrogation and tracking of the beacon by a number of radars is possible.

chain radar system—A radar system comprising a number of radars or radar stations located at various sites along a missile-flight path. These radar stations are linked together by data and communication lines for target acquisition, target positioning, and/or data-recording purposes. The target-acquisition link makes it possible for any radar to position any other radar on target.

chain reaction—A reaction that stimulates its own repetition. In a fission chain reaction a fissionable nucleus absorbs a neutron and fissions, releasing additional neutrons. These in turn can be absorbed by other fissionable nuclei, releasing still more neutrons. A fission chain reaction is self-sustaining when the number of neutrons released in a given time equals or exceeds the number of neutrons lost by absorption in nonfissioning material or by escape from the system.

challenger—*See* Interrogator.

chance failure—*See* Random Failure.

changer—A device that plays several phonograph records in sequence automatically. It consists of a turntable, an arm, and a record stacking and dropping mechanism.

channel—1. A portion of the spectrum assigned for the operation of a specific carrier and the minimum number of sidebands necessary to convey intelligence. 2. A single path for transmitting electric signals. (Note: The word "path" includes separation by frequency division or time division. Channel may signify either a one-way or two-way path, providing communication in either one direction only or in two directions.) *See also* Alternate Channel. 3. In electric computers, that portion of a storage medium which is accessible to a given reading station. 4. The path along which information, particularly a series of digits or characters, may flow. 5. In computer circulating storage, one recirculating path containing a fixed number of words stored serially by word. 6. An area, under the silicon dioxide of a planar surface, that has been changed from one type of conductivity to the opposite type. A channel is the conductive path between the source and drain in an IGFET. Generally, channels are undesirable in other instances. Thick oxides or heavily doped regions called channel stoppers are used to prevent channels. 7. A complete sound path. A single-channel, or monophonic system, has one channel. A stereophonic system has at least two full channels designated as left (A) and right (B). Monophonic material may be played through a stereo system; both channels will carry the same signal. Stereo material, if played on a monophonic system, will mix and emerge as a monophonic sound. 8. The conducting charge layer between source and drain induced by the applied gate voltage. The charge layer is holes in a p-type device, and electrons in n-types.

channel balance—Equal response on both left and right channels.

channel bank—The part of a carrier-multiplex terminal in which are performed the first step of modulation of the voice frequencies into a higher-frequency band and the final step in the demodulation of the received higher-frequency band into voice frequencies.

channel capacity—1. The maximum number of elementary digits that can be handled per unit time in a particular channel. 2. The maximum possible rate of information transmission through a channel at a specified error rate. It may be measured in bits per second or bauds. 3. The total number of individual channels in a system.

channel designator—A number assigned for reference purposes to a channel, tributary, or trunk. Also called channel sequence number.

channel diffusion stops—A narrow, doped region beside each sensing channel in a ccd that prevents excess charges generated within a particular light sensing site from spreading sideways.

channel effect—Current leakage over a surface path between the collector and emitter of some types of transistors.

channel frequency—The band of frequencies which must be handled by a carrier system to transmit a specific quantity of information.

channeling—The utilization of a modulation-frequency band for the simultaneous transmission from two or more communication channels in which the channel separation is accomplished by the use of carriers or subcarriers, each in a different discrete frequency band forming a subdivision of the main band. This covers a special case of multiplex transmission.

channel interval—The time allocated to a channel, including on and off time.

channelizing—The process of subdividing wideband transmission facilities for the purpose of putting many different circuits requiring comparatively narrow bandwidths on a single wideband facility.

channel pulse—A telemetering pulse that, by its time or modulation characteristics, represents intelligence on a channel.

channel pulse synchronization—Synchronization of a local-channel rate oscillator by comparison and phase lock with separate channel synchronizing pulses.

channel reliability—The percentage of time that the channel meets the arbitrary standards established by the user.

channel reversal—Shifting the outputs of a stereo system so the channel formerly heard from the left speaker now comes from the right and vice versa.

channel-reversing switch—A switch that reverses the connections of two speakers in a stereo system with respect to the channels, so that the channel heard previously from the right speaker is heard from the left and vice versa.

channel sampling rate — The number of times per second that individual channels are sampled. This is different from commutation rate, since it is possible for more than one channel to be applied to a given commutator input (with subcommutation).

channel selector—A switch or dial used for selecting a desired channel.

channel separation—1. In stereo the electrical or acoustical difference between the left and right channels. Inadequate separation can lessen the stereo effect; excessive separation can exaggerate it beyond natural proportions. 2. The degree to which the two signals in a stereo system are electrically isolated. Usually expressed as a ratio in decibels.

channel sequence number — *See* Channel Designator.

channel shifter — A radiotelephone carrier circuit by means of which one or two voice-frequency channels are shifted from normal channels to higher voice-frequency channels as a means of reducing cross talk between channels. At the receiving end, the channels are shifted back by a similar circuit.

channel stopper—In p-channel devices it is an n-type ring diffused around each transistor and connected to ground to isolate the device and prevent formation of parasitic devices in the field.

channel strip—An amplifier or other device having a sufficiently wide bandpass to amplify one television channel.

channel subcarrier—The channel required to convey telemetric data involving a subcarrier band.

channel synchronizing pulse separator — A device for separating channel synchronizing pulses from commutated signals.

channel-to-channel connection — A device for rapid data transfer between two computers. A channel adapter is available that permits the connections between any two channels on any two systems. Data are transferred at the rate of the slower channel.

channel-utilization index — In a computer, the ratio between the information rate (per second) through a channel and the channel capacity (per second).

channel wave—Any elastic wave propagated in a sound channel because of a low-velocity layer in the solid earth, the sea, or the atmosphere.

character—1. In electronic computers, one of a set of elementary symbols which may be collectively arranged in order to express information. These symbols may include the decimal digits 0 through 9, the letters A through Z, punctuation and typewriter symbols, and any other single symbol which a computer may read, store, or write. 2. One of a set of symbols used to present information on a display tube. 3. Part of a computer word that has a meaning in itself. For example, six bits recorded across a magnetic tape make up a character and signify a number or letter symbol. 4. A combination of holes punched in a line. 5. A letter, digit, or other symbol that is used as part of the organization, control or representation of data. A character is often in the form of a spatial arrangement or adjacent or connected strokes.

character boundary—In character recognition, the largest rectangle, having a side parallel to the document reference edge, each of the sides of which is tangent to a given character outline.

character check—Verification of the observance of rules for character formation.

character code—A special way of using a group of bits to represent a character. Different codes may be used in different equipment according to the internal design.

character crowding—The effect of reducing the time interval between subsequent characters read from tape. It is caused by a combination of mechanical skew, gap scatter, jitter, amplitude variation, etc. Also called packing.

character density—A measure of the number of recorded characters per unit of length or area.

character display tube—A form of cathode ray tube in which the cathode ray beam can be shaped either by electrostatic or electromagnetic deflection, or by passing the beam through a mask, into symbols or letters.

character emitter—In a computer, an electromechanical device that puts out coded pulses.

character generator—1. A unit that accepts input in the form of one of the alphanumeric codes and prepares the electrical signals necessary for its display in the proper position on a dot matrix, tv system, or crt. 2. That part of the display controller which draws alphanumeric characters and special symbols for the screen. A character is automatically drawn and spaced every time a character code is interpreted. 3. A hardware or software device which provides the means for formulating a character font and which also may provide some controlling function during printing.

characteristic—1. An inherent and measurable property of a device. Such a property may be electrical, mechanical, thermal, hydraulic, electromagnetic, or nuclear; and it can be expressed as a value for stated or recognized conditions. A characteristic may also be a set of related values (usually in graphical form). 2. The integral part of a logarithm to the base 10; also the power of 10 by which the significant digits of a floating point number are multiplied.

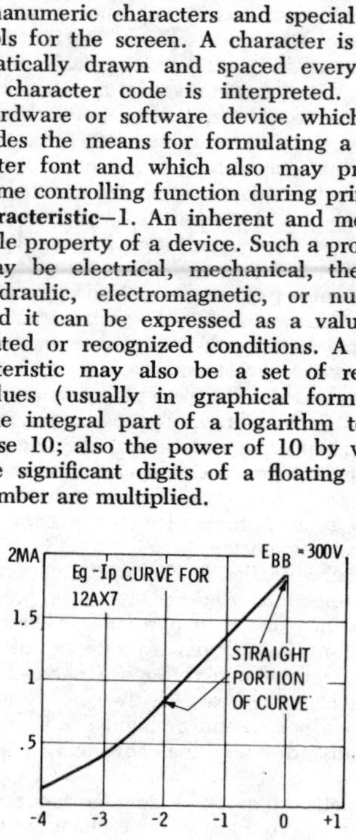

Characteristic curve.

characteristic curve—A graph plotted to show the relationship between changing values. An example would be a curve showing plate-current changes as the grid voltage varies.

characteristic distortion—1. Displacement of signal transitions due to the persistence of transients caused by preceding transitions. 2. Repetitive displacement or disruption peculiar to specific parts of a teletypewriter signal. The two types of characteristic distortion are line and equipment.

characteristic frequency—The frequency which can be easily identified and measured in a given emission.

characteristic impedance—Also called surge impedance. 1. The driving-point impedance of a line if it were of infinite length. 2. In a delay line, the value of terminating resistance which provides minimum reflection to the network input and output. 3. The ratio of voltage to current at every point along a transmission line on which there are no standing waves. 4. The square root of the product of the open- and short-circuit impedance of the line.

characteristic impedance of free space—The relationship between the electric and magnetic intensities of space due to the expansion of the impedance concept to electromagnetic fields.

characteristic spread—The range between the minimum and maximum values for a given characteristic that is considered normal in any large group of tubes or other devices.

characteristic telegraph distortion—Distortion which does not affect all signal pulses alike. Rather, the effect on each transition depends on the signal previously sent, because remnants of previous transitions or transients persist for one or more pulse lengths.

characteristic wave impedance—The ratio of the transverse electric vector to the transverse magnetic vector at the point it is crossed by an electromagnetic wave.

character (or byte) interleave – A technique in time-division multiplexing in which bytes of data are transmitted in one frame.

character reader—A computer input device which can directly recognize printed or written characters; they need not first be converted into punched holes in cards or paper or polarized magnetic spots.

character read-out systems—Photoelectrically controlled, alphanumeric reading devices that convert characters to audible or sorting signals which can be fed to a computer, electric typewriter, tapepunch or other machine.

character recognition—The automatic identification of graphic, phonic, or other characters. *See also* Magnetic Ink Character Recognition and Optical Character Recognition.

character sensing—To detect the presence of characters optically, magnetically, electrostaticallv etc.

character set—An ordered group of unique representations called characters, such as the 26 letters of the English alphabet, 0 and 1 of the Boolean alphabet, the signals in the Morse code alphabet, the 128 characters of the USASCII alphabet, etc.

character string—Two or more alphanumeric characters or special symbols (math, Greek, etc.) aligned in a textual format on the screen.

character subset—A selection from a character set of all characters having a specified common feature; for example, in the definition of a character set, the digits 0 through 9 are a character subset.

Charactron—Trade name of General Dynamics/Electronics for a specially constructed cathode-ray tube used to display alphanumeric characters and other special symbols directly on its screen.

charge—1. The electrical energy stored in a capacitor or battery or held on an insulated object. 2. The quantity of electrical energy in (1) above.

charge carrier—A mobile hole or conduction electron in a semiconductor.

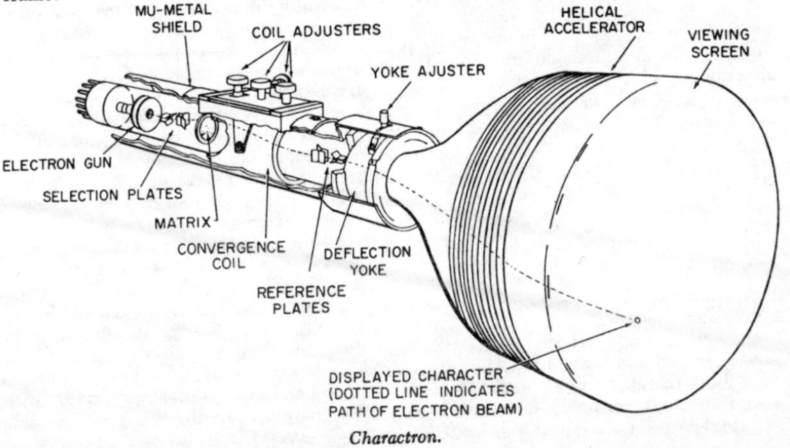

Charactron.

charge density—The charge per unit area on a surface, or charge per unit volume in space.

charged particle—An ion, an elementary particle that carries a positive or negative electric charge.

charge injection imaging device—*See* CID.

charger—A device used to convert alternating current into a pulsating direct current which can be used for charging a storage battery.

charge retention—The ability of a battery to hold its energy once it has been charged.

charge-storage tube—A storage tube which retains information on its surface in the form of electric charges.

charge transfer—The process in which an ion takes an electron from a neutral atom of the same type with a resultant transfer of electronic charge.

charging—The process of converting electrical energy to stored chemical energy.

charging current—The current that flows into a capacitor when a voltage is first applied.

charging rate—The rate of current flow used in charging a battery.

chaser—An array of elements similar to a ring except that as each successive element is switched to the "on" conditions the others remain on as well; when all stages are on, the next pulse turns them all off and the process generally repeats.

chassis – 1. A sheet-metal box, frame, or simple plate on which electronic components and their associated circuitry can be mounted. 2. The entire equipment (less cabinet) when so assembled. (*See also* Printed-Circuit Board.)

chassis ground—A connection to the metal structure that supports the electrical components which make up the unit or system.

chatter—1. A sustained rapid opening and closing of contacts due to variations in the coil current. 2. The vibration of a cutting stylus in a direction other than the direction in which it is driven.

chatter time—The interval of time from initial actuation of a contact to the end of chatter.

check—The partial or complete verification of the correctness of equipment operations, the existence of certain prescribed conditions, and/or the correctness of results.

check bit—A binary check digit.

check character—A character used to perform a check.

check digit—A digit carried along with a machine word and used to report information about the other digits in the word so that if a single error occurs, the check fails and an error alarm signal is initiated.

checkerboard—*See* Worst-Case Noise Pattern.

checkout—A series of operations and calibration tests used to determine the condition and status of a system or element of the system.

check point—In a computer routine, a point at which it is possible to store sufficient information to permit restarting the computation from that point.

check-point routine—A computer routine in which information for a check point is stored.

check problem—A problem which, when incorrectly solved, indicates an error in the programming or operation of a computer.

check register—A special register provided in some computers to temporarily store transferred information for comparison with a second transfer of the same information in order to verify that the information transferred each time agrees precisely.

check routine—A program the purpose of which is to determine whether a computer or a program is operating correctly.

checksum—In a computer, a summation of digits or bits summed according to an arbitrary set of rules and primarily used for checking purposes.

cheese antenna—An antenna with a cylindrical parabolic reflector enclosed by two plates perpendicular to the cylinder and so spaced that more than one mode can be propagated in the desired direction of polarization. It is fed on the focal line.

chelate—A molecule in which a central inorganic ion is covalently bonded to one or more organic molecules, with at least two bonds to each molecule; used as a laser dopant.

chemical deposition—The process of depositing a substance on a surface by means of the chemical reduction of a solution.

chemically deposited printed circuit—A printed circuit formed on a base by the reaction of chemicals alone. Dielectric, magnetic, and conductive circuits can be applied.

chemically reduced printed circuit—A printed circuit formed by chemically reducing a metallic compound.

chemisorption—*See* Adsorption.

CHIL—Abbreviation for current-hogging injection logic.

Child's law—Also known as the three-halves power equation. It states that the current in a thermionic diode varies directly with the three-halves power of the anode voltage and inversely with the square of the distance between electrodes.

chip—1. Also called thread. In mechanical recording, the material removed from the recording medium by the recording stylus as it cuts the groove. 2. In punched cards, a piece of cardboard removed in the punching process. 3. A single substrate on which all the active and passive circuit elements have been fabricated using one or all of the semiconductor techniques of diffusion, passivation, masking, photoresist, and epitaxial growth. A chip is not ready for use until packaged and provided with external connectors. The term is also applied to discrete capacitors and resistors which are small enough to be bonded to substrates by hybrid techniques. 4. A tiny piece of semiconductor material scribed or etched from a semiconductor slice on which one or more electronic components are formed. The total number of usable chips obtained from a wafer is the yield.

chip sets—The microprocessor chip in addition to RAMs, ROMs and interface i/o devices. The chip sets mounted on a board are also referred to as the CPU portion of the microcomputer. Also called microcontroller.

Chireix antenna—Resonant series-fed array of square loops with half-wave sides. The loops feed each other in cascade, corner to corner, and the antenna resembles a double zigzag. Also called Chireix-Mesny antenna.

chirp—1. An all-encompassing term for the various techniques of pulse expansion-pulse compression applied to pulse radar. A technique to expand narrow pulses to wide pulses for transmission, and to compress wide received pulses to the original narrow pulse width and waveshape. This improves the signal-to-noise ratio without degradation to the range resolution and range discrimination. 2. A colloquial expression for a coded pulse. In coding the pulse, the carrier frequency is increased in a linear manner for the duration of the pulse, and when the pulse is translated to an audio frequency, it sounds like a chirp. 3. A change in the pitch of code signals, generally due to poor regulation of the transmitter power supply.

chirp modulation—Swept-frequency modulation used in some radar and sonar equipment to increase the on-target energy and improve range resolution by making full use of the average power capability of the transmitter.

chirp radar—Radar in which a swept-frequency signal is transmitted, received after being returned from a target, and compressed in time to give a final narrow pulse called the chirp signal. This type of radar has high immunity to jamming and provides inherent rejection of random noise signals.

choke—1. An inductance used to impede the flow of pulsating direct current or alternating current by means of its self-inductance. 2. An inductance used in a circuit to present a high impedance to frequencies appreciably limiting the flow of direct current. Also called choke coil. 3. A groove or other discontinuity in a waveguide surface so shaped and dimensioned as to impede the passage of guided waves within a limited frequency range.

Chokes.

choke coil—Also called impedance coil. An inductor (reactor) used to limit or suppress the flow of alternating current without appreciable effect on the flow of direct current.

choke flange—A waveguide flange with a grooved surface; the groove is so dimensioned that the flange forms part of a choke joint.

choke-input filter—A power-supply filter in which a choke is the first element in series with the input current from the rectifier.

choke joint—1. A connector between two sections of transmission line in which the gap between sections to be connected is built out to form a series-branching transmission line carrying a standing wave, in which actual contact falls at or near a current minimum. 2. A joint for connecting two sections of waveguide together. Permits efficient energy transfer without the necessity of an electrical contact at the inside surface of the guide.

cholesteric phase — An arrangement of liquid crystal molecules that occurs only in optically active substances and is considered to be a twisted nematic phase with a helical structure. It consists of layers resembling the smectic phase but each layer has an order characteristic of a nematic phase.

chopped mode—A time-sharing method of displaying output signals of two or more channels with a single crt gun, in sequence, at a rate not referenced to the sweep.

chopper—1. A device for interrupting a current or a light beam at regular intervals. Choppers are frequently used to facilitate amplification. 2. An electromechanical switch for the production of modified square waves. The waves are of the same frequency as a driving sine wave and bear a definite relationship to it. 3. An electromechanical or electronic device used to interrupt a dc or low-frequency ac signal at regular intervals to permit amplification of the signal by an ac amplifier. It may also be used as a demodulator to convert an ac signal to dc.

4. A rotating shutter for interrupting an otherwise continuous stream of particles. Choppers can release short bursts of neutrons with known energies. Used to measure nuclear cross sections.

chopper amplifier—A circuit that amplifies a low-level signal after it has gone through a chopper.

chopper stabilization—1. The addition of a chopper amplifier to the regulator input circuitry of a regulated power supply in order to reduce output drift. 2. A method of improving the dc drift of an amplifier by utilization of chopper circuits.

chopper-stabilized amplifier—An amplifier configuration utilizing a carrier-type dc amplifier to reduce the effect of input offset and drift of a direct-coupled amplifier.

chopping—Removal by electronic means of one or both extremities of a wave at a predetermined level.

chord—A harmonious combination of tones sounded together through the use of one or more fingers on either or both hands. On chord organs a full chord is selected by depressing a single chord button.

chord organ—An organ with provision for playing a variety of chords, each produced by means of a single button or key.

chorus—A natural electromagnetic phenomenon in the vlf range. Probably originates in the exosphere. Also called "dawn chorus" because it sounds like birds at dawn. It generally consists of a multitude of rising tones, each tone rising from 1-2 kHz to 3-4 kHz and usually lasting 0.1 to 0.5 second.

Christiansen antenna—A radiotelescope composed of two interferometer arrays placed at right angles. It resembles a Mills cross antenna.

Christmas-tree pattern—1. *See* Optical Pattern. 2. A pattern resembling a Christmas tree, sometimes produced on the screen of a television receiver when the horizontal oscillator falls out of sync.

chroma—That quality which characterizes a color without reference to its brightness; that quality which embraces hue and saturation. White, black, and gray have no chroma.

chroma-clear raster—Also called white raster. Looks like a clear raster, but each of the three guns in the crt is operating under the influence of a color level determined by a white video signal. In this case, all tv set chroma circuits are working as though the tv set were receiving a color transmission of a completely white scene.

chroma control—A variable resistor which controls saturation by varying the level of chrominance signal fed to the demodulators of a color television receiver.

chromatic aberration—An effect which causes refracted white light to produce an image with colored fringes due to the various colors being bent at different angles.

chromaticity—1. The combination of the hue and saturation attributes of color. 2. A term quantitatively descriptive of a color, and dependent upon both hue and saturation, but without reference to brilliance.

chromaticity coordinate—The ratio of any one of the tristimulus values of a sample color to the sum of the three tristimulus values.

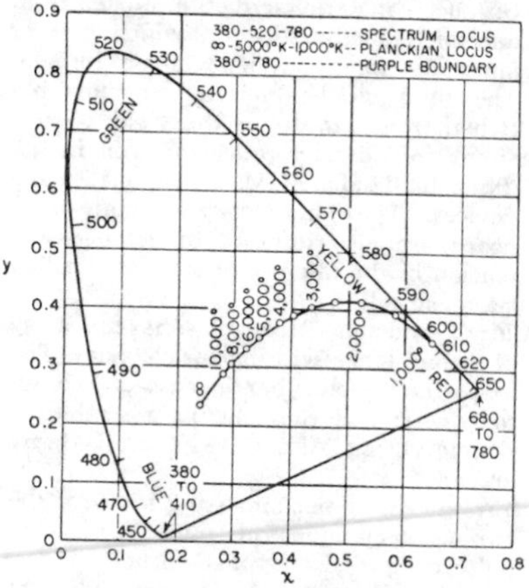

Chromaticity diagram.

chromaticity diagram—A plane diagram formed when any one of the three chromaticity co-ordinates is plotted against another.

chromaticity flicker—The flicker which results from fluctuation of the chromaticity only.

chrominance—Colorimetric difference between any color and a reference color of equal luminance, the reference color having a specified chromaticity. In standard color-television transmission, the specified chromaticity is that of the zero subcarrier.

chromatron—A color kinescope that has a single electron gun and whose color phosphors are laid out in parallel lines on its screen. The electron beam is directed to the correct phosphor by a deflection grid or wire grille near the face of the tube.

chrominance amplifier—The amplifier that separates the chrominance signal from the total video signal.

chrominance cancellation—A cancellation of the brightness variations produced by the chrominance signal on the screen of a monochrome picture tube.

chrominance-carrier reference—A continuous signal having the same frequency as the chrominance subcarrier, and a fixed phase with the color burst. The phase reference of carrier-chrominance signals for modulation or demodulation.

chrominance channel—In color television, a combination of circuits designed to pass only those signals having to do with the reproduction of color.

chrominance component—Either of the I and Q signals which add to produce the complete chrominance signal in NTSC systems.

chrominance demodulator—A demodulator used in color-television reception for deriving video-frequency chrominance components from the chrominance signal and a sine wave of the chrominance subcarrier frequency.

chrominance gain control—In red, green, and blue matrix channels, variable resistors which individually adjust the primary-signal levels. Used in color television.

chrominance modulator—A modulator used in color-television transmission for generating the chrominance signal from the video-frequency chrominance components and the chrominance subcarrier.

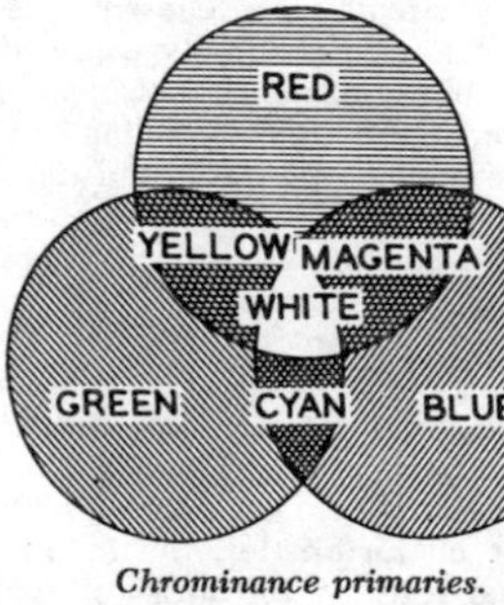

Chrominance primaries.

chrominance primary—One of two transmission primaries, the amounts of which determine the chrominance of a color. Chrominance primaries have zero luminance and are not physical.

chrominance signal—The chrominance-subcarrier sidebands added to a monochrome television signal to convey color information. The components of the chrominance signal represent hue and saturation but do not include luminance or brightness.

chrominance subcarrier—Also called color carrier. An rf signal which has a specific frequency of 3.579545 MHz and which is used as a carrier for the I and Q signals.

chrominance-subcarrier oscillator—In a color-tv receiver, a crystal-controlled oscillator which generates the subcarrier signal for use in the chrominance demodulators.

chrominance video signals—Output voltages from the red, green, and blue sections of a color-television camera or receiver matrix.

chromium dioxide – A type of recording-tape coating that produces very good quality at low recording speeds. Because of its magnetic properties, it requires a higher value of bias current in the recorder. The high performance inherent in chromium dioxide tape can only be realized in a tape machine having provision for a CrO_2 bias setting. On a standard recorder, the chrome tape will appear to have high frequency emphasis and may likely be difficult to erase. Chromium dioxide has a good dynamic range and a low noise level. Used in the cassette format with a suitable machine equipped with the Dolby system, it can make recordings that meet the best high-fidelity standards.

chronistor—A subminiature elapsed-time indicator that uses electroplating principles to totalize the operating time of equipment up to several thousand hours.

chronograph—An instrument for producing a graphical record of time as shown by a clock or other device.

chronometer—A portable timekeeper with compensated balance, capable of showing time with high precision and accuracy.

chronoscope—An instrument for measuring very small intervals of time.

cid—Abbreviation for charge injection imaging device. A matrix of electrodes on a silicon wafer that store photogenerated minority carriers in potential wells under each pair of electrodes. By removing the voltages on the electrodes, the charges are injected into the substrate and the current readout is proportional to the number of stored minority carriers.

CIE—Initials of the Commission Internationale de l'Eclairage or International Commission on Illumination.

CIE standard chromaticity diagram – A chromaticity diagram in which the x and y chromaticity coordinates are plotted in rectangular coordinates.

cinching – Longitudinal slippage between the layers in a tape pack as a result of acceleration or deceleration of the roll.

cipher telephony – A technique by which mechanical and/or electrical equipment is used for scrambling or unscrambling, or enciphering or decoding radio or voice messages.

ciphony—*See* Cipher Telephony.

circle cutter—A tool consisting of a center drill with an adjustable extension-arm cutter and used to cut holes in panels and chassis.

circle-dot mode—A method of storage of binary digits in a cathode-ray tube in which one kind of digit is represented by

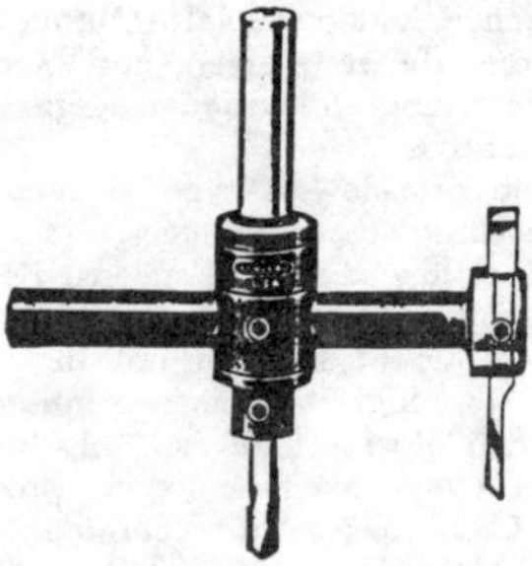

Circle cutter.

a small circle on the screen, and the other kind is represented by a similar circle with a concentric dot.

circle of confusion—The circular image of a point source due to the inherent aberrations in an optical system.

circlotron amplifier—A one-port, nonlinear cross-field high power microwave amplifier which uses a magnetron as a negative-resistance element, much as a maser uses an active material.

circuit—1. An electronic path between two or more points capable of providing a number of channels. 2. A number of conductors connected together for the purpose of carrying an electrical current. 3. The interconnection of a number of devices in one or more closed paths to perform a desired electrical or electronic function. Examples of simple circuits are high- or low-pass filters, multivibrators, oscillators, and amplifiers. 4. *See* Channel, 2.

circuit analyzer — Also called multimeter. Several instruments or instrument circuits combined in a single enclosure and used in measuring two or more electrical quantities in a circuit.

circuit breaker — 1. An automatic device which, under abnormal conditions, will open a current-carrying circuit without damaging itself (unlike a fuse, which must be replaced when it blows). 2. A device for interrupting a circuit under normal or abnormal conditions by means of separable contacts. 3. An electromagnetic device that opens a circuit automatically when the current exceeds a predetermined value. It can be reset by operating a lever or by other means.

circuit-breaker cascade system — A system wherein the protective devices are arranged in order of ratings such that those in series will coordinate and provide the required protection.

circuit capacity—The number of communication channels which can be handled by a given circuit at the same time.

circuit commutated turn-off time—The time interval between the instant when the principal current has decreased to zero after external switching of the principal voltage circuit, and the instant when a thyristor is capable of supporting a specified rate of rise of on-state voltage without turning on.

circuit dropout—A momentary interruption of a transmission because of the complete failure of a circuit.

circuit efficiency (of the output circuit of electron tubes)—Ratio of the power, at the desired frequency, delivered to a load at the output-circuit terminals of an oscillator or amplifier, to the power, at the desired frequency, delivered to the output circuit by the electron stream.

circuit element — Any basic constituent of a circuit except the interconnections.

circuit hole—On a printed-circuit board, a hole that lies partially or completely within the conductive area.

circuit noise—The noise brought to the receiver electrically from a telephone system, but not the noise picked up acoustically by the telephone transmitters.

circuit noise level—At any point in a transmission system, the ratio of the circuit noise at that point to some arbitrary amount of circuit noise chosen as a reference. This ratio is usually expressed in decibels above reference noise, abbreviated dBrn, signifying the reading of a circuit noise meter; or in adjusted decibels, abbreviated dBa, signifying the circuit noise meter reading adjusted to represent interfering effect under specified conditions.

circuit-noise meter—Also called noise-measuring set. An instrument for measuring the circuit-noise level. Through the use of a suitable frequency-weighting network and other characteristics, the instrument gives equal readings for noises of approximately equal interference. The readings are expressed in decibels above the reference noise.

circuit parameters—The values of the physical quantities associated with circuit elements—for example, the resistance (parameter) of a resistor (element), the amplification factor and plate resistance (parameters) of a tube (element), the inductance per unit length (parameter) of a transmission line (element).

circuit protection—Automatic protection of a consequence-limiting nature used to minimize the danger of fire or smoke, as well as the disturbance to the rest of the system which may result from electrical faults or prolonged electrical overloads.

circuit re-entrancy—*See* Re-Entrancy, 1.

circuit reliability—The percentage of time the circuit meets arbitrary standards set by the user.

Circuitron — A combination of active and passive components mounted in a single tube-type envelope and functioning as one or more complete operating stages.

circuits bonding jumper – The connection between portions of a conductor in a circuit to maintain required ampacity of the circuit.

circular antenna—A horizontally polarized antenna derived essentially from a half-wave antenna but having its elements bent into a circle.

circularly polarized loop vee—An airborne communications antenna that provides an omnidirectional radiation pattern for use in obtaining optimum near-horizon communications coverage.

circularly polarized wave—Applied usually to transverse waves. An electromagnetic wave for which the electric and/or magnetic field vector at a point describes a circle.

circular magnetic wave—A wave with circular magnetic lines of force.

circular mil—The universal term used to define cross-sectional areas. Equal to the area of a circle one mil (.001 inch) in diameter.

circular mil area—The square of the diameter of a round conductor measured in thousandths of an inch. The circular mil area of a braid is the sum of the circular mil area of each of the wires that make up the braid.

circular polarization—1. Polarization such that the vector representing the wave has a constant magnitude and rotates continuously about a point. 2. Simultaneous transmission of vertically and horizontally polarized radio waves.

circular scanning—Scanning in which the direction of maximum radiation generates a plane or a right circular cone with a vertex angle close to 180°.

circular trace—A cro time base produced by applying sine waves of the same frequency and amplitude, but 90° out of phase, to the horiozntal- and vertical-deflection plates of a cathode-ray tube. This results in a circular trace, and signals then give inward or outward radial deflections from the circle.

circular waveguide—A waveguide having a circular cross-sectional area.

circulating memory – 1. *See* Circulating Register. 2. A type of memory in which a data stream circulates in a loop. One example is a string of shift-register stages with the last output connected to the first input. At every clock pulse, a particular bit would be accessed as it passed a certain point in the circuit. Circulating memories also use other delay techniques, including electrical and acoustical delay lines.

circulating register—Also called circulating memory. A register (or memory) consisting of a means for delaying the information and a means for regenerating and reinserting it into the delaying means. This is accomplished as the information moves around a loop and returns to its starting place after a fixed delay.

circulating storage—A device using a delay line to store information in a train or pattern of pulses. The pulses at the output end are sensed, amplified, reshaped, and re-inserted into the input end of the delay line.

circulator—1. A microwave coupling device having a number of terminals so arranged that energy entering one terminal is transmitted to the next adjacent terminal in a particular direction. 2. An arrangement of phase shifters and waveguide or coax that distributes incoming signals among selected outputs. For example, a four port circulator will transfer a signal entering from port 1 to port 2. In turn, a signal entering port 2 will leave only by port 3. A signal entering port 3 will leave by port 4 and port 4 entering signal leaves by port 1. Can be used to isolate a transmitter and receiver when both are connected to the same antenna.

circumferential crimp—The type of crimp in which symmetrical indentations are formed in a barrel by crimping dies that completely surround the barrel.

Citizens band channels (Class D)—

Channel No.	*Frequency (MHz)*	*Channel No.*	*Frequency (MHz)*
1	26.965	21	27.215
2	26.975	22	27.225
3	26.985	*23	27.255
4	27.005	24	27.235
5	27.015	25	27.245
6	27.025	26	27.265
7	27.035	27	27.275
8	27.055	28	27.285
9	27.065	29	27.295
10	27.075	30	27.305
11	27.085	31	27.315
12	27.105	32	27.325
13	27.115	33	27.335
14	27.125	34	27.345
15	27.135	35	27.355
16	27.155	36	27.365
17	27.165	37	27.375
18	27.175	38	27.385
19	27.185	39	27.395
20	27.205	40	27.405

*Shared with Class-C radio control.

Citizens Radio Service – A radiocommunications service of fixed, land, and mobile stations intended for short-distance personal or business radiocommunications, radio signaling, and control of remote objects or devices by radio; all to the extent that these uses are not specifically prohibited by the FCC Rules and Regulations.

C³L—Abbreviation for complementary constant current logic.

clamp—*See* Clamping Circuit.

clamper—*See* Clamping Circuit.

clamping – The process that establishes a fixed level for the picture signal at the beginning of each scanning line.

clamping circuit—A circuit which adds a fixed bias to a wave at each occurrence of some predetermined feature of the wave. This is done to hold the voltage or current of the feature at (clamp it to) a specified fixed or variable level. (*See also* DC Restorer.)

clamping diode—A diode used to fix a voltage level at some point in a circuit.

clamp-on—*See* Clamp-On.

clapotis—A standing wave in which there is no horizontal motion of the crest. This phenomenon results from the interference caused by the reflection of a wave train from a barrier which the wave train approaches with its crest parallel to the barrier.

clapper—A hinged or pivoted armature.

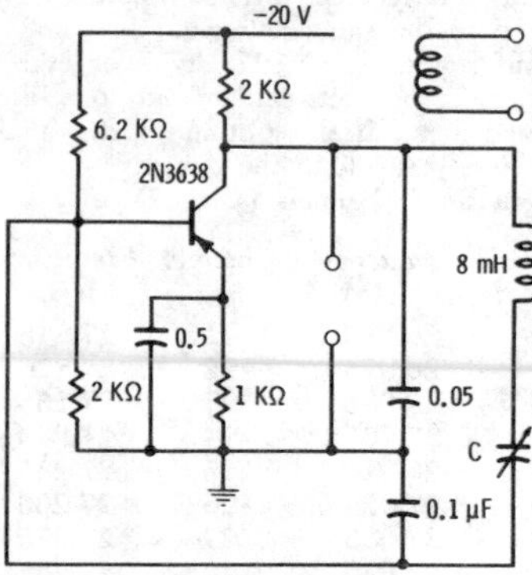

Clapp oscillator.

Clapp oscillator—A Colpitts-type oscillator using a series-resonant tank circuit for improved stability.

clarifier—A control on an ssb transceiver which enables adjustment of frequency so that the frequencies of the recovered audio signal will be essentially the same as the frequencies of the modulating signal fed to a distant transmitter.

Clark cell—An early standard cell that used an anode of mercury, a cathode of zinc amalgam, and an electrolyte containing zinc sulphate and mercurous sulphate. Its voltage is 1.433 at 15°C.

Class-A amplifier—1. An amplifier in which the grid bias and alternating grid voltage are such that plate current flows at all times. To denote that no grid current flows during any part of the input cycle, the suffix *1* is sometimes added to the letter or letters of the class identification. The suffix 2 denotes that grid current flows during part of the input cycle. 2. An amplifier where the output transistors or tubes are operating permanently on linear portions of their transfer characteristics. Efficiency is low, but a constant current is drawn from the power supply whatever the signal level. Usually recognized by the use of a single transistor or tube driving the loudspeaker. 3. Operation that implies biasing the tubes or transistors to the middle parts of their transfer characteristics so that the device is driven upwards on one half cycle and downwards on the other half cycle. In a push-pull power amplifier, one of the device pair is driven upwards on negative half cycles while its partner is driven downwards, the mode reversing on the positive half cycles. The stage thus draws a constant current at all drive levels within the dynamic range of the amplifier. A Class-A power amplifier has an efficiency of almost 50 percent.

Class-AB amplifier – 1. An amplifier in which the grid bias and alternating grid voltage are such that plate current flows for more than half but less than the entire electrical cycle. To denote that no grid current flows during any part of the input cycle, the suffix *1* is sometimes added to the letter or letters of the class identification. The suffix *2* denotes that grid current flows during part of the cycle. 2. One type of power amplifier engineered so that at low drive level the stage operates Class A, while at increasing drive level the mode changes to Class B.

Class-A0 emission—The incidental radiation of an unmodulated carrier wave from a station.

Class-A1 emission—A carrier wave (unmodulated by an audio frequency) keyed normally for telegraphy to transmit intelligence in the International Morse Code at a speed not exceeding 40 words per minute (the average word is composed of five letters).

Class-A2 emission—A carrier wave which is amplitude-modulated at audio frequencies not exceeding 1250 hertz. The modulated carrier wave is keyed normally for telegraphy to transmit intelligence in the International Morse Code at a speed not exceeding 40 words per minute, the average word being composed of five letters.

Class-A3 emission—A carrier wave which is amplitude-modulated at audio frequencies corresponding to those necessary for intelligible speed transmitted at the speed of conversation.

Class-A GFCI—A ground fault circuit interrupter that will trip when a fault current to ground is 5 milliamperes or more.

Class-A insulating material—A material or

combination of materials such as cotton, silk, and paper suitably impregnated, coated, or immersed in a dielectric liquid such as oil. Other materials or combinations of materials may be included if shown to be capable of satisfactory operation at 105°C.

Class-A modulator – A Class-A amplifier used for supplying the signal power needed to modulate the carrier.

Class-A operation—Operation of a vacuum tube with grid bias such that plate current flows throughout 360° of the input cycle.

Class-A signal area—A strong tv signal area, defined by the FCC as receiving a signal strength equal to or greater than approximately 2500 microvolts per meter for channels 2 through 6, 3500 microvolts per meter for channels 7 through 13, and 5000 microvolts per meter for channels 14 through 69.

Class-A station—A station in the Citizens Radio Service licensed to be operated on an assigned frequency in the 460- to 470-MHz band with input power of 60 watts or less.

Class-A transistor amplifier – An amplifier in which the input electrode and alternating input signal are biased so that output current flows at all times.

two transistors or tubes operate on positive and negative half cycles of the signal wave form. Each operates from a low initial current but this rises as the signal level increases. Usually recognized by the use of two transistors or tubes operating in antiphase to drive the loudspeaker. 3. Transistor or tube (power) amplifier whose biasing is adjusted such that the push-pull transistors (or tubes) operate at a low nodrive current (called quiescent current). When drive is applied the current in one of the pair rises while the partner is pushed into cutoff on one half cycle, the mode reversing on the other half cycle.

Class-B GFCI—A ground-fault circuit interrupter that will trip when a fault current to ground is 20 milliamperes or more.

Class-B insulating material—A material or combination of materials such as mica, glass fiber, asbestos, etc., suitably bonded. Other materials or combinations, not necessarily inorganic, may be included if shown to be capable of satisfactory operation at 130°C.

Class-B modulator – A Class-B amplifier used specifically for supplying the signal power needed to modulate a carrier.

Class-B operation—Operation of a vacuum

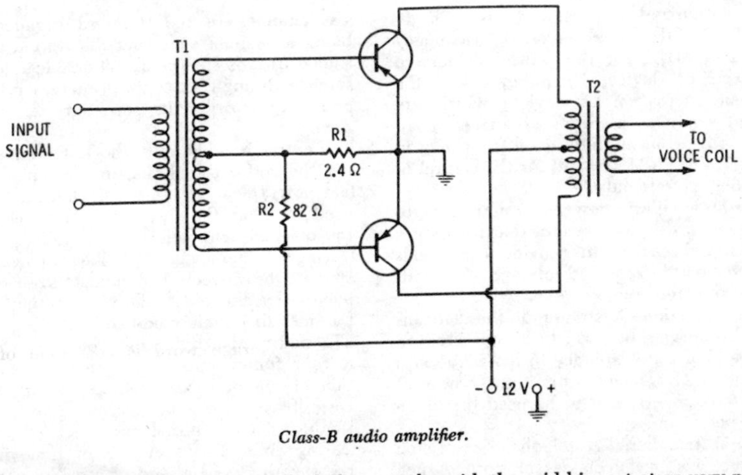

Class-B audio amplifier.

Class-B amplifier—An amplifier in which the grid bias is approximately equal to the cutoff value so that, when no exciting grid voltage is applied, the plate current will be approximately zero and will flow for approximately half of each cycle when an alternating grid voltage is applied. To denote that no grid current flows during any part of the input cycle, the suffix *1* is sometimes added to the letter or letters of the class identification. The suffix *2* denotes that grid current flows during part of the cycle. 2. An amplifier where

tube with the grid bias set at or very near cutoff, so that plate current flows for approximately the positive half of each cycle of the input signal.

Class-B station – A station in the Citizens Radio Service licensed to be operated on an authorized frequency in the 460- to 470-MHz band with input power of 5 watts or less.

Class-B transistor amplifier – An amplifier in which the input electrode is biased so that when no alternating input signal is applied, the output current is approximately zero, and when an alternating input signal is applied, the output current flows for approximately half a cycle.

Class-C amplifier – An amplifier in which the grid bias is appreciably beyond the cutoff point, so that plate current is zero, when no alternating grid voltage is applied, and plate current flows for appreciably less than half of each cycle when an alternating grid voltage is applied. To denote that no grid current flows during any part of the input cycle, the suffix *1* is sometimes added to the letter or letters of the class identification. The suffix *2* denotes that grid current flows during part of the cycle.

Class-C insulating material–Insulation consisting entirely of mica, porcelain, glass, quartz, or similar inorganic materials. Other materials or combinations of materials may be included if shown to be capable of satisfactory operation at temperatures over 220°C.

Class-C operation–Operation of a vacuum tube with grid bias considerably greater than cutoff. The plate current is zero with no input signal to the grid and flows for appreciably less than one-half of each cycle of the input signal.

Class-C station – A station in the Citizens Radio Service licensed to be operated on an authorized frequency in the 26.96- to 27.23-MHz band, or on the frequency 27.255 MHz, for the control of remote objects or devices by radio, or for the remote actuation of devices which are used solely as a means of attracting attention, or on an authorized frequency in the 72- to 76-MHz band for the control of model aircraft only.

Class-D auxiliary power–An uninterruptible (no-break) power unit that makes use of stored energy to provide continuous power within specified tolerances for voltage and frequency.

Class-D station–A station in the Citizens Radio Service licensed to be operated on an authorized frequency in the 26.965- to 27.405-MHz band, with input power of 5 watts or less, and to be used for radiotelephony only.

Class-D telephone–A telephone restricted to use in special classes of service such as fire alarm, guard alarm, and watchman services.

Class-F insulating material–A material or combination of materials such as mica, glass fiber, asbestos, etc., suitably bonded. Other materials or combinations of materials, not necessarily inorganic, may be included if shown to be capable of satisfactory operation at 155°C.

Class-H insulating material–A material or combination of materials such as silicone elastomer, mica, glass fiber, asbestos, etc., suitably bonded. Other materials or combinations of materials may be included if shown to be capable of satisfactory operation at 180°C.

Class-J oscilloscope–*See* J-Scope.

Class-O insulating material–An unimpregnated material or combination of materials such as cotton, silk, or paper. Other materials or combinations of materials may be included if shown to be capable of satisfactory operation at 90°C.

clavier–Any keyboard, either hand or foot operated.

clean room–An area in which high standards of control of humidity, temperature, dust and all forms of contamination are maintained.

clear–1. Also called reset. To restore a storage or memory device to a prescribed state, usually to zero. 2. Remove all components of a calculation in a calculator. 3. In a calculator, to "erase" the contents of a display, memory or storage register. 4. As used in security work, the term "clear" is synonymous with reset meaning that a latched circuit is restored to normal state.

clearance – The shortest distance through space between two live parts, between live parts and supports or other objects, or between any live part and grounded part.

clear channel–In the standard broadcast band, a channel such that the station assigned to it is free of objectionable interference through all of its primary service area and most of its secondary service area.

clear entry–Remove only the last number, not the entire calculation in a calculator.

clear entry/clear all–In a calculator a key used to clear the last entry or to clear the machine completely.

clearing–1. Removal of a flaw or weak spot in the dielectric of a metallized capacitor by the electrical vaporization of the metallized electrode at the flaw. 2. The ability of a lightning protector to interrupt follow current before the operation of circuit fuses or breakers. In the case of a simple gap, clearing frequently requires some external assistance.

clearing ends–The operation of removing the sheath from the end of a cable, eliminating all moisture, and checking for cosses, shorts, and grounds in preparation for testing.

clearing-out drop–A drop signal associated with a cord or trunk circuit and operated by ringing current to attract the operator's attention.

clear input–An asynchronous input to a flip-flop used to set the Q output to logic zero.

clear memory key–Removes what is stored in a memory register of a calculator.

clear raster–A raster free of snow such as

would be obtained in the absence of a video signal on either the cathodes or the grids of the three guns in the color crt (mostly a function of bias conditions).

clear terminal—*See* Reset Terminal.

click filter—A capacitor and resistor connected across the contacts of a switch or relay to prevent a surge from being introduced into an adjacent circuit. (*See also* Key-Click Filter.)

clipped-noise modulation—A clipping action performed to increase the bandwidth of a jamming signal. Results in more energy in the sidebands, correspondingly less energy in the carrier, and an increase in the ratio of average power to peak power.

clipper—A device the output of which is zero or a fixed value for instantaneous input amplitudes up to a certain value, but is a function of the input for amplitudes exceeding the critical value.

clipper amplifier—An amplifier designed to limit the instantaneous value of its output to a predetermined maximum.

clipper-limiter—Also called slicer. A device the output of which is a function of the instantaneous input amplitude for a range of values lying between two predetermined limits, but is approximately constant at another level for input values above the range.

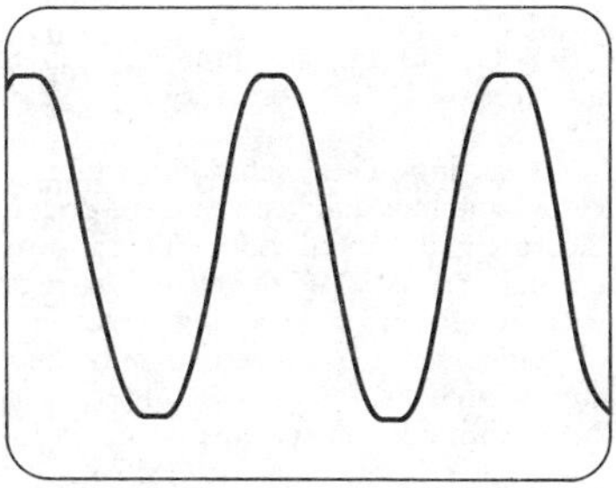

Clipping.

clipping—1. The loss of initial or final parts of words or syllables due to less-than-ideal operation of voice-operated devices. 2. Term used to express the clipping of the peaks of a waveform when an amplifier is driven beyond its power capacity. The flattening of the tips of a sine wave due to clipping. 3. Severe distortion caused by overloading the input of an amplifier. A sine wave signal waveform has a flat top and bottom at the peaks when clipping occurs.

clock—1. A pulse generator or signal waveform used to achieve synchronization of the timing of switching circuits and the memory in a digital computer system. It determines the speed of the cpu. 2. A timing device in a system: usually it provides a continuous series of timing pulses.

clocked—Pertaining to the type of operation in which gating is added to a basic flip-flop to permit the flip-flop to change state only when there is a change in the clocking input or an enabling level of the clocking input is present.

clocked flip-flop—A flip-flop circuit designed so that it is triggered only if trigger and clock pulses are present at the same time.

clocked R-S flip-flop—A flip-flop in which two conditioning inputs control the state the flip-flop assumes upon arrival of the clock pulse. If the S (set) input is enabled, the flip-flop assumes the 1 condition when clocked; if the R (reset) input is enabled, the flip-flop assumes the 0 condition when clocked. A clock pulse must be applied to change the state of the flip-flop.

clock frequency—In digital computers, the master frequency of periodic pulses that are used to schedule the operation of the computer.

clock input—That flip-flop terminal whose condition or change of conditions controls the admission of data through the synchronous inputs and thereby controls the output state of the flip-flop. The clock signal permits data signals to enter the flip-flop and, after entry, directs the flip-flop to change state accordingly.

clock pulse—A pulse used to gate information into a flip-flop operated in the synchronous mode. (In JK flip-flops, the clock pulse causes counting if the data inputs are both held at logic 1.)

clock rate—The rate at which a word or characters of a word (bits) are transferred from one internal computer element to another. Clock rate is expressed in cycles (in a parallel-operation machine, in words; (in a serial-operation machine, in bits) per second.

clock skew—Phase shift in a single clock distribution system in a digital circuit. It results from different delays in clock driving elements and/or different distribution paths.

clock stagger—1. Time separation of clock pulses in a multiphase clock system. 2. Voltage separation between the clock thresholds in a flip-flop.

clockwise-polarized wave — *See* Right-Handed Polarized Wave.

close coupling—Also called tight coupling. Any degree of coupling greater than critical coupling.

closed array—An array which cannot be extended at either end.

closed circuit—1. A complete electric circuit through which current may flow when a voltage is applied. 2. A program source that is not broadcast for general consumption, but is fed to remote monitoring units by wire.

closed-circuit communication systems – Certain communication systems which are entirely self-contained and do not exchange intelligence with other facilities and systems.

closed circuit jack–A jack which has its through circuits normally closed. Circuits are opened by inserting mating plug.

closed-circuit signaling–Signaling in which current flows in the idle conditions and a signal is initiated by increasing or decreasing the current.

closed-circuit television–1. A television system in which the television signals are not broadcast, but are transmitted over a closed circuit and received only by interconnected receivers. 2. Transmission and reception of video signals via wire carriers.

closed entry–A design that places a limit on the size of a mating part.

closed-entry contact–A female contact designed to prevent the entry of a device that has a cross-sectional dimension greater than that of the mating pin.

closed loop–1. A circuit in which the output is continuously fed back to the input for constant comparison. 2. In a computer, a group of indefinitely repeated instructions. 3. A system with feedback control in which the output is used to control the input. *See also* open-loop.

closed-loop bandwidth – The frequency at which the closed-loop gain drops 3 dB from its midband or dc value.

closed loop feedback–An automatic means of sensing speed variations and correcting to maintain close speed regulation.

closed-loop gain – 1. The response of a feedback circuit to a voltage inserted in series with the amplifier input. 2. The overall gain of an amplifier with an external negative-feedback loop.

closed-loop input impedance–The impedance looking into the input port of an amplifier with feedback.

closed-loop output impedance–The impedance looking into the output port of an operational amplifier with feedback.

closed-loop system – Automatic control equipment in which the system output is fed back for comparison with the input, for the purpose of reducing any difference between input command and output response.

closed-loop voltage gain–The voltage gain of an amplifier with feedback.

closed magnetic circuit–A circuit in which the magnetic flux is conducted continuously around a closed path through ferromagnetic materials.

closed routine–In a computer, a routine that is entered by basic linkage from the main routine rather than being inserted as a block of instructions within a main routine.

closed subroutine–In a computer, a subroutine not stored in the normal program sequence. Transfer is made from the program to the storage location of the subroutine, and then following execution of the subroutine, control is returned to the main program.

close memory–Part of a directly addressable computer memory which provides fast cycle time and is usually employed for frequently used accesses.

close-talking microphone–Also called noise-cancelling microphone. A microphone designed to be held close to the mouth of the speaker.

closing rating–In a relay, conditions under which the contact must close, with a prescribed duty cycle and contact life.

cloud absorption – Absorption of electromagnetic radiation as a result of water drops and water vapor in a cloud.

cloud attenuation–Reduction in microwave radiation intensity due largely to scattering, rather than absorption by clouds.

cloverleaf antenna – A nondirectional vhf transmitting antenna that consists of a number of horizontal four-element radiators arranged much like a four-leaf clover, stacked a half-wave apart vertically. These horizontal units are energized to give maximum radiation in the horizontal plane.

clutter–Confusing, unwanted echoes which interfere with the observation of desired signals on a radar display.

clutter gating–The technique which provides switching between moving-target indicators and normal video. This results in normal video being displayed in regions with no clutter, and moving-target indicator video being switched in only for the clutter areas.

CML–Abbreviation for current mode logic.

CMOS–*See* Complementary MOS.

cmrr–Abbreviation for common-mode rejection ratio.

C-network–A network composed of three impedance branches in series. The free ends are connected to one pair of terminals, and the junction points to another pair.

CO–*See* Cavity Oscillator.

coarse-chrominance primary – Also called the Q signal. A zero-luminance transmission primary associated with the minimum bandwidth of chrominance transmission and chosen for its relatively small importance in contributing to the subjective sharpness of the color picture.

coast–On a radar, a memory feature which, when activated, causes the range and/or angle systems to continue to move in the same direction and at the same speed as an original target was moving. Used to prevent lock-up to a stronger target if approached by the target being tracked.

coastal refraction—Bending of the path of a direct radio wave as it crosses the coast at or near the surface. It is caused by differences in electrostatic conditions between soil and water.

coast station—A land-based radio station in the maritime mobile service. It carries on communications with shipboard stations.

coated cathode—In a vacuum tube, a cathode that has been coated with compounds so as to increase its electron emission (e.g., an oxide-coated cathode).

coated filament — A vacuum-tube filament that has been coated with metal oxides to increase its electron emission.

coated tape—*See* Magnetic-Powder–Coated Tape.

coating—1. The magnetic layer, consisting of oxide particles held in a binder that is applied to the base film, used for magnetic recordings. 2. The magnetizable material on one surface of a recording tape which stores the (audio) signals when recording.

coating thickness — The thickness of the magnetic coating applied to the base film. Modern tape coatings range in thickness from 170 to 650 microinches, with a preponderance of coatings being approximately 400 microinches thick. In general, thin coatings give good resolution, at the expense of reduced output at long wavelengths; thick coatings give a high output at long wavelengths, at the expense of degraded resolution.

coaxial—Having a common axis.

coaxial antenna—An antenna comprised of a quarter-wavelength extension to the inner conductor of a coaxial line, and a radiating sleeve which in effect is formed by folding back the outer conductor of the coaxial line for approximately one-quarter wavelength.

coaxial cable—*See* Coaxial Line.

coaxial cavity — A cylindrical resonating cavity that has a central conductor in contact with its movable pistons or other reflecting devices. The conductor serves to pick up a desired wave in the microwave region.

coaxial diode—A diode that has the same outer diameter and terminations as a coaxial cable into which the diode is designed to be inserted.

coaxial-fed linear array—A beacon antenna having a uniform azimuth pattern.

coaxial filter—A passive, linear, essentially nondissipative network that transmits certain frequencies and rejects others.

coaxial line—Also called coaxial cable, coaxial transmission line, and concentric line. A transmission line in which one conductor completely surrounds the other, the two being coaxial and separated by a continuous solid dielectric or by dielectric spacers. Such a line has no external field and is not susceptible to external fields from other sources.

coaxial-line connector — A connection between two coaxial lines or between a coaxial line and the equipment.

coaxial-line frequency meter — A shorted section of coaxial line which acts as a resonant circuit and is calibrated in terms of frequency or wavelength.

coaxial relay — A type of relay used for switching high-frequency circuits.

coaxial speaker—1. A single speaker comprising a high- and a low-frequency unit plus an electrical crossover network. 2. A tweeter mounted on the axis of and inside the cone of a woofer.

coaxial stub—A short length of coaxial cable joined as a branch to another coaxial cable. Frequently a coaxial stub is short-circuited at the outer end, and its length is so chosen that a high or low impedance is presented to the main coaxial cable at a certain frequency range.

coaxial transistor — A diffused-base, alloy-emitter, epataxial mesa germanium semiconductor device with a bandwidth product up to 3 gigahertz capable of being operated at medium power.

coaxial transmission line—*See* Coaxial Line.

COBOL — Acronym for COmmon Business Oriented Language. Used to express problems of data manipulation and processing in English narrative form.

co-channel interference — Interference between two signals of the same type from transmitters operating on the same channel.

Cockcroft-Walton accelerator—A device for accelerating charged particles by the action of a high direct-current voltage on a stream of gas ions, in a straight insulated tube. The voltage is generated by a voltage-multiplier system consisting essentially of a number of capacitor pairs connected through switching devices (vacuum tubes). The particles (which are nuclei of an ionized gas, such as protons from hydrogen) gain energies of up to several million electronvolts from the single acceleration so produced. Named for the British physicists J. D. Cockcroft and E.T.S. Walton, who developed this machine in the 1930's.

codan — Acronym for carrier-operated device, antinoise. An electronic circuit that

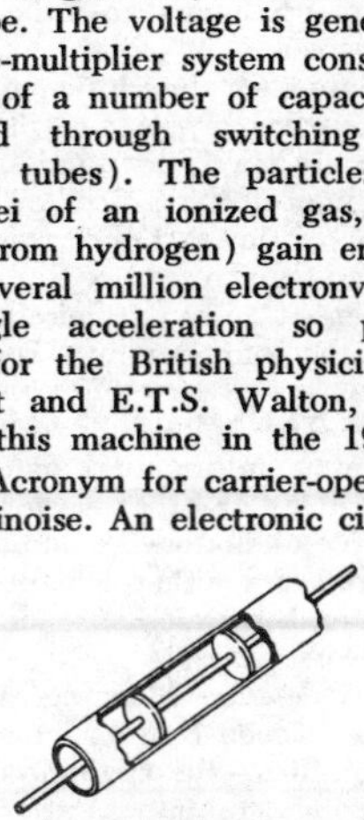

Coaxial line.

keeps a receiver inactive except when a signal is received.

codan lamp—A visual indication that a usable transmitted signal has been received by a particular radio receiver.

code — 1. A communications system in which arbitrary groups of symbols represent units of plain text of varying length. Codes may be used for brevity or security. 2. System of signaling by utilizing dot-dash-space, mark-space, or some other method where each letter or figure is represented by prearranged combinations. 3. System of characters and rules for representing information. 4. A system of symbols that represent data values and make up a special language that a computer can understand and use.

code character—One of the elements which make up a code and which represent a specified symbol or value to be encoded. Dot-dot-dot-dash is the Morse code character for the letter V.

code conversion—The changing of the bit grouping for a character in one code into the corresponding bit grouping in another code.

code converter—1. A device for translating one code to another. Examples: ASCII to EBCDIC, Gray to bcd, Hollerith to EBCDIC, etc. 2. A decision-making type of digital building block that converts information received at its inputs to another digital code which is transmitted at its outputs. (Also called encoder or decoder).

coded decimal digit — A decimal digit expressed in terms of four or more ones and zeroes.

coded passive reflector antenna—An object intended to reflect Hertzian waves and having variable reflecting properties according to a predetermined code for the purpose of producing an indication on a radar receiver.

coded program—A description of a procedure for solving a problem with a digital computer. It may vary in detail from a mere outline of the procedure to an explicit list of instructions coded in the machine's language.

code element—One of the finite set of parts of which the characters in a code may be composed.

code holes—The holes in perforated tape that represent information, as opposed to the feed holes or other perforations.

code-practice oscillator—An audio oscillator with a key and either headphones or a speaker, used to practice sending and receiving Morse code.

coder—1. A device which sets up a series of signals in code form. 2. A beacon circuit which forms the trigger-pulse output of a discriminator into a series of pulses and then feeds them to a modulator circuit. 3. A person who prepares instruction sequences from detailed flowcharts and other algorithmic procedures prepared by others, as contrasted with a programmer who prepares the procedures and flowcharts.

code ringing — In a telephone system, a method of ringing in which the number and/or duration of rings indicate which station on a party line is being called.

code set — The entire set of unique codes that represent specific characters. Different code sets are employed in different equipment.

code translation — In telephone operation, the changing of a directory code or number into a predetermined code that controls the selection of an outgoing trunk or line.

coding — 1. Converting program flowcharts into the language used by the computer. 2. The assignment of identification codes to transactions, such as a customer code number. 3. A method of representing characters within a computer. 4. Changing a communications signal into a form suitable for transmission or processing.

coding delay—Arbitrary time delay between pulse signals sent by master and slave transmitters.

coding disc—A disc with small projections that operate contacts to generate a predetermined code.

coding line—A single command or instruction that directs a computer to solve a problem usually written on one line.

codiphase radar—A radar system including a phased-array radar antenna and signal-processing and beam-forming techniques.

codistor—A multijunction semiconductor device which provides noise-rejection and voltage-regulation functions.

coefficient—1. The ratio of change under specified conditions of temperature, length, volume, etc. 2. A number (often a constant) that expresses some property of a physical system in a quantitative way.

coefficient of coupling—*See* Coupling Coefficient.

coefficient of performance of a thermoelectric cooling couple—The quotient of the net rate of heat removal from the cold junction divided by the electrical power input to the thermoelectric couple, assuming perfect thermal insulation of the thermoelectric arms.

coefficient of performance of a thermoelectric cooling device—The quotient of the rate at which heat is removed from the cooled body divided by the electrical power input to the device.

coefficient of performance of a thermoelectric heating couple—The quotient of the rate at which heat is added to the hot junction divided by the electrical power input to the thermoelectric couple, assuming perfect thermal insulation of the thermoelectric arms.

coefficient of performance of a thermoelectric heating device—The quotient of the rate at which heat is added to the heated body divided by the electrical power input to the device.

coefficient of reflection—The square root of the ratio of the power reflected from a surface to the power incident on the same surface.

coefficient of thermal expansion—The average expansion per degree over a specified temperature range, expressed as a fraction of the original dimension. The coefficient may be linear or volumetric.

coercive force—Symbolized by H. The magnetizing force that must be applied to a magnetic material in a direction opposite the residual induction in order to reduce the induction to zero.

coercivity—1. The property of a magnetic material measured by the coercive force corresponding to the saturation induction for the material. 2. A measure of the amount of applied magnetic field (of opposite polarity) that is necessary to restore a magnetized tape to a state of zero magnetism. High-coercivity tapes exhibit less tendency toward self-erasure and thus have enhanced high-frequency-response characteristics, but they require more current through the erase head for full erasure of a recorded signal.

cogging—1. Nonuniform angular velocity. The armature coil of a motor tends to speed up when it enters the magnetic field produced by the field coils, and to slow down when leaving it. This becomes apparent at low speeds, and the fewer the coils, the more noticeable it is. 2. In a motor the effect caused by improper ratio of stator to rotor slots for a particular speed. It is caused by a magnetic interaction between rotor and stator teeth. In dc motors it is caused by the rich ripple content of the power supply rectifier. The term "cogging torque" is usually applied to reluctance type synchronous motors. This is the maximum torque of such a synchronous motor before starting to cog or slip, as if a gear were slipping one tooth at a time.

coherence—1. The property of a set of waves by which their phases are completely predictable along an arbitrarily specified surface in space, also the relation between a set of sources by which the phases of their respective radiations are similarly predictable. 2. A term used to denote various forms of temporal or statistical phase correlations of electromagnetic fields at different spatial positions; the more extensive the correlations, the greater the coherence. (*See* Maser and Laser.)

coherent bundle — A fiber-optics bundle where each fiber maintains its relative location throughout the bundle. Thus, an image introduced at one end is transmitted to the other without being scrambled.

coherent carrier — A carrier, derived from a cw signal, the frequency and phase of which have a fixed relationship to the frequency and phase of the reference signal.

coherent carrier system—A transponder system in which the interrogating carrier is raised to a definite multiple frequency and retransmitted for comparison.

coherent detection—A method of deriving additional information from the phase of the carrier.

coherent detector—A detector that gives an output signal amplitude dependent on the phase rather than the strength of the echo signal. It is required for a radar display that shows only moving targets.

coherent display—In random-sampling oscilloscope technique, a plot of a group of samples in which the time sequence of signal events is maintained.

coherent echo—A radar echo that has relatively constant phase and amplitude at a given range.

coherent electroluminescence device — *See* Diode Laser.

coherent interrupted waves — Interrupted continuous waves occurring in wave trains in which the phase of the waves is maintained through successive wave trains.

coherent light — 1. A single frequency of light. Light having characteristics similar to a radiated radio wave that has a single frequency. 2. Light of but a single frequency that travels in intense, nearly perfect, parallel rays without appreciable divergence.

coherent light communications—Communications using amplitude or pulse-frequency modulation of a laser beam.

coherent light detection and ranging—*See* Colidar.

coherent oscillator — An oscillator within some radar sets which furnishes phase references for target returns during intervals between transmitter pulses, and has its output compared with the returns so that the echo becomes coherent video. Coherent video is applied to a cancellation circuit which eliminates nonmoving targets, and only moving targets are supplied to the indicator.

coherent-pulse operation — The method of pulse operation in which a fixed phase relationship is maintained from one pulse to the next.

coherent radar—A type of radar containing circuits that make possible comparison of the phase of successive received target signals.

coherent radiation—1. A form of radiation in which definite phase relationships exist between radiation at different positions in a cross section of the radiant beam. 2. Single-frequency energy such that there is reinforcement when portions of a signal coincide in phase and cancellation when they are in phase opposition.

coherent reference – The reference signal, usually of stable frequency, to which other signals are phase-locked to establish coherency throughout a system.

coherent system of units—Also called absolute system of units. A system of units in which the magnitude and dimensions of each unit are related to those of the other units by definite simple relationships in which the proportionality factors are usually chosen to be unity.

coherent transponder – A transponder, the output signal of which is coherent with the input signal. Fixed relations between frequency and phase of input and output signals are maintained.

coherent video – A video signal resulting from the combination of a radar echo signal with the output of a continuous-wave oscillator. After delay, the signal so formed is detected, amplified, and subtracted from the next pulse train to give a signal that represents only moving targets.

coherer—An early form of detector used in wireless telegraphy.

coil—1. A number of turns of wire wound around an iron core or onto a form made of insulating material, or one which is self-supporting. A coil offers considerable opposition to the passage of alternating current, but very little opposition to direct current. 2. A number of turns of wire used to introduce inductance into an electric circuit, to produce magnetic flux, or to react mechanically to a charging magnetic flux. In high-frequency circuits, a coil may be only a fraction of a turn. The electrical size of a coil is called inductance and is expressed in henrys. The opposition that a coil offers to alternating current is called impedance and is expressed in ohms. The impedance of a coil increases with frequency. Also called inductance and inductor.

coil dissipation—The amount of electrical power consumed by a winding. For most practical purposes, this equals the I^2R loss.

coil form – An insulating support of ceramic, plastic, or cardboard onto which coils are wound.

coil loading—As commonly understood, the insertion of coils into a line at uniformly spaced intervals. However, the coils are sometimes inserted in parallel.

coil neutralization—*See* Inductive Neutralization.

coil rating—*See* Input-Power Rating.

coil resistance—The total terminal-to-terminal resistance of a coil at a specified temperature.

coil serving—A covering, such as thread or tape, that serves to protect a coil winding from mechanical damage.

coil temperature rise – The increase in winding temperature above the ambient temperature when energized under specified conditions for a given period of time, usually that required to reach a stable temperature.

coil tube—A tubular coil form. (*See also* Spool.)

coincidence amplifier—An amplifier which produces no output unless two input pulses are applied simultaneously to the circuit.

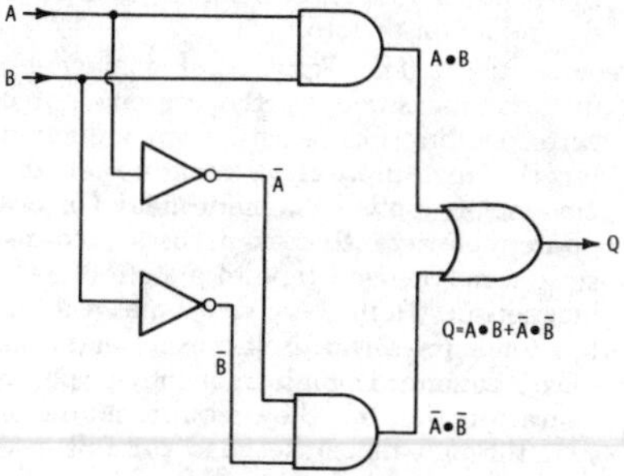

Coincidence circuit.

coincidence circuit—1. A circuit that produces a specified output pulse when and only when a specified number (two or more) or combination of input terminals receives pulses within a specified time interval. 2. A circuit that has an output signal only when all input signals are present. 3. An AND circuit.

coincidence counting—The use of electronic devices to detect when two or more pulses from separate counters occur within a given time interval. This is done to determine whether the pulses were produced by the same particle, for example in scintillation counting, or whether they correspond to the same event.

coincidence gate—A circuit with the ability to produce an output which is dependent upon a specified type or the coincident nature of the input.

coincident-current selection—The selection of a magnetic core, for reading or writing, by the simultaneous application of two or more currents.

cold—Idiomatic term generally used to describe electrical circuits that are disconnected from voltage sources and are at ground potential. Opposite of the term "hot."

cold cathode—A cathode the operation of which does not depend on its temperature being above the ambient temperature.

cold-cathode tubes — Tubes in which no external source is used for heating the cathode. These include tubes such as photoelectric cells, gas glow tubes, and mercury rectifiers.

cold emission—*See* Field Emission.

cold flow—Change of dimension or distortion, caused by sustained application of a force.

cold-pressure welding—A method of making an electrical connected in which the members to be joined are compressed to the plastic range of the metals.

cold rolling—Rolling a magnetic core alloy into the form of a rod so that the metallic grains are oriented in the long direction of the rod.

cold weld — A joint between two metals (without an intermediate material) produced by the application of pressure only.

colidar—Acronym for coherent light detection and ranging. An optical radar system that uses the direct output from a ruby laser source without further pulse modulation.

collate—To combine two or more similarly ordered sets of items to produce another ordered set. Both the number and the size of the individual items may differ from those of the original sets or their sums.

collating sequence—1. In digital computers, the sequence in which the characters acceptable to a computer are ordered. The British term is marshalling sequence. 2. An ordering assigned to a set of items, such that any two sets in that assigned order can be collated.

collator — 1. A device to collate sets of punched cards or other documents into a sequence. 2. A device for determining and indicating the coincidence or noncoincidence of two signals.

collector—1. In a transistor, the region into which majority carriers flow from the base under the influence of a reverse bias across the two regions. 2. The external terminal of a transistor that is connected to this region. 3. In certain electron tubes, an electrode to which electrons or ions flow after they have completed their function.

collector capacitance — Depletion-layer capacitance associated with the collector junction of a transistor.

collector-coupled logic—*See* Current-Sourcing Logic.

collector current—The direct current flowing in the collector of a transistor.

collector cutoff—The operating condition of a transistor when the collector current is reduced to the leakage current of the collector-base junction.

collector cutoff current—The minimum current that will flow in the collector circuit of a transistor with zero current in the emitter circuit.

collector-diffusion isolation — A technique for fabricaiton of bipolar IC's; the collector diffusion is used to isolate transistors on the same silicon chip electrically, thus reducing the number of photolithographic masking steps required. Abbreviated CDI.

collector efficiency—The ratio, usually expressed in percentage, of useful power output to final-stage power-supply power input of a transistor.

collector family—Set of transistor characteristic curves in which the collector current and voltage are variables.

collector junction—The semiconductor junction between the base and collector regions of a transistor. In normal transistor operation, it is reverse biased to collect carriers injected by the emitter (base-to-emitter) junction. In general, the collector junction is designed for a high breakdown voltage, and the emitter junction is designed for a high emitting efficiency.

collector resistance — Resistance in series with the collector lead in the common-T equivalent circuit of a transistor.

collector rings (slip rings) — Metal rings suitably mounted on an electric machine serving, through stationary brushes bearing thereon, to conduct current into or out of the rotating member.

collector transition capacitance — The capacitance across the collector-to-base transition region of a transistor.

collector voltage—The dc collector supply voltage applied between the base and collector of a transistor.

collimated light—Parallel light rays, as opposed to converging or diverging rays.

collimation—1. The process of adjusting an instrument so that its reference axis is aligned in a desired direction within a predetermined tolerance. 2. The process of making light rays parallel.

collimation equipment — Equipment designed specifically for aligning optical equipment.

collimation tower — A tower supporting a visual and a radio target used to check the electrical axis of an antenna.

collimator—An optical device that creates a beam made up of parallel rays of light, used in testing and adjusting certain optical instruments.

collinear array—An antenna array in which half-wave elements are arranged end to end on the same vertical or horizontal line.

color balance—The adjustment of electron-gun emissions to compensate for the difference in the light emitting efficiencies of the three phosphors on the screen of the color picture tube.

color-bar signal — A test signal used in

checking chrominance functions of color tv systems. Typically, it contains bars of six colors, yellow, cyan, green, magenta, red, and blue.

color breakup—Any fleeting or partial separation of a color picture into its display primary components because of a rapid change in the condition of viewing. For example, fast movement of the head, abrupt interruption of the line of sight, and blinking of the eyes are illustrations of rapid changes in the conditions of viewing.

color burst — Also called reference burst. Approximately nine cycles of the chrominance subcarrier added to the back porch of the horizontal blanking pedestal of the composite color signal and used in the color receiver as a phase reference for the 3.579545-MHz oscillator.

color carrier—*See* Chrominance Subcarrier.

color-carrier reference—A continuous signal of the same frequency as the color subcarrier and having a fixed phase relationship to the color burst. It is used for modulation at the transmitter and demodulation at the receiver.

colorcast—A color television broadcast.

color code—A system of colors for specifying the electrical value of a component part or for identifying terminals and leads. Also used to distinguish between cable conductors.

color coder—Also called color encoder. In a color-tv transmitter, that circuit or section which combines the camera signals and the chrominance subcarrier to form the transmitted color picture signal.

color coding—A system of identification of terminals and related devices through the use of colored markings.

color contamination—An error in color rendition due to incomplete separation of the paths carrying different color components of the picture. Such errors can arise in the optical, electronic, or mechanical portions of a color-television system.

color-coordinate transformation—Computation of the tristimulus values of colors in terms of one set of primaries from the tristimulus values of the same colors in another set of primaries. Such computation may be performed electrically in a color-television system.

color decoder — A section or circuit of a color-television receiver used for deriving the signals for the color-display device from the color-picture signal and the color burst.

color-difference signal — The signal produced when the amplitude of a color signal is reduced by an amount equal to the amplitude of the luminance signal. Color-difference signals are usually designated R – Y, B – Y, and G – Y. In a sense, I and Q signals are also color-difference signals because they are formed when specific proportions of R – Y and B – Y color-difference signals have been combined.

color edging—Spurious color at the boundaries of differently colored areas in a picture.

color encoder—*See* Color Coder.

color fidelity—The degree to which a color television system is capable of faithfully reproducing the colors in an original scene.

color flicker—The flicker which results from fluctuation of both the chromaticity and the luminance.

color fringing—Spurious chromaticity at the boundaries of objects in a color-tv picture. It can be caused by the change in relative position of the televised object from field to field, or by misregistration. Color fringing may cause small objects to appear separated into different colors.

colorimeter — An optical instrument designed to compare the color of a sample with that of a standard sample or a synthesized stimulus. (In a three-color colorimeter, the synthesized stimulus consists of three colors of contrast chromaticity but variable luminance.)

colorimetric — Pertaining to the measurement of color characteristics, particularly wavelength and primary-color content.

colorimetry — The technique of measuring color and interpreting the results.

color killer — A stage designed to prevent signals in a color receiver from passing through the chrominance channel during monochrome telecasts.

color match — The condition in which the two halves of a structureless photometric field look exactly the same.

color media—Transparent, colored materials that can be placed in front of an instrument to color the emitted light. These materials are often referred to as "gels" (for gelatin), but glass or plastic also may be used.

color mixture—Color produced by the combination of lights of different colors. The combination may be accomplished by successive presentation of the components, provided the rate of alternation is sufficiently high; or the combination may be accomplished by simultaneous presentation, either in the same or in adjacent areas, provided they are small enough and close enough together to eliminate pattern effects.

color oscillator — In a color-television receiver, the oscillator operating at the burst frequency of 3.579545 MHz. Its frequency and phase are synchronized by the master oscillator at the transmitter.

color phase — The difference in phase between a chrominance primary signal (I or Q) and the chrominance carrier reference.

color-phase alternation—Periodic changing of the color phase of one or more components of the color-television subcarrier between two sets of assigned values.

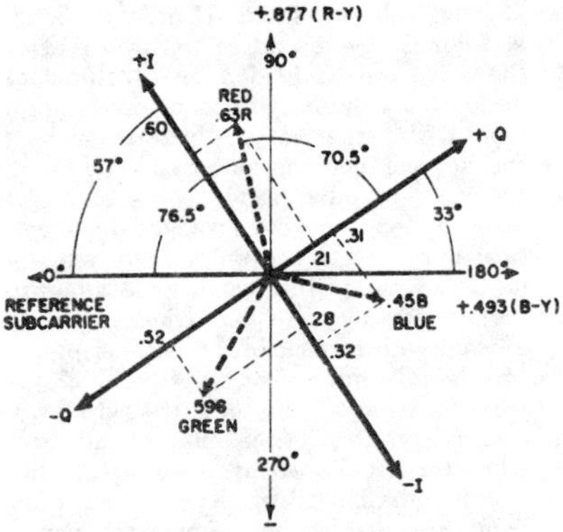

Color-phase diagram.

color-phase diagram — A vector diagram which denotes the phase difference between the color-burst signal and the chrominance signal for each of the three primary and complementary colors. This diagram also designates vectorially the peak amplitude of the chrominance signal for each of these colors, and the polarities and peak amplitudes of the in-phase and quadrature portions required to form these chrominance signals.

color-picture signal—1. A signal which represents electrically the three color attributes (brightness, hue, and saturation) of a scene. 2. A combination of the luminance and chrominance signals, excluding all blanking and synchronizing signals.

color picture tube—An electron tube that provides an image in color by scanning a raster and by varying the intensity at which it excites the phosphors on the screen to produce light of the chosen primary colors.

color primaries—In the color receiver, the saturated colors of definite hue and variable luminance produced by the receiver. These color primaries, when mixed in proper proportions, form other colors.

color purity — Freedom of a color from white light or any colored light not used to produce the desired color.

color-purity magnet — A magnet placed in the neck region of a color picture tube to alter the electron-beam path and thereby improve color purity.

color registration — The accurate superimposing of the red, green, and blue images that are used to form a complete color picture in a color television receiver.

color rendering index—A number that represents approximately the effect of a light source on the appearance of colored surfaces. Abbreviated CRI.

color sampling rate — In a color television system, the number of times per second that each primary color is sampled.

color saturation — The degree to which white light is absent in a particular color. A fully saturated color contains no white light. If 50% of the light intensity is due to the presence of white light, the color is said to have a saturation of 50%.

color sensitivity—The spectral sensitivity of a light-sensitive device such as a phototube or camera tube.

color signal—Any signal at any point in a color-television system, used for wholly or partially controlling the chromaticity values of a color-television picture. This is a general term encompassing many specific connotations, such as are conveyed by the words, "color-picture signal," "chrominance signal," "carrier-color sig-

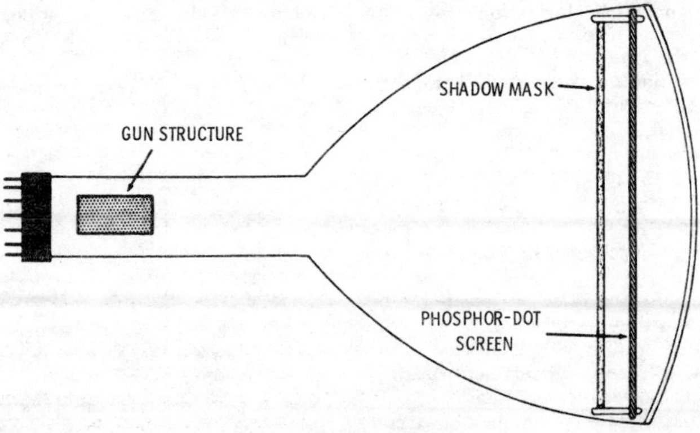

Color picture tube.

nal," "monochrome signal" (in color television), etc.

color subcarrier – A monochrome signal to which modulation sidebands have been added to convey color information.

color-sync signal–The series of color bursts (pulses of subcarrier-reference signal) applied to the back porch of the horizontal-sync pedestal in the composite video signal.

color-television receiver–A standard monochrome receiver to which special circuits have been added. Phosphors capable of glowing in the three primary colors are used on the special screen. By using these primary colors and mixing them to produce complementary colors, and by varying their intensity, it is possible to reproduce an image in somewhat the original colors.

color-television signal–The complete signal used to transmit a color picture. Included are horizontal-, vertical-, and color-sync components.

color temperature – The temperature to which a perfectly blackbody must be heated to match the color of the source being measured. Color-temperature measurements begin at absolute zero and are expressed in degrees Kelvin.

color temperature of a light source – The absolute temperature at which a blackbody radiator must be operated to have a chromaticity equal to that of the light source.

color transmission – The transmission of color-television signals which can be reproduced with different values of hue, saturation, and luminance.

color triad–One cell of a three-color, phosphor-dot screen of a phosphor-dot, color-picture tube. Each triad contains one dot of each of the three color-producing phosphors.

color triangle – A triangle drawn on a chromaticity diagram to represent the entire range of chromaticities obtainable when the 3 prescribed primaries are added. These are represented by the corners of the triangle.

Colpitts oscillator – A sinusoidal oscillator using a three-terminal active element, as a tube, transistor, etc., and a feedback loop containing a parallel LC circuit. The capacitance of the LC circuit consists of two capacitors, in series forming a voltage divider that serves to match the input and output impedance of the active device.

column–Also called place. In positional notation, a position corresponding to a given power of the radix. A digit located in any particular column is a coefficient of a corresponding power of the radix.

column-binary code–A punched-card code in which successive bits are represented by the presence or absence of punches in adjacent positions in successive columns rather than succesive rows. It is used with 36-bit-word computers, in which each group of three columns is used to represent a single word.

column speaker–1. A loudspeaker cabinet of long "column" shape. Usually the loudspeaker is at one end so that the rear of the drive unit is loaded by a column of air. Often columns are made from drain pipes. Also the name given to a long speaker cabinet containing several loudspeakers for public address work. 2. An array of loudspeakers arranged in a vertical line, having the property of spreading its radiation through a wide angle in the horizontal plane while keeping it in a beam with respect to the vertical plane.

com–*See* Computer-Output-Microfilm.

coma–1. An aberration of spherical lenses, occurring in the case of oblique incidence, when the bundle of rays forming the image is unsymmetrical. The image of a point is comet shaped, hence the name. 2. A cathode-ray tube image defect that makes the spot on the screen appear comet shaped when away from the center of the screen.

coma lobe–A side lobe that occurs in the radiation pattern of a microwave antenna when the reflector alone is moved to sweep the beam. The lobe appears because the feed is not always at the center of the reflector. This scanning method is used to eliminate the need for a rotary joint in the feed waveguide.

combinatorial logic – Digital circuitry in which the states of the outputs from a device depend only on the states of the inputs. *See also* Sequential Logic.

comb filter–A type of filter network that is, in effect, a multiple-bandpass design that passes only frequencies within a number of narrow bands, or provides outputs corresponding to each of its passbands. Comb filters are so named because their response characteristics have the appearance of a comb.

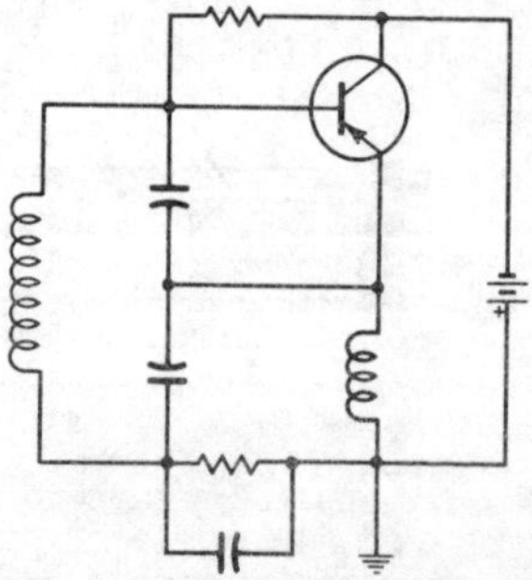

Colpitts oscillator.

comb generator – Usually a step-recovery diode circuit which converts a single frequency rf input into an rf output signal that contains a large number of spectral lines, each of which are harmonically related to the input frequency.

combination cable – A cable in which the conductors are grouped in combinations such as pairs and quads.

combination microphone – A microphone consisting of two or more similar or dissimilar microphones combined into one.

combination tones – Frequencies produced in a nonlinear device, such as in an audio amplifier having appreciable harmonic distortion.

combined-gate IC – A single IC chip in which several gate circuits are interconnected to form a more complex circuit.

combiner–A circuit for mixing video, trigger, and scan data from the synchronizer for the modulation of a link.

combiner circuit – One that combines the luminance and chroma channels with the sync signals in color-television cameras.

combustible–*See* Flammable.

come-back–A point in the stop band of a filter where a spurious response occurs beyond points at which there is proper attenuation. Come-backs usually occur at frequencies much higher than the pass-band frequencies, because of feed-through in parasitic elements.

command–1. In a computer, one or more sets of signals which occur as the result of an instruction. (*See also* Instruction, 1.) 2. An independent signal from which the dependent signals in a feedback-control system are controlled according to the prescribed system relationships. 3. A signal which initiates or triggers an action in the device that receives the signal.

command code–*See* Operation Code.

command control–A system whereby functions are performed as the result of a transmitted signal.

command destruct signal – A radio signal for destroying a missile in flight.

command guidance system–A missile guidance system in which both the missile and the target are tracked by radar. The missile is guided by signals transmitted to it while it is in flight. (*Also see* Command Link.)

command language – A computer source language that consists primarily of procedural operators, each of which is capable of invoking a function to be executed.

command link–The portion of a command guidance system used to transmit steering commands to the missile. (*See also* Command Guidance System.)

command module–*See* Module.

command net – A communication network which connects an echelon of command with some or all of its subordinate echelons for the purpose of command and control.

command reference–In a servo or control system, the voltage or current to which the feedback signal is compared. As an independent variable, the command reference exercises complete control over the system output.

comment–An expression which explains or identifies a particular step in a routine, but which has no effect on the operation of the computer in performing the instructions for the routine.

comment field–In a computer, an area in a record assigned for entry of explanatory comments about a program.

commercial test (measuring) equipment–Devices used as "working instruments" selected from suppliers' catalogs as suitable for the user's measuring needs and procured as standard ("off-the-shelf") items. This includes measuring devices installed, intact and in consoles. Devices procured from suppliers but modified by user-imposed specifications, in such a manner as to affect console performance, do not fall into this category.

commercial time-sharing–A type of computer use with remote terminal, interactive, and time-sharing characteristics. The user of the machine pays an amount determined by the time used.

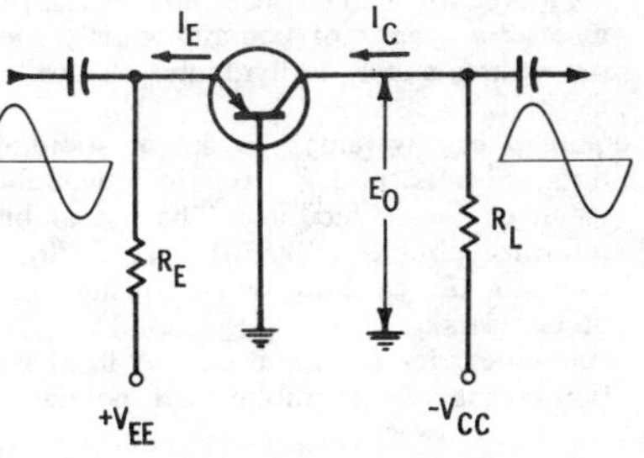

Common-base amplifier.

common-base amplifier – Also called a grounded-base amplifier. A transistor amplifier in which the base element is common to both the input and the output circuit. It is comparable to the grounded-grid configuration of a triode electron-tube amplifier.

common–1. Shared by two or more circuits. Used to designate the terminal of a three-terminal device that is shared by the *input* and *output* circuits. (Thus a transistor may be operated in a common-base configuration, a common-collector configuration, or a common-emitter configuration.) Vacuum tube connections may be characterized in a similar way, but *grounded* is normally used instead of *common*. 2. A point that acts as the reference potential for several circuits—a ground.

common-base circuit--A transistor circuit in which the base electrode is common to both input and output circuits.

common-base feedback oscillator – A common-base bipolar transistor amplifier with a feedback network between the collector (output) and the emitter (input) to produce oscillations at a desired frequency.

common-base transistor—Circuit configuration in which the base terminal is common to the input circuit and to the output circuit and in which the input terminal is the emitter terminal and the output terminal is the collector terminal.

common battery—A system of current supply in which all direct-current energy for a unit of a telephone system is supplied by one source in a central office or exchange.

common-battery office – A central office which supplies transmitter and signal current for its associated stations, and for signaling by the central office equipment, from a power source located in the central office.

common business oriented language – Specific language by which business data-processing procedures may be precisely described in a standard form. Intended not only as a means for directly presenting any business program to any suitable computer for which a compiler exists, but also as a means of communicating such procedures among individuals. Also called COBOL.

common bus system – A set of standard data, address, and control lines available to all computer modules. The use of bus interface circuits makes it possible for a user to tie in and communicate with other users.

common-carrier fixed station – A fixed station that is open to public correspondence.

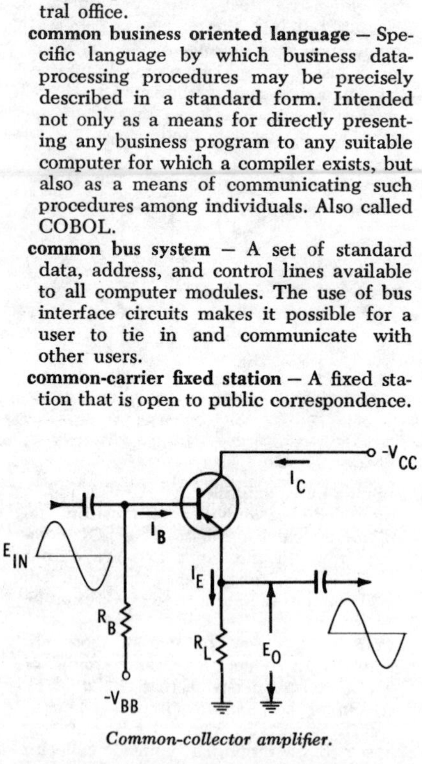

Common-collector amplifier.

common-collector amplifier—Also known as an emitter-follower and a grounded-collector amplifier. A transistor amplifier in which the collector element is common to both the input and the output circuit. This configuration is comparable to an electron-tube cathode follower.

common-collector transistor—Circuit configuration in which the collector terminal is common to the input circuit and to the output circuit and in which the input terminal is the base terminal and the output terminal is the emitter terminal.

common communications carrier – A company recognized by an appropriate regulatory agency as having a vested interest in furnishing communications services.

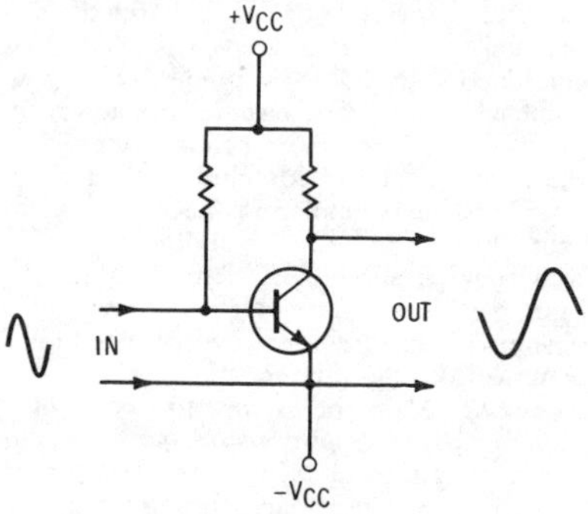

Common-emitter amplifier.

common-emitter amplifier – Also called grounded-emitter amplifier. A transistor amplifier in which the emitter element is common to both the input and the output circuit. This configuration is comparable to a conventional electron-tube amplifier.

common-emitter transistor – Circuit configuration in which the emitter terminal is common to the input circuit and to the output circuit and in which the input terminal is the base terminal and the output terminal is the collector terminal.

common language—A form of representing information which a machine can read and which is common to a group of computers and data-processing machines.

common mode – Signals that are identical with respect to both amplitude and time. Also used to identify the respective parts of two signals that are identical with respect to amplitude and time.

common-mode characteristics—The characteristics pertaining to performance of an operational amplifier where the inverting and noninverting inputs have a common signal.

common-mode gain—The ratio of the output voltage of a differential amplifier to the common-mode input voltage. The common-mode gain of an ideal differential amplifier is zero.

common-mode impedance input—The internal impedance between either one of the input terminals of a differential operational amplifier and signal ground.

common-mode input – That signal applied in phase (i.e., common mode) equally to both inputs of a differential amplifier.

common-mode input capacitance – The equivalent capacitance of both inverting and noninverting inputs of an operational amplifier with respect to ground.

common-mode input impedance—The open-loop input impedance of both inverting and noninverting inputs of an operational amplifier wtih respect to ground.

common-mode input voltage – The maximum voltage that can be applied simultaneously between the two inputs of a differential amplifier and ground without causing damage.

common-mode interference – Interference that appears between the terminals of the measuring circuit and ground.

common-mode output voltage—The output voltage of an operational amplifier resulting from the application of a specified voltage common to both inputs.

common-mode rejection – Also called in-phase rejection. A measure of how well a differential amplifier ignores a signal which appears simultaneously and in phase at both input terminals (called a common-mode signal). Usually and preferably stated as a voltage ratio, but more often stated in the dB equivalent of said ratio at a specified frequency—e.g., "120 dB at 60 hertz with a source impedance of 1000 ohms."

common-mode rejection in decibels – Twenty times the log of the common-mode rejection ratio.

common-mode rejection ratio—1. The ratio of the common-mode input voltage to the output voltage expressed in dB. The extent to which a differential amplifier does not provide an output voltage when the same signal is applied to both inputs. 2. The ratio of differential-mode gain to common-mode gain. 3. The ratio of the change of input offset voltage of an operational amplifier to the change in common-mode voltage producing it.

common-mode resistance – The resistance between the input- and output-signal lines and circuit ground. In an isolated amplifier, this is its insulation resistance. (Common-mode resistance has no connection with common-mode rejection.)

common-mode signal – The instantaneous algebraic average of two signals applied to a balanced circuit (i.e., two ungrounded inputs of a balanced amplifier), all signals referred to a common reference.

common-mode voltage—1. The amount of voltage common to both input lines of a balanced amplifier. Usually specified as the maximum voltage which can be applied without breaking down the insulation between the input circuit and ground. (Common-mode voltage has no connection with common-mode rejection.) 2. The voltage on both inputs of a differential input amplifier. 3. An undesirable signal picked up in a transmission line by both wires making up the circuit, to an equal degree, with respect to an arbitrary "ground."

common-mode voltage gain – The ratio of the ac voltage with respect to ground at the output terminal of an amplifier (or between the output terminals of an amplifier with differential outputs) to the common-mode input voltage.

common-mode voltage range—The range of voltage that may be applied to both inputs of an operational amplifier without saturating the input stage. This may limit the output capabilities in the voltage-follower connection.

common pool—A dedicated area of memory used as storage and shared by various processes.

common-user channels – Communication channels which are available to all authorized agencies for transmission of command, administrative, and logistic traffic.

common-user circuit – A circuit shared by two or more services, either concurrently or on a time-sharing basis. It may be a unilateral, bilateral, or joint circuit.

communal chained memory – A technique employed in dynamic storage allocation in a computer.

communication—1. The transmission of information from one point, person, or equipment to another. 2. The sensing of a measurement signal or phenomena for display, recording, amplification, transmission, computing or processing into useful information.

communication band—The band of frequencies due to the modulation (including keying) necessary for a given type of transmission.

communication channel—Part of a radio or wire circuit, or a combination of wire and radio which connects two or more terminals.

communication control character—A character the purpose of which is to control or facilitate data transmission over communication networks.

communication link—The physical means of connecting one location to another for the purpose of transmitting and receiving information.

communications common carriers – Companies that furinsh communications services to the public, regulated by the FCC or appropriate state agencies.

communications receiver – A receiver designed for reception of voice or code signals from stations operating in the communications service.

communications satellite—An orbiting space vehicle that actively or passively relays signals between communications stations.

communications security – The protection resulting from all measures designed to deny unauthorized persons information of value which might be derived from the possession and study of telecommunications or to mislead unauthorized persons in their interpretations of the results of such possession and study.

communication switch – A device used to execute repetitive sequential switching.

communication zone indicator – A device that indicates whether or not long-distance high-frequency broadcasts are successfully reaching their destination.

community antenna television—A television system that receives and retransmits television broadcasts. Microwave transmitters and coaxial cables are used to bring the television signals to subscribers in a community. Abbreviated catv.

community dial office – A small dial telephone office that serves an exchange area and that operates with no employees located in the building.

community television system—A receiving system by means of which television signals may be distributed over coaxial cables to homes in an entire community.

commutation—1. A mechanical process of converting the alternating current in the armature of direct-current generators into the direct-current generator output. 2. Sampling of various quantities in a repetitive manner, for transmission over a single channel. 3. The switching of currents back and forth between various paths as required for operation of some system or device. In particular, a switching of current to or from the appropriate armature coils of a motor or generator. The turning off of an active element at the correct time as in an inverter or power controller.

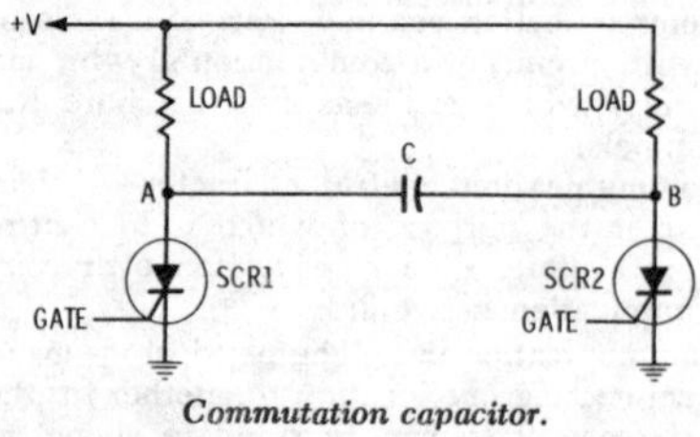

Commutation capacitor.

commutation capacitors – 1. Cross-connected capacitors in a thyratron inverter. They provide a path such that the start of conduction in one thyratron causes an extinguishing pulse to be applied to the alternate thyratron. Also used in inverter circuits employing semiconductor devices. 2. A specially designed capacitor used in the turn off (commutation) circuit of an SCR where it is subjected to exceedingly fast rise time pulses. Thus the capacitor must be capable of discharging large peak currents in very short periods of time.

commutation switch – A device used to carry out repetitive sequential switching.

commutator—1. The part of the armature to which the coils of a motor are connected. It consists of wedge-shaped copper segments arranged around a steel hub and insulated from it and from one another. The motor brushes ride on the outer edges of the commutator bars and thereby connect the armature coils to the power source. 2. Device used in a direct current generator to reverse the direction of an electric current and maintain a current flowing in one direction. 3. A switch or equivalent device that permits the reversal or exchange of external connections of a transducer to provide a desired sequencing of signals.

compactron—An electron tube based on a building-block concept which involves the standardizing of basic tube sections, diodes, triodes, pentodes; clipping them together as required; and sealing them in a single envelope.

compander—A combination consisting of a compressor at one point in a communication path to reduce the volume range of signals, followed by an expander at another point to improve the ratio of the signal to the interference entering the path between the compressor and expander.

companding—A process in which compression is followed by expansion. Companding is often used for noise reduction, in which case the compression is applied before the noise exposure and the expansion afterward.

compandor—*See* Compander.

companion keyboard – A remote keyboard connected by a multiwire cable to an ordinary keyboard and able to operate it.

comparator—1. A circuit which compares two signals and supplies an indication of agreement or disagreement. 2. In a computer, a circuit that determines whether the absolute difference between a data sample and the previous sample passed is greater than or equal to a redundancy criterion (which may be a tolerance or a limit). 3. A device that compares two inputs for equality. One type compares voltages and gives one of two outputs; less than, or greater than. Another type compares binary numbers and has three outputs: less than, equal to, or greater than. A third type compares phase or frequency, and gives a variable voltage depending on the relationship between the inputs. 4. A unit often found in audio showrooms, which by switch selection, will connect up a combination of speakers, amplifier, tuner, pick-up, tape player, etc. For comparing different types. 5. A circuit which compares two signals and provides a "difference" signal.

compare—A computer operation in which two quantities are matched for the purpose of discovering their relative magnitudes or algebraic values.

comparison—The examination of how two similar items of data are related. The comparison is usually followed by a decision.

comparison bridge—A type of voltage-comparison circuit resembling a four-arm electrical bridge. The elements are so arranged that if a balance exists in the circuit, a zero error signal is derived.

compatibility—1. That property of a color-television system which permits typical, unaltered monochrome receivers to receive substantially normal monochrome from the transmitted signal. 2. The property that makes possible use of a stereo system with a monophonic program source, or reproduction of a stereo program monophonically on a monophonic system. 3. The ability of one unit to be used with another without detrimental effect on the signal through mismatch. For example, a compatible pickup will play both mono and stereo records.

compatible IC—A hybrid IC in which the active circuit element is within the silicon planar integrated structure. A passive network, which may be separately optimized, is deposited onto its insulating surface to complete the IC device.

compatible monolithic integrated circuit — A device in which passive components are deposited by thin-film techniques on top of a basic silicon-substrate circuit containing the active components and some passive parts. *Also see* All-Diffused Monolithic Integrated Circuit.

compensated amplifier—A broad-band amplifier the frequency range of which is extended by the proper choice of circuit constants.

compensated-impurity resistor—A diffused-layer resistor into which are introduced additional n- and p-type impurities.

compensated-loop direction finder—A direction finder employing a loop antenna and a second antenna system to compensate for polarization error.

compensated semiconductor — A semiconductor in which one type of impurity or imperfection (donor) partially cancels the electrical effects of the other (acceptor).

compensated volume control—*See* Loudness Control.

compensating filter—A filter used to alter the spectral emission of an emulsion to a specified response to different wavelengths.

compensation—1. The controlling elements which compensate for, or offset, the undesirable characteristics of the process to be controlled in a system. 2. The shaping of an op-amp frequency response in order to achieve stable operation in a particular circuit. Some op amps are internally compensated while others require external compensation components in some circuits.

compensation signal—A signal recorded on the tape, along with the computer data and on the same track as the data; this signal is used during the playback of data to electrically correct for the effects of tape-speed errors.

compensation theorem—An impedance in a network may be replaced by a generator of zero internal impedence, the generated voltage of which at any instant is equal to the instantaneous potential difference produced across the replaced impedance by the current flowing through it.

compensator—1. In a direction finder, the portion which automatically applies to the direction indication all or part of the correction for the deviation. 2. An electronic circuit for altering the frequency response of an amplifier system to achieve a specified result. This refers to record equalization or loudness correction.

compile — To bring digital-computer programming subroutines together into a main routine or program.

compiler—1. An automatic coding system in a computer which generates and assembles a program from instructions written by a programmer. 2. A unit that converts computer programs written in higher-level languages, such as FORTRAN and BASIC, into the machine language of the computer. 3. Computer routine which translates symbolic instructions to machine instructions and replaces certain items of input with series of instructions, called subroutines.

compiler language — A computer language system consisting of various subroutines that have been evaluated and compiled into one routine that can be handled by the computer. FORTRAN, COBOL, and ALGOL are compiler language. Compiler language is the third level of computer language (*See* Machine Language, 3, for other levels).

compiler program — Software usually supplied by the manufacturer to convert an application program from compiler language to machine language.

compiling routine—A routine by means of which a computer can itself construct the program used to solve a problem.

complement—1. In an electronic computer, a number the representation of which is derived from the finite positional notation of another by one of the following rules:

True complement—Subtract each digit from 1 less than the base; then add 1 to the least significant digit and execute all required carries.

Base minus 1's complement — Subtract each digit from 1 less than the base (e.g., "9's complement" in the base 10, "1's complement" in the base 2, etc.)

2. To form the complement of a number. (In many machines, a negative number is represented as a complement of the corresponding positive number.) The binary opposite of a variable or function. The complement of 1 is 0 and the complement of 0 is 1; thus, for example, the complement of 011010 is 100101.

complementary colors—Two colors are complementary if, when added together in proper proportion such as by projection, they produce white light.

complementary constant current logic (C^3L)—A new high density approach to bipolar LSI that features switching speeds of 3 nanoseconds.

complementary MOS—Pertaining to n- and p-channel enhancement-mode devices fabricated compatibly on a silicon chip and connected into push-pull complementary digital circuits. These circuits offer low quiescent power dissipation and potentially high speeds, but they are more complex than circuits in which only one channel type (generally p channel) is used. Abbreviated CMOS.

complementary operator—The logic operator the result of which is the NOT of a given logic operator.

complementary push-pull—A power amplifier where the output transistors are of complementary polarities (i.e., pnp and npn). In some amplifiers of this kind the driven transistors also constitute a complementary pair.

complementary rectifier—Half-wave rectifying circuit elements which are not self-saturating rectifiers in the output of a magnetic amplifier.

complementary silicon controlled rectifier—A pnpn semiconductor device that is the polarity complement of the silicon controlled rectifier.

complementary-symmetry circuit — An arrangement of pnp and npn transistors that provides push-pull operation from one input signal.

complementary tracking—A system of interconnection of two or more devices in which one (the master) operates to control the others (the slaves).

complementary transistor amplifier—An amplifier that utilizes the complementary symmetry of npn and pnp transistors.

complementary transistor logic — A digital logic circuit configuration making use of a complementary transistor emitter coupled AND-OR gate. Basically a two-level diode gate using simultaneous npn and pnp action.

complementary transistors—Two transistors of opposite conductivity (pnp and npn) that are operated in the same functional unit.

complementary unijunction transistor — 1. An integrated semiconductor structure with characteristics similar to those of a unijunction transistor, but complementary to other unijunction transistors in the way that pnp transistors are complementary to npn transistors. Abbreviated CUJT. 2. A silicon planar device similar to a UJT except that the operating currents and voltages are of the opposite polarity. The electrical characteristics are stable, consistent and predictable over a wide temperature range. The CUJT will operate from a 5-volt supply and is therefore compatible with integrated circuits. Typically, the case is electrically connected to the substrate and must be isolated from the circuit.

complementary wave—A wave brought into existence at the ends of a coaxial cable or two-conductor transmission line, or any discontinuity along the line.

complementary wavelength — The wavelength of light of a single frequency. When combined with a sample color in suitable proportions, the wavelength matches the reference-standard light.

complement number — A number which, when added to another number, gives a sum equal to the base of the numbering system. For example, in the decimal system, the complement of 2 is 8. Complement numbers are used in some computer systems to facilitate arithmetic operations.

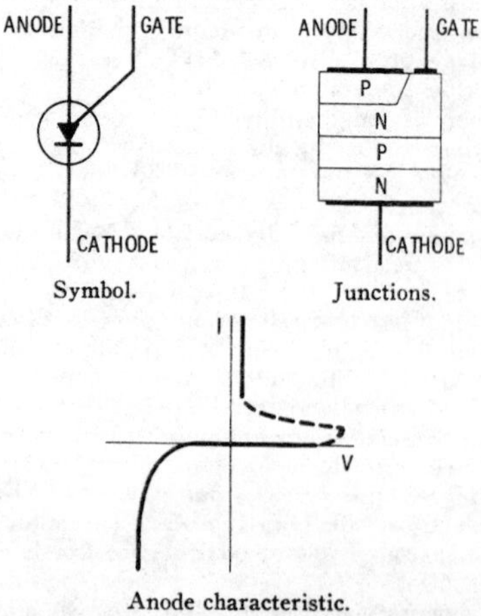

Symbol. Junctions.

Anode characteristic.

Complementary silicon controlled rectifier.

complement number system—A system of number handling in which operations are performed on the complement of the actual number. The system is used in some computers to facilitate arithmetic operations.

complementor—A circuit or device that produces a Boolean complement. A NOT circuit.

complete carry—A system of executing the carry process in a computer. All carries, and any other carries to which they give rise, are allowed to propagate to completion in this system.

complex components—Indivisible and nonrepairable components having more than one function.

complex function—An integrated device in which three or more circuits are integral to a single silicon chip. In addition, the circuits are interconnected on the chip itself to form some electronic function at a higher level of organization than a single circuit. The interconnection pattern for the function is predetermined by the fixed mask; no wiring discretion is available for yield purposes. The input and outputs of all the circuits are not normally exposed to the package terminals. An example of a complex function would be a full adder or a multibit serial shift register.

complex parallel permeability — The complex relative permeability measured under stated conditions on a core with the aid of a coil. The parameter characterizing the induction is the impedance of the coil when placed on the core, expressed as a parallel connection of reactance and resistance. The parameter characterizing the field strength is the reactance the coil would have if placed on a core of the same dimensions but with unity relative permeability, the distribution of the magnetic field being identical in both cases. The coil should have negligible copper losses.

complex permeability—Under stated conditions, the complex quotient of the moduli of the parallel vectors representing induction and field strength in a material. One of the moduli varies sinusoidally with time, and the component chosen from the other modulus varies sinusoidally at the same frequency.

complex series permeability—The complex relative permeability measured under stated conditions on a core with the aid of a coil. The parameter characterizing the induction in the core is the impedance of the coil when placed on the core, expressed as a series connection of reactance and resistance. The parameter characterizing the field strength is the reactance this coil would have if placed on a core of the same dimensions but with unity relative permeability. The coil should have negligible copper losses, etc.

complex steady-state vibration—A periodic vibration of more than one sinusoid. It includes repeating square waves, sawtooth waves, etc., because these waveforms can be expressed in terms of a Fourier series of sinusoidal terms.

complex target—A radar target made up of a number of reflecting surfaces that, taken together, are smaller in all dimensions than the resolution capability of the radar.

complex tone—A sound wave produced by the combination of simple sinusoidal components of different frequencies. A sound sensation characterized by more than one pitch.

complex wave—A periodic wave made up of a combination of several frequencies or several sine waves superimposed on one another.

complex-wave generator — A device which generates a nonsinusoidal signal having a desired repetitive characteristic and waveform.

compliance—1. The reciprocal of stiffness—i.e., the ability to yield or flex. 2. In a phono cartridge an indication of how easily the stylus can be deflected. It is the amount of lateral or vertical movement per unit of applied force (cm/dyne). In general, the higher the compliance the better is the low-frequency tracking at a given tracking force. Compliance can be measured in several ways—static and dynamic. 3. The mechanical and acoustical equivalent of capacitance. 4. The flexibility of a speaker cone's suspension. High compliance is important in a woofer for accurate reproduction of low-frequency signals of large amplitude.

compliance extension—A form of master/slave interconnection of two or more current-regulated supplies to increase their output-voltage range through series connection.

compliance range — The range of voltage needed to sustain a given constant current throughout a range of load resistances.

compliance voltage—The output voltage of a constant-current supply.

compliance voltage range—The output voltage range of a dc power supply operating in a constant-current mode.

component—1. An essential functional part of a subsystem or equipment. It may be any self-contained element with a specific function, or it may consist of a combination of parts, assemblies, accessories, and attachments. 2. In vector analysis, one of the parts of a wave, voltage, or current considered separately. 3. A self-contained element of a complete operating equipment which performs a function necessary to operation of the equipment. Normally interchangeable with "unit." 4. In high fidelity, a specialized item of equipment designed to do a particular part of the work in a sound system. 5. Any of the basic parts used in building electronic equipment, such as a resistor or capacitor, etc.

component density—The number of components contained in a given volume or within a given package or chip.

component layout — The physical arrangement of the components in a chassis or printed circuit.

component operating hours—A unit of measurement for the period of successful operation of one or more components (of a specified type) which have endured a given set of environmental conditions.

component population — The variety and number of components (transistors, resistors, transformers, etc.) necessary to perform the desired electrical function.

component stress—Those factors of usage or test, such as voltage, power, temperature, frequency, etc., which tend to affect the failure rate of component parts.

composite cable—A cable in which conductors of different gages or types are combined under one sheath.

composite circuit—A circuit which can be used simultaneously for telephony and direct-current telegraphy or signaling, the two being separated by frequency discrimination.

composite color signal — The color-picture signal plus all blanking and synchronizing signals. Includes luminance and chrominance signals, vertical- and horizontal-sync pulses, vertical- and horizontal-blanking pulses, and the color-burst signal.

composite color sync—The signal comprising all the sync signals necessary for proper operation of a color receiver. Includes the deflection sync signals to which the color sync signal is added in the proper time relationship.

composite conductor—Two or more strands of different metals, such as aluminum and steel or copper and steel, assembled and operated in parallel.

composite controlling voltage—The voltage of the anode of an equivalent diode, combining the effects of all individual electrode voltages in establishing the space-charge limited current.

composited circuit—A circuit which can be used simultaneously for telephony and for direct-current telegraphy or signaling, separation between the two being accomplished by frequency discrimination.

composite dialing—In telephone operations, a method of dialing between distant offices over one leg of a composite set.

composite filter—A filter with two or more sections.

composite guidance system — A guidance system using a combination of more than one individual guidance system.

composite picture signal — The television signal produced by combining a blanked picture signal with the sync signal.

composite video signal — The complete video signal. For monochrome, it consists of the picture signal and the blanking and

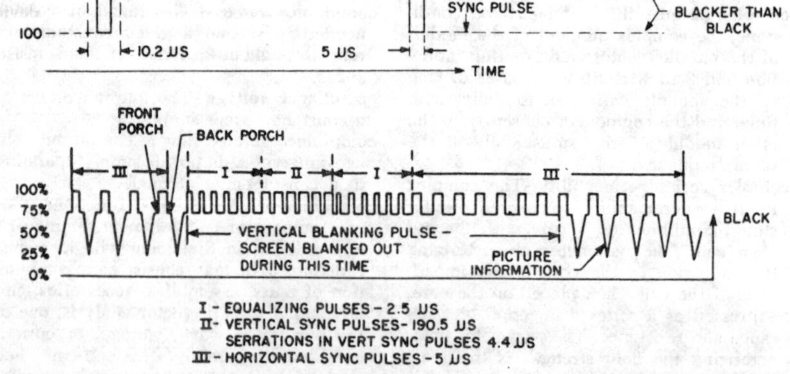

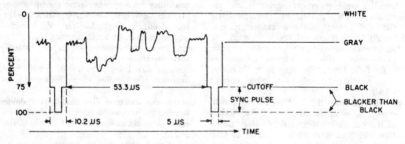

Composite picture signal.

synchronizing signals. For color, color-synchronizing signals and color-picture information are added.

composite wave filter—A combination filter consisting of two or more low-pass, high-pass, bandpass, or band-elimination filters.

composition resistor—*See* Carbon Resistor.

compound-connected transistor—Two transistors which are combined to increase the current amplification factor at high emitter currents. This combination is generally employed in power-amplifier circuits.

compound-filled transformer—A transformer which is contained within a case and in which the structural insulating material is supplemented by submergence in a solid or semisolid insulating material introduced into the case in a fluid state.

compound horn — An electromagnetic horn of rectangular cross section. The four sides of the horn diverge in such a way that they coincide with or approach four planes, with the provision that the two opposite planes do not intersect the remaining planes.

compound modulation—*See* Multiple Modulation.

compound-wound motor—A dc motor having two separate field windings. One, usually the predominant field, is connected in parallel with the armature circuit and the other is connected in series.

compress—To reduce some parameter of a signal as bandwidth, amplitude variation, duration, etc., while preserving its information content.

compressed-air loudspeaker—A loudspeaker that has an electrically actuated valve to modulate a stream of compressed air.

compression—1. A process in which the effective amplification of a signal is varied as a function of the signal magnitude, the effective gain being greater for small than for large signals. In television, the reduction in gain at one level of a picture signal with respect to the gain at another level of the same signal. 2. Electronic reduction of the dynamic range so that quiet sounds are raised and loud sounds lowered. The most common application is an "automatic" recording where it is important that all sounds recorded are made intelligible when played back. Also used where necessary to avoid overrecording and distortion, or to lift the signal level clear of background noise or hum.

compressional wave—In an elastic medium, a wave which causes a change in volume of an element of the medium without rotation of the element.

compression driver unit—A speaker driver unit that does not radiate directly from the vibrating surface. Instead, it requires acoustic loading from a horn which connects through a small throat to an air space adjacent to the diaphragm.

compression ratio—The ratio between the magnitude of the gain (or amplification) at a reference signal level and its magnitude at a higher stated signal level.

compressor — A transducer which, for a given amplitude range of input voltages, produces a smaller range of output voltages. In one important type of compressor, the envelope of speech signals is used to reduce their volume range.

compromise network—In a telephone system, a network used in conjunction with a hybrid coil to balance a subscriber's loop. The network is adjusted for an average loop length, an average subscriber's set, or both, and gives compromise (not precision) isolation between the two directional paths of the hybrid coil.

Compton diffusion — An elastic shock between a photon and an electron. The photon is diffused with a lesser energy and the electron acquires a kinetic energy equal to the energy decrease of the photons.

Compton effect—The elastic scattering of photons by electrons. Because the total energy and momentum are conserved in the collisions, the wavelength of the scattered radiation undergoes a change that depends on the scattering angle.

computational stability — The degree to which a computational process remains valid when subjected to such effects as errors or malfunctions.

computer—Any device capable of accepting information, applying prescribed processes to the information, and supplying the results of these processes; sometimes, more specifically, a device for performing sequences of arithmetic and logical operations; sometimes, still more specifically, a stored-program digital computer capable of performing sequences of internally stored instructions, as opposed to calculators on which the sequence is impressed manually (desk calculator) or from tape or cards (card programmed calculator).

computer access device input — A device that automatically routes to the computer all teletypewriter observation reports that are received in a standard format.

computer-aided design—*See* CAD.

computer code—Also called machine language. The code by which data are represented within a computer system. An example is a binary-coded decimal.

computer control — The parts of a digital computer that have to do with the carrying out of instructions in the proper sequence, the interpretation of each instruction, and the application of signals to the arithmetic unit and other parts in accordance with this interpretation.

computer control counter — 1. A counter that stores the next required address.

2. Any counter that provides information to the control unit.

computer diagnosis—The use of data processing systems for evaluation of raw data.

computer entry punch—A combination card reader and key punch used to enter data directly onto the memory drum of a computer.

computer interface—Peripheral equipment for attaching computer to scientific or medical instruments.

computer language—The method or technique used to instruct a computer to perform various operations. See machine language and high-level language.

computer-output-microfilm — Abbreviated com. A microfilm printer that will take output directly from the computer thus substituting for line printer or tape output.

computer-limited—Having to do with the condition in which the time required for computation is greater than the time required to read inputs and write outputs.

computer program—A series of instructions or statements prepared in a form acceptable to the computer, the purpose of which is to achieve a certain result. *See* Software.

computer terminal — Peripheral computer equipment for entering and retrieving data. Sometimes incorporates cathode-ray tube for display.

computer utility — A network of central computers linked through data communications facilities to remote terminal systems.

computer word—A sequence of bits or characters that is treated as a unit and that can be stored in one computer location. Same as machine word.

computing—Performing basic and more involved mathematical processes of comparing, adding, subtracting, multiplying, dividing, integrating, etc.

computing machine—An automatic device which carries out well defined mathematical operations.

concatenate—To link together or unite in a series.

concatenation — The joining or linking of sets or series.

concave—Curved inward.

concentrated-arc lamp—A type of low-voltage arc lamp having nonvaporizing electrodes sealed in an atmosphere of inert gas and producing a small, brilliant, incandescent cathode spot.

concentration gradient—A difference in carrier concentration (holes or free electrons) from point to point in a semiconductor.

concentrator—1. A device that feeds the signals from several data terminals into a single transmission line for input to a computer, or vice versa. 2. An analog or digital buffer switch used to reduce the required number of trunks. 3. A device for combining many low-speed data lines into one high-speed data line. 4. A device that uses hardware and software to perform computer communication functions. (A term first applied to telephone-switching systems which permitted greater economy in use of facilities by combining many phone circuits into one.)

concentric cable—*See* Coaxial Line.

concentric groove—*See* Locked Groove.

concentric-lay conductor—A conductor composed of a central core surrounded by one or more layers of helically laid wires. In the most common type of concentric-lay conductors, all wires are of the same size, and the central core is a single wire.

concentric line—*See* Coaxial Line.

concentric stranding—A method of stranding wire in which the final wire is built up in layers such that the inner diameter of a succeeding layer always equals the outer diameter of the underlying layer.

concentric-wound coil—A coil with two or more insulated windings which are wound one over the other.

concurrent processing — The ability of a computer to work on more than one program at the same time.

condensed mercury temperature—The temperature of a mercury-vapor tube, measured on the outside of the tube envelope, in the region where the mercury is condensing in a glass tube or at a designated point on a metal tube.

condenser—Obsolete term for capacitor.

condenser antenna—*See* Capacitor Antenna.

condenser microphone — *See* Electrostatic Microphone.

condenser speaker — *See* Electrostatic Speaker.

condenser tissue — Kraft paper of 0.002 inch or less nominal thickness used in the manufacture of capacitors with paper or paper/film dielectrics.

conditional—1. In a computer, subject to the result of a comparison made during computation. 2. Subject to human intervention.

conditional breakpoint instruction—A conditional jump instruction that causes a computer to stop if a specified switch is set. The routine then may be allowed to proceed as coded, or a jump may be forced.

conditional jump—Also called conditional transfer of control. An instruction to a computer which will cause the proper one of two (or more) addresses to be used in obtaining the next instruction, depending on some property of one or more numerical expressions or other conditions.

conditional transfer of control—*See* Conditional Jump.

condition code—In a computer, a limited group of program conditions such as carry, borrow, overflow, etc. which are pertinent to the execution of instructions. The codes are contained in a condition codes register.

conditioning — 1. Equipment modifications or adjustments required to provide matching of transmission levels and impedances or to provide equalization between facilities. 2. The addition of equipment to a leased voice-grade channel to provide minimum line characteristics necessary for data transmission.

Condor—A cw navigational system, similar to Benito, which automatically measures bearing and distance from a single ground station and displays them on a cathode-ray indicator. The distance is determined by phase comparison, and the bearing, by automatic direction finding.

conductance—Symbolized by G or g. 1. In an element device, branch, network, or system, the physical property that is the factor by which the square of an instantaneous voltage must be multiplied to give the corresponding energy lost by dissipation as heat or other permanent radiation, or by loss of electromagnetic energy from the circuit. 2. The real part of admittance. 3. Reciprocal of resistance and measured in mhos.

conducted heat—Thermal energy transferred by thermal conduction.

conducted interference—Any unwanted electrical signal conducted on the power lines supplying the equipment under test, or on lines supplying other equipment to which the one under test is connected.

conducted signals — Electromagnetic or acoustic signals propagated along wirelines or other conductors.

conductimeter—*See* Conductivity Meter.

conduction—The transmission of heat or electricity through or by means of a conductor.

conduction band—A partially filled energy band in which electrons can move freely, allowing the material to carry an electric current (with electrons as the charge carriers).

conduction current—The power flow parallel to the direction of propagation, expressed in mhos/meter.

conduction-current modulation — 1. Periodic variations in the conduction current passing a point in a microwave tube. 2. The process of producing such variations.

conduction electrons—The electrons which are free to move under the influence of an electric field in the conduction band of a solid.

conduction error—The error in a temperature transducer due to heat conduction between the sensing element and the mounting to the transducer.

conduction field—Energy which surrounds a conductor when an electric current is passed through the conductor, and which, because of the difference in phase between the electrical field and magnetic field set up in the conductor, cannot be detached from the conductor.

conductive gasket—A special highly resilient gasket used to reduce rf leakage in shielding which has one or more access openings.

conductive material—A material in which a relatively large conduction current flows when a potential is applied between any two points on or in a body constructed from the material. Metals and strong electrolytes are examples of conductors.

conductive pattern—The arrangement or design of the conductive lines on a printed-circuit board.

conductivity—1. The conductance between opposite faces of a unit cube of material. The volume conductivity of a material is the reciprocal of the volume resistivity. 2. The ability of a material to conduct current. The reciprocal of resistivity. 3. The ability to conduct or transmit heat or electricity.

conductivity meter—Also called conductimeter. An instrument that measures and/or records electrical conductivity.

conductivity modulation—The change in conductivity of a semiconductor as the charge-carrier density is varied.

conductivity-modulation transistor — A transistor in which the active properties are derived from minority-carrier modulation of the bulk of resistivity of the semiconductor.

conductor—1. A bare or insulated wire or combination of wires not insulated from one another, suitable for carrying an electric current. 2. A body of conductive material so constructed that it will serve as a carrier of electric current. 3. A material (usually a metal) that conducts electricity through the transfer of orbital electrons. 4. A material, such as copper or aluminum, which offers low resistance or opposition to the flow of electric current. 5. A medium for transmitting electrical current. A conductor usually consists of copper, aluminum, steel, silver or other material. 6. A solid, liquid or gas which offers little opposition to the continuous flow of electric current.

conduit—A tubular raceway for holding wires or cables designed and used expressly for this purpose.

conduit wiring—Wiring carried in conduits and conduit fittings.

cone—The diaphragm that sets the air in motion to create a sound wave in a direct-radiator loudspeaker. Usually it is conical in shape.

cone breakup—The inability of a speaker cone to work as a piston at high frequencies, the effect being that the cone is not under the complete control of the voice coil, certain parts of it moving in opposition to other parts like a "rippled rope." Responsible for uneven frequency response.

cone of nulls—A conical surface formed by directions of negligible radiation.

cone of silence—An inverted cone-shaped space directly over the aerial towers of some radio beacons. Within the cone, signals cannot be heard or will be greatly reduced in volume.

confetti—Flecks or steaks of color caused by tube noise in the chrominance amplifier. Because of its colors, confetti is much more noticeable than snow in a black-and-white picture. The chrominance amplifier is therefore cut off during a monochome program.

confidence—1. The likelihood, expressed in percent, that a statement is true. 2. The degree of assurance that the stated failure rate has not been exceeded.

confidence factor—The percentage figure expressing confidence level.

confidence interval—A range of values believed to include, with a preassigned degree of confidence, the true characteristic of the lot.

confidence level—1. The probability (expressed as a percentage) that a given assertion is true or that it lies within certain limits calculated from the data. 2. A degree of certainty.

confidence limits—Extremes of a confidence interval within which there is a designated chance that the true value is included.

configuration—1. The relative arrangement of parts (or components) in a circuit. 2. A listing of the names and/or serial numbers of the assemblies that make up an equipment.

confocal resonator—A wavemeter for millimeter wavelengths. It consists of two spherical mirrors that face each other; a change in the spacing between the mirrors affects the propagation of electromagnetic energy between them, making possible direct measurement of free-space wavelengths.

conformance error—The deviation of a calibration curve from a specified curve line.

confusion jamming—An electronic countermeasure by means of which a radar may detect a target, but the radar operator is denied accurate data regarding range, azimuth, and velocity of the target. This result is accomplished through amplification and retransmission of an incident radar signal with distortion to create a false echo. Also called deception jamming.

confusion reflector—A device that reflects electromagnetic radiation to create echoes for purposes of causing confusion of radar, guided missiles, and proximity fuses.

congestion—A condition in which the number of calls arriving at the various inputs of a communications network are too many for the network to handle at once and are subject to delay or loss. (The concept applies in an analogous way to any system in which arriving "traffic" can exceed the number of "servers.")

conical horn—A horn the cross-sectional area of which increases as the square of the axial length.

conical scanning—A form of scanning in which the beam of a radar unit describes a cone, the axis of which coincides with that of the reflector.

conjugate—Either of a pair of complex numbers that are mutually related in that their real parts are identical and the imaginary part of one is the negative of the imaginary part of the other, that is, if $a = x + iy$, then $a = x - iy$ is its conjugate.

conjugate branches—Any two branches of a network in which a driving force impressed on one branch does not produce a response in the other.

conjugate bridge—A bridge in which the detector circuit and the supply circuits are interchanged, compared with a normal bridge.

conjugate impedance—An impedance the value of which is the conjugate of a given impedance. For an impedance associated with an electric network, the conjugate is an impedance with the same resistance component and a negative reactive component of the original.

connected—A network is connected if, between every pair of nodes of the network, there exists at least one path composed of branches of the network.

connection—1. The attachment of two or more component parts so that conduction can take place between them. 2. The point of such attachment.

connection diagram—A diagram showing the electrical connections between the parts that make up an apparatus.

connector—1. A coupling device which provides an electrical and/or mechanical junction between two cables, or between a cable and a chassis or enclosure. 2. A device that provides rapid connection and disconnection of electrical cable and wire terminations. 3. A plug or receptacle which can be easily joined to or separated from its mate. Multiple contact connectors join two or more conductors with others in one mechanical assembly.

connector assembly—The combination of a mated plug and receptacle.

connector flange—A projection that extends from or around the periphery of a connector and incorporates provisions for mounting the connector to a panel.

connector receptacle—1. An electrical fitting with contacts constructed to be electrically connected to a cable, coaxial line, cord, or wire to join with another electrical connector mounted on a bulkhead, wall, chassis, or panel.

connect time—The total time required for establishing a connection between two points.

conoscope—An instrument for determining the optical axis of a quartz crystal.

consequent poles — Additional magnetic poles present at other than the ends of a magnetic material.

consol—*See* Sonne.

console—1. A cabinet for a radio or television receiver that stands on the floor rather than on a table. 2. Main operating unit in which indicators and general controls of a radar or electronic group are installed. 3. A part of a computer that may be used for manual control of the machine. 4. An array of controls and indicators for the monitoring and control of a particular sequence of actions, as in the checkout of a rocket, a countdown, or a launch.

console operator—A person who monitors and controls an electronic computer by means of a central control unit or console.

consonance—Electrical or acoustical resonance between bodies or circuits not connected directly together.

constant—An unvarying or fixed value or data item.

constant amplitude recording—In disc recording, a relationship between the modulations in the groove and the electrical signals making them so that the width of the groove (the excursions of the cutting stylus) is proportional to the amplitude, or power, of the signal. In playback, a similar relation between the record and the motion of the stylus so that the cartridge produces equal voltages regardless of frequency. Crystal and ceramic pickups have a constant amplitude characteristic on playback.

constant current—1. A current that does not undergo a change greater than the required precision of the measurement when the impedance of the generator is halved. 2. Having to do with a type of power-supply operation in which the output current remains at a preset value (within specified limits) while the load resistance varies, resulting in an output-voltage variation within the voltage range of the power supply.

constant-current characteristic—The relationship between the voltages of two electrodes, the current to one of them as well as all other voltages being maintained constant.

constant-current/constant-voltage supply—A power supply that behaves as a constant-voltage source for relatively large values of load resistance and as a constant-current source for relatively small values of load resistance. The crossover point between these two modes of operation occurs when the value of the critical load resistance equals the value of the supply voltage setting divided by the supply current setting.

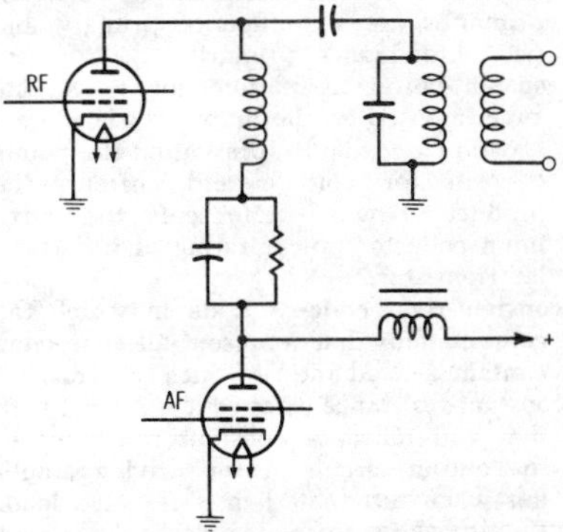

Constant current modulation.

constant-current modulation — Also called Heising modulation. A system of amplitude modulation in which the output circuits of the signal amplifier and carrier-wave generator or amplifier are directly and conductively coupled by a common inductor. The inductor has an ideally infinite impedance to the signal frequencies and therefore maintains the common plate-supply current of the two devices constant. The signal-frequency voltage thus appearing across the common inductor also modulates the plate supply to the carrier generator or amplifier, with corresponding modulation of the carrier output.

constant-current power supply — A regulated power supply which acts to keep its output current constant in spite of changes in load, line, or temperature. Thus, for a change in load resistance, the output current remains constant to a first approximation, while the output voltage changes by whatever amount necessary to accomplish this.

constant-current transformer — A transformer that automatically maintains a constant current in its secondary circuit under varying conditions of load impedance when supplied from a constant-potential source.

constant-delay discriminator – *See* Pulse Demoder.

constant-K filter–An image-parameter filter comprising a tandem connection of a number of identical prototype L-section filters. Each adjacent pair of L-sections together forms either a T- or π-network. The product of the series and shunt impedances is a constant that is independent of frequency.

constant-K network–A ladder network in which the product of its series and shunt impedances is independent of frequency within the range of interest.

constant-luminance transmission – A type of transmission in which the transmission primaries are a luminance primary and two chrominance primaries.

constant-power-dissipation line – A line superimposed on the output static characteristic curves and representing the points of collector voltage and current, the product of which represents the maximum collector power rating of a particular transistor.

constant-ratio code–A code in which the combinations that represent all characters contain a fixed ratio of ones to zeros.

constant-resistance network – A network that will reflect a constant resistance to the output circuit of the driving amplifier when terminated in a resistive load. Loudspeakers do not reflect a constant impedance; therefore, an amplifier does not see a constant resistance. This disadvantage may be somewhat compensated for by the use of negative feedback by the amplifier.

constant velocity recording – In disc recording, a relationship between the wiggles in the groove and the electrical signals making them, whereby the frequency of the signal determines the degree of excursion of the cutter. In playback, a similar relation between the recorded wiggles and the motion of the stylus so that the cartridge produces voltages that vary in strength, or amplitude, as the frequency in the groove varies. Magnetic cartridges have a constant velocity characteristic and must be equalized by special networks during playback.

constant voltage–1. Voltage that does not undergo a change greater than the required precision of the measurement when the impedance of the generator is doubled. 2. Having to do with a type of power-supply operation in which the output voltage remains at a preset value (within specified limits) while the output current is varied within the range of the power supply.

constant-voltage/constant-current crossover – Behavior of a power supply in which there is automatic conversion from voltage stabilization to current stabilization (and vice versa) when the output current reaches a preset value.

constant-voltage/constant-current (cv/cc) output characteristic–A regulated power supply which acts as a constant-voltage source for comparatively large load resistances and as a constant-current source for comparatively small load resistances.

constant-voltage power supply – A regulated power supply which acts to keep its output voltage constant in spite of changes in load, line, or temperature. Thus, for a change in load resistance, the output voltage of this type of supply remains constant to a first approximation, while the output current changes by whatever amount necessary to accomplish this. 2. A power supply capable of maintaining a fixed voltage across a variable load resistance and over a defined input voltage and frequency change. The output is automatically controlled to maintain constant the product of output current times load resistance.

constant-voltage transformer – A transformer delivering a fixed predetermined voltage over a limited range of input voltage variations (e.g. 95-125 volts).

consumer's reliability risk–The risk to the consumer that a product will be accepted by a reliability test when the reliability of the product is actually below the value specified for rejection.

contact–1. One of the current-carrying parts of a relay, switch, or connector that are engaged or disengaged to open or close the associated electrical circuits. 2. To join two conductors or conducting objects in order to provide a complete path for current flow. 3. The juncture point to provide the complete path.

contact-actuation time–The time required for any specified contact on the relay to function. When not otherwise specified, it is the initial actuation time. For some purposes, it is preferable to use either the final or effective actuation time.

contact arc – The electrical (current) discharge that occurs between mating contacts when the circuit is being disestablished.

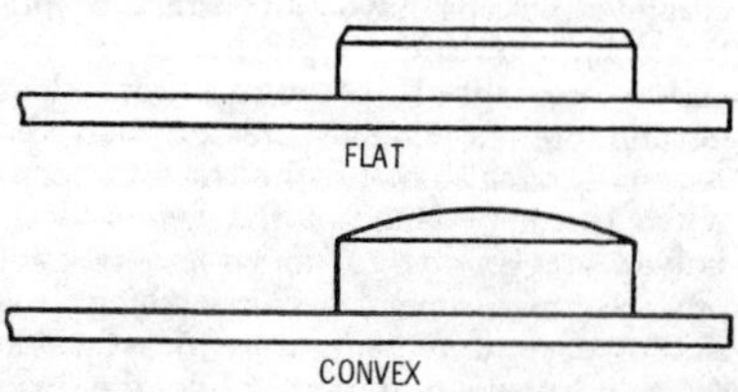

Contacts, 1 (side view).

contact area—The common area between two conductors or a conductor and a connector through which the flow of electricity takes place.

contact arrangement—1. The combination of contact forms that make up the entire relay-switching structure. 2. The number, spacing, and positioning of contacts in a connector.

contact bounce—1. The uncontrolled making and breaking of contact when the relay contacts are closed. 2. Internally caused intermittent and undesired opening of closed contacts, of a relay, caused by one or more of the following:

(a) Impingement of mating contacts.
(b) Impact of the armature against the coil core on pickup or against the backstop on dropout.
(c) Momentary hesitation, or reversal, of the armature motion during the pickup or dropout stroke.

contact chatter—The undesired vibration of mating contacts during which there may or may not be actual physical contact opening. If there is no actual opening but only a change in resistance, it is referred to as dynamic resistance, and it appears as "grass" on the screen of an oscilloscope having adequate sensitivity and resolution.

contact combination—1. The total assembly of contacts on a relay. 2. Sometimes used for contact form.

contact emf—A small voltage established whenever two conductors of different materials are brought into contact.

contact follow — The displacement of a stated point on the contact-actuating member following the initial closure of a contact.

contact force—1. The amount of force exerted by one of a pair of closed contacts on the other. 2. The force exerted by the moving mercury on a stationary contact or electrode in a mercury switch.

contact gap—Also called contact separation. The distance between a pair of mating relay contacts when they are open.

contact length—The length of travel of one contact while touching another contact during the assembly or disassembly of a connector.

contact load — The electrical power demands encountered by a contact set in any particular application.

contact microphone — A microphone designed to pick up mechanical vibrations directly from the sound source and convert them into corresponding electrical currents or voltages.

contact miss—Failure of a contact mating pair to establish the intended circuit electrically. This may be a contact resistance in excess of a specified maximum value.

contact modulator—A switch used to produce modified square waves having the same frequency as, and a definite phase relationship to, a driving sine wave. Also called electromechanical chopper.

contact noise—The random fluctuation of voltage across a junction through which current is flowing from one solid to another.

contactor—1. A device for the purpose of repeatedly establishing or interrupting an electric power circuit. 2. A heavy duty relay used to control electrical circuits.

contactor alarm—A signal calling attention to lowered pressure in a cable gas-pressure system.

contact potential—1. Also called Volta effect. The difference of potential that exists when two dissimilar, uncharged metals are placed in contact. One becomes positively charged and the other negatively charged, the amount of potential depending on the nature of the metals. 2. The potential difference between the contacting surfaces of two metals that have different work functions.

contact-potential difference — The difference between the work functions of two materials, divided by the electronic charge generated by them.

contact pressure—The amount of pressure holding a set of contacts together.

contact rating — The electrical power-handling capability of relay or switch contacts under specified environmental conditions and for a prescribed number of operations.

contact rectifier—A rectifier consisting of two different solids in contact. Rectification is due to the greater conductivity across the contact in one direction than in the other.

contact resistance—1. Total electrical resistance of a contact system, such as the resistance of a relay or a switch measured at the terminals. Usually this resistance is only a fraction of an ohm. 2. The ohmic resistance between the contacts of a switch connector or relay. It may be an extremely small value—typically in the milliohm range. Contact resistance is normally measured from terminal to terminal.

contact retainer—A device used to retain a contact in an insert or body; it may be either on the contact or in the insert.

contact retention—The minimum axial load a contact in a connector can withstand in either direction while remaining firmly fixed in its normal position in the insert.

contacts—In a relay, the current-carrying parts that engage or disengage to open or close electrical circuits.

contact separation—The maximum distance between the stationary and movable contacts when the circuit is broken.

contact spring — 1. A current-carrying spring to which the contacts are fastened. 2. A non-current-carrying spring that positions and tensions a contact-carrying member.

contact wetting—The coating of a contact surface with an adherent film of mercury.

contact wipe—The distance of travel (electrical engagement) of one contact during its engagement with or separation from another or during mating or unmating of the connector halves.

contaminated—Made radioactive by addition of a radioactive material.

content-addressed storage—*See* Associative Storage.

content indicator—A display device that indicates the content in a computer, and the program or mode in use.

contention—1. A condition that occurs on a multidrop communication channel when two or more locations attempt to transmit simultaneously. 2. Unregulated bidding for a line by multiple users.

contents—The information stored in any part of the computer memory.

Continental code—*See* International Morse Code.

continuity—1. A continuous path for the flow of current in an electric circuit. 2. In radio broadcasting, the prepared copy from which the spoken material is presented.

continuity test—An electrical test for determining whether a connection is broken.

continuity writer—In radio broadcasting, the person who writes the copy from which the spoken material is presented.

continuous carrier—A carrier over which transmission of information is accomplished by means which do not interrupt the carrier.

continuous commercial service—*See* CCS.

continuous data—Any set of data the information content of which can be ascertained continuously in time.

continuous-duty rating—The rating applied to equipment if operated for an indefinite length of time.

continuous load—A load where the maximum current is expected to continue for three hours or more.

continuously loaded cable — A submarine cable in which the conductors are continuously loaded.

continuous output power — The maximum power (in watts) that an amplifier will deliver from each channel (with all channels operating) without exceeding its rated harmonic distortion. Measured with a 1-kHz signal. Power ratings should include harmonic distortion unit, the load impedance (4, 8, or 16 ohms). For example: Continuous output power 40 W/40 W (at less than 1% harmonic distortion, into 8 ohm load).

continuous power—The power an amplifier is capable of delivering for at least 30 seconds with a sine-wave signal.

continuous power output—*See* Rated Output Power.

continuous rating—The rating that defines the load which can be carried for an indefinite length of time.

continuous recorder — A recorder that makes its record on a continuous sheet or web rather than on individual sheets.

continuous scan thermograph — Equipment for presenting a continuous scan image of thermal pattern (thermogram) of a patient or an object on a cathode-ray tube.

continuous spectrum—The spectrum which exhibits no structure and appears to represent a continuous variation of wavelength from one to the other.

continuous variable—A variable that may assume any value within a defined range.

continuous-wave radar—A system in which a transmitter sends out a continuous flow of radio energy to the target, which reradiates (scatters) the energy intercepted and returns a small fraction to a receiving antenna.

continuous waves — Abbreviated cw. Electromagnetic waves generated as a continuous train of identical oscillations. They can be interrupted according to a code, or modulated in amplitude, frequency, or phase in order to convey information.

continuous-wave tracking system—A tracking system which operates by keeping a continuous radio beam on a target and determining its behavior from changes in the antenna necessary to keep the beam on the target.

contour control system—In automatic control of machine tools, a system in which the cutting path of a tool is controlled along two or more axes.

contourograph—A device in which a cathode-ray oscilloscope is used to produce images that have a three-dimensional appearance.

contrahelical—In the wire and cable industry the term is used to mean the direction of a layer with respect to the previous layer. Thus it would mean a layer spiraling in an opposite direction from the preceding layer within a wire or cable.

CONTRAN—A computer-programming language in which instructions are written at a compiler level, thereby eliminating the need for translation by a compiling routine.

contrast—1. The actual difference in density between the highlights and the shadows. Contrast is not concerned with the magnitude of density, but only with the difference in densities. 2. Amplitude ratio between picture white and picture black. 3. Ratio between the maximum and minimum brightness values in a picture. 4. In optical character recognition, the differences between the color or shading of the printed material on a document and the background on which it is printed.

contrast control—A method of adjusting the contrast in a television picture by changing the amplitude of the video signal.

contrast range – The ratio between the whitest and blackest portions of a television image.

contrast ratio—Ratio of the maximum to the minimum luminance values in a television picture or a portion thereof.

control—Also called a control circuit. 1. In a digital computer, those parts which carry out the instructions in proper sequence, interpret each instruction, and apply the proper signals to the arithmetic unit and other parts in accordance with the interpretation. 2. Sometimes called a manual control. In any mechanism, one or more components responsible for interpreting and carrying out manually initiated directions. 3. In some business applications, a mathematical check. 4. In electronics, a potentiometer or variable resistor. 5. In an alarm system, any mechanism which sequences the interrogation of protected site units, resets latched alarms and performs similar functions.

control ampere turns—The magnitude and polarity of the control magnetomotive force required for operation of a magnetic amplifier at a specified output.

control block – A storage area through which information of a particular type required for control of the operating system is communicated among the parts of the system.

control card—In computer programming, a card containing input data or parameters for a specific application of a general routine.

control characteristic—1. A pilot of the load current of a magnetic amplifier as a function of the control ampere turns for various loads and at the rated supply voltage and frequency. 2. The relationship, usually shown by a graph, between the critical grid voltage and the anode voltage of a tube.

control circuit—*See* Control.

control circuits—In a digital computer, the circuits which carry out the instruction in proper sequence, interpret each instruction, and apply the proper commands to the arithmetic element and other circuits in accordance with the interpretation.

control-circuit transformer – A voltage transformer utilized to supply a voltage suitable for the operation of control devices.

control-circuit voltage – The voltage provided for the operation of shunt-coil magnetic devices.

control compartment—A space within the base, frame, or column of a machine used for mounting the control panel.

control counter—In a computer, a device which records the storage location of the instruction word to be operated on following the instruction word in current use.

control data—In a computer, one or more items of data used to control the identification, selection, execution, or modification of another routine, record file, operation, data value, etc.

control DATA or Control Data – A trademark and service mark of Control Data Corporation in respect to data processing equipment and related services.

control electrode—An electrode on which a voltage is impressed to vary the current flowing between other electrodes.

control field—In a sequence of similar items of computer information, a constant location where control information is placed.

control grid—The electrode of a vacuum tube, other than a diode, upon which a signal voltage is impressed to regulate the plate current.

control-grid bias—The average direct-current voltage between the control grid and cathode of a vacuum tube.

control-grid plate transconductance – The ratio of the amplification factor of a vacuum tube to its plate resistance, combining the effect of both into one term.

controlled avalanche—A predictable, nondestructive avalanche characteristic designed into a semiconductor device as protection against reverse transients that exceed its ratings.

controlled avalanche device – A semiconductor device that has very specific maximum and minimum avalanche-voltage characteristics and is also able to operate and absorb momentary power surges in this avalanche region indefinitely without damage.

controlled avalanche silicon rectifier – A silicon diode manufactured with characteristics such that, when operating, it is not damaged by transient voltage peaks.

controlled-carrier modulation—Also called variable-carrier or floating-carrier modulation. A modulation system in which the carrier is amplitude-modulated by the signal frequencies, and also in accordance with the envelope of the signal, so that the modulation factor remains constant regardless of the amplitude of the signal.

controlled rectifier—1. A rectifier employing grid-controlled devices such as thyratrons or ignitrons to regulate its own output current. 2. Also called an SCR (silicon controlled rectifier). A four-layer pnpn semiconductor which functions like a grid-controlled thyratron.

controller—1. An instrument that holds a process or condition at a desired level or status as determined by comparison of the actual value with the desired value. 2. A device or group of devices, which serves to govern in some predetermined manner, the electric power delivered to the apparatus to which it is connected.

controller function—Regulation, acceleration, deceleration, starting, stopping, reversing, or protection of devices connected to an electric controller.

control locus—A curve which shows the critical value of grid bias for a thyratron.

control panel—A panel having a systematic arrangement of terminals used with removable wires to direct the operation of a computer or punched-card equipment.

control point—A point which may serve as a reference for all incremental commands.

control-power disconnecting device—A disconnective device such as a knife switch, circuit breaker, or pullout fuse block used for the purpose of connecting and disconnecting, respectively, the source of control power to and from the control bus or equipment.

control program—A computer program that places another program and its environment in core memory in proper sequence and retains them there until it has finished operating.

control ratio—1. The ratio of the change in anode voltage to the corresponding change in critical grid voltage of a gas tube, with all other operating conditions maintained constant. 2. Also called programming coefficient. The required range in control resistance of a regulated power supply to produce a 1-volt change in output voltage. Expressed in ohms per volt.

control read only memory — Abbreviated CROM. A major component in the control block of some microprocessors. It is a ROM which has been microprogrammed to decode control logic.

control rectifier—A silicon rectifier capable of switching or regulating the flow of a relatively large amount of power through the use of a very small electrical signal. These solid-state devices can take the place of mechanical and vacuum tube switches, relays, rheostats, variable transformers and other devices used for switching or regulating electric power.

control register—In a digital computer, the register that stores the current instruction governing the operation of the computer for a cycle. Also called instruction register.

control section—*See* Control Unit.

control sequence—In a computer, the normal order of execution of instructions.

control tape—In a computer, a paper or plastic tape used to control the carriage operation of some printing output devices. Also called carriage tape.

control unit—1. That section of an automatic digital computer that directs the sequence of operations, interprets coded instructions, and sends the proper signals to the other computer circuits to carry out the instructions. Also called control section. 2. A preamplifier unit in an audio setup. Signals from audio sources, i.e., tuner, pickup microphone, are fed into it. Equalization (where necessary) is applied, then the signal is fed to the main amplifier. Volume and tone controls are usually incorporated together with any necessary program-selection switch.

control-voltage winding—The motor winding which is excited by a varying voltage at a time phase difference from the voltage applied to the fixed voltage windings of a servomotor.

control winding—In a saturable reactor, the winding used for applying a controlling magnetomotive premagnetization force to the saturable-core material.

convection—1. The motion in a fluid as a result of differences in density and the action of gravity. 2. The transfer of heat from a high-temperature region in a gas or a liquid as a result of movement of masses of the fluid.

convection cooling — A method of heat transfer which depends on the natural upward movement of the air warmed by the heat dissipated from the device being cooled.

convection current—The amount of time required for a charge in an electron stream to be transported through a given surface.

convection-current modulation — 1. The time variation in the magnitude of the convection current passing through a surface. 2. The process of producing such a variation.

convenience receptacle—An assembly consisting of two or more stationary contacts mounted in a small insulating enclosure that has slots to permit blades on attachment plugs to enter and make contact with the circuit.

convergence—The condition in which the electron beams of a multibeam cathode-ray tube intersect at a specified point.

convergence coil—One of the two coils associated with an electromagnet, used to obtain dynamic beam convergence in a color-television receiver.

convergence control—A variable resistor in

the high-voltage section of a color-television receiver. It controls the voltage applied to the three-gun picture tube.

convergence electrode – An electrode the electric field of which causes two or more electron beams to converge.

convergence magnet – A magnet assembly the magnetic field of which causes two or more electron beams to converge.

convergence phase control–A variable resistor or inductance for adjusting the phase of the dynamic convergence voltage in a color-tv receiver employing a three-gun picture tube.

convergence surface – The surface generated by the point at which two or more electron beams intersect during the scanning process in a multibeam cathode-ray tube.

conversational mode–A type of communication between a terminal and a computer in which there is a response from the computer for each entry from the terminal, and vice versa.

conversational operation–A type of operation similar to the interactive mode, except that the computer user must wait until a question is posed by the computer before interacting.

conversational system–*See* Interactive System.

conversion–1. The process of changing from one data-processing method or system to another. 2. The process of changing from one form of representation to another. 3. *See* Encode, 2.

conversion efficiency–1. The ratio of ac output power to the dc input power to the electrodes of an electron tube. 2. The ratio of the output voltage of a converter at one frequency to the input voltage at some other frequency. 3. In a rectifier, the ratio of dc output power to ac input power. 4. The ratio of maximum available luminous or radiant flux output to total input power.

conversion gain–1. The ratio of the intermediate-frequency output voltage to the input-signal voltage of the first detector of a superheterodyne receiver. 2. The ratio of the available intermediate-frequency power output of a converter or mixer to the available radio-frequency power input.

conversion loss–The ratio of available input power to available output power under specified test conditions.

conversion time– 1. The length of time required by a computer to read out all the digits in a given coded word. 2. The time required for a complete measurement by an analog-to-digital converter. In successive-approximation converters it ranges typically from 0.8 microseconds to 400 microseconds.

conversion transconductance – The magnitude of the desired output-frequency component of current divided by the magnitude of the input-frequency component of voltage when the impedance of the output external termination is negligible for all frequencies which may affect the result.

conversion transducer–One in which the signal undergoes frequency conversion. The gain or loss is specified in terms of the useful signal.

conversion voltage gain (of a conversion transducer)–With the transducer inserted between the input-frequency generator and the output termination, the ratio of the magnitude of the output-frequency voltage across the output termination to the magnitude of the input-frequency voltage across the input termination of the transducer.

convert–1. To change information from one form to another without changing the meaning, e.g., from one number base to another. 2. In computer terminology, to translate data from one form of expression to a different form.

converted data–The output from a unit which changes the language of information from one form to another so as to make it available or acceptable to another machine, e.g., a unit which takes information punched on cards to information recorded on magnetic tape, possibly including editing facilities.

converter–1. In a superheterodyne radio receiver, the section which converts the desired incoming rf signal into a lower carrier frequency known as the intermediate frequency. 2. A rotating machine consisting of an electric motor driving an electric generator, used for changing alternating current to direct current. 3. A facsimile device that changes the type of modulation delivered by the scanner. 4. Generally called a remodulator. A facsimile device that changes amplitude modulation to audio-frequency–shift modulation. 5. Generally called a discriminator. A device that changes audio-frequency – shift modulation to amplitude modulation. 6. A conversion transducer in which the output frequency is the sum or difference of the input frequency and an integral multiple of the local-oscillator frequency. 7. A device that accepts an input that is a function of maximum voltage and time, and converts it to an output that is a function of maximum voltage only. 8. *See* Shaft Position Encoder. 9. A device capable of converting impulses from one mode to another, such as analog to digital, or parallel to serial, or one code to another. 10. Device in a digital system that transforms information coded in one number system to its equivalent in another number system. Typically, conversion is either decimal-to-binary or binary-to-decimal.

converter tube – A multielement electron tube that combines the mixer and local-oscillator functions of a heterodyne conversion transducer.

converter unit—The unit of a radar system in which the mixer of a superheterodyne receiver and usually two stages of intermediate-frequency amplification are located. Performs a preamplifying operation.

converting—Changing data from one form to another to facilitate its transmission, storage, or manipulation of information.

convex—Curved outward.

Cook system—An early stereo-disc recording technique in which the two channels were recorded simultaneously with two cutters on different portions (bands) of a record as concentric spirals. The playback equipment consisted of two pickups mounted side by side so that each played at the correct spot on its own band.

Coolidge tube—An X-ray tube in which the electrons are produced by a hot cathode.

coordinate digitizer—A device that transcribes graphic information in terms of a coordinate system for subsequent processing.

coordinated indexing—1. In a computer, a system in which individual documents are indexed by descriptors of equal rank so that a library can be searched for a combination of one or more descriptors. 2. A computer indexing technique in which the coupling of individual words is used to show the interrelation of terms.

coordinated transpositions – Transpositions which are installed in either electric supply or communication circuits, or in both, for the purpose of reducing inductive coupling, and which are located effectively with respect to the discontinuities in both the electric supply and communication circuit.

coordinate system—A way by which a pair of numbers is associated with each point in a plane (or a triplet of numbers is associated with each point in three-dimensional space) without ambiguity.

coordination—A term describing the ability of the lower rating of two breakers in series to trip before the higher-rating one trips.

coordinatograph – A precision drafting instrument used in the preparation of artwork for mask making.

copper loss—*See* I^2R Loss.

copper oxide photocell – An early type of nonvacuum photovoltaic cell consisting of a layer of copper oxide on a metallic substrate, with a thin transparent layer of a conductor over the oxide. Light falling on the cell produces a small voltage between the substrate and the conducting layer. This type of cell is extensively used in exposure meters for cameras for it requires no external source of electric power.

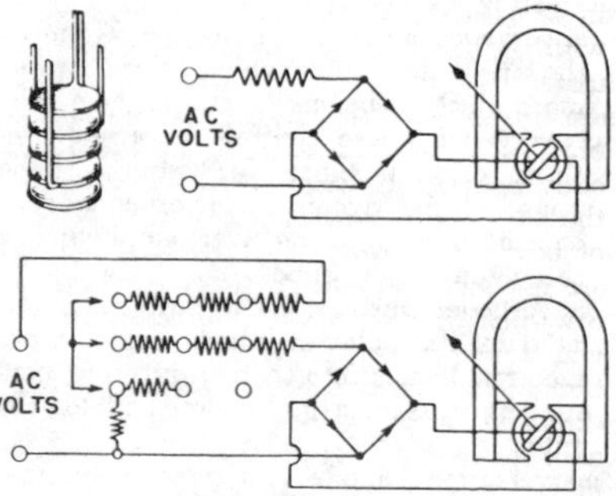

Copper-oxide rectifier.

copper-oxide rectifier—A metallic rectifier in which the rectifying barrier is the junction between metallic copper and cuprous oxide. A disc of copper is coated with cuprous oxide on one side, and a soft lead washer is used to make contact with the oxide layer.

copper-sulfide rectifier – A semiconductor rectifier in which the rectifying barrier is a junction between magnesium and copper sulfide.

copy—*See* Subject Copy.

copying telegraph—An absolute term for a facsimile system for the transmission of black-and-white copy only.

Corbino effect—A special case of the Hall effect that occurs when a disc carrying a radial current is placed perpendicularly into a magnetic field.

cord—One or a group of flexible insulated conductors covered by a flexible insulation and equipped with terminals.

cord circuit—A circuit, terminated in a plug at one or both ends, used at a telephone switchboard position in establishing connections.

cordless switchboard—A telephone switchboard in which manually operated keys are used to make connections.

cord sets—Portable cords fitted with any type of wiring device at one or both ends.

cordwood – 1. A sandwich-type construction wherein components lie in a vertical "cordwood" pattern between horizontal layers. 2. The technique of producing modules by bundling parts as closely as possible and interconnecting them into circuits by welding or soldering leads together.

cordwood module – A high-density circuit module in which discrete components are mounted between and perpendicular to two small, parallel printed-circuit boards to which their terminals are attached.

core—1. A magnetic material placed within a coil to intensify the magnetic field. 2. Magnetic material inside a relay or coil winding.

coreless-type induction heater — A device in which an object is heated by induction without being linked by a magnetic-core material.

core loss—Also called iron loss. Loss of energy in a magnetic core as the result of eddy currents, which circulate through the core and dissipate energy in the form of heat.

core memory—1. A magnetic type of memory made up of miniature toroids, each of which can be magnetized in one direction to represent a 0 and in the other direction to represent a 1. It is a permanent memory, since if the power is removed the stored information remains. They are characterized by low-cost storage and relatively slow memory operating speed. Core memories are nonvolatile, but have destructive readouts. 2. *See also* Internal Storage, 1.

core plane—A horizontal network of magnetic cores that contains a core common to each storage position.

core rope storage—Direct-access storage in which a large number of doughnut-shaped ferrite cores are arranged on a common axis, and sense, inhibit, and set wires are threaded through individual cores in a predetermined manner to provide fixed storage of digital data. Each core stores one or more complete words instead of a single bit.

core storage—In a computer, a form of high-speed storage that uses magnetic cores. 2. In a calculator, a storage register in which the contents will remain even after the machine has been switched off.

core store—A matrix of small magnetic rings or cores upon which electrical pulses may be stored. The presence of a pulse in a train is recorded by magnetizing a core, the absence of a pulse by leaving a core unmagnetized.

core transformer—A transformer in which the windings are placed on the outside of the core.

core-type induction heater — A device in which an object is heated by induction. Unlike the coreless type, a magnetic core links the inducing winding to the object.

core wrap—Insulation placed over a core before the addition of windings.

corner—1. An abrupt change in direction of the axis of a waveguide. 2. A neighborhood or point where a curve makes a sharp or discontinuous change of slope.

corner cut—A corner removed, for orientation purposes, from a card to be used with a computer.

corner effect—The rounding off of the attenuation versus frequency characteristic of a filter at the extremes (or corners) of the passband.

corner frequency — 1. The frequency at which the open-loop gain-versus-frequency curve changes slope. For a servo motor, the product of the corner frequency in radians per second and the time constant of the motor is unity. 2. The frequency at which the two asymptotes of the gain-magnitude curves of an operational amplifier intersect. 3. The upper frequency at which 3-dB attenuation occurs in a high-gain amplifier. A cornering circuit usually is introduced to attenuate the high-frequency signals before the natural phase shift of the amplifier becomes greater than 90 degrees. When properly designed, the cornering circuit prevents high-frequency oscillations in feedback amplifiers. The corner frequency is sometimes erroneously referred to as the cutoff frequency.

corner reflector—A reflecting object consisting of two (dihedral) or three (trihedral) mutually intersecting conducting surfaces. Trihedral reflectors are often used as radar targets.

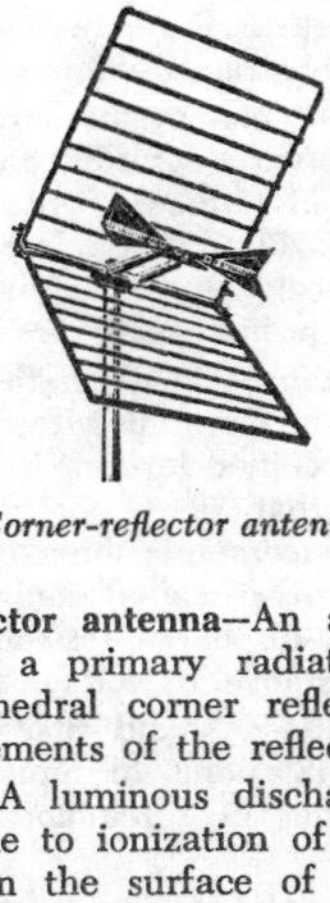

Corner-reflector antenna.

corner-reflector antenna—An antenna consisting of a primary radiating element and a dihedral corner reflector formed by the elements of the reflector.

corona—1. A luminous discharge of electricity, due to ionization of the air, appearing on the surface of a conductor when the potential gradient exceeds a certain value. 2. Any electrically detectable, field-intensified ionization that occurs in an insulating system but does not result immediately in catastrophic breakdown. (Corona always precedes dielectric breakdown.) 3. The ionization of gases about a conductor that results in a bluish-purple glow due to the voltage differential between a high-voltage conductor and the surrounding atmosphere. 4. A device used in an electrostatic copier to impart an electrical charge (in the dark) to the photoconductive material (zinc-oxide coated paper) to make it sensitive to the action of light.

corona discharge–A phenomena that occurs when an electric field is sufficiently strong to ionize the gas between electrodes and cause conduction. The effect is usually associated with a sharply curved surface, which concentrates the electric field at the emitter electrode. The process operates between an inception voltage and a spark breakdown voltage. These potentials and the current-voltage characteristics within the operating range are affected by the polarity of the corona electrodes as well as the composition and density of the gas in which the discharge occurs.

corona effect (of alternating current)–The effect produced when two wires, or other conductors having a great difference of voltage, are placed near each other.

corona endurance–Resistance to corona cutting.

corona extinction voltage–The voltage at which discharges preceded by corona cease as the voltage is reduced. The corona extinction voltage is always lower than the corona start voltage. Abbreviated cev.

corona failure–Failure due to corona degradation at areas of high voltage stress.

corona loss–A loss or discharge which occurs when two electrodes having a great difference of pressure are placed near each other. The corona loss takes place at the critical voltage and increases very rapidly with increasing pressure.

corona resistance–That length of time that an insulation material withstands the action of a specified level of field-intensified ionization that does not result in the immediate, complete breakdown of the insulation. Also called ionization resistance, brush-discharge resistance, slot-discharge resistance, or voltage endurance.

corona shield–A shield placed around a high-potential point to redistribute electrostatic lines of force and prevent corona.

corona start voltage–The voltage at which corona discharge begins in a given system. Abbreviated csv.

corona voltmeter–A voltmeter in which the peak voltage value is indicated by the beginning of corona at a known and calibrated electrode spacing.

correction – An increment which, when added algebraically to an indicated value of a measured quantity, results in a better approximation to the true value of the quantity.

corrective equalization–*See* Frequency-Response Equalization.

corrective maintenance–The maintenance performed on a nonscheduled basis to restore equipment to satisfactory condition.

corrective network – Also called shaping network. An electric network designed to be inserted into a circuit to improve its transmission or impedance properties, or both.

correed relay–A device consisting of a hermetically sealed reed capsule surrounded by a coil. It is used as a switching device in telephone equipment.

correlated characteristic – A characteristic known to be reciprocally related to some other characteristic.

correlation–1. The relationship, expressed as a number between minus one and plus one, between two sets of data, etc. 2. A relationship between two variables; the strength of the linear relationship is indicated by the coefficient of correlation. 3. A measure of the similarity of two signals.

correlation detection–A method of detection in which a signal is compared, point to point, with an internally generated reference. The output of such a detector is a measure of the degree of similarity of the input and reference signal. The reference signal is constructed in such a way that it is at all times a prediction, or best guess, of what the input signal should be at that time.

correlation direction finder–A satellite station separated from a radar to receive a jamming signal. By correlating the signals received from several such stations, the range and azimuth of many jammers may be obtained.

correlation distance–A term used in tropospheric propagation. The minimum spatial separation between antennas which will give rise to independent fading of the received signals.

Correlation Orientation Tracking and Range System–A system generally using a parabolic antenna for the analysis of a narrow band of radar energy for tracking and ranging purposes.

correlation tracking and ranging–A nonambiguous short-base-line, single-station cw phase comparison system measuring two direction cosines and a slant range from which space position can be computed.

correlation tracking and triangulation – A trajectory-measuring system composed of several antenna base lines separated by large distances and used to measure direction cosines to an object. From these measurements, the space position is computed by triangulation.

correlation tracking system–A system utilizing correlation techniques in which signals derived from the same source are correlated to derive the phase difference between the signals. This phase difference contains the system data.

corrosion—A chemical action which causes gradual destruction of the surface of a metal by oxidation or chemical contamination. Also caused by reduction of the electrical efficiency between the metal and a contiguous substance or to the disintegrating effect of strong electrical currents or ground-return currents in electrical systems. The latter is known as electrolytic corrosion.

cosecant-squared antenna — An antenna which emits a cosecant-squared beam. In the shaped-beam antenna used, the radiation intensity over part of its pattern in some specified plane (usually the vertical) is proportionate to the square of the cosecant of the angle measured from a specified direction in that plane (usually the horizontal).

cosecant-squared beam—A radar-beam pattern designed to give uniform signal intensity in echoes from distant and nearby objects. It is generated by a spun-barrel reflector. The beam intensity varies as the square of the cosecant of the elevation angle.

cosine law—The law which states that the brightness in any direction from a perfectly diffusing surface varies in proportion to the cosine of the angle between that direction and the normal to the surface.

cosmic noise—Radio static, the origin of which is due to sources outside the earth's atmosphere. The source may be similar to sunspots, or spots on other stars.

cosmic rays—Any rays of high penetrating power produced by transmutations of atoms in outer space. These particles continually enter the earth's upper atmosphere from interstellar space.

COS/MOS—*See* CMOS.

coulomb—1. The quantity of electricity which passes any point in an electric circuit in 1 second when the current is maintained constant at 1 ampere. The coulomb is the unit of electric charge in the mksa system. 2. The measure of electric charge, defined as a charge equivalent to that carried by 6.281×10^{18} electrons.

Coulomb's law—Also called law of electric charges or law of electrostatic attraction. The force of attraction or repulsion between two charges of electricity concentrated at two points in an isotropic medium is proportionate to the product of their magnitudes and is inversely proportionate to the square of the distance between them. The force between unlike charges is an attraction, and between like charges a repulsion.

coulometer — An electrolytic cell which measures a quantity of electricity by the amount of chemical action produced.

Coulter counter — An electronic cell-counting instrument operating on the ion-conductivity principle. Designed by J. R. Coulter. (*See also* Cell Counter.)

count—In radiation counters, a single response of the counting system.

countdown—A decreasing tally that indicates the number of operations remaining in a series.

counter—1. A circuit which counts input pulses. One specific type produces one output pulse each time it receives some predetermined number of input pulses. The same term may also be applied to several such circuits connected in cascade to provide digital counting. Also called divider. 2. In mechanical analog computers, a means for measuring the angular displacement of a shaft. 3. Sometimes called accumulator. A device capable of changing from one to the next of a sequence of distinguishable states upon receipt of each discrete input signal. 4. An arrangement of flip-flops producing a binary word that increases in value by one each time an input pulse is received. It may also be called a divider, since successive counter stages divide the input frequency by two. 5. A device capable of changing stages in a specified sequence upon receiving appropriate input signals; a circuit that provides an output pulse or other indication after receiving a specified number of input pulses.

counterbalance—A weight, usually adjustable, fitted at the pivot end of a pickup arm. It counters the weight of the pickup head and cartridge unit and allows adjustment of the stylus pressure to the desired value.

counter circuit—A circuit which receives uniform pulses representing units to be counted and produces a voltage in proportion to their frequency.

counterclockwise polarized wave—*See* Left-Handed Polarized Wave.

counter-countermeasures — Use of antijamming techniques and circuits designed to decrease the effectiveness of electronic countermeasure activities on electronic equipment.

counterelectromotive cell—A cell of practically no ampere-hour capacity used to oppose the line voltage.

counterelectromotive force — Abbreviated counter emf. A voltage developed in an inductive circuit by an alternating or pulsating current. The polarity of this voltage is at every instant opposite that of the applied voltage.

countermeasures—That part of military science dealing with the employment of devices and/or techniques intended to im-

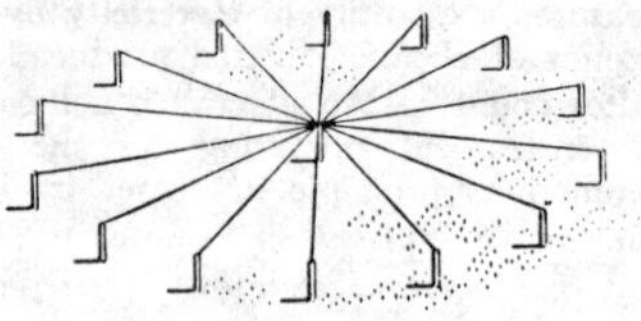

Counterpoise.

pair the operational effectiveness of enemy activity.

counterpoise—A system of wires or other conductors, elevated above and insulated from ground, forming a lower system of conductors of an antenna.

counters (Geiger and Scintillation)—Instruments used to detect ionizing radiations having very short wavelength (about one thousandth the wavelength of visible light). Natural sources of this radiation are radium, uranium isotopes, cosmic rays and ores in which these elements are present; man-contrived sources are the atomic bomb, nuclear reactors used for generating electric power, high-voltage radar crt's and X-ray machines.

counter tube—Also called radiation counter tube. An electron tube that converts an incident particle or burst of incident radiation into a discrete electric pulse. This is generally done by utilizing the current flow through a gas that is ionized by the radiation.

counting efficiency—In a scintillation counter, the ratio, under specified conditions, of the average number of photons or particles of ionizing radiation that produce counts to the average number of photons or particles incident on the sensitive area.

counting-rate meter—A device for indicating the time rate of occurrence of input pulses averaged over a time interval.

counting-type frequency meter—An instrument for measuring frequency. Its operation depends on the use of pulse-counting techniques to indicate the number and/or rate of recurring electrical signals applied to its input circuits.

counts—Clicking noises made by a radiation-detecting instrument in the presence of radiation. *See* Scintillation Counter.

counts per turn—The total number of code positions per 360° of encoder shaft rotation.

couple—Two or more dissimilar metals or alloys in electrical contact with each other that act as the electrodes of an electrolytic cell when they are immersed in an electrolyte.

coupled circuit — Any network containing only resistors, inductors (self and mutual), and capacitors, and having more than one independent mesh.

coupler—A passive device that divides an antenna signal to feed two or more receivers, or combines two or more antenna signals to feed a single down lead. A coupler provides some interset isolation and maintains an impedance match between the antenna and receiver.

coupling—The association of two or more circuits or systems in such a way that power may be transferred from one to another.

coupling angle—In connection with synchronous motors, the mechanical-degree relationship between the rotor and the rotating field.

coupling aperture — Also called coupling hole or coupling slot. An aperture, in the wall of a waveguide or cavity resonator, designed to transfer energy to or from an external circuit.

coupling capacitor—Any capacitor used to couple two circuits together. Coupling is accomplished by means of the capacitive reactance common to both circuits.

coupling coefficient—Also called coefficient of coupling. The degree of coupling that exists between two circuits. It is equal to the ratio between the mutual impedance and the square root of the product of the total self-impedances of the coupled circuits, all impedances being of the same kind.

coupling hole—*See* Coupling Aperture.

coupling loop—A conducting loop projecting into a waveguide or cavity resonator and designed to transfer energy to or from an external circuit.

coupling probe—A probe projecting into a waveguide or cavity resonator and designed to transfer energy to or from an external circuit.

coupling slot—*See* Coupling Aperture.

coupling transformer—A transformer that couples two circuits together by means of its mutual inductance.

covalent bond—A type of linkage between atoms. Each atom contributes one electron to a shared pair that constitutes an ordinary chemical bond.

coverage—A percentage of the completeness with which a braid or shield covers the surface of an underlying insulated conductor or conductors.

Covington and Broten antenna—A compound interferometer in which a long line source is adjacent to a two-element interferometer of comparable aperture, in the same straight line.

cpm—Abbreviation for cycles per minute.

C-power supply—A device connected in the circuit between the cathode and grid of a vacuum tube to apply grid bias.

cps—Abbreviation for cycles-per-second, an obsolete term. Replaced by term hertz, abbreviated Hz.

CPU—Abbreviation for central processing unit.

CPU portion—*See* Chip Sets.

crash-locator beacon—Airborne equipment consisting of various transmitters, collapsible antennas, etc., designed to be ejected from a downed aircraft and to transmit beacon signals to help searching forces to locate the crashed aircraft.

crater lamp—1. A glow-discharge type of vacuum tube the brightness of which is proportional to the current passing through the tube. The glow discharge takes place in a cup or crater rather than on a plate as in a neon lamp. 2. A gaseous lamp usually containing neon. Provides a point source of light that can be modulated with a signal.

crazing—Checking of an insulation material when it is stressed and in contact with certain solvents or their vapors.

credence—A measure of confidence in a radar target detection; generally it is proportional to the target-return amplitude.

credit balance indicator—In a calculator, warning light to indicate a negative answer.

creepage — The conduction of electricity across the surface of a dielectric.

creepage distance—The shortest distance between conductors of opposite polarities, or between a live part and ground, measured over the surface of the supporting material.

creepage path—The path across the surface of a dielectric between two conductors. Lengthening the creepage path reduces the possibility of arc damage or tracking.

creepage surface—An insulating surface that provides physical separation between two electrical conductors of different potential.

creep-controlled bonding — A method of diffusion bonding in which enough pressure is exerted to cause significant creep deformation at the joint interfaces. The method is characterized by use of intermediate and low unit loads for a period of hours.

creep distance—The shortest distance on the surface of an insulator between two electrically conductive surfaces separated by the insulator.

creep recovery—The change in no-load output occurring with time after removal of a load which had been applied for a specific period of time.

crest factor—The ratio of the peak voltage to the rms voltage of a waveform (with the dc component removed.)

crest value—Also called peak value. The maximum absolute value of a function.

crest voltmeter—A peak-reading voltmeter.

CRI—Abbreviation for color rendering index.

crimp—To compress or deform a connector barrel around a cable so as to make an electrical connection.

crimp contact—A contact whose back portion is a hollow cylinder to allow it to accept a wire. After a bared wire is inserted, a swedging tool is applied to crimp the contact metal firmly against the wire. An excellent mechanical and electrical contact results. A crimp contact often is referred to as a solderless contact.

crimping—A method of attaching a terminal, splice, or contact to a conductor through the application of pressure.

crimping tool—A device used to apply solderless terminals to a conductor.

"crippled leapfrog" test—In a computer, a variation of the "leapfrog" test in which the test is repeated from a single set of storage locations rather than from a changing set of storage locations.

critical angle—1. The maximum angle at which a radio wave may be emitted from an antenna and will be returned to the earth by refraction in the ionosphere. 2. The maximum angle of incidence for which light will be transmitted from one medium to another. Light approaching the interface at angles greater than the critical angle will be reflected back into the first medium.

critical area—*See* Elemental Area.

critical characteristic—A characteristic not having the normal tolerance to variables.

critical coupling — Also called optimum coupling. Between two circuits independently resonant to the same frequency, the degree of coupling which transfers the maximum amount of energy at the resonant frequency.

critical current—That current, at a specified temperature and in the absence of external magnetic fields, above which a material is normal and below which it is superconducting.

critical damping—The value of damping which provides the most rapid transient response without overshoot. Operation between underdamping and overdamping.

critical dimension — The dimension of a waveguide cross section that determines the cutoff frequency.

critical failure—A failure that causes a system to operate outside designated limits.

critical field—Also called cutoff field. Of a magnetron, the smallest theoretical value of a steady magnetic-flux density, at a steady anode voltage, that would prevent an electron emitted from the cathode at zero velocity from reaching the anode.

critical frequency—Also called penetrating frequency. The limiting frequency below which a magneto-ionic wave component is reflected by an ionospheric layer and above which the component penetrates the layer at vertical incidence. (*See also* Waveguide Cutoff Frequency.)

critical grid current – The instantaneous value of grid current in a gas tube when the anode current starts to flow.

critical grid voltage – The instantaneous value of grid voltage at which the anode current starts to flow in a gas tube.

critical high-power level – The radio-frequency power level at which ionization is produced in the absence of a control-electrode discharge.

critical inductance–In an inductor-input power supply filter, the minimum value of the input inductor needed to insure that the current drawn through the rectifier never goes to zero.

critical item–*See* Critical Part.

critical magnetic field–That field intensity below which at a specified temperature and in the absence of current, a material is superconducting and above which it is normal.

critical part–A part whose failure to meet specified requirements results in the failure of the product to serve its intended purpose. Also called critical item.

critical potential–*See* Ionization Potential.

critical race–*See* Race, 2.

critical rate of rise of off-state voltage–The minimum value of the rate of rise of principal voltage which will cause a semiconductor switching device to switch from the off-state to the on-state.

critical temperature–That temperature below which, in the absence of current and external magnetic fields, a material is superconducting and above which it is normal.

critical voltage–Also called cutoff voltage. In a magnetron, the highest theoretical value of steady anode voltage, at a given steady magnetic-flux density, at which electrons emitted from the cathode at zero velocity will fail to reach the anode.

critical wavelength–The free-space wavelength corresponding to the critical frequency.

cro – Abbreviation for cathode-ray oscilloscope.

CROM – Abbreviation for Control Read Only Memory.

Crookes dark space–*See* Cathode Dark Space.

Crosby system–A compatible multiplex fm stereo broadcast technique in which the right and left signals are combined in phase (sum signal) and transmitted on the main carrier, and also combined out-of-phase (difference signal) and transmitted on the subcarrier. The two signals are combined (matrixed) in the receiving apparatus to restore the right and left channels.

cross assembler – A symbolic language translator that runs on one type of computer to produce machine code for another type of computer. *See* Assembler.

crossbanding–The use of combinations of interrogation and reply frequencies such that either one interrogation frequency is used with several reply frequencies or one reply frequency is used with several interrogation frequencies.

crossbar switch–A switch having a number of vertical paths, a number of horizontal paths, and an electromagnetically operated mechanism for interconnecting any one vertical path with any one horizontal path.

crossbar switching system–A method of switching which, when directed by a common control unit, will select and close a path through a matrix arrangement of switches.

crossbar system–An automatic telephone switching system in which, generally, the selecting devices are crossbar switches. Common circuits select and test the switching paths and control the selecting mechanisms. The method of operation is one in which the switching information is received and stored by controlling mechanisms that determine the operation necessary to establish a connection.

cross beat – A spurious frequency that arises as a result of cross modulation.

crosscheck–To check a computation by two different methods.

cross color–The interference in a color-television receiver chrominance channel caused by cross talk from monochrome signals.

cross-correlation function–A measure of the similarity between two signals when one is delayed with respect to the other.

cross coupling – Unwanted coupling between two different communication channels or their components.

crossed-field amplifier–A high-power electron tube in which direct-current power is converted to microwave power by a combination of crossed electric and magnetic fields.

crossed-field device–An electronic device in which electrons from the cathode are influenced by a magnetic field that acts at right angles to the applied electric field. When electrons leave the cathode in a direction perpendicular to the magnetic field, this field causes a force to act at right angles to the electron motion. The electrons then spiral into orbit around the cathode instead of moving collinearly with the electric field. Most of the electrons gradually move toward the anode, giving up potential energy to the rf field as they interact with the anode slow-wave structure. The tube structure may be either cylindrical or linear.

crossed-pointer indicator – A two-pointer indicator used with instrument landing systems to indicate the position of an aircraft with respect to the glide path.

crossfire – Interfering current in one telegraph or signaling channel resulting from telegraph or signaling currents in another channel.

crossfoot—1. In a computer, to add or subtract numbers in different fields of the same punch card and punch the result into another field of the same card. 2. To compare totals of the same numbers obtained by different methods.

cross-hatching—In a printed-circuit board, the breaking up of large conductive areas where shielding is required.

cross magnetostriction – Under specified conditions, the relative change of dimension in a specified direction perpendicular to the magnetization of a body of ferromagnetic material when the magnetization of the body is increased from zero to a specified value (usually saturation).

cross modulation—A spurious response that occurs when the carrier of a desired signal intermodulates with the carrier of an undesired signal. This often happens in early stages of radio receivers particularly when strong signals from local stations drive these stages into nonlinear operation.

cross neutralization—A method of neutralization used in push-pull amplifiers. A portion of the plate-to-cathode ac voltage of each tube is applied to the grid-to-cathode circuit of the other tube through a neutralizing capacitor.

cross office switching time—The time required for connection of any input through the switching center to any selected output.

crossover—1. The point where two conductors that are insulated from each other cross. 2. A connection formed between two elements of a circuit by depositing conductive material across the insulated upper surface of another interconnection or element. 3. A point in an integrated or MOS circuit at which an interconnect pattern passes over another conductive part of a circuit but is insulated from it by a thin dielectric layer. *See also* Underpass.

crossover distortion—1. Distortion that occurs in a push-pull amplifier at the points of operation where the input signals cross over (go through) the zero reference points. 2. The type of distortion resulting from class-B push-pull power amplifiers owing to the lack of coincidence of the two transfer characteristics at the crossover point. The effect is reduced by applying a critical value of biasing to optimize the quiescent current and hence "linearize" the middle portion of the transfer characteristic. The situation is further improved by heavy negative feedback and by circuit design, such that the crossover distortion from hi-fi amplifiers is very small.

crossover frequency—1. As applied to electrical dividing networks, the frequency at which equal power is delivered to each of the adjacent frequency channels when all channels are terminated in the specified load. *See also* Transition Frequency. 2. A frequency at which other frequencies above and below it are separated. In a two-way speaker system, for instance, the crossover frequency is the point at which woofer and tweeter response is divided.

crossover network—1. An electrical filter that separates the output signal from an amplifier into two or more separate frequency bands for a multispeaker system. 2. A circuit (usually employing capacitors and coils) which feeds low notes to a low frequency speaker (woofer) and high notes to a high frequency speaker (tweeter). The crossover frequency is that at which frequency bands divide. Sometimes the audio spectrum is divided into more than two bands to drive more than two speakers.

crossover spiral—*See* Leadover Groove.

cross polarization—The component electric field vector normal to the desired polarization component.

cross-sectional area of a conductor—The summation of all cross-sectional areas of the individual strands in the conductor, expressed in square inches or more commonly in circular mils.

cross talk—1. Interference caused by stray electromagnetic or electrostatic coupling of energy from one circuit to another. 2. Undesired signals from another circuit in the same system. 3. Breakthrough of the signal from one channel to another by conduction or radiation. 4. Transient noise induced on a switching signal by interaction with other switching transitions. 5. Audio interference from one track of a stereo tape to another. Poor head alignment often causes this. 6. Leakage of recorded signal from one channel of a stereo device into the adjacent channel or channels. Cross talk between stereo channels impairs stereo separation; cross talk between reverse-direction track signal to be heard, backwards, during quieter parts of the desired program.

cross-talk coupling—Also called cross-talk loss. Cross coupling between speech communication channels or their component parts. Note: Cross-talk coupling is measured between specified points of the disturbing and disturbed circuit and is preferably expressed in decibels.

cross-talk level—The volume of cross-talk

energy, measured in decibels, referred to a base.

cross-talk loss—*See* Cross-Talk Coupling.

crossunder—1. A connection of two elements of a circuit by a conductive path deposited or diffused into a substrate. 2. A point in an integrated or MOS circuit at which there is a crossing of two conductive paths, one of which is built into the active substrate for interconnection.

crowbar — A term describing the action which effectively creates a high overload on the actuating member of the protective device. This crowbar action may be triggered by a slight increase in current or voltage.

crowbar circuit — An electronic switching system used to protect high-voltage circuits from damage caused by arc currents. The system places a momentary short across the circuit to be protected.

crowbar protection circuit — A protection circuit which by rapidly placing a low resistance across the output terminals of a power supply initiates action that reduces the output voltage to a low value.

crowbar voltage protector—A separate circuit which monitors the output of a power supply and instantaneously throws a short circuit (or crowbar) across the output terminals of the power supply whenever a preset voltage limit is exceeded. An scr is usually used as the crowbar device.

crt—Abbreviation for cathode-ray tube.

crt display — *See* Cathode-Ray-Tube Display.

cryogenic—Of or having to do with temperatures approaching absolute zero.

cryogenic device — A device intended to function best at temperatures near absolute zero.

cryogenic motor—A motor that operates at a temperature below −129°C and that uses a cryogenic fluid or gas to cool its windings and bearings.

cryogenics—1. The subject of physical phenomena at temperatures below about −50°C. More generally, the term is used to refer to methods for producing very low temperatures. Also called cryogeny. 2. The study of the behavior of matter at supercold temperatures.

cryogeny—*See* Cryogenics.

cryolectronics — Technology having to do with the characteristics of electronic components at cryogenic temperatures. The branch of electronics that is concerned with applications of cryogenics. A contraction of cryogenic electronics.

cryosar—A semiconductor device primarily intended for high-speed computer switching and memory applications. This device operates by the low-temperature avalanche breakdown produced by impact ionization of impurities.

cryosistor—A cryogenic semiconductor device in which the ionization between two ohmic contacts is controlled by means of a reverse-biased pn junction. After ionization, the device can act as a three-terminal switch, a pulse amplifier, an oscillator, or a unipolar transistor.

cryostat—A refrigerating unit such as that for producing or utilizing liquid helium in establishing extremely low temperatures (approaching absolute zero).

cryotron—A superconductive four-terminal device in which magnetic field, produced by passing a current through two input terminals, controls the superconducting-to-normal transition—and thus the resistance—between the two output terminals.

cryotronics—A contraction of "cryogenic electronics."

crypto—A prefix used to form words that pertain to the transformation of data to conceal its actual meaning, usually by conversion to a secret code.

crystal—1. A solid in which the constituent atoms are arranged with some degree of geometric regularity. In communication practice, a piezoelectric crystal, piezoelectric crystal plate, or crystal rectifier. 2. A thin slab or plate of quartz ground to a thickness that causes it to vibrate at a specific frequency when energy is supplied. It is used as a frequency-control element in radio-frequency oscillators.

crystal anisotropy—A force which directs the magnetization of a single-domain particle along a direction of easy magnetization. To rotate the magnetization of the particle, an applied magnetic field must provide enough energy to rotate the magnetization through a difficult crystal direction.

crystal audio receiver—Similar to the crystal video receiver except for the path direction bandwidth, which is audio rather than video.

crystal calibrator—A crystal-controlled oscillator used as a reference to check and set the frequency tuning of a receiver or transmitter.

crystal-can relay—A relay mounted in a can of a specific size and shape; called a crystal can because of its common usage as a mounting for quartz crystals used in frequency-control circuits.

crystal control—Control of the frequency of an oscillator by means of a specially designed and cut crystal.

crystal-controlled oscillator — *See* Crystal Oscillator.

crystal-controlled transmitter — A radio transmitter in which the carrier frequency is controlled directly by a crystal oscillator.

crystal cutter—A disc cutter in which the

mechanical displacements of the recording stylus are derived from the deformities of a crystal having piezoelectric propties.

crystal detector—A mineral or crystalline material which allows electrical current to flow more easily in one direction than in the other. In this way, an alternating current can be converted to a pulsating current.

crystal diode—A two-electrode semiconductor device that makes use of the rectifying properties of a pn junction (junction diode) or a sharp metallic point in contact with a semiconductor material (point-contact diode). Also called crystal rectifier, diode, and semiconductor diode.

crystal filter—A highly selective circuit capable of discriminating against all signals except those at the center frequency of a crystal, which serves as the selective element. Resonant mechanical section consists of ceramic discs which vibrate at the band of frequencies to be removed. 2. Electrically coupled, two-terminal electroacoustic resonators (crystals) in ladder and lattice configurations. Monolithic filters with quartz substrates are also called crystal filters.

crystal headphones—Headphones using Rochelle-salt or other crystal elements to convert audio-frequency signals into sound waves.

crystal holder—A case of insulating material for mounting a crystal. External prongs allow the crystal to be plugged into a suitable socket.

crystal imperfection — Any deviation in lattice structure from that of a perfect single crystal.

crystal lattice—A periodic geometric arrangement of points that correspond to the locations of the atoms in a perfect crystal.

crystal loudspeaker — A loudspeaker in which piezoelectric action is used to produce mechanical displacements. Also called piezoelectric loudspeaker.

crystal microphone — Also called piezoelectric microphone. A microphone which depends for its operation on the generation of an electric charge by the deformation of a body (usually crystalline) having piezoelectric properties.

crystal mixer—1. A mixer circuit with the frequency of the local oscillator being controlled by a crystal. Normally used in superheterodyne radio receivers. 2. A mixer that utilizes the nonlinear characteristic of a crystal diode to mix two frequencies. Frequently used in radar receivers to convert the received radar signal to a lower i-f signal by mixing it with a local-oscillator signal.

crystal operation—Operation using crystal-controlled oscillators.

crystal orientation — For MOS devices, terms <100> and <111> are commonly used. This refers to the angle with respect to crystal facets at which the silicon crystal is sliced. Each has a direct effect on MOS transistor characteristics.

crystal oscillator—Also called crystal-controlled oscillator. An oscillator in which the frequency of oscillation is controlled by a piezoelectric crystal.

crystal oven—A container, maintained at a constant temperature, in which a crystal

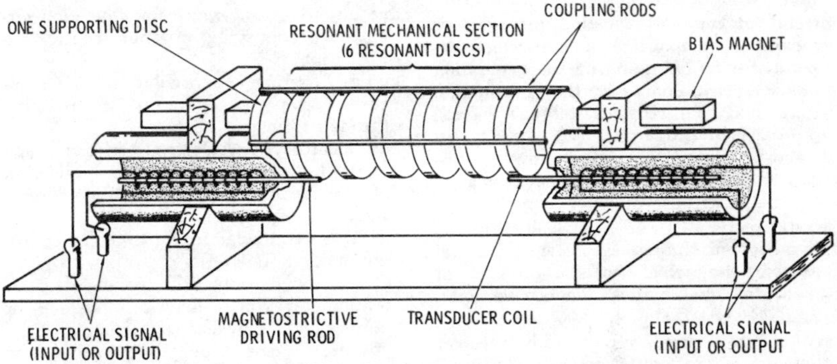

Crystal filter.

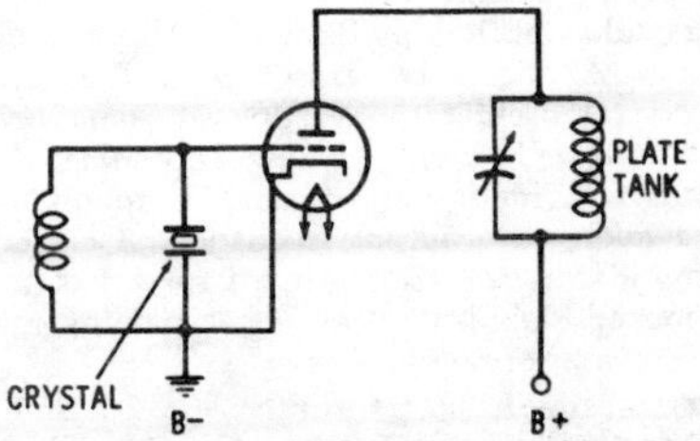

Crystal oscillator.

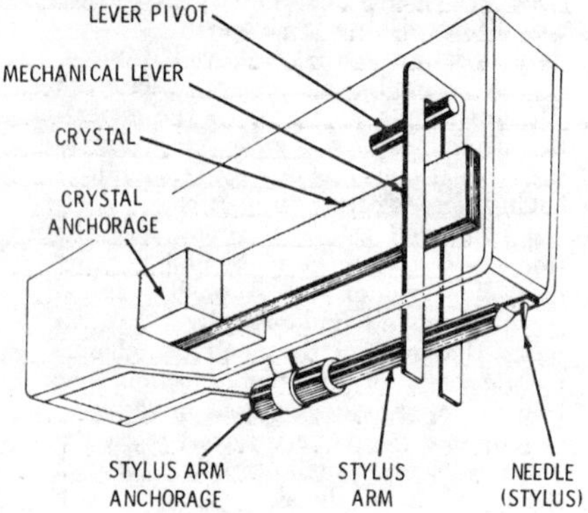

Crystal pickup.

and its holder are enclosed in order to keep their temperature constant and thereby reduce frequency drift.

crystal pickup – Also called piezoelectric pickup. A phonograph pickup which depends for its operation on the generation of an electric charge by the deformation of a body (usually crystalline) having piezoelectric properties.

crystal pulling–A method of growing crystals in which the developing crystal is gradually withdrawn from a melt.

crystal rectifier–1. An electrically conductive or semiconductive substance, natural or synthetic, which has the property of rectifying small radio-frequency voltages. 2. *See* Crystal Diode.

crystal set–A simple type of radio receiver having only a crystal-detector stage for demodulation of the received signal and no amplifier stages.

crystal shutter–A mechanical switch for shorting a waveguide or coaxial cable so that undesired rf energy is prevented from reaching and damaging a crystal detector.

crystal slab–A relatively thick piece of crystal from which crystal blanks are then cut.

crystal speaker – Also called piezoelectric speaker. A speaker in which the mechanical displacements are produced by piezoelectric action.

crystal-stabilized oscillator – Abbreviated xso. A microwave rf source that uses a crystal oscillator operating at some low frequency (usually below 150 MHz) to drive a multiplier or sampler to obtain a microwave output frequency. A crystal oscillator can also be used for injection-locked stabilization of a free-running microwave oscillator.

crystal-stabilized transmitter – A transmitter employing automatic frequency control, in which the reference frequency is the same as the crystal-oscillator frequency.

crystal transducer – *See* Piezoelectric Transducer.

crystal video receiver–A radar receiver consisting only of a crystal detector and video amplifier.

CSA–Abbreviation for Canadian Standards Association.

C-scope–A rectangular radar display in which targets appear as bright spots with azimuth indicated by the horizontal coordinate and elevation angle by the vertical coordinate.

csv–Abbreviation for corona start voltage.

CT-cut crystal–A natural quartz crystal cut to vibrate below 500 kHz.

CTL – Abbreviation for complementary transistor logic.

cubical antenna–An antenna array, the elements of which are positioned to form a cube.

cue–*See* Address.

cue circuit–A one-way communication circuit for conveying program control information.

cuing–Locating a particular spot on a recorded tape, preparatory to playing through from that spot.

cuff electrode–An electrode in the shape of the letter "C" designed for application of potentials to small circular bodies, such as peripheral nerves.

CUJT – Abbreviation for complementary unijunction transistor.

cup–A single mechanical section of a potentiometer which may contain one or more electrical resistance elements.

cup core–A core which forms a magnetic shield around an inductor. Usually a cylinder with one end closed. A center core inside the inductor is normally used and may or may not be part of the cup core.

cupping–Curvature of a recording tape in a direction perpendicular to the length of the tape.

cup washer–A washer formed with a recess in one side to retain compression springs or, on binding-post terminals, to prevent escape of connecting wire strands.

curie–A unit used for indicating the strength of radioactive sources in terms of the number of disintegrations per second in the source. One curie is equal to 3.7×10^{10} disintegrations per second.

Curie point – In ferroelectric dielectrics, the temperature or temperatures at which peak values of the dielectric constant occur. Also called Curie temperature.

curie temperature–Temperature in degrees centigrade at which a magnetized sample is completely demagnetized due to thermal agitation.

curing temperature–Temperature at which a material undergoes a curing process.

Curpistor—A subminiature constant-current tube containing two electrodes and filled with radioactive nitrogen.

current – The movement of electrons through a conductor. Current is measured in amperes, in milliamperes and microamperes.

current amplification—1. The ratio of the current produced in the output circuit of an amplifier to the current supplied to the input circuit. 2. In photomultipliers, the ratio of the signal output current to the photoelectric signal current from the photocathode.

current amplifier—A device designed to deliver a greater output current than its input current.

current antinode—Also called current loop. The point at which current is a maximum along a transmission line, antenna, or other circuit element having standing waves.

current attenuation—The ratio of the magnitude of the current in the input circuit of a transducer to the magnitude of the current in a specified load impedance connected to the transducer.

current-balance relay—A relay in which operation occurs when the magnitude of one current exceeds the magnitude of another current by a predetermined ratio.

current-carrying capacity – 1. The maximum current a conductor (or braid) can carry without heating beyond a safe limit. 2. The maximum current that can be continuously carried without causing permanent change in the electrical or mechanical properties of a device or conductor. (As applied to phone jacks, it refers to carrying current without interrupting the circuit.)

current-carrying rating – The current that can be carried continuously or for stated periodic intervals without impairment of the contact structure or interrupting capability.

current density—1. The amount of electric current passing through a given cross-sectional area of a conductor in amperes per square inch—i.e., the ratio of the current in amperes to the cross-sectional area of the conductor. 2. The ratio of current to surface area.

current echo—The signal which on a transmission line is reflected as the result of some discontinuity.

current feed—A method of feeding to a point where the current is a maximum (e.g., at the center of a half-wave antenna).

current flicker – Current surges resulting from momentary shorts that can occur within a solid-electrolytic capacitor. They may produce a leakage-current increase, a decrease, no change at all, or a catastrophic failure. Under certain conditions, current flicker can avalanche to cause a short, which under low-impedance circuit conditions results in catastrophic destruction of the capacitor.

current generator—A two-terminal circuit element with a terminal current independent of the voltage between its terminals.

current hogging—1. A condition in which one of several parallel logic circuits takes the largest share of the available current because it has a lower resistance than the other circuits have. 2. A condition that exists when several base-emitter junctions are driven from the same output and the input with the lowest base-emitter junction forward potential severely limits the drive current to the other transistor bases.

current-hogging injection logic – Abbreviated CHIL. A logic form which combines the input flexibility of current-hogging logic with the performance and packing density of injection logic.

current limiter— A device that detects current leakage and prevents potential shock hazard by minimizing or interrupting current flow.

current limiting (automatic)—An overload-protection mechanism which limits the maximum output current of a power supply to a preset value and automatically restores the output when the overload is removed. (*See also* Short-Circuit Protection.)

current-limiting fuse—A protective device which anticipates a dangerous short-circuit current and opens the circuit, precluding the development of the peak available current.

current-limiting reactor—A form of reactor intended for limiting the current that can flow in a circuit under short-circuit conditions.

current-limiting resistor – A resistor inserted into an electric circuit to limit the flow of current to some predetermined value. Usually inserted in series with a fuse or circuit breaker to limit the current flow during a short circuit or other fault, to prevent excessive current from damaging other parts of the circuit.

current limit-sense voltage – The voltage between the sense and limit terminals of a regulator which will cause current limiting.

current loop—*See* Current Antinode.

current margin – The difference between the steady-state currents flowing through a telegraph receiving instrument corresponding, respectively, to the two positions of the telegraph transmitter.

current-mode logic—Abbreviated CML. A nonsaturating logic circuit that employs the characteristics of a differential amplifier circuit in its design. Because it is nonsaturating, it is a very fast switching logic design. The gate input element is the base of a transistor with a separate transistor for each input.

current node—A point at which current is zero along a transmission line, antenna, or any other circuit element that has standing waves.

current noise—Also called excess noise. A low-frequency noise caused by current flowing in a resistor, particularly a carbon resistor. The amount varies widely with the type and construction of the resistor. This low-frequency noise is generally measurable only in the region below 100 kHz, and the noise power varies inversely with frequency.

current probe—A type of transformer usually having a snap-around configuration, used for measuring the current in a conductor.

current pump — A circuit that drives, through an external load circuit, an adjustable, variable, or constant value of current, regardless of the reaction of that load to the current, within rated limits of current, voltage, and load impedance.

current regulator—A device that functions to maintain the output current of a generator or other voltage source at a predetermined value, or varies the voltage according to a predetermined plan.

current relay—A relay that operates at a predetermined value of current. It can be an overcurrent relay, an undercurrent relay, or a combination of both.

current saturation—The condition in which the plate current of a vacuum tube cannot be further increased by increasing the plate voltage.

current-sensing resistor—A resistor of low value placed in series with the load to develop a voltage proportional to the output current. A regulated dc power supply regulates the current in the load by regulating the voltage across this sensing resistor.

current sensitivity—The current required to give standard deflection on a galvanometer.

current-sinking logic — Also called input-coupled logic. A logic form that requires that current flow out of the input of a circuit and back into the output of the preceding stage, which serves as a current sink instead of a source.

current-sourcing logic—Also called collector-coupled logic. A logic form where current flows from the output of a circuit and is forced into the input of a similar circuit to activate the circuit which drives.

current-stability factor—In a transistor, the ratio of a change in emitter current to a change in reverse-bias current between the collector and base.

current-transfer ratio—*See* Beta.

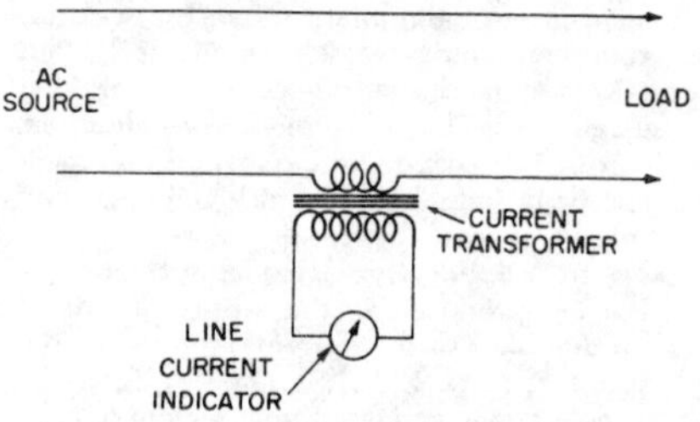

Current transformer.

current transformer — A transformer, intended for measuring or control purposes, designed to have its primary winding connected in series with a circuit carrying the current to be measured or controlled.

current-type telemeter — A telemeter in which the magnitude of a single current is the translating means.

cursor—1. A mechanically or electronically generated line which moves back and forth over another surface to delineate accurate readings. 2. A mechanical bearing line on a plan position indicator-type display for reading target bearing.

curtate—A portion of a punched card; it consists of adjacent punched rows.

curve tracer—An instrument capable of producing a display of one current or voltage as a function of a second voltage or current with a third voltage or current as a parameter.

customer set—*See* Subscriber Set.

Cutler antenna—A rear feed for a paraboloidal antenna reflector. It consists of a support waveguide with a terminating cavity containing two resonant slots, one on either side of the support waveguide, that face the reflector. Each slot is parallel to the broad faces of the feed waveguide.

Cutler feed—A resonant cavity, at the end of a waveguide, which feeds radio-frequency energy to the reflector of the antenna assembly of some airborne antennas.

cutoff—1. Minimum value of bias which cuts off, or stops, the flow of plate current in a tube. 2. The frequency above or below which a selective circuit fails to respond. 3. The frequency of transmission at which the loss exceeds by 10 decibels that observed at 1000 hertz. 4. The condition when the emitter-base junction of a transistor has zero bias or is reverse biased and there is no collector current. 4. The frequency at which the modulus of measured parameter has decreased to $1/\sqrt{2}$ of its low frequency value. (For a transistor, the cutoff frequency usually applies to the short-circuit small-signal forward current transfer ratio for either the common-base or common-emitter configuration.)

cutoff attenuator – A variable length of waveguide used below the cutoff frequency of the waveguide to introduce variable nondissipative attenuation.

cutoff current—Transistor collector current with no emitter current and normal collector-to-base bias.

cutoff field—*See* Critical Field.

cutoff frequency – 1. The frequency at which the gain of an amplifier falls below .707 times the maximum gain. 2. The frequency that marks the edge of the passband of a filter and the beginning of the transition to the stopband. 3. With respect to a line, the upper frequency limit, usually of a loaded transmission circuit, beyond which the attenuation rises very rapidly.

cutoff limiting – Keeping the output of a vacuum tube below a certain point by driving the control grid beyond cutoff.

cutoff voltage—1. The electrode voltage which reduces the dependent variable of an electron-tube characteristic to a specified value. (*See also* Critical Voltage.) 2. The voltage at which the discharge is considered complete. This need not be a very low voltage.

cutoff wavelength—The ratio of the velocity of electromagnetic waves in free space to the cutoff frequency of a waveguide.

cutout—1. An electrical device that interrupts the flow of current through any particular apparatus or instrument, either automatically or manually. 2. Pairs brought out of a cable and terminated at some place other than at the end of the cable.

cut-signal branch operation – In systems where radio reception continues without cutting off the carrier, the cut-signal branch operation technique disables a signal branch in one direction when it is enabled in the other to preclude unwanted signal reflections.

cutter—Also called mechanical recording head. An electromechanical transducer which transforms an electric input into a mechanical output (for example, the mechanical motions which a cutting stylus inscribes into a recording medium).

cut-through—The resistance of a solid material to penetration by an object under conditions of pressure, temperature, etc.

cut-through flow test—A test to measure the resistance to deformation of insulation subjected to heat and pressure.

cutting rate—The number of lines per inch the lead screw moves the cutting-head carriage across the face of a recording blank. Standard rates are 96, 104, 112, 128, 136, and greater in multiples of eight lines per inch. For microgroove recordings, 200 to 300 lines per inch are used.

cw—Abbreviation for: 1. Continuous wave. 2. Clockwise.

cw jamming—Transmission of a constant-amplitude, constant-frequency, unmodulated signal for the purpose of jamming a radar receiver by changing its gain characteristics.

cw reference signal—In color television, a sinusoidal signal used to control the conduction time of a synchronous demodulator.

cybernetics—1. The study of systems of control and communications in humans and animals, and in electrically operated devices such as calculating machines. 2. The comparative study of the control and intracommunication of information handling machines and nervous systems of animals and man in order to understand and improve communication. 3. The science that is concerned with the principles of communication and control, particularly as applied to the operation of machines and the functioning of organisms.

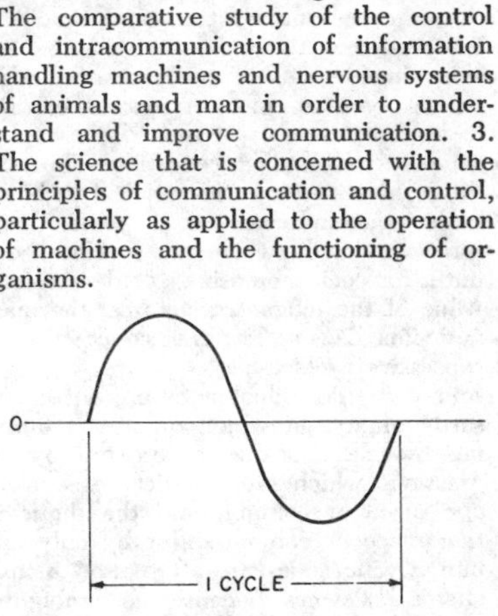

Cycle, 1.

cycle—1. The change of an alternating wave from zero to a negative peak to zero to a positive peak and back to zero. The number of cycles per second (hertz) is called the frequency. (*See also* Alternation.) 2. An off-on application of power. 3. A set of operations carried out in a predetermined manner. 4. In computer terminology, a regularly repeated sequence of operations, or the time required for one such sequence.

cycle counter – A mechanism or device used to record the number of times a specified cycle is repeated.

cycle criterion – In computer terminology, the number of times a cycle is to be repeated.

cycle index—1. In digital computer programming, the number of times a cycle has been executed. 2. The difference between the number of times a cycle has been executed and the number of times it is desired that the cycle be repeated.

cycle life—For rechargeable cells, the total number of discharge/charge cycles before the cell becomes inoperative.

cycle reset—To return a cycle index to its initial value.

cycle shift—In a computer, the removal of the digits of a number or characters from a word from one end of the number or word and their insertion, in the same sequence, at the other end.

cycle stealing — A memory cycle stolen from the normal CPU operation for a DMA operation.

cycle time—1. The interval of time between the occurrence of corresponding parts of successive cycles. 2. The length of time required to obtain information from a memory and then write information back into the memory. Also called read-write cycle time, since it is normally equal to the sum of the write time and the read time.

cycle timer — A controlling mechanism which opens or closes contacts according to a preset cycle.

cyclically magnetized condition—The condition of a magnetic material after being under the influence of a magnetizing force varying between two specific limits until, for each increasing (or decreasing) value of the magnetizing force, the magnetic-flux density has the same value in successive cycles.

cyclic code—Positional notation, not necessarily binary, in which quantities differing by one unit are represented by expressions which are identical except for one place or column, and the digits in that place or column differ by only one unit. Cyclic codes are often used in mechanical devices because no ambiguity exists at the changeover point between adjacent quantities.

cyclic decimal code—A four-bit binary code word only one digit of which changes state between one sequential code word and the next, and which translates to decimal numbers. It is categorized as one of a group of unit-distance codes.

cyclic memory—A memory that continuously stores information but provides access to any piece of stored information only at multiples of a fixed time called the cycle time.

cyclic shift—A shift in which the data moved out of one end of the storing register are re-entered into the other end, as in a closed loop.

cycling—1. A rhythmic variation, near the desired value, of the factor under control. 2. A periodic change from one value to another of the controlled variable in an automatic control system.

cycling vibration—Sinusoidal vibration applied to an instrument and varied in such a way that the instrument is subjected to a specified range of vibrational frequencies.

cycloconverter — A step-down static frequency converter that produces a constant or a precisely controllable output frequency from a variable-frequency ac power input. In general, the frequency ratio chosen is three to one or greater.

cyclogram — An oscilloscope display obtained by monitoring two voltages having a direct cyclic relationship to each other.

cyclograph—A device in which an electron beam moves in two directions, at right angles.

cycloidotron—A fast-wave crossed-field microwave tube similar to the cyclotron resonance tube.

cyclometer register—A set of four or five wheels numbered from 0 to 9 inclusive on their edges, and enclosed and connected by gearing so that the register reading appears at a series of adjacent digits.

cyclotron — 1. A device consisting of an evacuated tank in which positively charged particles (for example, protons, deuterons, etc.) are guided in spiral paths by a static magnetic field while being highly accelerated by a fixed-frequency electric field. 2. Type of accelerator of nuclear particles (protons or deuterons) that uses an oscillating electric field and a fixed magnetic field to accelerate the particles. *See* Accelerator.

cyclotron frequency — The frequency at which an electron traverses an orbit in a steady, uniform magnetic field and zero electric field. Given by the product of the electronic charge and the magnetic-flux density, divided by 2π times the electron mass.

cyclotron-frequency magnetron oscillations — Those oscillations having substantially the same frequency as that of the cyclotron.

cyclotron radiation — The electromagnetic radiation emitted by charged particles orbiting in a magnetic field. It arises from the centripetal acceleration of the particle moving in a circular orbit.

cyclotron resonance—The effect characterized by the tendency of charge carriers to spiral around an axis in the same direction as an applied magnetic field, with an angular frequency determined by the value of the applied field and the ratio of the charge to the effective mass of the charge carrier.

cylindrical-film storage—A computer storage device, each storage element of which is a short length of glass tubing with a thin film of nickel-iron alloy on its outer surface. Wires running through the tubing act as bit and sense lines, and conducting straps at right angles to the tubing function as word lines.

cylindrical reflector—A reflector which is part of a cylinder, usually parabolic.

cylindrical wave—A wave the equiphase surfaces of which form a family of coaxial cylinders.

Czochralski technique—A method of growing large single crystals by pulling them from a molten state. Usually used for growing single crystals of germanium and silicon.

D

D – Symbol for electrostatic flux density, deuterium, dissipation factor, or drain electrode.

DAC – Abbreviation for digital-to-analog converter.

d-a decoder—A device that changes a digital word to an equivalent analog value.

Dag – Trademark of Acheson Industries, Inc. Abbreviation for Aquadag.

daisy chain—In a computer, a bus line which is interconnected with units in such a way that the signal passes from one unit to the next in serial fashion. (The architecture of the Fairchild F-8 provides an example of daisy-chained memory chips. Each chip connects to its neighbors to accomplish daisy-chaining of interrupt priorities beginning with the chip closest to the CPU.)

Damon effect—The change in susceptibility of a ferrite, caused by a high rf power input.

damped natural frequency – 1. The frequency at which a system with a single degree of freedom will oscillate, in the presence of damping, after momentary displacement from the rest position by a transient force. 2. The rate of free oscillation of a sensing element in the presence of damping.

damped oscillation – The oscillation that occurs when the amplitude of the oscillating quantity decreases with time. If the rate of decrease can be expressed mathematically, the name of the function describes the damping. Thus, if the rate of decrease is expressed as a negative exponential, the system is said to be an exponentially damped system.

damped waves—Waves in which successive cycles at the source progressively diminish in amplitude.

dampen—1. To diminsh progressively in amplitude; usually said of waves or oscillations. 2. To deaden vibrations.

damper tube—The tube which conducts in the horizontal-output circuit of a television receiver when the current in the horizontal-deflecting yoke reaches its negative peak. This causes the sawtooth deflection current to decrease smoothly to zero instead of continuing to oscillate.

damper winding – 1. In electric motors, a permanently short-circuited winding, usually uninsulated, and arranged so that it opposes rotation or pulsation of the mag-

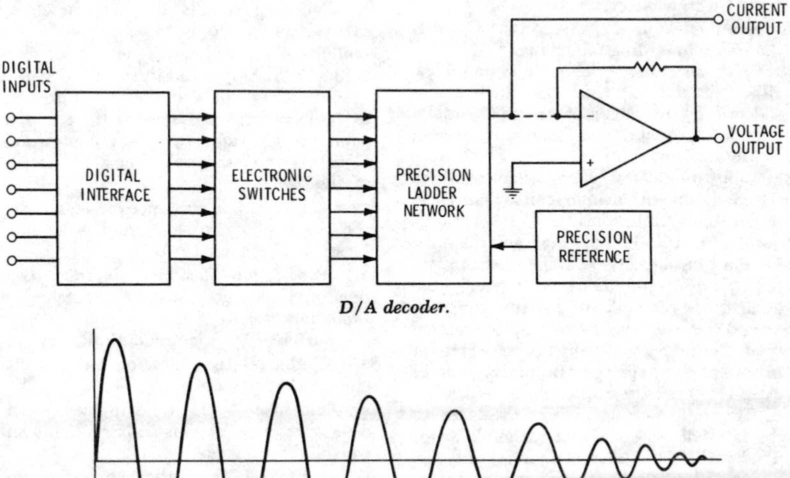

D/A decoder.

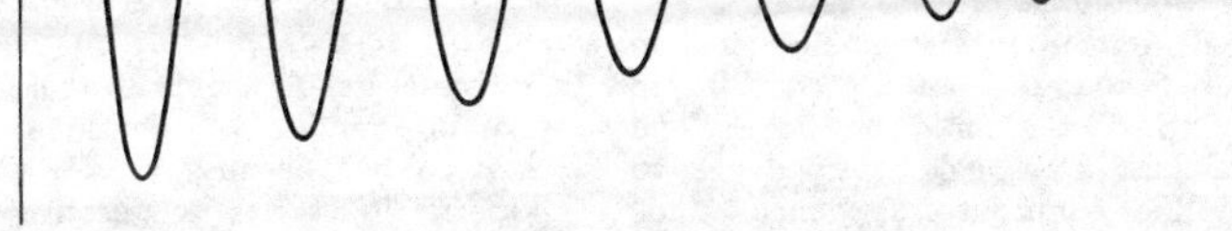

Damped oscillation.

netic field with respect to the pole shoes. 2. Also called amortisseur winding. A winding of copper bars or rods (squirrel cage) embedded in the pole face of synchronous motors and generators. Used as a starting winding on synchronous motors, acts as the squirrel cage of an induction motor. After the synchronous motor is up to speed, tends to prevent oscillations and hunting—a damping effect—thus, the name "damper winding."

damping—1. Reduction of energy in a mechanical or electrical system by absorption or radiation. 2. Act of reducing the amplitude of the oscillations of an oscillatory system, hindering or preventing oscillation or vibration, or diminishing the sharpness of resonance of the natural frequency of a system. 3. The dissipation of kinetic energy in a system by a controlled energy-absorbing medium. A system can be described as being either critically damped, overdamped, or underdamped. 4. The manner in which the pointer settles to its steady indication after a change in the value of the measured quantity. Two general classes of damped motion are: a. Periodic, in which the pointer oscillates about the final position before coming to rest; b. Aperiodic, in which the pointer comes to rest without overshooting the rest position. Sometimes referred to as overdamping. 5. The energy dissipating characteristic which, together with natural frequency, determines the upper limit of frequency response and the response time characteristics of a transducer.

damping coefficient — The ratio of actual damping to critical damping.

damping constant — The Napierian logarithm of the ratio of the first to the second of two values of an exponentially decreasing quantity separated by a unit of time.

damping diode — A vacuum tube which damps the positive or negative half-cycle of an ac voltage.

damping factor—1. For any underdamped motion during any complete oscillation, the quotient obtained by dividing the logarithmic decrement by the time required by the oscillation. 2. Numerical quantity indicating ability of an amplifier to operate a speaker properly. Values over 4 are usually considered satisfactory. 3. The ratio of rated load impedance to the internal impedance of an amplifier. 4. The ratio (larger to smaller) of the angular deviations of the pointer of an electrical indicating instrument on two consecutive swings from the equilibrium position. 5. *See* Decrement, 1. 6. The ratio of load or speaker impedance to the amplifier's output impedance. Thus the smaller the output impedance the greater the damping factor. The damping factor increases with increase of voltage negative feedback, and with the large amounts of feedback applied to transistor hi-fi amplifiers the source or output impedance can be as low as .1 ohm, giving a damping factor of 80 referred to an 8-ohm load.

Such a small value of output impedance is "seen" by the speaker as a virtual short-circuit (excluding the resistance of the speaker connecting cable) and this has the effect of damping the cone movement electromagnetically, thereby minimizing the speaker output at the resonance frequencies. The bass resonance is important from this respect, so the damping factor should remain desirably high down to 40 Hz or less.

damping magnet — A permanent magnet and a movable conductor such as a sector or disc arranged in such a way that a torque (or force) is produced which tends to oppose any relative motion.

damping ratio—1. The ratio of the degree of actual damping to the degree of damping required for critical damping. May be affected by changes in ambient temperature. 2. Of a galvanometer, the ratio, expressed as a positive number, of a given deflection to the next deflection in the opposite direction. The greater this ratio, the greater the degree of damping. The natural logarithm of this ratio is called the logarithmic decrement.

dancer arm—A device which senses tape tension, and signals the reel motor to take up or to supply tape.

Daniell cell—A cell having a copper electrode in a copper-sulfate solution and a zinc electrode in a diluted sulfuric acid or zinc-sulfate solution, with the two solutions separated by a porous partition. Generates an essentially constant electromotive force of about 1.1 volts.

daraf—The unit of elastance. It equals the reciprocal of capacitance and is actually farad spelled backward.

dark conduction—Residual electrical conduction in a photosensitive substance in total darkness.

dark current—*See* Electrode Dark Current.

dark discharge—In a gas, an electric discharge that has no luminosity.

dark-field disc—A disc used in the optical-electronic type of cell counter that controls light transmission.

dark resistance—The resistance of a photoelectric device in total darkness.

dark satellite—A satellite that does not give information to friendly ground stations, either because it is controlled or because it carries inoperative radiating equipment.

dark space—A nonluminous region of a glow-discharge tube.

dark spot—A phenomenon sometimes observed in a reproduced television image. It is caused by the formation of electron clouds in front of the mosaic screen in the transmitter camera tube.

dark spot signal—The signal existing in a television system while the television camera is scanning a dark spot.

dark-trace tube—A cathode-ray tube with a screen coated with a halide of sodium or potassium. The screen normally is nearly white, and whenever the electron beam strikes, it turns a magenta color which is of long persistence. The screen can be illuminated by a strong light source, so that the reflected image may be made intense enough to be projected.

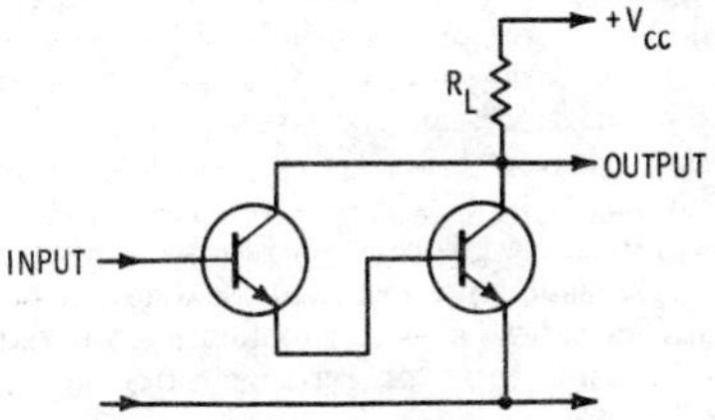

Darlington amplifier.

Darlington amplifier—Also called Darlington pair, double-emitter follower, or β multiplier. 1. A transistor circuit which, in its original form, consists of two transistors in which the collectors are tied together and the emitter of the first transistor is directly coupled to the base of the second transistor. Therefore, the emitter current of the first transistor equals the base current of the second transistor. This connection of two transistors can be regarded as a compound transistor with three terminals. 2. A two-transistor amplifier connected so that the amplification of the amplifier equals the product of the individual transistors' amplification. 3. A composite configuration of transistors which provides a high input impedance and a high degree of amplification.

Darlington-connected phototransistor — A phototransistor whose collector and emitter are connected to the collector and base, respectively, of a second transistor. The emitter current of the input transistor is amplified by the second transistor and the device has a very high sensitivity to light.

Darlington connection—A form of compound connection in which the collectors of two or more transistors are connected together and the emitter of one is connected to the base of the next. Two transistors connected in this way constitute a Darlington pair.

Darlington pair—*See* Darlington Amplifier.

D'Arsonval current — A high-frequency, low-voltage current of comparatively high amperage.

D'Arsonval galvanometer — A dc galvanometer consisting of a narrow rectangular coil suspended between the poles of a permanent magnet.

D'Arsonval instrument — *See* Permanent-Magnet Moving-Coil Instrument.

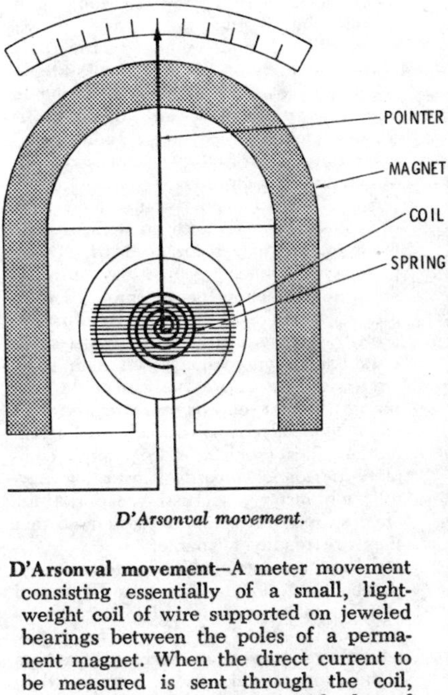

D'Arsonval movement.

D'Arsonval movement—A meter movement consisting essentially of a small, lightweight coil of wire supported on jeweled bearings between the poles of a permanent magnet. When the direct current to be measured is sent through the coil, its magnetic field interacts with that of the permanent magnet and causes the coil and attached pointer to rotate.

dash—Term used in radiotelegraphy. It consists of three units of a sustained transmitted signal followed by one unit during which no signal is transmitted.

dashpot—A device using a gas or liquid to absorb energy from or retard the movement of the moving parts of a circuit breaker or other electrical or mechanical device.

data—A general term used to denote any or all facts, numbers, letters, and symbols, or facts that refer to or describe an object, idea, condition, situation, or other factors. It connotes basic elements of information which can be processed or produced by a computer. Sometimes data are considered to be expressible only in numerical form, but information is not so limited.

data access arrangement—A protective connecting arrangement that serves as an interface between a customer-provided modem and the switched network.

data acquisition—The process by which events in the real world are translated to machine-readable signals. The term usually refers to automated systems in which sensors of one type or another are attached to machinery.

data acquisition system—1. A system in which a computer at a central computing facility gathers data from multiple remote locations. 2. System for recording data, usually in digital form, from several sources; can include computing functions.

data acquisition and conversion system—A method of processing analog signals and converting them into digital form for subsequent processing or analysis by computer or for data transmission.

data bank—A comprehensive collection of libraries of data. For example, one line of an invoice may form an item, a complete invoice may form a record, a complete set of such records may form a file, the collection of inventory control files may form a library, and the libraries used by an organization are known as its data bank. Synonymous with data base.

data base—The entire body of data that has to do with one or more related subjects. Typically, it consists of a collection of data files (such as a company's complete personnel records concerning payroll, job history, accrued vacation time, etc.) stored in a computer system so that they are readily available.

data base management—A systematic approach to storing, updating and retrieval of information stored as data items, usually in the form of records in a file, where many users, or even many remote installations, will use common data banks.

data block—Typically, all the data for one item that is entered into a computer for processing, or the computer output that results from processing. An example of an input data block is an individual shipping list, and an example of an output data block is a check to be sent.

data break—A facility that permits input/output transfers to take place on a cycle-stealing basis without disturbing execution of the program by a computer.

data bus—A wire or group of wires used to carry data to or from a number of different locations.

data code—A structured set of characters used to stand for the data items of a data element, for example, the numerals 1, 2, . . . 7 used to represent the data items Sunday, Monday, . . . Saturday.

data collection—In a computer, the transferring of data from one or more points to a central point. Also called data gathering.

data communications—1. The technology covering the transfer of data over relatively long distances. 2. Transmission of data in both directions between a central location (host computer) and remote locations (terminals) through communication lines. To facilitate this, interfaces such as modems, multiplexers, concentrators, etc. are required at each end of the lines.

data communications processor—A small computer used to control the flow of data between machines and terminals over communications channels. It may perform the functions of a concentrator, handshaking, and formating, but does not include long-term memory or arithmetic functions.

data compression—The process of reducing the number of recorded or transmitted digital data samples through the exclusion of redundant samples.

data control block — A control block through which is communicated to access routines the information required by them for storage and retrieval of data.

data conversion — The changing of data from one form of representation to another.

data display — Visual presentation of processed data by means of special electronic or electromechanical devices connected (either on or off line) with digital computers or component equipment. Although line printers and punch cards may display data, they usually are categorized as output equipment rather than as displays.

data distributor—An array of simple gates that can accept one or more input lines of data and route the data to specific outputs as determined by the levels of control inputs to the array.

data element—An element which converts data functions into a usable signal. (*See also* Element, 2.)

data flowchart—A flowchart that represents the path of data used in a problem and which defines the major phase of the processing as well as the various data media used.

data entry device—An electromechanical device to allow manual input of data to a display system. Examples of data entry devices are: alphanumeric keys, data tablet, function keys, joystick, mouse, and trackball.

data-flow diagram—An illustration having a configuration such that it suggests a certain amount of circuit operation.

data format — The structure and significance of data areas on a storage medium, without reference necessarily to the value of the data contained. Initial values, or limit values are considered part of a data format definition. The data format itself may have been specified by parameter values at system generation time.

data gathering—*See* Data Collection.

data generators – Specialized word generators in which the programming is designed to test a particular class of device, in which the pulse parameters and timing are adjustable, and in which selected words may be repeated, reinserted later in the sequence, omitted, etc.

data-handling capacity – The number of bits of information which may be stored in a computer system at one time. The rate at which these bits may be fed to the input either by hand or with automatic equipment.

data-handling system – Semiautomatic or automatic equipment used in the collection, transmission, reception, and storage of numerical data.

data hierarchy—A data structure, made up of sets and subsets, in which every subset of a set is of lower rank than the data of the set.

data item—The simplest type of information dealt with by a computer system (e.g., a name or employee number). A collection of data items constitutes a record (e.g., payroll information on one employee) and a number of related records constitute a file (e.g., payroll information on all employees of a company).

data link – Electronic equipment which permits automatic transmission of information in digital form.

data management – Those control-program functions that provide access to data sets, enforcement of data-storage conventions, and regulation of the use of input/output devices.

data organization—Any of the data-management conventions for determining the arrangement of a data set.

dataphone—1. A trademark of American Telephone and Telegraph Co. to identify the data sets made and supplied for use in the transmission of data over the regular telephone network. 2. A telephone equipped with a modem and appropriate switching for both voice and data transmission.

data pointer—A register holding the memory address of the data (operand) to be used by an instruction. Thus the register "points" to the memory location of the data.

data processing—The handling of information in a sequence of reasonable operations.

data-processing machines—A general name for a machine which can store and process numerical and alphabetical information.

data-processing system—A network of machine components that can accept information, process it in accordance with a plan, and produce the desired results.

data processor—1. An electronic or mechanical machine for handling information in a sequence of reasonable operations. 2. Any device that can perform operations on data.

data reduction—The process of converting a large quantity of information into a more manageable form, usually including a reduction in volume and a simplification of format.

data-reduction system – Automatic equipment employed to simplify the use and interpretation of a large amount of data gathered by instrumentation.

dataset—1. A circuit termination device used to provide interface between a circuit and terminal input/output equipment. 2. A modem. A device that converts the signals of a business machine to signals that are suitable for transmission over communication lines and vice versa. It may also perform other related functions.

data sheet—A compilation of terminal information on a specific device defining the electrical and mechanical characteristics of that device. Also called spec sheet.

data signaling rate—The data-transmission capacity, expressed in bits per second, of a set of parallel communication channels.

data sink—A communications device that can accept data signals from a transmission device. Also, it may check the received signals and originate signals for error control. (*See also* Data Source.)

data source – A communications device that can originate data signals for a transmission device and may accept error control signals. (*See also* Data Sink.)

data stabilization—Stabilization of a radar display with respect to a selected reference, regardless of changes in the attitude of the vehicle that carries the radar, as in azimuth-stabilized ppi.

data synchronizer—A device that controls and synchronizes the transmission of data between an input/output (i/o) device and the computer system.

data tablet—A data entry device consisting of a stylus and a graphic recorder with a coordinate grid similar to the number space of the screen. By pressing the stylus on the tablet, an interrupt is generated and the coordinates of the stylus are stored in special X-Y input registers. The registers are then read by the host computer. The tablet generally replaces the light pen for cursor and tracking symbol movements, and is used extensively in storage-tube display systems where light-pen tracking and identification are impossible.

data terminal—A common point at which data from various sources is collected and transferred; it may include or connect with several types of data-processing equipment.

data tracks—Positions of information storage on drum storage devices. Information storage on the drum surface is in the form of magnetized and nonmagnetized areas.

data transmission—The sending of information from one place to another or from one part of a system to another.

data-transmission equipment — The communications equipment used in direct support of data-processing equipment.

data transmission system — Means for transmitting dataphone, radio, etc. *See also* Telemetry.

data transmission utilization measure—In a data-transmission system, the ratio of the useful data output to the total data input.

data value—The information contained in data formats. Normally prepared for the generation of specific packages from generic packages.

dB—Abbreviation for decibel.

dBa—Abbreviation for decibels adjusted. Used in conjunction with noise measurements. The reference level is −90 dBm, and the adjustment depends on the frequency-band weighting characteristics of the measuring device.

dBj—A unit used to express relative rf signal levels. The reference level is zero dBj = 1000 microvolts. (Originated by Jerrold Electronics.)

dBk—Decibels referred to 1 kilowatt.

dBm—1. Abbreviation for decibels above (or below) one milliwatt. A quantity of power expressed in terms of its ratio to 1 milliwatt. 2. A term used to denote power level; 0 dBm is equal to 1 milliwatt across a 50-ohm circuit.

dB meter — A meter having a scale calibrated to read directly in decibel values at a specified reference level (usually one milliwatt equals zero dB).

dBRAP—Decibels above reference acoustical power, which is defined as 10^{-16} watt.

dBRN — Decibels above reference noise. This is a unit used to show the relationship between the interfering effect of a noise frequency, or band of noise frequencies, and a fixed amount of noise power commonly called reference noise. A tone of 1000 hertz having a power level of −90 dBm was selected as the reference noise power because it appeared to have negligible interfering effect and would permit the measurement of interfering effect in positive numbers.

dBV—The increase or decrease in voltage independent of impedance levels.

dBW—Decibels referred to 1 watt.

dBx—Decibels above the reference coupling. Reference coupling is defined as the coupling between two circuits that would be required to give a reading of 0 dBa on a two-type noise-measuring set connected to the disturbed circuit when a test tone of 90 dBa (using the same weighting as that used on the disturbed circuit) is impressed on the disturbing circuit.

dc—Abbreviation for direct current.

D cable—Two-conductor cable, each conductor having the shape of the capital letter D, with insulation between the conductors and between the conductors and the sheath.

dc amplifier—*See* Direct-Current Amplifier.

dc balance—An adjustment of circuitry to avoid a change in dc level when the gain is changed.

dc beta—The dc current gain of a transistor; the ratio of the collector current to the base current that caused it, measured at constant collector-to-emitter voltage.

dc block—A coaxial component employed to prevent the flow of dc or video along a transmission line while allowing the uninterrupted flow of rf. The structure is of a short section of coaxial line, having a capacitance in series with the center and/or outer conductor. The rf flows with negligible reflection or attenuation, while the video frequencies or dc are blocked.

dc breakdown—Voltage at which ionization occurs at a slowly rising dc voltage.

dcc—Abbreviation for double cotton-covered.

dc circuit breaker—A device used to close and interrupt a dc power circuit under normal conditions or to interrupt this circuit under fault or emergency conditions.

dc component—The average value of a signal. In television it represents the average luminance of the picture being transmitted, and in radar it is the level from which the transmitted and the received pulses rise.

dc continuity—A circuit in which an impressed dc current—a reading on a conventional ohmmeter applied across the terminals of a circuit with dc continuity—will result in a deflection of the meter.

dc coupled—The connection by a device which passes the steady-state characteristics of a signal and which largely eliminates the transient or oscillating characteristics of the signal.

dc dump—The withdrawal of direct-current power from a computer. This may result in loss of the stored information.

dc generator—A rotating electric device

for converting mechanical power into dc power.

dc inserter stage—A television transmitter stage that adds a dc component known as the pedestal level to the video signal.

dc leakage current—The relatively small direct current through a capacitor when dc voltage is impressed across it. Abbreviated dcl.

dcm—Abbreviation for dc noise margin.

dc noise—The noise arising when reproducing a magnetic tape which has been nonuniformly magnetized by energizing the record head with direct current, either in the presence or absence of bias. This noise has pronounced long-wavelength components which can be as much as 20 dB higher than those obtained from a bulk-erased tape.

dc noise margin—1. The difference between the normal applied logic levels and the threshold voltage of a digital integrated circuit. Abbreviated dcm. Noise margin is also called noise immunity. 2. The difference between the output voltage level of a driving gate and the input threshold voltage of a driven gate for both the "1" and the "0" states.

dc operating point—The dc values of collector voltage and current of a transistor with no signal applied.

dc overcurrent relay—A device which functions when the current in a dc circuit exceeds a given value.

dc patch bay—Specific patch panels provided for termination of all direct-current circuits and equipment used in an installation.

dc picture transmission—Transmission of the dc component of the television picture signal. This component represents the background or average illumination of the overall scene and varies only with the overall illumination.

dc plate resistance—The value or characteristic used in vacuum-tube computations. It is equal to the direct-current plate voltage divided by the direct-current plate current and is given the symbol R_p.

dc reclosing relay—A device which controls the automatic closing and reclosing of a dc circuit interrupter, generally in response to load-circuit conditions.

dc resistivity—The resistance of a body of ferromagnetic material having a constant cross-sectional area, measured under stated conditions by means of direct voltage, multiplied by the cross-sectional area, and divided by the length of the body.

dc restoration—The re-establishment, by a sampling process, of the dc and the low-frequency component which, in a video signal, have been suppressed by ac transmission.

dc restorer—Also called clamper or restorer. A clamping circuit which holds either amplitude extreme of a signal waveform to a given reference level of potential.

dc shift—An error in transient response, with a time constant approaching several seconds.

dc short—A coaxial component that provides a dc circuit between the center and outer conductors, while allowing the rf signal to flow uninterrupted. The unit has a high-impedance line shunted across the main coax line. This consequently makes the device frequency dependent.

dc signaling—A transmission method which utilizes direct current.

DCTL – Abbreviation for direct-coupled transistor logic.

dc transducer—A transducer capable of proper operation when excited with direct current. Its output is given in terms of direct current unless otherwise modified by the function of the stimulus.

dc transmission—Transmission of a television signal in such a way that the dc component of the picture signal is still present. This is done to maintain the true level of background illumination.

dcwv – Abbreviation for direct current working volts. The maximum continuous voltage which can be applied to a capacitor.

deac—A device employed in frequency-modulation receivers to de-emphasize the higher frequencies in the received signal so as to restore their proper relative amplitude. (*See also* De-Emphasis.)

deaccentuator—A network or circuit employed in frequency-modulated receivers to de-emphasize the higher frequencies in the received signal to restore their proper relative amplitude.

deactuate pressure—The pressure at which an electrical contact opens or closes as the pressure approaches the actuation level from the opposite direction.

dead—1. Free from any electric connection to a source of potential difference and from electric charge; having the same potential as that of the earth. The term refers only to current-carrying parts which are sometimes alive, or charged. 2. *See* Room Acoustics.

dead band—1. In a control system, the range of values through which the measurand can be varied without initiating an effective response. 2. Also called dead space, dead zone, or switching blank. A specified range of values in which the incoming signal can be altered without also changing the outgoing response.

deadbeat—Coming to rest without vibration or oscillation—i.e., the pointer which a highly damped meter or galvanometer moves to a new position without overshooting and vibrating about its final position.

deadbeat instrument – A voltmeter, ammeter, or similar device in which the movement is highly damped to bring it to rest quickly.

dead break–An unreliable contact made near the trip point of a relay or switch at low contact pressure. As a result, the switch does not actuate, even though the circuit is interrupted.

dead-center position–The place on the commutator of a dc motor or generator at which a brush would be placed if the field flux were not distorted by armature reaction.

dead end–1. In a sound studio, the end with the greater sound-absorbing characteristic. 2. In a tapped coil, the portion through which no current is flowing at a particular bandswitch position.

dead-end tower–An antenna or transmission-line tower designed to withstand unbalanced mechanical pull from all the conductors in one direction, together with the wind strain and vertical loads.

dead room–1. A room for testing the acoustic efficiency or range of electroacoustic devices such as speakers and microphones. The room is designed with an alsolute minimum of sound reflection, and no two dimensions of the room are the same. A ratio of 3 to 4 to 5 is usually employed (e.g., 15′ × 20′ × 25′). The walls, floor, and ceiling are lined with a sound-absorbing material. 2. *See* Anechoic.

dead short–A short circuit having minimum resistance.

dead space–1. An area or zone, within the normal range of a radio transmitter, in which no signal is received. 2. *See* Dead Band.

dead spot–1. A geographic location in which signals from one or more radio stations are received poorly or not at all. 2. That portion of the tuning range of a receiver where stations are heard poorly or not at all because of poor sensitivity.

dead time–1. The minimum interval, following a pulse, during which a transponder or component circuit is incapable of repeating a specified performance. 2. Any definite delay intentionally placed between two related actions to avoid overlap that could result in confusion or to permit another particular event, such as a control decision or switching event, to occur.

dead volume–The total volume of the pressure port cavity of a pressure transducer at the rest position (i.e., with no stimulus applied).

dead zone–*See* Dead Band, 2.

deafness–*See* Hearing Loss.

debicon–A high-efficiency microwave generator in which use is made of crossed-field effects.

de Brogile wavelength–The wavelength of radiation that corresponds to a photon the energy of which is one electron volt: 1.24 micron.

debug–1. To examine or test a procedure, routine, or equipment for the purpose of detecting and correcting errors. 2. To detect, locate, and remove mistakes from a program. Debugging programs are available that test for and isolate errors in another program.

debugging–1. Isolating and removing all malfunctions ("bugs") from a computer or other device to restore its operation. 2. A process of "shakedown operation" of each finished material which is performed prior to its being placed in use in order to exclude the early failure period. During debugging, "weak" elements are expected to fail and be replaced by elements of normal quality which are not subject to early failure.

debugging period–*See* Early-Failure Period.

debugging routines–Programs that aid in the isolation and correction of malfunctions and/or errors in a unit of equipment or another program.

debunching–Space-charge effect that tends to destroy the electron bunching in a velocity-modulation vacuum tube by spreading the beam due to mutual repulsion of the electrons.

Debye length–A theoretical length that describes the maximum separation at which a given electron is influenced by the electric field of a given positive ion. Also called Debye shielding distance or plasma length.

Debye shielding distance – *See* Debye Length.

decade–1. The interval between any two quantities having a ratio of 10:1. 2. A group or assembly of 10 units (e.g., a counter which counts to 10 in one column, or a resistor box which inserts resistance quantities in multiples of powers of 10).

decade band–A band having frequency limits related by the equation $f_b - f_a = 10$.

decade box–A special assembly of precision resistors, coils, or capacitors. It contains two or more sections, each having 10 times the value of the preceding section. Each section is divided into 10 equal parts. By means of a 10-position selector switch or equivalent arrangement, the box can be set to any desired value in its range.

decade counter–A logic device that has

10 stable states and may be cycled through these states by the application of 10 clock or pulse inputs. A decade counter usually counts in a binary sequence from state 0 through state 9 and then cycles back to 0. Sometimes referred to as a divide-by-10 counter.

decade resistance box—A resistance box containing two or more sets of 10 precision resistors.

Decade resistance box.

decade scaler—A decade counter, or scale-of-10 counter. A scaler with a factor of 10. It produces one output pulse for every 10 input pulses.

decalescent point — The temperature at which there is a sudden absorption of heat as the temperature of a metal is raised.

decametric waves—High-frequency band; 3 MHz to 30 MHz.

decay—1. Gradual reduction of a quantity. 2. The decrease in the radiation intensity of any radioactive material with respect to time. 3. In a storage tube, a change in magnitude or configuration of stored information by any cause other than erasing or writing.

decay characteristic—*See* Persistence Characteristic (of a Luminescent Screen).

decay constant — The probability that an atom will decay per unit of time. It may also be thought of as the fraction of a radioactive population of atoms that decay per unit of time.

decay distance—The distance between an area of wave generation and a point of passage of the resulting waves outside the area.

Decca — A British long-range hyperbolic navigational system that operates in the 70- to 130-kilohertz frequency band. It is a continuous-wave system in which the receiver measures and integrates the relative phase difference between the signal received from two or more synchronized ground stations. One master station and three slave stations are usually arranged in stat formation. Operational range is about 250 miles.

decelerated electrons — Electrons which, after traveling at a great rate of speed, strike a target, become quickly decelerated, and cause the target to emit X rays.

decelerating electrode — In an electron-beam tube, an electrode to which a potential is applied to slow down the electrons in the beam.

deceleration—The act or process of moving, or of causing to move, with decreasing speed; the state of so moving.

deceleration time—1. In a computer, the time interval between the completion of the reading or writing of a record on a magnetic tape and the time when the tape stops moving. 2. The time required to stop a motor, whether free running or with some braking means.

deception—Deliberate production of false or misleading echoes on enemy radar by the radiation of spurious signals synchronized to the radar or by the reradiation of the radar pulses from extraneous reflectors.

deception device—A device that works to make unfriendly signals either unusable or misleading.

deception jamming—*See* Confusion Jamming.

deci—Prefix meaning one-tenth (10^{-1}).

decibel—1. Abbreviated dB. The standard unit for expressing transmission gain or loss and relative power levels. The term "dBm" is used when a power of one milliwatt is the reference level. Decibels indicate the ratio of power output to power input:

$$\text{dB} = 10 \log_{10} \frac{P_1}{P_2}$$

One decibel is 1/10 of a bel. 2. A unit of change in sound intensity. One decibel is approximately the smallest change that the ear can perceive. Larger decibel increments reflect the fact that sound intensity must be squared in order for the ear to perceive a doubling of intensity. An increase in intensity is expressed as a + number of dB's, a decrease as a − value. No change in intensity is 0 dB and 0 is also used to indicate a starting point, from which changes are measured. 3. A unit used to measure and compare signal levels on a logarithmic scale.

decibel meter—1. Also called dB meter. An instrument for measuring the electric power level, in decibels, above or below an arbitrary reference level. 2. Sound-level indicator.

decibels above or below 1 milliwatt—The unit used to describe the ratio of the power at any point in a transmission system to a reference level of 1 milliwatt. The ratio expresses decibels above or below this reference level of 1 milliwatt.

decibels above or below 1 watt—A measure of power expressed in decibels to a reference level of 1 watt.

decibels above reference noise – An expression used to describe the ratio of the circuit noise level in a transmission system, at any point, to some arbitrarily chosen reference noise. The expression signifies the reading of a noise meter. Where the circuit noise meter has been adjusted to represent effect under specified conditions, the expression is in adjusted decibels.

decilog – A division of the logarithmic scale used for measuring the logarithm of the ratio of two values of any quantity. The number of decilogs is equal to 10 times the logarithm to the base 10 of the ratio. One decilog therefore corresponds to a ratio of $10^{0.1}$ (i.e. 1.25892+).

decimal—1. Pertaining to a characteristic or property involving a selection, choice, or condition in which there are 10 possibilities. 2. Pertaining to the number representation system with a radix of 10. 3. Pertaining to a system of numerical representation in which there are ten symbols, 0, 1, 2, 3, . . . 9.

decimal attenuator—A system of attenuators arranged so that a voltage or current can be reduced decimally.

decimal-binary switch—A switch by means of which a single input lead is connected to appropriate combinations of four output leads (representing 1, 2, 4, and 8) for each of the decimal-numbered settings of the associated control knob. For example, with the knob in position 7, the input lead would be connected to output leads 1, 2, and 4.

decimal code—A code in which each allowable position has one of ten possible states. The conventional number system with the base ten is a decimal code.

decimal-coded digit—One of ten arbitrarily selected patterns of 1's and 0's that are used to represent decimal digits.

decimal digit – One of the numbers 0 through 9 used in the number system with the base ten.

decimal encoder – An encoder in which there are ten output lines, one for each digit from 0 to 9, for each decade of decimal numbers.

decimal notation—The writing of quantities in the decimal numbering system.

decimal numbering system—The popular numbering system using the Arabic numerals 0 through 9 and thus having a base, or radix, of 10. For example, the decimal number 2345 can be derived in this way:

$$2000 + 300 + 40 + 5 = 2345$$

or:

$$2(10^3) + 3(10^2) + 4(10^1) + 5(10^0) = 2345.$$

In the decimal system, all numbers are obtained by raising the radix (total number of marks, or 10 in this system) to various powers.

decimal point—In a decimal number, the point which marks the place between integral and fractional powers of 10.

decimal-to-binary conversion—The mathematical process of converting a number written in the scale of 10 into the same number written in the scale of 2.

Decimal	*Binary*	*Decimal*	*Binary*
0	0	10	1010
1	1	11	1011
2	10	12	1100
3	11	13	1101
4	100	14	1110
5	101	15	1111
6	110	16	10000
7	111	32	100000
8	1000	64	1000000
9	1001	128	10000000

decimetric waves – 1. Electromagnetic waves having wavelengths between 0.1 and 1 meter. 2. Ultrahigh frequency band; 300 MHz to 3 GHz.

decineper—One-tenth of a neper.

decinormal calomel electrode—A calomel electrode containing a decinormal potassium chloride solution.

decision—In a computer, the process of determining further action on the basis of the relationship of two similar items of data.

decision box—On a flowchart, a rectangle or other symbol used to mark a choice or branching in the sequence of programming of a digital computer.

decision element—In computers or data-handling systems, a circuit which performs a logical operation—such as AND, OR, NOT, or EXCEPT on one or more binary digits of input information which represent "yes" or "no"—and expresses the result in its output.

decision table—A table of all contingencies that are to be considered in the description of a problem, together with the actions to be taken. Decision tables are sometimes used in place of flowcharts for problem description and documentation.

deck—1. In computer usage, a collection of cards, usually a complete set of cards punched for a definite purpose. 2. A term usually applied to a tape machine having no built-in power amplifiers or loudspeakers of its own, but intended rather for feeding a separate amplifier and speaker system, as in a component installation.

declination – The angular difference between the position of a compass needle and the true position of geographical north and south.

declinometer–Also called a compass declinometer. A device for measuring the direction of a magnetic field relative to astronomical or survey coordinates.

decode–1. In a computer, to obtain a specific output when specific character-coded input lines are activated. 2. To use a code to reverse a previous encoding. 3. To determine the meaning of characters or character groups in a message. 4. To determine the meaning of a set of pulses that describes an instruction, a command, or an operation to be carried out.

decoder – 1. A device for translating a combination of signals into one signal that represents the combination. It is often used to extract information from a complex signal. 2. In automatic telephone switching, a relay-type translator which determines from the office code of each call the information required for properly recording the call through the switching train. Each decoder has means, such as a cross-connecting field, for establishing the controls desired and readily changing them. 3. Sometimes called matrix. In an electronic computer, a network or system in which a combination of inputs is excited at one time to produce a single output. 4. A device that converts coded information into a more usable form, for example, a binary-to-decimal decoder. 5. A circuit which accepts coded input data and activates a specific output(s) in accordance with the code present at the input. 6. A circuit built into an fm tuner to enable it to translate stereo signal information into two matched audio outputs. 7. A means to extract and process recorded quadraphonic sound information from a complex signal into four matched outputs. 8. *See* Code Converter.

decoding – 1. The process of obtaining intelligence from a code signal. 2. In multiplex, a process of separating the subcarrier from the main carrier.

decoding matrix–A device for decoding many input lines into a single output line.

decoding network–A circuit made so that, when a particular combination of inputs is on, an output appears on one of a number of output lines.

decommutation–The process of recovering a signal from the composite signal previously created by a commutation process.

decommutator–Equipment for separating, demodulating, or demultiplexing commutated signals.

decoupler–A circuit for eliminating the effect of coupling in a common impedance.

decoupling–The reduction of coupling.

decoupling circuit–A circuit used to prevent interaction of one circuit with another.

decoupling network–A network of capacitors and chokes or resistors placed into leads which are common to two or more circuits, to prevent unwanted, harmful interstage coupling.

decoy – A reflecting object used in radar deception, having reflective characteristics of a target.

decrement – 1. Progressive diminution in the value of a variable quantity; also the amount by which a variable decreases. When applied to damped oscillations, it is usually called damping factor. 2. A specific part of an instruction word in some binary computers, thus a set of digits.

decremeter – An instrument for measurement of the logarithmic decrement (damping) of a wave train.

dedicated–To set apart for some special use. For example, a dedicated microprocessor is one that has been specifically programmed for a single application such as weight measurement by scale, traffic light control, etc. (ROMs by their very nature [read-only] are "dedicated" memories.)

dee–A hollow, D-shaped accelerating electrode in a cyclotron.

dee line–A structural member that supports the dee of a cyclotron and together with the dee forms the resonant circuit.

de-emphasis–1. Also called post-emphasis or post-equalization. Introduction of a frequency-response characteristic which is complementary to that introduced in preemphasis. 2. Reduction of the level of the higher audio frequencies during fm reception or tape replay, so that they compensate for the preemphasis (which see) applied to the transmission. This restores an overall uniform response, with improved signal-to-noise ratio.

de-emphasis network–A network inserted into a system to restore the preemphasized frequency spectrum to its original form.

de-energize–To disconnect a device from its power source.

deep discharge – The withdrawal of all available electrical energy before recharging a cell or battery.

deep space net–A combination of radar and communications stations in the U.S.,

Australia and South Africa so located as to keep a spacecraft in deep space under observation at all times.

defect—A condition considered potentially hazardous or operationally unsatisfactory and therefore requiring attention.

defect conduction—Hole conduction in the valence band in a semiconductor.

deferred entry—In a computer, an entry into a subroutine as a result of a deferred exit from the program that passes control to the subroutine.

deferred exit—In a computer, the transfer of control to a subroutine at a time controlled by the occurrence of an asynchronous event rather than at a predictable time.

defibrillator—An electronic device that applies a high, brief voltage potential to the heart by means of electrodes placed on the chest wall. The defibrillator is used to restore regular rhythm to a heart in ventricular fibrillation.

definite-purpose relay — A relay with some electrical or mechanical feature which distinguishes it from a general-purpose relay.

definition—1. The fidelity with which the detail of an image is reproduced. When the image is sharp (i.e., has definite lines and boundaries), the definition is said to be good. 2. The degree with which a communication system reproduces sound images or messages. 3. The fidelity with which the pattern edges in a printed circuit (conductors, inductors, etc.) are reproduced relative to the original master pattern.

deflecting coil—An inductor used to produce a magnetic field that will bend the electron beam a desired amount in the cathode-ray tube of an oscilloscope, television receiver, or television camera.

deflecting electrode — An electrode to which a potential is applied in order to deflect an electron beam.

deflecting torque—*See* Torque of an Instrument.

deflection—Movement of the electron beam in a cathode-ray tube as electromagnetic or electrostatic fields are varied to cause the light spot to traverse the face of the tube in a predetermined pattern.

deflection coil—One of the coils in the deflecting yoke.

deflection factor—*See* Deflection Sensitivity.

deflection plane—A plane perpendicular to the cathode-ray-tube axis and containing the deflection center.

deflection plates — Two pairs of parallel electrodes, the pairs set one forward of the other and at right angles to each other, parallel to the axis of the electron stream within an electrostatic cathode-ray tube. An applied potential produces an electric field between each pair. By varying the applied potential, this field may be varied to cause a desired angular displacement of the electron stream.

deflection polarity—The relationship between the direction of displacement of an oscilloscope trace and the polarity of the applied signal wave.

deflection sensitivity — Also called deflection factor. The displacement of the electron beam at the target or screen of a cathode-ray tube divided by the change in magnitude of the deflecting field. Deflection sensitivity is usually expressed in millimeters (or inches) per volt applied between the deflecting electrodes, or in millimeters (or inches) per ampere in the deflection coil.

deflection voltage—The voltage applied to the electrostatic plates of a cathode-ray tube to control the movement of the electron beam.

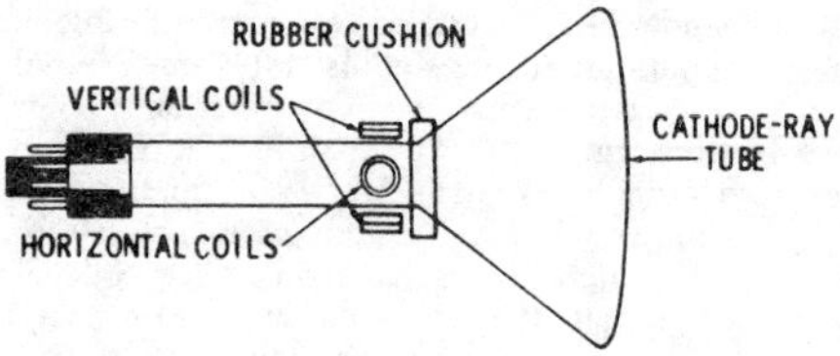

Deflection yoke.

deflection yoke—An assembly of one or more electromagnets for deflecting one or more electron beams.

defocus-dash mode—A method of storage of binary digits in a cathode-ray tube. Initially, the writing beam is defocused to excite a small circular area on the screen. For one kind of binary digit the beam remains defocused, and for the other kind of binary digit it is suddenly focused into a concentric dot, which traces out a dash on the screen during the interval of time before the beam is cut off and moved to the next position.

defocus-focus mode—A variation of the defocus-dash mode in which the focused dot is not caused to trace a dash.

defruiting — Method of eliminating asynchronous returns in radar beacon systems.

degassing—The process of driving out and exhausting the gases of an electron tube occluded in its internal parts.

degausser—1. Also called automatic degausser and bulk eraser. A device to clarify the color picture by means of coils within the set. The coils deactivate the magnetization which builds up around a color tv set when it is moved around, or when other electrical devices are brought too close to the receiver. 2. Any device for neutralizing magnetism, as in a recorder head or in a separate unit. Also called a "tape eraser" for use with a complete tape recording on its reel.

degaussing—Girdling a ship's hull with a web of current-carrying cable that sets up a magnetic field equal in value and opposite in polarity to that induced by the earth's magnetic field, thus rendering the ship incapable of actuating the detonator of a magnetic mine.

degeneracy—The condition in which two or more modes have the same resonant frequency in a resonant device.

degenerate modes—A set of modes having the same resonance frequency (or propagation constant). The members of a set of degenerate modes are not unique.

degenerate parametric amplifier – An inverting parametric device for which the two signal frequencies are identical and equal to one-half the frequency of the pump. (This exact but restrictive definition is often relaxed to include cases where the signals occupy frequency bands which overlap.)

degeneration—*See* Negative Feedback.

degradation—A gradual decline of quality or loss of ability to perform within required limits. The synonym "drift" is often used for electronic devices.

degradation failure—Failure of a device because a parameter or characteristic changes beyond some previously specified limit.

degree of current rectification—The ratio between the average unidirectional current output and the root-mean-square value of the alternating-current input from which it was derived.

degree of voltage rectification—The ratio between the average unidirectional voltage and the root-mean-square value of the alternating voltage from which it was derived.

deion circuit breaker – A circuit breaker built so that the arc that forms when the circuit is broken is magnetically blown into a stack of insulated copper plates, giving the effect of a large number of short arcs in series. Each arc becomes almost instantly deionized when the current drops to zero in the alternating-current cycle, and the arc cannot re-form.

deionization—The process by which an ionized gas returns to its neutral state after all sources of ionization have been removed.

deionization potential – The potential at which ionization of the gas within a gas-filled tube ceases and conduction stops.

deionization time—The time required for the grid of a gas tube to regain control after the anode current has been interrupted.

dekahexadecimal—*See* Sexidecimal Notation.

Dekatron—A cold-cathode counting tube.

delay—1. The time required for a signal to pass through a device or conductor. 2. The time interval between the instants at which any designated point in a wave passes any two designated points of a transmission circuit.

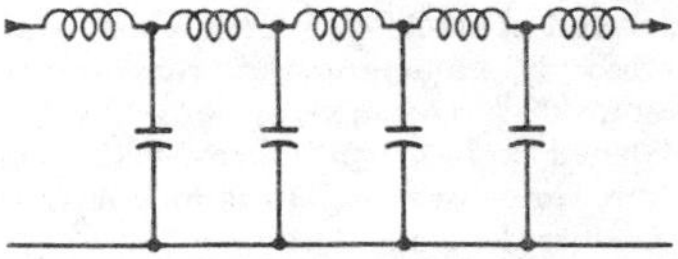

Lumped-constant line.

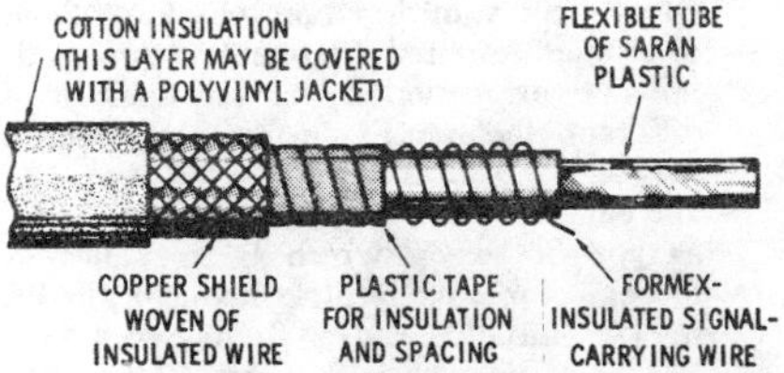

Early-model delay line.

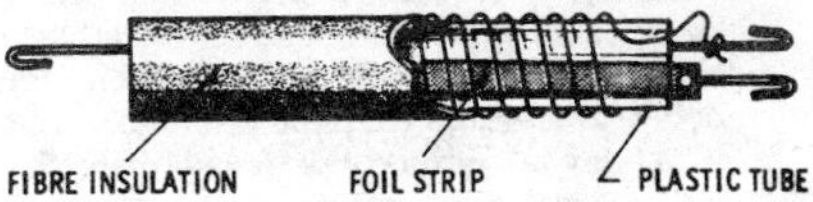

Modern delay line.

Delay lines, 1.

delay circuit—A circuit which delays the passage of a pulse or signal from one part of a circuit to another.

delay coincidence circuit—A coincidence circuit actuated by two pulses, one of which is delayed a specific amount with respect to the other.

delay counter—In a computer, a device that can temporarily delay a program a sufficient length of time for the completion of an operation.

delay distortion—1. Phase-delay distortion (i.e., departure from flatness in the phase delay of a circuit or system over the frequency range required for transmission, or the effect of such departure on a transmitted signal). 2. Also called envelope delay distortion or phase delay. Envelope distortion (i.e., departure from flatness in the envelope delay of a circuit or system over the frequency range required for transmission, or the effect of such departure on a transmitted signal). 3. The amount of variation in delay for various frequency components of the facsimile signal, usually expressed in microseconds from an average delay time. 4. The difference between the maximum and minimum phase delay within a specified band of frequencies.

delayed automatic volume control—Abbreviated delayed avc. An automatic volume-control circuit that acts only on signals above a certain strength. It thus permits reception of weak signals even though they may be fading, whereas normal automatic volume control would make the weak signals even weaker.

delayed avc—*See* Delayed Automatic Volume Control.

delayed contacts—Contacts which are actuated a predetermined time after the start of a (timing) cycle.

delayed ppi—A ppi (plan-position indicator) in which the initiation of the time base is delayed.

delayed repeater satellite — A satellite which stores information obtained from a ground terminal at one location and, upon interrogation by a terminal at a different location, transmits the stored message.

delayed sweep—1. In a cathode-ray tube, a type of sweep which is not allowed to begin for a while after being triggered by the initiating pulse. 2. A sweep that has been delayed either by a predetermined period or by a period determined by an additional independent variable.

delay equalizer—1. A device which adds delay at certain frequencies to a circuit in a way to reduce the delay distortion. 2. A corrective network which is designed to make the phase delay or envelope delay of a circuit or system substantially constant over a desired frequency range. 3. A network that introduces an amount of phase shift complementary to the phase shift in the circuit at all frequencies within the desired band.

delay/frequency distortion—That form of distortion which occurs when the envelope delay of a circuit or system is not constant over the frequency range required for transmissions.

delay line—1. A real or artificial transmission line or equivalent device designed to delay a signal or wave for a predetermined length of time. 2. A specially constructed cable used in the luminance channel of a color receiver to delay the luminance signal. 3. A sequential logic element that has one input channel and in which the state of an output channel at any instant is the same as the state of the input channel at the instant t — n, where n is a constant time interval for a given output channel (the input sequence undergoes a delay of n time units.) 4. A device that can cause the transmission of one unit of information to be retarded until another unit can synchronize with it. 5. A device capable of causing an energy impulse to be retarded in time from point to point, thus providing a means of storage by circulating intelligence-bearing pulse configurations and patterns. Examples of delay lines are material media such as mercury, in which sonic patterns may be propagated in time; lumped constant electrical lines; coaxial cables, transmission lines and recirculating magnetic drum loops.

delay-line memory—*See* Delay-Line Storage.

delay-line register—An acoustic or electric delay line in an electronic computer, usually one or an integral number of words in length, together with input, output, and circulation circuits.

delay-line storage—Also called delay-line memory. In an electronic computer, a storage or memory device consisting of a delay line and a means for regenerating and reinserting information into it.

delay multivibrator — A monostable multivibrator that produces an output pulse a predetermined time after it is triggered by an input pulse.

delay ppi—A radar indicator in which the start of the display sweep is delayed after the trigger so that distant targets are displayed on a short range scale that gives an expanded presentation.

delay relay—Also called time-delay relay. A relay in which there is a delay between the time it is energized or de-energized and the time the contacts open or close.

delay time—The amount of time one signal is behind (lags) another.

delay timer—A term sometimes used to designate a timer which is primarily used for energizing (or de-energizing) a load at the end of a timed period. (*See* Time Delay Relay.)

delay unit—The unit of a radar system in which pulses may be delayed a controllable amount.

deletion record — In a computer, a new record to replace or remove an existing record in a master file.

delimiter—1. In a computer, a character that limits a string of characters and therefore cannot be a member of the string. 2. Also called separator. A flag that separates and organizes items of data.

Dellinger effect—*See* Radio Fadeout.

delta—1. The Greek letter delta (Δ) represents any quantity which is much smaller than any other quantity of the same units appearing in the same problem. 2. In a magnetic cell, the difference between the partial-select outputs of the same cell in a one state and in a zero state. 3. Brain wave signals whose frequency is approximately 0.2 to 3.5 Hz. The associated mental state is usually a deep sleep, or a trancelike state.

delta circuit — A three-phase circuit in

which the windings of the system are connected in the form of a closed ring, and the instantaneous voltages around the ring equal zero. There is no common or neutral wire, so the system is used only for three-wire systems or generators.

delta connection—In a three-phase system, the terminal connections. So called because they are triangular like the Greek letter delta.

delta match—*See* Y Match.

delta matched antenna—Also called Y antenna. A single-wire antenna (usually one half-wavelength long) to which the leads of an open-wire transmission line are connected in the shape of a Y. The flared part of the Y matches the transmission line to the antenna. The top of the Y is not cut, giving the matching section its triangular shape of the Greek letter *delta*, hence the name.

delta matching transformer—An impedance device used to match the impedance of an open-wire transmission line to an antenna. The two ends of the transmission line are fanned out so that the impedance of the line gradually increases. The ends of the transmission line are attached to the antenna at points of equal impedance, symmetrically located with respect to the center of the antenna.

delta modulation—A means of encoding analog signals in control and communication systems. The output of the delta encoder is a single weighed digital pulse train which may be decoded at the receiving end to reconstruct an original analog signal.

delta network—A set of three branches connected in series to form a mesh.

delta pulse code modulation—A modulation system that converts audio signals into corresponding trains of digital pulses to provide greater freedom from interference during transmission over wire or radio channels.

delta tune—A control provided on some transceivers which permits tuning the receiving frequency slightly off the center to compensate for variations in transmitting frequency of other transceivers.

delta wave—A brain wave the frequency of which is below 9 hertz.

dem—Abbreviation for demodulator.

demagnetization—Partial or complete reduction of residual magnetism.

demagnetization curve — In the second quadrant of a hysteresis loop, the portion which lies between the residual induction point, B_r, and the coercive force point, H_c.

demagnetization effect—A decrease in internal magnetic field caused by uncompensated magnetic poles at the surface of a sample.

demagnetizer — A device which removes residual magnetism from recording or playback tape heads. This magnetism, if not removed, can introduce noise on recordings and cause high-frequency loss.

demagnetizing force—A magnetizing force applied in such a direction that it reduces the residual induction in a magnetized body.

demand factor—The ratio of the maximum demand of a power-distribution system to the total connected load of the system.

demarcation strip — A physical interface, usually a terminal board, between a business machine and a common carrier. *See also* Interface, 1.

Dember effect—Also known as the photo-diffusion effect. The production of a potential difference between two regions of a semiconductor specimen when one is illuminated. This phenomenon is related to the photoelectromagnetic effect, except there is no magnetic field. H. Dember discovered when an illuminated metal plate, producing electrons, is bombarded by other electrons from an outside source, the photoelectric emission increases because, in addition to photoelectrons, secondary electrons are also knocked out by bombardment.

demodulation—Also called detection. The operation on a previously modulated wave in such a way that it will have substantially the same characteristics as the original modulating wave.

demodulator—1. A device which operates on a carrier wave to recover the wave with which the carrier was originally modulated. 2. A facsimile device which detects an amplitude-modulated signal and produces the modulating frequency as a direct current of varying amplitude. This type of unit is used to provide a keying signal for a frequency-shift exciter unit for radio facsimile transmission. 3. A device which receives tones from a transmission circuit and converts them to electrical pulses, or bits, which may be accepted by a business machine. 4. Circuitry that plays back a CD-4 disc's four signals after reprocessing the base and carrier bands inscribed in each side of the record groove.

demodulator probe—A probe designed for use with an oscilloscope, for displaying modulated high-frequency signals.

deMorgan's theorem — A theorem which

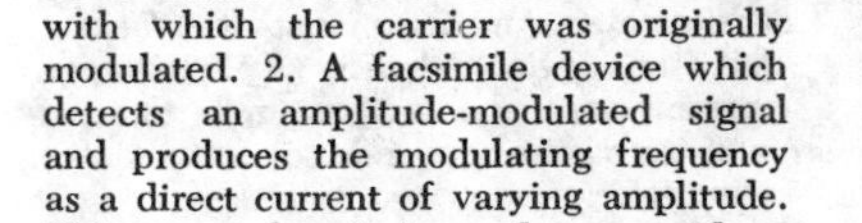

*R MAY BE INCREASED IN VALUE TO REDUCE CIRCUIT LOADING.

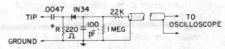

Demodulator probe.

states that the inversion of a series of AND implications is equal to the same series of inverted OR implications, or the inversion of a series of OR implications is equal to the same series of inverted AND implications. In symbols, $\overline{A \cdot B \cdot C} = \overline{A} + \overline{B} + \overline{C}$, or $\overline{A + B + C} = \overline{A} \cdot \overline{B} \cdot \overline{C}$.

demountable tube—A high-power electron tube having a metal envelope with porcelain insulation. Can be taken apart for inspection and for renewal of electrodes.

demultiplexer—1. A device used to separate two or more signals that were previously combined by a compatible multiplexer and are transmitted over a single channel. 2. A circuit that directs information from a single input to one of several outputs at a time in a sequence that depends on the information applied to the control inputs.

demultiplexing circuit – A circuit that is used to separate the signals that have been combined for transmission by multiplex.

denary band – A band having frequency limits with the ratio of $f_b/f_a = 10$.

dendrite—A semiconductor crystal with a heavily branched, treelike structure which grows from the nucleus as the metal becomes solidified.

dendritic growth—A technique of producing semiconductor crystals in long, uniform ribbons with optically flat surfaces.

dense binary code—A binary code in which all the possible states of the pattern are used.

densitometer—An instrument for measuring the optical density (photographic transmission, photographic reflection, visual transmission, and so forth) of a material.

density—1. A measure of the light-reflecting or transmitting properties of an area. 2. The mass per unit volume. The specific gravity of a body is the ratio of a density to the density of a standard substance. Water and air are commonly used as the standard substances. 3. Amount per unit cross-sectional area (e.g., current, magnetic flux, or electrons in a beam). 4. The logarithm of the ratio of incident to transmitted light. *See* Opacity.

density modulation – Modulation of an electron beam by varying the density of the electrons in the beam with time.

density packing—The number of magnetic pulses (representing binary digits) stored on tape or drum per linear inch on a single head.

density step tablet—A facsimile test chart consisting of a series of areas that increase in steps from a low value of density to a maximum value of density.

dentaphone—A device by which deaf persons can hear sounds via the teeth.

dentophonics—The technique of using electronics in broadcasting speech from the mouth. The principle is the same as that of a throat microphone, in which a transducer responds to sound energy transmitted through the tissues as a person speaks.

dependent linearity – Nonlinearity errors expressed as a deviation from a desired straight line of fixed slope and/or position.

depletion field-effect transistor (FET)–An active semiconductor device in which the main current is controlled by the depletion width of a pn junction.

depletion layer—Also called barrier layer. In a semiconductor, the region in which the mobile-carrier charge density is insufficient to neutralize the net fixed charge density of donors and acceptors.

depletion-layer capacitance – Also called barrier capacitance. Capacitance of the depletion layer of a semiconductor. It is a function of the reverse voltage.

depletion-layer rectification – Also called barrier-layer rectification. The rectification that appears at the contact between dissimilar materials, such as a metal-to-semiconductor contact or a pn junction, as the energy levels on each side of the discontinuity are readjusted.

depletion-layer transistor – Any of several types of transistors which rely directly for their operation on the motion of carriers through depletion layers (for example, a spacistor).

depletion-mode field-effect transistor—1. A FET which exhibits substantial device current (IDSS) with zero gate-to-source bias (VGS = 0V). 2. An MOS transistor normally on with zero gate voltage applied (channel formed during processing). A voltage of the correct polarity applied to the gate will force majority carriers from the channel, thus "depleting" it and turning the transistor off.

depletion-mode operation – The operation of a field-effect transistor such that changing the gate-to-source voltage from zero to a finite value decreases the magnitude of the drain current.

depletion region—The region, extending on both sides of a reverse-biased semiconductor junction, in which all carriers are swept from the vicinity of the junction; that is the region is depleted of carriers. This region takes on insulating characteristics and is capable of isolating semiconductor regions from each other. Depletion regions make planar bipolar integrated circuits possible. Also referred to as space-charge, barrier, or intrinsic region.

depolarization—The process of preserving the activity of a primary cell by the addition of a substance to the electrolyte.

This substance combines chemically with the hydrogen gas as it forms, thus preventing excessive buildup of hydrogen bubbles.

depolarize – To make partially or completely unpolarized.

depolarizer–A chemical used in some primary cells to prevent formation of hydrogen bubbles at the positive electrode.

deposited carbon–Resistive element made of a thin film of crystalline carbon or a carbon alloy sputtered onto a ceramic rod.

deposition–The application of a material to a substrate through the use of chemical, vapor, electrical, vacuum, or other processes.

depth finder–*See* Fathometer.

depth of cut–The depth to which the recording stylus penetrates the lacquer of a recording disc.

depth of heating–The depth at which effective dielectric heating can be confined below the surface of a material when the applicator electrodes are placed adjacent to only one surface.

depth of modulation–In a radio-guidance system obtaining directive information from the two spaced lobes of a directional antenna, the ratio of the difference in total field strength of the two lobes to the field strength of the greater lobe at a given point in space.

depth of penetration–The thickness of a layer extending inward from the surface of a conductor and having the same resistance to direct current as the whole conductor has to alternating current of a given frequency. *See also* Skin Depth.

depth sounder–*See* Fathometer.

de-*Q*–To reduce the *Q* of a tuned circuit, as generally applied to carrier-current transmission systems.

derate–To reduce the voltage, current, or power rating of a device to improve its reliability or to permit operation at high ambient temperatures.

derating–The reduction in rating of a device or component, especially the maximum power-dissipation rating at higher temperatures.

derating factor–The factor by which the ratings of component parts are reduced to provide additional safety margins in critical applications or when the parts are subjected to extreme environmental conditions for which their normal ratings do not apply.

derivative action–*See* Rate Action.

derivative control – Automatic control in which the rate of correction is determined by the rate at which the error producing it changes.

derived center channel – A monophonic composite signal derived from the sum or difference of the left and right stereo channels, often fed to an extra speaker to fill in an aural "hole" between the left and right speakers. The signal from a voltage-derived center-channel output must be fed to an external power amplifier before it can drive a speaker. A power-derived center channel can drive a speaker directly.

desensitization–1. The saturation of one component (an amplifier, for instance) by another so that the first cannot perform its proper function. 2. The reduction in receiver sensitivity due to the presence of a high-level off-channel signal overloading the radio-frequency amplifier or mixer stages, causing automatic gain control action.

desiccant–A substance used as a drying agent because of its affinity for water.

design-center rating–Values of operating and environmental conditions which should not be exceeded under normal conditions in a bogey electron device.

design compatibility – Electromagnetic compatibility achieved by incorporating in all electromagnetic radiating and receiving apparatus (including antennas) characteristics or features for elimination or rejection of undesired self-generated or external signals, for enhancement of operating capabilities in the presence of natural or man-made electromagnetic noise.

design engineer–An engineer who has been assigned to design a specific product for specific application.

design for maintainability–Those features and characteristics of design of an item that reduce requirements for tools, test equipment, facilities, spares, highly skilled personnel, etc., and improve the capability of the item to accept maintenance actions.

design-maximum rating–Values of operating and environmental conditions which should not be exceeded under the worst possible conditions in a bogey electron device.

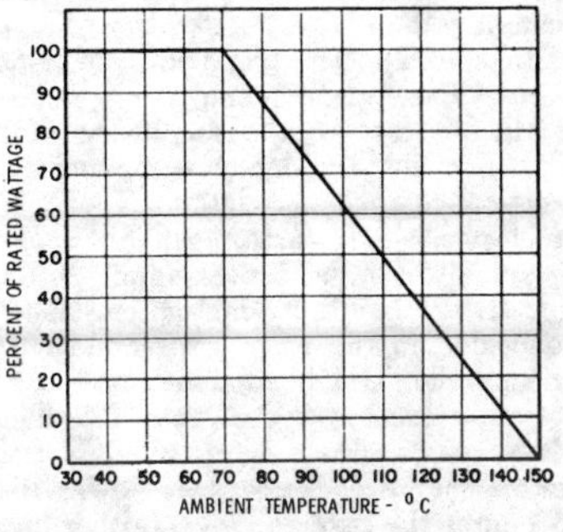

Derating curve.

design proof test—A test used to verify that a design specification meets the overall functional requirements of the finished product.

Desk-Fax—Trade name of Western Union Telegraph Co. for a small facsimile transceiver employed principally in short-line telegraph service.

desk-top computer — A computer that requires no formally prepared stored program entered in a special language. Most desk-top computers are operated like adding machines or mechanical calculators.

destaticization—Treatment of a material to minimize the accumulation of static electricity and, as a result, the amount of dust that adheres to the material because of such static charges.

destination register—In a computer, a register into which data is being placed.

Destriau effect — Sustained emission of light by suitable phosphor powders embedded in an insulator and subjected only to the action of an alternating electric field.

destructive readout—1. In a data-storage system, a readout that erases the stored information. 2. A characteristic of a memory. The memory is said to have a destructive readout if information retrieved from memory must be written back in immediately after it is used or else it is lost. A core memory has destructive readout. Computers with destructive readouts contain special circuits to write information back into memory after readout.

destructive-readout memory — *See* DRO Memory.

destructive test—Any test resulting in the destruction or drastic deterioration of the test specimen.

detail—A measure of the sharpness of a recorded facsimile copy or reproduced image. Generally related to the number of lines scanned per inch. Defined as the square root of the ratio between the number of scanning lines per unit length and the definition in the direction of the scanning line.

detail contrast—The ratio of the amplitude of the high-frequency components of a video signal to the amplitude of the reference low-frequency component.

detection—*See* Demodulation.

detectophone—An instrument for secretly listening in on a conversation. A high-sensitivity, nondirectional microphone is concealed in the room and connected to an amplifier and headphones or recorder remotely located. Sometimes the microphone feeds into a wired-wireless transmitter that broadcasts over power lines, to permit the listener to be farther away.

detector—1. A device for effecting the process of detection or demodulation. 2. A mixer or converter in a superheterodyne receiver; often referred to as a "first detector." 3. A device that produces an electrical output that is a measure of the radiation incident on the device. 4. A rectifier tube, crystal, or dry disc by which a modulation envelope on a carrier or the simple on-off state of a carrier may be made to drive a lower-frequency device.

detector balanced bias—A controlling circuit used in radar systems for anticlutter purposes.

detector circuit—That portion of a receiver which recovers the modulation signal from the rf carrier wave.

detector probe—A probe containing a high-frequency rectifying element such as a crystal diode or a tube. Used with an oscilloscope, vacuum-tube voltmeter, or signal tracer for recovering the modulation from a carrier.

detector quantum efficiency—The ratio of the number of carriers generated to the number of photons absorbed.

detent—1. A stop or other holding device, such as a pin, lever, etc., on a ratchet wheel. 2. Switch action typified by a gradual increase in force to a position at which there is an immediate and marked reduction in force.

deterministic signal—A signal the future behavior of which can be predicted precisely.

detune—To change the inductance and/or capacitance of a tuned circuit and thereby cause it to be resonant at other than the desired frequency.

detuning stub — A quarter-wave stub for matching a coaxial line to a sleeve-stub antenna. The stub tunes the antenna itself and detunes the outside of the coaxial feed line.

deuterium—Heavy hydrogen, so called because it weighs twice as much as ordinary hydrogen. The nucleus of heavy hydrogen is a deuteron.

deuteron—Also called deuton. The nucleus of an atom of heavy hydrogen (deuterium) containing one proton and one neutron. Deuterons are often used as atomic projectiles in atom smashers.

Deutsche Industrie Normenausschus—*See* DIN.

deviation—1. The difference between the actual and specified values of a quantity. 2. In fm transmissions and reception, the increase or decrease of signal carrier frequency from the nominal; also applied to drifting. Standard maximum deviation rating is ± 75kHz for fm radio. 3. A departure from specification requirements for which approval is obtained from the consumer prior to occurrence of the departure from specification requirements.

deviation absorption—Absorption that occurs at frequencies near the critical frequency. Occurs in conjunction with the slowing up of radio waves near the critical frequency, upon reflection from the ionosphere.

deviation distortion—Distortion caused by inadequate bandwidth, amplitude-modulation rejection, or discriminator linearity in an fm receiver.

deviation ratio—In frequency modulation, the ratio of the maximum change in carrier frequency to the highest modulating frequency.

deviation sensitivity – The smallest frequency deviation that produces a specified output power in fm receivers.

device—1. A single discrete conventional electronic part such as a resistor or transistor, or a microelectronic circuit. 2. Also called item. Any subdivision of a system. 3. A mechanical, electrical, and/or electronic contrivance intended to serve a specific purpose. 4. The physical realization of an individual electrical element in a physically independent body which cannot be further reduced or divided without destroying its stated function. This term is commonly applied to active devices. Examples are transistors, pnpn structures, tunnel diodes, and magnetic cores, as well as resistors, capacitors, and inductors. It is not, for example, an amplifier, a logic gate, or a notch filter.

device complexity—The number of circuit elements within an integrated circuit.

device independence—In a computer, the ability to request input/output operations without regard to the nature of the input/output devices.

Dewar flask – A container with double walls. The space between the walls is evacuated, and the surfaces bounding this space are silvered.

dewetted surface—A surface that was initially wetted, i.e., a surface on which the solder flowed uniformly.

dew point—The temperature at which condensation first occurs when a vapor is cooled.

df—Abbreviation for direction finder or dissipation factor.

df antenna—Any antenna combination included in a direction finder for obtaining the phase or amplitude reference of the received signal. May be a single or orthogonal loop, an adcock, or spaced differentially connected dipoles.

df antenna system—One or more df antennas and their combining circuits and feeder systems, together with the shielding and all electrical and mechanical items up to the receiver input terminals.

D flip-flop—A flip-flop the output of which is determined by the input that appeared one pulse earlier; for example, if a 1 appeared at the input, the output after the next clock pulse would be 1.

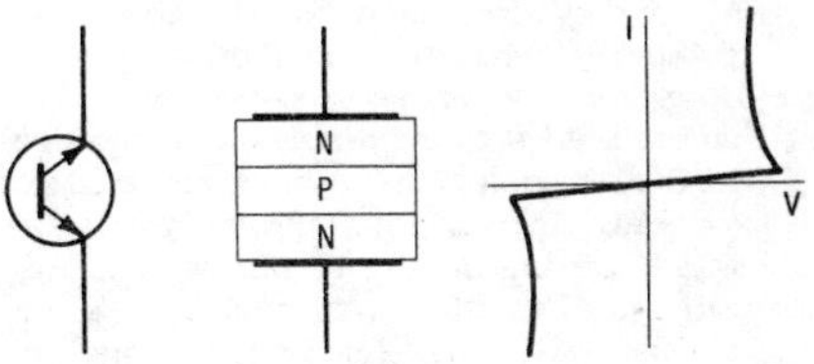

Symbol. Junctions. Characteristic.

Diac.

diac – 1. Two-lead alternating-current switch semiconductor. 2. *See* Three-Layer Diode. 3. A bidirectional breakdown diode which conducts only when a specified breakdown voltage is exceeded.

diagnostic—1. Having to do with the detection and isolation of a malfunction or error in a computer. 2. A message output by a compiler or assembler indicating that a computer program contains a mistake.

diagnostic function test—A program for testing overall system reliability.

diagnostic routine—An electronic-computer routine designed to locate a malfunction in the computer, a mistake in coding, or both.

diagnostics—Information on what tests a device failed and how they were failed, used to aid in troubleshooting.

diagnotor—In a computer, combined diagnostic and edit routine that questions unusual situations and makes note of the implied results.

diagonal horn antenna—A horn antenna all cross sections of which are square and in which the electric vector is parallel to a diagonal. The radiation pattern in the far field has almost perfect circular symmetry.

diagram – 1. Schematics, prints, charts, etc., or any other graphical representation, the purpose of which is to explain rather than to represent. 2. A schematic representation of a sequence of subroutines designed to solve a problem. 3. A coarser and less symbolic representation than a flowchart, frequently including descriptions in English words. 4. A schematic or logical drawing showing the electrical circuit or logical drawing showing the electrical circuit or logical arrangements within a component.

dial—1. A means for indicating the value to which a control knob has been set. 2. A calling device which generates the required number of pulses in a telephone set and thereby establishes contact with the party being called.

dial cable—*See* Dial Cord.

dial central office—A telephone or teletypewriter office at which is located the automatic equipment necessary for connecting two or more user stations together by wires for communication purposes.

dial cord—Also called dial cable. A braided cord or flexible wire cable connected to a tuning knob so that turning the knob will move the pointer or dial which indicates the frequency to which a radio receiver is tuned. Also used for coupling two shafts together mechanically.

dialer—A device that detects and reports emergencies by automatic dialing of telephone numbers. When an emergency is detected, the dialer usually begins playing a prerecorded tape containing the telephone number or numbers to be called (in the form of a series of pulses) and the emergency message. When the number has been dialed, the tape continues to play the prerecorded message.

dialing key—A dialing method in which a set of numerical keys instead of a dial is used to originate dial pulses. Generally, it is used in connection with voice-frequency dialing.

dial jacks—Strips of jacks associated with and bridged to a regular outgoing trunk jack circuit so that connections between the dial cords and the outgoing trunks can be made.

dial key—The key unit of the subscriber's cord circuit that is used to connect the dial to the line.

dial leg—The circuit conductor brought out for direct-current dial signaling.

dial light—A small pilot lamp which illuminates the tuning dial of a radio receiver.

diallyl phthalate — A thermosetting resin that has excellent electrical insulation properties.

dial pulse—Interruption in the direct current flowing through the loop of a calling telephone, produced by the opening and closing of the dial pulse springs of a calling telephone in response to the dialing of a digit. The current in the calling-line loop is interrupted as many times as there are units in the digit dialed —i.e., dialing of the digit 7 generates seven dial pulses (interruptions) in current flowing through the loop of the calling telephone.

dial pulsing — The transmission of telephone address information by the momentary opening and closing of a dc circuit a specified number of times, corresponding to the decimal digit which is dialed. This is usually accomplished, as with the ordinary telephone dial, by manual operation of a finger wheel.

dial register—*See* Standard Register of a Motor Meter.

dial telephone system—Telephone system in which telephone connections between customers are ordinarily established by electronic and mechanical apparatus, controlled by manipulations of dials operated by the calling parties.

dial tone—A hum or other tone employed in a dial telephone system to indicate that the line is not busy and that the equipment is ready for dialing.

dial-up—The use of a dial or push-button telephone for initiating a station-to-station call.

diamagnetic—1. Term applied to a substance with a negative magnetic susceptibility. 2. Bars of certain elements, such as zinc, copper, lead, and tin, when freely suspended in a magnetic field arrange themselves at right angles to the lines of force of the magnetic field, i.e., they are magnetized in the opposite direction to the magnetizing field. These elements are said to be diamagnetic.

diamagnetic material—A material which is less magnetic than air, or in which the intensity of magnetization is negative. There is no known material in which this effect has more than a very feeble intensity. Bismuth is the leading example of materials of this class.

diamagnetism—1. A phenomenon whereby the magnetization induced in certain substances opposes the magnetizing force. 2. The negative susceptibility exhibited by certain substances. The permeability of such substances is less than unity.

diamond antenna—Also called a rhombic antenna. A horizontal antenna having four conductors that form a diamond, or rhombus.

diamond lattice—The crystal structure of germanium and silicon (as well as a diamond).

diamond stylus — A phonograph pickup with a ground diamond as its point.

diapason—The unique fundamental tone color of organ music.

diaphragm—1. A flexible membrane used in various electroacoustic transducers for producing audio-frequency vibrations when actuated by electric impulses, or electric impulses when actuated by audio-frequency vibrations. 2. In electrolytic cells, a porous or permeable membrane, usually flexible, separating the anode and cathode compartments. 3. In waveguide technique, a thin plate, or plates, placed transversely across the waveguide, not completely closing it, and usually introducing a reactance component. *See also* Iris. 4. A sensing element consisting of a membrane placed between two volumes. The membrane is deformed by the pressure differential applied across it.

diathermal apparatus—Apparatus for generating heat in body tissue by high frequency electromagnetic radiation.

diathermy – The use of radio-frequency fields to produce deep heating in body tissues. The output of a powerful rf oscillator is applied to a pair of electrodes, known as pads, between which the portion of the body to be treated is placed. The body tissues thus become the dielectric of a capacitor, and dielectric losses cause heating of the tissues.

diathermy interference – A form of television interference caused by diathermy equipment, resulting in a horizontal herringbone pattern across the picture.

diathermy machine—A medical apparatus consisting of an rf oscillator frequently followed by rf amplifier stages, used to generate high-frequency currents that produce heat within some predetermined part of the body for therapeutic purposes.

dibit—A group of two bits. In four-phase modulation, each possible dibit is encoded in the form of one of four unique phase shifts of the carrier. The four possible states for a dibit are 00, 01, 10, and 11.

DIC—Dielectrically isolated integrated circuits. Also called DIIC.

dichroic mirror—A special mirror through which all light frequencies pass except those for the color which the mirror is designed to reflect.

dichroism—A property of an optical material which causes light of some wavelengths to be absorbed when the incident light has its electric-field vector in a particular orientation and not absorbed when the electric-field vector has other orientations.

dicing—The process of sawing a crystal wafer into blanks.

dictionary—In digital computer operations, a list of mnemonic code names together with the addresses and/or data to which they refer.

die—1. Sometimes called chip. A tiny piece of semiconductor material, broken from a semiconductor slice, on which one or more active electronic components are formed. (Plural: dice). 2. A portion of a wafer bearing an individual circuit or device cut or broken from a wafer containing an array of such circuits or devices.

die bounding – The method by which a semiconductor die, or chip, is attached to a mechanical support.

dielectric—1. The insulating (nonconducting) medium between the two plates of a capacitor. Typical dielectrics are air, wax-impregnated paper, plastic, mica, and ceramic. A vacuum is the only perfect dielectric. 2. A medium capable of recovering, as electrical energy, all or part of the energy required to establish an electric field (voltage stress). The field, or voltage stress, is accompanied by displacement or charging currents. 3. The insulating material between the metallic elements of an electromechanical component or any of a wide range of thermoplastics or thermosetting plastics.

dielectric absorption—1. Also called dielectric hysteresis (short-term effect), or dielectric soak (long-term effect). A characteristic of dielectrics which determines the length of time a capacitor takes to deliver the total amount of its stored energy. It manifests itself as the reappearance of potential on the electrodes after the capacitor has been discharged. Its magnitude depends on the charge and discharge time of the capacitor. 2. That property of an imperfect dielectric as a result of which all electric charges within the body of the material because of the application of an electric field are not returned to the field.

dielectric amplifier—An amplifier employing a device similar to an ordinary capacitor, but with a polycrystalline dielectric which exhibits a ferromagnetic effect.

dielectric antenna—An antenna in which a dielectric is the major component producing the required radiation pattern.

dielectric breakdown—An abrupt increase in the flow of electric current through a dielectric material as the applied electric field strength exceeds a critical value.

dielectric breakdown voltage—The voltage between two electrodes at which electric breakdown of the specimen occurs under prescribed test conditions. Also called electric breakdown voltage, breakdown voltage, or hi-pot.

dielectric capacity—The inductivity or specific inductive capacity of a substance, being its ability to convey the influence of an electrified body.

dielectric constant—The ratio of the capacitance of a capacitor with the given dielectric to the capacitance of a capacitor having air for its dielectric but otherwise identical. Also called permittivity, specific inductive capacity, or capacitivity.

dielectric current—The current flowing at any instant through the surface of an isotropic dielectric which is in a changing electric field.

dielectric dissipation—*See* Loss Tangent.

dielectric dissipation factor—The cotangent of the dielectric phase angle of a material.

dielectric fatigue—The property of some dielectrics in which the insulating quality decreases after a voltage has been applied for a considerable length of time.

dielectric guide—A waveguide made of a solid dielectric material through which the waves travel.

dielectric heating—A method of raising the temperature of a nominally insulating material by sandwiching it between two plates to which an rf voltage is applied. The material acts as a dielectric, and its internal losses cause it to heat up.

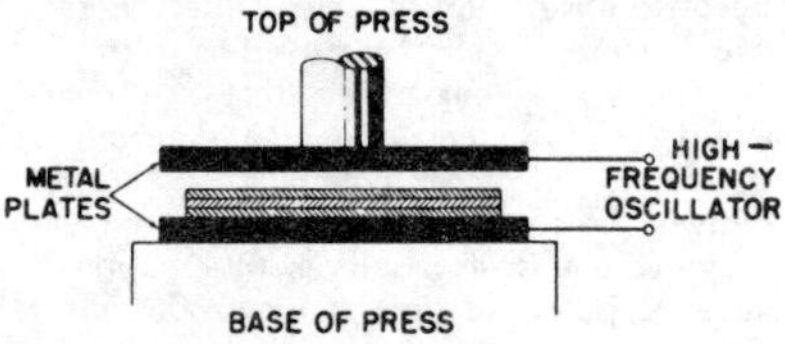

Dielectric heating.

dielectric hysteresis — Short-term effect of dielectric absorption (which see).

dielectric isolation—The electrical isolation of monolithic integrated circuit elements from each other by dielectric material rather than by reverse-biased pn junctions.

dielectric lens—A lens used with microwave antennas; it is made of dielectric material so that it refracts radio waves similar to the way an optical lens refracts light waves.

dielectric loss—The power dissipated by a dielectric as the friction of its molecules opposes the molecular motion produced by an alternating electric field.

dielectric loss angle—The complement of the dielectric phase angle (i.e., the dielectric phase angle minus 90°).

dielectric loss factor—The product of the dielectric constant of a material times the tangent of the dielectric loss angle. Also called dielectric loss index.

dielectric loss index—*See* Dielectric Loss Factor.

dielectric matching plate—In waveguide technique, a dielectric plate used as an impedance transformer for matching purposes.

dielectric mirror—A highly frequency-selective, multilayer dielectric reflector acting by partial reflection of light at the interface between materials of unequal refractive indices.

dielectric phase angle—The angular difference in phase between the sinusoidal alternating voltage applied to a dielectric and the component of the resultant alternating current having the same period.

dielectric phase difference—*See* Dielectric Loss Angle.

dielectric polarization—*See* Polarization, 3.

dielectric power factor—The cosine of the dielectric phase angle.

dielectric rating — Standard test voltages and frequencies above which failure occurs between specified points in a relay structure.

dielectric-rod antenna — An antenna in which propagation of a surface wave on a tapered dielectric rod produces an end-fire radiation pattern.

dielectric soak—Long-term effect of dielectric absorption (which see).

dielectric strength—The maximum voltage a dielectric can withstand without rupturing. Usually expressed as volts per mil. Also called: electric strength, breakdown strength, electric field strength, and insulating strength.

dielectric susceptibility—The ratio of the polarization in a dielectric to the electric intensity responsible for it.

dielectric tests—1. Tests which consist of the application of a voltage higher than the rated voltage for a specified time for the purpose of determining the adequacy against breakdown of insulating materials and spacings under normal conditions. 2. The testing of insulating materials by the application of a constantly increasing voltage until failure occurs.

dielectric waveguide—A waveguide constructed from a dielectric (nonconductive) substance.

dielectric wedge—A wedge-shaped piece of dielectric material used in one waveguide to match its impedance to that of another waveguide.

dielectric wire — A dielectric waveguide used for short-distance transmission of uhf radio waves between parts of a circuit.

difference—The signal energy representing the differences in information between the signals in two or more stereo channels. A difference signal is produced when stereo signals differing in electrical polarity or in intensity are mixed together in opposing polarity.

difference amplifier—*See* Differential Amplifier.

difference channel — In a stereophonic sound system, an audio channel that handles the difference between the signals in the left and right channels.

difference detector—A detector circuit in which the output is a function of the difference between the peak or rms amplitudes of the input waveforms.

difference frequency—1. A signal representing, in essence, the difference between the left and right sound channels of a stereophonic sound system. 2. One of the output frequencies of a converter. It is the difference between the two input frequencies.

difference in depth modulation—In directive systems employing overlapping lobes with modulated signals, a ratio obtained by subtracting from the percentage of modulation of the larger signal the percentage of modulation of the smaller signal and dividing by 100.

difference of potential – The voltage or electrical pressure existing between two points. It will result in a flow of electrons whenever a circuit is established between the two points.

difference signal—In a quadraphonic sound system, a signal arrived at by subtracting left or right back-channel signal from its respective front-channel signal. Left-front and left-back signals, if added to left-front minus left-back signals yield a left-front channel; if subtracted, they yield a left-back channel.

differential—1. A planetary gear system which adds or subtracts angular movements transmitted to two components and delivers the answer to a third. Widely used for adding and subtracting shaft movements in servo systems and for addition and subtraction in computing machines. 2. In electronics, the difference between two levels.

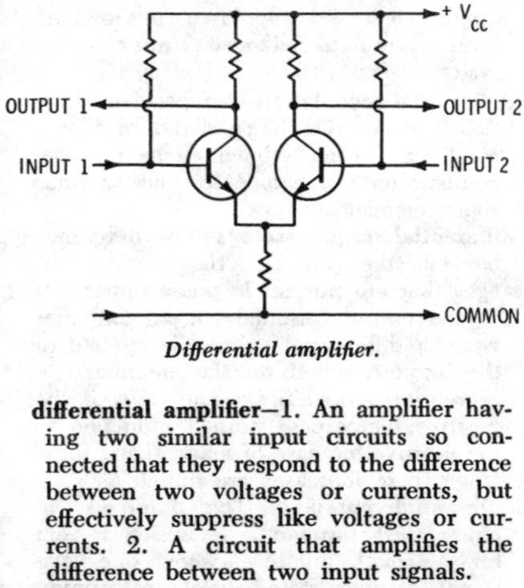

Differential amplifier.

differential amplifier—1. An amplifier having two similar input circuits so connected that they respond to the difference between two voltages or currents, but effectively suppress like voltages or currents. 2. A circuit that amplifies the difference between two input signals.

differential angle—The total angle from the operation to the releasing position in a mercury switch.

differential analyzer – A mechanical or electrical device primarily designed and used to solve differential equations.

differential capacitance – The derivative with respect to voltage of a capacitor charge characteristic at a given point on the curve.

differential capacitance characteristic – The function that relates differential capacitance to voltage.

differential capacitor—A variable capacitor having two similar sets of stator plates and one set of rotor plates. When the rotor is turned, the capacitance of one section is increased while the capacitance of the other section is decreased.

differential comparator—A circuit in which differential-amplifier design techniques are applied to the comparison of an input voltage with a reference voltage. When the input voltage is lower than the reference voltage, the circuit output is in one state; when the input voltage is higher than the reference voltage, the output is in the opposite state. Commonly used for pulse-amplitude detector circuits, a-d conversion, and differential receivers for data transmission in noisy environments over a twisted-pair line.

differential cooling—A lowering of temperature which takes place at a different rate at various points on an object or surface.

differential delay—The difference between the maximum and the minimum frequency delays occurring across a band.

differential discriminator—A discriminator that passes only pulses having amplitudes between two predetermined values, neither of which is zero.

differential duplex system—A duplex system in which the sent currents divide through two mutually inductive sections of the receiving apparatus. These sections are connected respectively to the line and to a balancing artificial line in opposite directions. Hence, there is substantially no net effect on the receiving apparatus. The received currents pass mainly through one section, or through the two sections in the same direction, and operate the apparatus.

differential flutter – Speed-change errors which occur at different magnitudes, frequencies, or phases across the width of a magnetic tape.

differential gain—The ratio of the differential output signal of a differential amplifier divided by the differential input signal causing that output.

differential gain control—Also called gain sensitivity control. A device for altering the gain of a radio receiver in accordance with an expected change of signal level, in order to reduce the amplitude differential between the signals at the receiver output.

differential galvanometer—A galvanometer having two similar but opposed coils, so that their currents tend to neutralize each other. A zero reading is obtained when the currents are equal.

differential gap—1. The difference between two target values, one of which applies to an upswing of conditions and the other to a downswing. 2. The span between on and off switching points. For example, a room thermostat set for 70° might switch the furnace on at 68° and off at 72°, resulting in a 4° differential. *See also* Dead Band.

differential gear—In an analog computer, a mechanism that relates the angles of rotation of three shafts. Usually it is designed so that the algebraic sum of the rotations of two shafts is equal to twice the rotation of the third. The device can be used for addition or subtraction.

differential generator—A synchro differential generator driven by a servo system.

differential impedance—The internal impedance observed between the input terminals of an operational amplifier.

differential input—1. An input circuit that rejects voltages which are the same at both input terminals and amplifies the voltage difference between the two input terminals. May be either balanced or floating and may also be guarded. 2. An input applied between two terminals of an operational amplifier, neither of which is at ground (earth) potential.

differential-input amplifier—An amplifier in which the output is ideally a function only of the difference between the signals applied to its two inputs, both signals being measured with respect to a common "low" or "ground" reference point.

differential-input capacitance—The capacitance between the inverting and noninverting input terminals.

differential-input impedance—1. The impedance between the inverting and noninverting input terminals of a differential amplifier. 2. The impedance measured between the + and the − input terminals of an operational amplifier.

differential-input measurement—Also called floating input. A measurement in which the two inputs to a differential amplifier are connected to two points in a circuit under test and the amplifier displays the difference voltage between the points. In this type of measurement, each input of the amplifier acts as a reference for the other, and ground connections are used only for safety reasons.

differential-input rating — The maximum differential input which may be applied between the two terminals of an operational amplifier.

differential-input resistance — The resistance between the inverting and noninverting input terminals of a differential amplifier.

differential-input voltage—The maximum voltage that can be applied across the input terminals of a differential amplifier without damaging the amplifier.

differential-input voltage range—The range of voltages that may be applied between input terminals without forcing the circuit to operate outside its specifications.

differential-input voltage rating—The maximum allowable signal that may be applied between the inverting and noninverting inputs of a differential amplifier without damaging the amplifier.

differential instrument—A galvanometer or other measuring instrument having two circuits or coils, usually identical, through which currents flow in opposite directions. The difference or differential effect of these currents actuates the indicating pointer.

differential keying—A method of obtaining chirp-free break-in keying of a cw transmitter by turning the oscillator on quickly before the keyed amplifier stage can pass any signal, and turning it off quickly after the keyed amplifier stage has cut off.

differential microphone—*See* Double-Button Carbon Microphone.

differential-mode gain—The ratio of the output voltage of a differential amplifier to the differential-mode input voltage.

differential-mode input—The voltage difference between the two inputs of a differential amplifier.

differential-mode signal—A signal that is applied between the two ungrounded terminals of a balanced three-terminal system.

differential modulation—A type of modulation in which the choice of the significant condition for any signal element is dependent on the choice for the previous signal element.

differential output voltage—The difference between the values of the two ac voltages that are present in phase opposition at the output terminals of an amplifier when a differential voltage is applied to the input terminals of the amplifier.

differential permeability—The ratio of the positive increase of normal induction to the positive increase of magnetizing force when these increases are minute.

differential phase—1. The difference in phase shift through a television system for a small, high-frequency sine-wave signal at two stated levels of a low-frequency signal on which the first signal is superimposed. 2. In a color tv signai, the phase change of the color subcarrier introduced by the overall circuit, measured in degrees as the picture signal on which it rides is varied from blanking to white level.

differential phase-shift keying—A modulation scheme in which the information is conveyed by changes in carrier phase during one interval relative to the preceding interval.

differential pressure—1. The difference in pressure between two pressure sources. 2. The difference between a reference pressure and a measured value of pressure.

differential pressure transducer – A pressure transducer that accepts simultaneously two independent pressure sources, and the output of which is proportional to the pressure difference between the sources.

differential protective relay–A protective device which functions on a percentage or phase angle or other quantitative difference of two currents or of some other electrical quantities.

differential relay–A relay with multiple windings that functions when the voltage, current, or power difference between the windings reaches a predetermined value. The power difference may result from the algebraic addition of the multiple inputs.

differential selsyn–A selsyn in which both the rotor and the stator have similar windings that are spread 120° apart. The position of the rotor corresponds to the algebraic sum of the fields produced by the stator and rotor.

differential stage–A symmetrical amplifier stage in which two inputs are balanced against each other so that when there is no input signal, or equal input signals, there is no output signal. An input-signal unbalance, including a signal to only one input, produces an output signal proportional to the difference between the input signals.

differential synchro–*See* Synchro Differential Generator *and* Synchro Differential Motor.

differential transducer–A device capable of simultaneously measuring two separate stimuli and providing an output proportionate to the difference between them.

differential transformer–Also called linear variable-differential transformer. 1. A transformer used to join two or more sources of signals to a common transmission line. 2. An electromechanical device which continuously translates displacement of position change into a linear ac voltage.

differential voltage–For a glow lamp, the difference between the breakdown and maintaining voltage.

differential voltage gain–1. The ratio of the change in output signal voltage at either terminal of a differential device to the change in signal voltage applied to either input terminal, all voltages measured to ground. 2. The ratio of the differential output voltage of an amplifier to the differential input voltage of the amplifier. If the amplifier has one output terminal, the differential voltage gain is the ratio of the ac output voltage (with respect to ground) to the differential input voltage.

differential winding–A coil winding so arranged that its magnetic field opposes that of a nearby coil.

differential-wound field–A type of motor or generator field having both series and shunt coils connected so they oppose each other.

differentiate–1. To distinguish. 2. To find the derivative of a function. 3. To deliver an output that is the derivative with respect to time of the input.

differentiating circuit–A circuit whose output voltage is proportional to the rate of change of the input voltage. The output waveform is then the time derivative of the input waveform, and the phase of the output waveform leads that of the input by 90°. An RC circuit gives this differentiating action. Also called differentiating network and differentiator.

differentiating network–*See* Differentiating Circuit.

differentiator–*See* Differentiating Circuit.

diffracted wave–A radio, sound, or light wave which has struck an object and been bent or deflected, other than by reflection or refraction.

diffraction – 1. The bending of radio, sound, or light waves as they pass through an object or barrier, thereby producing a diffracted wave. 2. The phenomenon whereby waves traveling in straight paths bend around an obstacle.

diffuse–To undergo or cause to undergo diffusion.

diffused-alloy transistor–Also called drift transistor. A transistor in which the semiconductor wafer is subjected to gaseous diffusion to produce a nonuniform base region, after which alloy junctions are formed in the same manner as for an alloy-junction transistor. It may also have an intrinsic region to give a pnip unit.

diffused-base transistor – Also called graded-base transistor. A type of tran-

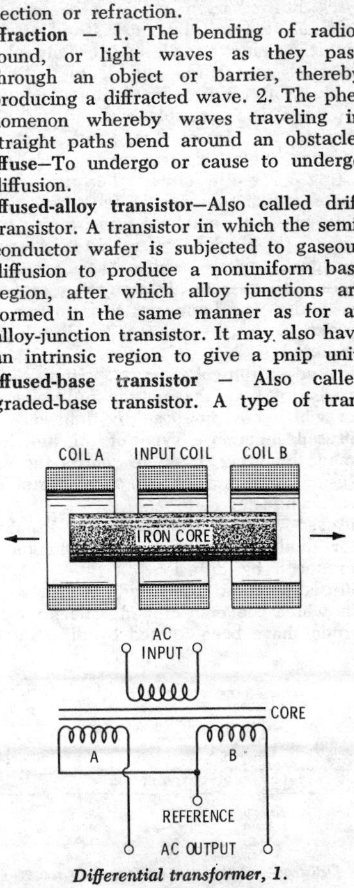

Differential transformer, 1.

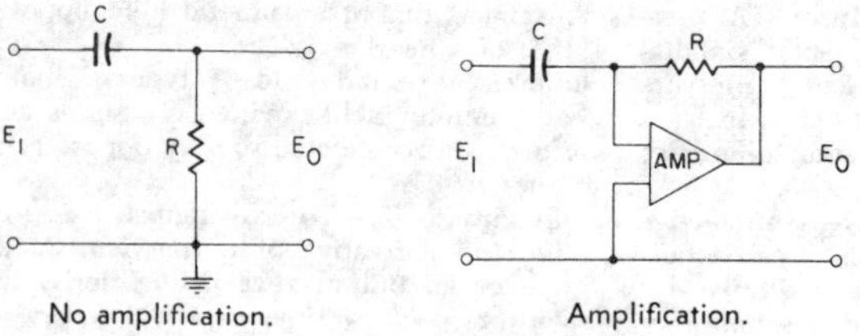

$$E_O = T\frac{dE_I}{dT} \quad (T=RC)$$

Differentiator.

sistor made by combining diffusion and alloy techniques. A nonuniform base region and the collector-to-base junction are formed by gaseous dissemination into a semiconductor wafer that constitutes the collector region. Then the emitter-to-base junction is formed by a conventional alloy process on the base side of the diffused wafer.

diffused device—A semiconductor device in which a base, usually of silicon, has successive layers of p and n characteristics diffused upon and into the base by means of a series of masks and around which p and n materials, usually phosphorous and boron, adhere to the base by gaseous diffusion in a high-temperature furnace. It is possible to build areas of resistance, capacitance, and "active" diodes and transistors into the base, creating an entire circuit. Performance is poor in the presence of radiation.

diffused-emitter-and-base transistor — Also called double-diffused transistor. A semiconductor wafer which has been subjected to gaseous dissemination of both n- and p-type impurities to form two pn junctions in the original semiconductor material.

diffused-emitter-collector transistor — A transistor both the emitter and collector of which are produced by diffusion.

diffused junction — Type of pn junction, made by using masks to control the diffusion of impurities into monocrystalline semiconductor material.

diffused-junction rectifier—A semiconductor diode in which the pn junction is produced by diffusion.

diffused-junction transistor — A transistor in which the emitter and collector electrodes have been formed by diffusion of

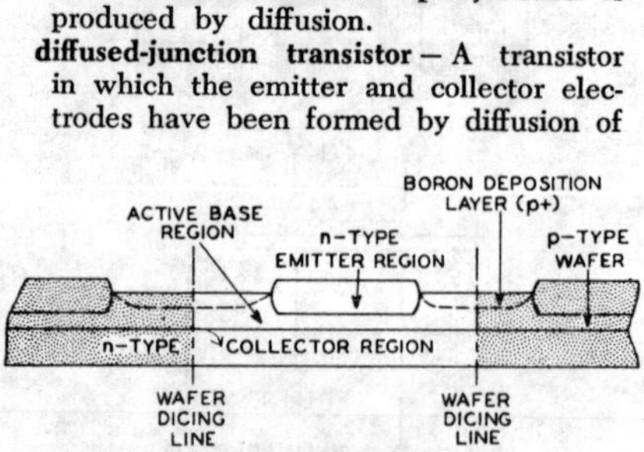

Diffused-junction silicon transistor wafer.

an impurity into the semiconductor wafer without heating.

diffused-layer resistor—A resistor formed by including an appropriate pattern in the photomask to define diffusion areas.

diffused mesa transistor—A transistor in which the collector-base junction is formed by gaseous diffusion, and the emitter-base junction is formed either by gaseous diffusion or by an evaporated metal strip. The collector-base junction is then defined by etching away the undesired parts of the emitter and base regions, thus producing a mesa.

diffused metal-oxide semiconductor — *See* DMOS.

diffused planar transistor — A transistor made by two gaseous diffusions, but in which the collector-base junction is defined by oxide masking. Junctions are formed beneath this protective oxide layer with the result that the device has lower reverse currents and good dc gain at low currents.

diffused sound—Sound which has uniform energy density, meaning the energy flux is equal in all parts of a given region.

diffused transistor—A transistor in which the emitter and collector junctions are both formed by diffusion.

diffused transmission—The total net transmission, by a medium or device, of light that is neither perfectly Lambertian nor parallel. Often used interchangeably with the term "gross transmission."

diffusion — 1. The movement of carriers from a region of high concentration to regions of lower concentration. 2. A thermally induced process in which one material permeates another. In silicon processing, doping impurities diffuse into the silicon at elevated temperatures to form the desired junctions. The same impurities penetrate silicon dioxide much more slowly, and therefore silicon dioxide on the surface of the silicon acts as a mask to determine the areas into which diffusion occurs. 3. Thermal process which distributes small amount of impurities in semiconductor materials according to precisely controlled patterns. 4. The process of adding impurities to a semiconductor material in order to affect its characteristics.

diffusion bonding—Formation of a metallurgical joint between similar or dissimilar metals by the process of interdiffusion of atoms across the joint interface in either the solid or liquid state. The term generally is applied to, but is not limited to, solid-state diffusion. The joining surfaces must be brought within atomic distances through the application of pressure.

diffusion capacitance—The capacitance of a forward-biased pn junction.

diffusion constant—The quotient of diffusion-current density in a homogeneous semiconductor, divided by the charge-carrier concentration gradient. It is equal to the drift mobility times the average thermal energy per unit charge of carriers.

diffusion current—1. The current produced when charges move by diffusion. 2. The flow of a particular type of carrier in a semiconductor due to a concentration difference in that type of carrier. Carriers will flow from an area of high concentration to an area of low concentration.

diffusion length—In a homogeneous semiconductor, the average distance the minority carriers move between generation and recombination.

diffusion process — Doping of a semiconductor material by injection of an impurity into the crystal lattice at an elevated temperature. Usually, the semiconductor crystal is exposed to a controlled surface concentration of dopants.

diffusion transistor—A transistor in which current depends on the diffusion of carriers, donors, or acceptors, as in a junction transistor.

diffusion under the epitaxial film — *See* DUF.

diffusion window—In a semiconductor, the opening etched through the oxide to permit the diffusion of the emitter and base.

digiralt—A system of high-resolution radar altimetry in which pulse-modulated radar and high-performance time-to-digital conversion techniques are combined.

digit—1. One of the symbols, 0, 1, 2, 3, 4, 5, 6, 7, 8, and 9, used in numbering in the scale of 10. One of these symbols, when used in a scale of numbering to the base *n*, expresses integral values ranging from *0* to $n-1$ inclusive. 2. A character used to represent a non-negative integer smaller than the radix, e.g., either 0 or 1 in binary notation. 3. In a dial telephone system, one of the successive series of pulses incoming from a dial for operation of a switching train.

digit absorbing selector — A dial switch that sets up and then falls back on the first of two digits dialed; it then operates on the next digit dialed.

digital — 1. Using numbers expressed in digits and in a certain scale of notation to represent all the variables that occur in a problem. 2. Of or pertaining to the class of devices or circuits in which the output varies in discrete steps (i.e., pulses or "on-off" operation). 3. Of or pertaining to an element or circuit the output of which is utilized as a discontinuous function of its input. 4. Circuitry in which data-carrying signals are restricted to either of two voltage levels, corresponding to logic 1 or 0.

digital circuit—A circuit which operates like a switch (it is either "on" or "off"), and can make logical decisions. It is used in computers or similar decision-making equipment. The more common families of digital integrated circuits (called "logic forms") are RTL, DTL, HTL, ECL, and TTL.

digital communication — The transmission of intelligence by the use of encoded numbers—usually uses binary rather than decimal number system.

digital communications—A system of telecommunications employing a nominally discontinuous signal that changes in frequency, amplitude, or polarity.

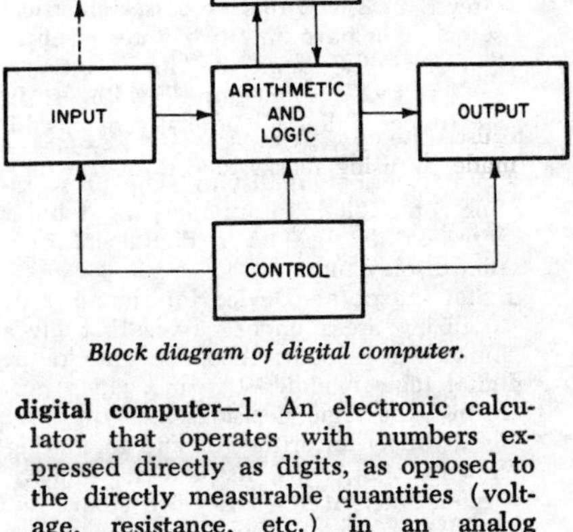

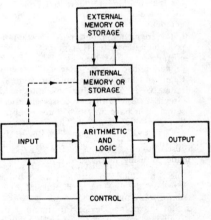

Block diagram of digital computer.

digital computer—1. An electronic calculator that operates with numbers expressed directly as digits, as opposed to the directly measurable quantities (voltage, resistance, etc.) in an analog computer. In other words, the digital computer counts (as does an adding machine); the analog computer measures a quantity (as does a voltmeter). 2. A computer that processes information in numerical form. Electronic digital computers generally use binary or decimal notation and solve problems by repeated high-speed use of the fundamental arithmetic processes of addition, subtraction, multiplication, and division. 3. A computer system in which circuit operation is based on specific signal levels. In a binary digital computer, there are two such signal levels, one at or near zero and the other at a defined voltage. 4. A device that performs sequences of arithmetic and logic operations on discrete data.

digital data—Data represented in discrete, discontinuous form, as contrasted with analog data represented in continuous form. Digital data are usually represented by means of coded characters (e.g., numbers, signs, symbols, etc.).

digital data-handling system — The electronic equipment which receives digital data, operates on them in a suitable manner, records them in a suitable manner on a suitable medium, and presents them directly to a computer or a display.

digital device — Typically, an IC that switches between two exclusive states or levels, usually represented by logical 1 or 0.

digital differential analyzer — A special-purpose digital computer that performs integration and that can be programmed for the solution of differential equations in a manner similar to that of an analog computer.

digital frequency monitor—A special-purpose digital counter that permits a train of pulses to pass through a gate for a predetermined time interval, counts them, and indicates the number counted.

digital IC—1. A switching type integrated circuit. 2. An IC that processes electrical signals that have only two states, such as "on" or "off," "high" or "low" voltages, or "positive" or "negative" voltages. In electronics, "digital" normally means binary or two-state.

digital information display—The presentation of digital information in tabular form on the face of a digital information display tube.

digital integrator—Device for summing or totalizing areas under curves that gives numerical readout. *Also see* Integrator.

digital logic modules—Circuits which perform basic logic decisions AND/OR/NOT; used widely for arithmetic and computing functions, flip-flops, half-adders, multivibrators, etc. *Also see* Logic System.

digitally programmable oscillator — A voltage-controlled oscillator designed to accept a digital tuning word instead of the usual analog signal. Internal digital-to-analog (d/a) converter circuits transform the digital input to an analog voltage. Tuning-curve linearization is usually accomplished through a digital memory. The frequency speed is primarily limited by the d/a circuits.

digital output — An output signal which represents the size of a stimulus or input signal in the form of a series of discrete quantities which are coded to represent digits in a system of numerical notation. This type of output is to be distinguished from one which provides a continuous rather than a discrete output signal.

digital phase shifter—A device which provides a signal phase shift by the application of a control pulse. A reversal of phase shift requires a control pulse of opposite polarity.

digital position transducer—A device that converts motion or position into digital information.

digital readout indicator — An indicator that reads directly in numerical form, as opposed to an analog indicator needle and scale.

digital rotary transducer—A rotating device utilizing an optical sensor that produces a serial binary output as a result of shaft rotation.

digital signal—An electrical signal with two states on or off, high or low, positive or negative, such as could be obtained from a telegraph key or two-position toggle switch. Digital normally means binary or two-state.

digital signals — Discrete or discontinuous signals whose various states are discrete intervals apart.

digital speech communications—Transmission of voice signals in digitized or binary form.

digital television—A television system in which reduction or elimination of picture redundancy is obtained by transmitting only the information needed to define motion in the picture, as represented by changes in areas of continuous white or black.

digital thermometer — Electronic temperature measuring device that reads and/or prints out numerically.

digital-to-analog conversion — The generation of analog (usually variable-voltage) signals in response to a digital code.

digital-to-analog converter—1. A comput-

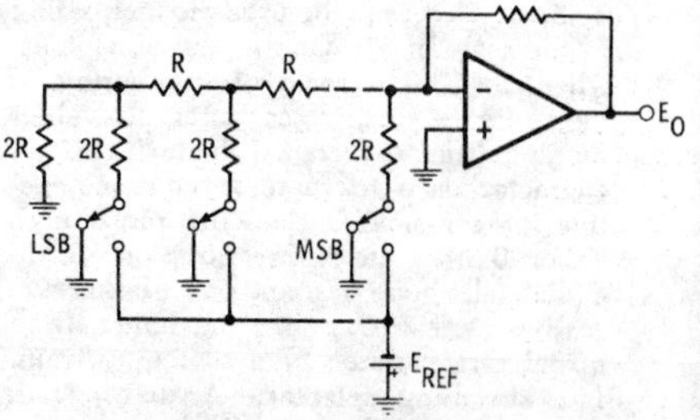

Digital-to-analog converter.

ing device that changes digital quantities into physical motion or into a voltage (i.e., a number output into turns of a potentiometer). 2. Abbreviated dac or d/a converter. A unit or device that converts a digital signal into a voltage or current whose magnitude is proportional to the numeric value of the digital signal. For example:

Digital input	Analog output
00101 (binary 5)	2 volts
01010 (binary 10)	4 volts
10100 (binary 20)	8 volts

digital transmission—A mode of transmission in which all information to be transmitted over the facility is first converted into digital form and then sent down the line as a stream of pulses. (Such transmission may imply a serial bit stream, but parallel forms are also possible.) When noise and distortion threaten to destroy the integrity of the pulse stream, the pulses are detected and regenerated.

digital voltmeter – 1. An indicator which provides a digital readout of measured voltage rather than a pointer indication. 2. An electronic instrument which converts an analog voltage of unknown magnitude into a digital display of known value.

digit compression – In a computer, any of several techniques used to pack digits.

digitize—To convert an analog measurement of a physical variable into a number expressed in digits in a scale or notation.

digitizer—A device which converts analog data into numbers expressed in digits in a system of notation.

digit selector—In a computer, a device for separating a card column into individual pulses that correspond to punched row positions.

digit-transfer bus—The main wire or wires used to transfer information (but not control signals) among the various registers in a digital computer.

digitron display—In a calculator, a type of display in which all digits appear in the same plane. Similar to mosaic lamp display.

diheptal base—Also called diheptal socket. A vacuum-tube base having 14 pins (such as the base of cathode-ray tube).

diheptal socket—*See* Diheptal Base.

DIIC – Dielectrically isolated integrated circuits. Devices isolated from each other by a layer of dielectric insulation, usually glass, rather than by the more conventional reverse-biased pn junction. This "insulated substrate" structure is far more radiation-resistant than junction

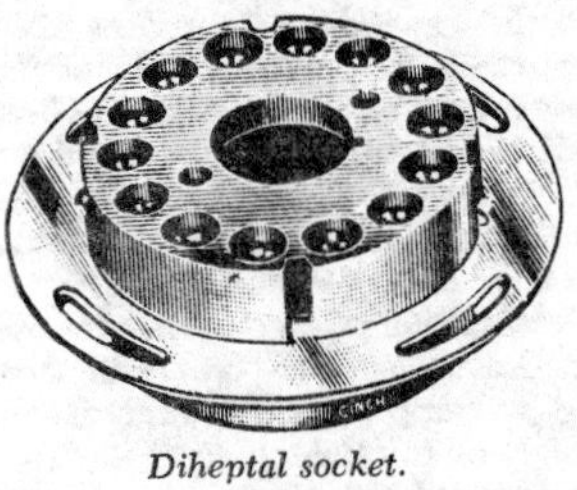

Diheptal socket.

isolated units, making DIICs valuable in military and aerospace applications.

dimensional stability – The ability of a body to maintain precise shape and size.

diminished-radix complement—*See* Radix-Minus-One Complement.

dimmer – A device for controlling the amount of light emitted by a luminaire. Common types employ resistance, autotransformer, magnetic amplifier, silicon controlled rectifier or semiconductor, thyratron, or iris control elements.

dimmer curve – The performance characteristic of a light dimmer expressed as a graph of the light output of a dimmer-controlled lamp versus the setting of the control in terms of an arbitrary linear scale of zero to ten.

DIN—The abbreviation for the association in West Germany that determines the standards for electrical and other equipment in that country. Deutsche Industrie Normenausschuss. Similar to the American USAS.

D-indicator—A radar indicator which combines Types B and C indicators. The signal appears as a bright spot, with azimuth angle as the horizontal coordinate and elevation angle as the vertical coordinate. Each horizontal trace is expanded vertically by a compressed time sweep to facilitate separation of the signal from noise and to give a rough range indication.

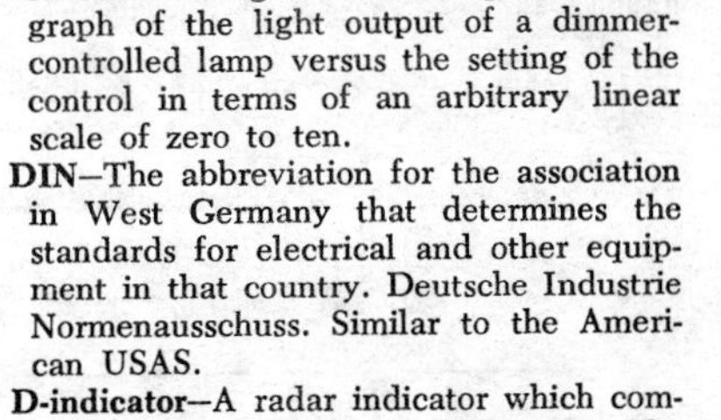

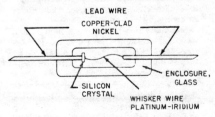

Diode, 2.

diode—1. An electron tube having two electrodes, a cathode and an anode. 2. *See* Crystal Diode. 3. A two-element electron tube or solid-state device. Solid-state diodes are usually made of either germanium or silicon and are primarily used for switching purposes, although they can also be used for rectification. Diodes are usually rated at less than one-half ampere. 4. A two-terminal, electronic device that will conduct electricity much more easily in one direction than in the other.

diode amplifier—A parametric amplifier that uses a special diode in a cavity. Used to amplify signals at frequencies as high as 6000 MHz.

diode assembly—A single structure of more than one diode.

diode characteristic—The composite electrode characteristic of a multielectrode tube, taken with all electrodes except the cathode connected together.

diode demodulator—A demodulator in which one or more semiconductor or electron-tube diodes are used to provide a rectified output that has an average value proportional to the original modulation. Also called diode detector.

diode logic—An electronic circuit using current-steering diodes in an arrangement such that the input and output voltages have relationships that correspond to AND or OR logic functions.

diode matrix—A two-dimensional array of diodes used for a variety of purposes such as decoding and read only memory.

diode mixer—A diode which mixes incoming radio-frequency and local-oscillator signals to produce an intermediate frequency.

diode modulator—A modulator in which one or more diodes are employed to combine a modulating signal with a carrier signal. It is used chiefly in low-level signalling because it has inherently poor efficiency.

diode pack—A combination of two or more diodes integrated into a solid block.

diode peak detector—A diode used in a circuit to indicate when audio peaks exceed a predetermined value.

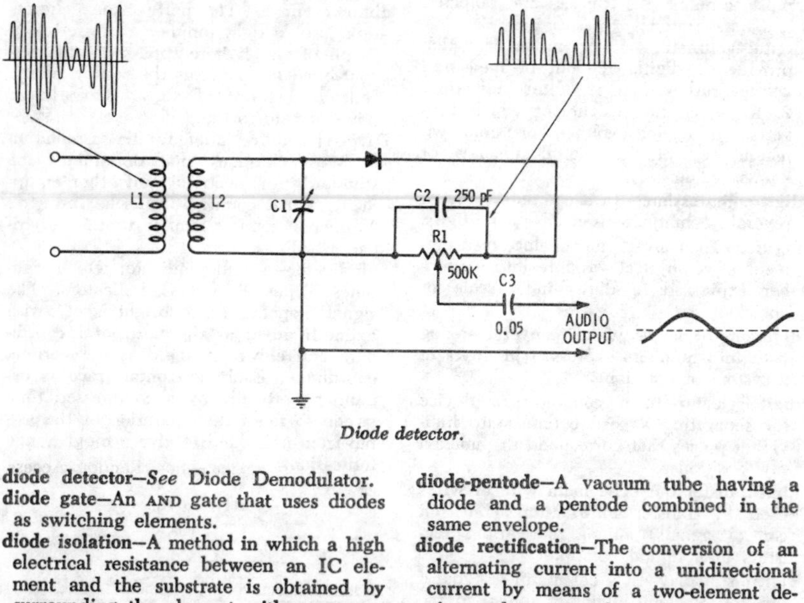

Diode detector.

diode detector—*See* Diode Demodulator.

diode gate—An AND gate that uses diodes as switching elements.

diode isolation—A method in which a high electrical resistance between an IC element and the substrate is obtained by surrounding the element with a reverse-biased pn junction.

diode laser—A pn junction semiconductor electron device which converts direct forward-bias electrical input (pump power) directly into coherent optical output power via a process of stimulated emission in the region near the junction. Called laser diode, injected laser, coherent electroluminescence device, semiconductor laser.

diode limiter—A circuit employing a diode and used to prevent signal peaks from exceeding a predetermined value.

diode-pentode—A vacuum tube having a diode and a pentode combined in the same envelope.

diode rectification—The conversion of an alternating current into a unidirectional current by means of a two-element device such as crystal, vacuum tube, etc.

diode switch—A diode in which positive and negative biasing voltages (with respect to the cathode) are applied in succession to the anode in order to pass and block, respectively, other applied waveforms within certain voltage limits. In this way, the diode acts as a switch.

diode transistor logic—Abbreviated DTL. A logic circuit that uses diodes at the input to perform the electronic logic function that activates the circuit transistor output. In monolithic circuits, the DTL diodes are a positive level logic AND function or a negative level OR function. The output transistor acts as an inverter to result in the circuit becoming a positive NAND or a negative NOR function.

diode-triode — A vacuum tube having a diode and triode combined in the same envelope.

diopter—The unit of optical measurement which expresses the refractive power of a lens or prism.

DIP—Abbreviation for dual in-line package.

dip—1. A drop in the plate current of a class-C amplifier as its tuned circuits are being adjusted to resonance. 2. The angle between the direction of the earth's magnetic field and the horizontal as measured in a vertical plane.

dip coating—1. A method of applying an insulating coating to a conductor by passing it through an applicator containing the insulating medium in liquid form. The insulation is then sized and passed through ovens to solidify. 2. The insulation so applied.

dip encapsulation — A type of conformal coating. An embedding process in which the insulating material is applied by immersion and without the use of an outer container. The coating conforms generally with the contour of the embedding part or assembly.

diplexer—A coupling unit which allows more than one transmitter to operate together on the same antenna.

diplex operation—The simultaneous transmission or reception of two messages from a single antenna or on a single carrier.

diplex radio transmission — Simultaneous transmission of two signals by using a common carrier wave.

diplex reception—The simultaneous reception of two signals having some feature in common—i.e., a single receiving antenna or a single carrier frequency.

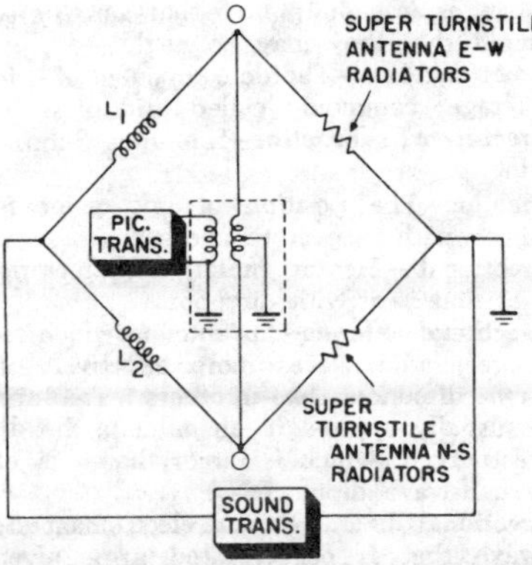

Diplexer circuit.

dipole—1. See Dipole Antenna. 2. A molecule which has an electric moment. For a molecule to be a dipole, the effective center of the positive charges must be at a different point from the center of the negative charges.

$$2L = 1/2 \text{ WAVELENGTH X }.95$$

$$L \text{ (FT.)} = \frac{234}{\text{FREQ. (MHz)}}$$

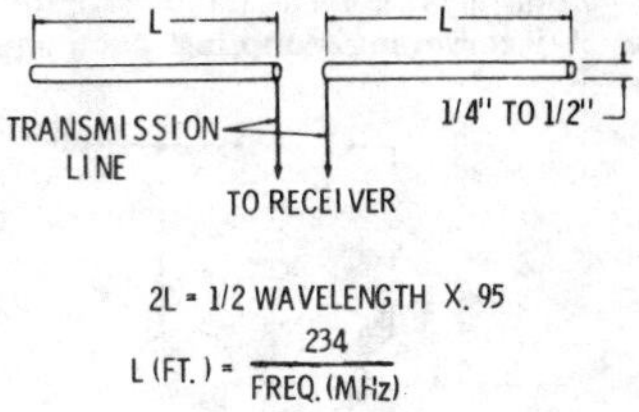

Dipole antenna.

dipole antenna — Also called dipole. A straight radiator usually fed in the center. Maximum radiation is produced in the plane normal to its axis. The length specified is the overall length.

dipole disc feed—An antenna, consisting of a dipole near a disc, used to reflect energy to the disc.

dipping—The process of impregnating or coating insulating materials or windings by the simple method of immersion in the liquid insulating material. A step in the process of treating insulating materials or electrical components by immersion in a liquid insulation, followed by draining and curing to provide increased electrical and mechanical protection.

dip soldering—1. The process of soldering component leads, terminals, and hardware to the conductive pattern on the "bottom" of a printed-circuit board by dipping that side into molten solder or floating it on the surface. 2. A process of joining metals, previously cleaned and fluxed, by immersing them wholly or partially into molten solder. The filling of the joint is by capillary attraction.

direct-access device — See Random-Access Device.

direct-acting recording instrument—An instrument in which the marking device is mechanically connected to or directly operated by the primary detector.

direct address—An address that specifies the location in a computer of an instruction operand.

direct addressing—This is the standard addressing mode in a computer. It is characterized by an ability to reach any point in main storage directly. Direct addressing is sometimes restricted to the first 256 bits in main storage.

direct capacitance — The capacitance between two conductors excluding stray capacitance that may exist between the two conductors and other conducting elements.

direct-coupled amplifier—1. A direct-current amplifier in which the plate of one stage is coupled to the grid of the next stage by a direct connection or a low-value resistor. 2. An amplifier in which the output of one stage is connected to the input of the next stage without the use of intervening coupling components.

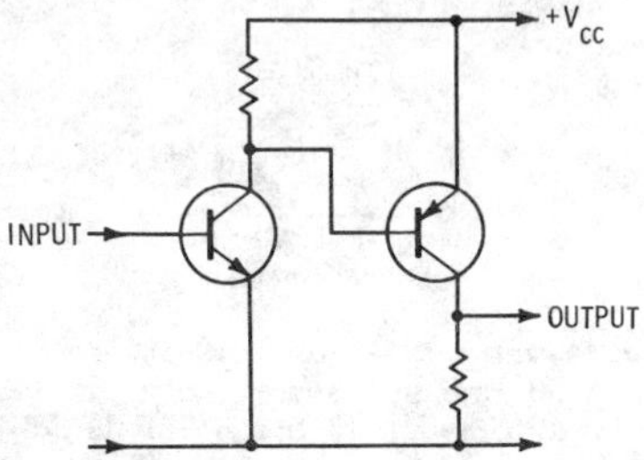

Direct-coupled amplifier.

direct-coupled transistor logic—A NOR-gate type of bipolar logic in which the output of one gate is coupled directly to the input of the succeeding gate. This form of logic evolved into resistor-transistor logic because of the difficulty in mass producing transistors within the close tolerances necessary for direct coupling. Abbreviated DCTL.

direct coupling—1. The association of two or more circuits by means of an inductance, a resistance, a wire, or a combination of these so that both direct and alternating currents can be coupled. 2. Interstage coupling or speaker coupling with no intervening transformer or capacitor. To the speaker it ensures that the damping factor remains high at low frequencies (but increasing power supply impedance at low frequencies can influence this), while direct coupling generally minimizes low-frequency phase shift and encourages enhanced bass performance.

direct current—1. Abbreviated dc. An essentially constant-value current that flows in only one direction. 2. A flow of continuous electric current in one direction as long as the circuit is closed (as opposed to alternating current). 3. A current that flows in one direction only in an electric circuit. It may be continuous or discontinuous. It may be constant or varying.

direct-current amplifier — Also called dc amplifier. An amplifier capable of boosting dc voltages. Resistive coupling only is generally employed between stages, but sometimes will be combined with other forms.

direct-current erasing head—A head which uses direct current in magnetic recording to produce the magnetic field required for erasure. Direct-current erasing is achieved by subjecting the medium to a unidirectional field. Such a medium is therefore in a different magnetic state from one erased by alternating current.

direct-current generator—A rotating machine that changes mechanical into electrical energy in the form of direct current. This is accomplished by commutating bars on the armature. The bars make contact with stationary brushes from which the direct current is taken.

direct-current restorer — The means by which a direct current or low-frequency component is reinserted after transmission. Used in a circuit incapable of transmitting slow variations, but capable of transmitting components of higher frequency.

direct-current transmission—Of television, that form of transmission in which a fixed setting of the controls makes any instantaneous value of signal correspond to the same value of brightness at all times.

direct digital control — Time sharing, or multiplexing, of a computer among many controlled loops.

direct distance dialing—A telephone exchange service which enables the telephone user to call other subscribers outside his local area without operator assistance. *See also* Area Code and Numbering Plan.

direct drive—A drive system used to rotate a turntable in which the platter is driven directly by the motor shaft at the exact speed required. These designs usually include electronic motor control.

direct-drive torque motor—A servoactuator which can be directly attached to the load it is to drive. It converts electrical signals directly into sufficient torque to maintain the desired accuracy in a positioning or speed control system.

direct electromotive force — A unidirectional electromotive force in which the changes in values are either zero or so small that they may be neglected.

direct grid bias—The dc component of grid voltage; commonly called grid bias.

direct-insert subroutine—*See* Open Subroutine.

direction—The position of one point in space with respect to another.

directional — Having radiative characteristics that vary with direction.

directional antenna—An antenna which radiates radio waves more effectively in some directions than in others. (The term is usually applied to an antenna the directivity of which is larger than that of a half-wave dipole.)

directional beam — An electromagnetic wave that is concentrated in a given direction.

directional coupler—A junction consisting of two waveguides coupled together in such a manner that a traveling wave in either guide will induce a traveling wave in the same direction in the other guide.

directional filter—A filter used to separate the two frequency ranges in a carrier system where one range of frequencies is used for transmission in one direction and another range of frequencies for transmission in the opposite direction. Also called directional separation filter.

directional gain—*See* Directivity Index.

directional homing—The procedure of following a path in such a way that the target is maintained at a constant relative bearing.

directional hydrophone — A hydrophone having a response that varies significantly with the direction of incidence of sound.

directional lobe—*See* Lobe.

directional microphone — 1. A microphone the response of which varies significantly with the direction of sound. (*See also* Unidirectional, Bidirectional, *and* Semidirectional Microphone.) 2. A microphone which is more sensitive to sounds coming from certain directions than to sounds coming from other directions.

directional pattern—Also called radiation pattern. A graphical representation of the radiation or reception of an antenna as a function of direction. Cross sections are frequently given as vertical and horizontal planes, and principal electric and magnetic polarization planes.

directional phase shifter—A passive phase-shifting device in which the phase change for transmission in one direction is different from the phase change for transmission in the opposite direction.

directional power relay—A device which functions on a desired value of power flow in a given direction, or upon reverse power resulting from arcback in the anode or cathode circuits of a power rectifier.

directional relay—A relay which functions in conformance with the direction of power, voltage, current, pulse rotation, etc. *See also* Polarized Relay.

directional separation filter — *See* Directional Filter.

direction angle—The angle between the antenna base line and a line connecting the center of the base line with the target.

direction cosine—The cosine of the angle between the base line and the line from the center of the base line to the target.

direction finder — Abbreviated df. Also called radio compass. Apparatus for receiving radio signals and taking their bearings in order to determine their points of origin.

direction finding—The principle and practice of determining a bearing by radio means, using a discriminating antenna system and a radio receiver so that the direction of an arriving wave, and ostensibly the direction or bearing of a distant transmitter, can be determined.

direction of lay—The lateral direction in which strands or the elements of a cable are wound over the top of the cable. Expressed as right- or left-hand lay, viewed as they recede from the observer.

direction of polarization—For a linearly polarized wave, the direction of the electrostatic field.

direction of propagation—At any point in a homogeneous, isotropic medium, the direction of the time-average energy flow. In a uniform waveguide, the direction of propagation is often taken along the axis. In a uniform lossless waveguide, the direction of propagation at every point is parallel to the axis and in the direction of time-average energy flow.

direction rectifier—A rectifier which supplies a direct-current voltage, the magnitude and polarity of which are determined by the magnitude and relative polarity of an alternating-current selsyn error voltage.

directive gain—In a given direction, 4π times the ratio of the radiation intensity to the total power radiated by the antenna.

directivity—1. The property that causes an antenna to radiate or receive more energy in some directions than in others. 2. The value of the directive gain of an antenna in the maximum-gain direction. 3. A tendency for some microphones to respond less strongly to sounds arriving from the sides and/or rear. Directional microphones are useful in discriminating on the basis of direction between wanted sounds (musical instruments) and unwanted sounds (audience noises). Directivity is typically graphed on a "polar

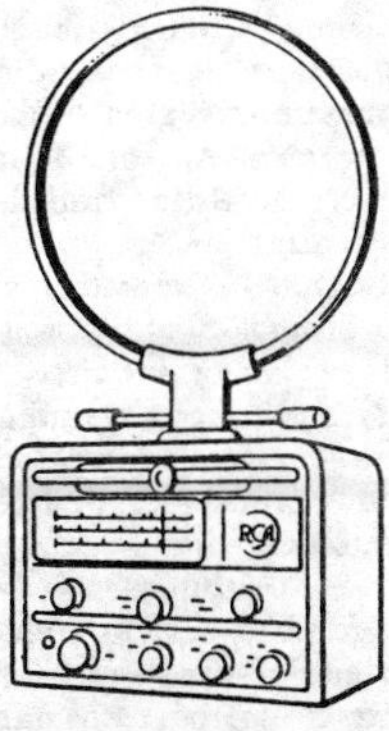

Direction finder.

pattern," and is thus classed as nondirectional (omnidirectional), bidirectional (figure-8), or unidirectional (cardioid), supercardioid or hyperdirectional.

directivity diagram of an antenna – The graphical representation of the gain of an antenna in the different directions of space.

directivity factor—1. In acoustics, the directivity factor is equivalent to directivity, as applied to an antenna. 2. Of a transducer used for sound emission, the ratio of the intensity of the radiated sound at a remote point in a free field on the principal axis to the average intensity of the sound transmitted through a sphere passing through the remote point and concentric with the transducer. 3. Of a transducer used for sound reception, the ratio of the square of the electromotive force produced in response to sound waves arriving in a direction parallel to the principal axis of the transducer to the mean square of the electromotive force that would be produced if sound waves having the same frequency and mean square pressure were arriving at the transducer simultaneously from all directions with random phase. The frequency should be specified in both cases.

directivity index – Also called directional gain. A measure of the directional properties of a transducer. It is the ratio, in decibels, of the average intensity or response over the whole sphere surrounding the projector or hydrophone to the intensity or response on the acoustic axis.

directivity of a directional coupler—Ratio of the power measured at the forward-wave sampling terminals with only a forward wave present in the transmission line, to the power measured at the same terminals when the forward wave reverses direction. This ratio is usually expressed in dB and would be infinite for a perfect coupler.

directivity of an antenna—The ratio of the maximum field intensity to the average field intensity at a given distance, implying a maximum value.

directivity pattern—A plot of the response of an electroacoustic transducer as a function of direction.

directivity signal—A spurious signal present in the output of any coupler because its directivity is not infinite.

direct light—Light from a luminous object such as the sun or an incandescent lamp, as opposed to reflected light.

directly grounded—*See* Solidly Grounded.

directly heated cathode—A wire, or filament, designed to emit the electrons that flow from cathode to plate. This is done by passing a current through the filament; the current heats the filament to the point where electrons are emitted. In an indirectly heated cathode, the hot filament raises the temperature of a sleeve around the filament; the sleeve then becomes the electron emitter.

direct lighting—A system of lighting which delivers a majority of light in useful directions without being deflected from the ceiling or walls. Any lamp equipped with a glass or metal reflector, arranged to reflect the light toward the object to be illuminated is classified as direct lighting.

direct material—A semiconductor material in which electrons move directly from the conduction band to the valence band to recombine with holes. The process of recombination conserves energy and momentum.

direct memory access – 1. Abbreviated DMA. In a computer, a method of gaining direct access to main storage to achieve data transfer without involving the CPU. The manner in which CPU is disabled while DMA is in progress differs in different models and some use several methods to accomplish DMA. 2. A mechanism that allows an input/output device to take control of the CPU for one or more memory cycles in order to write into memory or read from memory. The order of executing the program steps (instructions) remains unchanged.

direct numerical control—A system connecting a set of numerically controlled machines to a common memory for part program or machine program storage with provision for on-demand distribution of data to the machines. Direct numerical control systems have additional provisions for collection, display, or editing of part programs, operator instructions, or data related to the numerical control process.

director—1. A parasitic antenna element located in the general direction of the major lobe of radiation for the purpose of increasing the radiation in that direction. 2. Equipment in common-carrier telegraph-message switching systems, used to make cross-office selection and connection from an input-line to an output-line equipment in accordance with addresses in the message. 3. A telephone switch which translates the digits dialed into the directing digits actually used to switch the call. 4. Electromechanical equipment which is used to track a moving target in azimuth and angular height, and which, with the addition of other necessary information from an outside source such as a radar set or a range finder, continuously computes firing data and transmits them to the guns. 5. In a machine-tool or process control system, the part of the system that receives the command signals from a controller and converts and amplifies these signals to make them usable by the control devices in the machine or process.

direct pickup—Transmission of television images without resorting to an intermediate magnetic or photographic recording.

direct piezoelectricity—A name sometimes given to the piezoelectric effect in which an electric charge is developed on a crystal by the application of mechanical stress.

direct point repeater—A telegraph repeater in which the relay controlled by the signals received over a line sends corresponding signals directly into another line or lines without the use of any other repeating or transmitting apparatus.

direct radiative transition — A transition that involves photons alone. *See* Transition, 1.

direct-radiator speaker — A speaker in which the radiating element acts directly on the air instead of relying on any other element such as a horn.

direct recording — 1. The production of a visible record without subsequent processing, in response to received signals. 2. Analog recording in which continuous amplitude variations are recorded linearly through the use of ac bias.

direct-recording magnetic tape—A method of recording using a high-frequency bias in which the electrical input signal is applied to the recording head without alteration.

direct resistance-coupled amplifier — An amplifier in which the plate of one stage is connected either directly or through a resistor to the control grid of the next stage, with the plate-load resistor being common to both stages. Used to amplify small changes in direct current.

direct route—In wire communications, the trunks that connect two switching centers, regardless of the geographical path the actual trunk facilities may follow.

direct scanning—In this method, the entire subject is illuminated continuously but the television camera views only one portion of it at a time.

direct sound wave—A wave in an enclosure emitted from a source prior to the time it has undergone its first reflection from a boundary of the enclosure. Frequently a sound wave is said to be direct if it contains reflections that have occurred from surfaces within about 0.05 second after the sound was first emitted.

direct synthesizer—A frequency synthesizer producing an output frequency which is related to the reference frequency by the ratio of two integers. The primary advantage of this type synthesizer lies in its ability to change output frequencies at a moderately fast rate (typically in the microsecond range) and in a random way. Principal applications include frequency-agile radars, secure communication links, and electronic countermeasures.

direct voltage—Also called dc voltage. A voltage that forces electrons to move through a circuit in the same direction and thereby produce a direct current.

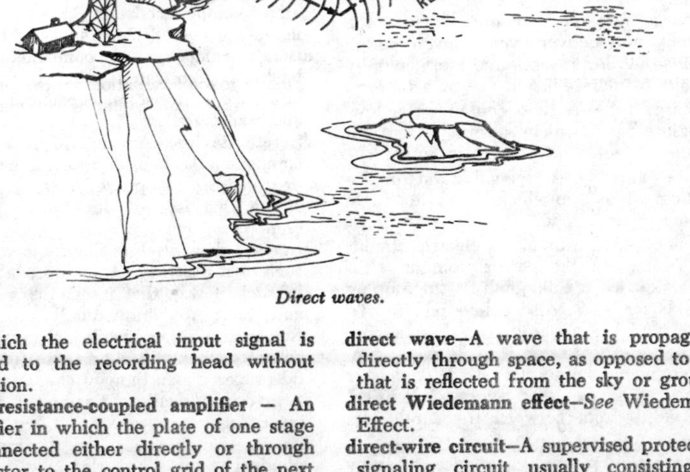

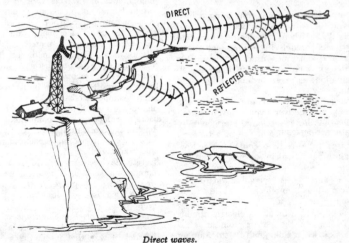

Direct waves.

direct wave—A wave that is propagated directly through space, as opposed to one that is reflected from the sky or ground.

direct Wiedemann effect—*See* Wiedemann Effect.

direct-wire circuit—A supervised protective signaling circuit usually consisting of one metallic conductor and a ground return and having signal-receiving equipment responsive to either an increase or a decrease in current.

direct writing galvanometer recorder—Recorder using a pen attached directly to a galvanometer movement for direct writing of signals of frequencies up to about 300 Hz.

disable—To prevent the passage of binary signals by application of the proper signal to the disable terminal of a device.

disc—1. A phonograph record. 2. The blank used in a recorder.

disc capacitor—A small disc-shaped capacitor with a ceramic dielectric, generally used for bypassing or for temperature compensation in tuned circuits.

disc files—A type of storage medium consisting of numbers of discs that rotate; each disc has a special coating for retaining stored information.

disc generator—A capacitive-charge type of voltage generator.

discharge—1. In a storage battery, the conversion of chemical energy into electrical energy. 2. The release of energy stored in a capacitor when a circuit is connected between its terminals.

discharge breakdown—Breakdown of a material as a result of degradation due to gas discharges.

discharge key—A device for switching a capacitor suddenly from a charging circuit to a load through which it can discharge.

discharge lamp—A lamp containing a low-pressure gas or vapor which ionizes and emits light when an electric discharge is passed through it. Fluorescent materials are sometimes used on the inside of the glass envelope to increase the illumination, as in an ordinary fluorescent lamp.

discharge rate—The amount of current a battery will deliver over a given period of time. A slower discharge rate generally results in more efficient use of a battery.

discharge tube—A tube containing a low-pressure gas which passes a current whenever sufficient voltage is applied.

discone antenna—A special form of biconical antenna in which the vertex angle of one cone is 180°.

disconnect—1. To break an electric circuit. 2. To remove the power from an electrical device (colloquially, "to unplug the device"). 3. Also called release. To disengage the apparatus used in a connection and to restore it to its ready condition when not in use. 4. A device or group of devices that removes electrical continuity from between the conductors of a circuit and the source of supply.

disconnecting means — A device whereby the current-carrying conductors of a circuit can be disconnected from their source of supply.

disconnector release—A device that disengages the apparatus used in a telephone connection to restore it to the condition in which it exists when not in use.

disconnect signal—A signal sent from one end of a trunk or subscriber line to indicate at the other end that the established connection should be released.

disconnect switch (motor circuit switch)—A switch intended for use in a motor branch circuit. It is rated in horsepower, and is capable of interrupting the maximum operating overload current of a motor of the same rating at the rated voltage.

discontinuous amplifier—An amplifier that reproduces an input waveform on some type of averaging basis.

disc pack—A set of magnetic discs that can be removed from a disc storage as one unit.

disc recorder—A recording device in which the sounds are mechanically impressed onto a disc—as opposed to a tape recorder, which impresses the sound magnetically on a tape.

discrete—1. An individual circuit component, complete in itself, such as a resistor, diode, capacitor, or transistor, and used as an individual and separable circuit element. 2. Pertaining to distinct elements, such as characters, or to representation by means of distinct elements. 3. A term applied to four channels when there are four electrically independent signals, as opposed to matrix. 4. A quad disc or record-playback method that keeps four signals separate, distinct, and independent from recording to playback.

discrete circuit—A circuit built from separate components that are individually manufactured, tested, and assembled.

discrete component—A component which has been fabricated prior to its installation (e.g., resistors, capacitors, diodes, and transistors).

discrete device — An individual electrical component such as a resistor, capacitor, or transistor, as opposed to an integrated circuit that is equivalent to several discrete components.

discrete element—An electronic element, such as a resistor or transistor, fabricated in such a way that it can be measured and transported individually.

discrete part—A separately packaged single circuit element supplying one fundamental property as a lumped characteristic in a given application. Examples: resistor, transistor, diode.

discrete sampling—The lengthening of individual samples so that the sampling process does not deteriorate the intelligence frequency response of the channel.

discrete thin-film component—An individually packaged electronic component having one or more thin films serving as resistive, conductive, and/or insulating elements. Resistors and potentiometers having thin-film metallic resistance elements are examples.

discretionary wiring—The use of a selective metalization pattern in the interconnection of large numbers of basic circuits on a slice of semiconductor material to form complex arrays. The metalization pattern connects only the "good" circuits on the wafers. Discretionary wiring requires a different interconnection pattern for each wafer.

discrimination—1. The difference between losses at specified frequencies, with the system or transducer terminated in specified impedances. 2. In a frequency-modulated system, the detection or demodulation of the imposed variations in the frequency of the carriers. 3. In a tuned circuit, the degree of rejection of unwanted signals.

discrimination ratio—The ratio of the width of the passband of a filter to the width of the stopband of the filter.

discriminator—1. A device in which amplitude variations are derived in response to frequency or phase variations. 2. A facsimile auxiliary device between the radio receiver and the recorder which converts an audio-frequency shift facsimile signal to an amplitude-modulated facsimile signal.

discriminator transformer — A transformer used in fm receivers to convert frequency changes directly to af signals.

discriminator tuning unit—A device which tunes the discriminator to a particular subcarrier.

disc-seal tube—Also called lighthouse tube or megatron. An electron tube with disc-shaped electrodes arranged in closely spaced parallel layers, to give a low interelectrode capacitance along with a high power output in the uhf region.

disc storage—1. Random-access auxiliary memory device in which information is stored on constantly rotating magnetic discs. 2. The storage of data on the surface of magnetic discs.

dish—1. A microwave antenna, usually shaped like a parabola, which reflects the radio energy leaving or entering the system. 2. A parabolic type of radio or radar antenna, roughly the shape of a soup bowl.

diskette—*See* Floppy Disk.

dislocation—In a crystal, a region in which the atoms are not arranged in the perfect crystal-lattice structure.

dispatcher—In a digital computer, the section which transfers the "words" to their proper destinations.

dispenser—A device that automatically distributes radar chaff from an aircraft.

disperse—In data-processing, to distribute grouped input items among a larger number of groups in the output.

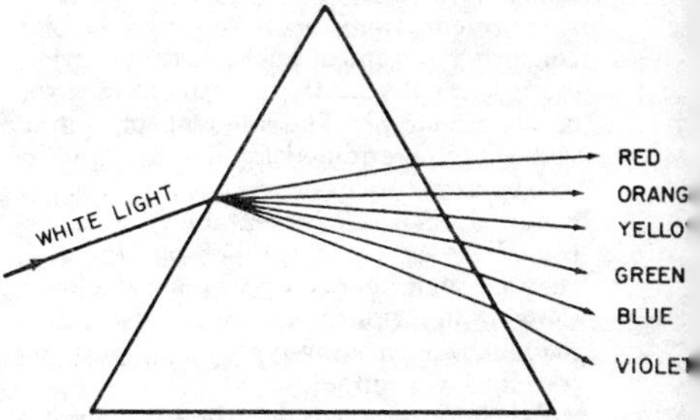

Dispersion, 3 (by a prism).

dispersion—1. Separation of a wave into its component frequencies. 2. Scattering of a microwave beam as it strikes an obstruction. 3. The property of an optical material which causes some wavelengths of light to be transmitted through the material at different velocities, and the velocity is a function of the wavelength. (This causes each wavelength of light to have a different refractive index.) 4. In a magnetostrictive delay line, the variation of delay as a function of frequency. 5. The frequency difference that can be analyzed in one sweep by a spectrum analyzer. Dispersion can be considered as that frequency width over which sampling can be performed, and is always equal to or less than the frequency range. 6. The extent to which a speaker distributes acoustical power widely and evenly into the listening area.

dispersive medium—A medium in which the phase velocity of a wave is related to the frequency.

displacement—The vector quantity representing change of position of a particle.

displacement current—A current which exists in addition to ordinary conduction

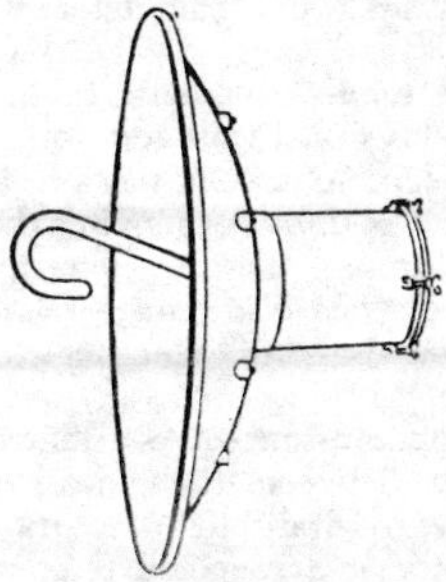

Dish antenna.

current in ac circuits. It is proportional to the rate of change of the electric field. The current at right angles to the direction of propagation determined by the rate at which the field energy changes.

displacement of porches—The difference in level between the front and back porch of a television signal.

displacement transducer—A device which converts mechanical energy into electrical energy, usually by the movement of a rod or an armature. The amount of output voltage is determined by the amount the rod or armature is moved.

display—1. Visual presentation of a received signal on a cathode-ray tube. 2. Row of digits across top of a calculator, showing input or final answer. Also called readout. In printing-type calculators referred to as printout.

display information processor—A computer used in a combat operations center to generate situation displays.

display loss—*See* Visibility Factor.

display modes—Each display mode, such as vector, increment, character, point, vector continue, or short vector specifies the manner in which points are to be displayed on the screen.

display primaries—Also called receiver primaries. The red, green, and blue colors produced by a color television receiver and mixed in proper proportions to produce other colors.

display-storage tube—A special cathode-ray tube with a long and controllable image persistence and high luminescence.

display unit—A device used to provide a visual representation of data.

display window—The width of the portion of the frequency spectrum presented on panoramic presentation, expressed in frequency units, usually megahertz.

disruptive discharge—The sudden, large current through an insulating medium when electrostatic stress ruptures the medium and thus destroys its insulating ability.

dissector—In optical character recognition, a mechanical or electronic transducer that sequentially detects the level of light in different areas of a completely illuminated sample space.

dissector tube—A camera tube having a continuous photocathode on which a photoelectric-emission pattern is formed. Scanning is done by moving the electron optical image of the pattern over an aperture. (*See also* Image Dissector, 1.)

dissipation—Loss of electrical energy as heat.

dissipation constant—A constant of proportionality between the power dissipated and the resultant temperature rise in a thermistor at a specified temperature.

dissipation factor—Symbolized by D. Ratio between the permittivity and conductivity of a dielectric. The reciprocal of the dissipation factor is the storage factor, sometimes called the quality factor. Abbreviated df.

dissipation line—A length of stainless-steel or Nichrome wire used as a noninductive impedance for termination of a rhombic transmitting antenna when power of several kilowatts must be dissipated.

dissonance—The formation of maxima and minima by the superposition of two sets of interference fringes from light of two different wavelengths.

dissymmetrical network—*See* Dissymmetrical Transducer.

dissymmetrical transducer—Also called dissymmetrical network. A transducer with unequal input and output image impedances.

distance mark—Also called range mark. A mark which indicates, on a cathode-ray screen, the distance from the radar set to a target.

distance-measuring equipment — Abbreviated DME. A radio navigational aid for determining the distance from a transponder beacon by measuring the time of transmission to and from it.

distance protection—The effect of a device operative within a predetermined electrical distance on the protected circuit to cause and maintain an interruption of power in a faulty circuit.

distance relay—1. A protective relay, the operation of which is a function of the distance between the relay and the point of fault. 2. A device which functions when the circuit admittance, impedance, or reactance increases or decreases beyond predetermined limits.

distance resolution—The ability of a radar to differentiate targets solely by distance measurement. Generally expressed as the minimum distance the targets can be separated and still be distinguishable.

distortion—1. Undesired changes in the waveform of a signal so that a spurious element is added. All distortion is undesirable. Harmonic distortion disturbs the original relationship between a tone and other tones naturally related to it. Intermodulation distortion (imd) introduces new tones caused by mixing of two or more original tones. Phase distortion, or nonlinear phase shift, disturbs the natural timing sequence between a tone and its related overtones. Transient distortion disturbs the precise attack and decay of a musical sound. Harmonic and imd distortion are expressed in percentages; phase distortion in degrees; transient distortion is usually judged from oscilloscope patterns. 2. Unwanted changes in the purity of sound being reproduced or in rf signals. In audio it generally implies intermodulation and/or harmonic distortion. These are derived from phase differences and/or amplitude distortion where the amplitude of the output does not bear the same proportion to the input at all frequencies.

distortion factor—*See* Harmonic Distortion.

distortion factor of a wave—The ratio of the effective value of the residue after the elimination of the fundamental to the effective value of the original wave.

distortionless line — A transmission line whose propagation constant is independent of frequency. (This is approached in a practical case by adjusting the line parameters, series inductance (l) shunt capacitance (c) series resistance (r) shunt conductance (g) so that $r/g = l/c$).

distortion meter — An instrument which measures the deviation of a complex wave from a pure sine wave.

distortion tolerance—Of a telegraph receiver, the maximum signal distortion which can be tolerated without error in reception.

distress frequency—A frequency reserved for distress calls, by international agreement. It is 500 kHz for ships at sea and aircraft over the sea.

distributed—Spread out over an electrically significant length, area, or time.

distributed amplifier—A multistage amplifier in which the high-frequency limitation, due to the input and output capacitances of the active element, is circumvented by making these capacitances the shunt elements of lumped-parameter delay lines. In this way the overall gain is the sum of the gains of the individual stages rather than the product, thus allowing amplification even when the individual gains are less than unity.

distributed capacitance—Also called self-capacitance. Any capacitance not concentrated within a capacitor, such as the capacitance between the turns in a coil or choke, or between adjacent conductors of a circuit.

distributed constants—Constants such as resistance, inductance, or capacitance that exist along the entire length or area of a circuit, instead of being concentrated within circuit components.

distributed-emission photodiode—A broadband photodiode for use in detecting modulated laser beams at millimeter wavelengths.

distributed inductance — The inductance along the entire length of a conductor, as distinguished from the inductance concentrated within a coil.

distributed network — An electrical-electronic device which for proper operation depends on physical size in comparison to a wavelength and physical configuration.

distributed parameter network—A network in which the parameters of resistance, capacitance, and inductance cannot be taken as being concentrated at any one point in space. Rather, the network must be described in terms of its magnetic and electric fields and the quantities related to the distributed constants of the network.

distributed paramp—A paramagnetic amplifier consisting essentially of a transmission line shunted by uniformly spaced, identical varactors. The varactors are excited in sequence by the applied pumping wave to give the desired traveling-wave effect.

distributed pole—A motor has distributed poles when its stator or field windings are distributed in a series of slots located within the arc of the pole.

distributing amplifier—An amplifier, either radio-frequency or audio-frequency, having one input and two or more isolated outputs.

distributing cable—*See* Distribution Cable, 1.

distributing frame—A structure for terminating permanent wires of a central office, private branch exchange, or private exchange, and for permitting the easy change of connections between them by means of cross-connecting wires.

distributing terminal assembly—A frame situated between each pair of selector bays to provide terminal facilities for the selector bank wiring and facilities for cross connection to trunks running to succeeding switches.

distribution—Also called frequency distribution. The number of occurrences of the particular values of a variable as a function of those values.

distribution amplifier—A power amplifier designed to energize a speech, music, or antenna distribution system. Its output impedance is sufficiently low that changes in the load do not appreciably affect the output voltage.

distribution cable—1. Also called distributing cable. A cable extended from a feeder cable for the purpose of providing service to a specific area. 2. In a system, the transmission cable from the distribution amplifier to the drop cable.

distribution center—In an alternating-current power system, the point at which control and rotating equipment is installed.

distribution coefficients—Equal-powered tristimulus values of monochromatic radiations.

distribution switchboard—A power switchboard used for the distribution of electrical energy at the voltage common for each distribution within a building.

distributor—1. *See* Memory Register. 2. The electronic circuitry which acts as an intermediate link between the accumulator and drum storage.

disturbance — An irregular phenomenon which interferes with the interchange of intelligence during transmission of a signal.

disturbed-one output—A "one" output of a magnetic core to which partial-write pulses have been applied since that core was last selected for writing.

disturbed-zero output—A "zero" output of a magnetic core to which partial-write pulses have been applied since that core was last selected for reading.

dither—1. An oscillation introduced for the purpose of overcoming the effects of friction, hysteresis, or clogging. 2. A small electrical signal deliberately injected into an electromechanical device for the purpose of overcoming static friction in the device. In a recording instrument it makes the indicator "ready to jump."

dithering—The application of intermittent or periodic acceleration forces sufficient to minimize the effect of static friction with a transducer, without introducing other errors.

divergence loss—The part of the transmission loss that is caused by the spreading of sound energy.

diverging lens—A lens that is thinner in the center than at the edges. Such a lens causes light passing through to spread out, or diverge.

diversity—A form of transmission using several modes, usually in space or in time, to compensate for fading or outages in any one of the modes. In the space diversity system, the same signal is sent simultaneously over several different transmission paths, which are separated enough so that independent propagation conditions can be expected. With time diversity, the same path may be used, but the signal is transmitted more than once, at different times. There are other forms of diversity, using different frequencies or different polarizations to provide the separate transmission modes.

diversity factor—The ratio of the sum of the individual maximum demands of the subdivisions of a system (or part of a system) to the maximum demand of the entire system (or part of the system).

diversity gain—The gain in reception as a result of the use of two or more receiving antennas.

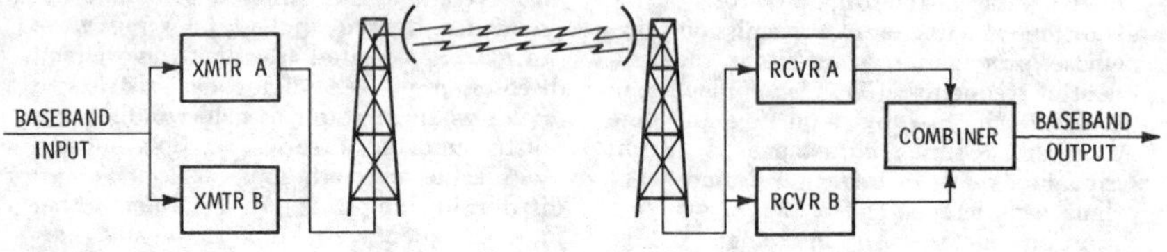

Frequency diversity.

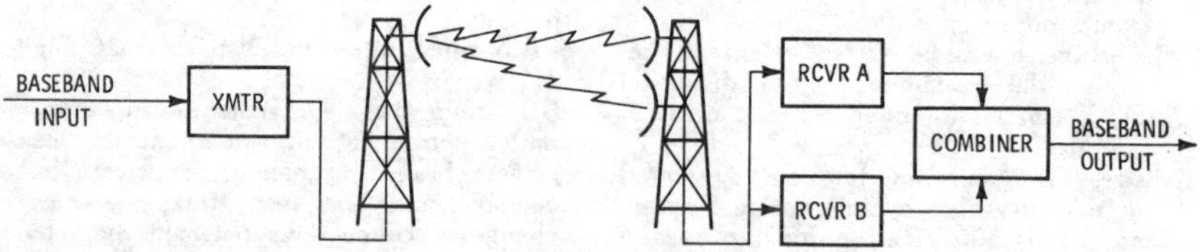

Space diversity.

Diversity reception.

diversity reception—A method of minimizing the effects of fading during reception of a radio signal. This is done by combining and/or selecting two or more sources of received-signal energy which carry the same intelligence but differ in strength or signal-to-noise ratio in order to produce a usable signal.

diverter pole generator — A compound-wound direct-current generator with the series winding of the diverter pole opposing the flux generated by the shunt-wound main pole; provides a close voltage regulation.

divide-by-N counter — A group of counter stages that can be programmed to divide an input frequency by any number up to N.

divide-by-16 counter—A logic device in which four flip-flops count from 0 through 15 and then recycle to 0. All 16 states of the combination of four flip-flops are used. Sometimes referred to as a hexadecimal counter.

divide-by-10 counter—*See* Decade Counter.

divide check—In a computer, an indicator which shows that an invalid division has occurred or has been attempted.

divided-carrier modulation—The process by which two signals are added so that they can modulate two carriers of the same frequency but 90° out of phase. The resultant signal will have the same frequency as the carriers, but its amplitude and phase will vary in step with the variations in amplitude of the two modulating signals.

divider—*See* Counter, 4.

dividing network—Also called speaker dividing network and crossover network. A frequency-selective network which divides the audio-frequency spectrum into two or more parts to be fed to separate devices such as amplifiers or speakers.

D layer—The lowest ionospheric layer. Its intensity is proportional to the height of the sun and is greatest at noon. Waves below approximately 3 MHz are absorbed by the D layer when it is present. High-angle radiation may penetrate the D layer and be reflected by the E layer.

DMA—Abbreviation for direct memory access.

DMOS—Abbreviation for diffused metal-oxide semiconductor. A process where n and p atoms are diffused through the same mask opening to give precise-sized narrow channels. Used on discrete field-effect transistors (not MOS ICs) for ultra-high gains and frequency performance. Its use of a larger area of active silicon makes it not too promising for MOS ICs.

DNL—*See* Dynamic Noise Limiter.

document reader—A general term referring to ocr or omr equipment which reads a limited amount of information (one to five lines). Generally operates from a predetermined format and is therefore more restricted in the location of information to be read. The forms involved are generally tab card size or slightly smaller or larger.

doghouse—A small enclosure located near the base of a transmitting-antenna tower and used to house antenna tuning equipment.

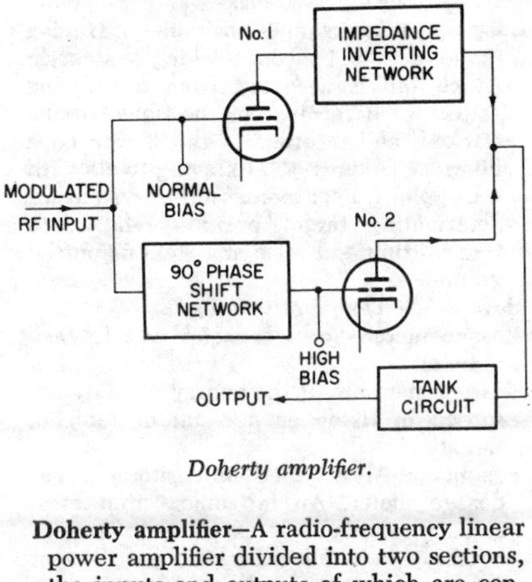

Doherty amplifier.

Doherty amplifier—A radio-frequency linear power amplifier divided into two sections, the inputs and outputs of which are connected by quarter-wave (90°) networks. As long as the input-signal voltage is less than half the maximum amplitude, section No. 2 is inoperative and section No. 1 delivers all the power to the load. The load presents twice the optimum impedance required for maximum output. At one-half the maximum input, section No. 1 is operating at peak efficiency but is beginning to saturate. Above this level, section No. 2 comes into operation and decreases the impedance presented to section No. 1. As a result, section No. 2 delivers more and more power to the load until, at maximum signal input, both sections are operating at peak efficiency and each section is delivering one-half the total output power.

dolby—1. A technique which increases the signal-to-noise ratio of a recording medium by raising the volume of quiet passages prior to recording, and lowering them to their original levels during playback. The lowering process automatically reduces any noise that was introduced as a result of the recording or playback processes. 2. Noise reduction circuit that boosts the recorded signal at the tape hiss frequencies for low levels and to reduce the boost progressively as the signal becomes large enough to mask the noise. (The dolby system has the important advantage that it is standardized and any dolby tape can be replaced accurately on any other dolby machine.)

dolby A—The original form of the dolby noise-reduction device, intended for professional use. It has four independently controlled noise-reduction channels, to increase signal-to-noise ratio at low, middle, high and very high frequencies.

dolby B—A simplified version of the original dolby A, intended primarily for use by nonprofessional recordists. Dolby B functions identically to the dolby A, but has only one controlled frequency band, which is effective primarily on tape hiss.

dolby system — A noise-reducing system whereby the recording signal is compressed and the replay signal expanded. This results in a reduction of background noise and is particularly designed for use with tape recorders.

dolly—A wheeled platform on which a television camera or other apparatus is mounted to give it wider mobility.

domain—In magnetic theory, that region of a magnetic material in which the spontaneous magnetization is all in one direction. In conventional magnetic-tape coatings, this corresponds to one oxide particle.

domestic induction heater—A home cooking utensil which is heated by induced currents within it. The unit contains a primary inductor, with the utensil itself acting as the secondary.

dominant mode—Also called fundamental mode or principal mode. In waveguide transmission, the mode with the lowest cutoff frequency. Designations for this mode are $TE_{0,1}$ and $TE_{1,1}$ for rectangular and circular waveguides, respectively.

dominant wave—The guided wave which has the lowest cutoff frequency. It is the only wave which will carry energy when the excitation frequency is between the lowest and the next higher cutoff.

dominant wavelength—Of a color sample, the wavelength of light that matches it in chromaticity when mixed with white light.

dominant wavelength of a color—The predominant wavelength of light in a color.

donor—Also called donor impurity. An impurity atom which tends to give up an electron and thereby affects the electrical conductivity of a crystal. Used to produce n-type semiconductors.

donor impurity—*See* Donor.

donut—*See* Land, 2.

doorknob tube—So called because of its shape. It is a vacuum tube designed for uhf transmitter circuits. It has a low electron-transit time and low interelectrode capacitance because of the close spacing and small size, respectively, of its electrodes.

dopant—An impurity added to a semiconductor to improve its electrical conductivity.

dope—To add impurities (called dopants) to a substance, usually a solid, in a controlled manner to cause the substance to have certain desired properties. For example, the number of electrical carriers in silicon can be increased by doping it with small amounts of other semimetallic elements. Ruby is aluminum oxide doped with chromium oxide.

doped junction—A semiconductor junction produced by the addition of an impurity to the melt during crystal growth.

doping—The addition of controlled amounts of impurities to a semiconductor to achieve a desired characteristic, such as to produce an n-type or p-type material. Common doping agents for germanium and silicon include aluminum, antimony, arsenic, gallium, and indium.

doping agent—An impurity element added to semiconductor materials used in crystal diodes and transistors. Common doping agents for germanium and silicon include aluminum, antimony, arsenic, gallium, and indium.

doping compensation—The addition of donor impurities to a p-type semiconductor or of acceptor impurities to an n-type semiconductor.

Doppler cabinet—A speaker cabinet in which either the speaker or a baffle board is rotated or moved to change the length of the sound path cyclically and thereby produce a vibrato effect mechanically.

Doppler effect—1. The observed change of frequency of a wave caused by a time rate of change of the effective distance traveled by the wave between the source and the point of observation. As the distance between a source of constant vibration and an observer diminishes or increases, the received frequencies are greater or less. 2. The apparent change in the frequency or radio wave reaching an observer, due either to motion of the source toward or away from the observer, to motion of the observer, or both.

Doppler radar—A radar unit that measures the velocity of a moving object by the shift in carrier frequency of the returned signal. The shift is proportionate to the velocity of the object as it approaches or recedes.

Doppler ranging (doran)—A cw trajectory-measuring system which utilizes the Doppler shift to measure the distance between a transmitter, missile, transponder, and several receiving stations. From these measurements trajectory data are computed. In contrast to a similar system, doran circumvents the necessity of continuously recording the Doppler signal by performing the distance measurements with four different frequencies simultaneously.

Doppler shift—1. The change in frequency of a wave reaching an observer or a system, caused by a change in distance or range between the source and the observer or the system during the interval of reception. It is due to the Doppler effect. 2. The change in frequency with which energy reaches a receiver when the source of radiation or a reflector of the radiation and the receiver are in motion relative to each other. The Doppler shift is used in many tracking systems.

Doppler velocity and position—1. Having to do with a beacon tracking system in which pulses are sent from a tracking station to a receiver in the object to be tracked, and returned to the station on a different frequency. 2. Having to do with a Doppler trajectory-measuring system for determining target position relative to transmitting and receiving stations on the ground.

doran—*See* Doppler Ranging.

dosage meter—*See* Dosimeter *and* Intensitometer.

dose—A measure of the energy actually absorbed in tissue as a result of ionizing radiation.

dosimeter—Also called intensitometer or dosage meter. An instrument that measures the amount of exposure to nuclear or X-ray radiation utilizing the ability of such radiation to produce ionization of a gas.

dot—1. *See* Button, 2. 2. Also called bubble. A symbol placed at the input of a

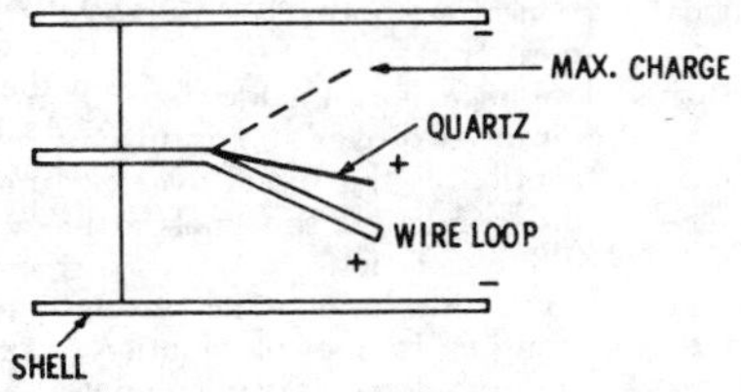

Expanded view of dosage indicator.

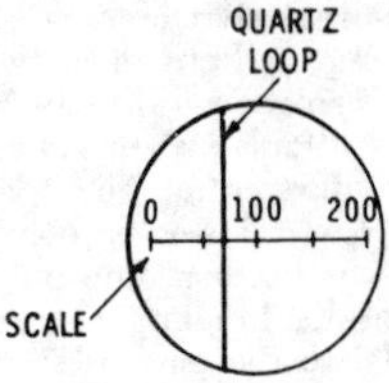

Dosage scale.

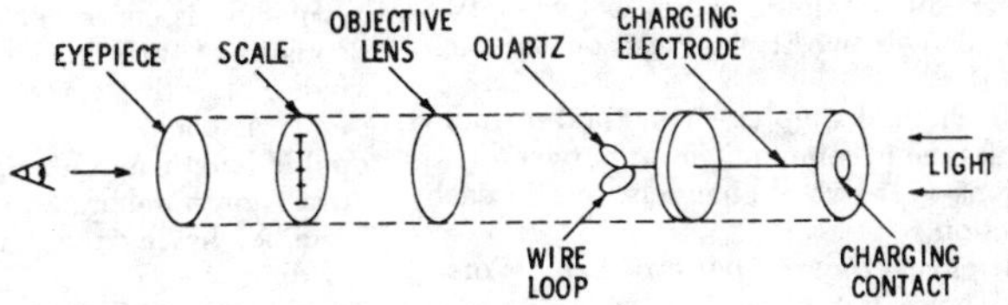

Dosimeter viewer.

Pocket dosimeter.

logic symbol to indicate that the active signal input is negative. The absence of a dot indicates a positive active signal.

dot AND—*See* Wired AND.

dot cycle—One cycle of a periodic alternation between two signaling conditions, each condition having unit duration. Thus, two-condition signaling consists of a dot, or marking element, followed by a spacing element. In teletypewriter applications, one dot cycle consists of a mark and a space. The speed of telegraph transmission sometimes is stated in terms of dot cycles per second, or dot speed (half the speed of transmission expressed in bauds).

dot encapsulation—A packaging process in which cylindrical components are inserted into a perforated wafer to form a solid block with interconnecting conductors on both surfaces joining the components.

dot generator—An instrument used in servicing color television receivers. It produces a pattern of white dots so that convergence adjustments can be made on the picture tube.

dot-matrix—A pattern of dots in a fixed area used for formulation of characters, such as 35 dots in a 5 × 7 pattern.

dot matrix display—A display format consisting of small light emitting elements arranged as a matrix. Various elements are energized to depict a character. A typical matrix is 5 × 7.

dot OR—*See* Wired OR.

dot pattern—Small dots of light produced on the screen of a color picture tube by the signal from a dot generator. If overall beam convergence has been obtained, the three color-dot patterns will merge into one white-dot pattern.

dot sequential—Pertaining to the association of the primary colors in sequence with successive picture elements of a color television system. (Examples: dot-sequential pickup, dot-sequential display, dot-sequential system, dot-sequential transmission.)

double-amplitude-modulation multiplier—A multiplier in which a carrier is amplitude modulated by one variable, and the modulated signal is again amplitude modulated by a second variable. The product of the two variables is obtained by applying the resulting double-modulated signal to a balanced demodulator.

double armature—An armature having two windings and commutators but only one core.

double-base diode—*See* Unijunction Transistor.

double-base junction transistor—Also called tetrode junction transistor. Essentially a junction triode transistor with two base connections on opposite sides of the central region of the transistor.

double-beam cathode-ray tube—A cathode-ray tube having two electron beams capable of producing on the screen two independent traces that may overlap. The beams may be produced by splitting the beam of one gun or by using two guns.

double-bounce calibration—A method of calibration used to determine the zero set error by using round-trip echoes. The correct range is the difference between the first and second echoes.

double-break contacts—A set of contacts in which one contact is normally closed and makes simultaneous connection with two other contacts.

double-break switch—A switch which opens the connected circuit at two points.

double bridge—*See* Kelvin Bridge.

double-button carbon microphone — Also called differential microphone. A microphone with two carbon-resistance elements or buttons, one on each side of a central diaphragm. They are connected in parallel to the current source in order to give twice the resistance change obtainable with a single button.

double-channel duplex—A method for simultaneous communication between two stations over two rf channels, one in each direction.

double-channel simplex—A method for nonsimultaneous communication between two stations over two rf channels, one in each direction.

double-checkerboard pattern—*See* Worst-Case Noise Pattern.

double-clocking—Incorrect setting of a flip-flop due to bounce in input signal.

double-conversion receiver — A receiver using a superheterodyne circuit in which the incoming signal frequency is converted twice, first to a high if and then to a lower one.

double-current generator—A machine which supplies both direct and alternating current from the same armature winding.

double-diffused epitaxial mesa transistor—*See* Epitaxial-Growth Mesa Transistor.

double-diffused transistor—A transistor in which two pn junctions are formed in the semiconductor wafer by gaseous diffusion of both p-type and n-type impurities. An intrinsic region can also be formed.

double diode—Also called duodiode. A vacuum tube or semiconductor having two diodes in the same envelope.

double-diode limiter—A type of limiter used to remove all the positive signals from a combination of positive and negative pulses or to remove all the negative signals from such a combination of positive and negative pulses.

double-doped transistor — A transistor formed by growing a crystal and successively adding p- and n-type impurities to the melt while the crystal is being grown.

double-doublet antenna—An antenna composed of two half-wave doublet antennas criss-crossed at their centers; one is made shorter than the other to give broader frequency coverage.

double emitter follower—*See* Darlington Amplifier.

double frequency-shift keying—A multiplex system in which two telegraph signals are combined and transmitted simultaneously by frequency shifting among four radio frequencies.

double-grip terminal—A solderless terminal with an extended flared barrel that permits a crimp to be made over the insulation of a wire as well as over the stripped portion.

double image—A television picture consisting of two overlapping images due to reception of the signal over two paths that differ in length so signals arrive at slightly different times.

double insulation—The insulation system resulting from a combination of functional and supplementary insulation.

double-junction photosensitive semiconductor—A semiconductor in which the current flow is controlled by light energy. It consists of three layers of a semiconductor material with electrodes connected to the ends of each.

double-length number—Also called double-precision number. An electronic computer number having twice the normal number of digits.

double local oscillator—An oscillator mixing system which generates two rf signals accurately spaced a few hundred hertz apart and mixes these signals to give the difference frequency which is used as the reference. This equipment is used in an interferometer system to obtain a detectable signal containing the phase information of an antenna pair, and the reference signal to allow removal of the phase data for use.

double-make contacts—A set of contacts in which one contact is normally open and makes simultaneous connection with two other independent contacts when closed.

double moding—Changing from one frequency to another abruptly and at irregular intervals.

double modulation—The process of modulation in which a carrier wave of one frequency is first modulated by a signal wave, and the resultant wave is then made to modulate a second carrier wave of another frequency.

double photoresist—A technique for eliminating pinholes in the photoresist coating during fabrication of microelectronic integrated circuits. The method may consist of two separate applications and exposures of photoresist emulsions of the same or different types.

double-play tape—Tape having half the thickness, and hence double the running time (for a given reel size) of standard 1½-mil tape.

double pole—A term applied to a contact arrangement to denote that it includes

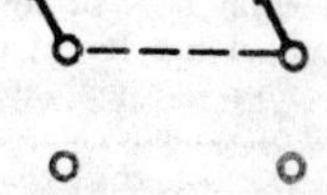

Double-pole, double-throw switch.

two separate contact forms (i.e., two single-pole contact assemblies).

double-pole, double-throw switch—Abbreviated dpdt. A switch that has six terminals and is used to connect one pair of terminals to either of the other two pairs.

double-pole, single-throw switch—Abbreviated dpst. A switch that has four terminals and is used to connect or disconnect two pairs of terminals simultaneously.

Double-pole, single-throw switch.

double-pole switch—A switch that operates simultaneously in two separate electric circuits or in both lines of a single circuit.

double precision—Having to do with the use of two computer words to represent one number.

double-precision arithmetic—The use of two computer words to represent a number. This is done where it is necessary to obtain greater accuracy than a single word of computer storage will provide.

double-precision number — *See* Double-Length Number.

double-pulsing station—A Loran station that receives two pairs of pulses and emits pulses at two pulse rates.

double pumping—A technique of pumping a laser for a relatively long time to store energy in subthreshold-level excited states, followed immediately by a very brief second pumping in which the threshold condition is exceeded in some region. This triggers laser oscillation throughout the entire active region and produces peak output powers several times larger than normally seen.

double rail—*See* Dual Rail.

double-rail logic—Pertaining to self-timing asynchronous circuits in which each logic variable is represented by two electrical lines that in combination can assume three meaningful states: zero, one, and undecided.

double screen—A three-layer screen consisting of a two-layer screen with the additional second long-persistence coating having a different color and different persistence from the first.

double-shield enclosure—A shielded enclosure or room the inner wall of which is partially isolated electrically from the outer wall.

double sideband—Amplitude-modulated intelligence which is transmitted at frequencies both above and below the carrier frequency by the audio-frequency value of the intelligence.

double-sideband transmitter—One which transmits not only the carrier frequency, but also the two sidebands resulting from modulation of the carrier.

double-spot tuning—Superheterodyne reception of a given station at two different local-oscillator frequencies. The local oscillator is adjusted either above or below the incoming signal frequency by the intermediate-frequency value.

double-stream amplifier — A microwave traveling-wave amplifier in which amplification occurs through interaction of two electron beams having different average velocities.

double-stub tuner—An impedance-matching device consisting of two stubs, usually fixed three-eighths of a wavelength apart, in parallel with the main transmission lines.

double superheterodyne reception — Also called triple detection. The method of reception in which two frequency converters are employed before final detection.

double-surface transistor—A point-contact transistor, the emitter and collector whiskers of which are in contact with opposite sides of the base.

doublet—The output voltage waveform of a delay line under linear operating conditions when the input to the line is a current step function.

doublet antenna—An antenna consisting of two elevated conductors substantially in the same straight line and of substantially equal length, with the power delivered at the center.

double tape mark—A delimiter consisting of two consecutive tape marks which is used to indicate the end of a volume or of a file set.

double throw—A term applied to a contact arrangement to denote that each contact form included is a breakmate.

double-throw circuit breaker—A circuit breaker by means of which a change in the circuit connections can be obtained by closing either of two sets of contacts.

double-throw switch—A switch which alternately completes a circuit at either of its two extreme positions. It is both normally open and normally closed.

double-track recorder—*See* Dual-Track Recorder.

double triode—*See* Duotriode.

doublet trigger—A trigger signal consisting of two pulses spaced by a fixed amount for coding.

double-tuned amplifier — An amplifier in which each stage utilizes coupled circuits having two frequencies of resonance for the purpose of obtaining wider bands than are possible with single tuning.

double-tuned circuit—A circuit in which two circuit elements are available for tuning.

double-tuned detector—A type of fm discriminator in which the limiter output transformer has two secondary windings, one tuned a certain amount above the center frequency and the other tuned an equal amount below the center frequency.

double-V antenna—Also called fan antenna. A modified single dipole which has a higher input impedance and broader bandwidth than an ordinary dipole.

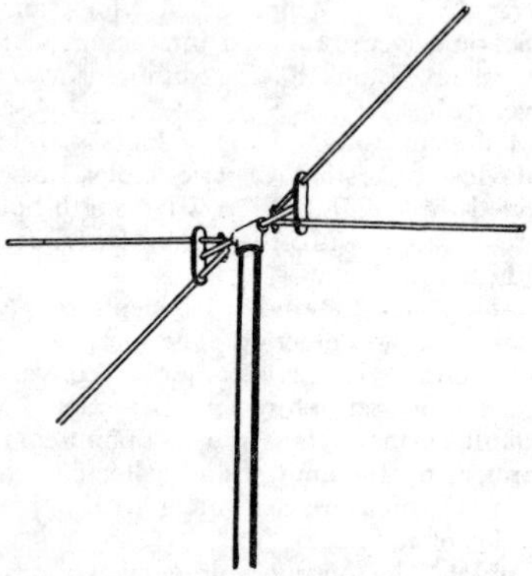

Double-V antenna.

double-winding synchronous generator—A synchronous generator which has two similar windings in phase with one another, mounted on the same magnetic structure but not connected electrically, designed to supply power to two independent external circuits.

doubling—The generation of large amounts of second-harmonic distortion by nonlinear motion of a loudspeaker cone.

doubly balanced modulator—A modulator circuit in which two Class-A amplifiers are supplied with modulating and carrier signals of equal amplitudes and opposite polarities. Carrier suppression takes place because the two amplifiers share a common plate circuit and only the sidebands appear at the output.

down lead—The wire that connects an antenna to a transmitter or receiver.

downtime – The period of time during which an equipment is malfunctioning or is not in operation because of electrical, mechanical, or electronic failure, but not because of a lack of work or the absence of an operator.

downward modulation – Modulation in which the instantaneous amplitude of the modulated wave is never greater than that of the unmodulated carrier.

dpdt—Abbreviation for double-pole, double-throw switch.

dpst—Abbreviation for double-pole, single-throw switch.

drag angle—A stylus cutting angle of less than 90° to the surface of the record. So called because the stylus drags over the surface instead of digging in. It is the opposite of dig-in angle.

drag cup—A nonmagnetic metal rotated in a magnetic field to generate a torque or voltage proportional to its speed.

drag-cup motor—A small, high-speed, two-phase, alternating-current electric motor having a two-pole, two-phase stator. The rotating element consists only of an extremely light metal cup attached to a shaft rotating on ball bearings. Reversal is accomplished by reversing the connections to one phase. Used in applications requiring quick starting, stopping, and reversal characteristics.

drag magnet—*See* Retarding Magnet.

drain—1. The current taken from a voltage source. 2. The working-current terminal (at one end of the channel in a FET) that is the drain for holes or free electrons from the channel. Corresponds to collector of bipolar transistor. 3. Terminal which receives carriers from the MOS channel.

drainage equipment—Equipment used to protect connected circuits from transients produced by the operation of protection equipment.

drain cutoff current—The current into the drain terminal of a depletion-type transistor with a specified reverse gate-to-source voltage applied to bias the device to the off state.

drain terminal—The terminal electrically connected to the region into which majority carriers flow from the channel.

drain wire – An uninsulated solid or stranded tinned copper wire which is placed directly under a shield. It touches the shield throughout the cable, and therefore may be used in terminating the shield to ground. It is completely necessary on spiral shielded cables because it eliminates the possibility of induction in a spiral shield.

D-region—The region of the ionosphere up to about 90 kilometers above the earth's surface. It is below the E-region.

dress—The exact placement of leads and components in a circuit to minimize or eliminate undesirable feedback and other troubles.

dressed contact—A contact that has a locking spring member permanently attached.

drift—1. Movement of carriers in a semiconductor as voltage is applied. 2. A change in either absolute level or slope of an input-output characteristic. 3. *See* Flutter, 1. 4. *See* Degradation. 5. An undesired change in one of the output parameters of a power supply (voltage, current, frequency, etc.) over a period of time. The change is unrelated to all other variables such as load, line, and environment. Drift is measured over a period of time by keeping all variables (such as line, load, and environment) constant. Specifications usually apply only after a warm-up period. 6. The angular displacement of an aircraft by the wind, generally expressed in degrees. 7. In a dc amplifier, the change in output with constant input, usually measured in terms of the dc input signal required to restore normal output; may be called out as microvolts or millivolts per hour. 8. A change in output attributable to any cause.

drift current—The flow of carriers in a semiconductor due to an electric field. In the same electric field, holes and electrons will flow in opposite directions due to their opposite charge.

drift mobility—The average drift velocity of carriers in a semiconductor per unit electric field. In general, the mobilities of electrons and holes are not the same.

drift space—1. In an electron tube, a region substantially free of alternating fields from external sources, in which relative repositioning of the electrons depends on their velocity distributions and the space-charge forces. 2. The distance between the buncher and catcher in a velocity-modulated vacuum tube.

drift speed—Average speed at which electrons or ions progress through a medium.

drift transistor—A type of transistor manufactured with a variable-conductivity base region. Such a base sets up an electric field which speeds up the carriers, thus reducing the transit time and improving high-frequency operation.

drift velocity—Net velocity of charged particles in the direction of the applied field.

drip loop—A loop formed in a transmission line at a point where it enters a building. Condensation of moisture and water that may form on the line will drip off at the loop and thus will not enter the building.

drip-proof motor—A motor in which the ventilating openings are such that foreign matter falling on the motor at any angle not exceeding 15° from the vertical cannot enter the motor either directly or indirectly.

driptight enclosure—An enclosure that is intended to prevent accidental contact with the enclosed apparatus and, in addition, is so constructed as to exclude falling moisture or dirt.

drive—*See* Excitation, 2.

drive belt—A belt used to transmit power from a motor to a driven device.

drive circuit—A circuit, usually a printed-circuit card or an encapsulated module, that converts an input pulse to the appropriate winding excitation sequence to produce one step of the motor shaft.

drive control—*See* Horizontal-Drive Control.

driven element—An antenna element connected directly to the transmission line.

driven sweep—A sweep signal triggered by an incoming signal only.

drive pattern—In a facsimile system, an undesired pattern of density variations that result from periodic errors in the position of the recording spot.

drive pin—In disc recording, a pin similar to the center pin but located at one side of it and used to prevent a disc record from slipping on the turntable.

drive pulse—A pulsed magnetomotive force applied to a magnetic cell from one or more sources.

driver—1. An electronic circuit which supplies input to another electric circuit. 2. A stage of amplification which precedes the power output stage. 3. In a radar transmitter, a circuit which produces a pulse to be delivered to the control grid of the modulator tube. 4. An element coupled to the output stage of a circuit to increase the power- or current-handling capability or fan-out of the stage; for example, a clock driver is used to supply the current necessary for a clock line. 5. A device in a logic family controlled with normal logic levels whose output has the capability of sinking or sourcing high current. The output may control a lamp, relay or a very large fanout of other logic devices; Also a device "driving" a higher output device or transistor by supplying power, voltage, or current to it.

driver element—An antenna array element that receives power directly from the transmitter.

driver stage—The amplifier stage preceding the power-output stage.

driving-point admittance—The complex ratio of the alternating current to the applied alternating voltage for an electron tube, network, or other transducer.

driving-point impedance—At any pair of terminals in a network, the driving-point impedance is the ratio of an applied potential difference to the resultant current at these terminals, all terminals being terminated in any specified manner.

driving power—The power supplied to the grid circuit of a tube where the grid swings positive and draws current for part of each cycle of the input signal.

driving-range potential—The voltage difference between the potential of the electrochemically more active anode and the less active protected metal or cathode. One example of driving potential is the electromotive force in a cathodic protection system that causes current between the protected structure (cathode) and the anode. The driving potential decreases as the electrodes become polarized.

driving signal—Television signals that time the scanning at the pickup point. Two kinds of driving signals are usually available from a central sync generator, one composed of pulses at the line frequency and the other of pulses at the field frequency.

driving spring—The spring driving the wipers of a stepping relay.

DRO memory – Destructive readout memory. A memory in which the contents of a storage location are destroyed in being read. Information must be rewritten after reading, if it is to be returned. An example of a DRO memory is the common computer core memory.

drone cone—An undriven speaker cone mounted in a bass-reflex enclosure.

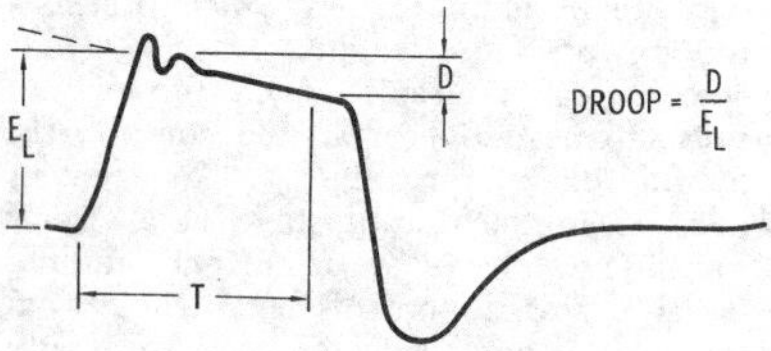

Droop.

droop—The decrease in mean pulse amplitude, expressed as a percentage of the 100% amplitude, at a specified time following the initial attainment of 100% amplitude.

drop—1. To develop a specified difference of potential between a pair of terminals as the result of a flow of current. 2. *See* Voltage Drop.

drop bar—A protective device used to ground a high-voltage capacitor when opening a door.

drop bracket transposition—Reversal of the relative positions of two parallel wire conductors while depressing one, so that the crossover is in a vertical plane.

drop cable—In a catv system, the transmission cable from the distribution cable to a dwelling.

drop channel—A type of operation in which one or more channels of a multichannel system are terminated (dropped) at some intermediate point between the end terminals of the system.

drop-in—The reading of a spurious signal of amplitude greater than a predetermined percentage of the nominal signal amplitude.

drop indicator—An indicator for signaling, consisting of a hinged flap normally held up by a catch. The catch is released by an electromagnet, allowing the flap to drop when a signal is received.

dropout—1. A momentary loss of volume or treble response due to a brief separation of the tape from the surface of the record or play head. A very slight separation causes a treble dropout; more severe loss of head-to-tape contact causes the whole signal to drop out. Dropouts can be caused by buckled or crinkled tape, lumps or pits in the magnetic coating, or detached clumps of oxide passing across the head surface. 2. Short pause in tape replay due to bad tape coating. 3. Momentary loss of signal in a transmission channel.

dropout error—An error, such as loss of a recorded bit, that occurs in recorded magnetic tape because of foreign particles on or in the magnetic coating or because of defects in the backing.

dropouts—Also called keys. Special images inserted at certain points in the array on a photomask used in the production of monolithic circuits.

drop-out value—The maximum value of current, voltage, or power which will de-energize a previously energized relay. (*See also* Hold Current *and* Pickup Value [Voltage, Current, or Power].)

dropping resistor—A resistor used to decrease a given voltage by an amount equal to the potential drop across the resistor.

drop relay—A relay activated by incoming ringing current to call the attention of an operator to the subscriber's line.

drop repeater—A microwave repeater station equipped for local termination of one or more circuits.

dropsonde—A parachute-carried radiosonde dropped from a high-flying aircraft to measure weather conditions and report them to the aircraft. It is used over water or other areas where ground stations cannot be maintained.

drop wire—A wire suitable for extending an open wire or cable pair from a pole or cable terminal to a building.

drum—A random-access auxiliary memory device in which information is stored on a revolving drum that is coated with a magnetic material.

drum controller—1. A device in which electrical contacts are made on the surface of a rotating cylinder or sector. 2. A device in which contacts are made by the operation of a rotating cam.

drum memory—A rotating cylinder or disc coated with magnetic material so that information can be stored in the form of magnetic spots.

drum parity—A parity error that occurs during the transfer of information to or from drums.

drum programmer—An electromechanical device that provides stored-program logic for control of a sequential operation such as batch processing or machine cycling. It ranks between relay and solid-state systems in the cost/complexity scale.

drum recorder—A facsimile recorder in

which the record sheet is mounted on a rotating cylinder.

drum speed—The number of revolutions per minute made by the transmitting or receiving drum of a facsimile transmitter or recorder.

drum storage—A storage device in which information is recorded magnetically on a rotating cylinder; a type of addressable storage associated with some computers.

drum switch—A switch in which the electrical contacts are made on pins, segments, or surfaces on the periphery of a rotating cylinder or sector, or by the operation of a rotating cam.

drum transmitter—A facsimile transmitter in which the copy is mounted on a rotating cylinder.

drum-type controller—1. A multicircuit timing device, with or without a motor, using a cylindrical carriage into which pins are inserted to program events. 2. A multicircuit timing device intended to be driven from an external rotary power source.

drunkometer—A device measuring the degree of alcoholic intoxication by analyzing the subject's breath.

dry—A condition in which the electrolyte in a cell is immobilized. The electrolyte may be either in the form of a gel or paste or absorbed in the separator material.

dry battery—Two or more dry cells arranged in series, parallel, or series-parallel within a single housing to provide desired voltage and current values.

dry cell—A voltage-generating cell having an immobilized electrolyte. The commonest form has a positive electrode of carbon and a negative electrode of zinc in an electrolyte of sal ammoniac paste.

dry circuit—1. A circuit in which the open-circuit voltage is 0.03 V or less and the current 200 mA or less. (The voltage is most important because at such a low level it is not able to break through most oxides, sulfides, or other films which can build up on contacting surfaces.) 2. A circuit over which voice signals are transmitted and which carries no direct circuit.

dry circuit contact—A contact that carries current but neither makes nor breaks while its load circuit is energized. Sometimes erroneously used if referring to low level.

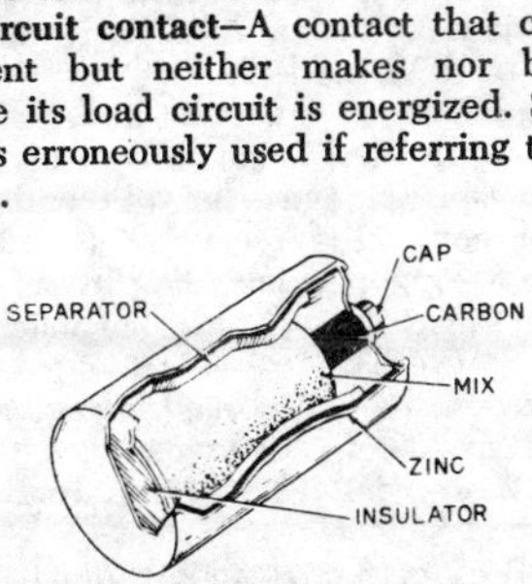

Dry cell.

dry-disc rectifier—A rectifier consisting of discs of metal and other materials in contact under pressure. Examples are the copper-oxide and the selenium rectifier.

dry-electrolytic capacitor—An electrolytic capacitor with a paste rather than liquid electrolyte. By eliminating the danger of leakage, the paste electrolyte permits the capacitor to be used in any position.

dry flashover voltage—The voltage at which the air surrounding a clean, dry insulator or shell completely breaks down between electrodes.

dry-reed contact—An encapsulated switch containing two metal wires that serve as the contact points for a relay.

dry-reed relay—A relay that consists of one or more capsules containing contact mechanisms that are generally surrounded by an electromagnetic coil for actuation. The capsule consists of a glass tube with a flattened ferromagnetic reed sealed in each end. These reeds, which are separated by an air gap, extend into the tube so as to overlap. When placed in a magnetic field, they are brought together and close a circuit.

dry shelf life—The length of time that a cell can stand without electrolyte before it deteriorates to a point where a specified output cannot be obtained.

dry-type forced-air-cooled transformer (class AFA)—A transformer which is not immersed in oil and which derives its cooling by the forced circulation of air.

dry-type self-cooled/forced-air-cooled transformer (class AA/FA)—A transformer which is not immersed in oil, and which has a self-cooled rating with cooling obtained by the forced circulation of air.

dry-type self-cooled transformer (class AA)—A transformer which is cooled by the natural circulation of air and which is not immersed in oil.

dry-type transformer—A transformer which is cooled by the circulation of air and which is not immersed in oil.

dsc—Abbreviation for double silk-covered.

D-scope—A radar display similar to a C-scope except that the blips extend vertically to give a rough estimate of the distance.

"D" service—FAA service pertaining to radio broadcast of meteorological information, advisory messages, and notices to airmen.

DT-cut crystal—A crystal cut to vibrate below 500 kHz.

DTL—Abbreviation for diode transistor logic.

D-type flip-flop—A flip-flop that, on the occurrence of the leading edge of a clock pulse, propagates to the 1 output whatever information is at its D (data) conditioning input prior to the clock pulse.

dual—Either of a pair of systems, circuit, etc. that are described by equations of the same form in which the same functional relationships hold provided that the dependent and independent dynamic variables are interchanged between these equations.

dual-beam oscilloscope—An oscilloscope in which the cathode-ray tube produces two separate electron beams that may be individually or jointly controlled.

dual capacitor—Two capacitors within a single housing.

dual-channel amplifier—An amplifier which has two channels independent of each other, but similar in design, construction, and output.

dual cone—Speaker unit containing a main cone for bass and middle frequencies and a smaller, stiffer inner cone for treble frequencies, sometimes called a "full range" speaker unit.

dual-diversity receiver—A radio receiver that receives signals from two different receiving antennas and uses whichever signal is the stronger at each instant to offset fading. In one arrangement, two identical radio-frequency systems, each with its own antenna, feed a common audio-frequency channel. In another arrangement, a single receiver is changed over from one antenna to the other by electronic switching at a rate fast enough to prevent loss of intelligibility.

dual-emitter transistor—A passivated pnp silicon planar epitaxial transistor with two emitters; used as a low-level chopper.

dual-frequency induction heater—A type of induction heater in which work coils operating at two different frequencies induce energy, either simultaneously or successively, to material within the heater.

dual-groove record—*See* Cook System.

dual in-line package—A type of housing for integrated circuits. The standard form is a molded plastic container about ¾ inch long and ⅓ inch wide, with two rows of pins spaced 0.1 inch between centers. This package is more popular than the flat pack or TO-type can for industrial use because it is relatively inexpensive and is easily dip-soldered into printed circuit boards. Abbreviated DIP.

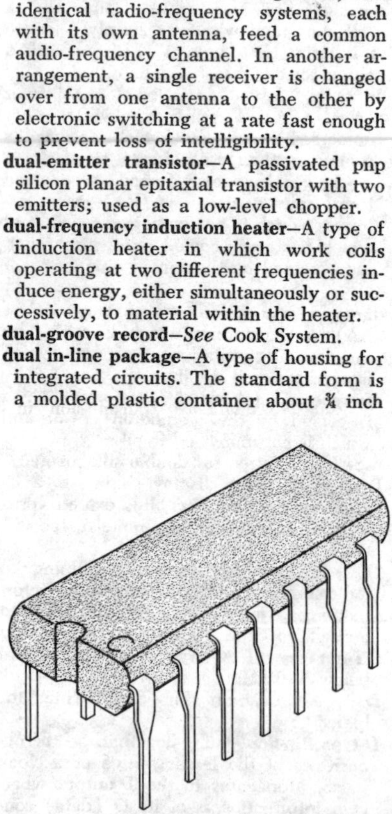

Dual inline package.

dual meter—A meter constructed so that two aspects of a circuit may be read simultaneously.

dual modulation—The use of two different types of modulation, each conveying separate intelligence, to modulate a common carrier or subcarrier wave.

dual operation—A logic operation the result of which is the negation of the result of an original operation when applied to the negation of its operands; for example, the OR operation is the dual of the AND operation. A dual operation is represented by writing 0 for 1 and 1 for 0 in the tabulated values of P, Q, and R for the original operation.

dual pickup—*See* Turnover Pickup.

dual rail—Pertaining to a method of transferring data in which the data and the complement of the data are available on different input or output lines or wires. Also called double rail.

dual slope converter—An integrating A/D converter in which the unknown signal is converted to a proportional time interval which is then measured digitally.

dual-tone multifrequency – A signaling method in which are employed set pairs of specific frequencies used by subscribers and PBX attendants, if their switchboard positions are so equipped, to indicate telephone address digits, precedence ranks, and end of signaling.

dual trace—A mode of operation in which a single beam in a cathode-ray tube is shared by two signal channels. *See also* Alternate Mode *and* Chopped Mode.

dual-track recorder – Normally a monophonic recorder where the recording head covers slightly less than half the width of a standard quarter-inch tape, making it possible to record one track on the tape in one direction and, after turning the reels over, a second track in the opposite direction. Known also as "half-track" or "two-track" recorder.

dual-use line—A communications link that normally is used for more than one mode of transmission (e.g., for voice and data).

dub—A copy of a recording (noun). To make a copy of a recording by recording on one machine what another machine is playing (verb). Also called re-recording.

dubbing—1. In radio broadcasting, the addition of sound to a prerecorded tape or disc. 2. Copying a tape from a previously recorded version.

duct—1. A protective pipe through which conductors or cables are run. 2. In microwave transmission, atmospheric conditions may cause radio waves to follow a narrower path than usual. The narrower path is called a duct. The presence of ducting sometimes causes unusual transmission because the transmission waves do not follow the intended path.

ducting—The trapping of an electromagnetic wave, in a waveguide action, between two layers of the earth's atmosphere, or between a layer of the atmosphere and the earth's surface.

DUF—Abbreviation for diffusion under the epitaxial film. A method for providing a low-resistance path between the active region of an IC transistor and the contact electrode at the surface. A region of high conductance is formed by selective diffusion in the required location prior to deposition of the epitaxial layer.

dummy—1. A simulating device that has no operating features. 2. A telegraphy network used to simulate a customer's loop for adjustment of a telegraph repeater. The dummy side of the repeater is the side toward the customer. 3. In a computer, an artificial address, instruction, or other unit of information inserted solely for the purpose of fulfilling such prescribed conditions as word length or block length without affecting operations.

dummy antenna—*See* Artificial Antenna.

dummy instruction—An artificial instruction or address inserted in a list of instructions to a computer solely to fulfill prescribed conditions (such as word length or block length) without affecting the operation.

dummy load—*See* Artificial Load.

dump — 1. Also called power dump. To withdraw all power from a computer, either accidentally or intentionally. 2. To transfer all or part of the contents of one section of a digital-computer memory into another section.

dump check—Checking a computer by adding all digits as they are dumped (transferred) to verify the sum to make sure no errors exist as the digits are retransferred.

dumping resistor—A resistor whose function is to discharge a capacitor or network for safety purposes.

Dunmore cell—*See* Lithium Chloride Sensor.

duodecal socket — A vacuum-tube socket having 12 pins. Used for cathode-ray tubes.

Duodecal socket.

duodecimal—1. Pertaining to a characteristic or property involving a selection, choice, or condition in which there are 12 possibilities. 2. Pertaining to the numbering system with a radix of 12.

duodiode—Also called dual diode. A vacuum tube or semiconductor having two diodes within the same envelope.

duodiode-pentode—An electron tube containing two diodes and a pentode in the same envelope.

duodiode-triode—An electron tube containing two diodes and a triode in the same envelope.

duolateral coil—*See* Honeycomb Coil.

duopole—An all-pass action with two poles and two zeros.

duotriode—An electron tube containing two triodes in the same envelope. Also called double triode.

duplex—1. The method of operation of a communication circuit in which each end can simultaneously transmit and receive. (Ordinary telephones are duplex. When used on a radio circuit, duplex operation requires two frequencies.) 2. Two-in-one, as two conductors with a common overall insulation or two telegraph transmission channels over one wire. 3. Two conductors twisted together, usually with no outer covering. This term has a double meaning and it is possible to have parallel wires and jacketed parallel wires, and still refer to them as duplex.

duplex artificial line—A balancing network simulating the impedance of the real line and distant terminal apparatus; it is employed in a duplex circuit for the purpose of making the receiving device unresponsive to outgoing signal currents.

duplex cable—A cable composed of two insulated stranded conductors twisted together. They may or may not have a common insulating covering.

duplex channel—A communication channel providing simultaneous transmission in both directions.

duplexer—A radar device which, by using the transmitted pulse, automatically switches the antenna from receive to transmit at the proper time.

duplexing assembly (radar)—*See* Transmit-Receive Switch.

duplex operation—Simultaneous operation of transmitting and receiving apparatus at two locations.

duplex system—A system with two distinct and separate sets of facilities, each of which is capable of assuming the system function while the other assumes a standby status. Usually both sets are identical in nature.

duplex tube—A combination of two vacuum tubes in one envelope.

duplicate—To copy in such a way that the result has the same physical form as the source. For example, to make a new punched card that has the same pattern of holes as an original punched card.

duplication check—A computer check in which the same operation or program is checked twice to make sure the same result is obtained both times.

duration control—A control for adjusting the time duration of reduced gain in a sensitivity-time control circuit.

during cycle—The interval while a timer is operating for its preset time period.

dust core—A pulverized iron core consisting of extremely fine iron particles mixed with a binding material for use in radio-frequency coils.

dust cover—A device specifically designed to cover the mating end of a connector so as to provide mechanical and/or environmental protection.

dust-ignitionproof motor — A totally enclosed motor, the enclosure of which is designed and constructed in a manner that will exclude ignitable amounts of dust or amounts which might affect the performance rating, and will not permit arcs, sparks, or heat otherwise generated or liberated inside the enclosure to cause ignition of exterior accumulations or atmospheric suspensions of a specific dust on or in the vicinity of the enclosure.

duty cycle—The amount of time a device operates, as opposed to its idle time. Applied to a device that normally runs intermittently rather than continuously.

duty cyclometer—A test meter which gives a direct reading of duty cycle.

duty factor—1. In a carrier composed of regularly recurring pulses, the product of their duration and repetition frequency. 2. Ratio of average to peak power. 3. Same as duty cycle except it is expressed as a decimal rather than a percentage. Usually calculated by multiplying pps × pulse width.

duty ratio—In a pulsed system, such as radar, the ratio of average to peak power.

dv/dt—The rate of change of voltage with respect to time. Proportional to current in a capacitor.

DX—1. Abbreviation for distance. 2. Reception of distant stations.

DX hound—An amateur who specializes in making distant contacts.

dyadic Boolean operator—A Boolean operator that has two operands. The dyadic Boolean operators are AND, equivalence, exclusion, exclusive OR, inclusion, NAND, NOR, and OR.

dyadic operation—An operation on two operands.

dynamic—Of, concerning, or dependent on conditions or parameters that change, particularly as functions of time.

dynamic acceleration—Acceleration in a constantly changing magnitude and direction, either simple or complex motion usually called vibration. Also measured in gravity units.

dynamic analogies—The similarities in form between the differential equations that describe electrical, acoustical, and mechanical systems that allow acoustical and mechanical systems to be reduced to equivalent electrical networks, which are conceptually simpler than the original systems.

dynamic behavior—The way a system or individual unit functions with respect to time.

dynamic braking—1. A system of braking of an electric drive in which the motor is used as a generator, and the kinetic energy of the motor and driven machinery is employed as the actuating means of exerting a retarding force. 2. A type of motor braking caused by current being applied to the windings after the power is shut off. This is accomplished either by self-excitation, (dc motors) or by separate excitation (ac motors).

dynamic characteristics — Relationship between the instantaneous plate voltage and plate current of a vacuum tube as the voltage applied to the grid is varied.

dynamic check—A check used to ascertain the correct performance of some or all components of equipment or a system under dynamic or operating conditions.

dynamic contact resistance—1. In a relay, a change in contact electrical resistance due to a variation in contact pressure on mechanically closed contacts. For example, during wiping motion of sliding contacts during make or prior to break. Also when contact members no longer actually open as in contact bounce, but members are still vibrating and varying the contact pressure and hence its resistance. 2. A varying contact resistance on contacts mechanically closed.

dynamic convergence — The condition where the three beams of a color picture tube come together at the aperture mask as they are reflected both vertically and horizontally. (*See also* Vertical Dynamic Convergence *and* Horizontal Dynamic Convergence.)

dynamic decay—In a storage tube, decay caused by an action such as that of ion charging.

dynamic demonstrator — A three-dimensional schematic diagram in which the components of the radio, television receiver, etc., are mounted directly on the diagram.

dynamic deviation—The difference between the ideal output value and the actual output value of a device or circuit when the reference input is changing at a specified constant rate and all other transients have expired.

dynamic dump—A dump performed while a program is being executed.

dynamic equilibrium of an electromagnetic system—1. The tendency of any electromagnetic system to change its configuration so that the flux of magnetic induction will be maximum. 2. The tendency of any two current-carrying circuits to maintain the flux of magnetic induction linking the two at maximum.

dynamic error—An error in a time-varying signal resulting from inadequate dynamic response of a transducer.

dynamic focus—The application of an ac voltage to the focus electrode of a color picture tube to compensate for the defocusing caused by the flatness of the screen.

dynamic magnetic field—A field the intensity of which is changing and the lines of force of which are expanding or contracting. Such change can be periodic or random. Unlike the static field, the dynamic field can transfer energy from one point to another without relative motion between the points.

dynamic memory—A type of semiconductor memory in which the presence or absence of an electrical charge represents the two states of a storage element.

dynamic microphone—*See* Moving-Coil Microphone.

dynamic MOS array—A circuit made up of MOS devices which requires a clock signal. The circuit must be tested at its rated (operating) speed. Known as clock-rate testing.

dynamic mutual-conductance tube tester—*See* Transconductance Tube Tester.

dynamic noise limiter—Abbreviated DNL. A compatible circuit designed primarily for use with tape recorders. It improves the effective signal-to-noise ratio during replay by selective filtering at low signal levels.

dynamic noise suppressor—An audio filter the bandpass of which is adjusted automatically to the signal level. At low signal levels filtering is highest, and at high signal levels all filter action is removed.

dynamic output impedance—*See* Output Impedance, 2.

dynamic pickup—A phonograph pickup the electrical output of which is the result of the motion of a conductor in a magnetic field.

dynamic plate impedance—The internal resistance to the flow of alternating current between the cathode and plate of a tube.

dynamic plate resistance—*See* AC Plate Resistance.

dynamic power—*See* Music Power.

dynamic printout—In a computer, a printout of data which occurs as one sequential operation during the machine run.

dynamic problem check—A dynamic check used to ascertain that the solution determined by an analog computer satisfies the given system of equations.

dynamic programming—A procedure used in operations research for optimization of a multistage problem solution in which a number of decisions are available at each stage of the process.

dynamic range—1. The difference, in decibels, between the overload level and the minimum acceptable signal level in a system or transducer. 2. The span of volume between the loudest and softest sounds, either in an original signal (original dynamic range) or within the span of a recorder's capability (recorded dynamic range). Dynamic range is expressed in decibels. *See* Signal-to-Noise Ratio. 3. The range of signal amplitudes from the loudest to the quietest that can be reproduced effectively by an equipment. Limited by the intrinsic noise of the amplifier and the ambient background noise level of the listening environment and by the power capacity of the amplifier and speaker system.

dynamic register—A memory in which the storage takes the form of capacitively charged circuit elements and therefore must be continually "refreshed" or recharged at regular intervals.

dynamic regulator—A transmission regulator in which the adjusting mechanism is in self-equilibrium at only one or a few settings and requires control power to maintain it at any other setting.

dynamic relocation—The ability to move computer programs or data from auxiliary memory to any convenient location in the memory. Normally the addresses of programs and data are fixed when the program is compiled.

dynamic reproducer — *See* Moving-Coil Pickup.

dynamic resistance—Incremental resistance measured over a relatively small portion of the operating characteristic of a device.

dynamic run—1. Also called dynamic test. The test performed on an instrument to obtain the overall behavior and to establish or corroborate specifications such as frequency response, natural frequency of the device, etc. 2. Test based on a time-interval measurement as, for example, the rise time or fall time of a pulse.

dynamic sequential control—A method of operation in which a digital computer can alter instructions or their sequence, as the computation proceeds.

dynamic shift register—A shift register in which information is stored by means of temporary charge storage techniques. The major disadvantage of this method is that loss of the information occurs if the clock repetition rate is reduced below a minimum value.

dynamic speaker—*See* Moving-Coil Speaker.

dynamic storage—1. A data-storage device in which the data are permitted to move or vary with time in such a way that the specified data are not always immediately available for recovery. Magnetic drums and discs are permanent dynamic storage; an acoustic delay line is a volatile dynamic storage. 2. Information storage using temporary charge storage techniques. It requires a clock repetition rate high enough to prevent loss of information.

dynamic storage allocation—A storage-allocation technique in which program and data locations are determined by criteria applied at the moment of need.

dynamic subroutine — In digital-computer programming, a subroutine which involves parameters (such as decimal point position) from which a properly coded subroutine is derived. The computer itself adjusts or generates the subroutine according to the parametric values chosen.

dynamic test—*See* Dynamic Run.

dynamic transfer-characteristic curve — A curve showing the variation in output current as the input current changes.

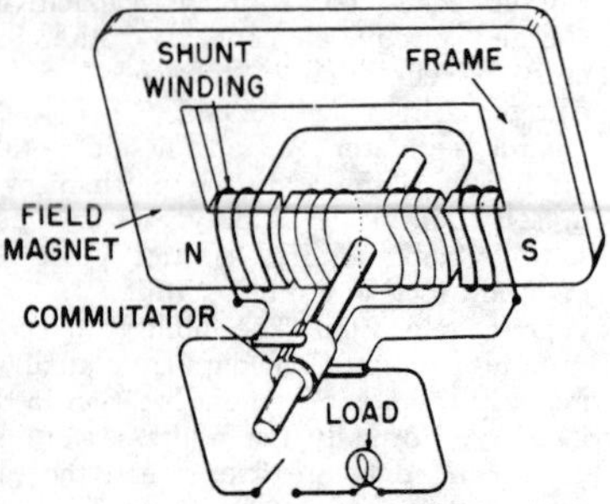

Dynamo.

dynamo—1. Normally called a generator. A machine that converts mechanical energy into electrical energy by electromagnetic induction. 2. In precise terminology, a generator of direct current—as opposed to an alternator, which generates alternating current.

dynamoelectric—Pertaining to the relationship between mechanical force and electrical energy or vice versa.

dynamometer—1. An instrument in which the force between a fixed and a moving coil provides a measure of current, voltage, or power. 2. Equipment designed to measure the power output of a rotating machine by determining the friction absorbed by a hand brake opposing the rotation.

dynamotor—Also called a rotary converter or synchronous inverter. A rotating device for changing a dc voltage to another value. It is a combination electric motor and dc generator with two or more armature windings and a common set of field poles. One armature winding receives the direct current and rotates (thus operating as a motor), while the others generate the required voltage (and thus operate as dynamos or generators).

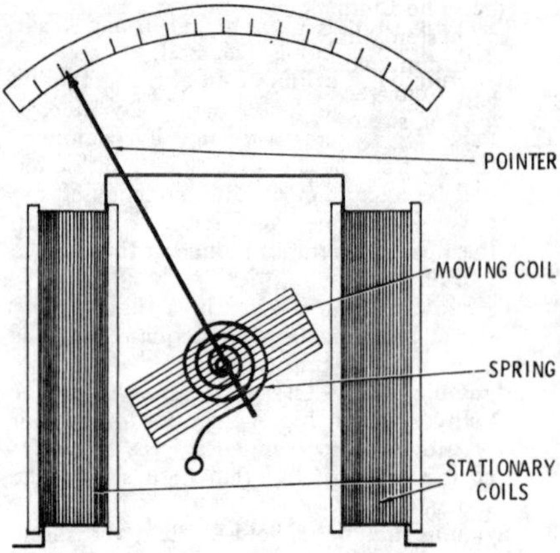

Dynamometer, 1.

dynaquad—A germanium pnpn semiconductor switching device which is base-controlled and has three terminals. Its operation is similar to that of a flip-flop circuit or latching relay.

dynatron—A type of vacuum tube in which secondary emission of electrons from the plate causes the plate current to decrease as the plate voltage increases, with the result that the device exhibits a negative-resistance characteristic. Used in oscillator circuits. Also called negatron. *See* Tetrode.

dynatron oscillator — A negative-resistance oscillator with negative resistance derived between the plate and cathode of a screen-grid tube operating such that secondary electrons produced at the plate are attracted to the higher-potential screen grid.

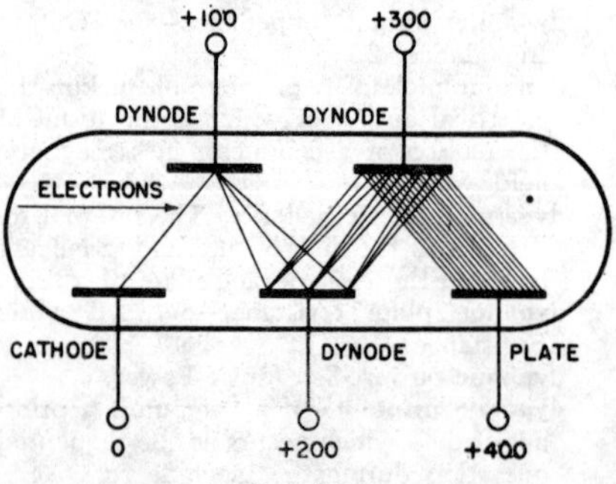

Dynode.

dyne—The fundamental unit of force in the cgs system that, if applied to a mass of one gram, would give it an acceleration of one cm/sec/sec.

dyne per square centimeter – The unit of sound pressure. One dyne per square centimeter was originally called a bar in acoustics, but the full expression is used in this field now because the bar is defined differently in other applications. Also called microbar.

dynistor – A nonlinear semiconductor having the characteristics of a small current flow as voltage is applied. As the applied voltage is increased a point is reached at which the current flow suddenly increases radically and will continue at this rate even though the applied voltage is reduced.

dynode—An electrode having the primary function of supplying secondary-electron emission in an electron tube.

E

E—1. Symbol for voltage or emitter. 2. Abbreviation for illumination.

E and M leads—In a signaling system, the output and input leads, respectively.

E- and M-lead signaling – Communications between a trunk circuit and a separate signaling unit by way of two leads, an M lead over which battery or ground signals are transmitted to the signaling equipment, and an E lead over which open or ground signals are received from the signaling unit.

early-failure period—The period of equipment life, starting immediately after final assembly, during which equipment failures initially occur at a higher than normal rate due to the presence of defective parts and abnormal operating procedures. Also called debugging period, burn-in period, or infant-mortality period.

early-warning radar—A radar which usually scans the sky in all directions, in order to detect approaching enemy planes and/or missiles at distances far enough away that interceptor planes can be in the air to meet their approach before they are near their target.

earphone—Also called receiver. An electroacoustic transducer intended to be placed in or over the ear.

earth – Term used in Great Britain for ground.

earth current—Also called ground current. 1. Current in the ground as a result of natural causes and affecting the magnetic field of the earth, sometimes causing magnetic storms. 2. Return, fault, leakage, or stray current passing from electrical equipment through the earth.

earthed—A British term meaning grounded.

earth ground—A connection from an electrical circuit or equipment to the earth through a water pipe or a metal rod driven into the earth. This connection reduces shock hazards from faulty equipment. Water pipes may no longer be reliable grounds because of the use of transite pipe, neoprene gaskets, and other nonconducting links. Any ground rods driven under the interior of a large building may gradually become ineffective because the building may drive the local water table down so far that the rod is essentially surrounded by dry soil.

earth inductor—*See* Generating Magnetometer.

earth-layer propagation—1. Propagation of electromagnetic waves through layers in the atmosphere of the earth. 2. Propagation of electromagnetic waves through layers below the surface of the earth.

earth oblateness—The slight departure from a perfect spherical shape of the shape of the earth and the form of its gravity field.

E-bend—*See* E-Plane Bend.

ebiconductivity – Conductivity induced as the result of electron bombardment.

ebmd – Abbreviation for electron-beam mode discharge.

EBS amplifier – Abbreviation for electron-bombarded semiconductor amplifier.

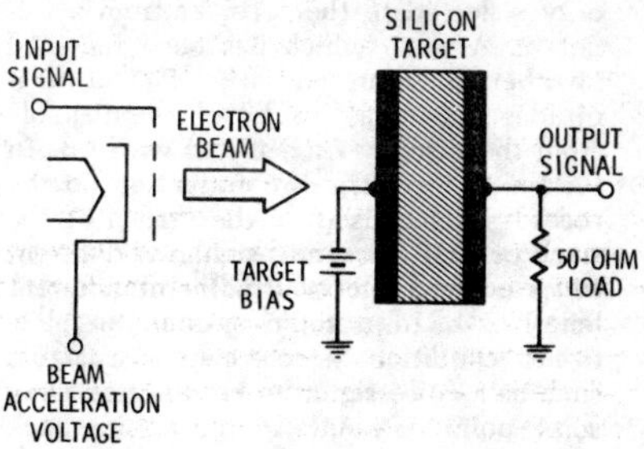

EBS amplifier.

ec—Abbreviation for enamel covered.

eccentric circle—*See* Eccentric Groove.

eccentric groove—Also called eccentric circle. An off-center locked groove for actuating the trip mechanism of an automatic record changer at the end of a recording.

eccentricity – In disc recording, the displacement of the center of the recording-groove spiral with respect to the record center hole.

Eccles-Jordan circuit – A flip-flop circuit consisting of a two-stage, resistance-coupled amplifier. Its output is coupled back to its input, two separate conditions of stability being achieved by alternately biasing the two stages beyond cutoff.

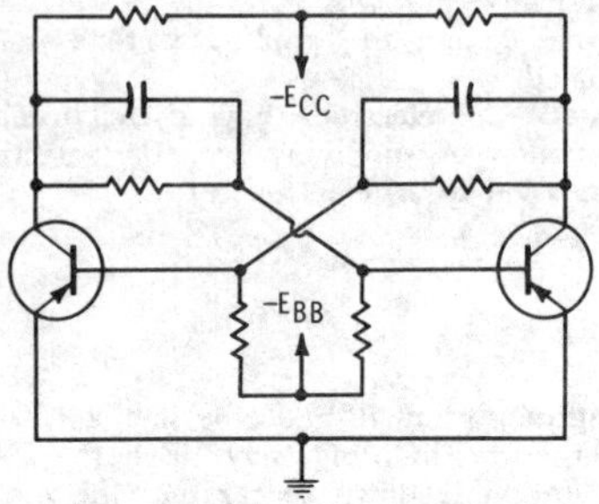

Eccles-Jordan multivibrator.

ECCM – Abbreviation for electronic counter-countermeasures. Retaliatory tactics used to reduce the effectiveness of electronic countermeasures.

ecdc–Abbreviation for electrochemical diffused-collector transistor. A pnp transistor in which all the mass of p material is etched off and replaced with metal, which acts as a heat sink. It is suitable for high-current, high-speed core driver and computer-memory applications.

echelon–One of a series of levels of accuracy of calibration, the highest of which is represented by an accepted national standard. There may be auxiliary levels between two successive echelons.

echo–1. In radar, that portion of the energy reflected to the receiver from a target. 2. A wave which has been reflected or otherwise returned with sufficient magnitude and delay to be distinguishable from the directly transmitted wave. 3. In facsimile, a multiple reproduction on the record sheet caused by the arrival of the same original facsimile signal at different times over transmission paths of different lengths. 4. In a radio system, an electronic condition which causes a signal such as a voice signal to be reflected from some point or points in the circuit back to the point of origination of the signal. 5. A delayed repetition (sometimes several rapid repetitions) of the original sound or signal. 6. In tape recording, this refers to a provision for picking up some of the sound from a play head while recording, and feeding it back to the record head to produce a rapidly periodic repetition of each sound. Correct echo-volume adjustment causes a "decay" of the repeated sounds to simulate acoustical reverberation.

echo area–Equivalent echoing area of a radar target (i.e., the relative amount of radar energy the target will reflect).

echo attenuation–In a four-wire (or two-wire) circuit equipped with repeater or mutliplex equipment in which the two directions of transmission can be separated from each other, the attenuation of the echo currents (which return to the input of the circuit under consideration) is determined by the ratio of the transmitted power, P1, to the echo power received, P2.

echo box – Also called phantom target. A device for checking the overall performance of a radar system. It comprises a resonant cavity which receives a portion of the pulse energy from the transmitter and retransmits it to the receiver as a slowly decaying transient. The time required for this transient response to decay below the minimum detectable level on the radar indicator is known as the ring time and is indicative of the overall performance of the radar set.

echo chamber–A reverberant room or enclosure used for adding hollow effects or actual echoes to radio or television programs.

echo check–A method of checking the accuracy of transmission of data in which the received data are returned to the sending end for comparison with the original data.

echo checking – A method of checking in which transmitted information is reflected back to the transmitting point and compared with what was sent.

echo depth sounder–*See* Fathometer.

echo depth sounding – A system of determining the ocean depth by producing a sound just below the water's surface and measuring the amount of time before the echo is reflected from the floor of the ocean.

echoencephalograph – Device using ultrasonic energy and echo-ranging technique to determine brain midline, hematoma, tumor, etc.

echoencephalography–Visualization of the fluid filled cavity of the brain by means of ultrasonic diagnostic devices. Since tumors are usually unilateral, they produce an asymmetry of the cerebral hemispheres and a movement of the midline toward one direction. Echoes reflected from the midline are therefore not equal bilaterally and a diagnosis is possible.

echoencephaloscope–An ultrasonic instrument for use in brain studies. A transducer that generates a series of ultrasonic pulses and detects the returning echoes is placed against the patient's head. Each pulse is displayed together with its associated echoes on a cathode-ray tube.

echo intensifier–A device, located at the target, that is used to increase the amplitude of the reflected energy to an abnormal level.

echo matching—Rotating an antenna to a position in which the pulse indications of an echo-splitting radar are equal.

echo ranging—Determination of both direction and distance of an underwater object from a vessel by ultrasonic radiation.

echo sounder—A sounding device used by ships to determine the depth of water. (*See also* Echo Depth Sounding.)

echo splitting—In certain radar equipment, the echo return is split and appears as a double indication on the screen of the radar indicator. This splitting is accomplished by special electronic circuits associated with the antenna lobe switching mechanism. When the two echo indications are of equal height, the target bearing is read from a calibrated scale.

echo suppression—1. A control used to disable a responder for a short interval of time so that reception of echoes of the interrogator pulse from nearby targets is prevented. 2. A circuit used to eliminate reflected waves.

echo suppressor — 1. A voice-operated device that is connected to a two-way telephone circuit to attenuate echo currents in one direction caused by telephone currents in the other direction. 2. In navigation, a circuit which desensitizes the equipment for a fixed period after the reception of one pulse, for the purpose of rejecting delayed pulses arriving from indirect reflection.

echo talker—A portion of the transmitted signal returned from a distant point to the transmitting source with sufficient time delay to be received as interference.

ECL—1. Abbreviation for emitter-coupled logic, a type of unsaturated logic performed by emitter-coupled transistors. Higher speeds may be achieved with ECL than are obtainable with standard logic circuits. 2. An IC logic family characterized by its very high speed of operation, low circuit density per chip and very high power dissipation when compared to other IC logic families.

ECM — Electronic countermeasures; methods of jamming or otherwise hindering the operation of enemy electronic equipment.

eco—Abbreviation for electron-coupled oscillator.

E-core—The laminated configuration resembling the capital letter *E* in some transformers and inductive transducers.

E-core.

eddy current clutch—A device that permits connection between a motor and a load by electrical (magnetic) means—no physical contact is involved. This method is also used for speed control by clutch "slippage."

eddy-current heating—Synonym for induction heating.

eddy-current loss—The core loss which results when a varying induction produces electromotive forces which cause a current to circulate within a magnetic material.

eddy currents — Also called Foucault currents. Those currents induced in the body of a conducting mass by a variation in magnetic flux.

edge effect—1. *See* Following Blacks, Following Whites, Leading Blacks, *and* Leading Whites. 2. Nonuniformity of electric fields between two parallel plates caused by an outward bulging of electric flux lines at the edges of the plates.

edge-triggered flip-flop—A type of flip-flop in which some minimum clock-signal rate of change is one necessary condition for an output change to occur.

edging — Undesired coloring around the edges of different-colored objects in a color television picture.

Edison base — Standard screw-thread base used for ordinary electric lamps.

Edison distribution system — A three-wire direct-current distribution system, usually 120 to 240 volts, for combined light and power service from a single set of mains.

Edison effect — Also called Richardson effect. The phenomenon wherein electrons emitted from a heated element within a vacuum tube will flow to a second element that is connected to a positive potential.

Edison storage cell—A storage cell having negative plates of iron oxide and positive plates of nickel oxide immersed in an alkaline solution. An open-circuit voltage of 1.2 volts per cell is produced.

E-display—In radar, a rectangular display in which targets appear as blips with distance indicated by the horizontal coordinate and elevation by the vertical coordinate.

edit—To arrange or rearrange output information from a digital computer before it is printed out. Editing may involve deleting undesired information, selecting desired information, inserting invariant symbols such as page numbers and typewriter characters, and applying standard processes such as zero suppression.

editing — The rearrangement of recorded material to provide a change of content or form, or for replacement of imperfect material. Usually accomplished by cutting and splicing the tape.

EDP – Abbreviation for electronic data processing.

EDP center–*See* Electronic Data-Processing Center.

EDPM – Abbreviation for electronic data-processing machine.

eeg – Abbreviation for electroencephalograph.

eeg electrode – Electrode that attaches to scalp for detecting brain waves.

effective acoustic center – The point from which the spherically divergent sound waves from an acoustic generator appear to diverge. Also called apparent source.

effective actuation time–The sum of the initial actuation time and the contact chatter intervals of a relay following such actuation.

effective address–The address that is actually used in carrying out a computer instruction.

effective ampere–That alternating current which, when flowing through a standard resistance, produces heat at the same average rate as one ampere of direct current flowing in the same resistance.

effective antenna length–The length which, when multiplied by the maximum current, will give the same product as the length and uniform current of an elementary electric dipole at the same location, and the same ratio field intensity in the direction of maximum radiation.

effective area–The effective area of an antenna in any specified direction is equal to the square of the wavelength multiplied by the power gain (or directive gain) in that direction, divided by 4π.

effective bandwidth–For a bandpass filter, the width of an assumed rectangular bandpass filter having the same transfer ratio at a reference frequency and passing the same mean-square value of a hypothetical current and voltage having even distribution of energy over all frequencies.

effective capacitance–The total capacitance existing between any two given points of an electric circuit.

effective conductivity – The conductance between the opposite parallel faces of a portion of a material having unit length and unit cross section.

effective confusion area–Amount of chaff whose radar cross-sectional area equals the radar cross-sectional area of a particular aircraft at a particular frequency.

effective current–That value of alternating current which will give the same heating effect as the corresponding value of direct current. For sine-wave alternating currents, the effective value is 0.707 times the peak value.

effective cutoff – *See* Effective Cutoff Frequency.

effective cutoff frequency–Also called effective cutoff. The frequency at which the insertion loss of an electric structure between specified terminating impedances exceeds the loss at some reference point in the transmission band.

effective facsimile band–A frequency band equal in width to the difference between zero frequency and the maximum keying frequency of a facsimile signal.

effective field intensity–Root-mean-square value of the inverse distance fields one mile from the transmitting antenna in all directions horizontally.

effective height–1. The height of the antenna center of radiation above the effective ground level. 2. In loaded or nonloaded low-frequency vertical antennas, a height equal to the moment of the current distribution in the vertical section divided by the input current.

effective irradiance to trigger–The minimum effective irradiance required to switch a light activated SCR from the off-state to the on-state.

effectively grounded–Grounded through a ground connection of sufficiently low impedance (inherent and/or intentionally added) so that fault grounds which may occur cannot build up voltages that are dangerous to connected equipment.

effective parallel resistance–The resistance leakage considered to be in parallel with a pure dielectric.

effective percentage modulation–For a single, sinusoidal input component, the ratio between the peak value of the fundamental component of the envelope and the direct-current component in the modulated conditions, expressed in percent.

effective radiated power–The product of the antenna power (transmitter power less transmission-line loss) times, either the antenna power gain, or the antenna field gain squared. Where circular or elliptical polarization is employed, the term "effective radiated power" is applied separately to the horizontal and vertical components of radiation. For allocation purposes, the effective radiated power authorized is the horizontally polarized component of radiation only. If specified for a particular direction, it is the antenna power gain in that direction only.

effective radius of the earth–A value used in place of the geometrical radius to correct the atmospheric refraction when the index of refraction in the atmosphere changes linearly with height. Under conditions of standard refraction, the effective radius is one and one-third the geometrical radius.

effective resistance–1. The average rate of dissipation of electric energy during a cycle divided by the square of the effective current. 2. The equivalent pure dc resistance which, when substituted for the winding of a motor being checked, will draw the same power. It is also equivalent to the impedance of a circuit having a capacitor connected in parallel with the winding and the capacitor adjusted to unity power factor for the circuit.

effective series resistance—A dropping resistance considered to be in series with an assumed pure capacitance.

effective sound pressure — The root-mean-square value of the instantaneous sound pressure at one point over a complete cycle. The unit is the dyne per square centimeter.

effective speed — The speed (less than rated) which can be sustained over a significant period of time and which reflects slowing effects of control codes, timing codes, error detection, retransmission, tabbing, hand keying, etc.

effective speed of transmission—The average rate over some specified time interval at which information is processed by a transmission facility. Usually expressed as average characters or average bits per unit time. Also called average rate of transmission.

effective thermal resistance—Of a semiconductor device, the effective temperature rise per unit power dissipation of a designated junction above the temperature of a stated external reference point under conditions of thermal equilibrium.

effective value—Also called the rms (root mean square) value. The value of alternating current that will produce the same amount of heat in a resistance as the corresponding value of direct current. For a sine wave, the effective value is 0.707 times the peak value.

effective wavelength—The wavelength corresponding to the effective propagation velocity and the observed frequency.

efficiency—1. Ratio of the useful output of a physical quantity which may be stored, transferred, or transformed by a device, to the total input of the device. 2. The ratio of the output power to the input power of a regulated power supply, expressed in percent. In the absence of statements to the contrary, it is assumed to be taken at nominal input and output levels and full load conditions. 3. As related to speaker systems, it is the relative ability to convert electrical energy into sound at a given volume level.

efficiency of a source of light—The ratio of the total luminous flux to the total power consumed. In the case of an electric lamp, it is expressed in lumens per watt.

efficiency of rectification—Ratio of direct-current power output to alternating-current power input of a rectifier.

EGO — Acronym for Eccentric orbit Geophysical Observatory.

ehf—Abbreviation for extremely high frequency.

E-H tee—A waveguide junction composed of a combination of E and H plane tee junctions which intersect the main guide at a common point.

E-H tuner—An E-H tee having two arms terminated in adjustable plungers. It is used for impedance transformation.

EIA—Abbreviation for Electronic Industries Association.

EIA interface — A set of signal properties (time duration, voltage, and current) specified by the Electronic Industries Association for business machine/data set connections.

eight-level code—A code in which eight impulses are utilized for describing a character. Start and stop elements may be added for asynchronous transmission. The term is often used to refer to the USASCII code.

E-indicator—A rectangular radar display in which the horizontal coordinate of a target blip represents range and the vertical coordinate represents elevation.

Einthoven string galvanometer—A moving-coil type of galvanometer in which the coil in a single wire suspended between the poles of a powerful electromagnet.

E-I pick-off—An assembly of transformer-like laminations, the output coils of which develop a voltage proportional to the displacement of a magnetic element from the neutral position for limited rotary as well as angular travel.

ekg—Abbreviation for electrocardiograph.

elastance—Symbolized by S. In a capacitor, the ratio of potential difference between its electrodes to the charge in the capacitor. It is the reciprocal of capacitance. The unit of measure is the daraf.

$$S\ (\text{daraf}) = \frac{V}{Q}$$

elastic collision—Collision resulting in no molecular excitation when the conservation of momentum and kinetic energy governs the energy transfer.

elastic limit—The maximum stress a solid can endure and still return to its unstrained state when the stress is removed.

elastic wave—A pure acoustic wave; a moving lattice distortion without a magnetic component.

elastivity—The resistance of an electrostatic field. It is the reciprocal of permittivity.

E-layer — One of the regular ionospheric layers with an average height of about 100 kilometers. This layer occurs during daylight hours, and its ionization is dependent on the sun's angle. The principal layer corresponds roughly to what was formerly called the Kennelly-Heaviside layer.

elbow—In a waveguide, a bend with a rela-

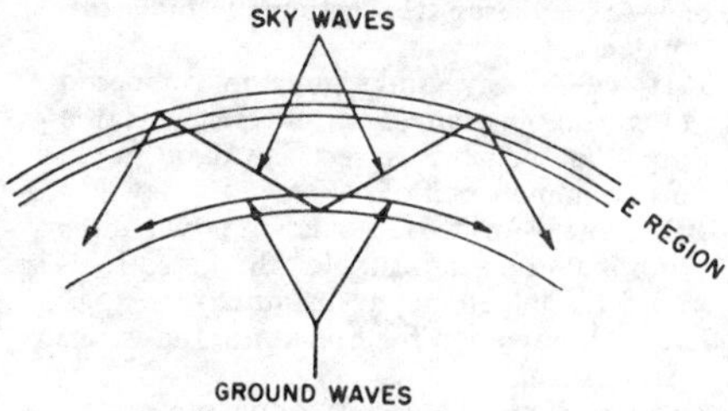

E-layer.

tively short radius and an angle normally of 90° but sometimes for acute angles down to 15°.

electra – A specific radionavigational aid that provides a number (usually 24) of equasignal zones. Electra is similar to sonne except that in sonne the equasignal zones as a group are periodically rotated in bearing.

electralloy–A nonmagnetic alloy frequently used in radio chassis.

electret–A permanently polarized piece of dielectric material produced by heating the material and placing it in a strong electric field during cooling. Some barium titanate ceramics can be polarized in this way, and so can carnauba waxes. The electric field of an electret corresponds somewhat to the magnetic field of a permanent magnet.

electric – Containing, producing, arising from, actuated by, or carrying electricity, or designed to carry electricity and capable of so doing. Examples: electric eel, energy, motor, vehicle, wave.

electrical–Related to, pertaining to, or associated with electricity but not having its properties or characteristics. Examples: electrical engineer, handbook, insulator, rating, school, unit.

electrical angle–A quantity that specifies a particular instant in a cycle of alternating current. One cycle is considered to be 360°, so a half cycle is 180° and a quarter cycle is 90°. If one voltage reaches its peak value a quarter cycle after another, the phase difference, or electrical angle between the voltages, is 90°.

electrical bail – A switch action in which, upon actuation of one station, the switch changes the contact position, electrically locks the switch in that position, and releases any station previously actuated.

electrical bias – An electrically produced force tending to move the armature of a relay toward a given position.

electrical boresight–The tracking axis as determined by an electrical indication, such as the null direction of a conical scanning or monopulse antenna system or the beam maximum direction of a highly directive antenna.

electrical center–The point approximately midway between the ends of an inductor or resistor. This point divides the inductor or resistor into two equal electrical values (e.g., voltage, resistance, inductance, or number of turns).

electrical coupling–Coupling discrete elements with either electrical conductors or reactances.

electrical charge–The excess on (or in) a body of one kind (polarity) of electricity over the other kind. A plus sign indicates that positve electricity predominates, and a minus sign indicates that negative electricity predominates. Symbol: Q or q.

electrical degree–A unit of time measurement applied to alternating current.

electrical discharge machining–Machining in which metal is removed by a controlled electrical spark in a dielectric.

electrical distance–The distance between two points, expressed as the length of time an electromagnetic wave in free space takes to travel between them.

electrical element–The concept in uncombined form of any of the individual building blocks from which electronic circuits are synthesized. Examples of basic electrical elements are insulation, conductance, resistance, capacitance, and inductance.

electrical erosion–The loss of contact material due to action of an electrical discharge.

electrical filter – Device for rejecting or passing a specific band of signal frequencies.

electrical forming–The application of electric energy to a semiconductor device in order to permanently modify the electrical characteristics.

electrical gearing–A term used to describe the action of a system in which the output shaft rotates at a different speed from the input shaft, the ratio being established by electrical means.

electrical glass insulation–Insulating materials made from glass fibers of varying diameters, lengths, compositions, etc., including yarns, rovings, slivers, cords, and sheets or mats, bounded or treated only as necessary to their manufacture.

electrical-impedance cephalography – A method of evaluating blood circulation in the brain by measuring changes in the impedance between two surface electrodes attached to the head. This impedance decreases when the blood volume in the brain increases. The technique is also known as rheoencephalography.

electrical inertia–Inductance that opposes any change in current flow through an inductor.

electrical initiation–Any source of electrical power used to start a function or sequence.

electrical interlocks—Switches mounted on contactors or other devices and operated by rods or levers. These interlocks open or close, depending on the open or closed position of the contractor or device with which they are associated, and are used to govern succeeding operations of the same or allied devices.

electrical length — Length expressed in wavelengths, radians, or degrees. Distance in wavelengths $\times 2\pi =$ radians; distance in wavelengths $\times 360 =$ degrees.

electrical load—A device (e.g., a speaker) comprising resistive and/or reactive components into which an amplifier, generator, etc., delivers power.

electrically connected — Joined through a conducting path or a capacitor as distinguished from being joined merely through electromagnetic induction.

electrically operated rheostat — A rheostat used to vary the resistance of a circuit in response to some means of electrical control.

electrically operated valve—A solenoid- or motor-operated valve used in vacuum, air, gas, oil, water, or similar lines.

electrically variable inductor—An inductor in which the inductance can be controlled by a current or a voltage. It is usually made in the form of a saturable reactor with two windings. One is called the signal, or tuned winding, corresponding to the ac or load winding of a power-handling saturable reactor; the other is the control winding and corresponds to the dc winding of the saturable reactor.

electrical noise — Unwanted electrical energy other than cross talk in a transmission system.

electrical radian — 57.296°, or $1/2\pi$ (1/6.28) of a cycle of alternating current or voltage.

electrical reset—A term applied to a relay to indicate that it is capable of being electrically reset after an operation.

electrical resistivity — The resistance of a material to passage of an electric current through it. Expressed as ohms (units of resistance) per mil foot or as microhms (millionths of an ohm) per centimeter cubed (cm^3) at a specified temperature.

electrical resolver—Special type of synchro having a single winding on the stator and two windings the axes of which are 90° apart on the rotor.

electrical scanning—Scanning accomplished through variation of the electrical phases or amplitudes at the primary radiating element of an antenna system.

electrical service entrance—A combination of intake wires and equipment including the service entrance wires, electric meter, main switch or circuit breaker and main distribution or service panel through which the supply of power enter the home.

electrical sheet—Iron or steel sheets from which laminations for electric motors are punched.

electrical system — The organized arrangement of all electrical and electromechanical components and devices in a way that will properly control the particular machine tool or industrial equipment.

electrical twinning—*See* Twinning.

electrical zero—A standard synchro position at which electrical outputs have defined amplitudes and time phase.

electric arc—A discharge of energy through a gas.

electric bell — An audible signaling device consisting of one or more gongs and an electromagnetically actuated striking mechanism.

electric brazing—A brazing (alloying) process in which the heat is furnished by an electric current.

electric breakdown voltage—*See* Dielectric Breakdown Voltage.

electric breeze or wind — The emission of electrons from a sharp point of a conductor which carries a high negative potential.

electric charge — 1. Electric energy stored on the surface of an object. 2. A property of electrons and protons. Similarly charged particles repel one another. Particles having opposite charges attract one another.

electric chronograph—A highly accurate apparatus for measuring and recording time intervals.

electric circuit—A continuous path consisting of wires and/or circuit elements over or through which an electric current can flow. If the path is broken at any point, current can no longer flow and there is no circuit.

electric contact — A separable junction between two conductors which is designed to make, carry, or break (in any sequence or singly) an electric circuit.

electric controller—A device which governs the amount of electric power delivered to an apparatus.

electric current—Electricity in motion. In the atoms of metallic substances, there are a number of "free" electrons or negatively charged particles which wander in the spaces between the atoms of the metal. The electron movement is normally without any definite direction and cannot be detected. The connection of an electric battery produces an electric field in the metal and causes the free electrons to move or drift in one direction, and it is this electron drift which constitutes an electric current. Electrons, being of negative polarity, are attracted to the positive terminal of the battery and so the actual direction of flow of electricity is from negative to positive, that is, opposite to the conventional direction usually adopted.

electric delay line—A delay line using properties of lumped or distributed capacitive and inductive elements. Can be used as a storage medium by recirculating the information-carrying signal.

electric dipole—Also called a doublet. A simple antenna comprising a pair of oppositely charged conductors capable of radiating an electromagnetic wave in response to the movement of an electric charge from one conductor to the other.

electric-discharge lamp—A sealed glass enclosure containing a metallic vapor or an inert gas through which electricity is passed to produce a bright glow.

electric displacement — *See* Electric-Flux Density.

electric-displacement density—*See* Electric-Flux Density.

electric eye—1. The layman's term for a photoelectric cell. 2. The cathode-ray, tuning-indicator tube used in some radio receivers.

electric field—1. The region about a charged body. Its intensity at any point is the force which would be exerted on a unit positive charge at that point. 2. A condition detectable in the vicinity of an electrically charged body such that forces act on other electric charges in proportion to their magnitudes. 3. Field of force which exists in the space around electrically charged particles. Lines of force are imagined to originate at the protons or positively charged particles and to terminate on electrons or negatively charged particles.

electric-field intensity—A measure of the force exerted at a point by a unit charge at that point.

electric-field strength — The magnitude of the electric field in an electromagnetic wave. Usually stated in volts per meter. (*See also* Dielectric Strength.)

electric-field vector—At a point in an electric field, the force on a stationary positive charge per unit charge. May be measured in either newtons per coulomb or volts per meter. This term is sometimes called the electric-field intensity, but such use of the word "intensity" is deprecated in favor of "field strength," since intensity denotes power in optics and radiation.

electric-filament lamp—A glass bulb either evacuated or filled with an inert gas and having a resistance element electrically heated to, and maintained at, the temperature necessary to produce incandescence.

electric filter—*See* Electric-Wave Filter.

electric-flux density — Also called electric-displacement density or electric displacement. At a point, the vector equal in magnitude to the maximum charge per unit area which would appear on one face of a thin metal plate introduced in the electric field at that point. The vector is normal to the plate from the negative to the positive face.

electric force—Electric field intensity measured in dynes.

electric furnace—A furnace in which electric arcs provide the source of heat.

electric generator—A machine that transforms mechanical power into electrical power.

electric governor-controlled series-wound motor—A series-wound motor having an electric speed governor connected in series with the motor circuit. The governor is usually built into the motor.

electric hygrometer—An instrument for indicating humidity by electric means. Its operation depends on the relationship between the electric conductance and moisture content of a film of hygroscopic material.

electric hysteresis—Internal friction in a dielectric material when subjected to a varying electric field (e.g., the paper or mica dielectric of a capacitor in an ac circuit). The resultant heat generated can eventually break down the dielectric and cause the capacitor to fail.

electrician—A person engaged in designing, making, or repairing electric instruments or machinery. Also, one who sets up an electrical installation.

electric image—The electrical counterpart of an object; i.e., the fictitious distribution of the same amount of electricity that is actually distributed on a nearby object.

electricity—The property of certain particles to possess a force field which is neither gravitational nor nuclear. The type of force field associated with electrons is defined as negative and that associated with protons and positrons as positive. The fundamental unit is the charge of an electron: 1.60203×10^{-19} coulomb. Electricity can be further classified as static electricity or dynamic electricity. Static electricity in its strictest sense refers to charges at rest, as opposed to dynamic electricity, or charges in motion. Static electricity is sometimes used as a synonym for triboelectricity or frictional electricity.

electric light—Light produced by an electric lamp.

electric lines of force—In an electric field, curves the tangents of which at any point give the direction of the fields at that point.

electric meter—A device that measures and registers the amount of electricity consumed over a certain period of time.

electric mirror—*See* Dynode.

electric moment—For two charges of equal magnitude but opposite polarities, a vector equal in magnitude to the product of the magnitude of either charge by the distance between the centers of the two charges. The direction of the vector is from the negative to the positive charge.

electric motor – A device which converts electrical energy into rotating mechanical energy.

electric network – A combination of any number of electric elements, having either lumped or distributed impedances, or both.

electric oscillations – The back-and-forth flow of electric charges whenever a circuit containing inductive and capacitance is electrically disturbed.

electric potential–A measure of the work required to bring a unit positive charge from an infinite distance or from one point to another (the difference of potential between two points).

electric precipitation – The collecting of dust or other fine particles floating in the air. This is done by inducing a charge in the particles, which are then attracted to highly oppositely charged collector plates.

electric probe–A rod inserted into an electric field during a test to detect dc, audio, or rf energy.

electric reset–A qualifying term indicating that the contacts of a relay must be reset electrically to their original positions following an operation.

electric shield–A housing, usually aluminum or copper, placed around a circuit to provide a low-resistance path to ground for high-frequency radiations and thereby prevent interaction between circuits.

electric strain gage – A device which detects the change in shape of a structural member under load and causes a corresponding change in the flow of current through the device.

electric strength – The maximum electric charge a dielectric material can withstand without rupturing. (*See also* Dielectric Strength *and* Insulating Strength.)

electric stroboscope–An instrument for observing or for measuring the speed of rotating or vibrating objects by electrically producing periodic changes in the intensity of light used to illuminate the object.

electric tachometer–A tachometer (rpm indicator) that utilizes voltage or electrical impulses.

electric telemeter–A system consisting of a meter which measures a quantity, a transmitter which sends the information to a distant station, and a receiver which indicates or records the quantity measured.

electric transcription – In broadcasting, a disc recording of a message or a complete program.

electric transducer–A device actuated by electric waves from one system and supplying power, also in the form of electric waves, to a second system.

electric tuning–A system by which a radio receiver is tuned to a station by pushing a button (instead of, say, turning a knob).

electric vector–A component of the electromagnetic field associated with electromagnetic radiation. The component is of the nature of an electric field. The electric vector is supposed to coexist with, but act at right angles to, the magnetic vector.

electric watch–A timepiece in which a battery replaces the mainspring as the prime energy source of the watch, and in which an electromagnet impels the balance wheel through a mechanical switching-contact arrangement.

electric wave–Another term for the electromagnetic wave produced by the back-and-forth movement of electric charges in a conductor.

electric-wave filter–Also called electric filter. A device that separates electric waves of different frequencies.

electrification–1. The process of establishing an excess of positive or negative charges in a material. 2. The process of applying a voltage to a component or device.

electroacoustic – Pertaining to a device (e.g., a speaker or a microphone) which involves both electric current and sound-frequency pressures.

electroacoustic device – One that employs phonon propagation or vibrations of a material's crystal lattice structure as the basic energy transport mechanism. Electrical energy is converted into acoustic energy by the material's piezoelectric properties.

electroacoustic transducer – A device that receives excitations from an electric system and delivers an output to an acoustic system, or vice versa. A speaker is an example of the first, and a microphone is an example of the second.

electroanalysis–The process of determining the quantity of an element or compound in an electrolyte solution by depositing the element or compound on an electrode by electrolysis.

electrobiology–The science concerned with electrical phenomena of living creatures.

electrobioscopy–The application of a voltage to produce muscular contractions.

electrocardiogram–Essentially an electromyogram of the heart muscle. All muscular activity in the body is characterized by the discharge of polarized cells, the aggregate current from which causes a voltage drop that can be measured on the skin. A changing emf will appear between electrodes connected to the arms, legs, and chest, which rises and falls with heart action such that the period of the resulting waveform is the time between heartbeats. Various positive and negative peaks within one cycle of this waveform have been lettered P, Q, R, S, and T, a notation which aids in subsequent analysis and diagnosis.

electrocardiograph – A medical instrument for detecting irregularities in the action of a human heart. It measures the changes in voltage occurring in the human body with each heartbeat. Abbreviated ekg.

electrocardiography – Recording and interpretation of the electrical activity of the heart. The voltage generated by the heart is picked up by surface electrodes on the limbs and chest, amplified, and applied to a strip-chart recorder.

electrocardiophonograph – An instrument that records heart sounds.

electrochemical deterioration—A process in which autocatalytic electrochemical reactions produce an increase in conductivity and in turn ultimate thermal failure.

electrochemical equivalent—The weight of an element, compound, radical, or ion involved in a specified electrochemical reaction during passage of a specified quantity of electricity such as a coulomb.

electrochemical junction transistor—A junction transistor produced by etching an n-type germanium wafer on opposite sides with jets of a salt solution such as indium chloride.

electrochemical recording – A recording made by passing a signal-controlled current through a sensitized sheet of paper. The paper reacts to the current and thereby produces a visual record.

electrochemical transducer—A device which uses a chemical change to measure the input parameter, and the output of which is a varying electrical signal proportional to the measurand.

electrochemical valve—Electric valve consisting of a metal in contact with a solution or compound, across the boundary of which current flows more readily in one direction than in the other direction and in which the valve action is accompanied by chemical changes.

electrochemistry – That branch of science concerned with reciprocal transformations of chemical and electrical energy. This includes electrolysis, electroplating, the charge and discharge of batteries, etc.

electrochromic display – A passive solid state display that is made from a material whose light-absorption properties are changed by an externally applied electric field. Ordinarily electrochromic materials do not absorb light in the visible range of the spectrum; so they are completely transparent. When a moderate electric field is applied, the material develops an absorption band in the visible spectrum and takes on a color that remains even after the electric field is removed and lasts from minutes to months. The color change can be reversed and the display returned to its original state when the polarity of the applied electric field is simply reversed.

electrocoagulation—The process of solidifying tissue by means of a high-frequency electrical current.

electrocution—Killing by means of an electric current.

electrode—1. In an electronic tube, the conducting element that does one or more of the following: emits or collects electrons or ions, or controls their movement by means of an electric field on it. 2. In semiconductors, the element that does one or more of the following: emits or collects electrons or holes, or controls their movements by means of an electric field on it. 3. In electroplating, the metal being plated. 4. A conductor by means of which a current passes into or out of a fluid or an organic material such as human skin; often one terminal of a lead. 5. A metallic conductor such as in an electrolytic cell, where conduction by electrons is changed to conduction by ions or other charged particles.

electrode admittance—The alternating component of the electrode current divided by that of the electrode voltage (all other electrode voltages maintained constant).

electrode capacitance—The capacitance between one electrode and all the other electrodes connected together.

electrode characteristic—The relationship, usually shown by a graph, between the electrode voltage and current, all other electrode voltages being maintained constant.

electrode conductance—The quotient of the in-phase component of the electrode alternating current divided by the electrode alternating voltage, all other electrode voltages being maintained constant. This is a variational and not a total conductance.

electrode current—Current passing into or out of an electrode.

electrode dark current—1. In phototubes, the component of electrode current that flows in the absence of ionizing radiation and optical photons. Also called dark current. 2. The current which flows in a photodetector when there is no incident radiation on the detector.

electrode dissipation—The power which an electrode dissipates as heat when bombarded by electrons and/or ions and radiation from nearby electrodes.

electrode drop—The voltage drop produced in an electrode by its resistance.

electrode impedance – The reciprocal of electrode admittance.

electrode inverse current—Current through a tube electrode in the direction opposite to that for which the tube was designed.

electrodeless discharge – A luminous discharge produced by means of a high-frequency electric field in a gas-filled glass tube that has no internal electrodes.

electrodeposition – Also called electrolytic deposition. The process of depositing a substance on an electrode by electrolysis, as in electroplating, electroforming, electrorefining, or electrotinning. *See also* Electroplating.

electrode potential–The instantaneous voltage on an electrode. Its value is usually given with respect to the cathode of a vacuum tube.

electrode reactance–The imaginary component of electrode impedance.

electrode resistance–The reciprocal of electrode conductance. It is the effective parallel resistance, not the real component of electrode impedance.

electrodermography–The recording of the electrical resistance of the skin, which is a sensitive indicator of the activity of the autonomic nervous system.

electrode voltage–The voltage between an electrode and the cathode or a specified point of a filamentary cathode. The terms "grid voltage," "anode voltage," "plate voltage," etc., designate the voltage between these electrodes and the cathode. Unless otherwise stated, electrode voltages are measured at the available terminals.

electrodialytic process–A process for producing fresh water by using a combination of electric current and two types of chemically treated membranes.

electrodynamic–Pertaining to electric current, electricity in motion, and the actions and effects of magnetism and induction.

electrodynamic braking–A method of stopping a tape-deck motor gently by the application of a predetermined voltage to the motors.

electrodynamic instrument–An instrument which depends for its operation on the reaction between the current in one or more moving coils and the current in one or more fixed coils.

electrodynamic machine–Electric generator or motor in which the output load current is produced by magnetomotive currents generated in a rotating armature.

electrodynamics–The science dealing with the various phenomena of electricity in motion, including interactions of currents with each other, with their associated magnetic fields, and with other magnetic fields.

electrodynamic speaker–A speaker consisting of an electromagnet called the field coil, through which a direct current flows.

electrodynamometer–An instrument for detecting or measuring an electric current by determining the mechanical reactions between two parts of the same circuit.

electroencephalogram – A waveform obtained by plotting brain voltages (available between two points on the scalp) against time. An electroencephalogram is not necessarily a periodic function, although it can be—particularly if the patient is unconscious. These voltages are of extremely low level and require recording apparatus which displays excellent noise rejection.

electroencephalograph (**eeg**) – An instrument for measuring and recording the rhythmically varying potentials produced by the brain by the use of electrodes applied to the scalp.

electroencephalography–Recording and interpretation of the electrical activity of the brain. Voltage (typically 50 microvolts) picked up by electrodes on the scalp is amplified and applied to a strip-chart recorder.

electroencephaloscope–An instrument for detecting brain potentials at many different sections of the brain and displaying them on a cathode-ray tube.

electroforming–1. Also called electrodeposition and electroplating. Making a metal object by using electrolysis to deposit a metal on an electrode. 2. Creating a pn junction by passing a current through point contacts on a semiconductor.

electrogastrogram–The graphic record that results from synchronous recording of the electrical and mechanical activity of the stomach.

electrograph–1. A plot, graph, or tracing made by means of the action of an electric current on sensitized paper or other material, or by means of an electrically controlled stylus or pen. 2. Equipment for facsimile transmission.

electrographic recording–Also called electrostatography. The producing of a visible record by using a gaseous discharge between two or more electrodes to form electrostatically charged patterns on an insulator. (*See also* Electrostatic Electrography.)

electrokinetics–The branch of physics concerned with electricity in motion.

electroless plating–1. A method of metal deposition by means of a chemical reducing agent present in the processing solution. The process is further characterized by the catalytic nature of the surface which enables the metal to be plated to any thickness. 2. A chemical process by which certain metals can be plated without electrical current. Tin may be plated onto copper in this manner.

electroluminescence – 1. Luminescence resulting from a high-frequency discharge through a gas or from application of an alternating current to a layer of phosphor. 2. Direct conversion of electrical energy into light energy in a liquid or solid; for example, photo emission as a result of electron-hole recombination in a pn junction. The standard abbreviation for the effect is written EL. (This process is not to be confused with the ordinary tungsten filament bulb, where there is an intermediate stage of heat making the process thermoluminescent.)

electroluminescent lamp – A lamp in the shape of a panel which is decorative as well as illuminative. It consists primarily of a capacitor having a ceramic dielectric with electroluminescent phosphor. The amount of illumination is determined by the voltage across the layer and by the frequency applied to it.

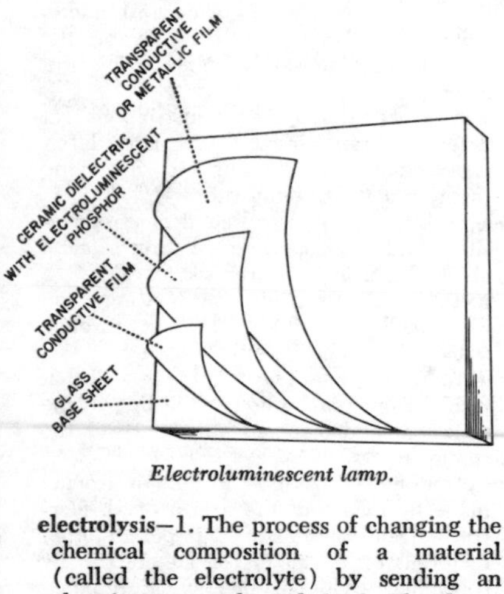

Electroluminescent lamp.

electrolysis–1. The process of changing the chemical composition of a material (called the electrolyte) by sending an electric current through it. 2. The decay of an underground structure by chemical action due to stray electrical currents.

electrolyte–1. A substance in which the conduction of electricity is accompanied by chemical action. 2. The paste which forms the conducting medium between the electrodes of a dry cell, storage cell, or electrolytic capacitor. 3. A substance which, when dissolved in a suitable liquid (often water), dissociates into ions, thus rendering the liquid electrically conducting. 4. The current-conducting substance (liquid or solid) between two capacitor electrodes at least one of which is covered by a dielectric film.

electrolytic–Pertaining to or made by electrolysis; deposited by electrolysis; pertaining to or containing an electrolyte.

electrolytic capacitor–1. A capacitor consisting of two conducting electrodes, with the anode having a metal oxide film formed on it. The film acts as the dielectric or insulating medium. The capacitor is operable in the presence of an electrolyte, usually an acid or salt. Generally used for filtering, bypassing, coupling, or decoupling. 2. A capacitor in which the dielectric is a film of oxide electrolytically deposited on a plate of aluminum or tantalum. The thinness of the film permits a high capacitance/volume ratio. The oxide acts as a dielectric in one direction only. The device is, therefore, polarized. (A nonpolarized electrolytic capacitor is, in effect, two polarized types in series with their like terminals connected together.)

electrolytic cell–In a battery, the container, two electrodes, and the electrolyte.

electrolytic conduction–The flow of current between electrodes immersed in an electrolyte. It is caused by the movement of ions from one electrode to the other when a voltage is applied between them.

electrolytic conductivity–Also called specific conductance. A measure of the ability of a solution to carry an electric current. Defined as the reciprocal of the resistance in ohms of a 1-cm cube of the liquid at a specified temperature. The units of specific conductance are the reciprocal ohm-cm (or mho/cm) and one millionth of this, micromho/cm. High-quality condensed steam and distilled or demineralized water have specific conductances at room temperatures as low as or lower than 1 micromho/cm).

electrolytic corrosion–*See* Corrosion.

electrolytic deposition – *See* Electrodeposition.

electrolytic dissociation–The breaking up of molecules into ions in a solution.

electrolytic interrupter–A device which is tilted to change the current through it.

electrolytic iron–Iron obtained by an electrolytic process. The iron possesses good magnetic qualities and is exceptionally free of impurities.

electrolytic potential–The difference in potential between an electrode and the immediately adjacent electrolyte, expressed in terms of some standard electrode difference.

electrolytic recording–A form of facsimile recording in which ionization causes a chemically moistened paper to undergo a change.

electrolytic rectifier–A rectifier consisting

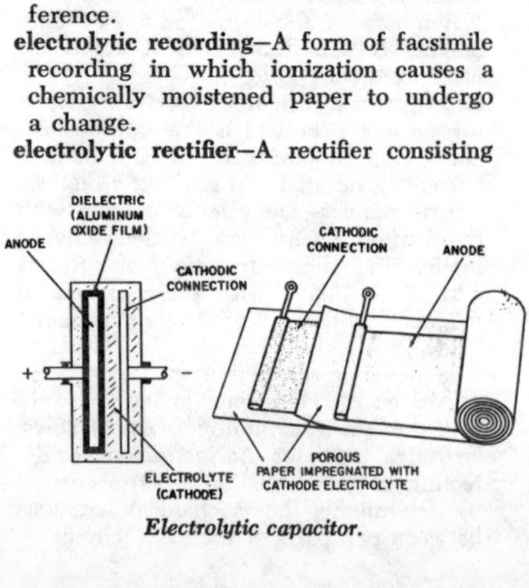

Electrolytic capacitor.

of metal electrodes in an electrolyte, in which rectification of alternating current is accompanied by electrolytic action. A polarization film formed on one of the electrodes permits current in one direction but not in the other.

electrolytic refining—The refining or purifying of metals by electrolysis.

electrolytic switch—A switch having two electrodes projecting into a chamber containing a precisely measured quantity of a conductive electrolyte, leaving an air bubble of predetermined width. When the switch is tilted from true horizontal, the bubble shifts position and changes the amount of electrolyte in contact with the electrodes, thereby changing the amount of current passed by the switch. Used as a leveling switch in gyro systems.

electrolyzer—An electrolytic cell that produces alkalies, metals, chlorine, or other allied products.

electromagnet—1. A temporary magnet consisting of a solenoid with an iron core. A magnetic field exists only while current flows through the solenoid. 2. A magnet, consisting of a solenoid with an iron core, which has a magnetic field existing only during the time of current flow through the coil.

electromagnetic—1. Having both magnetic and electric properties. 2. Pertaining to the mutually perpendicular electric and magnetic fields associated with the movement of electrons through conductors, as in an electromagnet. 3. Pertaining to the combined electric magnetic fields associated with radiation or with movements of charged particles.

electromagnetic amplifying lens—A system made up of a large number of waveguides symmetrically arranged with respect to an excitation medium so that they are excited with equal amplitude and phase in order to provide an effective gain in energy.

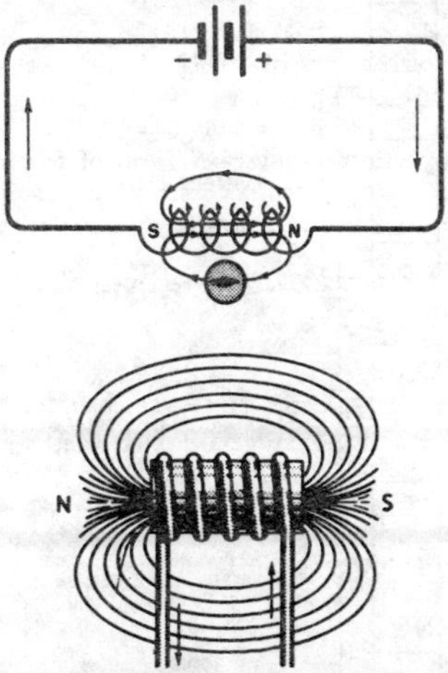

Electromagnet.

electromagnetic cathode-ray tube—A cathode-ray tube which uses electromagnetic deflection to deflect the electron beam.

electromagnetic communications—The electromagnetic wave conductor is space itself. The electromagnetic frequencies available today for communications fall into two categories: frequencies which form "wireless" communications (such as visual light of fairly high frequency), and frequencies man uses for wireless communications (such as radio, short wave, and microwave transmitting, of relatively lower frequencies). In communicating by radio, short wave, and microwave frequencies, translators similar in principle to those used in electrical communications are needed, although the equipment requirement increases.

electromagnetic compatibility—The ability of electronic communications equipment, subsystems, and systems to operate in their intended environments without suffering or causing unacceptable degradation of performance as a result of unintentional electromagnetic radiation or response. Abbreviated emc.

electromagnetic complex—The electromagnetic configuration of an installation, including all radiators of significant amounts of energy.

electromagnetic coupling—The mutual relationship between two separate but adjacent wires when the magnetic field of one induces a voltage in the other.

electromagnetic crack detector—An instrument for detecting hidden cracks in iron or steel objects by magnetic means.

electromagnetic deflection—The deflection of an electron stream by means of a magnetic field. In a television receiver, the magnetic field for deflecting the electron beam horizontally and vertically is produced by two pairs of coils, called the deflection yoke, around the neck of the picture tube.

electromagnetic deflection coil — A coil, around the neck of a cathode-ray tube, for deflecting the electron beam.

electromagnetic delay line—A delay line the operation of which is based on the time of propagation of electromagnetic waves through distributed or lumped capacitance and inductance.

electromagnetic energy—Forms of radiant energy such as radio waves, heat waves, light waves, X-rays, gamma rays, and cosmic rays.

electromagnetic environment—The rf field or fields existing in an area or desired in an area to be shielded.

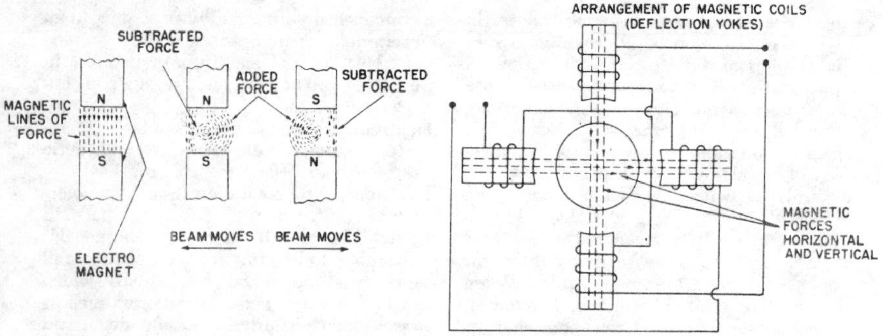

Electromagnetic deflection coil.

electromagnetic field—1. The field of influence produced around a conductor by the current flowing through it. 2. A rapidly moving electric field and its associated magnetic field. The latter is perpendicular to both the electric lines of force and their direction.

electromagnetic focusing—In a television picture tube, the focusing produced by a coil mounted on the neck. Direct current through the coil produces magnetic field lines parallel to the tube axis.

electromagnetic horn—A horn-shaped structure that provides highly directional radiation of radio waves in the 100-megahertz or higher frequency range.

electromagnetic induction — The voltage produced in a coil as the number of magnetic lines of force (flux linkages) passing through the coil changes.

electromagnetic inertia—1. The characteristic delay of a current in an electric circuit in reaching its maximum or zero value after application or removal of the source voltage. 2. The property of self-induction.

electromagnetic interference—Electromagnetic phenomena which, either directly or indirectly, can contribute to a degradation in performance of an electronic receiver or system. (The terms "radio interference," "radio-frequency interference," "noise," "emi," and "rfi" have been employed at various times in the same context.)

electromagnetic lens—1. An electron lens in which the electron beams are focused electromagnetically. 2. An electromagnet that produces a suitably shaped magnetic field for the focusing and deflection of charged particles in electron-optical systems.

electromagnetic mirror—A surface or region capable of reflecting radio waves, such as one of the ionized layers in the upper atmosphere.

electromagnetic oscillograph — An oscillograph in which a mechanical motion is derived from electromagnetic forces to produce a record.

electromagnetic pulse—Abbreviated emp. A reaction of large magnitude resulting from the detonation of nuclear weapons.

electromagnetic radiation—1. That form of energy which is characterized by transversely oscillating electric and magnetic fields and which propagates at velocity "c" in free space. At a sufficient distance from the source, the electric-field vector and the magnetic-field vector are at right angles to each other, forming a right-handed (coordinate) system. In an ionized medium, a longitudinal component may be present. 2. A form of power emitted from vibrating charged particles. A combination of oscillating electric and magnetic fields, electromagnetic radiation propagates through otherwise empty space

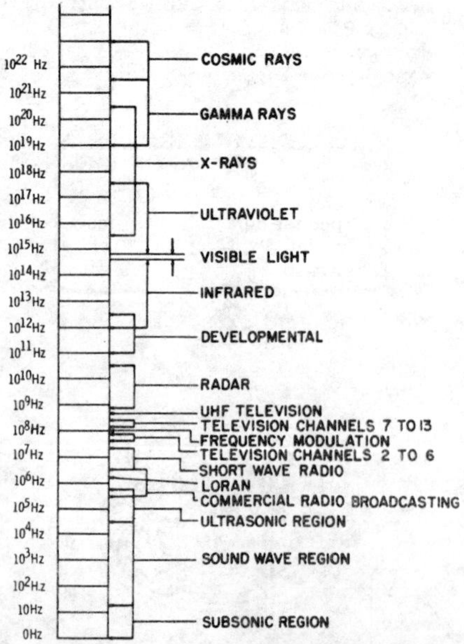

Electromagnetic radiation spectrum chart.

with the velocity of light. This (constant) velocity equals the alternation frequency multiplied by the wavelength; hence the frequency and wavelength are inversely proportional to each other. The spectrum of electromagnetic radiation is continuous over all frequencies. 3. Abbreviated emr. When discussing shielding describes radiation generated by an electrical means, ranging from a stationary magnetic or electrostatic field, to high frequencies changing fields and transmitted plane waves of radio frequency.

electromagnetic reconnaissance – Activity conducted to locate and identify potentially hostile sources of electromagnetic radiation, including radar, communication, missile-guidance, and air-navigation equipment.

electromagnetic relay—Device which opens or closes contacts by setting "moving" contacts against "fixed" contacts when current passes through an electromagnet. Current sets up a magnetic attraction between the core of the electromagnet and a hinged arm to the tip of which is attached the "moving" contact. The movement of the arm toward the core of the electromagnet brings "moving" and "fixed" contact together. When current is withdrawn, a spring returns the arm to its original position and the contacts separate.

electromagnetic repulsion – The repelling action between like poles of electromagnets.

electromagnetics – In physics, the branch concerned with the relationships between electric currents and their associated magnetic fields.

electromagnetic spectrum—A chart or graph showing the relationships among all known types of electromagnetic radiation classified by wavelengths. 2. The continuous range of frequencies, from 0.1 to 10^{22} hertz, of which a radiated signal is composed. Spectral dimensions are more conveniently described in terms of wavelength (Ängstroms) where one Ängstrom is equivalent to 10^{-7} mm. The electromagnetic spectrum includes radiofrequency waves, light waves, microwaves, infrared, X-rays (Roentgen rays) and gamma rays.

electromagnetic theory of light—The theory which states that electromagnetic and light waves have identical properties.

electromagnetic transduction – 1. Conversion of the measurand into the output induced in a conductor by a change in magnetic flux. 2. A wave produced by the oscillation of an electric charge. 3. A wave in which there are both electric and magnetic displacements. 4. A transverse wave associated with the transmission of electromagnetic energy.

electromagnetic-type microphones – Microphones in which the voltages are varied by an electromagnet (namely, ribbon or velocity, dynamic or moving-coil, and reluctance or moving-vane microphones).

electromagnetic unit—Abbreviated emu. A unit of electricity based primarily on the magnetic effect of an electric current. The fundamental centimeter-gram-second unit is the abampere. Now considered obsolete.

electromagnetic vibrator—A mechanical device for interrupting the flow of direct current and thereby making it a pulsating current. This is done where a circuit requires an alternating current to operate. A reed within the vibrator is alternately attracted to two electromagnets.

electromagnetic wave—1. The radiant energy produced by oscillation of an electric charge. It includes radio, infrared, visible and ultraviolet light waves, and X-, gamma, and cosmic rays. 2. A wave in which both electric and magnetic displacement are present.

electromagnetism – The magnetic field around a wire or other conductor when, and only when, current passes through it.

electromanometer – Instrument used for measuring pressure of gases or liquids by electronic methods.

electromechanical—Any device using electrical energy to produce mechanical movement.

electromechanical bell—A bell with a prewound spring-driven clapper which is tripped electrically to ring the bell.

electromechanical breakdown—A mechanical runaway that occurs when the mechanical restoring force fails to balance the electrical compressive force.

electromechanical chopper – *See* Contact Modulator.

electromechanical energy—Energy present in an induction coil or solenoid.

electromechanical frequency meter—A meter which uses the resonant properties of mechanical devices to indicate frequency.

electromechanical recorder—A device which transforms electrical signals into equivalent mechanical motion which is transferred to a medium by cutting, embossing, or writing.

electromechanical timer—Usually refers to a motor-driven timer, with or without an electrically operated clutch. Can also apply to pneumatic and thermal timers, or slow pull-in or drop-out relays.

electromechanical transducer – A device that transforms electrical energy into mechanical energy or vice versa. A speaker is an example of the first, and a microphone of the second.

electromechanics—That branch of electrical engineering concerned with machines producing or operated by electric currents.

electrometallurgy—That branch of science

concerned with the application of electrochemistry to the extraction or treatment of metals.

electrometer—1. An electrostatic instrument that measures a potential difference or an electric charge by the mechanical force exerted between electrically charged surfaces. 2. A dc vacuum-tube voltmeter with an extremely high input resistance, usually around 10^{10} megohms, as opposed to 10 megohms or less for a conventional type.

electrometer amplifier—An amplifier circuit having sufficiently low current drift and other noise components, sufficiently low amplifier input-current offsets, and adequate power and current sensitivities to be usable for measuring current variations of considerably less than 10^{-12} A.

electrometer tube—A vacuum tube having a very low control-electrode conductance, to facilitate the measurement of extremely small direct currents and voltages.

electromigration—A detrimental effect occurring in transistors employing aluminum metalization schemes. Electromigration of aluminum results from the mass transport of metal by momentum exchange between thermally activated metal ions and conducting electrons. When it occurs, the ideally uniform aluminum film reconstructs to form thin conductor regions and extrudedlike hillocks that may cause the transistor's destruction.

electromotive force — 1. Abbreviated emf. The force which causes electricity to flow when there is a difference of potential between two points. The unit of measurement is the volt. 2. Electrical pressure at the source. Not to be confused with potential difference which is the voltage developed across a resistance or impedance due to current flowing through it. Both are measured in volts. 3. Electric pressure which causes a current to flow in a circuit; it is the energy put into the circuit by the source per unit electric charge which it supplies to the circuit. The unit of emf is the volt, being the electromotive force required to cause a current of 1 ampere to flow in a resistance of 1 ohm. 4. The difference of electrical potential found across the terminals of a source of electrical energy, more precisely, the limit of the potential difference across the terminals of a source as the current between the terminals approaches zero.

electromotive series—A list of metals arranged in decreasing order of their tendency to pass into ionic form by losing electrons.

electromyogram — Classically, a waveform of the contraction of a muscle as a result of electrical stimulation. Usually the stimulation comes from the nervous system (normal muscular activity). The record of potential difference between two points on the surface of the skin resulting from the activity or action potential of a muscle.

electromyograph—An instrument for measuring and recording potentials generated by muscles.

electromyography—Recording and interpretation of the electrical activity of muscle tissue. Surface electrodes (for many muscle fibers) or needle electrodes (for one or a few fibers) provide a signal that is amplified and displayed on a cathode-ray tube. Abbreviated EMG.

electron—1. Also called negatron. One of the natural elementary constituents of matter; it carries a negative electric charge of one electronic unit and has approximately 1/1840th the mass of a hydrogen atom, or 9.107×10^{-28} gram. Electrons surround the positively charged nucleus and determine the chemical properties of the atom. Positive electrons, or positrons, also exist. 2. High-speed, negatively charged particle forming outer shell of an atom—smallest electric charge that can exist. 3. An electrically charged particle that orbits the nucleus of every atom and is responsible for the bonds between atoms. The electron has a charge of −1.

electronarcosis—1. The induction of unconsciousness by passage of a weak current through the brain. 2. Anesthesia induced by the passage of a precisely controlled electric current through the brain.

electron attachment—Process by which an electron is attached to a neutral molecule to form a negative ion. Often characterized by the attachment coefficient η, which is the number of attachments per centimeter of drift. Also characterized by the ratio $h = \sigma_a/\theta$, where σ_a is the attachment cross section and θ the total cross section.

electron avalanche — The chain reaction started when one free electron collides with one or more orbiting electrons and frees them. The free electrons then free others in the same manner, and so on.

electron band—A spectrum band composed of molecules that is usually found in the visible or the ultraviolet because of the electron transition taking place within the molecule.

electron beam—A narrow stream of electrons moving in the same direction under the influence of an electric or magnetic field.

electron-beam evaporation—An evaporation technique in which the evaporant is heated by electron bombardment.

electron-beam generator—A velocity-modulated generator, such as a klystron tube, used to generate extremely high frequencies.

electron-beam instrument — Also called a

cathode-ray instrument. An instrument in which a beam of electrons is deflected by an electric or magnetic field (or both). Usually the beam is made to strike a fluorescent screen so the deflection can be observed.

electron-beam machining — A process in which controlled electron beams are used to weld or shape a piece of material.

electron-beam magnetometer — An instrument that measures the intensity and direction of magnetic forces by the immersion of an electron beam into the magnetic field.

electron-beam mode discharge—A form of discharge produced by a perforated-wall hollow cathode operating under conditions of pressure, voltage, and geometry usually associated with the abnormal glow discharge.

electron-beam tube — An electron tube which depends for its operation on the formation and control of one or more electron beams.

electron-beam welding—The process of using a focused beam of electrons to heat materials to the fusion point.

electron-bombarded semiconductor—An amplifier consisting of an electron-gun modulation system, semiconductor target and output coupling network all within a glass or ceramic envelope. The semiconductor target is a pair of silicon diodes, each consisting of two metallic electrodes with a pn junction under the top contact. Amplifier operation is based on the fact that a modulated electron beam can control the current in a reverse-biased semiconductor junction. Abbreviated EBS amplifier.

electron-bombardment-induced conductivity—In a multimode display storage tube, a process by which the image on the surface of the cathode-ray tube is erased by the use of an electron gun.

electron charge — Also called elementary charge. The charge of a single electron. Its value is $1{,}60219 \times 10^{-19}$ coulomb. The fundamental unit of electrical charge.

electron-coupled oscillator — Abbreviated eco. A circuit using a multigrid tube in which the cathode and two grids operate as a conventional oscillator and the electron stream couples the plate-circuit load to the oscillator.

electron coupling—In vacuum (principally multigrid) tubes, the transfer of energy between electrodes as electrons leave one and go to the other.

electron diffraction camera—A special evacuated camera equipped with means for holding a specimen and bombarding it with a sharply focused beam of electrons. A cylindrical film placed around the specimen records the electrons which may be scattered or diffracted by it.

electron drift—The movement of electrons in a definite direction through a conductor, as opposed to the haphazard transfer of energy from one electron to another by collision.

electronegative—Having an electric polarity that is negative.

electron emission—The freeing of electrons into space from the surface of a body under the influence of heat, light, impact, chemical disintegration, or a potential difference.

electron flow—The movement of electrons from a negative to a positive point in a metal or other conductor, or from a negative to a positive electrode through a liquid, gas, or vacuum.

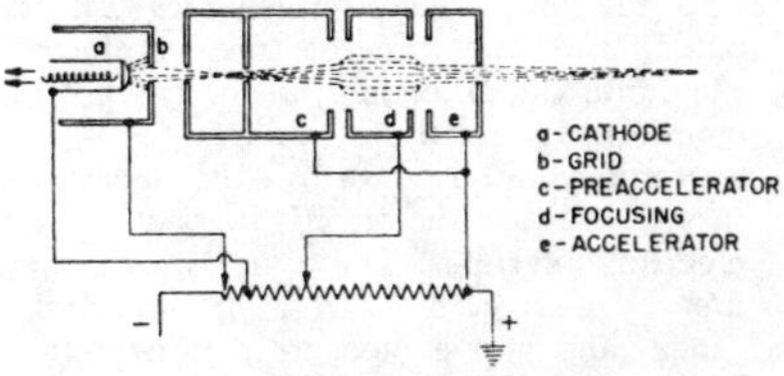

Electron gun.

electron gun—An electrode structure which produces and may control focus and may deflect and converge one or more electron beams.

electronic—1. Pertaining to that branch of science which deals with the motion, emission, and behavior of currents of free electrons, especially in vacuum, gas, or phototubes and special conductors or semiconductors. This is contrasted with electric, which pertains to the flow of large currents in metal conductors. 2. Of or pertaining to devices, circuits, or systems using the principle of electron flow through a conductor—for example, *electronic* control, equipment, instrument, circuit.

electronic autopilot—An arrangement of gyroscopes, electronic amplifiers, and servomotors for detecting deviations in the flight of an aircraft and applying the required corrections directly to its control cables.

electronic balance—Weighing balance that uses forces produced by known currents to balance unknown currents and, thereby, unknown weights very accurately to within parts of a microgram.

electronic "bug" — A keying system which converts the Morse signals from a hand key into correctly proportioned and spaced dots and dashes.

electronic calculator—Electronic device for arithmetic and logarithmic computations; may also include digital printer and computer.

electronic camouflage — Use of electronic

means, or exploitation of electronic characteristics, to reduce, submerge, or eliminate the radar-echoing properties of a target.

electronic circuit—A circuit containing one or more electron tubes, transistors, magnetic amplifiers, etc.

electronic commutator — A type of switch which provides a continuous switching or sampling of a number of circuits by means of a radial-beam electronic tube or electronic switching circuit.

electronic control — Also called electronic regulation. The control of a machine or condition by electronic devices.

electronic counter—An instrument capable of counting up to several million electrical pulses per second.

electronic counter-countermeasures—Efforts made to ensure effective use of electromagnetic radiation in spite of the use of countermeasures by an enemy.

electronic countermeasures — All measures taken to reduce the effectiveness of enemy electronic systems. There are two distinct areas: passive measures, or reconnaissance, and active measures, or jamming. Abbreviated ECM.

electronic countermeasures control—1. Collection and sorting of large quantities of data for the purpose of measuring and defining radar signals. 2. Examination of the data received in order to determine selection and switching of countermeasure devices with little or no time delay.

electronic coupling—The method of coupling electrical energy from one circuit to another through the electron stream in a vacuum tube.

electronic crowbar—An electronic switching device generally used in a power supply to divert a fault current from more delicate components until a fuse, circuit breaker, or the like has time to respond.

electronic data processing—Operations on data carried out mainly by electronic equipment. Abbreviated EDP.

electronic data-processing center—A place in which is kept automatically operated equipment, including computers, designed to simplify the interpretation and use of data gathered by instrumentation installations or information-collection agencies. Abbreviated EDP center.

electronic data-processing machine—Abbreviated EDPM. A machine or its device and attachments used primarily in or with an electronic data-processing system.

electronic data-processing system—Any machine or group of automatically intercommunicating machines capable of entering, receiving, sorting, classifying, computing and/or recording alphabetical of numerical accounting or statistical data (or all three) without intermediate use of tabulating cards.

electronic deception—Deliberate radiation, reradiation, alteration, absorption, or reflection of electromagnetic radiations in a manner intended to cause the enemy to obtain misleading data or false indications from his electronic equipment.

electronic device—A device in which conduction is principally by the movement of electrons through a vacuum, gas, or semiconductor.

electronic differential analyzer—A form of analog computer using interconnected electronic integrators to solve differential equations.

electronic digital computer — A machine which uses electronic circuitry in the main computing element to perform arithmetic and logical operations on digital data (i.e., data represented by numbers or alphabetic symbols) automatically, by means of an internally stored program of machine instruction. Such devices are distinguished from calculators, on which the sequence of instructions is externally stored and is impressed manually (desk calculators) or from tape or cards (card-programmed calculators).

electronic efficiency—The ratio of (a) the power at the desired frequency delivered by the electron stream to the oscillator or amplifier circuit to (b) the direct power supplied to the stream.

electronic flash—1. Also called strobe. The firing of special light-producing, high-voltage, gas-filled glass tubes with a high instantaneous surge of current furnished by a capacitor or bank of capacitors which have been charged from a high-voltage source (usually 450 volts or higher). 2. A device which upon command produces a pulse of luminous energy caused by a discharge of electrical energy through a gas. The term usually implies the use of a flashtube and associated power source and trigger circuit.

electronic flash tube—*See* Flash Tube.

electronic flash units—A small xenon-filled tube with metal electrodes fused into the ends. The gas flashes brilliantly when a capacitor is discharged through the tube.

electronic frequency synthesizer—A device which generates two or more selectable

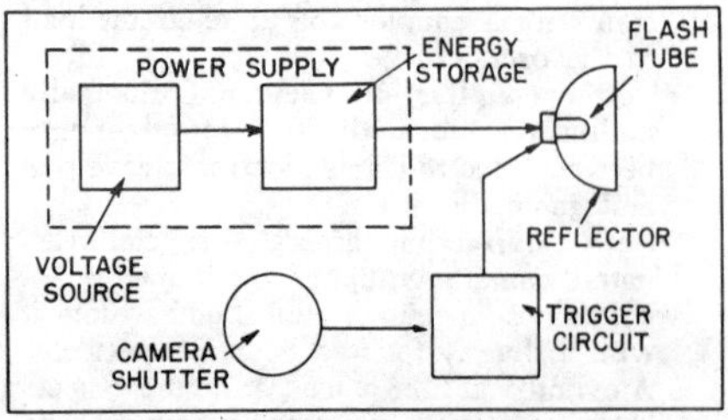

Electronic flash unit.

frequencies from one or more fixed-frequency sources.

electronic gate—A device in which diodes and/or transistors provide input-output relations that correspond to a Boolean-algebra function (AND, OR, etc.).

electronic heating — Also called high-frequency heating. A method of heating a material by inducing a high-frequency current in it or having the material act as the dielectric between two plates charged with a high-frequency current.

electronic industries — Industrial organizations engaged in the manufacture, design, development, and/or substantial assembly of electronic equipment, systems, assemblies, or the components thereof.

Electronic Industries Association (EIA)—A trade association of the electronics industry. Some of its functions are the formulation of technical standards, dissemination of marketing data, and the maintenance of contact with government agencies in matters relating to the electronics industry. The association was originally known as the Radio Manufacturers Association (RMA), and later as the Radio-Electronics-Television Manufacturers Association (RETMA).

electronic instrument — Any instrument which depends for its operation on the action of either one or more electron devices.

electronic intelligence — The technical and intelligence information derived from foreign noncommunications electromagnetic radiations emanating from other than nuclear detonations or radioactive sources.

electronic interference—Electrical or electromagnetic disturbances that result in undesired response in electronic equipment.

electronic jamming—Intentional radiation, reradiation, or reflection of electromagnetic energy for the purpose of reducing the effectiveness of enemy electromagnetic devices.

electronic keying — A method of keying whereby the dots and dashes are produced solely by electronic means.

electronic line scanning — Facsimile scanning in which a spot on a cathode-ray tube moves across the copy electronically while the record sheet or subject copy is moved mechanically in a perpendicular direction.

electronic microphone—A device which depends for its operation on the generation of a voltage by the motion of one of the electrodes in a special electron tube.

electronic mine detector—*See* Mine Detector.

electronic multimeter—A device employing the characteristics of an electron-tube circuit for the measurement of electrical quantities, at least one of which is voltage or current, or a single calibrated scale.

electronic music—The electronic generation and processing of audio signals or the electronic processing of natural sound, and the manipulation and arrangement of these signals via tape recorders into a finished musical composition.

electronic organ—The electronic counterpart of the pipe organ. All tones and tone variations such as vibrato, tremolo, etc., are produced by electronic circuits instead of by pipes.

electronic pacemaker—An electrical device, usually with electrodes planted in the myocardium, that performs the pacing function in a diseased heart no longer capable of pacing itself. Electronic pacemakers can receive power from implanted batteries, radio frequency signals, biological energy sources, etc.

electronic packaging—The coating or surrounding of an electronic assembly with a dielectric compound.

electronic part—A basic circuit element that cannot be disassembled and still perform its intended function. Examples of electronic parts are capacitors, connectors, filters, resistors, switches, relays, transformers, crystals, electron tubes, and semiconductor devices.

electronic photometer—A photometer with a photocell, phototransistor or phototube for measuring the intensity of light. Also called photoelectric photometer.

electronic power supply — A circuit which transforms electrical input energy—alternating or direct current—into output energy—alternating or direct current. (Sources operating on rotating machine principles, or deriving electrical power from other energy forms such as batteries and solar cells, are excluded.) Supplies covered by this definition fall into one of four groups: 1. Ac in, dc out—most common supplies. 2. Ac in, ac out—line regulators, variable frequency supplies. 3. Dc in, dc out—converters. 4. Dc in, ac out—inverters.

electronic products—Materials, parts, components, subassemblies, and equipment which employ the principles of electronics in performing their major functions. These products may be used as instruments and controls in communications, detection, amplification, computation, inspection, testing, measurement, operation, recording, analysis, and other functions employing electronic principles.

electronic profilometer—An electronic instrument for measuring surface roughness. The diamond-point stylus of a permanent-magnet dynamic pickup is moved over the surface being examined. The resultant variations in voltage are amplified, rectified, and measured with a meter calibrated to read directly in microinches of deviation from smoothness.

electronic raster scanning – Scanning by electronic means so that substantially uniform coverage of an area is provided by a predetermined pattern of scanning lines.

electronic reconnaissance–Search for electromagnetic radiations to determine their existence, source, and pertinent characteristics for electronic warfare purposes.

electronic rectifier–A rectifier using electron tubes or equivalent semiconductor elements as rectifying elements.

electronic regulation–*See* Electronic Control.

electronic relay–An electronic circuit that provides the functional equivalent of a relay, but has no moving parts.

electronics–1. The field of science and engineering concerned with the behavior of electrons in devices and the utilization of such devices. 2. Of or pertaining to the field of electronics, such as electronics engineer, course, laboratory, committee. 3. Name given to that branch of electrical engineering which deals with devices whose operation depends upon the movement of electrons in space as opposed to the movement of electrons in liquids or solid conductors, e.g., radio tubes, photoelectric cells, etc. It includes the study of radio, radar, television, sound films, and control of industrial processes.

electronic search reconnaissance–The determination of the presence, source, and significant characteristics of electromagnetic radiations.

electronic security – Protection resulting from measures designed to deny to unauthorized persons information of value that might be obtained by interception and analysis of noncommunications electromagnetic radiations.

electronic sky screen equipment–An electronic device for indicating the departure of a missile from a predetermined trajectory.

electronic sphygmomanometer–Device that measures and/or records blood pressure electronically.

electronic stethoscope–An electronic amplifier of sounds within a body. Its selective controls permit tuning for low heart tones or high pulmonary tones. It has an auxiliary output for recording or viewing audio patterns.

electronic stimulator–A device for applying electronic pulses or signals to activate muscles, or to identify nerves, or for muscular therapy, etc.

electronic surge arrestor–A device used to switch high-energy surges to ground so as to reduce the transient energy to a level that is safe for secondary protectors (e.g., zener diodes, silicon rectifiers, etc.).

electronic switch–1. A circuit element causing a start and stop action or a switching action electronically, usually at high speeds. 2. An electronic circuit used to perform the function of a high-speed switch. Applications include switching a cathode-ray oscilloscope back and forth between two inputs at such high speed that both input waveforms appear simultaneously on the screen.

electronic switching system – A telephone switching system in which is used a computer with a storage containing program switching logic. The output of the computer actuates reed or electronic switches that establish telephone connections automatically. Abbreviated ESS.

electronic thermal conductivity – The part of the thermal conductivity due to the transfer of thermal energy by means of electrons and holes.

electronic timer–1. A synchronizer, pulse generator, modulator, or keyer that originates a series of continuous control pulses at an unvarying repetition rate known as the pulse-recurrence frequency. 2. A timer using electronic circuits (either tube or transistor type) to control a time period, in place of a motor or other means.

electronic tube relay–A relay that employs electronic tubes as components.

electronic tuning–1. Altering the frequency of a reflex klystron oscillator by changing the repeller voltage. 2. Frequency changing in a transmitter or receiver by changing a control voltage rather than circuit components.

electronic voltmeter–Also called vacuum-tube voltmeter. A voltmeter which utilizes the rectifying and amplifying properties of electron tubes and their circuits to secure such characteristics as high input impedance, wide frequency range, peak-to-peak indications, etc.

electronic volt-ohmmeter–A device employing the characteristics of an electron-tube circuit for the measurement of voltage and resistance on a single calibrated scale.

electronic warfare–Military usage of electronics to reduce an enemy's effective use of radiated electromagnetic energy and to ensure our own effective use.

electronic waveform synthesizer–An instrument using electron devices to generate an electrical signal of a desired waveform.

electronic watch–A timepiece in which a battery replaces the mainspring, and semiconductor elements replace the mechanical switching-contact arrangement.

electron image tube–1. A cathode-ray tube having a photoemissive mosaic upon which an optical image is projected, and an electron gun to scan the mosaic and convert the optical image into corresponding electrical current. 2. A cathode ray tube that increases the brightness or size of an image, or forms a visible image from invisible radiation. The focal plane for the optical image is a large, light-sensitive cold-cathode. The emission from the cathode is first accelerated through a suitable lens system and then strikes a fluorescent screen, where an image is formed that is an enlarged and brightened reproduction of the original image.

electron lens—1. The convergence of the electrons into a narrow beam in a cathode-ray tube by deflecting them electromagnetically or electrostatically. So called because its action is analogous to that of an optical lens. 2. A system of deflecting electrodes or coils designed to produce an electric field that influences a beam of electrons in the same manner that a lens affects a light beam.

electron microscope—An instrument which uses an electron beam to penetrate thin samples of a material. It is possible to magnify images of the material on a screen or film up to 350,000 times.

electron multiplier — A vacuum tube in which electrons liberated from a photosensitive cathode are attracted successively to a series of electrodes called dynodes. In doing so, each electron liberates others by secondary emission and thereby greatly increases the number of electrons flowing in the tube.

electron-multiplier section—A section of an electron tube in which an electron current is amplified by one or more successive dynode stages.

electron optics—1. The branch of electronics concerned with the behavior of the electron beam under the influence of electrostatic and electromagnetic forces. 2. The control of free electron movement through the use of electric or magnetic fields, and use of this electron movement in research investigation of electronic diffraction phenomena, directly analogous to the control of light through the uses of lenses.

electron-pair bond—A valence bond formed by two electrons, one from each of two adjacent atoms.

electron paramagnetic resonance—A condition in which a paramagnetic solid subjected to two magnetic fields, one of which is fixed and the other normal to the first and varying at the resonance frequency, emits electromagnetic radiation associated with changes in the magnetic quantum number of the electrons.

electron-ray tube—1. Also called a "magic eye." A tube which indicates visibly on a fluorescent target the effects of changes in control-grid voltage applied to the tube. Used as a tuning indicator in receivers. 2. A type of recording-level indicator using a luminous display in a special tube. The display is typically like an "eye" with a keyhole in the middle, and maximum recording level corresponds to the closing-up of a slot at the bottom of the keyhole (largely superseded by meters in current-model recorders).

electron scanning — 1. The moving of an electron beam back and forth and/or up and down by deflecting the beam electromagnetically or electrostatically. 2. A deflection of a beam of electrons, at regular intervals, across a crt screen, according to a definite pattern.

electron spin—The twirling motion of an electron, independent of any orbital motion.

electron-stream potential—The time average of the difference in potential between a point in an electron stream and the electron-emitting surface.

electron-stream transmission efficiency — With respect to an electrode through which an electron stream passes, the ratio of the average stream current through the electrode to the stream current approaching the electrode. (In connection with multitransit tubes, the electron stream is considered to include only those electrons approaching the electrode for the first time.)

electron telescope—An apparatus for seeing through haze and fog. An infrared image is formed optically on the photoemissive mosaic of an electron-image tube and then made visible by the tube.

electron transit time—The time required for electrons to travel between two electrodes in a vacuum tube. This time is extremely important in tubes designed for ultrahigh frequencies.

electron tubes—Devices used to control the flow of electrons. They may be either gas filled, or partially or fully evacuated (vacuum). Common tubes include vacuum tubes, cathode ray tubes, phototubes, mercury vapor tubes, thyratrons and microwave tubes.

electron-tube static characteristic—The relationship between two variables of an electron tube, such as the voltage and current of an electrode with all other variables maintained constant.

electron unit—The unit of charge (nega-

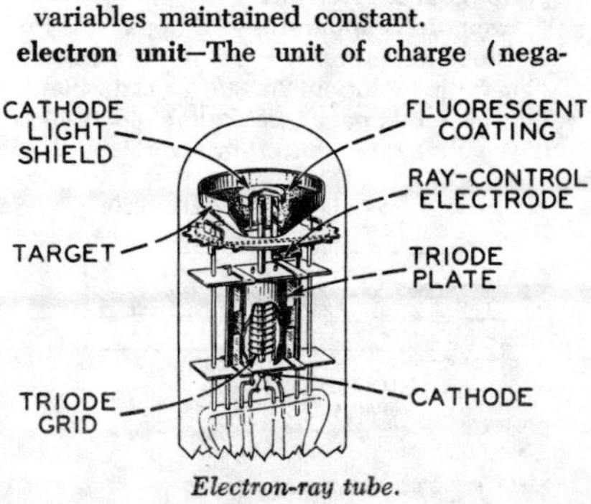

Electron-ray tube.

tive or positive) equal to the charge on an electron.

electronvolt—Abbreviated ev or eV. The amount of kinetic energy gained by an electron when it is accelerated through an electric potential difference of 1 volt. It is equivalent to 1.603×10^{-12} erg. It is a unit of energy or work, not of voltage.

electron-wave tube – An electron tube in which streams of electrons having different velocities interact and cause a progressive change in signal modulation along their length.

electro-oculography—Recording and interpretation of the voltages that accompany eye movements. Eye-position voltages from electrodes placed on the skin near the eye are amplified and applied to a strip-chart recorder.

electro-optical transistor—A transistor capable of responding in nanoseconds to both light and electrical signals.

electro-optic radar – A radar system in which electro-optic instead of microwave techniques and equipment are used to perform the acquisition and tracking operation.

electropad—The part of an electrocardiograph body electrode that makes contact with the skin.

electrophonic effect—The sensation of hearing produced when an alternating current of suitable frequency and magnitude from an external source is passed through an animal or human body.

electrophoresis—1. The movement of particles or ions in solution caused by applying an electric field, as reported by O. Lodge in 1886. 2. The migration of colloidal particles under the influence of an applied electrical field. A colloidal particle, such as a protein molecule, has large numbers of positive and negative radicals that act as if they were on the surface. Thus, since protein molecules carry electric charges, they will migrate when subjected to an electric field. The fractional nature of the net charge makes possible a wide variety of electrophoretic patterns at a given pH.

electrophoresis apparatus—An apparatus for causing migration of charged particles (ions) in solution in an electric field. Types include paper, cascading electrodes, high voltage, gel, and thin layer.

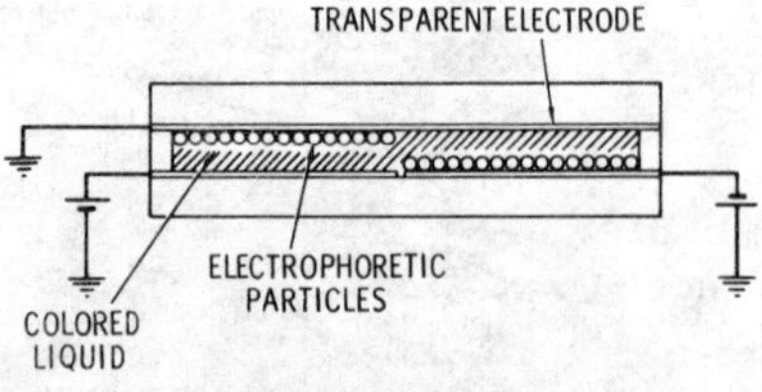

Electrophoretic display.

electrophoresis scanner—An instrument for reading bands on paper strips or gel, for the purpose of measuring particle movement due to electrophoresis.

electrophoretic display—A reflective display that offers a wide choice of colors and has a short-to-medium-term memory that consumes no power. The heart of the display is a suspension of charged pigment particles in a liquid of another color. The suspension, a layer typically 50 micrometers thick, is sandwiched between a pair of electrodes, one of which is transparent. When direct current of the right polarity is applied to the electrodes, the particles are pulled toward the transparent electrode thus displacing the contrasting liquid and showing their own coloration. When the polarity is reversed, they move to the other electrode and are hidden by the liquid.

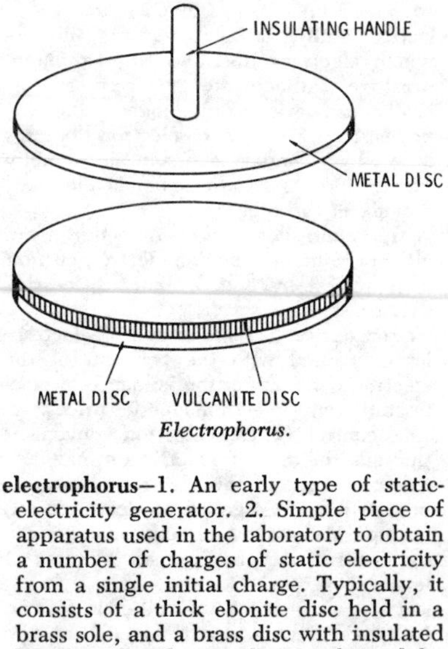

Electrophorus.

electrophorus—1. An early type of static-electricity generator. 2. Simple piece of apparatus used in the laboratory to obtain a number of charges of static electricity from a single initial charge. Typically, it consists of a thick ebonite disc held in a brass sole, and a brass disc with insulated handle. The ebonite disc is charged by rubbing with fur and the metal disc is brought near and allowed to pick up an induced charge which can be lifted and conveyed where required. 3. A device in which the electric field of an object that has been electrified by friction is used to induce charges in conductors.

electrophotographic process – The process in which images are formed by various electrical and photographic means. Examples are processes employing selenium-coated drums or zinc-oxide–coated paper.

electrophotography – A term referring to photographic process where electrical energy is used to make materials sensitive to light.

electrophotometer—An instrument using a

photoelectric sensor for colorimetric determinations.

electrophysiology—The science of physiology as related to electric reactions of the body.

electroplaques — Individual electricity-producing cells in eels and other electric fishes connected in series-parallel arrays, like miniature elements of a battery. They are usually thin waferlike cells, the two surfaces of which differ markedly.

electroplate—1. To deposit a metal on the surface of certain materials by electrolysis. 2. To effect the transfer of one metal to another by electrolysis.

electroplating—The electrodeposition of an adherent metal coating on a conductive object for protection, decoration, or other purposes. The object to be plated is placed in an electrolyte and connected to one terminal of a dc voltage source. The metal to be deposited is similarly immersed and connected to the other terminal. Ions of the metal provide transfer to the metal as they make up the current between the electrodes.

electropolishing—The process of producing a smooth, lustrous surface on a metal by making it the anode in an electrolytic solution and preferentially dissolving the minute protuberances.

electrorefining—The removal of impurities from a metal by electrolysis.

electroretinograph—An instrument for measuring the electrical response of the human retina to light stimulation.

electroretinography — Recording and interpretation of the voltage generated by the retina of the eye. An electrode fitted to a plastic contact lens is used to pick up voltage from the surface of the eyeball.

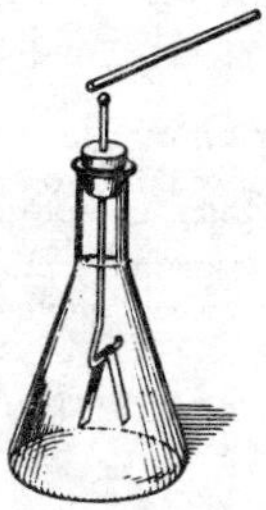

Electroscope.

electroscope — An electrostatic instrument for measuring a potential difference or an electric charge by means of the mechanical force exerted between electrically charged surfaces.

electrosection—A surgical cutting technique that makes use of an rf arc.

electrosensitive recording—The passage of electric current into a sheet of sensitive paper to produce a permanent record.

electroshock—A state of shock produced by passing an electric current through the brain. It is useful in the treatment of certain mental disorders.

electrospinograph—A device for detecting and recording electric signals of the spinal cord.

electrostatic—1. Pertaining to static electricity—i.e., electricity, or an electric charge, at rest. 2. Applied to loudspeakers and microphones (condenser type). An electrostatic force is used to activate the diaphragm. The charged diaphragm is suspended between two perforated plates. As an ac signal is applied to the outer plates, the diaphragm vibrates. 3. A form of electrical energy which has the capability of attracting and holding small particles having an opposite electrical charge.

electrostatic actuator—An apparatus comprising an auxiliary external electrode which permits known electrostatic forces to be applied to the diaphragm of a microphone for the purpose of obtaining a primary calibration.

electrostatic capacitor — Two conducting electrodes separated by an insulating material such as air, ceramic, mica, gas, paper, plastic film, or glass. These are generally high-impedance devices.

electrostatic charge — An electric charge stored in a capacitor or on the surface of an insulated object.

electrostatic component — The portion of radiation due to electrostatic fields.

electrostatic-convergence principle — The principle of electron-beam convergence through use of an electrostatic field.

electrostatic copier—A type of copier which employs the principles of photoconductivity and electrostatic attraction.

electrostatic coupling—Method of coupling by which charges on one surface influence those on another through capacitive action.

electrostatic deflection—The method of deflecting an electron beam by passing it between charged plates mounted inside a cathode-ray tube.

electrostatic electrography—The branch of electrostatography which produces a visible record by employing an insulating medium to form latent electrostatic patterns without the aid of electromagnetic radiation.

electrostatic electrophotography — That branch of electrostatography which produces a visible record by employing a photoresponsive medium to form latent electrostatic images with the aid of electromagnetic radiation.

electrostatic energy—The energy contained in electricity at rest, such as in the charge of a capacitor.

electrostatic field—The vector force field set up in the vicinity of nonmoving electrical

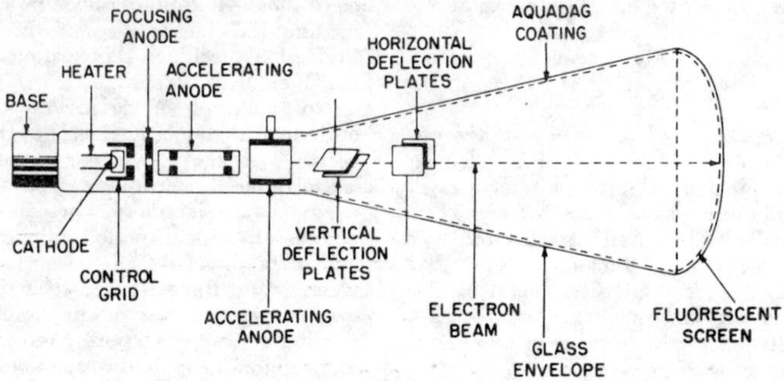

Electrostatic deflection.

charges. The strength of this static field at a point is defined as the force per unit charge on a stationary positive test charge, provided the test charge is so small that it does not disturb the original charge distribution.

electrostatic flux—The electrostatic lines of force existing between bodies at different potentials.

electrostatic focusing—The focusing of an electron beam by the action of an electric field.

electrostatic galvanometer — Galvanometer operated by the effects of two electric charges on each other.

electrostatic generator — A device for the production of electric charges by electrostatic action.

electrostatic headphones — A device held against the ear that reproduces incoming electrical signals as sound. It relies on changes in electrical charge across a diaphragm stretched between two perforated, polarized plates. All parts of the diaphragm experience equal force and the sound is inherently more linear.

electrostatic induction—1. The process of charging an object electrically by bringing it near a charged object. 2. Capacitive induction of interfering signals over an air gap separating an instrument (e.g., from its wiring or housing).

electrostatic instrument — An instrument which depends for its operation on the attraction and repulsion between electrically charged bodies.

electrostatic latent image — In an electrostatic copier, the invisible image formed on the zinc oxide coated paper by the action of light.

electrostatic loudspeaker—Loudspeaker in which the mechanical forces are produced by the action of electrostatic fields.

electrostatic memory — Also called electrostatic storage. A memory device in which information is retained by an electrostatic charge. A special type of cathode-ray tube is usually employed, together with associated circuitry.

electrostatic memory tube—Also called storage tube. An electron tube in which information is retained by electric charges.

electrostatic microphone—1. Also called capacitor microphone or condenser microphone. A microphone which contains a metal plate and a thin metal diaphragm set close together. A polarizing voltage is applied to the plates. The capacitance of the microphone is thus affected by movement of the diaphragm from air pressure waves. 2. A microphone whose transduction principle is based on the varying electrical charge across a sound-modulated capacitor.

electrostatic precipitation—The process of removing smoke, dust, and other particles from the air by charging them so that they can be attracted to and collected by a properly polarized electrode.

electrostatic process — A reproduction method in which image formation depends on electrical rather than chemical changes induced by light.

electrostatic recording — Recording by means of a signal-controlled electrostatic field.

electrostatic relay—A relay in which two or more conductors that are separated by insulating material move because of the mutual attraction or repulsion produced by electric charges applied to the conductors.

electrostatics—The branch of physics concerned with electricity at rest.

electrostatic separator — An apparatus in which a finely pulverized mixture of the materials to be separated is passed through the powerful electrostatic field between two electrodes.

electrostatic series — *See* Triboelectric Series.

electrostatic shield—A shield which prevents electrostatic coupling between circuits, but permits electromagnetic coupling.

electrostatic speaker—Also called capacitor or condenser speaker. A speaker in which

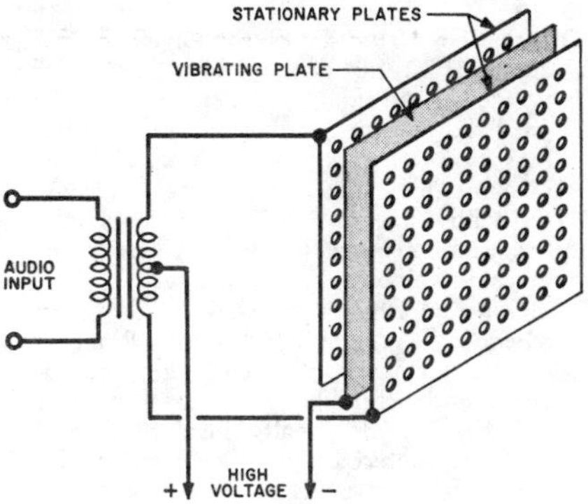

Electrostatic speaker.

the mechanical forces are produced by the action of electrostatic fields.

electrostatic storage – The storage of changeable information in the form of charged or uncharged areas usually on the screen of a cathode-ray tube.

electrostatic transducer–A transducer that consists of a capacitor, at least one plate of which can be set into vibration. Its operation depends on the interaction between its electric field and a change in its electrostatic capacity.

electrostatic tweeter – A speaker with a movable flat metal diaphragm and a nonmovable metal electrode capable of reproducing high audio frequencies. The diaphragm is driven by the varying high voltages applied across it and the electrode.

electrostatic unit–An electric unit based primarily on the dynamic interaction of electric charges. Defined as a charge which, if concentrated on a small sphere, would repel with a force of one dyne a similar charge one centimeter away in a vacuum.

electrostatic voltmeter – A voltmeter depending for its action on electrostatic forces. Its scale is usually graduated in volts or kilovolts.

electrostatography–The process of recording and reproducing visible patterns by the formation and utilization of latent electrostatic charge patterns.

electrostriction–A mechanical deformation caused by the application of an electric field to any dielectric material. The deformation is proportional to the square of the applied field. This phenomenon results from the induced dipole movement caused by the applied field, resulting in the mechanical distortion.

electrostrictive effect–The elastic deformation of a dielectric by an electrostatic field.

electrostrictive relay–A relay the operation of which is produced by an electrostrictive-dielectric actuator.

electrosurgery–The surgical use of electricity in such applications as dissection, coagulation, laser heating, laser welding, diathermy, desiccation of tumors, and hemostasis.

electrosurgical unit–An rf generator the output of which is applied to a blade or wire loop used instead of a conventional scalpel for surgical incision or excision.

electrotape–An electronic distance-measuring device.

electrotherapeutics–*See* Electrotherapy.

electrotherapy–1. Also known as electrotherapeutics. The medical science or use of electricity to treat a disease or ailment. 2. Equipment for applying electric current to the body for massage or heat treatment.

electrotherapy apparatus – Equipment for applying electric current to the body for massage or heat treatment.

electrothermal–The heating effect of electric current, or the electric current produced by heat.

electrothermal expansion element–An actuating element consisting of a wire strip or other shape and having a high coefficient of thermal expansion.

electrothermal recorder – A recorder in which heat produces the image on the recording medium in response to the received signals.

electrothermal recording–*See* Electrothermal Recorder.

electrothermic instrument – An instrument which depends for its operation on the heating effect of a current. Examples are the thermocouple, bolometric, hot-wire, and hot-strip instruments.

electrothermics–The branch of science concerned with the direct transformation of electric energy into heat.

electrowinning–The process by which metals are recovered from a solution by electrolysis.

element–1. One of the 104 known chemical substances that cannot be divided into simpler substances by chemical means. A substance the atoms of which all have the same atomic number (e.g., hydrogen, lead, uranium). 2. In a computer, the portion or subassembly which constitutes the means of accomplishing one particular function, such as the arithmetic element. 3. Any electrical device (such as an inductor, resistor, capacitor generator, line, or electron tube) with terminals at which it may be connected directly to other electrical devices. 4. The dot or dash of an International Morse character. 5. A radiator, either active or parasitic, that is part of an antenna. 6. The smallest portion of a televised picture that still retains the characteristics of the picture. 7. A portion of a part that cannot be renewed without destruction of the part.

8. A part of an integrated circuit which contributes directly to its electrical characteristics. An active element exhibits gain, as a transistor; a passive element does not have gain, such as a resistor or capacitor.

elemental area—*See* Picture Element, 2.

elemental semiconductor—A semiconductor containing only one element in the undoped state.

elementary charge—A natural unit or quantum into which both positive and negative charges appear to be subdivided. It is the charge on a single electron and has a value of about 4.77×10^{-10} electrostatic units.

element error rate—The ratio of the number of elements incorrectly received to the total number of elements sent.

elevation—The angular position perpendicular to the earth's surface.

elevation-position indicator — A radar display which simultaneously shows angular elevation and slant range of detected objects.

elevator leveling control — A positioning control used to align the platform of an elevator with the floor level of the building. Metal vanes are mounted in the elevator shaft at each floor level, and an oscillator is mounted on the elevator car. When the elevator is properly leveled, the metal vane is between the plate and the grid coils of the oscillator. A relay connected in the oscillator circuit now energizes. The contacts of this relay are connected in the motor-control circuit of the elevator so that the elevator stops in alignment with the floor level.

elf — Abbreviation for extremely low frequency.

eliminator—Also called a battery eliminator. A device operated from an ac or dc power line and used for supplying direct current and voltage to a battery-operated circuit.

E-lines—Contour lines of constant electrostatic field strength with respect to some reference base.

elliptically polarized wave—An electromagnetic wave the electric intensity vector of which describes an ellipse at one point.

elliptical polarization — Polarization in which the wave vector rotates in an elliptical orbit about a point.

elliptical stylus—*See* Biradial Stylus.

elliptic function—A mathematical function employed in obtaining the squarest possible amplitude response of a filter with a given number of circuit elements. The elliptic function has a Tchebychev response in both the passband and the stopband. The phase response and transient response of an elliptic-function filter are poorer than for any of the classical transfer functions.

elongation—Extension of the envelope of a signal as a result of the delayed arrival of certain of the multipath components.

e/m—The ratio of the electric charge to the mass for particles such as electrons and positive rays. For slow-moving electrons, the value of e/m is 1.77×10^{4} coulombs per gram. The value decreases with increasing velocity, however, because of an increase in effective mass.

embedding—A general term for all methods of surrounding or enclosing components and assemblies with a substantial thickness of electrically insulating solid or foam material that substantially fills the voids or interstices between parts.

embedment—The complete encasement of a part or assembly to some uniform external shape. A relatively large volume of a complete package consists of the embedment material.

embossed-foil printed circuits — A printed circuit formed by indenting the desired pattern of metal foil into an insulating base, and then mechanically removing the remaining unwanted raised portion.

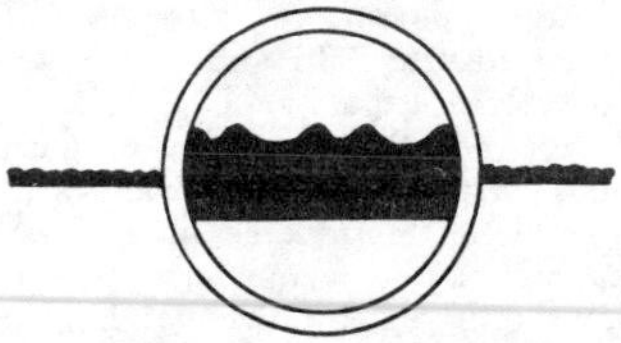

Embossed-groove recording.

embossed-groove recording — A method of recording sounds on discs or film strips by embossing sound grooves with a blunt stylus rather than by cutting into them with a sharp stylus. Embossing throws the material up in furrows on each side of the sound groove without actually removing any of the material in the disc or strip.

embossing stylus—A recording stylus with a rounded tip which forms a groove in the recording medium by merely displacing the material instead of removing it completely.

emc — Abbreviation for electromagnetic compatibility.

emergency communication — The transmission or reception of distress, alarm, urgent, or safety signals or messages relating to the safety of life or property, or the occasional operation of equipment to determine whether it is in working condition

emergency radio channel — Any radio frequency reserved for emergency use, particularly for distress signals.

emergency service — The radiocommunication service carrier used for emergency purposes.

emf–Abbreviation for electromotive force.

emg–Abbreviation for electromyography.

emi–Abbreviation for electromagnetic interference.

emission–1. The waves radiated into space by a transmitter. 2. The ejection of electrons from the surface of a material (under the influence of heat, for example).

emission characteristic – The relationship between the emission and the factor controlling it, such as temperature, voltage, or current of the filament or heater. This relationship is usually shown on a graph.

emission current–The current produced in the plate circuit of a tube when all the electrons emitted by the cathode pass to the plate.

emission efficiency – The rating of a hot cathode. Expressed in milliamperes per watt.

emission power – The time rate at which radiant energy is given off in all directions per unit surface area of a radiating body at a given temperature.

emission spectrum–The spectrum showing the radiation emitted by a substance, such as the light emitted by a metal when placed in an electric arc, or the light emitted by an incandescent filament.

emission types–The classification of modes of radio transmission adopted by international agreement. The am designations are:

Type A0. Unmodulated continuous-wave transmission.

Type A1. Telegraphy or pure continuous waves.

Type A2. Modulated telegraphy.

Type A3. Telephony.

Type A4. Facsimile.

Type A5. Television.

emission-type tube tester–Also called an English-reading tube tester. A tube tester for checking the electron emission from the filament or cathode. The indicating meter is generally calibrated to read "Good" or "Bad." The tester connects all elements such as the plate and screen, suppressor, and control grids together and uses them as an anode.

emission velocity–The initial velocity at which electrons emerge from the surface of a cathode, ranging from zero up to a few volts (attained by very few electrons). This effect accounts for the existence of virtual cathodes, and also for the shape of the cutoff region of plate current.

emissivity–The ratio of the radiant energy emitted by a radiation source to the radiant energy of a perfect (blackbody) radiator having the same area and at the same temperature and conditions.

emitron–A cathode-ray tube developed in England.

emitron camera–A British television camera tube resembling an iconoscope.

emittance–The power per unit area radiated by a source of energy.

emitter–1. An electrode within a transistor from which carriers are usually minority carriers; when they are majority carriers, the emitter is referred to as a majority emitter. 2. Of a transistor, a region from which charge carriers that are minority carriers in the base are injected into the base. 3. A device used on a punched-card machine to give timed pulses at regular intervals during the machine cycle.

emitter-base and collector-base junction–In a semiconductor the region where the base and collector, and the emitter and base meet. These junctions are defined on the surface of the chip as an oxide step.

emitter bias–The bias voltage applied to the emitter of a transistor.

emitter-coupled logic–Nonsaturated bipolar logic in which the emitters of the input logic transistors are coupled to the emitter of a reference transistor. The basic gate circuit employs a long-tailed pair. Abbreviated ECL.

emitter current–The direct current flowing in the emitter circuit of a transistor.

emitter follower–A transistor amplifier circuit configuration analogous to a vacuum-tube follower. The circuit is characterized by relatively high input impedance, low output impedance, and a voltage gain of less than unity.

emitter junction–A semiconductor junction normally biased in the low-resistance direction so that minority carriers are injected into the interelectrode region.

emitter resistance–The resistance in series with the emitter lead in the common-T equivalent circuit of a transistor.

emitter semiconductor–A junction normally biased in the low-resistance direction to inject minority carriers into an interelectrode region.

emitter voltage–The voltage between the emitter terminal and a reference point.

emp – Abbreviation for electromagnetic pulse.

emphasizer–A circuit or device that provides an intentional increase in signal strength at certain audio frequencies.

empire cloth–A cotton or linen cloth coated with varnish and used as insulation on coils and other parts of electrical equipment.

empirical–Based on actual measurement, observation, or experience, as opposed to theoretical determination.

emr–Abbreviation for electromagnetic radiation.

emu – Abbreviation for electromagnetic unit.

enable–To permit a circuit to be activated by the removal of a suppression signal.

enabling gate—A circuit which determines the start and length of a generated pulse.

enabling pulse—A pulse that opens a normally closed electric gate, or otherwise permits occurrence of an operation for which it is a necessary but not sufficient condition.

enameled wire—Wire coated with a layer of baked-enamel insulation.

encapsulating—Coating, by dipping, brushing, spreading, or spraying an electronic component or assembly. An encapsulated unit usually retains its original geometry.

encapsulating material—A composition primarily adapted for use on or around an electrical device to provide protection from the surrounding environment.

encapsulation—1. A protective coating of cured plastic placed around delicate electronic components and assemblies. It is identical to potting, except the cured plastic is removed from the mold. The plastic therefore determines the color and surface hardness of the finished part. The molds may be made of any suitable material. 2. An embedding process using removable molds or other techniques in which the insulating material forms the outer surfaces of the finished unit.

encased control—A self-contained motor speed/torque control completely housed in an enclosure. Switching, indicating and adjusting devices are provided on the outside of the enclosure. Unit portability, safety and component protection are leading assets of this design.

encipher—*See* Encode.

enciphered facsimile communications — Communications in which security is provided by mixing pulses from a key generator with the output of a facsimile converter. Plain text is recovered at the receiving terminal by subtracting identical key pulses. Unauthorized persons are unable to reconstruct the plain text unless they have an identical key generator and they know the daily key setting.

enclosed relay—A relay in which both the coil and the contacts are protected from the environment.

enclosed switch—Switch having internal parts protected by a housing. Enclosed switch can be dust proof, moisture proof, oil or contamination proof, or hermetically sealed.

enclosure—An acoustically designed housing or structure for a loudspeaker; also any cabinet for a component, electrical or electronic device.

encode—Also called encipher. 1. To use a code, frequently one composed on binary numbers, to represent individual characters or groups of characters in a message. 2. To change from one digital code to another. If the codes are greatly different, the process usually is called conversion. 3. To substitute letters, numbers, or characters, usually with the intention of hiding the meaning of the message except from persons who know the encoding scheme.

encoder—1. An electromechanical device that can be attached to a shaft to produce a series of pulses to indicate shaft position; when the output is differentiated, the device is an accurate tachometer. (It is fundamentally oriented to digital rather than analog techniques.) An encoder contains a disc with a printed pattern; as the disc rotates, it makes and breaks a circuit. The more make-and-break cycles per revolution, the better the resolution. 2. A digital-to-analog converter. 3. A unit that produces coded combinations of outputs from discrete inputs. 4. The process of converting an event such as a switch closure into a form suitable for transmission over a communication channel. 5. *See* Code Converter. 6. Circuitry in a quadriphonic sound sysctem that, by matrixing in the recording process, turns four signals into two for inscribing, stereostyle, on each wall of the record groove.

encoding—1. Translation of information from an analog or other easily recognized form to a coded form without a significant loss of information. 2. The process of converting an event such as a switch closure into a form suitable for transmission over a communication channel.

end-around carry—A computer operation in which the carried information from the left-most bit is added to the results of the right-most addition. It is used for 1's complement and 9's complement arithmetic.

end-around shift—In a computer, the movement of characters from one end of the register to the other end of the same register.

end bell—An accessory which is similar to a cable clamp and attaches to the back of a plug or receptacle. It serves as an adapter for the rear of connectors. Some angular end bells have built-in cable clamps. Angular end bells up to 90° are available.

end bracket—*See* End Shield.

end-cell rectifier—A small trickle-charge rectifier for maintaining the voltage of storage-battery end cells.

end cells—Cells that can be switched in series with a storage battery to maintain the output voltage of the battery when it is not being charged.

end distortion—A shifting of the ends of all marking pulses of start-stop teletypewriter signals from their proper positions relative to the beginning of the start pulse.

end effect—The capacitive effect at the ends

of a half-wave antenna. To compensate for this effect, a dipole is cut slightly shorter than a half wave.

end-fire array—A linear or cylindrical antenna having its direction of maximum radiation parallel to the long axis of the array.

end instrument—A device connected to one terminal of a loop and capable of converting usable intelligence into electrical signals or vice versa. Includes all generating, signal-converting, and loop-terminating devices at the transmitting and/or receiving location.

end item—A combination of products, parts, and/or materials that is ready for its intended use.

end mark—In a computer, a code or signal used to indicate the termination of a unit of information.

endodyne reception—A British term applying to reception of unmodulated code signals. A vacuum-tube circuit having a local oscillator whose frequency is slightly different from that of the carrier signal. Thus a beat signal in the audio range is produced.

end-of-block signal—A symbol or indicator that defines the end of a block of data.

end-of-file mark—In a computer, a code instruction indicating that the last record of a file has been read.

end of message—The end of data to be transmitted. It can be indicated by a special control code, as in the ASCII code set; by an absence of data for a specified time interval; or by a particular sequence of block gaps and data, as is done on magnetic tape.

end of tape—The point on a magnetic tape at which the system or operator is given a warning that the physical end of the tape is approaching. It is approximately 25 feet from the actual end of the tape on ½-inch computer tape and approximately 50 feet from the halt marker on ¼-inch tape.

end-of-tape marker—A marker placed on a magnetic tape to indicate the end of the permissible recording area. It may be a photo-reflective strip, a transparent section of tape, or a particular bit pattern.

end-of-transmission card—The last card of a message, used to signal the end of a transmission. It contains the same data as the header card, plus additional information for traffic analysis.

end-on armature—Of a relay, an armature which moves in the direction of the core axis, with the pole face at the end of the core and perpendicular to this axis.

end-on directional antenna—A directional antenna which radiates chiefly toward the line on which the antenna elements are arranged.

endoradiograph—Equipment for X-ray examination of internal organs and cavities by means of radiopaque materials.

endoradiosonde—Also called radio pill. A device for detecting and transmitting physiological data from the gastrointestinal tract or other inaccessible body cavities.

endothermic—A term describing a chemical reaction in which heat is absorbed.

end point—The shaft positions immediately before the first and after the last measurable change(s) in output ratio after wiper continuity has been established, as the shaft of a precision potentiometer moves in a specified direction.

end-point control—Quality control by means of continuous, automatic analysis. In highly automatic processes, the final product is analyzed, and if any undesirable variations are detected, the control system automatically brings about the necessary changes.

end-point sensitivity—The algebraic difference in electrical output between the maximum and minimum value of the measurand over which an instrument is calibrated.

end-point voltage—The terminal voltage of a cell below which equipment connected to it will not operate or should not be operated.

end resistance—The resistance of a precision potentiometer measured between the wiper terminal and an end terminal, with the shaft positioned at the corresponding end point.

end-resistance offset — In potentiometers, the residual resistance between a terminal and the moving contact, at a position corresponding to full rotation against that terminal.

end scale value—The value of the actuating electrical quantity that corresponds to end scale indication of an instrument. When zero is not at the end or at the electrical center of the scale, the higher value is taken. Certain instruments such as power-factor meters, ohmmeters, etc., are necessarily excepted from this definition.

end setting—In a potentiometer, the minimum resistance that is measured between one end of a potentiometer and the wiper, with the wiper mechanically positioned at that end.

end shield — 1. Frequently called end bracket or end bell. In a motor housing, the part that supports the bearing and also guards the electrical and rotating parts inside the motor. 2. In a magnetron, the shield that confines the space charge to the interaction space.

end spaces—In a multicavity magnetron, the two cavities at either end of the anode block which terminate all the anode-block cavity resonators.

end use—The way the ultimate consumer uses a device.

energize—To apply the rated voltage to a circuit or device, such as to the coil of a relay, in order to activate it.

energized—Also called alive, hot, and live. Electrically connected to a voltage source.

energized part—A part at some potential with respect to another part, or the earth.

energy—The capacity for performing work. A particle or piece of matter may have energy because it is moving or because of its position in relation to other particles or pieces of matter. A rolling ball is an example of the first; a ball at rest at the top of an incline, an example of the second.

energy conversion—The change of energy from one form to another, e.g., from chemical energy to electrical energy.

energy conversion devices—Devices including primary and secondary cells; fuel cells; photovoltaic systems; electrochemical energy converters; radiation conversion devices; thermionic converters; converters using solar, ionic, or nuclear energy sources; devices for creating a plasma in an interaction space between an emitter and a collector; electrostatic generators for creating an electrical output; organic and inorganic ion exchange and membrane devices; electronvolt energy devices; devices for direct conversion of fuel to electricity; and electrical-energy storage-unit devices capable of delivering a power output.

energy density—The ratio of the energy available from a cell to the weight or volume of the cell.

energy gap—The energy range between the bottom of the conduction band and the top of the valence band of a semiconductor.

energy level—A particular value of energy of a physical system, such as a nucleus, which the system can maintain for a reasonably long length of time. Systems on an atomic scale have only certain discrete energy levels and cannot occupy values between these levels.

energy-level diagram—A line drawing that shows the increase or decrease in electrical power as current intensities rise and fall along a channel of signal communications.

energy-measuring equipment — Equipment used to measure energy in electrical, electronic, acoustical, or mechanical systems.

energy of a charge — Represented by $E = \frac{1}{2}QV$, given in ergs, when the charge Q and the potential V are in electrostatic units.

energy product—The product of the magnetic flux density B in gauss times the magnetic field strength H in oersteds. Used as an index of magnet quality. The larger the maximum energy product, the smaller the required magnet for a given job.

energy-product curve—A curve obtained by plotting the product of the value of magnetic induction B and demagnetizing force H for each point of the demagnetization curve of a permanent magnetic material. Usually shown together with the demagnetization curve.

energy redistribution—A method of finding the duration of an irregularly shaped pulse by considering it as a power curve. The area under the curve can be represented by an equivalent rectangle of the same area and peak amplitude. The original-pulse duration is equal to the rectangle width.

energy state—The position and speed of an electron relative to the position and speed of other electrons in the same atom or adjoining atoms.

energy-variant sequential detection — A technique for sequential detection in which a fixed number of transmitted pulses of varying energy are received with a single (upper) threshold device.

engineered military circuit—1. Leased long lines of which only the station equipment, local loops, and reserved positions of interexchange channels are paid for continuously. The unreserved portions of leased long lines or interexchange channels are on a steady status and are placed in an operational status and paid for only when required by the command concerned. 2. A standby or on-call circuit that is engineered specifically to meet military criteria.

engineering — A profession in which a knowledge of the natural sciences is applied with judgment to develop ways of utilizing the materials and forces of nature.

English-reading tube tester—*See* Emission-Type Tube Tester.

enhanced carrier demodulation—An amplitude-demodulation system in which a synchronized local carrier of the proper phase is added to the demodulator. This has the effect of materially reducing the distortion produced in the demodulation process.

enhancement mode—1. An MOS transistor that is normally off with zero gate voltage applied. A gate voltage of the correct polarity attracts majority carriers to the gate area, thus "enhancing" it and forming a current-conducting channel. 2. A device type that is normally off with zero gate voltage. A threshold voltage is then required to turn the device on.

ensemble—A collection of sample functions of a random process, all of which start from the same zero time.

enhancement-mode field effect transistor—

An FET in which no device current flows (leakage only) when V_{GS} is zero volts. Conduction does not begin until V_{GS} reaches the threshold voltage.

enhancement-mode operation—The operation of a field-effect transistor such that changing the gate-to-source voltage from zero to a finite value increases the magnitude of the drain current.

enhancement MOS—A type of MOS transistor in which no current flows in the absence of an input control signal on the control terminal (called the gate) of the transistor. This reduces power dissipation (power dissipation occurs only when an input signal is present) and results in excellent logic state recognition (full OFF being one state and ON the other).

enterprise number—Unique telephone exchange number that permits the called party to be automatically billed for incoming calls. Also called "toll-free number."

entrance box—A metal box that houses overcurrent-protection devices and serves as the point of distribution for the various electrical circuits in a structure.

entrance cable—A cable by means of which electrical power is brought from an outside powerline into a building.

entropy—1. A measure of the unavailable energy in a thermodynamic system. 2. The unavailable information in a set of documents. 3. An inactive or static condition (total entropy).

entry—Each statement in a computer programming system.

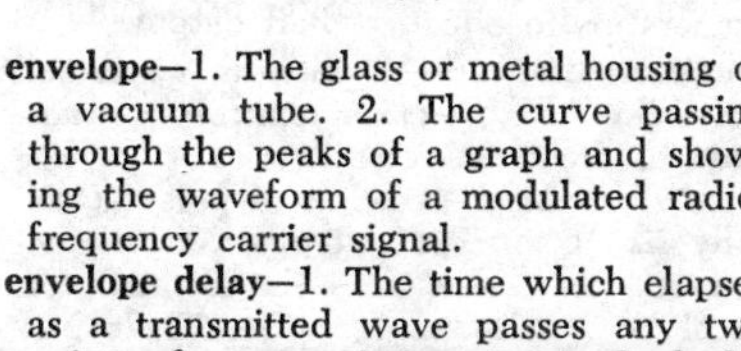

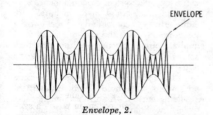

Envelope, 2.

envelope—1. The glass or metal housing of a vacuum tube. 2. The curve passing through the peaks of a graph and showing the waveform of a modulated radio-frequency carrier signal.

envelope delay—1. The time which elapses as a transmitted wave passes any two points of a transmission circuit. Such delay is determined primarily by the constants of the circuit and is measurable in milliseconds or microseconds. 2. Sometimes called time delay or group delay. The propagation time delay undergone by the envelope of an amplitude-modulated signal as it passes through a filter. Envelope delay is proportional to the slope of the curve of phase shift as a function of frequency. Envelope-delay distortion is introduced when the delay is not the same at all frequencies in the passband.

envelope-delay distortion—The distortion that occurs during transmission when the phase shift of a circuit or system is not constant over the frequency range.

environmental conditions—External conditions of heat, shock, vibration, pressure, moisture, etc.

environmentally sealed—Provided with gaskets, seals, potting, or other means to keep out contamination which might reduce performance.

environmental testing—The testing of a system or component under controlled environmental conditions, each of which tends to affect its operation or life.

environmentproof switch—A switch which is completely sealed to ensure constant operating characteristics. Sealing normally includes an "O" ring on the actuator shaft and fused glass-to-metal terminal seals or complete potting and an elastomer plunger-case seal.

eog—Abbreviation for electro-oculography.

ephemeris time – Astronomical time based on the motion of the earth around the sun during the tropical year 1900. Abbreviated ET.

episcotister—A device consisting of alternate opaque and transparent discs which rotate at a speed which interrupts light beams at an audio-frequency rate. It modulates the light beam used to excite a photoelectric element.

epitaxial – Pertaining to a single crystal layer on a crystalline substrate, oriented the same as the substrate. In certain semiconductor processes, an epitaxial layer is grown on a silicon substrate during the fabrication of transistors and integrated circuits.

epitaxial deposition—The growth of additional material, usually in a thin film, on a substrate. Often the added material has a crystal structure and orientation controlled by matching that of the substrate.

epitaxial device – A device constructed in such a manner that the crystalline structure of successive layers is oriented in the same direction as that of the original base material.

epitaxial film – 1. A film of single-crystal semiconductor material that has been deposited onto a single-crystal substrate. 2. Any deposited film, provided the orientation of its crystal is the same as that of the substrate material.

epitaxial growth—1. A semiconductor fabrication process in which single-crystal p or n material is deposited and grows on the surface of a substrate. Usually, this material has a different conductivity than the substrate. 2. Crystal growth obtained by depositing a film of monocrystalline semiconductor material on a monocrystalline substrate. 3. The process of producing an additional crystal layer of semiconductor material on a semiconductor substrate. The crystalline structure of the substrate is continued into the epitaxial layer; however, the impurity concentration can be made to differ greatly. 4. The deposition of a single-crystal film on the surface of a single-crystal substrate so that the crystal orientations of the two layers are alike.

epitaxial-growth mesa transistor—A transistor made by overlaying a thin mesa crystal over another mesa crystal.

epitaxial growth process — The process of growing a semiconductor material by depositing it in vaporized form on a semiconductor seed crystal. The deposited layer continues the single-crystal structure of the seed.

epitaxial layer — 1. A grown or deposited crystal layer with the same crystal orientation as the parent material and, in the case of semiconductor circuits of the same basic material as the original substrate. 2. A single-crystal p-type or n-type material deposited on the surface of a substrate. 3. A thin, precisely doped monocrystalline silicon layer grown on a heavily doped thick wafer, into which are diffused semiconductor junctions. In conventional processing of an integrated circuit the thick wafer is p doped and the exiptaxial layer is n doped.

epitaxial planar transistor—A transistor in which a thin collector region is eptiaxially deposited on a low-resistivity substrate, and the base and emitter regions are produced by gaseous diffusion with the edges of the junction under a protective oxide mask.

epitaxial process—The process of growing from the vapor phase a single-crystal semiconductor material with controlled resistivity and thickness.

epitaxial transistor—A transistor with one or more epitaxial layers.

epitaxy—The growth of a crystal on the surface of a crystal of another substance in such a way that the orientation of the atoms in the original crystal controls the orientation of the atoms in the grown crystal.

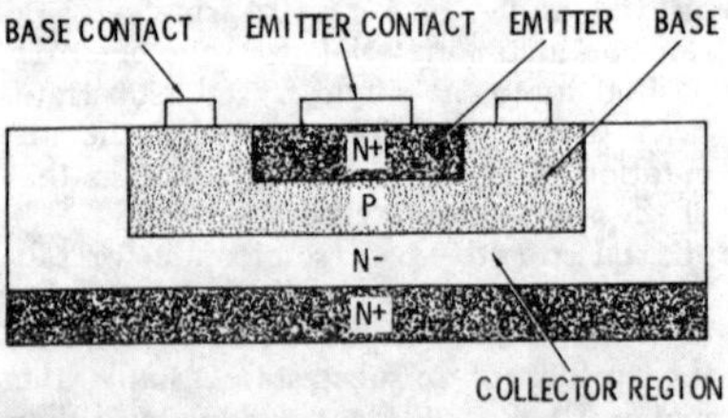

Epitaxial transistor (triple diffused).

E-plane—The plane of an antenna containing the electric field. The principal E-plane also contains the direction of maximum radiation.

E-plane bend — Also called E-bend. The smooth change in direction of the axis of a waveguide. The axis remains parallel to the direction of polarization throughout the change.

E-plane T-junction—Also called series T-junction. A waveguide T-junction in which the structure changes in the plane of the electric field.

epoxy—Pertaining to a family of thermosetting materials that are widely used for casting and potting and as adhesives.

epsilon—The greek letter E, or ϵ, frequently used to represent 2.71828, which is the base of the natural system of logarithms.

equal-energy source—A source of electromagnetic or sound energy that emits the same amount of energy at each frequency in the spectrum.

equal-energy white—The light produced by a source which radiates equal energy at all visible wavelengths.

equalization—The process of reducing the frequency and/or phase distortion of a circuit by the introduction of networks to compensate for the difference in attenuation and/or time delay at the various frequencies in the transmission band.

equalizer—1. A passive device designed to compensate for an undesired amplitude-frequency and/or phase-frequency characteristic of a system or component. 2. A series of connections made in paralleled, cumulatively compound direct-current generators to give the system stability.

equalizer circuit breaker—A breaker which serves to control or to make and break the equalizer or the current-balancing connections for a machine field, or for regulating equipment, in a multiple-unit installation.

equalizing current—A current circulated between two parallel-connected compound generators to equalize their output.

equalizing network—A network connected to a line to correct or control its transmission frequency characteristics.

equalizing pulses—A series of pulses (usually six) occurring at twice the line frequency before and after the serrated vertical tv synchronizing pulse. Their purpose is to cause vertical retrace to occur at the correct instant for proper interlace.

equal-loudness contours—*See* Fletcher-Munson Curves.

equation function — As applied to microelectronic circuitry, a combination of electronic elements or circuits capable of solving the electronic-counter portion of a mathematical or Boolean equation. In obtaining the solution, it performs the necessary function within an electronic or electromechanical system.

equation solver—A computer, usually of the analog type, designed to solve systems of linear simultaneous (nondifferential) equations or to find the roots of polynomials.

equilibrium brightness——The brightness of the viewing screen when a display storage tube is in a fully written condition.

equiphase surface—In a wave any surface over which the field vectors at the same instant are either in phase or 180° out of phase.

equiphase zone—In radionavigation, the region in space within which the difference in phase between two radio signals is indistinguishable.

equipment—1. An item having a complete function apart from being a substructure of a system. Sometimes called a set. 2. A general term referring to practically every part of an electrical system, including the parts consuming electrical energy. (Devices are also included in this category.)

equipment augmentation—1. Procuring additional automatic data procession equipment capability to accommodate increased workload within an established data system. 2. Obtaining additional automatic data procession equipment capability to extend an established data system to additional sites or locations.

equipment bonding jumper—The connection between two or more portions of the equipment grounding conductor.

equipment chain — A group of units of equipment that are functionally in series. The failure of one or more individual units results in loss of the function.

equipment characteristic distortion—A repetitive display or disruption peculiar to specific portions of a teletypewriter signal. Normally, it is caused by improperly adjusted or dirty contacts in the sending or receiving equipment.

equipment ground — A connection from earth ground to a noncurrent carrying metal part of a wiring installation of electric equipment. It reduces shock hazard and provides electrostatic shielding.

equipment life—The arithmetic mean of the cumulative operating times of identical pieces of equipment beginning with the time of acceptance by the ultimate consumer and ending when the equipment is no longer serviceable.

equipotential—A conductor having all parts at a single potential. The cathode of a heater-type tube is equipotential, whereas the filament is not because its voltage varies from one end to the other.

equipotential cathode — *See* Indirectly Heated Cathode.

equipotential line—An imaginary line in space having the same potential at all points.

equipotential surface—A surface or plane passing through all points having the same potential in a field of flow.

equisignal localizer—Also called tone localizer. A type of localizer in which lateral guidance is obtained by comparing the amplitudes of two modulation frequencies.

equisignal radio-range beacon — A radio-range beacon used for aircraft guidance. It transmits two distinctive signals, which are received with equal intensity only in certain directions called equisignal sectors.

equisignal surface — The surface formed around an antenna by all points which have a constant field strength (usually measured in volts per meter) during transmission.

equisignal zone—In radionavigation, the region in space within which the difference in amplitude between two radio signals is indistinguishable.

equivalence—A logic operator having the property that if P is a statement, Q is a statement, R is a statement, etc., then the equivalence of P,Q,R . . . is true if and only if all statements are true or all statements are false.

equivalent absorption—The rate at which a surface will absorb sound energy, expressed in sabins. Defined as the area of a perfect absorption surface that will absorb the same sound energy as the given object under the same conditions.

equivalent binary digits — The number of binary digits equivalent to a given number of decimal digits or other characters.

equivalent circuit—An arrangement of common circuit elements with electrical characteristics equivalent to those of a more complicated circuit or device.

equivalent circuit of a piezoelectric crystal unit—The electric circuit which has the same impedance as the unit in the frequency region of resonance. It is usually represented by an inductance, capacitance, and resistance in series, shunted by the direct capacitance between the terminals of the crystal unit.

equivalent component density—In circuits where discrete components are not readily identifiable, the volume of the circuit divided by the number of discrete components necessary to perform the same function.

equivalent conductance—The normal conductance of an atr tube in its mount, measured at its resonance frequency.

equivalent dark-current input—The incident luminous flux required to give an output current equal to the dark current.

equivalent differential input capacitance—The equivalent capacitance looking into the inverting or noninverting inputs of a differential amplifier with the opposite input grounded. *See also* Equivalent Differential Input Impedance.

equivalent differential input impedance—The equivalent impedance looking into the inverting or noninverting input, with the opposite input grounded and the operational amplifier operated in the linear amplification region.

equivalent differential input resistance – The equivalent resistance looking into the inverting or noninverting input of a differential amplifier with the opposite input grounded. *See also* Equivalent Differential Input Impedance.

equivalent diode—An imaginary diode consisting of the cathode of a triode or multigrid tube and a virtual anode to which is applied a composite controlling voltage of such a value that the cathode current would be the same as the current in the triode or multigrid tube.

equivalent four-wire system – A transmission system using frequency division to obtain full-duplex operation over only one pair of wires.

equivalent grid voltage—The grid voltage plus plate voltage divided by the mu of the tube.

equivalent height—The virtual height of an ionized layer of the ionosphere.

equivalent input noise current—The equivalent input noise current which would reproduce the noise seen at the output of an operational amplifier if all amplifier noise sources were set to zero, and the source impedances were large compared to the optimum source impedance.

equivalent input noise voltage—The equivalent input noise voltage which would reproduce the noise seen at the output of an operational amplifier if all amplifier noise sources and the source resistances were set to zero.

equivalent input offset current—The difference between the two currents flowing into the inverting and noninverting inputs of a differential amplifier when the output voltage is zero.

equivalent input offset voltage – The amount of voltage required at the input to bring the output to zero. Usually this voltage is adjustable to zero by using either a built-in or an external variable resistor (balance control).

equivalent input wideband noise voltage—The output noise voltage of a differential amplifier with the input shorted, divided by the dc voltage gain of the amplifier. This voltage is measured with a true rms voltmeter and is limited to the combined bandwidth of the amplifier and meter.

equivalent loudness—The intensity level of a sound relative to some arbitrary reference intensity, such as a 1000-hertz pure tone which is judged by the listeners to be equivalent in loudness.

equivalent network—A network which may replace another network without substantial change in the operation of the system.

equivalent noise conductance—The spectral density of a noise-current generator expressed in conductance units at a specified frequency.

equivalent noise input—In a photosensitive device, the value of incident luminous flux which produces an rms output current equal to the rms noise current within a specified bandwidth when the flux is modulated in a stated manner.

equivalent noise pressure—*See* Transducer Equivalent Noise Pressure.

equivalent noise resistance—A measure of the residual noise output of a potentiometer while the slider is being actuated. (The residual noise consists of active components in the form of self-generated voltages arising in the slider contact interface, and passive components in the form of ohmic contact resistance at the point of slider contact.)

equivalent noise temperature—The absolute temperature at which a perfect resistor with the same resistance as the component would generate the same noise as the component at room temperature.

equivalent periodic line – A periodic line that, when measured at its terminals or at corresponding section junctions, has the same electrical behavior at a given frequency as the uniform line with which it is compared.

equivalent permeability—The relative permeability which a component would have under specified conditions if it had the same reluctance as a component of the same shape and size but different materials.

equivalent plate voltage—The plate voltage plus mu times the grid voltage.

equivalent resistance—The concentrated or lumped resistance that would cause the same power loss as the actual small resistances distributed throughout a circuit.

equivalent series resistance – Abbreviated esr. In a circuit or component, the square root of the difference between the impedance squared and the reactance squared. All internal series resistance of the circuit or component treated as being concentrated in a single resistance at one point.

equivalent time—In random-sampling oscilloscope operation, the time scale associated with the display of signal events.

equivocation – In a computer, the conditional information contained in an input symbol given an output symbol, averaged over all input-output pairs.

erasable storage—Storage media in a computer which hold information that can be changed.

erase—To replace all the binary digits in a storage device by binary zeros. In a binary computer, erasing is equivalent to clearing, while in a coded decimal computer where the pulse code for decimal zero may contain binary ones, clearing leaves decimal zero while erasing leaves all-zero pulse codes in all storage locations.

erase head—A head on a tape recorder which applies a strong high-frequency alternating magnetic field to the tape so that earlier recordings may be "erased" as the tape runs past the head.

erasing speed—In charge-storage tubes, the rate of erasing successive storage elements.

erasure—A process by which a signal recorded on a tape is removed and the tape made ready for rerecording. This may be accomplished by ac erasure, where the tape is demagnetized by an alternating field which is reduced in amplitude from an initially high value, or by dc erasure, where the tape is saturated by applying a primarily unidirectional field.

E-region — The region of the ionosphere about 50 to 100 miles above the earth's surface.

E-register—The extension of the computer A-register for use in double-precision arithmetic or logic-shift operations.

erg—The absolute centimeter-gram-second unit of energy and work. The work done when a force of one dyne is applied through a distance of one centimeter.

E_r — E_y—The resultant color television signal when E_y is subtracted from the original full red signal.

erp — Abbreviation for effective radiated power. The amount of power radiated by an antenna which may be more or less than the power absorbed by it from the transmitter.

error—1. In mathematics, the difference between the true value and a calculated or observed value. A quantity (equal in absolute magnitude to the error) added to a calculated or observed value to obtain the true value is called a correction. 2. In a computer or data-processing system, any incorrect step, process, or result. In addition to the mathematical usage in the computer field, the term also commonly refers to machine malfunctions, or "machine errors," and to human mistakes, or "human errors."

error-correcting code — A code in which each acceptable expression conforms to specific rules of construction that also define one or more equivalent nonacceptable expressions, so that if certain errors occur, in an acceptable expression the result will be one of its equivalents and thus the error can be corrected.

error-correcting telegraph system — A system that employs an error-detecting code in such a way that any false signal initiates retransmission of the character incorrectly received.

error-correction routine—A series of computer instructions programmed to correct a detected error condition.

error detecting and feedback system — A system employing an error-detecting code and so arranged that a signal detected as being in error automatically initiates a request for retransmission of the signal detected as being in error.

error-detecting code—In a digital computer, a system of coding characters such that any single error produces a forbidden or impossible code combination.

error detector—That portion of an automatic control system that determines when the regulated quantity has deviated outside the dead zone.

error rate—1. A measure of quality of a digital circuit or equipment item. 2. The number of erroneous bits or characters in a sample; it is frequently taken as the number of errors per 100,000 characters.

error-rate damping—A damping method in which a signal proportional to the rate of change of error is added to the error signal for anticipatory purposes.

error signal—In an automatic control device, a signal whose magnitude and sign are used to correct the alignment between the controlling and the controlled elements.

error tape—A special tape developed and used for writing out errors in order to correct them by study and analysis after printing.

error voltage—A voltage which is present in a servo system when the input and output shafts are not in correspondence. The error voltage, which actuates the servo system, is proportional to the angular displacement between the two shafts.

Esaki diode—*See* Tunnel Diode.

escape character—A code-extension character used with one or more succeeding characters to form an escape sequence, which indicates that the interpretation of the succeeding characters is to be different.

escapement—A mechanical stepping device that stores energy over a predetermined time interval and delivers this stored energy at the end of said time interval to advance or move a camshaft a predetermined number of degrees.

escape velocity—The speed which a particle or larger body must attain in order to escape from the gravitational field of a planet or star.

E-scope—A radar display in which targets appear as blips, with distance indicated by the horizontal coordinate and elevation by the vertical coordinate.

escutcheon – A backing plate around an opening. Commonly the ornamental metal, wood, plastic, or other framework around a radio tuning dial, control knob, or other panel-mounted part in a radio receiver or television receiver, audio-frequency amplifier, etc.

esr–Abbreviation for equivalent series resistance.

ess–Abbreviation for electronic switching system.

established reliability–A quantitative maximum failure rate demonstrated under controlled test conditions in accordance with a military specification; usually expressed as percent failures per thousand hours of test.

are wound on the outer legs of the E, and the primary is on the center leg.

Ettingshausen effect – Analogous to the Hall effect. The different temperatures found on opposite edges of a metal strip which is perpendicular to a magnetic field and through which an electric current flows longitudinally.

eureka–The ground transponder of secondary radar system rebecca-eureka.

eutectic–1. An isothermal reversible reaction in which a liquid solution is converted into two or more intimately mixed solids on cooling, the number of solids formed being the same as the number of components in the system. 2. An alloy having the composition indicated by the

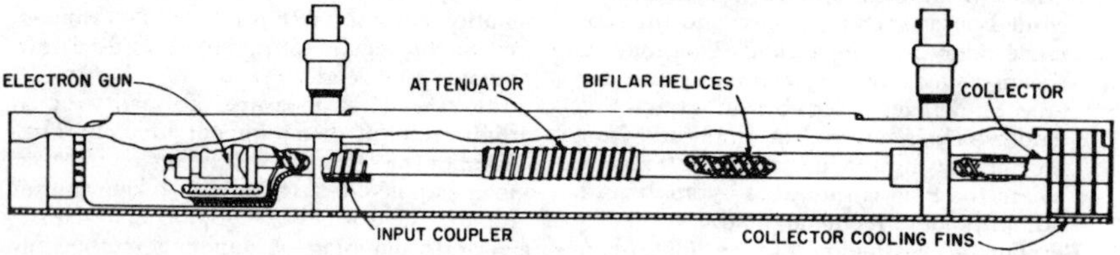

Estiatron.

Estiatron–A special type of electrostatically focused traveling-wave tube.

ET–*See* Ephemeris Time.

etchant–1. A chemical agent that can remove a solid material. For example, a highly selective etchant that acts on silicon dioxide, silicon, or both is employed in semiconductor processing to pattern the silicon surface for diffusion masking. 2. A solution used, by chemical reaction, to remove the unwanted portion of a conductive material bonded to a base. 3. A liquid solution of a chemical or combination of chemicals used to preferentially dissolve metal.

etched printed circuit–A type of printed circuit formed by chemically or electrolytically (or both) removing the unwanted portion of a layer of material bonded to a base.

etch factor–A ratio of etched depth to the lateral etch at the boundary of the interface of the photoresist and the substrate.

etching – The selective removal of unwanted material from a surface, usually by chemical means.

etching to frequency – Finishing a crystal blank to its final frequency by etching it in hydrofluoric acid.

ether–A hypothetical medium that pervades all space (including vacuum) and all matter and is assumed to be the vehicle for propagation of electromagnetic radiations.

E-transformer–A special form of differential transformer employing an E-shaped core. The secondaries of the transformer

eutectic point on an equilibrium diagram. 3. An alloy structure of intermixed solid constituents formed by a eutectic reaction. 4. Referring to an alloy or solid solution that has the lowest possible melting point, usually below that of its components.

eutectic alloy – A combination of two or more metals that has a sharply defined melting point and no plastic range.

eutectic bonding–Formation of a metallurgical joint in similar or dissimilar metals through the introduction of a thin film of another metal at the joint interface. Upon application of heat and moderate pressure, the intermediate film and the metals to be joined form a molten eutectic phase, which is then eliminated from the joint by thermal diffusion into the base metals.

eV (or ev)–Abbreviation for electronvolt.

evaporation of electrons – The cooling which occurs on the surface of a cathode during emission. It is analogous to the cooling of a liquid or solid as it evaporates.

evaporative deposition–The process of condensing a thin film of evaporated material upon a substrate. Evaporation usually is produced by heating a material in a high vacuum.

E-vector–A vector representing the electric field of an electromagnetic wave. In free space it is perpendicular to the direction of propagation.

even harmonic – Any harmonic that is an

even multiple (2, 4, 6, etc.) of the fundamental frequency.

event counter—Instrument that records and totalizes occurring events; can include time of occurrence of events.

E-wave — Designation for tm (transverse magnetic) wave, one of the two classes of electromagnetic waves that can be sent through waveguides.

EX—In a calculator, it is the abbreviation for exchange key. Interchanges the last entry with the preceding value in the calculator.

exalted-carrier receiver — A receiver that counteracts selective fading by maintaining the carrier at a high level at all times.

exalted-carrier reception—A method of receiving either amplitude- or phase-modulated signals in which the carrier is separated from the sidebands, filtered and amplified, and then recombined with the sidebands at a higher level prior to demodulation.

except gate—A gate in which the specified combination of pulses producing an output pulse is the presence of a pulse on one or more input lines and the absence of a pulse on one or more other input lines.

excess conduction — Conduction by excess electrons in a semiconductor.

excess electron — An electron introduced into a semiconductor by a donor impurity and available to promote conduction. An excess electron is not required to complete the bond structure of the semiconductor.

excess fifty—In a computer, a representation in which a number (N) is denoted by the equivalent of (N + 50).

excess meter — An electricity meter which measures and registers the integral, with respect to time, of those portions of the active power in excess of the predetermined value.

excess minority carriers—In a semiconductor, the number of minority carriers that exceed the normal equilibrium number.

excess modified index of refraction — *See* Refractive Modulus.

excess noise—Noise resulting from the passage of current through a semiconductor material. Also called current noise, bulk noise, and 1/f noise.

excess sound pressure—The total instantaneous pressure at a point in a medium containing sound waves, minus the static pressure when no sound waves are present. The unit is the dyne per square centimeter.

excess-three bcd—Excess-three binary coded decimal. Pertaining to a code based on adding 3 to a decimal digit and then converting the result directly to binary form. Use of this code simplifies the execution of certain mathematical operations in a binary computer that must handle decimal numbers.

exchange—To remove the contents of one storage unit of a computer and place it in a second, at the same time placing the contents of the second storage unit into the first.

exchange cable—A lead-covered, nonquadded, paper-insulated cable used in providing cable pairs between local subscribers and a central office.

exchange key—*See* EX.

exchange line—A line that joins a subscriber or switchboard to a commercial exchange.

exchange plant—Facilities used to serve the needs of subscribers as distinguished from facilities used for long-distance communication.

exchange register—*See* Memory Register.

exciplex—(From *exci*ted state com*plex.*) A chemical reaction occurring in certain lasing materials known as organic dyes and used for adjusting a laser so that it emits light in a color range from near ultraviolet to yellow.

excitation—1. Also called stimulus. An external force or other input applied to a system to cause it to respond in some specified way. 2. Also called drive. A signal voltage applied to the control electrode of an electron tube. 3. In electric or electromagnetic equipment, supplying with a potential, a charge, or a magnetic field. 4. The addition of energy to a system so as to transfer the system from its ground state to an excited state.

excitation anode—An auxiliary anode of a pool-cathode tube, used to maintain a cathode spot when the output current is zero.

excitation current—The resultant current in the shunt field of a motor when voltage is applied across the field.

excitation energy — The external electrical energy required for proper operation of a transducer.

excitation purity—Also called purity. The ratio between the distance from the reference point to the point representing the sample and the distance along the same straight line from the reference point to the spectrum locus or to the purple boundary, both distances being measured (in the same direction from the reference point) on the CIE chromaticity diagram.

excitation voltage—The voltage required for excitation of a circuit.

excited-field speaker—A speaker in which the steady magnetic field is produced by an electromagnet.

exciter—1. In a directional transmitting antenna system, the part connected directly to the source of power such as to the transmitter. 2. A crystal or self-excited oscillator that generates the carrier fre-

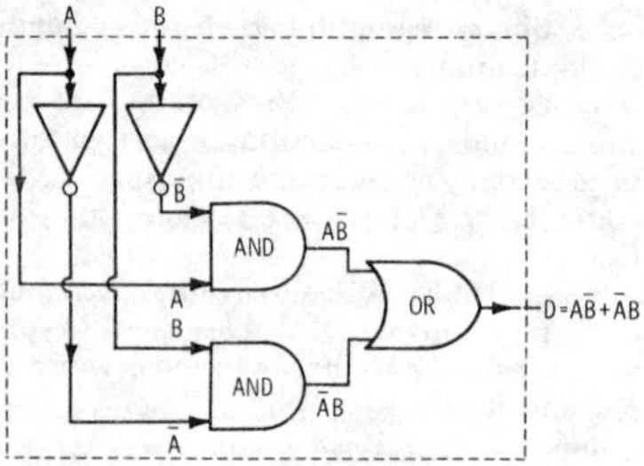

Exclusive OR.

quency of a transmitter. 3. A small, auxiliary generator that provides field current for an ac generator.

exciter lamp—1. A high-intensity incandescent lamp having a concentrated filament. It is used in making variable-area, sound-on-film recording and in reproducing all types of sound tracks on film, as well as in some mechanical television systems. 2. A light source used in a facsimile transmitter to illuminate the subject copy being scanned.

exciter or dc-generator relay — A device which forces the dc machine-field excitation to build up during starting, or which functions when the machine voltage has built up to a given value.

exciter response—In rotating electrical machinery, the rate of change of the main exciter voltage when the resistance in the main exciter field circuit is suddenly changed. The exciter response may be expressed in volts per second or by the numerical value obtained by dividing the volts per second by some designated value of voltage, such as the nominal collector-ring voltage.

exciting current — 1. The current which flows in the primary of a transformer when the secondary is open-circuited. This current produces a flux that generates a back emf equal to the applied voltage. 2. Also called magnetizing current. The current that passes through the field windings of a generator.

exciton—A mobile, electrically neutral, excited state of holes and electrons in a crystal. An example is a weakly bound electron-hole pair; when such a pair recombines, the energy yielded is the band gap reduced by the binding energy of the pair.

excitron—A type of rectifier tube used in applications with heavy power requirements and in power distribution systems. It has a single anode and a mercury-pool cathode and is provided with a means for maintaining a continuous cathode spot.

exclusion principle — The principle which states that if particles are considered to occupy quantum states, then only one particle of a given kind can occupy any one state. Particles differ in kind due to their direction of spin, momentum, orbit, etc.

exclusive OR—A function that is valid (its value is 1) if one and only one of the input variables is present. The exclusive OR applied to two variables is present, or 1, if the binary inputs are different. The term half-add is often applied to the exclusive OR with two input variables.

exclusive OR gate—A type of gate that produces an output when the inputs are the same, but not when they are different, i.e., AB or $\overline{AB}$.

excursion — A single movement away from the mean position in an oscillating or alternating motion.

execute—The process of interpreting an instruction and performing the indicated operation(s).

execution—The performance of an operation or instruction.

execution time—The time required for a computer to execute an instruction, usually several machine cycles.

executive routine—In computer operation, a set of coded instructions that controls loading and relocation of other routines, and in some cases employs instructions not known to the general programmer. Effectively, an executive routine is a non-hardware part of the computer itself, except that it is superimposed on all lower-level programs and instructional sets.

exercisers — Multioutput data generators, e.g., a "memory exerciser" produces, at different sets of output terminals, data-input words, addresses, and coincidence/complement signals for an error detector (discriminator). . . . as well as appropriate write and read commands. These instruments are actually small test systems, and might properly be classed with them.

exhaustion—The removal of gases from a space, such as the bulb of a vacuum tube, by means of vacuum pumps.

exit—In a computer, the means of halting a repeated cycle of operation in a program.

exosphere — The outermost region of the earth's atmosphere, where the atoms and molecules move in dynamic orbits under the influence of the gravitational field.

exothermic—A chemical reaction in which heat is produced.

expand—To spread out part or all of the trace of a cathode-ray display.

expandable gate—A logic gate the number of inputs to which can be increased by the simple addition of an expander block.

expanded contact—In a semiconductor, any pattern that has metalization crossing a diffused junction.

expanded-position-indicator display—An expanded display of a setcor from a plan-position-indicator presentation.

expanded scale meter—A meter in which the ratio of deflection per unit of applied energy becomes greater as the energy approaches a specified value.

expanded scope—A magnified portion of a cathode-ray-tube presentation.

expanded sweep—A preselected portion of a sweep, during which time the electron beam is speeded up in a cathode-ray tube.

expander—1. A transducer which produces a larger range of output voltages for a given amplitude range of input voltages. One important type expands the volume range of speech signals by employing their envelope. 2. A logic block that may be connected easily to an expandable gate to make a larger number of logic inputs available.

expander inputs—Gates used for increasing the number of logic-performing inputs.

expansion—1. A process in which the effective gain applied to a signal is increased for larger signals and decreased for smaller signals. 2. In facsimile transmission, an increase in the contrast between light and dark portions of the transmitted picture.

experimental model—An equipment model that demonstrates the technical soundness of a basic idea but does not necessarily have the same form or parts as the final design.

experimental station—Any station (except amateur) utilizing electromagnetic waves between 10 kHz and 3000 GHz in experiments, with a view toward the development of a science or technique.

exploring coil—*See* Magnetic Test Coil.

explosion-proof motor – A motor designed and constructed so as to withstand an internal explosion of a specified gas or vapor and to prevent the ignition of the specified gas or vapor surrounding the motor by sparks, flashes, or explosions of the specified gas or vapor inside the motor casing.

explosion-proof switch – A switch (UL listed) that can withstand an internal explosion of a specified gas without causing ignition of surrounding gases.

explosive atmosphere—The condition where air is mixed with dust, metal particles, or inflammable gas in such proportion that it is capable of igniting or exploding.

exponential—Pertaining to exponents or to an expression having exponents. A quantity that varies in an exponential manner increases by the square or some other power of a factor, instead of linearly.

exponential curve—A curve representing the variation of an exponential function.

exponential damping—Damping which follows an exponential law.

exponential decay – The decay of signal strength, radiation, charge, or some other quantity at an exponential rate.

exponential horn—A horn the cross-sectional area of which increases exponentially with axial distance.

exponential quantity – A single quantity which increases or decreases at the same rate as the quantity itself (e.g., the discharge current of a capacitor through a noninductive resistor).

exponential sweep—An electron-beam sweep which starts rapidly and slows down exponentially.

exponential transmission line—A two-conductor transmission line the characteristic impedances of which vary exponentially with the electrical length of the line.

exponential waveform—A waveform which is characterized by smooth curves but which possesses pulse properties because it contains numerous constituent frequencies. The exponential waveform undergoes a rate of amplitude change which is either inversely or directly proportional to the instantaneous amplitude.

exposure—A measure of the X- or gamma radiation at any point; it is used to describe the energy of the radiation field outside the body.

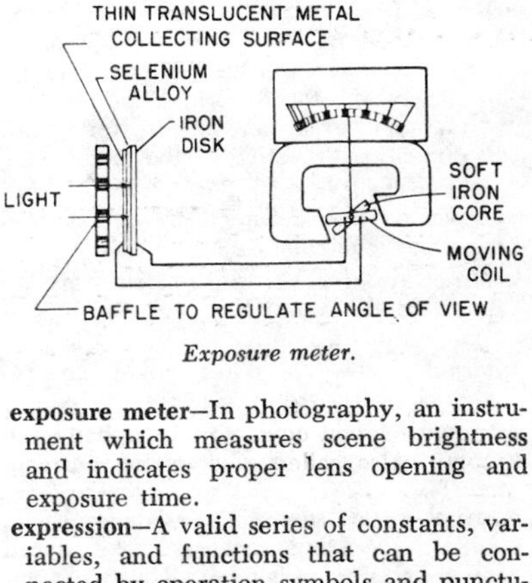

Exposure meter.

exposure meter—In photography, an instrument which measures scene brightness and indicates proper lens opening and exposure time.

expression—A valid series of constants, variables, and functions that can be connected by operation symbols and punctuated, if necessary, to describe a desired computation.

expression control—In an organ the control which regulates the over-all volume. Usually operated with the right foot.

extended addressing—An addressing mode in a computer that can reach any place in memory. *See also* Direct Addressing.

extended-cutoff tube – *See* Remote-Cutoff Tube.

extended-foil construction – A method of fabricating capacitors in which two foil electrodes, separated by a dielectric, are offset so that they may be wound with one foil extending from one end of the

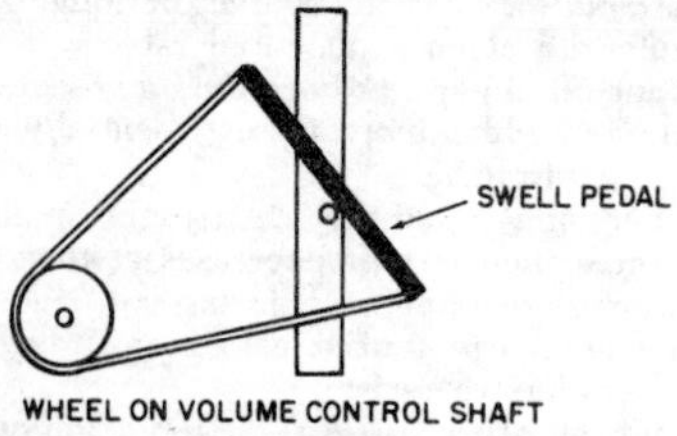

Expression control.

winding and the other foil extending from the opposite end.

extended-interaction tube – A microwave tube in which a moving electron stream and a traveling electric field interact in a long resonator. The bandwidth of such a tube is between that of a klystron and a traveling-wave tube.

extended octaves–In an organ, tones above or below the notes on the regular keyboard which can be sounded only when certain couplers are on.

extended play–Abbreviated ep. A 45-rpm record on a seven-inch disc. It provides eight minutes of playing time, instead of the five minutes of a standard 45-rpm disc.

extended speed range–An extension of the basic speed range of a motor by means of field voltage control. When operating in this range, the load must be reduced to maintain a constant horsepower output to prevent overheating of the motor.

extent – The physical positions on input-output devices occupied by or reserved for a particular data set.

external armature–A ring-shaped armature that rotates around the field magnets of a generator or motor.

external circuit–All wires and other conductors which are outside the source.

external control devices – All control devices mounted external to the control panel.

external critical damping resistance – The value of resistance that must be placed in series or in parallel with a galvanometer in order to produce the critically damped condition.

external device–An input or output device, such as a paper-tape reader, line printer, or magnetic tape recorder, under control of a computer.

external feedback–In a magnetic amplifier, the ampere-turns of auxiliary windings which assist in equalizing the load winding ampere-turns, thereby reducing the control ampere-turns required for control. External feedback may be degenerative to reduce the gain and improve the stability of an amplifier.

externally adjustable timer – A time delay which can be adjusted by varying the resistance of the timer resistor externally, usually by means of a screwdriver.

externally caused contact chatter – That chatter resulting from shock or vibration imposed on the relay by external action.

externally caused failure–A failure caused by an environment outside the design limitations, such as excessive loads, voltages, etc., resulting from operator error, accident, or failure of another part.

externally quenched counter tube–A radiation counter tube that requires an external circuit to prevent it from reigniting.

external memory–An auxiliary storage unit apart from the internal memory of a computer (e.g., punched or magnetic tape).

external *Q*–In a microwave tube, the reciprocal of the difference between the reciprocal of the loaded *Q* and the reciprocal of the unloaded *Q*.

external storage–Storage facilities separate from the computer itself but holding information in a form acceptable to the computer, e.g., magnetic tapes, magnetic wires, punched cards, etc.

extinction potential–The lowest value to which the plate voltage of a gaseous tube can be reduced without cutting off the flow of plate current.

extinguishing voltage–The voltage across a glow lamp when the lamp ceases to glow.

extra-class license – The highest classification of United States amateur license. Requirements include a code sending and receiving ability of 20 words per minute, a knowledge of advanced theory, and the holding of a general-or conditional-class license for two years.

extract–1. To remove from a set of items of information all those items that meet some arbitrary criterion. 2. In computer operations, to obtain specific digits from a stored word. 3. To form a new word from selected segments of given words.

extract instruction–In a digital computer, the instruction to form a new word by placing selected segments of given words side by side.

extractor–*See* Filter, 3.

extraneous emission – Any emission of a transmitter or transponder other than the carrier and those sidebands intentionally added to convey intelligence.

extraneous response – Any undesired response of a receiver, recorder, or other susceptible device, due to the desired signals, undesired signals, or any combination or interaction among them.

extraordinary wave – One of two components into which a sky wave is split in the ionosphere. When viewed below the ionosphere in the direction of propagation, it has clockwise or counterclockwise elliptical polarization, depending on whether the magnetic field of the earth has a positive or negative component in the same direction. The extraordinary wave is designated by the letter X and is sometimes called the X-wave. The other component is the ordinary wave, or O-wave.

extra play – Tape recording term. Originally, all recording tapes were 1½ mils (1/1000's of an inch) thick, and a 7-inch reel would accommodate about 1200 feet of this for a half-hour of continuous recording at 7½ ips. The later extra-play tapes are 1-mil thick, and allow for 45 minutes of recording, or 15 minutes "extra."

extrapolate – To estimate the value of a function for variables lying outside the range in which values of the function are known (e.g., to extend the graphs of the function beyond the plotted points).

extraterrestrial noise – Radio disturbances originating from sources other than those related to the earth; cosmic or solar noise.

extremely high frequency—Abbreviated ehf. The frequency band extending from 30 to 300 GHz.

extremely low frequency—A frequency below 300 hertz. Abbreviated elf.

extrinsic base resistance-collector capacitance product—The product of the base resistance and collector capacitance of a transistor. It is expressed in units of time, since it is an RC time constant and affects the high-frequency operation of a transistor.

extrinsic conductance—The conductance resulting from impurities or external causes.

extrinsic properties – The properties of a semiconductor, modified by impurities or imperfections within the crystal.

extrinsic semiconductor – A semiconductor with charge carrier concentration dependent upon impurities or other imperfections.

eyelet—A tubular metal piece having one end (and possibly the second) headed or rolled over at a right angle.

E-zone—In the making of frequency predictions, one of three zones into which the earth is divided to show the variations of the F_2 layer with respect to longitude. This zone includes Asia, Australia, the Philippines, and Japan.

F

F—1. Symbol for filament, fuse. 2. Abbreviation for Fahrenheit, farad.

f – 1. Abbreviation for femto (10^{-15}). 2. Symbol for focal length, frequency.

F− – *See* A−.

F+ – *See* A+.

fA – Abbreviation for femtoampere (10^{-15} ampere).

fabrication holes—*See* Pilot Holes.

fabrication tolerance—In the construction or assembly of an equipment or portion of an equipment, the maximum variation in the characteristics of a part which, considering the defined variations of the other parts in the equipment, will permit the equipment to operate within specified performance limits.

Fabry-Perot interferometer – A resonant cavity bounded by two end mirrors separated so as to produce interference for certain allowed optical frequencies. The enhancements and cancellations produced by the internal reflections are, in fact, optical standing waves. The structure can be used as a laser cavity, mode filter, or frequency selector.

face—1. A plane surface on a crystal which stands in a particular and invariable relation to the axes and planes of reference and to other faces. 2. Front, or viewing, surface of a cathode-ray tube. 3. The portion of a meter bearing the scale markings.

face bonding—A method of attaching active devices to thin-film passive networks. The semiconductor chips are provided with small mounting pads turned face down and bonded directly to the end of the thin-film conductors on the passive substrate. The term includes ultrasonic, solder-reflow, and solder-ball techniques.

faced crystal—A single or twinned mass of quartz bonded in part or entirely by the original crystal growth faces.

face-parallel cut – A Y-cut for a quartz crystal.

face-perpendicular cut – An X-cut for a quartz crystal.

facility—Anything used or available for use in providing communication service; a communications path.

facom—A long-distance measuring or radio-navigational system which derives information of distances by comparing the phases of received and locally generated signals. It is a base-line system operating in the low-frequency band and will work under adverse propagation and noise conditions at ranges of up to 3000 miles from the signal source.

facsimile—A process or the result of a process by which fixed graphic material including pictures or images is scanned and the information converted into electrical signal waves, which are used either locally or remotely to produce in record form a likeness (facsimile) of the subject copy.

facsimile broadcast station—A station licensed to transmit images of still objects for reception by the general public.

facsimile posting—The process of transferring, by a duplicating process, a printed line on a report to a ledger or other record sheet. These may be posted from a transaction listing previously prepared on an accounting machine. A data-processing function.

facsimile receiver—An apparatus that translates the facsimile signal into a reproduced image.

facsimile recorder—Apparatus which reproduces on paper the image transmitted by a facsimile system.

facsimile signal—The signal resulting from the scanning in a facsimile system.

facsimile-signal level—An expression of the maximum signal power or voltage created by the scanning of the subject copy as measured at any point in a facsimile system. According to whether the system employs positive or negative modulation, this will correspond to picture white or black, respectively. It may be expressed in decibels with respect to some standard value, such as 1 milliwatt or 1 volt.

facsimile system—An integrated assembly of the elements used for facsimile transmission and reception.

facsimile transmission—The transmission of signal waves produced by the scanning of a picture on a revolving drum by a photoelectric cell for reproduction in permanent form at the receiver.

facsimile transmitter — An apparatus employed to convert the subject copy into suitable facsimile signals. Usually, the copy is wrapped around a revolving drum and scanned by a photoelectric cell, which converts the darks and lights into corresponding signal amplitudes.

fade—The gradual lowering in amplitude of a signal.

fade chart—A graph of the null areas of an air-search radar antenna; it is used as an aid in estimating target altitude.

fade in — To increase the signal strength gradually in a sound or television channel.

fade out—1. The gradual decrease in signal strength in a sound or television channel. 2. The cessation or near cessation of radio-wave propagation through parts of the ionosphere due to a sudden atmospheric disturbance.

fader—A multiple-unit control used in radio for gradual changeover from one microphone or audio channel to another; in television, from one camera to another; and in motion-picture projection, from one projector to another.

fading—1. A drift in the level of received radio signals beyond intelligibility. It is often caused by changes in the upper atmosphere. 2. Deliberate slow reduction of signal level by means of the volume control.

fading margin—The number of decibels of attenuation which can be added to a specified radio-frequency propagation path before the signal-to-noise ratio of the channel falls below a specified minimum.

Fahnestock clip—A spring-type terminal to which a temporary connection can readily be made.

Fahrenheit temperature scale—A temperature scale in which the freezing point of water is defined as 32° and the boiling point as 212° under normal atmospheric pressure (760 mm of mercury).

fail hardover — Failure which results in a steady-state maximum system output with no input signal.

fail-safe circuit—A circuit that has an output state which indicates that either a

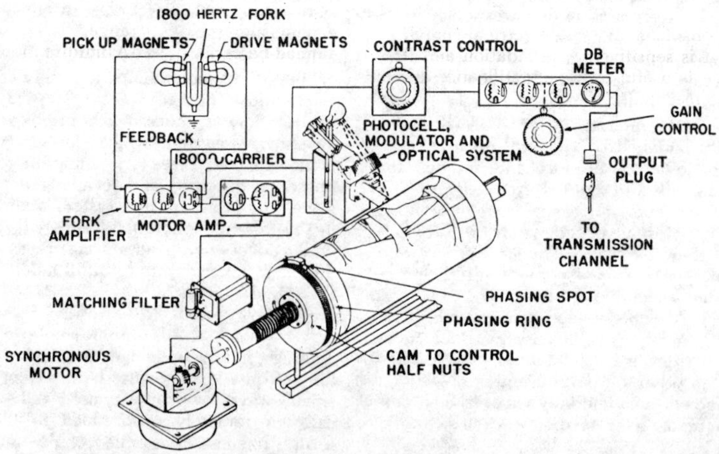

Facsimile transmitter.

circuit input or the circuit itself has failed. Finds circuit application in complex systems where self-healing subsystems exist. When a subsystem failure is detected, a back-up subsystem is automatically inserted.

fail-safe control—A system of remote control for preventing improper operation of the controlled function in event of circuit failure.

fail-safe operation—An electrical system so designed that the failure of any component in the system will prevent unsafe operation of the controlled equipment.

fail soft—An active failure which can be compensated for by the system or one where the operator can safely assume control.

fail-soft system—A system in a computer which continues to process data despite the failure of parts of the system. Usually accompanied by a deterioration in performance.

failure—The inability of a system, subsystem, component, or part to function in the required or specified manner.

failure activating cause—A stress or force (e.g., shock or vibration) that induces or activates a failure mechanism.

failure analysis—Examination of electronic parts to determine the cause of performance variations outside previously established limits, for the purpose of identifying failure modes and failure activating causes.

failure indicator—The observed characteristic that shows that an item is defective.

failure mechanism — 1. A structural or chemical defect, such as corrosion, a poor bond, or surface inversion, that causes failure. 2. *See* Failure Mode, 1.

failure mode—1. The manner in which a failure occurs, including the operating condition of the equipment or part at the time of the failure. Also called failure mechanism. 2. An electrical parameter that is sensitive to degradation and therefore is useful in the identification of failure mechanisms.

failure rate—Also called hazard. The average proportion of units failing per unit of time, normally expressed in percent per thousand hours. It is used in assessing the life expectancy of a device. To be meaningful, a statement of the failure rate must be accompanied by complete information regarding the testing conditions, failure criteria, parameters monitored, and confidence level.

failure unit—One failure in 10^9 device operating hours.

fall-in—In a synchronous motor, the point at which synchronous speed is reached.

fallouts—*See* Transistor Seconds.

fall time—1. The length of time during which a pulse is decreasing from 90% to 10% of its maximum amplitude. 2. A measure of time required for a circuit to change its output from a high level (1) to a low level (0). 3. The time required for the pointer of an electrical indicating instrument to move from a steady full-scale deflection to 0.1 (± a specified tolerance) of full scale when the instrument is short circuited.

false add—To form a partial sum; that is, to perform an addition without carries.

false alarm—A radar indication of a detected target even though one does not exist. It is caused by noise or interference levels that exceed the set threshold of detection.

false course—In navigation normally providing one or more course lines, a spurious additional course-line indication due to undesired reflections or maladjusted equipment.

false-echo device—A device for producing an echo different from that normally observed.

false statement—A statement for a zero in Boolean algebra.

fan—The volume of space energized periodically by a radar beam(s) as it repeatedly traverses an established pattern.

fan antenna—*See* Double-V Antenna.

fan beam—A field pattern having an elliptically shaped cross section in which the ratio of major to minor axes usually exceeds 3 to 1.

fan-in—1. The number of inputs that can be connected to a logic circuit. 2. The number of operating controls in a single device that individually or in combination result in the same output from the device.

fan-in circuit—A circuit that has many inputs feeding to a common point.

fan marker—A radio signal having a vertically directed fan beam which tells the pilot the location of his aircraft while flying along a radio range.

fanned-beam antenna—A unidirectional antenna so designed that transverse cross sections of the major lobe are almost elliptical.

fanning beam — A narrow antenna beam that is scanned repeatedly over a limited arc.

fanning strip — An insulated board, often made of wood, that spreads out the wires of a cable for distribution to a terminal board.

fan-out — The number of parallel loads within a given logic family that can be driven from one output mode of a logic circuit.

fan-out circuit—A circuit that has a single output point which feeds many branches.

farad—The capacitance of a capacitor in which a charge of 1 coulomb produces a change of 1 volt in the potential difference between its terminals. The farad is the unit of capacitance in the mksa systems.

faraday—A unit equal to the number of coulombs (96,500) required for an electrochemical reaction involving one electrochemical equivalent. In an electrolytic process, 1 gram equivalent weight of matter is chemically altered at each electrode for 1 faraday of electricity passed through the electrolyte.

Faraday cage—*See* Faraday Shield.

Faraday dark space—The relatively nonluminous region between the negative glow and the positive column in a glow-discharge–cold-cathode tube. (*See also* Glow Discharge.)

Faraday effect—1. The rotation of the plane of polarization of radio waves as they pass through the ionosphere in the earth's magnetic field. This effect produces a decoupling loss between linearly polarized antennas. 2. The rotation of the plane of polarization that occurs when a plane-polarized beam of light passes through certain transparent substances in a direction parallel to the lines of force of a strong magnetic field. This effect also governs the action of a ferrite rotator in a waveguide.

Faraday rotation—The apparent rotation of the plane of polarization of a linearly polarized wave as it passes through a medium (e.g., a ferrite material) that has a different propagation constant for each of the two component waves of opposite rotational sense.

Faraday screen—*See* Faraday Shield.

Faraday shield—Also called Faraday screen or Faraday cage. A network of parallel wires connected to a common conductor at one end to provide electrostatic shielding without affecting electromagnetic waves. The common conductor is usually grounded.

Faraday's laws—1. The mass of a substance liberated in an electrolytic cell is proportionate to the quantity of electricity passing through the cell. 2. When the same quantity of electricity is passed through different electrolytic cells, the masses of the substances liberated are proportionate to their chemical equivalents. 3. Also called the law of electromagnetic induction. When a magnetic field cuts a conductor, or when a conductor cuts a magnetic field, an electric current will flow through the conductor if a closed path is provided over which the current can circulate.

faradic current—An intermittent and nonsymmetrical alternating current like that obtained from the secondary winding of an induction coil.

faradmeter—An instrument for measuring electric capacitance.

far-end cross talk—Cross talk which travels along the disturbed circuit in the same direction as the signals in that circuit. To determine the far-end cross talk between two pairs, 1 and 2, signals are transmitted on pair 1 at station A, and the cross-talk level is measured on pair 2 at station B.

far field—The space beyond the near field of an antenna in which radiation is essentially confined to a fixed pattern, and power density along the axis of the pattern falls off inversely with the square of the distance.

far-field region—That region of the field of an antenna where the angular field distribution is essentially independent of the distance from the antenna.

far ir—The portion of the infrared spectrum that contains the longest wavelengths.

Farnsworth image-dissector tube—A special cathode-ray tube for use in television cameras.

fast-access storage—In a computer memory or storage, the section from which information may be obtained most rapidly.

fast automatic gain control — A radar agc method in which the response time is long compared to a pulse width, but short compared to the time on target.

fastener—A device used to secure a conductor (or other object) to the structure which supports it.

fast forward—1. The provision on a tape recorder permitting tape to be run rapidly through the recorder in the play direction, usually for search or selection purposes. 2. High-speed winding (shuttling) of tape from the supply reel onto the takeup reel.

fast-forward control—A tape-recorder control which permits running the tape through the machine rapidly in the forward direction.

fast groove—Also called fast spiral. In disc recording, an unmodulated spiral groove having a much greater pitch than the recorded grooves.

fast-operate–fast-release relay — A high-speed relay designed specifically for both short-operate and short-release times.

fast-operate relay—A high-speed relay designed specifically for short-operate but not short-release time.

fast-operate–slow-release relay—A relay designed specifically for short-operate and long-release times.

fast-release relay—A high-speed relay designed specifically for short-release but not short-operate time.

fast spiral—*See* Fast Groove.

fast time constant—An antijamming device used in radar video-amplifier circuits. It differentiates incoming pulses so that only the leading edges of the pulses are used.

fathometer—Also called depth finder, depth sounder, acoustic depth finder, echo depth

Fathometer.

sounder, or sonic depth sounder. A direct-reading device for determining the depth of water in fathoms or other units by reflecting sonic or ultrasonic waves from the ocean bottom. Also used to locate underwater bodies, such as schools of fish or sunken objects.

fatigue–The weakening of a material under repeated stress.

fault–A defect in a wire circuit due to unintentional grounding, a break in the line, or a crossing or shorting of the wires.

fault current–The current that may flow in any part of a circuit or amplifier under specified abnormal conditions.

fault electrode current–The peak current that flows through an electrode during a fault, such as an arc back or a load short circuit.

fault finder–A test set for locating troubles in a telephone system.

fault indicator–Equipment that provides an instantaneous alarm, both visual and audible, of failures detected in the various components of its assorted equipment.

Faure plate–A storage-battery plate, consisting of a conductive lead grid filled with active paste material.

fax–Abbreviation for facsimile.

fc–Abbreviation for footcandle.

FCC–Abbreviation for Federal Communications Commission.

F-display–Also called F-scan or F-scope. In radar, a rectangular display in which a target appears as a centralized blip when the radar antenna is aimed at it. Horizontal and vertical aiming errors are indicated by the horizontal and vertical displacement of the blip.

fdm – Abbreviation for frequency-division multiplex.

Federal Communications Commission–Abbreviated FCC. A board of seven commissioners appointed by the President under the Communications Act of 1934, having the power to regulate all interstate and foreign electrical communication systems originating in the United States.

feedback–1. In a transmission system or a section of it, the returning of a fraction of the output to the input. 2. In a magnetic amplifier, a circuit connection by which an additional magnetomotive force (which is a function of the output quantity) is used to influence the operating condition. 3. In a control system, the signal or signals returned from a controlled process to denote its response to the command signal. Feedback is derived from a comparison of actual response to desired response, and any variation is used as an error signal combined with the original control signal to help attain proper system operation. Systems employing feedback are termed closed-loop systems; feedback closes the loop. 4. Squeal or howl from speaker caused by speaker sound entering microphone of same recorder or amplifier. 5. The return of a portion of the output of a circuit or device to its input. With positive feedback, the signal fed back is in phase with the input and increases amplification, but may cause oscillation. With negative feedback, the signal is 180° out of phase with the input and decreases amplification but stabilizes circuit performance and tends to lower an amplifier's output impedance, improve signal stability and minimize noise and distortion.

feedback admittance–In an electron tube, the short-circuit transadmittance from the output electrode to the input electrode.

feedback amplifier–An amplifier that uses a passive network to return a portion of the output signal to modify the performance of the amplifier.

feedback attenuation–In the feedback loop of an operational amplifier, an attenuation factor by which the output voltage is attenuated to produce the input error voltage.

feedback control–1. A type of system control obtained when a portion of the output signal is operated upon and fed back to the input in order to obtain a desired effect. 2. An automatic means of sensing speed variations and correcting to maintain a constant speed or close speed regulation.

feedback control loop–A closed transmission path which includes an active transducer and consists of a forward path, a feedback path, and one or more mixing points arranged to maintain a prescribed relationship between the loop input and output signals.

feedback control signal – That portion of the output signal which is returned to the input in order to achieve a desired effect, such as fast response.

feedback control system–A control system comprising one or more feedback control loops; it combines the functions of the controlled signals and commands, tending to maintain a prescribed relationship between the two.

feedback cutter – An electromechanical transducer which performs like a disc cutter except it is equipped with an auxiliary feedback coil in the magnetic field.

Signals exciting the cutter are induced into the feedback coil, the output of which is fed back in turn to the input of the cutter amplifier. The result is a substantially uniform frequency response.

feedback loop—The components and processes involved in using part of the output as an input for correction or control of the operation of a system.

feedback oscillator—An oscillating circuit, including an amplifier, in which the output is coupled in phase with the input. The oscillation is maintained at a frequency determined by the parameters of the amplifier and the feedback circuits, such as LC, RC, and other frequency-selective elements.

feedback path—In a feedback control loop, the transmission path from the loop output signal to the loop feedback signal.

feedback regulator—A feedback control system which tends to maintain a prescribed relationship between certain system signals and other predetermined quantities.

feedback transfer function—In a feedback control loop, the transfer function of the feedback path.

feedback winding—In a saturable reactor, the control winding to which a feedback connection is made.

feeder—A conductor or group of conductors connecting two generating stations, two substations, a generating station and a substation or feeding point, a substation and a feeding point, or a transmitter and antenna.

feeder cable—1. A communication cable extending from the central office along a primary route (main feeder cable), or from a main feeder cable along a secondary route (branch feeder cable) and providing connections to one or more distribution cables. 2. Also called trunk cable. In a catv system, the transmission cable from the head end (signal pickup) to the trunk amplifier.

feed function—In automatic control of machine tools, the relative motion between the work and the cutting tool (excluding the motion provided for removal of material).

feed holes—A series of small holes in perforated paper tape which convey no information, but are solely for the purpose of engaging the feed pawls or sprocket which transports the tape over the sensing pins of various reading devices.

feed pitch—The distance between the centers of adjacent feed holes in a tape.

feed reel—On a tape recorder, the reel from which the tape unwinds while playing or recording.

feedthrough—1. The accidental or unintentional transfer of a signal from one track to another on a multitrack tape. 2. The use of special connectors to pass conductors through bulkheads or panels. Contacts can be male on one side and female on the other, or they can be male on both sides or female on both sides. Feedthrough connectors differ from rack-and-panel types in that connection can be made on both sides of the panel.

feedthrough capacitor—A feedthrough insulator that provides a desired value of capacitance between the feedthrough conductor and the metal chassis or panel through which the conductor is passing. Used chiefly for bypass purposes in uhf.

feedthrough insulator—A type of insulator which permits wire or cable to be fed through walls, etc., with minimum current leakage.

feed-thru connection—*See* Thru-Hole connection.

female—Pertaining to the recessed portion of a device into which another part fits.

female contact—A contact located in an insert or body in such a manner that the mating contact is inserted in the unit. This is similar in function to a socket contact.

femto—Prefix meaning 10^{-15}. Abbreviated f.

femtoampere—A unit of current equal to 10^{-15} ampere. Abbreviated fA.

femtovolt—A unit of voltage equal to 10^{-15} volt. Abbreviated fV.

fence—1. A line or system of early-warning radar stations. 2. A concentric steel fence placed around a ground radar transmitting antenna to act as an artificial horizon and suppress ground clutter that would otherwise mask weak signals returned from a target at a low angle.

Fermi level—The value of electron energy at which the Fermi distribution function is one half.

ferpic – *Fer*roelectric ceramic *pic*ture device. A sandwichlike structure made up of transparent electrodes, a photoconductive film, and a thin plate of fine-grained ferroelectric ceramic. The device stores images in the form of a variation of the birefringence of the ceramic plate (that is, as a variation in the way the plate transmits polarized light).

ferreed—An electromechanical switch that combines the rapid switching of bistable magnetic material with metallic contacts to produce output indications that persist as long as desired without further application of power.

ferret—An aircraft, ship, or vehicle especially equipped for ferret reconnaissance.

ferret reconnaissance—A form of reconnaissance which detects, locates, and analyzes enemy radars. It is a passive technique that listens for signals transmitted by enemy radars and thus cannot be jammed. The maximum effective range is limited only by the radio horizon. Ferret systems flown at altitudes of a few miles can accurately locate and analyze radars 300 miles away. Ferret is the only all-weather reconnaissance technique, and no camouflage system works against a properly designated ferret.

ferri- – Prefix indicating a material having a net dipole moment.

ferric oxide (Fe_2O_3) – A red, iron oxide coating for magnetic recording tapes.

ferrimagnetic amplifier–A microwave amplifier utilizing ferrite material in the coupling inductors and transformers.

ferrimagnetic limiter–A power limiter used to replace tr tubes in microwave systems. Its operation is based on a ferrimagnetic material, such as a piece of ferrite or garnet, that exhibits nonlinear properties.

ferrimagnetic materials – Those materials in which spontaneous magnetic polarization occurs in nonequivalent sublattices; the polarization in one sublattice is aligned antiparallel to the other.

ferrimagnetism–A type of magnetism that, as the magnetic moment of neighboring ions tend to align antiparallel, appears microscopically similar to antiferromagnetism. The fact that these moments may be of different magnitudes allows a large resultant magnetization that macroscopically resembles ferromagnetism.

ferristor–A two-winding ferroresonant magnetic amplifier that operates on a high carrier frequency.

ferrite – Also called ferrospinel. A powdered, compressed, and sintered magnetic material having high resistivity, consisting chiefly of ferric oxide combined with one or more other metals. The high resistance makes eddy-current losses extremely low at high frequencies. Examples of ferrite compositions include nickel ferrite, nickel-cobalt ferrite, manganese-magnesium ferrite, yttrium-ion garnet, and single-crystal yttrium-ion garnet.

ferrite bead–A magnetic device for storage of information. It is made of ferrite powder mixtures in the form of a bead fired on the current-carrying wires of a memory matrix.

ferrite circulator – A nonreciprocal microwave network which transmits power from one terminal to another in sequence. Can replace a conventional duplexer, provide isolation of transmitter from receiver, eliminate the requirement for an atr, and isolate the transmitter from antenna reflections.

ferrite core–A core made from iron and other oxides and usually shaped like a doughnut. It is used in circuits and magnetic memories and can be magnetized and demagnetized very rapidly.

ferrite-core memory–A magnetic memory in which read-in and read-out wires are threaded through a matrix of very small toroidal cores molded from a square-loop ferrite.

ferrite isolator–A device either in a waveguide or coax which allows power to pass through in one direction with very little loss, while the rf power in the reverse direction is absorbed. It is useful for maintaining signal source stability and eliminating long-line and frequency-pulling effects in all types of low-power microwave signal sources.

ferrite limiter–A passive low-power microwave limiter that provides an insertion loss of less than 1 dB, with minimum phase distortion, when operating in its linear range. It is used for protecting sensitive receivers from burnout and from blocking by a strong interfering signal.

ferrite phase-differential circulator–A combination microwave duplexer and load isolator that serves as a switching device between a radar antenna and the associated high-power radar magnetron and radar receiver.

ferrite-rod antenna–Also called ferrod or loopstick antenna. An antenna used in place of a loop antenna in a radio receiver. It consists of a coil wound around a ferrite rod.

ferrite rotator–A gyrator, composed of a ferrite cylinder surrounded by a ring-type permanent magnet, that is inserted in a waveguide to rotate the plane of polarization of electromagnetic waves that travel through the waveguide.

ferrites–1. Chemical compounds of iron oxide and other metallic oxides combined with ceramic material. They have ferromagnetic properties but are poor conductors of electricity. Hence they are useful where ordinary ferromagnetic materials (which are good electrical conductors) would cause too great a loss of electrical

Ferrite cores.

energy. 2. Ceramic structures made by mixing iron oxide (Fe_2O_3) with oxides, hydroxides, or carbonates of one or more of the divalent metals, such as zinc, nickel, manganese copper, cobalt, magnesium, cadmium or iron.

ferrite switch—A ferrite device that obstructs the flow of energy through a waveguide by causing a 90° rotation of the electric field vector.

ferroacoustic storage—A delay-line type of storage comprising a thin tube of magnetostrictive material, a central conductor that passes through the tube, and an ultrasonic driver transducer at one end of the tube.

ferrod—*See* Ferrite-Rod Antenna.

ferrodynamic instrument — An electrodynamic instrument in which the measuring forces are materially increased by the presence of ferromagnetic material.

ferroelectric—Pertaining to a phenomenon exhibited by certain materials in which the material is polarized in one direction or the other, or reversed in direction by the application of a positive or negative electric field of magnitude greater than a certain amount. The material retains the electric polarization unless it is disturbed. The polarization can be sensed by the fact that a change in the field induces an electromotive force which can cause a current.

ferroelectric converter — A device which generates high voltage when heat is applied to it. Its operation is based on the change in the dielectric constant or the permittivity of certain materials such as barium titanate when heated. This change reaches maximum at the Curie point.

ferroelectric crystal—A crystal which can be polarized in the opposite direction by applying an electric field weaker than the breakdown strength of the material.

ferroelectricity—A property of certain crystalline materials whereby they exhibit a permanent, spontaneous electric polarization (dipole moment) that is reversible by means of an electric field; the electric analog of ferromagnetism. Materials that show this effect are piezoelectric as well.

ferroelectric materials—Those materials in which the electric polarization is produced by cooperation action between groups or domains of collectively oriented molecules.

ferroelectrics—Pyroelectric materials the direction of polarization of which can be reversed by application of an electric field.

ferromagnetic—Pertaining to a phenomenon exhibited by certain materials in which the material is polarized in one direction or the other, or reversed in direction by the application of a positive or negative magnetic field of magnitude greater than a certain amount. The material retains the magnetic polarization unless it is disturbed. The polarization can be sensed by the fact that a change in the field induces an electromotive force, which can cause a current.

ferromagnetic amplifier—A parametric amplifier based on the nonlinear behavior of ferromagnetic resonance at high rf power levels. In one version, microwave pumping power is supplied to a garnet or other ferromagnetic crystal mounted in a cavity containing a strip line. A permanent magnet provides sufficient field strength to produce gyromagnetic resonance in the garnet at the pumping frequency. The input signal is applied to the crystal through the strip line, and the amplified output signal is extracted from the other end of the strip line. Sometimes incorrectly called a garnet maser, but the operating principle differs from that of the maser.

ferromagnetic material—1. A material having a specific permeability greater than unity, the amount depending on the magnetizing force. A ferromagnetic material usually has relatively high values of specific permeability and it exhibits hysteresis. The principal ferromagnetic materials are iron, nickel, cobalt, and certain of their alloys. 2. A paramagnetic material which exhibits a high degree of magnetizability.

ferromagnetic oxide parts—Parts, consisting primarily of oxides which display ferromagnetic properties.

ferromagnetic resonance—A condition under which the apparent permeability of a magnetic material reaches a sharp maximum at a microwave frequency.

ferromagnetics—The science that deals with the storage of information and the control of pulse sequences through use of the magnetic polarization properties of materials.

ferromagnetic tape — Tape made of magnetic material and used for winding closed cores for toroids and transformers.

ferromagnetism—1. A high degree of magnetism in ferrites and similar compounds. The magnetic moments of neighboring ions tend not to align parallel with each other. The moments are of different magnitudes, and the resultant magnetization can be large. 2. Strong magnetic property of such substances as iron, cobalt, nickel, and certain alloys. Ferromagnetic substances are essential to the construction of such pieces of equipment as speakers, transformers, electric generators, etc. 3. A property of certain metals, alloys, and compounds whereby below a certain critical temperature (the curie point) the magnetic moments of the atoms tend to align, giving rise to a spontaneous, permanent magnetism (dipole moment) that is reversible by means of a magnetic field.

ferromanganese—An alloy of iron and manganese.

ferrometer—An instrument for making permeability and hysteresis tests of iron and steel.

ferroresonance—Resonance associated with circuits in which at least one of the circuit elements is nonlinear and contains iron.

ferroresonant circuit—A resonant circuit in which one of its elements is a saturable reactor.

ferrospinel—A ceramiclike material containing iron and other elements combined with oxygen. A poor conductor of electricity, it is used in transformers, antenna loops, and television deflecting yokes. *See also* Ferrite.

ferrous—Composed of and/or containing iron. A ferrous metal exhibits magnetic characteristics, as opposed to a nonferrous metal such as aluminum which does not.

ferrule—A short tube.

ferrule resistor – A resistor having ferrule terminals for mounting in standard fuse clips.

FET – Abbreviation for field-effect transistor. 1. A transistor controlled by voltage rather than current. The flow of working current through a semiconductor channel is switched and regulated by the effect of an electric field exerted by electric charge in a region close to the channel called the gate. Also called unipolar transistor. A FET has either p-channel or n-channel construction. 2. A transistor whose internal operation is unipolar in nature. The metaloxide semiconductor FET (MOSFET) is widely used in integrated circuits because the devices are very small and can be manufactured with few steps. 3. A solid-state device in which current is controlled between source terminal and drain terminal by voltage applied to a nonconducting gate terminal.

fetch—To go after and return with things. In a microprocessor, the "object" fetched are instructions which are entered in the instruction register. The next, or a later step in the program, will cause the machine to execute what it was programmed to do with the fetched instructions. Often referred to as an instruction fetch.

fetal cardiotachometer—*See* Fetal Monitor.

fetal electrocardiograph—*See* Fetal Monitor.

fetal monitor—An instrument that displays or records the fetal electrocardiogram or other indication of heart action. In some instruments, the material electrocardiogram is recorded simultaneously. The instrument may be referred to more definitely as a fetal electrocardiograph, fetal cardiotachometer, or fetal phonocardiograph, depending on its primary purpose.

fetal phonocardiograph – An instrument which provides continuous instantaneous recording of beat-to-beat changes in the fetal heart rate. *See also* Fetal Monitor.

fetch—That portion of a computer cycle during which the location of the next instruction is determined. The instruction is taken from memory and modified if necessary, and it is then entered into the control register.

FET resistor – A field-effect transistor in which, generally, the gate is tied to the drain and the resultant structure is used in place of a resistor load for a transistor.

"f" factor—The slope of the straight line from which the nonlinearity of a displacement transducer is calculated in microvolts output per volt excitation per unit stimulus.

fiber—A tough insulating material, generally of paper and cellulose, compressed into rods, sheets, or tubes.

fiber metallurgy—The growing of superfine crystal whiskers whose characteristic is relatively great strength in their length to diameter ratio.

fiber needle—A playback point or phonograph needle made from fiber. Being softer than a metal or diamond needle, it is less scratchy; however, it has an extremely short life.

fiber-optic bundles—1. Assemblies of optical fibers. 2. Very fine transparent glass or plastic threads each of which transmits light.

fiber optics—Also called optical fibers or optical fiber bundles. An assemblage of transparent glass fibers all bundled together parallel to one another. The length of each fiber is much greater than its diameter. This bundle of fibers has the ability to transmit a picture from one of its surfaces to the other around curves and into otherwise inaccessible places with an extremely low loss of definition and light, by a process of total reflection.

fiber optics probe—A flexible probe made up of a bundle of fine glass fibers optically aligned to transmit an image, transmit light, or both.

fiber optics scrambler—Similar to a fiberscope except that the middle section of loose fiber is deliberately disoriented as much as possible, then potted and sawed. Each half is then capable of coding a picture, which can be decoded by the other half.

fiberscope—Optical glass fibers, when systematically arranged in a bundle, transmits a full color image that remains undisturbed when the bundle is bent. By mounting an objective lens on one end of the bundle, and an eyepiece at the other, the assembly becomes a flexible fiberscope that can be used to view ob-

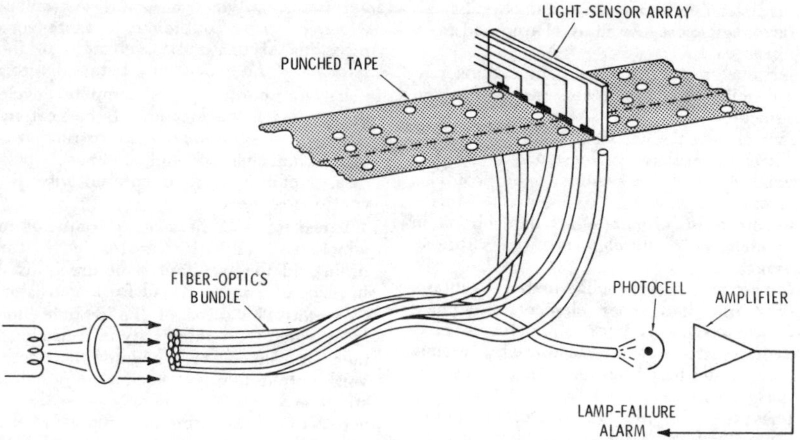

Fiber optics.

jects that would be inaccessible for direct viewing.

fidelity—1. The accuracy with which a system or portion of a system reproduces at its output the essential characteristics of the signal impressed on its input. 2. A measure of the exactness with which sound is duplicated or reproduced.

field—1. One half of a television image. With present U. S. standards, pictures are transmitted in two fields of 262½ lines each, which are interlaced to form 30 complete frames, or images, per second. 2. A general term referring to the region under the influence of some physical agency such as electricity, magnetism, or a combination produced by an electrically charged object, electrons in motion, or a magnet. 3. A group of characters in a computer which is treated as a single unit of information. 4. In each of a number of punch cards, a column or columns regularly used for a standard item of information. 5. That silicon area on a chip not used or occupied by active transistors.

field application relay—A device which automatically controls the application of the field excitation to an ac motor at some predetermined point in the slip cycle.

field circuit breaker—A device which functions to apply, or to remove, the field excitation of a machine.

filed coil—A coil of insulated wire wound around an iron core. Current flowing in the coil produces a magnetic field.

field control of speed—The varying of voltage applied to the field of a shunt-wound motor to control the motor's speed over the extended range.

field density—*See* Magnetic Induction.

field-discharge protection—A control function or device to limit the induced voltage in the field when the field current attempts to change suddenly.

field distortion – Distortion between the north and south poles of a generator due to the counterelectromotive force in the armature winding.

field-effect tetrode—A semiconductor device consisting basically of a thin n region adjacent to a similarly thin p region. Two contacts are made to the n side and two to the p side so that currents can be passed through each thin region parallel to the single junction. The two currents remain separate because reverse bias is maintained on the junction. A current in either side affects the resistance of the other side and hence the current in the other side.

field-effect transistor—A transistor in which current carriers (*holes* or electrons) are injected at one terminal (the *source*) and pass to another (the *drain*) through a channel of semiconductor material whose resistivity depends mainly on the extent to which it is penetrated by a *depletion region.* The depletion region is produced by surrounding the channel with semiconductor material of the opposite conductivity and reverse-biasing the resulting p-n junction from a control terminal (the

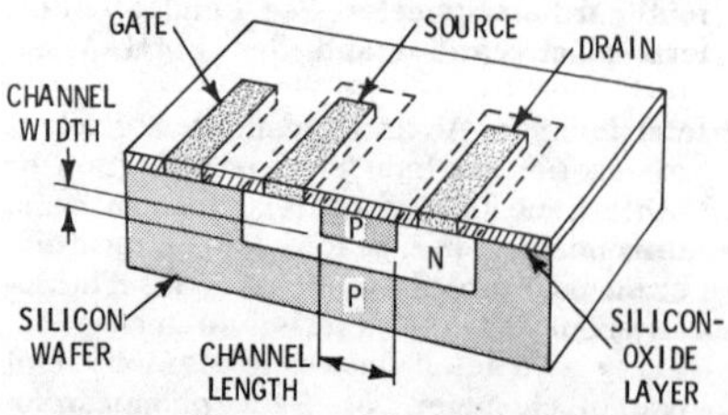

Field-effect transistor.

gate). The depth of the depletion region depends on the magnitude of the reverse bias. As the reverse-biased junction draws negligible current, the characteristics of the device are similar to those of a vacuum tube.

field-effect tube—A triode with its grid replaced by a nonintercepting control gate. A high positive voltage is applied to this gate in order to draw sufficient current from the cathode. The result is a strong concentration of the electric field at the gap between the gate and the cathode, producing an electron beam passing through the gate to the anode.

field-effect varistor – A passive, nonlinear, two-terminal semiconductor device that maintains a constant current over a wide range of voltage.

field emission – Also called cold emission. The liberation of electrons from a solid or liquid by application of a strong electric field at the surface.

field-enhanced photoelectric emission – Increased photoelectric emission resulting from the action of a strong electric field on the emitter.

field-enhanced secondary emission – Increased secondary emission resulting from the action of a strong electric field on the emitter.

field forcing—The effect of a control function or device which temporarily overexcites or underexcites the field of an electrical machine in order to increase the rate of change of flux.

field-free emission current – Also called zero-field emission. The electron current emitted by a cathode when the electric field at the surface of the cathode is zero.

field frequency—Also called field repetition rate. In television, the frame frequency multiplied by the number of fields contained in one frame. In the United States the field frequency is 60 per second, or twice the frame frequency.

field intensity—*See* Field Strength, 1.

field inversion—Also called "parasitic field turn-on." The creation of a channel between two nonassociated diffused beds in the field by voltages on conductors passing over.

field loss relay – *See* Motor-Field Failure Relay.

field magnet—An electromagnet or permanent magnet which produces a strong magnetic field in a speaker, microphone, phonograph pickup, generator, motor, or other electrical device.

field-neutralizing coil—A coil encircling the faceplate of a color picture tube. The current through it produces a magnetic field which offsets any effects of the earth's and other stray magnetic fields on the electron beams.

field-neutralizing magnet—Also called rim magnet. A permanent magnet mounted near the edge of the faceplate of a color picture tube to prevent stray magnetic fields from affecting the path of the electron beams.

field of view—The solid angle from which objects can be acceptably viewed, photographed, or otherwise detected.

field oxide—That portion of the oxide on an MOS device which is the thickest when measured perpendicular to the bulk silicon usually 12K to 20K Å.

field period—The time required to transmit one television field. In the United States, it is 1/60th of a second.

field pickup—Also called a remote or nemo. A radio or television program originating outside the studio.

field pole—A structure, made of magnetic material, on which may be mounted a field coil.

field relay—A device that functions on a given or abnormally low value or failure of machine field current, or on an excessive value of the reactive component of armature current in an ac machine, indicating abnormally low field excitation.

field-repetition rate—*See* Field Frequency.

field resistor—A component in which the resistance element is a thin layer of conductive material on an insulated form. The conductive material does not contain either binding or insulating material.

field rheostat – A variable resistance connected to the field coils of a motor or generator and used for varying the field current.

field ring – The part which supports the field of a dc or series-wound motor housing. The motor end shields are attached to the ends of the field ring.

field scan – In a television system, the downward excursion of an electron beam across the face of a cathode-ray tube, resulting in the scanning of alternate lines.

field selection—In a computer, the isolation of a particular data field with one computer word without isolating the word.

field sequential—Pertaining to the association of individual primary colors with successive fields in a color-television system (e.g., field-sequential pickup, display, system, transmission).

field-sequential color television – A color-television system in which the individual primary colors (red, blue, and green) are produced in successive fields.

field shield—A process whereby a conducting layer covers an entire MOS chip (except at transistor terminals) between the doped substrate and interconnecting conductors to control field inversion problems.

field-simultaneous system – A color-television system in which a complete full-color field is presented simultaneously as a unit.

The eyes then see a succession of full-color images rather than a succession of primary-color fields.

field strength—1. Also called field intensity. The value of the vector at a point in the region occupied by a vector field. In radio, it is the effective value of the electric-field intensity in microvolts or millivolts per meter produced at a point by radio waves from a particular station. Unless otherwise specified, the measurement is assumed to be in the direction of maximum field intensity. 2. The amount of magnetic flux produced at a particular point by an electromagnet or permanent magnet. 3. The strength of radio waves at a distance from a transmitting antenna, usually expressed in microvolts-per-meter. This is not the same as the strength of a radio signal at the antenna terminals of a receiver.

field-strength meter—A calibrated measuring instrument for determining the strength of radiated energy (field strength) being received from a transmitter.

field telephone—A durable, portable telephone designed for use in the field.

field weakening—The introduction of resistance in series with the shunt field of a motor to reduce the voltage and current and increase the motor speed.

field wire—A flexible insulated wire used in field-telephone and -telegraph systems.

FIFO (First In, First Out) buffer or shift register—A shift register with an additional control section that permits input data to "fall through" to the first vacant stage so that if there is any data contained, it is available at the output even though all the stages are not filled. In effect, it is a variable-length shift register whose length is always the same as the data stored therein.

figure-eight microphone—*See* Bidirectional Microphone.

figure of merit—1. The property or characteristic which makes a tube, coil, or other electronic device suitable for a particular application. It is a quality to look for in choosing a piece of equipment. 2. In a magnetic amplifier, the ratio of the power gain to the time constant. 3. For a thermoelectric material, the quotient of the square of the absolute Seebeck coefficient (α) divided by the product of the electrical resistivity (ρ) and the thermal conductivity.

filament—1. Also known as a filamentary cathode. The cathode of a thermionic tube, usually a wire or ribbon, which is heated by passing a current through it. 2. In tubes employing a separate cathode, the heating element. 3. A slender thread of material such as carbon or tungsten which emits light when raised to a high temperature by an electric current (as in an incandescent light bulb).

filamentary transistor—A conductivity-modulation transistor which is much longer than it is wide.

filament battery—The source of energy for heating the filament of a vacuum tube.

filament circuit—The complete circuit through which filament current flows.

filament current—The current supplied to a filament to heat it.

filament emission—The liberation of electrons when the filament in a vacuum tube is heated.

filament power supply—The source of power for the filament or heater of a vacuum tube.

filament resistance—The resistance (in ohms) of the filament of a vacuum tube or incandescent lamp.

filament rheostat—A variable resistance placed in series with the filament of a vacuum tube to regulate the filament current.

filament sag—The bending of a filament when it heats up and expands.

filament saturation—Also called temperature saturation. The condition whereby a further increase in filament voltage will no longer increase the plate current at a given value of plate voltage.

filament transformer—A transformer used exclusively to supply filament voltage and current for vacuum tubes.

filament voltage—The voltage value which must be applied to the filament of a vacuum tube to obtain the rated filament current.

filament winding—A secondary winding provided on a power transformer to furnish alternating filament voltage for one or more vacuum tubes.

file—1. A collection of related records—e.g., in inventory control, one line of an invoice containing data on the material, the quantity, and the price forms an item; a complete invoice forms a record; and the complete set of such records forms a file. 2. To insert an item into such a set.

file gap—On a data medium, an area intended to be used to mark the end of a file and, possibly, the start of another. A

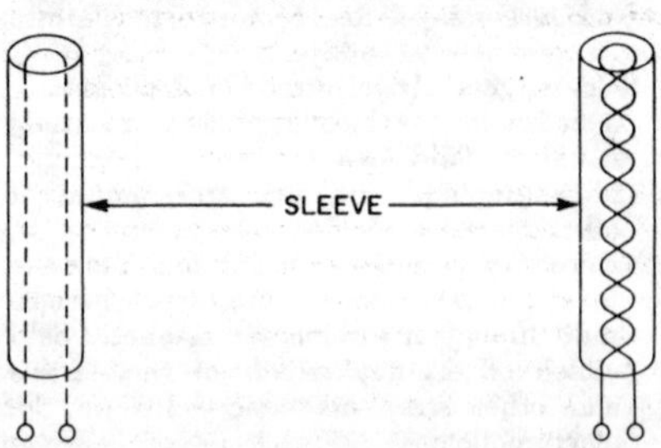

Filaments, 2.

file gap frequently is used for other purposes, in particular for indicating the end or beginning of some other group of data.

file layout—1. The organization and structure of data in a file, including the sequence and size of the components. 2. By extension, the description thereof.

file maintenance—The processing of a computer file in order to bring it up to date.

file management—An operating system facility for the manipulating of data files to and from secondary storage devices (usually disc files or magnetic tapes); it is used for building files, retrieving information from them, or modifying the information.

file-protection device—A ring that must be in place in the hub of a reel before data can be recorded on the tape contained by the reel. A reel of tape not provided with a file-protection device can be read but not written.

file section—That part of a file which is recorded on any one volume. The file sections may not have sections of other files interspersed.

file set—A collection of one or more related files, recorded consecutively on a volume set.

fill—1. The number of working lines in a particular cable or cable center. 2. The number of working lines as a percentage of the total pairs provided.

filler—1. In mechanical recording, the inert material of a recorded compound (as distinguished from the binder). 2. Nonconducting component cabeled with insulated conductors to impart roundness, flexibility, tensile strength, or a combination of all three, to the cable.

film badge—A type of dosimeter consisting of a badge containing a sensitized film which, when developed, gives an indication of the total dose of ionizing radiation to which the badge has been subjected.

film chain—An arrangement of a film projector or projectors and a cctv camera for transmitting moving pictures over a television system.

film integrated circuit—A circuit made up of elements that are films all formed in place upon an insulating substrate. Also called film microcircuit. To further define the nature of a film integrated circuit, additional modifiers may be prefixed. Examples are: thin-film integrated circuit; thick-film integrated circuit.

film microcircuit—*See* Film Integrated Circuit.

film pickup—A film projector combined with a television camera for telecasting scenes from a motion-picture film.

film reader—A computer input device which scans opaque and transparent patterns on photographic film and relays the corresponding information to the computer.

film recorder—A device that receives information from a computer and records it on photographic film.

film reproducer—An instrument which reproduces a recording on film.

film resistor—1. A fixed resistor the resistance element of which is a thin layer of conductive material on an insulated form. Some sort of mechanical protection is placed over this layer. 2. A resistor the characteristics of which depend on film rather than bulk properties.

film scanning—The process of converting movie film into corresponding electrical signals that can be transmitted by a television system.

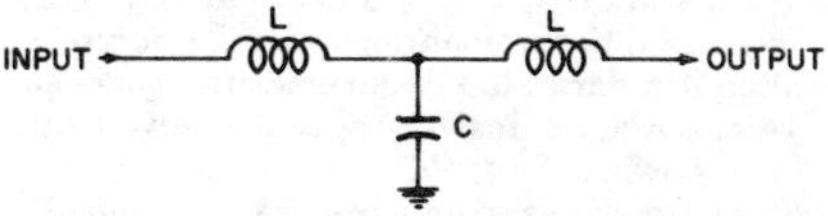

T-section.

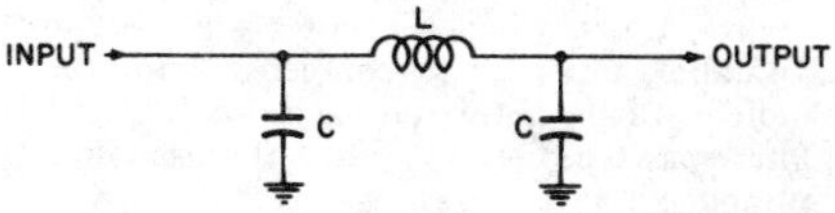

Pi section.

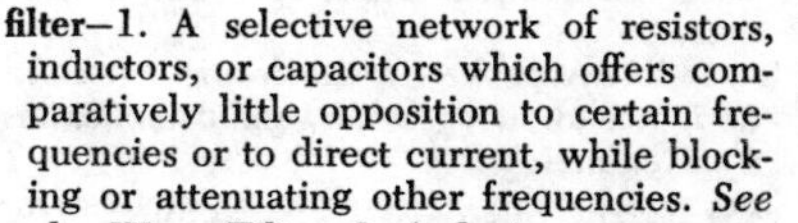

Ladder section.

Filters, 1.

filter—1. A selective network of resistors, inductors, or capacitors which offers comparatively little opposition to certain frequencies or to direct current, while blocking or attenuating other frequencies. *See also* Wave Filter. 2. A device or program that separates data, signals, or materials in accordance with specified criteria. 3. Also called extractor or mask. A machine word that specifies which parts of another machine word are to be operated on.

filter attenuation—A loss of power through a filter as a result of absorption in resistive materials, of reflection, or of radiation. Usually expressed in decibels.

filter attenuation band—Also called a filter stop band. A frequency band in which the attenuation constant is not zero if dissipation is neglected. In other words, a frequency band of attenuation.

filter capacitor—A capacitor used in a filter circuit. The term is usually reserved for electrolytic capacitors in a power-supply filter circuit.

filter center—In an aircraft control and warning system, a location at which information from observation posts is filtered for further dissemination to air-direction centers.

filter choke—Normally, an iron-core coil which allows direct current to pass while opposing the passage of pulsating or alternating current.

filter crystal or plate—A quartz plate or crystal used in an electrical circuit designed to pass energy only at certain frequencies.

filter discrimination – The difference between the minimum insertion loss at any frequency in the attenuation band of a filter, and the maximum insertion loss at any frequency in the transmission band of the filter. The loss is determined under the conditions of normal use of the filter.

filtered radar data—Radar data from which undesired returns have been removed by mapping.

filter-impedance compensator—An impedance compensator which is connected across the common terminals of electric-wave filters, when the latter are used in parallel, in order to compensate for the effects of the filters on each other.

filter passband – *See* Filter Transmission Band.

filter section—Any of various simple networks which may be connected in cascade to form a filter. The simplest is the half section, consisting of a series impedance (Z) followed by a shunt admittance (Y). A full section is either a tee network in which the shunt arm is Y and the series arms are Z/2, or a pi-network in which the series arm is Z and the shunt arms Y/2. Full sections, unlike half sections, have equal input and output impedances.

filter slot—A choke, in the form of a slot, designed to suppress unwanted modes in a waveguide.

filter stop band – *See* Filter Attenuation Band.

filter transmission band—Also called filter passband. A frequency band in which the attenuation constant is zero if dissipation is neglected. In other words, a frequency band of free transmission.

fin—A metal disc or a thin, projecting metal strip attached to a semiconductor to dissipate heat.

final actuation time—The time of termination of the chatter of a relay following contact actuation.

final amplifier—The stage which feeds the antenna in a transmitter.

final control element—The part of a control system that actually changes the amount of energy or fuel to the process. For example, in an industrial oven the final control element could be a valve that controls the amount of fuel reaching the burner.

final wrap—The outer layer of insulation around a coil, covering the saddle and splice insulation.

finder—In a telephone switching system, a name applied to the switch or relay group that selects the path which the call is to follow through the system.

fine-chrominance primary—Also called the I-signal. In the color television system presently standardized for broadcasting in the United States, the chrominance primary associated with the greater transmission bandwidth.

fine-tuning control – A receiver control which varies the frequency of the local oscillator over a small range to compensate for drift and permit fine adjustment to the carrier frequency of a station.

finger plethysmograph—An instrument for detecting and displaying changes in the volume of blood in the finger during the cardiac cycle. In some types, a light source and a photocell are placed on opposite sides of the finger; the volume of blood in the finger determines the amount of light reaching the photocell. In another type, the finger is placed between two electrodes. The increased blood volume during each contraction of the heart reduces the impedance of the finger, and the resulting change in the current between the electrodes is recorded or displayed on a cathode-ray tube.

finished blank – A crystal product after completion of all processes. It may also include the electrodes adherent to the crystal blank.

finishing—The process of repeated hand lapping and electrical testing by which a finished crystal blank is brought up to specifications.

finishing rate—Expressed in amperes, the rate of charge to which the charging current of a battery is reduced near the end of the charge to prevent excessive gassing and temperature rise.

finish lead—The lead connected to the finish, or outer end, of a coil.

finite—Having fixed and definite limits.

fins—Radia sheets or discs of metal attached to metal parts of a power tube or other component for the purpose of dissipating heat.

fin waveguide—A waveguide in which a thin longitudinal metal fin is placed to increase the range of wavelengths over which the waveguide can transmit signals efficiently. The method usually is used with circular waveguides.

fire—To change from a blocked condition, in which negligible current flows, to a saturated condition, in which heavy current flows.

fire-control equipment – Equipment that takes in target indications from optical or radar devices and, after calculating the motion of the target and firing vehicle, properties of air, etc., puts out directions of bearing, elevation, and timing for aiming and firing the guns.

fire-control radar—Radar employed for directing gunfire against the targets it observes.

fired tube (tr, atr, and pre-tr tubes)—The condition of a tube while a radio-frequency glow discharge exists at the resonant gap, resonant window, or both.

firing—1. In any gas- or vapor-filled tube, the ionization of the gas and the start of current flow. 2. The excitation of a device during a brief pulse. 3. In a magnetic amplifier, the transition from the unsaturated to the saturated state of the saturable reactor during the conducting or gating alternation. 4. An adjective modifying phase or time, to designate when firing occurs.

firing-angle—1. The electrical angle of the plate-supply voltage at which ionization of a gaseous tube occurs. 2. In a magnetic amplifier, the point on a sine-wave control voltage at which the control ampere-turns are sufficient to saturate the core. This is the point where the secondary winding (load) impedance drops to zero, and almost all of the supply voltage appears across the load.

firing point—The point at which the gas or vapor in a tube ionizes and current begins to flow.

firing potential—The controlled potential at which conduction through a gas-filled tube begins.

firmware—1. Programs or instructions that are stored in read-only memories; firmware is analogous to software in a hardware form. 2. The internal interconnections that permanently determine what functions a device or system can perform. Also called microprogram. 3. Part of a computer program that is incorporated, at least temporarily, as machine hardware. For example, instructions contained in a ROM.

first audio stage—The first stage in an audio amplifier.

first detector—Now called the mixer. In a superheterodyne receiver, the stage where the local-oscillator signal is combined with the modulated incoming radio-frequency signal to produce the modulated intermediate-frequency signal.

first Fresnel zone—In optics and radio communications, the circular portion of a wave front intersecting the line between an emitter and a more distant point where the resultant disturbance is being observed. The center intersects the front with the direct ray, and the radius is such that the shortest path from the emitter through the periphery to the receiving point is one-half wave longer than the ray.

first-generation computer—A computer in which vacuum-tube components are used.

first selector—The selector that immediately follows a line-finder in a switch train. It responds to the dial pulses that represent the first digit of the called telephone number.

fishbone antenna—An antenna consisting of a series of coplanar elements arranged in collinear pairs and loosely coupled to a balanced transmission line.

fishpaper—A tough fiber used in sheet form for insulating transformer windings from the core, field coils from field poles, or conductors from the armature.

fission—Also called atomic fission or nuclear fission. The splitting of an atomic nucleus into two parts. Fission reactions occur only with heavy elements such as uranium and plutonium and are accompanied by large amounts of radioactivity and heat.

fissionable—Capable of undergoing fission.

fission products—The elements which result from atomic fission. They may consist of more than forty different radioactive elements such as arsenic, silver, cadmium, iodine, barium, tin, cerium, and others.

fitting—An accessory, such as a locknut or bushing, to a wiring system. Its function is primarily mechanical rather than electrical.

five-layer device—A semiconductor, as a diac, triac, etc., in which there are four p-n junctions.

five-level code—A telegraph code in which five impulses are utilized for describing a character. For asynchronous transmission, start and stop elements may be added. A common five-level code is the Baudot code.

five-level start-stop operation — Simplex mode of teletypewriter operation. Each code character consists of five electrical units. The distributor unit of the machine makes a positive start and stop for the transmission of each character.

fix—A position determined without reference to any former position.

fixed bias—A constant value of bias voltage.

fixed capacitor—A capacitor designed with a definite capacitance that cannot be adjusted.

fixed composition (carbon composition resistor)—Resistive element consists of a carbon composition which is molded under extreme pressure, then enclosed in an insulating sleeve.

fixed crystal—A crystal detector with a nondefinite contact position.

fixed-cycle operation—1. A type of computer performance whereby a fixed amount of time is allocated to an operation. 2. Synchronous or clock-type arrangement in a computer in which events occur as a function of measured time.

fixed decimal—Calculator that is limited to established decimal category; can be preset for specified number of places in answer, or preset so that numbers are entered as they would be written.

fixed decimal point—Location of the decimal point in the display of a calculator chosen by a selector switch. For example, if the switch is set to position six on an eight-digit machine, the numbers between 99 and 0.001 can be used. In some machines no selector is provided, and a calculation like 123/456 yields the answer 0.28 instead of 0.2697368.

fixed echo—A stationary echo indication on a radar ppi display, indicating a fixed target.

fixed-frequency iff—A class of iff (identification friend or foe) equipment which responds immediately to every interrogation, thus permitting the response to be displayed on plan-position indicators.

fixed-frequency transmitter — A transmitter designed for operation on a single carrier frequency.

fixed-length record—Pertaining to a file in which all records are constrained to be of the same predetermined length. (Opposite of variable-length record.)

fixed logic—Circuit logic computers or peripheral devices that cannot be changed through operation of external controls. Connections must be physically changed to rearrange the logic.

fixed memory—1. A nondestructive-readout computer memory that is alterable only by mechanical means. 2. A memory into which information normally can be written only once. The ROM is a fixed program memory. Programs are usually stored in fixed memories.

fixed point—Pertaining to notation or a system of arithmetic in which all numeric quantities are expressed with a predetermined number of digits and the point is located implicitly at some predetermined position.

fixed-point arithmetic—1. Calculations in which the computing device is not concerned with the location of the point. An example is a slide-rule calculation, since the human operator must locate the decimal point. 2. A type of arithmetic in which all figures must remain within certain fixed limits.

fixed-point system—A system of notation in which a number is represented by a single set of digits and the position of the radix point is not numerically expressed. (*See also* Floating-Point System.)

fixed-program computer — *See* Wired-Program Computer.

fixed resistor—A resistor designed to introduce only a predetermined amount of resistance into an electrical circuit and not adjustable.

fixed screen—Application of a potential to a screen grid which is unaffected by other operating conditions within the tube.

fixed service—Any service communicating by radio between fixed points, except broadcasting and special services.

fixed station—1. A station in the fixed service. (A fixed station may, as a secondary service, transmit to mobile stations on its normal frequencies.) 2. A permanent station which communicates with other fixed stations.

fixed transmitter—A transmitter operated from a permanent location.

fixed-voltage winding—The motor winding which is excited by a fixed voltage.

fixer network or system—A combination of radio or radar direction-finding installations that, when operated in combination, can determine the position relative to the ground of an aircraft in flight.

fixturing—An assortment of electronic switches, wiring, black boxes, etc., to connect two systems, e.g., a test system and a board under test.

fl—Abbreviation for footlambert.

flag—1. A large sheet of metal or fabric for shielding television camera lenses from light. 2. In a computer, an indication that a particular operation has been completed and may be skipped by the program.

flag lines—Inputs to a microprocessor controlled by i/o devices and tested by branch instructions.

flag terminal—A type of solderless terminal in which the tongue projects from the side rather than the end of the terminal barrel.

flame-failure control—A system which automatically stops the fuel supply to a furnace if the pilot burner accidentally goes out.

flame microphone—A microphone in which the action of sound waves on a flame changes the resistance between two electrodes in the flame.

flame resistance—The characteristic of a material that prevents it from flaming when the source of heat is removed.

flame-resistant—*See* Flame-Retardant.

flame-retardant—Retarding ignition and the spread of flames, either inherently or because of special treatment. Also called flame-resistant.

flammability—The ability of a material to support combustion.

flammable—Term applied to material which readily ignites and burns when exposed to flames or elevated temperatures. Also called combustible.

flange—1. Also called waveguide flange. A fitting used at the end of a waveguide for making attachment to a microwave component or to another waveguide. 2. The side of a tape reel, which prevents the tape on the hub from slipping sideways off the "pie."

flange connector—A mechanical joint employing plane flanges bolted together in a waveguide.

flange coupling — A connection utilizing flanges not in mechanical contact between two parts of a waveguide, yet introducing no discontinuity in the flow of energy along the guide.

flange focus—The distance from the mounting flange or reference surface of a lens to the focal plane for a subject at infinity.

flanking effect—The effect on filter characteristics of connecting additional filters in parallel.

flap attenuator—A form of waveguide attenuator in which a variable amount of loss is introduced by insertion of a sheet of resistive material, usually through a nonradiating slot.

flare — An enlarged and distorted radar-screen target indication due to excessive brightness.

flare angle—The continuous change in cross section of a waveguide.

flare factor—A number expressing the degree of outward curvature of a speaker horn.

flash—Sometimes called hit. Momentary interference to a television picture, lasting approximately one field or less and of sufficient magnitude to totally distort the picture information. In general, this term is used only when the impairment is so short that the basic impairment cannot be recognized.

flashback voltage—The inverse peak voltage at which ionization takes place in a gas tube.

flasher—A device, generally a thermal or motor-driven switch, that rapidly and automatically lights and extinguishes electric lamps.

flasher relay—A self-interrupting relay.

flashing—The application of a high-frequency electromagnetic field to an electron tube through the envelope to flash its getter during evacuation.

flashing of a dynamo—The flashing or sparking which is likely to take place at the brushes of a commutator.

flash lamp—A device in which a large amount of stored electrical energy is converted into light by means of a sudden electrical discharge. The flash is obtained by storing electrical energy in a capacitor and allowing the capacitor to discharge through the lamp.

flash magnetization—Magnetization of a ferromagnetic object by an abrupt current impulse.

flashover—A disruptive discharge through air, around or over the surface of insulation, or between parts of different potential or polarity produced by the application of voltage, wherein the breakdown path becomes sufficiently ionized to maintain an electric arc.

flashover voltage—1. The highest value attained by any voltage impulse which caused a flashover. 2. The voltage at which insulation fails by discharge between electrodes across the insulation surface.

flashpoint of impregnant—The temperature to which a liquid or solid impregnant must be heated before it gives off sufficient vapor to form a flammable mixture.

flash pulsing—Transmission of short bursts of radiation at irregular intervals by a mechanically controlled keyer.

flash test—A method of testing insulation by momentarily applying a voltage much higher than the working voltage.

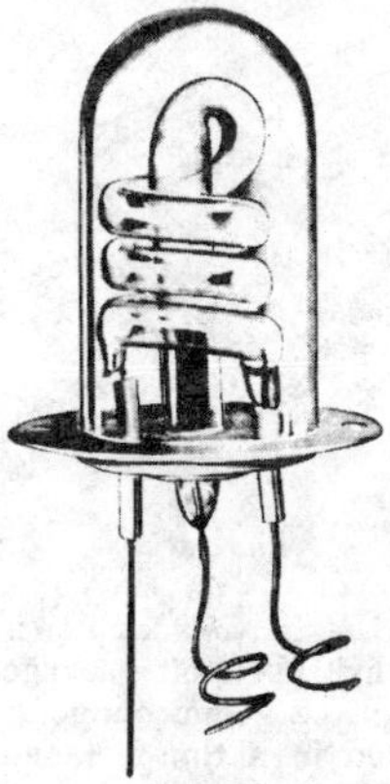

Flash tube.

flash tube—Also called electronic flash tube and photoflash tube. A gas-discharge tube for producing high-intensity, short-duration flashes of light. It consists of a glass tube bent in a U, a helix, or a combination of the two and filled with a rare gas. The tube has an anode, a cold cathode, and a trigger electrode. It is flashed by applying a high-voltage pulse to the trigger electrode.

flash welding—Welding in which an arc is first struck between the pieces to be welded. After the ends are thus heated, the weld is completed by bringing them together under pressure and cutting off the current.

flat—Having a slope of zero at all points, as a graph, curve, etc.

flat back paper—A flat kraft paper tape used in splicing electrical cable.

flat-compounded generator—A compound-wound generator in which the series field winding is adjusted so that the output voltage is virtually constant for currents between no load and full load.

flat fading—That type of fading in which all components of the received radio signal fluctuate in the same proportion simultaneously.

flat, flexible cable—*See* Tape Cable.

flat frequency response—The response of a system to a constant-amplitude function which varies in frequency is flat if the response remains within specified limits of amplitude, usually specified in decibels from a reference quantity.

flat leakage power (tr and pre-tr tubes)—The peak radio-frequency power transmitted through the tube after establishment of a steady-state radio-frequency discharge.

flat line—A radio-frequency transmission line or part of a line having a low standing wave ratio.

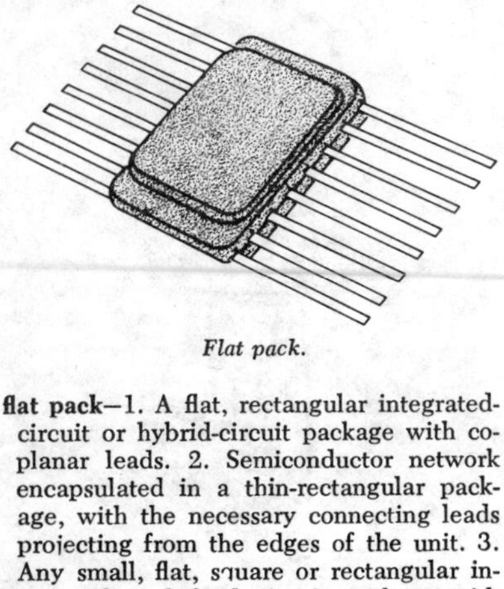

Flat pack.

flat pack—1. A flat, rectangular integrated-circuit or hybrid-circuit package with coplanar leads. 2. Semiconductor network encapsulated in a thin-rectangular package, with the necessary connecting leads projecting from the edges of the unit. 3. Any small, flat, square or rectangular integrated or hybrid circuit package with leads coming from the sides of the package in the same plane as the package.

flat response—Ability of a sound system to reproduce all tones (from the lowest to the highest) in their proper proportions. (For example, a specification of response within plus or minus one dB from 30 to 15,000 Hz would be considered "flat").

flat top—The horizontal portion of an antenna.

flat-top antenna—An antenna having two or more lengths of wire parallel to each other and to the ground.

flat-top response — Response characteristic in which a definite band of frequencies is transmitted uniformly.

flat-type armature — Of a relay, an armature which rotates about an axis perpendicular to that of the core, with the pole face on a side surface of the core.

flat-type relay—A relay having a flat-type armature.

flaw—In a material, any discontinuity that would be harmful to proper functioning of the material.

flaw detection—The process of using sonic or ultrasonic waves to locate imperfections in a solid material. This is done by transmitting the waves through the material and listening for reflections or variations in transmission when they strike an imperfection in the material.

F-layer—An ionized layer in the F-region, existing in the night hemisphere and in the weakly illuminated portion of the day hemisphere.

F_1-layer—One of the regular ionospheric layers at an average height of about 225 kilometers, which occurs during the daylight hours.

F_2-layer—The most useful of the ionospheric layers for radio-wave propagation. It is the most highly ionized and highest of the layers, having an average night height of 225 kilometers and a midday height of about 400 kilometers.

Fleming's rule—1. Also called the right-hand or left-hand rule. If the thumb and the first and second fingers are extended at right angles to one another, with the thumb representing the direction of the wire motion, the first finger representing the direction of magnetic lines of force (from the north pole to the south pole), and the second finger representing the direction of the current, then the right hand will give the correct relationships for a conductor in the armature of a generator, and the left hand will give the correct relationships for a conductor in the armature of a motor. This rule is applied to the so-called conventional current flow, which is the opposite of electron flow. 2. A rule stating that if the fingers of the right hand are placed around a current-carrying wire so that the thumb points in the direction of the conventional current, the fingers will point in the direction of the magnetic field.

Fleming valve—An early name for a diode, or two-electrode thermionic vacuum tube used as a detector.

Fletcher-Munson curves—Also called equal-loudness contours. A group of sensitivity curves showing the characteristics of the human ear for different intensity levels between the threshold of hearing and the threshold of feeling. The reference frequency is 1000 Hz.

Flewelling circuit—An early radio circuit in which one tube served as a detector, amplifier, and local oscillator.

flexible coupling—1. A device for connecting two shafts end to end so that they can be rotated even though not exactly aligned. 2. Mechanical connection be-

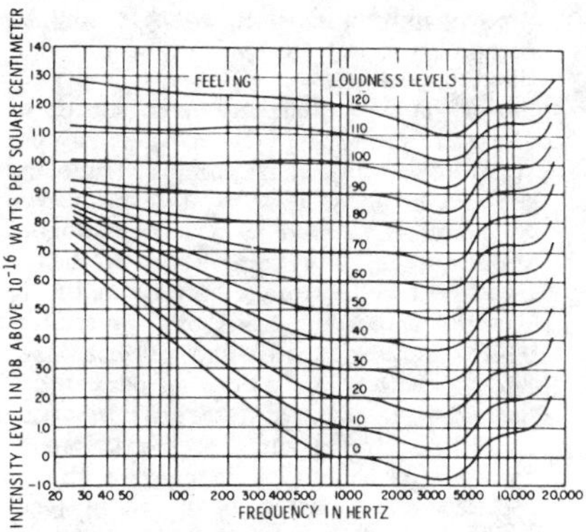

Fletcher-Munson curves.

tween two lengths of waveguide normally lying in a straight line and designed to allow a limited angular movement between axes.

flexible disk—See Floppy Disk.

flexible resistor—A wirewound resistor that looks like a flexible lead. It is made by winding *Nichrome* or any other type of resistance wire around asbestos or other heat-resistant cord. The wire is then covered with braided insulation, which is color coded to indicate the resistor value.

flexible shaft—A flexible core made up of layers of wire, which rotate inside a metal or rubber-covered flexible casing. The casing not only supports the core, but also acts as the bearing surface for the core.

flex life—A measure of the resistance of a conductor or other device to failure due to fatigue from repeated bending.

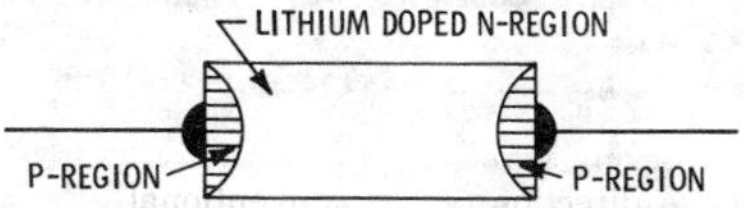

Flexode.

flexode—A flexible diode containing a junction that may be altered at will from a pn junction in one direction, to no junction at all, to a pn junction in the opposite direction. Thus the direction of easy current may be reversed without reversing the leads to the diode, and the resistance of the diode may be continuously varied from the back-resistance value to the forward-resistance value. It may be set to behave as a simple resistor, with the same value for both directions of current.

flicker—1. In television, the flickering produced in the picture when the field frequency is insufficient to completely synchronize the visual images. 2. In a regulated power supply, a phenomenon due primarily to sudden, minute changes of brief duration in the reference voltage or the input-stage junctions of the correction amplifier in the regulator. 3. Also called jitter or wobble. Noise in an amplifier, of higher frequency than drift, but lower than power-line or chopper-drive frequency noise.

flicker effect—Small variations in the plate current of a thermionic vacuum tube, believed to be due to random emission of positive ions by the cathode.

flicker photometer—A device for measuring the intensity of a light source. Illumination from the light source being measured and a standard light source are observed alternately in rapid succession. When the standard source is equal to the other, the flickering disappears.

flight control—Real-time calculations for the control of a vehicle in flight; includes stabilization, fuel monitoring, cruise control, etc.

flight path—A planned course for an airborne vehicle.

flight-path computer—A computer that includes all the functions of a course-line computer and also controls the altitude of an aircraft in accordance with a desired plan of flight.

flight-path deviation—The difference between the flight track of an aircraft and the actual flight path expressed in terms of either angular or linear measurement.

flight-path-deviation indicator—An instrument that provides a visual indication of deviation from a flight path.

flight track—The three-dimensional path in space actually traced by a vehicle.

Flinders bar—A bar of soft iron placed near a compass to correct errors due to variation of the vertical component of the earth's magnetism in different parts of the world.

flip-chip—1. An unencapsulated semiconductor device in which bead-type leads terminate on one face to permit flip (facedown) mounting of the device by contact of the leads to the required circuit interconnectors. 2. A mounting approach in which the chip (die) is inverted and connected directly to the substrate rather than using the more common wire bonding technique. Examples of this kind of flip chip mounting are beam lead and solder bump. 3. A generic term describing a semiconductor device having all terminations on one side in the form of bump contacts. After the surface of the chip has been passivated or otherwise treated, it is flipped over for attaching to a matching substrate.

flip-chip bonding—Method of interconnecting ICs in a circuit by bonding "bumps," located on the IC chip's back surface, to the circuit's conducting paths.

flip coil—A small coil used for measuring a magnetic field. When connected to a ballistic galvanometer or other instrument, it gives an indication whenever the magnetic field of the coil or its position in the field is suddenly reversed.

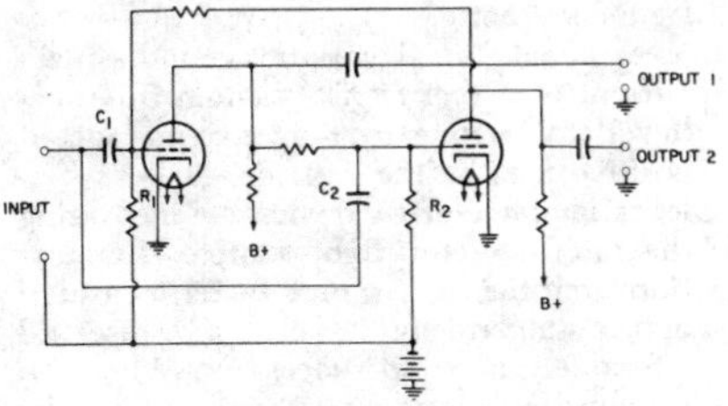

Flip-flop, 1.

flip-flop—1. Also called bistable multivibrator, Eccles-Jordan circuit, or trigger circuit. A two-stage multivibrator circuit having two stable states. In one state, the first stage is conducting and the second is cut off. In the other state, the second stage is conducting and the first stage is cut off. A trigger signal changes the circuit from one state to the other, and the next trigger signal changes it back to the first state. For counting and scaling purposes, a flip-flop can be used to deliver one output pulse for each two input pulses. 2. A similar bistable device with an input which allows it to act as a single-stage binary counter. 3. An electronic circuit having two static states and the ability to change from one state to the other on application of a signal in a special manner. 4. A type of digital circuit which can be in either of two states, depending both on the input received and on which state it was in when the input was received.

flip-flop circuit—An electronic circuit that has two conditions of permanent stability and a means for changing from one to the other in response to an external stimulus. *See also* Eccles-Jordan Circuit.

flip-flop equipment—An electronic or electromechanical device which causes automatic alternation between two possible circuit paths. The same term is often applied to any mechanical operation which is analogous to the principles of the flip-flop.

flip-flop key—Calculator that can display double its digital capacity in two steps by depressing flip-flop key.

flip-flop multivibrator—Also called start-stop multivibrator. A biased rectangular wave generator which operates for one cycle when a synchronizing trigger signal is applied.

flipover cartridge—A phonograph cartridge having separate needles for playing micro-groove and standard records. It may be turned to bring the proper needle into playing position.

float—1. To be connected to no source of electrical potential. (Often used with respect to a particular point). 2. To be maintained in a constant state of charge by being connected to a source of constant voltage, as a storage battery.

floated battery—A storage battery kept fully charged across the leads of a generator. The generator carries the normal load, and the battery assists during peaks.

floating—1. Keeping a storage battery connected in parallel with an electric supply to serve as a standby in case of supply failure and to assist in handling peak loads. 2. The condition of a device or circuit that is not grounded and not tied to any established potential.

floating address—*See* Symbolic Address.

floating-carrier modulation—*See* Controlled, Carrier Modulation.

floating charge—Continuous charging of a storage battery with a low current to keep the battery fully charged while idle or on light duty.

floating decimal—A calculator function that allows the user to calculate any decimal category; decimal may or may not be present; if present, automatically positions itself correctly in the answer.

floating-decimal arithmetic—*See* Floating-Point Arithmetic.

floating decimal point—Calculator entry that may contain the decimal point in any position. The number and decimal point will be properly positioned automatically when displayed.

floating grid—A vacuum-tube grid that is not connected to any circuit. It assumes a negative potential with respect to the cathode.

floating ground—A reference ground that is not earthed.

floating in—Decimal-point position need not be preset; numbers in a calculator are entered as they would be written.

floating input—1. An isolated input circuit not connected to ground at any point (the maximum permissible voltage to ground is limited by electrical design parameters of the circuit involved). It is understood that in a floating input circuit, both conductors are equally free from any reference potential, a qualification which limits the types of signal sources which can be operated floating. 2. *See* Differential-Input Measurement.

floating junction—A semiconductor junction through which no net current flows.

floating neutral—A circuit in which the voltage to ground is free to vary with circuit conditions.

floating out—Decimal point in a calculator is automatically aligned in the answer.

floating point – Pertaining to a form of number representation in which quantities are expressed in terms of a bounded number (mantissa) and a scale factor (characteristic or exponent) consisting of a power of the number base. For example, $127.6 = 0.1276 \times 10^3$ where the bounds are 0 and 1.

floating-point arithmetic—1. Computer handling of data in which the point is not always in the same position. Floating-point numbers are expressed in terms of digits and exponents. 2. In a digital computer, a form of arithmetic in which each number is represented by several significant digits, with an explicitly placed decimal point, multiplied by the base of the number system raised to a power, as for instance 6.3542×10^5. In computations of this kind the decimal point and exponent are adjusted automatically.

floating-point calculation—In a computer, a calculation taking into account the varying location of the decimal point (if base 10) or binary point (if base 2). The sign and coefficient of each number are specified separately.

floating-point routine – Coded instructions in proper sequence to direct a computer to perform a calculation with floating-point operation.

floating-point system—A system of numbering in which an added set of digits is used to denote the location of the radix point. (*See also* Fixed-Point System.)

floating potential—The dc voltage between an open-circuited terminal of a circuit and a reference point when a dc voltage is applied to the other circuit terminals as specified.

floating zero—In a machine-tool control system, the characteristic that allows the reference-point zero to be located readily anywhere along an axis of travel. Previously established reference points are eliminated from the control memory.

float switch—A switch actuated by a float on the surface of a liquid.

float-zone crystal—A crystal grown by passing a molten zone through a cylinder of material. No other material, with the possible exception of a gas, contacts the molten zone. When the crystal is grown in a vaccum, the term "vacuum float-zone crystal" is frequently used.

flock—Finely divided felt used on phonograph turntables, underneath microphone stands, or wherever a nonscratching surface is desired.

flood projection—In facsimile transmission, an optical method in which all of the subject to be transmitted is illuminated and the scanning spot is defined by an aperture between the subject and the light-sensitive device.

floppy disk—Also called diskette, flexible disk or minidisk. A large round nonrigid piece of magnetic tape. An oxide coated Mylar disk, 7.8 inches in diameter and 0.005 inch thick with a 1.5-inch hole in the center. The disk is contained in a flexible plastic envelope 8 inches square and 1/16 inch thick. It is coated on the inside with a soft material that permits easy rotation of the disk inside the envelope at 360 rpm. A slot in the envelope provides access for the read-write head to the disk surface.

flow—1. The passage of electrons (a current) through a conductor or through the space between electrodes. 2. A general term to indicate a sequence of events.

flow amplification—The rate of change of the flow in a specified load impedance, connected to a device, with respect to the change in the flow applied to the controls of the device.

flowchart—1. *See* Flow Diagram. 2. A graphic presentation of the major steps of work in process with accent on how the work flows through the process rather than on how the steps are done.

flowchart symbol—A symbol used on a flowchart to represent data, flow, equipment, or an operation.

flow diagram—Also called flowchart. A chart showing all the logical steps of a computer program. A program is coded by writing down the successive instructions that will cause the computer to perform the logical operations necessary for solving the problem, as represented on a flowchart.

flow direction—In flowcharting, the antecedent-to-successor relation between operations on a flowchart; it is indicated by arrows or other means.

flowed wax—A mechanical recording disc prepared by melting and flowing wax onto a metal base.

flowline—On a flowchart, a line that represents a connecting path between flowchart symbols, such as a line indicating transfer of data or control.

flowmeter—A device for measuring the rate of flow of liquids or gases.

flow soldering—Also called wave soldering. A method of soldering printed-circuit boards by moving them over a flowing wave of molten solder in a solder bath.

fluctuating current—A direct current that changes in value, but not at a steady rate.

fluctuation noise—*See* Random Noise.

fluctuation voltage – Small voltage variations in a thermionic tube due to thermal agitation, shot effect, flicker effect, etc.

fluid computer – A digital computer constructed totally from fluid logic elements. All logic functions are carried out by interaction between jets of air or liquid, and the device contains no moving parts or electronic circuits.

fluid damping—Damping obtained through displacement of a viscous fluid and the accompanying dissipation of heat.

fluidic—Of or pertaining to devices, systems, assemblies, etc., utilizing fluidic components.

fluidics—1. The branch of engineering and technology concerned with the design and production of logic elements, amplifiers, and the like, that depend for their operation on interactions between jets of fluid rather than on electrical phenomena. (While slower than electronic logic systems, fluid logic systems can operate in environments that (would damage electronic systems.) 2. The technology wherein sensing, control, information processing, and/or actuation functions are performed solely through utilizing fluid dynamic phenomena.

fluidized bed coating—A method of applying a resin coating to an article. The heated article is immersed in a dense-phase aerated bed of powdered resin, and then is heated in an oven to obtain a smooth, pinhole-free coating.

fluorescence — The emission of light (or other electromagnetic radiation of longer wavelengths) by a substance as a result of the absorption of some other radiation of shorter wavelengths only as long as the stimulus producing it is maintained. Luminescence persists for less than about 10^{-8} seconds after excitation is stopped.)

fluorescent—Having the property of giving off light when activated by electronic bombardment or a source of radiant energy.

fluorescent lamp — An electric discharge lamp in which a gas ionizes and produces radiation that activates the fluorescent material inside the glass tubing. The phosphors in the fluorescent material transform the radiant energy from the electric discharge into wavelengths giving more light (higher luminosity).

fluorescent material—A material that fluoresces readily when exposed to electron beams, X-rays, or other radiation.

fluorescent screen—The coating, on the face of a cathode-ray or television picture tube, which glows under electron bambardment.

fluorometer—An instrument for measuring fluorescence.

fluoroscope—An instrument with a fluorescent screen suitably mounted with respect to an X-ray tube, used for immediate indirect viewing of internal organs of the body, or internal structures in apparatus or masses of metals, by means of X-rays. A fluorescent image (really a kind of X-ray shadow picture) is produced.

fluoroscopy—The use in diagnosis, testing, etc., of a fluorescent screen activated by X-rays.

flush receptacle—A receptacle recessed into a wall, with only the plate extending beyond the surface.

flush-type instrument—An instrument designed to be mounted with its face projecting only slightly from the front of the panel.

flutter—1. Also called wow and drift. The frequency deviations produced by irregular motion of a turntable or tape transport during recording, duplication, or reproduction. The term "flutter" usually refers to relatively high cyclic deviations (for example, 10 hertz), and the term "wow" to relatively low ones (for example, a variation of once per turntable revolution). The term "drift" usually refers to a random rate close to zero hertz. 2. In communications, (a) distortion due to variations in loss resulting from simultaneous transmission of a signal at another frequency, or (b) a similar effect due to phase distortion. 3. Rapidly-repeated fluctuations in tape speed that introduce spurious burbling, quivering or shimmering variations in the pitch of the reproduced sound. 4. A fast change in pitch (about 10 Hz) caused by a change in the speed of a turntable specified as a percentage of the test frequency (usually 3000 or 3150 Hz) and figures below about 0.1% are good.

Flutter bridge.

flutter bridge—An instrument for measuring the irregularities in a constant-speed device such as a film, disc, or tape recorder.

flutter echo—1. A rapid succession of reflected pulses resulting from a single initial pulse. 2. A multiple echo in which the reflections occur in rapid succession. If periodic and audible, it is referred to as a musical echo.

flutter rate—The number of times per second the flutter varies.

flux — 1. A material used to promote the joining of metals in soldering. Rosin is widely used in soldering electronic parts. 2. Number of particles crossing a unit area per unit time. The common unit of flux is particles/cm²/sec. Integrated flux, after an exposure of time T, is equal to the total number of particles which have traversed a unit area during time T. 3. The number of photons that pass through a surface per unit time. Expressed in lumens or watts.

flux density—A measure of the strength of a wave; flux per unit area normal to the direction of the flux; number of photons passing through a surface per unit time per unit area. Expressed in watts/cm^2 or lumens/ft^2.

fluxgate—A magnetic azimuth-sensitive element of the fluxgate-compass system activated by the earth's magnetic field.

fluxgate compass—A gyrostabilized, remote-indicating compass and azimuth-control system used with automatic pilots.

fluxgraph — A machine that automatically plots on paper the magnetic field strength at various points in the vicinity of a coil.

flux guide—In induction heating, a magnetic material used for guiding the electromagnetic flux to the desired location or for confining it to definite regions.

flux linkage—Magnetic lines of force which link a coil of wire. Whenever the flux linkage changes, an emf is generated in the coil.

fluxmeter—An instrument used with a test coil for measuring magnetic flux. It consists usually of a moving-coil galvanometer in which the torsional control is either negligible or compensated.

flyback—1. The shorter of the two time intervals comprising a sawtooth wave. 2. Also called retrace. As applied to a cathode-ray tube, the return of the spot to its starting point after having reached the end of its trace. This portion of the wave is usually not seen because of blanking circuits or the shortage of time.

flyback checker — An instrument used to check flyback or other transformers or inductors for open windings or shorted turns.

flyback power supply — The power supply that generates the high dc voltage required by the second anode of a picture tube. This voltage is produced during the flyback period, the current in the horizontal-deflecting coils reversing and inducing

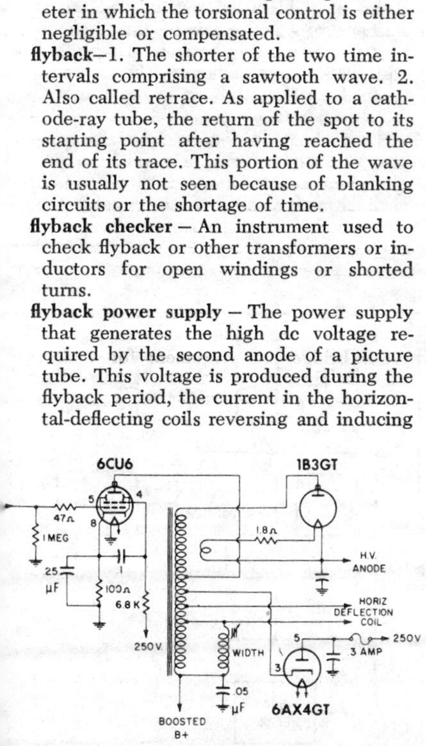

Flyback power supply.

a sharp pulse in the primary of the transformer supplying the deflection circuit. This pulse is stepped up by an autotransformer and rectified. After suitable filtering, it becomes a very high dc voltage.

flyback tester—An instrument that tests flyback transformers and sometimes also deflection yokes.

flyback time—The period during which the electron beam is returning from the end of a scanning line to begin the next line.

flyback transformer—Also called horizontal-output transformer. A transformer used in the horizontal-deflection circuit of a television receiver to provide the horizontal-scanning and accelerating-anode voltages for the cathode-ray tube. It also supplies the filament voltage for the high-voltage rectifier.

flycutter——An accessory used with a drill press to cut out large holes in metal or wood.

flying spot—A small, rapidly moving spot of light, usually generated by a cathode-ray tube, used to scan an image field for television transmission.

flying-spot scanner—1. Also called light-spot scanner. A television scanning device embodying a small beam which is moved over a scene or film and translates the highlights and shadows into electrical signals. 2. In optical character recognition, a device employing a moving spot of light to scan a sample space, the intensity of the transmitted or reflected light being sensed by a photoelectric transducer.

fly's-eye lens—A multiple lens made up of hundreds of small, closely spaced lenses. It forms many images of the same subject and eliminates the need for step and repeat techniques in the fabrication of microelectronic circuits.

flywheel effect—The maintaining of oscillations in a circuit in the intervals between pulses of excitation energy. The action is analogous to the rotation of a flywheel due to its stored mechanical energy.

flywheel synchronization — Automatic frequency control of a scanning system based on the average timing of the incoming sync signals rather than triggering of the scanning circuit by each pulse. It is used in high-sensitivity television receivers for fringe-area reception, in which noise pulses might otherwise trigger the sweep circuit prematurely.

flywheel tuning—A tuning-dial mechanism which uses a heavy flywheel on the control shaft for added momentum, to obtain a smoother tuning action.

fm—Abbreviation for frequency modulation.

fm/am—A system in which information subcarriers are frequency-modulated and are used to amplitude-modulate the carrier.

fm/am multiplier—A multiplier in which a carrier is modulated so that its frequency deviation from the center value is proportional to one variable, and its amplitude is proportional to another variable. The modulated carrier is then consecutively demodulated for fm and for am. The final output is proportional to the product of the two variables.

fm broadcast band—The band of frequencies extending from 88 to 108 MHz, which include those assigned to noncommercial educational broadcasting.

fm broadcast channel—A band of frequencies 200 kHz wide and designated by its center frequency. Channels for fm broadcast stations begin at 88.1 MHz and continue in steps of 200 kHz through 107.9 MHz.

fm broadcast station—A station employing frequency modulation in the fm broadcast band and licensed primarily for the transmission of radiotelephone emissions intended to be received by the general public.

fm discriminator—A device which converts frequency variations to proportional variations in the amplitude of an electrical signal. Discriminators may be of several basic types, such as pulse averaging, Foster-Seely, ratio detector, or phase-lock correlation detector.

fm discriminator (subcarrier) — The same as an fm discriminator except that it is used to convert subcarrier frequency variations into proportional voltage or current signals.

fm-fm—Frequency modulation of a carrier by one or more subcarriers that are themselves frequency modulated by information.

fm laser—A conventional laser with a phase modulator inside its Fabry-Perot cavity. It is characterized by a lack of noise resulting from random phase fluctuation in the various modes.

fm multiplex—*See* FM Stereo.

fm noise level—Residual frequency modulation of an aural transmitter as a result of disturbances in the frequency range between 50 and 15,000 Hz.

fm/pm — A system in which information subcarriers are frequency-modulated and used to phase-modulate the carrier.

fm radar—*See* Frequency-Modulated Radar.

fm receiver deviation sensitivity — The smallest frequency deviation that results in a specified output power.

fm recording (magnetic tape)—A method of recording in which the input signal modulates a voltage-controlled oscillator, the output of which is delivered to the recording head.

fm stereo—1. Also called fm multiplex. A means by which fm radio stations are able to transmit stereophonic program material to specially designed receivers and which is at the same time compatible with monophonic equipment. 2. Fm broadcasting in which two channels of sound are transmitted, offering a signal similar to the stereo available from records and tapes. To hear fm stereo requires either a stereo fm tuner or a monophonic fm tuner fitted with an fm stereo adapter. The technical means for transmitting fm stereo is known as multiplexing

fm stereophonic broadcast—The transmission of a stereophonic program by a single fm broadcast station utilizing the main channel and a subchannel to carry the

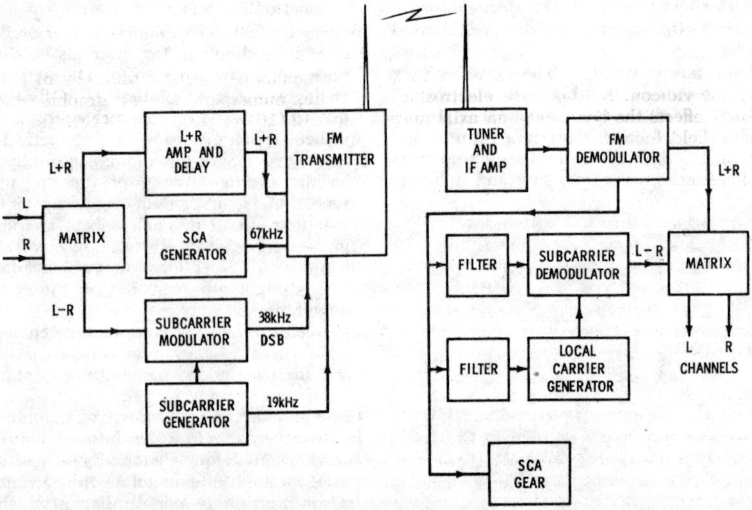

Fm stereophonic broadcast system.

signals required to produce the stereophonic effect.

f/number—In optical terminology, a number which describes a lens; ratio of focal length to lens diameter.

focal length—Symbolized by *f*. The distance from the principal focus (focus of parallel rays of light) to the surface of a mirror or the optical center of a lens.

focometer — An instrument for measuring the focal length of a lens or an optical system.

focus—1. The convergence of light rays or an electron beam at a selected point. 2. The sharp definition of a scanning beam in television receivers or optical systems.

focus control—On a television receiver, a potentiometer control used for fine focusing of the electron beam. The control varies the first-anode voltage of an electrostatic tube or the focus-coil current of a magnetic tube.

focusing — The process of controlling the convergence and divergence of an electron or light beam.

focusing anode—One of the electrodes used to focus the electron beam in a cathode-ray tube. As its potential changes, so does the electric field, thereby altering the path of the electrons.

focusing coil—The coil around the neck of a cathode-ray tube. It provides a magnetic field, parallel to the electron beam, for controlling the cross-sectional area of the beam on the screen.

focusing electrode—An electrode to which a potential is applied to control the cross-sectional area of the electron beam.

focusing magnet—A permanent magnet assembly that produces a magnetic field for focusing the electron beam in a cathode-ray tube.

focus projection and scanning—A method of magnetically focusing and electrostatically deflecting the electron beam in a hybrid vidicon. A transverse electrostatic field deflects the beam, and an axial magnetic field focuses the beam.

foldback — British term synonymous with talkback.

fold-back characteristic—*See* Current Liming (Automatic).

folded cavity — An arrangement used for producing a cumulative effect in a klystron repeater. This is done by making the incoming wave act in several places on the electron stream from the cathode.

folded-dipole antenna — An antenna comprising two parallel, closely spaced dipole antennas. Both are connected together at their ends, and one is fed at its center.

folded heater — A strand of bent, coated wire inserted into a cathode sleeve.

folded horn—1. An acoustic horn which is curled to permit more efficient use of the space it occupies. 2. A type of loudspeaker enclosure employing a horn-shaped passageway for aiding the bass response.

folding frequency—The frequency that is one-half the sampling rate when samples are made continuously at equal intervals.

foldover — A distorted television picture, which appears to overlap horizontally or vertically. It is due to nonlinear horizontal- or vertical-sweep circuits.

follow current—1. That line current which tends to follow a lightning discharge through an arrester to ground. 2. The current through a lightning protector from a connected steady-state power source which flows during and following the discharge of a surge or transient current.

follower—A circuit in which the output of a high-gain amplifier is fed directly back to its negative input. The input signal is reproduced without polarity reversal.

follower drive—Also called slave drive. A drive in which the reference input and operation are direct functions of a master drive.

follower with gain — A follower in which only a part of the output voltage is fed back in series opposition to the input signal. Hence, closed-loop gain greater than unity is obtained over the rated range of operation.

following blacks—Also called edge effect, trailing reversal, or trailing blacks. A picture condition in which the edge following a white object is overshaded toward black (i.e., the object appears to have a trailing black border).

following whites—Also called edge effect, trailing reversal, or trailing whites. A picture condition in which the edge following a black or dark gray object is shaded toward white (i.e., the object appears to have a trailing white border).

font—1. The characteristic style of a set of alphanumerics, e.g., Gothic. 2. An alphabetic, numeric, or other graphic shape, i.e., 3/16 gothic font, 1428E font, ocr (A) font, etc.

footcandle (fc)—The unit of illumination when the foot is taken as the unit of length. It is the illumination on a surface one square foot in area on which there is a uniformly distributed flux of one lumen,

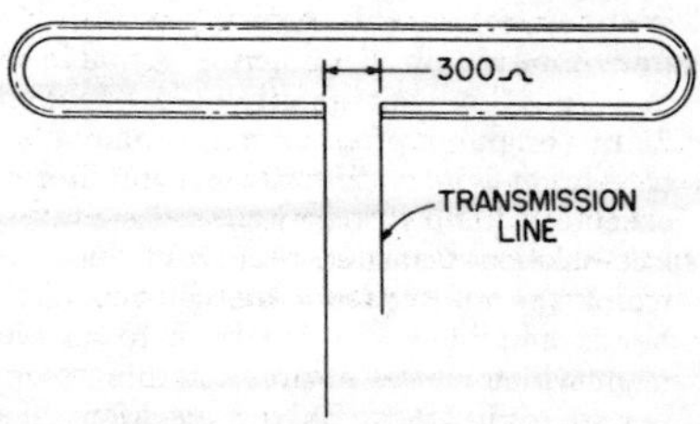

Folded-dipole antenna.

or the illumination produced on a surface all points of which are at a distance of one foot from a directionally uniform point source of one candela.

foot control – A foot-actuated start-stop switch, usually used for dictating and transcribing via tape.

footlambert—A unit of luminance equal to $1/\pi$ candle per square foot, or to the uniform luminance of a perfect diffusing surface emitting or reflecting light at the rate of one lumen per square foot.

footlambert (fl) – A unit of luminance (photometric brightness) equal to $1/\pi$ candela per square foot, or to the uniform luminance of a perfectly diffusing surface emitting or reflecting light at the rate of one lumen per square foot or to the average luminance of any surface emitting or reflecting light at that rate. The average luminance of any reflecting surface in footlamberts is, therefore, the product of the illumination in footcandles by the luminous reflectance of the surface.

foot-pound—A unit of measurement equivalent to the work of raising one pound vertically a distance of one foot.

forbidden band—The energy band lying between the conduction and valence bands. The energy difference across it determines whether a solid acts as a conductor, semiconductor, or insulator.

forbidden combination—A combination of bits or other representations that is invalid according to some criterion.

forbidden-combination check—A test, usually automatic, for the occurrence of a code expression that is not permissible. A self-checking code (or error-detecting code) uses code expressions such that errors result in a forbidden combination. A parity check uses a self-checking binary-digit code in which the total number of 1's (or 0's) in each permissible code expression is always even or always odd. A check may be made for either even parity or odd parity. A redundancy check makes use of a self-checking code that employs redundant digits called check digits.

forbidden energy gap—The energy range of a semiconductor between the bottom of the conduction band and the top of the valence band. Electrons cannot exist at energies within this range.

force—1. Any physical action capable of moving a body or modifying its motion. 2. In computer programming, manual intervention which directs the computer to execute a jump instruction.

force-balance transducer—A transducer in which the output from the sensing member is amplified and fed back to an element which causes a force-summing member to return it to its rest position. The magnitude of the signal fed back determines the output of the device, like the error signal in a servo signal.

forced coding – *See* Minimum-Access Programming.

force differential—The difference between the operating force and the release force of a momentary contact switch.

forced oscillation – In a linear constant-parameter system, the response to an applied driving force, excluding the transient which results from energy at the time the driving force is applied.

force factor (of an electromechanical or electroacoustic transducer)—1. The complex quotient of the force required to block the mechanical or acoustic system, divided by the corresponding current in the electrical system. 2. The complex quotient of the resultant open-circuit voltage in the electric system, divided by the velocity in the mechanical or acoustic system.

force-summing device—In a transducer the element directly displaced by the applied stimulus.

fore pump – An auxiliary vacuum pump used as the first stage in evacuating vacuum systems.

foreshortened addressing—A feature of control computers that makes it possible to use simpler instructions when addressing the computer; hence less of the available computer storage is used for this purpose.

fork oscillator—An oscillator in which a tuning fork is the frequency-determining element.

fork tines—The projecting ends of a tuning fork. When vibrated, they produce a constant frequency.

form—To apply a voltage to an electrolytic capacitor, semiconductor, or other component as part of a manufacturing process, in order to cause a desired change in its characteristics.

formant – 1. The particular frequency region in which the energy of a vowel sound is concentrated most strongly. 2. In an organ, an electrical circuit, the purpose of which is to alter the tone quality of sound amplified by it. A formant filter is applied to the entire output from a manual, rather than to individual tones.

formant filter—A waveshaping network used in an organ to modify the signal from the tone generator so it will assume the waveshape of the desired tone.

format—In a computer, a specified grouping of data to facilitate storage and movement of the data in the system. A given format may include control codes, record marks, block marks, and tape marks in a prearranged sequence. The format tells the operator or the system how the transfer, processing, and printing of data are to be controlled. The term format also describes the layout of characters on printed copy, which is directly related to the data format.

formatting—The arranging in a predefined order of code characters within a record.

form factor—1. Shape (diameter/length) of a coil. 2. Ratio of the effective value of a symmetrical alternating quantity to its half-period average value. 3. A figure of merit which indicates how much the current departs from pure dc or from a continuous, nonpulsating current. Unity represents pure dc. Values greater than one indicate an increasing departure from pure dc. A departure from unity form factor increases the heating effect in a motor and reduces brush life.

Formica—Trade name for a phenolic compound having good insulating qualities.

forming—The application of voltage to an electrolytic capacitor, electrolytic rectifier, or semiconductor device to produce a desired permanent change in electrical characteristics as a part of the manufacturing process.

formula translation—*See* FORTRAN.

form-wound coil—An armature coil that is formed or shaped over a fixture before being placed on the armature of a motor or generator. Any coil wound on a fixture or dummy form.

FORTRAN—1. Formula translation, a procedure-oriented computer language designed to be used with problems expressible in algebraic notation. There are several forms, FORTRAN II, FORTRAN IV, etc. 2. A computer-programming language designed mainly for scientific problems.

fortuitous conductor—Any conductor which may provide an unintended path for intelligible signals; for example, water pipe, wire or cable, metal structural members, and so forth.

fortuitous telegraph distortion — Distortion other than bias or characteristic. It occurs when a signal pulse departs from the average combined effects of bias and characteristic distortion for one occurrence. Since fortuitous distortion varies from one signal to another it must be measured by a process of elimination over a long period. It is expressed in a percentage of unit pulse.

forty-five/forty-five—Also called the Westrex system. A system of disc recording in which signals originating from two microphones are impressed on each side of a groove. The two sides are cut 45° from the surface of the record.

forty-five record—A 7-inch record with a 1½-inch center hole. It is recorded at 45 rpm and played at the same speed.

forty-four–type repeater — Type of telephone repeater used in a four-wire system. It employs two amplifiers and no hybrid arrangements.

forward—In or of the direction in which a nonlinear element, like a pn junction, conducts most easily.

forward-acting regulator — A transmission regulator which makes an adjustment without affecting the quantity that caused the adjustment.

forward-backward counter—A counter having both an add and subtract input and thus capable of counting in either an increasing or a decreasing direction.

forward bias—An external voltage applied in the conducting direction of a pn junction. The positive terminal is connected to the p-type region, and the negative terminal to the n-type region.

forward biased second breakdown—A local thermal runaway phenomenon in a semiconductor characterized by high local temperature and uneven current density. It is strongly a function of breakdown voltage and is affected by the structure used.

forward coupler — A directional coupler used for sampling incident power.

forward current—The current which flows across a semiconductor junction when a forward-bias voltage is applied.

forward gate current—The current into the gate terminal of an FET with a forward gate-to-source voltage applied.

forward gate-to-source breakdown voltage —The breakdown voltage between the gate and source terminals of an insulated-gate field effect transistor with a forward gate-to-source voltage applied and all other terminals short-circuited to the source terminal.

forward direction — The direction of easy current flow through a semiconductor device when a given voltage within the ratings of the device is applied. In a conventional rectifier or diode, the forward direction is from anode to cathode when the anode is at a positive voltage with respect to the cathode.

forward path—In a feedback control loop, the transmission path from the loop-actuating to the loop-output signal.

forward propagation by ionospheric scatter—A radiocommunication technique using the scattering phenomenon exhibited by electromagnetic waves in the 30- to 100-megahertz region when passing through the ionosphere at an elevation of about 85 kilometers.

forward propagation by tropospheric scatter — A method of communication by means of ultrahigh-frequency fm radio. It provides reliable multichannel telephone, teletype, and data transmission without line-of-sight restrictions or the necessity of using wire or cables.

forward recovery time—In a semiconductor diode, the time required for the current or voltage to arrive at a specified condition after instantaneous switching from zero or a specified reverse voltage to a specified forward voltage.

forward resistance – The resistance measured at a specified forward voltage drop or forward current in a rectifier.

forward scatter—1. Propagation of electromagnetic waves at frequencies above the maximum usable high frequency through use of the scattering of a small portion of the transmitted energy when the signal passes from a nonionized medium into a layer of the ionosphere. 2. A term referring collectively to very-high-frequency forward propagation by ionospheric scatter and ultrahigh-frequency forward propagation by tropospheric scatter communication techniques.

forward scattering—The reflected radiation of energy from a target away from the illuminating radar.

forward short-circuit–current amplification factor—In a transistor, the ratio of incremental values of output to input current when the output circuit is ac short-circuited.

forward-transfer function – In a feedback control loop, the transfer function of the forward path.

forward voltage—1. Voltage of the polarity that produces the larger current. 2. The voltage drop across a device after breaking over into conduction at some specified current.

forward voltage drop—The resultant voltage drop when current flows through a rectifier in the forward direction.

forward wave—In a traveling-wave tube, a wave with a group velocity in the same direction the electron stream moves.

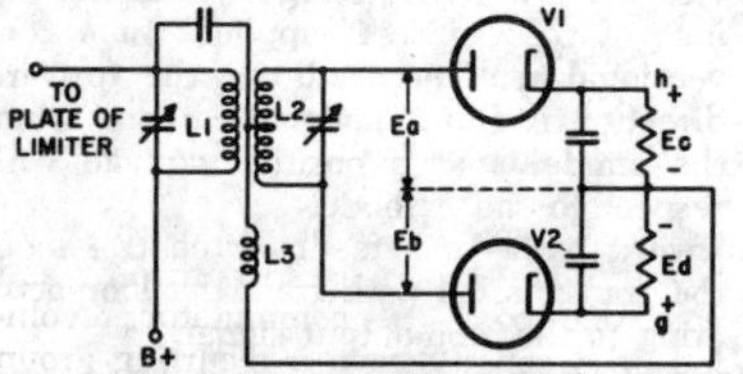

Foster-Seeley discriminator.

Foster-Seeley discriminator—A type of frequency discriminator that converts a frequency-modulated signal into an audio signal. It requires a limiter to prevent random amplitude variations of the fm signal from appearing in its output.

Foster's reactance theorem—The driving-point impedance of a finite two-terminal network composed of pure reactances is a reactance which is an odd rational function of frequency and which is completely determined, except for a constant factor, by assigning the resonant and antiresonant frequencies. In other words, the driving-point impedance consists of segments going from minus infinity to plus infinity (except that at zero, or infinite, frequency, a segment may start or stop at zero impedance). The frequencies at which the impedance is infinite are termed poles, and those at which it is zero are termed zeros.

Foucault currents—*See* Eddy Currents.

four-address code – An artificial language for describing or expressing the instructions carried out by a digital computer. In automatically sequenced computers, the instruction code is used for describing or expressing sequences of instructions. Each instruction word then contains a part specifying the operation to be performed, plus one or more addresses which identify a particular location in storage.

four-channel sound – *See* Quadraphonic Sound.

four-frequency diplex telegraphy – A method of frequency-shift telegraphy in which a separate frequency is used to represent each of the four possible signal combinations corresponding to two telegraph channels.

four-horn feed—A cluster of four rectangular horn antennas used as the radiating and receiving elements of parabolic or lens-type radar antennas. The four segments of the horn assembly define the four quadrants of information for direction to target sensing. Used on monopulse-type radar systems such as the AN/FPS-16.

fourier series—A mathematical analysis that permits any complex waveform to be resolved into a fundamental, plus a finite number of terms involving its harmonics.

Fourier transform—1. A mathematical relationship that provides a connection between information in the frequency domain and the time domain. The Fourier transform of correlation functions yields the power spectra. 2. A mathematical operation which decomposes a time varying signal into its complex frequency components (amplitude, and phase or real and imaginary components).

four-layer diode—1. A semiconductor diode

Four-layer diode.

that has three junctions with connections made only to the two outer layers that form the junctions. A Shockley diode is an example. 2. A pnpn two-terminal thyristor exhibiting a negative resistance characteristic in one direction. It has two stable states, an "off" state in which it displays a high series resistance and an "on" state in which the series resistance is quite low. Switching time for the 4-layer is in the nanosecond region. A very high ratio of hold current to switching makes it ideal for oscillator application.

four-layer transistor—A junction transistor that has four conductivity regions, but only three terminals. A thyristor is an example.

four-level laser—A type of laser that differs from a three-level type in that it has a terminal (lower level) for the laser transition which itself is an excited state of the system rather than ground level. Ordinarily, less energy is required to obtain the necessary population inversion in a four-level laser because the terminal level may be almost empty initially.

four-level system—A laser involving four electronic energy levels. The ground state (level 1) is pumped to level 4, from which the excited electrons make a downward transition to the upper laser level 3 (or metastable level 3). Then, stimulated transition to the lower laser level 2 occurs, followed by rapid decay to the ground state. The four-level system has the advantage that the pump level and ground state are isolated from the laser action.

four-pole network—*See* Two-Terminal–Pair Network.

four-quadrant multiplier — In analog computers, a multiplier in which operation is not restricted with regard to the signs of the input variables.

four-track recorder—*See* Track Configuration.

four-track recording — Also called quarter-track. On quarter-inch-wide tape, the arrangement by which four different channels of sound may be recorded on adjacent tracks. These may be recorded as four separate and distinct tracks (monophonic), or two related (stereo) pairs of tracks. By convention, tracks 1 and 3 are recorded in the forward direction of a given reel, and tracks 2 and 4 are recorded in the reverse direction.

four-track tape—Tape on which four separate sound paths are recorded. The use of four track's permits stereo in both directions of tape movement, or alternately, monophonic recording across four times the length of a given tape. Also known as quarter-track.

four-wire circuit—A two-way circuit with two paths. Each path transmits the electric waves in one direction only. The transmission paths may or may not employ four wires.

four-wire repeater — A telephone repeater used in a four-wire circuit. It has two amplifiers; one amplifies the telephone currents in one side of the four-wire circuit, and the other in the other side.

four-wire terminating set — A hybrid arrangement involving termination of four-wire circuits on a two-wire basis for interconnection with two-wire circuits.

fox message—A standard message which is used for testing teletypewriter circuits and machines because it includes all the alphanumerics on a teletypewriter, as well as most of the function characteristics such as space, figures shift, letters shift, etc. It is: THE QUICK BROWN FOX JUMPED OVER A LAZY DOG'S BACK 1234567 890———SENDING. The sending station's identification is inserted in the three space blanks which precede the word "sending."

fractional arithmetic units — Arithmetic units in a computer that is operated with the decimal point at the extreme left so that all numbers have a value less than 1.

fractional frequency offset—*See* Frequency Offset.

fractional-horsepower motor — Any motor having a continuous rating of less than one horsepower.

Frahm frequency meter—A meter that measures the frequency of an alternating current. It consists of a row of steel reeds, each with a different natural frequency. All are excited by an electromagnet fed with the current to be measured. The reed that vibrates is the one with a frequency corresponding most nearly to that of the current.

frame—1. In television, the total area occupied by the picture. In the United States, each frame contains 525 horizontal scanning lines, and 30 complete frames are shown per second. 2. One cycle of a recurring number of pulses. 3. In pam and pdm, one complete commutator revolution or sweep. In pcm, a recurring group of words which includes a single synchronizing signal. 4. The array of binary digits across the width of magnetic or paper tape. 5. The time period needed to transmit either bits or bytes of data along with parity and other control information.

frame frequency—1. The number of times per second the picture area is completely scanned (30 per second in the United States television system. 2. In a computer, the number of frames per unit time. 3. In telemetry, the number of times per second that a frame of pulses is sent or received.

frame grid—The grid of a vacuum tube consisting of a rigid welded frame on which tungsten wire is wound under tension, resulting in a firm precision structure that can be positioned accurately. It also allows the use of much finer grid wire, which reduces electron interception and power dissipation in the grid.

frame-grounding circuit – A conductor which is electrically bonded to the machine frame and/or to any conduction parts which are normally exposed to operating personnel. This circuit may further be connected to external grounds as may be required by applicable Underwriters code.

frame of reference—A set of points, lines, or planes used for defining space coordinates.

frame pulse synchronization—Synchronization of the local-channel rate oscillator by comparison and phase lock with the separate frame synchronizing pulses.

framer – A device for adjusting facsimile equipment so that the recorded elemental area bears the same relationship to the record sheet as the corresponding transmitted elemental area bears to the subject copy as the line progresses.

frame rate—*See* Frame Frequency.

frame roll—A momentary roll, or "flip-flop," of a television picture.

frame synchronization signal – In pam, a coded pulse or interval to indicate the start of the commutation frame period. In pcm, any signal used to identify a frame of data.

frame-synchronizing pulse—A recurrent signal that establishes each frame.

frame-synchronizing pulse separator—A circuit for separating frame-synchronizing pulses or intervals from commutated signals.

framing—1. Adjusting the picture to a desired position in the direction of line progression. 2. The process of selecting the bit groupings representing one or more characters from a continuous stream of bits.

framing bits—Also called sync bits. Noninformation-carrying bits used to make possible the separation of characters in a bit stream.

framing control—More often called centering control. A knob (or knobs) for centering and adjusting the height and width of a television picture.

framing magnet—*See* Centering Magnet.

Franklin antenna—A base-fed vertical antenna that is several wavelengths high and that gives broadside radiation as a result of the elimination of phase reversals by means of loading coils or wire folds.

Franklin oscillator – A two-terminal feedback oscillator using two tubes or transistors and having sufficient loop gain to permit extremely loose coupling to the resonant circuit.

Fraunhofer region—The region in which the energy from an antenna proceeds essentially as though coming from a source located in the vicinity of the antenna.

free electrons – Electrons which are not bound to a particular atom, but circulate among the atoms of a substance.

free energy—The available energy in a thermodynamic system.

free field—1. Theoretically, a field (wave or potential) which is free from boundaries in a homogeneous, isotropic medium. In practice, a field in which the effects of the boundaries are negligible over the region of interest. 2. A property of information-processing recording media which permits recording of information without regard to a preassigned or fixed field; e.g., information-retrieval-devices information may be dispersed in the record in any sequence or location.

free-field emission—Electron emission that occurs when the electric field at the surface of an emitter is zero.

free grid—A grid electrode that is left unconnected in a vacuum tube. Its potential exerts a control over the plate current.

free impedance—Also called normal impedance. The input impedance of a transducer when the load impedance is zero.

free magnetic pole—A magnetic pole so far from an opposite pole that it is free from the effect of the other pole.

free motional impedance—The complex remainder after the blocked impedance of a transducer has been subtracted from the free impedance.

free net—A net in which any station may communicate with any other station in the same net without first obtaining permission from the control station.

free oscillations—Commonly referred to as shock-excited oscillations. Oscillations that continue in a circuit or system after the applied force has been removed. The frequency of the oscillations is determined by the parameters of the system or circuit.

free-point tube tester—A tester instrument that permits transferring a tube from a circuit to a test panel at which either voltage or current measurement for any electrode of the tube is readily made by plugging a meter into appropriate jacks. Connections to the receiver are made by means of a cord and plug inserted into the socket from which the tube was removed.

free position—The initial position of the actuator of a momentary-contact switch when there is no external force (other than gravity) applied on the actuator, and the switch is in the specified position.

free progressive wave – Also called free wave. A wave free from boundary effects in a medium. In other words, there are no reflections from nearby surfaces. A free wave can only be approximated in practice.

free radicals—Atoms, ionized fragments of

atoms, or molecules which combine and release enormous amounts of energy.

free reel—The reel which supplies the magnetic tape on a recorder.

free-rotor gyro—A gyro the rotor of which is supported by a gas-lubricated spherical bearing.

free routing—A method of traffic handling in which a message is sent toward its destination over any available channel without dependence on a predetermined routing doctrine.

free-running frequency—The frequency at which a normally synchronized oscillator operates in the absence of a synchronizing signal.

free-running local synchronizer oscillator—A free-running oscillator circuit in the decommutator normally triggered by separated channel synchronizing pulses. It supplies substitute pulses for missing channel pulses.

free-running multivibrator — *See* Astable Multivibrator (Free-Running).

free-running sweep — A sweep operating without synchronizing pulses.

free sound field—A field in a medium free from discontinuities or boundaries. In practice it is a field in which the boundaries cause negligible effects over the region of interest.

free space—1. Empty space, or space with no free electrons or ions. It has approximately the electrical constants of air. 2. Having to do with a condition in which the radiation pattern of an antenna is not affected by surrounding objects such as the earth, buildings, vegetation, etc.

free-space field intensity — The radio-field intensity that would exist at a point in a uniform medium in the absence of waves reflected from the earth or other objects.

free-space loss — The theoretical radiation loss which would occur in radio transmission if all variable factors were disregarded.

free-space propagation—Electromagnetic radiation over a straight-line path in a vacuum or ideal atmosphere, sufficiently removed from all objects that affect the wave in any way.

free-space radar equation—The equation for determining the characteristic of a radar signal propagated between the radar set and a reflecting target in free space.

free-space radiation pattern—The radiation pattern of an antenna in free space, where there is nothing to reflect, refract, or absorb the radiated waves.

free-space transmission — Electromagnetic radiation over a straight line in a vacuum or ideal atmosphere sufficiently removed from all objects that affect the wave.

free speed—The angular speed of an energized motor under no-load conditions. (*See also* Angular Velocity.)

free wave—*See* Free Progressive Wave.

freewheeling circuit—A motor arrangement in which the field is shunted by a half-wave rectifier which discharges the energy stored in the field during the negative half cycles.

freewheeling diode—A rectifier diode connected across an inductive load to carry a current proportional to the energy stored in the inductance. It carries this current when no power is being supplied by the source to the load and until all the energy in the inductance has been dissipated or until the next voltage pulse is applied.

freeze-out—A short-time denial of a telephone circuit to a subscriber by a speech-interpolation system.

F-region — The region of the ionosphere above 100 miles.

freq—Abbreviation for frequency.

frequency—1. Symbolized by f. The number of recurrences of a periodic phenomenon in a unit of time. Electrical frequency is specified as so many hertz. Radio frequencies are normally expressed in kilohertz at and below 30,000 kilohertz and in megahertz above this frequency. 2. The number of complete cycles in one second of alternating current, voltage, electromagnetic or sound pressure waves. 3. Number of alternations or repetitions per second in any recurring action. In the case of alternating current and other forms of wave motion it is expressed in hertz.

frequency agility—The rapid and continual shifting of a radar frequency to avoid jamming by the enemy, reduce mutual interference with friendly sources, enhance echoes from targets, or provide necessary patterns of ecm (electronic countermeasures) or eccm (electronic counter-countermeasures) radiation.

frequency allocation — The assignment of available frequencies in the radio spectrum to specific stations, for specific purposes. This is done to yield maximum utilization of frequencies with minimum interference between stations. Allocations in the United States are made by the Federal Communications Commission.

frequency authorization—The document of power which legalizes the assignment of a frequency or a frequency band.

frequency-azimuth-intensity — Pertaining to a type of radar display in which frequency, azimuth, and strobe intensity are correlated.

frequency band—A continuous and specific range of frequencies.

frequency band of emission—The frequency band required for a specific type of transmission and speed of signaling.

frequency bias—A constant frequency purposely added to the frequency of a signal.

frequency changer — *See* Frequency Converter.

frequency-change signaling – A telegraph signaling method in which one or more particular frequencies correspond to each desired signaling condition of a telegraph code. The transition from one set of frequencies to the other may be either a continuous or a discontinuous change in frequency or in phase.

frequency-changing circuit–A circuit comprising an oscillator and a mixer and delivering an output at one or more frequencies other than the input frequency.

frequency channel–A continuous portion of the appropriate frequency spectrum for a specified class of emission.

frequency compensation–1. The technique of modifying an electronic circuit or device for the purpose of improving or broadening the linearity of its response with respect to frequency. 2. The compensation required in feedback amplifiers to ensure stability and prevent unwanted oscillations.

frequency constant – The number relating the natural vibration frequency of a piezoid (finished crystal blank) to its linear dimension.

frequency conversion – 1. The process of converting a signal to some other frequency by combining it with another frequency. 2. Of a heterodyne receiving system, converting the carrier frequency of a received signal from its original value to the intermediate frequency value in a superheterodyne receiver.

frequency converter–Also called frequency changer. A circuit, device, or machine that changes an alternating current from one frequency to another, with or without a change in voltage or number of phases. In a superheterodyne receiver, the oscillator and mixer first-detector stages together serve as a frequency converter.

frequency correction – Compensation, by means of an attenuation equalizer, for unequal transmission of various frequencies in a line.

frequency counter–An instrument in which frequency is measured by counting the number of cycles (pulses) occurring during a precisely established time interval.

frequency cutoff–The frequency at which the current gain of a transistor drops 3 dB below the low-frequency gain.

frequency demodulation – Removal of the intelligence from a modulated carrier.

frequency departure–The amount a carrier or center frequency deviates from its assigned value.

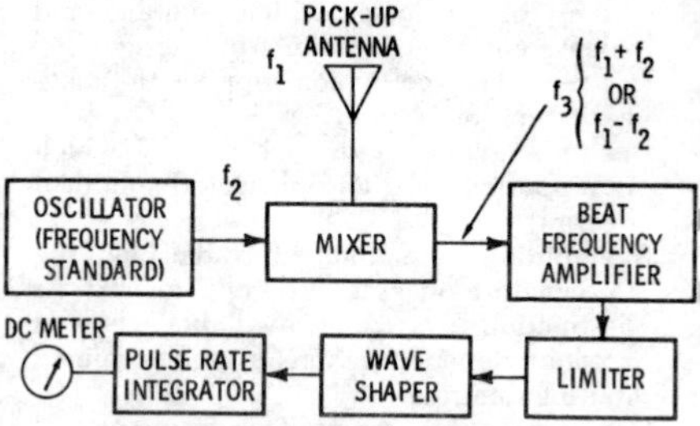

Frequency-deviation meter.

frequency deviation–1. In frequency modulation, the peak difference between the instantaneous frequency of the modulated wave and its carrier frequency. 2. A measure of the output frequency excursion around the carrier, caused by modulating the oscillator's tuning input which produces a frequency modulated output signal.

frequency-deviation meter–An instrument that indicates the number of hertz a transmitter has drifted from its assigned carrier frequency.

frequency discrimination–A term applied to the operation of selecting a desired frequency or frequencies from a spectrum of frequencies.

frequency discriminator–A circuit that converts a frequency-modulated signal into an audio signal.

frequency distortion–1. The distortion that results when all frequencies in a complex wave are not amplified or attenuated equally. 2. The unequal amplification of all frequencies over the passband of an amplifier. *Also see* Frequency Response.

frequency distribution–The number of occurrences of particular values plotted against those values.

frequency diversity–*See* Frequency-Diversity Reception.

frequency-diversity reception – Also called frequency diversity. The form of diversity reception which utilizes transmission at different frequencies.

frequency divider–1. A device delivering an output voltage which is at an integral submultiple or proper fraction of the input frequency. 2. A counter which has a gating structure added that provides an output pulse after a specified number of input pulses are received.

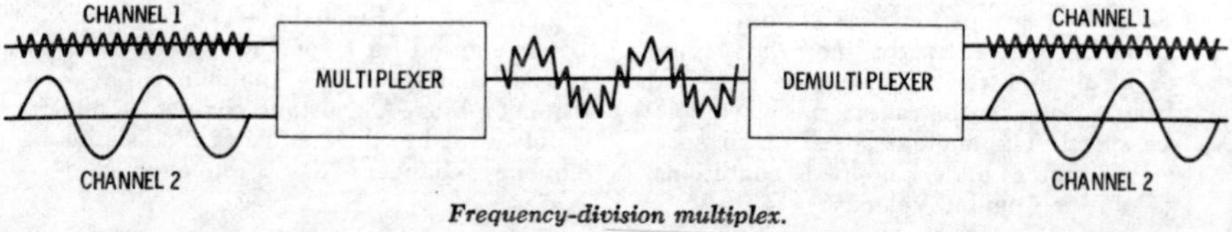

Frequency-division multiplex.

frequency-division data link—A data link in which frequency-division techniques are used for channel spacing.

frequency-division multiplex – 1. Abbreviated fdm. A device or process for transmitting two or more signals over a common path by sending each one over a different frequency band. 2. A multiplex system in which the available transmission frequency range is divided into narrower bands, each used for a separate channel.

frequency-division multiplexing – Taking the frequency spectrum of one leased line, and subdividing it into a series of lower frequency bands, each of which will transmit the data of an associated low-speed device.

frequency doubler—An electronic stage having a resonant output circuit tuned to the second harmonic of the input frequency. The output signal will then have twice the frequency of the input signal.

frequency-doubling transponder – A transponder that doubles the frequency of the interrogating signal before retransmission.

frequency drift – Any undesired change in the frequency of an oscillator, transmitter, or receiver.

frequency-exchange signaling—The method in which the change from one signaling condition to another is accompanied by a decay in amplitude of one or more frequencies and by a build-up in amplitude of one or more other frequencies.

frequency frogging—The interchanging of the frequency allocations of carrier channels to prevent singing, reduce cross talk, and correct for line slope. It is accomplished by having the modulation in a repeater translate a low-frequency group to a high-frequency group and vice versa.

frequency indicator—A device which shows when two alternating currents have the same phase or frequency.

frequency influence—In a measuring instrument other than a frequency meter, the change, expressed as a percentage of the full-scale value, in the indicated value as a result of a departure of the measured quantity from a specified reference frequency.

frequency interlace—In television, the relationship of intermeshing between the frequency spectrum of an essentially periodic interfering signal and the spectrum of harmonics of the scanning frequencies. Such relationship minimizes the visibility of the interfering pattern by altering its appearance on successive scans.

frequency keying—A method of keying in which the carrier frequency is shifted between two predetermined frequencies.

frequency-measuring equipment – Equipment for indicating or measuring the frequency or pulse-repetition rate of an electrical signal.

frequency meter—1. An instrument for measuring the frequency of an alternating current. 2. Instrument for measuring the repetition rate of a recurring phenomenon, as the cycles per second of a sinusoidal waveform.

frequency-modulated carrier-current telephony—A form of telephony in which a frequency-modulated carrier signal is transmitted over power lines or other wires.

frequency-modulated cyclotron – A cyclotron in which the frequency of the accelerating electric field is modulated in order to hold the positively charged particles in synchronism with the accelerating field despite their much greater mass at very high speeds.

frequency-modulated jamming—A jamming technique in which an rf signal of constant amplitude is varied in frequency about a center value to produce a signal that covers a band of frequencies.

frequency-modulated output—A transducer output which is obtained in the form of a deviation from a center frequency, where the deviation is proportional to the applied stimulus.

frequency-modulated radar—Also called fm radar. A form of radar in which the radiated wave is frequency-modulated. The range is measured by beating the returning wave with the one being radiated.

frequency-modulated transmitter – One in which the frequency of the wave is modulated.

frequency-modulated wave—A carrier wave whose frequency is varied by an amount proportionate to the amplitude of the modulated signal.

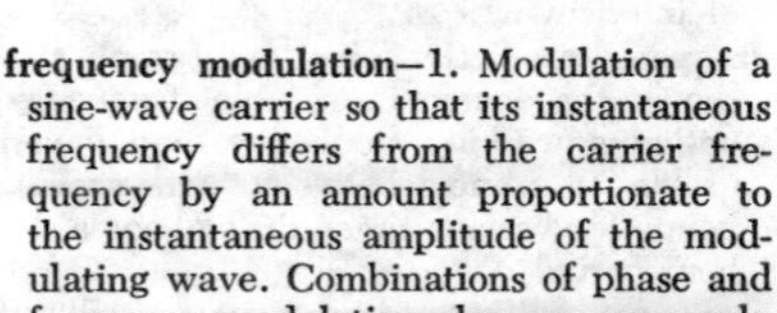

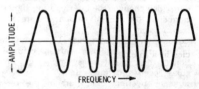

Frequency modulation.

frequency modulation—1. Modulation of a sine-wave carrier so that its instantaneous frequency differs from the carrier frequency by an amount proportionate to the instantaneous amplitude of the modulating wave. Combinations of phase and frequency modulation also are commonly referred to as frequency modulation. (*See also* Frequency-Modulated Wave.) 2. Abbreviated fm. One of three ways of modifying a sine wave signal to make it "carry" information. The sine wave or "carrier" has its frequency modified in accordance with the information to be transmitted. The frequency function of the modulated wave may be continuous or discontinuous. In the latter case, two or more particular frequencies may correspond to one significant condition.

frequency-modulation broadcast band—The band from 88 to 108 megahertz. It is divided into 100 channels, each 200 kilohertz in width, and set aside for frequency-modulated broadcasting.

frequency-modulation deviation—The peak difference between the instantaneous frequency of a modulated wave and the carrier or reference frequency.

frequency modulation-frequency modulation—A system in which frequency-modulated subcarriers are used to frequency modulate a second carrier.

frequency modulation-phase modulation — A system in which frequency-modulated subcarriers are used to phase-modulate a second carrier.

frequency monitor—An instrument for indicating the amount a frequency deviates from its assigned value.

frequency multiplex — A technique for the transmission of two or more signals over a common path. Each signal is characterized by a distinctive reference frequency or band of frequencies.

frequency multiplier—A device for delivering an output wave whose frequency is a multiple of the input frequency (e.g., frequency doublers and triplers).

frequency offset — The amount by which a frequency lies above or below a reference frequency. For example, if a frequency measures 1.000 001 MHz when compared against a reference frequency of 1.000,000 MHz, then its fractional frequency offset is 1 Hz/1 MHz, or 1 part in 10^6.

frequency offset transponder—A transponder which changes the interrogating signal frequency by a fixed amount before retransmission.

frequency output (transducer)—An output in the form of frequency which is a function of the applied measurand (e.g., angular speed and flow rate).

frequency overlap—That part of the frequency band which is shared as a result of interleaving.

frequency-prediction chart — A graph that shows the maximum usable frequency, optimum working frequency, and lowest usable frequency between two specific points for various times throughout a 24-hour period.

frequency pulling — A change in oscillator frequency due to a change in the load impedance.

frequency pushing — A source-frequency change caused by a change in electron flow within the source oscillator.

frequency range—1. In a transmission system, those frequencies at which the system is able to transmit power without attenuating it more than an arbitrary amount. 2. In a receiver, the frequency band over which the receiver is designed to operate, covering those frequencies the receiver will readily accept and amplify. 3. A designated portion of the frequency spectrum.

frequency record—A recording of various known frequencies at known amplitudes, usually for testing purposes.

frequency regulator—A regulator that maintains the frequency of the frequency-generating equipment at a predetermined value or varies it according to a predetermined plan.

frequency relay—A relay which functions at a predetermined value of frequency. It may be an overfrequency or underfrequency relay, or a combination of both.

frequency response—1. A measure of how effectively a circuit or device transmits the different frequencies applied to it. 2. The portion of the frequency spectrum which can be sensed by a device within specified limits of amplitude error. 3. A graphical characteristic showing relative signal levels at different frequencies with respect to a given reference level. A "flat" frequency response is one that has a uniform level at all frequencies within a given bandwidth. 4. A measure of the ability of a device to take into account follow or act upon a rapidly varying condition, e.g., as applied to amplifiers. 5. The measure of any component's ability to pass signals of different frequency without affecting their relative strengths. This is shown as a graph or "curve" which assumes input signals equally strong at all frequencies and plots their output intensities against a decibel scale. The ideal "curve" is a straight line. Frequency response may also be stated as a frequency range but with specified decibel limits, indicating the maximum deviations from flat response. For instance, 30 to 20,000 Hz ± 2 dB means the component will not change the relative intensities of any frequencies within that range by more than 2 dB above or 2 dB below the ideal zero dB (volume unchanged) point. 6. The range of frequencies over which an amplifier responds within defined limits of amplification (or signal output).

frequency-response analysis—The use of alternating or pulsating signals to excite a control system so that the response of the system to different frequencies can be ascertained to permit analysis of its operating characteristics.

frequency-response characteristic — The amount by which the gain or loss of a device varies with the frequency.

frequency-response curve—A graphical representation of the way a circuit responds to different frequencies within its operating range.

frequency-response equalization – Also called equalization or corrective equalization. The effect of all frequency-discrimination means employed in a transmission system to obtain the desired overall frequency response.

frequency run–A series of tests for determining the frequency-response characteristics of a transmission line, circuit, or device.

frequency-scan antenna – A radar antenna, similar to a phased-array antenna, in which scanning in one dimension is accomplished through frequency variation.

frequency scanning–A technique in which the output frequency is made to vary over a desired range at a specified rate.

frequency selectivity–The degree to which a transducer is capable of differentiating between the desired signal and signals or interference at other frequencies.

frequency-sensitive relay–A relay that operates only when energized with voltage, current, or power within specific frequency limits.

frequency-separation multiplier – A multiplying device in which each variable is split into low-frequency and high-frequency parts that are multiplied separately to obtain results that are added to give the required product. The system makes possible high accuracy and broad bandwidth.

frequency separator–The circuit which separates the horizontal-scanning from the vertical-scanning synchronizing pulses in a television receiver.

frequency shift–1. Pertaining to radioteletypewriter operation in which the mark and space signals are transmitted as different frequencies. 2. A change in the frequency of a radio transmitter or oscillator. 3. Pertaining to a modulation system in which one radio frequency represents picture black and another represents picture white; frequencies between the two limits represent shades of gray. 4. The frequency difference in a frequency-shift modulation system.

frequency-shift converter–A device which limits the amplitude of the received frequency-shift signal and then changes it to an amplitude-modulated signal.

frequency-shift indicator – In automatic code transmission, a device which designates mark and space by shifting the carrier back and forth between two frequencies instead of keying it on and off.

frequency-shift keying–1. Abbreviated fsk. A form of frequency modulation in which the modulating wave shifts the output frequency between predetermined values and the output wave has no phase discontinuity. 2. A method of modulating a carrier frequency. A binary "1" shifts the frequency above the center carrier frequency; a binary "0" shifts the frequency below the center carrier frequency. 3. A frequency-modulation method in which the frequency is made to vary at the significant instants; (1) by smooth transitions: the modulated wave and the change in frequency are continuous at the significant instants; (2) by abrupt transitions: the modulated wave is continuous, but the frequency is discontinuous at the significant instants.

frequency-shift telegraphy–Telegraphy by frequency modulation in which the telegraph signal shifts the frequency of the carrier between predetermined values. There is phase continuity during the shift from one frequency to the other.

frequency-shift transmission–A method of transmitting the mark and space elements of a telegraph code by shifting the carrier frequency slightly, usually about 800 Hz.

frequency-slope modulation–A method of modulation in which the carrier is swept periodically over the entire band. Modulation of the carrier with a communications signal changes the system bandwidth without affecting the uniform distribution of energy over the band. Thus, the desired information can be recovered from any part of the system bandwidth, and portions of the band that have interference can be filtered out without loss of desired information.

frequency spectrum – The entire range of frequencies of electromagnetic radiations.

frequency splitting–One condition of magnetron operation in which rapid alternation occurs from one mode of operation to another. This results in a similar rapid change in oscillatory frequency and consequent loss of power at the desired frequency.

frequency stability – The ability of electronic equipment to maintain the desired operating frequency.

frequency stabilization–The controlling of the center or carrier frequency so that it does not differ more than a prescribed amount from the reference frequency.

frequency standard–A stable low-frequency oscillator used for frequency calibration. It can generate a fundamental frequency of 50 to 100 kilohertz with a high degree of accuracy. Harmonics of this fundamental are then used as reference points for checking throughout the radio spectrum at 50- or 100-kilohertz intervals.

frequency swing–The instantaneous departure of the emitted wave from the center frequency when its frequency is modulated.

frequency synthesizer – 1. A frequency source of high accuracy generally charac-

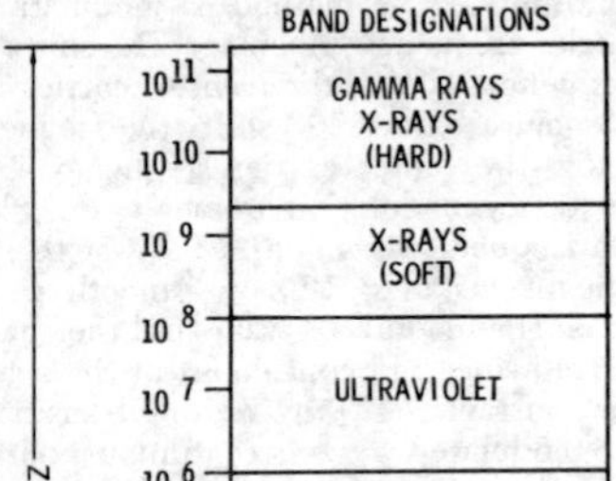

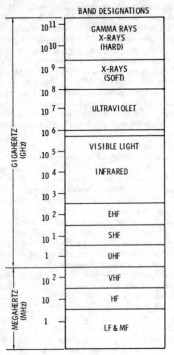

Frequency spectrum.

terized by the fact that the output frequency is composed of two components. The frequency steps, mostly decadic, are derived from the crystal-stabilized frequency standard and the variable frequency of a free-running oscillator which fills in between these steps. The simplest method of frequency synthesis is to derive standard frequencies from a crystal-controlled frequency by harmonic generation and frequency division. 2. A circuit capable of producing a multitude of output frequencies from a single input frequency. 3. An rf source that can provide by external command, any discrete and precise frequency within its range and resolution. The output signal is stabilized to a fixed-frequency reference, which may be internal or external to the synthesizer. Primary applications include automatic test equipment, electronic warfare, and communications systems. 4. A system utilizing the phase locked loop (pll) principle in conjunction with a programmable digital frequency divider to generate any of a number of discrete frequencies. (May replace a number of crystals or other timing elements with only one timing element.)

frequency-time-intensity – Pertaining to a type of radar display in which frequency, time, and strobe intensity are correlated.

frequency tolerance – The maximum permissible deviation with respect to the reference frequency of the corresponding characteristic frequency of an emission. Expressed in percent or in hertz.

frequency translation–The transfer *en bloc* of signals occupying a definite frequency band, such as a channel or group of channels, from one position in the frequency spectrum to another in such a way that the arithmetic frequency difference of the signals within the band is unaltered.

frequency tripler–An amplifier whose output circuit is resonant to the third harmonic of the input signal. The output frequency is three times the input frequency.

frequency-type telemeter – A telemeter which employs the frequency of a periodically recurring electric signal as the translating means.

frequency-wavelength relation – For radio waves, the frequency in hertz is equal to approximately 300,000,000 divided by the wavelength in meters. The wavelength in meters is equal to approximately 300,000,000 divided by the frequency in hertz or 300 divided by the frequency in megahertz.

Fresnel – A little-used unit of frequency equal to 10^{12} hertz.

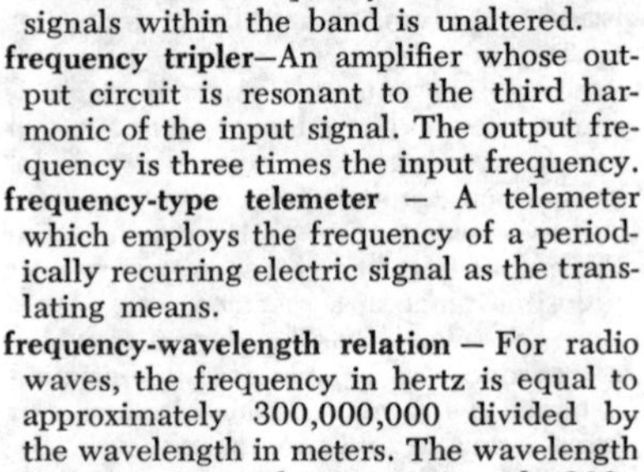

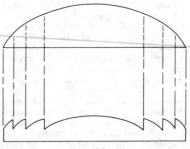

Fresnel lens.

Fresnel lens–A lens similar in action to a plano-convex lens but made thinner and lighter because of the presence of steps on the convex side. Often the flat side has a rough surface to diffuse the light slightly and thereby smooth the light beams.

Fresnel loss–*See* Surface Reflection.

Fresnel number–The square of the radius of a lens aperture divided by the product of the focal length and the wavelength. It provides a measure of the importance of diffraction in the image formed by the lens: a small Fresnel number indicates greater diffraction effects.

Fresnel region–The region between an antenna and the beginning of the Fraunhofer region.

Fresnel zone–1. An area selected in the aperture of a radiating system so that radiation from all parts of the system reaches some point at which it is desired at a common phase within 180°. 2. A circular zone about the direct path between a transmitter and a receiver at each radius that the distance from a point on this circle to the receiving point has a path length that is some multiple of a half-wavelength longer than the direct path.

frictional electricity—Electric charges produced by rubbing one material against another.

frictional error—Applied to pickups, the difference in values measured in percent of full scale before and after tapping, with the measurand constant.

frictional loss—The loss of energy due to friction between moving parts.

frictional machine—Also called a static machine. A device for producing frictional electricity.

friction effects—The difference in resistance or output between readings obtained prior to and immediately after tapping an instrument while applying a constant stimulus. Particularly applicable to potentiometric transducers.

friction error—A change in a reading originally taken in the absence of vibration that occurs after a transducer is tapped or dithered to remove internal friction.

friction-free calibration (transducer)—Calibration under conditions minimizing the effect of static friction often obtained by dithering.

friction tape—A fibrous tape impregnated with a sticky, moisture-resistant compound which provides a protective covering or insulation.

fringe—A unit of linear measurement equal to half the wavelength of thallium green light (approximately 0.01 mil). It is used in measuring the depth of diffusion in silicon.

fringe area—The area just beyond the limits of the reliable service area of a television transmitter. Signals are weak and erratic, requiring the use of high-gain directional receiving antennas and sensitive receivers for satisfactory reception.

fringe effect—The extension of the flux in a field beyond the edges of a gap, as electric flux at the edges of the plates of a capacitor or magnetic flux at the edges of an air gap in a magnetic circuit.

fringe howl—A squeal or howl heard when some circuit in a receiver is on the verge of oscillation.

frit—1. Metallic powders fused in a glass binder. 2. To melt and fuse together, as a set of electrical contacts that are subject to repeated discharges.

fritting—A type of contact erosion in which an electrical discharge makes a hole through the contact film and produces molten matter that is drawn through the hole by electrostatic forces and then solidifies and forms a conducting bridge.

frogging repeater—A carrier repeater that has provisions for frequency frogging to make possible the use of a single multipair voice cable without excessive cross talk.

Frohlich high-temperature breakdown theory—A thermal mechanism for breakdown in which electrons rather than ions carry the current. The necessary number of conduction electrons is produced by thermal excitation of the electrons in impurity and imperfection levels.

Frohlich low-temperature breakdown theory—Also called the high-energy criterion. Similar to the Von Hippel theory, except that an electron energy distribution is assumed. Only a few electrons on the high-energy tail of the distribution must gain the necessary critical energy.

front contact — A movable relay contact which closes a circuit when the associated device is operated.

front end—The section of a tuner or receiver which is used to select the desired station from either the am, fm, or tv band, and to convert the rf signal to if. To do its job properly, a front end requires a high-gain, low-noise rf stage, a mixer, and an oscillator. The degree to which a desired station can be received without interference and without adding noise is expressed by sensitivity, and signal-to-noise ratio.

front-end rejection—A dB expression of the relative ability of a receiver network to reject signals outside the tuned bandwidth.

front porch—In the composite television signal, that portion of the synchronizing signal (at the blanking or black level) preceding the horizontal-sync pulse at the end of each active horizontal line. The standard EIA signal is 1.27 microseconds in duration.

front projection—A system of picture enlargement using an opaque reflective screen. The projector and viewers are on the same side of the screen.

front-surface mirror—An optical mirror on which the reflecting surface is applied to the front of the mirror instead of the back.

front-to-back ratio — Also called front-to-rear ratio. 1. The ratio of power gain between the front and rear of a directional antenna. 2. Ratio of signal strength transmitted in a forward direction to that transmitted in a backward direction. For receiving antennas, the ratio of received-signal strength when the source is in the front of the antenna to the received-signal strength when the antenna is rotated 180°.

front-to-rear ratio—*See* Front-to-Back Ratio.

fruit—*See* Fruit Pulse.

fruit pulse—Also called fruit. Radar beacon system video display of a synchronous beacon return which results when several interrogator stations are located within the same general area. Each interrogator receives its own interrogated reply as well as many synchronous replies resulting from interrogation of the airborne transponders by other ground stations.

F-scan—*See* F-Display.

F-scope—*See* F-Display.

fsk—*See* Frequency-Shift Keying.

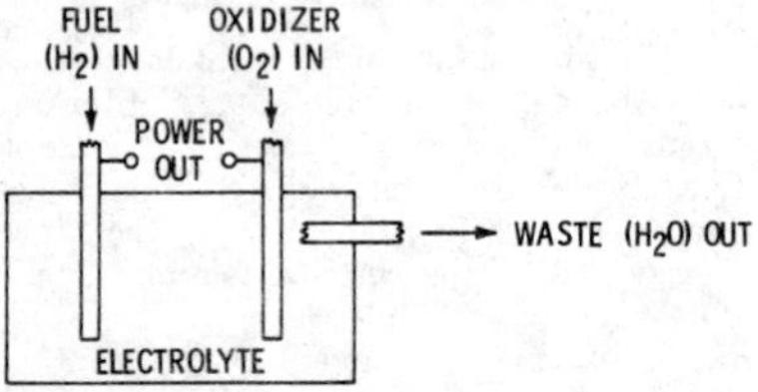

Fuel cell.

fuel cell—An electrochemical generator in which the chemical energy from the reaction of air (oxygen) and a conventional fuel is converted directly into electricity. A fuel cell differs from a battery in that it uses hydrocarbons (or some derivative such as hydrogen) for fuel, and it operates continuously as long as fuel and air are available.

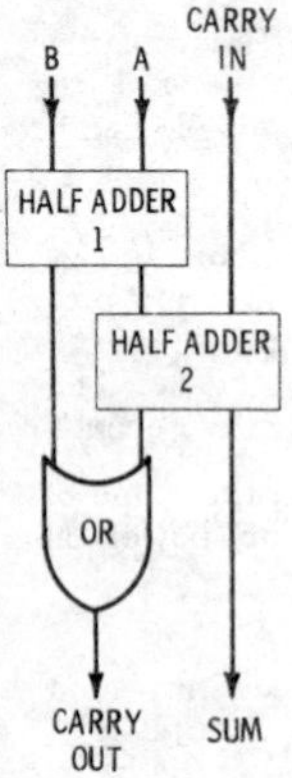

TRUTH TABLE

CARRY IN	B	A	CARRY OUT	SUM
0	0	0	0	0
0	0	1	0	1
0	1	0	0	1
0	1	1	1	0
1	0	0	0	1
1	0	1	1	0
1	1	0	1	0
1	1	1	1	1

Full adder.

full adder—A circuit that provides an output equal to the sum of three binary-digit inputs (two digits to be added and a carry digit from a previous stage). Sum and carry outputs are provided.

full duplex—1. A method of operation of a communication circuit in which each end can simultaneously transmit and receive. When used on a radio circuit, duplex operation requires two frequencies. 2. A data circuit which is capable of both sending and receiving data simultaneously.

full-duplex operation—Simultaneous operation in opposite directions in a telegraph system.

full excursion—The application of a measurand, in a controlled manner, over the entire range of a transducer.

fullhouse—A multichannel radio control system in which all controls work to allow the model to fly a complete flight pattern.

full load – The greatest load a piece of equipment is designed to carry under specified conditions.

full period allocated circuit—A communication link (allocated circuit) assigned exclusively for the use of previously defined users at two or more terminal points.

full-pitch winding – A type of armature winding in which the number of slots between the sides of the coil equals the pole-pitch measure in the slots.

full range speaker unit—*See* Dual Cone.

full scale—The total interval over which an instrument is intended to be operated. Also, the output from a transducer when the maximum rated stimulus is applied to the input.

full-scale cycle—The complete range of an instrument, from minimum reading to full scale and back to minimum reading.

full-scale error—The difference between the actual voltage or current that produces full-scale deflection of an electrical indicating instrument and the rated full-scale input of the instrument.

full-scale output—The algebraic difference in electrical output between the maximum and minimum values of measurand over which an instrument is calibrated. When the sensitivity slope is given by any other line than the end-point sensitivity, full scale expresses the algebraic difference, for the span of the instrument, which is calculated from the slope of the straight line from which nonlinearity is determined.

full-scale sensitivity—*See* Full-Scale Output.

full-scale value—The largest value of applied electrical energy which can be indicated on a meter scale. When zero is between the ends of the scale, the full-scale value is the arithmetic sum of the values of the applied electrical quantity corresponding to the two ends of the scale. On a suppressed meter the full-scale value is the largest value of applied electrical input less the smallest value of applied electrical energy input.

full-scale value of an instrument—The largest actuating electrical quantity which can be indicated on the scale; or, for an instrument having its zero between the ends of the scale, the sum of the values of the actuating electrical quantity corresponding to the two ends.

full subtractor – A device that can obtain the difference of two input bits, subtract a borrow from these bits, and provide difference and borrow outputs.

full-track recording–1. Defines the track width as essentially equal to the tape width. Applies to quarter-inch-wide (or less) tape only. 2. A recorded signal occupying the full width of a ¼-inch tape. *See* Track Configuration.

full-voltage starting – *See* Across-The-Line Starting.

full-wave rectification–The process of inverting the negative half cycle of current of an alternating input so that it flows in the same direction as the positive half cycle. A way to accomplish this is to use four diodes placed in a bridge configuration.

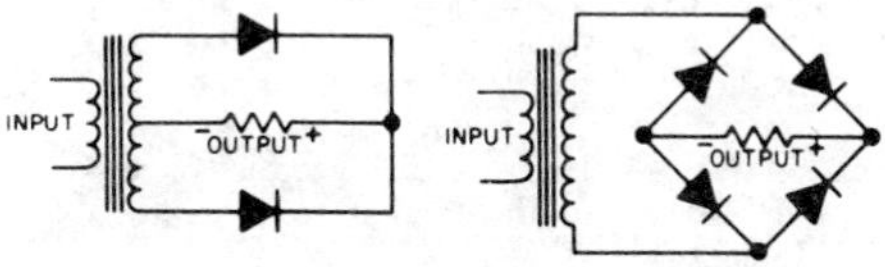

Full-wave center-tapped. Full-wave bridge.

Full-wave rectifier.

full-wave rectifier–A circuit, electron tube, or other device which uses both positive and negative alternations in an alternating current to produce direct current.

full-wave rectifier tube–A tube containing two sets of rectifying elements, to provide full-wave rectification.

full-wave vibrator – A vibrator having an armature that moves back and forth between two fixed contacts, so as to change the direction of the direct current flow through a transformer at regular intervals and thereby permit voltage step-up by the transformer. Used in battery-operated power supplies for mobile and marine radio equipment.

function–1. A quantity the value of which depends on the value of one or more other quantities. 2. A specific purpose of an entity, or its characteristic action.

functional blocks – Also called molecular electronic circuit. A more or less homogenous combination of several solid-state materials to perform a desired circuit function. The quartz crystal is often represented as a natural expression of the functional block combination of resistance, capacitance, and inductance.

functional device–*See* Integrated Circuit.

functional diagram–A diagram showing the functional relationships among the parts of a system.

functional electronic block – A fabricated device serving a complete electronic function, such as amplification, without other individual components or conducting wires except those required for input, power, and output.

functional insulation–The insulation necessary for the proper functioning of a device, and for basic protection against electrical shock hazard.

functional interface–An interface between the operating characteristics of equipment such as electrical power and signal characteristics, signal timing, and environmental coupling.

functional parts–Discrete items defined by functional characteristics and dimensions which are not repairable with the use of spare parts (e.g., resistors, capacitors, diodes, potted transformers, permanently sealed batteries, etc.).

functional testing – Testing to determine whether the device under test reacts correctly to inputs, qualitatively.

function characteristic – The relationship between the output ratio and the shaft position of a precision potentiometer.

function codes–Codes which appear in tape or cards to operate machine functions, such as carriage returns, space, shift, skip, tabulate, etc.

function digit–A coded instruction used in a computer for setting a branch order to link subroutines into the main program.

function generator–1. A device capable of generating one or more desired waveforms. 2. An electrical network that can be adjusted to make its output voltage (or current) a desired function of time. Used in conjunction with analog computers. 3. A character, conic, or vector generator in a display controller. Some display controllers have additional function generators such as sweep generators for displaying television type pictures on the display screen.

functioning time–In a relay, the time that elapses between energization and operation or between de-energization and release.

functioning value–In a relay, the value of applied voltage, current, or power at which operation or release occurs.

function keys–Data energy devices usually programmed to initiate or terminate a particular function or process in the graphic system. Function keys may or may not generate interrupts when depressed and/or released. Some systems will generate an interrupt upon depression and release, some just on depression, and others do not generate any interrupt but rather set a bit in a register. The latter type of function key must have the register periodically read by the host computer to determine when a key is depressed. The register bit is cleared when the key is released. Function keys may be mounted on the alphanumeric keyboard or on a separate box on the console.

function switch—1. A network or system having a number of inputs and outputs. When signals representing information expressed in a certain code are applied to the inputs, the output signals will represent the input information in a different code. 2. In adapters or control units, the switch which determines whether the system plays as a monophonic or stereophonic unit; it may parallel the speakers or cut out one or the other, switch amplifiers from one speaker to the other, reverse channels, etc.

function table—1. A table of values for a mathematical function. 2. A hardware device or a computer program which translates one representation of information into another. 3. A routine by means of which a computer can determine the value of a dependent variable from the values of independent variables. 4. A subroutine that can be used either to decode multiple inputs into a single output or to encode a single input into multiple outputs.

function unit—A device which can store a functional relationship and release it continuously or sporadically.

fundamental—*See* Fundamental Frequency.

fundamental component—The fundamental frequency component in the harmonic analysis of a wave.

fundamental frequency—The principal component of a wave; i.e., the component with the lowest frequency or greatest amplitude. It is usually taken as a reference.

fundamental group — In wire communications, a group of trunks by means of which each local or trunk switching center is connected to a trunk switching center of higher rank on which it "homes." The term also applies to groups that interconnect zone centers.

fundamental harmonic—The harmonic component with the lowest frequency.

fundamental mode—1. *See* Dominant Mode. 2. Of vibration, the mode having the lowest natural frequency.

funndamental piezoelectric crystal unit—A unit designed to use the lowest resonant frequency for a particular mode of vibration.

fundamental tone—1. In a periodic wave, the component corresponding to the fundamental frequency. 2. In a complex tone, the component tone of lowest pitch.

fundamental units — Units arbitrarily selected as the basis of an absolute system of units.

fundamental wavelength—The wavelength corresponding to the fundamental frequency. In an antenna, the lowest resonant frequency of the antenna alone, without inductance or capacitance.

fungusproof—To chemically treat a material, component, or unit to prevent the growth of fungus spores.

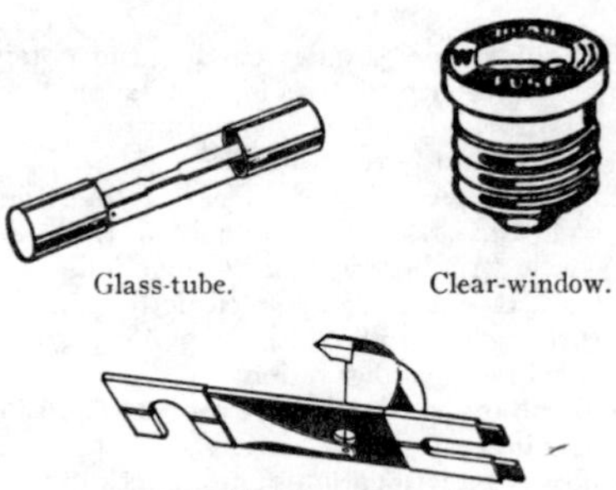

Glass-tube. Clear-window.

Grasshopper.

Fuses.

fuse—1. A protective device, usually a short piece of wire but sometimes a chemical compound, which melts and breaks the circuit when the current exceeds the rated value. 2. To equip with a fuse.

fuse alarm—A circuit which produces a visual and/or audible signal to indicate a blown fuse.

fuse block—An insulating base on which fuse clips or other contacts for holding fuses are mounted.

fuse box—An enclosed box containing fuse blocks and fuses.

fuse clips—Contacts on the fuse support for connecitng the fuse holder into the circuit.

fuse cutout—An assembly consisting of a fuse support and a fuse holder that may or may not include the fuse link.

fused disconnect—Generally an air-break switch with a fusing unit in the blade. Used for opening and closing high-voltage circuits.

fused junction—*See* Alloy Junction.

fused quartz—A glasslike insulating material having exceptional resistance to the action of heat and acid.

fused semiconductor—The junction formed by recrystallization on a base crystal from a liquid phase of one or more components and the semiconductor.

fuse filler—Material placed within the fuse tube to aid circuit interruption.

fuse holder—A device for supporting a fuse

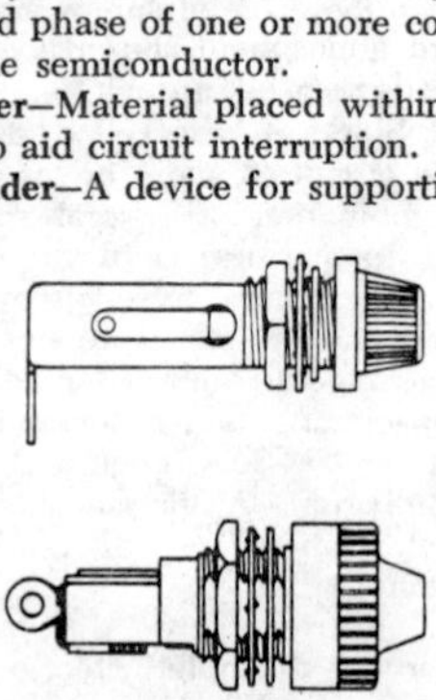

Fuse holders.

and providing connections for its terminals.

fuse link—In a fuse, the current-carrying portion which melts when the current exceeds a predetermined value.

Fusestat—Trade name for a time-delay fuse similar to a Fusetron. It has a sized base requiring a permanent socket adapter, which prevents insertion of a fuse or Fusetron of an incorrect rating.

Fusetron—A screw-plug fuse that permits up to 50% overload for short periods of time without blowing.

fuse tube—The insulated tube enclosing a fuse link.

fuse unit—An assembly consisting of a fuse link mounted in a fuse holder, which contains parts and materials essential to the operation of the fuse link.

fuse wire—A wire made from an alloy that melts at a relatively low temperature.

fusible-link-readout memory—A large semiconductor array of prediffused cells interconnected by a fixed metallization pattern that can be tailored to a particular need simply by burning out selected interconnections.

fusible resistor—A resistor designed to protect a circuit against overload by opening when the current drain exceeds the design limits.

fusible wire—A wire used in fire-alarm circuits. It is made of an alloy with a low melting point.

fusion—1. Also called atomic fusion or nuclear fusion. The melting of atomic nuclei, under extreme heat (millions of degrees), to form a heavier nucleus. The fusion of two nuclei of light atoms is accompanied by a tremendous release of energy. 2. Melting, usually as the result of interaction of two or more materials.

fuzz—An intentional distortion of the natural tone of an electric guitar. Also called fuzz tone.

fV—Abbreviation for femtovolt (10^{-15} volt).

G

G—1. Symbol for conductance, a grid of a vacuum tube, a generator, or ground. 2. Abbreviation for giga (10^9), or the gate of a field-effect transistor.

g—Also called G-force. Symbol for the acceleration of a free-falling body due to the earth's gravitational pull. Equal to 32.17 feet per second per second.

g-a (ground-to-air)—Communication with airborne objects from the ground.

GA coil—A coil wound with air spaces between its turns and layers to reduce the capacitance.

gage—Also spelled gauge. An instrument or means for measuring or testing. By extension, the term is often used synonymously with transducer.

gage pressure — 1. A differential pressure measurement using the ambient pressure as a reference. 2. A pressure in excess of a standard atmosphere at sea level (i.e., 14.7 pounds per square inch).

gage pressure transducer—A pressure transducer that uses ambient pressure as the reference pressure. The sensing element is normally vented to the ambient pressure.

gain—Also called transmission gain. 1. Any increase in power when a signal is transmitted from one point to another. Usually expressed in decibels. Widely used for denoting transducer gain. 2. The ratios of voltage, power, or current with respect to a standard or previous reading. 3. Any increase in the strength of an electrical signal, as takes place in an amplifier. Gain is measured in terms of decibels or number-of-times of amplification, as: 6 dB (a gain of 2) increases an input voltage to an output twice as large. 4. The change in source-drain current per unit change in gate voltage. Thus higher gain gives faster devices.

gain-bandwidth product — The product of the closed-loop gain of an operational amplifier and its corresponding closed-loop bandwidth. This product is often constant in operational amplifiers.

gain control—A device for varying the gain of a system or component.

gain function—A transfer function that relates either a pair of voltages or a pair of currents.

gain margin—The amount of gain change of an operational amplifier at 180° phase-shift angle frequency that would produce instability.

gain-sensitivity control — *See* Differential Gain Control.

gain stability—The extent to which the sensitivity of an instrument remains constant with time. (The property reported in specifications should be instability, which is the maximum change in sensitivity from the initial value over a stated period of time under stated conditions.)

gain-time control — *See* Sensitivity-Time Control.

galactic noise—All noise that originates in space as a result of radiation of celestial bodies other than the sun.

galena—A bluish-gray, crystalline form of lead sulfide often used as the crystal in a variable crystal detector.

galloping ghost—*See* Proportional Control, 2.

galvanic—An early term for current resulting from chemical action, as distinguished from electrostatic phenomena.

galvanic anode—A source of emf for cathodic protection provided by a metal less noble than the one to be protected (i.e., magnesium, zinc, or aluminum as used for cathodic protection of steel).

galvanic cell—An electrolytic cell capable of producing electric energy by electrochemical action.

galvanic corrosion — Accelerated electrochemical corrosion produced when one metal is in electrical contact with another more noble metal, both being in the same corroding medium, or electrolyte, with a current between them. (Corrosion of this type usually results in a higher rate of solution of the less noble metal and protection of the more noble metal.)

galvanic current—An electrobiological term for unidirectional current such as ordinary direct current.

galvanic series—A list of metals and alloys arranged in the order of their relative potentials (ability to go into solution) in a given environment. The table of potentials is arranged with the anodic, or least noble, metals at one end and the cathodic, or more noble, metals at the other. (For marine use, the potentials listed are related to a seawater environment.)

Galvanic Series

(Anodic)

Magnesium
Zinc
Cadmium
Steel or Iron
Cast Iron
Chromium-Iron (Active)
Lead-Tin Solders
Lead
Tin
Nickel
Brasses
Copper
Bronzes
Copper-Nickel Alloys
Monel
Silver
Graphite
Gold
Platinum

(Cathodic)

galvanizing—The coating of steel with zinc to retard corrosion.

galvanometer—An instrument for measuring an electric current. This is done by measuring the mechanical motion produced by the electromagnetic or electrodynamic forces set up by the current.

galvanometer constant — The factor by which a certain function of a galvanometer reading must be multiplied to obtain the current in ordinary units.

galvanometer lamp—The lamp that illuminates a movable mirror of the galvanometer in some spectrophotometers. The angle at which the light is reflected to the galvanometer scale depends on the amount of current through the galvanometer coil.

galvanometer recorder (for photographic recording) — A combination of a mirror and coil suspended in a magnetic field. A signal voltage, applied to the coil, causes a light beam from the mirror to be reflected across a slit in front of a moving photographic film.

galvanometer shunt—A resistor connected in parallel with a galvanometer to increase the range of the instrument. The resistor limits the current to a known fraction and thus prevents excessive current from damaging the galvanometer.

galvanometric controller—A temperature indicator which has been converted to a temperature controller. The indicator operates directly off the sensor's input signal. The controller portion detects the mechanical position of the pointer, and varies the output to keep the pointer at the desired temperature.

gamma — 1. A unit of magnetic intensity, equal to 10^{-5} oersted. 2. A number indicating the degree of contrast in a photograph, facsimile reproduction, or received television picture.

gamma correction—Introduction of a nonlinear output-input characteristic for the purpose of changing the effective value of gamma.

gamma ferric oxide—The magnetic constituent of practically all present-day tapes, in the form of a dispersion of fine acicular particles within the coating.

gamma radiation — A highly penetrating electromagnetic disturbance (photons) that emanates from the nucleus of an atom. This type of radiation travels in waveform much like X-rays or light, but it has a shorter wavelength of approximately 1 angstrom or 10^{-7} millimeter.

gamma rays — The emission from certain radioactive substances. They are electromagnetic radiations similar to X-rays, but with a shorter wavelength form about 10^{-12} to 10^{-9} centimeters.

gang—To mechanically couple two or more variable capacitors, switches, potentiometers, or other components together so they can be operated from a single control knob.

gang capacitor—Also called gang tuning capacitor. Two or more variable tuning capacitors mounted on the same shaft and controlled by a single knob, but each capacitor tuning a different circuit. Thus, more than one circuit can be tuned simultaneously by a single control.

gang control—Simultaneous control of sev-

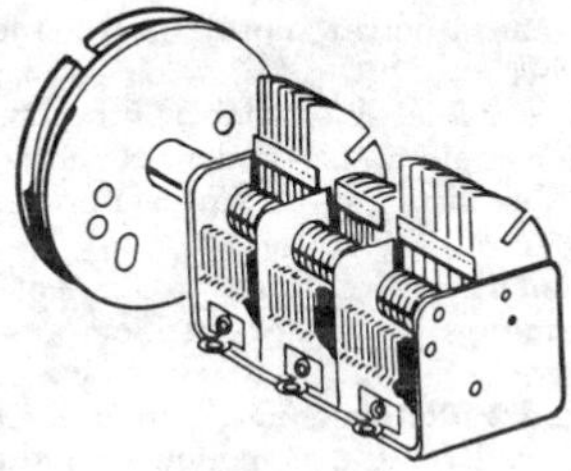

Gang capacitor.

eral similar pieces of apparatus with one adjustment knob or other control.

ganged tuning—Simultaneous tuning of two or more circuits with a single mechanical control.

gang punch—To punch information that is identical or constant into all of a group of punch cards.

gang switch – A number of switches mechanically coupled for simultaneous operation, but electrically connected to different circuits. In one common form, two or more rotary switches are mounted on the same shaft and operated by a single control.

gang tuning capacitor—*See* Gang Capacitor.

gap—1. In a magnetic circuit, the portion that does not contain ferromagnetic material (e.g., an air space). 2. The space between two electrodes in a spark gap. 3. The tiny space between the pole pieces of a record or playback head, across which the magnetic fields appear (when the tape is running) for transduction into alternating electrical signals. 4. A space between radar-antenna radiation lobes where the field strength is low, resulting in incomplete radar coverage. 5. A space in which the radiation fails to meet minimum coverage requirements, either because the space is not covered or because the minimum specified overlap is not obtained. 6. An interval of space or time used to indicate automatically the end of a word, record, or file of data on a tape.

gap arrester—A type of lightning arrester comprising a number of air gaps in series between metal cylinders or cones.

gap coding—1. A means for inserting periods of no transmission in a system in which transmission is normally continuous. The spacing and duration of the periods of silence from the code are variable. 2. Subdividing the response of a transponder into long and short groups of pulses (analogous to Morse code) for purposes of recognition.

gap depth—The dimension of the gap in a recording head, measured perpendicular to the surface of the head.

gap factor—In a tube that employs electron-accelerating gaps (a traveling-wave tube), the ratio of the maximum energy gained (expressed in volts) to the maximum gap voltage.

gap filler—1. A lightweight radar set used to fill in gaps in the coverage pattern of an early-warning radar net. 2. An auxiliary radar antenna used to fill in gaps in the pattern of the main radar antenna.

gap filling—Electrical or mechanical rearrangement of an antenna array, or the use of a supplementary array, to produce lobes where gaps previously occurred.

gap insulation—Insulation wound in a gap.

gap length—The dimension of the gap of a recording head, measured from one pole face to the other. In longitudinal recording, the gap length is the dimension of the gap in the direction of tape travel.

gap loss—The loss in output attributable to the finite gap length of the reproduce head. The loss increases as the wavelength decreases, amounting to approximately 4 dB when the wavelength is equal to twice the gap length, and subsequently increases rapidly toward a complete extinction of output when the wavelength is approximately equal to 1.15 times the gap length.

gap motor—A spark-gap drive motor.

gap scatter—In a computer, the deviation from true vertical alignment of the gaps of the magnetic readout heads for the several parallel tracks.

gap width—The dimension of the gap, measured in the direction parallel to the head surface and pole faces. The gap width of the record head governs the track width. The gap widths of reproduce heads are sometimes made appreciably less than those of the record heads to minimize tracking errors.

garbage—1. In a computer, a slang term for unwanted and meaningless information carried along in storage. Sometimes called hash. 2. Undecipherable or meaningless sequences of characters produced in computer output or retained within storage.

garbage in, garbage out—*See* Gigo.

garble—Faulty transmission, reception, or encoding which renders the message incorrect or unreadable.

garnet maser—*See* Ferromagnetic Amplifier.

garter spring—In facsimile, the spring fastened around the drum to hold the record sheet or copy in place.

gas—One of the three states of matter. An aeriform fluid having neither independent shape nor volume, but tending to spread out and occupy the entire enclosure in which it is placed. Gases are formed by heating a liquid above its boiling point.

gas amplification—Ratio of the charge collected to the charge liberated by the initial ionizing of the gas in a radiation counter.

gas-amplification factor – Ratio of radiant

or luminous sensitivites with and without ionization of the gas in a gas phototube.

gas cell—A cell the action of which is dependent on the absorption of gases by the electrodes.

gas cleanup—The tendency of many gas-filled tubes to lose their gas pressure and hence become inoperable. This occurs when the ions of gas are driven at high velocity into the metal parts or the glass envelope of the tube where they form stable compounds and are lost as far as the tube is concerned.

gas current—1. The current in the grid circuit of a vacuum tube when the gas ions within the tube are attracted by the grid. 2. A flow of positive ions to an electrode, the ions having been produced as a result of gas ionization by an electron current between other electrodes.

gas detector—An instrument used to indicate the concentration of harmful gases in the air.

gas diode—A tube having a hot cathode and an anode in an envelope containing a small amount of an inert gas or vapor. When the anode is made sufficiently positive, the electrons flowing to it collide with gas atoms and ionize them. As a result, the anode current is much greater than that for a comparable vacuum diode.

gas discharge device—A device utilizing the conduction of electricity in a gas, due to movements of electrons and ions produced by collision.

gas discharge display—A device containing an inert gas that gives off light when a high voltage is applied to break down (ionize) the gas. Gas discharge displays have a 7-segment format.

gas-discharge laser—*See* Gas Laser.

gas-electric drive—A self-contained power-conversion system comprising an electric generator driven by a gasoline engine. The generator in turn supplies power to the driving motor or motors.

gaseous discharge—The state of a gas or mixture of gases in which a conduction current can be maintained by ionization. The ionization results from collisions between electrons and atoms or molecules of the gas, the energy being furnished by an external source such as an electric field.

gaseous electronics—The field of study involving the conduction of electricity through gases and a study of all atomic-scale collision phenomena.

gaseous tube—An electronic tube into which a small amount of gas or vapor is introduced after the tube has been evacuated. Ionization of the gas molecules during operation of the tube affects its operating characteristics.

gaseous-tube generator — A power source comprising a gas-filled electron-tube oscillator and a power supply, plus associated controls.

gas-filled cable—A coaxial or other type of cable containing gas under pressure which serves as insulation and prevents moisture from entering.

gas-filled lamp — A tungsten-filament lamp containing nitrogen or an inert gas such as argon.

gas-filled radiation-counter tube — A gas tube used for the detection of radiation. It operates on the principle that radiation will ionize a gas.

gas-filled tube rectifier—A rectifier tube in which a unidirectional flow of electrons from a heated electrode ionizes the inert gas within the tube. In this way, rectification is accomplished.

gas focusing — Also called ionic focusing. The use of an inert gas to focus the electron beam in a cathode-ray tube. Beam electrons ionize the gas molecules, forming a core of positive ions along the beam path which tends to attract beam electrons, making the beam more compact.

gasket-sealed relay—A relay in an enclosure sealed with a gasket.

gas laser—A laser where the active medium is a discharge in a gas, vapor, or mixture within a glass or quartz tube that has a Brewster window at each end. This gas can be excited by a high-frequency oscillator or by a direct current between electrodes inside the tube to pump the medium so as to obtain a population inversion. Also called gas-discharge laser.

gas magnification—The increase in current through a phototube due to ionization of the gas within the tube.

gas maser—A maser in which the microwave electromagnetic radiation interacts with the molecules of a gas such as ammonia. Use is limited chiefly to highly stable oscillator applications, as in atomic clocks.

gas noise—Electrical noise produced by erratic motion of gas molecules in gas or partially evacuated vacuum tubes.

gas-phase laser—A continuous-wave device for general experimental work with coherent light. It employs a resonator made up of a fused-silica plasma tube 60 cm long having internal, multilayer, dielectric-coated confocal reflectors of optical-grade fused silica.

gas phototube—A phototube into which a quantity of gas has been introduced, usually to increase its sensitivity.

gas ratio—The ratio of the ion current in a tube to the electron current that produces it.

gassiness—The presence of unwanted gas in a vacuum tube, usually in relatively small amounts. It is caused by leakage from outside the tube or by evolution from its inside walls or elements.

gassing—1. Evolution of a gas from one or more electrodes during electrolysis. 2. The production of gas in a storage battery when the charging current is continued after the battery has been completely charged.

gassy—Having operating characteristics that are impaired as a result of an excessive amount of gas inside its envelope, as a vacuum tube.

gassy tube—*See* Soft Tube.

gaston—A modulator that produces a random-noise modulation signal from a gas tube. It may be attached to any standard aircraft communications transmitter to provide a counterjamming modulation.

gas tube – A partially evacuated electron tube containing a small amount of gas. Ionization of the gas molecules is responsible for the current flow.

gas-tube relaxation oscillator—A relaxation oscillator in which the abrupt discharge is provided by the breakdown of the gas in the tube.

gas X-ray tube—An X-ray tube in which electron emission from the cathode is produced by bombarding it with positive ions.

gate—1. A circuit having two or more inputs and one output, the output depending upon the combination of logic signals at the inputs. There are four gates, called AND, OR, NAND, and NOR. The definitions assume positive logic is used. In computer work, a gate is often called an AND circuit. 2. A signal used to trigger the passage of other signals through a circuit. 3. One of the electrodes of a field-effect transistor; it is analogous to the base of a transistor or the grid of a vacuum tube. Symbol: G. 4. An output element of a cryotron. 5. A circuit in which one signal (usually a square wave) switches another signal on or off. 6. To control the passage of a pulse or signal. 7. Voltage-actuated control terminal of an MOS transistor.

gate array – A geometric pattern of basic gates contained in one chip. It is possible to interconnect the gates during manufacture to form a complex function that may be used as a standard production.

gate circuit—A circuit that passes a signal only when a gating pulse is present.

gate-controlled switch – Also called gate turn-off switch. A three-junction, three-terminal, solid-state device, constructed very much like a silicon controlled rectifier except that it has a turn-off ability, which is controlled by a negative-current pulse applied to the gate.

gate controlled turn-on time—The time interval between the 10% rise of the gate pulse and the 90% rise of the principal current pulse during switching of a thyristor from the off-state to the on-state.

gate current—Instantaneous current flowing between the gate and cathode of a silicon controlled rectifier.

gate current for firing – Gate current required to fire a silicon controlled rectifier when the anode is at a fixed dc voltage with respect to the cathode and with the device at stated temperature conditions.

gated-beam detector—A single-stage fm detector using a gated-beam tube.

gated-beam tube – A five-element tube in which the electrons flow in a beam between the cathode and plate. A small increase in voltage on the limiter grid will cut off the plate current, and further increases will have a negligible effect on it.

gated buffer – A low-impedance inverting driver circuit that may be used as a line driver for pulse differentiation or in multivibrators.

gated flip-flop—One that has a steering circuit that prevents both flip-flop outputs from becoming "0" at the same time. Alternating-current input pulses must be used to prevent the flip-flop from oscillating. This situation can arise if both input lines are made high simultaneously.

gated sweep—Sweep in which the duration as well as the starting time is controlled to exclude undesired echoes from the indicator screen.

gated transistor – A transistor in which a gate electrode covers the emitter and collector junctions. This allows the application of an electric field at the surface of the base region.

gate electrode – A control electrode to which trigger pulses are applied.

gate equivalent circuit—A basic unit for describing relative digital circuit complexity. The number of gate equivalent circuits is that number of individual logic gates that would have to be interconnected to perform the same function.

gate generator—A circuit or device used to produce one or more gate pulses.

gate impedance—The impedance of a gate winding in a magnetic amplifier.

gate multivibrator – A rectangular-wave generator designed to produce a single positive or negative gate voltage upon being triggered, and then to become inactive until the arrival of the following trigger pulse.

gate nontrigger voltage – The maximum gate voltage that will not cause a thyristor to switch from the off-state to the on-state.

gate power dissipation – The power dissipated between the gate and cathode terminals of a silicon controlled rectifier.

gate-producing multivibrator—A rectangular-wave generator that produces a single positive or negative gate voltage only when triggered by a pulse.

gate pulse—A pulse that enables a gate circuit to pass a signal. The gate pulse generally has a longer duration than the signal to ensure time coincidence.

gate signal—That signal generated by some form of delay circuit required in connection with beam switching, automatic following, the application of agc to a selected echo, and many other purposes.

gate terminal – The terminal in an FET electrically connected to the electrode associated with the region in which the electric field, due to the control voltage, is effective.

gate trigger current—In a controlled rectifier, the minimum gate current, for a given anode-to-cathode voltage, required to switch the rectifier on.

gate trigger voltage—In a controlled rectifier, the gate voltage that produces the gate trigger current.

gate tube—A tube which does not operate unless two signal voltages, derived from two independent circuits, are applied simultaneously to two separate electrodes.

gate turn-off current—In a controlled rectifier, the minimum gate current, for a given collector current in the on state, required to cause the rectifier to switch off.

gate turn-off switch – *See* Gate-Controlled Switch.

gate turn-off voltage—In a controlled rectifier, the gate voltage required to produce the gate turn-off current.

gate voltage – 1. The voltage across the gate-winding terminals of a magnetic amplifier. 2. The instantaneous voltage between gate and cathode of a silicon controlled rectifier with anode open.

gate winding – The reactor winding that produces the gating action in a magnetic amplifier.

gating – 1. Selecting those portions of a wave which exist during certain intervals or which have certain magnitudes. 2. Applying a rectangular voltage to the grid or cathode of a cathode-ray tube, to sensitize it during the sweep time only. 3. Application of a specific waveform to perform electronic switching. *See also* Blanking.

gating circuit—A circuit that operates as a selective switch and allows conduction only during selected time intervals or when the signal magnitude is within specified limits.

gating pulse—A pulse which modifies the operation of a gate circuit.

gauge—*See* Gage.

gauge pressure—*See* Gage Pressure.

gauss—The centimeter-gram-second electromagnetic unit of magnetic induction. One gauss represents one line of flux (one maxwell) per square centimeter.

Gaussian distribution—Also called normal distribution. A density function of a population whch is bell-shaped and symmetrical and which is completely defined by two independent parameters, the mean and the standard deviation.

Gaussian function – A mathematical function used in designing a filter to pass a step function with zero overshoot and minimum rise time (similar to a Bessel-function filter).

Gaussian noise – Unwanted electrical disturbances or perturbations described by a probability density function that follows a normal law of statistics.

Gaussian random vibration – *See* Random Vibration.

Gaussian waveform—In pulse-compression systems, a waveform that produces very low transmitted side lobes.

gaussmeter – An instrument that provides direct readings of magnetic field density (flux density) by virtue of the interaction with an internal magnetic field.

Gauss's theorem – The summation of the normal component of the electric displacement over any closed surface is equal to the electric charge within the surface.

GCA – Abbreviation for ground-controlled approach.

GCI – Abbreviation for ground-controlled interception.

GCS – Abbreviation for gate controlled switch.

GCT or Gct—Abbreviation for Greenwich civil time.

G-display—Also called G-scan or G-scope. In radar, a rectangular display in which a target appears as a laterally centralized blip on which wings appear to grow as the target approaches. Horizontal and vertical aiming errors are indicated by horizontal and vertical displacement of the blip.

gear – An element shaped like a toothed wheel which engages one or more similar wheels. The energy transmitted can be stepped up or down by making the driven gears of different sizes.

geared synchro system—A system in which the transmitting and receiving synchros turn at a higher speed than the input and output shafts. Geared systems are generally used when a high degree of accuracy is required.

gearmotor—A train of gears and a motor used for reducing or increasing the speed of the driven object.

Geiger counter—Also called Geiger-Mueller or G-M counter. A radiation detector that uses a Geiger-Mueller counter tube, an amplifier, and an indicating device. The tube consists of a thin-walled gas-filled metal cylinder with a projecting electrode. Nuclear particles enter a window in the metal cylinder and temporarily ionize the gas, causing a brief pulse discharge.

These pulses, which appear at the projecting electrode, are amplified and indicated visibly or audibly. 2. Gas-chamber type radiation counter in which the chamber operates in avalanche region for high amplification and sensitivity.

Geiger-Mueller counter—*See* Geiger Counter.

Geiger-Mueller counter tube—A radiation-counter tube designed to operate in the Geiger-Mueller region.

Geiger-Mueller region—Also called Geiger region. The voltage interval in which the pulse size is independent of the number of primary ions produced in the initial ionizing event.

Geiger-Mueller threshold—Also called Geiger threshold. The lowest voltage at which all pulses produced in the tube by any ionizing event are of the same size regardless of the size of the primary ionizing event. This threshold is the start of the Geiger region where the counting rate does not substantially change with applied voltage.

Geiger region—*See* Geiger-Mueller Region.

Geiger threshold—*See* Geiger-Mueller Threshold.

Geissler tube—A gas-filled dual-electrode discharge tube that glows when electric current passes through the gas.

gel—A material composed of a solid held in a liquid.

genemotor—A type of dynamotor having two armature windings. One winding serves as the driving motor and operates from the vehicle battery. The other winding functions as a high-voltage dc generator for operation of mobile equipment.

general address—Group of characters included in the heading of a message to cause routing of the message to all addresses included in the general address category.

general class license—A license issued by the FCC to amateur radio operators who are able to send and receive code at the rate of 13 words per minute and who are familiar with general radio theory and practice. Holder enjoys all authorized amateur privileges except those reserved for higher license classes.

general-purpose computer—A computer designed to solve a wide variety of problems, the exact nature of which may have been unknown before the computer was designed.

general-purpose digital computer—A digital computer designed to solve a large variety of problems; that is, a computer that can be adapted to a large class of applications (as opposed to a computer designed specifically to control a manufacturing process). A typical general-purpose digital computer consists of four subsystems: (1) Input/output, which permits communication with the outside world; (2) memory, which stores data and instructions; (3) central processing unit (CPU), or arithmetic unit, which performs the arithmetic and data processing operations; and (4) control, which ties all of the subsystems together so that they operate in a fully automated way.

general-purpose motor—A motor of 200 hp or less and 450 rpm or more, rated for continuous operation, having standard ratings, and suitable for use without restriction to a particular application.

general-purpose relay—A relay that is adaptable to a variety of applications.

general rate—The amount of time taken by the creation of electron-hole pairs in a semiconductor.

general routine—A computer routine designed to solve a general class of problems, but when appropriate parametric values are supplied, it specializes in a specific problem.

generated noise—In potentiometric transducers, the noise which is attributable to causes such as the generation of emf when dissimilar metals are rubbed against each other, or the emf resulting from the thermocouple effects at points where dissimilar metals are joined.

generating electric field meter—Also called a gradient meter. A device for measuring the potential gradient at the surface of a conductor. A flat conductor is alternately exposed and then shielded from the electric field to be measured. The resultant current in the conductor is then rectified and used as the measure.

generating magnetometer—Also called earth inductor. A magnetometer which measures a magnetic field by the amount of emf generated in a coil rotated in the field.

generating station—An installation that produces electric energy from chemical, mechanical, hydraulic, or some other form of energy.

generating voltmeter—Also called a rotary voltmeter. A device which measures voltage. A capacitor is connected across the voltage, and its capacitance is varied cyclically. The resultant current in the capacitor is then rectified and used as a measure.

generation—The number of dubbing steps between a master recording and a given copy of the master. Thus, a second-generation dub is a copy of a copy of the original master.

generation data group—A collection of successive data sets that are historically related.

generation rate—The time rate of creation of electron-hole pairs in a semiconductor.

generator—1. Symbolized G. A rotating machine which converts mechanical energy into electrical energy. 2. An electronic device which converts dc voltage to alternating current of the desired frequency and wave-shape. 3. Any device that generates electricity. 4. In computer operation, a routine for producing specific routines from specific input parameters and skeletal coding. 5. A machine that converts mechanical energy into electric energy. In its commonest form, a large number of conductors are mounted on an armature that is rotated in a magnetic field produced by field coils. Also called dynamo. 6. A vacuum-tube oscillator or any other nonrotating device that generates an alternating voltage at a desired frequency when energized with dc power or low-frequency ac power. Such generators are used to produce large amounts of rf power, such as for high-frequency heating and ultrasonic cleaning. 7. A circuit that generates a desired repetitive or nonrepetitive waveform such as a pulse generator.

generator efficiency—1. In a generator, the ratio between the power required to drive the generator and the output power obtained from it. 2. In a thermoelectric couple, the ratio of the electrical power output to the thermal power input. It is an idealized efficiency assuming perfect thermal insulation of the thermoelectric arms.

generator field control—Regulation of the output voltage of a generator by control of the voltage that excites the field winding of the generator.

generator voltage regulator—A regulator which maintains or varies the voltage of a synchronous generator, capacitor motor, or direct-current generator at or within a predetermined value.

generic package—A collection of software items from which more than one installation-dependent specific package may be generated by a software manufacturing process. Generic packages may be in source-code form (generic source package) or in object-code form (generic object package).

gen-lock—A system of regenerating synchronizing pulses and subcarrier from a composite video source.

geodesic—The shortest line between two points on a given surface.

geometric distortion—In television, any geometric dissimiliarity between the original scene and reproduced image.

geometric mean—The square root of the product of two quantities.

geometric symmetry—Filter response in which there is mirror-image symmetry about the center frequency when frequency is plotted on a logarithmic scale. (This is the natural response of many electrical circuits.) *See also* Arithmetic Symmetry.

george box—An amplitude-sensitive device employed in an if amplifier. It rejects jamming signals of insufficient amplitude to operate its circuits; however, jamming signals having sufficient amplitude are not affected.

germanium—A brittle, grayish-white metallic element having semiconductor properties. Widely used in transistors and crystal diodes. Its atomic number is 32. Symbol, Ge.

Germanium diode.

germanium diode—A semiconductror diode in which a germanium crystal pellet is used as the rectifying element.

germanium transistor—A transistor in which germanium is the semiconducting material.

German silver—Usually called nickel silver. A silverish alloy of copper, zinc, and nickel.

getter—An alkali metal introduced into a vacuum tube during manufacture. It is fired after the tube has been evacuated, to react chemically with and eliminate any remaining gases. The getter then remains inactive inside the tube. The silvery deposit sometimes seen on the inside of the glass envelope is due to getter firing.

G force—*See* g.

g-g (ground-to-ground) – Communication between two points on the ground.

ghost—*See* Ghost Image.

ghost image—Also called ghost. An undesired duplicate image offset somewhat from the desired image as viewed on a television screen. It is due to a reflected signal traveling over a longer path and, hence, arriving later than the desired signal. It may be eliminated by the use of a directional antenna array which receives signals over only one path.

ghost mode—A waveguide mode in which there is a trapped field associated with an imperfection in the waveguide wall. A ghost mode can cause difficulty in a waveguide operated near the cutoff frequency of a propagation mode.

ghost pulse—*See* Ghost Signal.

ghost signal—Also called ghost pulse. An unwanted signal on the screen of a radar indicator. Echoes which experience multiple reflections before reaching the receiver are an example. The term also is applied to a reflected television signal (*see* Ghost Image).

giant grid—An extensive regional or national system of backbones and networks.

giant ties—*See* Interconnection, 1.

Gibson girl—A portable, hand-operated

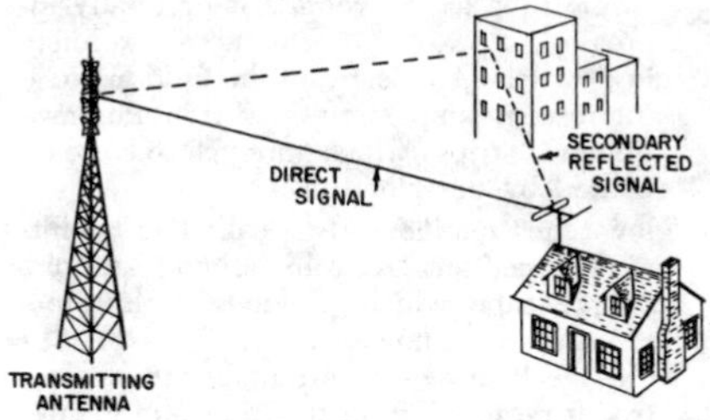

Ghost signal.

transmitter used by pilots forced down at sea.

giga—A prefix meaning one billion, or 10^9.

gigacycle—One kilomegacycle, or one billion cycles. An obsolete term. The currently preferred term is gigahertz.

gigahertz—A term for 10^9 cycles per second. Used to replace the more cumbersome and obsolete term kilomegacycle. Abbreviated GHz.

gigawatt—Abbreviated GW. One thousand megawatts (10^9 watts).

gigo—An acronym formed from the phrase Garbage In, Garbage Out. It is used to describe a computer the operation of which is suspect.

gigohm—One thousand megohms (10^9 ohms).

gilbert—A cgs unit of the magnetomotive force required to produce one maxwell of magnetic flux in a magnetic circuit of unit reluctance; 1 gilbert $= 10/4\pi$ ampere-turn.

gilbert per centimeter—The practical cgs unit of magnetic intensity. Gilberts per centimeter are the same as oersteds.

Gill-Morrell oscillator—A retarding-field oscillator in which the oscillation frequency depends not only on the electron-transit time within the tube, but also on the associated circuit parameters.

gill selector—A slow-acting telegraph sender and calling key for selective signaling.

gimbal—A mechanical frame having two perpendicularly intersecting axes of rotation.

gimmick—1. A capacitor with a value of a few picofarads, improvised by twisting together two insulated wires. 2. Length of twisted two-conductor cable, used as a variable capacity load, in which the capacity is varied by untwisting and separating the individual conductors.

gimp—A slang name given to the extremely flexible wire used in telephone cords and similar equipment. This wire cannot be directly soldered to, as it is a metallic cloth-type material.

Giorgi system—*See* MKSA Electromagnetic System of Units.

glass-ambient technology—The technique by which glass is applied directly to the surface of a semiconductor material. Typically, glass is placed on the surface of a microelectronic device by means of pyrolytic deposition, vapor deposition, or the firing of a glass powder to the surface.

glass electrode—In electronic pH measurement, an electrode used for determining the potential of a solution with respect to a reference electrode. The calomel type is the most common.

glassivation—1. A method of transistor passivation by a pyrolytic glass-deposition technique, whereby silicon semiconductor devices, complete with metal contact systems, are fully encapsulated in glass. 2. The deposition of glass on a chip to give protection to underlying device junctions. 3. A process in which a dielectric material is diffused over the entire wafer to provide mechanical and environmental protection for the circuits. Also called passivation.

glass-plate capacitor—A high-voltage capacitor in which the metal plates are separated by sheets of glass for the dielectric. The complete assembly is generally immersed in oil.

glass-to-metal seal—An airtight seal between glass and metal parts of an electron tube, made by fusing together a special glass and special metal alloy having nearly the same temperature coefficients of expansion.

glass tube—A vacuum or gaseous tube that has a glass envelope.

glide path—The approach path used by an aircraft making an instrument landing.

glide-path localizer—In an aircraft instrument-landing system, the part which indicates the altitude of the plane and creates a glide path for a blind landing.

glide-path transmitter—A transmitter that produces signals for vertical guidance of aircraft along an inclined surface that extends upward from the desired point of ground contact

glide slope—A radio beam used by pilots to determine the altitude of the aircraft during a landing.

glide-slope facility—A radio transmitting facility which provides the glide-slope signals.

glidetone—A device used in electronic music that produces a continuous shift in the frequency of an audio signal.

G-line—A round wire coated with a dielectric and used to transmit microwave energy.

glint—1. Also called glitter. A distorted radar-signal echo, which varies in amplitude from pulse to pulse because the beam is being reflected from a rapidly moving object such as an airplane propeller. 2. An electronic-countermeasures technique in which the scintillating, or flashing, effect of shuttered or rotating reflectors is used to degrade the tracking or seeking functions of an enemy weapons system.

glissando—A tone that changes smoothly from one pitch to another.

glitch—A form of low-frequency interference appearing as a narrow horizontal bar moving vertically through the television picture. This is also observed on an oscilloscope at the field or frame rate as an extraneous voltage pip moving along the signal at approximately the reference-black level.

glitter—*See* Glint.

glossmeter—A photoelectric instrument for determining the gloss factor of a surface (i.e., the ratio of light reflected in one direction to the light reflected in all directions).

glow discharge—A discharge of electricity through a gas in an electron tube. It is characterized by a cathode glow resulting from a space potential much higher than the ionization potential of the gas in the vicinity of the cathode.

glow-discharge microphone—A microphone in which the sound waves cause corresponding variations in the current forming a glow discharge between two electrodes.

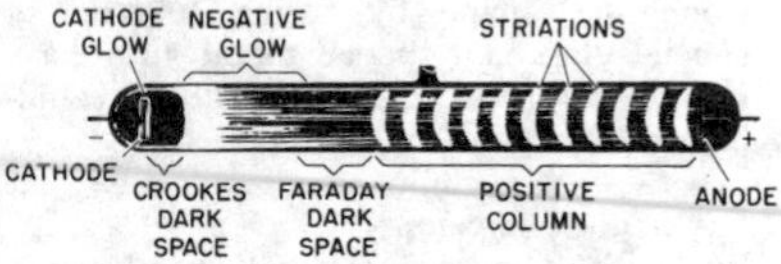

Glow-discharge tube.

glow-discharge tube—A gas tube that depends for its operation on the properties of a glow discharge.

glow-discharge voltage regulator—A gas tube used for voltage regulation. The resistance of the gas within the tube varies in step with the voltage applied across the tube.

glow lamp—1. A lamp containing a small amount of gas or vapor. Current between the two electrodes ionizes the gas and causes the lamp to glow but does not provide rectification. Neon gives a red-orange glow, mercury vapor a blue glow, and argon a purple glow. 2. Gas discharge tube serving as a concentrated source of light whose brightness varies in proportion to current flow. When an audio-frequency signal is combined with the lamp current, the brightness of the glow discharge varies according to the audio-frequency signal variations. 3. Glow-discharge types of tube whose light brightness is proportional to the current passing through the tube; used for photographic recording of facsimile signals.

glow potential—The voltage at which a glow discharge begins in a gas-filled electronic tube as the voltage is gradually increased.

glow switch—An electron tube used in some fluorescent-lamp circuits. It contains two bimetal strips which are closed when heated by the glow discharge.

glow-tube rectifier—Also called a point-plane rectifier. A cold-cathode gas-discharge tube which provides a unidirectional current flow.

glue-line heating—An arrangement of electrodes designed to heat a thin film of material having a high loss factor between alternate layers of materials having a low loss factor.

g_m—Symbol for the mutual conductance or transconductance of a vacuum tube.

G-M counter—Abbreviation for Geiger-Mueller counter.

G — Y signal—In color television, the green-minus-luminance signal representing primary green minus the luminance, or Y, signal. It is combined with a luminance, or Y, signal outside or inside the picture tube to yield a primary green signal.

GMT (or Gmt)—Abbreviation for Greenwich mean time.

gobo—A dark mat used to shield the lens of a television camera from stray lights.

gold-bonded diode—A semiconductor diode in which a preformed whisker of gold contacts an n-germination substrate as the junction is formed by millisecond electrical pulses.

gold doping—A technique used to control the lifetime of minority carriers in a diffused-mesa transistor. Gold is diffused into the base and collector regions to reduce the storage time.

gold-leaf electroscope—An apparatus comprising two pieces of gold leaf pointed at their upper ends and suspended inside a glass jar. When a charge is applied to the terminal connected to the leaves, they spread apart due to repulsion of the like charges on them.

Goldschmidt alternator—An early radio transmitter. It is a rotating machine employing oscillating circuits in connection with the field and the armature to introduce harmonics in the generated fundamental frequency. Interaction between the stator and rotor harmonics gives a cumulative effect and thereby provides very high radio frequencies.

goniometer—1. In a radio-range system, a device for electrically shifting the directional characteristics of an antenna. 2. An electrical device for determining the azimuth of a received signal by combining the outputs of individual elements of an antenna array in certain phase relationships.

go/no testing—Testing designed to show only whether the device under test passed or failed, with no indication of how it failed or why.

googol—In mathematics, the figure 1 followed by 100 zeros.

goto circuit—A circuit capable of sensing the direction of current. It can be used in majority logic circuits in which the output is either positive or negative, depending on whether the majority of the inputs is positive or negative.

goto pair—Two tunnel diodes connected in series in a way such that one is in the reverse tunneling region when the other is in the forwarded conduction region. This arrangement is used in high-speed gate circuits.

governed series motor—A motor used with teletypewriter equipment. It has a governor for regulating the speed.

governor—1. A motor attachment that automatically controls the speed at which the motor rotates. 2. The equipment which controls the gate or valve opening of a prime mover.

gpi—Abbreviation for ground-position indicator.

graceful degradation —A computer programming technique the purpose of which is to prevent catastrophic system failure by permitting the machine to operate, although in a degraded mode, in spite of failures or malfunctions in several integral units or subsystems.

graded-base transistor—*See* Diffused-Base Transistor.

graded filter—A power-supply filter in which the output stage of a receiver or audio amplifier is connected at or near the filter input so that the maximum available dc voltage will be obtained. The output stage has low gain; therefore, ripple is not too important.

graded insulation—A combination of insulation proportioned so as to improve the distribution of the electric field to which the combination is subjected.

graded-junction transistor—*See* Rate-Grown Transistor.

graded thermoelectric arm—A thermoelectric arm having a composition that changes continuously in the direction of the current.

gradient—The rate at which a variable quantity increases or decreases. For example, potential gradient is the difference of potential along a conductor or through a dielectric.

gradient meter—*See* Generating Electric Field Meter.

gradient microphone—A microphone in which the output rises and falls with the sound pressure (*See also* Pressure Microphone.)

gram—A unit of mass and weight in the metric system.

gramme ring—A ring-shaped iron armature around which the coils are wound. Each turn is tapped from the inside diameter of the ring to a commutator segment.

grandfather cycle—The period during which magnetic-tape records are retained before reusing so that records can be reconstructed in the event of loss of information stored on a magnetic tape.

granular carbon—Small particles of carbon used in carbon microphones.

granularity—A characteristic of the output data of a measuring instrument. The measure of granularity is the smallest increment of the output data when it is in a digital form. The smallest increment is also called least count.

graph—A pictorial presentation of the relationship between two or more variables.

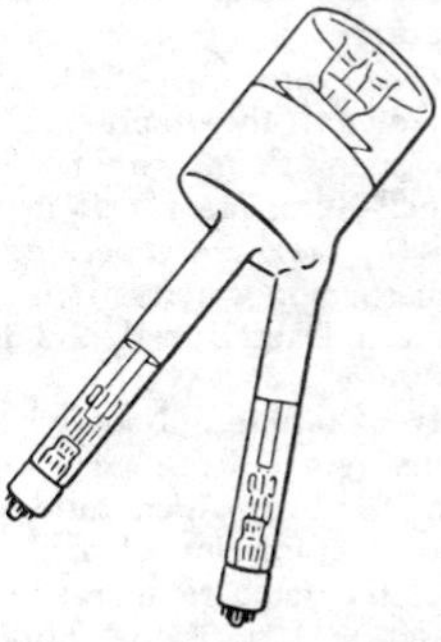

Graphechon.

graphechon—A specially designed electron memory tube, based on iconoscope principles, in which electrical signal information is stored and recovered at different scanning rates. It is used in radar and computer applications.

graphical analysis—The use of diagrams and other graphic methods to obtain operating data and answers to scientific or mathematical problems.

graphic instrument—*See* Recording Instrument.

graphics—1. In communications systems, an information mode in which intelligence is reproduced by use of a graphic system (a variation of facsimile). 2. Nonvoice analog information modes and devices such as facsimile and television.

graphic terminal—1. A cathode-ray-tube display. 2. An XY plotter.

graphite—A finely divided carbon used as a lubricant and in the construction of some carbon resistance elements. The most common use is in so-called lead pencils.

grass—The pattern produced by random noise on an A-scope; it appears as closely spaced, sharp, constantly moving pulses on the base line. *See also* Random Noise.

grasshopper fuse—A small fuse containing a spring which, upon release by the fusing wire, completes an auxiliary circuit to operate an alarm.

graticule—A calibrated screen placed in front of a cathode-ray tube for measurement purposes.

grating—A device for spreading out light or other radiation. It consists of narrow parallel slits in a plate or narrow parallel reflecting surfaces made by ruling grooves on polished metal. The slits or grooves break up the waves as they emerge. (*See also* Ultrasonic Space Grating *and* Ultrasonic Cross Grating.)

grating reflector—An antenna reflector consisting of an openwork metal structure that resembles a grating.

Gratz rectifier—An arrangement of two rectifiers per phase connected into a three-phase bridge circuit to provide full-wave rectification.

gravity—The force which tends to pull bodies toward the center of the earth, thereby giving them weight. (*See also* g.)

gravity cell—A primary cell in which two electrolytes are kept separated by differences in specific gravity. It is a modification of the Daniell cell and is now obsolete.

gray body—A radiating body whose spectral emissivity remains the same at all wavelengths. It is in constant ratio of less than unity to the radiation of a blackbody radiator at the same temperature.

Gray code—1. A positional binary number notation in which any two numbers whose difference is one are represented by expressions that are the same except in one place or column and differ by only one unit in that place or column. 2. A numeric code composed of a number of bits, assigned in such a way that only one bit changes at each increment (or decrement).

gray scale—A series of regularly spaced tones ranging from white to black through intermediate shades if gray used as a reference scale for control purposes in photography or tv.

great manual—In an organ, the keyboard normally used for playing the accompaniment to the melody. Also called the accompaniment manual or lower manual.

green-gain control—A variable resistor used in the matrix of a three-gun color television receiver to adjust the intensity of the green primary signal.

green gun—The electron gun the beam of which, when properly adjusted, strikes only the green phosphor dots in the color picture tube.

green restorer—A dc restorer used in the green channel of a three-gun color-television picture-tube circuit.

green video voltage—The signal voltage that controls the grid of the green gun in a three-gun picture tube.

Greenwich Civil Time—*See* Universal Time.

Greenwich mean time—Abbreviated GMT or Gmt. The mean solar time at the meridian of Greenwich (zero longitude). *See also* Universal Time.

grid—1. An electrode having one or more openings for the passage of electrons or ions. (*See also* Control Grid, Screen Grid, Shield Grid, Space-Charge Grid, *and* Suppressor Grid.) 2. An interconnected sys- in which high-voltage, high-capacity backbone lines overlay and are connected with networks of lower voltages. 3. A two-dimensional network consisting of a set of equally spaced parallel lines superimposed upon another set of equally spaced parallel lines so that the lines of one set are perpendicular to the lines of the other, thereby forming square areas. The intersections of the lines provide the basis for an incremental location system.

grid battery—Sometimes called a C-battery. A source of energy for supplying a bias voltage to the grid of a vacuum tube.

grid bearing—A bearing made with the reference line to grid north.

grid bias—Also called C-bias. A constant potential applied between the grid and cathode of a vacuum tube to establish an operating point.

grid-bias cell—A small cell used in a vacuum tube circuit to make the grid more negative than the cathode. It provides a voltage, but cannot supply an appreciable amount of current.

grid blocking—Blocking of capacitance-coupled stages in an amplifier because of an accumulated charge on the coupling capacitor as the result of current flow during the reception of large signals. 2. A method of keying a circuit by application of a negative grid voltage several times cutoff during key-up conditions; when the key is down, the blocking bias is removed, and normal current through the keyed circuit is restored.

grid cap—At the top of some vacuum tubes, the terminal which connects to the control grid.

grid capacitor—A capacitor in parallel with the grid resistor or in series with the grid lead of a tube.

grid-cathode capacitance—Capacitance between the grid and the cathode in a vacuum tube.

grid characteristic—The curve obtained by plotting grid-voltage values of a vacuum tube as abscissas against grid-current values as ordinates on a graph.

grid circuit—The circuit connected between the grid and cathode, and forming the input circuit of a vacuum tube.

grid-circuit tester—A tester designed to measure the grid resistance of vacuum tubes without discriminating between the type or polarity of impedance.

grid clip—A spring clip used for making a connection to the top-cap terminal of some vacuum tubes.

grid-conductance—The in-phase component of the alternating grid current divided by the alternating grid voltage, all other electrode voltages being maintained constant.

grid control—Control of the anode current of an electron tube by means of changes in the voltage between the control grid and cathode of the tube.

grid-controlled mercury-arc rectifier—A mercury-arc rectifier employing one or more electrodes exclusively for controlling start of the discharge.

grid-controlled rectifier—A triode mercury-vapor rectifier tube in which the grid determines the instant at which plate current starts to flow during each cycle, but does not determine how much current will flow.

grid-control tube—A mercury-vapor-filled thermionic vacuum tube an external grid control.

grid current—The current which flows in the grid-to-cathode circuit of a vacuum tube. It is usually a complex current made up of several currents having a variety of polarities and impedances.

grid detection—Detection by rectification in the grid circuit of a vacuum tube.

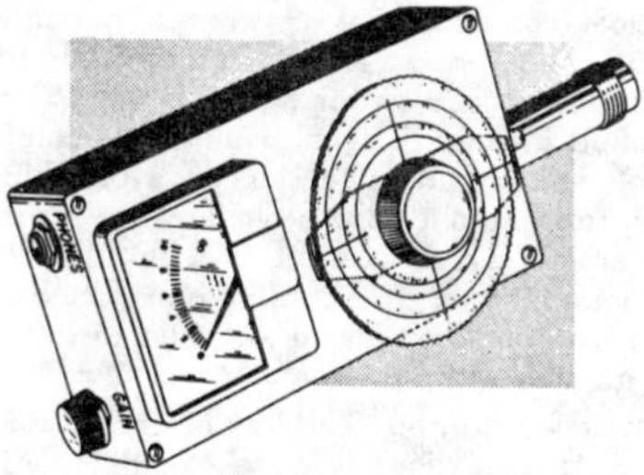

Grid-dip meter.

grid-dip meter—A multiple-range oscillator incorporating a meter in the grid circuit to indicate grid current. The meter is so named because its reading dips (reads a lower grid current) whenever an external resonant circuit is tuned to the oscillator frequency.

grid dissipation—The power lost as heat at the grid of a tube.

grid-drive characteristic—The relationship between the electrical or light output of an electron tube and the control-electrode voltage measured from cutoff.

grid driving power—The average product of the instantaneous value of grid current and the alternating component of grid voltage over a complete cycle.

grid emission – Electron or ion emission from the grid of an electron tube.

grid-glow tube—A glow-discharge, cold-cathode tube in which one or more control electrodes initiate the anode current, but do not limit it except under certain conditions.

gridistor – A field-effect transistor which uses the principle of centripetal striction and has a multichannel structure, combining advantages of both field-effect transistors and minority-carrier injection transistors.

grid leak—A high resistance connected across the grid capacitor or between the grid and cathode. It provides a direct-current path, to limit the accumulation of a charge on the grid.

grid-leak capacitor—A small capacitor connected in a vacuum-tube grid circuit, together with a resistor, to produce grid bias.

grid-leak detector—A triode or multielectrode tube in which rectification occurs because of electron current through a high resistance in the grid circuit. The voltage associated with this flow appears in amplified form in the plate circuit.

grid limiting—The use of grid-current bias derived from the signal, through a large series grid resistor, in order to cut off the plate current and consequently level the output wave for all input signals above a critical value.

grid locking—Faulty tube operation in which excessive grid emission causes the grid potential to become continuously positive.

grid modulation—Modulation produced by application of the modulating voltage to the control grid of any tube in which the carrier is present. Modulation in which the grid voltage contains externally generated pulses is called grid-pulse modulation.

grid neutralization—A method of neutralizing an amplifier. A portion of the grid-to-cathode alternating-current voltage is shifted 180° and applied to the plate-to-cathode circuit through a neutralizing capacitor.

grid north—An arbitrary reference direction used with the grid system of navigation.

grid-plate capacitance—The direct capacitance between the girl and plate of a vacuum tube.

grid-plate transconductance—Mutual conductance, which is the ratio of a plate-current change to the grid-voltage change that produces it.

grid-pool tank—A grid-pool tube having a heavy metal envelope somewhat resembling a tank in appearance.

grid-pool tube—A gas discharge tube having a mercury-pool cathode, one or more anodes, and a control electrode or grid to control the start of current. *See also* Excitron *and* Ignitron.

grid-pulse modulation—Modulation produced in an amplifier or oscillator by application of one or more pulses to a grid circuit.

grid pulsing—Method of controlling the operation of a radio-frequency oscillator. The oscillator-tube grid is biased so negatively that no oscillation occurs, even at full plate voltage, except when this negative bias is removed by application of a positive voltage pulse to the grid.

grid resistor—A general term that denotes any resistor in the grid circuit.

grid return – An external conducting path for the return of grid current to the cathode.

grid suppressor—A resistor, sometimes connected between the control grid and the external circuit of an amplifier, to prevent parasitic oscillations caused by stray-capacitance feedback.

grid swing—The total variation in grid-to-cathode voltage from the positive peak to the negative peak of the applied-signal voltage.

grid-to-cathode capacitance—The direct capacitance between the grid and cathode of a vacuum tube.

grid-to-plate capacitance—Designated C_{gp}. The direct capacitance between the grid and plate in a vacuum tube.

grid-to-plate transconductance – The mutual conductance, or ratio of plate-current to grid-voltage changes, in a vacuum tube.

grid voltage—The voltage between the grid and cathode of a tube.

grid-voltage supply—The means for supplying, to the grid of an electron tube, a potential which is usually negative with respect to the cathode.

grille cloth—A loosely woven fabric that is virtually transparent to sound, often stretched across the opening in a speaker enclosure to which the radiating side of loudspeaker is filled.

grommet—An insulating washer, usually of rubber or plastic, inserted through a hole in a chassis or panel to prevent a wire from touching the sides.

groove—In mechanical recording, the track inscribed in the record by a cutting or embossing stylus, including undulations or modulations caused by vibration of the stylus. In stereo discs, its cross section is a right-angled triangle, with each side at a 45° angle to the surface of the record; information is cut on both sides of the groove. In a long-playing record, groove dimensions could be: width 2.5 mils; depth 1 mil and pitch 250-350 groove revolutions per inch.

groove angle—In disc recording, the angle between the two walls of an unmodulated groove in a radial plane perpendicular to the surface of the recording medium.

groove shape—In disc recording, the contour of the groove in a radial plane perpendicular to the surface of the recording medium.

groove speed—In disc recording, the linear speed of the groove with respect to the stylus.

groove velocity—The speed with which the record groove moves under the cartridge. An lp record rotates at a constant 33⅓ rpm with grooves cut at diameters which decrease gradually from 11½ in to 4¾ in. Groove velocity therefore ranges from 20 in/s at the outside of the record to 8.3 in/s in the innermost groove.

gross information content—A measure of the total information, including redundant portions, contained in a message. It is expressed as the number of bits or hartleys necessary to transmit the message with specified accuracy by way of a noiseless medium without coding.

ground—1. A connection to the earth for conducting electrical current to and from the earth. 2. The voltage reference point in a circuit. There may or may not be an actual connection to earth, but it is understood that a point in the circuit said to be at ground potential could be connected to earth without disturbing the operation of the circuit in any way. 3. A point in an electrical system that has zero voltage. Usually, the chassis of an electrical component is at ground potential and thus serves as the return path for signals as well as for power circuits. The shield in coaxial signal cable is, or should be, at ground potential to avoid hum pickup. Ground also designates the earth, literally, which is used as a return path for radio waves from an antenna. In British terminology "earth" is used to designate all ground connections.

ground absorption—The loss of energy during transmission because of the radio waves dissipated to ground.

ground bus – A conductor, usually large-diameter wire, that connects a number of points to one or more ground electrodes.

ground check (base-line check)—1. A procedure followed prior to the release of a radiosonde in order to obtain the temperature and humidity correction for the radiosonde system. 2. Any instrumental check prior to the ground launch of an airborne experiment.

ground clamp—A clamp used for connecting a grounding conductor (ground wire) to a grounded object such as a water pipe.

ground clutter – The pattern produced on the screen of a radar indicator by undesired ground return.

ground conduit—A conduit used solely to

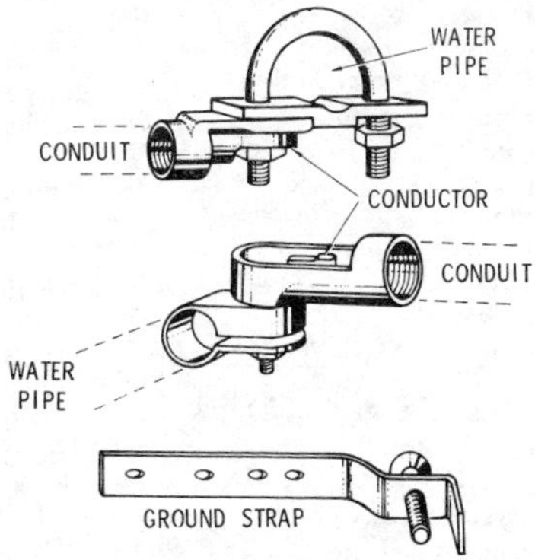

Ground clamps.

contain one or more grounding conductors.

ground controlled approach (GCA) – The radar system developed to give direction, distance and elevation along a fixed approach path to an airport. The ground controller at the radarscope communicates instructions to the pilot to direct the aircraft along the approach line.

ground-controlled interception – Abbreviated GCI. A radar system used for directing an aircraft to intercept enemy aircraft.

ground current – Current in the earth or grounding connection.

ground detector – An instrument or equipment that indicates the presence of a ground on a normally ungrounded system. 2. Device that indicates ground faults in electrical circuits.

ground dielectric constant – The dielectric constant of the earth at a given location.

ground distance – The great-circle component of distance from one point to another at mean sea level.

grounded – Connected to the earth, or to some conducting body in place of the earth.

grounded-base amplifier – *See* Common-Base Amplifier.

grounded cable bond–A cable bond used for grounding the armor and/or sheaths of cables.

grounded capacitance–In a system having several conductors, the capacitance between a given conductor and the other conductors when they are connected together and to ground.

grounded-cathode amplifier – The conventional amplifier circuit. It consists of a tube amplifier in which the cathode is at ground potential at the operating frequency. The input is applied between the control grid and ground, and the output load is between the plate and ground.

grounded circuit–A circuit in which one conductor or point (usually the neutral conductor or neutral point of transformer or generator windings) is intentionally grounded (earthed) either solidly or through a grounding device.

grounded-collector amplifier–*See* Common-Collector Amplifier.

grounded conductor–A conductor which is intentionally grounded, either directly or through a current-limiting device.

grounded-emitter amplifier–*See* Common-Emitter Amplifier.

grounded-gate amplifier–An FET amplifier circuit in which the gate electrode is connected to ground. The input signal is applied to the source electrode, and the output is taken from the drain electrode.

grounded-grid amplifier – An electron-tube amplifier circuit in which the control grid is at ground potential at the operating frequency. The input is applied between the cathode and ground, and the output load is between the plate and ground. The grid-to-plate impedance of the tube is in parallel with the load, instead of acting as a feedback path.

grounded-grid triode–A type of triode designed for use in a grounded-grid circuit.

grounded-grid triode circuit – A circuit in which the input signal is applied to the cathode and the output is taken from the plate. The grid is at rf ground and serves as a screen between the input and output circuits.

grounded-grid triode mixer – A triode in which the grid forms part of a grounded electrostatic screen between the anode and cathode. It is used as a mixer for centimeter wavelengths.

grounded outlet–An outlet equipped with a receptacle of the polarity type having, in addition to the current-carrying contacts, one ground contact which can be used for the connection of an equipment-grounding conductor.

grounded parts–Parts of a completed installation that are so connected that they are substantially at the same potential as the earth.

grounded-plate amplifier–Also called cathode follower. An electron-tube amplifier circuit in which the plate is at ground potential at the operating frequency. The input is applied between the control grid and ground, and the output load is between the cathode and ground.

grounded system–A system of conductors in which at least one conductor or point (usually the middle wire or neutral point of transformer or generator windings) is intentionally grounded, either solidly or through a current-limiting device.

ground environment–1. The environment surrounding and affecting a system or item of equipment that operates on the ground. 2. A system or part of a system that functions on the ground. 3. The aggregate of all ground-installed equipment that makes up a communications-electronics system, facility, station, set, etc. 4. The portion of an air-defense system that provides for the detection, surveillance, and control of airborne objects. It includes ground-based facilities and overwater facilities such as picket vessels and airborne early warning and control aircraft.

ground-equalizer inductors—Relatively low-inductance coils inserted in the circuit to one or more of the grounding points of an antenna to obtain a desired distribution of the current to the various points.

ground fault – An unintentional electrical path between a part operating normally at some potential to ground, and ground.

ground fault interrupter—*See* Current Limiter.

ground gating—The conversion of pam signals at a telemetry ground station to 50% duty-cycle signals.

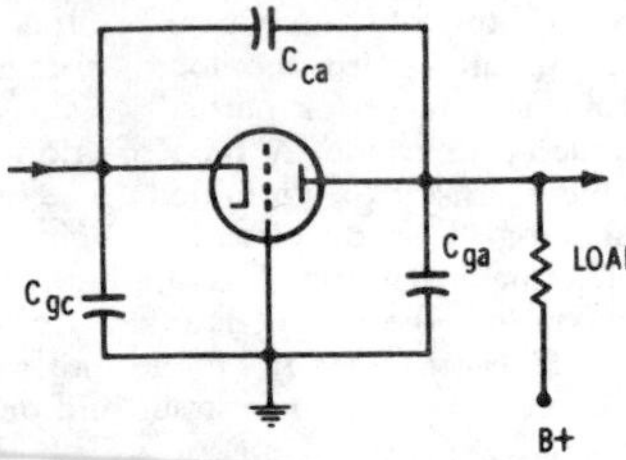

Grounded-grid amplifier.

ground grid—A system of grounding electrodes interconnected by bare cables buried in the earth to provide lower resistance than a single grounding electrode.

ground indication—An indication of the presence of a ground on one or more of the normally ungrounded conductors of a system.

grounding—1. Connecting to ground, or to a conductor which is grounded. 2. A means of referencing electrical circuits to the well-bonded equipotential surface.

grounding conductor—A conductor which, under normal conditions, carries no current, but serves to connect exposed metal surfaces to an earth ground, to prevent hazards in case of breakdown between current-carrying parts and the exposed surfaces. The conductor, if insulated, is colored green with or without a yellow stripe.

grounding connection—A connection used to establish a ground, consisting of a grounding conductor, a grounding electrode, and the earth surrounding the electrode.

grounding electrode—A conductor embedded in the earth and used for maintaining ground potential on conductors connected to it, or for dissipating into the earth any current conducted to it.

grounding plate—An electrically grounded metal plate on which a person stands in order to discharge any static electricity that may be picked up by his body.

grounding switch—A form of air switch for connecting a circuit or apparatus to ground.

grounding transformer—A transformer intended primarily for the purpose of providing a neutral point for grounding purposes.

ground insulation—The major insulation used between a winding and structural parts at ground potential.

ground junction—In a semiconductor, a junction formed during the growth of a crystal from a melt.

ground level—*See* Ground State.

ground loop – 1. An unwanted feedback condition in which power current in a single ground wire causes instability or errors. 2. A potentially detrimental condition produced when two or more points in an electrical system that are nominally at ground potential are connected by a conducting path. The term usually is applied when, because of improper design or by accident, unwanted noise signals are generated in the common return of relatively low-level signal circuits by the return currents or by magnetic fields produced by relatively high-power circuits or components. 3. The electrical path between two separate grounds.

ground lug—A lug for connecting a grounding conductor to a grounding electrode.

ground mat—A system of bare conductors, on or below the surface of the earth, connected to a ground or ground grid to provide protection from dangerous touch voltage.

ground noise—In recording and reproducing, the residual noise in the absence of a signal. It is usually caused by dissimilarities between the recording and reproducing media, but may also include amplifier noise such as from a tube or noise generated in resistive elements at the input of the reproducer amplifier system.

ground-noise margin—The voltage that may be applied at the ground connection of a logic circuit without causing the circuit to malfunction. It is usually measured by increasing the static ground voltage on a single gate until the logic fails to operate properly.

ground outlet – An electrical outlet equipped with a polarized receptacle that has, in addition to the current-carrying contacts, a grounded contact to which can be connected an equipment-grounding conductor.

ground plane—Copper or brass sheet used in interference testing to simulate missile, aircraft, or vehicle frame or skin so that

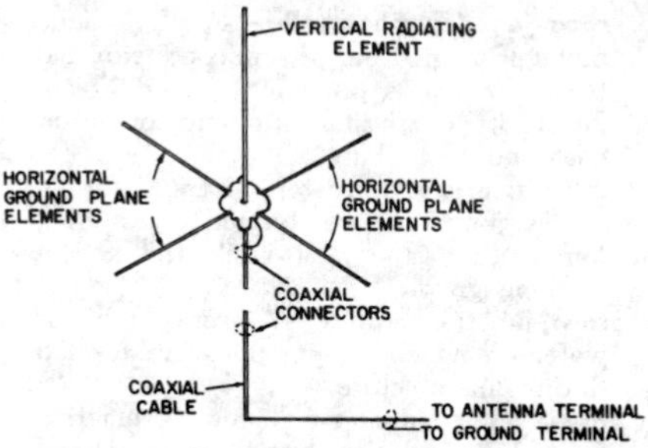

Ground-plane antenna.

actual installation and grounding conditions may be approximated.

ground-plane antenna—A vertical antenna combined with a turnstile element to lower the angle of radiation. It has a concentric base support and a center conductor that place the antenna at ground potential, even though located several wavelengths above ground.

ground plate—A plate of conductive material buried in the earth to serve as a grounding electrode.

ground-position indicator—Abbreviated gpi. A dead-reckoning computer, similar to an air-position indicator, with provision for taking drift into account.

ground potential—Zero potential with respect to ground or the earth.

ground protection—Protection of a circuit by means of a device that opens the circuit when a fault to ground occurs.

ground protective relay—A device which functions on failure of the insulation of a machine, transformer, or other apparatus to ground, or on flashover of a dc machine to ground.

ground range—In range measurements related to airborne radar, the distance on the surface of the earth between the object under consideration and a point directly below the aircraft that carries the radar.

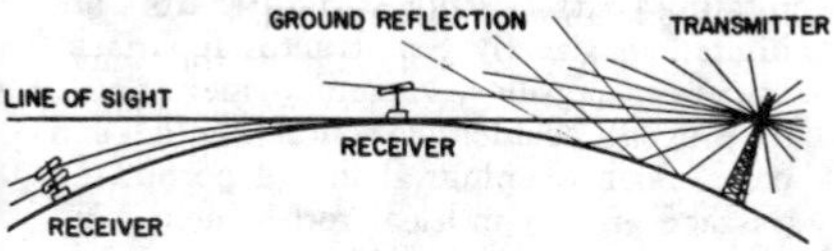

Ground-reflected waves.

ground-reflected wave—In a ground wave, the component reflected from the earth.

ground resistance—1. The opposition of the earth to the flow of current through it. Its value depends on the nature and moisture content of the soil; the material, composition, and physical dimensions of the connections to earth; and the electrolytic action present. 2. The ohmic resistance between a grounding electrode and a remote or reference grounding electrode so spaced that their mutual resistance is essentially zero.

ground return—1. In radar, the echoes reflected from the earth's surface and fixed objects on it. 2. A lead from an electronic circuit, antenna, or power line to ground.

ground-return circuit—A circuit which has a conductor (or two or more in parallel) between two points and which is completed through ground or the earth.

ground rod—A steel or copper rod driven into the earth to make an electrical contact with it.

ground-scatter propagation—Multihop ionospheric propagation of radio waves along other than the great-circle path between the transmitting and receiving points. Radiation from the transmitter is returned from the ionosphere to the surface of the earth, from which it is then scattered in many directions.

ground shift—The variation in signal amplitude at different grounding points because of a voltage drop along a ground line.

ground speed—In navigation, the speed of a vehicle with reference to ground.

ground state—The lowest energy level or state of an atom or atomic system; all other states of the system are called excited states. Also called ground level.

ground-support equipment — All ground equipment that is part of a complete weapons system and that must be furnished to ensure complete support of the weapons system.

ground system of an antenna—The portion of an antenna system that includes an extensive conducting surface, which may be the earth itself, and those parts of the antenna closely associated with that surface.

ground-to-air communication — One-way communication from ground stations to aircraft.

ground wave—A radio wave which travels along the earth's surface rather than through the upper atmosphere.

ground wire—A conductor leading to an electric connection with the earth.

group—In carrier telephony, a number of voice channels multiplexed together and treated as a unit. Commonly, a group contains twelve channels, each with a bandwidth of 4 kHz, frequency multiplexed and occupying the band from 60 to 180 kHz.

group busy tone—A high tone fed to the jack sleeves of an outgoing trunk group to serve as an indication that all trunks in the group are busy.

group delay—Also called envelope delay. The delay in transmission of information modulated on a carrier.

grouped-frequency operation—A method in

which different frequency bands are used for channels in opposite directions in a two-wire carrier system.

group frequency—The number of sets or groups of waves passing a given point in one second.

grouping—1. Nonuniform spacing between the grooves of a disc recording. 2. Periodic error in the spacing between recorded lines in a facsimile system.

grouping circuits—Circuits used to interconnect two or more positions of a switchboard so that one operator may handle the several positions from one operator's set.

group loop—A source of interference when a system is grounded improperly at several points.

group mark—A mark used to identify the beginning or the end of a set of data, which could include words, blocks, or other items.

group modulation—The process by which a number of channels, already separately modulated to a specific frequency range, are again modulated to shift the group to another range.

group velocity — 1. Of a traveling plane wave, the velocity of propagation of the envelope delay is approximately constant. It is equal to the reciprocal of the envelope delay per unit length. (Group velocity differs from phase velocity in a medium in which the phase velocity varies with the frequency.) 2. The velocity of the envelope of an electromagnetic wave as it travels in a medium, usually identified with the velocity of energy propagation.

Grove cell—A primary cell with a platinum electrode submerged in an electrolyte of nitric acid within a porous cup, surrounded by a zinc electrode in an electrolyte of sulfuric acid. This cell normally operates on a closed circuit.

growler — An electromagnetic device for locating short-circuited coils and for magnetizing or demagnetizing objects. So called because of the growling noise it makes when indicating a short circuit. It consists essentially of two field poles arranged as in a motor.

grown-diffused transistor — A transistor made by combining the diffusion and double-doped techniques. Suitable n- and p-type impurities are added simultaneously to the melt while the crystal is being grown. Subsequently, the base region is formed by diffusion as the crystal grows.

grown junction—The boundary between p- and n-type semiconducting materials. It is produced by varying the impurities during the growth of a crystal from the melt. Such junctions have strong rectifying properties, the forward current being obtained when p is positive to n.

grown-junction photocell — A photodiode made of a small bar of semiconductor material that has a pn junction at right angles to its length and an ohmic contact at each end of the bar.

grown-junction transistor—A transistor in which junctions are formed by adding impurities to the melt while the crystal is being grown.

grown-junction wafer — A semiconductor wafer on which pn junctions are formed during manufacture.

grown semiconductor junction—A junction formed during the growth of a crystal from a melt.

G-scan—*See* G-Display.

G-scope—*See* G-Display.

guard—A mechanism to terminate program execution (real or simulated) upon access to data at a specified memory location. (Used in debugging.)

guard arm—1. A crossarm placed across and in line with a cable to protect it from damage. 2. A crossarm placed over wires to prevent other wires from falling into them.

guard band—1. Also called interference-guard band. A frequency band left vacant between two channels to safeguard against mutual interference. 2. The unused chip surface area which by virtue of physical spacing serves to isolate functional elements in a printed circuit or an integrated circuit. Also refers to the consideration given instrumentation precision in electrical testing.

guard circle—An inner concentric groove on disc records. It prevents the pickup from being thrown to the center of the record and possibly damaged.

guarded input—An input that has a third terminal which is maintained at a potential near the input-terminal potential for a single-ended input, or near the mean input potential for the differential input. It is used to shield the entire input circuit.

guarded motor—An open motor in which all openings given direct access to live or rotating parts (except smooth shafts) are limited in size by the structural parts or by screens, grilles, expanded metal, etc., to prevent accidental contact with such parts. Such openings shall not permit the passage of a cylindrical rod ½ inch in diameter.

guarding—The introduction of conducting surfaces at critical points in a circuit to intercept and divert leakage currents that otherwise would cause undesired effects or measurement errors.

guard relay—A relay used in the linefinder circuit to prevent more than one linefinder from being connected to any line circuit when two or more line relays are operated simultaneously.

guard ring—1. A metal ring placed around

a charged terminal or object to distribute the charge uniformly over the surface of the object. 2. A ring-shaped electrode intended to limit the extent of an electric field, as, for instance, in elimination of the fringe effect at the edges of the plates of a capacitor.

guard shield—An internal floating shield surrounding the input section of an amplifier. Effective shielding results only when the absolute potential of the guard is stabilized relative to the incoming signal.

guard wire—A grounded wire used frequently where high-tension lines cross a thoroughfare. Should a line break, it will contact the guard wire and be grounded.

Gudden-Pohl effect—The momentary illumination produced when an electric field is applied to a phosphor previously excited by ultraviolet radiation.

guidance—Control of a missile or vehicle from within by a person, a preset or self-reacting automatic device, or a device that reacts to outside signals.

guidance system—A system which measures and evaluates flight information, correlates it with target data, converts the resultant into the parameters necessary to achieve the desired flight path, and communicates the appropriate commands to the flight-control system.

guidance tapes — Magnetic or paper tapes that are placed in a missile or computer and that contain previously entered information necessary for directing the missile to the selected target.

guide—In a tape recorder, a grooved or flanged pin or roller that guides the tape in a straight line between the reels and the heads, to keep it perfectly in line with the pole pieces.

guided ballistic missile—A ballistic missile which is guided during the powered portion of the trajectory and follows a free ballistic path during the remainder.

guided missile—An unmanned vehicle moving above the surface of the earth, the trajectory or flight path of which is capable of being altered by an external or internal mechanism.

guided propagation—A type of radio-wave propagation in which radiated rays are bent excessively by refraction in the lower layers of the atmosphere. This bending creates an effect much as if a duct or a waveguide has been formed in the atmosphere to guide part of the radiated energy over distances far beyond the normal range.

guided spark—An electrical discharge between two electrodes that has its path guided or constrained by the presence of a dielectric material or a gas jet.

guided wave—A wave in which the energy is concentrated near a boundary (or between substantially parallel boundaries) separating materials of different properties. The direction of propagation is parallel to the boundary.

guide pin—A pin or rod that extends beyond the mating faces of a connector in such a way that it guides the closing or mating of the connector and ensures proper engagement of the contacts.

guide wavelength—*See* Waveguide Wavelength.

Guillemin line—A special type of artificial transmission line or pulse-forming network used in radar sets to control the duration of the pulses. It generates a nearly square pulse for use in high-level pulse modulation.

guillotine capacitor—A translatory motion tuning capacitor consisting of a pair of stators and a sliding plunger in place of a rotor.

gun-directing radar—Radar used for directing antiaircraft or similar artillery fire.

gunn diode—A tiny wafer of n-type gallium arsenide consisting of a thin active layer of n-type gallium arsenide grown on a low resistivity substrate of the same material. The substrate is bonded to the anode terminal of the encapsulation and the other face of the wafer has an evaporated cathode contact connected by a bonded gold wire. The diode has no pn junction and cannot be used for rectification. When a few volts dc are applied to make the anode positive with respect to the cathode, the current which flows is dc with superimposed pulses.

gunn effect—Current oscillations that occur at an rf rate when an electric field of about 3000 V/cm is applied to a short (0.005 inch or less) specimen of n-type gallium arsenide. This effect takes place because electrons under the influence of sufficiently high fields are transferred from high- to low-mobility valleys in the conduction band of GaAs.

gunn oscillator—An oscillator in which the active element is a gunn diode operating in the negative resistance mode. This type of oscillator is one of the simplest means of generating microwave signals as only a microwave-tuned circuit, gunn diode, bias network and low-voltage power supply are required. Presently available units are restricted to operation above 4 GHz, and have dc to rf conversion efficiencies of less than 10%.

gutta-percha — A natural vegetable gum, similar to rubber, used principally as insulation for wires and cables.

guy wire—A wire used to brace the mast or tower of a transmitting or receiving antenna system.

gyrator—A two-port circuit element that exhibits a 180° differential, or nonreciprocal phase shift. The gyrator circuit symbol indicates that an rf signal transmitted

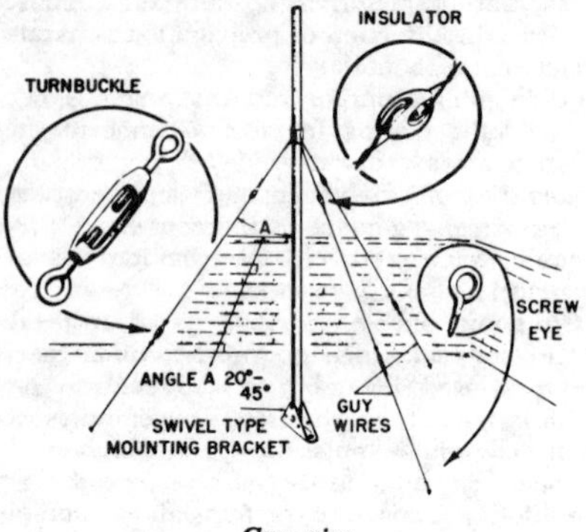

Guy wire.

from port 1 to port 2 will undergo a 180° phase shift relative to an rf signal transmitted in the reverse direction.

gyro—Abbreviation for gyroscope.

gyrocompass—*See* Gyroscope.

gyrofrequency—The natural frequency at which charged particles rotate around the lines of force of the earth's magnetic field. For electrons, it is 700 to 1600 kilohertz; for ions, it is in the audio-frequency range.

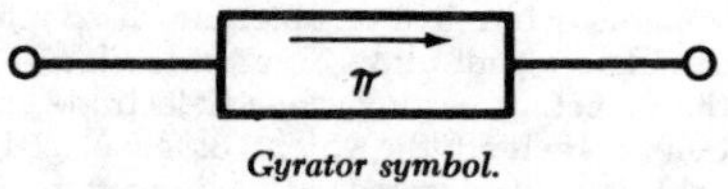

Gyrator symbol.

gyromagnetic—The magnetic properties of rotating electric charges, such as electrons spinning within atoms.

gyromagnetic effect—The change in the angular momentum of a body as a result of being magnetized, arising as a result of the fact that the magnetic moments of its electrons are associated with their spins or orbital angular momentum.

gyropilot—*See* Autopilot.

gyroscope—A rotating device the axle of which will maintain a constant direction, even though the earth is turning under it. It consists of a wheel mounted so that its spinning axis is free to rotate around either of two other axes perpendicular to itself and to each other. When its axle is pointed north, it can be used as a gyrocompass. Abbreviated gyro.

gyroscopic action—An action that causes a mass to turn on an axis perpendicular to the applied torque and to the axis of spin.

gyrostabilized platform—*See* Stable Platform.

H

H—1. A radar air-navigation system using an airborne interrogator to measure the distance from two ground responder beacons. (*See also* Shoran.) 2. Symbol for heater, magnetic field strength, or henry.

hairpin pickup coil—A hairpin-shaped, single-turn coil for transferring uhf energy.

hairpin tuning bar — A sliding hairpin-shaped metal bar inserted between the two halves of a doublet antenna to vary its electrical length.

halation—1. Distortion seen as blurred images and caused by reflection of the image rays off the back of a fluorescent screen that is too thick. 2. The spreading of light in a photographic emulsion outside the intended area of exposure by reflection from the rear surface of the material supporting the emulsion; this is distinguished from the diffusion which takes place within the emulsion layer.

half-add—In a computer, an operation that is performed first in carrying out a two-step binary addition. It consists of addition of corresponding bits in two binary numbers, with any carry information being ignored. *See also* Exclusive OR.

half-adder — 1. A circuit that will accept two binary input signals and produce corresponding sum and carry outputs. So called because, above the first order, two half-adders per order are required when adding two quantities. 2. Building-block circuit used in digital computers. A combination of logic gates adds two bits and delivers an answer—two bits called "sum" and "carry." Half-adders can be combined to add numbers of any length. Two half-adders make up a full-adder. 3. A logic element that adds two input bits, but does not have provision for adding in the carry from a previous addition. *See also* Full-Adder.

half cell—An electrode, submerged in an electrolyte, for measuring single electrode potentials.

half cycle—The time interval required for the operating frequency to complete one half, or 180°, of its cycle.

half-duplex—1. A communication system in which information can be transmitted in either direction, but only in one direction at a time. 2. In communications, pertaining to an alternate independent transmission made in one direction at a time.

half-duplex circuit—A circuit which permits one-direction electrical communications between stations. Technical arrangements may permit operation in either direction, but not simultaneously. Therefore, this term is qualified by one of the following suffixes: s/o for send only; r/o for re-

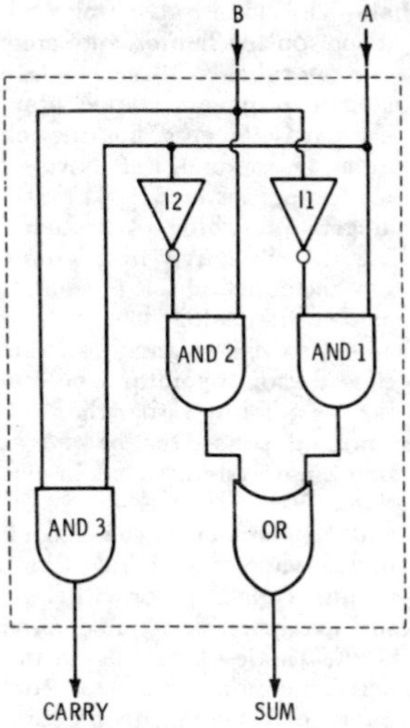

TRUTH TABLE

B	A	CARRY	SUM
0	0	0	0
0	1	0	1
1	0	0	1
1	1	1	0

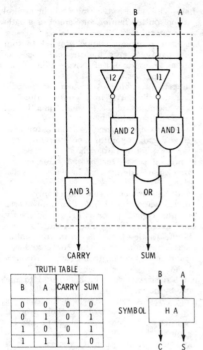

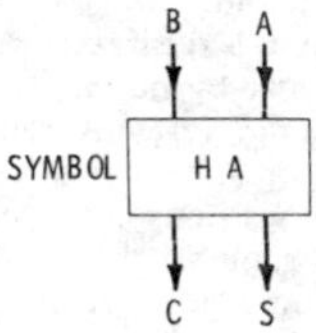

Half adder.

ceive only; s/r for send or receive. Abbreviated hdx.

half-duplex operation—A duplex telegraph system capable of operating in either direction, but not in both simultaneously.

half-duplex repeater—A duplex telegraph repeater provided with interlocking arrangements which restrict the transmission of signals to one direction at a time.

half life—The time interval used to measure the rate of decay of radioactive material. In the first half-life, the amount of radioactive material left unchanged is one-half the original amount; in the next half-life interval, half of the remaining amount, or one-fourth of the original amount remains. Thus, by determining the remaining radioactivity of a fossil and comparing it with the half-life of the material, scientists can estimate the age of the fossil. The half-life of various materials varies greatly—from millionths of a second to billions of years.

half-nut—A feed nut which engages half the circumference or less of a lead screw, so that it can be withdrawn from the lead screw to stop the lateral scanning movement.

half-power frequency—Either a high frequency or a low frequency at which the output of an amplifier, network, transducer, etc., falls to one half (−3dB) of its maximum or nominal response.

half-power point — On an amplitude response characteristic or other curve of the magnitude of a network quantity versus frequency, distance, angle, or other variable, the point that corresponds to half the power of a neighboring point having maximum power.

half-power width of a radiation lobe—In a plane containing the direction of the maximum of the lobe, the full angle between the two directions in that plane in which the radiation intensity is one half the maximum value of the lobe.

half-shift register—1. A logic circuit that consists of a gated input storage element with or without an inverter. 2. A logic device equivalent to half of a full master-slave flip-flop.

half step—*See* Semitone.

half tap — A bridge that can be placed across conductors without disturbing their continuity.

half-time emitter — A device that produces synchronous pulses midway between the row pulses of a punched card.

half-tone characteristic — In facsimile, the fidelity of the recorded density shadings in comparison with the original transmitted subject copy. Also used to express the relationship between the facsimile signal and the subject or recorded copy.

half-track recorder — *See* Dual-Track Recorder.

half-track tape—Also called two-track tape. Quarter-inch magnetic tape on which half the width of the tape is used for one sound path. Such a tape provides stereo in one direction of tape travel, or mono sound in both directions.

half wave — A wave with an electrical length of half a wavelength.

half-wave antenna—An antenna having an electrical length equal to half the wavelength of the signal being transmitted or received.

half-wave dipole—A straight, ungrounded antenna measuring substantially one-half wavelength.

half-wave rectification—1. The production of a pulsating direct current by passing only half the input cycle of an alternating current. The other half is blocked by the rectifier. 2. The process of blocking the negative half cycle of current of an alternating input. This is accomplished by a single diode.

half-wave rectifier — A rectifier utilizing only one half of each cycle to change alternating current into pulsating direct current.

half-wave transmission line — A piece of transmission line having an electrical length equal to half the wavelength of the signal being transmitted or received.

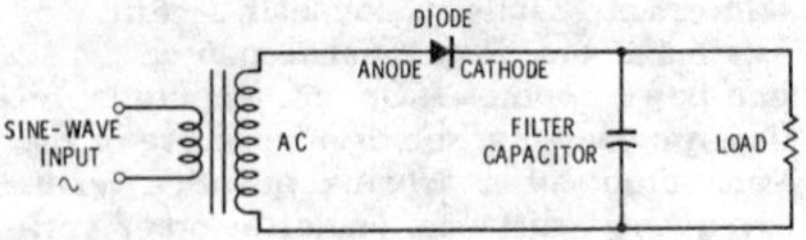

Half-wave rectifier.

half-wave vibrator—A vibrator used mainly in battery-operated mobile power supplies. It has only one pair of contacts, and supplies an intermittent unidirectional current at its output (usually connected to a half-wave rectifier).

half word—A nonbroken sequence of bits or characters that makes up half a computer word and that can be addressed as a unit.

Hall constant—The constant of a proportionality in the equation for a current-carrying conductor in a magnetic field. The constant is equal to the transverse electric field (Hall field) divided by the product of the current density and the magnetic field strength. The sign of the majority carrier can be inferred from the sign of the Hall constant.

Hall effect—In a current-carrying semiconductor bar located in a magnetic field that is perpendicular to the direction of the current, the production of a voltage perpendicular to both the current and the magnetic field.

Hall-effect modulation — Use of a Hall-effect multiplier as a modulator to produce an output voltage proportional to the product of two input voltages or currents.

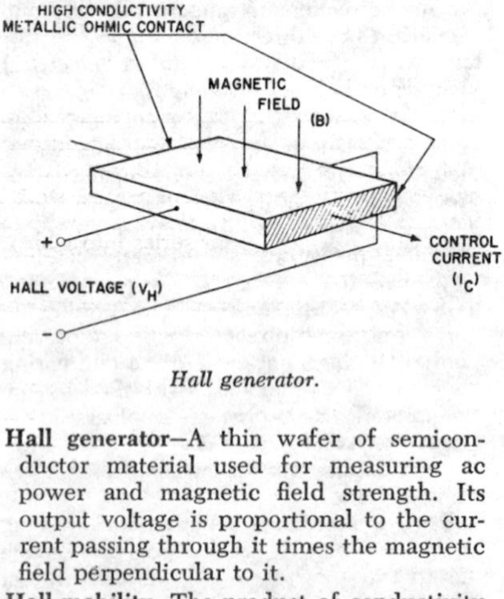

Hall generator.

Hall generator—A thin wafer of semiconductor material used for measuring ac power and magnetic field strength. Its output voltage is proportional to the current passing through it times the magnetic field perpendicular to it.

Hall mobility—The product of conductivity and the Hall constant for a conductor or semiconductor. It is a measure of the mobility of the electrons or holes in a semiconductor.

halo—The undesirable ring of light around a spot on the fluorescent screen of a cathode-ray tube.

halogen—A general name applied to four chemical elements, fluorine, chlorine, bromine, and iodine, that have similar chemical properties.

halogen quenching—A method of quenching the discharge in a counter tube by the introduction of a small quantity of one of the halogens.

ham—Also called amateur. Slang for a licensed radio operator who operates a station as a hobby rather than a business.

Hamming code—One of the error correction code systems used in data transmission.

hand capacitance — The capacitance introduced when one's hand is brought near a tuning capacitor or other insufficiently shielded part of a tuned circuit.

Handie-Talkie—Trade name of the Motorola Communications Div. for a two-way radio small enough to be carried in one's hand.

hand receiver — An earphone held to the ear by hand.

hand reset—A relay in which the contacts must be reset manually to their original positions after normal conditions are resumed.

handset — A telephone-type receiver and transmitter mounted on a single frame.

handset telephone — *See* Hand Telephone Set.

handshaking—1. A process in which predetermined arrangements of characters are exchanged by the receiving and transmitting equipment to establish synchronization. 2. The exchange of predetermined signals between machines connected by a communications channel to assure each that it is connected to the other. May also include the use of passwords and codes by an operator. 3. A colloquial term which describes the method used by a modem to establish contact with another modem at the other end of a telephone line. Often used interchangeably with buffering and interfacing, but with a fine line of difference in which handshaking implies a direct package-to-package connection regardless of functional circuitry.

hand telephone set—Also called a handset telephone. A telephone set having a handset and a mounting which supports the handset when not in use.

hangover — 1. Also called tailing. The smeared or blurred bass notes reproduced by a poorly damped speaker or one mounted in an improperly vented enclosure. 2. In television, overlapping and blurring, in the direction opposite to subject motion, of successive frames as a result of improper transient response. 3. In facsimile, distortion that occurs when the signal changes from maximum to minimum at a slower rate than required, with the result that there is tailing on the lines in the copy.

hangup—A condition in which the central processor of a computer is trying to perform an illegal or forbidden operation or in which it is continually repeating the same routine.

hard—1. Indicating an electron tube that has been evacuated to a high degree. 2. Indicating X-rays of relatively high penetrating power.

hard copy—1. Typewritten or printed characters on paper produced by a computer at the same time information is copied or converted into machine language that is not easily read by a human. 2. A printed copy of a machine output.

hard-copy printer – An automatic device, sometimes resembling a typewriter, which produces intelligible symbols in a permanent form.

hard-drawn copper wire—Copper wire that is not annealed after work hardening during drawing, thus providing increased tensile strength.

hardened links—1. Transmission links for which special construction or installation is necessary to assure a high probability of survival under nuclear attack. 2. Passive protection to aid survival.

hard firing—A condition in which the gate signal of an SCR is several times the dc triggering current and in which the rise time of the gate current is short relative to the turn-on time.

hardline – The intelligence link between two objects, consisting of a wire or wires, as opposed to a radio or radar link.

hard magnetic materials—Magnetic, materails that are not easily demagnetized.

hardness—Referring to X-rays, the quality which determines their penetrating ability. The shorter the wavelength, the harder and hence more penetrating they are.

hardness tester—Equipment for determining the force required to penetrate the surface of a solid.

hard rubber – A material formerly widely used for insulation. It is formed by vulcanizing rubber at high temperature and pressure to give it the desired hardness.

hard solder – Solder composed principally of copper and zinc. It must be red-hot before it will melt. Hard soldering is practically equivalent to brazing.

hard tube—A high-vacuum electronic tube.

hardware—1. Mechanical, magnetic, electrical, or electronic devices; physical equipment. (Contrasted with software.) 2. Particular circuits or functions built into a system. 3. The physical components of a computer or a system. (Software is the term used to describe the programs and instructions for a computer.

hardwire—A colloquialism meaning a circuit evidencing dc continuity.

hard-wired logic – A group of solid-state logic modules mounted on one or more circuit boards and interconnected by electrical wiring. The logic control functions are determined by the way in which the modules are interconnected. (As contrasted with a programmable controller or microprocessor in which the logic is in program form.)

hard X-rays—Highly penetrating X-rays as distinguished from less penetrating, or soft, X-rays.

harmful interference—Any radiation or any induction which disrupts the proper functioning of an electromagnetic system.

harmonic—A sinusoidal wave having a frequency that is an integral multiple of the fundamental frequency. For example, a wave with twice the frequency of the fundamental is called the second harmonic.

harmonic anaylsis—1. A method of identifying and evaluating the harmonics that make up a complex waveform of voltage, current, or some other varying quantity. 2. The expression of a given function as a series of sine and cosine terms that are approximately equal to the given function, such as a Fourier series.

harmonic analyzer—Also called harmonic-wave analyzer. A mechanical or electronic device for measuring the amplitude and phase of the various harmonic components of a wave from its graph.

harmonic antenna—An antenna the electrical length of which is an integral multiple of a half wavelength.

harmonic attenuation – Elimination of a harmonic frequency by using a pi network and tuning its shunt resistances to zero for the frequency to be eliminated.

harmonic component—Of a periodic quantity, any one of the simple sinusoidal quantities of the Fourier series into which the periodic quantity may be resolved.

harmonic content—1. The degree of distortion in the output signal of an amplifier. 2. The components remaining after the fundamental frequency has been removed from a complex wave.

harmonic conversion transducer—A conversion transducer in which the useful output frequency is a multiple or submultiple of the input frequency.

harmonic detector—A voltmeter circuit that measures only a particular harmonic of the fundamental frequency.

harmonic distortion—1. The production of harmonic frequencies at the output by the nonlinearity of a transducer when a sinusoidal voltage is applied to the input. The amplitude of the distortion is usually

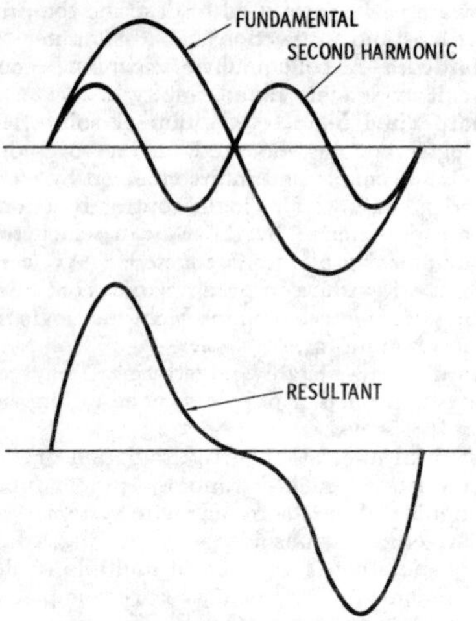

Harmonic content, 1.

a function of the amplitude of the input signal. 2. The voltages of harmonics resulting from amplitude distortion expressed as a percentage of the voltage of the fundamental. A common measurement is total harmonic distortion (thd) where the fundamental of a very low distortion sine-wave test signal is removed by a steep notch filter. The summed harmonics that remain are then measured as a voltage and expressed as a percentage (or dB value) of the voltage of the fundamental at the required test power of the amplifier. (When measured in this manner the term thd is not really correct since the distortion also has the amplifier noise added to it within the test passband. The correct term is distortion factor.) 3. The sum of all signals in an output which are multiples of the input signal frequencies (harmonics). Their intensities are expressed as a percentage of the total output intensity.

harmonic filter – A combination of inductance and capacitance tuned to an undesired harmonic to suppress it.

harmonic generator – A vacuum tube or other generator operated so that it generates strong harmonics in the output.

harmonic interference – Interference between radio stations because harmonics of the carrier frequency are present in the output of one or more stations.

harmonic-leakage power (tr and pre-tr tubes)—The total radio-frequency power transmitted, through the fired tube in its mount, at other than the fundamental frequencies generated by the transmitter.

harmonic motion—Back and forth motion, such as that of a pendulum, in which the distance on one side of equilibrium always equals the distance of the other side; the acceleration is toward the point of equilibrium and directly proportional to the distance from it. Graphically, harmonic motion is represented by a sine wave.

harmonic oscillator—1. A circuit in which the oscillating frequency of the active device and the output frequency are not the same. For example, in a push-push configuration, each transistor oscillates at f_o but the output is combined to provide 2 f_o. 2. An oscillator whose output is very nearly a sine wave and whose output amplitude and frequency are very nearly constant.

harmonic producer – A tuning-fork–controlled oscillator used to provide carrier frequencies for broad-band carrier systems. It is capable of producing odd and even harmonics of the fundamental tuning-fork frequency.

harmonic ringing—A system of selectively signaling several parties on a subscriber's line. The different rings are produced by currents which are harmonics of several fundamental frequencies.

harmonic selective ringing—Selective ringing which employs currents of several frequencies and ringers, each tuned mechanically or electrically to the frequency of one of the ringing currents, so that only the desired ringer responds.

harmonic series of sounds – A series in which each basic frequency in it is an integral multiple of a fundamental frequency.

harmonic telephone ringer—A ringer which responds only to alternating current within a very narrow frequency band. A number of such ringers, each responding to a different frequency, are used in one type of selective ringing.

harmonic-wave analyzer – *See* Harmonic Analyzer.

harness – Wires and cables arranged and tied together so they can be connected or disconnected as a unit.

hartley—In computers, a unit of information content equal to one decimal decision, or the designation of one of ten possible and equally likely values or states of anything used to store or convey information. One hartley equals $\log_2 10$ (3.23) bits.

Hartley oscillator—An oscillator in which a parallel-tuned tank circuit is connected between the grid and plate of an electron tube or between the base and collector of a junction transistor, the inductive element of the tank having an intermediate tap at the cathode or emitter potential.

hash—1. Electrical noise generated within a receiver by a vibrator or a mercury-vapor rectifier. (*See also* Grass, Garbage.) 2. A completely random interfer-

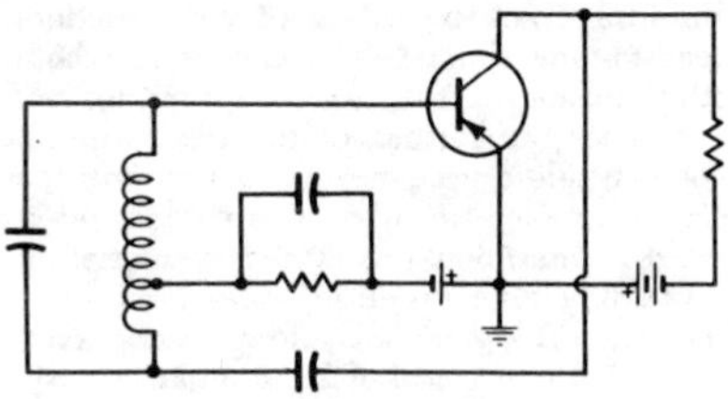

Hartley oscillator.

ing signal usually caused by arcing and occasionally by natural environmental disturbances.

hash total – In a computer, a total for checking purposes. It is determined by adding all the digits or all the numbers in a particular field in a batch of unit records, with no attention paid to the meaning or significance of the total.

hat–To arrange a fixed number of symbols or groups of symbols in a random sequence, as if they had been drawn from a hat.

hatted code–A randomized code consisting of an encoding section. The plain-text groups are arranged in a significant order, accompanied by their code groups arranged in a random order.

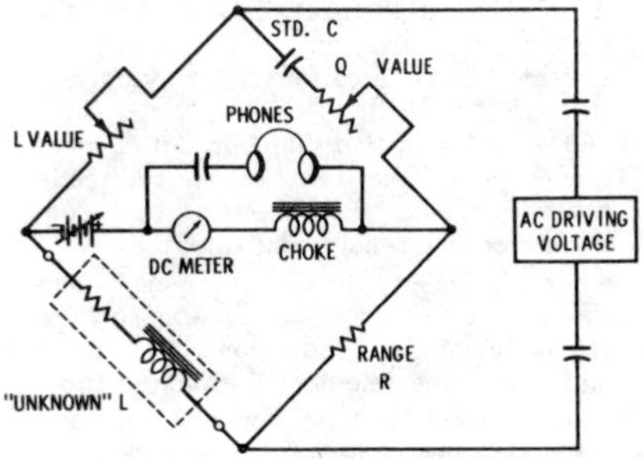

Hay bridge.

Hay bridge – A four-arm, alternating-current bridge used for measuring inductance in terms of capacitance, resistance, and frequency. The arms adjacent to the unknown impedance are nonreactive resistors, and the opposite arm is composed of a capacitor in series with a resistor (unlike the Maxwell bridge, where it is in parallel). Usually the bridge is balanced by adjustment of the resistor, which is also in series with the capacitor and one of the nonreactive arms. The balance depends upon the frequency.

hazard–*See* Failure Rate.

Hazeltine neutralizing circuit – An early form of neutralized radio-frequency amplifier circuit.

H-beacon – A nondirectional homing beacon with an output power of 50 to 2000 watts.

H-bend – Also called H-plane bend. In waveguide technique, a smooth change in the direction of the axis of the waveguide. Throughout the change, the axis remains perpendicular to the direction of polarization.

HCD–Abbreviation for hot-carrier diode.

H-display–Also called H-scan. In radar, a B-display modified to indicate the angle of elevation. The target appears as two closely spaced blips which approximate a short, bright line that slopes in proportion to the sine of the angle of target elevation.

head–1. A device that reads, records, or erases data on a storage medium. For example, a small electromagnet used to read, write, or erase data on a magnetic drum or tape, or the set of perforating reading or marketing devices used for punching, reading, or printing on paper tape. 2. In a tape recorder, any device intended to change the magnetic state of the tape. Specialized types of heads are used for erasing the tape, recording a signal on it, and playing back the signal from it. In many recorders, the recording and playback functions are both performed (at different times) by the same head.

head alignment – Positioning the record-playback head on a tape recorder so that its gap is perpendicular to the tape.

head amplifier–An audio-frequency amplifier mounted on or near the sound head of a motion-picture projector to amplify the extremely weak output of the phototube.

head degausser – A special demagnetizer with elongated pole pieces enabling them to be brought into proximity to head surfaces for elimination of the built-up magnetic "charge" that develops over a period of time as a result of asymmetrical electrical input signals.

head demagnetizer–Device for eliminating any magnetism built up in a recording head.

header–The part of a sealed component or assembly that provides support and insulation for the leads passing through the walls.

header card–A card containing information about the data in other cards that follow.

header record – A computer input record that contains common, constant, or identifying information for other records that follow.

head gap – A space inserted intentionally into the magnetic circuit of a magnetic-recorder head to force or direct the flux into the recording medium.

head guy–A messenger cable and attachments placed so they pull toward the pole line.

heading—The direction of a ship, aircraft, or other object with reference to true, magnetic, compass, or grid north.

headlight—An aircraft radar antenna small enough to be housed in the wing, like an automobile headlight. The beam operates like a searchlight.

headphones—Also called a head receiver or phone. 1. A device held against the ear and having a diaphragm which vibrates according to current variations. It reproduces the incoming electrical signals as sound. Thus, the headphone permits private listening to a receiver amplifier or other device. 2. Small sound reproducers, superficially resembling miniature loudspeakers, set in a suitable frame for wearing about the head and listening to by close coupling to the ears. Recent headphones, improved greatly in fidelity, have become increasingly popular among audiophiles for private listening without disturbing others, as well as to prevent outside noises from interfering with the listening. Headphones are available in mono or stereo.

head receiver—*See* Headphone.

head room – The "safety margin" that is normally provided between the "maximum recording level" as indicated on a recorder level indicators, and the actual point of severe tape overload. Most good recorders provide 6 to 8 dB of head room above the indicated zero-VU or normal maximum indicated recording level, to allow for the inability of the needle of the VU meter needle to respond fully to sudden, intense bursts of signal energy.

Headset.

headset—1. A headphone (or a pair of headphones) and its associated headband and connecting cord. 2. Small portable telephone receivers, usually in pairs, with a connecting clamp to support the phones against the ears, for operators of receiving equipment.

headshell—The end of a pickup arm where the cartridge fits. Sometimes bonded to the arm, though often detachable.

head stack—A group of two or more heads mounted in a single unit, used to provide multiple-track recording or reproduction.

head-to-tape contact—The degree to which the surface of the magnetic coating approaches the surface of the record or replay heads during normal operation of a recorder. Good head-to-tape contact minimizes separation loss and is essential in obtaining high resolution.

hearing aid – A small audio reproducing system for the hard of hearing. It consists of a microphone, amplifier, battery, and earphone and is used to increase the sound level normally received by the ear.

hearing loss – Also called deafness. The hearing loss of an ear at a specified frequency—i.e., the ratio, expressed in decibels, of its threshold of audibility to the normal threshold.

hearing loss for speech—The difference in decibels between the speech levels at which the average normal ear and the defective ear, respectively, reach the same intelligibility. It is often arbitrarily set at 50%.

heart pacer—*See* Pacemaker.

heat aging—A test used to indicate the relative resistance of various insulating materials to heat degradation.

heat coil—A protective device that grounds or opens a circuit, or both, when the current rises above a predetermined value. A mechanical element moves when the fusible substance that holds it in place is heated above a certain point by current through the circuit.

heater—1. Also called filament. An element that supplies the heat to an indirectly heated cathode. 2. A resistor that converts electrical energy into heat.

heater biasing—Application of a dc potential to the heater of a vacuum tube to eliminate diode conduction between it and some other element within the tube.

heater current – The current flowing through a heater in a vacuum tube.

heater voltage – The voltage between the terminals of a heater.

heater-voltage coefficient – In a klystron, the frequency change per volt of heater voltage change when the reflector voltage is adjusted for the peak of a reflector voltage mode.

heat-eye tube—A cathode-ray tube powered by a midget generator. It is used as an infrared instrument that can "see" in the dark.

heat gradient—The difference in temperature between two parts of the same object.

heating effect of a current – Assuming a constant resistance, the amount of heat produced by the current through it. It is proportionate to the square of the current.

heating element—The wirewound resistor,

terminals, and insulating supports used in electric cooking and heating devices.

heating pattern—In induction or dielectric heating, the distribution of temperature in a load or charge.

heating station—In induction or dielectric heating, the work coil or applicator and its associated production equipment.

heat loss—The loss due to conversion of part of the electric energy into heat.

heat of emission—Additional heat energy that must be supplied to an electron-emitting surface to keep its temperature constant.

heat of radioactivity – Heat generated by radioactive disintegration.

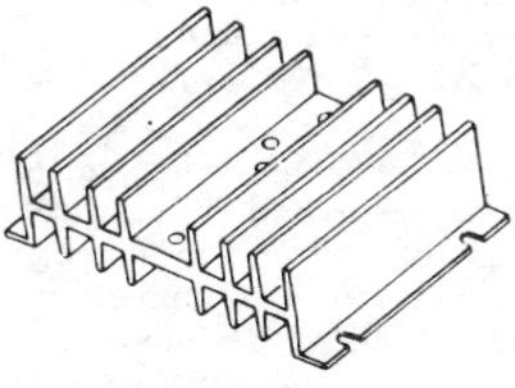

Heat sink.

heat sink—1. A mounting base, usually metallic, that dissipates, carries away, or radiates into the surrounding atmosphere the heat generated within a semiconductor device. The package of the device often serves as a heat sink, but, for devices of higher power, a separate heat sink on which one or more packages are mounted is required to prevent overheating and consequent destruction of the semiconductor junction. 2. A mass of metal that is added to a device for the absorption or transfer of heat away from critical parts.

heat waves—Infrared radiation similar to radio waves but of a higher frequency.

heat-writing recorder—A type of strip-chart recorder in which a heated stylus writes on a strip of chemically treated paper. The paper is discolored by the heat, and the path followed by the pen over the surface of the paper is thus made visible.

Heaviside - Campbell mutual - inductance bridge – A Heaviside mutual-inductance bridge in which one inductive arm contains a separate inductor that is included in the bridge arm during the first of a pair of measurements and is short-circuited during the second. The balance is independent of frequency. (*See also* Heaviside Mutual-Inductance Bridge.)

Heaviside layer—Also called the Kennelly-Heaviside layer. The region of the ionosphere that reflects radio waves back to earth.

Heaviside mutual-inductance bridge – An alternating-current bridge normally used for the comparison of self- and mutual-inductances. Each of the two adjacent arms contains self-inductance, and one or both of them have mutual inductance to the supply circuit. The other two arms normally are nonreactive resistors. The balance is independent of frequency.

heavy hydrogen—Another term for deuterium (H^2) or tritium (H^3).

hecto—A prefix meaning 100.

hectometric wave – An electromagnetic wave between the wavelength limits of 100 and 1000 meters corresponding to the frequency range 300 kHz to 3 MHz.

heelpiece—Part of a relay magnetic structure at the end of a coil, opposite the armature. It generally supports the armature and complete the magnetic path between it and the core of the coil.

Hefner lamp—A standard source that gives a luminous intensity of 0.9 candlepower.

height control—In a television receiver, the adjustment which determines the amplitude of the vertical-scanning pulses and hence the height of the picture.

height finder—A radar which measures the altitude of an airborne object.

height input—Information regarding target height received by a computer from a height finder and relayed by way of a ground-to-ground data link or telephone.

height overlap coverage – A region of height-finder coverage within which there is duplicated coverage from adjacent height finders of other radar stations.

height-position indicator—A radar display which simultaneously shows the angular-elevation slant range and height of objects.

height-range indicator—A cathode-ray tube from which altitude and range measurements of airborne objects may be viewed.

Heising modulation—*See* Constant-Current Modulation.

helical—Spiral-shaped.

helical antenna—1. Also called a helical-beam antenna. A spiral conductor wound around a circular or polygonal cross section. The axis of the spiral normally is

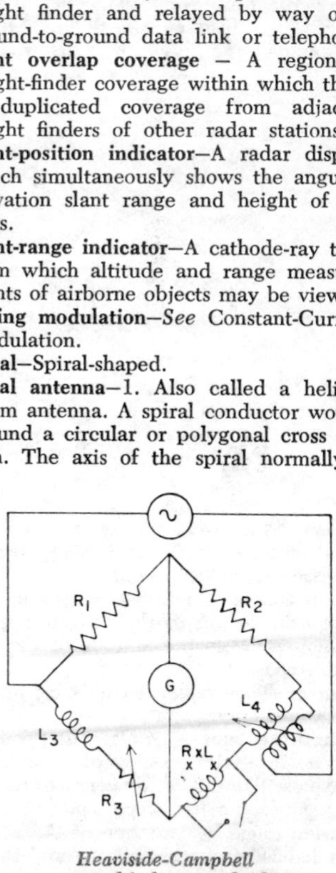

Heaviside-Campbell mutual-inductance bridge.

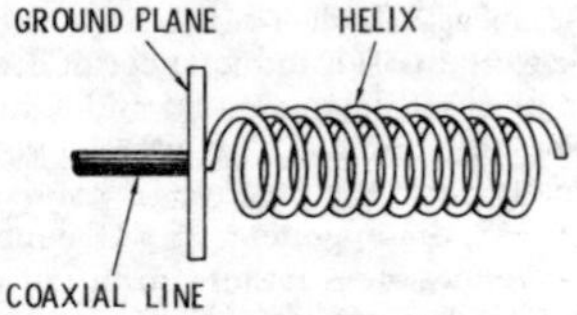

Helical antenna.

mounted parallel to the ground and fed at the adjacent end. The radiation produced has approximately a circular polarization and is confined mainly to a single lobe located along the axis of the spiral. 2. An antenna made of wire wound as a coil, usually on a Fiberglas rod and with the wire usually within Fiberglas.

helical-beam antenna – *See* Helical Antenna.

helical potentiometer—A precision potentiometer which requires several turns of the control knob to move the contact arm from one end of the spiral-wound resistance element to the other end.

helical scanning – 1. Radar scanning in which the rf beam describes a distorted spiral motion. The antenna rotates about the vertical axis while the elevation angle rises slowly from zero to 90°. 2. Method of facsimile scanning in which the elemental area sweeps across the copy in a spiral motion.

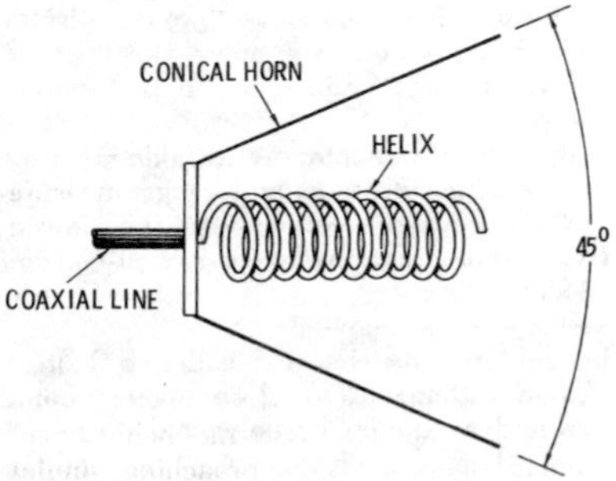

Helicone.

helicone – A circularly polarized antenna that produces a low side-lobe level. It consists of a helix excited in the axial mode and placed inside a conical horn. The axial length of the helix is approximately equal to the altitude of the truncated cone.

helionics—The conversion of solar heat to electric energy.

helitron oscillator – An electrostatically focused, low-noise, backward-wave microwave oscillator. The frequency of the output signal can be swept rapidly through a wide range by variation of the voltage applied between the cathode and the associated rf circuit.

helium tight—*See* Hermetic.

helix—A spiral.

helix recorder—A recorder in which helical scanning is used.

Helmholtz coil – A phase-shifting network used for determining the range in certain types of radar equipment. It consists of fixed and movable coils. The phase is kept constant at the input, but may be continually shifted from 0° to 360° at the output.

Helmholtz resonator – An acoustic enclosure with a small opening which causes the enclosure to resonate. The frequency at which it does depends on the geometry of the resonator.

hemimorphic—Terminated at the two ends by dissimilar sets of faces.

HEM wave – *See* Hybrid Electromagnetic Wave.

henry—The cgs electromagnetic unit of inductance or mutual inductance. The inductance of a circuit is 1 henry when a current variation of 1 ampere per second induces 1 volt. Abbreviated H.

heptode – A vacuum tube that contains seven electrodes: an anode; a cathode; a control electrode; and four additional electrodes, usually grids.

hermaphroditic connector—1. A connector in which both mating contacts are exactly alike at their mating face. 2. Either of a pair of coaxial connectors whose mating faces are alike. 3. A connector design in which pin and socket contacts are arranged in a balanced manner such that both mating connectors are identical. The contacts may also be hermaphroditic and arranged as male and female contacts, as for pins and sockets. Hermaphroditic contacts may also be used in a manner such that one half of each contact mating surface protrudes beyond the connector interface and both mating connnectors are identical.

hermaphroditic contact—A contact designed so that it is neither pin or socket and can be mated with any other contact of the same design.

hermetic—Pertaining to permanent sealing, by fusion, soldering, or other means, to prevent the transmission of gases. Also called helium tight, leak tight, and vacuum tight.

hermetically sealed relay—A relay in a gas-tight enclosure which has been completely sealed by fusion or comparable means to ensure a low rate of gas leakage over a long period of time.

hermetic seal—1. An airtight seal between two parts of a container, such as between the can and header of a metal component package. 2. A mechanical or physical closure that is impervious to moisture or gas, including air. Usually pertains to an envelope or enclosure containing electronic components or parts, or to a header.

herringbone pattern – Television interference seen as one or more horizontal bands of closely spaced V- or S-shaped lines.

hertz–Abbreviated Hz. A unit of frequency equal to one cycle per second.

Hertz antenna – An antenna system which does not depend for its operation on the presence of ground. Its resonant frequency is determined by its distributed capacitance, which varies according to its physical length.

Hertz effect–The ionization and spark discharge produced by ultraviolet radiation.

Hertzian oscillator–A type of oscillator for producing ultrahigh-frequency oscillations. It consists of two metal plates or other conductors separated by an air gap. The capacitor formed has such a small capacitance that ultrahigh-frequency oscillations can occur.

Hertzian waves–Electromagnetic waves of frequencies between 10 kHz and 30,000 GHz. Radio waves.

hertz-matching loran–*See* Low-Frequency Loran.

Hertz vector–A vector which specifies the electromagnetic field of a radio wave. Both the electric and the magnetic intensities can be specified in terms of it.

heterodyne–Also called beat. To mix two frequencies together in order to produce two other frequencies equal to the sum and difference of the first two. For example, heterodyning a 100-kHz and a 10-kHz signal will produce a 110-kHz (sum frequency) and a 90-kHz (difference frequency) signal.

heterodyne conversion transducer (converter)–A conversion transducer in which the output frequency is the sum or difference of the input frequency and an integral multiple of the frequency of another wave.

heterodyne detection – Detection (or conversion) by mixing two signals together to generate the intermediate frequency in a superheterodyne receiver or to make cw signals audible.

heterodyne detector–A detector that converts an incoming rf signal to an audible tone by heterodyning. It incorporates a local oscillator (called a beat-frequency oscillator).

heterodyne frequency–The sum or difference frequency produced by combining two other frequencies.

heterodyne frequency meter – *See* Heterodyne Wavemeter.

heterodyne oscillator – An oscillator which produces a desired frequency by combining two other frequencies (e.g., two radio frequencies to produce an audio frequency, or the incoming and local-oscillator frequencies to produce the intermediate frequency of a superheterodyne receiver).

heterodyne principle–*See* Heterodyne.

heterodyne reception–Also called beat reception. Reception by combining a received high-frequency wave with a locally generated wave in a nonlinear device to produce sum and difference frequencies at the output.

heterodyne repeater–A radio repeater in which the incoming radio signals are converted to an intermediate frequency, amplified, and reconverted to another frequency band before being transmitted over the next repeater section.

heterodyne-type frequency meter – An instrument for measuring frequency by producing a zero difference frequency (zero beat) between the signal under test and an internally generated signal.

heterodyne wavemeter–A wavemeter employing the heterodyne principle to compare the frequency being measured with a frequency being generated in a calibrated oscillator circuit.

heterodyne whistle–A steady squeal heard in a radio receiver when the signals from stations having nearly equal frequencies beat together.

heterodyning–*See* Heterodyne.

heterogeneity–A state or condition of being unlike in nature, kind, or degree.

heterogeneous–Composed of different materials (opposite of homogeneous).

heterosphere–The portion of the upper atmosphere in which the relative proportions of oxygen, nitrogen, and other gases are unfixed and radiation particles and micrometeroids are mixed with the air particles.

heuristic – Pertaining to exploratory problem-solving methods in which solutions are discovered through evaluation of the progress made toward the final result (as opposed to algorithmic methods).

heuristic program–A set of computer instructions that simulate the behavior of human operators in approaching similar problems.

hexadecimal – Number system using 0, 1, ...A,B,C,D,E,F to represent all the possible values of a 4-bit digit. The decimal equivalent is 0 to 15. Two hexadecimal digits can be used to specify a byte.

hexadecimal counter – *See* Divide-by-16 Counter.

hexadecimal number system – A number system having as its base the equivalent of the decimal number sixteen.

hex inverter–A group of six logic inverters contained in a single package.

hexode – A vacuum tube containing six electrodes: an anode; a cathode; a control electrode; and three additional electrodes, usually grids.

hf–Abbreviation for high frequency.

HH beacon—A nondirectional radio homing beacon with a power output of 2000 watts or more.

HIC – Abbreviation for hybrid integrated circuit.

hierarchy—A series of items classified according to rank or order.

hi-fi—*See* High Fidelity.

high band—Television channels 7-13, covering a frequency range of 174-216 MHz.

high boost—*See* High-Frequency Compensation.

high-contrast image – A picture in which strong contrast between light and dark areas is visible. Intermediate values, however, may be missing.

high definition—The condition of a reproduced television or facsimile image in which it contains sufficient accurately reproduced elements for the picture details to approximate those of the original scene.

high-energy materials – Also called hard magnetic materials. Magnetic materials having a comparatively high energy product, e.g., materials used for permanent magnets.

higher-level language—A programming language that closely resembles natural language. A statement in a higher-level language will produce many machine-language instructions. The higher-level languages are usually independent of the computer.

higher-order language – A programming language that is independent of the computer. Usually, it resembles natural languages, and a compiler is required for translation into machine language. Examples are FORTRAN and ALGOL.

highest probable frequency – Abbreviated hpf. An arbitrarily chosen frequency value 15% above the F_2-layer muf (maximum usable frequency) for the radio circuit. For the E-layer, the hpf is equal to the muf.

high fidelity—1. Popularly called hi-fi. The characteristic which enables a system to reproduce sound as nearly like the original as possible. 2. Reproduction of audio so perfect that listeners hear exactly what they would have heard if present at the original performance. Also called hi-fi.

high-fidelity receiver—A radio receiver capable of receiving and reproducing, without noticeable distortion, the original modulation impressed on the carrier waves.

high filter—An audio circuit designed to remove undesired high-frequency noise from the program material. Such noise includes record scratch, tape hiss, am whistles, etc.

high frequency – Abbreviated hf. The frequency bands from 3 to 30 MHz (100 meters to 10 meters).

high-frequency alternator – An alternator capable of generating radio-frequency carrier waves.

high-frequency band – The band of frequencies extending from 3 to 30 MHz.

high-frequency bias—In a tape recorder, a sinusoidal voltage that is mixed with the signal being recorded to better the linearity and dynamic range of the recorded signal. In practice, the bias frequency is three to four times the highest information frequency to be recorded.

high-frequency carrier telegraphy—Carrier telegraphy with the carrier currents above the frequencies transmitted over a voice telephone channel.

high-frequency compensation – Also called high boost. An increase in the amplification of the high frequencies with respect to the low and middle frequencies within a given band of frequencies.

high-frequency heating – *See* Electronic Heating.

high-frequency induction heater or furnace —An induction heater or furnace using frequencies much higher than the standard 60 hertz.

high-frequency resistance—Also called rf or ac resistance. The total resistance offered by a device in a high-frequency ac circuit. This includes the dc and all other resistances due to the effects of the alternating current.

high-frequency treatment—Therapeutic use of intermittent and isolated trains of heavily damped oscillations having a high frequency and voltage and a relatively low current.

high-frequency trimmer—A trimmer capacitor that is used to calibrate the high-frequency end of the tuning range in a superheterodyne receiver.

high-frequency unit—*See* Tweeter.

high-frequency welding – *See* Radio-Frequency Welding.

high level—In digital logic, the more positive of the two binary-system logic levels. *See also* Low Level, Negative Logic, and Positive Logic.

high-level detector—A linear power detector with a voltage-current characteristic that may be treated as a straight line or two intersecting lines.

high-level firing time – The time required to establish a radio-frequency discharge in a switching tube after radio-frequency power is applied.

high-level language – Programming language that generates machine codes from problem or function-oriented statements. FORTRAN, COBOL and BASIC are three commonly used high-level languages. A single functional statement may translate into a series of instructions or subroutines in machine language, in contrast to a low-level (assembly) language in which statements translate on a one-for-one basis. 2. A problem-oriented programming language as distinguished from a machine-oriented programming language. The former's instruction approach is closer to the needs of the problems to be handled than the language of the machine on which they are to be implemented.

high-level modulation—A system in which the modulation is introduced at a point where the power level approximates the output power.

high-level, radio-frequency signal (tr, atr, and pre-tr tubes)—A radio-frequency signal with sufficient power to fire the tube.

high-level vswr (switching tubes) — The voltage standing-wave ratio due to a fired tube in its mount, located between a generator and the matched termination in a waveguide.

highlight—The brightest portion of a reproduced image.

high-mu tube—A vacuum tube with a high amplification factor.

high-noise-immunity logic — Abbreviated HNIL. A special type of logic designed specifically to provide very high resistance to electrical noise. Sometimes called HTL (high threshold logic).

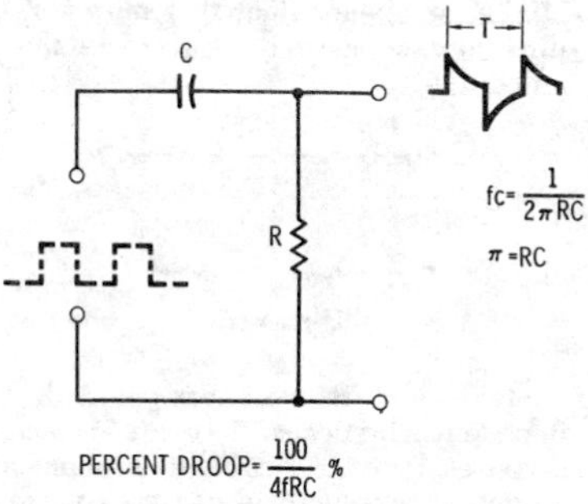

High-pass RC filter.

high-pass filter—1. A wave filter having a single transmission band extending from some critical, or cutoff, frequency other than zero, up to infinite frequency. 2. A filter which, above a critical frequency, allows the unrestricted passage of high-frequency signals. Reciprocally, a bass-cut filter. 3. A filter passing frequency components above some limited frequency and rejecting components below that limit.

high-performance equipment — Equipment having sufficiently exacting characteristics to permit their use in trunk or link circuits.

high-potential test—A test for determining the breakdown point of insulating materials and spacings. It consists of applying a voltage higher than the rated voltage between two points or between two or more windings. However, it is not a test of conductor insulation.

high-power silicon rectifiers — A group of rectifiers with continuous ratings exceeding 50 average amperes per section in a single-phase, half-wave circuit.

high *Q*—Having a high ratio of reactance to effective resistance. The factor determining the efficiency of a reactive component.

high-rate discharge — The storage-battery discharge equivalent to the heaviest possible duty in service.

high-recombination rate contact — A semiconductor-to-semiconductor or metal-to-semiconductor contact at which thermal equilibrium charge-carrier concentrations are maintained substantially independent of current density.

high-resistance joint — A faulty union of conductors or conductor and terminal. The result is less current flow and a drop in voltage at the union.

high-resistance voltmeter—A voltmeter having a resistance considerably higher than 1000 ohms per volt. As a result, it draws very little current from the circuit being measured.

high-speed bus—*See* Memory Register.

high-speed carry—In a computer, a type of carry in which: 1. A carry into a column results in a carry out of that column, because the sum without carry in that column is 9. 2. Instead of a normal adding process, a special process is used which takes the carry at high speed to the actual column where it is added. Also called standing-on-nines carry.

high-speed dc circuit breaker — A device which starts to reduce the current in the main circuit in 0.01 second or less, after the occurrence of the dc overcurrent or the excessive rate of current rise.

high-speed excitation system—An excitation system that can change its voltage rapidly in response to a change in the field circuit of the excited generator.

high-speed printer—A printer that has a speed of operation compatible with the speed of computation and data processing so that it may operate on-line.

high-speed reader—A reading device that can be connected to a computer so as to operate on-line without seriously slowing the operation of the computer.

high-speed relay—A relay designed specifically for short-operate or short-release time, or both.

high-speed storage—*See* Rapid Storage.

high-speed telegraph transmission—Transmission of code at higher speeds than are possible with hand-operated keys.

high tension—Lethal voltages, on the order of thousands of volts.

high-tension magneto – A self-contained generator in which the required high potential is generated directly; no induction coil is needed.

high-threshold logic – Abbreviated HTL. Logic with a high noise margin, used primarily in industrial applications. It closely resembles DTL, except that in HTL a reverse-biased emitter junction is used as a threshold element operating as a zener diode. A typical noise margin is 6 volts with a 15-volt supply.

high-vacuum phototube–A phototube that is highly evacuated so that its electrical characteristics are essentially unaffected by gaseous ionization. In a gas phototube, some gas is intentionally introduced.

high-vacuum rectifier–A vacuum-tube rectifier in which conduction is entirely by electrons emitted from the cathode.

high-vacuum tube–An electron tube that is highly evacuated so that its electrical characteristics are essentially unaffected by gaseous ionization.

high-velocity scanning–The scanning of a target with electrons of such velocity that the secondary-emission ratio is greater than unity.

high voltage – The accelerating potential that speeds up the electrons in a beam of a cathode-ray tube.

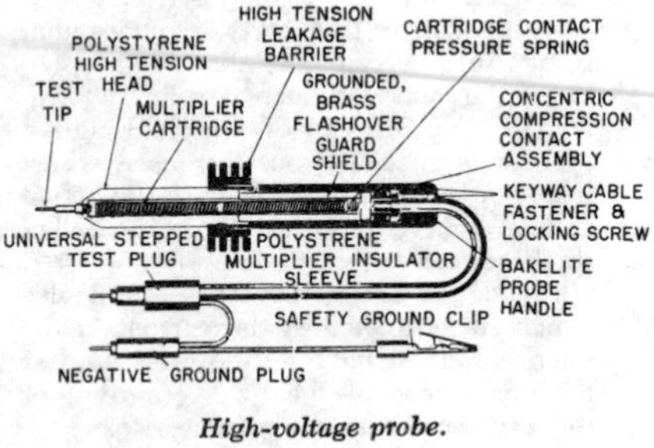

High-voltage probe.

high-voltage probe – A probe with a high internal resistance, for measuring extremely high voltages. It is used with a voltmeter having an internal resistance of 20,000 ohms per volt or more.

hill-and-dale recording – *See* Vertical Recording.

hinge – A joint in a relay that permits movement of the armature relative to the stationary parts of the relay structure.

hinged-iron ammeter – A moving-iron ammeter in which the fixed portion of the magnetic circuit is placed around the conductor to measure the current through it.

HIPERNAS – Acronym for high performance navigation system. A self-compensated, pure-inertial guidance system.

hipot – 1. *See* Dielectric Breakdown Voltage. 2. Contraction of high potential. Commonly refers to a device used for testing insulation breakdown or leakage, with high voltages. High potting is the verb. 3. High-potential voltage applied across a conductor to test the insulation.

hiss – 1. Random noise characterized by prolonged sibilant sounds in the audio-frequency range. 2. The primary background noise in tape recording, stemming from circuit noise in the playback amplifiers or from residual magnetism of the tape.

histogram – 1. A graphical representation of a frequency distribution by a series of rectangles which have for one dimension a distance proportional to a definite range of frequencies, and for the other dimension a distance proportional to the number of frequencies appearing within the range. 2. A description of one (or all) parameters, showing distribution, standard deviation, mean-value failure limits, and sample lot size for all samples within the lot.

hit–1. *See* Flash. 2. Momentary surge of voltage on a transmission channel.

hit-on-the-fly printer–A mechanical printer in which the printing head is in continual motion.

hits–Momentary line disturbances which could result in mutilation of characters being transmitted.

H-lines–Imaginary lines that represent the direction and strength of magnetic flux on a diagram.

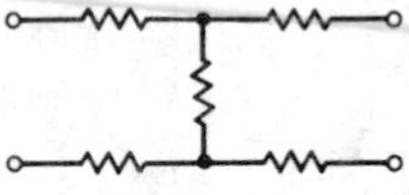

H-network.

H-network – A network composed of five impedance branches. Two are connected in series between an input terminal and an output terminal and two are connected between another input terminal and another output terminal. The fifth is connected from the junction points of the two branches.

HNIL – Abbreviation for high-noise-immunity logic.

hog horn–A microwave feed horn shaped so that the input energy from the waveguide approaches from the same direction as the horn opening.

hold – Opposite of clear. 1. To maintain storage elements in charge storage tubes at equilibrium potentials by electron bombardment. 2. To retain the information contained in one storage location of a computer after copying the information into another storage location, as opposed to clearing or erasing the information.

hold control–In a television receiver, the adjustment which controls the frequency of the vertical- or horizontal-scanning pulses and hence the stability of the picture.

hold current – Also called the electrical hold value. The minimum current which will keep the contact springs energized in a relay.

hold electrode—In a mercury switch, the electrode that remains in contact with the mercury pool while the circuit is being closed or opened.

holding anode—In a mercury-arc rectifier, a small auxiliary anode that maintains the ionization while the main anode current is zero.

holding beam – A diffused beam of electrons for regenerating the charges retained on the dielectric surface of an electrostatic memory or storage tube.

holding circuit—Also called a locking circuit. An alternate operating circuit which, when completed, maintains sufficient current in a relay winding to keep the relay energized after the initial current has ceased.

holding coil—A separate relay coil which keeps the relay energized after the original current has been removed.

holding current—1. That value of average forward current (with the gate open) below which a silicon controlled rectifier returns to the forward blocking state after having been in forward conduction. 2. The minimum current that must pass through a device such as a silicon controlled rectifier, thyratron, neon glow tube, etc., to maintain it in a conducting condition. 3. The minimum current that must be passed through the coil of a relay in order to keep it activated.

holding gun—In a storage tube, the source of electrons constituting the holding beam.

holding time—The total time a trunk or circuit is in use on a call, including both operator's and user's time.

hold lamp—An indicating lamp that stays lighted while a telephone connection is being held.

hold mode—In integrators or other charge-storage circuits, a condition or time interval in which input(s) are removed and the circuit is commanded (or expected) to maintain a constant output.

hold-off voltage—The maximum voltage an electronic flash tube will stand without self-flashing. Normal hold-off voltage is reduced at the end of lamp life and in the presence of high temperatures or rf fields.

holdover – The condition which occurs when a lightning-protector gap continues to conduct follow current.

holdover voltage—The steady-state voltage at which a gap just fails to clear a given value of follow current.

hold time – 1. In resistance welding, the time that is allowed for the weld to harden. 2. The length of time after the clocking of a flip-flop that data must remain unchanged. Also called release time.

hole—1. In the electronic valence structure of a semiconductor, a mobile vacancy which acts like a positive electronic charge with a positive mass. 2. In a semiconductor, the term used to describe the absence of an electron; has the same electrical properties as an electron except that it carries a positive charge.

hole conduction—Conduction occurring in a semiconductor when electrons move into holes under the influence of an applied voltage and thereby create new holes. The apparent movement of such holes is toward the more negative terminal, and is hence equivalent to a flow of positive charges in that direction.

hole current—Conduction in a semiconductor when electrons move into holes, creating new holes. The holes appear to move toward the negative terminal, giving the equivalent of positive charges flowing to the terminal.

hole density—In a semiconductor, the density of holes in an otherwise full band.

hole-electron pair—A positive charge carrier (hole) and a negative charge carrier (electron), considered together as one entity.

hole injection – The production of mobile vacancies in an n-type semiconductor when a voltage is applied to a sharp metal point in contact with the surface of the material.

hole injector—A pointed metallic device for injecting holes into an n-type semiconductor.

hole-in-the-center effect—Also called hole-in-the-middle effect. The lower volume or absence of sound between the left and right speakers of a stereo system.

hole mobility – The ability of a hole to travel easily through a semiconductor.

hole site—The area on a computer punch card or paper tape where a hole may or may not be punched. It can be a form of binary stoarge in which a hole represents a 1 and the absence of a hole represents a 0.

hole storage factor (K'_s)—In a transistor, the excess stored charge (when the transistor is in saturation) per unit excess base current. Excess base current is defined as the amount of current supplied to the base in excess of the current required to just keep the transistor in saturation.

hole trap—A semiconductor impurity which can trap holes by releasing electrons into the conduction or valence bands.

Hollerith—Pertaining to a particular type of code or punched card utilizing 12 rows per column and usually 80 columns per card.

Hollerith code – A code based on the punching of holes in cards at specified locations. From one to three punches may be made in each column of the card and up to 80 columns may be punched in each card. Each column corresponds to one character; the specific character is determined by the number and location of the punches in that column.

hollow-cathode tube–A gas discharge tube with a hollow cathode closed at one end. Almost all the radiation is from the cathode glow within the hollow cathode.

hollow core–A plain ferrite core having a center hole for mounting purposes.

hologram–1. A recording of the two-dimensional intensity distribution of the interference pattern produced by the interaction of two or more monochromatic waves that have phases derived from the same source. One of the waves is reconstructed when a replica of the other wave is diffracted from the hologram. 2. An interference pattern recorded on photographic film or similar media. This pattern is created by directing two beams of coherent light into the film. One, called the reference beam, strikes the film directly. The second, called the object beam, bounces off, or passes through the test specimen, then strikes the film. The interaction of these two beams makes up the interference pattern called a hologram. To "decode" the swirls and dots of the pattern and create a visible image, a coherent light beam is directed onto the hologram.

holography–1. The optical recording of an object wave formed by the resulting interference pattern of two mutually coherent, component light beams. A coherent beam is first split into two component beams, one of which irradiates the object, the second of which irradiates a recording medium. The diffraction or scattering of the first wave by the object forms the object wave which proceeds to and interferes with the second coherent beam, or reference wave at the medium. The resulting pattern is a three-dimensional record (hologram) of the object wave. 2. The recording of an object wave (usually optical) in such a way that an identical wave can subsequently be reconstructed. Whereas a conventional photograph records only the intensity of the light incident on it, a hologram records both the amplitude and phase. The additional phase information is contained in an interference pattern which is formed from the object wave and a reference wave.

home loop – An operation involving only those input and output units associated with the hole terminal.

home-on-jam–A radar feature that permits angular tracking of a jamming source.

hometaxial-base transistor – A transistor manufactured by a single-diffusion process so that both the emitter and collector junctions are formed in a uniformly doped silicon slice. The homogeneously doped base region that results is free from accelerating fields in the axial (collector-to-emitter) direction; such fields could cause undesirable high flow and destroy the transistor.

homing–1. Approaching a desired point by maintaining some indicated navigational parameter constant (other than altitude). 2. In missile guidance, the use of radiation from a target to establish a collision course.

homing adapter – A device used with an aircraft radio receiver to produce aural and/or visual signals which indicate the direction of a transmitting radio station.

homing antenna–A type of directional-antenna array used for pinpointing a target.

homing beacon–A radio transmitter which emits a distinctive signal for determining bearing, course, or location.

homing device – 1. An automatic device that moves or rotates in the correct direction without first having to go to the end of its travel in the opposite direction. 2. A radio device that guides an aircraft to an airport or transmitter site.

homing guidance – A missile-guidance system in which the missile steers itself toward a target by means of a self-contained mechanism (infrared detectors, radar, etc.). It is activated by some distinguishing characterism of the target. Homing guidance may be active, semiactive, or passive.

homing relay – A stepping relay that returns to a specified starting position prior to each operating cycle.

homing station–A radionavigational aid incorporating direction-finding facilities.

homodyne reception–Also called zero-beat reception. A system of reception using a locally generated voltage at the carrier frequency.

homogeneity–The state or condition of being similar in nature, kind, or degree.

homogeneous–Of the same nature (the opposite of heterogeneous).

homologous field – A field in which the lines of force in a given plane all pass through one point (e.g., the electric field between two coaxial charged cylinders).

homopolar – Electrically symmetrical–i.e., having equally distributed charges.

homopolar generator – A dc generator in which all the poles presented to the armature are of the same polarity, so that the armature conductor always cuts the magnetic lines of force in the same direction. A pure direct current can thus be produced without commutation.

homopolar magnet—A magnet with concentric pole pieces.

homosphere—That part of the atmosphere which is made up mostly of atoms and molecules found near the earth's surface, and retaining the same relative proportions of oxygen, nitrogen, and other gases throughout.

homotaxial—A term coined by RCA from "homogeneous" and "axial" to describe a single-diffused transistor with a base region of homogeneous resistivity silicon in the axial direction (emitter-to-collector).

honeycomb coil—Also called duolateral or lattice-wound coil. A coil with the turns wound crisscross to reduce the distributed capacitance.

honeycomb winding—A method of winding a coil with crisscross turns to minimize distributed capacitance.

hood—A shield placed over a cathode-ray tube to eliminate extraneous light and thus make the image on the screen appear more clearly.

hookswitch—A switch that is located within the supporting structure on which a telephone handset rests when it is not in use. When the handset is lifted, the switch closes the telephone circuit or loop.

hook transistor – A transistor having four alternating p-type and n-type layers, with one layer floating between the base layer and the collector layer. This arrangement gives high emitter-input current gains. The pnpn transistor has a p-type floating layer, while an npnp transistor has an n-type floating layer.

hookup—1. Method of connection between the various units in a circuit. 2. The diagram of connections used.

hookup wire—1. The wire used in coupling circuits together. It may be solid or stranded, and is usually tinned and insulated No. 18 or 20 soft-drawn copper. 2. Wire used for point-to-point connection within electronic equipment, usually carrying low voltages and currents.

hop—An excursion of a radio wave from the earth to the ionosphere and back. It is usually expressed as single-, double-, and multihop. The number of hops is called the order of reflection.

hopoff – In a potentiometer, the sudden jump in resistance as the contact is rotated over the junction of two resistance slopes. The magnitude of the hopoff is dependent on the ratio of the slopes and on the junction blending characteristic.

horizon—An apparent or visible junction of earth and sky as seen on or above the earth. It bounds the part of the earth's surface that can be reached by the direct wave of a radio station. The distance to the horizon is affected by atmospheric refraction.

horizon distance—The space between the farthest visible point and the transmitter antenna. It is the distance over which ultra-high-frequency transmission can be received under ordinary conditions with an unelevated receiving antenna.

horizontal—1. Perpendicular to the direction of gravity. 2. In the direction of or parallel to the horizon. 3. On a level.

horizontal angle of deviation—The horizontal angle between the great circle path from the transmitter to the receiver and the direction of departure or arrival of the wave along the line of propagation.

horizontal axes—The three horizontal axes of crystallographic reference.

horizontal blanking – Cutting off the electron beam between successive active horizontal lines during retrace.

horizontal blanking pulse – A rectangular pedestal in the composite television signal. It occurs between active horizontal lines and cuts off the beam current of the picture tube during retrace.

horizontal centering control – In a television receiver or cathode-ray oscilloscope, the adjustment for moving the entire display back and forth.

horizontal-convergence control—In a color television receiver, the control which adjusts the amplitude of the horizontal dynamic convergence voltage.

horizontal-deflecting electrodes—A pair of electrodes that move the electron beam from side to side on the screen of a cathode-ray tube employing electrostatic deflection.

horizontal-discharge tube—A vacuum tube used in the horizontal-deflection circuit to discharge a capacitor and thereby form the sawtooth scanning wave. (*See also* Discharge Tube).

horizontal-drive control—In an electromagnetically deflected television receiver, the control which adjusts the ratio of the pulse amplitude to the linear portion of the scanning-current wave.

horizontal dynamic convergence – Convergence of the three electron beams in a color picture tube at the aperture mask during scanning of a horizontal line.

horizontal field-strength diagram—A representation of the field strength in a horizontal plane and at a constant distance from an antenna. Unless otherwise specified, the plane passes through the antenna.

horizontal frequency—*See* Line Frequency, 1.

horizontal hold control—A synchronization control which varies the free-running frequency of the horizontal deflection oscillator so it will be in step with the scanning frequency at the transmitter.

horizontal hum bars – Broad, horizontal, moving or stationary bars, alternately dark and light, that extend over an entire television picture. They are caused by interference at approximately 60 Hz or a harmonic of 60 Hz.

horizontal-linearity control—In a television receiver, the control for adjusting the width at the left side of the screen.

hor:zontal line frequency – *See* Line Frequency, 1.

horizontal lock—The circuit that maintains horizontal synchronization in a television receiver.

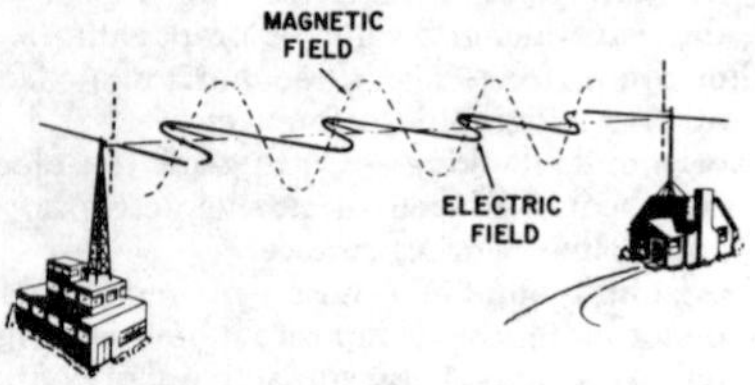

Horizontally polarized wave.

horizontally polarized wave—A linearly polarized wave with a horizontal electric-field vector.

horizontal-output transformer—*See* Flyback Transformer.

horizontal parabola control – *See* Phase Control, 1.

horizontal polarization—1. Transmission in which the electrostatic field leaves the antenna in a horizontal plane. Elements of the transmitting and receiving antennas likewise are horizontal. Horizontal polarization is standard for television in the United States. 2. Transmission of radio waves whose undulations vary horizontally with respect to the earth. (Horizontally polarized antennas are used mainly for base-to-base transmission.)

horizontal repetition rate—Also called horizontal scanning frequency. The number of horizontal lines per second (15,750 hertz in the United States).

horizontal resolution – 1. The number of picture elements in a horizontal scanning line that can be distinguished. 2. The capability of a tv system to resolve detail in a horizontal direction across the screen. The higher the resolution number, the sharper the picture will be.

horizontal retrace—The line that would be seen on the screen, while the spot is returning from right to left, if retrace blanking were not used.

horzontal ring-induction furnace – A furnace for melting metal. It comprises an open trough or melting channel, a primary inductor winding, and a magnetic core which links the melting channel to the primary winding.

horizontal-scanning frequency – *See* Horizontal Repetition Rate.

horizontal sweep—Movement of the electron beam from left to right across the screen or the scene being televised.

horizontal-sync discriminator – A circuit employed in the flywheel method of synchronization to compare the phase of the horizontal-sync pulses with that of the horizontal-scanning oscillator.

horizontal-sync pulse – The rectangular pulses which occurs above the pedestal level between each active horizontal line. They keep the horizontal scanning at the receiver in step with that at the transmitter.

horn—1. Also called an acoustic horn. A tubular or rectangular enclosure for radiating or receiving acoustic waves. 2. A primary element consisting of a section of metal waveguide in which one or both of the cross-sectional dimensions increase toward the open end. (*See also* Horn Antenna.)

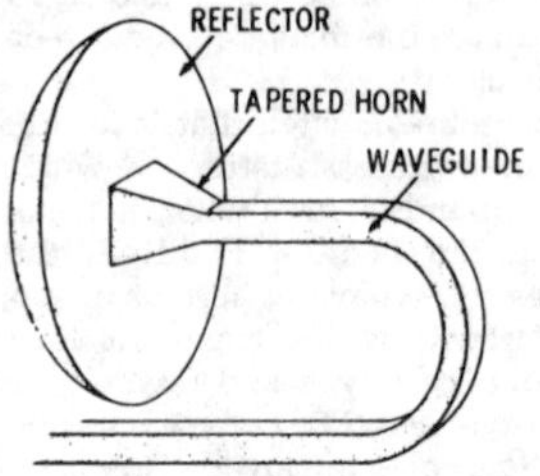

Horn antenna.

horn antenna—Also called a horn. A tubular or rectangular microwave antenna which is wider at the open end and through which radio waves are radiated into space.

horn arrester—A lightning arrester that has a spark gap with upward-projecting diversion horns of thick wire. When the arc is formed, it travels up the gap, and is extinguished upon reaching the widest part of the gap.

horn cutoff frequency—A frequency below which an exponential horn will not function correctly, because it fails to provide for proper expansion of the sound waves.

horn gap—A type of spark gap with divergent electrodes.

horn-gap switch—A form of air switch with arcing horns.

horn loading – A method of coupling a speaker diaphragm to the listening space by an expanding air column having a small throat and large mouth.

horn mouth—An open-ended, metallic device for concentrating energy from a waveguide and directing this energy into space.

horn speaker—A speaker in which a horn couples the radiating element to the medium.

horn throat—The narrow end of a horn.

horsepower – Abbreviated hp. A unit of power, or the capacity of a mechanism to do work. It is the equivalent of raising 33,000 pounds one foot in one minute, or 550 pounds one foot in one second. One horsepower equals 746 watts.

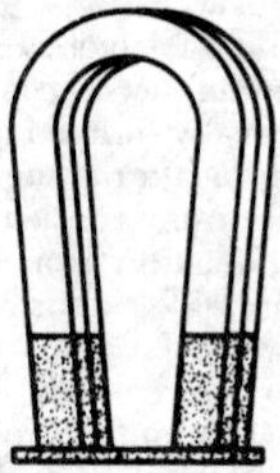

Horseshoe magnet.

horseshoe magnet—A permanent magnet or electromagnet shaped like a horseshoe or U to bring the two poles close together.

host computer—That part of the hardware which processes interrupts and data from the various data-entry devices. Display list creation and updating are accomplished in the computer. In some graphics systems, the host computer is a minicomputer with memory and attached peripherals, while in other systems it is a part of the display controller. In either configuration, the host computer is programmable.

hot – 1. Connected, alive, energized; pertains to a terminal or any ungrounded conductor. 2. Not grounded. 3. Strongly radioactive. 4. Excited to a relatively high energy level. 5. Idiomatic term generally used to describe conductors carrying an electrical charge.

hot-carrier diode – A diode in which a closely controlled metal-semiconductor junction provides virtual elimination of charge storage. The device has extremely fast turn-on and turn-off times, excellent diode forward and reverse characteristics, lower noise characteristics, and wider dynamic range. Abbreviated HCD.

hot carriers—In barrier diodes, carriers that have energies greater than those that are in thermal equilibrium with the metal. Thus, electrons that cross the junction from semiconductor to metal must be energetic enough to surmount the barrier. Therefore electrons that are energetic enough to cross the junction are called "hot electrons."

hot cathode—Also called thermionic cathode. A cathode which supplies electrons by thermionic emission. (As opposed to a cold cathode, which has no heater.)

hot-cathode tube – Also called thermionic tube. Any electron tube containing a hot cathode.

hot-cathode X-ray tube – A high-vacuum-ray tube in which a hot rather than cold cathode is used.

hot electrons—*See* Hot Carriers.

hot-electron triode – A solid-state, evaporated-thin-film structure that is directly equivalent to a triode vacuum tube.

hot plate—Electrically heated flat surface, sometimes combined with auxiliary equipment such as a magnetic stirrer.

hot spot—The point of maximum temperature on the outside of a device or component.

hot stamping – Method of imprinting letters, numbers, and symbols, with a heated die.

hot tip dip—A process of passing a bare wire through a bath of molten tin to provide a coating.

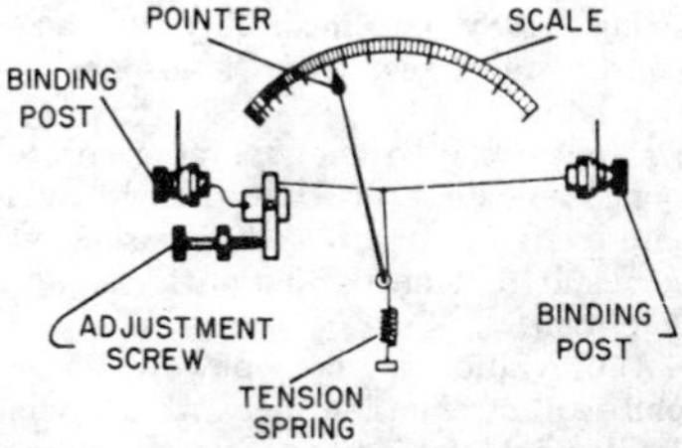

Hot-wire ammeter.

hot-wire ammeter—Also called thermal ammeter. An ammeter in which the expansion of a wire moves a pointer to indicate the amount of current being measured. The current flows through the wire and changes its length in proportion to I^2. Instability because of wire stretching, and the lack of ambient temperature compensation make the hot-wire ammeter commercially unsatisfactory.

hot-wire anemometer—An instrument that measures the velocity of wind or a gas by its cooling effect on an electrically heated wire.

hot-wire instrument—1. An electrothermic instrument operated by expansion of a wire heated by the current it is carrying. 2. A measuring device or transducer whose operation depends either on the expansion of a wire due to its being heated by an electric current or on the change in electrical resistance on the part of a wire that is heated or cooled.

hot-wire microphone – A microphone in which the cooling or heating effect of a sound wave changes the resistance of a hot wire and thus the current through it.

hot-wire relay—A form of linear-expansion time-delay relay in which the longitudinal expansion of a wire, when heated, provides the mechanical motion to open or close contacts. The time required to heat the wire constitutes the delay.

hot-wire transducer – A unilateral transducer in which the cooling or heating effect of a sound wave changes the resistance of a hot wire and thus the current.

housekeeping – In a computer routine, those operations, such as setting up constants and variables for use in the program, that contribute directly to the proper operation of the computer but not to the solution of the problem.

howl – An undesirable prolonged wail produced in a speaker by elecrtic or acoustic feedback.

howler – 1. An electromechanical device which produces an audio-frequency tone. 2. A unit by which the operator at a telephone test desk may apply a high tone of varying loudness to a line to call a subscriber's attention to the fact that his receiver is off the hook.

howling – System instability caused by acoustic feedback from loudspeaker to microphone.

howl repeater – In the operation of telephone repeaters, a condition in which more energy is returned than is sent, with the result that an oscillation is set up on the circuit.

hp–Abbreviation for horsepower.

H-pad–An attenuation network in which the elements are arranged in the form of the letter ***H***.

h-parameters–*See* Hybrid Parameters.

h-particle – The positive hydrogen ion or proton resulting from bombardment of the hydrogen atom by alpha rays or fast-moving positive ions.

hpf–Abbreviation for highest probable frequency.

H plane–The plane in which the magnetic field of an antenna lies. It is perpendicular to the E plane. The principal H plane of an antenna is the H plane that also contains the direction of maximum radiation.

H-plane bend–*See* H-Bend.

H-plane T-junction–Also called shunt T-junction. A waveguide T-junction in which the structure changes in the plane of the magnetic field.

H-scan–*See* H-Display.

HTL – Abbreviation for high-threshold logic.

hub–1. On a control panel or plugboard, a socket or receptacle into which an electrical lead or plug wire may be connected for the purpose of carrying signals. 2. The narrow "spindle" around which the tape is wound on a reel or in a cassette.

hue – Often used synonymously with the term "color," but does not include gray. It is the dominant wavelength–i.e., the one which distinguishes a color as red, yellow, etc. Varying saturations may have the same hue.

hue control–On a color television receiver, the operating control that changes the hue (color) of the picture.

hull potential – The voltage difference between a reference electrode and an immersed metallic hull, or the bonded underwater metallic appendages of a nonmetallic hull.

hum–1. In audio-frequency systems, a low-pitched droning noise consisting of several harmonically related frequencies. It results from an alternating-current power supply, ripple from a direct-current power supply, or induction from exposure to a power system. By extension, the term is applied in visual systems to interference from similar sources. 2. A pattern produced on a facsimile record sheet when a signal at the power-line frequency or a harmonic of the power-line frequency is mixed with or modulates the facsimile signal. 3. A continuous low-frequency interference caused by inadvertent pickup of 60-Hz or 120-Hz energy from nearby ac power sources. Most likely to originate in devices (like microphones) requiring substantial amplification. 4. A background tone caused by improper shielding of audio components or inadequate filtering of line voltage entering the equipment.

human engineering–The science and art of developing machines for human use, giving consideration to the abilities, limitations, habits, and preferences of the human operator. 2. The determination of man's capabilities and limitations as they relate to the equipment or systems he will use, and the application of this knowledge to the planning, design, and testing of man-machine combinations to obtain optimum performance, operability, reliability, efficiency, safety, and maintainability. 3. The study of the behavioral properties of man in interaction with machines, and of total man-machine systems; the structuring of man-machine systems to enhance system performance.

hum-balancing pot – A potentiometer usually placed across the heater circuit. Its arm is grounded so that the heater voltage is balanced with respect to ground.

hum bar–A dark band extending across the picture. It is caused by excessive 60-hertz hum (or harmonics) in the signal applied to the picture-tube input.

hum bucking–The introduction of a small amount of voltage, at the power-line frequency, into a circuit to cancel unwanted power.

hum-bucking coil – A coil wound around the field coil of a dynamic speaker and connected in series opposition with the voice coil. In this way, any hum voltage induced in the field coil will be induced in the voice coil in the opposite direction and buck, or cancel, the effects of the hum.

humidity—An indication of the water-vapor content of a gas mixture.

humidity transducer — A layer of hygroscopic (moisture-absorbing) substance deposited between two metal electrodes. These electrodes establish electrical contact with the hygroscopic chemical, which serves as a resistance element. Since the chemical coating tends to absorb moisture from the surrounding air, its resistance decreases as the humidity increases. In this manner, humidity variations are converted to resistance variations.

hum loop — A condition arising from the connection of two or more "grounds" to an amplifier system whereby circulating currents of low value at power-line frequency and harmonics are added to the program signals, causing hum to appear in the background.

humming—A sound produced by transformers having loose laminations or by magnetostriction effects in iron cores. The frequency of the sound is twice the power-line frequency.

hum modulation—Modulation of an rf signal or detected af signal by hum. This type of hum is heard only when the receiver is tuned to a station.

hunting—1. Continuous, cyclical searching by a control system for a desired or ideal value. Rapid hunting usually is termed oscillation; slower cycling is called birddogging. 2. Movements of a selector from terminal to terminal until an idle one is found.

hv—Abbreviation for high voltage.

H-vector — A vector which represents the magnetic field of an electromagnetic wave. In free space, it is perpendicular to the E-vector and the direction of propagation.

H-wave—A mode in which electromagnetic energy can be tranmsitted in a waveguide. An H-wave has an electric field perpendicular to the length of the waveguide, and a magnetic field parallel as well as perpendicular to the length.

hybrid—1. An electronic circuit that contains both vacuum tubes and transistors. 2. A mixture of thin-film and discrete integrated circuits. 3. A computer that has both analog and digital capabilities. 4. *See* Hybrid Junction. 5. A transformer or combination of transformers or resistors that affords paths to three branches, A, B and C, so arranged that A can send to C, and B can receive from C, but A and B are effectively isolated. 6. A mixture or combination of two different technologies.

hybrid balance—A measure of the degree of balance between two impedances connected to two conjugate sides of a hybrid set. Given by the formula for return loss.

hybrid circuit—1. A circuit which combines the thin-film and semiconductor technologies. Generally, the passive components are made by thin-film techniques, and the active components by semiconductor techniques. The active devices are attached to the thin-film passive components by a suitable bonding process. 2. Also called two-wire–four-wire terminating set. In telephone transmission circuits, a circuit for interconnecting two-wire and four-wire circuits through a differential balance or bridge circuit in which the two sides of the four-wire circuit form conjugate arms. 3. Any circuit made by using a combination of the following component manufacturing technologies: monolitic IC, thin film, thick film and discrete component. 4. An integrated microelectronic circuit in which each component is fabricated on a separate chip or substrate, interconnected by means of lead wires so that each component can be independently optimized for performance.

hybrid coil—Also called bridge transformer. A single transformer which has, effectively, three windings and which is designed to be connected to four branches of a circuit so as to render these branches conjugate in pairs.

hybrid computer—1. A computer that results from the interconnection of an analog computer and a digital computer, plus conversion equipment, each contributing its special advantages to an assigned part of the solution of a class of complex problems. 2. A computer that combines both analog and digital equipment for purposes of solving problems that cannot be adequately or economically handled by either type of computer operating independently. The term hybrid computer does not denote the use of some analog equipment to preprocess data that is then converted to digital form and subsequently entered into a conventional digital computer. Rather, there is usually a continual flow of data in both directions between analog and digital equipment. 3. A computer for data processing in which both analog and discrete representations of data are used.

hybrid electromagnetic wave—Abbreviated HEM wave. An electromagnetic wave having components of both electric- and magnetic-field vectors in the direction of propagation.

hybrid integrated circuit — 1. Abbreviated HIC. An integrated circuit combining parts made by a number of techniques, such as diffused monolithic portions, thin-film elements, and discrete devices. 2. An arrangement consisting of one, or more, integrated circuits in combination with one, or more, discrete devices. Alternatively, the combination of more than one type of integrated circuit into a single integrated component. 3. A composite of either monolithic integrated circuits or discrete semiconductor device circuits, in a unit-packaging configuration. 4. The physical realization of electronic circuits or subsystems from a number of extremely small circuit elements electrically and mechanically interconnected on a substrate.

hybrid junction—1. Also called hybrid-T or magic-T. A waveguide arrangement with four branches. When they are properly terminated, energy is transferred from any one branch into two of the remaining three branches. In common usage, this energy is divided equally between the two. 2. Also called bridge hybrid, and hybrid. A transformer, resistor, or waveguide circuit or device that has four pairs of terminals so arranged that a signal entering at one terminal pair will divide and emerge from the two adjacent terminal pairs, but will be unable to reach the opposite terminal pair.

hybrid microcircuit — A microcircuit in which thin-film or diffusion techniques are combined with separately attached semiconductor chips to form the circuit.

hybrid network—A nonhomogeneous communication network required to operate with signals of dissimilar characteristics (such as analog and digital modes).

hybrid parameters — Also called h-parameters. The resultant parameters of an equivalent transistor circuit when the input current and output voltage are selected as independent variables.

hybrid ring—Also called a "rat race." A hybrid junction commonly used as an equal power divider. It consists of a reentrant line (waveguide) to which four side arms are connected. The line is of the proper electrical length to sustain standing waves.

hybrids — A particular type of circuit or module consisting of a combination of two or more integrated circuits, or one integrated circuit and discrete elements.

hybrid set—Two or more transformers interconnected to form a network having four pairs of accessible terminals. Four impedances may be connected to the four terminals, so that the branches containing them may be made conjugate in pairs.

hybrid-T—*See* Hybrid Junction.

hybrid thin-film circuit — A microcircuit formed by attaching discrete components and semiconductor devices to networks of passive components and conductors that have been vacuum deposited on glazed ceramic, sapphire, or glass substrates.

hybrid transformer—*See* Hybrid Coil.

hybrid-type circuit—*See* Multichip Circuit.

hydroacoustics—The generation of acoustic energy from the flow of fluids under pressure.

hydroacoustic transducer — A transducer that produces high-level acoustic energy from the flow of high-pressure fluid.

hydroelectric—The production of electricity by water power.

hydrogen electrode — A platinum electrode covered with platinum black, around which a stream of hydrogen is bubbled. The hydrogen electrode furnishes a standard against which other electrode potentials can be compared.

hydrogen lamp — A special light source, used in some spectrophotometers, which produces invisible light energy. It is used in finding the light-energy frequency of test solutions.

hydrogen thyratron—A thyratron containing hydrogen.

hydrolysis—The chemical decomposition of a substance in the presence of water. Usually, it is considered in the sense of chemical degradation of insulating materials under the influence of heat or pressure and in contact with moisture (for example, hydrolysis of polyester films and coatings).

hydromagnetics—*See* Magnetohydrodynamics.

hydromagnetic waves—Waves in which the energy oscillates between the magnetic field energy and kinetic energy of the hydrodynamic motion, the reservoirs being the self-inductance of the conductive matter and the mass inertia of the moving fluid.

hydrometer—An instrument for measuring the specific gravity of a liquid such as the electrolyte of a storage battery. It contains a graduated float which indicates the specific gravity by the amount of liquid displaced.

hydrophone—An electroacoustic transducer which responds to waterborne sound waves and delivers essentially equivalent electric waves.

hydrostatic pressure—*See* Static Pressure.

hygrometer—An instrument that measures the relative humidity of the atmosphere.

hygroscopic—Readily absorbing and retaining moisture from the atmosphere. The opposite term is nonhygroscopic.

hygrostat—A device that closes a pair of contacts when the humidity reaches a prescribed level.

hyperacoustic zone—The region in the upper atmosphere, above about 60 miles, in which the distance between the air molecules is roughly equal to the wavelength of sound, so that sound is transmitted with less efficiency than at lower levels. Sound waves cannot be propagated above this zone.

hyperbola—1. A curve that is the locus of points having a constant difference of distance from two fixed points. 2. In hyperbolic guidance systems, a path along which the difference between the arrival times of pulses from two transmitters is constant. (*See also* Hyperboiic Guidance System).

hyperbolic error–The error in an interferometer system arising from the assumption that the directions of the wavefronts incident at two antennas of a base line are parallel, whereby the equiphase path is a cone. Mathematically the equiphase path is a hyperbola.

hyperbolic grind–A shape of tape playback and record heads. It permits good head contact and better response at high frequencies.

hyperbolic guidance system–A method of guidance in which sets of ground stations transmit pulses from which a hyperbolic path can be derived to give range and course information for steering. (*See also* Hyperbola.)

hyperbolic head–A recording head whose pole-piece surfaces (when viewed from the edge of the tape) are shaped like the graph of a mathematical hyperbolic function. This shape offers a good compromise between intimate tape contact at the gap and proximity to the tape of the rest of the pole-piece face (the latter is necessary for good low-frequency response).

hyperbolic horn – A horn in which the equivalent cross-sectional radius increases according to a hyperbolic law.

hyperbolic navigation system – A method of radionavigation (e.g., loran) in which pulses transmitted by two ground stations are received by an aircraft or ship. The difference in arrival time from each station is a measure of the difference in distance between the aircraft or ship and each station. This distance is plotted on one of many hyperbolic curves on a map. A second reading from another pair of stations (or from the same master and a different slave) establishes another point on a different hyperbolic curve. The intersection of the two curves gives the position of the aircraft or ship.

hyperdirectional–*See* Shotgun.

hyperfrequency waves–Microwaves having wavelengths in the range from 1 centimeter to 1 meter.

hypersensor – A single-component, resettable circuit breaker that operates as a majority-carrier tunneling device. It is used to provide overcurrent or overvoltage protection of integrated circuits.

hypersonic–Having five or more times the speed of sound.

hyspersyn motor – A synchronous motor which combines the desirable features of the induction, hysteresis, and dc excited synchronous motor, resulting in high efficiency and power factor. It possesses the vigorous starting torque of an induction motor, the synchronization torque of a hysteresis motor, and the stiffness of an externally dc excited synchronous motor.

hysteresigraph–A device for experimentally presenting or recording the hysteresis loop of a magnetic specimen.

hysteresis–1. The amount the magnetization of a ferrous substance lags the magnetizing force because of molecular friction. 2. A type of oscillator behavior where multiple values of the output power and/or frequency correspond to given values of an operating parameter. 3. The temporary change in the counting-rate–vs–voltage characteristic of a radiation-counter tube (caused by previous operation). 4. The difference between the response of a unit or system to an incrasing and a decreasing signal. 5. A form of nonlinearity in which the response of a circuit to a particular set of input conditions depends, not only on the instantaneous values of those conditions, but also on the immediate past (recent history) of the input and output signal. Hysteretical behavior is characterized by inability to "retrace" exactly on the reverse swing a particular locus of input/output conditions. 6. The lag in the response of an instrument or process when a force acting on it changes abruptly. 7. The property of a magnetic material by virtue of which the magnetic induction for a given magnetizing force depends upon the previous conditions of magnetization.

hysteresis brake–*See* Hysteresis Clutch.

hysteresis clutch – Also called hysteresis

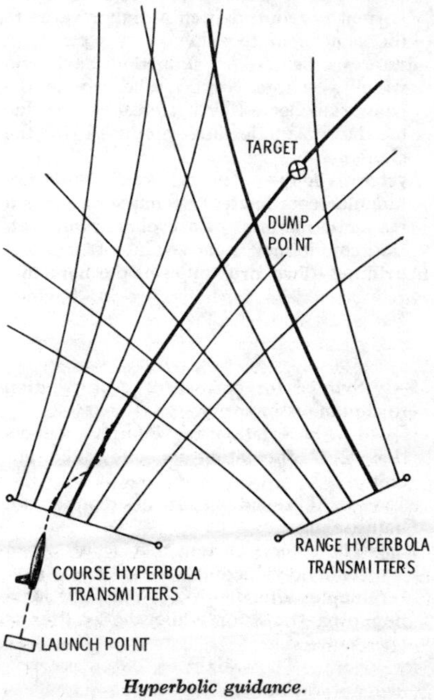

Hyperbolic guidance.

brake. A proportional torque-control device which employs the hysteresis effect in a permanent-magnet rotor to develop its output torque. It is capable of synchronous driving or continuous slip, provided heat can be removed, with almost no torque variation at any slip differential. Its control-power requirement is small enough for vacuum-tube or transistorized drive.

hysteresiscope—An instrument used to obtain hysteresis loops on a cathode-ray oscilloscope screen without the need for specially prepared ring samples. It is used in the inspection of magnetic material.

hysteresis curve—1. A curve showing the relationship between a magnetizing force and the resultant magnetic flux. 2. A graph showing the amount of magnetism imparted to a magnetizable material as the result of a varying magnetic field. This coincides with the variations of the applied field only through a relatively narrow range between zero magnetism and saturation, but the addition of a bias allows an audio signal to be recorded on a magnetic tape within this linear range, for minimum distortion.

hysteresis distortion — Distortion of waveforms in circuits containing magnetic components. It is due to the hysteresis of the magnetic cores.

hysteresis error—The difference in the reading obtained on a measuring instrument containing iron when the current is increased to a definite value and when the current is reduced from a higher value to the same definite value.

hysteresis heater—An induction device in which a charge (or a muffle around the charge) is heated by hysteresis losses due to the magnetic flux produced in the charge.

hysteresis loop—A curve (usually with rectangular coordinates) which shows, for a magnetic material in a cyclically magnetized condition, for each value of the magnetizing force, two values of the magnetic

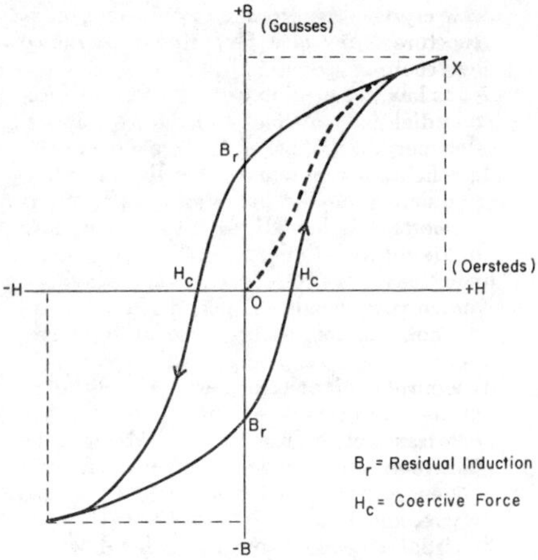

Hysteresis loop.

induction; one when the magnetizing force is increasing, the other when it is decreasing.

hystoroscope — An instrument used to observe, measure, and record the magnetic characteristics of both easy and hard axes of magnetic materials.

hysteresis loss—The power expended in a magnetic material as a result of magnetic hysteresis.

hysteresis meter—An instrument for determining the hysteresis loss in a ferromagnetic material. It measures the torque produced when the test specimen is placed in a rotating magnetic field or is rotated in a stationary magnetic field.

hysteresis motor — A synchronous motor without salient poles or direct-current excitation. It is started by the hysteresis losses innduced in its secondary by the revolving field.

Hz — Abbreviation for hertz, meaning cycles-per second (of any periodic phenomenon).

I

I—1. Symbol for current. 2. Abbreviation for luminous intensity.

IC — 1. Abbreviation for internal connection. 2. Abbreviation for integrated circuit.

icbm — Abbreviation for intercontinental ballistic missile.

I_{CBO}—The reverse current that occurs when a specific dc voltage is applied in the nonconducting direction to the collector junction of a transistor while the emitter is open-circuited.

ice loading—The weight of ice an antenna can accumulate without being damaged.

iconoscope — A camera tube in which a beam of high-velocity electrons scans a photoemissive mosaic capable of storing an electrical charge pattern.

icw—Abbreviation for interrupted continuous wave.

ideal bunching — A theoretical condition where bunching of the electrons in a velocity-modulated tube would give an infinitely large current peak during each cycle.

ideal capacitor—A capacitor having a single-valued transferred-charge characteristic.

ideal crystal—A crystal having no mosaic structure and capable of X-ray reflection in accordance with the Darwin-Ewald-Prins law.

ideal dielectric—A dielectric in which all the energy required to establish an electric field in the dielectric is returned to the source when the field is removed. (A perfect dielectric must have zero conductivity. Also, all absorption phenomena must be lacking. A vacuum is the only known perfect dielectric.)

ideal-noise diode—A diode that has an infinite internal impedance and in which the current exhibits full shot-noise fluctuations.

ideal transducer—Theoretically, any linear passive transducer which—if it dissipated no energy and, when connected to a source and load, presented its combined impedance to each—would transfer maximum power from source to load.

ideal transformer — A hypothetical transformer which would neither store nor dissipate energy. Its self-inductances would have a finite ratio and unity coefficient of coupling, and its self- and mutual impedances would be pure inductances of infinitely great value.

I demodulator—A demodulator circuit the inputs of which are the chrominance signal and the signal from the local 3.58-MHz oscillator. The output of this demodulator is a video signal representing color in the televised scene. The Q demodulator is similar except that its input from the local oscillator is shifted 90°.

identification—1. In radar, determining the identity of a displayed target (i.e., which one of the blips in the display represents the target). 2. In a computer, a code number or code name which uniquely identifies a record, block, file, or other unit of information.

identification beacon—A code beacon used for positively identifying a particular point on the earth's surface.

identification, friend or foe—A system using radar transmissions to which equipment carried by friendly forces automatically responds, for example, by emitting pulses, thereby distinguishing themselves from enemy forces. It is the primary method of determining the friendly or unfriendly character of aircraft and ships by other aircraft or ships and by ground forces employing radar-detection equipment and associated identification, friend or foe units. Abbreviated iff.

identifier—A symbol the purpose of which is to identify, indicate, or name a body of data.

identify—In a computer, to attach a unique code or code name to a specific unit of information.

idiochromatic—Having photoelectric properties characteristic of the pure crystal itself and not due to foreign matter.

I-display—In radar, a display in which a target appears as a circle when the radar antenna is pointed directly at it. The radius of the circle is proportionate to the target distance. When the antenna is not pointing at the target, only a segment of the circle appears. Its length is inversely proportionate to the magnitude of the pointing error, and the segment points away from the direction of error.

idle characters — Control characters interchanged by a synchronized transmitter and receiver to maintain synchronization during a nondata period.

idler—1. A rubber-tired wheel that transfers power from a phonograph motor to the turntable rim. 2. An intermediate drive wheel, usually with a rubber or neoprene "tire," which transfers rotational energy from a driven wheel to a third wheel. Often used for speed reduction between a drive motor and capstan shaft. *See also* Pinch Wheel.

idler drive—A drive system used to rotate a turntable, which consists of a driveshaft that is turned by the motor pulley, and which drives the inside rim of the turntable platter.

idler frequency—A sum or difference frequency, other than the input, output, or pump frequencies, generated within a parametric device and requiring specific circuit consideration to achieve the desired performance of the device.

idler pulley—A pulley used only for tightening a belt or changing its direction. The shaft does not drive any other part.

idle time—That portion of available time during which the hardware is not in use.

idle-trunk lamp—A signal lamp that indicates the outgoing trunk with which it is associated is not busy.

idling current — Also called quiescent current. The zero-signal power supply current drawn by a circuit or by a complete amplifier.

IDT — Abbreviation for interdigital transducer.

IEC—Abbreviation for integrated electronic component.

IEEE—Abbreviation for Institute of Electrical and Electronic Engineers. A professional organization of scientists and engineers whose purpose is the advancement of electrical engineering, electronics, and allied branches of engineering and science. (The IEEE resulted from the merger of the IRE and the AIEE.)

if — Abbreviation for intermediate frequency.

if amplifier — *See* Intermediate-Frequency Amplifier.

if canceler—In radar, a moving-target-indicator canceler operating at intermediate frequencies.

iff—*See* Identification, Friend or Foe.

IFRU—*See* Interference-Rejection Unit.

if strip—*See* Intermediate-Frequency Strip.

IGFET — Abbreviation for insulated-gate field-effect transistor. Though a less popular term than MOS, it more precisely defines devices made by various MOS processes.

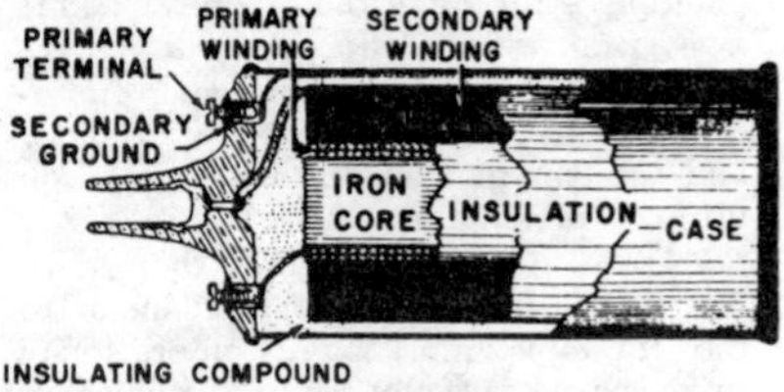

Ignition coil.

ignition coil — An iron-core transformer which converts a low direct voltage to the 20,000 volts or so required to produce an ignition spark in gasoline engines. It has an open core, a heavy primary winding connected to the battery or other source through a vibrating armature contact, and a secondary winding with many turns of fine wire.

ignition control—Control of the instant that static current begins to flow in the anode circuit of a gas tube.

ignition interference — Noise produced by sparks or other ignition discharges in a car, motor, or furnace ignition, or by equipment with loose contacts or connections.

ignition reserve—In a gasoline engine, the difference between the available voltage and the required ignition voltage.

ignition voltage—In a gasoline engine, the peak voltage required to produce a spark across the plug electrodes.

ignitor discharge—In switching tubes, a dc glow discharge between the ignitor electrode and a suitably located electrode. It is used to facilitate radio-frequency ionization.

ignitor electrode—An electrode (which is partly immersed in the mercury-pool cathode of an ignitron) used to initiate conduction at the desired points in each cycle.

ignitor firing time—In switching tubes, the interval between application of a dc voltage to the ignitor electrode and start of current flow.

ignitor interaction—In a tr, pre-tr, or attenuator tube, the difference between the insertion loss measured at a specified level of ignitor current and that measured at zero ignitor current.

ignitor leakage resistance—In a switching tube, the insulation resistance measured between the ignitor electrode terminal and the adjacent rf electrode in the absence of an ignitor discharge.

ignitor oscillation—A relaxation type of oscillation in the ignitor circuit of a tr, pre-tr, or attenuator tube.

ignitor voltage drop — In switching tubes, the dc voltage between the cathode and anode at a specified ignitor current.

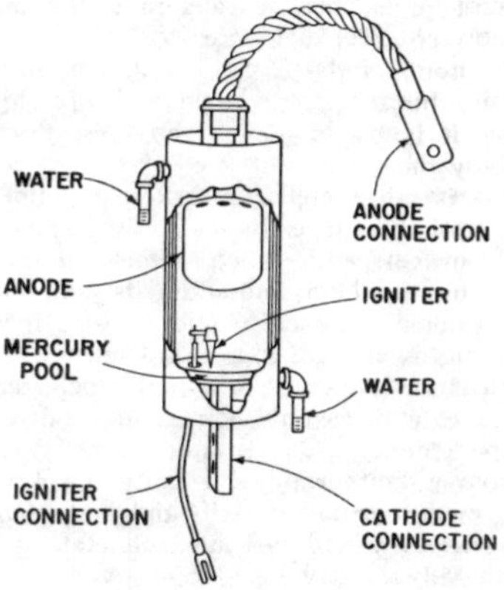

Ignitron.

ignitron—A type of mercury-pool rectifier which has only one anode. The arc is started for each cycle of operation by an ignitor which dips into the mercury pool. The mercury pocl serves as the cathode of the rectifier. The ignitron is characterized by the ability to withstand tube currents several times as high as rated values for a few cycles.

ignore—In a computer, a character code indicating that no action is to be taken.

IGY — Abbreviation for international geophysical year.

IHF—The Institute of High Fidelity, the official association of the manufacturers and certain related organizations in the high-fidelity field.

IHFM—Abbreviation for Institute of High-Fidelity Manufacturers, an association of manufacturers which publishes ratings and standards for high-fidelity equipment.

illegal character—A character or combination of bits that does not have validity according to some criterion; for example, a character that is not a member of a specified alphabet.

illuminance—The density of luminous flux at a given distance from the center of a source. It is equal to the total flux divided by the surface area over which it is uniformly spread. The units of illuminance are lumens/cm^2, lumens/ft^2, etc. One lumen/ft^2 is the same as a ft-candle; one lumen/cm^2 is the same as one phot;

and one lumen/meter^2 is equal to one lux, or one meter-candle.

illuminant-C—The reference white of color television—i.e., light which most nearly matches average daylight.

illuminate—1. To expose to light. 2. In radar, to strike with a radar signal so that reflection returns the signal to the source for interpretation.

illumination—1. The light flux incident on a unit projected area; it is the photometric counterpart of irradiance and is expressed in foot-candles. 2. The density of the luminous flux incident on a surface; it is the quotient of the luminous flux by the area of the surface, when the latter is uniformly illuminated.

illumination control—A photorelay circuit that turns on artificial lighting when natural illumination decreases below a predetermined level.

illumination sensitivity — The output current of a photosensitive device divided by the incident illumination at constant electrode voltages.

illuminometer—A portable photometer for measuring the illumination on a surface.

ILO—*See* Injection-Locked Oscillator.

ILS — Abbreviation for instrument landing system.

image—1. The instantaneous illusion of a picture on a flat surface. 2. The unused one of the two groups of sidebands generated in amplitude modulation. 3. A spatial distribution of some physical property (e.g., radiation, electric charge, conductivity, or reflectivity) made to correspond with another distribution of the same or another physical property.

image admittance—Reciprocal of image impedance.

image antenna—The imaginary counterpart of an actual antenna. For mathematical purposes it is assumed to be located below the ground and symmetrical with the actual antenna.

image-attenuation constant—The real part of the transfer constant.

image converter—1. A solid-state optoelectric device capable of changing the spectral characteristics of a radiant image. Examples of such changes are infrared-to-visible and X-ray-to-visible. 2. An electron tube that employs electromagnetic radiation to produce a visual replica of an image produced on its cathode. Electrons ejected from the photosensitive cathode by the incident radiation are accelerated to and focused upon a fluorescent phosphor screen, thus forming the visual replica. Image converters can be used in the infrared, ultraviolet, and X-ray regions as well as in the visible. An example of an infrared-sensitive image converter is the snooperscope.

image-converter tube—*See* Image Tube.

image dissection — An optical, mechanical, or electronic process, or a combination of such processes, in which an optical image is divided into discrete segments prior to being photographed, recorded, transmitted, or processed in some other way.

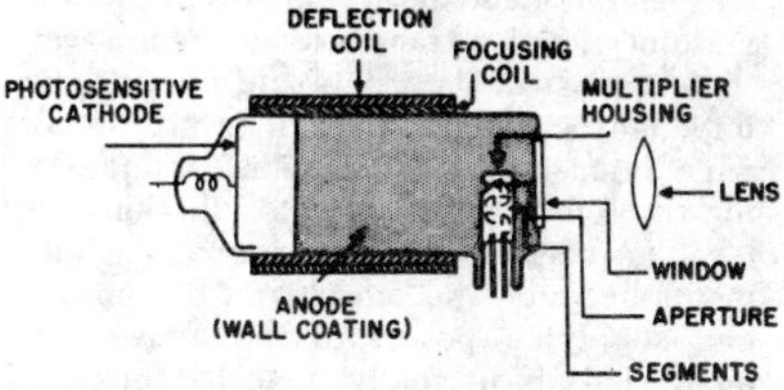

Image dissector, 1.

image dissector — 1. A television camera tube in which the image is swept past an aperture in a series of 525 interlaced lines thirty times per second. Instead of a beam scanning the image, the entire image is scanned past the aperture, which "dissects" the image—hence, the name. (*See also* Dissector Tube.) 2. In ocr, a mechanical or electronic transducer that detects in sequence the light levels in different areas of a completely illuminated sample space.

image distortion—Failure of the reproduced image in a television receiver to resemble the original scene scanned by the camera.

image effect—An effect produced on the field of an antenna as the electromagnetic waves are reflected from the earth's surface.

image-enhancing equipment—An elaborate device, often involving a computer in which a photograph is scanned by a point of light, the amplitude of the electrical signal being modified electronically before being re-recorded on another film.

image force—The force on a change due to that charge or polarization which it induces on neighboring conductors or dielectrics.

image frequency—In heterodyne frequency converters, an undesired input frequency capable of producing the selected frequency by selecting one of the two sidebands produced by beating. The word "image" implies the mirrorlike symmetry of signal and image frequencies about the beating oscillator frequency or intermediate frequency, whichever is higher.

image-frequency rejection ratio—Of a superheterodyne receiver, the ratio of the response at the desired frequency to the response at the image frequency.

image iconoscope—An iconoscope in which greater sensitivity is obtained by separating the function of charge storage from that of photoelectric emission. An optical image is projected on a continuous photosensitive screen, and the electron emission from the back of this screen is focused electromagnetically onto a mosaic screen that is scanned by an electron beam as in the original emitron cathode-ray tube. The British term is super-emitron.

image impedance—The impedances which will simultaneously terminate all inputs and outputs of a transducer in such a way that at each of its inputs and outputs the impedances in both directions are equal.

image intensifier—1. A system for increasing the sensor response to a radiation pattern or image by interposing active elements between the sensor and the image, and supplying power to the active element. This is normally done by focusing the scene to be imaged on the photocathode of the tube, giving rise to a photoelectron pattern corresponding to the optical image. This pattern is accelerated and focused onto a phosphor which emits light to reproduce a visual image of the scene. 2. Device used in X-ray techniques for brightening the fluoroscopic image several hundred times and reducing radiation exposure. 3. An electronic tube equipped with a light-sensitive electron emitter at one end and a phosphor screen at the other end; an electron lens inside the tube relays the image. These devices are used in astronomy for photographing very faint celestial objects.

image interference – In a receiver, a response due to signals of a frequency removed from the desired signal by twice the intermediate frequency.

image-interference ratio—In a superheterodyne receiver, the effectiveness of the preselector in rejecting signals at the image frequency.

image orthicon—A camera tube in which a photoemitting surface produces an electron image and focuses it on one side of a separate storage target. The opposite side of the target is then scanned by low-velocity electrons to produce the output.

image phase constant—The imaginary part of the transfer constant.

image ratio—In a heterodyne receiver, the ratio of the image-frequency signal input to the desired signal input for identical amplitude outputs.

image rejection—1. The suppression of image-frequency signals in a superheterodyne receiver. 2. The rejection by the tuner of a signal at the image (second-channel) frequency, corresponding to the tuned (real) frequency plus twice the intermediate frequency when the local oscillator is working above the signal frequency or minus twice the intermediate frequency when the local oscillator is working below the signal frequency.

image-rejection ratio—The ratio (in decibels) of the signal required for a 30-dB snr to that required for the same ratio but at the image frequency. An increase in front-end selectivity increases the ratio.

image-reject mixer—A combination of two balanced mixers and associated hybrid circuits designed for separation of the image channel from the signal channels normally present in a conventional mixer. The arrangement makes possible image rejection of up to 30 dB without the use of filters.

image response—The response of a heterodyne receiver to a signal that is separated by twice the intermediate frequency from the frequency to which the receiver is tuned. Unless there is some preselection, images will cause spurious unwanted re-

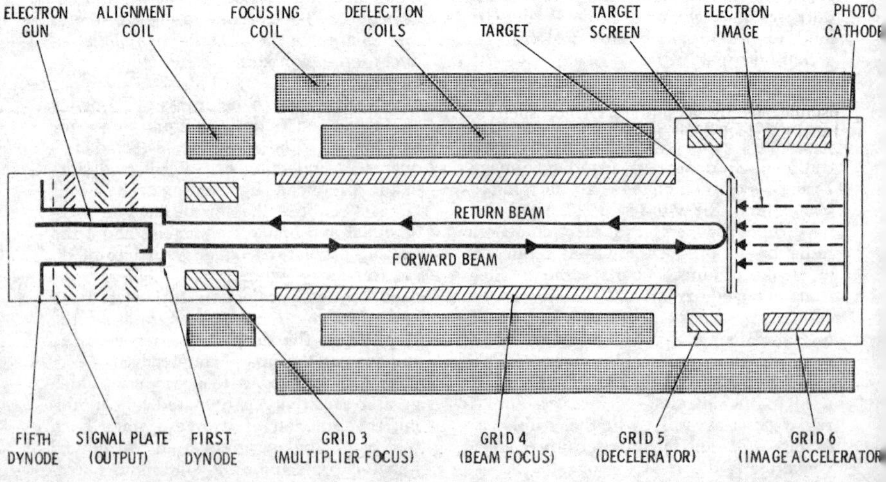

Image orthicon.

sponses when the spectrum occupied by a signal is greater than twice the frequency of the first if stage of the receiver.

image-transfer constant—*See* Transfer Constant.

image tube—Also called an image-converter tube. An electron tube which reproduces on its fluorescent screen an image of an irradiation pattern incident on its photosensitive surface.

imd—See Intermodulation Distortion.

imitative deception — The transmission of messages in the enemy's communication channels for the purpose of deceiving him.

immediate access — The ability of a computer to put data in storage or remove them from storage without delay.

immediate-access store—A store the access time of which is negligible compared to other operating times.

immediate addressing — In a computer, a mode of addressing where the operand contains the value to be operated on, and no address reference is required.

immediate data—Data which immediately follows an instruction in a memory, and is used as an operand by that instruction.

immersion plating—A method of metal deposition that depends on a galvanic displacement of the metal being plated by the substrate. Thickness of the plating is limited to 10 to 50 millionths of an inch.

immersion pyrometer — An instrument for determining molten-steel temperature and normally consisting of a platinum-platinum rhodium bimetal thermocouple junction and a recording device for transposing the millivoltage into degrees of temperature.

immittance—A term that denotes both impedance and admittance. It is commonly applied to transmission lines, networks, and certain kinds of measuring instruments.

IMOS—*See* Ion-Implanted MOS.

impact excitation—The starting of damped oscillations by a sudden surge, such as by a spark discharge.

impact modulator amplifier—A fluidic device in which the impact plane position of two opposed streams is controlled to alter the output.

impact predictor—A device which can determine, in real time, the point on the earth's surface where a ballistic missile will impact if thrust is instantaneously terminated.

IMPATT diode — (Impact avalanche and transmit time diode.) 1. A pn-junction diode operated with heavy back bias so that avalanche breakdown occurs in the active region. To prevent burnout, the device is so constructed that the active region is very close to a good heat sink. For the same reason, the bias supply must be a constant-current type. 2. A device whose negative resistance characteristic is produced by a combination of impact avalanche breakdown and charge-carrier transit-time effects.

Avalanche breakdown occurs when the electric field across the diode is high enough for the charge carriers (holes or electrons) to create electron-hole pairs. With the diode mounted in an appropriate cavity, the field patterns and drift distance permit microwave oscillations or amplification.

IMPATT oscillator—An oscillator in which the active element is an IMPATT diode operating in a negative resistance mode. Dc to rf conversion efficiencies are normally less than 20%. Present devices operate above 5 GHz.

impedance—The total opposition (i.e., resistance and reactance) a circuit offers to the flow of alternating current at a given frequency; the ratio of the potential difference across a circuit or element of a circuit to the current through the circuit or element. It is measured in ohms, and its reciprocal is called admittance. Symbol: Z.

impedance angle—Angle of the impedance vector with respect to the resistance vector. Represents the phase angle between voltage and current.

impedance bridge—A device for measuring the combined resistance and reactance of a component part of a circuit.

impedance characteristic—A graph of impedance versus frequency of a circuit or component.

impedance coil—A coil in which its inductive reactance is used to hinder the flow of alternating current in or between circuits.

impedance compensator — An electric network used with a line or another network to give the impedance of the combination a certain characteristic over a desired frequency range.

impedance coupling—A method of coupling

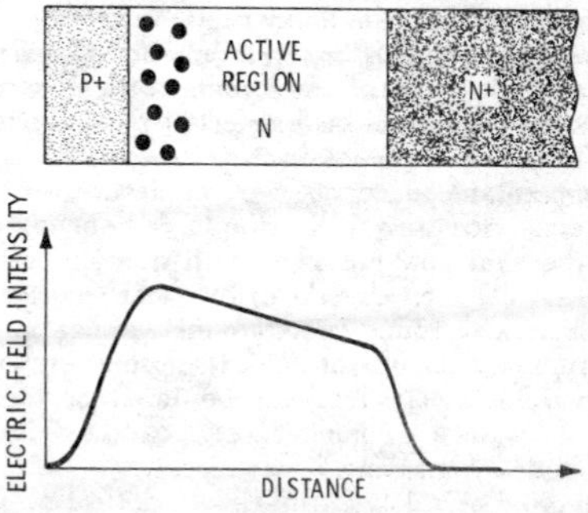

IMPATT diode.

using an impedance as the coupling device, common to both the primary and secondary circuits. This type of coupling is usually limited to audio systems, where high gain and limited bandpass are required.

impedance drop—The vector sum of the resistance drop and the reactance drop. (For transformers, the resistance drop, the reactance drop, and the impedance drop are, respectively, the sum of the primary and secondary drops reduced to the same terms. They are usually expressed in percent of the secondary-terminal voltage.)

impedance ground — An earth connection made through an impedance of predetermined value usually chosen to limit the current of a short-circuit to ground.

impedance irregularities—Breaks or abrupt changes which occur in an impedance-frequency curve when unlike sections of a transmission line are joined together or when there are irregularities on the line.

impedance matching — The connection across a source impedance of another impedance having the same magnitude and phase angle. If the source is a transmission line, reflection is thereby avoided.

impedance-matching transformer—A transformer used to match the impedance of a source and load.

impedance plethysmograph—An instrument used to detect the increased blood volume in the tissues of the body during a contraction of the heart. *See also* Electrical-Impedance Cephalography *and* Finger Plethysmograph.

impedance transformer—A transformer that transfers maximum energy from one circuit to another.

impedance triangle—A diagram consisting of a right triangle. The sides are proportional to the resistance and reactance in an ac circuit, with the hypotenuse representing the impedance.

imperfect dielectric—A dielectric in which part of the energy required to establish its electric field is converted into heat instead of being returned to the electric system when the field is removed.

imperfection (of a crystalline solid)—Any deviation in structure from an ideal crystal (one which is perfectly periodic in structure and contains no foreign atoms).

implantable pacemaker—A miniature pulse generator surgically implanted beneath the skin and provided with output leads that connect directly to the heart muscle. The electrodes may contact either the outer wall of the heart muscle (myocardial electrodes) or the inner surface of the heart chamber (endocardiac electrodes).

implied AND — Also called dot AND or wired AND. A logic element in which the combined outputs are true if and only if all outputs are true.. (Sometimes improperly called dot OR or wired OR.)

implied OR—Also called wired OR. A logic element in which the combined outputs are true if one or more of the outputs is true.

implode—The inward bursting of a picture tube due to its high vacuum.

impregnant—A substance, usually a liquid, used to saturate the paper dielectric of a capacitor and replace the air between its fibers, thereby increasing the dielectric strength and the dielectric constant of the capacitor.

impregnated coils — Coils that have been permeated with an electric grade varnish or other protective material to protect them from mechanical vibration, handling, fungus, and moisture.

impregnating—Complete filling of even the smallest voids in a component or closely packed assembly of parts. Low-viscosity compounds, usually liquids, are used. The process is frequently accomplished by a vacuum process, where all air is removed before introducing the impregnating material. Typical examples of impregnating are the filling of capacitors or transformer windings.

impregnation—The process of coating the insides of coils and closely packed electronic assemblies by dipping them into a liquid and letting it solidify.

impressed voltage—The voltage applied to a circuit or device.

improvement threshold—A characteristic of fm radio receivers which determines the minimum rf signal power required to overcome the inherent thermal noise. For increasing values of rf power above this point, an improvement of signal-to-noise ratio is obtained.

impulse — A pulse that begins and ends within so short a time that it may be regarded mathematically as infinitesimal. The change produced in the medium, however, is generally of a finite amount. (*See also* Pulse.)

impulse bandwidth — The area divided by the height of the voltage-response selectivity as a function of frequency. It is used in the calculation of broad-band interference.

impulse-driven clock—An electric clock in which the hands are moved forward at regular intervals by current impulses from a master clock.

impulse excitation—Also called shock excitation. A method of producing oscillatory current in which the duration of the impressed voltage is relatively short compared with that of the current produced.

impulse frequency—The number of pulse periods per second generated by the dial-pulse springs in a telephone as they rapidly open and close in response to the dialing of a digit.

impulse generator – 1. Also called surge generator. An electric apparatus that produces high-voltage surges for testing insulators and for other purposes. 2. A device that generates a broad energy spectrum by means of a very narrow impulse. Usually generated by discharging a short coaxial or waveguide transmission line. The pulses are discrete and regularly spaced, and are generally variable at a repetition rate from a few pulses per second to a few thousand pulses per second. The output of an impulse generator is specified as the rms equivalent of the peak voltage in dB above 1 microvolt per megahertz. 3. An oscillator circuit that generates electric impulses for synchronizing purposes in a television system. 4. A circuit, typically using a step-recovery diode, used to convert a sinusoidal input to a voltage impulse output. The basic circuit block in both SRD multiples and comb generators.

impulse noise—Noise due to disturbances having abrupt changes and of short duration. (These noise impulses may or may not have systematic phase relationships. The noise is characterized by nonoverlapping transient disturbance. The same source may produce impulse noise in one system and random noise in a different system.)

impulse-noise generator – Equipment for generating repetitive pulses which provide random noise signals uniformly spread over a wide band of frequencies.

impulse period—*See* Pulse Period.

impulse ratio—The ratio of the flashover, sparkover, or breakdown voltage of an impulse to the crest value of the power-frequency flashover, sparkover, or breakdown voltage.

impulse relay – 1. A relay that stores enough energy from a brief impulse to complete its operation after the impulse ends. 2. A relay that can distinguish between different types of impulses, operating on long or strong impulses, and not operating on short or weak ones. 3. An integrating relay.

impulse response of a room—The time sequence of signals received at some point in a room due to a sound pulse generated at some other point in the room. It defines the arrival of a sound that has traversed the direct path between source and microphone and the arrivals of the various reflections.

impulse sealing—A heat-sealing technique in which a pulse of intense thermal energy is applied to the sealing area for a very short time, followed immediately by cooling. It is usually accomplished by using an rf heated metal bar which is cored for water cooling, or is of such a mass that it will cool rapidly at ambient temperatures.

impulse separator – Normally called sync separator. In a television receiver, the circuit that separates the synchronizing impulses from the video information in the received signal.

impulse speed—The rate at which a telephone dial mechanism makes and breaks the circuit to transmit pulses.

impulse strength—A measure of the ability of insulation to withstand voltage surges on the order of microseconds in duration.

impulse timer—A timing device electrically powered by a synchronous motor, featuring a mechanical stepping device which enables it to advance a predetermined number of degrees within a predetermined time interval, controlling a multiple number of circuits. Said circuits are controlled by individual cams which program their activity.

impulse train—*See* Pulse Train.

impulse transmission—The form of signaling used principally to reduce the effects of low-frequency interference. Impulses of either or both polarities are employed for transmission, to indicate the occurrence of transitions in the signals.

impulse-transmitting relay – A relay in which a set of contacts closes briefly when the relay changes from the energized to the de-energized position, or vice versa.

impulse-type telemeter—A telemeter which employs the characteristics of intermittent electric signals, other than their frequency, as the translating means.

impurity—A material such as boron, phosphorus, or arsenic added to a semiconductor such as germanium or silicon to produce either p-type or n-type material. Impurities that provide free electrons are called donors and cause the semiconductor material to be n-type. Impurities that accept electrons are called acceptors and cause the material to be p-type.

impurity density—The amount of impurity material diffused into a certain volume of semiconductor material used in manufacturing semiconductor devices.

impurity ions – An alien, electrically charged atomic system in a solid; an ion substituted for a constituent atom or ion in a crystal lattice, or located in an interstitial site in the crystal.

impurity level—The energy level existing in a substance because of impurity atoms.

inaccuracy—1. The difference between the input quantity applied to a measuring instrument and the output quantity indicated by that instrument. The inaccuracy of an instrument is equal to the sum of its instrument error and its uncertainty. 2. The term sometimes used to indicate the deviation from an indicated or recorded value or the measure of conformity to an accepted standard.

inactive leg—Within a transducer, an electrical element which does not change its electrical characteristics with the applied stimulus. Applied specifically to elements that complete a Wheatstone bridge in certain transducers.

in-band signaling—The transmission of signaling tones at some frequency or frequencies within the channel normally used for voice transmission.

incandescence—1. The state of a body with such a high temperature that it gives off light. 2. The generation of light caused by passing an electric current through a wire filament. The resistance of the filament to the current causes the filament to heat up and emit radiant energy, some of which is in the visible range.

incandescent lamp – An electric lamp in which electric current flowing through a filament of resistance material heats the filament until it glows.

INCH—Acronym for integrated chopper. It is a device designed to operate as a chopper, commutator, modulator, demodulator, or mixer, depending on circuit requirements.

inching—*See* Jogging.

incidence angle—The angle between an approaching light ray or emission and the perpendicular (normal) to the surface in the path of the ray.

incidental fm—Also called residual fm. 1. The short-term jitter or undesired fm deviation of a local oscillator. It limits resolution when it approaches the if bandwidth in magnitude. 2. Peak-to-peak variations of a carrier frequency caused by external variations not a part of normal action of the carrier tuned circuits.

incident field intensity—The field strength of a down-coming sky wave, not including the effects of earth reflections at the receiving location.

incident power—The product of the outgoing current and voltage traveling from a transmitter down a transmission line to an antenna.

incident wave – In a medium of certain propagation characteristics, a wave which strikes a gap in the medium or strikes a medium having different propagation characteristics.

incipient failure—A degradation failure in its beginning stages.

inclination – The angle which a line, surface, or vector makes with the horizontal.

inclinometer—An instrument for measuring the magnetic inclination of the earth's magnetic field. It uses a magnetic needle that pivots vertically to indicate the inclination.

inclusive AND—A logic element the output of which is true if all inputs are true, all inputs are false, or all inputs but one are false.

inclusive NAND—A logic element the output of which is true if one and only one of the inputs is false.

incoherent – Denotes the lack of a fixed phase relationship between two waves. If two incoherent waves are superposed, interference effects cannot last longer than the individual coherent times of the waves.

incoherent detection – Detection wherein the information contained in the phase of the carrier is discarded.

incoherent scattering – The disordered change in their direction of propagation when radio waves encounter matter.

incoming selector—In a telephone central office, a selector associated with trunk circuits from another central office.

incomplete sequencer relay – A device which returns the equipment to the normal, or off, position and locks it out if the normal starting, operating, or stopping sequence is not properly completed within a predetermined time.

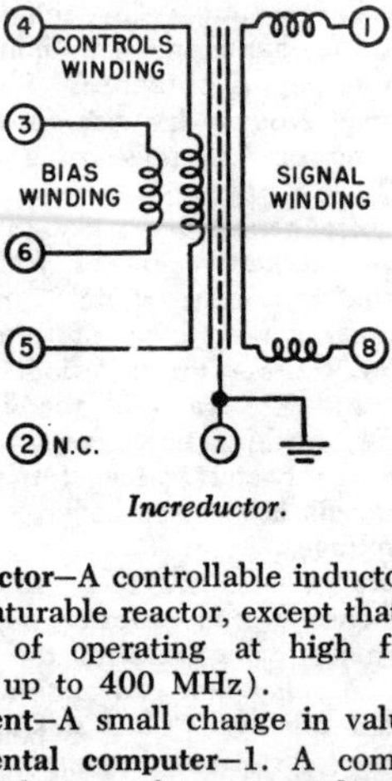

Increductor.

increductor—A controllable inductor similar to a saturable reactor, except that it is capable of operating at high frequency (e.g., up to 400 MHz).

increment—A small change in value.

incremental computer—1. A computer in which the use of incremental representation of data is predominant. 2. A special-purpose computer designed specifically to process changes in the variables as well as absolute values of the variables.

incremental digital recorder—A magnetic tape recorder that advances the tape across the recording head step by step, as in a punched-tape recorder. It is used for economical and reliable recording of an irregular flow of data.

incremental frequency shift—A method by which incremental intelligence may be superimposed on other intelligence by shifting the center frequency of an oscillator a predetermined amount.

incremental hysteresis loss – Losses in a magnetic material that has been subjected to a pulsating magnetizing force.

incremental induction—One half the algebraic difference between the maximum and minimum magnetic induction at a point in a material which has been subjected simultaneously to a polarizing and a varying magnetizing force.

incremental integrator—A digital integrator modified so that the output signal is maximum negative, zero, or maximum positive when the value of the output is negative, zero, or positive, respectively.

incremental permeability—1. Ratio of the cyclic change in magnetic induction to the corresponding cyclic change in magnetizing force when the mean induction is other than zero. 2. The ratio of incremental change in flux density to the incremental change in magnetizing force at any point on the hysteresis loop.

incremental sensitivity – The smallest change that can be detected by a particular instrument in a quantity under observation.

incremental tuner—A television tuner in which antenna, rf amplifier, and rf oscillator inductors are continuous or in small sections connected in series, Rotary switches, connected to taps on the inductors, provide the portion of total inductance required for a channel, or short-circuit all remaining inductance except that required for the channel.

independent failure—A failure which has no significant relationship to other failures in a given device and can occur without interaction with other component parts in the equipment.

independent load contacts—Contacts which can control electrical loads that must be isolated from the timer clutch solenoids and motor circuit.

independent variable.—One of several voltages and current chosen arbitrarily and considered to vary independently.

index counter—An odometer-type cumulative-digit indicator for "keeping score" on the amount of tape that has passed through a tape machine. The counter is generally driven by the takeup-reel turntable and thus registers rotation rather than tape footage, although the accuracy is generally good enough to allow for locating specific recorded segments according to previously noted index counter numbers.

indexed address—An address that is altered by the content of an index register before or during the execution of a computer instruction.

indexing—In a computer, a technique of address modification that is often implemented by means of index registers.

indexing slots—*See* Polarizing Slots.

index of cooperation—In rectilinear scanning or recording, the product of the total length of a line and the number of lines per unit length.

index of modulation—The modulation factor.

index of refraction—Ratio of the speeds of light or other radiation in two different materials. This determines the amount the ray will be refracted or bent when passing from one material to the other, such as from air to water.

index register—In a computer, a register that holds a quantity which may be used for modifying addresses or for other purposes, as directed by the program.

indicating demand meter – A meter equipped with a scale over which a pointer is advanced to indicate maximum demand.

indicating fuse—A protective device placed in a telephone circuit to provide visual and audible indication of a fault in the line. It consists of a fuse, pilot lamp, relay, and buzzer. When a line fault blows the fuse, the lamp lights and the buzzer sounds.

indicating instrument – An instrument which visually indicates only the present value of the quantity being measured.

indicating lamp—A lamp which indicates the position of a device or the condition of a circuit.

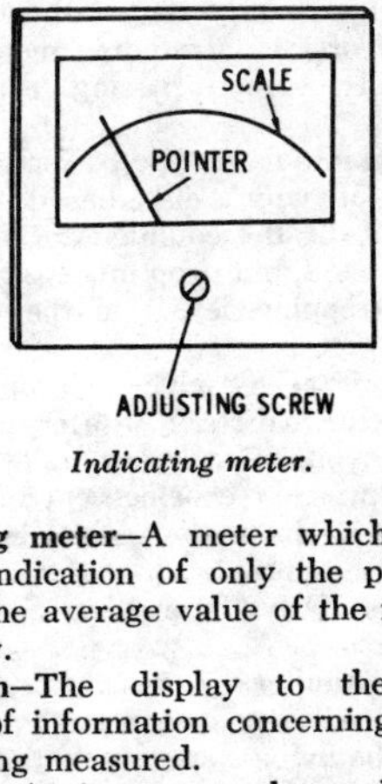

Indicating meter.

indicating meter—A meter which gives a visual indication of only the present or short-time average value of the measured quantity.

indication—The display to the human senses of information concerning a quantity being measured.

indicator—An instrument that makes information available, but does not store it.

indicator gate—A rectangular voltage waveform applied to the grid or cathode circuit of an indicator cathode-ray tube to sensitize or desensitize the tube during the desired portion of the operating cycle.

indicator tube—An electron-beam tube which conveys useful information by the variation in cross section of the beam at a luminescent target.

indirect-acting recording instrument – An instrument in which the marking device is actuated by raising the level of measurement energy of the primary detector. This is done mechanically, electrically, electronically, photoelectrically, or by some other intermediate means.

indirect address–An address in a computer instruction that indicates a location where the address of the referenced operand is to be found.

indirect addressing–A method of computer cross reference in which one memory location indicates where the correct address of the main fact can be found.

indirect light–Light from an object which has no self-luminous properties. Instead, it reflects light from another source.

indirect lighting–A system of lighting where all the light is directed to the ceiling or walls, which in turn reflect it to the objects to be illuminated.

indirectly controlled variable – A variable that is related to and influenced by the directly controlled variable but is not directly measured for control.

indirectly heated cathode. – Also called equipotential or unipotential cathode. A cathode which is heated by an independent heater.

indirectly heated thermistor–A thermistor which incorporates, as part of its composite structure, an electrical heater. A thermistor the body temperature of which in use is significantly higher than the temperature of its surrounding medium as a result of current passing through its heater.

indirect material–A type of semiconductor material in which electrons do not drop directly from the conduction band to the valence band, but drop in steps as a result of the trapping levels in the forbidden gap.

indirect piezoelectricity – The production of a mechanical strain in a crystal by applying a voltage to it (as opposed to the more common piezoelectric effect of applying a strain to the crystal in order to produce a voltage).

indirect radiative transition–*See* Transition, 1.

indirect scanning – A television technique used in early mechanical systems, and today in the flying-spot scanning of films. A small beam of light is moved across the subject and then reflected to a battery of phototubes.

indirect synthesizer–A synthesizer employing phase-locked loops, digital dividers, and high-Q varactor-tuned oscillators. The discrete output frequencies are not limited to integer ratios of the reference frequency. Frequency step size or increments, are primarily determined by the digital dividers. Switching speed between discrete output frequencies is usually limited by the phase-lock circuits (applications include automatic test and satellite communications systems).

indirect wave–A wave reaching a given reception point by a path from the transmitting point other than the direct-line path between the two (e.g., a sky wave received after deflection from the ionosphere layers).

individual gap azimuth–In a magnetic-tape record or reproduce head stack, the angle of an individual gap relative to a line perpendicular to the precision milled mounting pads in a plane parallel to the surface of the tape.

individual line–A subscriber line that serves one main station and optional additional stations connected to the line as extensions; the line is not arranged for discriminatory ringing with respect to the stations.

indoor antenna–Any receiving antenna located inside a building but outside the receiver.

indoor transformer – A transformer which must be protected from the weather.

induced–Produced by the influence of an electric or magnetic field.

induced charge – An electrostatic charge produced in one object by the electric field surrounding a nearby object.

induced current–1. The current that flows in a conductor which is moved perpendicularly to a magnetic field, or which is subjected to a magnetic field of varying intensity. The former takes place in an induction-motor rotor; the latter, in the secondary winding of a transformer. 2. In induction heating, the current that flows in a conductor when a varying electromagnetic field is applied.

induced electromotive force – Represented by E and is proportional to the rate of change of magnetic flux through the circuit ($d\phi/t$).

induced environment – The temperatures, vibrations, shocks, accelerations, pressures, and other conditions imposed on a system due to the operation or handling of the system.

induced failure–A failure that is basically caused by a condition or phenomenon external to the item that fails.

induced voltage–The voltage produced in a conductor when the conductor is moved up and down through the magnetic field of a second conductor, or when the field varies in intensity and cuts across the first conductor. Even though there is no mechanical coupling between the two conductors, the one producing the field will produce a voltage in the other.

inductance–1. Property of a circuit that tends to oppose any change of current because of a magnetic field associated with the current itself. Whenever an electric current changes in value—rises or falls—in a circuit, its associated magnetic field changes, and when this links with the conductor itself an emf is induced which tends to oppose the original current change. Self-inductance is the full name for this, but the term inductance only is usually used. The unit of inductance is the "henry." When a current changing at the rate of 1 ampere per second induces a voltage of 1 volt, the inductance of the circuit is 1 henry. 2. *See also* coil.

inductance bridge—An instrument, similar to a Wheatstone bridge, for measuring an unknown inductance by comparing it with a known inductance.

inductance coil—*See* Inductor.

inductance-tube modulation—A method of modulation employed in frequency-modulated transmitters. An oscillator control tube acts as a variable inductance in parallel with the tank circuit of the radio-frequency oscillator tube. As a result, the oscillator frequency varies in step with the audio-frequency voltage applied to the grid of the oscillator control tube.

induction—1. The establishment of an electric charge or a magnetic field in a substance by the proximity of an electrified source, a magnet, or a magnetic field. 2. The setup of an electromotive force and current in a conductor by variation of the magnetic field affecting the conductor.

induction brazing — The electric brazing process in which heat is produced by an induced current.

induction coil—A device for changing direct current into high-voltage alternating current. Its primary coil contains relatively few turns of heavy wire; and its secondary coil, wound over the primary, contains many turns of fine wire. Interruption of the direct current in the primary by a vibrating-contact arrangement induces a high voltage in the secondary.

induction compass—A compass in which the indications are produced by the current generated in a coil revolving in the magnetic field of the earth.

induction-conduction heater—A heating device through which electric current is conducted but is restricted by induction to a preferred path.

induction density—*See* Flux Density.

induction factor—In an alternating current circuit, the ratio between that element of the current which does no work and the total strength of the current.

induction field—1. That portion of the electromagnetic field of a transmitting antenna which acts as if it were permanently associated with the antenna, and into which energy is alternately stored and removed. 2. The electromagnetic field of a coil carrying alternating current, responsible for the voltage induced by that coil in itself or in a nearby coil.

induction frequency converter—A slip-ring induction machine driven by an external source of mechanical power. Its primary circuits are connected to a source of electric energy having a fixed frequency. The energy delivered by its secondary circuits is proportionate in frequency to the relative speed of the primary magnetic field and the secondary member.

induction furnace—A furnace heated by electromagnetic induction.

induction hardening—The process of hardening the surface of a casting by heating it above the transformation range by electrical induction, followed by rapid cooling.

induction heating—The method of producing heat by subjecting a material to a variable electromagnetic field. Internal losses in the material then cause it to heat up.

induction instrument—An instrument operated by the reaction between the magnetic flux set up by one or more currents in fixed windings and the currents set up by electromagnetic induction in movable conductive parts.

induction motor—1. An alternating-current motor in which the primary winding (usually the stator) is connected to the power source and induces a current into a polyphase secondary or squirrel-cage secondary winding (usually the rotor). 2. A motor which runs asynchronously; that is, not in step with the alternations of the alternating current.

induction-motor meter.—A meter containing a rotor that moves in reaction to a magnetic field and the currents induced into it.

induction noise — The noise — other than thump, flutter, cross fire or cross talk—produced when two circuits are inductively coupled together.

induction-resistance welding—Welding in which electromagnetic induction alone causes the heating current to flow in the parts being welded.

induction-ring heater—A core-type induction heater adapted principally for heating round objects. The core is open or can be taken off to facilitate linking the charge.

induction speaker—A speaker in which the current that reacts with the steady magnetic field is induced into the moving member.

induction-voltage regulator—A device having a primary winding in shunt, and a secondary winding in series with a circuit for gradually adjusting the voltage or the phase relation of the circuit by changing the relative position of the exciting and series windings of the regulator.

inductive—Pertaining to inductance or to the inducing of a voltage through mutual or electrostatic induction.

inductive circuit—A circuit with more inductive than capacitive reactance.

inductive coordination — Location, design, construction, operation, and maintenance of electric supply and communication systems in a manner that prevents inductive interference.

inductive coupling—1. The association of one circuit with another through inductance common to both. When used without modifying words, the term commonly refers to coupling by means of mutual inductance, whereas coupling by means of self-inductance common to both circuits is called direct inductive coupling. 2. In inductive-coordination practice, the interrelation of neighboring electric supply and communication circuits resulting from electric and/or magnetic induction.

inductive feedback—The transfer of energy from the output circuit to the input circuit of an amplifying device through an inductor or inductive coupling.

inductive interference—1. Interference produced in communication systems by induced voltages within the system. 2. Effect arising from the characteristics and inductive relations of electric supply and communication systems of such character and magnitude as would prevent the communication circuit from rendering service satisfactorily and economically if methods of inductive coordination were not applied.

inductive kick—The voltage, many times higher than the impressed voltage, produced by the collapsing field in a coil when the current through it is abruptly cut off.

inductive load—Also called lagging load. A load that is predominantly inductive, so that the alternating load current lags behind the alternating voltage of the load.

inductively coupled circuit — A network with two meshes having only mutual inductance in common.

inductive microphone—*See* Inductor Microphone.

inductive neutralization—Also called shunt or coil neutralization. A method of neutralizing an amplifier, whereby the equal and opposite susceptance of an inductor cancels the feedback susceptance caused by interelement capacitance.

inductive pickup—Signals generated in a circuit or conductor due to mutual inductance between it and a disturbing source.

inductive post—A metal post or screw extended across a waveguide parallel to the E field to act as inductive susceptance in parallel with the waveguide for purposes of tuning or matching.

inductive reactance—The opposition to the flow of alternating or pulsating current by the inductance of a circuit. It is measured in ohms, and its symbol is X_L. It is equal to 2π times the frequency in hertz times the inductance in henrys.

inductive transducer—A transducer in which changes in inductance convey the stimulus information.

inductive transduction—The conversion of the measurand into a change in the self-inductance of a single coil.

inductive tuning—A method of tuning a radio by moving a core into and out of a coil to vary the inductance.

inductive winding—A coil through which a varying current is sent to give it an inductance.

inductive window—A conducting diaphragm extended into a waveguide from one or both sidewalls to act as an inductive susceptance in parallel with the waveguide.

inductor—1. Also called inductance or retardation coil. A conductor used for introducing inductance into an electric circuit. The conductor is wound into a spiral, or coil, to increase its inductive intensity. 2. A passive fluidic element which, because of fluid inertness, has a pressure drop that leads flow by essentially 90°. 3. *See* coil.

inductor microphone—Also called inductive microphone. A. microphone in which the sound waves move a conductor back and forth, cutting magnetic lines of force and producing an electrical output of the same frequency and proportional to the amplitude of the sound waves.

inductor-type synchronous motor — A type of synchronous motor having field magnets that are fixed in magnetic position relative to the armature conductors, the torques being produced by forces between the stationary poles and salient rotor teeth. Such motors usually have permanent-magnet field excitation, are built in fractional-horsepower frames, and operate at low speeds (300 revolutions per minute or less).

Inductosyn—An extremely precise transducer based upon the magnetic circuit of a conductor deposited on glass for stability, and operated at a relatively high frequency. Extremely accurate, but requires much auxiliary equipment (Farrand Controls Inc.).

industrial-grade IC — Typically, an IC whose performance is guaranteed over the temperature range from 0 to 70°C.

industrial radio services—Radiocommunication services essential to, operated by, and for the sole use of those enterprises which require radiocommunications in order to function efficiently.

industrial television – Television used for remote viewing of manufacturing or assembling processes, usually over cables rather than through the air.

industrial timer–A timing device, impulse or constant-speed type, used in industrial applications other than the appliance industry.

industrial tube–A vacuum tube designed for industrial electronic equipment.

inelastic collision–Collision resulting in excitation of a molecule.

inertance–Acoustical equivalent of inductance.

inert gas–*See* Noble Gas.

inertia–The tendency of an object at rest to remain at rest, or of a moving object to continue moving in the same direction and at the same speed, unless disturbed by an outside force. Resulting from mass and inhibiting change in velocity. Important in pickup mechanics.

inertial guidance/navigation–A self-contained system for navigation where position can be computed by knowing a craft's starting point, and where it has been. Changes in acceleration are detected by gyroscopes for direction and attitude and by accelerometers for velocity. These signals are integrated to determine resulting velocity and distance. The system needs no outside reference and cannot be jammed.

inertial navigation–A guidance technique in which air-frame acceleration is first measured and then integrated twice with respect to time in order to determine the distance traveled. External aids such as radio and radar are not necessary. The acceleration or deceleration of the air frame is measured continuously with accelerometers oriented in some convenient frame of reference, usually corresponding to the earth's north-south, east-west coordinates.

inertia relay–A relay having added weights or other modifications that increase its moment of inertia and either slow it or cause it to continue in motion after the energizing force is removed.

inertia switch–A switch capable of sensing acceleration, shock, or vibration. It is designed to actuate upon an abrupt change in velocity.

inertia welding – A forge-welding process in which stored kinetic energy is released as frictional heat when two parts are rubbed together under the proper conditions.

infant mortality–The occurrence of premature catastrophic-type failures at a rate substantially greater than that observed during life prior to wearout.

infant-mortality period – *See* Early-Failure Period.

inferential – The kind of instrumentation, especially its signal source, in which there is sampling of an entirely different quantity from the one of interest, upon assuming that they vary in perfect proportion. Linearity or a perfectly repeatable relationship between the two is inferred, for the sake of a more convenient signal-source arrangement.

infinite–Boundless; having no limits whatsoever.

infinite baffle–1. A loudspeaker enclosure in which the speaker's rear sound waves are completely absorbed or dissipated. 2. A loudspeaker mounting where ideally there is no path of air between the back and front of the speaker diaphragm. An infinite baffle improves the forward radiation of sound at low frequencies and preferably should be a very large plane surface like the wall of a room or a screen of very rigid material (e.g., ¾-inch wood) on which a loudspeaker is mounted. (In practice truly infinite baffles are rarely accomplished except in sealed boxes, but these give rise to problems of resonance.) 3. An airtight speaker enclosure containing a bass speaker with very low open-air resonance, plus a sealed midrange speaker and tweeter.

infinite-baffle speaker system – A speaker where the bass driver is located in an almost airtight enclosure.

infinite-impedance detector–A detector circuit in which the load is a resistor connected in parallel with an rf bypass capacitor between the cathode and ground. Since the grid is always negative with respect to the cathode, the tube presents an infinite impedance to the input.

infinite line–A transmission line with the same characteristics as an ordinary line that is infinitely long.

infinite resolution–The capability of a device to provide continuous output over its entire range.

infinitesimal – Immeasurably small; approaching zero.

infinity – 1. A hypothetical amount larger than any assignable amount. 2. A number larger than any number a computer can store in any register. 3. Any distance of a subject from a lens for which the image no longer moves when the subject moves along the optical axis.

inflection point–The point where a curve changes direction.

infobond–An automated system of point-to-point wiring on the back of a two-sided printed wiring board (the components are on the front, or other side). The No. 38 AWG copper wire used is solder-bonded to terminations by an automatic soldering gun.

information—1. In computing, the basic data and/or program entered into the system. 2. That property of a signal or message whereby it conveys something meaningful and unpredictable to the recipient, usually measured in bits.

information bits – In telecommunications, those bits originated by the data source and not used for error control by the data-transmission system.

information center – A facility specifically designed for storing, processing, and retrieving information to be disseminated at regular intervals, on demand, or selectively, according to the needs of users.

information channel—The transmission and intervening equipment involved in the transfer of information in a given direction between two terminals. An information channel includes the modulator and demodulator and any error-control equipment irrespective of its location, as well as the backward channel, when provided.

information feedback system—In telecommunications, an information-transmission system in which an echo check is employed to verify the accuracy of the transmission.

information gate – A circuit that permits information or data pulses to pass when the circuit is triggered by an external source.

information handling – The storing and processing of information and its transmission from the source to the user. Information handling excludes the creation and use of information.

information rate—In computers, the minimum number of binary digits per second required to specify the source messages.

information rate changer – A device that speeds up the playback of tape-recorded speech without pitch change or deterioration of characteristic resonances. This is accomplished by rotating the playback head in the direction of tape travel.

information retrieval—A method for cataloging vast amounts of data related to one field of interest so that any part or all of this data can be called out at any time with accuracy and speed.

information-retrieval system—A system for locating and selecting on demand certain documents or other graphic records relevant to a given information requirement from a file of such material.

information separator—A control character used to identify a logical boundary of information. The name of the separator is not necessarily indicative of what it separates.

information theory—The branch of learning that deals with the likelihood of accurate transmission of messages subject to transmission failure, distortion, and noise.

infra—(prefix) Below; beneath; less than.

infradyne receiver—A superheterodyne receiver the intermediate frequency of which is made higher than the signal frequency in order to obtain high selectivity.

infrared – 1. Pertaining to or designating those radiations, such as are emitted by a hot body, with wavelengths just beyond the red end of the visible spectrum. These wavelengths are longer than those of visible light and shorter than those of radio waves. 2. That section of the electromagnetic spectrum, invisible to the eye, lying between wavelengths of 770 and about 1,000,000 A. Thermography utilizes waves in this region for recording changes in temperature.

infrared communications set – The collection of components necessary to operate a two-way electronic system in which infrared radiation is used to carry intelligence.

infrared counter-countermeasures – Action taken to employ infrared radiation equipment and systems in spite of enemy measurces to counter their use.

infrared countermeasures—Action taken to reduce the effectiveness of enemy equipment employing infrared radiation.

infrared detector – A transducer which is sensitive to invisible infrared radiation (wavelength between 0.75 and 1000 microns), usually using a semiconductor, thermocouple bolometer, or pneumatic (pressure) device to detect the radiation.

infrared emitter—*See* Infrared Light-Emitting Diode.

infrared-emitting diode—A semiconductor device with a semiconductor junction in which infrared radiant flux is nonthermally produced when a current flows as a result of applied voltage.

infrared guidance—A system using infrared heat resources for reconnaissance of targets or for navigation.

infrared homing—A type of missile homing in which the guidance system tracks the target from the infrared radiation it emits.

infrared light – Light rays just below the red end of the visible spectrum.

infrared light-emitting diode—Also called infrared emitter. An optoelectronic device containing a semiconductor pn junction that emits radiant energy in the 0.78-μm to 100-μm wavelength region when forward-biased.

infrared optics—Lenses, prisms, and other optical elements for use with infrared radiation (radiation with a wavelength between 0.75 and 1000 microns).

infrared radiation—Invisible radiation with wavelengths in the range between 7800

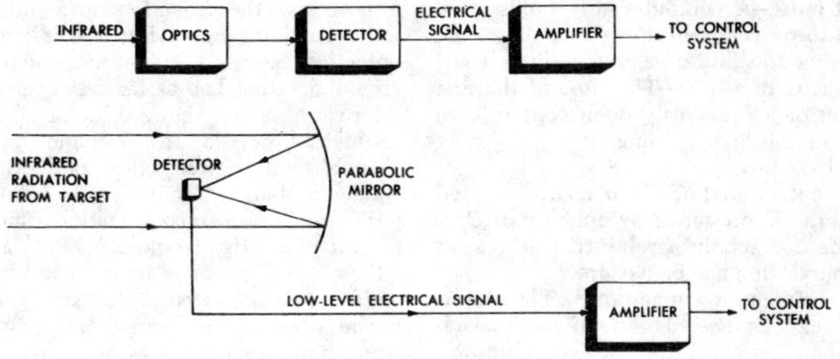

Infrared guidance.

angstroms (red) and about 1,000,000 angstroms (microwaves).

infrared sources – Emitters of radiation with a wavelength between 0.75 and 1000 microns.

infrared spectrum – That portion of the electromagnetic spectrum between the wavelengths of 0.75 and 1000 microns.

infrared waves – Also called black light. Invisible waves longer than the longest visible red light waves but shorter than radio-frequency waves.

infrared window – A region of relatively high transmission in the infrared-frequency range.

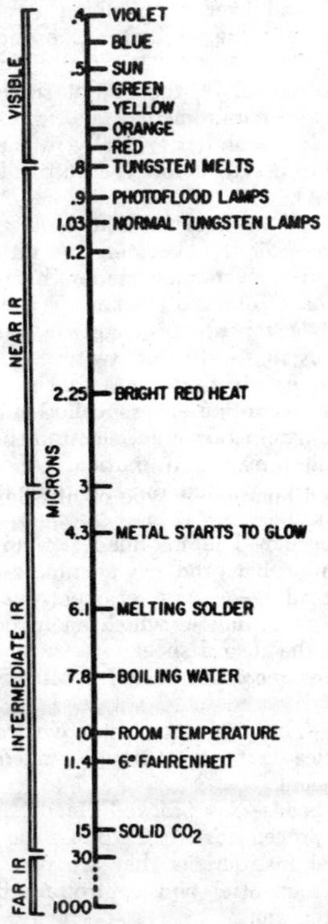

Infrared spectrum.

infrasonic–Pertaining to frequencies below the range of human hearing, hence below about 15 hertz. Formerly called subsonic.

infrasonic frequency–A frequency below the audio range. Infrasonic vibrations can be felt but not heard. Replaces the obsolete term subsonic frequency.

inharmonic frequency–A frequency which is not a rational multiple of another frequency.

inherent delay–Delay between the insertion of information into a unit and presentation of the information at the output. For example, a delay inserted into the crt vertical amplifier of pulse analyzers to allow the leading edge of the signal triggering the sweep to be seen.

inherent interference–A type of electromagnetic interference generated within a receiver by thermal agitation, shot effect, and nonlinear impedance.

inherited error–In a computer, the error in the initial values, especially that error accumulated from prior steps in a step-by-step integration.

inhibit–To prevent an action from taking place or data from being accepted by applying an appropriate signal (generally a logic 0 in positive logic) to the proper input.

inhibit gate–A circuit that provides an output when the inputs differ and only for one of these combinations. This function can be implemented directly or by using any previous gate, plus an inverter.

inhibiting input–A computer gate input which, if in its prescribed state, prevents any output which might otherwise occur.

inhibiting signal–A signal the presence of which prevents an operation from taking place.

inhibition gate–A gate circuit used as a switch and placed in parallel with the circuit it is controlling.

inhibitor–In a digital computer, a logic circuit that clamps a specified output to the zero level when energized. Also called inhibition gate.

inhibit pulse—A computer drive pulse that tends to prevent certain drive pulses from reversing the flux of a magnetic cell.

initial actuation time—The time of the first closing of a previously open contact of a relay or the first opening of a previously closed contact.

initial contact chatter—That chatter caused by vibration produced by opening or closing the contacts in a relay themselves, as by contact impact in closure.

initial differential capacitance—The differential capacitance of a nonlinear capacitor when the capacitor voltage is zero.

initial drain—The current supplied at nominal voltage by a cell or battery.

initial element—*See* Primary Detector.

initial erection—The mode of operation of a vertical gyro in which the gyro is being erected or slaved initially. The initial erection rate is usually relatively fast.

initial failure—The first failure that occurs in use.

initial inverse voltage—Of a rectifier tube, the peak-inverse anode voltage immediately following the conducting period.

initial ionizing event—Also called primary ionizing event. An ionizing event that initiates a tube count.

initialization—The process in which information (memory locations for data and results, tolerances, limits, etc.) is supplied to a computer prior to the running of a program.

initialize—To set counters, switches, and addresses to their starting values at the beginning of a computer routine or at prescribed points in the routine.

initializing—1. The preliminary steps in arranging those instructions and data in a computer memory that are not to be repeated. 2. Setting flip-flops to known states prior to testing.

initial permeability—Permeability at a field density approaching zero.

initial program loader—The procedure that results in loading of the initial part of an operating system or other program so that the program can then proceed under its own control.

initial reversible capacitance—In a nonlinear capacitor, the reversible capacitance at a constant bias voltage of zero.

initial-velocity current — A current which flows between an electrode, such as the grid of a vacuum tube, and its cathode as a result of electrons thrown off from the cathode because of heat alone. Their velocity is sufficient to allow the electrons to reach the grid unaided by an accelerating field.

injected laser—*See* Diode Laser.

injection grid — A vaccum-tube grid that controls the electron stream without causing interaction between the screen and control grids. In some superheterodyne receivers, the injection grid introduces the oscillator signal into the mixer stage.

injection laser—1. An optical oscillator or amplifier that has as its active medium a forward-biased semiconductor diode in which a population inversion has been established between the conduction and valence bands. Radiation is emitted in the process of recombination across the band gap. High-frequency modulation of the output beam can be achieved by modulating the input current. Usually, the optical resonator is formed by cleaving or polishing opposite faces of the diode crystal. Typical dimensions of the device are 0.1 mm × 0.1 mm × 0.5 mm. 2. A semiconductor diode, carrying a high current in the forward direction. Radiation is produced as electrons recombine with holes in the junction region. For coherent emission, the current density must exceed a threshold commonly about 10,000 amp/sq cm for gallium arsenide diodes. 3. A solid-state semiconductor device with at least one pn junction capable of emitting coherent or stimulated radiation under specified conditions. Incorporates a resonant optical cavity.

injection-locked oscillator — Abbreviated ILO. A freerunning microwave oscillator that is stabilized by injecting a reference signal into the oscillator's resonant circuitry. The required injected signal level is determined by the output signal characteristic requirements (i.e., noise, stability, etc.) and is typically in the range of 70 to 30 dB below the output level of the ILO.

injection luminescent diode — A gallium-arsenide diode, operating in either the laser or noncoherent mode, that can be used as a source of visible or near-infrared light for use in triggering such devices as light-activated switches.

injector—An electrode on a spacistor.

ink-mist recording — Also called ink-vapor recording. In facsimile, electromechanical recording in which particles of an ink mist are deposited directly onto the record sheet.

ink recorder—The ink-filled pen or capillary tube that produces a graphic record.

ink recording—A type of mechanical facsimile recording in which an inked helix marks the record sheet.

ink-vapor recording—*See* Ink-Mist Recording.

inleads—Those portions of the electrodes of a device that pass through an envelope or housing.

in-line heads—*See* Stacked Heads.

in-line procedures—In COBOL, the procedural instructions that are part of the main sequential and controlling flow of the program.

in-line processing—The processing of data

in random sequence not subject to preliminary sorting or editing.

in-line subroutine—A subroutine that is inserted directly into the linear operational sequence. Such a subroutine must be recopied at each point in a routine where it is needed.

in-line tuning—The method of tuning the intermediate-frequency strip of a superheterodyne receiver in which all the intermediate-frequency amplifier stages are made resonant to the same frequency.

inorganic electrolyte—A solution that conducts electricity due to the presence of ions of substances not of organic origin.

in phase—Two waves of the same frequency that pass through their maximum and minimum values of like polarity at the same instant, are said to be in phase.

in-phase portion of the chrominance signal—That portion of the chrominance signal having the same phase as, or exactly the opposite phase from, that of the subcarrier modulated by the I-signal. This portion of the chrominance signal may lead or lag the quadrature portion by 90 electrical degrees.

input—1. The current, voltage, power, or other driving force applied to a circuit or device. 2. The terminals or other places where current, voltage, power, or driving force may be applied to a circuit or device. 3. Data to be processed. 4. The process of transferring data from an external computer storage to an internal storage. 5. Terminals, jacks, or receptacles provided for the introduction of an electrical input signal into a tape recorder or other electrical component.

input admittance—1. The reciprocal of input impedance. 2. The admittance between the input terminals with the outputs shorted together.

input area—In a computer, the area of internal storage into which data from external storage is transferred.

input bias current—The current that must be supplied to each input of an IC op amp to assure proper biasing of the differential-input-stage transistors. In specification sheets, this term refers to the average of the two input bias currents. 2. One half the sum of the separate currents entering the two input terminals of a balanced amplifier.

input block—In a computer, a section of the internal storage reserved for receiving and processing input data.

input capacitance—The capacitance at the input terminals of a device.

input channel—A channel through which a state is impressed on a device or logic element.

input common-mode range—The maximum input that can be applied to either input of an operational amplifier without causing damage or abnormal operation.

input common-mode rejection ratio. – 1. The ratio of the change in input voltage to the corresponding change in output voltage, divided by the open-loop voltage gain. 2. The ratio of the full differential voltage gain to the common-mode voltage gain.

input device—The device or set of devices through which data are brought into another device.

input equipment—The equipment that introduces information into a computer.

input error voltage—The error voltage appearing across the input terminals of an operational amplifier when a feedback loop is applied around the amplifier.

input extender—A high-speed diode array used in a logic circuit when increased fan-in capability is required.

input gap—Also called buncher gap. In a microwave tube, the gap where the initial velocity modulation of the electron stream occurs.

input impedance – 1. The impedance a transducer presents to a source. 2. The effective impedance "seen looking into" the input terminals of an amplifier; circuit details, signal level, and frequency must be specified.

input impedance of a transmission line—The impedance between the input terminals with the generator disconnected.

input offset current—The difference between the input bias currents flowing into each input of an IC op amp, when the output of the op amp is at zero volts.

input offset voltage – That voltage which must be applied between the input terminals of an operational amplifier, through two equal resistances, to obtain zero output voltage.

input-output – Abbreviated i/o. 1. The transmission of information from an external source to the computer or from the computer to an external source. 2. A general term applied to the equipment used in communicating with a computer and the data involved in the computer.

input-output bound – *See* Input-Output Limited.

input-output limited—Pertaining to a system or condition in which the time taken by input and output operation exceeds the time for other operations. Also called input-output bound.

input-power rating—Also called coil rating. A statement of the allowable voltage, current, or power to the actuating element of a relay beyond which unsatisfactory performances will occur.

input process—1. The process in which a device receives data. 2. The transmission of data from peripheral equipment or external storage to internal storage.

input recorder–Any device which makes a record of an input electrical signal.

input register–In a computer, the register of internal storage able to accept information from outside the computer at one speed and supply the information to the computer calculating unit at another, usually much greater, speed.

input resonator–The buncher resonator in a velocity-modulated tube. It modifies the velocity of the electrons in the beam.

input sensitivity – The input signal level that will result in rated output of a piece of amplifying equipment. In preamplifiers, it is the signal that gives the rated voltage output of the preamplifier; in power amplifiers, the signal that gives the rated power output. (In preampliers, the phono sensitivity is commonly 1 millivolt; high-level inputs, such as tape and tuner, are commonly 250 millivolts. In power amplifiers, common values are between 0.5 and 1.0 volt.)

input transformer – A transformer that transfers energy from an alternating-voltage source to the input of a circuit or device. It usually provides the correct impedance match, also.

input uncertainty–In an operational amplifier, the algebraic sum of all the factors, including environmental and time effects, that contribute to the nonideal behavior of the input circuit.

input unit–In a computer, the unit that takes information from outside the computer into the computer.

input voltage drift–The change in output voltage of an operational amplifier divided by the open-loop gain, the quotient expressed as a function of temperature or time.

input voltage offset–The dc potential difference between the two inputs of a differential amplifier when the potential between the output terminals is zero.

input winding–*See* Signal Winding.

inquiry–1. The withdrawal of stored information from an electronic data processing system by interrogating the contents of the storage of a computer. 2. A technique for initiating the interrogation of the contents of the storage of a computer.

inquiry station–A remote terminal from which an inquiry may be sent over a wire line to a computer.

inquiry unit–A device used to extract a quick reply to a random question regarding information in a computer storage.

inrush current–In a solenoid or coil, the steady-state current drawn from the line with the armature in its maximum open position.

insert core–An iron core used generally for adjusting an inductor to a fixed frequency. It consists of a threaded metal insert molded or cemented into one or both ends of the core.

insert earphones–Small earphones which fit partially inside the ear.

insertion gain–The gain resulting from the insertion of a transducer in a transmission system is the ratio of the power delivered to that part of the system following the transducer to the power delivered to that same part before insertion. (If more than one component is involved in the input or output, the particular component used must be specified. This ratio is usually expressed in decibels.)

insertion loss–The difference between the power received at the load before and after the insertion of apparatus at some point in the line.

insertion phase shift–The change in phase of an electric structure when inserted into a transmission system.

insertion switch–A process by which information is inserted into a computer by the manual operation of switches.

inside lead–*See* Start Lead.

inside spider–A flexible device placed inside a voice coil to center it with the pole pieces of a speaker.

inspection chamber–In a spectrophotometer, the part in which the solution to be tested is placed for analysis.

inspectoscope–An instrument for viewing quartz crystals, while they are immersed in oil, to determine mechanical faults, the approximate direction of the optical axis, and regions of optical twinning.

instability–1. The measure of the fluctuations or irregularities in the performance of a device, system, or parameter. 2. An undersired change that occurs over a period of time and that is not related to input, operating conditions, or load.

instantaneous automatic gain control – A portion of a radar system that automatically adjusts the gain of an amplifier for each pulse so that there is a substantially constant output-pulse peak amplitude with different input-pulse peak amplitudes. The circuit is capable of acting during the time in which a pulse is passed through the amplifier.

instantaneous companding – Companding which varies the effective gain in response to instantaneous values of the signal wave.

instantaneous contacts–Contacts which are actuated immediately when a starting signal is applied to a timer.

instantaneous disc–A blank recording disc that can be played back on a phonograph immediately after being cut on a recorder.

instantaneous frequency – The rate at which the angle of a wave changes when the wave is a function of time. If the angle is measured in radians, the frequency in hertz is the rate of change of the angle divided by 2π.

instantaneous overcurrent relay – Also called rate-of-rise relay. A device which functions instantaneously on an excessive value of current, or on an excessive rate of current rise, thus indicating a fault in the apparatus of the circuit being protected.

instantaneous power–The power at the points where an electric circuit enters a region. It is equal to the rate at which the circuit is transmitting electrical energy into the region.

instantaneous power output–The rate at which energy is delivered to a load at a particular instant.

instantaneous readout–Readout by a radio transmitter at the instant the information to be transmitted is computed.

instantaneous recording – A recording intended for direct reproduction without further processing.

instantaneous sampling–The process of obtaining a sequence of instantaneous values of a wave. These values are called instantaneous samples.

instantaneous sound pressure – The total instantaneous pressure at a certain point, minus the static pressure at that point. The most common unit is the microbar.

instantaneous speech power – The rate at which the speaker is radiating sound energy at any given instant.

instantaneous value – The magnitude, at any particular instant, of a varying value.

instruction – 1. Information which, when properly coded and introduced as a unit into a digital computer, causes the computer to perform one or more of its operations. All instruction commonly includes one or more addresses. 2. A binary code applied to a logic circuit to affect its mode of operation. 3. A statement that specifies an operation and the values or locations of its operands. In this context, the term "instruction" is preferable to the terms "command" or "order," which are sometimes used synonymously. 4. A set of bits that defines a computer operation, and is a basic command understood by the CPU. It may move data, do arithmetic and logic functions, control i/o devices, or make decisions as to which instructions to execute next.

instructional constant–Also called pseudo instruction. In a computer, data stored in the program or instructional area which will be used only as a test constant.

instruction code–The list of symbols, names, and definitions of the instructions which are intelligible to a given computer or computing system.

instruction counter–*See* Control Counter.

instruction cycle–The process of fetching an instruction from memory and executing it.

instruction deck–A set of punched cards containing a symbolic coded program to be read into a computer.

instruction fetch–*See* Fetch.

instruction length–The number of words needed to store an instruction. It is one word in most computers, but some will use multiple words to form one instruction. Multiple-word instructions have different instruction execution times depending on the length of the instruction.

instruction modification–A change in the operation-code portion of a computer instruction or command such that, if the routine containing the instruction or command is repeated, the computer will perform a different operation.

instruction register – In a computer the register that temporarily stores the instruction currently being performed by the control unit of the computer.

instruction set – A means of describing computer capability. It consists of a listing of all the instructions the computer can execute. An instruction usually specifies an operation such as "add," and the data on which the operation is to be performed.

instruction storage–The storage medium that contains basic machining instructions in coded form.

instruction time – The time required to fetch an instruction from memory and then execute it.

instruction word–*See* Word.

instrument – A device capable of measuring, recording, and/or controlling.

instrument approach–A blind landing–i.e., solely by navigational instruments, without visual reference to the terrain.

instrument-approach system – In navigation, a system furnishing vertical and horizontal guidance to aircraft during descent. Touchdown requires some other guidance.

instrumentation – The use of devices to measure the values of varying quantities, usually as part of a system for keeping the quantities within prescribed limits.

instrument chopper – A vibrating switch used for modulating, demodulating, and switching dc or low-frequency ac information in instrumentation. It is driven synchronously from an ac or pulsating-dc source. The driven switching circuit is designed for low-level (0- to 10-volt) signal information.

instrument error–The inaccuracy of an instrument.

instrument flight–A blind flight–i.e., one in which the pilot controls the path and altitude of the aircraft solely by instrument.

instrument lamp—A light that illuminates or irradiates an instrument.

instrument-landing station—A special radio station for aiding in landing aircraft.

instrument-landing system — Abbreviated ILS. A radionavigation system intended to aid aircraft in landing. It provides lateral and vertical guidance, including distance from the landing point. Consists of four ground radio transmitting stations at and in the vicinity of an airport, which radiate direction and position signals to approaching aircraft. The signals are received on an instrument in the aircraft and alert the pilot to any deviation from the safe approach path to the correct touchdown point.

instrument multiplier—*See* Voltage-Range Multiplier.

instrument relay—A relay which operates on the principles employed in such electrical measuring instruments as the electrodynamometer, iron-vane, and D'Arsonval meters.

instrument shunt—An internal or external resistor connected in parallel with the circuit of an instrument to extend its current range.

instrument switch—A switch disconnecting an instrument or transferring it from one circuit or phase to another.

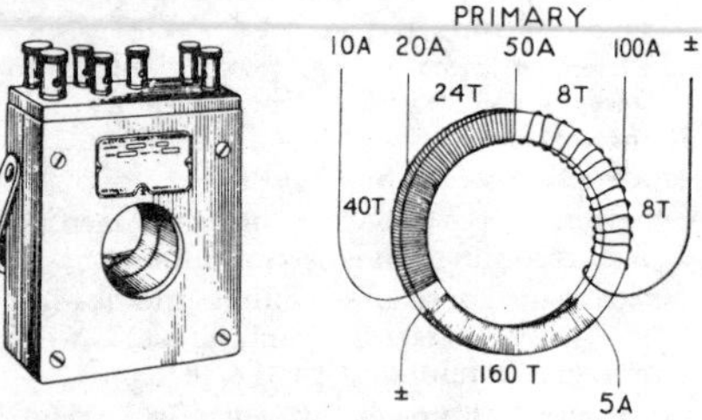

Instrument transformers.

instrument transformer — A transformer which reproduces, in its secondary circuit, the primary current (or voltage) with its phase relationship substantially preserved, suitable for utilization in measurement, control, or protective devices.

instrument-transformer correction factor—The factor by which a wattmeter reading must be multiplied to correct for the effect of the instrument-transformer ratio correction factor and phase angle.

insulated—Separated from other conducting surfaces by a nonconductive material offering a high, permanent resistance to the passage of current and disruptive discharge.

insulated carbon resistor—A carbon resistor encased in fiber, plastic, or other insulation.

insulated clip — A clip terminating in an insulated eye through which flexible cords or wires may be run and supported.

insulated enclosure — A special shielded enclosure design providing insulation against weather or providing maximum temperature stability. Usually prefabricated as an exterior building panel in modular construction.

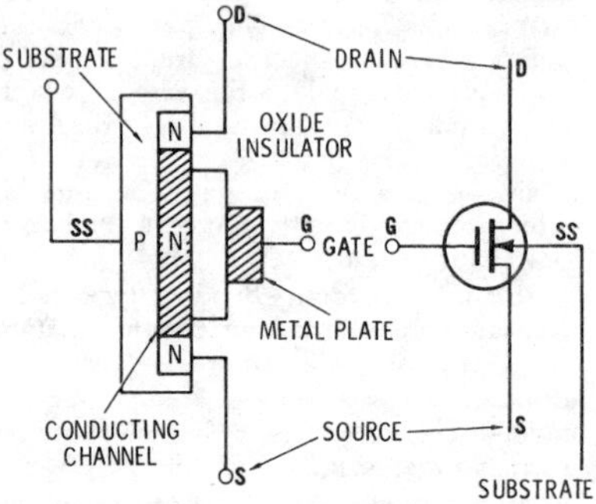

Insulated gate field-effect transistor.

insulated-gate field-effect transistor — In general, any field-effect transistor that has an insulated gate regardless of the fabrication process. Abbreviated IGFET.

insulated-substrate monolithic circuit—An integrated circuit which may be either an all-diffused device or a compatible structure so constructed that the components within the silicon substrate are insulated from one another by a layer of silicon dioxide, instead of the reverse-biased pn junctions used for isolation in other techniques.

insulated terminals — Solderless terminals provided with an insulated sleeve over the barrel to prevent a short circuit.

insulated wire—A conductor covered with a nonconductive material.

insulating material—1. A material on or through which essentially no current will flow. It is used to confine the flow of current within a conductor or to eliminate the shock hazard of a bare conductor. 2. Any composition primarily adapted for preventing the transfer of electricity therethrough, the useful properties of which depend on its chemical composition, or atomic arrangement.

insulating strength — The measure of the ability of an insulating material to withstand electrical stress without breaking down. It is defined in terms of the voltage per unit thickness necessary to initiate a disruptive discharge and usually is measured in volts per centimeter. *See also* Dielectric Strength *and* Electric Strength.

insulating tape — Tape that is wrapped around joints in insulated wires or cables. It is impregnated with an insulating material and covered with adhesive on one side.

insulating varnish – A varnish applied to coils and windings to improve their insulation (and, at times, their mechanical rigidity).

insulation – 1. A nonconductive material that prevents the leakage of electricity from a conductor, provides mechanical spacing or support, or protects against accidental contact. 2. The use of a material that passes negligible current to surround or separate a conductor to prevent loss of current.

insulation rating – The dielectric-strength and insulation-resistance values required to ensure satisfactory performance.

insulation resistance–1. The resistance offered by an insulating material to the flow of current resulting from an impressed dc voltage. 2. The ratio of the voltage applied between two electrodes in contact with a specific insulator to the total current between the electrodes. 3. The ratio of the direct-current voltage applied to the terminals of a capacitor and the resulting leakage current after the initial charging current has ceased. It includes both the volume and surface resistance. Industrial specifications usually call for a certain minimum value (several thousand megohms) determined with a specific voltage applied.

insulation resistivity–The insulation resistance per unit volume of insulation.

insulation system–All of the insulation materials used to insulate a particular electrical or electronic product.

insulator–1. A material in which the outer electrons are tightly bound to the atom and are not free to move. Thus, there is negligible current through the material when a voltage is applied. The resistivity is greater than 10^8 ohm-cm and generally decreases when the temperature rises. 2. A nonconducting substance such as porcelain, plastic, glass, rubber, etc.

insulator arcing ring – A circular or oval metal part placed at one or both ends of an insulator to prevent current from arcing over and damaging it and/or the conductor.

insulator arc-over–The flow of power current over an insulator in the form of an arc following a surface discharge.

insulazing–*See* Surface Insulation.

insulectrics–The science encompassing insulating materials in electrical insulation.

integer – A whole number, not fractional or mixed (e.g., 1, 2, 9, 100, etc.–not 1.1 or ½).

integral-cavity, reflex-klystron oscillator – A reflex-klystron oscillator in which tuning is accomplished by changing the physical dimensions of the resonant cavity. It is usually referred to as a diaphragm- or grid-gap–tuned klystron, since a flexible diaphragm is used to change the cavity dimension, i.e., the gap between the cavity grids.

integral circuit packages–Microcircuits assembled from discrete components and all circuits created essentially in an active or passive substrate.

integral contact–Current-carrying member of jack, switch, or relay. Usually a flat, flexible spring or other conducting member having no separate contacts attached at point of mating.

integral-external–cavity reflex oscillator – A reflex-klystron oscillator in which a fixed internal cavity is tightly coupled to a permanently attached external cavity. Tuning is achieved by varying a reactance probe in the external cavity.

integral-horsepower motor–A motor which is built into a frame and has a continuous rating of one horsepower.

integral resistor–An internal or external resistor preconnected to the electrical element and forming an integral part of the cup assembly to provide a desired electrical characteristic of a precision potentiometer.

integrated–A type of design in which two or more basic components or functions are physically, as well as electrically, combined–usually on one chassis, such as an integrated amplifier.

integrated amplifier – An amplifier which embodies in a common housing the preamplifier and control section and the power amplifier. Some early amplifiers of large power were in two separate units, one the control unit with preamplifiers and the other the power amplifier. (*See also* Amplifier.)

integrated circuit–Abbreviated IC. 1. A combination of interconnected circuit elements inseparably associated on or within a continuous substrate. (*See also* Monolithic Integrated Circuit, Multichip Integrated Circuit, Film Integrated Circuit *and* Hybrid Integrated Circuit.) To further define the nature of an integrated circuit, additional modifiers may be prefixed. Examples are: dielectric isolated monolithic integrated circuit, beam-lead monolithic integrated circuit, or silicon chip-tantalum thin-film hybrid integrated circuit. 2. Any electronic device in which both active and passive elements are contained in a single package. The term frequently is used for circuits other than those containing semiconductors; for example, microwave designers consider many types of waveguide assemblies to be integrated circuits. 3. A small chip of solid material (generally a semiconductor) upon which, by various techniques, an array of active and/or passive components has been fabricated and interconnected to form a functioning circuit. Integrated circuits, which are generally encapsulated with only input, output, power supply, and control terminals accessible, offer great advantages in terms of small size, economy, and reliability. 4. The physical realization of a number of electrical circuit elements inseparably associated on or within a continuous body of semiconductor material to perform the function of a circuit. 5. An electronic device containing several elements, active or passive, that perform all or part of a circuit function. 6. An interconnected array of conventional components—transistors, diodes, capacitors, and resistors—fabricated *in-situ* within and on a single crystal of semiconductor material with the capability of performing a complete electronic circuit function. Also called functional device. 7. An electronic circuit containing transistors, diodes, resistors, and perhaps capacitors and photocells, along with interconnecting electrical conductors processed and contained entirely within a single chip of silicon.

integrated-circuit array – Multiple integrated circuits formed on a common substrate and electrically interconnected during fabrication.

is especially significant; more specifically, that portion of the art dealing with integrated circuits.

integrated electronic system – *See* Integrated Circuit.

integrated equipment components – Integrated-circuit chips which contain a complete logic function. Abbreviated IECs.

integrated microcircuit – *See* Integrated Circuit.

integrated morphology – The structural characterization of an electronic component in which the identity of the current or signal modifying areas, patterns, or volumes has become lost in the integration of electronic materials, in contrast to an assembly of devices performing the same function.

integrated transducers – Semiconductor components that change the form of energy (e.g., piezoelectric devices, photogenerators, thermistors, etc.) and that are integrated into multifunction chips.

integrated voltage regulator – An integrated structure that serves as the reference, error amplifier, and shunt elements for shunt voltage regulation. A resistive voltage divider connected to its input provides adjustment of the output voltage. Abbreviated IVR.

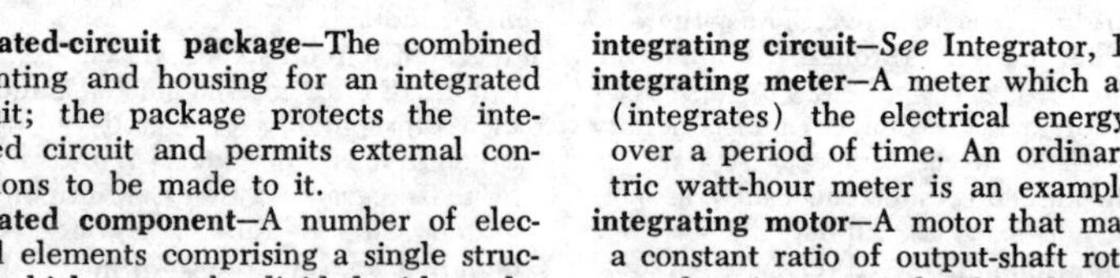

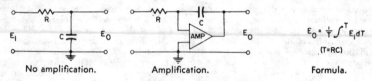

Integrating circuits.

integrated-circuit package—The combined mounting and housing for an integrated circuit; the package protects the integrated circuit and permits external connections to be made to it.

integrated component—A number of electrical elements comprising a single structure which cannot be divided without destroying its stated electronic function.

integrated data processing – A method of transforming disjointed and repetitive paperwork tasks into a correlated and mechanized production of information for any purpose.

integrated electronic component – An assembly that consists of several integrated circuits interconnected on a single chip of silicon to provide a complete electronic function with a circuit content greater than 10 equivalent gates. Abbreviated IEC.

integrated electronics – That portion of electronic art and technology in which the interdependence of material, device, circuit, and system-design considerations

integrating circuit—*See* Integrator, 1.

integrating meter—A meter which adds up (integrates) the electrical energy used over a period of time. An ordinary electric watt-hour meter is an example.

integrating motor—A motor that maintains a constant ratio of output-shaft rotational speed to input signal. Thus the angle of rotation of the shaft is proportional to the time integral of the input signal.

integrating photometer – A photometer which, with a single reading, indicates the average candlepower from a source in all directions or at all angles in a single plane.

integrating relay—A relay that sums up the inputs of voltage or current supplied to it and opens or closes its contacts in response to the input so integrated.

integrating-sphere densitometer—A photoelectric instrument that measures the density of motion-picture film or its sound track.

integration time—That time during which all electrons formed by impinging photons are gathered in a potential well under an energized electrode.

integrator—1. A device with an output proportionate to the integral of the input signal. 2. In certain digital machines, a device that numerically approximates the mathematical process of integration. 3. A device that determines on a continuous basis the total value of a quantity being measured, usually as a function of time. 4. Device for summing or totalizing counts, areas under curves, etc.

intelligence bandwidth — The total audio (or video) frequency bandwidths of one or more channels.

intelligence sample—Part of a signal taken as evidence of the quality of the whole.

intelligence signal—Any signal which conveys information (e.g., voice, music, code, or television).

intelligent terminal—Also called smart terminal. A programmable remote terminal with preprocessing capability. This requires memory and logic capability. For example an intelligent crt (cathode ray tube with typewriter) terminal features insertion or deletion of lines and individual characters to occur in the remote terminal. A data terminal containing a minicomputer to reduce the data transmitted and to expand the data received.

intelligibility—*See* Articulation.

intensifier electrode—Also called post-accelerating electrode. In some types of electrostatic cathode-ray tubes, an electrode which permits additional acceleration of the electron beam after it has been deflected. This electrode permits greater intensity of the trace without materially reducing the deflection sensitivity of the tube.

intensifying screen — A thin fluorescent screen placed next to a photographic plate to increase the effect of radiation on the plate.

intensitometer—Also called a dosage meter or dosimeter. An instrument that estimates the amount of X-ray radiation, for determining the duration of exposure during X-ray pictures or therapy.

intensity—1. The strength of a quantity. 2. The relative strength, or amplitude, of electric, magnetic, or vibrational energy. 3. The brilliance of an image on the screen of a cathode-ray tube. 4. The strength of light or other electromagnetic energy being radiated or reflected per steradian.

intensity control — Used with cathode-ray tubes to control the intensity of the electron beam and hence the amount of light generated by the fluorescent screen. Generally, the grid bias of the tube is regulated.

intensity level—Ratio of the intensity of the sound to a reference intensity of a free plane wave of 1 microwatt per square centimeter under normal conditions. Commonly expressed in decibels.

intensity modulation—The process and/or effect of varying the electron-beam current in a cathode-ray tube, resulting in varying brightness or luminance of the trace. (*See also* Z-Axis Modulation.)

intensity of radiation—The radiant energy emitted in a specified direction per unit time, per unit area of surface, per unit solid angle.

interaction—The effects two or more parts, components, etc., have on each other while each is performing a function.

interaction cross talk—Cross talk resulting from mutual coupling between two paths by means of a third path. For example, if a signal on pair 1 is coupled to pair 2, and then coupled from pair 2 to pair 3, where it is measured, it is known as interaction cross talk.

interaction loss (of a transducer) — Expressed in decibels, it equals 20 times the logarithm (to the base *10*) of the scalar value of the reciprocal of the interaction factor.

interaction space—In an electronic tube, the region where the electrons interact with an alternating electromagnetic field.

interactive graphics—The use of a large-screen, high-precision crt and its associated circuitry—usually linked to a large-scale computer system through a small control computer—on which both alphanumeric and vector data are displayed and manipulated. By using the data entry device, two- and three-dimensional geometric designs can be created, deleted, and modified in real time to achieve desired results.

interactive processing mode — A type of process in which the programmer or computer user can communicate with the machine and modify its operation as ideas are suggested by the results of the computation.

interactive system—A system in which it is possible for the human user or the device serviced by the computer to communicate directly with the operating program. For human users, this arrangement is termed a conversational system.

interaxis error — The deviation from 90° perpendicularity of one set of resolver windings when excitation is applied to one of the other windings. For rotor interaxis error one stator winding is excited; for stator interaxis error one rotor winding is excited.

interbase current — In a junction tetrode transistor, the current that flows from one base connection to the other through the base region.

interbase resistance — Resistance between Base 2 and Base 1 of a UJT measured at a specified interbase voltage with IE = 0.

interblock space—*See* Interrecord Gap.

intercarrier noise suppression—The means of suppressing the noise resulting from increased gain when a high-gain receiver with automatic volume control is tuned between stations. The suppression circuit automatically blocks the audio-frequency output of the receiver when there is no signal at the second detector.

intercarrier sound system—A television receiving system in which use of the picture carrier and the associated sound-channel carrier produces an intermediate frequency equal to the difference between the two carrier frequencies. This intermediate frequency is frequency-modulated in accordance with the sound signal.

intercellular massage—The ultrasonic stimulation of body cells. Sometimes called micromassage.

intercepting—Routing of a call or message placed for a disconnected or nonexistent destination to an operator position or a specially designated terminal or machine answering device.

intercepting trunk—A trunk to which a call made to a vacant number, a changed number, or a line out of order is connected so that action may be taken by an operator.

intercept receiver—Also called search receiver. A specially calibrated receiver which can be tuned over a wide frequency range in order to detect and measure enemy rf signals.

intercept service—In a telephone system, a service provided to subscribers whereby calls to disconnected stations or dead lines are either routed to an intercept operator for explanation or the calling party receives a distinctive tone signal to indicate that he has made such a call.

intercept tape—A tape used for temporary storage of messages intended for trunk channels and tributary stations in which there is equipment or circuit trouble.

intercept trunk—*See* Intercepting Trunk.

intercharacter space — In telegraphy, the space between characters of a word. It is equal to three unit lengths.

intercom—*See* Intercommunication System.

intercommunication apparatus—Equipment and systems for paging and intercommunication within a building, including audio, bell systems, pillow systems, and pocket page systems.

intercommunication system—Also called intercom. A two-way communication system without a central switchboard, usually limited to a single vehicle, building, or plant area. Stations may or may not be equipped to originate a call, but can answer any call.

interconnecting wire—Wires used for connections between subassemblies, panels, chassis, and remotely mounted devices. Does not necessarily apply to the internal connections of these units.

interconnection—1. Also called tie line. A transmission line connecting two electric systems or networks and permitting energy to be transferred in either direction. Larger interconnections are often called interties, giant ties, or regional interconnections. 2. Also called intraconnection. The physical wiring between components (outside a module), between modules, between units, or between larger portions of a system or systems.

interconnection diagram — Diagram showing the identity of all units in a piece of electronic equipment and the connections between them.

interconnections (microelectronic)—Those conductors and connections which are not in continuous integral contact with the substrate or circuit elements of an integrated circuit.

interconnection system—The electrical and mechanical interconnection of any one or all of the six levels of interconnections generally common to electronic equipment. The six levels of interconnection are: intramodule, module to motherboard, intramotherboard, motherboard to back panel, backpanel wiring, and input/output.

interdigital magnetron—A magnetron with anode segments around the cathode. Alternate segments are connected together at one end, and remaining segments at the opposite end.

interdigital transducer—Abbreviated IDT. A number of interleaved metal electrodes whose width and spacing is equal and uniform throughout the transducer pattern. When a harmonic voltage is applied to the transducer terminals, the IDT pattern excites a periodic electric field which penetrates into the piezoelectric substrate. The substrate responds by periodically expanding and contracting in unison with these fields. With the proper choice of substrate orientation, this piezoelectric excitation gives rise to surface acoustic waves which propagate in the two directions normal to the IDT electrodes. The electric field can be produced by applying signals of opposite potential to two parallel metal electrodes formed in films deposited on the surface of the crystal. As the field is applied, the alternating signals send an acoustic wave across its surface and this wave is reconverted to an electric signal at a second pair of similar electrodes at the other end of the crystal.

In practice, the transducer consists of several pairs of interleaved parallel electrodes which gives the transducer its name. The fingers, and the space between them, must be related to the size of a wave of the frequency desired. In certain applications the transducer fingers are less than a thousandth of a millimeter wide.

interelectrode capacitance – The capacitance between one electron-tube electrode and the next electrode toward the anode.

interelectrode coupling – Capacitive feedback from the plate of a tube to the grid. In triodes, this limits the maximum amplification possible without starting oscillation.

interelectrode leakage–The undesired current which flows between elements not normally connected in any way.

interelectrode transit time – The time required for an electron to travel between two electrodes.

interelement capacitance–The capacitance caused by the pn junctions between the regions of a transistor and measured between the external leads of the transistor.

interexchange channel–A channel connecting two different exchange areas.

interface–1. A point or device at which a transition between media, power levels, modes of operation, etc., is made. 2. The two surfaces on the contact sides of mating connectors that face each other when mated. 3. A common boundary between two or more items. May be mechanical, electrical, functional, or contractual 4. A common aspect at the boundary between two systems involving intersystem communication–e.g., the interaction between research and development, basic and applied science, or engineering and systems development. 5. The physical and space boundary surrounding the system, subsystem, equipment, or component, through which all environmental and operational stimuli essential to the device or affecting its proper operation must propagate or interact with other related devices or structures. 6. The hardware for linking two units of electronic equipment, for example, a hardware component to link a computer with its input (or output) device. 7. The means of connection between two logic elements, often elements that belong to two different "families."

interface circuit–A circuit that links one type of logic family with another or with analog circuitry.

interface resistance – *See* Cathode Interface.

interface unit–A device that translates incoming signals that are incompatible with the electrical characteristics of the computer without changing the information content. Also translates outgoing signals for the benefit of associated equipment that is designed to different electrical standards.

interfacial connection–In a printed-circuit board, a conductor that connects conductive patterns on opposite faces of the base.

interference–Any electrical or electromagnetic disturbance, phenomenon, signal, or emission, man-made or natural, which causes or can cause undesired response, malfunctioning, or degradation of the electrical performance of electrical and electronic equipment.

interference blanker–A device used with two or more pieces of radio or radar equipment to permit simultaneous operation without confusion of intelligence, or used with a single receiver to suppress undesired signals.

interference eliminator–A device designed for the purpose of reducing or eliminating interference.

interference fading – Fading produced by different wave components traveling slightly different paths in arriving at the receiver.

intererence filter–A device added between a source of man-made interference and a radio receiver to attenuate or eliminate noise signals. It generally contains a combination of capacitance and inductance.

interference guard band–*See* Guard Band.

interference pattern – 1. The resultant space distribution of pressure, particle velocity, or energy flux when progressive waves of the same frequency and kind are superimposed. 2. The pattern produced on a radar scope by interference signals.

interference prediction–Estimation of the interference level of a particular item of equipment with respect to its future electromagnetic environment.

interference-rejection unit–A tunable filter or wave trap capable of being adjusted to reject any frequency within the if passband of a receiver while allowing the remainder of the passband curve to remain intact. It is adjusted to reject an interference signal and thus constitutes a form of antijamming. Abbreviated IFRU.

interference source suppression – Techniques applied at or near a source of radiation to reduce its emission of undesired signals.

interference spectrum–The frequency distribution of the jamming interference.

interferometer–An apparatus that shows interference between two or more wave trains coming from the same luminous area, and also compares wavelengths with observable displacements of reflectors or other parts.

interferometer homing – A homing guidance system in which the direction of the target is determined by comparing the phase of the echo signal as received at more than one antenna.

interferometer system—A method of determining the azimuth of a target through use of an interferometer to compare the signal phases at the output terminals of two antennas receiving a common signal from a distant source.

interior label—In a computer, a magnetically recorded sequence added to a tape to identify the contents.

interior-wiring-system ground—The ground connection to one of the current-carrying conductors of an interior wiring system.

interlace—1. In a computer, to assign successive storage location numbers to physically separated storage locations on a magnetic drum. This serves to reduce access time. 2. To transmit different interrogation modes on successive sweeps.

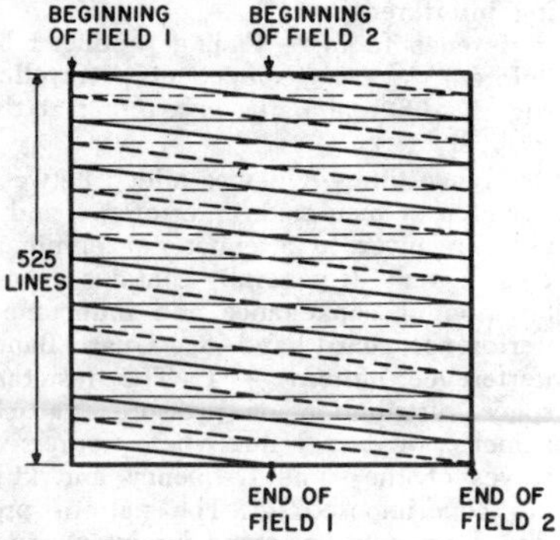

Interlaced scanning.

interlaced scanning—Also called line interlace. A system of scanning whereby the odd- and even-numbered lines of a picture are transmitted consecutively as two separate fields. These are superimposed to create one frame, or complete picture, at the receiver. The effect is to double the apparent number of pictures and thus reduce flicker.

interlace factor—A measure of the degree of interlace of normally interlaced fields.

interlace operation — A type of computer operation in which data can be read out of or copied into the memory without causing interference to the other activities of the computer. *See also* Interrupt *and* Time Sharing, 1.

interlacing—A method of scanning used in television, in which each picture is divided into two or more complete sets of interlacing lines to reduce flicker.

interleave—1. In a computer, to insert segments of one program into another program so that the two programs can be executed essentially simultaneously. 2. *See* interlace.

interleaving—Placing between. For example, in the transmission of a composite color signal, the bands of energy of the chrominance signal are interleaved with, or placed between, those of the luminance signal.

interlock—A device actuated by the operation of some other device with which it it directly associated, to govern succeeding operations of the same or allied devices. Interlocks may be either electrical or mechanical.

interlock circuit — A circuit in which a given action cannot occur until after one or more other actions have taken place. The interlocking action is generally obtained through the use of relays.

interlocking—The forcing of a voltage of one frequency to be in step with a voltage of another frequency.

interlock relay—A relay in which one armature cannot move or its coil be energized unless the other armature is in a certain position.

interlock switch—A safety switch that de-energizes a high-voltage supply when a door or other access cover is opened.

intermediate current—The range of current (milliamperes) at which formulation of carbonaceous material may significantly affect contact resistance.

intermediate frequency—1. Abbreviated if. A frequency to which a signal wave is shifted locally as an intermediate step in transmission or reception. 2. The fixed frequency resulting from heterodyning (i.e., "beating" or modulating to develop the sum or difference frequency signal) the incoming signal with a signal from the local oscillator. The if used in fm tuners is commonly 10.7 MHz and in am tuners is 455 kHz. The if signal is amplified in the if channel, and it is here where most of the selectivity is introduced by tuned bandpass transformers and/or crystal or ceramic filters.

intermediate-frequency amplifier — An amplifier tuned to a fixed frequency, or capable of single-control tuning over a range of frequencies, for the purpose of selecting one of the frequency components generated in a mixer circuit.

intermediate-frequency harmonic interference — Interference caused in superheterodyne receivers by the radio-frequency circuit accepting harmonics of the intermediate-frequency signal.

intermediate-frequency interference ratio—*See* Intermediate-Frequency Response Ratio.

intermediate-frequency jamming — A form of jamming in which two cw signals are transmitted at frequencies separated by an amount equal to the center frequency of the if amplifier in the radar receiver.

intermediate-frequency response ratio —

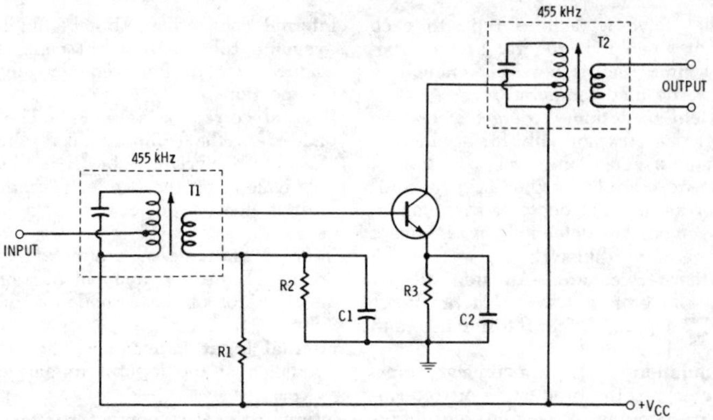

Intermediate-frequency amplifier.

Also called intermediate-frequency interference ratio. In a heterodyne receiver, the ratio of intermediate-frequency signal input at the antenna to the desired signal input for identical outputs.

intermediate-frequency strip—A subassembly containing the intermediate-frequency stages in a receiver.

intermediate-frequency transformer — A transformer designed for use in the intermediate-frequency amplifier of a superheterodyne receiver.

intermediate - frequency — transformer - lead color code — Transformer leads in many radio receivers are identified by the following standard EIA colors:

Blue—plate.	Green—grid or diode.
Red—B+.	Black—grid return.

Green-black—second diode (full-wave transformers only).

intermediate horizon — A screening object (such as a hill, mountain, ridge, building, etc.) similar to the radar horizon, but nearer to the radar site. For example, a distant mountain range might be the radar horizon on a given azimuth while a nearer, lower ridge might screen a valley between it and the mountain range; the ridge would be an intermediate horizon.

intermediate means — All system elements needed to perform distinct operations in the measurement sequence between the primary detector and the end device.

intermediate-range ballistic missile—A tacticle missile or rocket weapon with a range of 200 to 1500 miles.

intermediate repeater — A repeater used other than at the end of a trunk or line.

intermediate state—The partial superconductivity that occurs when a magnetic field of appropriate strength is applied to a sphere of material below its critical temperature (i.e., the temperature below which the material superconducts if no magnetic field were present).

intermediate storage — The portion of the computer storage facilities in which information in the processing stage usually is stored.

intermediate subcarrier—A carrier used for modulating a carrier or another intermediate subcarrier. It also may have been modulated by one or more subcarriers.

intermediate switching region—In a relay, an area between low level (including dry circuit) and power switching (including full rated load) where the contact arc does not destroy deposits which are byproducts of the switching function.

intermediate trunk distributing frame — A frame in which are mounted terminal blocks for connecting linefinders and first selectors.

intermittent—1. Occurring at intervals. 2. Electrical connections when conducting paths alternately open and close at some essentially uncontrolled rate. (Intermittents are undesirable since continuous connections are normally required.)

intermittent current—A unidirectional current that is interrupted at intervals.

intermittent defect—A defect that depends on variable conditions in a circuit. Hence it is not present at all times.

intermittent duty — Operation for specified alternate intervals of load and no-load; load and rest; or load, no-load, and rest.

intermittent-duty rating — The output rating of a device operated for specified intervals rather than continuously.

intermittent-duty relay — A relay which must be de-energized at intervals to avoid excessive temperature, or a relay that is energized at regular or irregular intervals, as in pulsing.

intermittent pulsing — The transmission of short bursts of radiation at irregular intervals.

intermittent rating — The permissible output of a piece of apparatus when it is operated for alternate periods of load and rest that have a definite ratio to each other, or when it is run for a stated period of time that is not long enough to produce the final temperature.

intermittent reception—A defect where the receiver operates normally for a while, at regular or irregular intervals.

intermittent scanning – One or two 360° scans of an antenna beam at irregular intervals to make detection by intercept receivers more difficult.

intermittent-service area—An area still receiving the ground wave of a broadcast station, but subject to interference and fading.

intermodulation—1. In a nonlinear transducer element, the production of frequencies corresponding to the sums and differences of the fundamentals and harmonics of two or more frequencies transmitted through the transducer. 2. An error form that occurs in chopper-stabilized amplifiers when a beat component forms between the chopper drive frequency and normal signals which have frequencies near that of the chopper.

intermodulation distortion—1. Nonlinearity characterized by the appearance of frequencies in the output equal to the sums and differences of integral multiples of the component frequencies present in the input signal. (Harmonics are usually not included.) 2. Abbreviated imd. The introduction of unwanted signal energy as the result of interaction between two or more simultaneously reproduced tones, causing a smearing or veiling of the sound. All recording and amplifying equipment produces a certain amount of intermodulation distortion, but it can be held to sufficiently low levels to be below the threshold of audibility.

intermodulation frequencies—The sum and difference frequencies generated in a nonlinear element.

intermodulation interference—The combination-frequency tones produced at the output by a nonlinear amplifier or network when two or more sinusoidal voltages are applied at the input. Generally expressed as the ratio of the root-mean-square voltage of one or more combination frequencies to that of one of the parent frequencies measured at the output.

intermodulation noise—Noise introduced in the channel of interest by signals being transmitted in other channels.

internal arithmetic—Any computations performed by the arithmetic unit of a computer, as distinguished from those performed by peripheral equipment.

internal calibration—Calibration by an internal voltage source (provided with the instrument) rather than an external standard.

internal connection—Abbreviated IC. In a vacuum tube, a base-pin connection designed not to be used for any circuit connections.

internal correction voltage – The voltage added to the composite controlling voltage of an electron tube. It is the voltage equivalent of those effects produced by initial electron velocity, contact potential, etc.

internal graticule—A graticule the rulings of which are a permanent part of the inner surface of the cathode-ray tube faceplate.

internal input impedance—The actual impedance at the input terminals of a device.

internally caused contact chatter – That chatter resulting from the operation or release of the relay. It may be classified as initial, armature-impact, armature-rebound, or armature-hesitation chatter.

internally stored program—A sequence of instructions (program) stored inside a computer in the same storage facilities as the computer data, as opposed to being stored externally on punched paper tape, pin boards, etc.

internal magnetic recording—Storage of information within the material itself, such as used in magnetic cores.

internal memory—Also called internal storage. The total memory or storage which is automatically accessible to a computer. It is an integral physical part of the computer and is directly controlled by it.

internal output impedance—The actual impedance at the output terminals of a device.

internal resistance—The effective series resistance in a source of voltage.

internal storage—1. Storage facilities in a computer forming an integral physical part of and directly controlled by the computer. Also called main memory and core memory. 2. The total storage automatically available to the computer.

international broadcast station—A station licensed for transmission of broadcast programs for international public reception. By international agreement, frequencies are allocated between 6000 and 26,600 kHz.

international call sign—The identifying letters and numbers assigned to a radio station in accordance with the International Telecommunications Union. The first character, or the first two, identify the nationality of the station.

international code signal—A code, adopted by many nations for international communication, in which combinations of letters are used in lieu of words, phrases, and sentences.

international communication service – A telecommunication service between offices or stations (including mobile) belonging to different countries.

international control station—A fixed station in the Fixed Public Control Service, directly associated with the International Fixed Public Radio Communication Service.

international coulomb — The quantity of electricity passing any section of an electric circuit in one second when the current is one international ampere. One international coulomb equals 0.99985 absolute coulomb.

international farad—The capacitance of a capacitor when a charge of one international coulomb produces a potential difference of one international volt between the terminals. One international farad equals 0.99952 absolute farad.

International Fixed Public Radiocommunication Service — A fixed service, the stations of which are open to public correspondence, intended to provide radiocommunication between the United States or its territories and foreign points.

international henry—The inductance which produces an electromotive force of one international volt when the current is changing at a rate of one international ampere per second. One international henry equals 1.00018 absolute henrys.

international joule—The energy required to transfer one international coulomb between two points having a potential difference of one international volt. One international joule equals 1.00018 absolute joules.

International Morse code—Also called Continental code. A system of dot-and-dash signals used chiefly for international radio and wire telegraphy. It differs from American Morse code in certain code combinations only.

international ohm—The resistance at 0°C of a column of mercury of uniform cross section 106.300 centimeters in length and with a mass of 14.4521 grams. One international ohm equals 1.00048 absolute ohms.

International Radio Consultative Committee—Abbreviated CCIR. An international committee which studies technical operating and tariff questions pertaining to radio, broadcast television, and multichannel video transmissions, and issues recommendations. It reports to the International Telecommunications Union.

international radio silence — Three-minute periods of radio silence, commencing 15 and 45 minutes after each hour, on a frequency of 500 kHz only. During this time all radio stations are supposed to listen on that frequency for distress signals of ships and aircraft.

international radium standard—A standard of radioactivity, consisting of 21.99 milligrams of pure radium chloride.

international system (of electrical and magnetic units)—A system for measuring electrical and magnetic quantities by using four fundamental quantities. Resistance and current are arbitrary values that correspond approximately to the absolute ohm and the absolute ampere. Length and time are arbitrarily called centimeter and second. The international system of electrical units was used between 1893 and 1947. By international agreement, it was discarded on January 1, 1948, in favor of the mksa (Giorgi) system.

international telecommunication service — A telecommunication service between offices or stations in different states, or between mobile stations that are not in the same state or that are subject to regulation by different states.

International Telecommunication Union—The United Nations specialized agency that deals with telecommunications. Its purpose is to provide standardized communications procedures and practices, including frequency allocation and radio regulations on a worldwide basis.

International Telegraph Consultative Committee — Abbreviated CCIT. An international committee responsible for studying technical operating and tariff questions pertaining to telegraph and facsimile and issuing recommendations. It reports to the International Telecommunications Union.

International Telephone Consultative Committee — Abbreviated CCIF. An international committee responsible for studying and issuing recommendations regarding technical operations and tariff questions pertaining to ordinary telephones; carrier telephones; and music, picture, television, and multichannel telegraph transmission over wire line. It reports to the International Telecommunication Union.

international temperature scale—A temperature scale adopted in 1948 by international agreement. Between the boiling point of oxygen (−182.97°C) and 630.5°C it is based upon the platinum resistance thermometer. From 630.5°C to 1063.0°C it is based on the platinum rhodium thermocouple, and above 1063.0°C on the optical pyrometer.

international volt — The voltage that will produce a current of one international ampere through a resistance of one international ohm. One international volt equals 1.00033 absolute volts.

international watt — The power expended when one international ampere flows between two points having a potential difference of one international volt. One international watt equals 1.00018 absolute watts.

interoffice trunk – The telephone channel between two central offices.

interphase transformer – An autotransformer or a set of mutually coupled reactors used with three-phase rectifier transformers to modify current relationships in the rectifier system and thereby cause a greater number of rectifier tubes to carry current at any instant.

interphone – A telephone communication system wholly contained within an aircraft, ship, or activity.

interphone system–An intercommunication system like that in an aircraft or other mobile unit.

interpolation – The process of finding a value between two known values on a chart or graph.

interpole–A small auxiliary pole placed between the main poles of a direct-current generator or motor to reduce sparking at the commutator.

interposition trunk – A trunk connecting two positions of a large switchboard so that a line on one position can be connected to a line on the other position.

interpreter – 1. A punch-card machine which will read the information conveyed by holes punched in a card and print its translation in characters arranged in specified rows and columns on the card. 2. A computer executive routine by which a stored program expressed in pseudocode is translated into machine code as a computation progresses, and the indicated operation is performed by means of subroutines as they are translated.

interpreter code–A computer code which an interpretive routine can use.

interpretive programming–The writing of computer programs in a pseudo machine language, which the computer precisely converts into actual machine-language instructions before performing them.

interpretive routine–Computer routine designed to transfer each pseudocode and, using function digits, to set a branch order that links the appropriate subroutine into the main program.

interrecord gap – Also called interblock space. The space between records on magnetic tape caused by delays involved in starting and stopping the tape motion. This gap is used to signal that the end of a record has been reached.

interrogation – The triggering of one or more transponders by transmitting a radio signal or combination of signals.

interrogation signal–A pulsed or cw signal emitted to initiate a reply signal from a transponder or responder.

interrogation suppressed time delay–The overall fixed time that elapses between transmission of an interrogation and reception of the reply to this interrogation at zero distance.

interrogator–Also called challenger. A radio transmitter used to trigger a transponder.

interrogator-responser – A combined radio transmitter and receiver for interrogating a transponder and displaying the replies.

interrupt–1. In a computer, a break in the normal flow of a system or routine such that the flow can be resumed from that point at a later time. The source of the interrupt may be internal or external. 2. A method of stopping a process and identifying that a certain condition exists. In graphic systems, interrupts can originate from data entry devices, the display list, the host computer, the refresh clock, and display error conditions. When an interrupt occurs, the host computer and display refresh cease until the interrupt is answered and processed. At that time, the host computer will restart the refresh —usually from where it was halted. If a new display list is to be presented, the display starts at the beginning of the list.

interrupted continuous waves–Continuous waves that are interrupted at an audio-frequency rate.

interrupt enable–*See* Interrupt Mask.

interrupter–1. A magnetically operated device used for rapidly and periodically opening and closing an electric circuit in doorbells and buzzers and in the primary circuit of a transformer supplied from a dc source. 2. A device used to produce interrupted ringing cycles. It also may be employed with the release alarm to start signal alarm circuits of the switching equipment and thereby provide timed delay in the sounding of a failure alarm. 3. An electrical, electronic, or mechanical device that periodically interrupts a continuous current to produce pulses.

interrupter contacts–On a stepping relay, an additional set of contacts operated directly by the armature.

interrupting capacity – The maximum power in the arc that can be interrupted by a circuit breaker or fuse without the occurrence of restrike or violent failure. Rated in volt-amperes for ac circuits and in watts for dc circuits.

interrupting rating – Conditions under which the contact of a relay must interrupt with a prescribed duty cycle and contact life.

interrupting time–In a circuit breaker, the interval between the energizing of the trip coil and the interruption of the circuit, at the rated voltage.

interrupt mask–A mechanism which allows the program to specify whether or not interrupt request will be accepted. Also called interrupt enable.

interrupt request–A signal to a computer

that temporarily suspends the normal sequence of a routine and transfers control to a special routine. Operation can be resumed from this point later. Ability to handle interrupts is very useful in communication applications where it allows the microprocessor to service many channels.

interstage—Between stages.

interstage coupling – Coupling between stages.

interstage punching – A system in which only the odd-numbered rows are punched in the British standard card.

interstage transformer—A transformer that couples two stages together.

interstation noise suppression – Canceling of the noise which occurs when a high-gain radio receiver with automatic volume control is tuned between stations.

interstitial site—A position that is inside a crystal lattice but is not one of the proper sites ordinarily occupied by the atoms of the crystalline material. Impurity ions of the proper size can occupy such positions in a lattice that is otherwise regular.

intersymbol interference—1. In a transmission system, extraneous signal energy during one or more keying intervals which tends to hinder the reception of the signal in another keying interval. 2. The disturbance which results from this condition.

interties—*See* Interconnection, 1.

intertoll trunk—A trunk linking toll offices in different telephone exchanges.

interval calibration—*See* Step Calibration.

interval circuit—A circuit which is energized during timing only. This can be accomplished by using a timer with interval contacts; or by using a timer with delayed contacts in series with the start switch, or one with instantaneous contacts in series with delayed contacts.

interval contacts – In a timer, contacts which are actuated only for the duration of the preset time interval.

interval timer—1. A device for measuring the time interval between two actions. 2. A timer which switches electrical circuits on or off for the duration of the preset time interval.

interword space—In telegraphy, the space between words or coded groups. It is equal to seven unit lengths.

intonation – The slight modification of pitch, or frequency, that makes a note sound flat or sharp compared with the natural frequency of the note played.

intracardiac—Pertaining to instruments the pickup element of which is inserted through a vein directly into the heart chambers.

intraconnections (microelectronics)—Those conductors and connections which are in continuous integral contact with the substrate or circuit elements of an integrated circuit.

intrinsically safe – Incapable of releasing sufficient electrical energy to ignite a specific atmospheric mixture under normal conditions or such abnormal conditions as accidental damage to any part of the equipment or wiring insulation, failure of electrical components, application of overvoltage, or improper adjustment or maintenance operations.

intrinsic-barrier diode – A pin diode in which a thin region of intrinsic material separates the n-type and p-type regions.

intrinsic-barrier transistor—A pnip or npin transistor in which a thin region of intrinsic material separates the collector from the base.

intrinsic brightness—The luminous intensity measured in a given direction per unit of apparent (projected) area when viewed from that direction.

intrinsic characteristics—Characteristics of a material that depend on the material itself and do not result from impurities.

intrinsic coercive force – The magnetizing force that, when applied to a magnetic material in a direction opposite to that of the residual induction, reduces the intrinsic induction to zero.

intrinsic coercivity—The measurement (in oersteds) of the force required to reduce the intrinsic induction of the magnetized material to zero.

intrinsic concentration—In a semiconductor, the number of minority carriers that exceeds the normal equilibrium number.

intrinsic conduction—In an intrinsic semiconductor, the conduction associated with the directed movement of electron-hole pairs under the influence of an electric field.

intrinsic contact potential difference—The true potential difference between two spotlessly clean metals in contact.

intrinsic electric strength—The characteristic electric strength of a material.

intrinsic flux—The product of the intrinsic flux density and the cross-sectional area of a uniformly magnetized sample of material.

intrinsic flux density—In a sample of magnetic material, the excess of the normal flux density over the flux density in a vacuum for a given magnetizing field strength. In the cgs system, the intrinsic flux density is numerically equal to the difference between the ordinary flux density and the magnetizing field strength.

intrinsic hysteresis loop – A curve that shows intrinsic flux density as a function of magnetizing field strength, when the magnetizing field is cycled between equal values of opposite polarity. Hysteresis is indicated by the fact that the ascending and descending portions of the curve do not coincide.

intrinsic induction – The excess magnetic induction produced in a magnetic material by a given magnetizing force, over the induction that would be produced by the same magnetizing force in a vacuum.

intrinsic-junction transistor–*See* Intrinsic-Region Transistor.

intrinsic layering–The method of separating two conductive semiconductor regions by a region of near-intrinsic semiconductor material.

intrinsic material–A semiconductor material in which there are equal numbers of holes and electrons, i.e., no impurities.

intrinsic mobility – The mobility of electrons in an intrinsic semiconductor, or in a semiconductor having a very low concentration of impurities.

intrinsic noise–Noise that is due to the device or transmission path and is independent of modulation.

intrinsic permeability – Ratio of intrinsic normal induction to the corresponding magnetizing force.

intrinsic properties – The semiconductor properties which are characteristic of the pure, ideal crystal.

intrinsic *Q* – *See* Unloaded *Q* (Switching Tubes).

intrinsic region–*See* Depletion Region.

intrinsic-region transistor – Also called intrinsic-junction transistor. A four-layer transistor with an intrinsic region between the base and collector. Examples are npin, pnip, npip, and pnin transistors.

intrinsic semiconductor–1. A semiconductor in which some hole and electron pairs are created by thermal energy at room temperature, even though there are no impurities in it. 2. A semiconductor with substantially the same electrical properties as those of the ideal crystal.

intrinsic standoff ratio–In a UJT the difference between the emitter voltage at the peak point with a specified interbase voltage and the forward voltage drop of the emitter junction, divided by the voltage on Base 2 with respect to Base 1.

intrinsic temperature range–The temperature range at which impurities or imperfections within the crystal do not modify the electrical properties of a semiconductor.

Invar–An alloy containing 63.8% iron, 36% nickel, and 0.2% carbon. Has a very low thermal coefficient of expansion. Used primarily as resistance wire in wirewound resistors.

inverse beta–The transistor gain that results when the emitter and collector loads are physically reversed in the operation of a circuit.

inverse common base – Transistor circuit configuration in which the base terminal is common to the input circuit and to the output circuit and in which the input terminal is the collector terminal and the output terminal is the emitter terminal.

inverse common collector – Transistor circuit configuration in which the collector terminal is common to the input circuit and to the output circuit and in which the input terminal is the emitter terminal and the output terminal is the base terminal.

inverse common emitter–Transistor circuit configuration in which the emitter terminal is common to the input circuit and to the output circuit and in which the input terminal is the collector terminal and the output terminal is the base terminal.

inverse electrical characteristics – In a transistor, those characteristics obtained when the collector and emitter terminals are interchanged and the transistor is then tested in the normal manner.

inverse electrode current – The current flowing through an electrode in the opposite direction from that for which the tube was designed.

inverse feedback–*See* Negative Feedback.

inverse-feedback filter–A tuned circuit at the output of a highly selective amplifier having negative feedback. The feedback output is zero for the resonant frequency, but increases rapidly as the frequency deviates.

inverse Fourier transform–A mathematical operation that synthesizes a time-domain signal from its complex spectrum components. If a time-domain signal is Fourier-transformed and then inverse Fourier-transformed, the original time function is reconstructed.

inverse limiter–A transducer with a constant output for inputs of instantaneous values within a specified range. Above and below that range, the output is linear or some other prescribed function of the input.

inverse networks – Any two two-terminal networks in which the product of their impedances is independent of frequency within the range of interest.

inverse neutral telegraph transmission–A form of transmission in which zero-current intervals are used as marking signals, and current pulses of either polarity are used as spacing signals.

inverse-parallel connection – *See* Back-to-Back Circuit.

inverse peak voltage – The peak instantaneous voltage across a rectifier tube during the nonconducting half-cycle.

inverse photoelectric effect–The transformation of the kinetic energy of a moving electron into radiant energy, as in the production of X-rays.

inverse piezoelectric effect–Contraction or

expansion of a piezoelectric crystal under the influence of an electric field.

inverse ratio – The seesaw effect whereby one value increases as the other decreases or vice versa.

inverse-square law—The strength of a field, or the intensity of radiation, decreases in proportion to the square of the distance from its source.

inverse time—A qualifying term applied to a relay, indicating that its time of operation decreases as the magnitude of the operating quantity increases.

inverse voltage – The effective voltage across a rectifier tube during the half-cycle when current does not flow.

inverse Wiedemann effect – *See* Wiedemann Effect.

inversion—1. The bending of a radio wave because the upper part of the beam is slowed down as it travels through denser air. This may occur when a body of cold air moves in under a moisture-laden body of air. 2. The producing of inverted or scrambled speech by beating an audio-frequency signal with a fixed band of the resultant beat frequencies. The original low audio frequencies then become high frequencies and vice versa.

inversion layer—A layer of doped semiconductor material that has changed to the opposite type, such as a p layer at the surface of an n-doped region. Surface inversion layers may be the result of surface ions or dopant gettering by surface passivation material or the action of induced electric fields. *See also* Channel, 6.

inverted amplifier—An amplifier stage containing two vacuum tubes. The control grids are grounded, and the driving excitation is applied between the cathodes. The grids then serve as a shield between the input and output circuits. Thus, the output-circuit capacitance is greatly reduced.

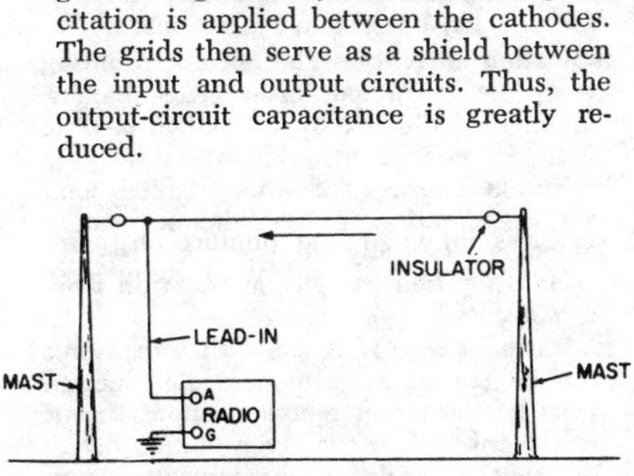

Inverted-L antenna.

inverted-L antenna—An antenna consisting of one or more horizontal wires with a vertical wire connected at one end.

inverted speech—*See* Scrambled Speech.

inverter—1. A circuit which takes in a positive signal and puts out a negative one, or vice versa. 2. A device that changes alternating current to direct current or vice versa. It frequently is used to change 6-volt or 12-volt direct current to 110-volt alternating current. 3. A device that accepts an input that is a function of the maximum voltage and changes it into an output that is a function of both the maximum voltage and time. 4. A circuit with

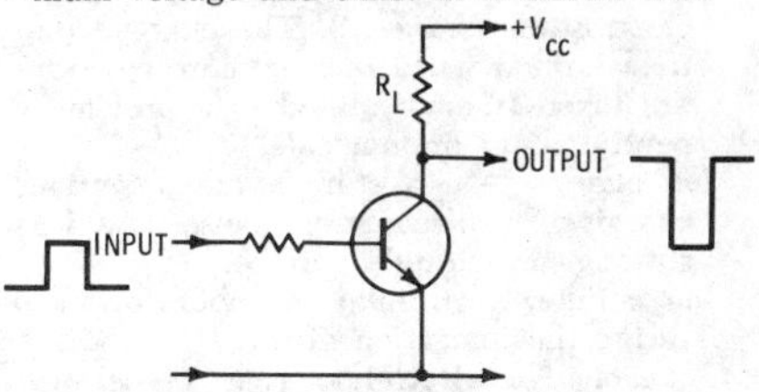

Inverter, 4.

one input and one output, and its function is to invert the input. When the input is high, the output is low, and vice versa. The inverter is sometimes called a NOT circuit, since it produces the reverse of the input. 5. A device or circuit that complements a Boolean function.

inverter circuit—*See* NOT Circuit.

inverting amplifier—An amplifier in which the output polarity is the opposite of the input polarity. Such an amplifier obtains negative feedback through a connection from the output to the input, and with high gain it is widely used as an operational amplifier.

inverting connection—The closed-loop connection of an operational amplifier when the forward gain is negative for dc signals. A 180° phase shift.

inverting input—A differential-amplifier input terminal for which the output signal is opposite in phase to the input signal (180° phase shift).

inverting parametric device—A parametric device the operation of which depends essentially upon three frequencies—a harmonic of the pump frequency, and two signal frequencies—of which the higher signal frequency is the difference between the pump harmonic and the lower signal frequency.

invister—A high-frequency, high-transconductance unipolar structure made by means of lateral diffusion.

inward-outward dialing system—A dialing system by means of which calls within the local exchange area may be dialed directly to or from base private branch exchange telephone stations without the assistance of an operator at the base private branch exchange.

i/o—Abbreviation for input-output.

i/o device—1. Input/output device. A card reader, magnetic tape unit, printer, or similar device that transmits data to or receives data from a computer or secondary storage device. 2. Input/output equipment used to communicate with a computer.

ion – 1. An electrically charged atom or group of atoms. Positively charged ions have a deficiency of electrons, and negatively charged ions have surplus electrons. 2. An atom or molecule with an electrostatic charge. 3. The charged particle formed when one or more electrons are taken from or added to a previously neutral atom or molecule.

ion charging—In a storage tube, spurious charging or discharging caused by ions striking the storage surface.

ion counter—A tubular chamber for measuring the ionization of air.

ion-exchange electrolyte cell—A fuel cell that uses the reaction of hydrogen with oxygen from the air. It is similar to the standard hydrogen-oxygen fuel cell, except that an ion-exchange membrane replaces the liquid electrolyte. Operation is at atmospheric pressure and room temperature.

ionic focusing—Focusing the electron beam in a cathode-ray tube by varying the filament voltage and temperature to change the electrostatic focusing field automatically produced by the accumulation of positive ions in the tube.

ionic-heated cathode – A cathode that is heated primarily by bombardment with ions.

ionic-heated cathode tube – An electron tube containing an ionic-heated cathode.

ionic tweeter—A type of speaker in which a varying electrostatic field activates a mass of air ionized by a high-voltage radio-frequency field. Ionic speakers are capable of extremely extended high-frequency response (up to 100 kHz or so) because of the extreme lightness of the ionic "diaphragm."

ion implantation—1. A method of semiconductor doping in which impurities that have been ionized and accelerated to high velocity penetrate the semiconductor surface and became deposited in the interior. 2. A processing step by which standard p-channel diffused MOS devices are made directly compatible with TTL/DTL logic. It is a highly controllable process that allows the adjusting of gate threshold voltages and also allows the fabrication of both enhancement-mode and depletion-mode transistors on the same chip. 3. A process that uses accelerated atoms to implant source and drain regions in metal-oxide semiconductors. It offers higher speed and lower threshold voltages and can also be used with PMOS, NMOS, and CMOS. 4. A process that shoots a purified, kinetically accelerated stream of ions at a room- or elevated-temperature substrate target. Impacting ions penetrate the target to a depth dependent upon the accelerating voltage. Although the impacting ions cause damage to the crystalline structure of the substrate, annealing during or subsequent to ion implantation at relatively low temperatures (considerably lower than used for diffusion processing) repairs most of the damage.

ion-implanted MOS—Abbreviated IMOS. A method for doping substrates with a stream of ionized dopant atoms. Ions are electrically shot into the substrate instead of diffusing atoms at high temperatures.

ionization—1. The dissociation of an atom or molecule into electrons and/or ions. 2. The state of an insulator in which it facilitates the passage of current because of the presence of charged particles (usually induced artificially). 3. The electrically charged particles produced by high-energy radiation (such as light or ultraviolet rays) or by the collision of particles during thermal agitation. 4. The formation of ions.

ionization arcover – 1. Formation of an electrical arc between terminals or contacts as a result of ionization of the adjacent air or gas. 2. Formation of an arc between the terminals of a satellite antenna as the satellite passes through the ionized regions of the ionosphere.

ionization chamber—1. An enclosure containing two or more electrodes between which an electric current may pass when the gas within is ionized. The current is a measure of the total number of ions produced in the gas by externally induced radiations. 2. A chamber containing a gas through which ionizing particles pass. A voltage is applied across the chamber so as to collect the ions produced and permit the ion current to be measured.

ionization current – The current resulting from the movement of electric charges, under the influence of an applied electric field, in an ionized medium.

ionization energy—Sometimes called ionization potential. The minimum amount of energy (usually expressed in electronvolts) required to eject an electron from a molecule.

ionization gage—1. A gage that measures the degree of a vacuum in an electron tube by the amount of ionization current in the tube. 2. A type of radiation detector that depends on the ionization produced in a gas by the passage of a charged particle through it. One of the best known is the Geiger-Mueller counter. Cloud chambers and spark chambers can also be included in this category.

ionization-gage tube—An electron tube that measures low gas pressure by the amount of ionization current produced.

ionization potential – 1. The energy, expressed in electronvolts, needed to remove one electron from a neutral atom or molecule in its ground state. 2. The amount of energy for a particular kind of atom required to remove an electron from the atom to infinite distance. (Usually expressed in volts.)

ionization pressure — An increase in the pressure in a gaseous discharge tube due to ionization of the gas.

ionizing radiation — Radiant electromagnetic energy and high-energy particles that cause the division of a substance into parts carrying positive and negative charges. While high-energy particles can directly ionize substances, electromagnetic radiation sets in motion charged particles, which then produce ions.

ionization resistance — *See* Corona Resistance.

ionization time — 1. The time interval between the initiation and the establishment of conduction in a gas tube at some stated voltage drop for the tube. 2. The elapsed time to achieve normal glow after a voltage greater than the breakdown voltage is applied to a glow lamp.

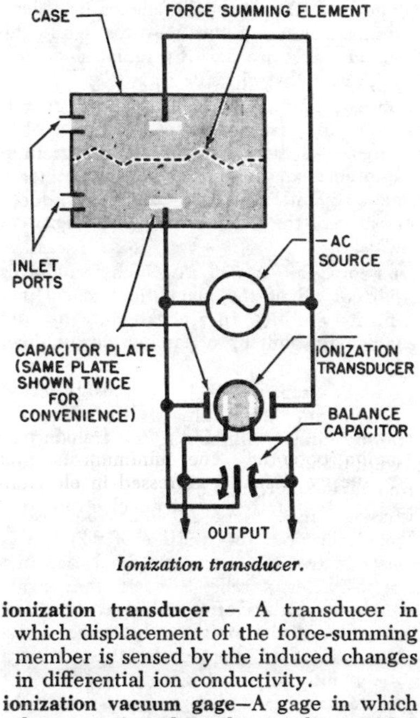

Ionization transducer.

ionization transducer — A transducer in which displacement of the force-summing member is sensed by the induced changes in differential ion conductivity.

ionization vacuum gage—A gage in which the operation depends on the positive ions produced in a gas by electrons as they accelerate between a hot cathode and another electrode in a vacuum. It ordinarily covers a pressure range of 10^{-4} to 10^{-10} mm of mercury.

ionize—To free an electron from an atom or molecule (e.g., by X-ray bombardment) and thus transform the atom or molecule into a positive ion. The freed electron attaches itself to another atom or molecule, which then becomes a negative ion.

ionized layers—Layers of increased ionization within the ionosphere. They are responsible for absorption and reflection of radio waves and are important for communication and for tracking satellites and other space vehicles.

ionizing event—Any interaction by which one or more ions are produced.

ionizing radiation—Electromagnetic radiation or particle radiation having sufficient energy to dislodge electrons from atoms or molecules, thereby producing ions.

ion migration—Movement of the ions produced in an electrolyte by application of an electric potential between electrodes.

ionophone — A high-frequency speaker in which the audio-frequency signal modulates an rf supply to maintain an arc in the mouth of a quartz tube. The resultant modulated wave acts directly on the ionized air under pressure and thus creates sound waves.

ionosphere—The part of the earth's outer atmosphere where sufficient ions and electrons are present to affect the propagation of radio waves.

ionospheric disturbance — The variation in the state of ionization of the ionosphere beyond the normal observed random day-to-day variation from average value for the location, date, and time of day under consideration.

ionospheric D scatter meteor burst—A phenomenon in which the penetration of meteors through the D region of the ionosphere affects ionospheric scatter communications.

ionospheric error—Also called sky error. In navigation, the total systematic and random error resulting from reception of the navigational signal after it has been reflected from the ionosphere. It may be due to variations in the transmission path, uneven height of the ionosphere, or uneven propagation within the ionosphere.

ionospheric prediction—The forecasting of ionospheric conditions and the preparation of radio propagation data derived from it.

ionospheric scatter—*See* Forward Scatter.

ionospheric storm—An ionospheric disturbance associated with abnormal solar activity and characterized by wide variations from normal, including turbulence in the F-region and increases in absorption. Often the ionization density is decreased and the virtual height is increased. The effects are most marked in high magnetic latitudes.

ionospheric wave—Also called a sky wave.

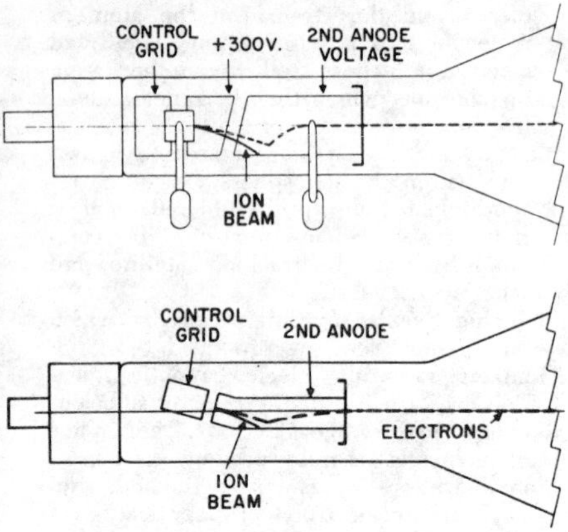

Ion trap.

A radio wave that is propagated by way of the ionosphere.

ion sheath—A positive-ion film which forms on or near the grid of a gas tube and limits its control action.

ion spot—1. In camera or image tubes, the spurious signal resulting from bombardment of the target or photocathode by ions. 2. On a cathode-ray-tube screen, an area where the luminescence has been deteriorated by prolonged bombardment with negative ions.

ion trap—Also called a beam bender. An electron-gun structure and magnetic field which diverts negative ions to prevent their burning a spot in the screen, but permits electrons to flow toward the screen.

I_p—Symbol for the plate current of a vacuum tube.

I-phase carrier—Also called in-phase carrier. A carrier separated in phase by 57° from the color subcarrier.

ips — Abbreviation for inches per second. Used for specifying the speed of a tape travelling past the heads of a tape recorder. The most common speeds are 1⅞ ips (4.75 cm/s); 3¾ ips (9.5 cm/s) and 7½ ips (19 cm/s).

ir — Abbreviation for: 1. Interrogator response. 2. Infrared. 3. Insulation resistance.

IRAC—Acronym for Interdepartmental Radio Advisory Committee. It is composed of representatives of eleven government agencies: The FCC; Army; Navy; Air Force; Maritime Commission; and the Treasury, State, Commerce, Agriculture, Interior, and Justice Departments.

IR compensation — A control device that compensates for voltage drop due to current flow.

IR drop—The voltage produced across a resistance (R) when there is a current (I) through the resistor.

IRE—Abbreviation for Institute of Radio Engineers, an organization now merged with AIEE. *See* IEEE.

i (intrinsic) region—In silicon, a pure region of the group-IV element (i.e., having neither excess holes or excess electrons, and therefore having very high resistivity).

iridescence — The rainbow exhibition of colors, usually caused by interference of light of different wavelengths reflected from superficial layers in the surface of a material.

iris—Also called diaphragm. In a waveguide, a conducting plate (or plates) which is very thin (compared with the wavelength) and occupies part of the cross section of a waveguide. When only a single mode can be supported in the waveguide, an iris appears substantially as a shunt admittance.

iris diaphragms—A simple mechanism used to vary the diameter of an aperture. Consists of a number of thin, arc-shaped, metal blades that surround the aperture, each blade having a lower stud at one end and an upper stud at the other end. The lower studs fall into holes in a fixed ring surrounding the aperture, while the upper studs are held by radial slots in a rotatable control ring.

iron constantan—A combination of metals used in thermocouples, thermocouple wires, and thermocouple lead wires. Constantan is an alloy of copper, nickel, manganese, and iron. The iron wire is the positive, the constantan the negative wire.

iron-core coil—A coil in which iron forms part or all of the magnetic circuit, linking its winding. In a choke coil, the core is usually built up of laminations of sheet iron.

iron-core transformer — A transformer in which iron forms part or all of the magnetic circuit, linking the transformer windings.

iron loss—*See* Core Loss.

iron-vane instrument—An indicating instrument the operating portion of which consists of two iron bars, one fixed, one pivoted, placed parallel to each other inside a signal coil. Current through the coil magnetizes the bars in the same direction, and they repel each other, causing the pointer to pivot against the force of a hairspring. A damping vane may be used to slow the movement of the pointer. Deflection is the same for ac or dc; the meter does not have polarity. The instrument has a nonlinear scale, and readings below ⅕ scale are extremely difficult to make. Because of inductance effects, use of this type of meter is limited to power-line frequencies.

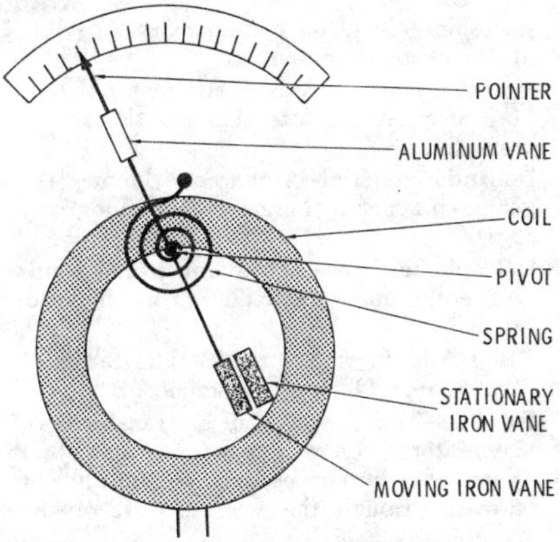

Iron-vane instrument.

irradiance – The incident radiated power per unit area of a surface; the radiometric counterpart of illumination, usually expressed in watts/cm^2.

irradiation—1. The application of X-rays, radium rays, or other radiation. 2. The amount of radiant energy per unit area received during a given time interval. This term is used in radiological therapy as well as in describing accidental exposure to radiation. It also can be used to denote radiant energy that ultimately passes through the skin to supply power to implanted electronic devices.

irregularity—A change from normal.

I-scan—A radar display in which a target appears as a complete circle when the radar antenna is correctly pointed at it, the radius of the circle being proportional to the target distance.

I-scope—*See* I-scan.

I-signal – Also called the fine-chrominance primary. A signal formed by the combination of +.74 of an R–Y signal and −.27 of a B–Y signal. One of the two signals used to modulate the chrominance subcarrier, the other being the Q signal.

ISM equipment – A Federal Communications Commission designation for industrial, scientific, and medical equipment capable of causing interference.

isobar—1. On meteorological maps, a line denoting places having the same atmospheric pressure at a given time. 2. One of a group of atoms or elements having the same atomic weights but different atomic numbers.

isochromatic – Having the same color as with the lines of the same tint in the interference figures of anisotropic crystals.

isochrone—On a map or chart, a line joining points associated with a constant time difference in the reception of radio signals.

isochrone determination—A radio location in which a position line is determined by the difference in transit times of signals along two paths.

isochronous circuits – Circuits having the same resonant frequency.

isochronous multiplexer – A multiplexer which can interleave two time-independent data streams into one higher-speed stream independent of the master timing control required by a synchronous multiplexer.

isoclinic line—*See* Aclinic Line.

isodynamic lines – On a magnetic map, lines passing through points of equal strength of the earth's magnetic field.

isoelectric—Uniformly electric throughout, or having the same electric potential, and therefore producing no current.

isoelectronic—Having the same number of electrons outside the nucleus of the atom.

isolated—Utterly cut off from; refers to that condition in which a conductor, circuit, or device is not only insulated from another (or others), but the two are mutually unable to engender current, emf, or magnetic flux in each other. As commonly used, insulation is associated predominantly with direct current, whereas isolation implies additionally a bulwark against ac fields.

isolated amplifier—A differential amplifier in which the input-signal lines are conductively isolated from the output-signal lines and chassis ground.

isolating diode—A diode that passes signals in one direction through a circuit but blocks signals and voltages in the opposite direction.

isolating switch—A switch intended for isolating an electric circuit from the source of power. It has no interrupting rating and is intended to be operated only after the circuit has been opened by some other means.

isolation—1. Electrical or acoustical separation between two locations. 2. The technique for producing a high electrical resistance between an integrated-circuit component and the substrate in which it is formed.

isolation amplifier—An amplifier employed to minimize the effects of the following circuit on the preceding circuit.

isolation diffusion—A technique for separation of the individual components within a monolithic silicon n structure; p diffused isolation zones are used to form pn junctions that act as reverse-biased diodes. The transistors are double diffused; that is, they are processed by two diffusion steps after the isolation diffusion.

isolation network—A network inserted into a circuit or transmission line to prevent interaction between circuits on each side of the insertion point.

isolation transformer – A transformer designed to provide magnetic coupling (flux coupling) between one or more pairs of isolated circuits, without introducing significant coupling of any other kind between them—i.e., without introducing either significant conductive (ohmic) or significant electrostatic (capacitive) coupling.

isolator ferrite—A microwave device which allows rf energy to pass through in one direction with very little loss but absorbs rf power in the other direction.

isolith — An integrated circuit of components formed on a single silicon slice, but with the various components interconnected by beam leads and with circuit parts isolated by removal of the silicon between them.

isomer — One of two or more substances composed of molecules having the same kinds of atoms in the same proportions but arranged differently. Hence, the physical and chemical properties are different. Isomers which do not have the same molecular weights are called polymers.

isoplanar — A bipolar fabrication process that replaces conventional planar P+ isolation diffusion with an insulating oxide to provide isolation between active elements of a silicon IC. Circuit elements can be fabricated in less space than conventional isolation techniques with improved speed and power performance.

isopulse system — In adaptive communications, a pulse coding system in which special inserted pulses indicate the number of information pulses that are transmitted.

isostatic—Being subjected to equal pressure from every side.

isothermal region — The stratosphere considered as a region of uniform temperature.

isotones—A group of atoms the nuclei of which have the same number of neutrons.

isotope—A species of matter the atoms of which contain the same number of protons as some other species, but a different number of neutrons. The atomic numbers of isotopes are identical, but the mass numbers (atomic weights) differ. *See* Radioisotope.

isotropic — 1. Having properties with the same values along axes in all directions. 2. Term applied to substances certain of whose properties are manifest in every direction, e.g., electrical conductivity in metals.

isotropic antenna—Also called unipole. A hypothetical antenna radiating or receiving equally in all directions. A pulsating sphere is a unipole for sound waves. In the case of electromagnetic waves, unipoles do not exist physically, but represent convenient reference antennas for expressing directive properties of actual antennas.

isotropic gain of an antenna—*See* Absolute Gain of an Antenna.

isotropic magnet—A magnetic material having no preferred axis of magnetic characteristics.

isotropic material—A material having the same magnetic characteristics along any axis.

isotropic radiator—A radiator which sends out equal amounts of energy in all directions.

I^2R—Power in watts expressed in terms of the current (I) and resistance (R).

I^2R loss—The power lost in transformers, generators, connecting wires, and other parts of a circuit because of the current flow I through the resistance R of the conductors.

item—A general term denoting one of a number of similar units, assemblies, objects, etc.

iterations per second—The number of approximations per second in iterative division in a computer; the number of times a cycle of operation can be repeated in one second.

iterative array — In a computer, a large number of identical, interconnected processing modules used, with appropriate driver and control circuits, to perform simultaneous parallel operations.

iterative division—In computers, a method of performing division by use of addition, subtraction, and multiplication operations. A quotient of specified precision is obtained by a series of progressively closer approximations.

iterative filter—A four-terminal filter that provides iterative impedance.

iterative impedance — An impedance that, when connected to one pair of terminals of a four-terminal transducer, causes the same value of impedance to appear between the other two terminals. The iterative impedance of a uniform transmission line is equal to the characteristic impedance of the line. In a symmetrical four-terminal transducer, the iterative impedances for the two pairs of terminals are equal and the same as the image impedances and the characteristic impedance.

iterative process—The calculating of a desired result by means of a repeating cycle of operations wihch comes closer and closer to the desired result.

iterative routine—A computer routine composed of repetitive computations, so that the output of every step becomes the input of the succeeding step.

ITU—Abbreviation for International Telecommunication Union.

itv—Abbreviation for industrial television.

i-type—Intrinsic semiconductor.

IVR — Abbreviation for integrated voltage

J

J—Abbreviation for joule.

jack—1. A socket to which the wires of a circuit are connected at one end, and into which a plug is inserted at the other end. 2. A type of two-way, or more, concentric contact socket for carrying audio signals. 3. A receptacle into which a mating connector may be plugged. 4. The receptacle that accepts a plug, specifically a phone plug.

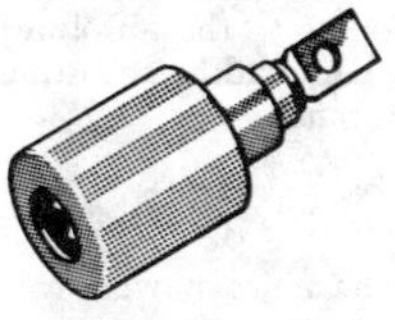

Jack.

jacket—Pertaining to wire and cable, the outer sheath which protects against environment and may also provide additional insulation.

jack panel—An assembly composed of a number of jacks mounted on a board or panel.

jackscrew—A screw attached to one half of a two-piece, multiple-contact connector and used to draw both halves together and to separate them.

jaff—Slang for the combination of electronic and chaff jamming.

jag—In facsimile, distortion caused in the received copy by a momentary lapse in synchronism between the scanner and recorder.

jam—1. In punch-card machines, a condition in the card feed which interferes with the normal travel of the punch cards through the machine. 2. To interfere electronically with the reception of radio signals.

jam input—The presetting or loading of a counter, using inputs provided for the purpose. Also, the establishment of a desired logic state or logic line by the direct application of the appropriate voltage level to the line, regardless of the outputs of other devices connected to it.

jammer—An electronic device for intentionally introducing unwanted signals into radar sets to render them ineffective.

jammer band — The radio-frequency band where the jammer output is concentrated. It is usually the band between the points where the intensity is 3 dB down from maximum.

jammer finder — Also called burnthrough. Radar which attempts to obtain the range of the target by training a highly directional pencil beam on a jamming source.

jammers tracked by azimuth crossings—Semiautomatic strobe processing and tracking that permits automatic detection and tracking on the basis of azimuth information obtained from the jamming signals emanating from an airborne vehicle.

jamming—1. The intentional transmission of radio signals in order to interfere with the reception of signals from another station. 2. Interference with hostile radio or radar signals for the purpose of deceiving or confusing the operator. It may be accomplished by saturating a receiver with sufficient noise to prevent detection and location of a target, or by deceiving the operator with intentionally misleading signals or false echoes without his knowing that such signals are present. Also called active ECM.

jamming effectiveness—The jamming-to-signal ratio—i.e., the percentage of information incorrectly received in a test message.

JAN specification—Joint Army-Navy specification. The forerunner of present Military Specifications, which are generally superseded by the designation MIL.

J-antenna—A half-wave antenna fed at one end by a parallel-wire, quarter-wave section having the configuration of a *J*.

J-carrier system—A broad-band carrier system which provides 12 telephone channels and utilizes frequencies up to about 140 kilohertz by means of four-wire transmission on a single open-wire pair.

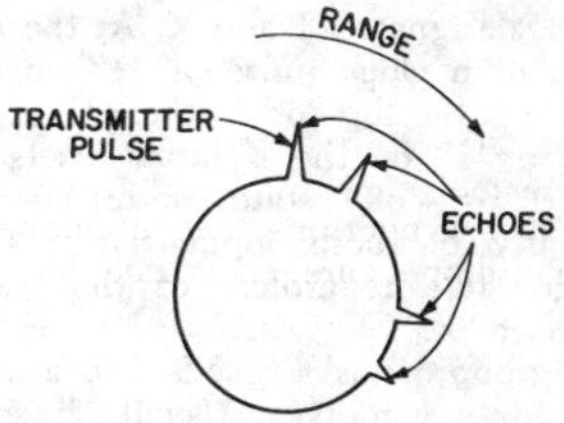

J-display.

J-display—Also called J-scan. In radar, a modified A-display in which the time base is a circle and the target signal appears as a radial deflection from it.

JEDEC—Acronym for Joint Electron Device Engineering Council.

JETEC—Acronym for Joint Electron Tube Engineering Council.

jewel bearing—A natural or synthetic jewel, usually sapphire, used as a bearing for a pivot or other moving parts of a delicate instrument.

jezebel—A system for the detection and classification of submarines.

JFET—Abbreviation for junction field-effect transistor.

JHG—Abbreviation for Joule heat gradient.

jitter—1. Instability of a signal in either its amplitude, its phase, or both. The term is applied especially to signals reproduced on the screen of a cathode-ray tube. The term "tracking jitter" describes minor variations in the pointing of an automatic tracking radar. 2. In facsimile, raggedness in the received copy caused by erroneous displacement of recorded spots in the scanning direction. 3. An aberration of a repetitive display, indicating instability of the signal or of the oscilloscope. May be random or periodic, and is usually associated with the time axis. 4. A loss of synchronization caused by electrical or mechanical malfunctions. 5. *See* Flicker, 3 *and* Fortuitous Telegraph Distortion.

jittered pulse recurrence frequency—The random variation of the pulse-repetition period. Provides a discrimination capability against repeater-type jammers.

JK flip-flop—1. A flip-flop with two conditioning inputs (J and K) and one clock input. If both conditioning inputs are disabled prior to a clock pulse, the flip-flop does not change condition when a clock pulse occurs. If the J input is enabled and the K input is disabled, the flip-flop will assume the 1 condition upon arrival of a clock pulse. If the K input is enabled and the J input is disabled, the flip-flop will assume the 0 condition when a clock pulse arrives. If both the J and K inputs are enabled prior to the arrival of a clock pulse, the flip-flop will complement, or assume the opposite state, when the clock pulse occurs. 2. A flip-flop having two inputs, designated J and K. At the application of a clock pulse, a "1" on the J input sets the flip-flop to the "1" or "on" state; a "1" on the K input resets it to the "0" or "off" state; and "1" simultaneously on both inputs causes it to change state regardless of the state it had been in.

job—A group of tasks specified as a unit of work for a computer. Usually by extension, a job includes all necessary programs, linkages, files, and instructions for the operating system.

job library—A related series of user-identified, partitioned data sets that serve as the primary source of load modules for a given job.

job statement — A control statement that identifies the start of a series of job control statements for a single job.

job step—The carrying out of a computer program explicitly identified by a job control statement. The execution of several job steps may be specified by a job.

jogging—Also called inching. Quick and repeated opening and closing of a motor starting circuit to produce slight movements of the motor.

Johnson counter—A counter composed of an N-stage shift register with the complement of the last stage returned to the input. It normally has 2N states through which it cycles. It has the distinguishing characteristic that only one stage changes state at each count. Also called Mobius Counter or twisted-ring counter. *See* Ring Counter, 2.

Johnson noise—1. Also called thermal noise. The noise generated by any resistor at a temperature above absolute zero. It is proportionate to the absolute temperature and the bandwidth, according to the following formula:

$$N = KTB$$

where,

N is the noise power in watts,
K is Boltzmann's constant, or 1.38047×10^{-23}
T is the absolute temperature in degrees Kelvin,
B is the bandwidth in hertz.

2. A frying or sizzling sound produced by thermal agitation voltages generated in amplifier circuits. It usually occurs in the input circuit (or front end) of an amplifier.

joined actuator—A multiple breaker such that when one pole trips, all trip, but whereas the faulted pole is trip-free, the other poles may be kept maintained by a restraining actuator.

joint—A connection between two or more conductors.

joint circuit — A communication link in which there is participation by the elements of more than one service, through control, operations, management, etc.

joint communications—The common use of communication facilities by two or more services of the same country.

joint use—The simultaneous use of pole, line, or plant facilities by two or more kinds of utilities.

Jones plug—A type of polarized connector designed in the form of a receptacle and having several contacts.

Josephson effect — The phenomenon described by Brian Josephson to explain the action of currents through and voltages across hairlike gaps in superconductors. On the basis of theoretical considerations, it is predicted that if two superconductors would be brought close enough together, a current could be made to flow across the gap between them. Under certain conditions, a voltage appears across the gap, and high-frequency radiation emanates from it. This predicted radiation would have a frequency precisely equal to 2eV/h, where V is the measured voltage across the gap.

joule—1. The work done by a force of 1 newton acting through a distance of 1 meter. The joule is the unit of work and energy in the mksa system. 2. The energy required to transport 1 coulomb between two points having a potential difference of 1 volt. The joule is 10^7 ergs. The kilowatt-hour is 3.6×10^6 joules.

Joule effect—In a circuit, electrical energy is converted into heat by an amount equal to I^2R. Half of this heat flows to the hot junction and the other half to the cold junction.

Joule heat—The thermal energy produced as a result of the Joule effect.

Joule heat gradient—The rate at which the thermal heat produced by the Joule effect increases or decreases. Abbreviated JHG.

Joule's law of electric heating—The amount of heat produced in a conductor is proportional to the resistance of the conductor, the square of the current, and the time.

joystick—1. A control device consisting of a handle with freedom of motion in all directions of a plane, connected to potentiometers or other control devices through suitable linkage permitting natural human input of positioning or other information. The term is derived from the joystick of aircraft. 2. A data-entry device used to manually enter coordinate values in special X-, Y-, and Z-input registers. The device consists of a vertically mounted stick or column which can be moved and twisted. When it is moved backward or forward or sideways, coordinate values are stored in the X- and Y-input registers. The Z-input register is varied whenever the joystick is twisted clockwise and counterclockwise. These registers must be scanned by the host computer since joysticks normally do not generate interrupts when they are activated. Usually the joystick is used to move a cursor and/or tracking symbol on the face of the crt screen.

J-scan—*See* J-Display.

J-scope—Also called Class-J oscilloscope. A cathode-ray oscilloscope that presents a J-display.

J/S ratio—A ratio, normally expressed in dB, of the total interference power to the signal-carrier power in the transmission medium at the receiver.

juice—Slang for electric current.

jump—1. To cause the next instruction to be selected from a specified storage location in a computer. 2. A deviation from the normal sequence of execution of instructions in a computer. 3. A departure from the normal one-step incrementing of the program counter. By forcing a new value (address) into the program counter the next instruction can be fetched from an arbitrary location (either further ahead or back). (For example, a program jump can be used to go from the main program to a subroutine, from a subroutine back to the main program, or from the end of a short routine back to the beginning of the same routine to form a loop.)

jumper—1. A short length of wire used to complete a circuit temporarily or to bypass a circuit. 2. A direct electrical connection, which is not a portion of the conductive pattern, between two points in a printed circuit.

junction—1. A connection between two or more conductors or two or more sections of a transmission line. 2. A contact between two dissimilar metals or materials (e.g., in a rectifier or thermocouple). 3. A region of transition between p- and n-type semiconductor material. The controllable resultant asymmetrical properties are exploited in semiconductor devices. There are diffused, alloy, grown, and electrochemical junctions.

junction barrier—The opposition to the diffusion of majority carriers across a pn junction due to the charge of the fixed donor and acceptor ions.

junction battery—A nuclear type of battery in which radioactive strontium 90 irradiates a silicon pn junction.

junction box—A box for joining different runs of raceway or cable, plus space for connecting and branching the enclosed conductors.

junction capacitor — A capacitor in which the capacitance is that of a reverse-biased pn junction.

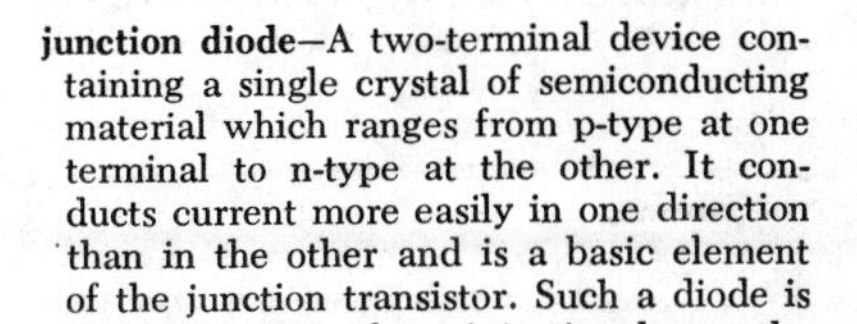

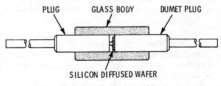

Junction-diode assembly.

junction diode—A two-terminal device containing a single crystal of semiconducting material which ranges from p-type at one terminal to n-type at the other. It conducts current more easily in one direction than in the other and is a basic element of the junction transistor. Such a diode is the basic part of an injection laser; the region near the junction acts as a source of emitted light. When fabricated in a suitable geometrical form, the junction diode can be used as a solar cell.

junction FET—*See* Junction Field Effect Transistor.

junction field-effect transistor—A transistor

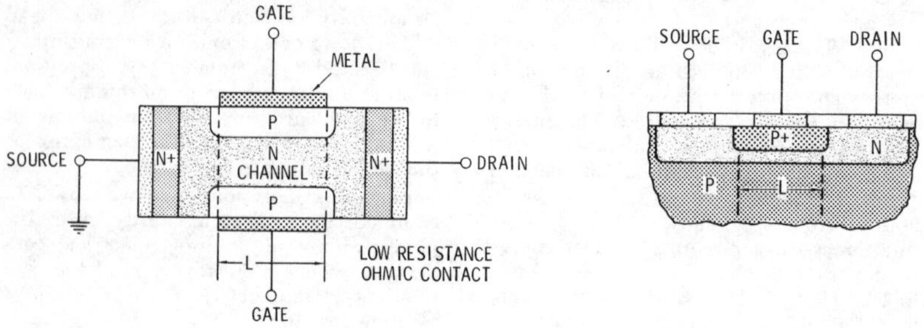

Junction field-effect transistor.

made up of a gate region diffused into a channel region. The gate element is a region of semiconductor material (ordinarily, the substrate) insulated by a pn junction from the channel, which is material of opposite polarity. When a control voltage is applied to the gate, the channel is depleted or enhanced, and the current between source and drain is thereby controlled. There is no current when the channel is "pinched off." Abbreviated JFET.

junction filter—A combination of a high-pass and a low-pass filter which is used to separate frequency bands for transmission over separate paths. For example, junction filters are used to separate voice and carrier frequencies at the junction between open-wire and cable so that the carrier frequencies and voice frequencies can be sent over nonloaded and vf loaded cable pairs, respectively.

junction loss—In telephone circuits, that part of the repetition equivalent that can be assigned to interaction effects originating at trunk terminals. *See* Repetition Equivalent.

junction point—*See* Node, 1

junction pole—Pole at the end of a transposition section of an open-wire line or the pole common to two adjacent transposition sections.

junction station—A microwave relay station that joins a microwave radio leg or legs to the main, or through, route.

junction transistor — A transistor having three alternate sections of p-type or n-type semiconductor material. *See also* Pnp Transistor *and* Npn Transistor.

junction transposition — Transposition located at the junction pole between transposition sections of an open-wire line.

junctor—In crossbar systems, a circuit extending between frames of a switching unit and terminated in a switching device on each frame.

justification—The act of adjusting, arranging, or shifting digits to the left or right so that they conform to a prescribed pattern.

justify—To align data about a particular reference.

just-operate value — Also called drop-out value. The measured functioning value at which a particular relay releases.

just-release value—The measured functioning value at which a particular relay releases.

just scale—A musical scale formed by three consecutive trians (those in which the highest note of one is the lowest note of the other), each having the ratio 4:5:6 or 10:12:15.

jute—Cordage fiber (such as hemp) saturated with tar and used as a protective layer over cable.

jute-protected cable — A cable having its sheath covered by a wrapping of tarred jute or other fiber.

K

K—1. Symbol for cathode or dielectric constant. 2. Abbreviation for Kelvin or kilo. 3. Abbreviation for luminosity factor. 4. In a calculator, a fixed number (a constant) that can be used repetitively.

k—Abbreviation for kilo.

kA—Abbreviation for kiloampere.

Karnaugh map—A display of a truth table in a way such that reduction (simplification) of a Boolean expression is facilitated. It consists of a rectangular or square array (depending on the number of variables) of "locations" the coordinates of which correspond to truth-table inputs.

K-band—A radio frequency band extending from 11 to 36 GHz and having wavelengths of 2.73 to 0.83 cm.

kc—Abbreviation for kilocycle. Now obsolete. Replaced by kHz.

K-carrier system—A broad-band carrier system which provides 12 telephone channels and utilizes frequencies up to about 60 kHz by means of four-wire transmission on cable facilities.

K-display—Also called K-scan. Modification of a type-A scan, used for aiming a double-lobe system in bearing or elevation. The entire range scale is displaced toward the antenna lobe in use. One signal appears as a double deflection from the range and relative scales. The relative amplitudes of these two pips indicate the amount of error in aiming the antenna.

keep-alive anode — An auxiliary electrode that maintains a dc discharge in a mercury-pool tube. It has the disadvantage of reducing the peak inverse voltage rating

keep-alive circuit—In a tr or anti-tr switch, a circuit for producing residual ionization in order to reduce the time for full ionization when the transmitter fires.

keep-alive voltage—A dc voltage that maintains a small glow discharge within one of the gap electrodes of a tr tube. This allows the tube to ionize more rapidly when the transmitter fires thus preventing damage to the receiver.

keeper—A magnetic conductor placed over the ends of a permanent magnet to protect it against being demagnetized.

Kel-f — Polymonochlorotrifluoroethylene — used as a high-temperature insulation (−55°C to +135°C).

Kelvin balance—An instrument for measuring current. This is done by sending it through a fixed and a movable coil attached to one arm of a balance. The resultant force between the coils is then compared with the force of gravity acting on a known weight at the other end of the balance arm.

Kelvin bridge — Also called a double or Thomson bridge. A seven-arm bridge for comparing the resistances of two 4-terminal resistors or networks. Their adjacent potential terminals are spanned by a pair of auxiliary resistance arms of known ratio, and they are connected in series by a conductor joining their adjacent current terminals.

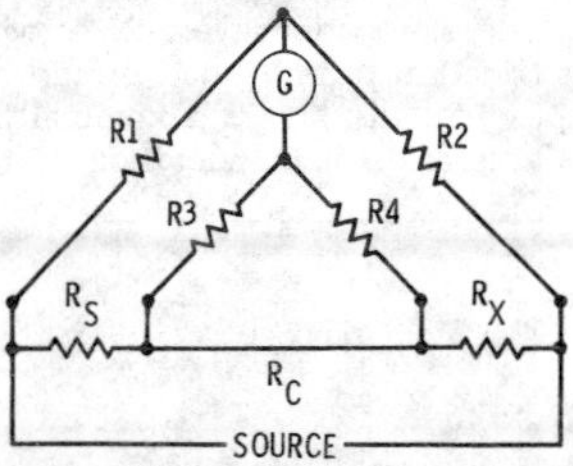

Kelvin bridge.

Kelvin scale—Also called absolute scale. A temperature scale using the same divisions as the Celsius scale, but with the zero point established at absolute zero (≅ −273°C) theoretically the lowest possible temperature.

Kendall effect—A spurious pattern or other distortion in a facsimile record. It is caused by unwanted modulation produced by transmission of a carrier signal. Such modulation appears as a rectified baseband that interferes with the lower sideband of the carrier.

Kennelly-Heaviside layer — *See* Heaviside Layer.

kenopliotron—A diode-triode vacuum tube within one envelope. The anode of the diode also serves as the cathode of the triode.

kenotron—1. Also called a value tube. A term used primarily in industrial and X-ray fields for a hot-cathode vacuum tube. 2. A high-vacuum, high-voltage thermionic diode, used primarily as a high-voltage rectifier.

keraunophone—A radio circuit device for audibly demonstrating the occurrence of distant lightning flashes.

kernel—A line within a current-carrying conductor along which the magnetic intensity due to the current is zero.

Kerr cell—A container filled with a transparent material that, when subjected to a strong electric field, exhibits double refraction. Because the two polarized components of an incident light beam have different phase velocities in the medium, this device can rotate the plane of polarization. When placed between crossed polarizers, the Kerr cell, usually containing nitrobenzene, can act as an extremely high-speed shutter for light beams.

Kerr effect—1. An electro-optical effect in which certain transparent substances become double refracting when subjected to an electric field perpendicular to a beam of light. 2. The conversion of plane into elliptically polarized light when reflected from the polished end of a magnet.

keV—Abbreviation for kiloelectronvolt.

key—1. A hand-operated switching device for switching one or more parts of a circuit. It ordinarily consists of concealed spring contacts and an exposed handle or push button. 2. A projection which slides into a mating slot or groove so as to guide two parts being assembled and assure proper polarization.

keyboard—In a calculator, keys for digits 0 through 9, plus additional keys for various functions, such as add, multiply, divide, subtract, clear, memory, etc.

keyboard computer—A computer whose input employs a keyboard, e.g., an electric typewriter.

keyboard perforator—A mechanism that

punches a paper tape from which messages are automatically transmitted by a transmitter distributor. The keyboard is similar to that of a typewriter and can be operated by any trained typist after a few hours' instruction. As each key is depressed, the tape is punched with corresponding code symbols.

keyboard send/receive—A combination teletypewriter transmitter and receiver with transmission capability from a keyboard only.

key cabinet—A case installed on a customer's premises and providing facilities so that different lines to the control office can be connected to various telephone stations. It has signals that indicate originating calls and busy lines.

key click — A transient signal sometimes produced when the key of a radiotelegraph transmitter is opened or closed. The transient is heard in a speaker or headphone as a click.

key-click filter—Also called a click filter. A filter that attenuates the surges produced each time the keying circuit of a transmitter is opened or closed.

keyed agc—Abbreviation for keyed automatic gain control.

keyed automatic gain control—Abbreviated keyed agc. A television automatic gain control in which the agc tube is kept cut off except when the peaks of the positive horizontal-sync pulse act on its grid. The agc voltage is therefore not affected by noise pulses occurring between the sync pulses.

keyed clamp—A clamping circuit in which a control signal determines the time of clamping.

keyed interval — In a periodically keyed transmission system, an interval that starts from a change in state and has a length equal to the shortest time between changes in state.

keyed rainbow generator—A color television test instrument which displays the individual colors of the spectrum, separated by black bars, on the picture tube.

keyed rainbow signal—A 3.563795-MHz (3.56-MHz) continuous sine-wave signal from a color-bar generator that is pulsed on and off. This signal creates a series of different color bars on the screen of the color picture tube. A typical pulse rate (for 10 color bars) is 12 times per 1 horizontal line.

keyer—1. In telegraphy a device which breaks up the output of a transmitter or other device into the dots and dashes that are used in the code. 2. A radar modulator.

keyer adapter—A device that detects a modulated signal and produces a dc output signal whose amplitude varies in accordance with the modulation. In radio facsimile transmission, it is used to provide the keying signal for a frequency-shift exciter unit.

keying—The forming of signals, such as those employed in telegraph transmission, by an abrupt modulation of the output of a director an alternating-current source (e.g., by interrupting it or by suddenly changing its amplitude, frequency, or some other characteristic).

keying chirps—Sounds accompanying code signals when the transmitter is unstable and shifts slightly in frequency each time the sending key is closed.

keying frequency—In facsimile, the maximum number of times a second a black-line signal occurs while scanning the subject copy.

keying wave—Also called marking wave. The emission that takes place in telegraphic communication while the information portion of the code characters is being transmitted.

keyless ringing—A type of machine ringing on a manual switchboard. Ringing is started automatically when the calling plug is inserted into the jack of the called line.

key pulse—A telephone signaling system in which numbered keys are depressed instead of a dial being turned.

key pulsing—A switchboard arrangement using a nonlocking keyset for the transmission of pulse signals corresponding to the key depressions.

key-pulsing signal—The signal which indicates a circuit is ready for pulsing, in multifrequency and direct-current key pulsing.

key punch—A keyboard machine for manually punching information into paper tape or cards.

keyshelf—A shelf on which are mounted the keys by means of which the operator of a manual telephone switchboard performs switching of one or more of the switchboard circuits.

key station—The master station from where a network radio or television program originates.

keystone distortion—1. The distortion produced when a plane target area not normal to the average direction of the beam is scanned rectilinearly with constant-amplitude sawtooth waves. 2. A type of geometrical distortion that brings about a trapezoidal display of a nominally rectangular picture in a television system.

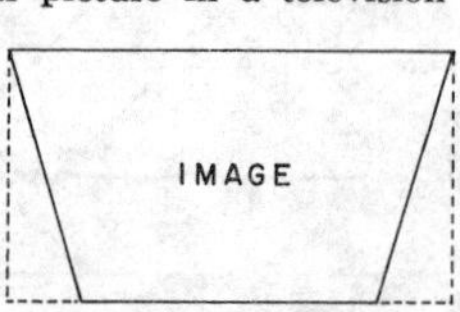

Keystone distortion.

keystone shaped—Wider at the top than at the bottom, or vice versa.

keystoning—The keystone-shaped scanning pattern produced when the electron beam in the television camera tube is at an angle with the principal axis of the tube. (*See also* Keystone Shaped.)

keyswitch—In an organ, the switch which is closed to allow a tone from the tone generator to sound when a key is depressed.

keyway — The mating slot or groove in which a key slides.

kHz—Abbreviations for kilohertz.

kickback—The voltage developed across an inductance by the sudden collapse of the magnetic field when the current through the inductance is cut off.

kickback power supply — *See* Flyback Power Supply.

kick-sorter—British term for pulse-height analyzer.

kidney joint—A flexible joint, or air-gap coupling, located in the waveguide and near the transmitting-receiving position of certain radars.

Kikuchi lines—A series of spectral lines obtained by the scattering of electrons, when an electron beam is directed against a crystalline solid. The pattern may be interpreted to yield information on the structure of the crystal and its mechanical perfection.

killer circuit—1. The vacuum tube or tubes and associated circuits in which are generated the blanking pulses used to temporarily disable a radar set. 2. In a transponder, a logic circuit that kills replies to side-lobe interrogations.

killer pulse—*See* Suppression Pulse.

kilo—Abbreviated k. A prefix representing 10^3, or 1000.

kiloampere—1000 amperes. Abbreviated kA.

kilobit—1000 bits.

kilocycle (kc)—1000 cycles. Generally interpreted as meaning 1000 cycles per second. Obsolete term, replaced by kilohertz (kHz).

kiloelectronvolt — 1000 electronvolts. The energy acquired by an electron that has been accelerated through a voltage difference of 1000 volts. Abbreviated keV.

kilogauss—1000 gausses.

kilogram (unit of mass)—The mass of a particular cylinder of platinum-iridium alloy, called the international prototype kilogram, which is preserved in a vault at Sevres, France, by the International Bureau of Weights and Measures.

kilohertz—1000 hertz. Abbreviated kHz.

kilohm—One thousand ohms. Abbreviated k or kohm.

kilohmmeter—A meter designed for measuring resistance in kilohms.

kilomega (kM)—Obsolete prefix for giga (G), representing 10^9, or 1,000,000,000.

kilomegacycle—Now called gigahertz. One billion cycles per second.

kilometer—One thousand meters, or approximately 3280 feet.

kilometric waves—British term for electromagnetic waves between 1000 and 10,000 meters in length.

kilosecond—1000 seconds.

kilovar—One reactive kilovolt-ampere, or 1000 reactive volt-amperes. Abbreviated kvar.

kilovar-hour — 1000 reactive volt-ampere-hours.

kilovolt—1000 volts. Abbreviated kV.

kilovolt-ampere — 1000 volt-amperes. Abbreviated kVa.

kilovoltmeter — A voltmeter which reads thousands of volts.

kilowatt—1000 watts. Abbreviated kW.

kilowatt hour—The equivalent energy supplied by a power of 1000 watts for one hour. Abbreviated kWh.

kine—Slang term for kinescope recording.

kine-klydonograph—An instrument that records the current-time characteristics of a lightning stroke. The instrument records a series of Lichtenberg figures in a manner similar to that of the field gradient recorder.

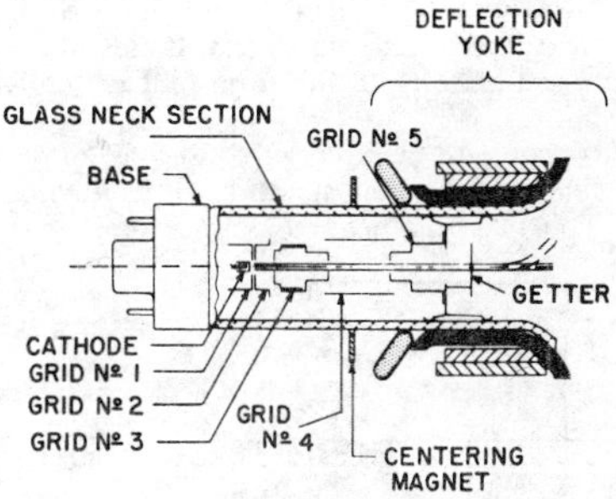

Kinescope, 1.

kinescope—1. In television receivers, the cathode-ray tube in which the electrical signals are translated into a visible picture on a luminescent screen. 2. A film recording made from a television program on a picture tube and used as a permanent record or for subsequent rebroadcasting.

kinescope recorder—A camera which photographs television images directly from the picture tube onto motion-picture film.

kinetic energy—Energy which a system possesses by virtue of its motion.

Kirchhoff's laws—1. The current flowing to a given point in a circuit is equal to the current flowing away from that point. 2. The algebraic sum of the voltage drops in any closed path in a circuit is equal to

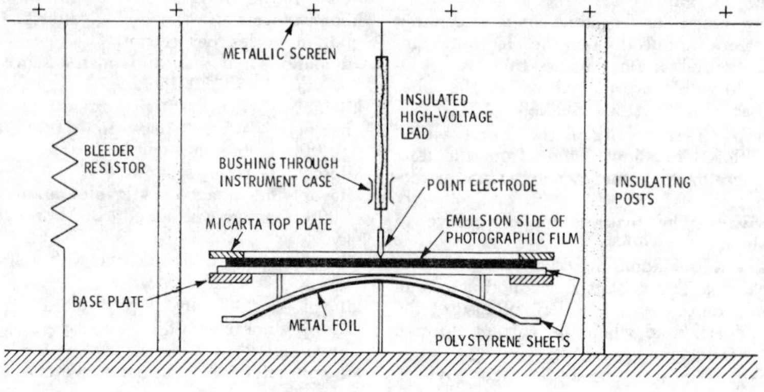

Klydonograph gradient recorder.

the algebraic sum of the electromotive forces in that path. (Laws 1 and 2 are also called laws of electric networks.) 3. At a given temperature, the emissive power of a body is the same as its radiation-absorbing power for all surfaces.

kit—A prepared package of parts with instructions for assembly and/or wiring a component or chassis (also a small accessory item).

klydonograph—A field gradient recording instrument that registers voltages on photographic film in the form of Lichtenberg figures.

klystron—An electron tube used as an oscillator or amplifier at ultrahigh frequencies. The electron beam is velocity-modulated (periodically bunched) to accomplish the desired results.

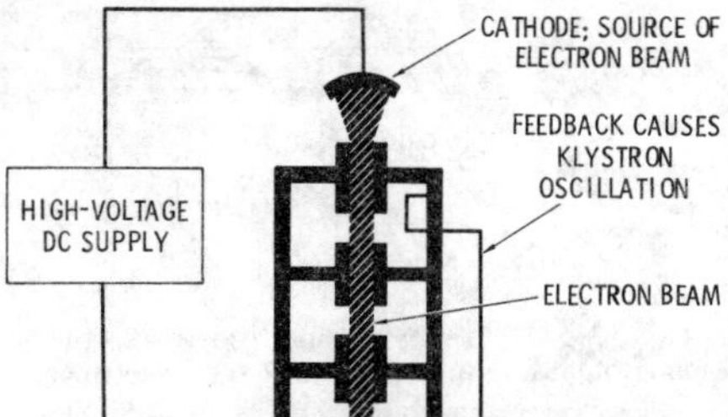

Klystron.

klystron control grid—An electrode which controls the emission, or beam current, of a klystron or other velocity-modulated tube.

klystron frequency multiplier—A two-cavity klystron that has the output cavity tuned to a multiple of the fundamental frequency.

klystron generator—A klystron tube used as a generator. Its cavity feeds energy directly into a waveguide.

klystron oscillator—An oscillator employing a klystron tube to generate radio-frequency power.

klystron repeater—A klystron tube operated an an amplifier and inserted directly into a waveguide in such a way that incoming waves velocity-modulate the electron stream emitted from a heated cathode. A second cavity converts the energy of the electron clusters into waves of a much higher amplitude and feeds them into the outgoing guide.

kM — Abbreviation for kilomega (an obsolete term).

kMc—Abbreviation for kilomegacycle. Now replaced by gigahertz.

knee—1. An abrupt change in direction between two relatively straight segments of a curve, such as the region of a magnetization curve near saturation or the top bend of a vacuum-tube characteristic curve. 2. A section between two comparatively straight segments of a curve in which the magnitude of curvature, although of the same sign, is relatively high.

knife-edge diffraction — In radio-wave propagation, an effect by which the atmospheric attenuation of a signal is reduced when the signal is diffracted as it passes over a sharp obstacle such as a mountain ridge.

knife-edge pointer (of a meter)—End of pointer is flattened and turned edgewise so smallest dimension or edge is seen. Usually used with mirror-backed scales to eliminate parallax and increase the accuracy of reading.

knife switch—A form of air switch in which a moving element is sandwiched between two contact clips. The moving element is usually a hinged blade; when it is not, it is removable.

knob – A round, polygonal, or pointer-shaped part which is fastened to one end of a control shaft so that the shaft can be turned more easily. The knob sometimes indicates the degree of rotation also.

knocker—A term used with some fire-control radars to indicate a subassembly comprising synchronizing and triggering circuits. It drives the rf pulse-generating equipment in the transmitter, and also synchronizes the cycle of operation with the transmitted pulse in range units and indicators.

knockout—A removable portion in the side of a box or cabinet. During installation it can be readily taken out with a hammer, screwdriver, or pliers so the raceway, cables, or fittings can be attached.

knot—One nautical mile (6,080.20 feet, or 1.15 statute miles) per hour.

kohm—Abbreviation for kilohm.

Kooman antenna—A vertical array of horizontal full-wave dipoles that are driven by transposed two-conductor line and backed by a parasitic reflecting curtain or horizontal dipoles.

Kovar – An iron-nickel-cobalt alloy with a coefficient of expansion similar to that of glass and silicon and thermal characteristics similar to those of alumina. It is used as a material for headers and in glass-to-metal seals.

kraft paper—Relatively heavy, high-strength sulfate paper used for electrical insulating material. (Capacitor tissue is kraft paper of normal thickness equal to 0.002 inch or less.)

K-scan—*See* K-Display.

K-series—A series of frequencies in the X-ray spectrum of an element.

KSR—(Keyboard send-receive set.) A combination transmitter and receiver with transmission capability from the keyboard only (teletypewriter term). Refers to a terminal device (teletype or similar) having only a keyboard for sending and a printer for receiving, i.e., no paper or magnetic tape equipment.

kurtosis—The degree of curvature of the peak of a probability curve.

kV—Abbreviation for kilovolt.

kVa—Abbreviation for kilovolt-ampere.

kvar—Abbreviation for kilovar.

kW—Abbreviation for kilowatt.

kWh—Abbreviation for kilowatt-hour.

kymograph—An instrument for recording wavelike oscillations of varying quantities for medical studies.

L

L—Symbol for coil or inductance.

label—1. A code name used to identify or classify a name, term, phrase, or document. 2. One or more characters that serve to identify an item of data. 3. A numerical value or a memory location in the programmable system of a computer. The specific absolute address is not necessary since the intent of the label is a general destination. Labels are a requisite for jump and branch instructions.

label group—A collection of continuous label sets of the same label type.

labile oscillator—A local oscillator the frequency of which is remote-controlled by a signal received from a raido or over a wire.

laboratory power supply – A regulated dc source having (a) less than 10-kV output at up to 500 W. (b) Output adjustable over a wide range, usually down to zero. (c) Regulation on the order of ±0.01% static line and load.

labyrinth—Speaker enclosure with absorbing air chambers at the rear to eliminate acoustic standing waves. A mazelike construction extends the air column. Resonances are tamed by heavy damping material.

labyrinth loudspeaker – Loudspeaker mounted in an acoustic baffle having air chambers designed to prevent acoustic standing waves.

lacquer disc – Also called cellulose-nitrate disc. A mechanical recording disc, usually made of metal, glass, or paper and coated with a lacquer compound often containing cellulose nitrate.

lacquer master—*See* Lacquer Original.

lacquer original—Also called lacquer master. An original recording made on a lacquer surface to be used as a master.

lacquer recording—Any recording made on a lacquer medium.

ladder attenuator—A series of symmetrical sections used in signal generators and other devices where voltages and currents must be reduced in known ratios. They are designed so that the required ratio of voltage loss per section is obtained with image-impedance operation. The impedance between any junction point and

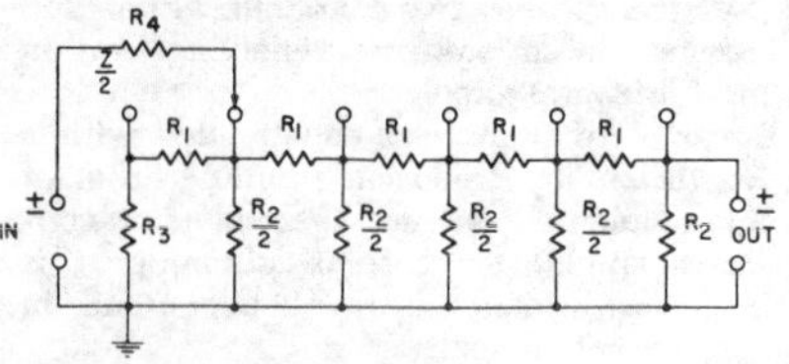

Ladder attenuator.

common ground in a ladder attenuator is half the image impedance.

ladder diagram—A diagram that shows actual component symbols and the basic wiring configuration of a relay logic circuit (as opposed to a logic diagram).

ladder network – Also called series-shunt network. A network composed of H-, L-, T-, or pi-networks connected in series.

LAFOT—Coded weather broadcasts issued by the U. S. Weather Bureau for the Great Lakes region. They are broadcast every six hours by Marine radiotelephone broadcasting stations on their assigned frequencies.

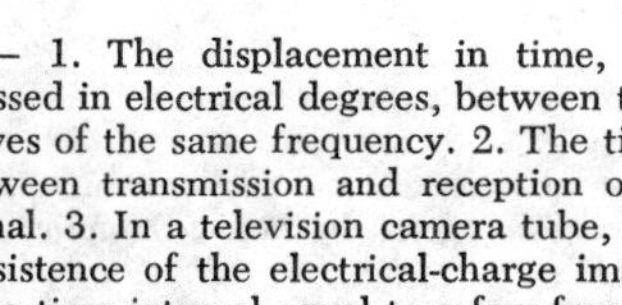

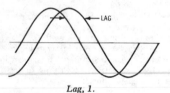

Lag, 1.

lag – 1. The displacement in time, expressed in electrical degrees, between two waves of the same frequency. 2. The time between transmission and reception of a signal. 3. In a television camera tube, the persistence of the electrical-charge image for a time interval equal to a few frames. 4. A time difference between the occurrence of two events.

lagged-demand meter – A meter in which there is a characteristic time lag, by either mechanical or thermal means, before maximum demand is indicated.

lagging current—The current flowing in a circuit that is mostly inductive. If the circuit contains only inductance, the current lags the applied voltage by 90°. Because of the characteristics of an inductance, the current does not change direction until after the corresponding voltage does.

lagging load – A predominantly inductive load—i.e., one in which the current lags the voltage.

lag-lead (lead-lag)—A circuit the response of which includes lag components and their derivatives.

lambda—Greek letter λ, used to designate wavelength measured in meters.

lambda wave – An electromagnetic wave propagated over the surface of a solid that has a thickness comparable to the wavelength of the wave.

lambert—A unit of luminance (photometric brightness) equal to $1/\pi$ candela per square centimeter and, therefore, equal to the uniform luminance of a perfectly diffusing surface emitting or reflecting light at the rate of one lumen per square centimeter. The lambert also is the average luminance of any surface emitting or reflecting light at the rate of one lumen per square centimeter. For the general case, the average must take account of variation of luminance with angle of observation, also of its variation from point to point on the surface considered.

Lambert's law of illumination—The illumination of a surface on which the light falls normally from a point source is inversely proportional to the square of the distance of the surface from the source. If the normal to the surface makes an angle with the direction of the rays, the illumination is proportional to the cosine of that angle.

laminated—Made of layers.

laminated contact—A switch contact made up of a number of laminations, each making contact with an opposite conducting surface.

laminated core – An iron core for a coil, transformer, armature, etc. It is built up from laminations to minimize the effect of eddy currents. The sheet iron or steel laminations are insulated from each other by surface oxides or by oxides and varnish.

laminated record—A mechanical recording medium composed of several layers of material (normally a thin face of material on each side of a core).

lamination—A single stamping of sheet material used in building up a laminated object such as the core of a power transformer.

Lamont's law—The permeability of steel at any flux density is proportional to the difference between the saturation value of the flux density and its value at the point in question. This law is only approximately accurate and is not true for the initial part of the magnetization curve.

lamp—A device for producing light.

lamp bank—An arrangement of incandescent lamps commonly used as a resistance load during electrical tests.

lamp cord – 1. A twin conductor, either twisted or parallel, used for connecting floor lamps and other electric appliances to wall outlets. 2. Flexible stranded conductor cord, rubber or plastic insulated, used in wiring of lamps, household fans, and similar appliances. Not subject to hard usage.

lamp holder—A lamp socket.

lamp jack – Special electronic electromechanical component having a frame which holds a lamp and has the contact springs and terminals for applying power to the lamp. Used extensively in jack panels and other types of telephone equipment as a visual-indicating signal device.

Lampkin oscillator – A variation of the Hartley oscillator. Its distinguishing feature is that an approximate impedance match is effected between the tank and grid-cathode circuits.

lamp receptacle – A device that supports an electric lamp and connects it to a power line.

land—1. The surface between two adjacent grooves of a recording disc. 2. Also called boss, pad, terminal point, blivet, tab, spot, donut. In a printed-circuit board, the conductive area to which components or separate circuits are attached. It usually surrounds a hole through the conductive material and the base material.

Landau damping—The damping of a space-charge wave by electrons moving at the phase velocity of the wave.

landing beacon—The radio transmitter that produces a landing beam for aircraft. (*See also* Landing Beam.)

landing beam—A highly directive radio signal projected upward from an airport to guide aircraft in making a landing during poor visibility.

landline – A telegraph or telephone line passing over land, as opposed to submarine cables.

landline facilities – Domestic communications common-carrier's facilities that are within the continental United States.

landmark beacon—Any beacon other than an airport or airway beacon.

land mobile service – A radio service in which communication is between base station and land mobile stations or between land mobile stations.

land mobile station—A two-way mobile station that operates solely on land.

land radio positioning station—A station in the radio positioning service, not intended to be operated while in motion.

land return – Radiation reflected from nearby land masses and returned to a radar set as an echo.

lands—Bonding points used in the manufacture of microelectronic circuits.

land station—A permanent, or fixed, station.

land transportation radio services – Radiocommunication services, the transmitting facilities of which include fixed, land, or mobile stations, operated by and for the sole use of certain land transportation carriers.

Langevin ion—An electrified particle produced in a gas by an accumulation of ions on dust particles or other nuclei.

Langmuir dark space – The nonluminous region surrounding a negatively charged probe inserted into the positive column of a glow or arc discharge.

language—1. A set of computer symbols, with rules for their combination. They form a code to express information with fewer symbols and rules than there are distinct expressible meanings. 2. A system for representing information and communicating it between people, or between people and machines. 3. A definition of the elements and syntax within which a computer program must be encoded.

language converter—A data-processing device designed to change one form of data, i.e., microfilm, strip chart, etc., into another (puch card, paper tape, etc.).

language translation – The process performed by an assembler, compiler, or other routine that accepts statements in one language and converts them to equivalent statements in another language.

L-antenna – An antenna consisting of an elevated horizontal wire to which a vertical lead is connected at one end.

lanyard—A device which is attached to certain quick-disconnect connectors and which permits uncoupling and separation of connector halves by a pull on a wire or cable.

lap—1. A rotating plate covered with liquid abrasive, used for grinding quartz crystals. 2. A fire-resistant, untwisted, ribbonlike form of asbestos felt made from slivers of asbestos fiber blended with cotton or other organic fibers. Used as a wrapping on wire and cable.

lap dissolve—In motion pictures or television, simultaneous transition in which one scene is faded down and out while the next scene is faded up and in.

lapel microphone—A microphone worn on the user's clothing.

lap joint—The connecting of two conductors by placing them side by side so that they overlap.

Laplace's law – The strength of the magnetic field at any given point due to any element of a current-carrying conductor is directly proportional to the strength of the current and the projected length of the element, and is inversely proportional to the square of the distance of the element from the point in question.

Laplace transform—A mathematical substitution the use of which permits the solution of a certain type of differential equation by algebraic means.

lapping—Bringing quartz crystal plates up to their final frequency by moving them over a flat plate over which a liquid abrasive has been poured.

lap winding – An armature winding in

which opposite ends of each coil are connected to adjoining segments of the commutator so that the windings overlap.

lap wrap—Tape wrapped around an object in an overlapping condition.

large-scale integrated circuit — An integrated circuit which contains 100 gates or more on a single chip, resulting in an increase in the scope of the function performed by a single device.

large-scale integration — Abbreviated LSI. 1. The simultaneous achievement of large-area circuit chips and optimum density of component packaging for the express purpose of cost reduction by maximization of the number of system interconnections made at the chip level. 2. Monolithic digital ICs with a typical complexity of 100 or more gates or gate-equivalent circuits. The number of gates per chip used to define LSI depends on the manufacturer. The term sometimes describes hybrid ICs built with a number of MSI or LSI chips. 3. The physical realization of a microelectronic circuit fabricated from a single semiconductor integrated circuit having circuitry equivalent to more than 100 individual gates. 4. Monolithic integrated circuits of very high density. Such circuits typically have on a single chip the equivalent of about two hundred to several thousand simple logic circuits.

large-signal characteristics—The characteristics of an amplifier when rated (full) output signals are produced.

large-signal dc current gain—The dc output current of a transistor with the dc output circuit shorted, divided by the dc input current producing the dc output current.

large-signal power gain — The ratio of the ac output power to the ac input power under specified large-signal conditions. Usually expressed in decibels (dB).

large-signal, short-circuit, forward-current transfer ratio—In a transistor, the ratio under specified test conditions of a change in output current to the corresponding change in input current.

Larmor orbit—The path of circular motion of a charged particle in a uniform magnetic field. The motion of the particle is unimpeded in the direction of the magnetic field, but motion perpendicular to the direction of the field is always accompanied by a force perpendicular to the direction of motion and the field.

laryngaphone—Also called a throat microphone. A microphone applied to the throat of a speaker to pick up voice vibrations directly. It is very useful in noisy locations because it picks up only the speaker's voice—no outside noises.

LASCR — 1. Light-activated silicon controlled rectifier. A pnpn device in which incident light performs the function of gate current; three of the four semiconductor regions are available for circuit connections. A photoswitch. 2. A semiconductor device that is triggered into conduction when the light falling on the base-collector photodiode junction exceeds a given threshld level. Operation of the LASCR is similar to the SCR, with the major difference being that an external resistance between gate and cathode (in addition to bias voltage and current) determines light sensitivity. A positive electrical signal applied to the gate can be used to trigger the LASCR, as well as to modify the light sensitivity.

LASCS — 1. Light-activated silicon controlled switch; it is similar to an LASCR, except that all four semiconductor regions are accessible. 2. A semiconductor device that combines the LASCR and the PSPS. Having four terminals, the LASCS can be triggered by light positive signals (at the gate terminal) and negative signals (at the anode gate terminal).

laser—1. A device for transforming incoherent light of various frequencies of vibration into a very narrow, intense beam of coherent light. The name is derived from the initial letters of "Light Amplification by Stimulated Emission of Radiation." In the emission of ordinary light the molecules or atoms of the source emit their radiation independently of each other, and consequently there is no definite phase relationship among the vibrations in the resultant beam. The light is incoherent. The laser, by means of an optical resonator, forces the atoms of the material of the resonator to radiate in phase. The emitted radiation is stimulated by the excitation of atoms to a higher energy level by means of energy supplied to the device. In the microwave

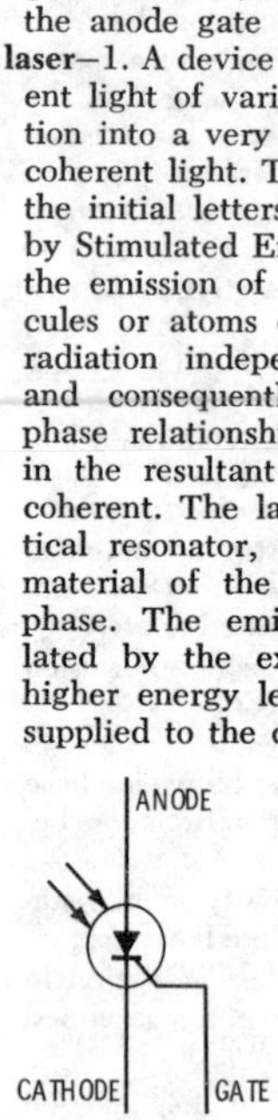

Symbol.

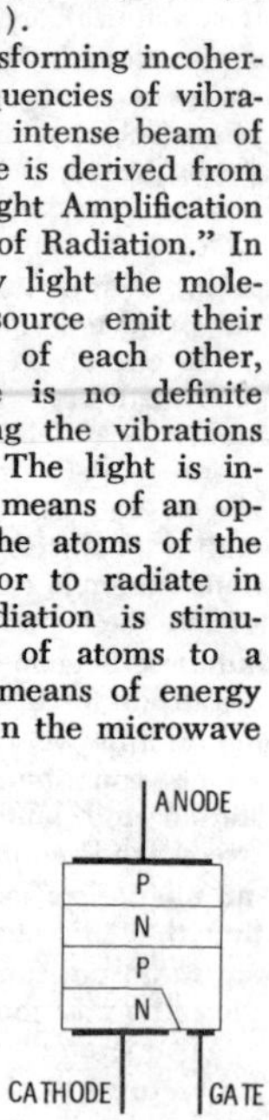

Junctions.

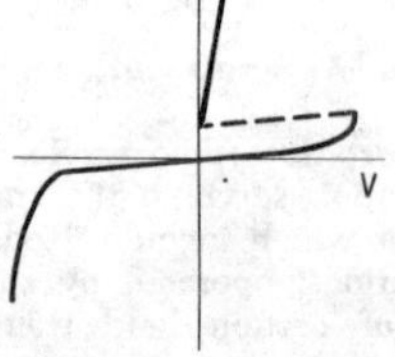

Anode characteristic.

LASCR.

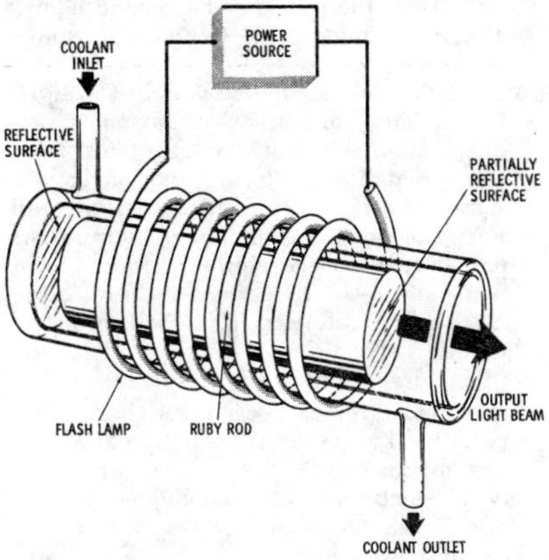

Laser.

region, the corresponding device is called a maser, and hence the laser is also known as a light maser. 2. A device for producing light by emission of energy stored in a molecular or atomic system when stimulated by an input signal. 3. A mechanically designed semiconductor junction which will optically pump (amplify light) short pulses of high-energy coherent radiation.

laser cavity—An optical resonant and hence mode-selecting low-loss structure in which laser action occurs through the build-up of electromagnetic field intensity upon multiple reflection.

laser diode—*See* Diode Laser.

laser ranger—A device similar to conventional radar but using high intensity light rather than microwaves.

laser welding – A method of welding in which material heating is accomplished by concentration of a beam of coherent light on the area until fusion of the materials takes place.

lasing—The phenomenon that certain materials exhibit when the threshold condition for self-sustaining photon emission has been achieved.

latch—1. A feedback loop used in a symmetrical digital circuit (such as a flip-flop) to retain a state. 2. A simple logic storage element. The most basic form consists of two cross-coupled logic gates that store a pulse applied to one logic input until a pulse is applied to the other input; thus, the complementary information is stored in the latch. 3. A name commonly used to refer to a flip-flop (usually a "D" type) when used for data storage, as opposed to counting and logic functions.

latching—A technique for storing an event such as the momentary breaking of a perimeter circuit. The fact that the event has occurred will be available until the latched circuit has been reset. *See* Alarm Hold.

latching current – The minimum value of principal current required to maintain a thyristor in the "on" state after switching from the "off" state to the "on" state has occurred and the trigger signal has been removed.

latching relay—A relay with contacts that lock in either the energized or de-energized position, or both, until reset either manually or electrically.

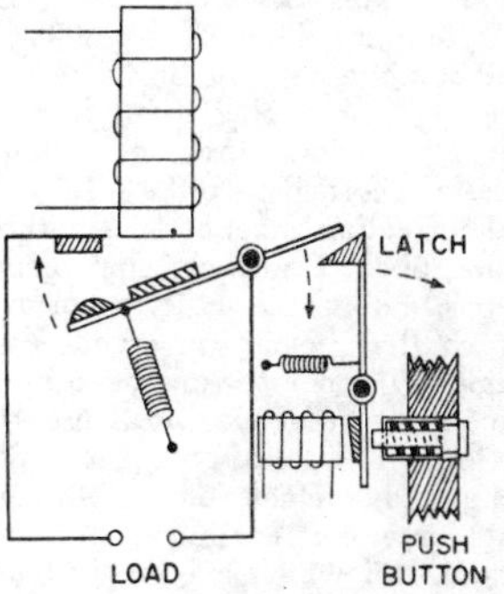

Latch-in relay.

latch-in relay—Also called locking relay. A relay with contacts which remain energized or de-energized until reset manually or electrically.

latch mode – A mode of operation for a storage circuit in which all encoder contact closures, even momentary ones, are latched "on."

latch-up—1. A condition in which the collector voltage in a given circuit does not return to the supply voltage when a transistor is switched from saturation to cutoff. Instead, the collector finds a stable operating point in the avalanche region of the collector characteristics. 2. An unintended stable circuit mode which will not revert to a previous intended circuit mode after removal of a stimulus such as a spurious signal or radiation. The effect is usually caused by parasitic circuit elements. 3. The characteristic of some op amps to remain in positive or negative saturation after their maximum differential input voltage is exceeded. 4. The switching of an electronic circuit to an unintended mode by improper voltage application.

latch voltage—The effective input voltage at which a flip-flop changes states.

late contacts—In a relay, contacts that open or close after other contacts when the relay operates.

latency—1. In a serial storage computer system, the time necessary for the desired storage location to appear under the drum heads. 2. In computers, the time required to establish communication with a specific storage location, not including transfer time, i.e., access time less word time. 3. A state of seeming inactivity, such as that occurring between the instant of stimulation and the beginning of response.

latent image—A stored image (e.g., the one contained in the charged mosaic capacitance in an iconoscope).

lateral chromatic aberration — Aberration which affects the sharpness of images off the axis. This occurs because different colors produce different magnifications.

lateral compliance—The force required to move the reproducing stylus from side to side as it follows the modulation on a laterally recorded record.

lateral-correction magnet — In a three-gun picture tube, an auxiliary component used for positioning the blue beam horizontally so that beam convergence will be obtained. It operates on the principle of magnetic convergence and is used in conjunction with a set of pole pieces mounted on the focus element of the blue gun.

lateral forced-air cooling — A method of heat transfer which employs a blower to produce side to side circulation of air through or across the heat dissipators.

lateral recording—A mechanical recording in which the groove modulation is perpendicular to the direction of motion of the recording medium and parallel to its surface.

lattice—1. In navigation, a pattern of identifiable intersection lines placed in fixed positions with respect to the transmitters that establish them. 2. The geometrical arrangement of atoms in a crystalline material.

lattice network — A network composed of four branches connected in series to form a mesh. Two nonadjacent junction points serve as input terminals, and the remaining two as output terminals.

lattice structure—In a crystal, a stable arrangement of atoms and their electron-pair bonds.

lattice-wound coil—*See* Honeycomb Coil.

launch complex — The entire launch, control, and support system required for launching rockets.

launching—The transferring of energy from a coaxial cable or shielded paired cable in a waveguide.

lavalier microphone — A microphone with acoustical and vibration-isolation properties suiting it to use for speech pickup from a position on the speaker's chest. Lavalier mikes are fitted with a band or strap for hanging around the neck, and are frequently used when it is important that the mike not be conspicuously visible (as to a tv audience). The use of this mike frees the speaker's hands and allows a certain amount of freedom to move about.

LAWEB—Weather bulletins issued every 6 hours. They are given in layman's language from ship and shore positions along the Great Lakes during the navigation season. Part 1 is from land stations, and Part 2 is from a ship four or more miles off shore with the ship's position given.

lawn mower—1. In facsimile, a term often used when referring to a helix-type recorder mechanism. 2. A type of rf preamplifier used with a radar receiver.

law of electric charges—Like charges repel; unlike charges attract. (*See also* Coulomb's Law.)

law of electromagnetic induction—*See* Faraday's Laws, 3.

law of eletrcomagnetic systems — Every electromagnetic system tends to change its configuration so that the flux of magnetic induction will be a maximum.

law of electrostatic attraction — *See* Coulomb's Law.

law of magnetism—Like poles repel; unlike poles attract.

law of normal distribution—The Gaussian law of the frequency distribution of any normal, repetitive function. It describes the probability of the occurrence of deviants from the average.

law of reflection—The angle of reflection is equal to the angle of incidence—i.e., the incident, reflected, and normal rays all lie in the same plane.

laws of electric networks—*See* Kirchhoff's Laws, 1 and 2.

lay—Pertaining to wire and cable, the axial distance required for one cable conductor or conductor strand to complete one revolution about the axis around which it is cabled.

layer—The consecutive turns of a coil lying in a single plane.

layer-to-layer adhesion—The tendency for adjacent layers of recording tape in a roll to adhere, particularly after prolonged storage under conditions of high temperature and/or humidity.

layer-to-layer signal transfer—The magnetization of a layer of tape in a roll by the field from a nearby recorded layer. The magnitude of the induced signal tends to increase with storage time and temperature, and to decrease after the tape is unwound. These changes are a function of the magnetic instability of the oxide.

layer winding — A coil-winding method in which adjacent turns are placed side by side and touch each other. Additional layers may be wound over the first and are usually separated by sheets of insulation.

layout—1. Diagram indicating the positions of parts on a chassis or panel. 2. The ac-

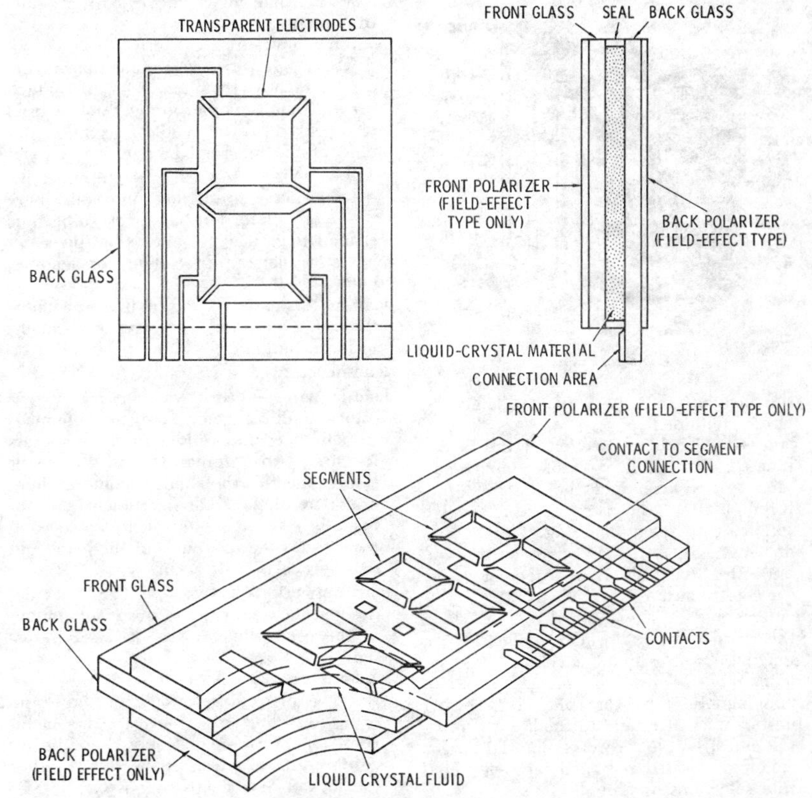

Liquid-crystal display.

tual positions of the parts themselves. 3. The topological arrangement of conductors and components in the design of integrated circuits. It precedes the artwork.

L-band—A radio-frequency band of 390 to 1550 MHz and corresponding wavelengths of 77 to 19 cm.

L-carrier system—A telephone carrier system employed on coaxial-cable systems and microwave line-of-sight and tropospheric-scatter radio systems. It occupies a frequency band extending from 68 kHz to over 8 MHz.

LCD – Abbreviation for liquid crystal display. A 7-segment (typically) display device consisting basically of a liquid crystal hermetically sealed between two glass plates. One type of LCD (dynamic scattering) depends upon ambient light for its operation, while a second type depends upon a backlighting source. The readout is either dark characters on a dull white background or white on a dull black background. LCDs have very low power requirements.

LC product – Inductance (L) in henrys multiplied by capacitance (C) in farads.

L/C ratio—Inductance in henrys divided by capacitance in farads.

LCS—Abbreviation for loudness-contour selector.

L-display—Also called L-scan. A radar display in which the target indication appears as two horizontal blips, one extending to the right and one to the left from a vertical time base. Azimuth pointing error is indicated by relative blip amplitude, and range is indicated by the position of the signal along the base line.

lead—1. A wire to or from a circuit element. 2. To precede (the opposite of lag).

lead-acid cell—Also called lead cell. A cell in an ordinary storage battery. It consists of electrodes (plates) immersed in an electrolyte of dilute sulfuric acid. The electrodes contain certain lead oxides that change their composition as the cell is charged or discharged.

lead cell—*See* Lead-Acid Cell.

lead-covered cable – A cable with a lead sheath. The sheath offers protection from

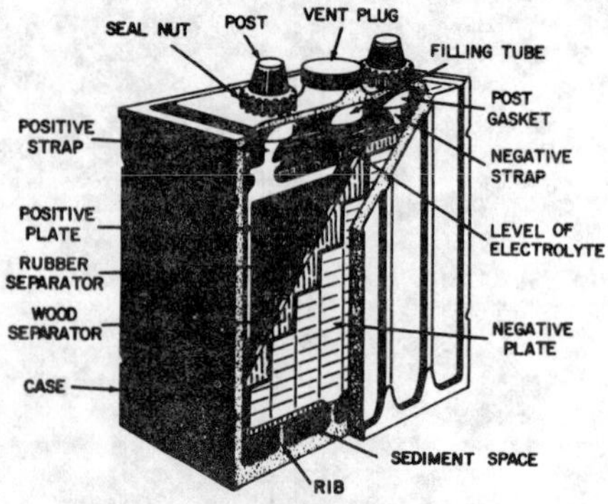

Lead-acid cell.

the weather and mechanical damage to wires contained.

leader—1. The blank section at the beginning of a tape. 2. Tough, nonmagnetic tape spliced ahead of the recorded material on a tape that is expected to receive rough or frequent handling. Usually has one matte-finished surface for writing on, and often available in a variety of colors for coding purposes.

leader cable—A navigational aid in which the path to be followed is defined by a magnetic field around a cable.

leader tape—Also called timing tape. Plain nonmagnetic tape for splicing to either end of magnetic tape to facilitate threading and preserve recorded material, or for splicing between recorded tapes to separate selections or provide pauses.

lead frame—A metal frame that holds the leads of a plastic encapsulated package (DIP) in place before encapsulation and is cut away after encapsulation.

lead-in – The conductor that provides the path for rf energy between the antenna and the radio/television receiver or transmitter.

leading blacks—Also called edge effect. In a television picture, the condition where the edge preceding a white object is overshaded toward black (i.e., the object appears to have a preceding, or leading, black border).

leading current—Current that reaches maximum before the voltage that produces it does. A leading current flows in any predominantly capacitive circuit.

leading edge – That transition of a pulse which occurs first.

leading-edge pulse time—The time required by a pulse to rise from its instantaneous amplitude to a stated fraction of its peak amplitude.

leading ghost—A twin image appearing to the left of the original in a televised picture.

leading load – A predominantly capacitive load—i.e., one in which the current leads the voltage.

lead-in groove—Also called a lead-in spiral. A blank spiral groove around the outside of the record. Its pitch is usually much greater than the other grooves and is used to quickly lead the needle into the beginning of the recorded groove.

leading whites—Also called edge effect. In a television picture, the condition where the edge preceding a black object is shaded toward white (i.e., the object appears to have a preceding, or leading, white border).

lead-in insulator – A tubular insulator through which cables or wires are brought inside a building.

lead-in spiral—*See* Lead-in Groove.

lead-in wires – Wires which carry current into a building (e.g., from an antenna).

lead-length compensation—In dc ammeters for use with external shunts, the leads that connect to the shunt become an integral part of the total instrument. An adjustable resistor is often included to compensate for the resistance of the leads and to improve overall accuracy.

lead network—A network, either ac or dc, designed to provide error-rate damping in the controlling device of a servo system.

lead-out groove – Also called a throw-out spiral. A blank spiral groove on the inside of a recording disc, next to the label. It is generally much deeper than the recording groove and is connected to either the locked or eccentric groove.

lead-over groove – Also called a crossover spiral. On disc records containing several selections, the groove in which the needle travels as it crosses from one selection to the next.

lead polarity of transformer – Also called polarity. A designation of the relative instantaneous directions of currents in the leads of the transformer. Primary and secondary leads are said to have the same polarity when, at a given instant, the current enters the primary lead in question and leaves the secondary lead in question in the same direction as though the two leads formed a continuous circuit.

lead screw—1. In a recording, a threaded rod which guides the cutter or reproducer across the surface of a disc. 2. In facsimile, a threaded shaft which moves the scanning mechanism or drum lengthwise.

lead time—In the display of a random sampling oscilloscope, the interval represented that occurs immediately before trigger recognition.

leaf insulator—Leaf-spring-shaped insulator located in a switch stack adjacent to a contact spring or actuator spring to keep that spring from making electrical contact with an adjacent spring or other metallic surface.

leak—A condition that causes current to be shunted away from its destination through a low resistance.

leakage — 1. Undesired flow of electricity over or through an insulator. 2. The portion not utilized most effectively in a magnetic field (e.g., at the end pieces of an electromagnet).

leakage coefficient—Ratio of total to useful flux produced in the neutral section of a magnet.

leakage current — 1. An undesirable small value stray current which flows through (or across the surface of) an insulator or the dielectric of a capacitor. 2. A current which flows between two or more electrodes in a tube other than across the interelectrode space. 3. Current prior to switching at a specified voltage. 4. Undesirable flow of current through or over a surface of an insulating material or insulator. 5. All currents, including capacitively coupled currents, which may be conveyed between energized parts of a circuit and ground or other parts.

leakage flux—The flux which does not pass through the air gap, or useful part, of the magnetic circuit.

leakage inductance—A self-inductance due to the leakage flux generated in the winding of a transformer.

leakage power — In tr and pre-tr tubes, the radio-frequency power transmitted through a fired tube.

leakage radiation—Spurious radiation in a transmitting system — i.e., radiation from other than the system itself.

leakage reactance — The reactance represented by the difference in value between two mutually coupled inductances when their fields are aiding and then opposing.

leakage resistance—The normally high resistance of the path over which leakage current flows.

leakance—The reciprocal of insulation resistance.

leaktight—*See* Hermetic.

leaky—Usually applied to a capacitor in which the resistance has dropped so far below normal that objectionably high leakage current flows.

leaky waveguide—A waveguide with a narrow longitudinal slot, permitting a continuous energy leak.

leaky waveguide antenna—An antenna constructed from a long waveguide with radiating elements along its length. It has a very sharp pattern.

leapfrogging—The process of phasing, or delaying, the ranging pulse of a tracking radar in order to move, or shift (on the scope presentation) the tracking gate (at target blip) past the target blip from another radar.

leapfrog test — A computer check routine using a program that calls for performing a series of arithmetical or logical operations on one section of memory locations, transferring to another section, checking correctness of transfer, and repeating the series of operations. Eventually, all storage positions are checked by this process.

least maximum deviation—A manner of expressing nonlinearity as a deviation from a straight line for which the deviations for proportional or normal linearity are minimized.

least mechanical equivalent of light — The radiant power that is contained in one lumen at the wavelength of maximum visibility. It is equal to 1.46 milliwatts at a wavelength of 555 millimicrons.

least significant bit—1. The digit with the lowest weighting in a binary number. Abbreviated LSB. 2. In a system in which a numerical magnitude is represented by a series of binary (i.e., two-valued) digits, that digit (or "bit") that carries the smallest value or weight.

least significant digit — Abbreviated LSD. The rightmost digit of a number; the digit that has the lowest place value in a number.

least voltage coincidence detection—A system that provides protection against interfering signals by blocking all signals except those having a pulse-repetition frequency the same as or some exact multiple of the radar-set pulse-repetition frequency. Abbreviated LVCD.

Lecher line—*See* Lecher Wire.

Lecher oscillator—A device for producing standing waves on two parellel wires called Lecher wires.

Lecher wire—A type of transmission line used to measure wavelength, consisting of a pair of wires whose electrical length is adjustable. If a source of radio frequency is coupled to one end of the line and the line is adjusted until a set of standing waves is formed, the wavelength may be determined by measurement of the distance between adjacent nodes.

Leclanche cell—1. Type of dry cell comprising a positive carbon pole contained in a porous vessel filled with manganese dioxide, the whole assembly standing in a container of an ammonium chloride solution which also contains the negative zinc pole. The electromotive force generated by a cell of this type is approximately 1.5 volts. 2. An ordinary dry cell. It is a primary cell with a positive electrode of carbon and a negative electrode of zinc in an electrolyte of sal ammoniac and a depolarizer of manganese dioxide.

LED — Abbreviation for light-emitting diode. A pn junction semiconductor device specifically designed to emit light when forward biased. This light can be one of several visible colors—red, amber yellow, or green—or it may be infrared and thus invisible. (The schematic symbol for a LED is similar to the symbol for a conventional diode except that the arrows are added to indicate light emission.) Electrically, a LED is similar to a conventional diode in that it has a relatively low forward voltage threshold. Once this threshold is exceeded, the junction has a low impedance and conducts current readily. This current must be limited by an external circuit, usually a resistor. The amount of light emitted by a LED is proportional to the forward current over a broad range, thus it is easily controlled, either linearly or by pulsing. The LED is extremely fast in its light output response after the application of forward current. Typically, the rise and fall times are measured in nanoseconds.

ledger balance – A facility used with message switching equipment to ensure that no messages are lost within the center. It involves comparing the number of addresses received with the number of addresses transmitted.

left-handed polarized wave – Also called counterclockwise-polarized wave. An elliptically polarized transverse-electromagnetic wave in which the electric intensity vector rotates counterclockwise (looking in the direction of propagation).

left-hand rule—*See* Fleming's Rule, 1.

left-hand taper—The greater resistance in the counterclockwise half of the operating range of a rheostat or potentiometer than in the clockwise half (looking from the shaft end).

left (or right) signal—The electrical output of a microphone or combination of microphones placed so as to convey the intensity, time, and location of sound originating predominantly to the listener's left (or right) of the center of the performing area.

left (or right) stereophonic channel—The left (or right) signal as electrically reproduced in the reception of fm stereophonic broadcasts.

leg—A section or branch of a component or system (e.g., one of the windings of a transformer).

legend—A table of symbols or other data placed on a map, chart, or diagram to assist the reader in interpreting it.

Lenard rays – Cathode rays that emerge from a special vacuum tube through a thin glass window or metallic foil.

Lenard tube – An electron tube in which the beam can be taken through a section of the wall of the evacuated enclosure.

length of a scanning line—1. The length of the path traced by the scanning or recording spot as it moves from line to line. 2. On drum-type equipment, the circumference of the drum. 3. The spot speed divided by the scanning-line frequency.

lens – 1. An optical device which focuses light by refraction. 2. An electrical device which focuses microwaves by refraction or diffraction. 3. An acoustic device which concentrates sound waves by refraction. 4. An electronic optical device which focuses electrons.

lens antenna—A microwave antenna with a dielectric lens placed in front of the dipole or horn radiator so that the radiated energy is concentrated into a narrow beam.

lens disc—A television scanning disc having a number of openings arranged in a spiral, with a lens set into each opening.

lens speed—The amount of light a lens will pass. It is equal to the focal length divided by the diameter of the lens.

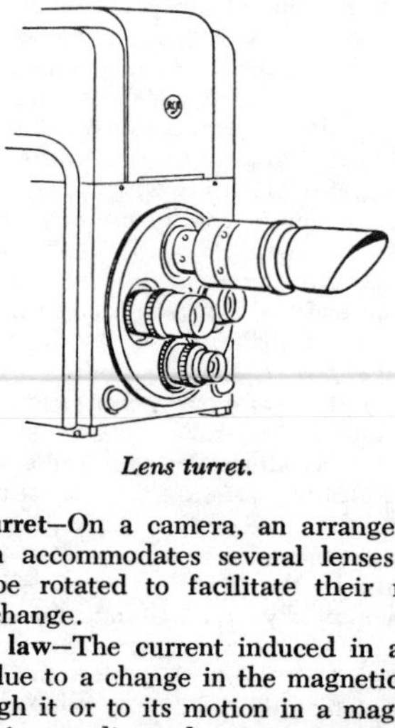

Lens turret.

lens turret—On a camera, an arrangement which accommodates several lenses and can be rotated to facilitate their rapid interchange.

Lenz's law—The current induced in a circuit due to a change in the magnetic flux through it or to its motion in a magnetic field is so directed as to oppose the change in flux or to exert a mechanical force opposing the motion. If a constant current flows in a primary circuit A and if by motion of A or the secondary circuit, B, a current is induced in B, the direction of the induced current will be such that, by its electromagnetic action on A, it tends to oppose the relative motion of the circuits.

Lepel discharger – A quenched spark gap used in early radiotelegraph transmitters employing shock excitation.

letters shift—In the Baudot code, a control character following which all characters are interpreted as being in the group containing letters (lower case).

let-through current—The current that actually passes through a circuit breaker under short-circuit conditions.

level—1. The magnitude of a quantity in

relation to an arbitrary reference value. Level normally is stated in the same units as the quantity being measured (e.g., volts, ohms, etc.). However, it may be expressed as the ratio to a reference value (e.g., dB—as in blanking level, transmission level, etc.). 2. A voltage that remains constant over a long period of time. 3. In describing codes or characters, a bit or element. 4. The intensity of an electrical signal.

level above threshold — Also called sensation level. The pressure level of a sound in decibels above its threshold of audibility for the individual listener.

level compensator — 1. An automatic gain control which minimizes the effect of amplitude variations in the received signal. 2. A device that automatically controls the gain in telegraph-receiving equipment.

level indicator—A device for showing visually the level of the audio signal, as a means of establishing the optimum amount of signal being fed to the tape.

level translator—A circuit that accepts digital input signals at one pair of voltage levels and delivers output signals at a different pair of voltage levels. For example, a circuit to "translate" the −0.8-V "zero" and 1.6-V "one" of ECL to −0.8-V "zero" and −4.2-V "one" suitable for COS/MOS.

level-triggered flip-flop—A flip-flop that responds to the voltage level rather than the rate of change of an input signal.

lever switch — Commonly referred to as a key lever or lever key. A hand-operated switch for rapidly opening and closing a circuit.

Lewis antenna—A microwave scanning antenna consisting of a lensed flat horn that tapers to a narrow rectangular opening across which a waveguide feed can be moved to scan the beam. The horn is folded by the incorporation of a 45° reflecting strip, and the thin rectangular end is formed into a circular annulus, around which the feed can be rotated. The deformed parallel-plate region that results has a conical shape with the feed circle as base.

Leyden jar—The original capacitor. It consists of metal foil sheets on the inside and outside of a glass jar. The foil serves as the plates and the glass as the dielectric.

lf — Abbreviation for low frequency—(i.e., between 30 and 300 kHz).

LFM — A vhf fan-type marker. It is low powered (5 watts) and has a range of only 10 miles or less.

library—1. A collection, usually stored on magnetic tape, of computer programs or subroutines for special purposes. 2. A group of standard, proven computer routines that can be incorporated into larger routines.

LIC—Abbreviation for linear integrated circuit.

Lichtenberg figure camera — Also called klydonograph or surge-voltage recorder. A device for indicating the polarity and approximate crest value of a voltage surge by the appearance and dimensions of the Lichtenberg figure produced on a photographic plate or film. The emulsion coating of the plate or film contacts a small electrode coupled to the circuit in which the surge occurs. The film is backed by an extended plane electrode.

lie detector—Also called a polygraph. An electronic instrument which measures the blood pressure, temperature, heart action, breathing, and skin moisture of the human body. Abrupt or violent changes in these variables indicate the subject is not telling the truth.

life—The expected number of full excursions over which a transducer would operate within the limits of the applicable specification.

life test—The test of a component or unit under the conditions which approximate, or simulate by acceleration, a normal lifetime of use. The test is performed to determine life expectancy or reliability throughout a predetermined life expectancy.

lifetime — The average time interval between the introduction and recombination of minority carriers in a semiconductor.

lifter—In a tape recorder, a movable rod or guide which draws the tape away from the heads during fast-forward or rewind modes, to eliminate needless head wear. Lifters work automatically on most machines in either high-speed mode.

lifting magnet—A powerful electromagnet used on the end of a crane to lift iron and steel objects. They can be dropped

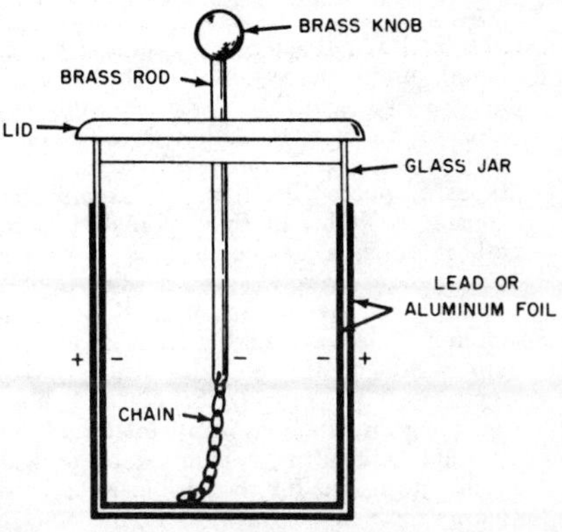

Leyden jar.

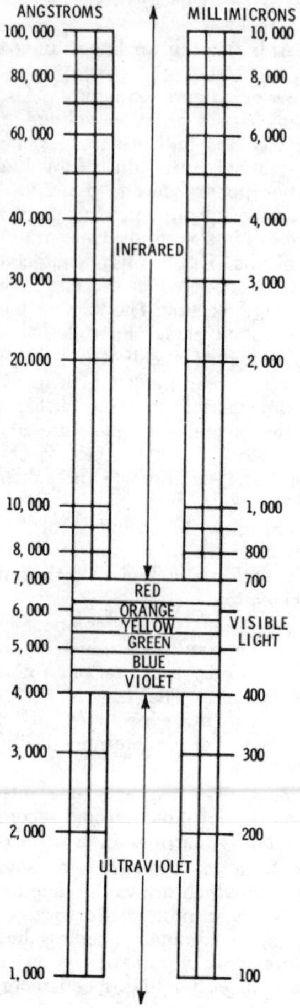

Light, 1 (spectrum).

instantly by merely cutting off the current.

light—1. Radiant energy within the wavelength limits perceptible by the average human eye (roughly, between 400 and 700 millimicrons). Although ultraviolet and infrared emissions will excite some types of photocells, they are usually not considered light. 2. In combination with other terms, a device used as a source of luminous energy (e.g., a pilot light). 3. Radiant energy transmitted by wave motion with wavelengths from about 0.3 μm to 30 μm; this includes visible wavelengths (0.38 μm to 0.78 μm) and those wavelengths, such as ultraviolet and infrared, which can be handled by optical techniques used for the visible region. In more restricted usage, radiant energy within the limits of the visual spectrum.

light-activated silicon controlled rectifier—*See* LASCR.

light-activated silicon controlled switch – *See* LASC.

light-activated switch—A semiconductor diode which is triggered into conduction by light irradiation of a light-sensitive part of the semiconductor pellet.

light amplifier – A solid-state amplifier using photoconductive and electroluminescent films.

light-beam cathode-ray-tube recorder—Recorder using electron beam to make multiple traces on crt screen. Traces are reflected from a fixed plane mirror onto moving photosensitive paper via an optical system.

light-beam galvanometer—A modified form of the D'Arsonval meter movement in which a small mirror is cemented to a moving coil mounted in the field of a permanent magnet. Current through the coil causes the coil to be deflected angularly, and the mirror reflects a beam of light onto a moving strip of photographic paper. The developed chart shows the waveform of the current through the coil.

light-beam instrument – An instrument in which a beam of light is the indicator.

light-beam oscillograph—Recorder using a mirror on a galvanometer to achieve recording response to 5 kHz.

light-beam pickup – A phonograph pickup utilizing a beam of light as a coupling element of the transducer.

light chopper—A device for interrupting a light beam. It is frequently used to facilitate amplification of the output of a phototube on which the beam strikes.

light current – The current that flows through a photosensitive device, such as a phototransistor or a photodiode, when it is exposed to illumination or irradiance.

light-dimming control—A circuit, often employing a saturable reactor, used to control the brightness of the lights in theaters, auditoriums, etc.

light-emitting diode—Abbreviated LED. A pn junction that emits light when biased in the forward direction.

light flux—*See* Luminous Flux, 1.

light gun – A photoelectric cell used by computer operators to take specific actions in assisting and directing computer operation. So called because of its gun-like case.

lighthouse tube – An ultrahigh-frequency electron tube shaped like a lighthouse and having disc-sealed planar elements. *See also* Disc-Seal Tube.

lighting outlet – An outlet for direct connection of a lamp holder, lighting fixture, or pendant cord terminating in a lamp holder.

light level – The amount of (or intensity of) light falling upon a subject.

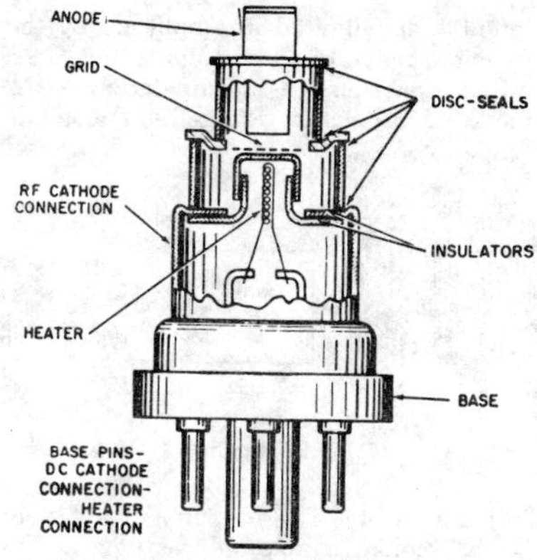

Cutaway view of lighthouse tube.

light load—A fraction of the total load the device is designed to handle.

light meter—An electronic device that contains a photosensitive cell and calibrated meter for the measurement of light levels.

light microsecond—The unit for expressing electrical distance. It is the distance over which light travels in free space in one microsecond (i.e., about 983 feet, or 300 meters).

light modulation—Variation in the intensity of light, usually at audio frequencies, for communications or motion-picture sound purposes.

light modulator—The device for producing the sound track on a motion-picture film. It consists of a source of light, an appropriate optical system, and a means for varying the resulting light beam (such as a galvanometer or light valve).

light negative – Having a negative photoconductivity when subjected to light.

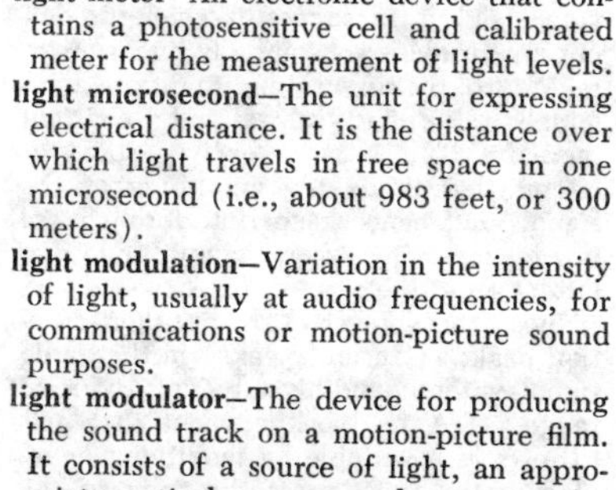

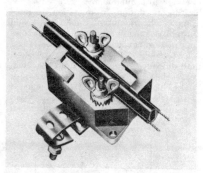

Lightning arrester.

lightning arrester – A device to prevent damage to electrical equipment by transient overvoltages whether from lightning or switching. Spark gaps which can only be bridged by voltages above those used in the equipment allow the higher voltages to be discharged to ground.

lightning generator – A generator of high-voltage surges (e.g., for testing insulators).

lightning rod—A rod projecting above the highest point on a structure and connected to ground in such a way that it can carry a lightning discharge to ground.

lightning surge—A transient disturbance in an electric circuit caused by lightning.

lightning switch—A switch for connecting a radio antenna to ground during electrical storms.

light pen—1. A light-sensitive device used with a computer-operated crt display for selecting a portion of the display for action by the computer. 2. A photosensor placed in the end of a penlike probe. It is used in conjunction with a crt (cathode-ray-tube) display for drawing, erasing, or locating characters. Operation is by comparison of the time it senses a light pulse to the scanning time of the display. 3. A hand-held data-entry device used only with refresh displays. It consists of an optical lens and photocell, with associated circuitry, mounted in a wand. Most light pens have a switch on the barrel which makes the pen sensitive to light from the screen. An activated light pen, when pointed at a vector or character on the screen, will generate an interrupt. It is then possible to identify the vector or character since the display stopped refreshing when the item was drawn that caused the interrupt. The most common uses of light pens are light-button selection and tracking. 4. A pencillike tube with a photocell at its lower end. The photocell is connected by wires to the control circuitry of the display it is being used with. The visible data on the display are produced by a sequential scanning action, and any lighted dot or segment is actually turning on and off very rapidly.

As a result of the scanning nature of the display, the control electronics knows at all times exactly what dot or portion has been reached in the scanning sequence. Thus, when the light pen touches a lighted point of the display, the pen's photocell produces an electrical signal that coincides in time with that specific display point. Effectively, then the light pen signal uniquely identifies that location on the display.

light pipe – 1. A bundle of transparent fibers which can transmit light around corners with small losses. Each fiber transmits a portion of the image through its length, reflection being caused by the lower refractive index of the surrounding material, usually air. 2. Transparent matter that usually is drawn into a cylindrical or conical shape through which light is channeled from one end to the other by total internal reflections. Optical fibers are examples of light pipes.

light positive – Having positive photoconductivity—i.e., increasing in conductivity when subjected to light.

light ray—A very thin beam of light.

light relay – A photoelectric device that opens or closes a relay when the intensity of a light beam changes.

light sensitive—Exhibiting a photoelectric effect when irradiated (e.g., photoelectric emission, photoconductivity, and photovoltaic action).

light-sensitive Darlington amplifier – Two stages of transistor amplification in one light-detector device. Darlingtons give higher gain than single transistors.

light-sensitive tube – A vacuum tube that changes its electrical characteristics with the amount of illumination.

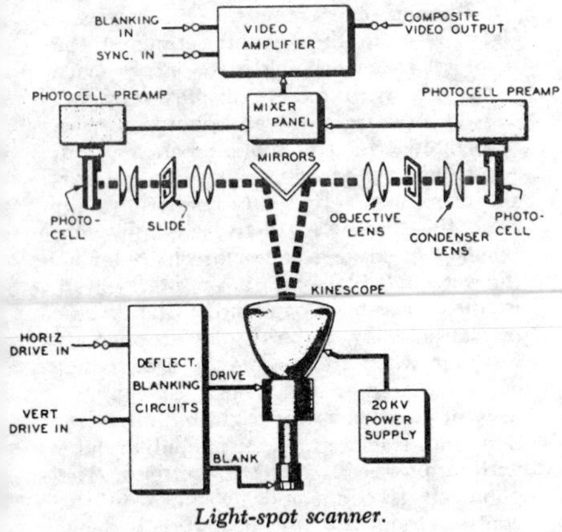

Light-spot scanner.

light-spot scanner—Also called a flying-spot scanner. A television camera in which the source of illumination is a spot of light that scans the scene to be televised. The picture signal is generated in a phototube, which picks up light either transmitted through the scene or reflected from it.

light valve—A device the light transmission of which can be varied in accordance with an externally applied electrical quantity such as voltage, current, an electric or magnetic field, or an electron beam.

light-year—The distance traveled by light in one year, or about 5,880,000,000,000 miles.

limit bridge—A form of Wheatstone bridge used for rapid routine production testing. Conformity with tolerance limits, rather than exact value, is determined.

limited signal—In radar, a signal that is intentionally limited in amplitude by the dynamic range of the system.

limited space-charge accumulation – A mode of oscillation for gallium arsenide diodes.

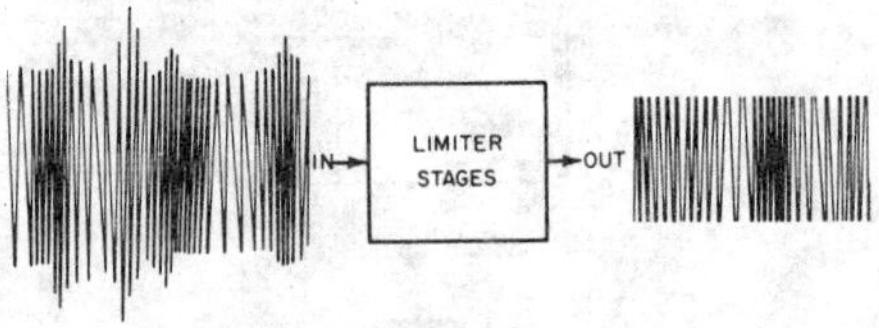

Limiter action.

limited stability—The property of a system which remains stable only as long as the input signal falls within a particular range.

limiter—1. A device in which some characteristic of the output is automatically prevented from exceeding a predetermined value—e.g., a transducer in which the output amplitude is substantially linear (with regard to the input) up to a predetermined value and substantially constant thereafter. 2. A radio-receiver stage or circuit that limits the amplitude of the signals and hence keeps interfering noise low by removing excessive amplitude variations from the signals. 3. A device that reduces the intensity of very-short-duration peaks (transient peaks) in the audio signal without audibly affecting dynamic range. 4. A feedback element that acts to restrain a variable by modifying or replacing the function of the primary element when predetermined conditions have been reached.

limiting—The restricting of the amplitude of a signal so that interfering noise can be kept to a minimum.

limiting resolution—In television, a measure of the resolution usually expressed in terms of the maximum number of lines per picture height discriminated on a test chart.

limit of error—An accuracy index that indicates the expected maximum deviation of the measured value from the true value if all of the factors causing deviations act simultaneously and in the same direction.

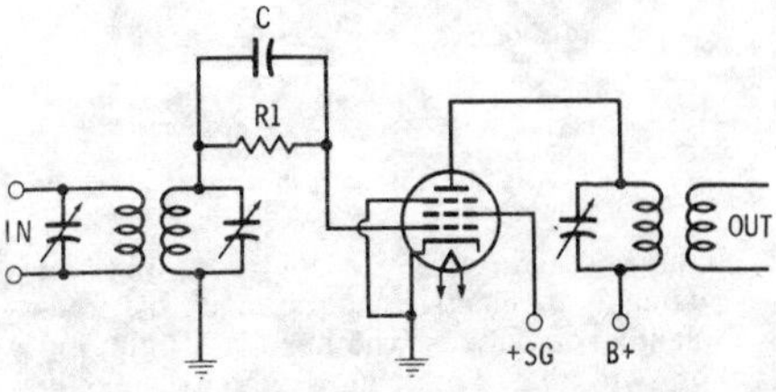

Limiter circuit.

limit ratio—The ratio of peak value to limited value.

limits—The minimum and maximum values specified for a quantity.

limit switch—1. A mechanically operated contact-making or -breaking device mounted in the path of a moving object and actuated by its passage. 2. An electromechanical device that uses changes in mechanical motion to control electrical circuits. It functions as the interlocking link between a mechanical motion and an electrical circuit.

line—1. In television, a single trace of the electron beam from left to right across the screen. The present United States standard is based on 525 lines to a complete picture. 2. A conductor of electrical energy. 3. The path of the moving spot in a cathode-ray tube. 4. A term used interchangeably for maxwell. 5. A row of actual or potential holes at right angles to the direction in which a punched tape advances. Line width is measured in terms of the maximum number of holes permissible, excluding the sprocket hole. 6. The interconnection between two electrical devices. Usually used with reference to a long run of interconnecting cable, as from a microphone to its tape-recorder input.

line advance—Also called line feed. The distance between the centers of the scanning lines.

line amplifier—1. An amplifier that supplies a program transmission line or system with a signal at a specified level. 2. An amplifier, usually remotely powered, used in a trunk line in a distribution system to increase the strength of the signal in order to drive an additional length of cable. Also called line stretcher.

line and trunk group—A group consisting of four-wire line circuits, incoming trunks from private automatic branch exchanges, and intertoll trunk groups.

linear—Having an output that varies in direct proportion to the input.

linear acceleration—The rate of change in linear velocity.

linear accelerator—A device for speeding up charged particles such as protons. It differs from other accelerators in that the particles move in a straight line instead of in circles or spirals.

linear accelerometer — A transducer for measuring linear accelerations.

linear actuator — An actuator which produces mechanical motion from electrical energy.

linear amplification—Amplification in which the output is directly proportional to the input.

linear amplifier—1. An amplifier the output signal of which is always an amplified replica of the input signal. 2. Amplifier whose gain is constant for a wide variation in amplitude of input signal—i.e., output signal is proportional to input signal.

linear array—An antenna array in which the elements are equally spaced and in a straight line.

linear circuit — 1. A circuit in which the output voltage is approximately directly proportional to the input voltage; this relationship generally exists only over a limited range of signal voltages and often over a limited range of frequencies. 2. A circuit whose output is an amplified version or a predetermined variation of its input.

linear control—A rheostat or potentiometer having uniform distribution of graduated resistance along the entire length of its resistance element.

linear detection — Detection in which the output voltage is substantially proportionate to the input voltage over the useful range of the detector.

linear detector — A detector that produces an output signal directly proportionate in amplitude to the variations in amplitude (for am transmission) or frequency (for fm transmission) of the rf input.

linear differential transformer — A type of electromechanical transducer that converts physical motion into an output voltage, the phase and amplitude of which are proportional to position. *See also* Linear Motion Transducer.

linear distortion — Amplitude distortion in which the output- and input-signal envelopes are not proportionate, but no alien frequencies are involved.

linear electrical parameters of a uniform line—Frequently called the linear electrical constants. The series resistance and inductance, and the shunt conductance and capacitance, per length of a line.

linear electron accelerator — An evacuated metal tube in which electrons are accelerated through a series of small gaps (usually cavity resonators in the high-frequency range). The gaps are so spaced that, at a specific excitation frequency, the electrons gain additional energy from the electric field as they pass through successive gaps.

linear feedback-control system — A feedback-control system in which the relationship between the pertinent measures of the system signals are linear.

linear integrated circuit—An integrated circuit the output of which remains proportional to the input level. Generally the term is taken to mean an analog IC, such as a voltage regulator, comparator, sense amplifier, driver, etc., as well as a linear amplifier. The operation of the circuit can be made nonlinear by connecting the basic linear amplifier to external circuit elements that have thresholds or other nonlinear characteristics. Abbreviated LIC.

linearity – 1. The relationship existing between two quantities when a change in a second quantity is directly proportionate to a change in the first quantity. 2. Deviation from a straight-line response to an input signal. 3. The ability of a meter to provide equal angular deflections proportional to the applied current. Usually expressed as a percent of the full-scale deflection 4. The relationship between the actual electrical energy input and the deflection of a meter pointer, as referenced to a theoretical straight line. Linearity is often confused with tracking.

linearity control–A control that adjusts the variation of scanning speed through the trace interval.

linearity error–The deviation of a calibration curve from a specified straight line.

linear logarithmic intermediate-frequency amplifier – An amplifier used to avoid overload or saturation as a protection against jamming in a radar receiver.

linearly polarized wave–At a point in a homogeneous isotropic medium, a transverse electromagnetic wave the electric field vector of which lies along a fixed line.

linear magnetostriction–Under stated conditions, the relative change of length of a ferromagnetic object in the direction of magnetization when the magnetization of the object is increased from zero to a specified value (usually saturation).

linear mobility – The synchronized incremental mobility of functionally transitional electrons in a semiconductor.

linear modulation – Modulation in which the amplitude of the modulation envelope (or the deviation from the resting frequency) is directly proportionate to the amplitude of the modulating wave at all audio frequencies.

linear modulator–A modulator in which the modulated characteristic of the output wave is substantially linear with respect to the modulating wave for a given magnitude.

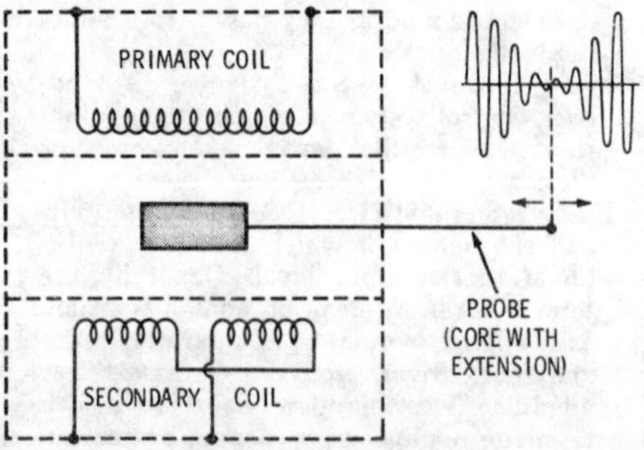

Linear motion transducer.

linear motion transducer–An instrumentation component that translates straight-line (linear) mechanical motion into an ac analog which is usable as a feedback signal for control or display. A transformer-type device in which a movable magnetic core is displaced axially by the moving component being monitored. When the core is moved in one direction from the center of its stroke, the output voltage is in phase with the excitation voltage, and when the core is moved in the opposite direction from the center, the output voltage is 180° out of phase. At the center, the output voltage is (virtually) zero. In either direction from the center, the voltage increases as a precise linear function of probe displacement. Thus the output signal has two basic analog components; phase relationship with the excitation voltage, indicating the direction of travel; and voltage amplitude, indicating the length of travel.

linear polarization – The polarization of a wave radiated by an electric vector that does not rotate but that alternates so as to describe a line. Normally the vector is oriented either horizontally or vertically.

linear power amplifier–A power amplifier in which the output voltage is directly proportionate to the input voltage.

linear programming–In computers, a mathematical method of sharing a group of limited resources among a number of competing demands. All decisions are interlocking because they must be made under a common set of fixed limitations.

linear pulse amplifier – A pulse amplifier which maintains the peak amplitudes of the input and output pulses in proportion.

linear rectification–The production, in the rectified current or voltage, of variations that are proportionate to variations in the input-wave amplitude.

linear rectifier–A rectifier with the same output-current or -voltage waveshape as that of the impressed signal.

linear scan–A radar beam which traverses only one arc or circle.

linear scanning–Scanning in which a radar beam generates only one arc or circle.

linear sweep–In a television receiver, the movement of the spot across the screen at a uniform velocity during active scanning intervals.

Linearsyn–A linear displacement pickoff of the differential-transformer type consisting of a coil assembly and a movable magnetic core. Linear velocity units of high-coercive-force permanent magnetic cores which induce sizable dc voltages while moving concentrically within shielded coils; the voltage varies linearly with the core velocity. (Sanborn Co.)

linear taper – A potentiometer which

changes the resistance linearly as it is rotated through its range.

linear time base—In a cathode-ray tube, the time base in which the spot moves at a constant speed along the time scale. This type of time base is produced by application of a sawtooth waveform to the horizontal-deflection plates of a cathode-ray tube.

linear transducer — 1. A transducer for which the pertinent measures of all the waves concerned are related by a linear function (e.g., a linear algebraic differential, or integral equation). 2. A transducer having its output at any given frequency proportional to the received input.

linear variable-differential transformer — *See* Differential Transformer.

linear varying parameter network — A linear network in which one or more parameters vary with time.

linear velocity transducer — A transducer which produces an output signal proportionate to the velocity of single-axis translational motion between two objects.

line-a-time printing — A type of computer output in which an entire horizontal row of characters is printed at the same time. *See also* Line Printer.

line balance—1. The degree to which the conductors of a transmission line are alike in their electrical characteristics with respect to each other, other conductors, and ground. 2. Impedance equal to that of the line at all frequencies (e.g., in terminating a two-wire line).

line-balance converter — A device used at the end of a coaxial line to isolate the outer conductor from ground.

line characteristic distortion—Distortion experienced in teletypewriter transmission when the presence of changing current transitions in the wire circuit affects the lengths of the received signal impulses.

line circuit — In a telephone system, the relay equipment associated with each station connected to a dial or manual switchboard. The term is also applied to a circuit for interconnecting an individual telephone and a channel terminal.

line coordinate — In a matrix, a symbol (normally at the side) identifying a specific row of cells and, in conjunction with a column coordinate, a specific cell.

line cord—Also called a power cord. A two-wire cord terminating in a two-prong plug at the end that goes to the supply, and connected permanently into a radio receiver or other appliance at the other end.

line-cord resistor — An asbestos-enclosed, wirewound resistance element incorporated into a line cord along with the two regular wires. It lowers the line voltage to the correct value for the series-connected tube filaments and pilot lamps of a universal ac/dc receiver.

line diffuser — A circuit used to produce small vertical oscillations of the spot on the screen of a television monitor or receiver to make the line structure of the image less noticeable to an observer close to the screen.

line driver — 1. An integrated circuit designed for transmitting logic information through long lines (normally at least several feet in length). 2. A buffer circuit with special output characteristics (i.e., high current and/or low impedance) suitable for driving logic lines longer than normal interconnection length (greater than a few feet). It may have complementary (push-pull) outputs to work with the differential inputs of a line receiver. *See* Line Receiver.

line drop — A voltage loss occurring between any two points in a power or transmission line. Such a loss, or drop, is due to the resistance, reactance, or leakage of the line. An example is the voltage drop between a power source and load when the line supplying the power has excessive resistance for the amount of current.

line-drop signal—A signal associated with a subscriber line on a manual switchboard.

line-drop voltmeter compensator—A device using a voltmeter to enable it to indicate the voltage at some distant point in the circuit.

line equalizer — An inductance and/or capacitance inserted into a transmission line to correct its frequency-response characteristics.

line-equipment balancing network — A hybrid network designed to balance filters, composite sets, and other line equipment.

line-fault protection—A means of eliminating or reducing the effect of faults which occur on a transmission line such as a telephone circuit. Such faults include momentary losses of transmission due to signal outages and high noise levels.

line feed—*See* Line Advance.

line fill—The ratio of the number of main telephone stations connected to a line to the nominal main-station capacity of the same line.

line filter—1. A device containing one or more inductors and capacitors. It is inserted between a transmitter, receiver, or appliance and the power line to block noise signals. In a radio receiver, it prevents powerline noise signals from entering the receiver. In other appliances, it prevents their own electrical noises from entering the power line. 2. A filter associated with a transmission line. In some applications, line filter may imply a filter used to separate the speech frequencies.

In other applications, it may imply directional separation, etc.

line-filter balance—A network designed to maintain phantom-group balance when one side of the group is equipped with a carrier system. Since the network must balance the phantom group for voice frequencies only, its configuration is much simpler than the filter it balances.

linefinder—1. A switching mechanism that locates a calling telephone line among a group and connects it to a trunk, selector, or connector. 2. An electromechanical device that automatically line-feeds the platen of a printer to a predetermined line on a printed form.

linefinder shelf—A group (usually 20) of linefinders with the equipment required for routing the dial pulses from any of its associated calling telephones to a selector or connector.

linefinder switch—In a telephone system, an automatic switch for seizing the selector apparatus that provides the dial tone transmitted to the calling party.

line-focus tube—An X-ray tube in which the focal spot is roughly a line.

line frequency — 1. Also called horizontal line frequency or horizontal frequency. In television, the number of times per second the scanning spot crosses a fixed vertical line in the picture in one direction, including vertical-return intervals. 2. The frequency of the supply voltage.

line-frequency regulation — The change in output (current voltage or power) of a regulated power supply for a specified change in line frequency.

line group—The frequency spectrum occupied by a group of carrier channels as they are applied to a transmission facility.

line hit—An electrical interference causing the introduction of spurious signals on a circuit.

line hydrophone—A directional hydrophone consisting of a single straight-line element, an array of adjacent electroacoustic transducing elements in a straight line, or the acoustic equivalent of such an array.

line impedance—The impedance measured across the terminals of a transmission line.

line input—Input channel of an amplifier designed to accept signal at a given level from a line at a specific impedance, usually 600 ohms.

line interlace—*See* Interlaced Scanning.

line leakage — Resistance existing through the insulation between the two wires of a telephone-line loop.

line lengthener—A device for altering the electrical length of a waveguide or transmission line, but not its physical length or other electrical characteristics.

line level—1. The intensity of the signal at a certain position on a transmission line. 2. Based roughly on the "standardized" signal intensity sent over a telephone line, this refers to any audio singal having a maximum intensity of between ½ and 1½ volts. Typically, this is the signal level put out by audio components which do not require preamplification (tuners, for instance).

line load — Usually a percentage of maximum circuit capability to reflect actual use during a period of time (e.g., peak-hour line load).

line loop—An operation performed over a communication line from an input unit at one terminal to output units at a remote terminal.

line loop resistance — The metallic resistance of the line conductors between an individual telephone set and the dial central office.

line loss — The total of the various energy losses in a transmission line.

line microphone—A directional microphone consisting of a single straight-line element, an array of adjacent electroacoustic transducing elements in a straight line, or the acoustical equivalent of such an array.

line noise—Noise originating in a transmission line.

line of force—In an electric or magnetic field, an imaginary line in the same direction as the field intensity at each point. Sometimes called a maxwell when used as a unit of magnetic flux.

line of propagation—The path over which a radio wave travels through space.

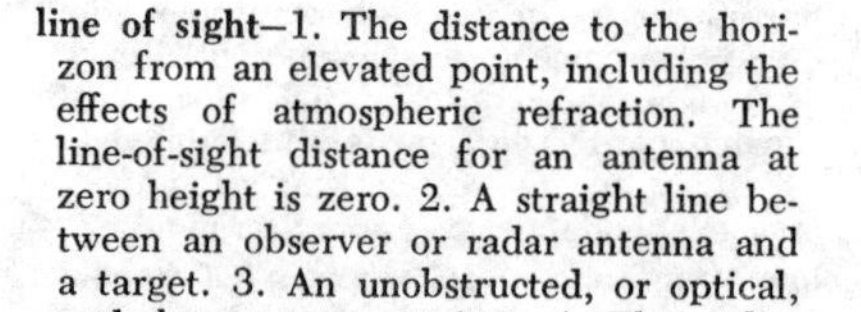

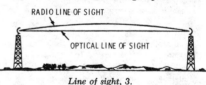

Line of sight, 3.

line of sight—1. The distance to the horizon from an elevated point, including the effects of atmospheric refraction. The line-of-sight distance for an antenna at zero height is zero. 2. A straight line between an observer or radar antenna and a target. 3. An unobstructed, or optical, path between two points. 4. The radio-propagation characteristic of a microwave.

line-of-sight coverage—The maximum distance for transmission above the highest usable frequency. Radio waves at those frequencies do not follow the curvature of the earth and are not reflected from the ionosphere, but go off into space and are lost.

line-of-sight stabilization—In shipboard or airborne radar, compensating for the roll and pitch by automatically changing the elevation of the antenna in order to keep the beam pointed at the horizon.

line of travel—The path followed by an electromagnetic wave from one point to another.

line oscillator—An oscillator in which the resonant circuit is a section of transmission line an integral number of quarter wavelengths in electrical length.

line output—Output channel of an amplifier designed to deliver signal at a given level to a line at a specific impedance, usually 600 ohms.

line pad—In radio broadcasting, a pad inserted between the final program amplifier and the transmitter to ensure a constant load on the amplifier.

line printer—In computers, a high-speed printer which produces an entire line at one time. All characters of the alphabet are contained around the rim of a continuously rotated disc, and there are as many discs as there are characters in the line. The computer momentarily stops the discs at the right characters for each line, and stamps out an impression in a fraction of a second.

line pulsing—A method of pulsing a transmitter by charging an artificial line over a relatively long period, and then discharging it through the transmitter tubes at a shorter interval determined by the line characteristics.

line receiver—A circuit to receive signals from a line, usually driven by a line driver and having features such as differential input, Schmitt trigger, and the like. *See also* Line Driver.

line regulation—1. The change in output (current, voltage, or power) of a regulated power supply for a specified change in line voltage. It may be stated as a percentage of the specified output and/or an absolute value. 2. Percent change in output voltage at constant junction temperature for a specified change in input voltage. This determines output accuracy of a regulator for changes in input voltage.

line relay—A relay activated by the signals on a line.

line-sequential color-television system — A color-television system in which the individual lines of green, red, and blue are scanned in sequence rather than simultaneously.

lines of force—In electric and magnetic fields, the electric and magnetic forces of repulsion or attraction which are taken to follow certain imaginary lines radiating from the electric charge or the magnetic pole. (It is assumed that any unit electric charge or unit magnetic pole placed in the appropriate field will be acted upon so as to move in the direction of these imaginary lines.)

line spectrum—The spectrum of a periodic, discrete signal consisting of one or more frequencies. For example, a square wave is characterized by a fundamental and odd-order harmonics.

line speed—The maximum rate at which signals may be sent over a given channel, usually expressed in bauds, or bits per second.

line-stabilized oscillator—An oscillator in which the frequency is controlled by using one section of a line as a sharply selective circuit.

line stretcher—1. A section of rigid coaxial line with telescoping inner and outer conductors that permit the section to be conveniently lengthened or shortened. 2. *See* Line Amplifier, 2.

line supervision—A means of determining that a transmission line is functional.

line switching—Also called circuit switching. A communications switching system which completes a circuit from sender to receiver at the time of transmission, as opposed to message switching.

line transformer—A transformer inserted into a system for such purposes as isolation, impedance matching, or additional circuit derivation.

line triggering—Triggering from the power-line frequency.

line unit—An electric device used in sending, receiving, and controlling the impulses of a teletypewriter.

line voltage—The voltage level of the main power supply to the equipment.

line-voltage regulator—A device that counteracts variations in the power-line voltage and delivers a constant voltage to the connected load.

linguistic—Pertaining to language or its study, including its origin, structure, phonetics, etc.

link — 1. A transmitter-receiver system connecting two locations. 2. In a digital computer, the part of a subprogram that connects it with the main program. 3. An interconnection. 4. In automatic switching, a path between two units of switching apparatus within a central office.

linkage—1. A measure of the voltage that will be induced in a circuit by magnetic flux. It is equal to the flux times the number of turns linked by the flux. 2. A mechanical arrangement for transferring motion in a desired manner. It consists of solid pieces with movable joints. 3. In a computer, a technique used to provide interconnections for entry and exit of a closed subroutine to or from the main routine.

link circuit—A closed loop used for coupling purposes. It generally consists of two coils, each having a few turns of wire, connected by a twisted pair of wires or by other means, with each coil placed over, near, or in one of the two coils to be coupled.

link coupling—Inductive coupling between circuits. A coil in one circuit acts as the primary, and a coil in the second circuit as the secondary.

link fuse—An unprotected fuse consisting of a short, bare wire between two fastenings.

link neutralization — Neutralization of a tuned radio-frequency amplifier by means of an inductive coupling loop from the output to the input.

link transmitter—In broadcasting, a booster for a remote pickup or from studio to main transmitter.

lin-log receiver—A radar receiver in which the amplitude response is linear for small-amplitude signals and logarithmic for large ones.

lip microphone — A sensitive microphone which is placed in contact with the lip.

liquid-borne noise—Undesired sound characterized by fluctuations of pressure of a liquid about the static pressure as a mean.

liquid cooling—The use of a circulating liquid to cool components or equipment that heat up during operation.

liuqid-crystal display—*See* LCD.

liquid crystals — Liquids that are doubly refracting and that display interference patterns in polarized light.

liquid-filled capacitor — A capacitor in which a liquid impregnant occupies substantially all of the case volume not required by the capacitor element and its connections. Space may be allowed for the expansion of the liquid under temperature variations.

liquid fuse unit—A fuse unit in which the fuse link is immersed in a liquid or the arc is drawn into the liquid when the fuse link melts.

liquid-impregnated capacitor—A capacitor in which a liquid impregnant is dominantly contained within the foil and paper winding, but does not occupy substantially all of the case volume not required by the capacitor element and its connections.

liquid laser—A laser in which the active material is in the liquid state. Present types employ a chelated rare-earth ion dissolved in an organic liquid.

liquid rheostat—A rheostat consisting of metal plates immersed in a conductive liquid. The resistance is changed by raising or lowering the plates or the liquid level to vary the area of the plates contacting the liquid.

liser—A microwave oscillator of very high spectrum purity. Its emission consists of right circularly polarized waves of two different cavity resonant frequencies.

Lissajous figures — Patterns produced on the screen of a cathode-ray tube when sine-wave signal voltages of various amplitude and phase relationships are applied to the horizontal- and vertical-deflection circuits simultaneously.

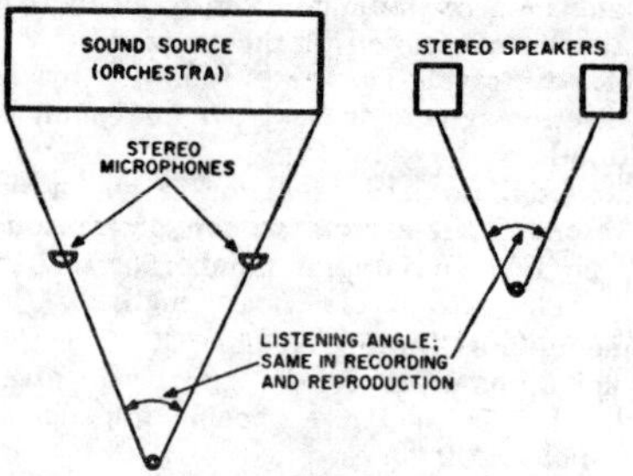

Listening angle.

listening angle—The enclosed angle between the listener and the two speakers of a stereo reproducing system.

listening sonar—*See* Sonar.

lithium—An alkali metal used in the construction of photocells and batteries.

lithium chloride sensor—A hygroscopic element that has fast response, high accuracy, and good long-term stability, and whose resistance is a function of relative humidity. Also called Dunmore cell.

litz wire—Also called Litzendraht wire. A conductor composed of a number of fine, separately insulated strands which are woven together so that each strand successively takes up all possible positions in the cross section of the entire conductor. Litz wire gives reduced skin effect; hence, lower resistance to high-frequency currents.

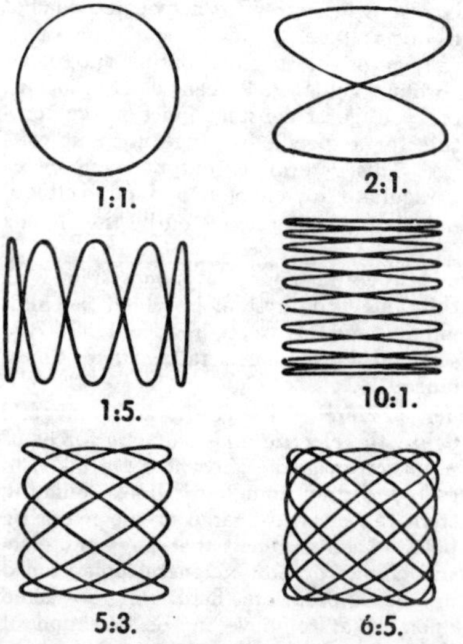

Lissajous figures.

live—A term applied to a circuit through which current is flowing. *See also* Energized.

live cable test cap – A protective cap placed over the end of a cable to insulate the cable and seal its sheath.

live end—The end of a radio studio where the reflection of sound is greatest.

live room—A room with a minimum of sound-absorptive material such as drapes, upholstered furniture, rugs, etc. Because of the many reflecting surfaces, any sound produced in the room will have a long reverberation time.

L-network—A network composed of two impedance branches in series. The free ends are connected to one pair of terminals, and the junction point and one free end are connected to another pair.

load—1. The power consumed by a machine or circuit in performing its function. 2. A resistor or other impedance which can replace some circuit element. 3. The power delivered by a machine. 4. A device that absorbs power and converts it into the desired form. 5. The impedance to which energy is being supplied. 6. Also called work. The material heated by a dielectric or induction heater. 7. In a computer, to fill the internal storage with information obtained from auxiliary or external storage. 8. The resistance or impedance that the input of one device offers to the output of another device to which it is connected. *See* Termination. 9. The circuit or transducer (e.g., speaker) connected to the output of an amplifier. The source (e.g., pickup) is loaded by the amplifier's input impedance (*See* Input Impedance).

load and go—In a computer, an operation and compiling technique in which the pseudo language is converted directly to machine language and the program is then run without the creation of an output machine-language program.

load balance—*See* Load Division.

load cell—1. Transducer that measures an applied load by a change in its properties, such as a change in resistance (strain-gage load cell), pressure (hydraulic load cell), etc. 2. A device that produces an output signal proportional to the applied weight or force.

load circuit—The complete circuit required to transfer power from a source to a load (e.g., an electron tube).

load-circuit efficiency—In a load circuit, the ratio between its input power and the power it delivers to the load.

load-circuit power input—The power delivered to the load circuit. It is the product of the alternating component of the voltage across the load circuit and the current passing through it (both root-mean-square values), times their power factor.

load coil—Also called a work coil. In induction heaters a coil which, when energized with an alternating current, induces energy into the item being heated.

load curve—A curve of power versus time —i.e., the value of a specified load for each unit of the period covered.

load divider—A device for distributing power.

load division—Also called load balance. A control function that divides the load in a prescribed manner between two or more power sources supplying the same load.

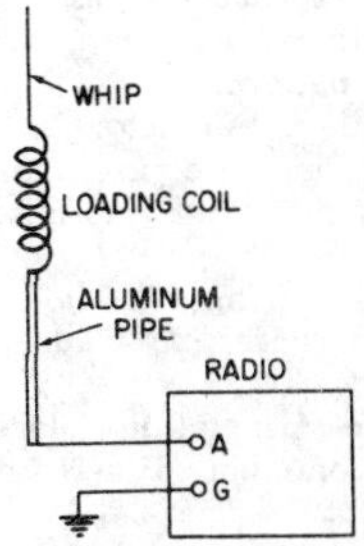

Center-loaded antenna.

loaded antenna—1. An antenna to which extra inductance or capacitance has been added to change its electrical (but not its physical) length. 2. An antenna employing a loading coil at its base or above its base to achieve the required electrical length using physically shorter elements.

loaded applicator impedance—In dielectric heating, the complex impedance measured at the point of application with the load material properly positioned for heating and at the specified frequency.

loaded impedance—In a transducer, the impedance at the input when the output is connected to its normal load.

loaded line—A transmission line that has lumped elements (inductance or capacitance) added at uniformly spaced intervals. Loading is used to provide a given set of characteristics to a transmission line.

loaded motional impedance—*See* Motional Impedance.

loaded *Q*—Also called the working *Q*. The *Q* of an electric impedance when coupled or connected under working conditions. 2. In bandpass filters, a quantity that defines the percentages of 3-dB bandwidths. Numerically, it is equal to the center frequency divided by the 3-dB bandwidth.

loader—1. A program in a minicomputer that takes a program from some input or storage device and places it in memory at some address. Programs loaded by a loader are then ready to run. 2. A program that places a binary (machine language) program into successive core memory locations for execution. Some can replace relocatable addresses with absolute addresses.

load factor—The ratio of average power demand, over a stipulated period of time, to the peak or maximum demand for that same interval.

load impedance—The impedance which the load presents to a transducer.

loading—1. In communication practice, the insertion of reactance into a circuit to improve its transmission characteristics. 2. Placing some material at the front or rear of a speaker so as to change its acoustic impedance and thus alter its radiation pattern.

loading coil—An inductor inserted into a circuit to increase its inductance and thereby improve its transmission characteristics.

loading-coil spacing—The line distance between the successive loading coils of a line.

loading disc—A metal disc placed on top of a vertical antenna to increase its natural wavelength.

loading error—The error introduced when more than negligible current is drawn from the output of a device.

loading factor—A number that represents the relative capacity of a gate for driving other gates or the relative load presented by a gate to the gate that drives it.

loading noise — Any unwanted signal caused by fluctuating contact resistance between the slider and the wire or film in a potentiometric transducer when current is drawn from the instrument.

loading routine—In a computer, a routine which, when in the memory, is able to read other information into the memory from cards, tape, etc.

load isolator—A waveguide or coaxial device (usually ferrite) which provides a good energy path from a signal source to a load, but provides a poor energy path for reflections from a mismatched load back to the signal source.

load life—The ability of a device to withstand its full power rating over an extended period of time, usually expressed in hours.

load line—1. A straight line drawn across a series of plate-current-plate-voltage characteristic curves on a graph to show how plate current changes with grid voltage when a specified plate-load resistance is used. 2. A line drawn on the collector characteristic curves of a transistor on which the operating point of the transistor moves as the collector current changes. Called a load line because the slope of the line depends on the value of the collector load resistance.

load matching—In induction and dielectric heaters, adjustment of the load-circuit impedance so that the desired energy will be transferred from the power source to the load.

load-matching network—An electrical impedance network inserted between the source and the load to provide for maximum transfer of energy.

load-matching switch—In induction and dielectric heaters, a switch used in the load-matching network to alter its characteristics and thereby compensate for a sudden change in the load characteristics (such as in passing through the Curie point).

load mode — In some variable-word-length computers, data transmission in which certain delimiters are moved with the data (in contrast with move mode).

load point—Sometimes called BOT (beginning of tape). The point on a magnetic tape at which writing and reading begin. It is indicated by a reflective marker placed on the tape.

load regulation—1. (For a constant-current supply.) The change in the steady-state value of the output dc current due to a change in load resistance from a short-circuit current (zero resistance) to a value which results in the maximum rated output voltage. 2. (For a constant-voltage supply.) The change in the steady-state value of the output dc voltage due to a change in load resistance from an open-circuit condition (infinite resistance) to a value which results in the maximum rated output current.

loadstone—*See* Lodestone.

load-transfer switch—A switch for connecting either a generator or a power source to one load circuit or another.

lobe—Also called directional, radiation, or antenna lobe. One of the areas of greater transmission in the pattern of a directional antenna. Its size and shape are determined by plotting the signal strength in various directions. The area with the greatest signal strength is known as the major lobe, and all others are called minor lobes.

lobe frequency—The number of times a lobing pattern is repeated per second.

lobe half-power width—In a plane containing the direction of the maximum energy of the lobe, the angle between the two directions in that plane about the maximum in which the radiation intensity is one-half the maximum value of the lobe.

lobe penetration—The penetration of the radar coverage of a station which is not limited by the pulse-repetition frequency, scope limitations, or the screening angle at the azimuth of penetration.

lobe switching—A form of scanning in which the maximum radiation or reception is periodically switched to each of two or more directions in turn.

lobing—The formation of maxima and minima at various angles of the vertical-plane antenna pattern by the reflection of energy from the surface surrounding the radar antenna. These reflections reinforce the main beam at some angles and tend to cancel it at other angles, producing fingers of energy.

local action—In a battery, the loss of otherwise usable chemical energy by currents which flow within regardless of its connections to an external circuit.

local battery—1. A battery made of single dry cells located at the subscriber's station. 2. A battery that actuates the recording instruments at a telegraph station (as distinguished from the battery that furnishes current to the line). 3. A telephone circuit power source usually in the form of dry cells, and located at the customer's end of the line.

local-battery telephone set — A telephone set that obtains transmitter current from a battery or other current supply circuit individual to the telephone set. The signaling current may be obtained from a local hand-operated generator or from a central power source.

local cable—A handmade cable form for circuit terminations at an attendant's switchboard, at unit equipment, and at other locations where wiring is routed inside the section or unit.

local central office—A central office arranged for termination of subscriber lines and provided with trunks for making connections to and from other central offices.

local channel—1. In private line services, that portion of a through channel within an exchange which is provided to connect the main station with an interexchange channel. 2. A standard broadcast channel in which several stations, with powers not in excess of 1000 watts daytime or 250 watts nighttime, may operate.

local control—Control of a radio transmitter directly at the transmitter, as opposed to remote control.

localizer—A radio facility which provides signals for guiding aircraft onto the center line of a runway.

localizer one-course line—A vertical line passing through a localizer. Indications of opposite sense are received on either side of the line.

localizer station—A ground radionavigation station that provides signals for the lateral guidance of aircraft with respect to the center line of a runway.

local oscillator—1. Also called beat oscillator. An oscillator used in a superheterodyne circuit to produce a sum or difference frequency equal to the intermediate frequency of the receiver. This is done by mixing its output with the received signal. 2. The oscillator whose output is mixed with the incoming signal in superheterodyne receivers to produce an intermediate frequency for signal processing (i.e., filtering, amplifying, detecting, etc.).

local-oscillator injection — An adjustment used to vary the magnitude of the local-oscillator signal that is coupled into the mixer.

local-oscillator radiation—Radiation of the fundamental or harmonics of the local oscillator of a superheterodyne receiver.

local-oscillator tube — The vacuum tube which provides the local-oscillator signal in a superheterodyne receiver.

local program—A program originating at and released through only one broadcast station.

local side — The connections from a data terminal to input-output devices.

local trunk—A trunk between local and long-distance switchboards, or a trunk between local and private branch exchange switchboards.

location—A unit-storage position in the main or secondary storage of a computer.

location counter—In the control section of a computer, a register which contains the address of the instruction currently being executed.

lock—1. To terminate the processing of a magnetic tape in a manner such that the contents of the tape are no longer accessible. 2. To synchronize or become synchronized with; follow or control precisely, as in frequency, phase, motion, etc., used with on, onto, in, etc.

locked groove — Also called concentric groove. The blank and continuous groove in the center, near the label, of a disc record. It prevents the needle from traveling farther inward, onto the label.

locked-rotor current — In a motor, the steady-state current taken from the line while the rotor is locked and the rated voltage (and frequency in alternating-current motors) is applied to the motor.

locked-rotor torque — Also called static torque. The minimum torque a motor will develop at rest, for all angular positions of the rotor, when the rated voltage is applied at rated frequency.

lock-in—1. The term used when a sweep oscillator is in synchronism with the applied sync pulses. 2. The shifting and automatic holding of one or both of the frequencies of two oscillating systems which are coupled together, so that the two frequencies have the ratio of two integral numbers.

lock-in amplifier—A form of synchronous detector having a balanced amplifier in which the signal is applied to the grids of two tubes as the control signal is applied to their plates or another grid. The difference of their output currents is then measured.

locking—1. Controlling the frequency of an oscillator by means of an applied signal of constant frequency. 2. Automatic following of a target by a radar antenna. 3. Pertaining to code extension characters that indicate a change in the interpretation of an unspecified number of subsequent characters.

locking circuit—*See* Holding Circuit.

locking-in—In two oscillators which are coupled together, the shifting and automatic holding of one or both of their frequencies so that the two frequencies are in synchronism (i.e., have the ratio of two integral numbers).

locking-out relay—An electrically operated hand or electrically reset device which functions to shut down and hold an equipment out of service on the occurrence of abnormal conditions.

locking relay—*See* Latch-in Relay.

lock-on—1. The instant at which radar begins to track a target automatically. 2. Signifies that a tracking or target-seeking system is continuously and automatically tracking a target in one or more coordinates.

lock-on range—The range from a radar to its target at the instant when lock-on occurs.

lockout—1. In a telephone circuit controlled by two voice-operated devices, a condition in which excessive local circuit noise or continuous speech from either or both subscribers results in the inability of one or both subscribers to get through. 2. *See* Receiver Lockout System.

lockout—Mechanical function whereby not more than one switch station can be fully depressed simultaneously — for switches with interlock, nonlock or push-lock/push-release mechanical functions.

lockup—Electromechanical function whereby switch stations are immobilized by operation (either actuating or releasing) of a solenoid (sometimes manually). Actuated switch stations cannot be released, and unactuated switch stations cannot be actuated until lockup is released. Lockup release can be accomplished locally or remotely.

lock-up relay—A relay that is locked in the energized position magnetically or electrically rather than mechanically.

loctal base—*See* Loktal Base.

lodar—Also called lorad. A direction finder which compensates for night effect by observing the distinguishable ground- and sky-wave loran signals on a cathode-ray oscilloscope and positioning a loop antenna to obtain a null indication of the more suitable components.

lodestone—Also spelled loadstone. A natural magnet consisting chiefly of a magnetic oxide of iron called magnetite.

Lodge antenna — A counterpoise antenna that consists of a vertical dipole provided with horizontal top and bottom plates or screens, the lower plate or screen being spaced from the ground. Other versions include the bow-tie antenna, in which two narrow triangular plates are connected at their smaller ends by a coil, and the umbrella antenna, in which the end plates are conical and made up of wires.

Loftin-White circuit—A type of direct-coupled amplifier.

log—1. A listing of radio stations and their frequency, power, location, and other pertinent data. 2. A record of the station with which a radio station has been in communication. Amateur radio operators, as well as all commercial operators, are required by law to keep a log. 3. At a broadcast station, a detailed record of all programs broadcast by the station. 4. At a broadcast transmitter, a record of the meter readings and other measurements required by law to be taken at regular intervals. 5. Abbreviation for logarithm.

logarithmic amplifier — An amplifier the output of which is a logarithmic (as opposed to linear) function of its input.

logarithmic curve—A curve on which one coordinate of any point varies in accordance with the logarithm of the other coordinate of the point.

logarithmic decrement—1. For an exponentially damped alternating current, the natural logarithm of the ratio of the first to the second of two successive amplitudes having the same polarity. 2. *See* Damping Ratio, 2.

logarithmic fast time constant—A constant false alarm rate system in which a logarithmic intermediate-frequency amplifier is followed by a fast-time-constant circuit.

logarithmic horn—A horn the diameter of which varies logarithmically with the length.

logarithmic scale—A scale on which the various points are plotted according to the logarithm of the number with which the point is labeled.

logarithmic scale meter—A meter having deflections proportional to the logarithms of the applied energies.

logger—An instrument that automatically scans certain quantities in a controlled process and records readings of the values of these quantities for future record.

logic—1. The science dealing with the basic principles and applications of truth tables, switching, gating, etc. 2. *See* Logical Design. 3. Also called symbolic logic. A mathematical approach to the solution of complex situations by the use of symbols to define basic concepts. The three basic logic symbols are AND, OR, and NOT. When used in Boolean algebra, these symbols are somewhat analogous to addition and multiplication. 4. In computers and information-processing networks, the systematic method that governs the operations performed on the information, usually with each step influencing the one that follows. 5. The systematic plan that defines the interactions of signals in the design of a system for automatic data processing. Circuitry designed to enhance separation between a matrix disc's recovered channels in a quadriphonic sound system. Such circuits monitor the four decoded channels and adjust the decoder dynamically to favor the main, or dominant channel.

logical choice—In a computer, the correct decision where alternatives or different possibilities are open.

logical comparison—In computers, the consideration of two items to obtain a yes (1) if they are the same with regard to some characteristic, or a no (0) if they are not.

logical decision—1. In a computer, the operation by which alternative paths of flow are selected depending on intermediate program data. 2. The ability of a computer to make a choice between alternatives; basically, the ability to answer yes or no to certain fundamental questions concerning equality and relative magnitude.

logical design—1. The preplanning of a computer or data-processing system prior to its detailed engineering design. 2. The synthesizing of a network of logical elements to perform a specified function. 3. The result of 1 and 2 above, frequently called the logic of the system, machine, or network.

logical diagram—In logical design, a diagram that represents logical elements and the interconnection between them, without necessarily including construction or engineering details.

logical element—In a computer or data-processing system, the smallest building blocks which operators can represent in an appropriate system of symbolic logic. Typical logical elements are the AND gate and the "flip-flop."

logical flow chart—A detailed, graphical presentation of work flow in its logic sequence; often, the built-in operations and characteristics of a given machine, with types of operations indicated by symbols.

logically equivalent circuits—Logic circuits that perform the same function even though details of the circuits differ.

logical 1—The opposite of logical 0.

logical operations—Those operations considered to be nonarithmetical, such as selecting, searching, sorting, matching, and comparing.

logical state—Signal levels in logic devices are characterized by two stable states—the logical *one* state and the logical *zero* state. The designation of the two states is chosen arbitrarily. Commonly the logical one state represents an "on" signal and the zero state represents an "off" signal.

logical threshold voltage—At the output of a logic device, the voltage level at which the following logic device switches states.

logical 0—One of the two values of a binary digit signal. If the signal is a voltage, logical 0 usually is the lower of the two voltage levels.

logic array—An integrated device in which 50 or more circuits are integral to a single silicon chip. In addition, the circuits are interconnected on the chip to form some electronic function at a higher level of organization than a single circuit. Logic arrays are constructed by the unit cell method in which a simple circuit (or function) is repeated many times on a slice. The interconnection pattern for converting groups of cells into large functions is determined after cell probe tests are completed. Each interconnection pattern may be unique to a single slice. In general, logic arrays are characterized by multiple levels of metallization to effect the large-scale function.

logic circuit—1. A circuit (usually electronic) that provides an input-output relationship corresponding to a Boolean-algebra logic function. 2. An electronic device or devices used to govern a particular sequence of operations in a given system.

logic devices—Digital components that perform logic functions. They can gate or inhibit signal transmission in accordance with the application, removal, or combination of input signals.

logic diagram—1. In logical design, a diagram representing the logical elements and their interconnections, but not necessarily their construction or engineering details. 2. A pictorial representation of interconnected logic elements using standard symbols to represent the detailed functioning of electronic logic circuits. The logic symbols in no way represent the types of electronic components used, but represent only their functions.

logic element—A device that performs a logic function; a gate or flip-flop, or in some cases a combination of these devices treated as a single entity.

logic function—1. A means of expression of a definite state or condition in a magnetic-amplifier, relay, or computer circuit. 2. One of the Boolean-algebra func-

LOGIC FUNCTION	LOGIC SYMBOL	BOOLEAN EXPRESSION
AND	A, B	AB
OR	A, B	A+B
NAND	A, B	$\overline{AB}$
NEGATED-INPUT OR	A, B	$\overline{A}+\overline{B}$
AND/NOR	A, B, C, D	$\overline{AB+CD}$
NOR	A, B	$\overline{A+B}$
NEGATED-INPUT AND	A, B	$\overline{A}\,\overline{B}$
INVERTER	A	$\overline{A}$

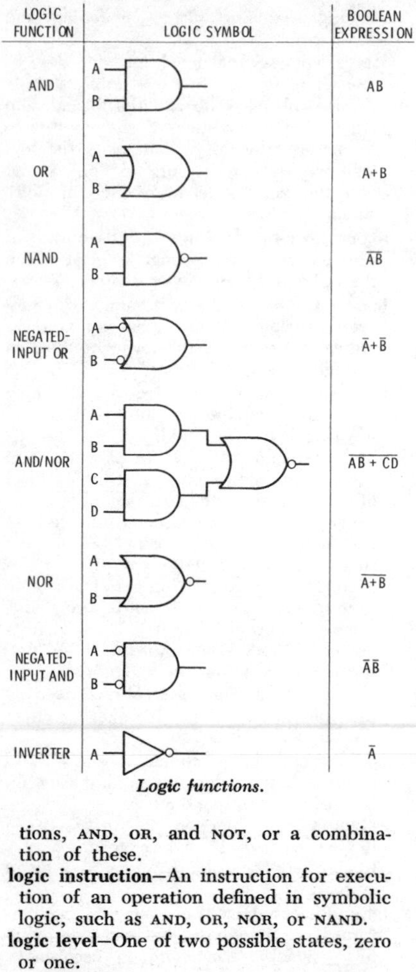

Logic functions.

tions, AND, OR, and NOT, or a combination of these.

logic instruction—An instruction for execution of an operation defined in symbolic logic, such as AND, OR, NOR, or NAND.

logic level—One of two possible states, zero or one.

logic module—Circuit element comprising logic gates (AND, OR, NOT) and variations (NAND, NOR). Used in the design of binary arithmetic circuits, Boolean logic arrangements, etc.

logic swing—The difference in voltage between the voltage levels for 0 and 1 in a binary logic circuit.

logic switch—A diode matrix or other switching arrangement that can direct an input signal to a selected one of several outputs.

logic symbol—A symbol used to represent a logical element on a graph.

logic system—1. A group of interconnected logic elements that act in combination to perform a relatively complex logic function. 2. Programming-recording system constructed of solid-state modules based on series of binary logic (go/no go) components.

log/linear preamplifier—A preamplifier the electrical output of which is proportional to the common logarithm of the input, but which can be switched to a different mode wherein the output is proportional to the input. Any amplifier the transfer characteristic or calibration curve of which can be arbitrarily switched from the y = A log x to the form y = Ax (y is the output and x is the input).

log-periodic antenna—A type of directional antenna that achieves its wideband properties by geometric iteration. The radiating elements and the spacing between elements have dimensions that increase logarithmically from one end of the array to the other so that the ratio of element length to element spacing remains constant.

log receiver—A receiver in which the output amplitude is proportionate to the logarithm of the input amplitude.

log scan—A spectrum display in which the frequency axis is calibrated logarithmically.

loktal base—Also spelled loctal. An eight-pin base for small vacuum tubes. It is designed so that it locks the tube firmly in a corresponding eight-hole socket. Unlike in other tubes, the pins are sealed directly into the glass envelope.

loktal tube—Also spelled loctal. A vacuum tube with a loktal base.

long base-line system—A system in which the distance separating ground stations approximates the distance to the target being tracked.

long-distance loop—A line that connects a subscriber's station directly to a long-distance switchboard.

long-distance navigational aid—A navigational aid usable beyond radio line of sight.

long-distance xerography—A facsimile system at the receiving terminal of which a lens projects a cathode-ray image onto the selenium-coated drum of a xerographic copying machine; a cathode-ray scanner is employed at the microwave transmitting terminal.

longevity—The normal operating lifetime of a piece of equipment, usually considered to be the period in which its failure rate is acceptably low and essentially constant.

longitudinal chromatic abberration—A lack of sharpness in the image because different colors come to a focus at different distances from the lens.

longitudinal circuit—A circuit in which one conductor is a telephone wire (or two or more wires in parallel) and the return is through the earth or any other conductors except those taken with the original wire or wires to form a metallic telephone circuit.

longitudinal coils—The coils that are used in transmission lines to suppress longitudinal currents.

longitudinal current—A current which flows in the same direction in both wires of a parallel pair. The earth is its return path.

longitudinal magnetization — In magnetic recording, magnetization of the recording medium in a direction essentially parallel to the line of travel.

longitudinal parity—Parity associated with bits recorded on one track in a data block to indicate whether there is an even or odd number of bits in the block.

longitudinal redundance—In a computer, a condition in which the bits in each track or row of a record do not total an even (or odd) number. The term is generally used to refer to records on magnetic tape, and a system can have either odd or even longitudinal parity.

longitudinal wave—A wave in which the direction of displacement at each point in the medium is perpendicular to the wave front, as, for example, in a sound wave propagated in air.

long-line effect—An effect that occurs when an oscillator is coupled to a transmission line with a serious mismatch. Oscillation may be possible at two or more frequencies, and the oscillator jumps from one side of these frequencies to another as its load changes.

long-lines engineering — Engineering with the objective of developing, modernizing, or expanding long-haul, point-to-point communications facilities that use radio, microwave, or wire circuits.

long-persistence screen—A fluorescent and phosphorescent screen in which the light intensity of its spots does not immediately die out after the beam has moved on. The generation of light at the instant the electron beam strikes the screen is due to fluorescence; the light that persists after the beam has moved on is due to phosphorescence. The time of persistence varies with the type of tube employed and the coating of the screen.

long-play record — Abbreviated lp record. Also called a microgroove record. A 10- or 12-inch record or transcription with finely cut grooves which give it a long playing time.

long-pull magnet — An electromagnet designed to exert a practically uniform pull, for an extended range of armature movement. It consists of a conical plunger moving up and down inside a hollow core.

long-range navigation — A long-range electronic navigation system which uses the time divergence of pulse-type transmission from two or more fixed stations.

long-range radar—A radar installation capable of detecting targets 200 or more miles away.

long-reach mike—*See* Shotgun.

long shunt—A shunt field connected across the series field and the armature, instead of directly across the armature alone, of a motor or generator.

long-tailed pair — A two-tube circuit in which decreased plate current through one tube results in increased plate current through the other tube, and vice versa.

long-term stability (or long-term instability) — The slow changes in average frequency arising from changes in an oscillator. Statements of long-term stability for quartz oscillators often term this characteristic "aging rate" and specify it as "parts per day" (fractional frequency change over 24 hours). For cesium standards, this term commonly refers to the total fractional frequency drift for the life of the cesium beam tube.

long throw—A method of speaker design in which the woofer moves freely through long excursions, providing excellent low-frequency response with low distortion.

long wave—Wavelengths longer than about 1000 meters. They correspond to frequencies below 300 kHz.

long-wire antenna—An antenna that has a length greater than one-half wavelength at the operating frequency.

look ahead—1. A feature of the CPU of a computer which allows the machine to mask an interrupt request until the following instruction has been completed. 2. A feature of adder circuits and alu's which allow these devices to look ahead to see that all carrys generated are available for addition.

lookthrough—1. In jamming, sporadic interruption of the emission for extremely short periods in order to monitor the victim signal. 2. When a set is being jammed, the monitoring of the desired signal during lulls in the jamming signals.

loom—A flexible nonmetallic tubing placed around insulated wire for protection.

loop—1. A complete electrical circuit. 2. In a computer, a series of instructions being carried out repeatedly until a terminal condition prevails. 3. In automatic control, the path followed by command signals, which direct the actions to be performed, and feedback signals, which are returned to the command point to indicate what is actually happening. *See also* Closed Loop, 1. 4. *See* Mesh *and* Antinodes. 5. A length of tape having its ends spliced together to form an endless loop. Frequently used by film and radio/tv sound departments for prolonged backgrounds of continual or repetitive sound effects. The loop is now the basis of the 8-track cartridge format. 6. A combination of one or more interconnected instruments arranged to measure or control a process variable, or both.

loop actuating signal—The signal derived from mixing the loop-input and loop-feedback signals.

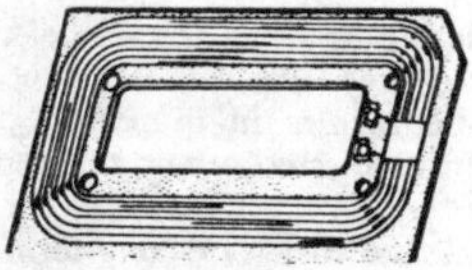

Loop antenna.

loop antenna—An antenna used in radio direction-finding apparatus and in some radio receivers. It consists of one or more loops of wire.

loopback test—A test in which signals are looped from a test center through a data set or loopback switch and back to the test center for measurement. *See also* Bussback.

loop circuit – A communication circuit which more than two parties share. In teletypewriter application, all machines print all data entered on the loop.

loop control—The maintaining of a specified loop of material between two sections of a machine by automatically adjusting the speed of at least one of the driven sections.

loop dialing—A return-path dialing method in which pulses are sent out over one side of the interconnecting line or trunk and returned over the other side. This arrangement is limited to short-haul traffic.

loop difference signal – The output signal from a summing point of a feedback control loop. It is a specific type of loop actuating signal produced by a particular loop input signal applied to that summing point.

loop error—The desired value minus the actual value of the loop output signal.

loop error signal—The loop actuating signal, when it is the loop error.

loop feedback signal—The signal derived from the loop output signal and fed back to the mixing point for control purposes.

loop feeder—A feeder which follows along a circuit and distributes the voltage more evenly at different points.

loop gain—1. The total usable power gain of a carrier terminal or two-wire repeater. Because any closed system tends to sing (oscillate), its usable gain may be less than the sum of the enclosed amplifier gains. The maximum usable gain is determined by—and cannot exceed—the losses in the closed path. 2. The increase in gain observed when the feedback path of an amplifier is opened, but with all circuit loads intact. 3. In an operational amplifier circuit, the product of the transfer characteristics of all of the elements (active or passive) encountered in a complete trip around the loop, starting at any point and returning to that point.

looping—Repetition of instructions at delayed speeds until a final value is determined (as in a weight scale indication). The looped repetitions are usually frozen into a ROM memory location and then jumped when needed. Looping also occurs when the CPU of a computer is in a wait condition.

loop input signal—An external signal applied to a feedback control loop.

loop-mile—The length of wire in a mile of two-wire line.

loop output signal – The controlled signal extracted from a feedback control loop.

loop pulsing—Also called dial pulsing. The regular, momentary interruption of the direct-current path at the sending end of a transmission line.

loop resistance—*See* Line Loop Resistance.

loop return signal—The signal returned, via a feedback control loop, to a summing point in response to a loop input signal applied to that summing point. The loop return signal is a specified type of loop input signal and is subtracted from it.

Loopstick antenna – *See* Ferrite-Rod Antenna.

loop test—A method of locating a fault in the insulation of a conductor when the conductor can be arranged to form part of a closed circuit.

loop transfer function – The transfer function of the transmission path. It is formed by opening and properly terminating a feedback loop. (*See also* Transfer Function.)

loose coupler – An obsolete tuning system consisting of two coils, one inside the other. Coupling was varied over a wide range by sliding one coil over the other.

loose coupling—Also called weak coupling. In resonant systems, a degree of coupling that is considerably less than critical coupling. Hence, there is very little transfer of energy.

LORAC – A trademark of Seismograph Service Corp. for a specific navigation system that determines a position fix by the intersection of lines of position. Each line is defined by the phase angle between two heterodyne beat-frequency waves. One wave is the beat frequency between two cw signals from two widely spaced transmitters. The other is a reference wave of the same frequency and is obtained by deriving the heterodyne beat of the same two cw signals at a fixed location and transmitting it, via a second radio-frequency channel, to the receiver being located.

lorad—*See* Lodar.

loran—A contraction of long-range navigation. A navigation aid sending out pulses at radio frequencies between 1800 and 2000 kHz. It defines lines of position which are based on the differences in travel time between radio waves from a master and a slave station. Airborne equipment utilizes a picture tube and measuring circuits to determine the time differences and relate them to time-difference lines drawn on a map. A navigation fix is established by the intersection of two or more lines of position.

loran C – A hyperbolic navigation system that relies on the stability of electromagnetic propagational characteristics at a frequency of 100 kHz. Each loran-C station operates in a pulse mode in such a way that no two stations can be received simultaneously. During the transmission period of the master station, both slave stations remain silent. After a fixed delay time, slave X will transmit a similar pulse. After a second delay, slave Y will transmit a pulse.

loran D–A tactical loran system that employs the coordinate converter of loran C and can be operated in conjunction with inertial systems on board aircraft, without dependence on ground facilities and without radiation of rf energy that could reveal the location of the aircraft.

loran line–A line of position on a loran chart where each line is the locus of points, the distances from two fixed stations of which differ by a constant amount.

loran station–A radionavigational land station transmitting synchronized pulses.

loran tables–Tables giving terrestrial coordinates (latitude and longitude) of loran lines of position and values of the sky-wave correction.

Lorentz force – The force exerted by an electric field and a magnetic field on a moving electric charge.

Lorentz force equation–An equation relating the force on a charged particle to its motion in an electromagnetic field.

lorhumb line–In navigation, a course line in a lattice; the derivation of one coordinate from the other is always equal to the ratio of the difference between coordinates at the beginning and ending of the course line.

loss–1. A decrease in power suffered by a signal as it is transmitted from one point to another. It is usually expressed in decibels. 2. Energy dissipated without accomplishing useful work.

loss angle–The complement of the phase angle of an insulating material.

losser–A circuit having less power in the output as compared to the power applied to the input. This term is particularly applicable to mixers; a crystal mixer is a losser.

losser circuit – A resonant circuit having sufficient high-frequency resistance to prevent sustained oscillation at the resonant frequency.

Lossev effect – The resultant radiation when charge carriers recombine after being injected into a forward-biased pn or pin junction.

loss factor–The characteristic which determines the rate at which heat is generated in an insulating material. It is equal to the dielectric constant times the power factor.

loss index – The product of the dielectric constant and the loss tangent. It is a measure of the power loss per cycle and receives extensive use in microwave and dielectric-heating applications.

lossless line–A theoretically perfect line–i.e., one that has no loss and hence transmits all the energy fed to it.

loss modulation – *See* Absorption Modulation.

loss tangent–Also called dielectric dissipation factor. The decimal ratio of the irrecoverable to the recoverable part of the electrical energy introduced into an insulating material by the establishment of an electric field in the material.

lossy – 1. Insulating material which dissipates more than the usual energy. 2. Of, like, or made of an insulating material capable of damping out an unwanted mode of oscillation while having little effect on a desired mode.

lossy attenuator – A length of waveguide made from some dissipative material and used to deliberately introduce transmission loss.

lossy line – A transmission line designed with high attenuation.

lot–A group of similar components which have been either all manufactured in a continuous production run from homogeneous raw materials under constant process conditions or assembled from more than one production run and submitted for random sampling and acceptance testing. Specifically, a quantity of material all of which was manufactured under identical conditions, and assigned an identifying lot number.

lot tolerance percent defective–A process average that will be rejected 90% of the time. Abbreviated LTPD.

loudness–1. A measure of the sensitivity of human hearing to the strength of sound. Scaled in sones, it is an overall single evaluation resulting from calculations based on several individual band-index values. 2. Generally synonymous with volume, which is the intensity of perceived sound.

loudness compensation–A variety of equalization applied to a signal according to its volume, in order to compensate for the tendency of the ear to change frequency response at different listening levels.

loudness contour – A curve showing the sound pressure required at each frequency to produce a given loudness sensation to a typical listener.

loudness-contour selector – Abbreviated as LCS. A circuit that alters the frequency response of an amplifier so that the characteristics of the amplifier will more closely match the requirements of the human ear.

loudness control—Also called compensated volume control. A combined volume and tone control which boosts the bass frequencies at low volume to compensate for the inability of the ear to respond to them. Some loudness controls provide similar compensation at the treble frequencies.

loudness level—The sound-pressure level of a 1000-Hz tone judged by a listener to be as loud as the sound under consideration. It is measured in decibels relative to .0002 microbar.

loudspeaker—*See* Speaker.

loudspeaker dividing network – *See* Dividing Network.

loudspeaker impedance – *See* Speaker Impedance.

loudspeaker system—*See* Speaker System.

loudspeaker voice coil—*See* Speaker Voice Coil.

louver—The grill of a speaker.

love wave—A dispersive or frequency-sensitive surface effect which also decays exponentially with depth. Love waves consist of particle motion in the plane of the surface and normal to the direction of wave propagation. These are generated when a substrate is covered wtih a film of solid material, whose thickness is considerably smaller than the wavelength.

low-angle radiation – Radiation that proceeds at low angles above ground.

low band—Television channels 2 through 6 covering frequencies between 54 and 88 MHz.

low-capacitance contacts – A type of contact construction providing small capacitance between contacts.

low-capacitance probe—A test probe with very low capacitance. It is connected between the input of an oscilloscope and the circuit under observation.

low-corner frequency – The frequency at which the output of a resolver is 3 dB below the midfrequency value and the phase shift is 45°.

low-definition television—A television system employing less than 200 scanning lines per frame.

low-energy circuit—Also called a dry circuit. A circuit application that functions at low voltage (i.e., approximately 10 volts or less) and low current (i.e., approximately 1 mA or less).

low-energy criterion – *See* Von Hippel Breakdown Theory.

low-energy material – *See* Soft Magnetic Material.

lower manual – *See* Accompaniment Manual.

lower pitch limit—The minimum frequency at which a sinusoidal sound wave will produce a pitch sensation.

lower sideband—The lower of two frequencies or groups of frequencies produced by the amplitude-modulation process.

lowest effective power—The minimum product of the antenna input power in kilowatts and the antenna gain required for satisfactory communication over a particular radio route.

lowest useful high frequency – In radio transmission, the lowest frequency effective at a specified time for ionospheric propagation of radio waves between any two points, excluding frequencies below several megahertz. It is determined by such factors as absorption, transmitter power, antenna gain, receiver characteristics, type of service, and noise conditions.

low filter—An audio circuit designed to remove low-frequency noises from the program material. Such noises include turntable rumble, tone-arm resonance, etc.

low frequency—Abbreviated lf. The band of frequencies extending from 30 to 300 kHz (10,000 to 1000 meters).

low-frequency compensation—A technique for extending the low-frequency response of a broad-band amplifier.

low-frequency impedance corrector – An electric network designed to be connected to a basic network, or to a basic network and a building network, so that the combination will simulate, at low frequencies, the sending-end impedance, including the dissipation, of a line.

low-frequency induction heater or furnace —A heater or furnace in which the charge is heated by inducing a current at the power-line frequency through it.

low-frequency loran – A modification of standard loran, with operation in the frequency range of approximately 100 to 200 kHz, for increased range over land and during daytime. Whereas the enve-

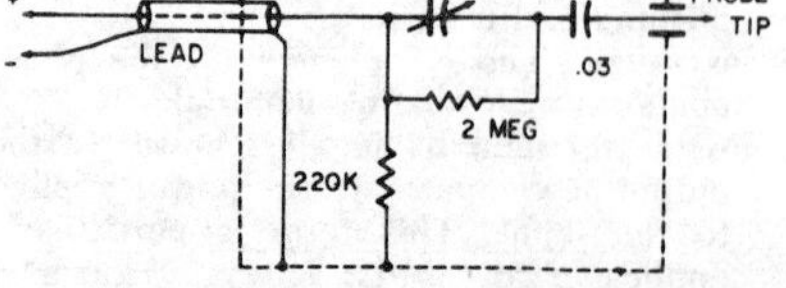

Low-capacitance probe circuit.

lopes of the pulses are matched, to obtain a line of position with ordinary loran, the hertz within the pulses are matched to provide a much more accurate fix with low-frequency loran. Also called hertz-matching loran.

low-frequency padder – In a superheterodyne receiver, a small adjustable capacitor connected in series with the oscillator tuning coil. During alignment it is adjusted to obtain correct calibration of the circuit at the low-frequency end of the tuning range.

low level—In digital logic, the more negative of the two binary-system logic levels. *See also* High Level. Negative Logic, *and* Positive Logic.

low-level contacts – Contacts that control only the flow of relatively small currents in relatively low-voltage circuits—e.g., alternating currents and voltages encountered in voice or tone circuits, direct current on the order of microamperes, and voltages below the softening voltages on record for various contact materials (that is, 0.080 volt for gold, 0.25 volt for platinum, and the like). Also defined as the range of contact electrical loading where there can be no electrical (arc transfer) or thermal effects and where only mechanical forces can change the conditions of the contact interface.

low-level modulation—The modulation produced in a system when the power level at a certain point is lower than it is at the output.

low-level radio-frequency signal (tr, atr, and pre-tr tubes)—A radio-frequency signal with insufficient power to fire the tube.

low-level signal – Very small amplitudes serving to convey information or other intelligence. Variations in signal amplitude are frequently expressed in microvolts.

low-loss insulator—An insulator with high radio-frequency resistance and hence slight absorption of energy.

low-loss line—A transmission line with relatively low losses.

low-noise tape—Magnetic tape with a signal-to-noise ratio 3 to 5 dB better than conventional tapes, making it possible to record sound – especially music with a wide frequency range – at reduced tape speeds without incurring objectionable background noise (hiss) and with little compromise of fidelity.

low-pass filter – 1. A filter network which passes all frequencies below a specified frequency with little or no loss but discriminates strongly against higher frequencies. 2. Wave filter having a single transmission band extending from zero frequency up to some critical or cutoff frequency not infinite. 3. Filter which passes all frequencies below a certain cutoff point and attenuates all frequencies above that point.

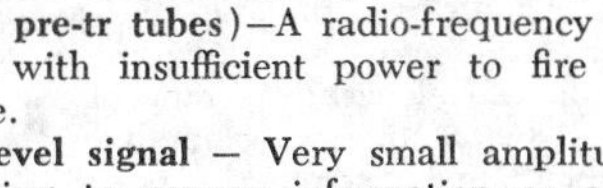

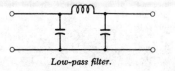

Low-pass filter.

low-print tape—Special magnetic recording tape significantly less susceptible to print-through (the transfer of signal from one layer to another), which results when tape is stored for long periods of time. These tapes are especially useful for speech "master recording" (making an original recording from which copies will be made) on professional-quality equipment.

low-rate discharge – The withdrawal of a small current from a cell or battery for long periods of time.

low tension—British term for low voltage. Generally refers to the heater or filament voltage.

low-velocity scanning—The scanning of a target with electrons having a velocity below the minimum required to give a secondary-emission ratio of unity.

low-voltage protection – *See* Undervoltage Protection.

L-pad—A volume control that has practically the same impedance at all settings. It consists essentially of an L-network in which both elements are adjusted simultaneously.

L + R, L − R—The sum and difference signals of the two stereo channels; the L + R signal combines the signals of both channels in phase; the L − R signal combines them out of phase. By combining ing in suitable circuitry, L + R and L − R can be added to obtain 2L, the signal from one channel; L − R can be subtracted from L + R to obtain 2R, the signal from the other channel.

lp record – Abbreviation for long-playing record.

LSA diode – A solid-state diode operating in an oscillator in the Limited Space-charge Accumulation mode at frequencies that produce millimeter wavelengths.

LSB—Abbreviation for least significant bit.

L-scan—*See* L-Display.

LSD – Least significant digit—Number at the extreme right of a group of numbers. Example: 6937. Digit 7 is the LSD.

LSI—Abbreviation for large-scale integration.

LTPD—Abbreviation for lot tolerance percent defective.

lug—A device soldered or crimped to the end of a wire lead to provide an eye or fork that can be placed under the head of a binding screw.

lumen – Abbreviated lm. 1. The unit of luminous flux. It is equal to the flux through a unit solid angle (steradian) from a uniform point source of one candela (candle) or to the flux on a unit surface all points of which are at unit distance from a uniform point source of one candela. 2. A unit of light emitted from a point light source of one candle through a unit solid angle. All lamps are rated in their output in lumens of light.

lumen-hour–A unit of the quantity of light delivered in one hour by a flux of one lumen.

lumen-second – The unit quantity of light equal to one lumen of luminous flux emitted for one second.

luminaire–A complete lighting unit comprising a lamp or lamps and the parts required to distribute the light, position and protect the lamps, and connect the lamps to the power supply.

luminance–1. The luminous flux emitted, reflected, or transmitted from the source. Usual units are the lumen per steradian per square meter, candle per square foot, meter-lambert, millilambert, and foot-lambert. Formerly known as brightness. 2. In color television, the photometric quantity of light radiation. 3. The amount of light emitted or scattered by a surface, usually measured in foot-lamberts. One footcandle falling upon a perfectly diffusing white surface with no loss would produce one foot-lambert. 4. The luminous flux unit solid angle emitted per unit projected area of a source.

luminance channel – In a color television system, any path intended to carry the luminance signal. The luminance channel may also carry other signals such as the color carrier, which may or may not be used.

luminance channel bandwidth–The bandwidth of the path intended to carry the luminance signal.

luminance flicker – The flicker resulting from fluctuation of the luminance only.

luminance primary – The one of three transmission primaries whose amount determines the luminance of a color.

luminance signal–Also called Y signal. The signal in which the amplitude varies with the luminance values of a televised scene. It is part of the composite color signal and is made up of 0.30 red, 0.59 green, and 0.11 blue. The luminance signal is capable of producing a complete monochromatic picture.

luminescence–1. The absorption of energy by matter, and its subsequent emission as light. If the light is emitted within 10^{-8} second after the energy is absorbed, the process is known as fluorescence. If the emission takes longer, the process is called phosphorescence. 2. The emission of light of certain wavelengths or limited regions of the spectrum in excess of that due to incandescence and the emissivity of the surface. This property is not exhibited by all materials. 3. Emission of light due to any other cause than high temperature (incandescence). 4. A general term which is applied to the production of light, either visible or infrared, by the direct conversion of some other form of energy. The general term is then subdivided to denote the particular energy conversion involved (i.e., thermoluminescence, cathodoluminescence, photoluminescence, and electroluminescence).

luminescence threshold–Also called threshold of luminescence. The lowest radiation frequency that will excite a luminescent material.

luminescent–Any material which will give off light but not heat when energized by an external source such as a stream of electrons or radiant energy.

luminescent screen – A screen which becomes luminous in those spots excited by an electron beam. The screen of a cathode-ray tube is the best-known example.

luminosity–Ratio of luminous flux to the corresponding radiant flux at a particular wavelength. It is expressed in lumens per watt.

luminosity coefficients–The constant multipliers for the respective tristimulus values of any color. The sum of the three products is the luminance of the color.

luminosity curve – A distribution curve showing luminous flux per element of wavelength as a function of wavelength.

luminosity factor–For radiation of a particular wavelength, the ratio of the luminous flux at that wavelength to the corresponding radiant flux. It is expressed in lumens per watt. Abbreviated K.

luminous – An adjective used to indicate the production of light, e.g., "luminous source," to distinguish from electrical sources, etc. It is sometimes used before "intensity" or "flux."

luminous efficiency–The ratio of luminous flux to radiant flux. It is usually expressed in lumens per watt of radiant flux and should not be confused with the term "efficiency" when applied to a practical source of light. The latter is based on the power supplied to the source–not the radiant flux from the source. For energy radiated at a single wavelength, luminous efficiency is synonymous with luminosity.

luminous emittance – The luminous flux emitted per unit area of the source; normal units of measurement are the lumen/cm^2 and the lumen/ft^2.

luminous flux – 1. Also called light flux. Time rate of flow of light (the total visible energy produced by a source per unit time). Usually measured in lumens. 2.

Radiant power weighted at each wavelength in accordance with the ability of the eye to perceive it.

luminous intensity—The ratio of luminous flux emitted by a source to the solid angle in which the flux is emitted. It is normally measured in terms of the lumen per steardian, which is the same as one candela.

luminous sensitivity—1. In a phototube, the output current divided by the incident luminous flux at a constant electrode voltage. (The term "output current" here does not include the dark current.) 2. The sensitivity of an object to light from a tungsten-filament lamp operating at a color temperature of 2870°K. This definition is permissible, since luminous sensitivity is not an absolute characteristic but depends on the spectral distribution of the incident flux.

luminous transmittance – The ratio of the luminous flux transmitted by an object to the incident luminous flux.

lumped – Concentrated in single, discrete elements rather than being distributed throughout a system.

lumped constant—The single quantity of a circuit property that is electrically equivalent to the total of that property distributed in a coil or circuit.

lumped-constant elements—Distinct electrical units smaller than a wavelength. They are calibrated and used, in conjunction with other electrical/electronic equipment, for controlling voltage and current.

lumped-constant oscillator – A microwave power-generation circuit that is realized by using discrete circuit elements such as inductors and capacitors.

lumped-constant-tuned heterodyne-frequency meter – A device for measuring frequency. Its operation depends on the use of a tuned electrical circuit consisting of lumped inductance and capacitance.

lumped impedance – Impedance concentrated in a component, as distinguished from impedance due to stray or distributed effects.

lumped inductance – Inductance concentrated in a component, as opposed to stray or distributed inductance.

lumped loading – Inserting uniformly spaced inductance coils along the line, since continuous loading is impractical.

lumped parameter—Any circuit parameter which—for purposes of analysis—can be considered to represent a single inductance, capacitance, resistance, etc., throughout the frequency range of interest.

Luneberg lens—A type of lens used to focus radiated ultrahigh-frequency electromagnetic energy to increase the gain of an antenna.

lux—A practical unit of illumination in the metric system. It is equivalent to the meter-candle and is the illumination on a one square-meter area on which there is a uniformly distributed flux of one lumen.

Luxemburg effect—A nonlinear effect in the ionosphere. As a result, the modulation on a strong carrier wave will be transferred to another carrier passing through the same region.

luxmeter—A type of illumination photometer employing a variable aperture and the contrast principle.

LVCD—Abbreviation for Least Voltage Coincidence Detection.

M

M – 1. Symbol for mutual inductance. 2. Abbreviation for the prefix mega or meg meaning one million.

m—Abbreviation for the prefix milli-.

mA—Abbreviation for milliampere.

MA—Abbreviation for megampere.

mach—Unit of speed measurement equal to speed of sound. Sound travels 759 mph at sea level; it decreases about 2% for every thousand feet of altitude.

machine-available time – In a computer, power-on time less maintenance time.

machine code—Code which is acceptable to the computing device on which the software is to be executed without further modification. *Also see* Machine Language.

machine cycle—The shortest complete process or action that is repeated in order. The minimum length of time in which the foregoing can be performed.

machine error—A deviation from the correct result in computer-processed data as a result of equipment failure.

machine hardware—Circuits contained in the five parts of the computer: input, output, control, storage, and arithmetic sections; any other circuit is "off-line."

machine independent – Pertaining to procedures or programs created without regard to the actual devices with which they will be used.

machine instruction – An instruction that can be recognized and executed by a machine.

machine language—1. Also called computer code. A set of symbols, characters, or signs and the rules for combining them to convey instructions or data to a computer. 2. Information recorded in a form a computer may use without prior translation.

3. A series of bits written as such for instruction of computers. The first level of computer language; *see* Assembly Language (second level) *and* Compiler Language (third level). 4. Sets of numeric instructions that control the functioning of a computer. These numeric instructions execute the logic functions of the computer's logic circuits. For example, the instructions 010003 might tell the computer to clear the A register. To overcome the problems inherent in working with long strings of numbers, computer programs are more often written in assembly language.

machine ringing—Telephone ringing that is started either mechanically or by an operator and that continues automatically until the call is either answered or abandoned.

machine run—In a computer, the performance of one or more machine routines which are linked to form one operating unit.

machine sensible—Information in a form such that it can be read by a specific machine.

machine thermal relay—A device which functions when the temperature of an ac machine armature, or of the armature or other load-carrying winding or element of a dc machine, or converter, power rectifier, or power transformer (including a power-rectifier transformer) exceeds a predetermined value.

machine word—*See* Computer Word.

macroassembler — A computer assembly program that translates alphanumeric language into machine language.

macro—A symbol which the assembler expands into one or more machine language instructions, relieving the programmer of having to write out frequently occurring instruction sequences.

macrocode—1. A coding system which assembles groups of computer instructions into single code words and which therefore requires interpretation or translation so that an automatic computer can follow it. 2. An instruction in a source language which is equivalent to a specified sequence of machine instruction.

macrocommand—A computer program entity formed by a string of standard, but related, commands which are put into effect by means of a single macrocommand. Any group of frequently used commands can be combined into a macrocommand. The many become one.

macroelement—An ordered set of two or more elements used as one data element with one data use identifier. For example, the macroelement data could be the ordered set of the data elements year, month, or day.

macroinstruction—In a computer, a source-language instruction that is equivalent to a specified sequence of machine instructions.

macroprogram—A computer program in the form of a sequence of instructions written in a source language.

macroprogramming — In a computer, the writing of machine-procedure statements in terms of macrocode instructions.

macroscopic—Large enough to be observed by the unaided eye.

macrosonics—The utilization of high amplitude sound waves for performing functions such as cleaning, drilling, emulsification, etc.

madistor — A semiconductor device that makes use of the effects of a magnetic field on a plasma current. The strength of a magnetic field determines the conductivity of the madistor material. A small change in the magnetic field produces a larger change in the madistor current.

MADT transistor—A microalloy diffused-base transistor.

magamp—Abbreviation for magnetic amplifier.

magic eye—*See* Electron-Ray Tube.

magic-T—*See* Hybrid Junction, 1.

magnaflux—A magnetic method of determining surface and subsurface defects in metals.

magnal base—An 11-pin base used on cathode-ray tubes.

Magnal socket.

magnal socket—An 11-pin socket used with cathode-ray tubes.

magnesium—An alkaline metal the compounds of which are sometimes used for cathodes.

magnesium anode—A bar of magnesium, buried in the earth, connected to an underground cable to prevent cable corrosion due to electrolysis.

magnesium cell—A primary cell the negative electrode of which is made of magnesium or one of its alloys.

magnesium-copper-sulfide rectifier — A dry disc rectifier consisting of magnesium in contact with copper sulfide.

magnesium-silver chloride cell—A reserve primary cell that becomes activated when water is added.

Magnesyn—Trade name for a device made by Bendix Aviation Corp. It is a portion of a repeater unit consisting of a two-pole permanent-magnet rotor within a three-phase, two-pole, delta-connected stator. The rotor carries the indicating pointer and is free to rotate in any direction.

magnet—A body that has the property of attracting or repelling magnetic materials. In its natural form it is called a lodestone. It may also be produced by permanently magnetizing a piece of iron or steel. A temporary magnet—called an electromagnet—is produced by passing a current through a coil surrounding a piece of iron or steel; the magnetism persists only while current is flowing. When suspended freely, a magnet will turn and align its poles with the north and south magnetic poles of the earth. (*See also* Electromagnetism *and* Permanent Magnet.)

magnet brake—A friction brake controlled electromagnetically.

magnetic—Pertaining to magnetism.

magnetic aging—The normal change in the metallurgical change in the material. The term also applies when the metallurgical changes are accelerated by an increase or decrease in temperature. When used in reference to core loss, this term, unless otherwise modified, implies an increase in loss. When used in reference to permeability or remanence, the term, in a positive sense, indicates a decrease in these quantities.

magnetic air gap—The air space, or nonmagnetic portion, of a magnetic circuit.

magnetically damped—A form of damping achieved by moving a metal vane through a magnetic field. This motion induces currents in the vane which set up magnetic fields opposing those of the stationary magnets, thus tending to bring the pointer to rest. This type of damping is found in many quality moving-iron and dynamometer-type instruments.

magnetic amplifier—1. A device in which one or more saturable reactors are used, either alone or with other circuit elements, to obtain power gain. Often called magamp. 2. A device in which a control signal applied to a system of saturable reactors modulates the flow of an alternating current in a power circuit.

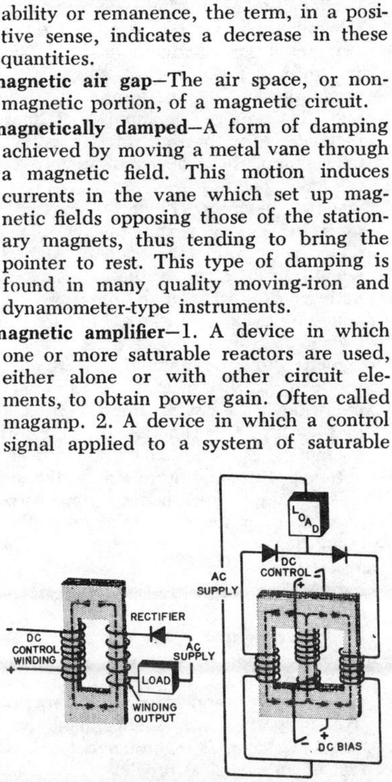

Magnetic amplifiers.

magnetic analysis—The separation of a stream of electrified particles by a magnetic field, in accordance with their mass, charge, or speed. This is the principle of the mass spectrograph.

magnetic anisotropy—The dependence of the magnetic properties of some materials on direction.

magnetic-armature speaker—Also called a magnetic-armature loudspeaker or magnetic speaker. A speaker comprising a ferromagnetic armature actuated by magnetic attraction.

magnetic azimuth—Azimuth measured from magnetic north.

magnetic bearing—The position in which an object is pointing with respect to the earth's magnetic north pole. It is expressed in degrees clockwise from that pole.

magnetic bias—A steady magnetic field applied to the magnetic circuit of a relay.

magnetic biasing—The superimposing of another magnetic field on the signal magnetic field of a tape while a magnetic recording is being made.

magnetic blowout—A magnet for establishing a field where an electrical circuit is broken. The field lengthens the arc and thus helps to extinguish it.

magnetic blowout switch — A switch that contains a small permanent magnet which provides a means of switching high dc loads. The magnet deflects the arc to quench it.

magnetic braking — Application of the brakes by magnetic force. The current for exciting the electromagnets is derived either from the traction motors acting as generators, or from an independent source.

magnetic bremsstrahlung—*See* Synchrotron Radiation.

magnetic brush development—A type of development in which the material that forms the image is carried to the field of the electrostatic image by ferromagnetic particles that act as carriers under the influence of a magnetic field.

magnetic card—A card on which data can be stored by selective magnetization of portions of a flat magnetic surface.

magnetic cartridge — *See* Variable-Reluctance Pickup.

magnetic character—A character imprinted with ink having magnetic properties. These characters are unique in that they can be read directly by both humans and machines.

magnetic circuit—The path of the flux as it travels from the north pole, through the circuit components, and back to the north pole. In a generator, the magnetic-circuit components include the field yoke, field pole pieces, air gap, and armature core. The magnetic circuit may be compared to an electrical circuit, with magnetomotive force corresponding to voltage, flux lines to current, and reluctance to resistance.

magnetic circuit breaker—A circuit breaker that depends on the response of an electromagnetic coil for operation.

magnetic coated disc—A magnetic disc-recording medium consisting of a coat of magnetizable material over a nonmagnetic base.

magnetic coating thickness – *See* Coating thickness.

magnetic coil—The winding of an electromagnet.

magnetic compass—A device for indicating direction. It consists of a magnetic needle which pivots freely and points toward the earth's north magnetic pole.

magnetic conduction current—The rate of flow of magnetism through a magnetized body.

magnetic contactor – A contactor actuated electromagnetically.

magnetic controller—An electric controller that contains devices operated by means of electromagnets.

magnetic-convergence principle—The obtaining of beam convergence through the use of a magnetic field.

magnetic core—1. A configuration of magnetic material which is, or is intended to be, placed in a rigid special relationship to current-carrying conductors and the magnetic properties of which are essential to its use. For example, it may be used to concentrate an induced magnetic field as in a transformer, induction coil, or armature; to retain a magnetic polarization for the purpose of storing data; or for its nonlinear properties as in a logic element. It may be made of such material as iron, iron oxide, or ferrite in such shapes as wires, tapes, toroids, or thin film. 2. The ferrous material in the center of an electromagnet.

magnetic core storage—A type of computer storage which employs a core of magnetic material surrounded by a coil of wire. The core can be magnetized to represent a binary 1 or 0.

magnetic course—A course in which the reference line is magnetic north.

magnetic cutter—A cutter in which the mechanical displacement of the recording stylus is produced by the action of magnetic fields.

magnetic cycle—The sequence of changes in the magnetization of an object corresponding to one cycle of the alternating current producing the magnetization.

magnetic damping—The damping of a mechanical motion by means of the reaction between a magnetic field and the current generated in a conductor moving through that field. The resistance of this conductor converts excess kinetic energy to heat.

magnetic deflection—The moving of the electron beam by means of a magnetic field produced by a coil around the neck of the cathode-ray tube. Linear motion is produced by a sawtooth current through the coil.

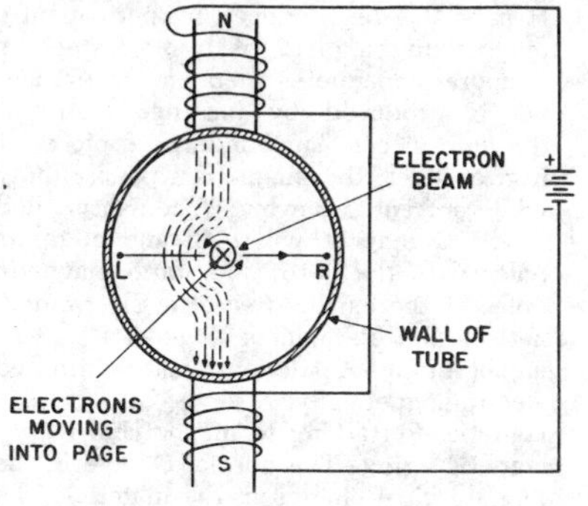

Magnetic deflection.

magnetic delay line – A computer delay line in which magnetic energy is propagated consisting in essence of a metallic medium along which the velocity of propagation of magnetic energy is small when compared to the speed of light. The storage of information, which is usually in a binary form, is accomplished by recirculation of wave patterns.

magnetic density—The number of lines of magnetic force passing through a magnet or magnetic field per unit area of cross section.

magnetic detecting device—A device for detecting cracks in iron or steel. This is done by introducing magnetic particles, which are attracted to the opposing magnetic poles created at the break.

magnetic device—Any device actuated electromagnetically.

magnetic diode—A magnetic-sensitive semiconductor device that (like indium-antimony devices) has an internal resistance that varies as a function of an external magnetic field. Electrical signals may be obtained through alteration of the magnetic field, and, hence, nonelectrical quantities can be converted into electrical quantities.

magnetic dip—Also called magnetic inclination. The angle of the magnetic field of the earth with the horizontal at a particular location.

magnetic dipole—A pair of equal-strength north and south magnetic poles spaced close together, and so small that its directive properties are independent of its size and shape. It is the magnetic equivalent of an electrical dipole.

magnetic dipole antenna—A loop antenna

that radiates an electromagnetic wave when electric current circulates in the loop.

magnetic direction indicator—Abbreviated MDI. An instrument that provides a compass indication, which it obtains electrically from a remote gyrostabilized magnetic compass (or its equivalent).

magnetic disc—A flat circular plate with a magnetic surface on which data can be stored by selective magnetization of portions of the flat surface.

magnetic disc storage—A device or system in which information is stored in the form of magnetic spots arranged on the surfaces of magnetically coated discs to represent binary data.

magnetic displacement—Magnetic flux density or magnetic induction.

magnetic drum—A storage device consisting of a rapidly rotating cylinder, the surface of which can be easily magnetized and which will retain the magnetization. Information is stored in the form of magnetized spots (or no spots) on the drum surface.

magnetic-drum receiving equipment — Radar developed for detection of targets beyond line of sight using ionospheric reflection and very low power. To distinguish radar signals from ground noise, a noise-free standard system is employed whereby the amplitude and characteristics are reproduced and preserved for comparison with returned signals. Fluctuations in the amplitude of the returned signals indicate the position and velocity of the target accurately to within 10 miles.

magnetic electron multiplier—One type of electron multiplier in which the paths of the emitted secondary electrons are controlled by an externally applied magnetic field.

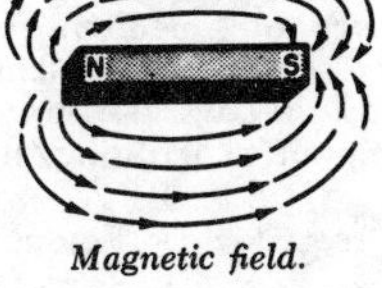

Magnetic field.

magnetic field — An area where magnetic forces can be detected around a permanent magnet, natural magnet, or electromagnet.

magnetic field strength—*See* Magnetizing Force.

magnetic figures—A pattern showing the distribution of a magnetic field. It is made by sprinkling iron filings on a nonmagnetic surface in the field.

magnetic flip-flop — A bistable amplifier using one or more magnetic amplifiers. The two stable output levels are determined by appropriate changes in the control voltage or current.

magnetic flux—The magnetic induction in a material. An electromotive force will be induced in a conductor placed in a magnetic field whenever the magnitude of the flux changes.

magnetic flux density—*See* Magnetic Induction.

magnetic focusing—The focusing of an electron beam by the action of a magnetic field.

magnetic freezing—In a relay, sticking of the armature to the core due to residual magnetism.

magnetic gap—The nonmagnetic part of a magnetic circuit.

magnetic gate—A gate circuit used in a magnetic amplifier.

magnetic head—In magnetic recording, a transducer that converts electric variations into magnetic variations for storage on magnetic media, or for reconverting such stored energy into electric energy. Stored energy can also be erased by this method.

magnetic heading—The heading of an aircraft with reference to magnetic north.

magnetic hysteresis—In a magnetic material, the property by virtue of which the magnetic induction for a given magnetizing force depends on the previous conditions of magnetization.

magnetic hysteresis loop—A closed curve that shows the induction of magnetization in a magnetic substance as a function of the magnetization force for a complete cycle of the magnetization force.

magnetic hysteresis loss—The power expended in a magnetic material, as a result of magnetic hysteresis, when the magnetic induction is cyclic.

magnetic inclination—*See* Magnetic Dip.

magnetic induction—Also called magnetic flux density. The flux per unit area perpendicular to the direction of the flux. The cgs unit of induction is called the gauss (plural, gausses) and is defined by the equation:

$$B = \frac{d\phi}{dA}$$

magnetic ink—Visible ink containing magnetic particles. When printed on a document (e.g., a bank check), the ink can be read by a magnetic character sensor and also by humans.

magnetic-ink character recognition — Recognition by a machine of characters printed with magnetic ink.

magnetic instability—The property of a magnetic material that leads to variations in the residual flux density of a recording tape with time, mechanical flexing, or changes in temperature.

magnetic integrated circuit—The physical realization of one or more magnetic elements inseparably associated to perform all or at least a major portion of its intended function.

magnetic-latch relay—*See* Polarized Double-Biased Relay.

magnetic leakage — Passage of magnetic flux outside the path along which it can do useful work.

magnetic lens—An apparatus that uses a nonuniform magnetic field to focus beams of rapidly moving electrons or ions in a cathode-ray or other tube.

magnetic line of force—In a magnetic field, an imaginary line which has the direction of the magnetic flux at every point.

magnetic materials—Materials that show magnetic properties. Ferromagnetic materials are more strongly magnetic than paramagnetic materials.

magnetic memory — A computer memory (or any portion) in which information is stored in the form of magnetism.

magnetic memory plate—A magnetic memory that consists of a ferrite plate containing a grid of small holes through which the read-in and read-out wires are threaded. Printed wiring applied directly to the plate may replace conventionally threaded wires, making possible mass production of plates with a high storage capacity.

magnetic microphone—*See* Variable-Reluctance Microphone.

magnetic mine—An underwater mine that detonates when near the steel hull of a ship. This is accomplished by relays, which redistribute the magnetic field in the mine when it is near the ship.

magnetic modulator—Also called a magnettor. A modulator employing a magnetic amplifier as the modulating element.

magnetic moment—Ratio of the maximum torque exerted on a magnet, to the magnetizing force of the field in which the magnet is situated.

magnetic needle—The magnetized needle used in a compass. When freely suspended, it will point to the earth's magnetic north and south poles.

magnetic north—The direction indicated by the north-seeking end of the needle in a magnetic compass.

magnetic phase modulator—A ferrite-core delay line in which the delay is varied by an external magnetic field.

magnetic pickup—*See* Variable-Reluctance Pickup.

magnetic-plated wire—A wire with a nonmagnetic core and a plated surface of ferromagnetic material.

magnetic poles—Those portions of a magnet toward which the lines of flux converge. All magnets have two poles, called north and south, or north-seeking and south-seeking, poles.

magnetic pole strength—The magnetic moment of a magnetized body divided by the distance between the poles.

magnetic potential difference—The line integral of magnetizing force between two points in a magnetic field.

magnetic-powder–coated tape—Also called coated tape. A tape consisting of a coating of uniformly dispersed powdered ferromagnetic material on a nonmagnetic base.

magnetic printing—Also called magnetic transfer, and cross talk. The permanent transfer of a recorded signal from a section of one magnetic recording medium to another section of the same or a different medium when they are brought near each other.

magnetic probe—A loop-type conductor for detecting the presence of static, audio, or rf magnetic fields.

magnetic radio bearing—*See* Radio Bearing.

magnetic recorder—Equipment incorporating an electromagnetic transducer and a means for moving a magnetic recording medium past the transducer. Electric signals are recorded in the medium as magnetic variations. (*See also* Magnetic Recording.)

magnetic recording—Recording audio frequencies by magnetizing areas of a tape or wire. The magnetized tape or wire is played back by passing it through a reproducing head. Here the magnetized areas are reconverted into electrical energy, which headphones or speakers then change back into sound.

magnetic recording head—In magnetic recording, a magnetic head that transforms electric variations into magnetic variations for storage on a magnetic medium such as tape or discs.

magnetic recording medium—A wire, tape, cylinder, disc, or other magnetizable material which retains the magnetic variations imparted to it during magnetic recording.

magnetic reproducer — Equipment which

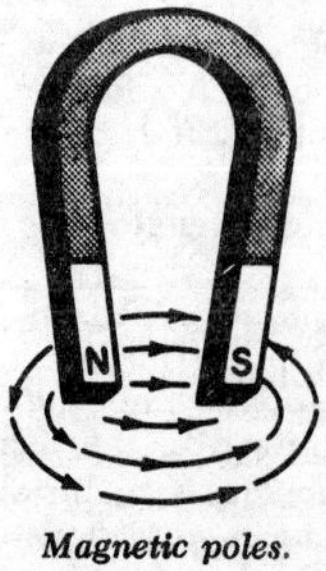

Magnetic poles.

picks up the magnetic variations on magnetic recording media and converts them into electrical variations.

magnetic reproducing head – In magnetic recording, the head that converts the magnetic variations into electric variations.

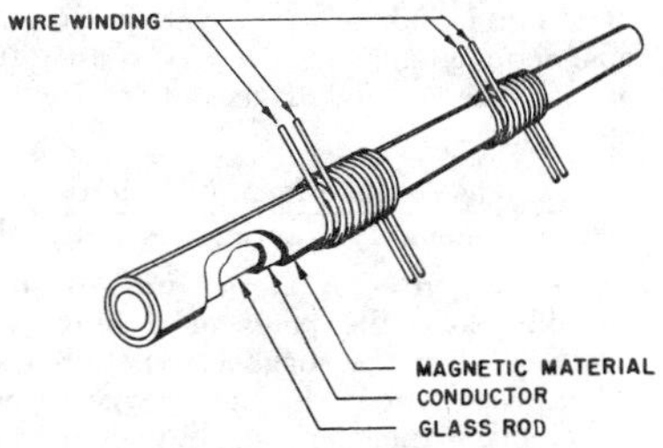

Magnetic rod.

magnetic rod–A square-loop switching and storage element for digital systems. It consists of a silver-coated glass rod upon which a thin layer of iron and nickel is deposited. Conductors are wound around the rod for the drive, sense, enable, and inhibit currents.

magnetics–The branch of science concerned with the laws of magnetic phenomena.

magnetic saturation–In an iron core, the point where application of a further increase in magnetizing force will produce little or no increase in the magnetic lines of force.

magnetic separator–An apparatus for separating powdered magnetic ores from nonmagnetic ores, or iron filings and other small iron objects from nonmagnetic materials. An electromagnet is employed which deflects the magnetic materials from the path taken by the nonmagnetic materials.

magnetic shield–A sheet or core of iron enclosing instruments or radio parts to protect them from stray magnetic fields. The shield provides a convenient path for the magnetic lines of force and thus diverts them from the component being protected.

magnetic shift register–A register in which magnetic cores are used as binary storage elements. By means of pulses, the pattern of binary digital information can be shifted one position to the left or right in the register.

magnetic shunt–A piece of iron used during instrument calibration to divert a portion of the magnetic lines of force passing through an air gap in the instrument.

magnetic sound–Sound recording in which magnetic impulses are stored in a ferric emulsion bonded to plastic or film.

magnetic speaker–A speaker in which acoustic waves are produced by mechanical forces resulting from magnetic reaction.

magnetic starter–A starter actuated electromagnetically.

magnetic storage–Any storage system in a computer which makes use of the magnetic properties of materials to store information.

magnetic storm – A disturbance in the earth's magnetic field. It is associated with abnormal solar activity and is capable of disrupting both radio and wire transmission.

magnetic strain gage–Also spelled gauge. An instrument for measuring strain in rails or other structural members that bend only microscopically under a normal load. It does this by determining the change in reluctance of a magnetic circuit having a movable armature.

magnetic susceptibility–Ratio of the intensity of magnetization to the corresponding value of the magnetizing force.

magnetic tape – The recording medium used in tape recorders. A paper or plastic tape on which a magnetic emulsion (usually ferric oxide) has been deposited. The most common width for home recorders is one-quarter inch. Some tapes are made of a magnetic material and hence need no magnetic-emulsion coating.

magnetic-tape core–A toroidal core made by winding a strip of thin magnetic-core material around a form. A toroidal winder is then used to wind coils around the core.

magnetic-tape reader–A computer device capable of converting the information recorded on magnetic tape into corresponding electric pulses.

magnetic-tape storage–A storage system based on the use of magnetic spots (bits) on metal or coated-plastic tape; the spots are arranged so that the desired code is read out as the tape travels past the read-write head.

magnetic test coil–Also called search or exploring coil. A coil which is connected to a suitable device to measure a change in the magnetic flux linked with it. The flux linkage may be changed by either moving the coil or varying the magnitude of the flux.

magnetic thin film–A layer of magnetic material, usually less than 10,000 angstroms thick. In electronic computers, magnetic thin films may be used for logic or storage elements.

magnetic transducer–*See* Variable-Reluctance Transducer.

magnetic transfer–*See* Magnetic Printing.

magnetic transition temperature – Also called the Curie point. In a ferromagnetic material, the point where its transition to paramagnetic seems to be complete as its temperature is raised.

magnetic tubes of flux—Regions in space the sides of which are everywhere tangent to the magnetic induction and the ends of which may meet to form closed rings.

magnetic units – Ampere-turn, gauss, gilbert, line of force, maxwell, oersted, and unit magnetic pole are examples of magnetic units—i.e., those used in measuring magnetic quantities.

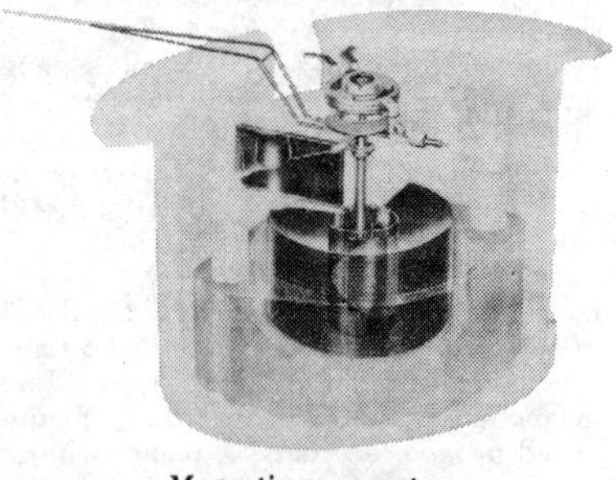

Magnetic-vane meter.

mangetic-vane meter—Also called a moving-vane meter. A meter for measuring alternating current. It contains a metal vane which pivots inside a coil. Alternating current flows through the coil and sets up magnetic forces which rotate the vane and attached pointer in proportion to the value of the current.

magnetic variometer – An instrument for measuring the differences in a magnetic field with respect to space or time.

magnetism—A property possessed by certain materials by which these materials can exert mechanical force on neighboring masses of magnetic materials and can cause voltages to be inducted in conducting bodies moving relative to the magnetized bodies.

magnetite – A mineral which exists in a magnetized condition in its natural state. It consists chiefly of a magnetic oxide of iron.

magnetization curve—A curve plotted on a graph to show successive states during magnetization of a ferromagnetic material. A normal magnetization curve is a portion of a symmetrical hysteresis loop. A virgin magnetization curve shows what happens to the material the first time it is magnetized.

magnetization intensity—At any point in a magnetized body, the ratio between the magnetic moment of the element of volume surrounding the point, and an infinitesimal volume.

magnetize—To make magnetic.

magnetizing current—The current through the field windings of a generator. Also called exciting current.

magnetizing field strength – The instantaneous strength of the magnetic field to which a sample of magnetic material is subjected.

magnetizing force – Also called magnetic field strength. The magnetomotive force per unit length at any given point in a magnetic circuit. In the cgs system the unit is the oersted. It is defined by the quotient:

$$\frac{\text{Magnetomotive force in gilberts}}{\text{Length in centimeters}}$$

magnet keeper – A bar of iron or steel placed across the poles of a horseshoe magnet when the magnet is not in use. The keeper prevents the magnet from becoming demagnetized, by completing the magnetic circuit to keep the flux from leaking off.

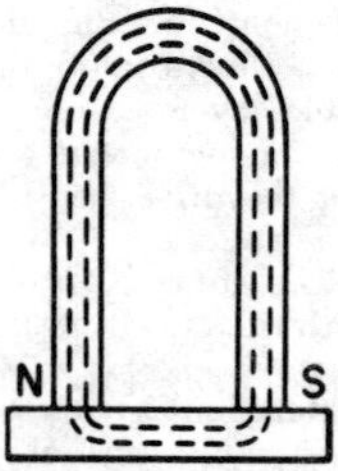

Magnet keeper.

magnet meter—Also called a magnet tester. An instrument for measuring the magnetic flux produced by a permanent magnet. It usually comprises a torque-coil or moving-magnet magnetometer with a particular arrangement of pole pieces.

magneto—1. An ac generator for producing ringing signals. 2. An ac generator for producing the ignition voltage in some gasoline engines.

magnetodiode—A high-sensitivity magnetosensitive semiconductor that operates on the principle of controlled lifetime of injected carriers by an external magnetic field.

magnetoelastic coupling – Energy transfer between elastic and spin-wave modes of propagation. The strength of the effect varies with the material used.

magnetoelastic energy—In a crystal, energy of interaction between elastic (or lattice) strains and applied magnetization. If a ferromagnetic crystal is placed in a magnetic field, or if a magnetic field is varied about the crystal, the lattice structure of the crystal is distorted (magnetostriction occurs) and changes the amount of potential energy in the lattice. The energy associated with this change is magnetoelastic.

magnetoelastic wave—A hybrid of spin and elastic waves.

magnetoelectric generator—An electric generator with permanent-magnet field poles.

mangetoelectric surface waves—Extensions of bulk magnetoelastic waves phenomena, which combines acoustic and spin wave properties. The spin waves, created by oscillations of the angle between adjacent atomic moments in a ferromagnetic solid, have high dispersive characteristics which depend on a dc magnetic field. A strain of a crystal lattice can affect the magnetic moments, resulting in a coupling between acoustic and spin waves. This coupling is most effective when wavelengths and frequencies of the two waves are comparable.

magnetoelectric transducer – A transducer which measures the emf generated by the movement of a conductor relative to a magnetic field.

magnetofluiddynamics—*See* Magnetohydrodynamics.

magnetofluidmechanics – *See* Magnetohydrodynamics.

magnetogasdynamics – *See* Magnetohydrodynamics.

magnetograph – A magnetometer that provides a continuous record of the changes occurring in the earth's magnetic field.

magnetohydrodynamic gyroscope—A gyroscope in which a rotating magnetic field drives a conducting fluid (e.g., mercury) around the closed path formed between the inner surface of a magnetic sleeve and the outer surface of a concentric magnetic cylinder.

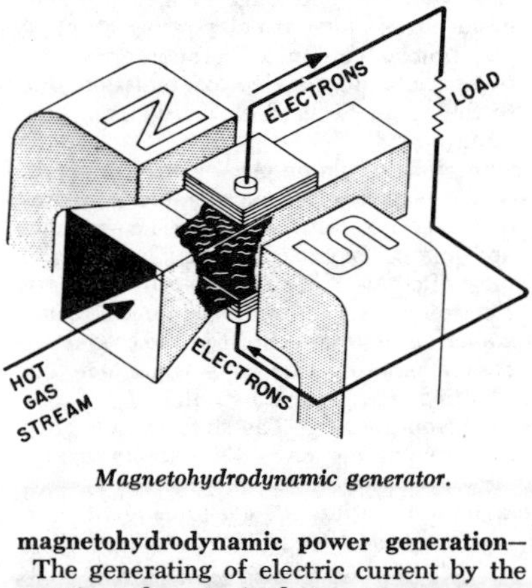

Magnetohydrodynamic generator.

magnetohydrodynamic power generation—The generating of electric current by the motion of an ionized gas.

magnetohydrodynamics—Abbreviated mhd. The study of the interaction between magnetic fields and electrically conductive fluids and gases. Also called magnetogasdynamics, hydromagnetics.

magnetoionic—Pertaining to the combined effect of atmospheric ionization and the magnetic field of the earth on electromagnetic wave propagation.

magnetoionic duct—A duct along the geomagnetic lines of force which exhibit waveguide characteristics for radio-wave propagation between conjugate points on the earth's surface.

magnetoionic wave component – Either of the two elliptically polarized wave components into which a linearly polarized wave in the ionosphere is separated by the earth's magnetic field.

magnetometer—An instrument for measuring the intensity or direction (or both) of a magnetic field (or component) in a particular direction.

magnetomotive force—The force by which a magnetic field is produced, either by a current flowing through a coil of wire or by the proximity of a magnetized body. The amount of magnetism produced in the first method is proportional to the current through the coil and the number of turns in it. The cgs unit of magnetomotive force is called the gilbert.

magnetoplasmadynamic generator – A device that generates an electric current by shooting an ionized gas (plasma) through a magnetic field.

magnetoresistance—The change in electrical conductivity of a material when a magnetic field is applied. This change is quite pronounced in materials with a high carrier mobility.

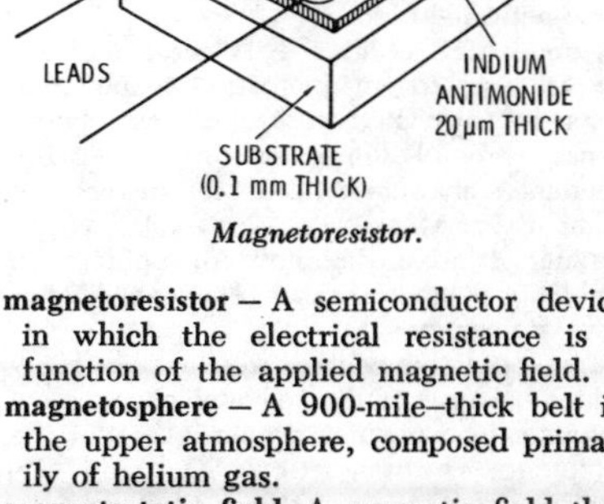

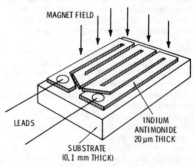

Magnetoresistor.

magnetoresistor – A semiconductor device in which the electrical resistance is a function of the applied magnetic field.

magnetosphere – A 900-mile-thick belt in the upper atmosphere, composed primarily of helium gas.

magnetostatic field—A magnetic field that is neither moving nor changing direction. Such a field could be produced by a stationary magnetic pole or by a constant current flowing in a stationary conductor.

magnetostriction—That property of certain

ferromagnetic metals, such as nickel, iron, cobalt, and manganese alloys, which causes them to shrink or expand when placed in a magnetic field. Conversely, if subjected to compression or tension, the magnetic reluctance changes, thus making it possible for a magnetostrictive wire or rod to vary a magnetic field in which it may be placed. This is true for lateral as well as longitudinal strains.

magnetostriction microphone – A microphone in which the deformation of a magnetostrictive material generates the required voltages.

magnetostriction oscillator—An oscillator in which the frequency is determined by the characteristics of the magnetostrictive element that inductively couples the plate circuit to the grid circuit.

magnetostriction speaker – A speaker in which the mechanical displacement is derived from the deformation of a magnetostrictive material.

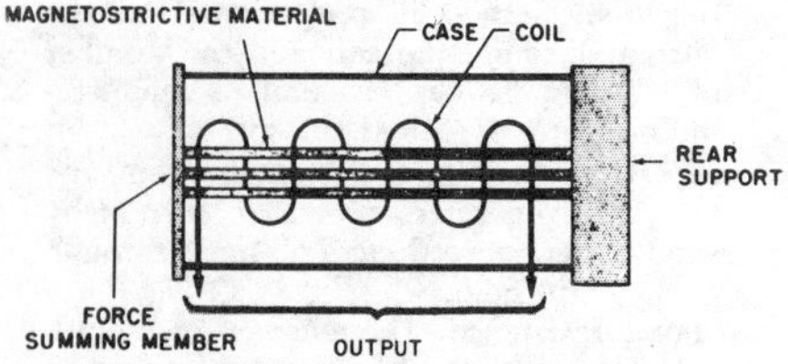

Magnetostriction transducer.

magnetostriction transducer—A transducer comprising an element of magnetostrictive material inside a coil, and a force-summing member attached to one end of the element. Current flows through the coil, and the magnetic field around it expands and contracts the element to move the member back and forth.

magnetostrictive delay line – A delay line made of nickel or certain other materials which become shorter when placed in a magnetic field.

magnetostrictive filter – A filter network which uses the magnetostrictive phenomena to form high-pass, low-pass, band-pass, or band-elimination filters. The impedance characteristic is the inverse of that of a crystal.

magnetostrictive oscillator—An oscillator in which a magnetostrictive element controls the frequency.

magnetostrictive relay – A relay that functions because of dimensional changes occurring in a magnetic material under the influence of a magnetic field.

magnetostrictive resonator – A ferromagnetic rod which can be excited magnetically so that it will resonate (vibrate) at one or more desired frequencies.

magnetostrictive transducer—Sensor using contraction or expansion of an iron or nickel rod due to a magnetic field. Magnetostrictive elements are used in magnetostriction oscillators.

magneto switchboard exchange—A manual telephone exchange arranged so that calling and clearing by the subscribers and operators are done by means of magnetoelectric generators.

magneto telephone—A telephone equipped with a magneto (a hand-driven, two-pole, ringing-signal generator). Although obsolete for home and business telephones, it is still used in many field applications.

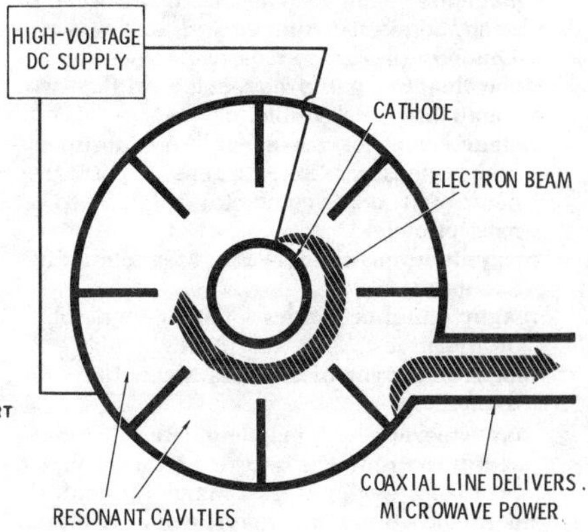

Magnetron.

magnetron—An electric tube used to generate high power output in the uhf and shf bands. The basis of its operation is the interaction of electrons with the electric field of a circuit element in crossed steady electric and magnetic fields to generate alternating-current power output.

magnetron effect—In a thermionic vacuum tube, the reduced electron emission due to the magnetic field of the filament current.

magnetron oscillator—An electron tube in which electrons are accelerated by a radial electric field between the cathode and one or more anodes and by an axial magnetic field that provides a high-energy electron stream to excite the tank circuits.

mangetron pulling—A shift in the frequency of a magnetron due to a change in the standing waves on the rf lines.

magnetron pushing—The shift in frequency of a magnetron caused by faulty modulator operation.

magnetron rectifier—A gas-tube rectifier in which the electron stream is controlled by an optical element or instrument instead of by heated electrodes.

magnet steel—A special steel used in permanent magnets because of its high retentivity. In addition to steel, it also contains tungsten, cobalt, chromium, and manganese.

magnet strip – Sheet or foil aluminum (either bare or insulated) used as a conductor in electric windings. Copper is also used sometimes.

magnet tester—*See* Magnet Meter.

magnettor—*See* Magnetic Modulator.

magnet wire – An insulated copper wire used for winding the coils of electromagnets.

magnification ratio—The ratio of the size of an image produced by a lens to the size of the source.

magnified sweep – In an oscilloscope, a sweep the time per division of which has been decreased by amplification of the sweep waveform rather than by changing the time constants used to generate it.

Magnistor—A saturable reactor for controlling electrical pulses or sine waves having frequencies of 100 kHz to 30 MHz and power levels ranging from microwatts to tens of watts.

magnitude—1. Size; the quantity assigned to one unit so that it may be compared with other units of the same class (i.e., the ratio of one quantity to another). 2. The size of a quantity irrespective of its sign. For example +10 and −10 have the same magnitude.

magnitude-controlled rectifier—A type of rectifier circuit in which a thyratron is used as the rectifying element. The load current is controlled by varying the basis on the grid of the thyratron.

mag-slip—A British term for synchro (i.e., a synchronous device such as the selsyn, autosyn, motortorque, and generator).

mail box—A set of computer locations in a common RAM storage area reserved for data addressed to specific peripheral devices as well as other microprocessors in the immediate environment. (Such an arrangement enables the coordinator CPU and the supplementary microprocessors to transfer data among themselves in an orderly fashion with minimal hardware.)

mailer—A cassette or very small reel of tape sold in a cardboard box printed with name and address blanks, for sending through the mail.

main anode – The anode which conducts the load current.

main bang—The transmitted pulse of a radar system.

main bonding jumper—The connection between the grounded circuit conductor and the equipment grounding conductor at the service.

main control unit—Transmitter or receiver controls for energizing, adjustment, etc., of the transmitter or receiver, but not for operating it while on the air.

main distributing frame – In a telephone central office, a distributing frame, on one section of which terminate permanent outside lines, and on another section of which terminate the subscriber-line multiple cabling, trunk multiple cabling, etc., used for associating any outside line with any desired terminal in such a multiple or with any other outside line. The main distributing frame usually carries the central-office protective devices and serves as a test point between line and office.

main exciter – An exciter which supplies energy for the field excitation of another exciter.

main frame—1. Also called central processing unit. The central processor of the computer system. It contains the main storage, arithmetic unit, and special register groups. 2. The CPU of a computer plus the input/output unit and memory. As distinguished from peripheral equipment.

main gap – The conduction path between the principal cathode and anode.

main memory—*See* Internal Storage, 1.

main power—Power supplied to a complete system from a line.

mains—Interior wires extending from the service switch, generator bus, or converter bus to the main distribution center.

main service panel – The main electrical switch or circuit breaker and the circuit panel box which houses the circuit breakers of fuses for a branch circuit.

main station – 1. A telephone station that has a distinct call number and a direct connection to a central office. 2. With respect to leased lines for customer equipment, the main point of interfacing of such equipment with the local loop.

main storage – Usually the fastest storage device of a computer and the one from which instructions are executed.

main sweep—The longest-range scale available on some fire-control radars.

maintained contact switch—A switch which remains in a given condition until actuated to another condition, which is also maintained until further actuation.

maintained switch – A switch which remains at the operated circuit condition when the actuating force is removed. It returns to the normal circuit condition, or moves to another position, when actuated a second time.

maintaining voltage—The voltage across a glow lamp after breakdown occurs.

maintenability—The probability of restoring a system to its specified operating conditions within a specified total down time, when maintenance action is initiated under stated conditions.

maintenance—1. All procedures necessary to keep an item in, or restoring it to, a serviceable condition, including servicing, repair, modification, modernization, overhaul, inspection, etc. 2. *See* File Maintenance.

maintenance time—Time used for preventive and corrective maintenance of hardware.

major-apex face—In a natural quartz crystal, any one of the three larger sloping faces extending to the apex (pointed) end. The other three are called the minor-apex faces.

major cycle—1. In a memory device which provides serial access to storage positions, the time interval between successive appearances of a given storage position. 2. A number of minor cycles.

major defects—Those defects usually responsible for the failure of a component to function in its intended manner.

major face—Any one of the three larger sides of a hexagonal quartz crystal.

major failure—1. A noncritical failure that can degrade the system performance due to cumulative tolerance buildup. 2. A malfunction in a system which, while not causing complete breakdown, causes it to function uselessly.

major items – Items and components of communications electronics equipment which are designed to perform a specific function (e.g., radio, radar sets, transmitters, receivers, modulators, amplifiers, and assemblies) and for which the procurement and supply lead times are such that the items must be programmed and procured to be available at a given time. A programming term.

majority—A logical operator with the property that if P, Q, R, etc., are statements, the majority of P, Q, R, . . . is true if more than half the statements are true, and false if half or fewer are true.

majority carrier—The predominant carrier in a semiconductor. Electrons are the majority carrier in n-type semiconductors, since there are more electrons than holes. Likewise, holes are the majority carrier in p-types, since they outnumber the electrons.

majority-carrier contact—An electrical contact across which the ratio of majority-carrier current to applied voltage is substantially independent of the voltage polarity but the ratio of minority-carrier current to applied voltage is not.

majority emitter—The transistor electrode from which the majority carriers flow into the interelectrode region.

majority gate—A logic element the output of which is true if more than half of the inputs are true, but the output of which is false for other input conditions.

major lobe—The antenna lobe in the direction of maximum energy radiation or reception.

major loop – A continuous network composed of all the forward elements and the primary feedback elements of a feedback control system.

major relay station—A tape relay station to which two or more trunk circuits are connected in order to provide an alternate route or to meet command requirements.

make—The closing of a relay, key, or other contact.

make-before-break – The action of closing a switching circuit before opening another associated circuit.

make-before-break contacts—Double-throw contacts so arranged that the moving contact establishes a new circuit before disrupting the old one.

make-before-break switch – *See* Shorting Switch.

make-break contacts—Also called continuity-transfer contact. A contact form of a relay, in which one contact closes connection to another contact and then opens its prior connection to a third contact.

make-break electrode—In a mercury switch, that electrode which serves the function of making and breaking contact with the mercury pool to close or open the electrical circuit.

make contact—A contact that closes a circuit upon operation of the device of which the contact is a part. *See also* Normally Open Contact.

make percent—In pulse testing, the length of time, in comparison with the duration of the test signal, that a circuit stands closed.

make-up time – That portion of available time used for reruns made necessary by malfunctions or mistakes that occurred during a previous operating time.

male—Adapted so as to fit into a matching hollow part.

male contact—A contact located in the insert or body of a connector in such a way that the mating portion of the contact extends into the female contact. It is similar in function to a pin contact.

malfunction—*See* Error, 2.

mandrel test—A test used to determine the flexibility of insulation. In it a wire, with or without previous stretch, is wrapped around a mandrel.

manganin—A alloy wire used in precision wirewound resistors because of its low temperature coefficient of resistance.

manipulated variable – The one variable (condition, quantity, etc.) of a process that is being controlled. The process can be controlled through manipulations of this variable.

man-made interference—A type of electromagnetic interference generated by electric motors, communication and broadcast transmitters, fluorescent lighting, and other electrical and electronic systems that radiate spurious signals.

man-made noise—High-frequency noise signals caused by sparking in an electric circuit.

man-made static—High-frequency noise signals created by sparking in an electric circuit. When picked up by radio receivers, it causes buzzing and crashing sounds from the speaker.

manometer—A gage for measuring the pressure of gases. It contains a column of incompressible liquid. The amount the liquid is displaced indicates the magnitude of the pressure causing the displacement.

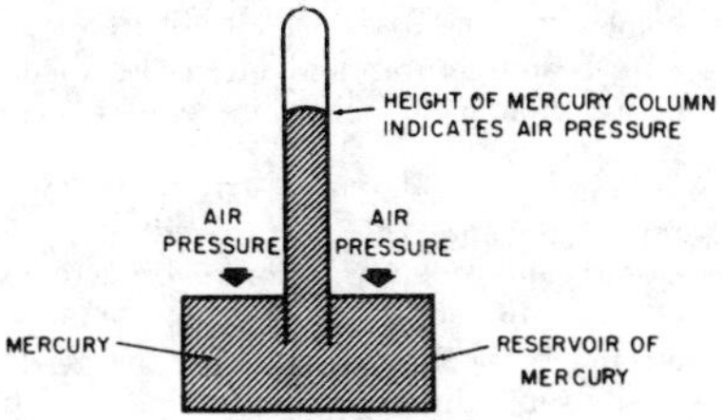

Manometer.

manpack—Also called packset. A portable radio-transmitting/receiving device which can be carried easily in a harness by a man.

manual—1. Hand operated. 2. In an organ, a group of keys played with the hand. In two-manual organs the upper manual, also referred to as the solo or swell manual, is normally used to play the melody. The lower manual, also referred to as the accompaniment manual or great manual, is normally used to play the accompaniment.

manual central office—A central office of a manual telephone system.

manual control—1. The opening or closing of switches by hand. 2. The direction of a computer by means of manually operated switches.

manual controller – An electric controller in which all but its basic functions are performed by hand.

manual dimmer – A dimmer in which the only linkage between the control lever and the moving electrical contact that conducts the electrical power is mechanical.

manual direction finder—A radio compass that uses a rotatable loop which is operated manually.

manual exchange—A telephone exchange in which the lines are connected to a switchboard and interconnections are controlled by an operator.

manual input—1. The entry of data into a system by hand at the time of processing. 2. Direct computer entry by means of manual intervention or drum entry of manual data through card machines.

manual operation—Data processing in a system by means of direct manual techniques.

manual preset timer—A manual start timer, the cycle of which is initiated by turning a pointer to the desired setting.

manual rate-aided tracking—Radar tracking of individual targets by means of circuits that compute the velocity from manually inserted position fixes.

manual reset—A qualifying term used to indicate that a relay may be reset manually after an operation.

manual ringing – A method of ringing a telephone. The key must be held down for the ringing to continue. Nor does it stop when the receiver is lifted off the hook, unless the caller releases the key.

manual single play—A turntable operation in which the records are played one at a time, and the arm is placed on the record and removed again manually.

manual start timer – An interval timer on which each cycle must be manually started at the timer.

manual switch—A switch which is actuated by an operator.

manual switchboard – A telephone switchboard in which the operator makes connections manually with plugs and jacks or with keys.

manual switching—A characteristic of a circuit breaker permitting manual opening and closing of the circuit by operation of the actuator.

manual telegraphy – Telegraphy in which an operator forms the individual characters of the alphabet in code.

manual telephone set—A set not equipped with a dial for securing the number to be called. Instead, lifting the receiver alerts the switchboard operator, who then connects the caller to the person being called.

manual telephone system – A system in which telephone connections between customers are ordinarily established manually by telephone operators, in accordance with orders given verbally by the calling parties.

manual tuning – Rotation by hand of a knob on a radio receiver to tune in a desired station.

manufacturing holes—*See* Pilot Holes.

manuscript – A form of storage medium, such as programming charts, in which is contained raw information in a sequential form suitable for translation.

Marconi antenna – An antenna system in which one end of the signal source is connected to a radiating element and the other end is connected to ground.

margin—1. The difference between an actual operating point and the point or condition where a failure to operate properly will occur. 2. Also called range or printing range. In telegraphy, the interval between limits on a scale, usually arbitrary, in which printing is error-free. 3. The distance that the electrode foil is indented from the edge of the dielectric when a capacitor is wound.

marginal checking – Also called marginal testing. Preventive maintenance in which certain operating conditions (e.g., supply voltage or frequency) are varied from normal in order to locate defects before they become serious.

marginal relay—A relay with a small margin between its nonoperative current value (maximum current applicable without operation) and its operative value (minimum current that operates the relay).

marginal testing—*See* Marginal Checking.

margin-punched card – A card in which holes representing data are punched only on the border, with the center left free for written or printed information.

marine broadcast station—A coastal station which regularly broadcasts the time and meteorological and hydrographic information.

marine radiobeacon station – A radionavigation land station, the emissions from which are used to determine the bearing or direction of a ship in relation to the marine radiobeacon station.

maritime mobile service—The radio service in which ships communicate with each other or with coastal and other land stations on specified frequencies.

maritime radionavigation service—A radio service intended to be used for the navigation of ships.

mark—1. In telegraphy, the closed-circuit condition—i.e., the signal that closes the circuit at the receiver to produce a click of the sounder or to print a character on a teletypewriter. 2. The presence of signal. A mark impulse is equivalent to a binary 1. 3. *See* Flag, 1.

mark and space impulses—In neutral operation of a teletypewriter system, the mark impulse is the closed-circuit signal, and the space impulse is the open-circuit signal. In other than neutral operation, the mark impulse is the circuit condition that produces the same result in the terminal equipment that a mark impulse produces in neutral operation. Similarly, the space impulse is the circuit condition that produces the same result in the terminal equipment that a space impulse produces in neutral operation.

marker—Also called marker beacon. A radio navigational aid consisting of a transmitter that sends a signal to designate the small area around and above it.

marker antenna—The transmitting antenna for a marker beacon.

marker beacon—*See* Marker.

marker generator—An rf generator that injects one or more pips of specific frequency onto the response curve of a tuned circuit being displayed on the screen of a cathode-ray oscilloscope.

marker pip—The inverted V (Λ) or spot of light used as a frequency index mark in cathode-ray oscilloscopes for alignment of tv sets. It is produced by coupling a fixed-frequency oscillator to the output of a sweep-driven signal generator.

marker thread—A colored thread laid parallel and adjacent to the strands of an insulated conductor which identifies the wire manufacturer and often the specification under which the wire is constructed.

marking-and-spacing intervals – In telegraphy, the intervals corresponding to the closed and open positions, respectively, of the originating transmitting contacts.

marking bias—Bias that affects the results in the same direction they are affected by marking current.

marking current—The magnitude and polarity of line current when the receiving mechanism is in the operated condition.

marking pulse—The signal interval during which the selector unit of a teletypewriter is operated.

marking wave—Also called keying wave. In telegraphy, the emission while the active portions of the code characters are being transmitted.

mark sense—To mark a position on a punch card, using a special pencil that leaves an electrically conductive deposit for later conversion to machine punching.

mark sensing – A technique for detecting special pencil marks entered in special places on a card and automatically translating the marks into punched holes.

mark-to-space transition—The change from a marking impulse to a spacing impulse.

marshalling sequence – *See* Collating Sequence, 1.

maser—1. Acronym for Microwave Amplification by Stimulated Emission of Radiation. A low-noise microwave amplifier in which a signal is boosted by changing the energy level of a gas or crystal (commonly, ammonia or ruby, respectively). 2. A means of focusing a stream of particles, which concentrates only on the high energy particles. These are passed into a resonator which is resonating at the radiation frequency of the particles. The particles are in this state raised to a strong oscillation, and can be used for control purposes. By reducing the flow of particles to the resonator, to maintain oscillations, it can be used as an amplifier. (There are many other applications). 3. Device for amplifying a microwave frequency signal by "stimulated emission of radiation"—i.e., the weak microwave signal causes electrons in an atom to change orbit in such a manner as to emit an amplified signal of the same frequency as the weak signal.

mask—1. A frame mounted in front of a television picture tube to limit the viewing area of the screen. 2. A device (usually a thin sheet of metal which contains an open pattern) used to shield selected portions of a base during a deposition process. 3. A device used to shield selected portions of a photosensitive material during photographic processing. 4. A logical technique in which certain bits of a word are blanked out or inhibited. 5. Template used to etch circuit patterns on semiconductor wafers. Images of the circuit patterns are produced on glass or metal photographically. The mask is then used to control the diffusion process, plus metallization.

masked diffusion—The use of a mask pattern to obtain selective impregnation of portions of a semiconductor material with impurity atoms.

masking—The process by which a sound is made audible by the addition of a second sound called the masking sound. The unit of measurement is usually the decibel.

masking audiogram—A graphical representation of the amount of masking by a noise. It is plotted in decibels as a function of the frequency of the masked tone.

mask microphone—A microphone designed for use inside an oxygen or other respiratory mask.

mask set — A set of plates, usually glass, which are used to transfer a device topology in sequence to a wafer during fabrication.

Masonite — Trade name of The Masonite Corp. Fiberboard made from steam-exploded wood fiber. Its highly compressed forms are used for panels in electrical equipment.

mass—1. The quantity of matter in an object. It is equal to the weight of a body divided by the acceleration due to gravity. 2. The bulk of matter though not necessarily equal to its weight. A mechanical unit whose electrical analog is inductance.

mass data—A larger amount of data than can be stored in the central processing unit of a computer at any one time.

mass-memory unit—A drum or disc memory that provides rapid-access bulk storage for messages being held until outgoing channels are available.

mass radiator—A spark radiator which generates a low-level, broad-band signal extending into and above the ehf band. Arcing occurs between fine metal particles suspended in a liquid dielectric.

mass spectrometer—An instrument that permits rapid analysis of chemical compounds. It consists of a vacuum tube into which a small amount of the gas to be studied is admitted. The gas is ionized by the electrons emitted from the cathode and speeded up by an accelerating grid. An electric field draws the ions out of the ionizing chamber. They are then sent through electric and magnetic fields that sort them according to their ratios of mass to charge.

mass spectrum—The spectrum obtained by deflecting a beam of electrons with an electric or magnetic field as they emerge from a tube containing a small quantity of the gas being investigated. The amount a particle is deflected depends on the ratio of its mass to its atomic charge. Hence, every element has a characteristic mass-spectrum line.

mast — The pole on which an antenna is mounted.

master—1. The mold from which other disc recordings are cast. It is made by electroforming from a disc recording, and is a "negative" of the disc (i.e., has ridges instead of grooves). 2. An original, or first special copy, of a recorded performance from which other copies may be made. 3. An original recording, made directly from recording microphones. A disc master is the lacquer original, usually cut from a tape from which stampers are made for vinyl pressings. 4. An element of a system that controls or initiates the action or responses of the other elements of the system.

master brightness control—In a color television receiver, a variable resistor that adjusts the bias level on all three guns in the picture tube at the same time.

master clock—1. In a computer, the primary source of timing signals. 2. A very accurate timer with an absolute time reference, providing controlled power to drive slave or auxiliary timers and display units. May also provide correcting pulses for slave devices.

master contactor—A device which is generally controlled by a master element or equivalent, and the necessary permissive and protective devices, and which serves to make and break the necessary control circuits to place an equipment into operation under the desired conditions, and to take it out of operation under other or abnormal conditions.

master control—1. In a studio, a central point from which sound or television programs are switched to one or more destinations. 2. An application-oriented routine usually applied to the highest level of a subroutine hierarchy.

master die—A substrate that contains unconnected active and passive elements in a predetermined pattern. Connection pads for each element and subelement are provided, and a variety of circuits may be obtained by appropriate choices of intraconnection patterns.

master drawing – A drawing showing the dimensional limits or grid locations applicable to any or all parts of a printed circuit including the base.

master drive–A drive that determines the reference input for one or more follower drives.

master element–The initiating device, such as a control switch, voltage relay, float switch, etc., which serves either directly or through such permissive devices as protective and time-delay relays to place an equipment in or out of operation.

master file–In a computer, a file of relatively more permanent information that is usually updated periodically.

master gain control–1. An amplifier control that permits adjusting the gain of two or more channels simultaneously. 2. Control of overall gain of an amplifying system as opposed to varying the gain of several individual inputs.

master instruction tape–A computer magnetic tape on which are recorded all programs for a system of runs.

master oscillator–1. In a transmitter, the oscillator that establishes the carrier frequency of the output. 2. An oscillator that controls or provides modulator drive frequencies for a number of channels or channel groups.

master oscillator-power amplifier–Abbreviated mopa. An oscillator followed by a radio-frequency buffer-amplifier stage.

master pattern – An accurately scaled pattern which is used to produce the printed circuit within the accuracy specified in the master drawing.

master reticle – A reticle plate properly aligned in a reticle frame and sealed in place. The master reticle is inserted into a photorepeater to make the final-sized photomask. Used in the production of monolithic circuits.

master routine–*See* Subroutine, 2.

master scheduler – The control-program function that responds to operator commands, initiates requested actions, and returns information that is requested or required; the overriding medium for control of the use of the computing system.

master-slave–A binary element consisting of two independent storage stages arranged with a definite separation of the clock function to enter information to the master stage and to transfer it to the slave stage.

master-slave flip-flop – A circuit that contains two flip-flops, a master and a slave. The master flip-flop receives information on the leading edge of a clock pulse, and the slave (output) flip-flop receives information on the trailing edge of the clock pulse.

master slice–A silicon wafer containing 30 or more groups of components. The elements can be interconnected with paths of aluminum to form desired circuits. The wafer is then diced into single devices.

master stamper–A master from which phonograph records are pressed.

master station – 1. The radio station to which the emission of other stations of a synchronized group are referred. 2. In a hyperbolic navigation system, the one of a pair of transmitting stations that controls the transmission of the other (slave) station in the pair, and maintains the time relationship between the pulses emitted from the two stations.

master switch – A switch located electri-

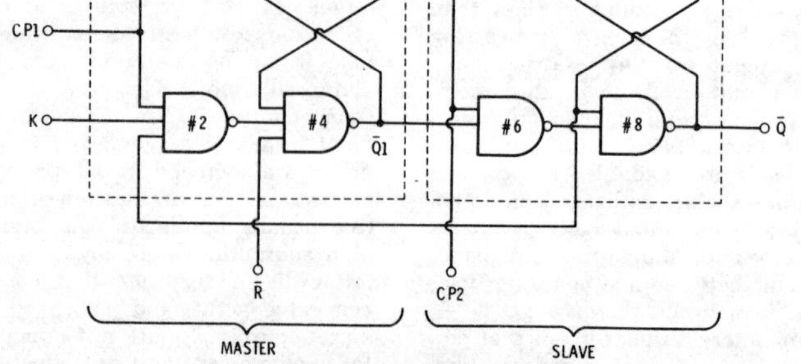

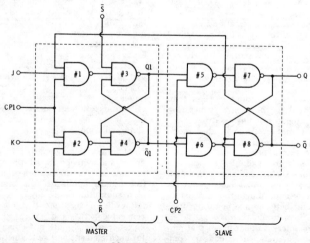

Master-slave JK flip-flop.

cally ahead of a number of other switches.

master synchronization pulse—A pulse distinguished from other telemetering pulses by its different amplitude and/or duration and used to indicate the end of a sequence of pulses.

master tv system—A combination of components for providing multiple tv set operation from one antenna.

master wafer — A processed semiconductor wafer with unconnected active and passive circuit elements located in a standardized pattern. Different integrated circuits may be synthesized by using various interconnection paths.

MAT—Acronym for microalloy transistor.

match—1. The similarity or equality of one thing to another. 2. To compare two or more items of data for identity.

matched filter—An optimum filter for separating a waveform of known shape from random perturbing noise.

matched impedance—The condition that exists when two coupled circuits are adjusted so that the impedance of one circuit equals the impedance of the other.

matched load—A device used to terminate a transmission line or waveguide so that all the energy from the signal source will be absorbed.

matched power gain — That power gain which is obtained when the impedance of the load is matched to the effective output impedance of the amplifier to which the load is connected.

matched pulse intercepting — A technique used in intercepting calls on party lines in a terminal-per-line office. A ground pulse is matched in time with the particular ringing frequency of the intercepted station.

matched symmetrical transistor—A special case of the bidirectional transistor in which not only are the requirements for a symmetrical transistor met, but actual matching specifications are also given. *See also* Symmetrical Transistor.

matched termination — A termination that causes no reflected wave at any transverse section of the transmission line. Its impedance is equal to the characteristic impedance of the line.

matched transmission line—A transmission line along which there is no wave reflection.

matched waveguide — A waveguide along which there is no reflected wave.

matching — 1. Connecting two circuits or parts together with a coupling device in such a way that the maximum transfer of energy occurs between the two circuits, and the impedance of either circuit will be terminated in its image. 2. The connection of a component's output to an input which provides the recommended value of load or "termination impedance."

matching diaphragm—A window consisting of an aperture (slit) in a thin piece of metal, placed transversely across a waveguide; used as a matching device. The orientation of the slit (whether parallel to the long or short dimensions of the waveguide) determines whether it is respectively capacitive or inductive.

matching impedance—The impedance value that must be connected to the terminals of a signal-voltage source for proper matching.

matching plate — In waveguides, a diaphragm used for matching.

matching stub—A device placed on a radio-frequency transmission line to vary its electrical length and hence its impedance.

matching transformer—A transformer used for matching impedances.

mate—Joining of two connector halves, or of a cable to a connector.

material system—The designation of the number of basic metals (e.g., silver-antimony-telluride) making up thermoelectric materials.

mathematical check—A check making use of mathematical identities or other properties.

mathematical logic—Also called symbolic logic. Exact reasoning concerning non-numerical relations by using symbols that are efficient in calculation.

mathematical model—The general characterization of a process, object, or concept in terms of mathematics that enables the relatively simple manipulation of variables to be accomplished in order to determine how the process or concept would behave in different situations.

matrix—1. A coding network or system in a computer. When signals representing a certain code are applied to the inputs, the output signals are in a different code. 2. In electronic computers, any logical network the configuration of which is a rectangular array of intersections of its input-output leads, with elements connected at some of these intersections. The network usually functions as an encoder or decoder. 3. A computer network or system in which only one input is excited at a time and produces a combination of outputs. 4. In a color tv circuit, the section that combines the I, Q, and Y signals and transforms them into individual red, green, and blue signals which are applied to the picture-tube grids. 5. A rectangular array of scalar quantities, usually numbers or letters used to represent numbers. 6. An orderly two-dimensional array. An arrangement of circuit elements, such as wires, relays, diodes, etc., which can transform a digital code from one type to another. 7. A rectangular array of a given set of numbers, letters, or symbols or a combination of them. Rectangular array means that the elements are arranged into definite rows and columns. 8. The terminology applied to the several methods for encoding four channels onto two channels for later recovery back to four channels. Also referred to as 4-2-4. The actual electronics used to encode into two channels or decode back to four are known as matrixing electronics. 9. A rectangular array of elements, in cross-match fashion. Used to describe memory organization, character formation, diode layouts, and so forth.

matrixer—Also called matrix unit. A device which transforms the color coordinates, usually by electrical or optical means.

matrixing electronics—*See* Matrix.

matrix life test—A test in which each test condition has two components. For example in transistor matrix life testing each life-test condition is represented by an ambient temperature and a power dissipation. At each stress level of temperature, there are several dissipations; and at each stress level of dissipation, there are several temperatures. The test conditions are placed into blocks or groups.

matrix line printer—A printer in which each character is composed of a matrix of dots.

matrix storage—Storage in which the elements are arranged in such a way that access to any location requires the use of two or more coordinates, as, for example, in cathode-ray-tube storage and core storage.

matrix unit—*See* Matrixer.

matter — Any physical entity — i.e., having mass.

Matteucci effect—The ability of a twisted ferromagnetic wire to generate a voltage as its magnetization changes.

max—Abbreviation for maximum.

maxima/minima—In radar, regions of maximum and minimum return from the transmitted pulse caused by additive and subtractive combinations of the direct and reflected wave. A plot of these data is usually known as a null pattern or fade chart.

maximum—Abbreviated max. The highest value occurring during a stated period.

maximum average power output—In television, the maximum radio-frequency output power, averaged over the longest repetitive modulation cycle.

maximum deviation sensitivity — Under maximum system deviation, the smallest signal input for which the output distortion does not exceed a specified limit.

maximum dissipation—The maximum average power a device can dissipate during operation while still remaining within published life specifications.

maximum frequency of oscillation — The highest frequency at which a transistor or vacuum tube will oscillate.

maximum keying frequency—In a facsimile system, the frequency (in hertz) equal to half the number of critical areas of the subject copy scanned per second.

maximum modulating frequency—In a facsimile system, the maximum scanning frequency process that can be transmitted without degrading the recorded copy.

maximum output—The highest average output power into a rated load, regardless of distortion.

maximum overshoot—The maximum amplitude deviation from the average of the steady-state values that exist immediately before and after the transient.

maximum peak plate current—The highest instantaneous plate current a tube can safely carry.

maximum percentage modulation — The highest percentage of modulation permitted in a transmitter without producing excessive harmonics in the modulating frequency.

maximum permeability—The highest permeability reached as induction or magnetization is increased.

maximum power transfer theorem — The maximum power will be absorbed by one network from another joined to it at two terminals, when the impedance of the receiving network is varied, if the impedances (looking into the two networks at the junction) are conjugates of each other.

maximum record level—In direct recording, the amount of record-head current required to produce 3-percent third-harmonic distortion of the reproduced signal at the record-level set frequency. Such distortion must result from magnetic-tape saturation, not from electronic circuitry.

maximum response speed — The maximum pulse frequency that can be applied to a stepper motor at random and result in synchronized steps. The motor must not miss in step while operating within this range.

maximum retention time — The maximum time interval between writing into a storage element of a charge storage tube, and reading an acceptable output.

maximum sensitivity — The smallest signal input that produces a specified output.

maximum signal level — In an amplitude-modulated system, the level corresponding to copy black or copy white—whichever has the higher amplitude.

maximum sound pressure — For any given cycle of a periodic wave, the maximum absolute value of the instantaneous sound pressure. The most common unit is the microbar.

maximum storage time—In a storage tube, the length of time after writing during which an acceptable output can be read. (*See also* Maximum Usable Viewing Time.)

maximum system deviation – In a frequency-modulation system, the greatest permissible deviation in frequency.

maximum torque–*See* Pull Out Torque.

maximum undistorted output–Also called maximum useful output. The maximum power an amplifier can deliver without producing excessive harmonics.

maximum usable frequency – Abbreviated muf. In radio transmission by ionospheric reflection, the highest frequency that can be transmitted by reflection from regular ionized layers.

maximum usable viewing time–Also called maximum storage time. The length of time during which the visible output of a storage tube can be viewed, without rewriting, before a specified decay occurs.

maximum useful output–*See* Maximum Undistorted Output.

maximum writing rate–The maximum spot speed which produces a line of specified density on a photographic negative or on the screen of a cathode-ray tube.

Maxterm form–A function expanded into a product of sums, such as (A + B) (C – D) (E + F).

maxwell–The cgs electromagnetic unit of magnetic flux, equal to 1 gauss per square centimeter, or one magnetic line of force.

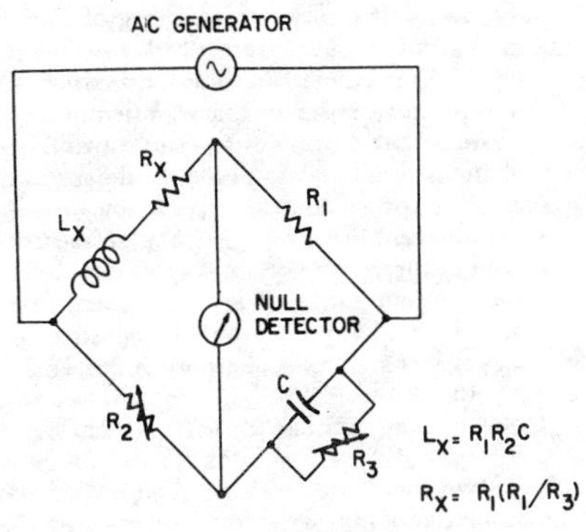

Maxwell bridge.

Maxwell bridge–A four-arm ac bridge normally used for measuring inductance in terms of resistance and capacitance (or capacitance in terms of resistance and inductance). One arm has an inductor in series with a resistor, and the opposition arm has a capacitor in parallel with a resistor. The other two arms normally are nonreactive resistors. The balance is independent of frequency.

Maxwell inductance bridge–A four-arm ac bridge for comparing inductances. Two adjacent arms have inductors, and the other two arms usually have nonreactive resistors. The balance is independent of frequency.

Maxwell mutual-inductance bridge–A four-arm ac bridge for measuring mutual inductance in terms of self-inductance. Mutual inductance is present between the supply circuit and the arm which includes one coil of the mutual inductor. The other three arms normally are nonreactive resistors. The balance is independent of frequency.

Maxwell's equations – Fundamental equations, developed by J. C. Maxwell, for expressing radiation mathematically and describing the condition at any point under the influence of varying electric and magnetic fields.

Maxwell's law–A movable portion of a circuit will always travel in the direction that gives maximum flux linkages through the circuit.

Maxwell triangle–A graph that defines the chromaticity values of a color in terms of three coordinates.

maxwell-turn–A unit of magnetic linkage equal to one magnetic line of force passing through one turn of a circuit.

mayday–International distress call for radiotelephone communication. It is derived from the French *m'aidez,* meaning "help me."

mc–Abbreviation for megacycle–an obsolete term superseded by MHz (megahertz).

MCM–*See* Monte Carlo Method.

McNally tube – A velocity-modulated vacuum tube that produces low-power uhf oscillations. It is used as a local oscillator in some radar receivers.

mcw–Abbreviation for modulated continuous wave.

m-derived filter–A type of constant-k filter in which the constant-k elements are multiplied by the factor m or a function of m. Generally, the m-derived filter has more elements and provides sharper cutoff and more uniform attenuation in the pass region It may provide high-pass, low-pass, bandpass, or band-stop filtering action.

m-derived L-section filter–A reactance network derived from the prototype L-section filter, so that the image-transfer coefficient and one image impedance are changed, but the other image impedance is left unchanged.

MDI–Abbreviation for magnetic direction indicator.

M-display–*See* M-Scan.

Meacham-bridge oscillator–A crystal oscillator in which the crystal forms one arm in a bridge so as to obtain effective multiplication of the actual Q of the crystal.

meaconing–The interception and rebroadcast of beacon signals. They are rebroadcast on the received frequency to confuse enemy navigation. As a result, aircraft or ground stations are given inaccurate bearings.

mean carrier frequency—The average carrier frequency of a transmitter (corresponding to the resting frequency in a frequency-modulated system).

mean charge—The arithmetic mean of the transferred charges corresponding to a given capacitor voltage, as determined from a specified alternating-charge characteristic curve.

mean charge characteristic — The function giving the relation of mean charge to capacitor voltage.

meander line—A transmission-line-matching section the electrical length of which is dependent on frequency. The characteristics of a meander line are determined primarily by the width of the structure, the spacing betwen adjacent turns, and the angle of the line with respect to the ground plane. If the turns have sufficient separation that there is no space coupling between adjacent turns, the meander line becomes a simple length of transmission line.

mean free path—1. The average distance which sound waves travel between successive reflections in an enclosure. 2. The average distance between collisions of atomic particles, which may be further specified according to type of collision (e.g., elastic, inelastic).

mean life—1. Also called average life. In a semiconductor, the time taken by injected excess carriers to recombine with others of the opposite sign. 2. A measure of the probability that a part or equipment will function satisfactorily during its constant-failure–rate period. It is unrelated to longevity.

mean power of a radio transmitter—Power supplied to the antenna transmission line by a transmitter during normal operation, averaged over a time sufficiently long compared with the period of the lowest frequency encountered in the modulation. A time of 1/10 second during which the mean power is greatest will be selected normally.

mean pulse time—The arithmetic mean of the leading-edge and trailing-edge pulse times.

means of communications — The medium (i.e., electromagnetic or sound waves, visual messenger) by which a message is conveyed from one person or place to another.

mean time between failures—Abbreviated MTBF. The average length of time between successive system failures. It is the reciprocal of the sum of the failure rates of all components and connection in the system.

mean time to failure—Abbreviated MTTF. In a piece of equipment, its measured operating time divided by its total number of failures during that time. Normally this measurement is made between the early-life and wearout failures.

mean time to first failure — Abbreviated MTTFF. A special case of MTBF, where T is the accumulated operating time to first failure of a number of devices (failures).

mean time to repair — The total effective maintenance time during a given time interval divided by the total number of failures during the interval.

measurand—Also called stimulus. The physical quantity, force, property, or condition measured by an instrument.

measurement—The determination of the existence or magnitude of a variable. Measuring instruments include all devices used directly or indirectly for this purpose.

measurement component — Those parts or subassemblies used primarily for the construction of measurement apparatus, excluding screws, nuts, insulated wire, or other stable materials.

measurement device—An assembly of one or more basic elements with other components needed to form a self-contained unit for performing one or more measurement operations. Included are the protecting, supporting, and connecting, as well as functioning parts.

measurement energy—The energy required to operate a measurement device or system. Normally it is obtained from the measurand or the primary detector.

measurement equipment — Any assemblage of measurement components, devices, apparatus, or systems.

measurement inverter — *See* Measuring Modulator.

measurement range of an instrument—That part of the total range of measurement through which the accuracy requirements are to be met.

measurement voltage divider — Also called voltage-ratio or volt box. A combination of two or more resistors, capacitors, or other elements arranged in series so that the voltage across any one is a definite, known fraction of the voltage applied to the combination (provided the current drain at the tap point is negligible or taken into account). The term "volt box" is usually limited to resistance voltage dividers intended for extending the range of direct-current potentiometers.

measuring modulator—Also called measurement inverter or chopper. An intermediate means of modulating a direct-current or low-frequency alternating-current input in a measurement system to give a proportionate alternating-current output, usually as a preliminary to amplification.

mechanical bail – Switch action in which, upon actuation of one station, the switch changes the contact position, mechanically locks the switch in that position, and releases any station previously actuated.

mechanical bandspread—A vernier tuning dial or other mechanical means of lengthening the rotation of a control knob. This permits more precise tuning in crowded short-wave bands.

mechanical bias—A mechanical force tending to move the armature of a relay toward a given position.

mechanical compliance—The displacement of a mechanical element per unit of force, expressed in centimeters per dyne. It is the reciprocal of stiffness and is analogous to capacitance.

mechanical coupling – *See* Acoustic Coupling.

mechanical damping – Mechanical resistance, generally associated with the moving parts of a cutter or reproducer.

mechanical damping ring—A loose member mounted on a contact spring for the purpose of reducing contact chatter.

mechanical differential analyzer – A form of analog computer in which interconnected mechanical surfaces are used for solving differential equations.

mechanical filter—1. *See* Mechanical Wave filter. 2. Mechanical resonators coupled by mechanical means. Piezoelectric or magnetostrictive transducers are used to convert electrical and mechanical energy at input and output. Resonators are bars, discs or electrode pairs; coupling elements are rods, wires or nonelectroded regions. (*See* Monolithic Filter.)

mechanical impedance—The complex ratio of the effective force acting on a specified area of an acoustic medium or mechanical device to the resulting effective linear velocity through or of that area, respectively. The unit is newton-sec/m or the mks mechanical ohm. (In the cgs system, the unit is the dyne-sec/cm or the mechanical ohm.)

mechanical joint—A joint made by clamping cables or other conductors together mechanically rather than by soldering them.

mechanical life—The maximum number of complete cycles through which a device may be actuated without electrical or mechanical failure.

mechanically timed relay—A relay that is mechanically timed by such means as a clockwork, escapement, bellows, or dashpot.

mechanically tuned oscillator—Any oscillator that is specifically designed for frequency tuning by mechanical means. Typically a cavity or discrete element.

mechanical ohm—The magnitude of a mechanical resistance, reactance, or impedance for which a force of 1 dyne produces a linear velocity of 1 centimeter per second (dyne-sec/cm). When expressed in newton-sec/m it is called the mks mechanical ohm.

mechanical overtravel—The shaft travel of a precision potentiometer between each end point (or limit of theoretical electrical electrical travel) and its adjacent corresponding limit of total mechanical travel.

mechanical phonograph—A phonograph the playback stylus of which drives the diaphragm of an acoustic pickup that radiates acoustic energy without any further amplification.

mechanical phonograph recorder – Also called mechanical recorder. Equipment that converts electric or acoustic signals into mechanical motion and cuts or embosses it into a medium.

mechanical reactance – The magnitude (size) of the imaginary component of mechanical impedance.

mechanical reader – A reader that senses characters on a perforated tape by means of a contact closure caused by each hole.

mechanical recorder—*See* Mechanical Phonograph Recorder.

mechanical recording head—*See* Cutter.

mechanical rectifier – A rectifier in which its action is done mechanically (e.g., by making and breaking the electrical circuit at the correct times with a rotating wheel or vibrating reed).

mechanical register—An electromechanical device which records or indicates a count.

mechanical reproducer – *See* Phonograph Pickup.

mechanical resistance—The real part of the mechanical impedance. The cgs unit is the mechanical ohm. The mks unit is the mks mechanical ohm.

mechanical scanning—An obsolete type of scanning in which a rotating device, such as a disc or mirror, breaks up a scene into a rapid succession of narrow lines for conversion into electrical impulses.

mechanical shock – Shock which occurs when the position of a system is significantly changed in a relatively short time in a nonperiodic manner. It is characterized by suddenness and large displacements which develop significant internal displacements within the system.

mechanical television system—A television system which uses mechanical scanning.

mechanical tilt—1. Tilt of the mechanical axis of an antenna. 2. The angle of this tilt is shown by the tilt indicator dial.

mechanical transducer—A device that transforms mechanical energy directly into acoustical energy.

mechanical transmission system—An assembly of gears, etc., for transmitting mechanical power.

mechanical tuning range – The frequency range of oscillation of a klystron that is obtainable by tuning mechanically while keeping the reflector voltage optimized for the peak of the reflector-voltage mode.

mechanical tuning rate—In a klystron, the frequency change per degree of rotation of the tuning apparatus while oscillation is maintained on the peak of the reflector-voltage mode.

mechanical wave filter – Also called mechanical filter. A filter that separates mechanical waves of different frequencies.

mechanical waveform synthesizer – A device which mechanically generates an electrical signal with the desired waveform.

mechanism of failure—*See* Failure Mode, 1.

mechanized assembly—The joining together of elements by operators using semiautomatic equipment as contrasted to fully automatic assembly.

MECL—*See* Current-Mode Logic.

median—The middle, or average, value in a series (e.g., in the series 1, 2, 3, 4, and 5, the median is 3).

medical amplifier—Amplifier designed for receiving medical and biological signals (EEG, ECG, etc.) and increasing their magnitude.

medical diathermy—The production of heat in body tissues for therapeutic purposes by high-frequency currents that are insufficiently intense to destroy tissues or to impair their vitality. Diathermy has been used in treating chronic arthritis, bursitis, fractures, gynecologic diseases, sinusitis, and other conditions.

medical electronics – The branch of electronics concerned with its therapeutic or diagnostic applications in the field of medicine.

medical sonic applicator – An electromechanical transducer designed for the local application of sound for therapeutic purposes, for example in the treatment of muscular ailments.

medium frequency—The band of frequencies between 300 kHz and 3 MHz (100 to 1000 meters).

medium-power silicon rectifiers – Rectifiers with maximum continuous rating of 1 to 50 average amperes per section in a single-phase, half-wave circuit.

medium-scale integration – 1. Integrated circuits that function as simple, self-contained logic systems, such as decade counters or five-bit shift registers. Such chips may contain up to 100 gates. Abbreviated MSI. 2. The accumulation of several circuits (usually less than 100) on a single chip of semiconductor. 3. The physical realization of a microelectronic circuit fabricated from a single semiconductor integrated circuit having circuitry equivalent to more than 10 individual gates.

meg—Abbreviation for megohm.

mega—Abbreviated M. Prefix denoting 10^6 (one million).

megabar – The absolute unit of pressure equal to one million bars.

megabit—A unit equal to one million binary digits.

megacycle (mc)—One million cycles—obsolete term replaced by megahertz. (MHz.)

megahertz—One million hertz. Abbreviated MHz.

megampere—One million amperes. Abbreviated MA.

megatron—A tube having a high power output at high frequencies, but very low interelectrode capacitances because its electrodes are arranged in parallel layers. *See also* Disc-Seal Tube.

megavolt—One million volts. Abbreviated MV.

megavolt-ampere – One million volt-amperes. Abbreviated MVA.

megawatt—One million watts. Abbreviated MW.

megawatt-hour – One million watt-hours. Abbreviated MWh.

Megger—A high-range ohmmeter having a built-in hand-driven generator as a direct voltage source, used for measuring insulation resistance values and other high resistances. Also used for continuity, ground, and short-circuit testing in general electrical power work.

megohm – Abbreviated meg. One million ohms.

megohm-farads—*See* Megohm-Microfarad.

megohm-microfarad—A term used to indicate the insulation resistance of capacitors. It is equal to the product of the insulation resistance in megohms and the capacitance in microfarads. For larger high-voltage capacitors, megohm-farads are used.

megohm sensitivity—The resistance in megohms which must be placed in series with a galvanometer in order that an applied emf of 1 volt shall produce the standard deflection. If the resistance of the galvanometer coil itself is neglected, the number representing the megohm sensitivity is equal to the reciprocal of the number representing the current sensitivity.

Meissner effect – The sudden loss of magnetism in superconductors as they are cooled below the temperature required for superconductivity. As a result, they become diamagnetic—i.e., the self-induced magnetization opposes the applied magnetic field to such an extent that there is no longer a magnetic field.

Meissner oscillator—An oscillator in which the grid and plate circuits are inductively coupled through an independent tank circuit, which determines the frequency.

mel—A unit of pitch. A simple 1000-hertz tone, 40 dB above a listener's threshold, produces a pitch of 1000 mels. The pitch of any sound that is judged by the listener to be *n* times that of a 1-mel tone is *n* mels.

melodeon—A broad-band, panoramic counter-measures receiver. It displays all types of received electromagnetic radiation as vertical pips on a frequency-calibrated crt indicator.

melt—Molten semiconductor material from which are drawn the basic single-crystal ingots.

meltback process—The making of junctions by melting a correctly doped semiconductor and allowing it to refreeze.

meltback transistor — A grown transistor produced by melting the tip of a double-doped pellet. Junctions are formed when the tip recrystallizes.

melting channel—The restricted portion of the charge in a submerged horizontal ring induction furnace. The induced currents are concentrated here to effect high energy absorption and thereby melt the charge.

melt-quench transistor—A junction transistor made by quickly cooling a melted-back region.

membrane potential—The electric potential that exists across the two sides of a membrane.

memistor — A nonmagnetic memory device composed of a resistive substrate in an electrolyte. When the device is used in an adaptive system, a dc signal deposits copper from an anode on the substrate, thus reducing the resistance of the substrate. Reversing the current reverses the process, increasing the resistance of the substrate.

memory — 1. The equipment and media used to hold machine-language information in electrical or magnetic form. Usually, the word "memory" means storage within a control system, whereas "storage" is used to refer to magnetic drums, MOS devices, discs, cores, tapes, punched cards, etc., external to the control system. Either term means collecting and holding pertinent information until it is needed by the computer. 2. The tendency of a material to return to its original shape after having been deformed. 3. Any device or circuit capable of storing a digital word or words. 4. The component of a computer, control system, guidance system, instrumented satellite or the like designed to provide ready access to data or instructions previously recorded so as to make them bear upon an immediate problem.

memory address register—The CPU register in a computer, which holds the address of the memory location being accessed.

memory addressing modes—The method of specifying the memory location and an operand. Common addressing modes are direct, immediate, relative, indexed, and indirect. (These modes are important factors in program efficiency.)

memory buffer register—In a computer, a register in which a word is stored as it comes from memory (reading) or just prior to its entering memory (writing).

memory capacity—*See* Storage Capacity.

memory cell—A single storage element of a memory, together with the associated circuits for inserting and removing one bit of information.

memory cycle—In a computer, an operation consisting of reading from and writing into memory.

memory dump—In a computer, a process of writing the contents of memory consecutively in such a form that it can be examined for computer or program errors.

memory fill—In a computer, the placing of a pattern of characters in the memory registers not in use in a particular problem to stop the computer if the program, through error, seeks instructions taken from forbidden registers.

memory hierarchy — A set of computer memories with differing sizes and speeds and usually having different cost-performance ratios. A hierarchy might consist of a very high-speed, small semiconductor memory, a medium-speed core memory and a large, slow-speed core.

memory light — In a calculator, indicates there is a number in the memory.

memory + and − keys — In a computer, direct access to the memory for storing numbers. On machines without these, the memory has to be addressed first and the working register + and − keys used to store a number in the memory.

memory protection — *See* Storage Protection.

memory register—1. Also called high-speed bus, distributor, or exchange register. In some computers, a register used in all data and instruction transfers between the memory, the arithmetic unit, and the control register. 2. A register in a calculator in which the contents can be added to or subtracted from without being recalled, or which can be recalled for other operations. The contents are available for repeated recall until the register is cleared.

memory relay — A relay in which each of two or more coils may operate independent sets of contacts, and another set of contacts remains in a position determined by the coil last energized. The term is sometimes erroneously used for polarized relay.

memory timer — A timing device wherein

the cycle duration is infinitely variable within the specified overall cycle time and having the ability, once the cycle is selected, to repeat this cycle any number of times by a simple mechanical actuation of the time shaft.

memory unit—That part of a digital computer in which information is stored in machine language, using electrical or magnetic techniques.

menu—A list of options or functions which are displayed on a crt. The items in the menu are composed of light buttons for light-pen selection and/or character strings which are alphanumeric key and/or function key selectable.

mercuric-oxide–cadimum cell – A primary-cell electrochemical system. Its primary advantage is its long shelf life in the fully charged condition and its operation at low temperatures, far below 0°C. Nominal cell voltage is about 0.9 volt.

mercury – A silver-white metal that becomes a liquid above −38.87° Centigrade. In addition to thermometers, it is used in switches and many electronic tubes. When it is vaporized, mercury ionizes readily and conducts electricity.

mercury arc—A cold-cathode arc through ionized mercury vapor. A very bright bluish-green glow is given off.

mercury-arc converter – A frequency converter using a mercury-vapor tube.

mercury-arc rectifier—Also called mercury-vapor or simply mercury rectifier. A diode rectifier tube containing mercury vapor. The mercury vapor is ionized by the voltage across the tube, and a much greater current can flow. There is only a small voltage drop in the tube during conduction.

mercury barometer – An instrument for measuring atmospheric pressure.

mercury battery—A type of battery characterized by extremely uniform output throughout its life. Employs a zinc-powder anode; the cathode is mercuric-oxide powder and graphite powder.

mercury cell—A primary cell with a zinc anode and a mercuric-oxide cathode in a potassium-hydroxide electrolyte. This cell offers a higher capacity than the alkaline or zinc-carbon cells and a much flatter voltage-discharge characteristic.

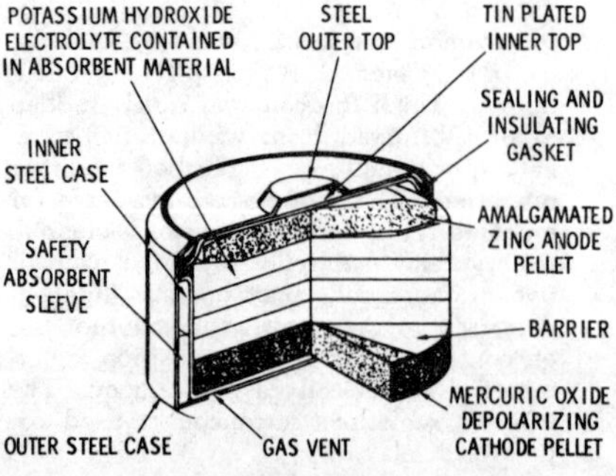

Mercury cell.

mercury-contact relay – A relay in which the contacts are mercury.

mercury delay line—*See* Mercury Memory.

mercury displacement relay – A relay in which the displacement of mercury, such as caused by a solenoid-actuated plunger, results in a mercury-to-mercury electrical contact.

mercury-hydrogen spark-gap converter – A spark-gap generator in which the source of radio-frequency power is the oscillatory discharge of a capacitor through an inductor and a spark gap. The latter comprises a solid electrode and a pool of mercury in hydrogen.

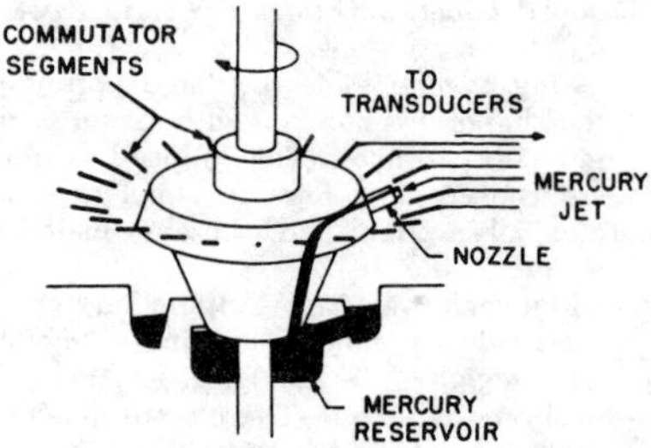

Mercury-jet scanning switch.

mercury-jet scanning switch—A commutating switch in which a stream of mercury performs the switching between the common circuit and those to be sampled.

mercury memory—Also called mercury storage and mercury delay line. Delay lines using mercury as the medium for storage of a circulating train of waves or pulses.

mercury-motor meter – A motor meter in which a portion of the rotor is immersed in mercury, which directs the current through conducting portions of the rotor.

mercury-pool cathode – The cathode of a gas tube consisting of a pool of mercury. An arc spot on the pool emits electrons.

mercury rectifier—*See* Mercury-Arc Rectifier.

mercury relay—A relay in which the energized coil pulls a magnetic plunger into a tube containing mercury. The plunger moves the mercury in order to make connection between the contacts.

mercury storage—*See* Mercury Memory.

mercury switch – An electric switch comprising a large globule of mercury in a metal or glass tube. Tilting the tube causes the mercury to move toward or away from the electrodes to make or break the circuit.

mercury tank—In a computer, a container

of mercury holding one or more delay lines for storing information.

mercury-vapor lamp – A glow lamp in which mercury vapor is ionized by the electric current, producing a bluish-green luminous discharge.

mercury-vapor rectifier – *See* Mercury-Arc Rectifier.

mercury-vapor tube – A tube containing mercury vapor, which when ionized allows conduction and produces a luminous glow.

mercury-wetted-contact relay – A special form of reed relay. It consists of a glass-encapsulated reed, with its base mounted in a pool of mercury and the other end arranged so as to move between two sets of stationary contacts. By capillary action, the mercury flows up the reed to coat the movable and stationary contact surfaces, thus assuring mercury-to-mercury contact during make.

merge – To produce a single sequence of items ordered according to some rule (i.e., arranged in some orderly sequence), from two or more sequences previously ordered according to the same rule, without changing the size, structure, or total number of the items.

mesa—In certain transistors, the raised area (somewhat resembling a geological mesa) left when semiconductor material is etched away to allow access to the base and collector regions.

mesa diffusion – Technique used to manufacture semiconductors having diffused pn junctions. A single base region is diffused over the entire wafer, and an acid is used to etch a mesa configuration for the transistor elements.

mesa isolation—A process for isolating IC's in which transistors, resistors, and other components are fabricated before isolation.

mesa structure – A semiconductor whose structure is moundlike. During processing, material is etched away from the original chip in order to produce the final shape.

mesa transistor – A transistor that is produced by chemically etching away a transistor chip formed by either a double-diffused or diffused-alloy process. When the etching process is complete, the base and emitter regions appear as plateaus above the collector region. One result of the mesa construction is a reduction in the collector-base capacitance as a result of lowering the junction area.

mesh—Sometimes called a loop. A set of branches forming a closed path in a network—provided that if any branch is omitted, the remaining branches do not form a closed path.

mesh beat—*See* Moiré, 1.

mesh current—The current assumed to exist over all cross sections of a closed path in a network. It may be the total current in a branch included in the path, or a partial current which, when combined with the others, forms the total current.

meson—A particle which weighs more than an electron but generally less than a proton. Mesons can be produced artificially as well as by cosmic radiation. They are so unstable that they disintegrate in millionths of a second.

message – 1. An ordered selection of an agreed set of symbols for the purpose of communicating information. 2. The original modulating wave in a communication system. 3. An arbitrary amount of information whose beginning and end are defined or implied. 4. One or more blocks of data that contain the total information to be transmitted. 5. A group of characters that have a meaning when taken together and that always are handled as a group.

message circuit—A long-distance telephone circuit used in providing regular long-distance or toll service to the general public, as opposed to a circuit used for private-line service.

message exchange—A device used between a communications line and a computer to perform certain comunications functions and free the computer for other tasks.

message interpolation – Insertion of data between syllables or during speech pauses on a busy voice channel without noticeably affecting the voice transmission.

message switching—The process of receiving a message, storing it until a suitable outgoing circuit and station are available, and then sending it on toward its destination.

metadyne—British term for amplidyne. A direct-current machine used for voltage regulation or transformation. It has more than two brushes for each pair of holes.

metal—A material that has high electrical and thermal conductivity at normal temperatures.

metal-base transistor—A transistor with a base of a thin metal film sandwiched between two n-type semiconductors, with the emitter doped more heavily than the base to give it a high electron-current–to–hole-current ratio.

metal detector—Also called metal locator. An electronic device for detecting concealed metal objects.

metal-etched mask – A mask formed by chemically etching openings in a metal film or plate where it is not protected by photoresist or other chemically resistant material.

metal film resistor – An electronic component in which the resistive element is an extremely thin layer of metal alloy vacuum-deposited on a substrate.

metal gate—Refers to the use of aluminum as gate conductor instead of silicon or refractory metals.

metal-insulator-silicon—*See* MIS.

metalization—The deposition of a thin-film pattern of conductive material onto a substrate to provide interconnection of electronic components or to provide conductive contacts (pads) for interconnections.

metalized capacitor — A self-healing fixed capacitor. A thin film of metal is vacuum-deposited directly on the dielectric, which can be paper or plastic. When a breakdown occurs, the metal film around it immediately burns away.

metalized resistor—A fixed resistor in which the resistance element is a thin film of metal deposited on the surface of a glass or ceramic substrate.

metalizing — Applying a thin coating of metal to a nonmetallic surface. This may be done by chemical deposition or by exposing the surface to vaporized metal in a vacuum chamber.

metallic circuit — A circuit in which the earth itself is not used as ground.

metallic insulator—A shorted quarter-wave section of transmission line, which acts as an electrical insulator at the transmitted frequency.

metal-nitride-oxide semiconductor—A process using a layer of nitride between the metal gate contact and the oxide protective layer.

metallic noise—Weighted noise current in a metallic circuit at a given point when the circuit is terminated at that point in the nominal characteristic impedance of the circuit.

metallic rectifier—A rectifier in which the asymmetrical junction between dissimilar solid conductors presents a high resistance to current flow in one direction and a low resistance in the opposite direction.

metallic-rectifier cell—An elementary rectifying device having only one positive electrode, negative electrode, and rectifying junction.

metallic-rectifier stack—A single structure made up of one or more metallic-rectifier cells.

metal locator—*See* Metal Detector.

metal on glass mask—An optical mask comprising a glass substrate selectively covered by a thin opaque metal layer; a type of photomask.

metal master—*See* Original Master.

metal negative—*See* Original Master.

metal-nitride-oxide semiconductor — An MOS structure with a layer of silicon nitride added to offset the permeability of silicon dioxide by contaminating sodium ions. In suitably fabricated devices, it is possible to generate a relatively nonvolatile memory element by charge storage at the nitride-oxide interface.

metal-oxide resistor—A type of film resistor in which the material deposited on the substrate is tin oxide, which provides good stability.

metal-oxide semiconductor — Abbreviated MOS. 1. A field-effect transistor in which the gate electrode is isolated from the channel by an oxide film. 2. A capacitor in which semiconductor material forms one plate, aluminum forms the other plate, and an oxide forms the dielectric. 3. A circuit in which the active region is a metal-oxide-semiconductor sandwich. The oxide acts as the dielectric insulator between the metal and the semiconductor. 4. A process which results in a structure of metal over silicon oxide over silicon. By appropriate topology this generates field effect transistors, capacitors or resistors. 5. Metal-oxide-semiconductor, referring to a field-effect transistor (MOSFET) that has a metal gate insulated by an oxide layer from the semiconductor channel. A MOSFET is either enhancement-type (normally turned off) or depletion-type (normally turned on). MOS also refers to integrated circuits that use MOSFET's (virtually all enhancement-type). 6. Technology that employs field-effect transistors having a metal or conductive electrode which is insulated from the semiconductor material by an oxide layer of the substrate material. Whereas bipolar devices permit current to flow in only one direction, MOS devices permit bidirectional current flow.

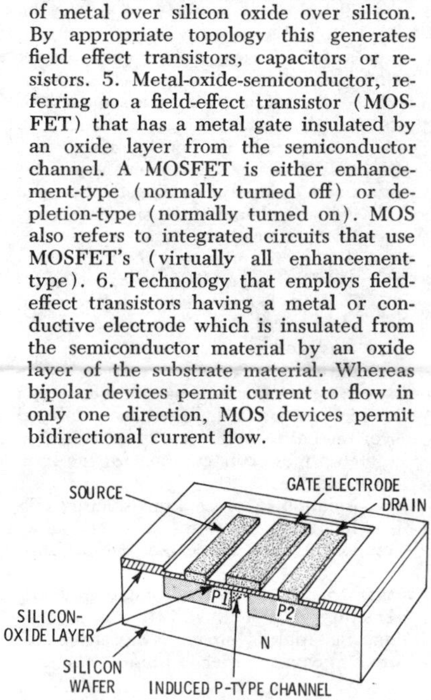

Metal-oxide-semiconductor transistor.

metal-oxide semiconductor field-effect transistor — Abbreviated MOSFET. A device consisting of diffused source and drain regions on either side of a p or n channel region, and a gate electrode insulated from the channel by silicon oxide. When a control voltage is applied to the gate, the channel is converted to the same type of semiconductor as the source and drain. This eliminates part of the pn junction and permits current to be established between the source and drain. Functionally, the main difference between a MOSFET and a bipolar transistor is that the source and drain of the MOSFET are interchangeable, unlike the emitter and collector of the bipolar transistor.

metal-tank mercury-arc rectifier – A mercury-arc rectifier in which its anodes and mercury cathode are enclosed in a metal chamber.

metal–thick-nitride semiconductor – A device similar to an MTOS device except that a thick layer of silicon nitride or silicon nitride-oxide is used instead of a layer of oxide.

metal–thick-oxide semiconductor–A device in which the oxide outside the desired active gate area is made much thicker so that problems with unwanted parasitic devices are reduced. Abbreviated MTOS.

metal tube–A vacuum tube with a metal envelope. The electrode leads pass through glass beads fused into the metal housing.

metamers–Lights of the same color but of different spectral energy distribution.

meteorological aids service–A radio service in which emissions consist of special signals intended solely for meteorological uses.

meteorological radar station–A station, operating in the meteorological aids service, that employs radar and is not intended for operation while in motion.

meter–1. Any measuring device; specifically, any electrical or electronic measuring instrument. 2. In the metric system of measurement, the unit of length equal to 39.37 inches, 3.281 feet, or 1.094 yards.

meter ampere–A measure of the strength of a radio transmitter. It is equal to the antenna height in meters times the maximum antenna current in amperes.

meter-candle–*See* Lux.

meter-correction factor – The factor by which a meter reading must be multiplied to compensate for meter errors in order to obtain the true reading.

meter display–A display in which one or more pointer instruments give the indications.

meter-kilogram-second-ampere system of units–*See* MKSA Electromagnetic System of Units.

meter rating–A manufacturer's designation used to indicate the operating limitations of the meter. The full-scale marking on a meter scale does not necessarily correspond to the rating of the meter.

meter resistance–The resistance of a meter as measured at the terminals at a given reference temperature. When applied to rectifier-type meters, the frequency and waveshape of the applied energy, as well as the indicated value at which the measurement is to be made, must be specified.

meter-type relay – A meter movement in which the armature function is performed by a contact-bearing pointer.

metrechon–A storage tube used in scanning converters (e.g., in radar and industrial tv).

metric system – The decimal system of weights and measures, used extensively by scientists.

metric waves–Waves with lengths between 1 and 10 meters (corresponding to frequencies between 300 and 30 MHz, respectively).

MeV–Usually pronounced M-E-V. Abbreviation for million electronvolts. 1 MeV equals 4.45×10^{-20} kilowatt hours, or 1.6020×10^{-6} erg.

MEW–Abbreviation for microwave early warning.

mf–1. Abbreviation for medium frequency. 2. Alternate abbreviation for microfarad.

mfd–Alternate abbreviation for microfarad.

mH–Abbreviation for millihenry.

mhd – Abbreviation for magnetohydrodynamics.

mho – The unit of conductance of a conductor when a potential difference of 1 volt between its ends maintains an unvarying current of 1 ampere. The mho is the unit of conductance in the mksa system. This term has now been replaced by siemens.

MHz–Abbreviation for megahertz.

MIC–1. Abbreviation for monolithic integrated circuit. 2. *See* Microwave Integrated Circuit.

mica–A transparent mineral which can be split into thin sheets. Because of its excellent insulating and heat-resisting qualities, it is used to separate the plates of capacitors and to insulate electrode elements in vacuum tubes.

mica capacitor–A fixed capacitor employing mica sheets as the dielectric material between adjacent plates. The complete units are usually encased in molded *Bakelite*.

micr–Abbreviation for magnetic ink character recognition. Machine recognition of characters printed with magnetic ink. (Contrast with ocr.)

micro–1. In the metric system, a prefix meaning one millionth (1/1,000,000). 2. A prefix meaning something very small.

microalloy diffused transistor – A transistor fabricated by etching emitter and collector pits into heavily doped germanium, depositing epitaxial layers of intrinsic germanium on both sides of the wafer, and adding electrodes that are plated and electrolayered in the pits.

microalloy transistor – A high-frequency transistor in which the emitter and collector are alloyed to a slight depth on opposite sides of the germanium base material by an electrochemical process. Abbreviated MAT.

microammeter – A meter, having a scale that reads in microamperes, for measuring extremely small currents.

microampere–One millionth of an ampere.

microbar – A unit of pressure commonly used in acoustics. One microbar is equal to one dyne per square centimeter. (Originally the term "bar" denoted dyne per square centimeter. Therefore, to avoid confusion it is preferable to use microbar or "dynes per square centimeter" when speaking of sound pressures.)

micro B-display–A B-scope in which range and azimuth data are so expanded that only a small portion of the area under surveillance is presented, and, because of the degree of expansion, distortion of the presentation is negligible.

microcircuit–1. A small circuit with a high equivalent-circuit-element density, and considered as a single part, composed of interconnected elements on or within a single substrate, that performs an electronic-circuit function. (According to this definition, such structures as printed wiring boards, circuit-card assemblies, and modules composed exclusively of discrete electronic parts are not considered to be microcircuits.) 2. An integrated circuit.

microcircuit module – An assembly of microcircuits or of microcircuits and discrete parts, designed to perform an electronic-circuit function or functions and constructed in such a manner that it is considered to be a single entity for the purposes of specification, testing, commerce, and maintenance.

microcircuit wafer–A microwafer carrying one or more circuit functions such as a flip-flop or gate. Integrated-circuit chips may be bonded to deposited conductors.

microcode–A computer-coding system that includes suboperations, such as multiplication and division, that ordinarily are inaccessible in programming. A list of very small program steps.

microcoding–In a computer, a system of coding that uses suboperations not ordinarily accessible in programming.

microcomponents – 1. Those components smaller than existing components by several orders of magnitude. 2. An assembly of very small, interconnected discrete components – active or passive – which forms an electronic circuit. Interconnection of the various leads is by soldering or welding. Microcomponents use no substrates.

microcomputer–1. A general-purpose computer composed of standard LSI components built around a central processing unit (CPU). The CPU (or microprocessor) is program-controlled featuring arithmetic and logical instructions, and general-purpose parallel i/o bus. The CPU is contained on a single chip or a small number of chips, usually not greater than four. Generally intended for dedicated applications, the microcomputer also includes any number of ROMs and RAMs (for instruction and data storage) and in some cases, one or more i/o devices. The simplest microcomputer consists of one CPU chip and one ROM. 2. A computer whose major sections–CPU, control, timing, and memory–are each contained on a single, integrated-circuit chip, or, at most, a few chips. An LSI computer.

microcontroller–*See Chipsets.*

microdensitometer–An instrument used in spectroscopy to measure lines in a spectrum by light transmission measurement.

microelectrode – An extremely small electrode. Some microelectrodes are small enough to contact a single biological cell.

microelectronic circuit–Discrete electrical components assembled and connected in extremely small and compact form.

microelectronic device–An alternate term for "integrated circuit."

microelectronics – 1. Also called microsystems electronics. The entire body of electronic art which is connected with or applied to the realization of electronic systems from extremely small electronic parts. 2. *See* Integrated Circuit. 3. All techniques for the manufacture of extremely small electronic circuits generally including all types of silicon integrated circuits, thin-film circuits, and thick-film circuits.

microelement – A resistor, capacitor, or transistor, diode, inductor, transformer, or other electronic element or combination of elements mounted on a small ceramic wafer (0.01 inch thick and about 0.3 inch square). Individual microelements are stacked, interconnected, and potted to form micromodules.

microelement wafer–A microwafer carrying one or more components or a simple network. The network can consist, for example, of several thin-film resistors deposited directly on the wafer.

microfarad–One millionth of a farad. Abbreviated mfd, mf, or μF.

microfaradmeter–*See* Capacitance Meter.

microfiche–A form of microfilming which reduces printed material 24 or more times and reproduces it on 4″ × 6″ film cards. When inserted into a reader, each page is enlarged to its original size.

microflash lamp–A lamp that emits radiation pulses having a duration of approximately one microsecond.

microgroove–In disc recording, the groove width of most long-play and 45-rpm records. Normally it is .001 inch, or about half as wide as the groove on a 78-rpm record.

microgroove record – *See* Long-Play Record.

microhenry—One millionth of a henry.

microhm—One millionth of an ohm.

microinstruction—A very simple instruction (typically a register-to-register copy). Also called elementary operation, cycle or function.

microlock – A phase-lock-loop system for transmitting and receiving information. Because the system reduces bandwidth drastically, it is used as a radar beacon for tracking, or to provide telemetering data.

micrologic—A group of high-speed, low-powered integrated logic building blocks primarily intended to be used in building the logic section of a digital computer.

micrologic elements – Semiconductor networks used in computer and other critical circuits.

micromanipulators – Devices that provide means for accurately moving miniscule tools over and on to the surface of a microscopic object.

micromassage—*See* Intercellular Massage.

micromho—One millionth of a mho.

micromicro – An obsolete prefix meaning one-millionth of a millionth, or 10^{-12}. Now called pico.

micromicrofarad—Obsolete term for 10^{-12} farad. Now called picofarad.

micromicrowatt – Obsolete term for 10^{-12} watt. Now called picowatt.

microminiature lamp – Any incandescent lamp, usually rated in the milliwatt range, that operates on 3 volts or less. Diameters range from 0.01 to 0.06 inch.

microminiaturization—1. The producing of microminiature electronic circuits from individual miniature solid-state and other nonthermionic components. 2. A relative degree of miniaturization resulting in an equipment or assembly volume an order of magnitude smaller than that existing in subminiature equipment. 3. The technique of packaging a microminiature part of an assembly composed of elements radically different in shape and form. Electronic parts are replaced by active and passive elements, through use of fabrication processes such as screening, vapor-deposition diffusion, and photoetching.

micromodule – 1. A tiny ceramic wafer made from semiconductive and insulative materials. It is capable of functioning as either a transistor, resistor, capacitor, or other basic component. 2. A microcircuit constructed of a number of components (e.g., microwafers) and encapsulated to form a block that is still only a fraction of an inch in any dimension.

micron—1. A unit of length equal to 10^{-6} meter. 2. A unit used in the measurement of very low pressures. It is equivalent to 0.001 mm (10^{-6} meter) of mercury at 32°F.

microphone – 1. An electroacoustic transducer which responds to sound waves and delivers essentially equivalent electric waves. A device for converting sound waves or sound-producing vibrations (as from the strings of a guitar) into corresponding electrical impulses. Microphones may use as transducing elements crystal or ceramic chips, ribbons, moving coils, or capacitors, and different recording applications may call for different transducers as well as for different directional patterns and impedances.

microphone amplifier—Also called a microphone preamplifier. An audio-frequency amplifier that boosts the output of a microphone before the signal reaches the main audio-frequency amplifier.

microphone boom—A movable crane from which a microphone is suspended.

microphone button—The resistance element of a carbon microphone. It is button-shaped and filled with carbon particles.

microphone cable – A shielded cable for connecting a microphone to an amplifier.

microphone mixer – An audio mixer that feeds the output from two or more microphones into a single input to an audio amplifier. The output from each microphone is adjustable by individual controls on the mixer.

microphone preamplifier—*See* Microphone Amplifier.

microphone stand—A stand that holds a microphone the desired distance above the floor or a table.

microphone transformer – An iron-core transformer used for coupling certain microphones to an amplifier or transmission line.

microphonics – 1. The generation of an electrical noise signal by mechanical motion of internal parts within a device. 2. Electrical disturbance (noise) due to mechanical disturbances of circuit elements. 3. A form of noise interference arising from the tendency for vibrations of certain objects to be converted into corresponding electrical signals. A microphonic device will cause a "bong" or "bing" in the signal when subjected to jarring.

microphonism—1. The production of noise as a result of mechanical shock or vibration. 2. The quasiperiodic voltage output of a tube produced by mechanical resonance of its elements as a result of mechanical impulse excitation. 3. The periodic voltage output of a tube produced by mechanical resonances of its elements as a result of sustained mechanical excitation. 4. The output voltage of a tube acting as an electrical transducer of mechanical energy.

microphonograph—A device which amplifies and records weak sounds; used in training the deaf to speak.

microphonoscope – A binaural stethoscope using a membrane in the chest piece to accentuate the sound.

microphotograph – A small picture of a large subject. The microfilming of a check or other document produces a microphotograph.

microprocessor–1. The control and processing portion of a small computer or microcomputer, that can be built with LSI MOS circuitry usually on one chip. Like all computer processors, microprocessors can handle both arithmetic and logic data in a bit-parallel fashion under control of a program. But they are distinguished both from a minicomputer processor by their use of LSI with its lower power and costs and from other LSI devices (except calculator chips) by their programmable behavior. 2. A computer contained on as few as three chips, which functions as central processor for executing instructions, a volatile memory for storing data, and an interface unit through which data and instructions are transmitted.

microprogram – 1. A computer program written in the most basic instructions or subcommands that can be executed by the computer. Frequently, it is stored in a read-only memory. *See also* Firmware, 2. A special-purpose program, stored in a fixed memory, that is initiated by a single instruction in a system's main program. For example, one instruction in the main program may initiate a stored microprogram of 6 or 7 instructions needed to execute the single main program instruction. 3. In a computer a subelement of a conventional program built up of a sequence of even smaller operations called microinstructions. Each microinstruction is further subdivided into a collection of micro-operations carried out in one basic machine cycle. (For example, the computer program consists of a sequence of instructions that are carried out in a specific order. Each instruction consists of a routine of one or more steps. This sequence of computer machine cycles necessary to execute a single instruction is called a microprogram.)

microprogramming–1. The setting up of basic suboperations for a computer to handle, after which the programmer combines them, and they are presented to the computer again in a higher-level program. For example, if a computer has only basic instructions for addition, subtraction, and multiplication, the instruction for division would be defined by microprogramming. 2. A method of operating the control part of a computer where each instruction is broken into several small steps (microsteps) that form part of a microprogram. 3. A method of organizing a general-purpose computer to perform desired functions, using instructions stored in a control array.

microradiometer – Also called a radio micrometer. A thermosensitive detector of radiant power. It consists of a thermopile supported on and connected directly to the moving coil of a galvanometer.

microsecond–One millionth of a second.

microstrip–A microwave transmission component in which a single conductor is supported above a ground plane. Also called stripline.

microsyn–A precise and sensitive pick-off device for converting angular displacement within a small range to an electrical signal.

microsystems electronics – *See* Microelectronics.

microvolt–One millionth of a volt.

microvoltmeter – A highly sensitive voltmeter, which measures millionths of a volt.

microvolts per meter–The potential difference in microvolts developed between an antenna system and ground, divided by the distance in meters between the two points.

microvolts/meter/mile – One method of stating the field strength of a radiated field. Radiation from industrial heating equipment, for example, must be suppressed so that the radiated field strength does not exceed 10 microvolts per meter at a distance of 1 mile from the source.

microwafer–A basic microcircuit building block generally made of beryllia, alumina, or glass. Terminations on the edges are usually of gold on top of chromium, with a heavy nickel overlay for welding.

microwatt–One millionth of a watt.

microwave–A term applied to radio waves in the frequency range of 1000 megahertz and upward. Generally defines operations in the region where distributed constant circuits enclosed by conducting boundaries are used instead of the conventional lumped-constant circuit components.

microwave amplification by stimulated emission of radiation–Amplification by a low-noise radio-frequency amplifier in which an input signal stimulates emission of energy stored in a molecular or atomic system by a microwave power supply.

microwave discriminator – A tuned cavity which converts a frequency-modulated microwave signal into an audio or video signal.

microwave early warning – Abbreviated MEW. A high-power, long-range, early-warning radar. It has numerous indicators that give high resolution and large traffic-handling capacity.

microwave filter—A filter built into a microwave transmission line to pass desired frequencies but reject or absorb all other frequencies.

microwave frequencies—Frequencies of approximately 1000 MHz and above.

microwave integrated circuit—1. An electronic circuit fabricated by microelectronic techniques and capable of operating at frequencies above one gigahertz. Either hybrid or monolithic integrated circuit technology may be employed. 2. A hybrid type of construction in which thick- or thin-film technology is used to lay out a pattern of conducting lines on a ceramic substrate, and uncased active devices are then bonded to the conductor pattern.

microwave power transmission—A method of transmitting power through space from a microwave transmitting antenna to a remotely located receiving antenna.

microwave radio relay — The relaying of long-distance telephone calls and television broadcast programs by means of highly directional high-frequency radio waves that are received and sent on from one booster station to another.

microwave refractometer — A device for measuring the refractive index of the atmosphere at microwave frequencies, usually in the 3-cm region.

microwave region—The portion of the electromagnetic spectrum between the far infrared and conventional radio-frequency portion. Commonly regarded as extending from 1000 (30 cm) to 300,000 (1 mm) megahertz.

microwave relay system—A series of ultrahigh-frequency radio transmitters and receivers comprising a system for handling communications (usually multichannel).

microwaves — 1. Radio frequencies with such short wavelengths that they exhibit some of the properties of light. Their frequency range is from 1000 MHz up. (Microwaves are preferred in point-to-point communications because they are easily concentrated into a beam.) 2. Short electromagnetic waves located between the television transmission and infrared frequency regions. For communication purposes, microwaves offer considerable appeal as they can be focused and directed like light and can be manipulated like electricity providing a practical means of transmitting information great distances without the use of wires.

microwave tube—The source of power for generating microwave frequencies. Primary microwave generating tubes are klystrons, magnetrons and traveling wave tubes. Other microwave devices, include masers, parametric amplifiers and backward wave oscillators.

middle marker — In an instrument landing system, a marker located on a localizer course line, about 3500 feet from the approach end of the runway.

middle-side system—*See* Mitte-Seite Stereo System.

midrange — The frequency range between bass and treble. Speaker systems often use a midrange unit which typically operates between about 400 and 3000 Hz.

migration--The movement of some metals, notably silver, from one location to another as a result of a plating action that takes place in the presence of moisture and an electrical potential.

mike—Slang for microphone.

mil—One thousandth of an inch. Used in the United States for measuring wire diameter and tape thickness.

MIL—Abbreviation for military. Pertains to nation's armed forces, including its army, navy, and air force. Specifically, the armed forces of the United States.

military grade IC—Typically, an IC whose performance is guaranteed over the temperature range from −55° to +125°C.

Miller bridge—A type of bridge circuit for measuring the amplification factor of vacuum tubes.

Miller capacitance—Feedback capacitance caused by gate metal overlapping source and drain regions.

Miller effect—The increase in the effective grid-to-cathode capacitance of a vacuum tube because the plate induces a charge electrostatically on the grid through grid-to-plate capacitance.

Miller oscillator—A crystal-controlled oscillator in which the crystal oscillates at its parallel resonant frequency due to the connection of negative resistance across its plates.

milli—Abbreviated m. Prefix meaning one thousandth (1/1000, or 10^{-3}).

milliammeter — An electric current meter calibrated in milliamperes.

milliampere — Abbreviated mA. One one-thousandth (.001) of an ampere.

millihenry — Abbreviated mH. One one-thousandth (.001) of a henry.

millilambert—A unit of brightness equal to one one-thousandth (.001) of a lambert.

millimaxwell — One-thousandth of a maxwell.

millimeter waves — Electromagnetic radiation in the frequency range of 30 to 500 gigahertz with corresponding wavelengths of 10 millimeters to 0.6 millimeter.

millimicro—Obsolete prefix for nano, representing 10^{-9}.

millimicron—A unit of length equal to one ten-millionth of a centimeter (10^{-7} cm), of one one-thousandth of a micron.

milliohm — One one-thousandth (.001) of an ohm.

milliroentgen — A unit of radioactive dose

equal to one one-thousandth of a roentgen.

millisecond – One-thousandth of a second. Abbreviated ms.

millitorr–One-thousandth of a torr.

millivolt–One-thousandth of a volt. Abbreviated mV.

millivoltmeter–A sensitive voltmeter calibrated in millivolts.

millivolts per meter – The potential difference in millivolts developed between an antenna system and ground, divided by the distance in meters between the two points.

milliwatt – One one-thousandth of a watt. The reference level used for dB measurements.

Mills antenna–A combination of two independent fan-beam antennas placed at right angles in a cross formation, with a common center and common pencil-beam volume of low gain. Antennas are combined by switching from output phase addition to phase opposition at a constant rate to secure the angular resolution of the pencil-beam components.

mine detector–Also called electronic mine detector. An electronic device that indicates the presence of metallic or nonmetallic explosive mines under the ground or under water.

miniature lamp–1. A small, filament-type lamp with an operating voltage less than 60 volts. 2. Very small tungsten lamps that are used where space is limited. They are sometimes called "grain-of-wheat" lamps.

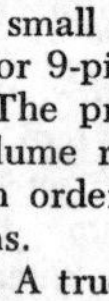

Miniature tube.

miniature tube–A small electron tube usually having a 7- or 9-pin base.

miniaturization – The process of reducing the minimum volume required by equipments or parts in order to perform their required functions.

minicomputer – 1. A true digital computer similar to its larger predecessors but reduced in size by modern electronic packaging by eliminating certain optional features, and by setting a limit on the amount of high-speed memory available. Physically, it weighs approximately 50 lbs and has about the same dimensions as a stereo radio. It has no special operating requirements (voltage, air conditioning), being able to operate in the office, home or factory. The computational speed is equivalent to computers costing ten times as much. A character can be retrieved from memory in one microsecond and the addition of two numbers takes three microseconds. Data can be transmitted in and out of the computer at a million characters per second. The major limitations of a minicomputer are the maximum memory size of 64,000 characters (bytes) as compared to several hundred thousand characters for larger computers, and the speed with which simultaneous data transfers can be made between high-speed data storage devices and the minicomputer. These limitations restrict the minicomputer system from applications involving data bases with hundreds of millions of characters. 2. A loosely used term for describing any general-purpose digital computer in the low-to-moderate price range. The approximate ceiling price used to define a minicomputer is subject to wide interpretation.

minidisk–*See* Floppy Disk.

minigroove–A recording having more lines per inch than the average 78-rpm phonograph record but not as many as the extended play, long playing, or microgroove records.

minimum-access programming–Also called minimum-latency programming, or forced coding. Programming a digital computer so information is obtained from the memory in the minimum waiting time.

minimum-access routine–A computer routine coded in a way such that the actual waiting time to obtain information from a serial memory is much less than the expected random-access waiting time.

minimum detectable signal – The signal level which just exceeds the threshold.

minimum discernible signal – Abbreviated mds. In a receiver, the smallest input signal that will produce a discernible signal at the output. The smaller the signal required, the more sensitive the receiver.

minimum firing power – The lowest radio-frequency power that will initiate a radio-frequency discharge in a switching tube at a specified ignitor current.

minimum flashover voltage – The crest value of the lowest voltage impulse of a given waveshape and polarity which causes flashover.

minimum-latency programming–*See* Minimum-Access Programming.

minimum-phase network – A network for which the phase shift at each frequency equals the minimum value determined solely by the attenuation-frequency characteristic.

minimum reject number – A number that defines the maximum number of rejects allowed for each sample given for each LTPD. Abbreviated MRN.

minimum reliable current – As applied to relay contacts, the range at which there is insufficient energy under arcing conditions at a mating contact surface to ensure good contacting for the kind of contact material, shape, and forces employed.

minimum resistance – The resistance of a precision potentiometer, measured between the wiper terminal and any terminal with the shaft positioned to give a minimum value.

minimum shift frequency – The minimum frequency at which a flip-flop can be operated in a shift-register application.

minimum-signal level – In facsimile, the level corresponding to the copy white or copy black signal, whichever is the lower.

minimum starting voltage – The minimum voltage of rated frequency applied to the control-voltage winding of a servomotor necessary to start the rotor turning at no-load conditions with rated voltage and frequency on the fixed-voltage winding.

minimum toggle frequency—The minimum frequency at which a flip-flop can be toggled. In a typical device, the maximum toggle frequency usually is about 20 percent higher.

minitrack – A satellite tracking system which uses a miniature pulse-type telemeter and a precise directional antenna system, with phase-comparison tracking techniques.

minor apex face—In a quartz crystal, any one of the three smaller sloping faces near but not touching the apex (pointed end). The other three are called the major apex faces.

minor bend—A rectangular waveguide bent so that one of its longitudinal axes is parallel to its narrow side throughout the length of the bend.

minor cycle – Also called word time. In a digital computer using serial transmission, the time required to transmit one word or the space between words.

minor face—One of the three smaller sides of a hexagonal quartz crystal.

minor failure—A failure which has no significant effect on the satisfactory performance of a system.

minority carrier – The less predominant carrier in a semiconductor. Electrons are the minority carriers in p-type semiconductors, since there are fewer electrons than holes. Likewise, holes are the minority carriers in n-types, since they are outnumbered by the electrons.

minority emitter—An electrode from which minority carriers flow into the interelectrode region.

minor lobe—Any lobe of an antenna radiation pattern other than the major lobe.

minor loop – A continuous network composed of both forward and feedback elements, which is only part of the overall feedback control system.

minor relay station – A tape relay station that has tape relay responsibility but does not provide an alternate route.

Minter stereo system – A stereo recording technique for producing the right and left channels. The two program channels are combined additively and recorded with a monophonic cutter. A 25-kHz note is also recorded and is frequency-modulated by the two channels combined subtractively. The sum and difference signals are then matrixed (combined) to produce the right and left channels.

mirror galvanometer—A suspended-coil instrument which, instead of using a pointer to indicate the reading, employs a light beam reflected from a mirror attached to the moving coil.

mirror galvanometer oscillograph – An instrument that photographs the deflection of a light spot from a mirror attached to a moving coil. Used for recording small current variations.

mirror-reflection echoes – Multiple-reflection echoes produced when a radar beam is reflected from a large, flat surface (such as the side of an aircraft carrier) and strikes nearby targets.

mirror scale—Meter scale with a mirror arc used to align the eyeball perpendicular to the scale when taking a reading. By eliminating this human error in reading, accuracy can be improved by half.

MIS—Technology wherein a silicon dioxide layer is formed on a single crystal silicon substrate and a polysilicon conductive is formed on the oxide. This layer is etched to form the electrode pattern and then doped with phosphorous to create the desired conductivity. Abbreviation for metal-insulator-silicon.

misfire—Failure of a mercury-pool-cathode

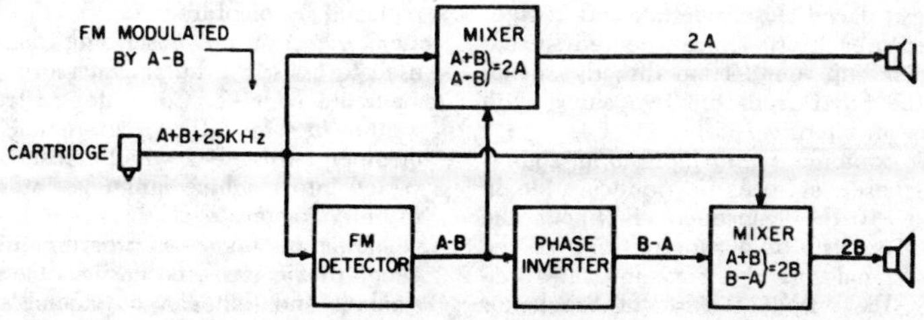

Minter stereo-disc playback.

tube to establish an arc between the main anode and cathode during a scheduled conducting period.

mismatch – A condition whereby the coupled impedances inhibit optimum power transference or the source or output signal fails to match the amplifier sensitivity or the load's (i.e., speaker) power capacity.

mismatch factor–*See* Reflection Factor, 1.

mismatch loss – The ratio between the power a device would absorb if it were perfectly matched to the source, and the power it actually does absorb.

mistake–*See* Error.

mistor–A magnetic-field sensing device the resistance of which increases with an increase in magnetic-field intensity.

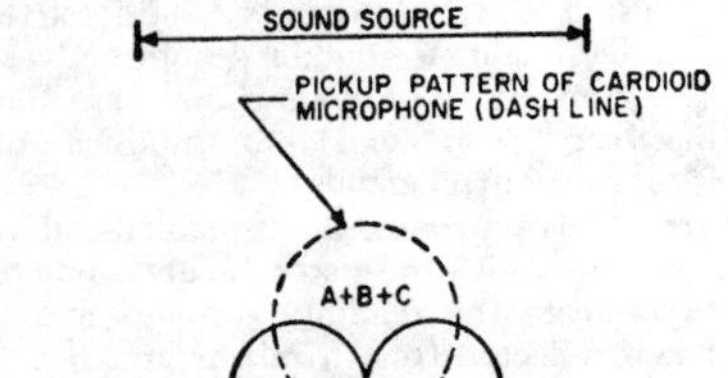

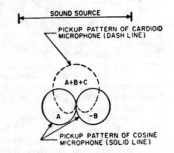

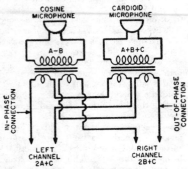

Mitte-seite stereo system.

mitte-seite stereo system–German for middle-side system. A technique of stereo pickup in which two directional microphones, placed close together and at right angles to each other, are oriented so that one picks up sound from directly ahead, and the other from the two sides, with maximum intensity.

mix–To combine two or more input signals in a transducer so as to produce a single output. If the transducer is linear the output consists of a superposition of the input signals. If the transducer is nonlinear, the output consists of the heterodyne products of the input signals.

mixed-base notation–A number system in which a single base, such as 10 in the decimal system, is replaced by two number bases, such as 2 and 5, used alternately.

mixed calculation–In a calculator, calculation involving more than one arithmetic mode. Saves time by permitting calculations to be handled as a single problem.

mixed highs – In color television, the method of reproducing very fine picture detail by transmitting high-frequency components as part of the luninance signal for achromatic reproduction.

mixer–1. In a sound transmission, recording or reproducing system, a device having two or more inputs (usually adjustable) and a common output. The latter combines the separate input signals linearly in the desired proportion to produce an output signal. 2. A circuit that generates output frequencies equal to the sum and difference of two input frequencies. 3. A device for blending two or more signals for special effects.

mixer tube–An electron tube which, when supplied with voltage or power from an external oscillator, performs the frequency-conversion function of a heterodyne-conversion transducer.

mixing–1. Combining two or more signals –e.g., the outputs of several microphones, or the received and local-oscillator signals in a superheterodyne receiver. 2. Blending of two or more signals for special effects, while exercising individual control over the volume of each.

mixing amplifier–An amplifier which combines several signals, each with a different amplitude and waveshape, into a composite signal.

mixing point–In a block diagram of a feedback control loop, a symbol indicating that the output is a function of the inputs at any instant.

mksa electromagnetic system of units–Also called the Giorgi system. A system in which the fundamental units are the meter kilogram, second, and ampere.

MLM–*See* Multilayer Metalization.

mm–Abbreviation for millimeter.

mmf – Abbreviation for micromicrofarad; replaced by picofarad.

mmfd–Abbreviation for micromicrofarad; replaced by picofarad.

mnemonic–A term describing something used to assist the human memory.

mnemonic code–1. Computer instructions written in a form the programmer can remember easily, but which must be converted into machine language later. 2. A memory jogger.

mnemonic language–A programming language that is based on easily remembered symbols and that can be assembled into machine language by the computer.

mnemonic operation codes—Computer instructions that are written in a meaningful notation, for example: ADD, MPY, STO.

mnemonic symbol—A symbol chosen so that it assists the human memory; for example, the abbreviation MPY used for "multiply."

MNOS—Abbreviation for metal-nitride-oxide-semiconductor.

mobile radio service—Radio service between a fixed location and one or more mobile radio stations, or between mobile stations.

mobile receiver—The radio receiver in an automobile, truck, or other vehicle.

mobile-relay station—A type of base station in which the base-station receiver automatically turns on the base-station transmitter which then retransmits all signals received by the base-station receiver. Such a station is used to extend the range of mobile units and requires two frequencies for operation.

mobile service—A service of radiocommunication between mobile and land stations, or between mobile stations.

mobile station – A radio station intended for use while in motion or during halts at unspecified points. Included are hand- and pack-carried units.

mobile telemetering—Electric telemetering between moving objects, where interconnecting wires cannot be used.

mobile transmitter—A radio transmitter installed and operated in a vessel, land vehicle, or aircraft.

mobile unit—A tape recorder or playback machine that receives its power from a large movable device such as an automobile, yacht or airplane. Mobile units are generally mounted in or under the control panel of the mobile device, frequently with easy removal facilities which, if taken advantage of, can minimize the probability of theft.

mobility—Symbolized by the Greek letter *mu* (μ). The ease with which carriers move through a semiconductor when they are subjected to electric forces. In general, electrons and holes do not have the same mobility in a given semiconductor. Also, their mobility is higher in germanium than in silicon.

mobius counter—*See* Johnson Counter.

mod/demod – Abbreviated form of modulating and demodulating.

mode—1. One of several types of electromagnetic wave oscillation that may be sustained in a given resonant system. Each type of vibration is designated as a particular mode and has its own particular electric- and magnetic-field configurations. 2. One of several methods of exciting a resonant system. The term has also been used to describe the existence of a number of different input voltages which allow operation of a klystron at the same frequency. 3. A computer system of data representations (e.g., the binary mode). 4. The most frequent value in a series of measurements or operations.

mode coupling – The interaction or exchange of energy between similar modes.

mode filter—A waveguide filter designed to separate waves of the same frequency but of different transmission modes.

mode hopping – The random shifting of laser output energy from one mode to another for short durations, usually in the microsecond range. The modes are generally closely spaced.

mode jump—A change in the mode of magnetron operation from one pulse to the next. Each mode represents a different frequency and power level.

modem – 1. Acronym for MOdulator/DEModulator. A device that transforms a typical two-level computer signal into a form suitable for transmission over the telephone network (for example, conversion of a two-level signal into a two-frequency sequence of signals). 2. A device that modulates and demodulates signals transmitted over communication facilities. 3. A device that performs modulation in the form of signal conversion, interfacing computers or computer peripheral equipment to the telephone line. Instead of trying to send a logic "0" or "1" dc voltage level over phone lines where voltage transients or noise pulses could be interpreted as false signals at the other end of the line and where transformer coupling is used, the modem changes the logic "0" or "1" into pulse audio tones. The tones travel over the phone line and enter a companion modem at the other end of the line and are converted back into "ones" and "zeros" to properly interface and communicate with a computer or computer peripheral equipment. 4. A modulator-demodulator, whose primary function is to convert input digital data to a form compatible with basically analog transmission lines. Modulation is the D-A conversion process; demodulation is the reverse, whereby transmitted analog signals are reconverted to digital data compatible with the receiving data-handling equipment. In the process of carrying out this primary function, the modem must be compatible with the data communications equipment at its digital interface as well as with the telephone line at its analog interface.

mode number—1. In a reflex klystron, the number of whole cycles that a mean-speed electron remains in the drift space. 2. The number of radians of phase shift resulting from going once around the anode of a magnetron divided by 2π.

mode of failure—*See* Failure Mode.

mode of resonance—A form of natural electromagnetic oscillation in a resonator. It is characterized by an unvarying field pattern.

mode of transmission propagation — 1. A form of guided-wave propagation characterized by a particular field pattern that intersects the direction of propagation. The field pattern is independent of its position along the waveguide. For a uniconductor waveguide, it also is independent of frequency. 2. In a vibrating system, the state which corresponds to one of the resonant frequencies.

mode of vibration—The pattern formed by the movement of the individual particles in a vibrating body (e.g., a piezoelectric crystal). This pattern is determined by the stresses applied to the body, the properties of the body, and the boundary conditions. The three common modes of vibration are flexural, extensional, and shear.

mode purity—The freedom of an atr tube from undesirable mode conversion while the tube is in its mount.

mode separation—In an oscillator, the difference in frequency between resonator modes of oscillation.

mode shift—In a magnetron, the change in mode during a pulse.

mode skip—Failure of a magnetron to fire during each successive pulse.

mode transducer—Also called mode transformer. A device that transforms an electromagnetic wave from one mode of propagation to another.

mode transformer—*See* Mode Transducer.

modification — The physical alteration of a system, subsystem, etc., for the purpose of changing its designed capabilities or characteristics.

modified constant-voltage charge — The charging of a storage battery in which the voltage of the charging circuit is held substantially constant, but a fixed resistance is inserted in the battery circuit, producing a rising voltage characteristic at the battery terminals as the charge progresses.

modified index of refraction — Also called modified refractive index. In the troposphere, the index of refraction at any height increased by h/a, where h is the height above sea level and a is the mean geometrical radius of the earth. When the index of refraction in the troposphere is horizontally stratified, propagation over a hypothetical flat earth through an atmosphere with the modified index of refraction is substantially equivalent to propagation over a curved earth through the real atmosphere.

modified refractive index—*See* Modified Index of Refraction.

modifier—A device that alters an instruction but does not change the form of energy (for example, the changing of electrical input signals into electrical output signals).

moding—A defect of magnetron oscillation in which the magnetron oscillates in one or more undesired modes.

modular—1. Made up of modules. 2. Dimensioned according to a prescribed set of size increments.

modulate — To vary the amplitude, frequency, or phase of a wave by impressing one wave on another wave of constant properties.

modulated amplifier—In a transmitter, the amplifier stage where the modulating signal is introduced and modulates the carrier.

modulated-beam photoelectric system — An intrusion-detector system which provides reliable beam ranges of several thousand feet. This is done by interrupting the light beam at the source with a rotating punched or slotted disc. In this way, the phototube output is converted to an ac signal, which can then be easily amplified.

modulated carrier—A radio-frequency carrier in which the amplitude or frequency has been varied by, and in accordance with, the intelligence to be conveyed.

modulated continuous wave — Abbreviated mcw. A wave in which the carrier is modulated by a constant audio-frequency tone. In telegraphy service, the carrier is keyed to produce the modulation.

modulated light—Light whose intensity has been made to vary in accordance with variations in an audio-frequency or code signal.

modulated oscillator—An oscillator the output frequency of which is varied by an input signal.

modulated signal generator — A device which produces an output signal that may be changed in amplitude and/or frequency according to a desired pattern. It is calibrated in units of both power (or voltage) and frequency.

modulated stage — The radio-frequency stage to which the modulator is coupled and in which the continuous wave (carrier wave) is modulated, in accordance with the system of modulation and the characteristics of the modulating wave.

modulated wave—A carrier wave in which the amplitude, frequency, or phase varies in accordance with the intelligence signal being transmitted.

modulating amplifier using variable reactance — A very-high-frequency electron-beam parametric amplifier with bandpass characteristics that are independent of gain, and which is unconditionally stable.

modulating electrode — In a cathode-ray

tube, an electrode to which a potential is applied to control the magnitude of the beam current.

modulating signal—*See* Modulating Wave.

modulating wave—Also called modulating signal, or simply signal. A wave which varies some characteristic (i.e., frequency, amplitude, phase) of the carrier.

modulation – 1. The process of modifying some characteristic of a wave (called a carrier) so that it varies in step with the instantaneous value of another wave (called a modulating wave or signal). The carrier can be a direct current, an alternating current (provided its frequency is above the highest frequency component in the modulating wave), or a series of regularly repeating, uniform pulses called a pulse chain (provided their repetition rate is at least twice that of the highest frequency to be transmitted). 2. The controlled variation of frequency, phase and/or amplitude of a carrier wave of any frequency in order to transmit a message.

modulation capability—The maximum percentage of modulation possible without objectionable distortion.

modulation code – A code used to cause variations in a signal in accordance with a predetermined scheme; normally used to alter or modulate a carrier wave to transmit data.

modulation distortion—Distortion occurring in the radio-frequency amplifier tube of a receiver when the operating point is at the bend of the grid-voltage–plate-current characteristic curve. As a result, the plate-current changes are greater on positive than on negative half-cycles. The effect is equivalent to an increase in the percentage of modulation.

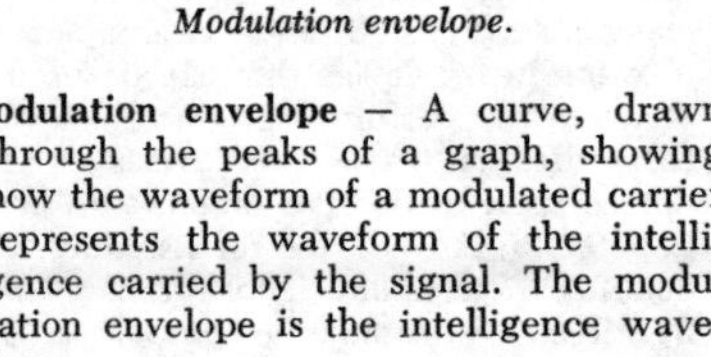

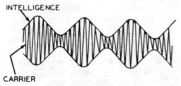

Modulation envelope.

modulation envelope – A curve, drawn through the peaks of a graph, showing how the waveform of a modulated carrier represents the waveform of the intelligence carried by the signal. The modulation envelope is the intelligence waveform.

modulation factor—In an amplitude-modulated wave, the ratio of half the difference between the maximum and minimum amplitudes to the average amplitude. This ratio is multiplied by 100 to obtain the percentage of modulation.

modulation frequency – That signal which causes the output frequency of an oscillator to be modulated.

modulation index – In frequency modulation with a sinusoidal modulating wave, the ratio of the frequency deviation to the frequency of the modulating wave.

modulation meter—Also called modulation monitor. An instrument for measuring the degree of modulation (modulation factor) of a modulated wave train. It is usually expressed in percent.

modulation monitor – *See* Modulation Meter.

modulation noise—Also called noise behind the signal. The noise caused by the signal, but not including the signal.

modulation percentage – The modulation factor multiplied by 100.

modulation plan – The arrangement by which the individual channels or groups of channels are modulated to a final frequency allocation.

modulation ratio—For an electrically modulated source, the number obtained by dividing the percentage of radiation modulation by the percentage of current modulation.

modulation rise—An increase in the modulation percentage. It is caused by a nonlinear tuned amplifier, usually the last intermediate-frequency stage of a receiver.

modulator – 1. A device that effects the process of modulation. 2. In radar, a device for generating a succession of short energy pulses which cause a transmitter tube to oscillate during each pulse. 3. An electrode in a spacistor.

modulator crystal—A crystal which is used to modulate a polarized light beam by the use of Pockel's effect. Useful as a modulator in laser systems.

modulator driver—A transmitter circuit that produces a pulse to be delivered to the control grid of the modulator stage.

modulator glow tube – A cold-cathode recorder tube used for facsimile and sound-on-film recording. It provides a modulated, high-intensity, point-of-light source.

modulator stage – The last stage through which the signal that modulates the radio-frequency wave is passed.

module—1. A unit in a packaging scheme displaying regularity and separable repetition. It may or may not be separate from other modules after initial assembly. Usually all major dimensions are in accordance with a prescribed series of dimensions. 2. A packaging concept in which identical forms are used. 3. A complete subassembly of a larger system combined in a single package.

modulo-n counter—A counter with n unique states. *See also* Programmable Counter.

modulus—An integer designating the number of states through which a counter sequences during each cycle. Abbreviated mod.

moiré—1. A wavy or satiny effect produced by the converegnce of lines. Usually appears as a curving of the lines in the horizontal wedges of the test pattern and is most pronounced near the center, where the lines forming the wedges converge. A moiré pattern is a natural optical effect when converging lines in the picture are nearly parallel to the scanning lines. To a degree this effect is sometimes due to the characteristics of color picture tubes and of image-orthicon pickup tubes (in the latter, termed "mesh beat"). 2. A coarse pattern of shading that occurs in a facsimile system when half-tone material is scanned.

moisture absorption—The amount of moisture (in percentage) that an insulation will absorb. The figure should be as low as possible when the insulation is to be used in a moist environment.

moisture repellent—Having properties such that moisture will not penetrate.

moisture resistance—The ability of a material not to absorb moisture, either from the air or from being immersed in water.

moisture resistant — Having characteristics such that exposure to a moist atmosphere will not readily lead to a malfunction.

mol—Abbreviation for molecular weight.

mold—In disc recording, a metal part derived from a master by electroforming. It is a positive of the recording (i.e., it has grooves similar to those of a recording and can thus be played).

molded capacitor — A capacitor that has been encased in a molded plastic insulation.

molectronics—Acronym for molecular electronics.

molecular circuit—1. *See* Monolithic Integrated Circuit. 2. An electronic circuit demonstrating a measurable input and output, where the portions performing various functions such as resistance or capacitance are indistinguishable as discrete areas.

molecular circuitry — *See* Morphological Circuitry.

molecular clock—A device for time measurement based on an electromagnetic oscillation of extremely stable frequency from a beam of the rotation spectrum, the vibration or inversion of a particular molecule, and a counting device.

molecular electronics—Abbreviated molectronics. The science of making a single block of matter perform the function of a complete circuit. This is done by merging the function with a material, using solid-state functional blocks.

molecular integrated circuit—An integrated circuit such that the identity and location of specific electric elements cannot be determined even by microscopic disassembly of the material of which the circuit is formed. In contrast to a conventional microelectronic circuit, the molecular integrated circuit can be defined only by function, which in turn can be described only by mathematical models and incremental circuit representation.

molecular technique—A practice method of causing a single piece of material or a crystal to provide a complete circuit function.

molecular weight—The sum of the weights of all the atoms in a molecule, expressed relative to the arbitrary weight of atomic oxygen ($O = 16$). Abbreviated mol.

molecule — In any substance, the smallest particle that still retains the physical and chemical characteristics of that substance. A molecule consists of one or more atoms of one or more elements. Sometimes two entirely different substances may have similar chemical elements, but their atoms will be arranged in a different order.

molybdenum—A metallic element (chemical symbol, *Mo;* atomic number, 42; atomic weight, 95.95) sometimes used for the grid and plate electrodes of vacuum tubes.

momentary-contact switch — A mercury switch designed to make contact for only a brief transitory interval while the mercury moves from one extreme position to another.

momentary switch—A switch which returns to its normal circuit condition when the actuating force is removed.

monaural—*See* Monophonic.

monaural recorder — Literally, a tape recorder intended for listening with one ear only; however, in popular usage refers to single-channel recorders, as distinguished from multichannel (stereophonic, binaural, etc.) types. More correctly but less universally called monophonic recorder.

monitor—1. "To listen . . ." to a communication service, without disturbing it, to determine its freedom from trouble or interference. 2. A device (e.g., a receiver, oscilloscope, teleprinter, etc.) used for checking signals. 3. A software package or a hardware device that can be used to measure the performance of a system or the utilization of specific devices. 4. Any device used to observe or measure a parameter. 5. Any device for listening incidentally to an audio signal that is primarily directed to some other purpose at that moment. A monitor loudspeaker is used for auditioning a recording or radio program incidentally to its committal to tape or its broadcast. 6. The operator of a television monitoring system who selects one out of several camera images for broadcasting.

monitored fast forward – A feature in a cartridge deck where the playback amplifier is left on at low volume during fast forward so the user can hear the program running through at the faster speed, to spot or cue up to the desired program.

monitor head – A playback head that is separate from the record head, enabling the recordist to listen to what is coming off the tape a fraction of a second after it has been recorded. Without a monitor head, a tape must be recorded to its end and then rewound and replayed before the recordist can evaluate the tape.

monitoring – Observing the characteristics of transmitted signals as they are being transmitted.

monitoring amplifier – A power amplifier used primarily for evaluation and supervision of a program.

monitoring key–A key that, when operated, permits an attendant or operator to listen on a telephone circuit without causing appreciable impairment of transmission on the circuit.

monitoring radio receiver–A radio receiver for checking the operation of a transmitting station.

monitor systems–Programs that supervise other programs and keep computers functioning efficiently with a minimum of assistance from human operators.

monkey chatter–So called because of the garbled speech or music heard along with the desired program. This interference occurs when the side frequencies of an adjacent-channel station beat with the signal from the desired station.

monobrid–A method of manufacturing an integrated circuit by using more than one monolithic chip within the same package.

monobrid circuit–An integrated circuit using a combination of monolithic and multichip techniques by means of which a number of monolithic circuits or a monolithic device in combination with separate diffused or thin-film components are interconnected in a single package.

monochromatic–1. Pertaining to or consisting of a single color. 2. Radiation of a single wavelength.

monochromatic emissivity–*See* Total Emissivity.

monochromaticity–The degree of response to one color.

monochromatic light–Light consisting of just one wavelength. No light is *completely* monochromatic. The closest approach is particular lines in the mercury 198 spectrum excited in a discharge tube with no electrodes.

monochromatic sensitivity – The response of a device to light of a given color only.

monochromator–An instrument used to isolate narrow portions of the spectrum by making use of the dispersion of light into its component colors.

monochrome–Also called black-and-white in referring to television. Having only one chromaticity–usually achromatic, or black and white and all shades of gray.

monochrome channel–In a color television system, any path intended to carry the monochrome signal (although it may carry other signals also).

monochrome channel bandwidth – The bandwidth of the path that carries the monochrome signal.

monochrome signal–1. In a monochrome television transmission, the signal wave that controls the luminance values in the picture. 2. In a color-television transmission signal wave, the portion with major control of luminance–whether displayed in color or monochrome.

monochrome television–Also called black-and-white television. Television in which the final reproduced picture is monochrome. That is, it has only shades of gray between black and white.

monochrome transmission – Also called black-and-white transmission. In television, the transmission of a signal wave for controlling the luminance–but not the chromaticity–values in the picture.

monoclinic–A crystal structure in which two of the three axes are perpendicular to the third, but not to each other.

monocord switchboard–A local-battery telephone switchboard in which each line terminates in a single jack and plug.

monocrystalline–Material made up of a single continuous crystal.

monoergic–A type of emission in which the particles or radiations are produced with a small energy spread (i.e., a "line spectrum").

monofier–A complete master oscillator and power amplifier system contained in a single evacuated envelope. It is equivalent electrically to a stable low-noise oscillator, an isolator, and a two- or three-cavity klystron amplifier.

monogroove stereo – Also called single-groove stereo. A stereo recording in which both channels are contained in one groove.

monolithic–1. Existing as one large, undifferentiated whole. 2. Refers to the single slice of silicon substrate on which an integrated circuit is built; hence, monolithic integrated circuit. 3. Elements or circuit formed within a single semiconductor substrate.

monolithic circuit–A monolithic semiconductor integrated circuit has all circuit components manufactured in or on top of a single crystal semiconductor material. Interconnections between the component parts within a given circuit are made by means of metallization patterns, and the individual parts are not separable from the complete circuit. Hence, the monolithic circuit is often referred to as a "fully integrated circuit."

monolithic filter—A filter whose operations are based on the use of deposited electrode pairs acting as shear or thickness-mode resonators separated by nonelectrode regions which act as acoustic coupling elements. Entire filter is on a single quartz or ceramic substrate.

monolithic IC – *See* Monolithic Integrated Circuit.

monolithic integrated circuit—1. Abbreviated MIC. An integrated circuit the elements of which are formed *in situ* upon or within a semiconductor substrate with at least one of the elements formed within the substrate. (*See also* Integrated Circuit, 1.) To further define the nature of a monolithic integrated circuit, additional modifiers may be prefixed. Examples are: pn junction isolated monolithic integrated circuit, dielectric isolated monolithic integrated circuit, or beamlead monolithic integrated circuit. 2. The physical realization of electronic circuits or subsystems from a number of extremely small circuit elements inseparably associated on or within a continuous body or a thin film of semiconductor material. 3. An IC that is fabricated completely on a single chip and which contains no thin film, or discrete components. (As opposed to hybrid IC.) 4. A complete electronic circuit fabricated as an inseparable assembly of circuit elements in a single small structure. It cannot be divided without permanently destroying its intended electronic function.

monolithic microcircuit – *See* Monolithic Integrated Circuit.

monophonic—Also called by the older term "monaural." Pertaining to audio information on one channel (e.g., as opposed to binaural or stereophonic). Monophonic and monaural are usually, although not necessarily, associated with a one-speaker system.

monophonic recorder – *See* Monaural Recorder.

monopinch – Antijam application of the monopulse technique in which the error signal is used to provide discrimination against jamming signals.

monopole—A stub antenna fed against a ground plane.

monopole antenna—A vertical antenna with a voltage node at the lower end and a current node at the top, that is 0.25 wavelength at the operating frequency.

monopulse—A method of determining azimuth and elevation angles simultaneously.

monopulse radar—Radar using a receiving antenna system having two or more partially overlapping lobes in the radiation pattern. Sum and difference channels in the receiver compare the amplitudes or phases of the antenna outputs.

monopulse tracking – A form of telemetry tracking which compares the phase of the carrier received simultaneously at four points on a plane (two elevation and two azimuth references) and positions the antenna so that all four signals are in phase, or equidistant from the source of radiation. A variation of the monopulse tracking technique is the scan-coded or single-channel monopulse tracking system.

monorange speaker—A speaker that provides the full spectrum of audio frequencies.

monoscope – Also called phasmajector or monotron. An electron-beam tube in which the picture signal is generated by scanning an electrode, parts of which have different secondary-emission characteristics.

monostable—1. A term used to describe a circuit that has one permanently stable state and one quasi-stable state. An external trigger causes the circuit to undergo a rapid transition from the stable state to the quasi-stable state, where it remains for a time and then spontaneously returns to the stable state. 2. A type of multivibrator that has one stable state. The integrated-circuit version usually includes input gating and sometimes a Schmitt trigger. Also called single-shot or one-shot. 3. A system that has an at-rest bias causing one output condition consistently, until appropriate input signaling occurs.

monostable multivibrator—1. Also called one-shot multivibrator, single-shot multivibrator, or start-stop multivibrator. A circuit having only one stable state, from

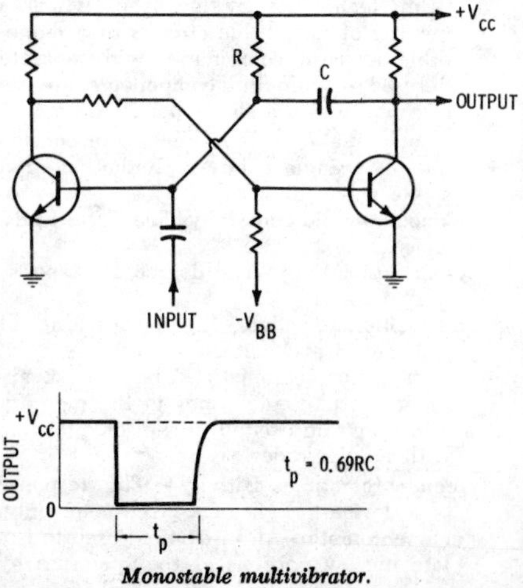

Monostable multivibrator.

which it can be triggered to change the state, but only for a predetermined interval, after which it returns to the original state. 2. A multivibrator having one stable and one semistable condition. A trigger is used to drive the unit into the semistable state where it remains for a predetermined time before returning to the stable condition.

monostatic reflectivity—The characteristic of a reflector which reflects energy only along the line of the incident ray (e.g., a corner reflector).

monotone—A single musical tone unvaried in pitch.

monotonicity — A measure of the ability of a converter, (either d/a or a/d) to produce an output in response to a continuously increasing input signal (a/d) or count (d/a) without decreasing in value or skipping codes.

monotron—*See* Monoscope.

Monte Carlo method — Abbreviated MCM. 1. A computer technique in which a number of possible models under study are mathematically constructed from constituents selected at random from representative populations. 2. Any procedure that involves statistical sampling techniques in order to obtain an approximate solution of a mathematical or physical problem.

mopa—Acronym for master oscillator-power amplifier.

morphological circuitry—Also called molecular circuitry. A circuit made from a material in which the molecular structure has been arranged to perform a certain electrical function.

Morse code—A system of dot-and-dash signals developed by Samuel F. B. Morse and now used chiefly in wire telegraphy.

Morse sounder—A telegraph receiving instrument that produces a sound at the beginning and end of each dot and dash. From these sounds, a trained operator can interpret the message.

Morse telegraphy—Telegraphy in which the Morse code or its derivative is used—specifically, the International (also called Continental) or American Morse code.

MOS — Abbreviation for metal-oxide-semiconductor.

mosaic — The light-sensitive surface of an iconoscope or other television camera tube. In one form, it consists of millions of tiny silver globules on a sheet of ruby mica. Each globule is treated with cesium vapor to make it sensitive to light. The globules retain a positive charge when bombarded with light and absorb electrons from a scanning beam in proportion to the amount of light they have received.

mosaic detector—A device in which a number of active elements are arranged in an array. It is generally used as an imaging device.

mosaic-lamp display — A form of display made up of seven straight line segments which can be used to form all digits. An individual lamp' behind each segment lights to form each digit.

MOS capacitor—A capacitor in which a silicon-oxide dielectric layer and a metal top electrode are deposited on a conducting semiconductor region that acts as the bottom electrode.

MOS device — Semiconductor component (typically, in an IC) formed by metal-oxide semiconductor process. In an MOS device, current can flow in either of two directions, whereas in a bipolar semiconductor current flows in only one direction.

MOSFET — Abbreviation for metal-oxide-semiconductor field-effect transistor.

MOS insulated-gate field-effect transistor —A semiconductor device consisting of two electrodes (source and drain) diffused into a silicon substrate and separated by a finite space, thus forming a majority-carrier conducting channel. A metal electrode (gate) is placed above and insulated from the channel.

MOS monolithic IC—A single-chip IC, consisting largely of interconnected unipolar active-device elements or MOS field-effect devices. This class of circuits results in higher equivalent parts density.

most significant digit—Abbreviated MSD. 1. Number at the extreme left of a group of numbers. Example, 6,937. Digit 6 is

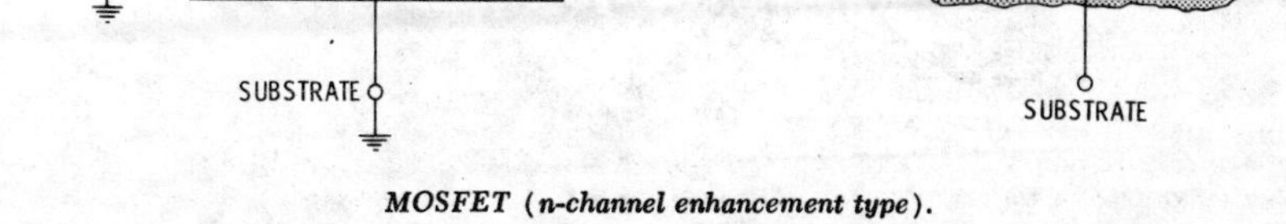

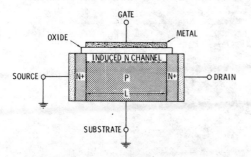

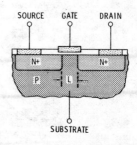

MOSFET (n-channel enhancement type).

the MSD. Does not apply to number 0 when in the MSD position. 2. In a binary number the digit with the highest weighting.

mother—In disc recording, a mold electroformed from the master.

motherboard—A relatively large piece of insulating material on which components, modules, or other electronic subassemblies are mounted and interconnections made by welding, soldering, or other means, using point-to-point or matrix wire, or circuitry fabricated integrally with the board.

mother crystal – The quartz crystal found in nature. It has the characteristic geometric design of a crystal (i.e., flat faces at definite angles to each other), but all or some of the faces may be worn because of abrasion with stones or other objects.

motional impedance – Also called loaded motional impedance. The complex remainder after the blocked impedance of a transducer has been subtracted from the loaded impedance.

motion detector—A device used in security systems which reacts to any movement on a cctv monitor by automatically setting off an alarm when the monitor is not manned.

motion frequency – 1. The natural frequency of a servo system. 2. The frequency at which a given servo tends to oscillate.

motion-picture pickup—A television camera or technique for televising scenes directly from motion-picture film.

motor – A device that moves an object. Specifically, a machine that converts electric energy into mechanical energy.

motor board – Also called tape-transport mechanism. The platform or assembly of a tape recorder on which the motor (or motors), reels, heads, and controls are mounted. It includes parts of the recorder other than the amplifier, preamplifier, speaker, and case.

motorboating—1. Interference heard as the characteristic "putt-putt" made by a motorboat. It is due to self-oscillation, usually pulsating, in an amplifier below or at a low audio frequency. 2. A generally periodic, relatively low-frequency pulse disturbance of the output voltage of a regulated power supply, frequently line- or load-dependent, unstable, and significantly large, as in the case of oscillations.

motor circuit switch – *See* Disconnect Switch.

motor converter—A device for converting an alternating current to a pulsating direct current. It consists of an induction motor to which an ac supply is connected. The armature of the induction motor is linked mechanically to the armature of a synchronous converter, which is connected to a dc circuit.

motor-driven relay—A relay in which the contacts are actuated by the rotation of a motor shaft.

motor effect—The repulsion force exerted between adjacent conductors carrying currents in the opposite direction.

motor element—That portion of an electroacoustic receiver that converts energy from the electrical system into mechanical energy.

motor-field control—The method of controlling the speed of a motor by changing the magnitude of its field current.

motor-field failure relay – A relay which functions to disconnect the motor armature from the line in the event of loss of field excitation. Also called field loss relay.

motor-field induction heater—An induction heater in which the inducing winding typifies that of a rotary or linear induction motor.

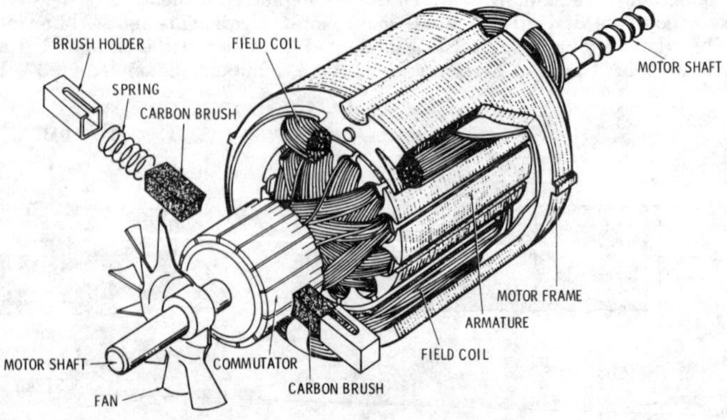

Motor.

motor-generator set – A motor-generator combination for converting one kind of electric power to another (e.g., alternating current to direct current). The two are mounted on a common base and their shafts are coupled together.

motorized lens–A camera lens fitted with a small electric motor which can focus the lens, open the diaphragm or in the case of a zoom lens, change the focal length, all by remote control.

motor junction (conduit) box–An enclosure on a motor for the purpose of terminating a conduit run and joining the motor to power conductors.

motor meter–A meter comprising a rotor, one or more stators, and a retarding element which makes the speed of the rotor proportionate to the quantity being measured (e.g., power or current). A register, connected to the rotor by suitable gearing, counts the revolutions of the rotor in terms of the total.

motor-operated sequence switch–A multicontact switch which fixes the operating sequence (or the major devices) during starting and stopping, or during other sequential switching operations.

motor-run capacitor–A capacitor which is left in the auxiliary motor winding, and which is in parallel with the main winding to obtain a higher power factor and efficiency.

motor-start capacitor–A capacitor which is in the circuit only during the starting period of a motor. The capacitor and its auxiliary winding are disconnected automatically by a centrifugal switch or other device when the motor reaches a predetermined speed, after which the motor runs as an induction motor.

motor starter – A device arranged to start an electric motor and accelerate it to normal speed; a motor starter has no running position other than fully on. It is a combination of all the switching means required to start and stop the motor, and it incorporates suitable overload protection.

mount – The flange or other means by which a switching tube, or a tube and cavity, are connected to a waveguide.

mount structure–The essential elements of a vacuum tube except the envelope.

mouth of a horn – The end having the larger cross section.

M-out-of-N code–A type of fixed-weight binary code in which M of the N digits are always in the same state.

movable contact–The one of a pair of contacts that is moved directly by the actuating system.

movement differential – The distance or angle from the operating position to the releasing position of a momentary contact switch.

move mode–In some variable-word-length computers, data transmission in which certain delimiters are not moved with the data (as opposed to load mode).

moving-coil galvanometer–A galvanometer in which the moving element is a suspended or pivoted coil.

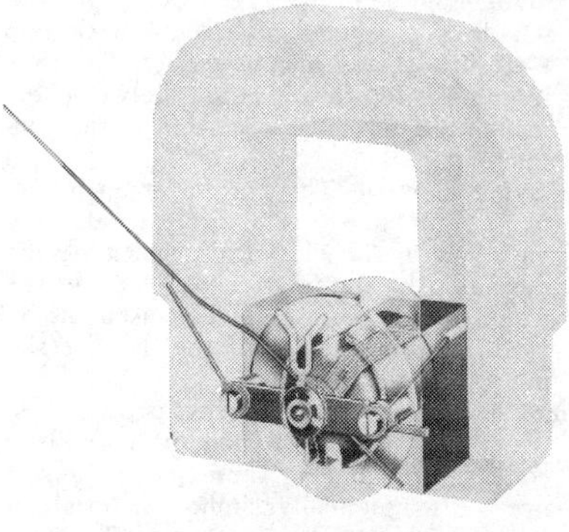

Moving-coil meter.

moving-coil meter – A meter in which a coil pivots between permanent magnets.

moving-coil microphone – 1. Also called a dynamic microphone. A moving-conductor microphone in which the diaphragm is attached to a coil positioned in a fixed magnetic field. The sound waves strike the diaphragm, moving it, and hence the coil, back and forth. An audio-frequency current is induced in the moving coil in the magnetic field and coupled to the amplifier. 2. A microphone using a permanent magnet and a vibrating coil or ribbon as its transducing system.

moving-coil pickup–Also dynamic pickup or reproducer. A phonograph pickup in which a conductor or coil produces an electric output as it moves back and forth in a magnetic field.

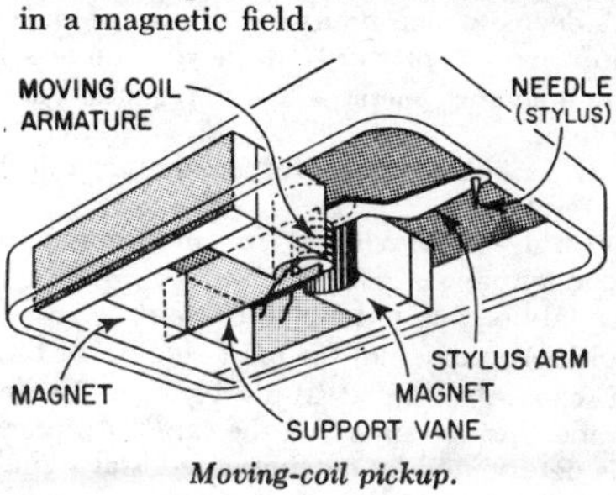

Moving-coil pickup.

moving-coil speaker–Also called a dynamic speaker. A speaker in which the moving diaphragm is attached to a coil, which is conductively connected to the source of electric energy and placed in a constant magnetic field. The current through the coil interacts with the magnetic field, causing the coil and diaphragm to move back and forth in step with the current variations through the coil.

moving-conductor microphone – A microphone which produces its electric output from the motion of a conductor in a magnetic field.

moving-conductor speaker – A speaker in which a conductor is moved back and forth in a steady magnetic field. The cone is moved by the reaction between the magnetic field and the current in the conductor.

moving element–The portion of an instrument that moves as a direct result of a variation in the electrical quantity being measured by the instrument. One-half the weight of the springs (when used) is included in the weight of the moving element.

moving-iron instrument–An instrument in which the current in one or more fixed coils acts on one or more pieces of soft iron or magnetically similar material, at least one of which is movable. The various forms of this instrument (plunger, vane, repulsion, attraction, repulsion-attraction) are distinguished chiefly by their mechanical construction. Otherwise, the action is the same.

moving-magnet instrument–An instrument in which a movable permanent magnet aligns itself in the field produced by another permanent magnet and an adjacent coil or coils carrying current, or by two or more current-carrying coils.

moving-magnet magnetometer–A magnetometer in which the torques act on one or more permanent magnets, which can turn in the field to be measured. Some types use auxiliary magnets (gaussian magnetometer); others use electric coils (sine or tangent galvanometer).

moving-target indicator–Abbreviated mti. A device which limits the display of radar information primarily to moving targets.

moving-vane meter – *See* Magnetic-Vane Meter.

MRN – Abbreviation for minimum reject number.

mr/min – Abbreviation for milliroentgens per minute.

ms–Abbreviation for millisecond.

MSB–Abbreviation for most significant bit.

M-scan–Also called M-display. An A-scan radar display in which the target distance is determined by moving a pedestal signal along the base line until it coincides with the horizontal position of the target-signal deflection. The control that moves the pedestal is calibrated in distance.

MSD – Abbreviation for most significant digit.

msec–Abbreviation for millisecond.

MSI–Abbreviation for medium-scale integration.

M-S stereo system–*See* Mitte-Seite Stereo System.

MTBF–Abbreviation for mean time between failures.

mti–Abbreviation for moving-target indicator.

MTNS – Abbreviation for metal-thick-nitride semiconductor.

MTOS–Abbreviation for metal–thick-oxide semiconductor.

MTTF–Abbreviation for mean time to failure.

MTTFF – Abbreviation for mean time to first failure.

M-type backward-wave oscillator–A cross-field injected-beam oscillator. The electrons in this device interact with an rf wave traveling backward or opposite to the electron beam. It is efficient, broadband, and can be voltage-tuned. It is also insensitive to load variations.

mu–English spelling for the Greek letter μ.

mux–Abbreviation for multiplex.

μ–Greek letter mu. 1. Symbol for amplification factor. 2. Symbol for permeability. 3. Abbreviation for prefix micro-. 4. Abbreviation for micron.

μA–Abbreviation for microampere.

μH–Abbreviation for microhenry.

μmho–Abbreviation for micromho.

$\mu\Omega$–Abbreviation for micro-ohm.

μs–Abbreviation for microsecond.

μV–Abbreviation for microvolt.

μW–Abbreviation for microwatt (10^{-6} watt).

mu-circuit–In a feedback amplifier, the circuit which amplifies the vector sum of the input signal and the feedback portion of the output signal in order to generate the output signal.

muf – Abbreviation for maximum usable frequency.

mu-factor–Ratio of the changes between two electrode voltages, assuming the current and all other electrode voltages are maintained constant–i.e., it is a measure of the relative effect which the voltages on two electrodes have on the current in the circuit of a specified electrode.

Muller tube – A thermionic vacuum tube having an auxiliary cathode or grid connected internally to the main cathode through a high-value resistor.

multiaddress – Pertaining to computer instructions which specify two or more addresses.

multianode tank–*See* Multianode Tube.

multianode tube – Also called multianode tank. An electron tube having two or more main anodes and a single cathode.

multiaperture reluctance switch – A two-aperture ferrite storage core that may be used to provide a nondestructive-readout memory for a computer.

multiband antenna–An antenna usable at more than one frequency band.

multicasting – Broadcasting a stereo program by using two fm stations. Two fm receivers are required.

multicavity magnetron – A magnetron in which the circuit has more than one cavity.

multicellular horn–A cluster of horns with juxtaposed mouths lying in a common surface. The cluster controls the directional pattern of the radiated energy.

multichannel radio transmitter – A radio transmitter having two or more complete radio-frequency portions capable of operating on different frequencies, either individually or simultaneously.

multichannel R/C–A radio-control installation that employs tuned reeds to supply several control functions. The basic carrier frequency remains the same, but different tones make possible a number of control channels.

multichip circuit–A microcircuit in which discrete, miniature active electronic elements (transistors and/or diode chips) and thin-film or diffused passive components or component clusters are interconnected by thermocompression bonds, alloying, soldering, welding, chemical deposition, or metalization.

multichip integrated circuit–An integrated circuit the elements of which are formed on or within two or more semiconductor chips that are separately attached to a substrate. *See also* Integrated Circuit. 2. Hybrid integrated circuit which includes two or more SIC, MSI, or LSI chips. 3. An electronic circuit in which two or more semiconductor wafers that contain single elements or simple circuits are interconnected and encapsulated in a single package to give a more complex circuit.

multichip microcircuit–A microcircuit the elements of which are formed on or within two or more semiconductor chips that are attached separately to a substrate.

multicoupler–A device for connecting several receivers to one antenna and properly matching their impedances.

multielectrode tube–An electron tube containing more than three electrodes associated wtih a single electron stream.

multielement parasitic array–An antenna consisting of driven dipoles and parasitic elements arranged to produce a highly directive beam.

multiemitter transistor – A transistor that has more than one emitter. Used mainly in logic circuits.

multifrequency transmitter–A radio transmitter capable of operating on two or more selectable frequencies, one at a time, using preset adjustments of a single radio-frequency portion.

multifunction–Pertaining to an integrated device containing two or more circuits integral to a single silicon chip. In addition, each circuit has all inputs and outputs available at terminals for testing and interconnection with other packages. Typical multifunction devices are quadruple gates and dual flip-flops.

multifunction array radar – An electronic scanning radar which will perform target detection and identification, tracking, discrimination, and some interceptor missile tracking on a large number of targets simultaneously and as a single unit.

multigun tube–A cathode-ray tube having more than one electron gun. Used in color television receivers and multiple-presentation oscilloscopes.

multihop propagation–The bouncing of radio waves from the ionosphere to increase their range.

multilayer circuit–Three or more conductive patterns interspersed with layers of insulating base and laminated into sandwiches, with interconnection between layers provided by plate-through holes.

multilayer interconnection pattern–A technique used for the interconnection of arrays performing large, complicated electronic functions. This technique involves the use of alternating films of insulating and conducting materials as a precondition for the realization of multiple interconnection planes.

multilayer metalization–1. A method used to increase the component density of a particular monolithic integrated circuit. Two or more layers of interconnecting metalization are stacked on top of the chip; the layers are separated by a thin dielectric (insulating) film except at the desired contact points. 2. Abbreviated MLM. An integrated circuit fabricating technique that makes economically feasible more complex monolithic logic subsystems or arrays: LSI, MLM permits signal paths to be routed most efficiently and to cross without interaction. As a result, integrated circuit chips can be smaller or can contain more circuitry.

multilayer printed circuit – A type of printed circuitry wherein 2 to 14 or more printed circuit layers are fabricated as a complete assembly.

multimeter–Electronic device for measuring resistance and ac or dc current and voltage. *See also* Circuit Analyzer.

multimode operation–The operation of a laser sufficiently above threshold so as to stimulate more than one mode. If the modes are not degenerate, the output pattern will contain measurable power at more than one frequency.

multimoding–The simultaneous generation of many frequencies instead of one discrete frequency.

multipactor–A high-power, high-speed microwave switching device in which an rf electric field drives a thin cloud of electrons back and forth between two parallel plane surfaces located in a vacuum. The device can be used as a switch in a waveguide or for high-speed switching of pulses in other microwave systems.

multipass sort — A computer program for sorting more data than can be contained by the internal computer storage. Intermediate storage, such as disc, tape, or drum, is required.

multipath cancellation—In effect, complete cancellation of signals because of the relative amplitude and phase differences of the components arriving over separate paths.

multipath delay—A form of phase distortion occurring most often in high-frequency layer-refracted or reflected signals, and also in vhf scattered signals. The existence of more than one signal path between transmitter and receiver causes the signal components to reach the receiver at slightly different times, causing echoes or ghosting.

multipath distortion/reception — Owing to reflection of the vhf fm signal by large buildings, hills, etc., a receiver sometimes receives not only the direct signal but also a reflected signal slightly later due to the greater path distance traveled. This results in high-order harmonic distortion and impairment to the stereo separation and quality.

multipath effect — The arrival of radio waves at slightly different times because all components do not travel the same distance.

multipath reception—Reception in which the radio signal from the transmitter travels to a receiver antenna by more than one route, usually because it is reflected from obstacles. The result is seen as ghosts in a tv picture.

multipath transmission — The phenomenon where the signals reach the receiving antenna from two or more paths and usually have both amplitude and phase differences. It may cause jitter in facsimile (in Europe it is called echo).

multiple—1. A group of terminals arranged in parallel to make a circuit or group of circuits accessible at any of several points to which a connection can be made. 2. To render a circuit accessible as in (1) above by connecting it in parallel with several terminals.

multiple access—1. The use of a communication satellite by more than one pair of ground stations at a time. Ideally it lets many independent ground stations use an active repeater without interfering with each other. 2. Pertaining to a computer system in which a number of on-line communication channels provide concurrent access to the common system.

multiple accumulating registers—In a computer, additional internal storage capacity which can contain loading, storing, adding, subtracting, and comparing factors up to four computer words in length.

multiple-address code—A computer instruction code that includes more than one address.

multiple-address instruction — In a computer, an instruction that contains more than one address.

multiple-aperture core — A magnetic core that has two or more holes through which may be passed wires and around which there may be magnetic flux. Such cores may be used for nondestructive reading.

multiple break—In an electrical circuit, an interruption at more than one point.

multiple-break contacts—A contact arrangement such that a circuit is opened in two or more places.

multiple-chip circuit — *See* Hybrid Integrated Circuit.

multiple-contact switch—A switch in which the movable contact can be set to any one of several fixed contacts.

multiple course—One of a number of lines of position defined by a navigational system. Any one of these lines may be selected as a course line.

multiple jacks—A series of jacks that appear on different panels and that have their tips, rings, and sleeves, respectively, connected in parallel.

multiple-length number—In a computer, a quantity or expression that occupies two or more registers.

multiple modulation — Also called compound modulation. A succession of modulation processes in which the modulated wave from one process becomes the modulating wave for the next.

multiple pileup—Also called multiple stack. An arrangement of contact springs composed of two or more pileups.

multiple processing — Configuring two or more processors in a single system, operating out of a common memory. This arrangement permits execution of as many programs as there are processors.

multiple programming — In computer programming, simultaneous execution of two or more arithmetical or logical operations.

multiple punching—The punching of more than one hole in the same column on a card by executing more than one key stroke.

multiple-purpose tester—A single test instrument having several ranges, for measuring voltage, current, and resistance.

multiple-reflection echoes—Returned echoes that have been reflected from an object in the radar beam. Such echoes give a false bearing and range.

multiple sound track—Two or more sound tracks printed side by side on the same medium, containing the same or different material, but meant to be played at the same time (e.g., those used for stereophonic recording).

multiple stack—*See* Multiple Pileup.

multiple switchboard—A manual telephone switchboard in which each subscriber line is connected to two or more jacks so that they are within reach of more than one operator.

multiple tube counts—Spurious counts induced in a radiation counter by previous tube counts.

multiple-tuned antenna — A low-frequency antenna with one horizontal section and several tuned vertical sections.

multiple-twin quad — Quad cable in which the four conductors are arranged in two twisted pairs, and the two pairs twisted together.

multiple-unit semiconductor device — A semiconductor device having two or more sets of electrodes associated with independent carrier systems. It is implied that the device has two or more output functions which are independently derived from separate inputs (e.g., a duotriode transistor).

multiple-unit steerable antenna — An antenna unit composed of a number of stationary antennas, the major composite lobe of which is electrically steerable.

multiple-unit tube—Also called a duodiode, duotriode, diode-pentode, duodiode-triode, duodiode-pentode an triode-pentode. An electron tube containing two or more groups of electrodes associated with independent electron streams within one envelope.

multiplex—1. To carry out several functions simultaneously in an independent but related manner. 2. To interleave or transmit two or more messages simultaneously on a single channel. 3. A technique for transmitting two or more signals on the same carrier frequency. (*See* fm stereo) 4. Abbreviated mux. Transmission of two or more channels on a single carrier so that they may be recovered independently by a tuner. In fm stereo, transmission of left-plus-right (sum) signal and left-minus-right (difference) signal on main carrier and subcarrier, respectively. The multiplex decoder in the tuner recovers independent left and right stereo channels from the multiplex signal. 5. To combine two or more electrical signals into a single, composite signal. This may be done on a frequency basis (frequency-division multiplexing) or on a time basis (time-division multiplexing).

multiplex adapter—A circuit incorporated in an fm tuner or receiver to permit two-channel, or stereophonic, reception from a station transmitting multiplex broadcasts. The other audio channel is produced by demodulating the transmitted subcarrier.

multiplex code transmission—The simultaneous transmission of two or more code messages in either or both directions over the same transmission path.

multiplexer—1. A device for accomplishing simultaneous transmission of two or more signals over a common transmission medium. 2. An analog or linear device for selecting one of a number of inputs and switching its information to the output; the output voltage follows the input voltage with a small error. 3. A digital device that can select one of a number of inputs and pass the logic level of that input on to the output. Information for input-channel selection usually is presented to the device in binary weighted form and decoded internally. The device acts as a single-pole mutiposition switch that passes digital information in one direction only. 4. A device that will interleave (time-division) or simultaneously transmit (frequency-division) two or more messages on the same communications channel. 5. A device which uses several communication channels at the same time and transmits and receives messages and controls the communication lines. This device itself may or may not be a stored-program computer. 6. A device which can combine several low-speed inputs into one high-speed output. Multiplexers can also function in reverse, a process called "demultiplexing."

multiplexing — 1. A method of signaling characterized by the simultaneous and/or sequential transmission and reception of multiple signals over a communication channel with means for positively identifying each signal. The signaling may be accomplished over a wire path or radio carrier or combination of both. 2. The division of a transmission facility into two or more channels either by splitting the frequency band transmitted by the channel into narrower bands, each of which is used to constitute a distinct channel (frequency-division multiplex) or by alloting this common channel to several different information channels, one at a time (time-division multiplexing).

multiplex operation — Simultaneous transmission of two or more messages in either or both directions over the same transmission path.

multiplex printing telegraphy—The form of printing telegraphy that uses a line circuit to transmit one character (or one or more pulses of a character) for each of two or more independent channels.

multiplex radio transmission—The simultaneous transmission of two or more signals over a common carrier wave.

multiplex stereo—A system of broadcasting

both channels of a stereo program on a single carrier. This is commonly done by modulating an ultrasonic subcarrier — either with the signals of one of the stereo channels, or by a difference signal composed of the two channels combined out of phase, or subtractively.

multiplex telegraphy—Telegraphy employing multiplex code transmission.

multiplex transmission — The simultaneous transmission of two or more signals within a single channel. Multiplex transmission as applied to fm broadcast stations means the transmission of facsimile or other signals in addition to the regular broadcast signals.

multiple x-y recorder—A recorder that plots a number of independent charts simultaneously, each showing the relation of two variables, neither of which is time.

multiplication point—A mixing point the output of which is obtained by multiplication of its inputs.

multiplier—1. A device in which the output represents the product of the magnitudes represented by the two or more input signals. 2. A series resistor placed in a voltmeter to increase the voltage range.

multiplier phototube—Phototube with one or more dynodes between its photocathode and the output electrode. The electron stream from the photocathode is reflected off each dynode in turn, with secondary emission adding electrons to the stream at each reflection. Also called photomultiplier.

multiplier-quotient register—In a computer, a register used for operations that involve multiplication and division.

multiplier resistor—*See* Multiplier, 2.

multiplier traveling-wave photodiode — A device in which increased sensitivity is obtained by combining the construction of a traveling-wave tube with that of a multiplier phototube.

multiplier tube—A vacuum tube in which sequential secondary emission from a number of electrodes is used to obtain increased output current. The electron stream is reflected from the first electrode of the multiplier to the second, and so on.

multiplying factor—The number by which the reading of a meter must be multiplied to obtain the true value.

multipoint circuit—A circuit interconnecting several locations, wherein information transmitted is available at all locations simultaneously.

multipolar—Having more than one pair of magnetic poles.

multiposition relay — A relay having more than one operate or nonoperate position, e.g., a stepping relay.

multiprocessing—1. The simultaneous or interleaved execution of two or more programs or instruction sequences by a computer or computer network. It may be accomplished by multiprogramming, parallel processing, or both. 2. The operation of more than one processing unit within a single system. Separate processors may take over communications or peripheral control, for example, while the main processor continues program execution. 3. A complex technique which permits more than one computer program to be executed simultaneously on more than one processor in a shorter period of time than is required for the overall sequentially organized task for which time may not be available.

multiprocessor—A computer capable of executing one or more programs with two or more processing units under integrated control of programs or devices.

multiprogramming — A method in which many programs are interleaved or overlapped so that they can be operated on within the same interval of time. This technique is the basis for time-shared operation.

multirate meter—A meter which registers at different rates or on different dials at different hours of the day.

multi-rf-channel transmitter—A radio transmitter having two or more complete radio-frequency portions that may be operated on different frequencies either individually or simultaneously.

multisegment magnetron — A magnetron with an anode divided into more than two segments, usually by parallel slots.

multispeed motor—A motor which can be operated at any one of two or more definite speeds, each practically independent of the load.

multistage tube — An X-ray tube in which the cathode rays are accelerated by multiple ring-shaped anodes, each at a progressively higher potential.

multistate noise—Noise consisting of erratic switching that is generated within a device at various sharply defined levels of applied current.

multitrack magnetic system — A magnetic recording system in which its medium has two or more tracks.

multitrack recording system — A recording system in which the medium has two or more recording paths, which may carry the same or different material but are played at the same time.

multiturn potentiometer — A potentiometer that must be rotated more than one turn for the slider to travel the complete length of the resistive element.

multiunit tube—An electron tube containing within one glass or metal envelope, two or more groups of electrodes, each associated with separate electron streams.

multivibrator — A relaxation oscillator in which the in-phase feedback voltage is

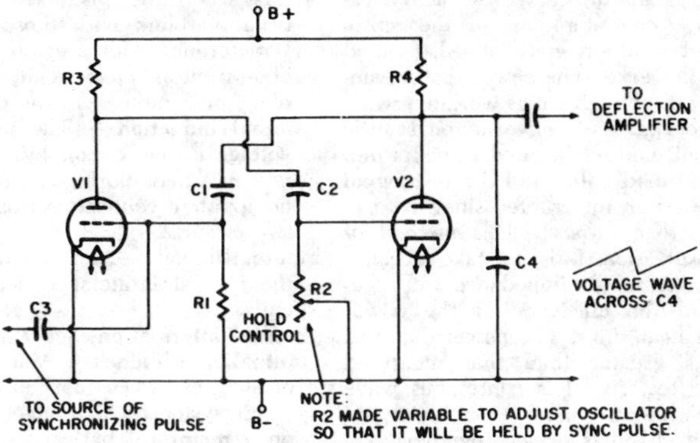

Multivibrator (asymmetrical).

obtained from two electron tubes or transistors. Typically, their outputs are coupled through resistive-capacitive elements. The time constants of the coupling elements determine the fundamental frequency, which may be further controlled by an external voltage. When such a circuit is normally in a nonoscillating state and a trigger signal is required to start a cycle of operation, it is called a one-shot, flip-flop, or start-stop multivibrator.

multivoltage control—A method of controlling the voltage of an armature by successive impression of a number of substantially fixed voltages on the armature. The voltages are usually obtained from multicommutator generators common to a group of motors.

multiwire — An automated interconnection system in which 33 AWG polymide-insulated magnet wire is first laid down on an adhesive-coated epoxy glass board. Then terminations are formed by drilling through the wire and board and electroplating the sides of the hole. The resulting tubelets are then used for components insertion and soldering. Components and wiring are on the same side of the board.

Mumetal—A metallic alloy with high permeability and a low hysteresis loss. It is excellent for magnetic shielding.

Mumetal shield—A cone-shaped covering made of Mumetal. Placed over the flared portion of a picture tube, it acts as a shield to prevent outside magnetic fields from affecting the alignment of the electron beams in the tube.

Munsell color system—A system of surface-color specification based on perceptually uniform color scales for the three variables; Munsell hue, Munsell value, and Munsell chroma. For an observer of normal color vision, adapted to daylight, and viewing the specimen when illuminated by daylight and surrounded with a middle gray to white background, the Munsell hue, value and chroma of the color correlate well with the hue, lightness, and saturation of the perceived color.

Munsell system — A color-specification system used principally in photography and color printing. It is based on sample cards containing the hue scale in five principal and five intermediate hues, and the brilliance scale in ten steps ranging from black to white. These represent visual, not physical, intervals.

Munsell value—In the Munsell system of object-color specification, the dimension which indicates the apparent luminous transmittance or reflectance of the object on a scale having approximately equal perceptual steps under the usual conditions of observation.

Murray loop test—A method of localizing a fault in a cable. This is done by replacing two arms of a Wheatstone bridge with a loop formed by the cable under test and a good cable connected to the far end of the defective cable.

MUSA antenna—Acronym for Multiple-Unit Steerable Antenna. It consists of several stationary antennas, the composite major lobe of which is electrically steerable.

mush winding—A type of winding in an ac machine. The conductors are placed one by one in prepared slots and the end connections are separately insulated.

musical cushion—A musical selection added at the end of a program which is running short or over. By playing it slowly or cutting out portions, the program director can make the program come out on time.

musical echo—*See* Flutter Echo, 2.

musical quality—*See* Timbre.

music power — 1. The short-term power available from an amplifier for the reproduction of program material. The music power output exceeds the rms power rating to a greater or lesser extent. Its measurement is standardized by the Institute of High Fidelity (IHF) and represents a practical means of stating the actual capabilities of an amplifier for the reproduction of program material. Also called dynamic power. 2. The power that an amplifier (with both channels working, when stereo) is capable of delivering on a music type signal. Output is rated in watts into a specified load value and the test signal is sometimes an interrupted sinewave signal (i.e., tone bursts). This method of power expression fails to take account of the power supply impedance and regulation and the efficiency of the power transistor heat sinks. The power obtained is always greater than that measured when the test signal is continuous wave (i.e., steady-state sine-wave signal) and with both channels driven together.

must-operate value – A specified functioning value at which all relays meeting the specification must operate.

must-release value–The specified operating value at which all relays that meet a certain specification must release.

muting – Muffling or deadening a sound. *Also see* Quieting.

muting circuit–1. A circuit which cuts off the receiver output when the rf carrier reaching the first detector is at or below a predetermined intensity. 2. A circuit for making a receiver insensitive while its associated transmitter is on.

muting switch–A switch used with automatic tuning systems to silence the receiver while it is being tuned.

mutual capacitance–Capacitance between two conductors with all other conductors connected together and to a grounded shield. (Does not apply to a shielded single conductor.)

mutual conductance – *See* Transconductance.

mutual-conductance meter – *See* Transconductance Meter.

mutual-conductance tube tester–*See* Transconductance Tube Tester.

mutual impedance–Between any two pairs of network terminals (all other terminals being open), the ratio of the open-circuit potential at either one of the two pairs, to the current applied to the other pair of network terminals.

mutual inductance – The property that exists between two current-carrying conductors when the magnetic lines of force from one link with those of the other. It determines, for a given rate of change of current in one circuit, the electromotive force induced in the other.

mutual induction – The production of a voltage in one circuit by a changing current in a neighboring circuit, even though no apparent connection exists between the two circuits.

mutual inductor–An inductor for changing the mutual inductance between two circuits.

mutual information–*See* Transinformation.

mutual interference – Man-made interference from two or more electrical or electronic systems which affects these systems on a reciprocal basis.

mV–Abbreviation for millivolt.

MV–Abbreviation for megavolt.

MVA–Abbreviation for megavolt-ampere.

MW–Abbreviation for megawatt.

MWh–Abbreviation for megawatt-hour.

Mycalex–Trade name of the Mycalex Corp. for mica bonded with glass. It has a low power factor at high temperatures, and is a good insulator at all frequencies.

Mylar–Trade name of E. I. duPont de Nemours and Co., Inc., for a highly durable, transparent plastic film of outstanding strength. It is used as a base for magnetic tape and as a dielectric in capacitors.

Mylar capacitor–A capacitor in which Mylar film, either alone or in combination with paper, is the dielectric.

myocardial electrodes – *See* Implantable Pacemaker.

myoelectric signals–Complex pulse potentials of 10 to 1000 microvolts with durations between 1 and 10 milliseconds. As generated by a muscular effort, recorded from the body surface.

myograph–A recorder of forces of muscular contraction. (*See also* Electromyograph.)

myokinesimeter–An apparatus for measuring the response of a muscle to stimulation by electric current.

myophone–An instrument for making audible the sound of a muscle when it is contracting.

myriametric waves – Very-low frequency band; 3 kHz to 30 kHz (100 km to 10 km).

N

N – Symbol for number of turns, or the northseeking pole of a magnet.

n–Abbreviation for the prefix nano- (10^{-9}).

nA–Abbreviation for nanoampere.

NAB – Abbreviation for National Association of Broadcasters.

NAB curve – 1. The standard playback equalization curve set by the National Association of Broadcasters. 2. In tape recording, this is used to describe anything for which standards have been established by the National Association of

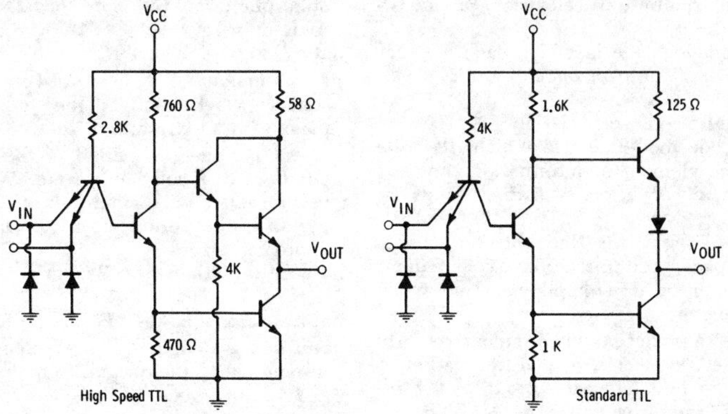

NAND gates.

Broadcasters. to ensure complete interchangeability from one recorder to another of the same speed.

nail-head bond—An alternate term for ball bond.

NAND—A function of A and B that is true if either A or B is false.

NAND circuit—A circuit in which the normal output of an AND circuit in inverted; a NOT-AND circuit.

NAND gate – 1. A combination of a NOT function and an AND function in a binary circuit that has two or more inputs and one output. The output is logic 0 only if all inputs are logic 1; it is logic 1 if any input is logic 0. With logic of the opposite polarity, this type of gate becomes a NOR gate. 2. A multiple-control device that evidences the simultaneous appearance of all positive-pressure control signals with a single zero-output signal. With the loss of one or more control signals, the zero NAND output signal ceases. The NAND gate can also be defined as a completed AND or NOT AND device. 3. An AND circuit that delivers an inverted output signal.

nano—Abbreviated n. Prefix meaning one-billionth (10^{-9}).

nanoampere – One-thousandth of a microampere (10^{-9} ampere). Abbreviated nA.

nanocircuit—An integrated microelectronic circuit in which each component is fabricated on a separate chip or substrate so that independent optimization for performance can be achieved.

nanofarad—One-billionth of a farad, equal to 10^{-9} farad, 0.001 microfarad, and 1000 picofarads. Abbreviated nF.

nanohenry – One-thousandth of a microhenry, equal to 10^{-9} henry and 1000 picohenrys. Abbreviated nH.

nanosecond – One-billionth of a second (10^{-9} second). Light travels approximately 1 foot in 1 nanosecond. Abbreviated ns.

nanovolt – One-thousandth of a microvolt (10^{-9} volt). Abbreviated nV.

nanovoltmeter – A voltmeter sufficiently sensitive to give readings in thousandths of microvolts.

Naperian logarithm—Also called hyperbolic or natural logarithm. A logarithm to the base 2.7128.

napier—*See* Neper.

narrow band – 1. A band whose width is greater than 1% of the center frequency and less than one-third octave. 2. Pertaining to a communication channel of less than voice grade.

narrow-band amplifier – An amplifier designed for optimum operation over a narrow band of frequencies.

narrow-band axis – In phasor representation of the chrominance signal, the direction of the coarse-chrominance primary of a color tv system.

narrow-band fm—Abbreviation for narrow-band frequency modulation.

narrow-band fm adapter – An attachment which converts an a-m communications receiver to fm.

narrow-band frequency modulation – Abbreviated nbfm. Frequency modulation which occupies only a small portion of the conventional fm bandwidth. Used mainly for two-way voice communication by police, fire, taxicabs, and amateurs.

narrow-band interference—Sharply tunable interference, having a spectrum that is small compared with the bandwidth of the measuring instrument. (Sine-wave carriers—both modulated and unmodulated—are good examples.)

narrow-sector recorder – A radio direction finder with which atmospherics are received from a limited sector related to the position of the antenna. The antenna is usually rotated continuously, and the bearing of the atmospherics recorded automatically.

NARTB—Abbreviation for National Association of Radio and Television Broadcasters.

n-ary code—A code in which each element can be any one of *n* distinct kinds or values.

n-ary pulse-code modulation – A type of pulse-code modulation in which the code for each element of information can consist of any one of *n* distinct kinds or values.

NASA—Acronym for National Aeronautics and Space Administration. The federal agency charged with all scientific space missions.

National Association of Broadcasters—Abbreviated NAB. The official association of the radio and television broadcasting industry. Formerly called the NARTB.

National Association of Radio and Television Broadcasters—Abbreviated NARTB. A name used for a number of years by an association of broadcasters. In 1958 the name was changed back to National Association of Broadcasters, an earlier title.

National Electrical Manufacturers Association—Abbreviated NEMA. An organization of manufacturers of electrical products.

National Electrical Code – Abbreviated NEC. A recognized authority on safe electrical wiring. The code is used as a standard by federal, state, and local governments in establishing their own laws, ordinances, and codes on wiring specifications.

National Television System Committee—Abbreviated NTSC. A committee organized in 1940 and comprising all United States companies and organizations interested in television. Between 1940 and 1941, it formulated the black and white television standards and between 1950 and 1953, the color television standards that were approved by the Federal Communications Commissicn.

natural antenna frequency – The lowest resonant frequency of an antenna operated without external inductance or capacitance.

natural frequency – 1. The frequency at which a system with a single degree of freedom will oscillate from the rest position when displaced by a transient force. Sometimes used synonymously with damped natural frequency. 2. The lowest resonant frequency of a circuit or component without adding inductance or capacitance.

natural frequency of an antenna—The lowest resonant frequency of an antenna with no added inductance or capacitance.

natural interference – Electromagnetic interference caused by natural terrestrial phenomena (atmospheric interference) or by natural disturbances outside of the atmosphere of the earth (galactic and solar noise).

natural logarithm—*See* Naperian Logarithm.

natural magnet—Magnetic ore (e.g., a lodestone) which exhibits the property of magnetism in its natural state.

natural period—The period of the free oscillation of a body or system. When the period varies with amplitude, the natural period is the period when the amplitude approaches zero.

natural radiation—*See* Background Radiation.

natural resonance—*See* Periodic Resonance.

natural wavelength—The wavelength corresponding to the natural frequency of an antenna or circuit.

Navaglobe – A long-distance navigational system of the continuous-wave, low-frequency type. Bearing information is provided by amplitude comparison.

navaids—navigational aids. The electronic facilities provided on or in the immediate vicinity of an airport so that a pilot using compatible airborne equipment can execute the instrument approach or approaches authorized for the airport.

navar—A coordinated series of radar air-navigation and traffic-control aids utilizing transmissions at wavelengths of 10 and 60 centimeters. In an aircraft, it provides distance and bearing from a given point, display of other aircraft in the vicinity, and commands from the ground. On the ground, it provides a display of all aircraft in the vicinity, as well as their altitudes and identities, plus means for transmitting certain commands.

navarho – A continuous-wave, low-frequency navigation system that provides simultaneous bearing and range information over long distances.

Navascreen—A system for the display and computation of air-traffic-control data based on information from radar and other sources.

navigation—The process of directing an airplane or ship toward its destination by determining its position, direction, etc.

navigational parameter—In a navigational aid, a visual or aural output having a specific relationship to navigational coordinates.

navigation beacon—A light, radio, or radar beacon that provides navigational aid to ships and aircraft.

nbfm – Abbreviation for narrow-band frequency modulation.

nc—1. Abbreviation for normally closed. 2. Abbreviation for no connection (in a vacuum-tube base).

NC—Abbreviation for numerical control.

n-channel – A device constructed on a p-type silicon substrate whose drain and source components are of n-type silicon.

n-channel FET—An FET the resistive bar of which is an n-type semiconductor.

n-channel MOS—An MOS (unipolar) transistor in which the working current consists of negative (n) electrical charges,

N-conductor concentric cable – A cable composed of an insulated central conducting core with (N – 1) tubular stranded conductors laid over it concentrically and separated by layers of insulation.

n-cube—Also called n-dimensional cube or n-variable cube. In switching theory, a term used to indicate two n – 1 cubes having corresponding points connected.

NC value—*See* Noise-Criteria Value.

n-dimensional cube—*See* n-cube.

N-display—Also called N-scan, or N-scope. A radar display similar to the K-display. The target appears as a pair of vertical deflections (blips) from the horizontal time base. Direction is indicated by the related amplitude of the vertical deflections. Target distance is determined by moving a pedestal signal (the control of which is calibrated in distance) along the base line until it coincides with the horizontal position of the vertical deflections.

near-end cross talk—In a disturbed telephone channel, cross talk propagated in the opposite direction from the current in the disturbing channel. The terminal where the near-end cross talk is present is ordinarily near, or coincides with, the energized terminal of the disturbing channel.

near field—1. The acoustic radiation field close to the speaker or some other acoustic source. 2. The electromagnetic field within a distance of 1 wavelength from a transmitting antenna.

near infrared – 1. Name applied to the spectral region primarily comprising wavelengths between 3 and 30 microns. 2. That nonvisible radiant energy with wavelengths nearest the red end of the visible spectrum.

NEC—Abbreviation for National Electrical Code.

necessary bandwidth—For a given class of emission, the minimum value of the occupied bandwidth sufficient to ensure the transmission of information at the rate and the quality required for the system employed, under specified conditions. Emissions useful for the good functioning of the receiving equipment—for example, the emission corresponding to the carrier of reduced carrier systems—shall be included in the necessary bandwidth.

neck—The narrow tubular part of the envelope of a cathode-ray tube; it extends from the funnel to the base and houses the electron gun.

needle – 1. A probe used on stacks of punched cards. 2. *See* Stylus, 1.

needle chatter—*See* Needle Talk.

needle drag—*See* Stylus Drag.

needle electrode – Small instrument used for subcutaneous electrical recording and stimulating.

needle gap—A spark gap having needlepoint electrodes.

needle pressure—*See* Stylus Force.

needle scratch—*See* Surface Noise.

needle talk—Audible sounds from a record player pickup in the vicinity of the stylus.

needle test points—Sharp steel probes connected to test cords for making contacts with conductors.

neg—Abbreviation for negative.

negate—To invert the value of a function or variable; that is, to change it to 1 if it is 0 or to 0 if it is 1. The symbol for negation is a superscript bar.

negated-input OR gate—A gate that has a high output when one or both inputs are low. A negated-input OR gate is identical to a NAND gate in function.

negation – The NOT operation in Boolean algebra.

negative – 1. Abbreviated neg. Less than zero. 2. The opposite of positive (e.g., negative resistance, transmission, feedback, etc.). 3. A terminal or electrode having an excess of electrons. Electrons flow out of the negative terminal of a voltage source, toward a positive source.

negative acceleration – A relative term often used to indicate that, referenced to zero acceleration, a negative electrical output will be obtained from a linear accelerometer when its sensitive axis is oriented normal to the surface of the earth, and a given minus reference point along the center of the seismic mass.

negative acknowledgement character—Abbreviated NAK. A communication control character transmitted from a receiving point as a negative response to a sender. It may also be used as an accuracy control character.

negative bias—In a vacuum tube, the voltage which makes the control grid more negative than the cathode.

negative booster – A booster used with a ground-return system to reduce the difference of potential between two points to the grounded return. It is connected in series with a supplementary insulated feeder extending from the negative bus of the generating station or substation to a distant point on the grounded return.

negative charge—A condition in a circuit when the element in question retains more than its normal quantity of electrons.

negative conductor—A conductor connected to the negative terminal of a source of supply. Such a conductor is frequently used as an auxiliary return circuit in a system of electric traction.

negative-effective-mass amplifiers and generators—A class of solid-state devices for broad-band amplification and generation of microwave energy. Operation of these devices is based on the phenomenon by which the effective masses of charge carriers in semiconductors become negative when their kinetic energies are sufficiently high.

negative electricity—The type of electricity possessed by a body that has an excess of electrons.

negative electrode—Thte electrode from which the forward current flows.

negative electron—Also called negatron. An electron, distinguished from a positive electron or positron.

negative feedback—Also called degeneration, inverse feedback, or stabilized feedback. A process by which a part of the output signal of an amplifying circuit is fed back to the input circuit. The signal fed back is 180° out-of-phase with the input signal; therefore, the amplification is decreased, and distortion reduced.

negative-feedback amplifier—An amplifier in which negative feedback is employed to improve the stability or frequency response, or both.

negative ghost—A ghost which has the opposite shading from that of the original image (i.e., is black when the image is white, and vice versa).

negative glow—The luminous glow between the cathode and Faraday dark spaces in a glow-discharge, cold-cathode tube (*See also* Glow Discharge.)

negative-grid generator — A conventional oscillator circuit in which oscillation is produced by feedback from the plate circuit to a grid which is normally negative with respect to the cathode and which is designed to operate without drawing grid current at any time.

negative ground — The negative battery terminal of a vehicle is connected to the body and frame.

negative image — A televised picture in which the whites appear black and vice versa. It is due to the picture signal having a polarity opposite to that of a normal signal.

negative impedance—Also called negative resistance when there is no inductance or capacitance in the circuit. A characteristic of certain electrical devices or circuits—instead of increasing, the voltage decreases when the current is increased, and vice versa.

negative ion—An atom with more electrons than the normal number. Thus, it has a negative charge.

negative-ion generator—A device that bombards air molecules with electrons so that molecules are ionized negatively.

negative light modulation—In television, the process whereby a decrease in initial light intensity causes an increase in the transmitted power.

negative logic—A form of logic in which the more positive voltage level represents logic 0 and the more negative level represents logic 1.

negative modulation—In an a-m television system, that modulation in which an increase in brightness corresponds to a decrease in transmitted power.

negative modulation factor—The maximum negative departure of the envelope of an a-m wave from its average value, expressed as a ratio. This rating is used whenever the modulation signal wave has unequal positive and negative peaks.

negative phase-sequence relay — A relay which functions in conformance with the negative phase-sequence component of the current, voltage, or power of the circuit.

negative picture phase—For a television signal, the condition in which an increase in brilliance makes the picture-signal voltage swing in a negative direction from the zero level.

negative plate—The grid and active material connected to the negative terminal of a storage battery. When the battery is discharging, electrons flow from this terminal, through the external circuit, to the positive terminal.

negative resistance—A resistance which exhibits the opposite characteristic from normal—i.e., when the voltage is increased across it, the current will decrease instead of increase.

negative-resistance magnetron—A magnetron operated so that it acts as a negative resistance.

negative-resistance oscillator—An oscillator produced by connecting a parallel-tuned resonant circuit to a two-terminal negative-resistance device (one in which an increase in voltage results in a decrease in current). Dynatron and transitron oscillators are examples.

negative-resistance region—That operating region in which an increase in applied voltage results in a current decrease.

negative-resistance repeater—A repeater in which gain is provided by a series or shunt negative resistance, or both.

negative temperature coefficient — The amount of reduction in the value of a quantity, such as capacitance or resistance, for each degree of increase in temperature. (*See also* Temperature Coefficient.)

negative terminal—In a battery of other voltage source, the terminal having an excess of electrons. Electrons flow from it, through the external circuit, to the positive terminal.

negative torque—A torque developed in opposition to the normal torque of a motor. This may occur at starting—common to pole motors—or at some speed below the nameplate rpm. This causes "cusps" or "saddles" in the graphed torque curves.

negative-transconductance oscillator – An electron-tube oscillator the output of which is coupled back to the input without phase shift, the phase condition for oscillation being satisfied by the negative transconductance of the tube. A transitron oscillator is an example.

negative transmission—The modulation of the picture carrier by a picture signal with a polarity such that the sync pulses occur in the blacker-than-black region and a decrease in initial light intensity increases the transmitted power.

negatron—1. An electron. *See also* negative electron. 2. A four-electrode vacuum tube having a negative-resistance characteristic. 3. *See* Dynatron.

NEMA—Abbreviation for National Electrical Manufacturers Association.

NEMA standards—Specifications adopted as standard by the National Electrical Manufacturers Association.

nematic liquid—An organic chemical used in liquid crystal displays (LCDs) that simultaneously exhibits liquid and crystal properties. It is transparent until the threadlike molecules are disturbed by an electric field; then it takes on a milky-white appearance.

nematic phase—An arrangement of parallel-oriented liquid crystal molecules. The nematic phase is turbid but much less viscous than the smectic phase, and the molecules are not arranged in layers.

nemo—*See* Field Pickup.

neon—An inert gas used in neon signs and in some electron tubes. It produces a bright red glow when ionized. Symbol, *Ne;* atomic number 10; atomic weight, 20.183.

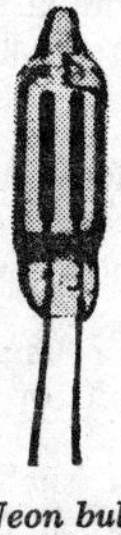

Neon bulb.

neon bulb—A glass envelope filled with neon gas and containing two or more insulated electrodes. The tube will not conduct until the potential difference between two electrodes reaches the firing, or ionization, potential, and will remain conductive until the voltage is reduced to the extinction level.

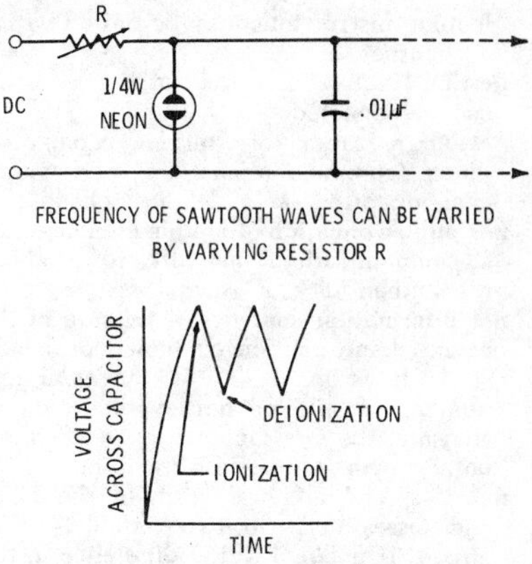

Neon-bulb oscillator.

neon-bulb oscillator—A simple relaxation oscillator in which a capacitor charges to the ionization potential of the neon gas within a bulb. Ionization rapidly depletes the charge on the capacitor, and a new charge cycle begins.

neon indicator tube—A cold-cathode tube containing neon and designed to visually indicate a potential difference or field.

neon oscillator—*See* Neon-Bulb Oscillator.

neper—Also called napier. The fundamental division of a logarithmic scale for expressing the ratio between two currents, powers, or voltages. The number of nepers denoting such a ratio is the natural (Naperian) logarithm of the square root of this ratio. One neper equals 0.8686 bels, or 8.686 decibels. Expressed as a formula:

$$N = \log_{\epsilon} \sqrt{\frac{E_1}{E_2}}$$

where,

N is the number of nepers denoting their ratio,

E_1 and E_2 are the two voltages.

Nernst effect—The effect whereby a potential difference is developed across a heated metal strip placed perpendicular to a magnetic field.

Nernst-Ettinghausen effect—A thermomagnetic effect that occurs in certain pure crystals, where a temperature difference is produced in a direction perpendicular to an applied magnetic field and a longitudinal electric current.

Nernst lamp—An electric lamp consisting of a short, slender rod of zirconium oxide that is heated to incandescence by a current.

nesting—A term used to indicate that a subroutine in a computer is enclosed inside a larger routine, but is not necessarily part of the outer routine. A series of looping instructions may be nested within each other.

nesting level—The number of times nesting can be repeated.

net—Organization of stations capable of direct communications on a common channel, often on a definite schedule.

net authentification—Identification used on a communications network to establish the authenticity of several stations.

net information content—A measure of the essential information contained in a message. It is expressed as the minimum number of bits or hartleys required to transmit the message with specified accuracy over a noiseless medium.

net loss—The algebraic sum of the gains and losses between two terminals of a circuit. It is equal to the difference in the levels at these points.

net reactance—The difference between the capacitive and inductive reactance in an ac circuit.

network—1. A combination of electrical elements. 2. An interconnected system of transmission lines that provides multiple connections between loads and sources of generation. 3. An organization of stations with a capability for intercommunication, although not necessarily on the same channel. 4. Two or more interrelated circuits.

network analog — The expression and solution of a mathematical relationship between variables through the use of a circuit or circuits to represent those variables.

network analysis—The obtaining of the electrical properties of a network (e.g., its input and transfer impedances, responses, etc.) from its configuration, parameters, and driving forces.

network analyzer—1. A group of electric-circuit elements which can readily be connected to form models of electric networks. From corresponding measurements on the model, it is then possible to infer the electrical quantities at various points on the prototype system. 2. Also called network calculator. An analog device designed primarily for simulating electrical networks.

network calculator—An analog device designed primarily for simulating electric networks.

network constant—Any one of the resistance, inductance, mutual-inductance, or capacitance values in a circuit or network. When these values are constant, the network is said to be linear.

network filter—A transducer for separating waves on the basis of their frequency.

network master relay—A relay that closes and trips an alternating-current, low-voltage network protector.

network phasing relay — A relay which functions in conjunction with a master relay to limit closure of the network protector to a predetermined relationship between the voltage and the network voltage.

network relay—A form of relay (e.g., voltage, power, etc.) used in the protection and control of alternating-current, low-voltage networks.

network synthesis—The obtaining of a network from prescribed electrical properties such as input and transfer impedances, specified responses for a given driving force, etc.

network transfer function — A frequency-dependent function, the value of which is the ratio of the output to the input voltage.

network transformer — A transformer suitable for use in a vault to feed a variable-capacity system of interconnected secondaries.

Neuman's law—Mutual inductance is a constant for a given relative physical position of coils, and independent of the fact that the current flows in one or the other coil, and of frequency, current, and phase.

neuristor—1. A two-terminal active device with some of the properties of neurons (e.g., propagation that suffers no attenuation and has a uniform velocity, and a refractory period). 2. A device that is essentially an active transmission line designed to propagate signals without attenuation. It is an electrical analog of a nerve fiber.

neuroelectricity—The minute electric voltage generated by the nervous system.

neuron—The basic building block of the human nervous system. A specialized body cell that can conduct and code an electrical pulse.

neutral—In a normal condition; hence, neither positive nor negative. A neutral object has its normal number of electrons —i.e., the same number of electrons as protons.

neutral circuit—A teletypewriter circuit in which current flows in only one direction. The circuit is closed during the marking condition and open during the spacing condition.

neutral conductor—That conductor of a polyphase circuit or of a single-phase three-wire circuit which is intended to have a potential such that the potential differences between it and each of the other conductors are approximately equal in magnitude and are also equally spaced in phase.

neutral ground—A ground connection to the neutral point or points of a circuit, transformer, rotating machine, or system.

neutralization—The nullifying of voltage feedback from the output to the input of an amplifier through the interelectrode impedance of the tube. Its principal use is in preventing oscillation in an amplifier. This is done by introducing, into the input, a voltage equal in magnitude but opposite in phase to the feedback through the interelectrode capacitance.

neutralize—In an amplifier stage, to balance out the feedback voltage due to grid-plate capacitance, thus preventing regeneration.

neutralized radio-frequency stage—A stage in which a circuit is added for the purpose of feeding back, in the opposite phase, an amount of energy sufficient to cancel the energy that would otherwise cause oscillation.

neutralizing capacitor—A capacitor, usually variable, employed in a radio receiving or transmitting circuit to feed a portion of the signal voltage from the plate circuit back to the **grid circuit**.

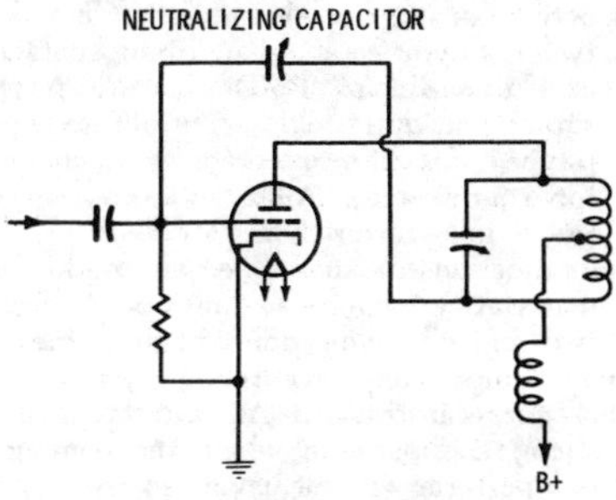

Neutralizing capacitor.

neutralizing circuit—In an amplifier circuit, the portion that provides an intentional feedback path from plate to grid. This is done to prevent regeneration.

neutralizing indicator—An auxiliary device (e.g., a lamp or detector coupled to the plate tank circuit of an amplifier) for indicating the degree of neutralization in an amplifier.

neutralizing tool — Also called a tuning wand. A small screwdriver or socket wrench, partly or entirely nonmetallic, for making neutralizing or aligning adjustments in electronic equipment. (*See also* Alignment Tool.)

neutralizing voltage—The ac voltage fed from the grid circuit to the plate circuit (or vice versa). It is deliberately made 180° out of phase with and equal in amplitude to the ac voltage similarly transferred through undesired paths (usually the grid-to-plate interelectrode capacitance).

neutral operation — The system whereby marking signals are formed by current impulses of one polarity, either positive or negative, and spacing signals are formed by reducing the current to zero or nearly zero.

neutral point—The point which has the same potential as the point of junction of a group of equal nonreactive resistances connected at their free ends to the appropriate main terminals or lines of the system.

neutral relay—Also called a nonpolarized relay. A relay in which the armature movement does not depend on the direction of the current in the controlling circuit.

neutral transmission—Transmission of teletypewriter signals in such a way that a mark is represented by current on the line, and a space is represented by the absence of current. By extension to tone signaling, a method of signaling involving two states, one of which represents both a space and the absence of signaling. Also called unipolar transmission.

neutral zone—A range in the total control zone in which a controller does not respond to changes in the controlled process; a dead zone, usually in the middle of the control range.

neutrino—An atomic particle with essentially no mass and no charge postulated to explain the conversation of energy and momentum when a radioactive atom emits an electron.

neutrodyne—An amplifier circuit used in early tuned radio-frequency receivers. It was neutralized by the voltage fed back by a capacitor.

neutron—One of the three elementary particles (the electron and proton are the other two) of an atom. It has approximately the same mass as the hydrogen atom, but no electric charge. It is one of the constituents of the nucleus (the portion is the other one).

neutron flux—A term used to express the intensity of neutron radiation, usually in connection with the operation of a reactor.

newton—In the mksa system, the unit of force that will impart an acceleration of 1 meter per second per second to a mass of 1 kilogram.

nF—Symbol for nanofarad.

n-gate thyristor—A thyristor in which the gate terminal is connected to the n-region nearest the anode and which is normally switched to the on state by applying a negative signal between the gate and anode terminals.

nibble—A sequence of 4 bits operated upon as a unit. *Also see* Byte.

Nichrome—Trade name of Driver-Harris Co. for an alloy of nickel and chromium used extensively in wirewound resistors and heating elements.

nickel-cadmium cell — The most widely used rechargeable sealed cell, nickel-cad-

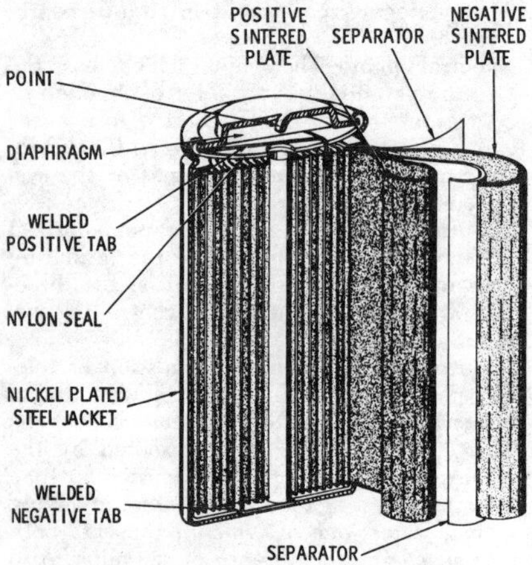

Nickel-cadmium cell.

mium gives a flat voltage discharge characteristic, nominal cell voltage of 1.25 volts, and good low-temperature operation. A cell with a nickel and oxide positive electrode and a cadmium negative electrode. The plates are wrapped with a separator between them, and are immersed in a potassium-hydroxide electrolyte.

nickel-oxide film diode—A solid-state diode made of nickel-oxide film. It may be switched from off to on by applying a 300-volt low-current pulse for 10 μsec, and may be switched from on to off by applying a 30-volt high-current pulse for 10 μsec. In effect it works like a dipole flip-flop.

nickel silver—*See* German Silver.

nif — Abbreviation for noise-improvement factor.

ni junction—A semiconductor junction between n-type and intrinsic materials.

nines complement—1. A decimal digit that yields 9 when added to another decimal digit. 2. Pertaining to an arithmetic method of negating a decimal number so that subtraction can be performed by using addition techniques. The nines-complement negation of a number results from subtracting each decimal digit individually from 9. End-around carry must be used in performing a nines-complement addition.

Nipkow disc—A round plate used for scanning small elementary areas of an image in correct sequence for a mechanical television system. It has one or more spirals of holes around its outer edge, and successive openings are positioned so that rotation of the disc provides the scanning.

NIPO—Acronym for Negative Input, Positive Output.

nit—1. In a computer, a choice among events that are equally probable. One nit equals 1.44 bits. 2. Abbreviated nt. The unit of luminance (photometric brightness) equal to one candela per square meter.

Nixie tube—A glow tube which converts a combination of electrical impulses into a visual display. (*See also* Numerical-Readout Tube.)

Nixie-tube display — A form of display composed of a number of neon-filled tubes, each containing wires in the shape of digits zero to nine, with separate wires for each digit. On signal, the correct wires are energized, causing the neon gas around the wires to glow.

n-level logic—Pertaining to a collection of gates connected in such a way that not more than n gates appear in series. (Computer term).

NMOS—Pertaining to MOS devices made on p-type silicon substrates in which the active carriers are electrons that flow between n-type source and drain contacts. The opposite of PMOS, where n-type source and drain regions are diffused into p-type substrate to create an n-channel for conduction. NMOS is from two to three times faster than PMOS.

n-n junction—In an n-type semiconducting material, a region of transition between two regions having different properties.

no—Abbreviation for normally open.

no-address instruction—An instruction specifying an operation which the computer can perform without having to refer to its storage or memory unit.

noble—Chemically inert. A term often used to describe metals such as gold, platinum, etc.

noble gas—Also called inert gas or rare gas. One of the chemically inert gases, including helium, neon, argon, krypton, and xenon.

noctovision—A television system employing invisible rays (usually infrared) for scanning purposes at the transmitter. Hence, no visible light is necessary.

nodal points—1. Two points on the axis of a lens such that a ray entering the lens

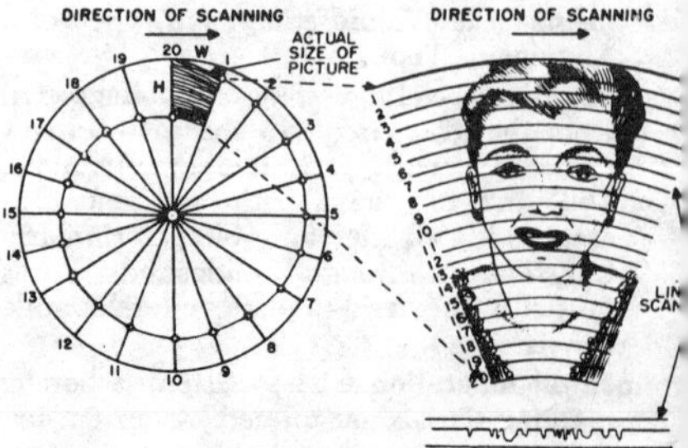

Nipkow's television system.

in the direction of one leaves as if from the other and parallel to the original direction. 2. *See* Node, 2.

nodal-point keying—Keying an arc transmitter at a point that is essentially at ground potential in the antenna circuit.

node—1. Also called junction point, branch point, or vertex. A terminal of any branch of a network, or a terminal common to two or more branches. 2. Also called modal point. The point, line, or surface in a standing-wave system where some characteristic of the wave field has essentially zero amplitude. The appropriate modifier should be used with the word "node" to signify the type that is intended (e.g., pressure node). 3. Provide data entry/exit points in computers and terminals, and switch or process data. Smart terminals and smart programmable controllers —the ones based on microprocessors—also have true processing capability and qualify as nodes. Each node is potentially capable of performing application-oriented tasks.

nodules—Clusters of oxide particles which protrude above the surface of magnetic tape.

noise—1. Any unwanted disturbance within a dynamic electrical or mechanical system (e.g., undesired electromagnetic radiation in a transmission channel or device). 2. Any unwanted electrical disturbance or spurious signal which modifies the transmitting, indicating, or recording of desired data. 3. In a computer, extra bits or words which have no meaning and must be ignored or removed from the data at the time it is used. 4. Random electrical variations generated internally in electronic components. 5. (acoustics) A potentially distracting or distorting sensory pattern. 6. Any sounds that are not harmonically related to the signal and are added subsequently by recording or playback equipment. (Includes hiss, crackles, hum, rumblings, ticks, pops, etc.)

noise analysis—Determination of the frequency components that make up the noise being studied.

noise analyzer—An instrument used for determining the amplitude versus frequency characteristics of noise.

noise bandwidth—The width of the equivalent rectangular power gain versus frequency response of a network. It is the width of the rectangle the area of which equals that of the actual power gain versus frequency response and the height of which equals the maximum available power gain at a reference frequency f_o. It does not include any spurious responses. Any responses other than those in the useful channel are accounted for by a degraded noise figure.

noise behind the signal – S
Noise.

noise blanker—A circuit used in
receivers just before the detector
mize ignition noise.

noise-canceling microphone – *See* C
Talking Microphone.

noise clipper—A circuit that automatically or manually clips the noise from the output of a receiver.

noise-criteria value – A measure of background noise; it is a single overall value determined from the greatest level of sound pressure of several individual frequency bands. Sometimes values are stated for each band. Also called NC value.

noise-current generator—A current generator in which the output is a random function of time.

noise diode – A standard electrical-noise source consisting of a diode operated at saturation. The noise is due to random emission of electrons.

noise equivalent bandwidth – The useful bandwidth of a thermistor bolometer for various frequencies in the input radiation is ¼t (where t is the time constant of the bolometer).

noise equivalent power—The value of radiation which produces, in a detector, an rms signal-to-noise ratio of unity. It is usually measured with a black-body radiation source at 500 K and a bandwidth of 1 or 5 hertz. The modulation rate varies with the type of detector and is usually between 10 and 1000 hertz.

noise factor—Also called noise figure. 1. For a given bandwidth, the ratio of total noise at the output, to the noise at the input. 2. A number expressing the amount by which a receiver falls short of equaling the theoretical optimum performance.

noise figure—1. The ratio of the total noise power at the output of an amplifier to that portion of the total output noise power attributable to the thermal agitation in the resistance of the signal source. Abbreviated nf. *See also* Noise Factor. 2. In a network with a generator connected to its input terminal, the ratio of the available signal-to-noise-power ratio at the signal-generator terminals (weighted by the network bandwidth) to the available signal-to-noise-power ratio at the output. 3. The common logarithm of the ratio of the input signal-to-noise ratio to the output signal-to-noise ratio.

noise filter – A combination of electrical components which prevent extraneous signals from passing into or through an electronic circuit.

noise grade – A number that defines the noise at a particular location relative to the noise at other locations throughout the world.

of the insen-
spurious or
or noise. 2.

a receiver,
-noise ratio
ratio. (The
broad sense
lse demodula-

ming.

noise killer – An electric network inserted into a telegraph circuit (usually at the sending end) to reduce interference with other communication circuits.

noise level—1. The strength of extraneous audible sounds at a given location. 2. The strength of extraneous signals in a circuit. (Noise level is referred to a specified base and usually measured in decibels.)

noise limiter – A circuit that cuts off all noise peaks stronger than the highest peak in the received signal. In this way, the effects of strong atmospheric or man-made interference are reduced.

noise margin – 1. The extraneous-signal voltage amplitude that can be added algebraically to the noise-free worst-case input level of a logic circuit before deviation of the output voltage from the allowable logic levels occurs. In this application, the term "input" generally refers to logic input terminals, ground reference terminals, or power supply terminals. 2. The difference between the operating voltage and the threshold voltage of a binary logic circuit.

noise-measuring set—*See* Circuit-Noise Meter.

noise pulse – Also called noise spike. A spurious signal of short duration and of a magnitude considerably in excess of the average peak value of the ordinary system noise.

noise quieting – The ability, usually expressed in decibels, of a receiver to reduce background noise in the presence of a desired signal.

noise ratio—The ratio of the available noise power at the output of a circuit divided by the noise power at the input.

noise-reducing antenna system – A receiving-antenna system so designed that only the antenna proper can pick up signals. It is placed high enough to be out of the noise-interference zone, and is connected to the receiver with a shielded cable or twisted transmission line that is incapable of picking up signals.

noise source—A device employed to generate random noise (e.g., photomultiplier and gaseous-discharge tubes).

noise spike—*See* Noise Pulse.

noise suppression—1. The ability of a radio receiver to materially reduce the noise output when no carrier is being received. 2. A means of reducing surface noise during reproduction of a phonograph record.

noise suppressor – 1. A circuit which reduces high-frequency hiss or noise. It is utilized primarily with old phonograph records. 2. In a receiver circuit, the portion which reduces noise automatically when no carrier is being received. 3. Any device intended to reduce the audibility of background noise.

noise temperature—1. At a pair of terminals and a specific frequency, the temperature of a passive system having an available noise power per unit bandwidth equal to that of the actual terminals. 2. The temperature at which the thermal noise available from a device equals the total fluctuation noise actually available from the device.

noise-voltage generator – A voltage generator the output of which is a random function of time.

noisy mode—In a computer, a floating-point arithmetic procedure associated with normalization. In this procedure digits other than zero are introduced in the low-order positions during the left shift.

no-load current—*See* Exciting Current, 1.

no-load losses—The losses in a transformer when it is excited at the rated voltage and frequency, but is not supplying a load.

nominal band—In a facsimile-signal wave, the frequency band equal to the width between the zero and the maximum modulating frequency.

nominal bandwidth—1. The difference between the nominal upper and lower cutoff frequencies of a filter. It may be expressed in octaves, in hertz, or as a percentage of the passband center frequency. 2. The maximum band of frequencies, inclusive of guard bands assigned to a channel.

nominal horsepower—The rated power of a motor, engine, etc.

nominal impedance—The impedance of a circuit under normal conditions. Usually it is specified at the center of the operating-frequency range.

nominal line pitch—The average separation between the centers of adjacent lines in a raster.

nominal line width – 1. In television, the reciprocal of the number of lines per unit length in the direction of line progression. 2. In facsimile transmission, the average separation between centers of adjacent scanning or recording lines.

nominal power rating of a resistor – The power which a resistor can dissipate continuously at a specified ambient temperature and for a stipulated length of time without excessive resistance drift.

nominal value – The stated or specified value as opposed to the actual value.

nominal voltage – The voltage of a fully charged storage cell when delivering rated current.

nomograph–*See* Alignment Chart.

nomography–A method of representing the relation between any number of variables graphically on a plane surface such as a piece of paper.

nonblinking–The ability of a digital meter display to appear steady and free of blinking. Usually achieved by storing the reading during the same period.

nonbridging–A contact transfer in which the movable contact leaves one contact before touching the next.

nonbridging contacts – A contact arrangement in which the opening of one set of contacts occurs before the closing of another set.

noncoherent radiation–Radiation in which the waves are out of phase with respect to space and/or time.

noncombustible–*See* Nonflammable.

nonconductor–An insulating material–i.e., one through which no electric current can flow.

noncorrosive flux–A flux that does not contain acid and other substances which might corrode the surfaces being soldered.

nondegenerate amplifier–A parametric amplifier in which the pumping frequency is considerably higher than twice the signal frequency. The output is at the input signal frequency. This type of amplifier has negative impedance characteristics, and therefore is capable of oscillation.

nondestructive readout–A type of memory operation or device in which reading out does not cause stored data to be lost from the memory. Such memories, therefore, do not require a write operation immediately after each read operation, as do destructive types.

nondeviated absorption–1. Absorption that occurs without any appreciable slowing up of waves. 2. Normal sky-wave absorption.

nondirectional–*See* Omnidirectional.

nondirectional antenna – *See* Omnidirectional Antenna.

nondirectional microphone – *See* Omnidirectional Microphone.

nondissipative stub – A lossless length of waveguide or transmission line coupled into the sides of a waveguide.

nonequivalence element – A logic element having an action that represents the Boolean connective exclusive OR.

nonerasable storage–In a computer, a storage medium which cannot be erased and reused, e.g., punched cards or perforated paper tape.

nonferrous – Not made of or containing iron.

nonflammable–Also called noncombustible. Term applied to material which will not burn when exposed to flame or elevated temperatures, e.g., asbestos, ceramics, and structural metals.

nonhoming tuning system–A motor-driven automatic-tuning system in which the motor starts in the direction of the previous rotation. If this direction is incorrect for the new station, the motor reverses, after turning to the end of the dial, then proceeds to the desired station.

noninductive–Having practically no inductance.

noninductive capacitor – A capacitor in which the inductive effects at high frequencies are reduced to a minimum. Foil layers are offset during winding so that an entire layer of foil (projecting at either end) is connected together for contact-making purposes. A current then flows laterally rather than spirally around the capacitor, and the inductive effect is minimized.

noninductive circuit–A circuit in which the inductance is reduced to a minimum or negligible value.

noninductive load–A load that has no inductance. It may consist entirely of resistance or capacitance.

noninductive resistor–A wirewound resistor with little or no self-inductance.

noninductive winding–A winding in which the magnetic fields produced by its two parts cancel each other and thereby provide a noninductive resistance.

noninverting connection – The closed-loop connection of an operational amplifier when the forward gain is positive for a dc signal (0° phase shift).

noninverting input–A differential-amplifier input terminal for which the output signal has the same phase as the input signal.

noninverting parametric device – A parametric device the operation of which depends essentially upon three frequencies, a harmonic of the pump frequency and two signal frequencies, of which one is the sum of the other plus the pump harmonics. Such a device can never provide gain at either of the signal frequencies. It is said to be noninverting because, if either of the two signals is moved upward in frequency, the other will move upward in frequency.

nonionizing radiation – Radiation which does not produce ionization (e.g., infrared, ultraviolet, and visible light).

nonlinear–Having an output that does not rise or fall in direct proportion to the input.

nonlinear capacitor–A capacitor that has a nonlinear mean-charge characteristic or peak-charge characteristic or a reversible capacitance that varies with bias voltage.

nonlinear coil–A coil with an easily saturable core. Its impedance is high at low or zero current, and is low when enough current flows to saturate the core.

nonlinear distortion—1. Distortion that occurs when the output does not rise and fall directly in proportion to the input. The input and output values need not be of the same quantity—e.g., in a linear detector, these values are the signal voltage at the output and the modulation envelope at the input. 2. Distortion which occurs because the transmission properties of a system are dependent upon the instantaneous magnitude of the transmitted signal. NOTE: Nonlinearity distortion gives rise to amplitude and harmonic distortion, intermodulation, and flutter. 3. The generation of unwanted signals as a result of nonlinear operations on a desired signal or group of signals. (Harmonic distortion is one well-known example; if the main signal is a sine wave of frequency f, the unwanted signals appear as sine waves at 2f, 3f, and so on.)

nonlinear feedback control system — Feedback control system in which the relationships between the pertinent measures of the system input and output signals cannot be adequately described by linear means.

nonlinear network — A network (circuit) not specifiable by linear differential equations with time as the independent variable.

nonlinear optical effects—Optical phenomena that can be observed only with the use of directional, nearly monochromatic light beams, such as those produced by a laser. Examples are optical mixing, harmonic generation, and the stimulated raman effect.

nonloaded *Q*—Also called basic *Q*. The value of the *Q* of an electric impedance without external coupling or connection.

nonmagnetic — Material which is not attracted by a magnet and cannot be magnetized (e.g., paper, plastic, tin, glass). In a strict sense, having a permeability equal to that of air or 1.

nonmagnetic armature shim — Also called nonmagnetic armature stop. A nonmagnetic member separating the pole faces of the core and armature in the operated position, used to reduce and stabilize the pull from residual magnetism in release.

nonmagnetic steel—A steel alloy that contains about 12% manganese, and sometimes a small quantity of nickel. It is practically nonmagnetic at ordinary temperatures.

nonmetallic sheathed cable—Two or more rubber- or plastic-covered conductors assembled in an outer sheath of nonconducting fibrous material that has been treated to make it flame- and moisture-resistant.

nonmultiple switchboard — A manual telephone switchboard in which each subscriber line is connected to only one jack.

nonphysical primary—A color primary represented by a point outside the area of the chromaticity diagram enclosed by the spectrum locus and the purple boundary.

nonplanar network—A network that cannot be drawn on a plane without crossing branches.

nonpolar crystals—Crystals having the property that each lattice point is identical.

nonpolarized electrolytic capacitor — An electrolytic capacitor which can be connected without regard to polarity. This is possible because the dielectric film is formed on both electrodes.

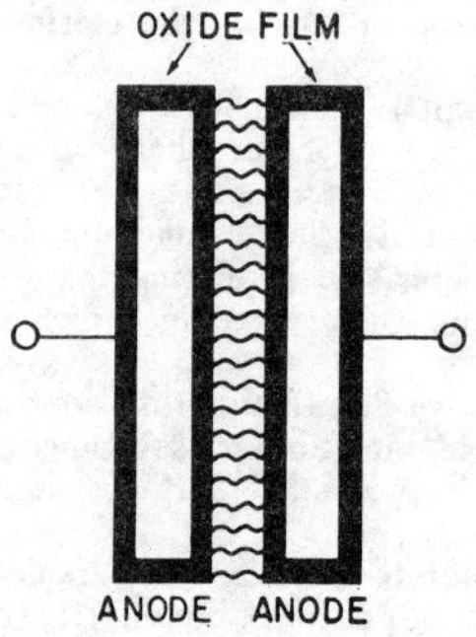

Nonpolarized electrolytic capacitor.

nonpolarized relay—*See* Neutral Relay.

nonrenewable fuse unit—A fuse unit that cannot be readily restored for service after operation.

nonreset timer—A timer which cannot be reset by electrical means.

nonresonant line—A transmission line with a natural resonant frequency different from that of the transmitted signal.

nonreturn-to-zero—A method of writing information on a magnetic surface in which the current through the write-head winding does not return to zero after the write pulse.

nonsaturated color — A color that is not pure—i.e., one that has been mixed with white or its complementary color.

nonsaturated logic—A type of logic circuit in which short delay times are achieved by preventing transistors from saturating.

nonshorting switch — Switch in which the width of the movable contact is less than the distance between contact clips so that one circuit is broken before another is completed.

nonsimultaneous transmission — Usually, transmission in which data can be moved in only one direction at a time by a device or facility.

nonsinusoidal wave—Any wave that is not a sine wave. It therefore contains harmonics.

nonstorage camera tube—A television camera tube in which the picture signal is always in proportion to the intensity of the illumination on the corresponding area of the scene being televised.

nonsynchronous—Not related in frequency, speed, or phase to other quantities in a device or circuit.

nonsynchronous starting torque—The maximum load torque with which motors can start. However, the motor will not come to a synchronous speed.

nonsynchronous vibrator — A vibrator that interrupts a direct-current circuit at a frequency unrelated to the other circuit constants. It does not rectify the resulting stepped-up alternating voltage.

nontrip-free circuit breaker—A breaker that can be maintained closed by manual override action while a tripping condition persists.

nonuniform field — A field in which the scalar (or vector) at that instant does not have the same value at every point in a given region.

nonvolatile memory — A memory whose stored data is undisturbed by removal of operating power. Core memories are of this type.

nonvolatile storage—Storage media which retain information in the absence of power and which will make the information available when power is restored.

nonwirewound trimming potentiometer—A trimming potentiometer characterized by the continuous nature of the surface area of the resistance element to be contacted. Contact is maintained over a continuous unbroken path. The resistance is achieved by using material compositions other than wire, such as carbon, conductive plastic metal film, and cement.

no-operation instruction—An instruction for a computer to do nothing but process the next instruction in sequence. Also called blank instruction or waste instruction. *See also* Skip.

NOR—A function of A and B that is true if both A and B are false.

NOR device—1. A device that has its output in the logic 1 state if and only if all of the control signals assume the logic 0 state. 2. An OR circuit that delivers an inverted output signal.

NOR element—A gate circuit having multiple inputs and one output that is energized only if all inputs are zero.

NOR gate—An OR gate followed by an inverter to form a binary circuit in which the output is logic 0 if any of the inputs is logic 1 and is logic 1 only if all the inputs are logic 0. With the opposite logic polarity, this type of gate is a NAND gate.

norm—1. The mean or average. 2. A customary condition or degree.

normal condition—The de-energized condition of a relay.

normal contact—A contact which in its normal position closes a circuit and permits current to flow.

normal distribution—The most common frequency distribution in statistics. The probability curve is bell-shaped, and the greatest probability occurs at the arithmetical average (i.e., at the top of the curve). The probability of occurrence of a particular value is shown by the areas between two abscissa values on the curve. *See also* Gaussian Distribution.

normal electrode — A standard electrode used for measuring electrode potentials.

normal failure period—That period of time during which an essentially constant failure rate exists.

normal impedance—*See* Free Impedance.

normal induction—The limiting induction, either positive or negative, in a magnetic material that is under the influence of a magnetizing force that varies between two extremes.

normal induction curve — The curve obtained by plotting B (induction) against H (magnetizing force), starting from a totally demagnetized state.

normalization—The transforming of signals to a common basis—e.g., adjusting two signals, representing the same spoken word but differing in loudness, to the same loudness.

normalize—1. To adjust the representation of a quantity so that it lies within a prescribed range. 2. In a computer, to shift all digits of a word to the right or left to accommodate the maximum number of digits.

normalized admittance — The reciprocal of normalized impedance.

normalized impedance—An impedance divided by the characteristic impedance of a waveguide.

normalized plateau slope—The slope of the substantially straight portion of the counting-rate-versus-voltage characteristic of a radiation counter tube, divided by the quotient of the counting rate and the voltage at the Geiger-Mueller threshold.

normal linearity—A manner of expressing linearity as the deviation from a straight line in terms of a given percentage of the output at a certain stimulus value, usually the full-scale value.

normally closed—Symbolized by nc. Designation applied to the contacts of a switch or relay when they are connected so that the circuit will be completed when the switch is not activated or the relay coil is not energized.

normally closed contact—Also called break contact. A contact pair which is closed when the coil of a relay is not energized.

normally closed switch – A switch that passes current until actuated.

normally open—Symbolized by no. Designation applied to the contacts of a switch or relay when they are connected so that the circuit will be broken when the switch is not activated or the relay coil is not energized.

normally open contact – A contact pair which is open when the coil of a relay is not energized.

normally open switch—A switch that must be actuated to pass current.

normal mode—The expected or usual operating conditions, such as the voltage that occurs between the two input terminals of an amplifier.

normal-mode interference – Interference that appears between the terminals of a measuring circuit.

normal mode of vibration – A characteristic distribution of vibration amplitudes among the parts of the system, each part of which is vibrating freely at the same frequency. Complex free vibrations are combinations of these simple vibration forms.

normal-mode voltage—1. The actual signal voltage developed by a transducer. 2. The difference voltage between two input-signal lines.

normal operating period—The time interval between debugging and wear-out.

normal permeability—Ratio of the normal induction to the corresponding magnetizing force. In the cgs system, the flux density in a vacuum is numerically equal to the magnetizing force.

normal position—The position of the relay contacts when the coil is not energized.

normal propagation – The phenomenon of passing radio waves through space when atmospheric and/or ionospheric conditions are such as to permit the passage with little or no difficulty.

normal record level – In direct recording, the amount of record-head current required to produce 1% third-harmonic distortion of the reproduced signal at the record-level set frequency. Such distortion must result from magnetic-tape saturation, not from electronic circuitry.

normal-stage punching – In a computer, a card-punching system in which only the even-numbered rows are punched on the British standard card.

north pole—1. In a magnet, the pole from where magnetic lines of force are considered to leave the magnet; the north-seeking pole. 2. A pole that attracts the south-seeking pointer of a field compass.

Norton's theorem—The current in any impedance Z_R, connected to two terminals of a network, is the same as though Z_R were connected to a constant-current generator the generated current of which is equal to the current that flows through the two terminals when these terminals are short-circuited, the constant-current generator being in shunt with an impedance equal to the impedance of the network, looking back from the terminals in question.

NOT AND circuit—An AND gating circuit which inverts the pulse phase.

notation – A manner of representing numbers. Some of the more important notation scales are:

Base	*Name*
2	binary
3	ternary
4	quaternary, tetral
5	quinary
8	octonary, octal
10	decimal
12	duodecimal
16	hexadecimal, sexidecimal
32	duotricenary
2.5	biquinary

notch – 1. A rectangularly shaped depression that extends below the sweep line of the radar indicator in some equipment. 2. On a graph (a graph of frequency response, in particular), a point where a curve dips sharply and returns equally sharply to its original value.

notch antenna—An antenna that forms a pattern by means of a notch or slot in a radiating surface. Its characteristics are similar to those of a properly proportioned metal antenna and may be evaluated with similar techniques.

notch filter—An arrangement of electronic components designed to attenuate or reject a specific frequency band with sharp cutoff at either end. *See also* Band-Reject Filter.

notch gate—The early and late gate display on the range crt. It appears as a negative deflection of the range base line equal in width to the early and late gates.

notching—A term indicating that a predetermined number of separate impulses are required to complete operation of a relay.

notching circuit—A control circuit used in a cathode-ray oscilloscope to expand portions of the displayed image.

NOT circuit—A binary circuit with a single output that is always the opposite of the single input. Also called inverter circuit.

NOT device—A device which has its output in the logic 1 state if and only if the control signal assumes the logic 0 state. The NOT device is a single-input NOR device.

note—1. The pitch, duration, or both of a tone sensation. 2. The sensation itself. 3. The vibration causing the sensation. 4. The general term when no distinction is desired between the symbol, the sensation, and the physical stimulus. 5. A single musical tone. The notes of the musical scale are referred to by letters running alphabetically from A to G for the white keys. A black key may be called a sharp of the note below it or a flat of the note above it. Each note has a frequency exactly one-half that of the corresponding note in the next higher octave.

NOT gate—1. An inhibitory circuit equivalent to the logical operation of negation (mathematical complement). The output of the circuit is energized only when its single input is not energized and there will be no output if the input is energized. 2. A single-control device that evidences the appearance of a positive-pressure control signal with a single zero output signal. With no positive control signal, the zero NOT output signal ceases. A NOR gate with a single control connection performs as a NOT gate.

NOT majority—A logic function in which the output is false if more than half the number of inputs are true; otherwise the output is true.

noval base — A nine-pin glass base for a miniature electron tube. For orientation of the tube into its socket, the spacing between pins 1 and 9 is greater than between the other pins.

novar—A beam-power tube that has a nine-pin base.

novice license—A class of amateur license. Its requirements are the easiest of all, but novice operation is limited. Transmitters must be crystal controlled, and the maximum permissible plate input power is 75 watts. Novice licensees are restricted to certain frequencies (3.70-3.75, 7.15-7.20, 21.10-21.25, and 145-147 MHz). A novice license is valid for two years only, and it is not renewable.

noys — A measure of perceived noisiness; the noys scale is linear.

npin transistor—An npn transistor with a layer of high-purity germanium added between the base and collector to extend the frequency range.

npip transistor—An intrinsic-region transistor in which the intrinsic region is located between two p-regions.

n-plus-one address instruction — In a computer, a multiple-address instruction in which one address serves to specify the location of the next instruction of the normal sequence to be executed.

npnp transistor—A hook transistor with a p-type base, n-type emitter, and a hook collector.

npn semiconductor — Double junction formed by sandwiching a thin layer of p-type material between two layers of n-type material of a semiconductor.

npn transistor—A transistor with a p-type base and an n-type collector and emitter. In construction, it consists of a thin layer of p-type semiconductor material sandwiched between two layers of n-type semiconductor material.

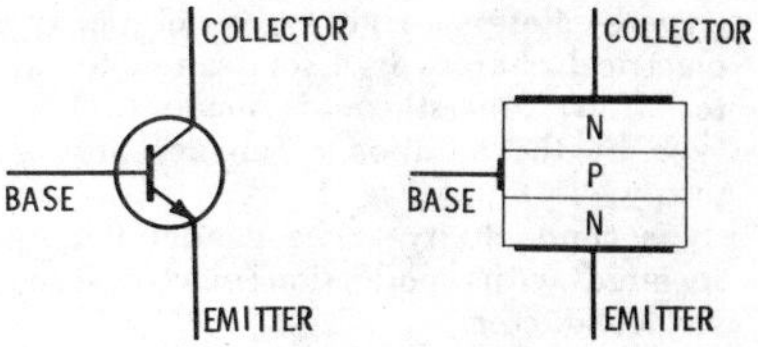

Symbol. Junctions.
Npn transistor.

npo body (negative positive zero)—A designation referring to a class of capacitors in the temperature-compensating series that have an essentially variant dielectric constant over a specified temperature range (commonly +25°C to +85°C).

np semiconductor—Material in a region of transition between n- and p-type material.

N-quadrant — One of the two quadrants where the N-signal is heard in an A-N radio range.

n-region—Also called n-zone. The region where the conduction-electron density in a semiconductor exceeds the hole density.

nrz — Abbreviation for nonreturn to zero. Data transmission and recording in which the transition represents the data, and the state remains unchanged until the next transition transmits more data on that line.

ns — Abbreviation for nanosecond (10^{-9} sec).

N-scan—*See* N-Display.

N-scope—*See* N-Display.

N-signal—A dash-dot signal heard in either a bisignal zone or an N-quadrant of a radio range.

NSOS—Same as PSOS, except it denotes n-channel devices.

***n*-terminal network**—A network with *n* accessible terminals.

***n*th harmonic** — A harmonic whose frequency is *n* times the frequency of the fundamental component.

NTSC — National Television System Committee, named after an all-industry engineering group who developed U.S. color-television specifications. Now describes the American system of color telecasting.

NTSC signal — A 3.579545-MHz (3.58-MHz) signal, the phase of which is varied with the instantaneous hue of the televised color, and the amplitude of which is varied with the instantaneous saturation of the color—as specified by the National Television System Committee.

NTSC triangle — On a chromaticity diagram, a triangle which defines the gamut of color obtainable through the use of phosphors.

n-type – Refers to an excess of negative electrical charges in a semiconductor material. Natural silicon is made to be p-type by the addition of an acceptor impurity.

n-type conductivity – The conductivity associated with conduction electrons in a semiconductor.

n-type crystal rectifier – A crystal rectifier in which forward current flows when the semiconductor is more negative than the metal.

n-type material—1. A crystal of pure semiconductor material to which has been added an impurity (an electron donor such as arsenic or phosphorous) so that electrons serve as the majority charge carriers. 2. A quadrivalent semiconductor material, with electrons as the majority charge carriers, that is formed by doping with donor atoms.

n-type region – Portion of semiconductor material containing a small number of dopant atoms that have an extra (free) electron in their outer orbit. The n region is a source of mobile negative charges.

n-type semiconductor – An intrinsic semiconductor in which the conduction-electron density exceeds the hole density. By implication, the net ionized impurity concentration is a donor type.

nuclear battery – 1. Also called an atomic battery. A battery which converts nuclear energy into electrical energy. 2. Direct-conversion secondary cell or battery using radioisotopes as the energy source.

nuclear bombardment – The shooting of atomic particles at nuclei, usually in an attempt to split the atom in order to form a new element.

nuclear clock—A clock in which the time base is provided by the mean time interval between two successive disintegrations in a radioactive source.

nuclear energy—Also called atomic energy or power. The energy released in a nuclear reaction when a neutron splits the nucleus of an atom into smaller pieces (fission), or when two nuclei are joined together under millions of degrees of heat (fusion).

nuclear fission—*See* Fission.

nuclear fusion—*See* Fusion, 1.

nuclear magnetic resonance – The flipping over of a particle such as a proton as the result of the application of an alternating magnetic field at right angles to a steady magnetic field in which the particle is placed.

nuclear pile—*See* Nuclear Reactor.

nuclear reaction – The reaction, accompanied by a tremendous release of energy, when the nucleus of an atom is split into smaller pieces (fission) or when two or more nuclei are joined together under millions of degrees of heat (fusion).

nuclear reactor—Also called an atomic reactor, nuclear pile, or reactor. A device which can sustain nuclear fission in a self-supporting chain reaction.

nucleation—The occurrence in an existing phase or state of a new phase or state.

nucleonics—The application of nuclear science in physics, chemistry, astronomy, biology, industry, and other fields.

nucleus—The core of an atom. It contains most of the mass and has a positive charge equal to the number of protons it contains. Its diameter is about one ten-thousandth that of the atom. Except for the ordinary hydrogen atom, the nuclei of all other atoms consist of protons and neutrons tightly locked together.

null—1. A balanced condition which results in zero output from a device or system. 2. To oppose an output which differs from zero so that it is returned to zero. 3. The minimum output amplitude (ideally, zero) in direction-finding systems where the amplitude is determined by the direction from which the signal arrives or by the rotation in bearing of the system's response pattern. 4. In a computer, a lack of information as opposed to a zero or blank for the presence of no information.

null balance—A condition in which two or more signals are summed and produce a result that is essentially zero.

null detection – A method of making df measurements. The antenna is turned to the point where the received signal is weakest. The true bearing of the signal is then found by noting the antenna direction and using a correction factor.

null detector—An apparatus that senses the complete balance, or zero-output condition, of a system or device.

null-frequency indicator—A device that indicates frequency by heterodyning two electrical signals together to give a zero-beat indication.

null indicator – A device that indicates when current, voltage, or power is zero.

null method – Also called zero method. A method of measurement in which the circuit is balanced, to bring the pointer of the indicating instrument to zero, before a reading is taken (e.g., in a Wheatstone bridge, or in a laboratory balance for weighing purposes).

null-spacing error—In a resolver, the difference between 180° and the angle between null positions of the output winding with respect to one input winding.

number—An abstract mathematical symbol for expressing a quantity. In this sense, the manner of representing the number is immaterial. Take 26, for example. This is its decimal form—but it could be expressed as binary 011010 and still mean the same. Some common numbering systems are: binary (base *2*), quinary (base *5*), octonary or octal (base *8*), and decimal (base *10*).

number of overshoots—The total number of times that the transient response goes to a minimum or maximum value that is outside a specified range.

number of scanning lines—The ratio of the scanning frequency to the frame frequency.

number system—Any system for the representation of numbers. (*See also* Positional Notation.)

numeric—In computers, consisting entirely or partially of digits as distinguished from alphabetic composition.

numerical analysis—The providing of convenient methods for finding useful solutions to mathematical problems and for obtaining useful information from available solutions that are not expressed in easily handled forms.

numerical control — 1. Descriptive of systems in which digital computers are used for the control of operations, particularly of automatic machines (e.g., drilling or boring machines, wherein the operation control is applied at a discrete point in the operation or process). Contrasted with process control, in which control is applied continuously. 2. Abbreviated nc. The technique of controlling a machine or process through the use of command instructions in coded numerical form.

numerical control system — A system controlled by direct insertion of numerical data at some point. The system must automatically interpret at least some portion of the data.

numerical data—Data expressed in terms of a set of numbers or symbols that can assume only discrete values or configurations.

numerical-readout tube—A gas-filled, cold-cathode, digital-indicator tube having a common anode and containing stacked metallic elements in the form of numerals. When a negative voltage is applied to one of them, the element then becomes the cathode of a simple gas-discharge diode.

numeric coding—A system of abbreviation used in the preparation of information for machine acceptance in which all information is reduced to numerical quantities, in contrast to alphabetic coding.

nutating feed—In a tracking radar, an oscillating antenna feed that produces an oscillating deflection of the beam without changing the plane of polarization.

nutation field—The time-variant, three-dimensional field pattern of a directional or beam-producing antenna having a nutating feed.

nuvistor – An electron tube in which all electrodes are cylindrical and are closely spaced, one inside the other, in a ceramic envelope.

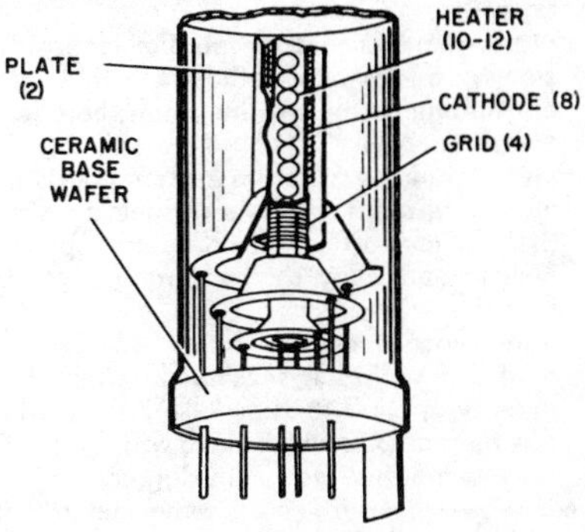

Nuvistor.

nV—Abbreviation for nanovolt.

n-variable cube—*See* N-Cube.

nW – Abbreviation for nanowatt (10^{-9} watt).

Nyquist diagram – A plot, in rectangular coordinates, of the real and imaginary parts of factor $\mu\beta$ for frequencies from zero to infinity in a feedback amplifier—where μ is the amplification in the absence of feedback, and β is the fraction of the output voltage superimposed on the amplifier input.

Nyquist interval—The maximum separation in time which can be given to regularly spaced instantaneous samples of a wave of a given bandwidth for complete determination of the waveform of the signal. Numerically it is equal in seconds to one-half the bandwidth.

Nyquist rate—Of a channel, the maximum rate at which independent signal values can be transmitted over the specified channel without exceeding a specified amount of mutual interference.

Nyquist's theorem—*See* Sampling Theorem.

n-zone—*See* N-Region.

O

OAO—Abbreviation for Orbiting Astronomical Observatory.

object code—Compiler or assembler output that is itself executable machine code or that can be processed to produce executable machine code.

object language—The computer language in which the output from a compiler or assembler is expressed.

object module—A module that serves as the output of an assembler or compiler and the input to a linkage editor.

object program – A computer program which is the output of an automatic coding system. The object program may be a machine-language program ready for execution, or it may be in an intermediate language.

oblique-incidence transmission—Transmission by means of a radio wave that travels obliquely up to the ionosphere and down again.

oboe—A radar navigation system consisting of two ground stations, which measure the distance to an airborne transponder beacon and relay this information to the aircraft.

obsolescence-free—Not likely to become outdated within a reasonable time. Frequently applied to tube testers and other test instruments which have provisions for accommodating new developments.

occlude—To absorb—e.g., some metals will occlude gases, which must be driven out before the metals can be used in the electrodes or supports of a vacuum tube.

occluded gas—Gas that has been absorbed by a material (e.g., by the electrodes, supports, leads, and insulation of a vacuum tube).

occupied bandwidth—The frequency bandwidth such that, below its lower and above its upper frequency limits, the mean powers radiated are each equal to 0.5% of the total mean power radiated by a given emission.

ocr – Abbreviation for optical character recognition. 1. A technology that enables a machine to automatically read and convert typewritten, machine-printed, mark-sensed or hand-printed characters into electrical impulses for processing by a computer. Ocr devices are designed to read man-readable data and convert it directly to a computer language. 2. The machine recognition of printed or written characters based on inputs from photoelectric transducers. (Contrast with micr.)

octal—*See* Notation.

octal base—An eight-pin tube base (although unneeded pins are often omitted without changing the position of the remaining pins). An aligning key in the center of the base assures correct insertion of the tube into the socket.

octal, binary coded—Pertaining to a binary-coded number system with the radix 8, in which the natural binary values of 0 through 7 are used to represent octal digits with values from 0 to 7.

octal digit—One of the symbols 0, 1, 2, 3, 4, 5, 6, and 7 when used in numbering in the scale of 8.

octal fraction—A shorthand expression of the binary contents of a half word.

octal fractional—A quantity less than a whole number referenced to a numbering system that has the radix eight.

octal loading program—A computer utility program with provision for making changes in programs and tables that are in core memory or drum storage, reading in words that are coded in octal notation on punched cards or tape.

octal numbering system—A numbering system based on powers of eight. This system is used extensively in computer work because it is derived simply from binary numbering.

octave – 1. The interval between two sounds having a basic frequency ratio of two—or, by extension, the interval between any two frequencies having a ratio of 2:1. The interval, in octaves, between any two frequencies having a ratio of 2:1. The interval, in octaves, belogarithm to the base *10*) of the frequency ratio. 2 The musical interval between two pitches whose fundamental frequencies differ by a ratio of 2 to 1. It encompasses eight successive notes on the diatonic musical scale. Thus, from A 440 Hz to A 880 Hz is an interval of one octave.

octave band – A band of frequencies the limits of which have the ratio 2 to 1.

octave-band pressure level—Also called octave pressure level. The pressure level of a sound for the frequency band corresponding to a specified octave.

octave pressure level – *See* Octave-Band Pressure Level.

octode—An eight-electrode electron tube containing an anode, a cathode, a control electrode, and five additional grids.

octonary—*See* Notation.

octonary signaling—A mode of communication in which information is represented by the presence and absence, or plus and minus variations, of eight discrete levels of a parameter of the signaling medium.

odd-even check—An automatic computer check in which an extra digit is carried along with each word, to determine whether the total number of 1's in the word is odd or even, thus providing a check for proper operation. *See* Parity Check.

odd harmonic—Any frequency that is an odd multiple of the fundamental frequency (e.g., 1, 3, 5, etc.).

odd-line interlace – The double-interlace system in which, since there are an odd number of lines per frame, each field contains a half line. In the 525-line television frame used in the United States, each field contains 262.5 lines.

odograph—Automatic electronic map tracer used in military vehicles for map making and land navigation. It automatically plots, on an existing map or on cross-sectional paper, the exact course taken by the vehicle. This is done by phototubes and thyratrons, which transfer the indication of a precision magnetic compass onto a plotting unit actuated by the speedometer drive cable. A pen then traces the course.

oersted—In the cgs electromagnetic system, the unit of magnetizing force equal to $1000/4\pi$ ampere-turns per meter. At any point in a vacuum, the magnetic intensity in oersteds is equal to the force in dynes exerted on a unit magnetic pole at that point.

off-center display—A ppi display, the center of which does not correspond to the position of the radar antenna.

off-delay—A circuit that retains an output signal some definite time after the input signal is removed.

off-ground — The voltage above or below ground at which a device is operated.

offhook—A condition that occurs when the telephone handset is lifted from its mounting, thus causing the hookswitch to operate (close) and closing the loop to the central office. The offhook condition indicates a busy condition to incoming calls.

offhook service—Priority telephone service in which a connection from caller to receiver is afforded by removing the phone from its cradle or hook.

off-limit contacts—Contacts on a stepping relay used to indicate that the wiper has reached the limiting position on its arc and must be returned to normal before the circuit can function again.

off-line equipment — Peripheral equipment that is not in direct communication with the computer central processing unit.

off-line operation—1. A control system in which the computer does not respond immediately or directly to the events of the controlled process; some time may elapse between the occurrence of an event and the reaction of the computer to it. 2. In a computer system, operation of peripheral equipment independent from the central processor, e.g., the transcribing of card information to magnetic tape, or of magnetic tape to printed form. 3. Operation performed while the computer is not engaged in monitoring or controlling a process or operation.

off-line system — A kind of teleprocessing system in which there must be human operations between the original recording functions and the ultimate data-processing function.

off-line unit—In a computer, the input/output device or auxiliary equipment not under direct control of the central processing unit.

off-normal contacts—Relay contacts that assume one condition when the relay is in its normal position and the reverse condition for any other position of the relay.

off-period—That portion of an operating cycle during which an electron tube or semiconductor is nonconducting.

offset — 1. The measure of unbalance between halves of a symmetrical circuit. Generally caused by differences in transistor betas or in values of biasing resistors. 2. The change in input voltage necessary to cause the output voltage in a linear amplifier circuit to be zero. 3. In digital circuits, the dc voltage on which a signal is impressed. 4. The difference between the desired value or condition and the value or condition actually attained.

offset angle — In lateral-disc reproduction, the smaller of the two angles between the projections, into the plane of the disc, of the vibration axis of the pickup stylus and the line connecting the vertical pivot (assuming a horizontal disc) of the pickup arm with the stylus point.

offset current — The difference in current into the two inputs of an operational amplifier required to bring the output voltage to zero.

offset error—The error by which the transfer function fails to pass through the origin, referred to the analog axis. This is adjustable to zero in available a/d converters.

offset stacker — In a computer, a card stacker having the ability to stack cards selectively under machine control so that they protrude from the balance of the stack, thus giving physical identification.

offset voltage—The difference in voltage at the two inputs of an operational amplifier required to bring the output voltage to zero.

off-target jamming—Employment of a jammer away from the main units of the force. This is done to prevent the enemy from monitoring the jamming signals and using them to poinpoint the location of the force.

OGO—Abbreviation for Orbiting Geophysical Observatory.

ohm — Symbolized by the Greek letter *omega* (Ω). The unit of resistance. It is defined as the resistance, at 0°C, of a uniform column of mercury 106.300 cm long and weighing 14.451 grams. One ohm is the value of resistance through which a potential difference of one volt will maintain a current of one ampere.

ohmic contact — Between two materials, a contact across which the potential difference is proportionate to the current passing through it.

ohmic heating – The energy imparted to charged particles as they respond to an electric field and make collisions with other particles. The name was chosen due to the similarity of this effect to the heat generated in an ohmic resistance due to the collisions of the charge carriers in their medium.

ohmic resistance–Resistance to direct current.

ohmic value–The resistance in ohms.

ohmmeter–A direct-reading instrument for measuring electric resistance. Its scale is usually graduated in ohms, megohms, or both. (If the scale is graduated in megohms, the instrument is called a megohmmeter; if the scale is calibrated in kilohms, the instrument is a kilohmmeter.)

ohmmeter zero adjustment – In an ohmmeter, a potentiometer or other means of compensating for the drop in battery voltage with age. Usually a knob is rotated until the meter pointer is at zero on the particular scale being used.

Ohm's law–The voltage across an element of a dc circuit is equal to the current in amperes through the element, multiplied by the resistance of the element in ohms. Expressed mathematically as $E = I \times R$. The other two equations obtained by transposition are $I = E/R$ and $R = E/I$.

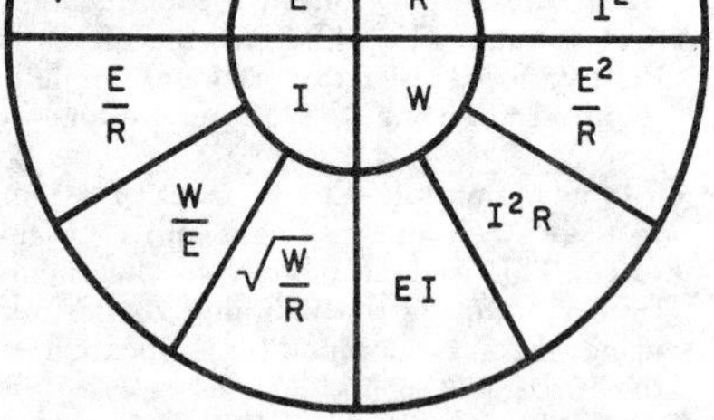

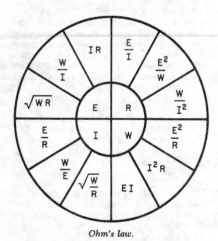

Ohm's law.

ohms per square–A unit of measurement of resistance by topological shape. A resistor topology can be considered to consist of continuous squares. The value of a resistor is equal to the number of squares times the ohms per square.

ohms per volt–A sensitivity rating for voltage-measuring instruments (the higher the rating, the more sensitive the meter). On any particular range, it is obtained by dividing the resistance of the instrument (in ohms) by the full-scale voltage value of that range.

oil–A prefix used with a device which interrupts an electrical circuit; it indicates that the interruption occurs in oil.

oil circuit breaker – A circuit breaker in which the interruption occurs in oil to suppress the arc and prevent damage to the contacts.

oiled paper – Paper that has been treated with oil or varnish to improve its insulating qualities.

oil-filled cable – Paper-insulated, lead-sheathed cable into which high-grade mineral oil is forced under pressure, saturating the insulation. Main object is to prevent moisture and gases from entering. Also easier to detect flaws due to leakage, as the oil is kept under constant pressure at all times.

oil fuse cutout–An enclosed fuse cutout in which all or part of the fuse support is mounted in oil.

oil-immersed forced-oil–cooled transformer (class FOA or FOW)–A transformer having its core and coils immersed in oil and cooled by the forced circulation of its oil through an external heat exchanger. If the heat exchanger utilizes forced circulation of air, the transformer is class FOA; if it utilizes forced circulation of water, the transformer is class FOW.

oil-immersed self-cooled forced-air–cooled transformer (class OA/FA) – A transformer having its core and coils immersed in oil and having a self-cooled rating with cooling obtained by the natural circulation of air over the cooling surface and a forced-air–cooled rating with cooling obtained by the forced circulation of air over this same cooling surface.

oil-immersed self-cooled transformer (class OA)–A transformer having its core and coils immersed in oil, the cooling being effected by the natural circulation of air over the cooling surface. (Smooth tank surfaces are generally sufficient to cool small transformers, but for about 25 kVA and larger, a supplementary heat-dissipating surface is provided in the form of external fins, tubes, or tubular radiators.)

oil-immersed transformer – A transformer the core and coils of which are immersed in oil.

oil-immersed water-cooled transformer (class OW) – A transformer having its core and coils immersed in oil, the cooling being effected by the natural circulation of oil over the water-cooled surface. The water-cooled surface is provided by water flowing through copper cooling coils immersed in the oil and located inside the main transformer tank, through which the oil flows by thermosiphon action.

oil switch–A switch in which the interruption of the circuit occurs in oil to suppress the arc and prevent damage to the contacts.

Omega – A long-range hyperbolic navigation system which transmits interrupted cw signals from which phase differences are extracted. Transmissions occur from two or more locations on a single carrier frequency utilizing time-sharing techniques.

omnibearing – The bearing indicated by a navigational receiver on transmissions from an omnirange.

omnibearing converter–An electromechanical device which combines an omnirange signal with aircraft heading information to furnish electrical signals for operating the pointer of a radio magnetic indicator. (*See also* Omnibearing Indicator.)

omnibearing indicator – An omnibearing converter to which a dial and pointer have been added.

omnibearing line–One of an infinite number of straight, imaginary lines radiating from the geographical location of a vhf omnirange.

omnibearing selector – Instrument capable of being set manually to any desired omnibearing or its reciprocal in order to control a course-line deviation indicator.

omniconstant – A calculator that can add consecutively in any steps of any predetermined size, raising the power of any number in consecutive steps.

omnidirectional–Also called nondirectional. All-directional; not favoring any one direction. Having no particular direction of maximum emission or sensitivity. An omnidirectional speaker is one that theoretically radiates equally in all directions.

omnidirectional antenna–Also called nondirectional antenna. An antenna producing essentially the same field strength in all horizontal directions and a directive vertical radiation pattern.

omnidirectional hydrophone – A hydrophone having a response that is essentially independent of the angle of arrival of the incident sound wave.

omnidirectional microphone – Also called nondirectional microphone. A microphone that responds to a sound wave from almost any angle of arrival.

omnidirectional radio range – Radio aid to air navigation using a transmitter that radiates throughout 360° azimuth, providing aircraft with a direct indication of the bearing of the transmitter.

omnidirectional range – Also called omnirange. A radio facility providing bearing information to or from such facilities at all azimuths within its service area.

omnidirectional range station–In the aeronautical radionavigational service, a land station that provides a direct indication of the bearing (omnibearing) of that station from an aircraft.

omnigraph – An instrument that produces Morse-code messages for instruction purposes. It contains a buzzer circuit which is usually actuated by a perforated tape.

omnirange–*See* Omnidirectional Range.

omr–Abbreviation for optical mark recognition.

on-call channels–Similar to allocated channels except that full-time exclusive use of the channel is not warranted.

on-course curvature – In navigation, the rate at which the indicated course changes with respect to the distance along the course path.

on-course signal–The monotone radio signal which indicates to the pilot that he is neither too far to the right nor to the left of the radio beam being followed.

on-delay–A circuit that produces an output signal some definite time after an input signal is applied.

on-demand system – A system from which the desired information or service is available at the time of request.

ondograph–An instrument for drawing alternating voltage waveform curves. It employs a capacitor which is momentarily charged to the amplitude at a particular point on the curve and then discharged through a recording galvanometer. This is repeated at intervals of about once every hundred cycles with the sample taken a little farther along on the waveform each time.

ondoscope–A glow-discharge tube used on an insulating rod to indicate the presence of high-frequency rf near a transmitter. The radiation ionizes the gas in the tube, and the visible glow indicates the presence of rf.

one-input terminal–Called the set terminal. The terminal which, when triggered, will put a flip-flop in the one (opposite of starting) condition.

one-many function switch – A function switch in which only one input is excited at a time and each input produces a combination of outputs.

one output–*See* One State.

one-output terminal – The terminal which produces an output of the correct polarity to trigger a following circuit when a flip-flop circuit is in the one condition.

ones complement – 1. Having to do with arithmetic that provides a method of negating a binary number so that binary subtraction can be performed with the techniques of addition. The ones complement of a binary number is obtained by complementing all bits in that number. In ones-complement addition, end-around carry must be used. 2. A binary digit that when added to another binary digit yields

1; the inverse binary state of any given bit.

one-shot—1. A circuit that produces an output signal of fixed duration when an input signal of any duration is applied. 2. *See* Monostable.

one-shot multivibrator – *See* Monostable Multivibrator.

one state – Also called one output. In a magnetic cell, the positive value of the magnetic flux through a specified cross-sectional area, determined from an arbitrarily specified direction. (*See also* Zero State.)

one-third–octave band—A frequency band in which the ratio of the extreme frequencies is 1.2599.

one-to-partial-select ratio—In a computer, the ratio of a *1* output to a partial-select output.

one-to-zero ratio—In a computer, the ratio of a *1* output to the *0* output.

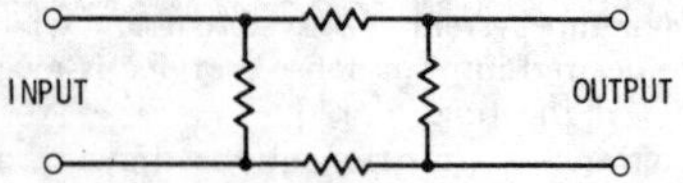

O-network.

O-network – A network composed of four impedance branches connected in series to form a closed circuit. Two adjacent junction points serve as input terminals, and the remaining two as output terminals.

one-way communication – Applied to certain radiocommunication or intercommunications systems where a message is transmitted from one station to one or more receiving stations that have no transmitting apparatus.

one-way repeater—*See* Repeater.

on-hook—A condition that exists when the telephone handset is on its mounting, thus keeping the hookswitch open. The on-hook condition opens the dc loop, indicating that calls can be accepted.

on line—1. Pertaining to a computer that is actively monitoring or controlling a process or operation, or to an operation performed by such a computer. 2. Pertaining to equipment or devices under direct control of the central processing unit. 3. Pertaining to a user's ability to interact with a computer.

on-line data reduction – The processing of information as rapidly as the information is received by the computing system or as rapidly as it is generated by the source.

on-line operation—1. An operation carried on within the main computer system (e.g., computing and writing results onto a magnetic tape, printed report, or paper tape). 2. Operation in which information concerning a controlled process is fed directly into a computer, and the computer exercises direct control on the basis of this information. Since the computer reacts without appreciable time lag, results are said to be in real time. 3. A type of system application in which the input data to the system is fed directly from the measuring devices and the computer results obtained during the process of the event; e.g., a computer receives data from wind tunnel measurements during a run, and the computations of dependent variables are performed during the run, enabling a change in the conditions so as to produce particularly desirable results.

on-line system—1. A teleprocessing system in which the input enters the computer directly from the originating point and/or in which the output is transmitted directly to the point at which it is used. 2. A telegraph system that involves transmitting directly into the system.

on-off control—A simple control system in a switch in which the device being controlled is either fully on or fully off and no operating positions are available.

on-off keying—Keying in which the output of a source is alternately transmitted and suppressed to form signals.

on-off ratio—Ratio of the duration (on) of a pulse to the space (off) between successive pulses.

on-off switch—*See* Power Switch.

on-off tests—The switching of various suspected interference sources on and off while the victim receiver is monitored so that the actual source can be identified.

on period—That portion of an operating cycle during which an electron tube is conducting.

on-state voltage – The maximum voltage when a thyristor is in the on state.

on the air—Transmitting.

on-the-fly printer—A high-speed line printer in which continuously rotating print wheels and fast-acting hammers are used so that the successive letters in a line of text are printed so rapidly that they appear to be printed simultaneously.

opacimeter – Also called turbidimeter. A photoelectric instrument for measuring the turbidity (amount of sediment) of a liquid. It does this by determining the amount of light that passes through the liquid.

opacity—The degree of nontransparency of a substance—i.e., its ability to obstruct, by absorption, the transmission of radiant energy such as light. Opacity is the reciprocal of transmission.

op amp—Abbreviation for operational amplifier.

op code—Abbreviation for operation code.

open—A circuit interruption that results in an incomplete path for current flow (e.g., a broken wire, which opens the path of the current).

open-center display – A ppi display on which zero range corresponds to a ring around the center of the display.

open circuit—A circuit which does not provide a complete path for the flow of current. (*See also* Open.)

open-circuit impedance—The driving-point impedance of a line or four-terminal network when the far end is open.

open-circuit jack – A jack that has its through circuit(s) normally open. Circuits are closed by inserting mating plug.

open-circuit parameters—The parameters of an equivalent circuit of a transistor that are the result of selecting the input current and output current as independent variables.

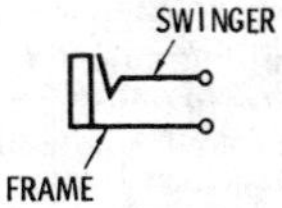

Open-circuit jack.

open-circuit signaling—Signaling in which no current flows under normal (i.e., inoperative) conditions.

open-circuit voltage – The voltage at the terminals of a battery or other voltage source when no appreciable current is flowing.

open collector output – A feature of some semiconductor memories which permits the formation of larger arrays by wiring several units together.

open core—A core fitting inside a coil but having on external return path. The magnetic circuit thus has a long path through air.

open-delta connection – Two single-phase transformers connected so that they form only two sides of a delta, instead of the three sides with three transformers in a regular delta connection.

open-ended – Pertaining to a system or process that is receptive to augmentation.

open-fuse cutout—An enclosed fuse cutout in which the fuse support and fuse holder are exposed.

open loop—1. Pertaining to a control system that does not provide self-correcting action for errors in the desired operational condition. 2. Having to do with an operational amplifier that does not have feedback.

open-loop bandwidth – Without feedback, the frequency limits at which the voltage gain of the device drops off 3 dB below the gain at some reference frequency.

open-loop control system—A control system in which there is no self-correcting action, as there is in a closed-loop system.

open-loop differential voltage gain – *See* Differential Voltage Gain.

open-loop gain—The ratio of the (loaded) output of an amplifier without any feedback to its net input at any frequency. Usually implies voltage gain.

open-loop output impedance—The complex impedance seen looking into the output terminals of an operational amplifier with no external feedback and in the linear-amplification region. In closed-loop operation, the output impedance is equal to the open-loop impedance divided by the loop gain. If the open-loop impedance is not more than a few hundred ohms and the loop gain is high enough for good gain accuracy and stability, the closed-loop impedance will be on the order of an ohm or less, which can be neglected in most applications.

open-loop output resistance—The resistance looking into the output terminal of an operational amplifier operating without feedback and in the linear-amplification region.

open-loop system – A control system that does not have a means of comparing input and output for control purposes.

open-loop voltage gain – The ratio of the output signal voltage of an operational amplifier to the differential input signal voltage producing it with no feedback applied.

open magnetic circuit—A magnet that has no closed external ferromagnetic circuit and does not form a complete conducting circuit itself (e.g., a permanent magnet ring interrupted by an air gap).

open motor – A motor having ventilating openings which permit passage of external cooling air over and around the windings of the machine. When such motors have an internal fan to aid the movement of ventilating air, they may be referred to as self-ventilated.

open-phase protection—The effect of a device operating on the loss of current in one phase of a polyphase circuit to cause and maintain the interruption of power in the circuit.

open-phase relay—A relay which functions when one or more phases of a polyphase circuit open and sufficient current is flowing in the remaining phase or phases.

open plug—A plug designed to hold jack springs in their open position.

open-reel—Also called reel-to-reel. Used to designate any tape format in which the running tape is wound onto a separate takeup reel. As distinguished from cartridges and cassettes, where the takeup reel is either nonexistent (cartridge) or is included in the tape package (cassette).

open relay—A relay not having an enclosure.

open routine—In a computer, a routine that

it is possible to insert directly into a larger routine without a linkage or calling sequence.

open subroutine – Also called direct-insert subroutine. A subroutine inserted directly into a larger sequence of instructions. Such a subroutine is not entered by a jump instruction; hence, it must be recopied at each point where it is needed.

open temperature pickup – A temperature transducer in which its sensing element is directly in contact with the medium whose temperature is being measured.

open wire – A conductor separately supported above the earth's surface.

open-wire circuit – A circuit made up of conductors separately supported on insulators.

open-wire loop–The branch line on a main open-wire line.

open-wire transmission line–A transmission line formed by two parallel wires. The distance between them, and their diameters, determine the surge impedance of the transmission line.

operand–The quantity that is affected, manipulated, or operated upon.

operate current–The minimum current required to trip all the contact springs of a relay.

operate time–1. The time that elapses, after power is applied to a relay coil, until the contacts being checked have operated (i.e., first opened in a normally closed contact, or first closed in a normally open contact). 2. The phase of computer operation during which an instruction is being executed.

operate-time characteristic – The relation between the operate time of an electromagnetic relay and the operate power.

operating angle–The electrical angle (portion of a cycle) during which plate current flows in an amplifier or an electronic tube. Class-A amplifiers have an operating angle of 360°; Class-B, 180° to 360°; and Class-C, less than 180°.

operating code – Abbreviated op code. Source statement which generates machine codes after assembly.

operating conditions–Those conditions, excluding the variable measured by the device, to which a device is subjected.

operating cycle–The complete sequence of operations required in the normal functioning of an item of equipment.

operating frequency – The rated ac frequency of the supply voltage at which a relay is designed to operate.

operating life – The minimum length of time over which the specified continuous and intermittent rating of a device, system, or transducer applies without change in performance beyond the specified tolerance.

operating-mode factor–A failure-rate modifier determined by the type of equipment environment, e.g., laboratory computer, nose-cone compartment, etc.

operating overload – The overcurrent to which electric apparatus is subjected in the course of the normal operating conditions that it may encounter.

operating point – Also called quiescent point. On a family of characteristic curves of a vacuum tube or transistor, a point the coordinates of which describe the instantaneous electrode voltages and currents for the operating conditions under consideration.

operating position – The operator-attended terminal of a communications channel. It is usually used in its singular sense (e.g., a radio-operator's position, a telephone-operator's position), even when there is more than one operating position.

operating power–The power actually reaching a transmitting antenna.

operating range–*See* Receiving Margin.

operating ratio–*See* Availability, 1.

operating system – Software that controls the carrying out of computer programs and that may provide scheduling, debugging, input/output control, accounting, compilation, storage assignment, data management, and related services.

operating temperature–The temperature or range of temperatures at or over which a device is expected to operate within specified limits of error.

operating temperature range–The interval of temperatures in which a component device or system is intended to be used, specified by the limits of this interval.

operating time–The time period between turn-on and turn-off of a system, subsystem, component, or part during which the system, etc., functions as specified. Total operating time is the summation of all operating-time periods.

operating voltages–The direct voltages applied to the electrodes of a vacuum tube under operating conditions.

operation–1. In mathematics, the determination of a quantity from two or more other quantities according to some rule. Examples are addition and multiplication in conventional arithmetic, and the AND and OR operations in Boolean algebra. 2. A specific action which a computer will perform whenever an instruction calls for it (e.g., addition, division).

operational amplifier–1. An amplifier that performs various mathematical operations. Application of negative feedback around a high-gain dc amplifier produces a circuit with a precise gain characteristic that depends on the feedback used. By the proper selection of feedback components, operational amplifier circuits can be used to add, subtract, average, integrate, and differentiate. An operational amplifier can have a single-input–single-output, differential-input–single-output, or differential-input - differential - output configuration. Also called op amp. 2. Generally, any high-gain amplifier whose gain and response characteristics are determined by external components. 3. An exceptionally versatile linear amplifier (usually a linear IC) used extensively in control, computation and measurement applications. 4. A stable, high-gain direct-coupled amplifier that depends on external feedback from the output to the input to determine its functional characteristics. The term operational is derived from the fact that the first op amps were employed to perform addition, multiplication, division, integration, and differentiation, among other mathematical operations, in analog computers.

operational differential amplifier—An operational amplifier which utilizes a differential amplifier in the first stage.

operational readiness—The probability that, at any point in time and under stated conditions, a system will be either operating satisfactorily or ready to be placed in operation.

operational reliability—*See* Achieved Reliability.

operation code (computer)—1. Also called the operation part of an instruction. The list of operation parts occurring in an instruction code, together with the names of the corresponding operations (e.g., "add," "unconditional transfer," "add and clear," etc.). 2. The part of a computer instruction word which specifies, in coded form, the operation to be performed.

operation decoder—In a computer, circuitry which interprets the operation-code portion of the machine instruction to be executed and sets other circuitry for its execution.

operation number—In a computer, a number that designates the position of an operation or its equivalent subroutine in the sequence of operations that makes up a routine. A number that identifies each step in a program stated in symbolic code.

operation part—In a computer instruction, the part that specifies the kind of operation, but not the location of the operands. (*See also* Instruction Code.)

operation register—In a computer, the register that stores the operation-code portion of an instruction.

operations research—The solving of large-scale or complicated problems that arise from an operation in which a major decision must be based on solutions of relationships involving a large number of variables.

operation time — The amount of time required by a current to reach a stated fraction of its final value after voltages have been applied simultaneously to all electrodes. The final value is conventionally taken as that reached after a specified length of time.

operator—1. Any person who operates, adjusts, and maintains equipment—specifically, a computer, radio transmitter, or other communications equipment. 2. A symbol that designates a mathematical operation, such as +, −, etc.

opposition—The phase relationship between two periodic functions of the same period when the phase difference between them is half of a period.

optical ammeter—An electrothermic instrument for measuring the current in the filament of an incandescent lamp. It normally uses a photoelectric cell and indicating instrument to compare the illumination with that produced by a current of known magnitude in the same filament.

optical axis—The Z-axis of a crystal.

optical character recognition—Identification of printed characters by a machine using light-sensitive devices. *See also* OCR.

optical communications — Communications through the use of beams of visible, infrared, or ultraviolet radiation, or, over much longer distances, through the use of laser beams. *See also* Coherent Light Communications.

optical coupler — Also called optical isolation. A device that couples signals from one electronic circuit to another by means of electromagnetic radiation, usually infrared, or visible. A typical optical coupler uses a light-emitting diode (LED) to convert the electrical signal of the primary circuit into light and a phototransistor in the secondary circuit to reconvert the light signal back into an electrical signal.

optical damping—A damping ratio slightly less than unity in which the overshoot is less than the specified uncertainty of the instrument.

optical horizon — The locus of points at which straight lines extended from a given point are tangent to the surface of the earth.

optical isolator—A photon-coupled device in which an electrical signal is converted into light that is projected through an insulating interface and reconverted to an electrical signal. *See also* Optical Coupler.

optical mark recognition—Abbreviated omr. An information-processing technology that converts data into another medium for computer input. This is accomplished by the presence of a mark in a given position, each position having a value known to the computer and which may or may not be understandable to humans.

optical maser—*See* Laser.

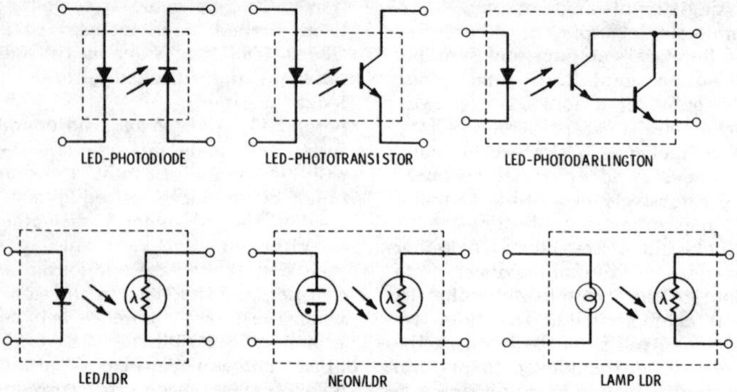

Optical isolator (opto isolator).

optical mode—In a crystal lattice, a mode of vibration that produces an oscillating dipole.

optical pattern—Also called Christmas-tree pattern. In mechanical recording, the pattern observed when the surface of the record is illuminated by a light beam of essentially parallel rays.

optical pumping—The process of changing the number of atoms or atomic system in a set of energy levels as a result of the absorption of light incident on the material. This process causes the atoms to be raised from certain lower to certain higher energy levels, and it may cause a population inversion between certain intermediate levels.

optical pyrometer—A temperature-measuring device comprising a comparison source of illumination, together with some convenient arrangement for matching this source—either in brightness or color—against the source whose temperature is to be measured. The comparison is usually made by the eye.

optical scanner — A computer-system input device that reads a line of printed characters and produces a corresponding electronic signal for each character.

optical sound—A system of recording and reproducing sound by using modulated light areas at the side of motion-picture film.

optical spectrometer — An instrument with an entrance slit, a dispersing device, and one or more exit slits, with which measurements are made at selected wavelengths within the spectral range, or by scanning over the range. The quantity detected is a function of radiant power.

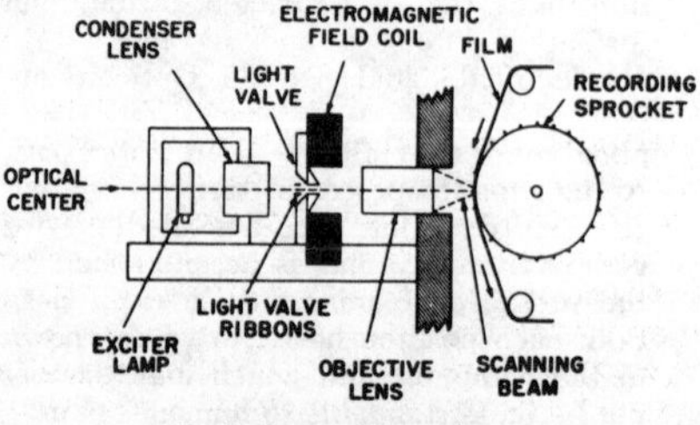

Optical sound recorder.

optical twinning – A defect occurring in natural quartz crystals. The right quartz and left quartz both occur in the same crystal. This generally results in small regions of unusable material, which are discarded when the crystal is cut up.

optics – The branch of science concerned with vision—i.e., the nature and propagation of electromagnetic radiation in the infrared, visible, and ultraviolet regions.

optimization—The continual adjustment of a process for the best obtainable set of conditions.

optimize—To arrange instructions or data in storage in such a way that a minimum of machine time is required for access when instructions or data are called out.

optimum bunching – The bunching condition required for maximum output in a velocity-modulation tube.

optimum coupling—*See* Critical Coupling.

optimum damping—The value of damping that permits fast response with some overshoot; this value is about 65 percent of critical damping.

optimum load – The value of load impedance which will transfer maximum power from the source to the load.

optimum plate load – The ideal plate-load impedance for a given tube and set of operating conditions.

optimum reliability—The value of reliability that yields a minimum total successful mission cost.

optimum working frequency – The frequency at which transmission by ionospheric reflection can be expected to be most effectively maintained between two specified points and at a certain time of day. (For propagation by way of the F_2 layer, the optimum working frequency is often taken as being 15% below the monthly median value of the F_2 maximum usable frequency for a specified path and time of day.)

optocoupler—A light source (input) and a light detector (output) where both the light source and detector are housed in a single package, sealed against outside light. An electrical signal applied to the light source changes the amount of light emitted. The emitted light falls upon, and is collected by, the detector. These input electrical signals are thus "coupled" to the output. From the output, the signals perform normal electronic functions, such as driving amplifiers, triggering a thyristor power supply, or switching logic levels.

optoelectronic integrated circuit—An integrated component that uses a combination of electroluminescence and photoconductivity in the performance of all or at least a major portion of its intended function.

optoelectronics — 1. Technology dealing with the coupling of functional electronic blocks by light beams. 2. Circuitry in which solid-state emitters and detectors of light are involved.

optoelectronic transistor—A transistor that has an electroluminescent emitter, a transparent base, and a photoelectric collector.

optoisolator—A coupling device consisting of a light sensor. Used for voltage and noise isolation between input and output while transferring the desired signal.

optophone—A photoelectric device that converts light energy into sound energy. Thus, a blind person by using a selenium cell and a circuit for converting the resulting signals into sounds of corresponding pitch can "read" by ear.

orange peel—A term applied to the surface of a recording blank which resembles an orange peel. Such a surface has a high background noise.

orbit—1. The path of a body or particle under the influence of a gravitational or other force. For instance, the orbit of a celestial body is its path relative to another body around which it revolves. 2. To go around the earth or other body in an orbit.

orbital electron—An electron which is visualized as moving in an orbit around the nucleus of an atom or molecule—as opposed to a free electron.

OR circuit—*See* OR Gate.

order—In computer terminology, the synonym for instruction, command, and—loosely—operation part. These three usages, however, are losing favor because of the ambiguity between them and the more common meanings in mathematics and business.

ordering—In a computer, the process of sorting and sequencing.

orders of logic—A measure of the speed with which a signal can propagate through a logic network (commonly referred to as orders-of-logic capability).

order tone—A tone sent over trunks to indicate the trunk is ready to receive an order, and to the receiving operator that it is about to arrive.

OR device—A device the output of which is logical zero if and only if all the control signals are logical zero.

ordinary wave – Sometimes called the O-wave. One of the two components into which the magnetic field of the earth divides a radio wave in the ionosphere. The other component is called the extraordinary wave, or X-wave. When viewed below the ionosphere in the direction of propagation, the ordinary wave has counterclockwise or clockwise elliptical polarization, depending on whether the magnetic field of the earth has a positive or negative component, respectively, in the same direction.

ordinate—1. The vertical line, or one of the coordinates drawn parallel to it, on a graph. 2. The vertical scale on a graph.

organ—In a computer subassembly, the portion which accomplishes some operation or function (e.g., arithmetic organ).

OR gate—1. Also called an OR circuit. A gate that performs the function of logical "inclusive or." It produces an output whenever any one (or more) of its inputs is energized. 2. A logic circuit which requires that at least one input be in the on state to drive the output into the on state.

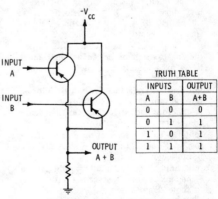

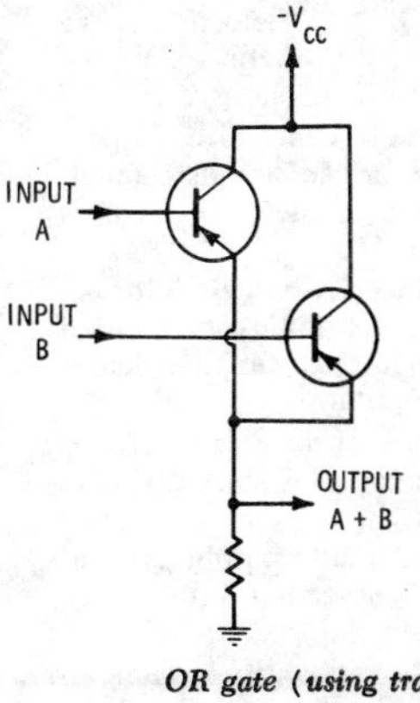

TRUTH TABLE

INPUTS		OUTPUT
A	B	A+B
0	0	0
0	1	1
1	0	1
1	1	1

OR gate (using transistors).

orient—To position or otherwise adjust with respect to some reference point (e.g., to orient an antenna for best reception).

orientation—An adjustment of the time, relative to the start transition, that teletypewriter receiving apparatus starts selection.

oriented — In crystallography, a crystal in which the axes of its individual grains are aligned so that they have directional magnetic properties.

orifice – An opening or window–specifically, in a side or end wall of a waveguide or cavity resonator, an opening through which energy is transmitted.

original lacquer–An original disc recording made on a lacquer surface for the purpose of producing a master.

original master–Also called metal master or metal negative. In disc recording, the master produced by electroforming from the face of a wax or lacquer recording.

O-ring–A circular piece of material with a round cross section; it effects a seal under pressure.

orthicon–Also called an orthiconoscope. A camera tube in which a beam of low-velocity electrons scans a photoemissive mosaic capable of storing and electrical-charge pattern. It is more sensitive than the iconoscope.

orthiconoscope–*See* Orthicon.

orthocode – An arrangement of black and white bars that resembles a piano keyboard and that can be read by an electric-eye device.

orthocore–A completely closed flux memory device designed to almost duplicate the geometry of the ferrite core memory, eliminate the wiring of memory cores, and provide a plurality of wires through the memory element. The concept involves the formation of a group of plastic rods around a suitable wiring array.

orthogonal antennas–A pair of radar transmitting and receiving antennas, or a single antenna for transmitting and receiving, designed to permit detection of a difference in polarization between the transmitted and returned energy.

orthogonal axes – Axes which are perpendicular to each other. In an instrument, these axes usually coincide with its axes of symmetry.

osciducer–A transducer in which information pertaining to the stimulus is provided in the form of deviation from the center frequency of an oscillator.

oscillate–To repeat a cycle of motions or to pass through a cycle of state with strict periodicity.

oscillating current–An alternating current –specifically, one that changes according to some definite law.

oscillating quantity–A quantity which alternately increases and decreases in value, but always remains within finite limits–e.g., the discharge of current from a capacitor through an inductive resistance (provided the inductance is greater than the capacitance times the resistance squared).

oscillating transducer – A transducer in which information pertaining to the stimulus is provided in the form of deviation from the center frequency of an oscillator.

oscillation–1. The state of a physical quantity when, in the time interval under consideration, the value of the quantity is continually changing in such a manner that it passes through maxima and minima (e.g., oscillating pendulum, oscillating electric current, and oscillating electromotive force). 2. Fluctuations in a system or circuit, especially those consisting of the flow of electric currents alternately in opposite directions; also, the corresponding changes in voltages. 3. *See* Hunting, 1.

oscillator – 1. An electronic device which generates alternating-current power at a frequency determined by the values of certain constants in its circuits. An oscillator may be considered an amplifier with positive feedback, with circuit parameters that restrict the oscillations of the device to a single frequency. 2. Something that oscillates. In particular, a self-excited electronic circuit whose output voltage or current is a periodic function of time. 3. A generator of an alternating signal, continuous, sinusoidal, or pulsed.

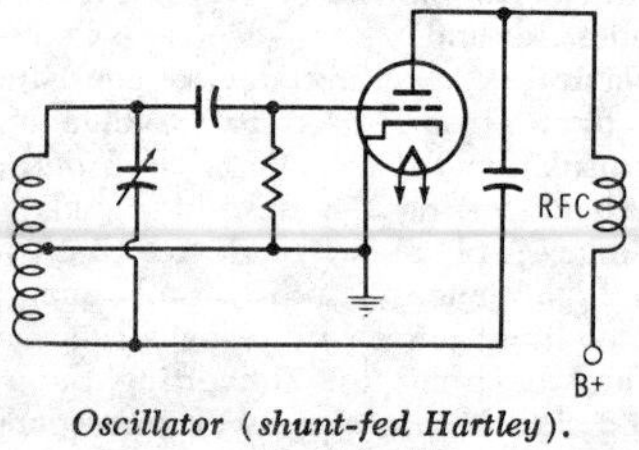

Oscillator (*shunt-fed Hartley*).

oscillator circuit–*See* Oscillator.

oscillator coil – A radio-frequency transformer that provides the feedback required for oscillation in the oscillator circuit of a superheterodyne receiver or in other oscillator circuits.

oscillator harmonic interference – Interference caused in a superheterodyne receiver by the interaction of incoming signals with harmonics (usually the second harmonic) of the local oscillator.

oscillator-mixer–first-detector – A single stage which, in a superheterodyne receiver, combines the functions of the local oscillator and the mixer–first-detector. It usually employs a pentagrid converter tube.

oscillator padder–An adjustable capacitor placed in series with the oscillator tank circuit of a superheterodyne receiver. It is used to adjust the tracking between the oscillator and preselector at the low-frequency end of the tuning coil.

oscillator radiation–The amount of voltage available across the antenna terminals of a receiver (or at a distance), traceable to any oscillators incorporated in the receiver.

oscillatory circuit—A circuit containing inductance and/or capacitance and resistance, so arranged or connected that a voltage impulse will produce a current which periodically reverses.

oscillatory current—A current which periodically reverses its direction of flow.

oscillatory discharge – Alternating current of gradually decreasing amplitude which, under certain conditions, flows through a circuit containing inductance, capacitance, and resistance when a voltage is applied. (*See also* Damped Waves.)

oscillatory surge—A surge which includes both positive and negative polarity values.

oscillistor—A semiconductor bar which is subjected to a magnetic field and a direct current, and which generates oscillations believed to be due to diffusion of ions toward the surface of the semiconductor as a result of the magnetic field.

oscillogram – 1. The recorded trace produced by an oscillograph. 2. A photograph of the luminous trace or image produced by an oscilloscope.

oscillograph – An instrument primarily for producing a record of the instantaneous values of one or more rapidly varying electrical quantities as a function of time, or of another electrical or mechanical quantity.

oscillograph recorder—A form of mechanical oscillograph in which the waveform is traced on a moving strip of paper by a pen.

oscillograph tube—*See* Oscilloscope Tube.

oscillography—The art and practice of utilizing the oscillograph.

oscillometer—An instrument for measuring oscillations (periodic variations) of any kind.

oscilloscope – An instrument in which the horizontal and vertical deflection of the electron beam of a cathode-ray tube are, respectively, proportional to a pair of applied voltages. In the most usual application of the instrument the vertical deflection is a signal voltage and the horizontal deflection is a linear time base.

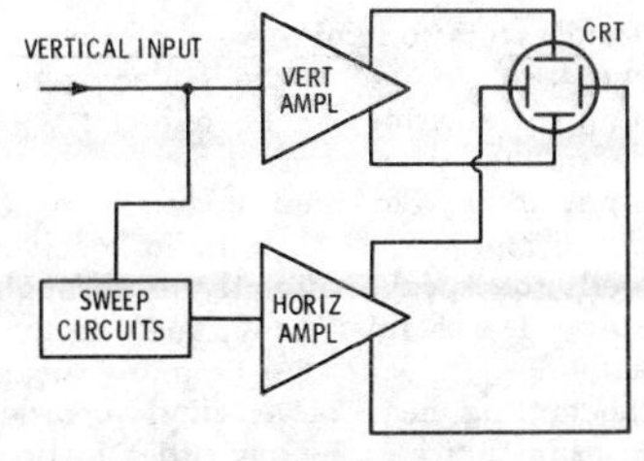

Oscilloscope.

The electron beam strikes the fluorescent screen of the cathode-ray tube and temporarily produces a visible pattern or wave form of some fluctuating electrical quantity such as voltage. The pattern is employed to reveal the detailed variations in rapidly changing electric currents, potentials, or pulses.

oscilloscope differential amplifier—A device that amplifies and displays the voltage difference that eixsts at every instant between signals applied to its two inputs.

oscilloscope tube—Also called oscillograph tube. A cathode-ray tube that produces a visible pattern which is the graphical representation of electric signals. The pattern is seen as a spot or spots, which change position in accordance with the signals.

OSO—Abbreviation for Orbiting Solar Observatory.

O-type backward-wave oscillator—A wideband, voltage-tunable microwave oscillator that uses a fundamental or space harmonic with phase and group velocity of different signs.

outage—Loss of signal in a channel, usually the result of a dropout or a hit.

outconnector—In a flowchart, a connector indicating a point at which a flowline is broken to be continued at another point.

outdoor antenna – A receiving antenna located on an elevated site otuside a building.

outdoor transformer – A transformer of weatherproof construction.

outer marker—In an instrument-landing system, a marker located on a localizer course line at a recommended distance (normally about 4½ miles) from the approach end of the runway.

outgassing—A phenomenon in which a substance in a vacuum spontaneously releases absorbed and occluded constituents as vapors or gases.

outlet—1. The point where current is taken from a wiring system. 2. Convenience receptacle used for supplying power in the home, shop, or laboratory from power-company mains.

outline drawing—A drawing showing approximately overall shape, but no detail.

out of phase—Two or more waveforms that have the same shape, but do not pass through corresponding values at the same instant.

out-of-service jack—A jack, associated with a test jack, into which a shorted plug may be inserted to remove a circuit from service.

outphaser—In electronic organs, a circuit that changes a sawtooth wave to something approaching a square wave by adding to the sawtooth a second sawtooth of twice the frequency and half the amplitude in reverse phase, thus cancelling the even harmonics.

outphasing–In electronic organs, a term applied to a method sometimes used for producing certain voices. Special circuitry, placed between the keying-system output and the formant filters, either adds or subtracts harmonics or subharmonics of the tone-generator signal.

output–1. The current, voltage, power, or driving force delivered by a circuit or device. 2. The terminals or other places where the circuit or device may deliver the current, voltage, power, or driving force. 3. Information transferred from the internal to the secondary or external storage of a computer. 4. The electrical quantity, produced by a transducer, which is a function of the measurand. 5. Useful energy delivered.

output amplifier – A circuit that energizes high-power-level devices upon application of a low-power-level input signal.

output axis – The axis around which the spinning wheel of a gyroscope precesses after the wheel has received an input.

output block–1. In a computer, a portion of the internal storage reserved for holding data which is to be transferred out. 2. A block of computer words treated as a unit and intended to be transferred from internal storage to an external location. 3. A block used as an output buffer.

output capability – The intensity of the strongest signal that a device can put out without exceeding certain limits of overload distortion.

output capacitance – 1. Of an n-terminal electron tube, the short-circuit transfer capacitance between the outut terminal and all other terminals, except the input terminal, connected together. 2. The shunt capacitance at the output terminal of a device.

output capacitive loading – The maximum capacitance that can be placed on the output of an operational amplifier at unity gain without increasing the phase shift to the point of inducing oscillation. The limiting value increases in direct proportion to the closed-loop gain.

output capacity–The number of loads that can be driven by the output of a circuit.

output device – The part of a machine which translates the electrical impulses representing data processed by the machine into permanent results such as printed forms, punched cards, and magnetic writing on tape.

output equipment – Equipment that provides information in visible, audible, or printed form from a computer.

output gap–An interaction gap with which usable power can be extracted from an electron stream.

output impedance–1. The impedance measured at the output terminals of a transducer with the load disconnected and all impressed driving forces (including those connected to the input) taken as zero. 2. Also called dynamic outut impedance. The impedance presented by a power supply to the load. It is calculated from the ratio of the change in output voltage (at the prescribed terminals) to the change in load current causing the change. The impedance is specified from dc to a stated maximum ac. 3. The impedance a device presents to its load.

output indicator–A meter or other device that indicates variations in the signal strength at the output circuits.

output limit–The maximum output signal available when an operational amplifier is operated in the saturation region.

output load current–The maximum current that the amplifier will deliver to, or accept from, a load. This rating includes the amount, however small, which is caused to flow in the feedback loop.

output meter–An alternating-current voltmeter that measures the signal strength at the output of a receiver or amplifier.

output-meter adapter–A device that can be slipped over the plate prong of the output tube of a radio receiver to provide a conventional terminal to which an output meter can be connected during alignment.

output offset voltage–1. The difference between the dc voltages at the two output terminals (or at the output terminal and ground in an amplifier that has one output) when both input terminals are grounded. 2. The output voltage of a negative-feedback op-amp circuit when the input voltage to the circuit is zero. An ideal op amp has zero output offset voltage.

output port–In a fluidic device, the port at which the output signal appears.

output power–1. The power which a system or component delivers to its load. 2. The maximum amount of power, limited by clipping or a specified value of distortion, that an amplifier is capable of delivering to a load of given value.

output resistance–The small-signal ac resistance of an operational amplifier seen looking into the output with no feedback applied and the output dc voltage near zero.

output saturation voltage–The lowest voltage level to which the collector of the output transistor can be reduced without degrading circuit performance.

output stage–The final stage in any electronic equipment. In a radio reeciver, it feeds the speaker directly or through an output transformer. In an audio-frequency amplifier, it feeds one or more speakers, the cutting head of a sound recorder, a transmission line, or any other load. In a transmitter, it feeds the antenna.

output transformer—A transformer used to couple the output stage of an amplifier to a load.

output tube – A power-amplifier tube designed for use in an output stage.

output unit—A computer unit that transfers data from the computer to an external device or from internal storage to external storage.

output voltage—The maximum output voltage which an amplifier will develop in the linear operating region (i.e., before the onset of saturation).

output winding—The winding of a saturable reactor, other than a feedback winding, through which power is delivered to the load.

outside lead—*See* Finish Lead.

overall electrical efficiency (induction- and dielectric-heating usage) – Ratio of the power absorbed by the load material, to the total power drawn from the supply lines.

overall loudness level—A measure of the response of human hearing to the strength of a sound. It is scaled in phons and is an overall single evaluation calculated for the levels of sound pressure of several individual bands.

overall thermoelectric generator efficiency —The ratio of electrical power output to thermal power input to the thermoelectric generator.

overall ultrasonic system efficiency – The acoustical power output at the point of application, divided by the electrical power input into the generator.

overbiasing—A setup procedure whereby the bias current of a tape recorder is adjusted to slightly beyond the point which produces maximum or peak output from the tape. Overbiasing reduces differences due to imperfect uniformity from one tape to another, at the cost of some high-frequency response.

overbunching – The condition where the buncher voltage of a velocity-modulation tube is higher than required for optimum bunching of the electrons.

overcompounding—In a compound-wound generator, use of sufficient series turns to raise the voltage as the load increases, in order to compensate for the increased line drop. In a motor, overcompounding makes it run faster as the load increases.

overcoupled circuit – A tuned circuit in which the coupling is greater than the critical coupling. The result is a broadband response characteristic.

overcurrent—In a circuit, the current which will cause an excessive or even dangerous rise in temperature in the conductor or its insulation.

overcurrent protection—*See* Overload Protection.

overcurrent protective device—A device operative on excessive current which causes and maintains the interruption of power in the circuit.

overcutting—In disc recording, the cutting through of one groove into an adjacent one when the level becomes excessive.

overdamping – Any periodic damping greater than the amount required for critical damping. *See also* Aperiodic Damping.

overdriven amplifier – An amplifier stage designed to distort the input-signal waveform by permitting the grid signal to drive it beyond cutoff or even into plate-current saturation.

overflow—1. The condition occurring whenever the result of an arithmetic operation exceeds the capacity of the number representation in a digital computer. 2. The carry digit arising from (1) above. 3. The generation of a quantity that exceeds the capacity of the computer storage facility. 4. That portion of the result of an operation that exceeds the capacity of the intended unit of storage. 5. In a calculator, when a number that is entered, or the answer to a calculation, exceeds the capacity of the working register, the excess digits "overflow" the register. In most machines, an overflow signal (either audible or visible) indicates when overflow occurs.

overflow indicator – 1. A bistable trigger that changes state on the occurrence of overflow in the computer register with which it is associated. The overflow indicator may be interrogated and/or restored to the original state. 2. In a calculator, a warning light or illuminated letter E, to indicate existence of overflow condition.

overflow position—In a computer, an extra register position in which the overflow digit is developed.

overflow storage – Additional storage provided in a store and forward switching center of a computer to prevent the loss of messages (or parts of messages) offered to a completely filled line store.

overhang—1. In terms of printed wiring, the inverted shelf of plating formed when conductor material is selectively removed from under the plating. The measurement between the base copper and the plating (generally refers to one side of the conductor only; thus, for a conductor width

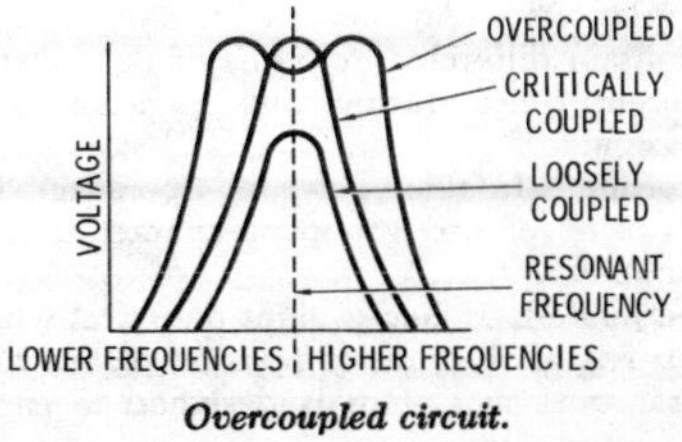

Overcoupled circuit.

reduction, one must take two times the overhang). 2. The critical dimension by which the stylus overreaches the center spindle of a turntable when an offset tone arm is mounted for minimum tracking error.

overhead line—A conductor carried on elevated poles (e.g., telephone or telegraph wires).

overinsulation—In a coil, the insulating material placed over a wire which is brought from the center over the top or bottom wall.

overlap—1. The amount by which the effective height of the scanning facsimile spot exceeds the nominal width of the scanning line. When the spot is rectangular, overlap may be expressed as a percentage of the nominal width of the scanning line. 2. To perform one operation at the same time that another operation is being performed; for example, to carry on input/output operations at the same time that instructions are being executed by the central processing unit.

overlapping contacts—Combinations of two sets of contacts actuated by a common means, each set closing in one of two positions, and so arranged that the contacts of one set open after the contacts of the other set have been closed.

overlap radar—Long-range radar which is located in one sector but also covers a portion of another sector.

overlay—In a computer, the technique of using the same blocks of internal storage for different routines during different stages of a problem, e.g., when one routine is no longer needed in internal storage, another routine can be placed in that storage location.

overlay load module — A computer load module that has been divided into overlay segments and has been provided by the linkage editor with information that permits the desired loading of segments to be implemented by the overlay supervisor when requested.

overlay supervisor — A routine for control of the sequencing and positioning of computer-program segments in limited storage during their execution.

overlay transistor—A transistor containing a large number of emitters connected in parallel to provide maximum power amplification at extremely high frequencies.

overlay zone—An area of memory used to contain different programs or data at different times during the operation of a system.

overline — In teletypewriter operation, the printing of one group of characters over another.

overload—1. A load greater than that which an amplifier, other component, or a whole transmission system is designed to carry. It is characterized by waveform distortion or overheating. 2. An electrically operated counter that indicates the number of times all trunks are busy between the various telephone office units. 3. In an analog computer, a condition within a computing element or at its output that causes a significant computing error as a result of the saturation of one or more parts of the computing element. 4. The amount beyond the specified maximum magnitude of the measurand which, when applied to a transducer, does not cause a change in performance beyond specified tolerance.

overload capacity — The level of current, voltage, or power beyond which a device will be ruined. It is usually higher than the rated load capacity.

overload level—The level at which a system, component, etc., ceases to operate satisfactorily and produces signal distortion, overheating, damage, etc.

overload margin — In an amplifier, the safety margin prior to the onset of overload to avoid clipping on transients. This also enhances the reproduction, giving it a "smoother" quality in many cases.

overload operating time — The length of time a system, component, etc., may be safely subjected to a specified overload current.

overload protection—1. A device which automatically disconnects the circuit whenever the current or voltage becomes excessive 2. Effect of a device operative on excessive current but not necessarily on short circuit, to cause and maintain the interruption of current flow to the device governed. Also called overcurrent protection.

overload recovery time—The time required for an amplifier to regain its ability to amplify within stated specification limits after distortion of the output voltage amplitude by the application of a specified input voltage exceeding the rated amplitude.

overload relay—A relay designed to operate when its coil current rises above a predetermined value.

overmodulation — Modulation greater than 100%. Distortion occurs because the carrier is reduced to zero during certain portions of the modulating signal.

overpotential—Also called overvoltage. A voltage greater than the normal operating voltage of a device or circuit.

overpressure — Pressure greater than the full-scale rating of a pressure transducer.

overpunch—Also called zone punch. A hole punched in one of the three top rows of a punch card and which, in combination with a second hole in one of the nine lower rows, identifies an alphabetic or special character.

overrange—The situation where the signal

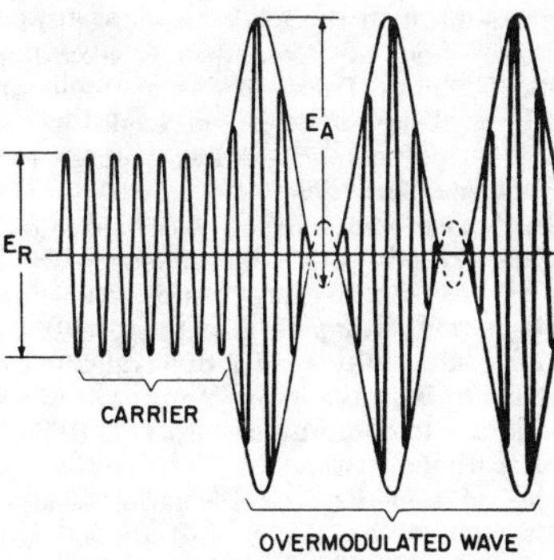

Overmodulation.

being measured exceeds the full-scale value of a digital panel meter. Digital panel meters are designed to indicate this condition with a blinking display or the actuation of an appropriate symbol.

override—To manually or otherwise deliberately overrule an automatic control system or circuit and thereby render it ineffective.

overscanning—In a cathode-ray tube, the deflection of the electron beam beyond the normal limits of the screen.

overscan recovery—A characteristic of differential comparators that states the time required for the amplifier to recover to within some amount of voltage after a return to the screen of a crt.

overshoot—1. The initial transient response, which exceeds the steady-state response, to a unidirectional change in input. 2. Amplitude of the first maximum excursion of a pulse beyond the 100% amplitude level expressed as a percentage of this 100% amplitude. 3. A transient rise beyond regulated output limits, occurring when the ac power input is turned on or off, and for line or load step changes. 4. The amount which the indicator travels beyond its final steady deflection when a new constant value of the measured quantity is suddenly applied to the instrument. The overtravel and deflection are determined in angular measure, and the overshoot is expressed as a percentage of the change in steady deflection. 5. Reception of microwave signals at an unintended location because of an unusual atmospheric condition that sets up variations in the index of refraction.

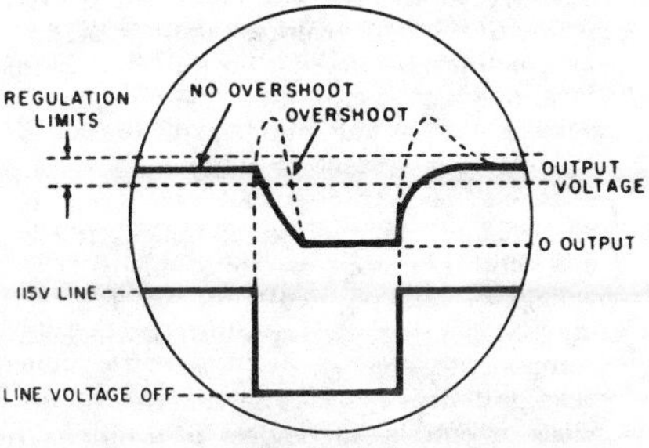

Overshoot, 3.

overshoot distortion—*See* Overthrow Distortion.

overtemperature protection—A thermal relay or other protective device which turns off the power automatically in the event of the occurrence of an overtemperature condition.

over-the-horizon radar — A type of high-powered radar used to "see" over the horizon by means of scatter propagation.

over-the-horizon transmission—*See* Scatter Propagation.

overthrow distortion—Also called overshoot distortion. The distortion that occurs in a signal wave when the maximum amplitude of the signal wavefront exceeds the steady-state amplitude.

overtone—A component of a complex tone having a pitch higher than that of the fundamental component. The term "overtone" has frequently been used in place of "harmonic," the *n*th harmonic being called the (*n*-1)th overtone. There is, however, ambiguity sometimes in the numbering of components of a complex sound when the word "overtone" is employed. Moreover, the word "tone" has many different meanings, so that it is preferable to employ terms which do not involve "tone" whenever possible.

overtone crystal — A quartz crystal cut so that it will operate at a harmonic of its fundamental frequency or at two frequencies simultaneously, as in a synthesizer.

overvoltage—1. The amount by which the applied voltage in a radiation-counter tube exceeds the Geiger-Mueller threshold. 2. *See* Overpotential.

overvoltage relay—A relay designed to operate when its coil voltage rises above a predetermined value.

Ovshinsky effect — The characteristic of a special thin-film solid-state switch that has identical response to both positive and negative polarities so that current can be made to have the same magnitude in both directions.

O-wave—*See* Ordinary Wave.

Owen bridge—A four-arm, alternating-current bridge for measuring self-inductance in terms of capacitance and resistance. One arm, adjacent to the unknown inductor, comprises a capacitor and resistor in series. The arm opposite the unknown consists of a second capacitor, and the fourth arm is a resistor. Usually the bridge is balanced by adjusting the resistor in series with the first capacitor, and also the resistor in series with the inductor. The balance is independent of frequency.

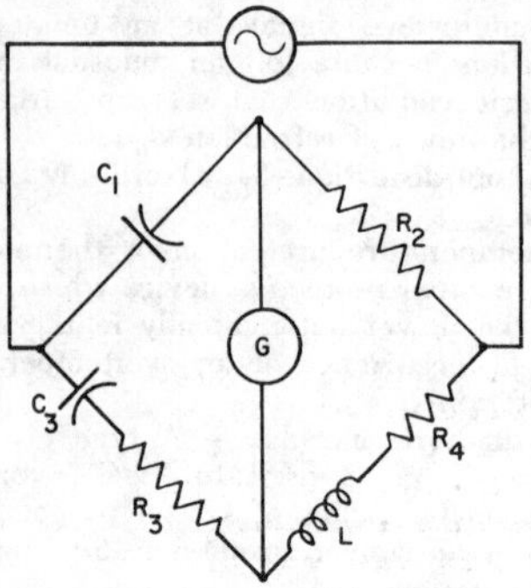

$C_3R_4 = C_1R_2$ $L = C_1R_3R_2$

Owen bridge.

oxalizing–*See* Surface Insulation.

oxidation – 1. Commonly known as rust when ferrous material is involved. The increase in oxygen or in an acid-forming element or radical in a compound. 2. The process of combining with oxygen. More generally, the process by which atoms lose valence electrons or begin to share them with more electronegative atoms. 3. The reaction of oxygen on a compound. Usually detected by a change in the appearance or feel of the surface or by a change in physical properties or both.

oxide–In magnetic recording, microscopic particles of ferric oxide dispersed in a liquid binder and coated on a recording-tape backing. These oxides are magnetically hard–i.e., once magnetized, they remain so permanently unless exposed to a strong magnetic field.

oxide breakdown voltage – That voltage which exceeds gate oxide dielectric breakdown, causing a gate-to-substrate short.

oxide-coated cathode – A cathode that has been coated with oxides of alkaline-earth metals to improve its electron emission at moderate temperatures. Also called Wehnelt cathode.

oxide isolation–Electrical isolation of a circuit element by a layer of silicon oxide formed between the element and the substrate.

oximeter – An instrument that determines the degree of oxygenation of the blood (e.g., in the ear lobe). It uses a photoelectric cell.

ozone–An extremely reactive form of oxygen, normally occurring around electrical discharges and present in the atmosphere in small but active quantities. It is faintly blue and has the odor of weak chlorine. In sufficient concentrations, it can break down certain rubber insulations under tension (such as a bent cable).

ozone-producing radiation – Ultraviolet energy shorter than about 220 nanometers, which decomposes oxygen (O_2), thereby producing ozone (O_3). Some ultraviolet sources generate energy at 184.9 nanometers, which is particularly effective in producing ozone.

P

P–Symbol for permeance.

p – Abbreviation for power (combining form, as in pf for power factor), plate of an electron tube, or the prefix pico- (10^{-12}).

pA–Abbreviation for picoampere.

PABX–Abbreviation for Private Automatic Branch Exchange. Has the same usage as a PBX except that calls within the system are completed automatically by dialing. An attendant at an attendant's board is required to route and complete incoming calls from the central office. Stations within the system are connected to the central office by dialing directly, or they are made to go through the attendant as company policy dictates.

pacemaker – An electronic instrument for starting and/or maintaining the heartbeat. The instrument is essentially a pulse generator with its output applied either externally to the chest or internally to the heart muscle. In cases requiring long-term application, the device is surgically implanted in the body, and its electrodes contact the heart directly. Also called pacer.

pacer–*See* Pacemaker.

pack – 1. In computer programming, to combine several fields of information into one machine word. 2. To compress data in a storage medium by taking advantage of known characteristics of the data in such a way that the original data can be recovered; e.g., to compress data in a storage medium by making use of bit or byte locations that would otherwise go unused.

package count–The number of packaged circuits in a system or subsystems.

packaged magnetron–An integral structure comprising a magnetron, its magnetic circuit, and its output matching device.

packaging–The physical process of locating, connecting, and protecting devices, components, etc.

packaging density–1. The number of devices or equivalent devices in a unit volume of a working system. 2. In a computer, the number of units of information per dimensional unit.

packing—Excessive crowding of carbon particles in a carbon microphone. The abnormal pressure of the particles lowers their resistance. As a result, the current increases excessively and fuses some of the particles together, further lowering the resistance and raising the current. Packing causes the sensitivity of the microphone to decrease.

packing density—1. In a digital computer, the number of units of desired information contained within a storage or recording medium. 2. The amount of digital information recorded along the length of a tape measured in bits per inch (bpi).

packing factor – The number of pulses or bits of information that can be written on a given length of magnetic surface.

pack unit – A term applied to a compact, combination radio transmitter/receiver that can be carried or strapped on the back. Some pack units are popularly known as walkie-talkies.

pad—1. A transducer capable of reducing the amplitude of a wave without introducing appreciable distortion. 2. A device inserted into a circuit to introduce transmission loss or to match impedances. 3. A metal electrode that is connected to the output of a diathermy machine and placed on the body over the region being treated. 4. *See* Land, 2.

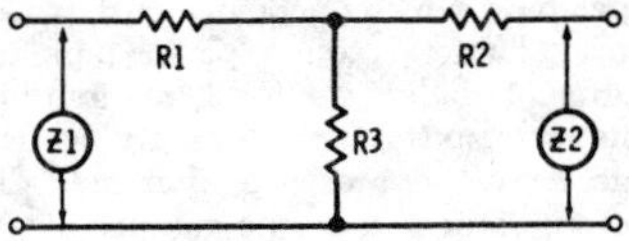

Pad, 2.

pad character – A character introduced to consume time while a function (usually mechanical, such as carriage return, form eject, etc.) is being accomplished.

padder—A series capacitor inserted into the oscillator tuning circuit of a superheterodyne receiver to control the calibration at the low-frequency end of the tuning range.

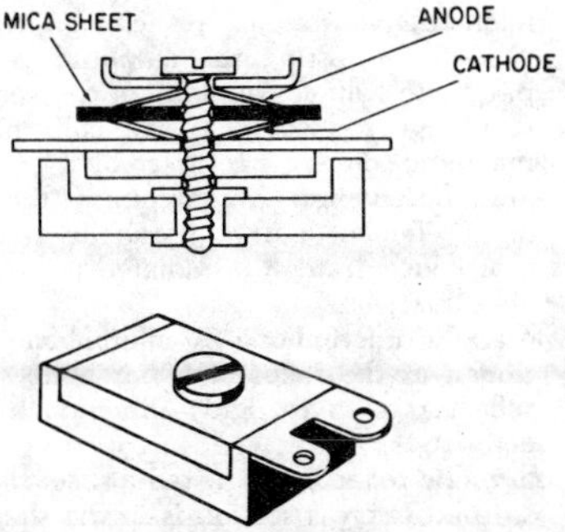

Padder capacitor.

pad electrode—One of a pair of electrode plates between which a load is placed for dielectric heating.

page—A natural grouping of memory locations by higher order address bits. In an 8-bit microprocessor, $2^8 = 256$ consecutive bytes often may constitute a page. Then words on the same page only differ in the lower-order 8 address bits.

page printer—A high-speed unit that prints characters one at a time to full page format.

paging—Methods for locating and exchanging segments to and from the main computer memory.

paging system—Communications system for summoning individuals (doctors, nurses, hospital personnel) or making public announcements.

paint—1. Vernacular for a target image on a radarscope. 2. To draw vectors when the beam is unblanked. 3. To shade the interior of a closed graphical image with diagonal lines, cross hatch, points, etc.

pair—In electric transmission, two like conductors employed to form an electric circuit.

paired cable—A cable in which all of the conductors are arranged in the form of twisted pairs, none of which is arranged with others to form quads.

pairing—In television, the imperfect interlace of lines comprising the two fields of one frame of the picture. Instead of being equally spaced, the lines appear in groups of two—hence the name.

PAL—(Phase Alternation Line.) Pertaining to a color television system in which the subcarrier derived from the color burst is inverted in phase from one line to the next in order to minimize hue errors that may occur in color transmission.

Palmer scan – A combination of circular and conical scans. The beam is swung around the horizon at the same time the conical scan is performed.

pam – Abbreviation for pulse-amplitude modulation.

pam/fm—Frequency modulation of a carrier by pulses which, in turn, are modulated by data.

pam/fm/fm – Frequency modulation of a carrier by subcarriers modulated by pulses which, in turn, are modulated by data.

pan – 1. To move a television or movie camera slowly up and down or across a scene to secure a panoramic effect. 2. To move the camera up and down, or back and forth, in order to keep it trained on a moving object.

pan and tilt—An accessory upon which a camera is mounted to facilitate movement (panning and tilting) by the operator or by a remote control unit.

pancake coil—A coil shaped like a pancake, usually with the turns arranged in a flat spiral.

panel—1. An electrical switchboard or instrument board. 2. A mounting plate of metal or insulation for the controls and/or other parts of equipment.

panel layout—The general physical layout of an electrical control panel with all relays, disconnect switch, control transformer, terminal strip, etc.

panning—*See* Pan.

panoramic adapter – An attachment used with a search receiver to provide, on an oscilloscope screen, a visual presentation of the band of frequencies extending above and below the center frequency to which the search receiver is tuned.

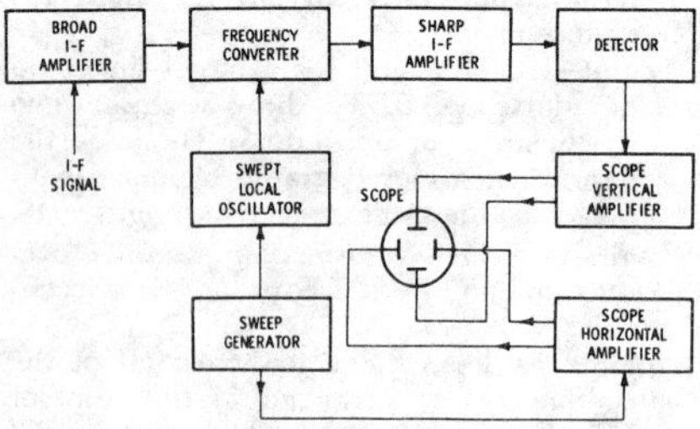

Panoramic adapter.

panoramic display—A display that shows at the same time all the signals received at different frequencies.

panoramic presentation—A presentation of signals as intensity pips (vertical deflections) along a line. The horizontal distance along the line represents frequency.

panoramic radar – A nonscanning radar which transmits signals omnidirectionally over a wide beam.

panoramic receiver – A radio receiver that displays, on the screen of a cathode-ray tube, the presence and relative strength of all signals within a wide frequency range. Used in communications for monitoring a wide band, locating open channels quickly, indicating intermittent signals or interference, and monitoring a frequency-modulated transmitter.

panoramic sonic analyzer – A heterodyne-type instrument which separates the frequency components of a complex waveform and displays them on an oscillographic screen, indicating both frequency and magnitude.

pan-range—Intensity-modulated A-type indication with slow vertical sweep applied to video. Stationary targets give solid vertical deflection, and moving targets give broken vertical deflection.

pantography – A system for transmitting and automatically recording radar data from an indicator to a remote point.

paper capacitor—A fixed capacitor consisting of two strips of metal foil separated by oiled or waxed paper or other insulating material, the whole rolled together into a compact roll. The foil strips can be staggered so that one strip projects from each end, or tabs can be added. The connecting wires are attached to the strips or tabs.

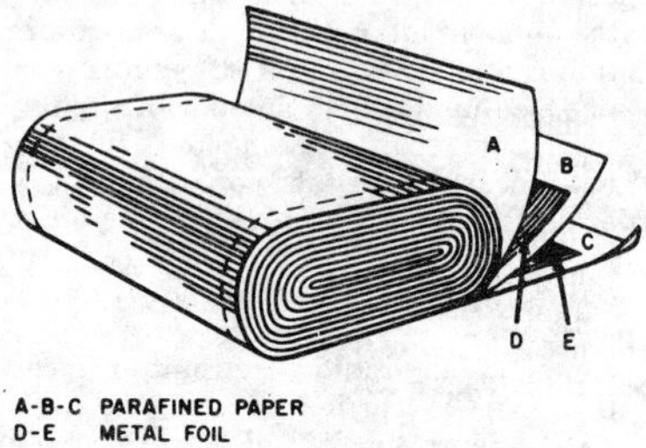

Paper capacitor.

paper electrophoresis – Analytical instrument for a technique in which ions migrate along a strip of porous filter paper saturated with an electrolyte when a potential gradient is applied across the length of the strip. It is used to identify ion types in analysis of serums, proteins, biochemicals, inorganic ions, rare earths, etc.

paper-tape punch—A device that places binary characters on a paper tape in the form of holes punched in appropriate channels on the tape. A binary one is indicated by the presence of a hole, and a zero is indicated by the absence of a hole.

paper-tape reader – A device that senses and translates holes punched in a tape into electrical signals.

par—*See* Precision Approach Radar.

parabola—Locus of points equidistant from a fixed point and a straight line.

parabola controls—Sometimes called vertical-amplitude controls. Three controls in a color television receiver employing the magnetic-convergence principle. They are used for adjusting the amplitude of the parabolic voltages applied, at the vertical-scanning frequency, to the coils of the magnetic-convergence assembly.

parabolic antenna—An antenna with a radiating element and a parabolic reflector that concentrates the radiated power into a beam.

parabolic microphone—A microphone positioned at the focus of a parabolic sound reflector to give highly directional characteristics.

parabolic reflector—A metallic sheet formed so that its cross section is in the shape of a cylindrical parabola. The antenna elements are placed along the line that runs through the focal point of the parabola, parallel to the leading edge of the reflecting sheet.

parabolic-reflector microphone – A microphone employing a parabolic reflector for improved directivity and sensitivity.

paraboloid – A reflecting surface of paraboloidal shape (the shape of a surface formed by rotating a parabola about its axis of symmetry).

paraboloidal reflector – A hollow concave reflector which is a portion of a paraboloid of revolution.

paraffin–A vegetable wax having insulating properties.

parallax–An optical illusion which makes an object appear displaced when viewed from a different angle. Thus, a meter pointer will seem to be at different positions on the scale, depending from which angle it is read. To eliminate such errors, the eye should be directly above the meter pointer.

parallel–1. Also called shunt. Connected to the same pair of terminals, so that the current can branch out over two or more paths. 2. In electronic computers, the simultaneous transmission of, storage of, or logical operations on a character or other subdivision of a word, using separate facilities for the various parts. 3. Indicating a type of computer in which several operations are performed on the same or different data at once.

parallel access–The process of taking information from or placing information into computer storage where the time required for such access depends on simultaneously transferring all elements of a word from a given storage location. Also called simultaneous access.

parallel adder – A conventional technique for adding where two multibit numbers are presented and added simultaneously in parallel.

parallel addition – A form of addition in which the computer operates simultaneously on each set of corresponding digits of two numbers.

parallel arithmetic unit–In a computer, a unit in which separate equipment operates (usually simultaneously) on the digits in each column.

parallel buffer–An electronic device (magnetic cores or flip-flops) used for temporary parallel storage of digital data.

parallel circuit–A circuit in which all positive terminals are connected to a common point, and all negative terminals are connected to a second common point. The voltage is the same across each element in the circuit.

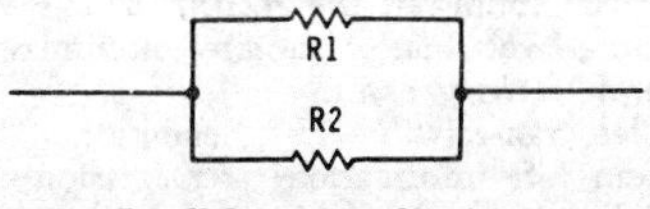

Parallel-connected resistors.

parallel computer – A computer in which the digits or data lines are handled at the same time by separate units.

parallel connection–Also called shunt connection. Connection of two or more parts of a circuit to the same pair of terminals, so that current divides between the parts—as contrasted with a series connection, where the parts are connected end to end so that the same current flows through all.

parallel cut–A Y-cut in a crystal.

parallel digital computer – A computer in which the digits are handled in parallel. Mixed serial and parallel machines are frequently called serial or parallel according to the way arithmetic processes are performed. For example, a parallel digital computer handles decimal digits in parallel, although the bits which comprise a digit might be handled either serially or in parallel.

parallel feed–Also called shunt feed. Application of a dc voltage to the plate or grid of a tube in parallel with an ac circuit, so that the dc and ac components flow in separate paths.

parallel gap welding–A method of resistance welding in which both electrode tips are in close proximity to each other, being separated by a small gap or insulating material, approach the work from the same direction, and contact only one of the two materials being welded.

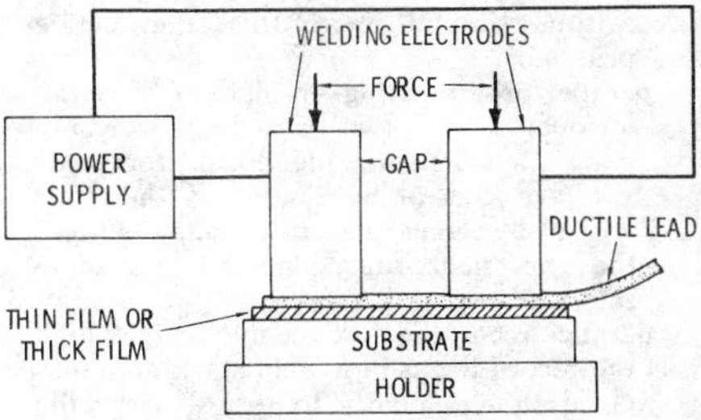

Parallel gap welding.

paralleling reactor–A reactor for correcting the division of load between parallel-connected transformers with unequal impedance voltages.

parallel load–*See* Shift, 2.

parallelogram distortion–In camera or image tubes, a form of distortion which amounts to a skewing of the reproduced image laterally across the crt face.

parallel operation–1. Also called master/slave operation. The connecting of two or more power supplies so that their outputs are tied together, permitting the accumulated flow of current from all units to a

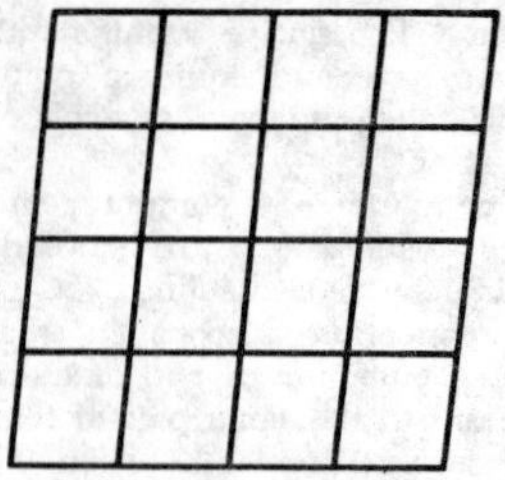

Parallelogram distortion.

common load. In regulated power supplies, interconnections other than the output terminals themselves may be required. For example, the amplifiers of all units but one may be made inoperable, and this single amplifier would control all regulating elements. 2. Pertaining to the manipulation of information within computer circuitry, in which the digits of a word are transmitted simultaneously on separate lines. Faster than serial operation, but requires more equipment.

parallel output—An output arrangement in which two or more bits, channels, or digits are available simultaneously.

parallel-plate oscillator—A push-pull, ultrahigh-frequency oscillator circuit that uses two parallel plates as the main frequency-determining elements.

parallel-plate waveguide—A pair of parallel conducting planes for propagating uniform cylindrical waves that have their axes normal to the plane.

parallel processing – In a computer, the processing of more than one program at a time through more than one active processor.

parallel programming—A method of parallel operation of two or more power supplies in which the feedback terminals (voltage control terminals) of the units are also connected in parallel. Often, these terminals are connected to a separate programming source.

parallel recording – A technique in which the record heads in a head stack are energized simultaneously to record a specific set of bits.

parallel resonance—In a circuit comprising inductance and capacitance connected in parallel, the steady-state condition that exists when the current entering the circuit from the supply line is in phase with the voltage across the circuit.

parallel-resonant circuit—An inductor and capacitor connected in parallel to furnish a high impedance at the frequency to which the circuit is resonant.

parallel-rod oscillator – An ultrahigh-frequency oscillator circuit in which the tank circuits are formed by parallel rods or wires.

parallel-rod tank circuit – A tank circuit

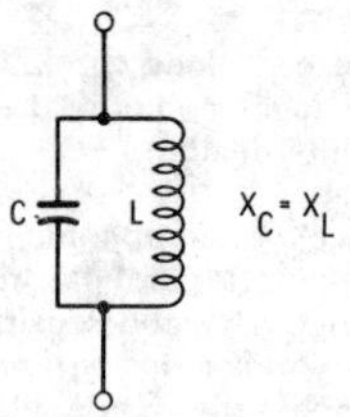

Parallel-resonant circuit.

consisting of two parallel rods connected at their far ends. This is done to provide the small values of inductance and capacitance in parallel required for ultrahigh-frequency circuits.

parallel-rod tuning—A tuning method sometimes used at ultrahigh frequencies. The transmitter, receiver, or oscillator is tuned by sliding a shorting bar back and forth on two parallel rods.

parallel search storage—A type of computer storage in which one or more parts of all storage locations are queried at the same time. *See also* Associative Storage.

parallel-series circuit – Also called shunt-series circuit. Two or more parallel circuits connected together in series.

parallel shift—*See* Shift, 2.

parallel splice—A device in which two or more conductors are joined and lie parallel and adjacent to each other.

parallel storage – Computer storage in which characters, words, or digits are accessed simultaneously.

parallel-T network—Also called twin-T network. A network composed of separate T-networks (usually two), the terminals of which are connected in parallel.

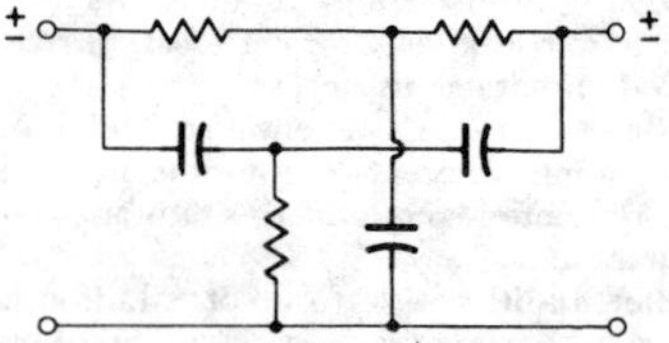

Parallel-T network.

parallel-T oscillator—An RC sine-wave oscillator which provides phase inversion at one discrete frequency and is so connected that positive feedback results only when phase inversion occurs.

parallel transfer—Data transfer where all characters of a word are transferred simultaneously over a set of lines.

parallel-tracking arm—Pickup system which allows phonograph cartridge to track on the true radius of the record, as the recording was made, thereby minimizing lateral tracking error.

parallel transmission—In a computer, the system of information transmission in which the characters of a word are transmitted (usually simultaneously) over separate lines, as contrasted with social transmission.

parallel-wire line—A transmission line consisting of two wires a fixed distance apart.

parallel-wire resonator—A resonator circuit consisting of two parallel wires connected at one end to the oscillator tube or transistor. The other end is short circuited and can be adjusted for the desired frequency.

Paraloc—A phase-shift amplifier the positive feedback of which results in an oscillation that is frequency-modulated by the change in a sensed variable. The sensor is one arm of a resistance bridge.

paramagnetic—Having a magnetic permeability greater than that of a vacuum but less than that of ferromagnetic materials. Unlike the latter, the permeability of paramagnetic materials independent of the magnetizing force.

paramagnetic amplifier—A parametric amplifying device in which a nonlinear element (varactor diode) is pumped at twice the frequency of the signal to be amplified. Amplification is obtained because the pumped state of the element can be returned to the normal state by an input signal of relatively low power, thus releasing the excess energy to an external circuit at the signal frequency.

paramagnetic material—A material having a permeability which is slightly greater than that of a vacuum and which is approximately independent of the magnetizing force.

paramagnetism—Magnetism that involves a permeability somewhat greater than unity.

parameter—1. A constant or element, the value of which characterizes the behavior of one or more variables associated with a given system. 2. A measured value which expresses performance. 3. A test variable used as an arbitrary constant. 4. A variable that is given a constant value for a specific purpose or process.

parameter extraction—A technique that reduces the bandwidth required to transmit a given data sample by means of an information-describing irreversible transformation. These transformations are considered irreversible in that, while they provide useful descriptions of the input signal, they so distort the signal that it is impossible to reconstruct the original waveform.

parameter format — The structure and significance of information which must be assigned values before a generic package may be generated.

parameter tags—Constants used by several programs.

parameter value — The actual information which is assigned to a parameter format.

parametric amplification—A means of amplifying optical waves whereby an intense coherent pump wave is made to interact with a nonlinear optical crystal to produce amplification at two other optical wavelengths.

parametric amplifier—A low-noise device for amplifying signals in the uhf and microwave regions of the electromagnetic spectrum. The essential element of the amplifier is a semiconductor crystal (varactor). By its very nature the diode contributes very little noise, unlike an amplifier utilizing electron tubes, where the random nature of the electron emission from the hot cathode produces large fluctuations in the current, called shot noise. Thus the signals to be amplified can be of extremely low levels.

parametric converter—An inverting or noninverting parametric device used to convert an input signal at one frequency into an output signal at a different frequency.

parametric device—A device the operation of which depends essentially upon the time variation of a characteristic parameter usually understood to be a reactance.

parametric down-converter—A parametric converter in which the output signal is at a lower frequency than the input signal.

parametric excitation—A term referring to the method of exciting and maintaining oscillation in either an electrical or mechanical dynamic system, in which excitation results from a periodic variation in an energy storage element in a system such as a capacitor, inductor, or spring constant.

parametric frequency converter — A frequency converter that utilizes the variation of the reactance parameter of an energy-storage element for frequency conversion.

parametric modulator — A modulator that utilizes the variation in the reactance parameter of an energy-storage element to produce modulation.

parametric oscillator — A device using a parametric amplifier inside a resonant optical cavity to generate a frequency-tunable coherent beam of light from an intense laser beam of fixed frequency. It is tuned by varying the phase-matching properties of the nonlinear material.

parametric testing—Testing based on reasonably precise measurements of voltages or currents.

parametric up-converter—A parametric converter in which the output signal is at a higher frequency than the input signal.

parametron—A digital circuit element utilizing the principle of parametric excitation. It is essentially a resonant circuit with a nonlinear reactive element which oscillates at half the driving frequency. The oscillation can be made to represent a binary digit by the choice between two stationary phases π radians apart.

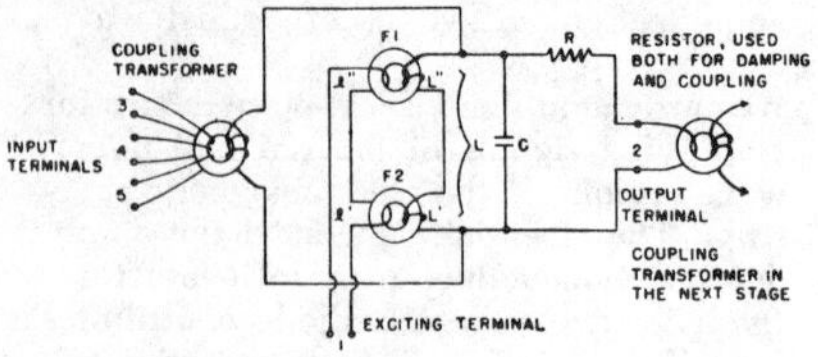

Magnetic types.

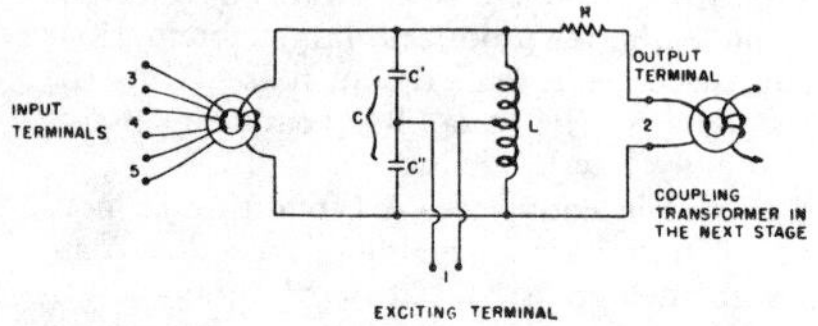

Capacitive types.

Parametrons.

paramistor—A digital logic-circuit module containing several parametron elements.

paraphase amplifier – An amplifier which converts a single input signal into two out-of-phase signals for driving a push-pull stage.

parasite—Current in a circuit, due to some unintentional cause such as inequalities of temperature or of composition; particularly troublesome in electrical measurement.

parasitic – An undesired low- or high-frequency signal in an electronic circuit.

parasitic antenna—An antenna that is excited by radiation from other antennas rather than by electrical connection with them.

parasitic array—An antenna array containing one or more elements not connected to the transmission line.

parasitic components—In a monolithic integrated circuit, the capacitors and diodes which are formed between the planned circuit elements and the substrate during processing. The circuit design must allow for the functional effects of these parasitic components.

parasitic element – 1. Also called passive element. An antenna element (i.e., reflector, director, etc.) not connected to the transmission line or to any driven element. A parasitic element affects the gain and directivity pattern of an antenna, and also acts on a driven element by absorbing and returning energy from it. In a dipole reflector combination, the reflector is the parasitic element. 2. An undesirable but inherent element in a circuit, such as wire resistance, core losses, winding capacitance, or leakage inductance.

parasitic field turn-on—*See* Field Inversion.

parasitic oscillation—1. An undesired, self-sustaining oscillation at a frequency other than the operating frequency. Parasitic oscillations occur chiefly in vacuum-tube circuits. 2. Any undesired oscillation in an oscillator or amplifier stage.

parasitics—Parasitic oscillations.

parasitic suppressor—A parallel resistance, or a parallel combination of inductance and resistance, inserted into a grid or plate circuit to suppress parasitic oscillations.

pard – In an electronic power conversion unit, the Periodic And Random Deviation of the output dc voltage, current, or power from its average value, with all external operational and environmental parameters maintained constant. The load impedance in particular must be held constant. Perturbations in the output which are induced by load impedance changes are dynamic load regulation or transient response. Pard may be defined as rms pard or peak-to-peak pard.

parent population – Prototype or initial group of the articles under consideration.

parity—A method of checking the accuracy of binary numbers. An extra bit, called a parity bit, is added to a number. If even parity is used, the sum of all 1's in the number and its corresponding parity bit is always even. If odd parity is used, the sum of the 1's and the parity bit is always odd.

parity bit—1. An additional bit used with a computer character or electronic channel data processor to provide a check for accuracy. 2. A binary digit appended to an array of bits to make the sum of all the bits always odd or always even.

parity check—A check that tests whether the number of ones (or zeros) in an array of binary digits is odd, or even. Synonymous with odd-even check.

parity tree—A group of exclusive OR gates that can be used to check a number of input bits for either odd or even parity. Parity trees are used both to check and generate parity wherever a redundant bit is added to a word in order to check for error.

part—1. The smallest subdivision of a system. 2. An item which cannot ordinarily be disassembled without destruction.

part failure—A breakdown that cannot be repaired and which ends the life of a part.

part-failure rate—The number of occasions, during a specified time period, on which a given quantity of identical parts will not function properly.

partial—1. A physical component of a complex tone. 2. A component of a sound sensation that can be distinguished as a simple tone which cannot be further analyzed by the ear and which contributes to the character of the complex sound.

partial carry—In parallel addition, a technique involving temporary storage of some or all of the carries instead of allowing them to propagate immediately.

partial dial tone—A high tone that notifies a calling party that he has not completed dialing within a specified period of time, or that not enough digits have been dialed.

partial motor—A motor sold with rotor and stator only—no end bells and no containing frame. Also called a shell-type motor.

partial node—The place in a standing-wave system at which some characteristic of the wave field has a minimum amplitude other than zero.

partial-read pulse—In a computer, any one of the applied currents which cause selection of a core for reading.

partial-select output — In a computer, the voltage response of an unselected magnetic cell produced by the application of partial-read pulses or partial-write pulses.

partial-write pulse—In a computer, any one of the applied currents which cause a core to be selected for writing.

particle — An infinitesimal subdivision of matter—e.g., a molecule, atom, or electron.

particle accelerator—Any device for accelerating charged particles to high energies (e.g., cyclotron, betatron, Van de Graaff generator, linear accelerator, etc.).

particle orientation—The process by which acicular particles are positioned so that their longest dimensions tend to be parallel. Orientation is accomplished in magnetic tape by the combined effects of the sheer force applied during the coating process and a magnetic field applied to the coating while it is still fluid.

particle velocity—The velocity of a given infinitesimal part of a sound wave. The most common unit is centimeter per second.

partitioning—Also called segmenting. In a computer, subdividing a large block into smaller, more conveniently handled subunits.

partition noise—A noise caused in an electron tube by random fluctuations as the electron stream divides between the electrodes. It is more pronounced in pentodes and tetrodes than in triodes.

part 95 rules — FCC rules and regulations governing the Citizens Radio Service.

part programmer — One who translates the physical operations for machining a part into a series of mathematical steps and then prepares the coded computer instructions for those steps.

parts density — The number of parts in a unit volume.

party line—A telephone line serving more than one subscriber, with discriminatory ringing for each.

Paschen's law—The sparking potential between two terminals in a gas is proportional to the pressure times the spark length. For a given voltage, this means the spark length is inversely proportionate to the pressure.

pass—One cycle of processing of a body of data.

passband — 1. The band of frequencies which will pass through a filter with essentially no attenuation. 2. The frequency range in which a filter is intended to pass signals.

passband ripple—In a filter, the difference, in dB, between the minimum loss point and the maximum loss point in a specified bandwidth.

pass element—An automatic variable resistance device, either a vacuum tube or power transistor, in series with the source of dc power. The pass element is driven by the amplifier error signal to increase its resistance when the output needs to be lowered or to decrease its resistance when the output must be raised. *See also* Series Regulator.

passivate—To treat the surface of a semiconductor with a relatively inert material in order to protect it from contamination.

passivation — 1. The growth of an oxide layer on the surface of a semiconductor to provide electrical stability by isolating the transistor surface from electrical and chemical conditions in the environment. This reduces reverse-current leakage, increases breakdown voltages, and raises the power dissipation rating. 2. *See* Glassivation.

passive—1. An inert component which may control, but does not create or amplify energy. 2. Pertaining to a general class of device that operates on signal power alone.

passive acoustic monitoring — The use of microphones and ancillary equipment to provide surveillance by monitoring the sounds in a protected premise.

passive communication satellite — A communication satellite which simply reflects a signal without amplification. In essence, it is a radio mirror. It requires a large reflecting surface and large, high-powered, complex ground stations.

passive component—A nonpowered component generally presenting some loss (expressed in dB) to a system.

passive decoder — A device that is set so that only one specific reply code will pass a decoder and give an output from one decoder for display.

passive detection—Detection of a target by reception of signals emitted by the target rather than by means involving a signal source independent of the target.

passive device—1. A device which exhibits no transistance. It has no gain or control and does not require any input other than a signal to perform its function. Examples of passive devices are conductors, resistors, and capacitors. 2. A component that does not provide rectification, amplification, or switching, but reacts to voltage and current; e.g., resistor, capacitor.

passive electric network – An electric network with no source of energy.

passive element – 1. A parasitic element. 2. A circuit element with no source of energy (e.g., a resistor, capacitor, inductor, etc.).

passive film circuit – A thin- or thick-film circuit network consisting entirely of passive circuit elements and interconnections.

passive homing system—A guidance system based on the sensing of energy radiated by the target. (*See also* Active Homing *and* Homing Guidance.)

passive network—A network with no source of energy.

passive pull-up – A gate output circuit in which the charging current for a load capacitance is obtained through a resistor.

passive reflector—A reflector often used on microwave relay towers to change the direction of a microwave. This permits convenient location of transmitter, repeater, and receiver equipment on the ground rather than at the tops of towers.

passive satellite – A satellite that reflects, without amplification, communications signals from one ground station to another.

passive sonar—*See* Sonar.

passive substrate – A substrate that may serve as a physical support and thermal sink for a thick- or thin-film integrated circuit but does not exhibit transistance. Examples of passive substrates are glass, ceramic, alumina, etc.

passive system—A system that emits no energy and therefore does not reveal its position or existence.

passive tracking system—Usually a system that tracks by reflected radiation from some external source, or by the jet emission of the vehicle (e.g., optical systems, use of commercial radio or television, reflection and infrared systems).

passive transducer—A transducer that does not require any local source of energy other than the received energy.

paste—In batteries, the medium, in the form of a paste or jelly, containing an electrolyte. It is positioned adjacent to the negative electrdoe of a dry cell. In an electrolytic cell, the paste serves as one of the conducting plates.

pa system—Abbreviation for public-address system.

patch—1. To connect circuits together temporarily with a special cord known as a patch cord. 2. In a computer, to make a change or correction in the coding at a particular location by inserting transfer instructions at that location and by adding elsewhere the new instructions and the replaced instructions. This procedure is usually used during checkout. 3. The section of coding so inserted.

patch board—A board or panel where circuits are terminated in jacks for patch cords.

patch cord – Sometimes called an attachment cord. A short cord with a plug or a pair of clips on one end, for conveniently connecting two pieces of sound equipment such as a phonograph and tape recorder, an amplifier and speaker, etc.

Patch cord.

patching—Connecting two lines or circuits together temporarily by means of a patch cord.

patching jack—A jack for interconnection of circuit elements.

patch panel—In a computer, a panel that contains means for changing circuit configurations; usually, it consists of receptacles into which jumpers can be inserted.

path—1. In navigation, an imaginary line connecting a series of points in space and constituting a proposed or traveled route. 2. *See* Channel, 2, 4, 5, 6, *and* 7.

path attenuation—The power loss between transmitter and receiver resulting from all causes.

patient monitor – A system of instruments that permits remote monitoring at a central location in a hospital of such quantities as heart rate, blood pressure, temperature, etc.

pattern – 1. The means of specifying the character of a wave in a guide. This is done by showing the loops of force existing in the guide for that wave. The pattern identifies the order and mode of the wave and the cross-sectional shape of the guide. 2. A geometrical figure representing the directional qualities of an antenna array.

pattern definition — The accuracy, relative to the original artwork, with which pattern edges are reproduced in integrated-circuit elements.

pattern recognition—In a computer, the examination of records for certain code-element combinations.

pattern-sensitive fault—A fault that appears in response to some particular data pattern.

pause control—A feature of some tape recorders making it possible to temporarily stop the movement of the tape without switching the machine from the "play" or "record" position. (Essentially for a tape recorder used for dictation and generally helpful for editing purposes.)

pax — Acronym for private automatic exchange. An automatic system used exclusively for interoffice dial communications and having no trunks to the central office.

pay television — Also called subscription television. A system whereby viewers must insert coins or record cards into a decoding device in order to view a television program that has been deliberately scrambled to prevent unpaid viewing.

P-band—A radio-frequency band extending from 225 to 390 MHz and having a wavelength from 133.3 to 76.9 cm.

pbx — Abbreviation for private branch exchange. A manual telephone system located on the premises of a business and requiring an attendant to complete all calls. It is usually owned by the telephone company and is equipped with trunks to a telephone-company central office.

pC — 1. Abbreviation for picocoulomb. 2. Abbreviation for picocurie.

p-channel device—A device constructed on an n-type silicon substrate, whose drain and source components are of p-type silicon. *See also* PMOS.

p-channel FET—An FET in which the "resistive bar" is of p-type semiconductor material.

pcm—1. Abbreviation for pulse-code modulation. 2. Abbreviation for punched card machine.

pcm/fm—Frequency modulation of a carrier by pulse-code-modulated information.

pcm/fm/fm — Frequency modulation of a carrier by subcarrier(s) which is (are) frequency modulated by pulse-code-modulated information.

pcm level—The number by which identification of a given quantized-signal subrange may be made.

pcm/pm—Phase modulation of a carrier by pulse-code-modulated information.

P-display—*See* Plan-Position Indicator.

pdm—Abbreviation for pulse-duration modulation.

pdm/fm—Frequency modulation of a carrier by pulses which are modulated in duration by information.

pdm/fm/fm — Frequency modulation of a carrier by subcarrier(s) which is (are) frequency modulated by pulses which are time duration modulated by information.

pdm/pm — Phase modulation of a carrier by pulses which are duration modulated by information.

peak—Also called crest. 1. A momentary high amplitude level occurring in electronic equipment. 2. A momentarily high volume level during a radio program. It causes the volume indicator at the studio or transmitter to swing upward. 3. The maximum instantaneous value of a quantity. 4. To increase or sharpen the peaks of a waveform. 5. To broaden the frequency response of an amplifier by including inductors in its coupling networks so as to cancel the input and output capacitances of its active elements.

peak alternating gap voltage—In a microwave tube, the negative of the line integral of the peak alternating electric field, taken along a specified path across the gap.

peak amplitude—The maximum deviation (e.g., of a wave) from an average or mean position.

peak anode current—The maximum instantaneous value of an anode current in an electron tube.

peak cathode current (fault)—The highest instantaneous value of a nonrecurrent pulse of cathode current occurring under fault conditions.

peak cathode current (steady-state) — The maximum instantaneous value of a periodically recurring cathode current.

peak cathode current (surge)—The highest instantaneous value of a randomly recurring pulse of cathode current.

peak-charge characteristic — The function giving the relation of one half the peak-to-peak value of transferred charge in the steady state to one half the peak-to-peak value of a specified symmetrical alternating voltage applied to a nonlinear capacitor.

peak current — 1. The maximum current which flows during a complete cycle. 2. Maximum amplitude of current an ionized device can pass without permanent change in breakdown ratings or published life specifications.

peak discharge energy — The maximum amount of energy a device can withstand during operation without permanent change in breakdown ratings or published life specifications.

peak distortion—The largest total distortion of signals noted during a period of observation.

peak electrode current—The maximum instantaneous current that flows through an electrode.

peak-envelope power of a radio transmitter—The average power supplied to the antenna transmission line by a transmitter during one radio-frequency cycle at the highest crest of the modulation envelope, taken under conditions of normal operation.

peak flux density—The maximum flux density in a magnetic material.

peak forward anode voltage — The maximum instantaneous anode voltage in the direction the tube is designed to pass current.

peak forward-blocking voltage—The maximum instantaneous value of repetitive positive voltage that may be applied to the anode of an SCR with its gate circuit open.

peak forward drop—The maximum instantaneous voltage drop measured when a tube or rectifier cell is conducting forward current, either continuously or during transient operation.

peaking—Adjusting a component so as to increase the response of a circuit at a desired frequency or band of frequencies.

peaking circuit—A circuit capable of converting an input wave into a peaked waveform.

peaking control—In a television receiver, a fixed or variable resistor-capacitor circuit which controls the negative shape of the pulses originating at the horizontal oscillator. This is done to assure a linear sweep.

peaking network—A type of interstage coupling network used to increase the amplification at the upper end of the frequency range. It consists of an inductance effectively in series (series peaking network) or shunt (shunt peaking network) with a parasitic capacitance.

peaking resistor—A resistor placed in series with the charging capacitor of the vertical sawtooth generator. By adding a negative peaking pulse to the sawtooth voltage, it creates the waveform required to produce a linear sawtooth current in the yoke.

peaking transformer—A transformer operated in such a way that its core is saturated in one direction or the other for most of a single ac cycle, with the result that the secondary voltage waveform is sharply peaked at each flux reversal. Sharpness of the peaking is enhanced by an approximately rectangular hysteresis loop in the core.

peak inverse anode voltage—The maximum instantaneous anode voltage in the direction opposite from that in which the tube is designed to pass current.

peak inverse voltage—The peak ac voltage which a rectifying cell or pn junction will withstand in the reverse direction.

peak level — The maximum instantaneous level that occurs during a specific time interval (i.e., in acoustics, the peak sound pressure level).

peak limiter—A device which automatically limits the magnitude of its output signal to approximate a preset maximum value by reducing its amplification when the instantaneous signal magnitude exceeds a preset value.

peak load—The maximum electrical power load consumed or produced in a stated period of time. It may be the maximum instantaneous load or the maximum average load over a designated interval of time.

peak magnetizing force — The upper or lower limiting value of a magnetizing force.

peak-or-valley readout memory — A circuit in which the output remains at the condition corresponding to the most positive (least negative) or vice versa input signal since the circuit was set to initial conditions, until reset to those conditions.

peak plate current—The maximum instantaneous current passing through the plate circuit of a tube.

peak point—The point on the characteristic curve of a tunnel diode corresponding to the lowest voltage in the forward direction for which the differential conductance is zero.

peak-point emitter current—The maximum emitter current that can flow without allowing a UJT to go into the negative-resistance region.

peak power—1. The mean power supplied to the antenna of a radio transmitter during one radio-frequency cycle at the highest crest of the modulation envelope. 2. The maximum power of the pulse from a radar transmitter. Since the resting time of a radar transmitter is longer than its operating time, the average power output is much lower than the peak power. 3. Maximum instantaneous audio power available from a power amplifier.

peak power output—1. The output power averaged over the radio-frequency cycle having the maximum peak value that can occur under any combination of signals transmitted. 2. Maximum instantaneous power output from any power amplifier. Usually related to the saturation power of the amplifier.

peak pulse amplitude—The maximum absolute peak value of the pulse, excluding unwanted portions such as spikes.

peak pulse power—The maximum power of a pulse, excluding spikes.

peak response—The maximum response of a system to an input.

peak signal level — An expression of the maximum instantaneous signal power or voltage as measured at any point in a facsimile transmission system. This includes auxiliary signals.

peak sound pressure—The maximum absolute value of instantaneous sound pressure for any specified time interval. The most common unit is the microbar.

peak spectral emission—The wavelength at which the radiation from a lamp has the highest intensity.

peak speech power—The maximum instantaneous speech power over the time interval considered.

peak to peak—The algebraic difference between the positive and negative maximum values of a waveform.

peak-to-peak amplitude—The amplitude of an alternating quantity, measured from positive peak to negative peak.

peak-to-peak voltmeter—A voltmeter which indicates the overall difference between the positive and negative voltage peaks.

peak value—Also called crest value. The maximum instantaneous value of a varying current, voltage, or power. For a sine wave, it is equal to 1.414 times the effective value of the sine wave.

peak voltage—The maximum value present in a varying or alternating voltage. This value may be either positive or negative.

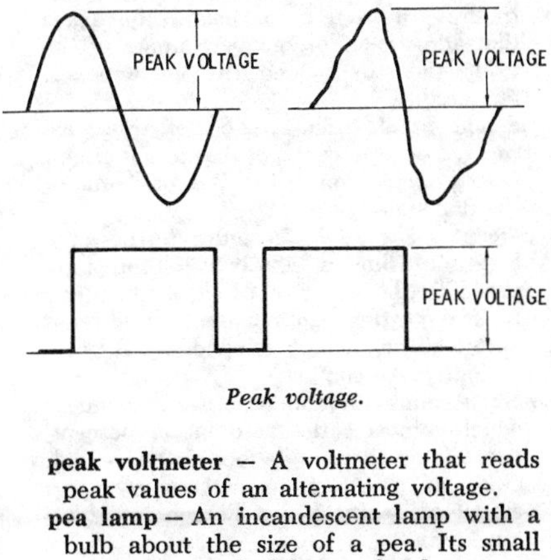

Peak voltage.

peak voltmeter — A voltmeter that reads peak values of an alternating voltage.

pea lamp — An incandescent lamp with a bulb about the size of a pea. Its small size makes it ideal for use by doctors, on instrument panels, and in small flashlights.

pedal clavier—In an organ, the pedal keyboard which supplies the bass accompaniment for the other manuals.

pedal keyboard—*See* Pedal Clavier.

pedestal — 1. A substantially flat-topped pulse which elevates the base level for another wave. 2. The base of a radar antenna.

pedestal level—*See* Blanking Level.

pedestal pulse — A square-wave pulse or gate on which a video signal or sweep voltage may be superimposed.

peek-a-boo—In a computer, a method of determining the presence or absence of holes in identical locations on punched cards by placing one card on top of another. (*See also* Batten System.)

peel-strength adhesion—*See* Bond Strength.

peg-count meters — In telephone practice, meters or registers used to indicate the number of trunks tested, circuits passed busy, test failures, and repeated tests completed.

pellicle—A thin membrane that has the capability of splitting beams, polarizing light, and reflecting images with few or no optical side effects.

Peltier coefficient—The quotient of the rate of Peltier heat absorption by the junction of two dissimilar conductors divided by the current through the junction. The Peltier coefficient of a couple is the algebraic difference between either the relative or absolute Peltier coefficients of the two conductors making up the couple.

Peltier effect — The production or absorption of heat at the junction of two metals when current is passed through the junction. Reversing the direction of the current changes a production of heat to an absorption, and vice versa.

Peltier electromotive force—1. The component of voltage produced by a thermocouple after being heated by the Peltier effect at the junction of the different metals. It adds to the Thomson electromotive force to produce the total voltage of the thermocouple. 2. The boundary emfs produced across the junctions of two different metals, associated with the heating and cooling effects of the two junctions.

Peltier heat—The thermal energy absorbed or produced as a result of the Peltier effect.

pen centering—An electrical or mechanical adjustment by which an oscillograph pen is positioned to channel center.

pencil beam—A radar beam in which the energy is confined to a narrow cone.

pencil-beam antenna—A unidirectional antenna in which those cross sections of the major lobe perpendicular to the maximum radiation are approximately circular.

pencil tube—A small tube designed for operation in the ultrahigh-frequency band and used as an oscillator or rf amplifier.

pendant—The type of plug and/or receptacle that is not mounted in a fixed position or attached to a panel or side of equipment.

pendant station—A push-button station suspended from overhead and connected by means of flexible cord or conduit, but supported by a separate cable.

pendulous accelerometer – A device which measures linear accelerations by means of a restrained unbalanced mass. Two pivots and jewels support the unbalanced gimbal, and the torsion bar functions as the spring.

penetrating frequency – *See* Critical Frequency.

penetration depth—1. In induction heating, the effective depth of the induced current. The skin effect causes this to be nearer the surface with high frequencies than with low frequencies. 2. The extent to which an external magnetic field penetrates a superconductor.

Penning discharge—A type of discharge in which electrons are forced to oscillate between two opposed cathodes and are prevented from going to the surrounding anode by the presence of a magnetic field. It is sometimes referred to as a pig discharge because the device producing it was first used as an ionization gage called the Penning Ionization Gage.

Penning Ionization Gage—*See* Penning Discharge.

pen position – An electrical or mechanical adjustment by which an oscillograph pen is positioned to any desired amplitude grid mark on the chart to represent zero signal.

pent—Abbreviation for pentode.

pentagrid converter—A pentagrid tube used as a combination oscillator and mixer in a superheterodyne receiver.

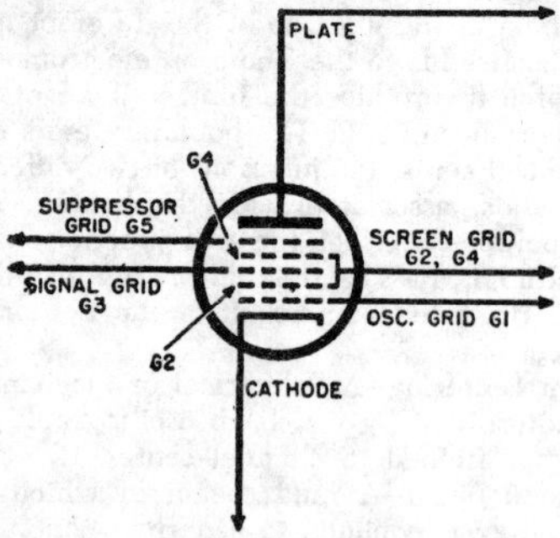

Pentagrid converter.

pentagrid mixer—A pentagrid tube used to mix the rf and local-oscillator signals in a superheterodyne receiver.

pentagrid tube – An electron tube having five grids, plus an anode and a cathode.

pentatron – A five-electrode vacuum tube that provides push-pull amplification with a single tube. It has one cathode, two grids, and two anodes. In effect, it is two tubes in one.

pentode—A five-electrode electron tube containing an anode, a cathode, a control electrode, and two grids.

pentode field-effect transistor – A five-lead transistor with three gates. It can be connected like a pentode if each of the gates is supplied from an independent bias source.

pentode transistor—A point-contact transistor in which there are four point-contact electrodes; the body serves as a base, and there are three emitters and one collector.

perceived noise level – An empirical measure that includes allowance for the subjective reaction of people to noise in the various frequency ranges. It is expressed in decibels (PNdB).

percentage-differential relay—A differential relay which functions when the difference between two quantities of the same nature exceeds a fixed percentage of the smaller quantity. This term includes relays formerly known as ratio-balance relays, biased relays, and ratio-differential relays.

percentage of meter accuracy—The ratio of the actual meter reading to the true reading, expressed as a percent.

percentage modulation – *See* Percent of Modulation.

percentage ripple – *See* Percent of Ripple Voltage.

percentage sync—Ratio of the amplitude of the synchronizing signal to the peak-to-peak amplitude of the picture signal between blanking and reference white level, expressed in percent.

percentage timer—A repeat-cycle nonreset timer with fixed cycle length, having a dial adjustment of the percentage of the cycle time for which the contacts are operated.

percent break – The period of time, expressed as a percentage, that a dial circuit stands open compared to the total time of the dial signals.

percent make – 1. In pulse testing, the length of time a circuit is closed compared to the duration of the test signal. 2. The portion (in percent) of a pulse period during which telephone-dial pulse springs make contact.

percent-modulation meter – An instrument which indicates the modulation percentage of an amplitude-modulated signal, either on a meter or a cathode-ray tube.

percent of deafness—*See* Percent of Hearing Loss.

percent of harmonic distortion—A measure of the harmonic distortion in a system or component. It is equal to 100 times the ratio of the square root of the sum of the squares of the root-mean-square harmonic voltages (or currents), to the root-mean-square voltage (or current) of the fundamental.

percent of hearing—At a given frequency, 100 minus the percent of hearing loss at that frequency.

percent of hearing loss – Also called percent of deafness. At a given frequency, 100 times the ratio of the hearing loss in decibels, to the number of decibels between the normal threshold levels of audibility and feeling.

percent of modulation—1. In am, the ratio of half the difference between the maximum and minimum amplitudes of a wave to the average amplitude, expressed in percentage. 2. In fm and tv audio transmission, the ratio of the actual frequency swing, to the frequency swing defined as 100% modulation, expressed in percentage. For fm broadcast stations, a frequency swing of ±75 kHz is defined as 100% modulation. For television, it is ±25 kHz.

percent of ripple voltage—Ratio of the effective (root-mean-square) value of the ripple voltage to the average value of the total voltage, expressed in percent.

percent of syllabic articulation—*See* Syllable Articulation.

percent ripple (rms)—The ratio of the effective (rms) value of the ripple voltage to the average value of the total voltage, expressed in percent. The new term is pard.

perceptron—A system capable of—either in theory or in practice—performing knowledgeable functions such as recognition, classification, and learning. These functions may exist as mathematical analyses, computer programs, or "hardware."

percussion – Musical sounds characterized by sudden or sharp transients. Organ percussion is achieved by causing the tone to start to decay the instant it is played rather than waiting until the key is released.

percussive arc welding – A process where the surfaces to be welded are held at a fixed gap while rf energy is applied. This ionizes the air gap between the two surfaces, causing the air gap to become a conductor. When the air gap is conductive, a capacitor bank dumps a controlled amount of energy into the system for a controlled time period. This results in an electric arc which scarifies the surfaces to be welded and heats them to welding temperature. As the pulse from the capacitor bank decays, a mechanical system drives the two hot surfaces together, consummating the weld.

perfect dielectric—Also called ideal dielectric. A dielectric in which all the energy required to establish the electric field in it is returned to the electric system when the field is removed. A perfect dielectric has zero conductivity and exhibits no absorption phenomena. A vacuum is the only known perfect dielectric.

perforated tape—*See* Punched Tape.

perforator – In telegraphy, a device that punches code signals into paper tape for application to a tape transmitter.

performance—Degree of effectiveness of operation.

performance characteristic—A characteristic measurable in terms of some useful denominator—e.g., gain, power output, etc.

period—1. The time required for one complete cycle of a regular, repeating series of events. 2. The time between two consecutive transients of the pointer or indicating means of an electrical indicating instrument in the same direction through the rest position. Sometimes called "periodic time."

periodic—Repeating itself regularly in time and form.

periodic antenna—An antenna in which the input impedance varies as the frequency does (e.g., open-end wires and resonant antennas).

periodic current – Oscillating current, the values of which recur at equal time intervals.

periodic damping—Also called underdamping. Damping in which the pointer of an instrument oscillates about the final position before coming to rest. The point of change between periodic and aperiodic damping is called critical damping.

periodic duty—Intermittent duty where the load conditions recur at regular intervals.

periodic electromagnetic wave—A wave in which the electric field vector is repeated in detail—either at a fixed point, after a lapse of time known as the period; or at a fixed time, after the addition of a distance known as the wavelength.

periodic electromotive force—An oscillating electromotive force that repeats its sequence of values over equal intervals of time.

periodicity—The variations in the insulation diameter of a transmission cable that result in reflections of a signal, when its wavelength or a multiple thereof is equal to the distance between two diameter variations.

periodic law—When chemical elements are arranged in the ascending or descending order of their atomic number, their properties will occur in cycles.

periodic line—A line consisting of identical, similarly oriented sections, each section having nonuniform electrical properties.

periodic pard—Pertains to that portion of the total pard, in an electronic power supply, the frequency of which is identical or harmonically related to the input frequency and/or intentionally internally generated signal frequencies. This phenomenon is frequently referred to as "ripple."

periodic pulse train—A pulse train made up of identical groups of pulses repeated at regular intervals.

periodic quantity – An oscillating quantity in which any value it attains is repeated at equal time intervals.

periodic rating – The load which can be carried for the alternate periods of load and rest specified in the rating without exceeding the specified heating limits.

periodic resonance—Also called natural resonance. Resonance in which the applied agency maintaining the oscillation has the same frequency as the natural period of oscillation of a system.

periodic time—*See* Period *and* Period of an Underdamped Instrument.

periodic vibration – A vibration having a regularly recurring waveform, e.g., sinusoidal vibration.

periodic wave – Wave in which the displacement has a periodic variation with time, distance, or both.

period of an underdamped instrument – Also called periodic time. The time required, following an abrupt change in the measurand, for the pointer or other indicating means to make two consecutive transits in the same direction through the rest position.

peripheral—Having to do with a device by means of which a computer communicates to the outside world. Auxiliary memories, such as tape, disc, and drum, may also be considered to be peripheral devices.

peripheral control unit – An intermediary control device which links a peripheral unit to the central processor, or, in the case of offline operation, to another peripheral unit.

peripheral device – Any instrument or machine that enables a computer to communicate with the outside world or that otherwise aids the operation of the computer, but does not form part of the basic installation.

peripheral electron – Also called a valence electron. One of the outer electrons of an atom. Theoretically, it is responsible for visible light, thermal radiation, and chemical combination.

peripheral equipment—1. In a data processing system, any unit of equipment, distinct from the central processing unit, which may provide the system with outside communication. 2. Equipment that is external to and not a part of the central processing instrumentation. Includes such equipment as tape punches and readers, magnetic tape or disc storage units, digital printers, graphic recorders, and typewriters.

peripheral processor—A general term for a laser computer associated with a large machine. Among the functions may be multiplexing, data formating, concentrating, polling, and the handling of simple routines to increase the capacity of a communications channel or to relieve the main (often called "host") computer.

peripheral transfer – The transmission of data between two peripheral units.

permalloy – A high-permeability magnetic alloy composed mainly of iron and nickel.

permanent echo – Signal received and displayed by a radar, indicating reflections from fixed objects.

permanent-field synchronous motor—A type of synchronous motor in which the member carrying the secondary laminations and windings also carries permanent-magnet field poles that are shielded from the alternating magnetic flux by the laminations. It behaves as an induction motor when starting but runs at synchronous speed.

permanently connected – Connected to a supply circuit by way of fixed electrical conductors.

permanent magnet – A piece of hardened steel or other magnetic material which has been so strongly magnetized that it retains the magnetism indefinitely.

permanent-magnet centering – Vertical or horizontal shifting of a television picture by means of magnetic fields from permanent magnets mounted around the neck of the picture tube.

permanent-magnet focusing – Focusing of the electron beam in a television picture tube by means of one or more permanent magnets located around the neck.

permanent-magnet material—Ferromagnetic material which, once having been magnetized, resists external demagnetizing forces (i.e., requires a high coercive force to remove the magnetism).

permanent-magnet, moving-coil instrument —Also called D'Arsonval instrument. An

Permanent magnet moving-coil type meter.

instrument in which a reading is produced by the reaction between the current in a movable coil or coils and the field of a fixed permanent magnet.

permanent-magnet, moving-iron instrument —Also called polarized-vane instrument. An instrument in which a reading is produced by an iron vane as it aligns itself in the magnetic field produced by a permanent magnet and by the current in an adjacent coil of the instrument.

permanent-magnet speaker—A moving-conductor speaker in which the steady magnetic field is produced by a permanent magnet.

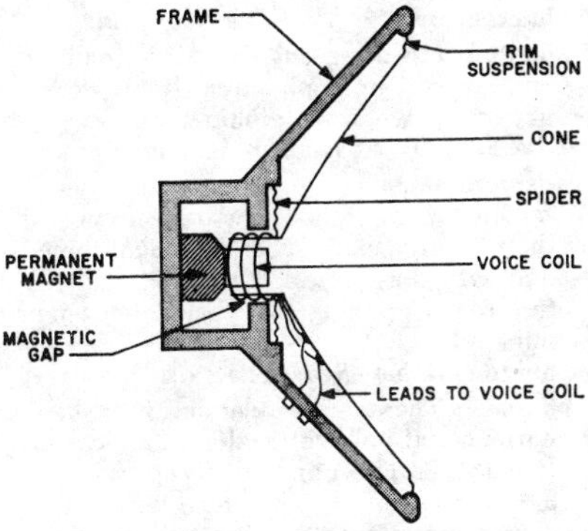

Permanent-magnet speaker.

permanent-magnet stepping motor—A type of motor in which a permanent magnet serves as the rotor. Current is switched sequentially through different stator coils, and the rotor aligns itself with the energized stator poles.

permanent magnistor – A saturable reactor which has the properties of memory and the ability to handle appreciable power.

permanent memory—A type of storage device that retains data intact when the computer has been shut down.

permanent-memory computer—A computer in which the stored information remains intact, even after the power has been turned off.

permanent set – The deformation that remains in a specimen after it has been stressed in tension for a definite time interval and released for a definite time interval.

permanent storage—A computer storage device which retains the stored data indefinitely.

permatron—A thermionic gas diode, the discharge of which is controlled by an external magnetic field. It is used mainly as a controlled rectifier and functions like a thyratron.

permeability – Symbolized by the Greek letter *mu* (μ). The measure of how much better a given material is than air as a path for magnetic lines of force. (Air is assumed to have a permeability of 1.) It is equal to the magnetic induction (B) in gausses, divided by the magnetizing force (H) in oersteds.

permeability tuning – A method of tuning a circuit by moving a magnetic core into or out of a coil to vary its inductance.

permeameter—An apparatus for determining the magnetizing force and flux density in a test specimen. From these values, the normal induction curves or hysteresis loops can then be plotted and the magnetic permeability computed.

permeance – The reciprocal of reluctance. Through any cross section of a tubular portion of a magnetic circuit bounded by lines of force and by two equipotential surfaces, permeance is the ratio of the flux to the magnetic potential difference between the surfaces under consideration. In the cgs system, it is equal to the magnetic flux (in maxwells) divided by the magnetomotive force (in gilberts).

permissive control device—Generally a two-position, manually operated switch which in one position permits the closing of a circuit breaker or the placing of an equipment into operation, and in the other position prevents the circuit breaker or the equipment from being operated.

permittivity—*See* Dielectric Constant.

permutation modulation—Proposed method of transmitting digital information by means of band-limited signals in the presence of additive white Gaussian noise. Pulse-code modulation and pulse-position modulation are considered simple special cases of permutation modulation.

permutation table – In computers, a table for use in the systematic construction of code groups. It may also be used in the correction of garbles in groups of code text.

peroxide of lead – A lead compound that forms the principal part of the positive plate in a charged lead-acid cell.

perpendicular magnetization – In magnetic recording, magnetization that is perpendicular to the line of travel and parallel to the smallest cross-sectional dimension of the medium. Either single– or double–pole-piece magnetic heads may be used.

persistence—The length of time a phosphor dot glows on the screen of a cathode-ray tube before going out—i.e., the length of time it takes to decay from initial brightness (reached during fluorescence) until it can no longer be seen.

persistence characteristic (of a luminescent screen)—Also called the decay characteristic. The relationship (usually shown by a graph) between the time a luminescent screen is excited and the time it emits radiant power.

persistence of vision – The phenomenon whereby the eye retains an image for a short time after the field of vision has disappeared.

persistent current–A curernt that is magnetically induced and flows undiminished in a superconducting material or circuit.

persistor–A bimetallic circuit used for storage or readout in a computer. It is operated near absolute zero, and changes from a resistive to a superconductive state at a critical current value.

persistron–A device in which electroluminescence and photoconductivity are combined into a single panel capable of producing a steady or persistent display with pulsed signal input.

persuader–In a storage tube, an element that directs secondary emission toward the electron-multiplier dynodes.

perveance–The space-charge-limited cathode current divided by the three-halves power of the anode voltage in a diode.

petticoat insulator–An insulator having an outward flaring lower part that is hollow to increase the length of the surface leakage path and keep part of the path dry at all times.

pF–Abbreviation for picofarad.

pg–Abbreviation for power gain.

p-gate thyristor – A thyristor in which the gate terminal is connected to the p-region nearest the cathode and which is normally switched to the on-state by applying a positive signal between the gate and cathode terminals.

pH–A measure of the degree of acidity or alkalinity of a solution. In a neutral solution the pH value is 7. In acid solutions it ranges from 0 to 7, and in alkaline solutions it ranges from 7 to 14.

phanotron–A term used primarily in industrial electronics to mean a hot-cathode gas diode.

phantastron – An electronic circuit of the multivibrator type which is normally used in the monostable form. It is a stable trigger generator in this connection and is used in radar systems for gating functions and sweep-delay functions.

phantom channel – In a stereo system, an electrical combination of the left and right channels fed to a third, centrally located speaker.

phantom circuit – A superimposed circuit derived from two suitably arranged pairs of wires called side circuits. Each pair of wires is a circuit itself, and at the same time acts as one conductor of the phantom circuit.

phantom-circuit loading coil – A loading coil that introduces the desired amount of inductance into a phantom circuit and a minimum amount into the constituent side circuits.

phantom-circuit repeating coil–A repeating coil used at a terminal of a phantom circuit in the terminal circuit extending from the midpoints of the associated side-circuit repeating coils.

phantom coil – A coil originally used in a phantom circuit for impedance matching. Now, generally, any coil, side or phantom, in a phantom circuit. When the term is used, the meaning should be made clear.

phantom group–1. A group of four open-wire conductors suitable for the derivation of a phantom circuit. 2. Three circuits which are derived from simplexing two physical circuits to form a phantom circuit.

phantom OR and AND–*See* wired OR.

phantom repeating coil – A side-circuit repeating coil or a phantom-circuit repeating coil, when discrimination between these two types is not necessary.

phantom signals–Signals appearing on the screen of a cathode-ray-tube indicator; their cause cannot readily be determined, and they may be due to circuit fault, interference, propagation anomalies, jamming, etc.

phantom target–*See* Echo Box.

phase–1. The angular relationship between current and voltage in alternating-current circuits. 2. The number of separate voltage waves in a commercial alternating-current supply (e.g., single-phase, three-phase, etc.). Symbolized by the Greek letter *phi* (ϕ) 3. In a periodic function or wave, the fraction of the period which has elapsed, measured from some fixed origin. If the time for one period is represented as 360° along a time axis, the phase position is called phase angle.

phase advancer – A phase modifier which supplies leading reactive volt-amperes to the system to which it is connected. Phase advancers may be either synchronous or nonsynchronous.

phase angle–1. Of a periodic function, the angle obtained by multiplying the phase by 2π if the angle is to be expressed in radians, or 360 for degrees. 2. The angle between the vectors representing two periodic functions that have the same frequency. 3. The phase difference, in degrees, between corresponding stages of progress of two cyclic operations.

phase-angle correction factor–That factor by which the reading of a wattmeter or watthour meter operated from the secondary of a current or potential transformer, or both, must be multiplied to correct for the effective phase displacement of current and voltage due to the measuring apparatus.

phase-angle measuring relay – Also called out-of-step protective relay. A device which functions at a predetermined phase angle between two voltages or currents, or between voltage and current.

phase-angle meter—*See* Phase Meter.

phase angle of a current transformer—The angle between the primary-current vector and the secondary-current vector reversed. This angle is conveniently considered as positive when the reversed secondary-current vector leads the primary-current vector.

phase angle of a potential (voltage) transformer—The angle between the primary-voltage vector and the secondary-voltage vector reversed. This angle is conveniently considered as positive when the reversed secondary-voltage vector leads the primary-voltage vector.

phase anomaly—A sudden irregularity in the phase of an lf of vlf signal.

phase balance — In a chopper, the phase-angle difference between positive and negative halves of the square wave; the difference in degrees between 180° and the measured angle between square-wave midpoints.

phase-balance current relay—*See* Reverse-Phase Current Relay.

phase-balance relay — A relay which functions by reason of a difference between two quantities associated with different phases of a polyphase circuit.

phase center (center of radiation) — Pertaining only to antenna types which have radiation characteristics such that while they are radiating energy, one can observe the antenna from a distance of many wavelengths and see the energy radiating from a point within the antenna array. The position of a point-source radiator which would replace the antenna and produce the same far-field phase contour.

phase characteristic—A graph of phase shift versus frequency, assuming sinusoidal input and output.

phase-comparison tracking system—A system which provides target-trajectory information by the use of cw phase-comparison techniques.

phase-compensation network — A network used to provide closed-loop stability in an operational amplifier. No greater than 12dB/octave rolloff of the open-loop gain is allowed.

phase conductors—Those conductors other than the neutral conductor of a polyphase circuit.

phase constant—The imaginary component of the propagation constant. For a traveling plane wave at a given frequency, the rate in radians per unit length, at which the phase lag of a field component (for the voltage or current) increases linearly in the direction of propagation.

phase control — 1. Also called horizontal parabola control. One of three controls for adjusting the phase of a voltage or current in a color television receiver employing the magnetic-convergence principle. Each control varies the phases of the sinusoidal voltages applied, at the horizontal-scanning frequency, to the coils of the magnetic-convergence assembly. 2. A technique for proportional control of an output signal by conduction only during certain parts of the cycle of the ac line voltage. 3. A method of regulating a supply of alternating current by use of a switching device such as a thyristor, by varying the point in each ac cycle or half-cycle at which the device is switched on.

phase-controlled rectifier—A rectifier circuit in which the rectifying element is a thyratron having a variable-phase, sine-wave grid bias.

phase correction—The process of keeping synchronous telegraph mechanisms in substantially correct phase relationship.

phase corrector — A network designed to correct for phase distortion.

phased array—A group of simple radiating elements arranged over an area called an aperture. A beam (or beams) can be formed by superposition of the radiation from all the elements, and the direction of the beam can be adjusted by varying the relative phase of the signal applied to each element or by varying the frequency of the main oscillator.

phase delay—1. In the transfer of a single frequency wave from one point to another in a system, the delay of part of the wave identifying its phase. 2. The insertion phase shift (in cycles) divided by the frequency (in cycles per second, or hertz). *See also* Delay Distortion, 2.

phase-delay distortion—The difference between the phase delay at one frequency and the phase delay at a reference frequency.

phase detector—1. A tv circuit in which a dc correction voltage is derived to maintain a receiver oscillator in sync with some characteristic of the transmitted signal. 2. A circuit which detects both the magnitude and the sign of the phase angle between two sine-wave voltages or currents.

phase deviation—In phase modulation, the peak difference between the instantaneous angle of the modulated wave and the angle of the carrier.

phase difference—The time in electrical degrees by which one wave leads or lags another.

phase discriminator—*See* Phase Detector.

phase distortion—The alteration of a complex waveform produced as it passes through a network or transducer whose phase shift is a function of frequency. *See also* Phase-Frequency Distortion.

phase-distortion coefficient—In a transmission system, the difference between the maximum and minimum transit times for frequencies within a specified band.

phase equalizer — A circuit employed to neutralize the effect of phase-frequency distortion in a particular range of frequencies.

phase-frequency distortion — Also called phase distortion. Distortion that occurs when the phase shift is not directly proportionate to the frequency over the range required for transmission.

phase inversion — The condition whereby the output of a circuit produces a wave of the same shape and frequency but 180° out of phase with the input.

phase inverter — 1. A stage that functions chiefly to change the phase of a signal by 180°, usually for feeding one side of a following push-pull amplifier. 2. *See* Vented Baffle. 3. A network or device such as a paraphrase amplifier, which produces two output signals that differ in phase by half a cycle.

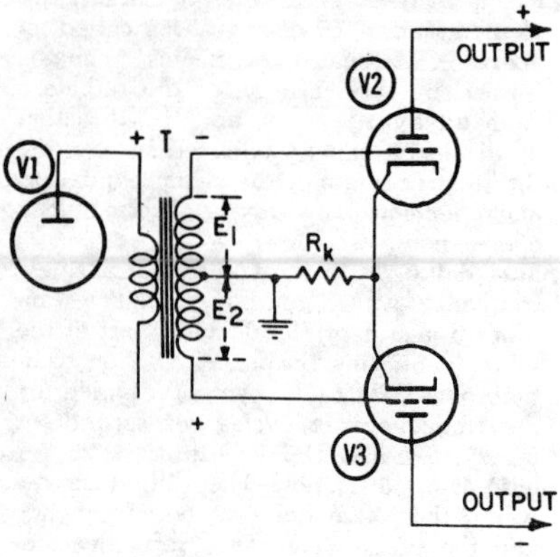

Phase inverter, 1.

phase localizer—An airfield runway localizer in which lateral guidance is obtained by comparing the phases of two signals.

phase lock—The technique of making the phase of an oscillator signal follow exactly the phase of a reference signal by comparing the phases between the two signals and using the resultant difference signal to adjust the frequency of the reference oscillator.

phase-locked loop — 1. A communications circuit in which a local oscillator is synchronized in phase and frequency with a received signal. 2. A closed-loop electronic servomechanism the output of which locks onto and tracks a reference signal. Phase lock is accomplished by comparing the phases of the output signal (or a multiple of it) and the reference signal. Any phase difference between these signals is converted into a correction voltage that causes the phase of the output signal to change so that it tracks the reference.

phase magnet—Also called trip magnet. A magnetically operated latch used to phase a facsimile transmitter or recorder.

phase margin—1. A safety factor in phase shift. When the loop gain is 1.0 or more and the phase shifts total 180°, instability will occur. The amount that the total phase shift is less than 180° is called the phase margin. 2. The additional amount of phase shift of the output signals in an operational amplifier at the open-loop unity-gain crossover frequency that would produce instability.

phase meter—Also called phase-angle meter. An instrument for measuring the difference in phase between two alternating quantities of the same frequency.

phase modifier — A device that supplies leading or lagging volt-amperes to the system to which it is connected.

phase-modulated transmitter — A transmitter, the output of which is a phase-modulated wave.

phase-modulated wave — A wave whose phase angle has been caused to deviate from its original (no-signal) angle by an amount proportional to the modulating signal amplitude.

phase modulation — 1. Abbreviated pm. Modulation in which the angle of a sine-wave carrier deviates from the original (no-signal) angle by an amount proportional to the instantaneous value of the modulating wave. Phase and frequency modulation in combination are commonly referred to as "frequency modulation." 2. Method of modulation in which the amplitude of the modulated wave remains constant, while varying in phase with the amplitude of the modulating signal. A phase-modulated wave is electrically identical to a modified frequency-modulated wave and vice versa. 3. Phase of the audio-modulating signal is varied in accordance to the superimposed intelligence. Unlike amplitude modulation, in both

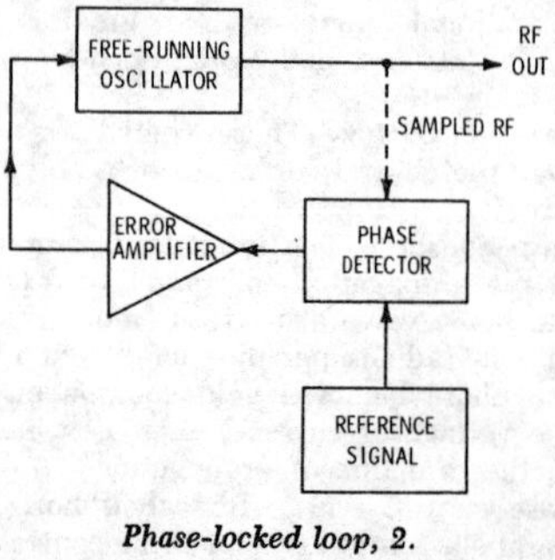

Phase-locked loop, 2.

lation, the average energies of the modulated signals are the same.

phase modulator – A circuit which modulates the phase of a carrier signal.

phase multiplier—A device that multiplies the frequency of signals used for phase comparison so that phase differences may be measured to a higher degree of resolution.

phase noise – A measure of the random phase instability of a signal.

phase offset—The difference between voltage and current in an ac power line with a lagging or leading power factor.

phase-propagation ratio—In wave propagation, the propagation ratio divided by its magnitude. Expressed as a unit vector of the same angle as the propagation ratio.

phaser—1. A device for adjusting facsimile equipment so that the recorded area bears the same relationship to the record sheet as the corresponding transmitted area bears to the subject copy in the direction of the scanning line. 2. A microwave ferrite phase shifter that employs a longitudinal magnetic field along a rod or rods of ferrite in a waveguide.

phase-recovery time (tr and pre-tr tubes)—The time required for a fired tube to deionize to such a level that a specified phase shift is produced in the low-level radio-frequency signal transmitted through the tube.

phase resonance—Also called velocity resonance. Resonance in which the angular phase difference between the fundamental components of the oscillation or vibration and the applied agency is 90°.

phase-response characteristic – The phase displacement versus frequency properties of a network or system.

phase reversal—A 180° change in phase (or one half cycle) such as a wave might undergo upon reflection under certain conditions.

phase-reversal protection – In a polyphase circuit, the interruption of power whenever the phase sequence of the circuit is reversed.

phase-reversal switch—A switch used on a stereo amplifier or in a speaker system to shift the phase 180° on one channel.

phase-rotation relay – *See* Phase-Sequence Relay.

phase-sensitive amplifier—A servoamplifier the outut signal polarity or phase of which is dependent upon the polarity or phase relationship between an error (input) voltage and a reference voltage.

phase-sensitive detector—A system that produces a dc output signal in response to an ac input signal of a defined frequency equal to the frequency of ac reference signal. The dc output is proportional to both the amplitude of the ac input signal and the cosine of its phase angle relative to that of the reference signal. Used as synchronous rectifiers in chopper dc amplifiers and for the accurate measurement of small ac signals obscured by noise.

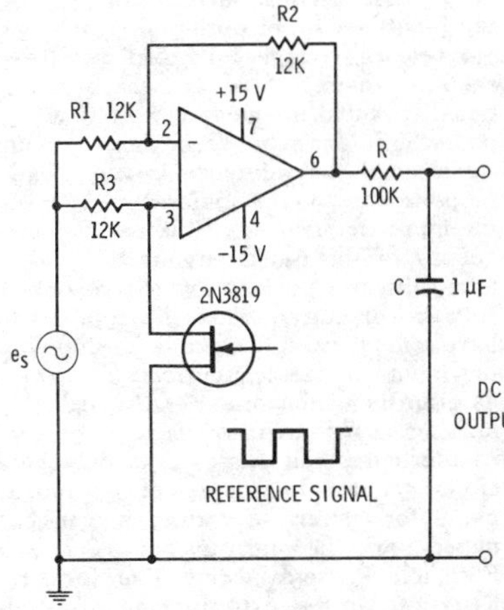

Phase-sensitive detector.

phase-sequence indicator—A device that indicates the sequence in which the fundamental components of a polyphase set of potential differences, or currents, successively reach some particular value (e.g., their maximum positive value).

phase-sequence relay – Also called phase-rotation relay. A relay which functions according to the order in which the phase voltages successively reach their maximum positive values.

phase-sequence voltage relay – A device which functions upon a predetermined value of polyphase voltage in the desired phase sequence.

phase shift—1. The difference between corresponding points on input and output signal waveshapes (not affected by a magnitude) expressed as degrees lead or lag. 2. A change in the phase of a periodic quantity. 3. The changing of phase of a signal as it passes through a filter. A delay in time of the signal is referred to as phase lag and in normal networks, phase lag increases with frequency, producing a positive envelope delay (*See* Envelope Delay). It is possible for an output signal to experience a time shift ahead of the input signal and this is called phase lead. The phase shift is always dependent on frequency.

phase-shift circuit—A network which shifts the phase of one voltage with respect to another voltage of the same frequency.

phase-shift discriminator – A circuit that produces an output proportional to the phase difference between two input signals. When used as an fm demodulator, the input is a set of push-pull signals and a reference voltage 90° displaced from each of them. These are taken across tuned circuits in such a way that the phase difference between the push-pull signals and the reference is very nearly proportional to the difference between the input frequency and the resonant frequency of the tuned circuits.

phase shifter—A device in which the output voltage (or current) may be adjusted to have some desired phase relationship with the input voltage (or current).

phase-shifting transformer — Also called a phasing transformer. A transformer connected across the phases of a polyphase circuit to provide voltages of the proper phase for energizing varmeters, var-hour meters, or other instruments. (*See also* Rotatable Phase-Adjusting Transformer.)

phase-shift keying—A form of phase modulation in which the modulating function shifts the instantaneous phase of the modulated wave between predetermined discrete values.

phase-shift microphone — A microphone the directional properties of which are provided by phase-shift networks.

phase-shift oscillator — An oscillator in which a network having a phase shift of an odd multiple of 180° (per stage) at the oscillation frequency is connected between the output and input of an amplifier. When the phase shift is obtained by resistance-capacitance elements, the circuit is called an RC phase-shift oscillator.

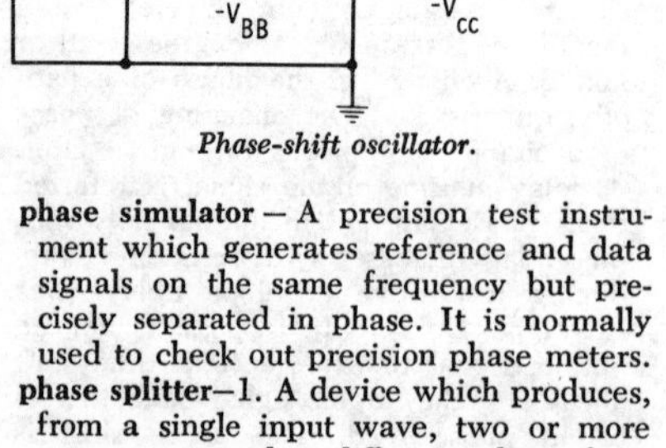

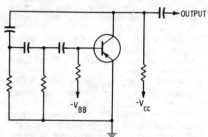

Phase-shift oscillator.

phase simulator — A precision test instrument which generates reference and data signals on the same frequency but precisely separated in phase. It is normally used to check out precision phase meters.

phase splitter—1. A device which produces, from a single input wave, two or more output waves that differ in phase from one another. 2. In color television, the stage which takes I and Q signals from demodulators, produces four signals, positive and negative I and Q, and feeds them to the matrix. 3. A circuit which generates out of an ac input signal two equal-amplitude outputs, one of which is 180° out of phase with the other, i.e., one is the other inverted. The dc levels may not be identical.

phase-tuned tube (**tr tubes**)—A fixed tuned broad-band tr tube in which the phase angle through it and the reflection it introduces are kept within limits.

phase undervoltage relay—A relay which is tripped by the reduction of one phase voltage in a polyphase circuit.

phase velocity—1. The velocity at which a point of constant phase is propagated in a progressive sinusoidal wave. 2. The velocity with which a point where there exists an electromagnetic wave of a certain fixed phase, moves through space in the direction of propagation of the wave.

phase–versus–frequency response characteristic — A graph or other tabulation of the phase shift occurring, in an electrical transducer, at several frequencies within a band.

phasing—1. Causing two systems or circuits to operate in phase or at some desired difference from the in-phase condition. 2. Adjusting a facsimile-picture position along the scanning line. 3. In stereo application, the establishment of the correct relative polarity in the connection between amplifier output and speakers so that one speaker tends to reinforce rather than cancel the output of the other (particularly evident at low frequencies).

phasing capacitor — A capacitor used in a crystal-filter circuit for neutralizing the capacitance of the crystal holder.

phasing line—In facsimile, the portion of the scanning line set aside for the phasing signal.

phasing pulse—A short pulse or signal employed for phasing the recorder with the transmitter in a television or facsimile system.

phasing signal—In facsimile, a signal used for adjusting the position of the picture along the scanning line.

phasing transformer — *See* Phase-Shifting Transformer.

phasitron — A tube designed to produce a frequency-modulated audio signal, which is induced by a varying field from a magnet placed around the glass envelope of the tube.

phasmajector—*See* Monoscope.

phasor—An entity which includes the concepts of magnitude and direction in a reference plane.

pH electrode—Transducer sensitive to hydrogen ion concentration. The sensor comprises a thin-walled glass membrane (glass electrode) or spongy platinum exposed to gaseous hydrogen (hydrogen electrode) or platinum exposed to quinhydron (quinhydrone electrode), all of which develop an electric force proportional to the hydrogen-ion concentration of a solution when immersed in the solution.

phenolic material—Any one of several thermosetting plastic materials available which may be compounded with fillers and reinforcing agents to provide a broad range of physical, electrical, chemical, and molding properties.

Phillips gate—A vacuum gage in which gas pressure is determined by measuring the current in a glow discharge.

Phillips screw—A screw with an indented cross in its head, instead of the conventional slot. It must be removed or inserted with a special screwdriver, also called a Phillips.

phi polarization — In an electromagnetic wave, the state in which the E vector of the wave is tangential to the lines of latitude of some given spherical frame of reference.

pH meter — An instrument used with a probe to determine the alkalinity or acidity of a solution.

phon—The unit for measuring the apparent loudness level of a sound. Numerically equal to the sound-pressure level, in decibels relative to 0.0002 microbar, of a 1000-hertz tone that is considered by listeners to be equivalent in loudness to the sound under consideration.

phone—*See* Headphone.

phone jack—1. Also called a telephone jack. A jack designed for use with phone plugs. 2. Receptacle having two or more through circuits. May also have shunt circuits and/or isolated switching circuits. Used for extending circuits through mating plugs. Phone jacks are short or long types, depending upon physical dimensions.

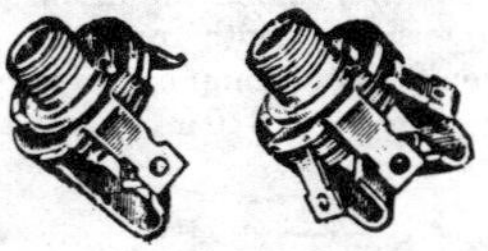

Phone jack.

phonemes—The minimal set of shortest segments of speech which, if substituted one for another, convert one word to another.

phone plug—Also called telephone plug. A plug used with headphones, microphones, and other audio equipment. It is a male connecting device (almost always connected to a cable) which connects with a phone jack. Consists usually of finger and handle which comprise the through circuit, terminals, insulators and handle. A cable clamp may or may not be part of a phone plug design.

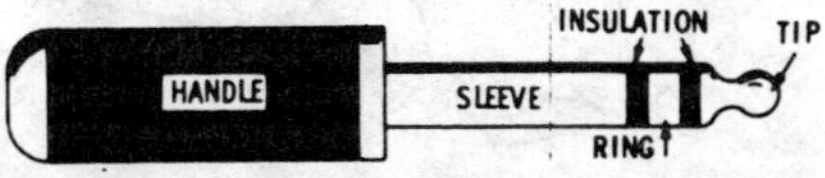

Phone plug.

phonetic alphabet — A list of standard words, one for each letter in the alphabet. It is used for distinguishing the letters in a spoken radio or telephone message. The list reads:

ALFA	NOVEMBER
BRAVO	OSCAR
CHARLIE	PAPA
DELTA	QUEBEC
ECHO	ROMEO
FOXTROT	SIERRA
GOLF	TANGO
HOTEL	UNIFORM
INDIA	VICTOR
JULIET	WHISKEY
KILO	X-RAY
LIMA	YANKEE
MIKE	ZULU

phonocardiogram—A graphic recording of the sounds produced by the heart and its associated parts (e.g., its mitral or aortic valves).

phonocardiograph — An instrument for recording sounds of the heart on a strip chart.

phonocardiography—The recording and interpretation of the sounds of the heart. A typical instrument for this purpose consists of a microphone, an amplifier, a cathode-ray tube or strip-chart recorder, and sometimes a loudspeaker or headset.

phono cartridge—The means by which the stylus movements are converted into an electrical signal. Various versions of magnetic (moving iron, magnet, or coil), ceramic-crystal, capacitive (electret), and strain-gauge devices are in use.

phonocatheter—A catheter-microphone combination that is inserted through the artery into the heart. It picks up inner cardiac sounds.

phonoelectrocardioscope—A dual-beam oscilloscope which develops both ecg signals and heart-sound signals.

phonograph—An instrument for reproducing sound. It consists of a turntable on which the grooved medium containing the impressed sound is placed, a needle that rides in the groove, and an electrical (formerly mechanical) amplifying system for taking the minute vibrations of the needle and converting them into electrical (formerly mechanical) impulses that drive a speaker.

phonograph oscillator—An rf oscillator circuit, the output of which is modulated by a phonograph pickup and sent through space to a receiver. Thus, no wires to the receiver are needed.

phonograph pickup—Also called mechanical reproducer, pickup, or phono pickup. A mechanoelectric transducer which is actuated by modulations present in the groove of a recording medium and transforms this mechanical input into an electric output.

phono jack—A jack designed to accept a phono plug. Receptacle having two through circuits (coaxial oriented) primarily intended for connecting audio signals between phonograph and amplifiers. Now widely used for many other types of signal, including occasionally rf.

phonon—1. A lattice vibration with which a discrete amount (quantum) of energy is associated. Some thermal and electrical properties of the lattice are theoretically treated in terms of electron-phonon interactions. 2. Quantum of thermal energy used to help describe thermal vibration of a crystal lattice. 3. Sharply tuned radiation of superhigh-frequency sound waves.

phonons—Packets of sound energy vibrating in a solid at ultrahigh frequencies—so high that the energy is commonly thought of as heat.

phono pickup—*See* Phonograph Pickup.

phono plug—A plug used at the end of a shielded conductor for feeding af signals to a mating phono jack on an audio preamplifier or amplifier.

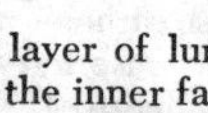

Phono plug.

phonoselectroscope—A stethoscopic device which suppresses low frequencies (characteristic of the normal heart function) to permit detection of higher-frequency sounds.

phosphor—1. A layer of luminescent material applied to the inner face of a cathode-ray tube. During bombardment by electrons it fluoresces, and after the bombardment, it phosphoresces. 2. A material that emits light when excited (energized) by radiant energy.

phosphor-dot faceplate—The glass viewing screen on which the trios of color phosphor dots are mounted in a three-gun picture tube.

phosphor dots — Minute particles of phosphor on the vewing screen of a picture tube. On a tricolor picture tube, the red, green, and blue phosphor dots are placed on the viewing screen in a pattern of dot triads—a phosphor dot of each color forming one-third of the triad.

phosphorescence—1. The emission of light from a substance after excitation has been removed. (*See also* Afterglow.) 2. The emission of light from a source that is delayed by over 10^{-8} second, following excitation.

phosphors — Chemical substances that exhibit fluorescence when excited by ultraviolet radiation, X-rays, or an electron beam. The amount of visible light is proportional to the amount of excitation energy. If fluorescence decays slowly after the exciting source is removed, the substance is said to be phosphorescent.

phosphor trio—In the phosphor screen of a tricolor kinescope, closely spaced triangular groups of three phosphor dots accurately deposited in interlaced positions.

phot (pt)—The unit of illumination when the centimeter is taken as the unit of length; it is equal to one lumen per square centimeter.

photobiology—The study of the effects on living matter (or substances derived therefrom) of electromagnetic radiation extending from the ultraviolet through the visible light spectrum into the infrared. The conversion of electromagnetic energy into chemical energy. Photosynthesis is an important branch of photobiological investigation.

photocathode—An electrode which releases electrons when exposed to light or other suitable radiation. Used in phototubes, television camera tubes, and other light-sensitive devices.

photocathode tube response—The photoemission current resulting from a specified luminous flux from a tungsten lamp filament at a color temperature of 2870°K when the flux is filtered by a specified blue filter.

photocathode luminous sensitivity. — The quotient of photoelectric emission current from the photocathode divided by the incident luminous flux. The measurement is made under specified conditions of illumination, usually with radiation from a tungsten-filament lamp operated at a color temperature of 2870°K. The cathode is

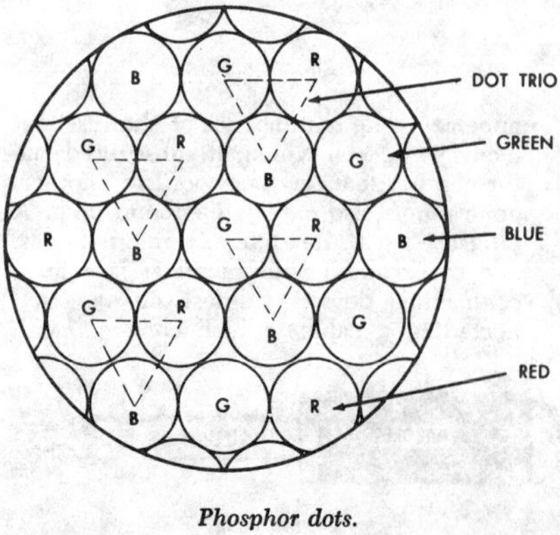

Phosphor dots.

usually illuminated by a collimated beam at normal incidence.

photocathode radiant sensitivity—The quotient of the photoelectric emission current from the photocathode divided by the incident radiant flux. It is usually measured at a given wavelength under specified conditions of irradiation with a collimated beam at normal incidence.

photocell—*See* Photoelectric Cell.

photochemical radiation—Energy in the ultraviolet, visible, and infrared regions used to produce chemical changes in materials.

photochromic—Pertaining to a single-crystal inorganic material used as a display and storage element. The material can be from one of several families of materials, such as fluorides or titanates. The display color and storage time are determined by the amount and kind of doping of the material.

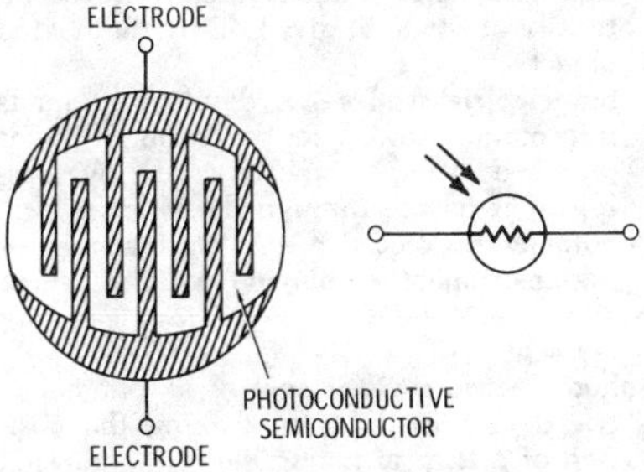

Photoconductive cell.

photoconductive cell—A photoelectric cell, the electrical resistance of which varies inversely with the intensity of light that strikes its active material.

photoconductive effect—The change of electrical conductivity of a material when exposed to varying amounts of radiation.

photoconductive material—Material having a high resistance in the dark, and a low resistance when exposed to the light.

photoconductivity—1. The greater electrical conductivity shown by some solids when illuminated. The incoming radiation transfers energy to an electron, which then takes on a new energy level (in the conduction band) and contributes to the electrical conductivity. 2. The ability of a material to hold a charge of electricity (insulator) in the absence of light yet act as a conductor of electricity when exposed to light.

photoconductor—1. A passive, high-impedance device composed of thin single-crystal or polycrystalline films of compound semiconductor materials. When the sensitive surface is illuminated its resistance decreases and hence its conductivity increases. 2. A light-sensitive resistor whose resistance decreases with increase in light intensity, when illuminated. Consists of a thin single crystal or polycrystalline films of compound semiconductor substances.

photocurrent—The difference between light current and dark current in a photodetector.

photodetector—1. A device that senses incident illumination. 2. Any device that utilizes the photoelectric effect to detect the presence of light.

photodielectric effect—The change in the dielectric constant and loss of a material when illuminated. The effect is observed only in phosphors that show photoconductivity during luminescence.

photodiffusion effect—*See* Dember Effect

photodiode—A pn semiconductor diode designed so that light falling on it greatly increases the reverse leakage current, so that the device can switch and regulate electric current in response to varying intensity of light.

Photodiode symbol.

photoelasticity – Changes in the optical properties of transparent isotropic dielectrics subject to stress.

photoelectric – Pertaining to the electrical effects of light or other radiation—i.e., emission of electrons, generation of a voltage, or a change in electrical resistance upon exposure to light.

photoelectric absorption—Conversion of radiant energy into photoelectric emission.

photoelectric cathode—A cathode the primary function of which is photoelectric emission.

photoelectric cell—Also called photocell. A cell, such as a photovoltaic or photoconductive cell, the electrical properties of which are affected by illumination. The term should not be used for a phototube, which is a vacuum tube and not a cell.

photoelectric colorimeter – A colorimeter which uses a photoelectric cell and a set of color filters to determine, by the output current for each filter, the chromaticity coordinates of light of a given sample.

photoelectric conductivity – The increased conductivity exhibited by certain crystals when struck by light (e.g., a selenium cell).

photoelectric constant – A quantity which, when multiplied by the frequency of the radiation causing the emission, gives (in centimeter-gram-second units) the voltage absorbed by the escaping photoelectron.

The constant is equal to h/e, where h is Planck's constant and e is the electronic charge.

photoelectric control—The control of a circuit or piece of equipment in response to a change in incident light impinging on a photosensitive device.

photoelectric counter—A device that registers a count whenever an object breaks the light beam shining on its phototube or photocell. An amplifier then boosts the minute energy to register on a mechanical or other type of counter.

photoelectric current—The stream of electrons emitted from the cathode of a phototube under the influence of light.

photoelectric cutoff control—A photorelay circuit used in machines for cutting long strips of paper, cloth, metal, or other material accurately into predetermined lengths or at predetermined positions.

photoelectric effect—1. The transfer of energy from incident radiation to electrons in a substance. This phenomenon includes photoelectric emission of electrons from the surface of a metal, the photovoltaic effect, and photoconductivity. 2. Interaction between radiation and matter resulting in the absorption of photons and the consequent generation of mobile charge carriers.

photoelectric electron-multiplier tube — A vacuum phototube that employs secondary emissions to amplify the electron stream emitted from the illuminated photocathode.

photoelectric emission – Electron emission due directly to the incidence of radiant energy on the emitter.

photoelectric flame-failure detector—An industrial electronic control employing a phototube and amplifier to actuate an electromagnetic or other valve that cuts of the fuel flow when the fuel-consuming flame is extinguished and light no longer falls on the phototube.

photoelectric inspection—Quality control of a product by means of a phototube, light-beam system, and associated electronic equipment.

photoelectric intrusion detector—A burglar-alarm system in which interruption of a light beam by an intruder reduces the illumination on a phototube and thereby closes an alarm circuit.

photoelectric liquid-level indicator—A level indicator in which the rising liquid interrupts the beam of light in a photoelectric control system.

photoelectric material – Any material that will emit electrons when illuminated in a vacuum (e.g., barium, cesium, lithium, potassium, rubidium, sodium, and strontrium).

photoelectric multiplier — A phototube in which the primary photoemission current, prior to being extracted at the anode, is multiplied many times.

photoelectric phonograph pickup—A phonograph reproducing device consisting essentially of a light source, a jewel stylus to which a very thin mirror is attached, and a selenium cell that picks up light reflected from the mirror. Sidewise movements of the stylus in the record groove cause the amount of reflected light to vary, and accordingly the resistance of the selenium cell. The light source is fed by a radio-frequency oscillator rather than from the power line, to eliminate 60-hertz flicker from the light beam.

photoelectric photometer—1. A photometer which incorporates a phototube or photoelectric cell for measurements of light. 2. *See* Electronic Photometer.

photoelectric pickup — A transducer that transforms a change in light into an electric signal.

photoelectric pyrometer—An instrument for measuring high temperatures from the intensity of the light given off by the heated object.

photoelectric reader — A device that reads information stored in the form of holes punched in paper tape or cards, by sensing light passed through the holes.

photoelectric recorder — An optical recording instrument employing a light source and phototube for the basic measuring element.

photoelectric register control—A photoelectric device used for controlling the position of a strip of paper, cloth, metal, etc., with respect to the machine through which it is being passed.

photoelectric relay—Also called light relay. A relay combined with a phototube (and amplifier if necessary), so arranged that changes in incident light on the phototube cause the relay contacts to open or close.

photoelectric scanner—A light source, lens system, and one or more phototubes in a single, compact housing. It is mounted a few inches above a moving surface, where it actuates control equipment when the amount of light reflected from the surface changes.

photoelectric sensitivity—Also called photoelectric yield. The rate at which electrons are emitted from a metal per unit radiant flux at a given frequency.

photoelectric smoke detector—A photoelectric instrument used to measure the density of smoke and to sound an alarm when a predetermined smoke density is exceeded.

photoelectric sorter — An industrial-electronic control employing a light beam, phototube, and amplifier to sort objects according to color, size, shape, or other characteristics.

photoelectric threshold—The quantum energy just sufficient to release photoelectrons from a given surface. The corresponding frequency is the critical, or threshold, frequency.

photoelectric timer – An electronic instrument that automatically turns off an X-ray machine or other photographic device when the film reaches the correct exposure.

photoelectric transducer—A transducer that converts changes in light energy into electrical changes.

photoelectric tube—*See* Phototube.

photoelectric work function – The energy required to transfer electrons from a given metal to a vacuum or other adjacent medium during photoelectric emission. It is sometimes expressed as energy in ergs or joules per unit of emitted charge, and sometimes as energy per electron in electron volts.

photoelectric yield—*See* Photoelectric Sensitivity.

photoelectromagnetic effect – *See* Dember Effect.

photoelectromotive force – Electromotive force caused by photovoltaic action.

photoelectrons—The electrons emitted from a metal in the photoelectric effect.

photoemissive – Capable of emitting electrons when under the influence of light or other radiant energy.

photoemissive effect—The emission of electrons as a result of incident radiation.

photoemissive detector – An electron tube in which the anode current varies with the intensity of light incident on the cathode.

photofabrication – The production of precise shapes in metal or other substances by recording a precise photographic image on the surface of the substance and etching away the unprotected areas by chemical or electrical means.

photoflash – A means of firing expendable flashbulbs with an instantaneous surge of current supplied by two or more 1½-volt, single-cell batteries, or by the discharge of a capacitor which has been charged to full capacity by a medium-voltage battery.

photoflash bulb – 1. An oxygen-filled glass

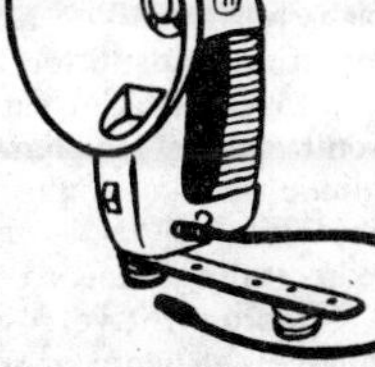

Photoflash.

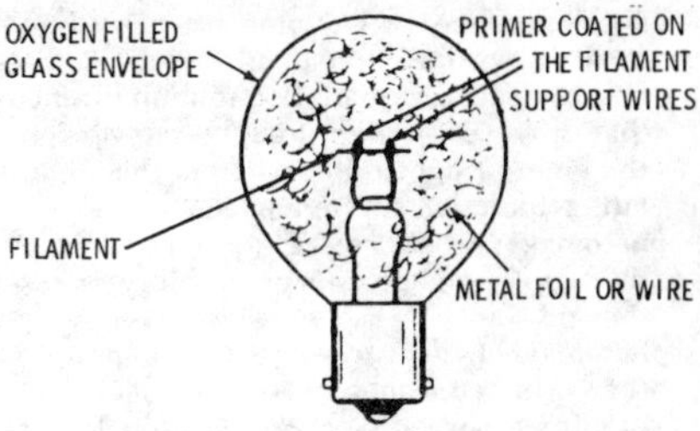

Photoflash bulb.

bulb containing a metal foil or wire. A surge of current heats the metal to incandescence, and a brilliant flash of light is produced when the wire burns in the oxygen. 2. A glass bulb packed with a highly inflammable mixture of zirconium wool and glass. The mixture is ignited and produces an instantaneous flash when a small powder-filled primer cap in the base of the bulb is set off by a small hammer.

photoflash tube—*See* Flash Tube.

photoflood lamp – An incandescent lamp employing excess voltage to give brilliant illumination. Used in television and photography, it has a life of only a few hours.

photogalvanic cell—A cell which generates an electromotive force when light falls on either of the electrodes immersed in an electrolyte.

photogenerator – A semiconductor-junction device that emits light when pulsed.

photoglow tube – A gas-filled phototube used as a relay. This is done by making the operating voltage so high that ionization and a glow discharge occur, accompanied by considerable current flow, when certain illumination is reached.

photographic writing speed – A figure of merit used to describe the ability of a particular combination of camera, film, oscilloscope, and phosphor to record a high-speed trace. It expresses the maximum single-event spot velocity (usually in centimeters per microsecond) that can be recorded on film as an image just discernible to the eye.

photoionization—Ionization occurring in a gas as a result of visible light or ultraviolet radiation.

photo-island grid—A photosensitive surface in the storage-type Farnsworth dissector tube used with television cameras. It comprises a thin, finely perforated (about 400 holes per square inch) sheet of metal.

photojunction battery—A nuclear-type battery in which the radioactive material, promethium 147, irradiates a phosphor which converts nuclear energy into light. The light is then converted to electrical energy by a small silicon junction.

photoluminescence – Luminescence stimulated by visible light or ultraviolet radiation. A special case of electroluminescence that involves a light-to-light conversion, the emitted light differing from the stimulating radiation in frequency.

photomagnetic effect – The direct effect of light on the magnetic susceptibility of certain substances.

photomagnetoelectric effect – The production in a semiconductor of an electromotive force normal to both an applied magnetic field and to a photon flux of proper wavelength.

photomask – 1. A photographic template through which images are transferred by light onto a photoresist coating. 2. A transparent plate slightly larger than a silicon slice, containing numerous tiny opaque spots, used in the planar diffusion process as a shadowmask over a slice coated with photoresist to expose the surface of the slice to acid in desired spots in a later step.

photometer–1. An instrument for measuring the intensity of a light source or the amount of illumination, usually by comparison with a standard light source. 2. A device used to compare the luminous intensities of two sources by comparing the illuminance they produce.

photometric–Related to measurements of light.

photometry–The techniques for measuring luminous flux and related quantities (e.g., luminous intensity, illuminance, luminance, luminosity, etc.).

phototmultiplier pulse-height resolution–A measure of the smallest change in the number of electrons emitted during a pulse from the photocathode that can be discerned as a change in output-pulse height.

photomultiplier tube–*See* Multiplier Phototube.

photon–1. A quantum of electromagnetic energy. The equation is hv, where h is Planck's constant and v is the frequency associated with the photon. 2. A quantum of light; used to help describe characteristics of light in conjunction with the wave theory of light. 3. A quantum of electromagnetic energy carried in small amounts by the energy and moving with the speed of light. Optical photons have energies corresponding to wavelengths between 120 and 1800 nanometers. 4. The smallest unit of radiant energy.

photon-coupled isolator–A circuit-coupling device consisting of an infrared emitter diode-coupled to a photon detector over a short shielded light path, which provides extremely high circuit isolation.

photon coupling – Coupling between circuits by a beam of light.

photonegative – Having a negative photoconductivity–hence, decreasing in conductivity (increasing in resistance) under the action of light. Selenium sometimes exhibits this property.

photo-optic memory – A memory that uses an optical medium for storage. For example, a laser might be used to record on photographic film.

photoparametric diode–A pill-sized device for simultaneously detecting and amplifying optical energy modulated at microwave frequencies.

photophone–A device for converting variations in light intensity into sound.

photopositive–Having a positive photoconductivity–hence, increasing in conductivity (decreasing in resistance) under the action of light. Selenium ordinarily has this property.

photoradiometer–Instrument for measuring the intensity and penetrating power of radiation.

photorelay circuit–A form of on-off control actuated by a change of illumination.

photoresist – 1. A solution that when exposed to ultraviolet light becomes extremely hard and resistant to etching solutions that dissolve materials such as silicon dioxide. 2. A liquid plastic which hardens into a tough acid-resistant solid when exposed to ultraviolet light. 3. A chemical substance rendered insoluble by exposure to light. By means of a photoresist, a selected pattern can be imaged on a metal. The unexposed areas are washed away and are then ready for etching by acid or doping to make a microcircuit. 4. A material which, when properly applied to a variety of substrates, becomes sensitive to portions of the electromagnetic spectrum, and when properly exposed and developed, masks portions of the substrate with a high degree of integrity.

photoresistive or photoconductive transduction – Conversion of the measurand into a change in the resistance of a semiconductor material (by changing the illumination incident on the material).

photoresistor – A semiconductor resistor which, when illuminated, drops in resistance.

photosensitive – Capable of emitting electrons when struck by light rays.

photosensitive field-effect transistor–A special unipolar field-effect transistor (FET) structure that is positioned on a header to receive illumination transmitted through a lens in the top of the header can. It combines the circuit and device characteristics of a photodiode and a high-impedance low-noise amplifier.

photosensitive recording–Recording by the exposure of a photosensitive surface to a signal-controlled light beam or spot.

photosensitive semiconductor–A semiconductor material in which light energy controls the current-carrier movement.

photosphere—The outermost luminous layer of the gaseous body of the sun.

photoswitch—A solid-state device that functions as a high-speed power switch activated by incident radiation. *See also* LASCR *and* LASCS.

phototelegraphy—In facsimile, the process of sending photographs over a wire.

phototransistor—A junction transistor with its base exposed to light through a lens in the housing. The collector current increases as the light intensity increases, because of the amplification of the base current by the transistor structure. The device may have only collector and emitter leads, or it may also have a base lead.

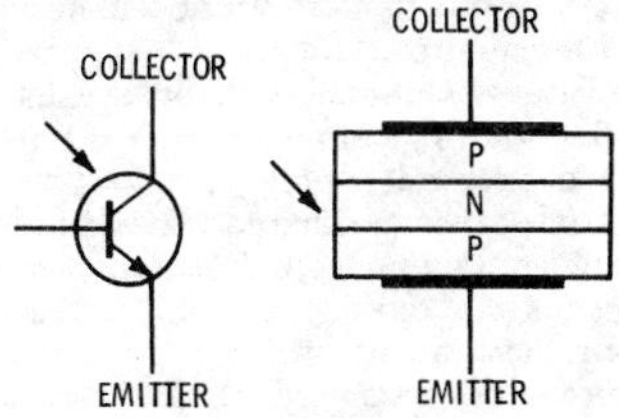

Phototransistor.

phototube—Also called photoelectric tube. An electron tube containing a photocathode. Its output depends on the total photoelectric emission from the irradiated area of the photocathode.

phototube bridge circuit—A circuit in which a phototube is one arm of a bridge circuit. With such a circuit, a balanced condition (no signal output) can be reached under either a black-signal or white-signal condition, depending on the impedance adjustments in the other arms.

phototube relay — An electrical relay in which the action of a beam of light on a phototube operates mechanical devices such as counters and safety controls.

photovaristor—A varistor in which the current-voltage relation may be modified by illumination. Cadmium sulfide and lead telluride exhibit such properties.

photovoltaic—Capable of generating a voltage when exposed to visible or other light radiation.

photovoltaic cell — A self-generating semiconductor device which converts light into electrical energy.

photovoltaic converter—A device for converting light to electric energy by means of the photovoltaic effect.

photovoltaic effect — The generation of a voltage (or an electric field) in a material that is illuminated with radiation of a suitable wavelength.

photovoltaic transduction — Conversion of the measurand into a change in the voltage generated when a junction of dissimilar materials is illuminated.

photox cell—A type of photovoltaic cell in which a voltage is generated between a copper base and a film of cuprous oxide during exposure to visible or other radiation.

photran—A triode pnpn-type switch of the reverse-blocking pnpn-type switch class. The photran provides optical triggering in addition to standard gate-terminal triggering.

phototronic cell — A type of photovoltaic cell in which a voltage is generated in a layer of selenium during exposure to visible or other radiation.

physical configuration — The arrangement of materials, components, and wires on a structural assembly for physical strength, conservation of size, and minimal electrical interaction.

physiological patient monitor—A device for automatically measuring and/or record-

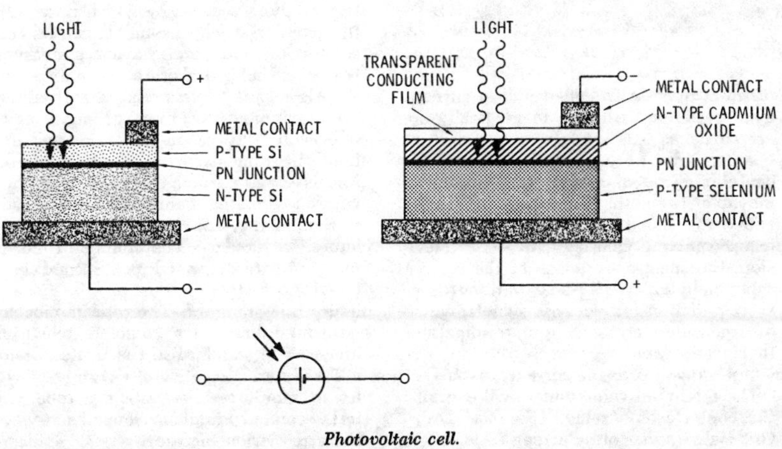

Photovoltaic cell.

ing one or several physiological variables and responses of a patient, including heart potential, blood flow, blood pressure, pulse rate, respiration rate, temperature, etc.

pi—The Greeek letter π. It designates the value of the ratio of the circumference of a circle to its diameter, approximately 3.14159.

pickoff—A device which produces a signal output, generally a voltage as a function of the angle between two gimbals, or between a gimbal and a base.

pickup—1. A device that converts a sound, scene, or other form of intelligence into corresponding electric signals (e.g., a microphone, television camera, or phonograph pickup). 2. The minimum current, voltage, power, or other value which will trip a relay. 3. Interference from a nearby circuit or system.

pickup arm—*See* Tone Arm.

pickup cartridge—The removable portion of a pickup arm. It contains the electromechanical transducing elements and the reproducing stylus.

Pickup cartridge.

pickup current—Also called pull-in current. That current at which a magnetically operated device starts to operate.

pickup head—The end of the pickup arm containing the cartridge. It is often removable from the arm and is usually called the "head shell."

pickup spectral characteristic — In television, the spectral response of the camera tube including the optical parts. It converts radiation into electric signals, which are measured at the output terminals of the pickup tube.

pickup value (voltage, current, or power) —The minimum value which will energize the contacts of a relay. (*See also* Drop-Out Value *and* Hold Current.)

pickup voltage—That voltage at which a magnetically operated device starts to operate.

pico—Prefix meaning 10^{-12}. (Formerly micromicro.) Abbreviated p.

picoammeter—A sensitive ammeter that indicates current values in picoamperes.

picoampere—One-millionth of a microampere (10^{-12} ampere). Abbreviated pA.

picocoulomb — One-millionth of a microcoulomb (10^{-12} coulomb). Abbreviated pC.

picocurie — One-millionth of a microcurie (10^{-12} curie). Abbreviated pC.

picofarad — Abbreviated pF. 10^{-12} farad; formerly micromicrofarad. Called puff in England.

picosecond — A micromicrosecond (10^{-12} second); one-thousandth of a nanosecond. Abbreviated ps.

picowatt — One-millionth of a microwatt (10^{-12} watt). Abbreviated pW. Formerly called micromicrowatt.

pictorial wiring diagram—A wiring diagram containing actual sketches of components and clearly showing all connections between them.

picture black—Also called black signal. In facsimile, the signal produced at any point by the scanning of a selected area of subject copy having maximum density.

picture brightness — A measure of the brightness of the highlights in a television picture, usually measured in foot-lamberts.

picture carrier—Also called luminance carrier. The carrier frequency 1.25 MHz above the lower frequency limit of a standard NTSC television signal. In color television, this carrier is used for transmitting luminance information; and the chrominance subcarrier, which is 3.579545 MHz higher in frequency, transmits the color information.

picture element—1. In facsimile, that portion of the subject copy which is seen by the scanner at any instant. It can be considered a square area having dimensions equal to the width of the scanning line. 2. Also called elemental area, critical area, scanning spot, or recording spot. In television, any segment of a scanning line, the dimension of which along the line is exactly equal to the nominal line width. The area being explored at any instant in the scanning process.

picture frequency — The number of complete pictures scanned per second in a television system.

picture line standard—The total number of horizontal lines in a complete television image. The standard in the United States is 525 lines.

picture monitor—A cathode-ray tube and its associated circuits arranged for viewing a television picture.

picture signal—In television, the signal resulting from the scanning process.

picture-signal amplitude – The difference between the white peak and the blanking level of a video signal.

picture-signal polarity—The polarity of the signal voltage which represents a dark area of a scene given with respect to the signal voltage representing a light area. Expressed as black negative or black positive.

picture size—The usable viewing area on the screen of a television receiver, measured in square inches.

picture-synchronizing pulse—*See* Vertical-Synchronizing Pulse.

picture transmission—Electric transmission of a shaded (halftone) picture.

picture transmitter—*See* Visual Transmitter.

picture tube—*See* Kinescope, 1.

picture-tube brightener – An accessory added to an aging picture tube to increase the image brightness and thereby extend its useful life. When the cathode emission is subnormal, the brightener raises the filament voltage and thereby increases the electron emission from the cathode.

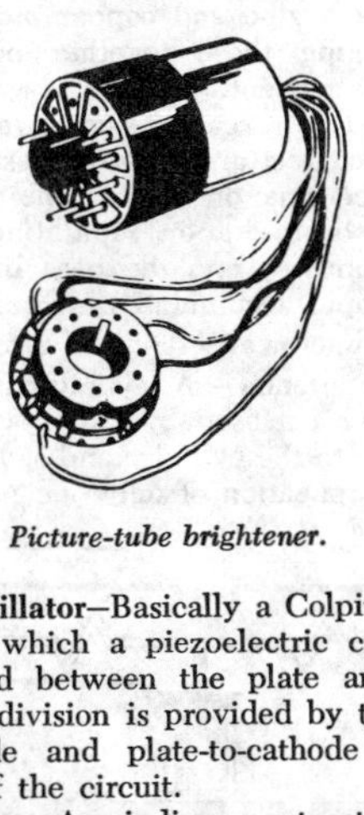

Picture-tube brightener.

Pierce oscillator—Basically a Colpitts oscillator in which a piezoelectric crystal is connected between the plate and grid. Voltage division is provided by the grid-to-cathode and plate-to-cathode capacitances of the circuit.

pie winding—A winding constructed from individual washer-shaped coils called pies.

piezodielectric—Pertaining to a change in dielectric constant under mechanical stress.

piezoelectric—The property of certain crystals, which: 1. Produce a voltage when subjected to a mechanical stress. 2. Undergo mechanical stress when subjected to a voltage.

piezoelectric accelerometer – Basically a crystalline material which, when force is applied, generates a charge. Through the incorporation of a mass in direct contact with the crystal, an acceleration transducer is produced.

piezoelectric axis—In a crystal, one of the directions in which tension or compression will develop a piezoelectric charge.

piezoelectric crystal—1. A piece of natural quartz or other crystalline material capable of demonstrating the piezoelectric effect. A quartz crystal, when ground to certain dimensions, will vibrate at a desired radio frequency when placed in an appropriate electric circuit. 2. A substance that has the ability to become electrically polarized and has strong piezoelectric properties in it, so cut as to emphasize the coupling to some distinct mechanical mode of the crystal. Used as an electromechanical transducer.

piezoelectric crystal cut – The orientation of a piezoelectric crystal plate with respect to the axes of the crystal. It is usually designated by symbols—e.g., GT, AT, BT, CT, and DT identify certain quartz-crystal cuts having very low temperature coefficients.

piezoelectric crystal element – A piece of piezoelectric material cut and finished to a specified shape and orientation with respect to the crystallographic axes of the material.

piezoelectric crystal plate – A piece of piezoelectric material cut and finished to specified dimensions and orientation with respect to the crystallographic axes of the material and having two essentially parallel surfaces.

piezoelectric crystal unit—A complete assembly comprising a piezoelectric crystal element mounted, housed, and adjusted to the desired frequency, with means for connecting it into an electric circuit. Such a device is commonly employed for frequency control or measurement, electric

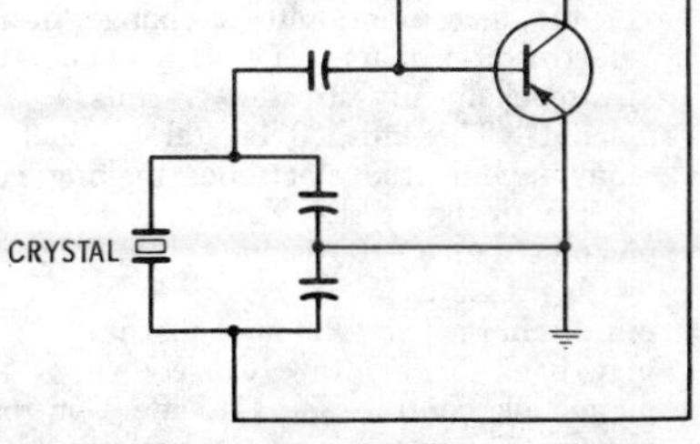

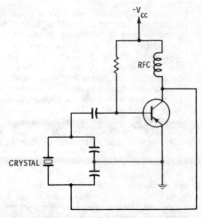

Pierce oscillator.

wave filtering, or interconversion of electric and elastic waves.

piezoelectric device—1. A substance which generates an electric voltage when bent, squeezed, or twisted. 2. Conversely, when a voltage is applied, it will twist, bend, expand, or contract.

piezoelectric effect—1. The mechanical deformity of certain natural and synthetic crystals under the influence of an electric field. This effect is used in high-precision oscillators and certain high-frequency filters. 2. The property of certain natural and synthetic crystals to produce a voltage when subjected to mechanical stress (compression, expansion, twisting, etc.).

piezoelectricity—The phenomenon whereby certain crystalline substances, such as barium titanate, generate electrical charges when subjected to mechanical deformation. The effect was first noticed by the Curie brothers in the 1880's as a result of extensive study of the symmetry of crystalline materials. The reverse effect also occurs.

piezoelectric loudspeaker – *See* Crystal Loudspeaker .

piezoelectric material – A material which generates an electrical output when subjected to a mechanical stress. (The word is derived from the Greek *piezein*, meaning to squeeze or press.)

piezoelectric microphone—*See* Crystal Microphone.

piezoelectric oscillator—A crystal-oscillator circuit in which the frequency is controlled by a quartz crystal.

piezoelectric pickup—*See* Crystal Pickup.

piezoelectric pressure gage—An apparatus for measuring or recording very high pressures. The pressure is applied to quartz discs or other piezoelectric crystals. The resultant voltage, after amplification, is then measured or is recorded with an oscillograph.

piezoelectric speaker—*See* Crystal Speaker.

piezoelectric transducer – Also called ceramic or crystal transducer. A transducer that depends for its operation on the interaction between the electric charge and the deformation of certain asymmetric crystals having piezoelectric properties.

piezoelectric transduction—The conversion of the measurand into a change in the electrostatic charge or voltage generated by mechanically stressed crystals.

piezoid – The finished crystal product. It may include the electrodes making contact with the crystal blank.

piezoresistance – Resistance that changes with pressure.

pig discharge—*See* Penning discharge.

piggyback—*See* Voltage Corrector.

piggyback control—*See* Cascade Control.

piggyback twistor – An electrically alterable, nondestructive-readout information-storage device that consists of a thin, narrow tape of magnetic material wound spirally around a fine copper conductor. A second similar tape is wrapped on top of the first to sense the stored information. A binary digit is stored at the intersection of a copper strap and a pair of these twistor wires.

pigtail—1. Either a wire attached for terminating purposes to a shield, or a conductor extending from a small component. 2. The termination of a capacitor winding to its lead. 3. The disc-shaped head, at the end of a lead, that is attached to a capacitor winding.

pigtail splice – A splice made by tightly twisting the bared ends of parallel conductors together.

pile – 1. A nuclear reactor. So called because early reactors were piles of graphite blocks and uranium slugs. 2. (Voltaic) Invented by Volta in 1800, it was the first primary battery known to the modern world. The pile consisted of an arrangement of pairs of discs of copper and zinc, each pair separated by a disc of moistened pasteboard. Later Volta arranged a series of cups filled with brine, each containing a zinc and copper plate, and by connecting these together obtained an electric current.

pileup—1. Also called stack. On a relay, a set of contact arms, assemblies, or springs fastened one on top of the other with layers of insulation separating them. 2. A departure from the base line because of a rapid accumulation of pulses.

pill—A microwave stripline termination.

pillbox antenna – A cylindrical parabolic reflector enclosed by two plates perpendicular to the cylinder and spaced to permit propagation of only one mode in the

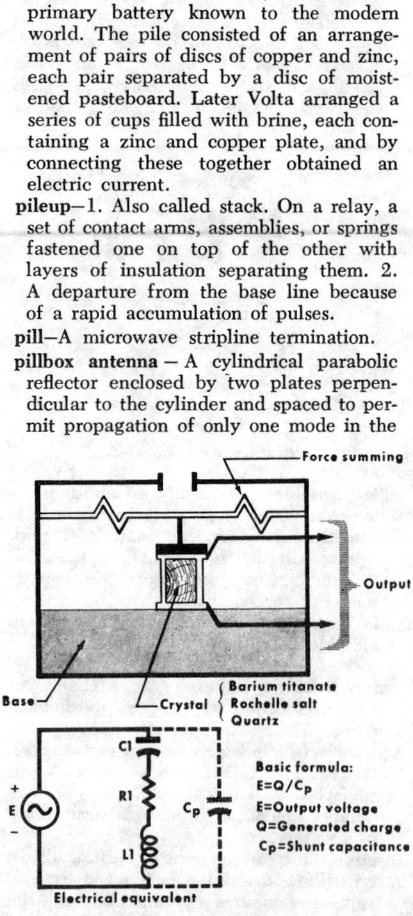

Piezoelectric transducer.

desired direction of polarization. It is fed on the focal line.

pilot—1. In a transmission system, a signal wave, usually a single frequency, transmitted over the system to indicate or control its characteristics. 2. In a tape relay, instructions appearing in a routing line, relative to the transmission or handling of that message.

pilot cell—The storage-battery cell selected because its temperature, voltage, and specific gravity are assumed to be those of the entire battery.

pilot channel—A very narrow-band channel over which a single frequency, the pilot frequency, is transmitted to operate trouble alarms or automatic level regulators, or both.

pilot contacts—Contacts, the opening, closing, or transfer of which govern an operation of relays or similarly controlled devices.

pilot holes — Also called manufacturing holes or fabrication holes. Holes on a printed-circuit board used as guides during manufacturing operations. Mounting holes in a part are sometimes used as pilot holes.

pilot lamp—A light that indicates whether a circuit is energized.

pilot light—1. A light which, by means of position or color, indicates whether a control is functioning. 2. A light that indicates which of a number of normal conditions of a system or device exists. It is unlike an alarm light, which indicates an abnormal condition. The pilot light is also known as a monitor light.

pilot regulator—A device for maintaining the receiving level of a carrier-derived circuit constant under varying attenuation conditions of the transmission line.

pilot spark—A low-power preliminary spark used in a gas-discharge tube to produce an ionized path for the large main spark discharge.

pilot subcarrier—A subcarrier used as a control signal in the reception of compatible fm stereophonic broadcasts.

pilot tone—1. A single frequency sent over a narrow channel to cause an alarm or automatic control to operate. 2. The signal included with the stereo information, and applied in a subchannel on the fm signal, which is used to reform the suppressed subcarrier and synchronize the switching of the detectors in the stereo decoder. Pilot-tone frequency is 19kHz (broadcast at the equivalent of 10% fm modulation), and doubling this (38kHz) gives the subcarrier frequency. The pilot tone also activates the stereo indicator light switching.

pilot wire — An auxiliary conductor used with remote measuring devices or for operating apparatus at a distance.

pilot-wire regulator—An automatic device for controlling adjustable gains or losses associated with transmission circuits to compensate for transmission changes caused by temperature variations, the control usually depending upon the resistance of a conductor or pilot wire having substantially the same temperature conditions as the conductors of the circuit being regulated.

pi mode—A mode of magnetron operation in which the fields of successive anode openings facing the interaction space differ inphase by pi radians.

pin—Also called a prong or base pin. A terminal on a connector, plug, or tube base.

pinboard—A perforated board into which pins may be inserted manually to control the operation of equipment.

pinch effect—1. The result of an electromechanical force that constricts, and sometimes momentarily ruptures, a molten conductor which is carrying a high-density current. 2. In the reproduction of lateral recordings, the pinching of the reproducing stylus tip twice each cycle. This is due to a decrease in the groove angle cut by the recording stylus during the swing from a negative to a positive peak. 3. The self-contraction of a plasma column carrying a large current due to the interaction of this current with the magnetic field it produces. The current required for such an effect is on the order of 10^5 amperes. If the current is in the form of a short pulse, a radially imploding shock wave is generated.

pinch-off—1. In a field-effect transistor, a condition in which the gate bias causes the depletion region to extend completely across the channel, with a resulting cessation of drain current. 2. The voltage required to stop majority current flow is a FET.

pinch-off voltage—The gate voltage of a field-effect transistor that blocks the current for all source-drain voltages below the junction breakdown value. Pinch-off occurs when the depletion zone completely fills the area of the device.

pinch resistor—A monolithic silicon resistor derived from a p-base diffusion resistor by superimposing an n+ emitter diffusion over the resistor. As a result, the surface of the p resistor reverts to n-type material, and a narrow "pinched" resistor is left under the n+ diffusion. A pinch resistor features 10,000 ohms per square, but it is limited to low voltages, and it has a high temperature coefficient.

pinch roller—An idling roller used to press a tape against the capstan to cause advance. A friction clamp is used with the roller, as a brake to stop perforated tape.

pinch wheel—1. *See* Pressure Roller. 2. *See* Idler.

pin connection—Connections made to the base of pins in a vacuum tube. They are identified by the following abbreviations: NC, no connection; IS, internal shield; IC, internal connection (not an electrode connection); P, plate; G, grid; SG, screen grid; SU, suppressor; K, cathode; H, heater; F, filament; RC, ray-control electrode; TA, target.

pin contact—A male-type contact, usually designed to mate with a socket or female contact. It is normally connected to the "dead" side of a circuit.

pincushion distortion—Distortion which results in a monotonic increase in radial magnification in the reproduced image away from the axis of symmetry of the electron optical system. The four sides of the raster are curved inward, leaving the corners extending outward.

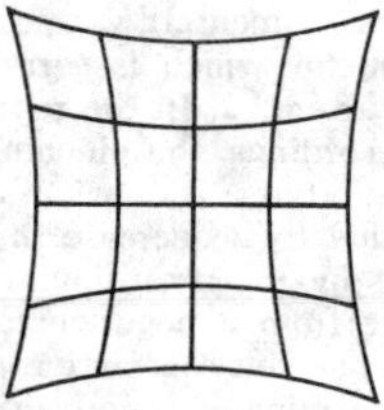

Pincushion distortion.

pin diode—A diode made by diffusing the semiconductor with p dopant from one side and n dopant from the opposite side with the process so controlled that a thin intrinsic region separates the n and p regions. The storage time of the pin diode is long enough that it cannot rectify at microwave frequencies. Instead, it behaves as a variable resistor with its value controlled by a dc bias current. Therefore, this type of diode is well suited for use as a variable microwave attenuator.

pin-diode attenuator—A two-port network consisting of two or more pin diodes controlled by a driver circuit. At microwave frequencies, the diodes act as a small value of capacitance shunted by a resistance that can be varied over a range of about 2 to 10,000 ohms through control of the bias current by the driver circuit.

pine-tree array—An array of dipole antennas aligned vertically (termed the radiating curtain), behind which and approximately a quarter wavelength away is a parallel array of dipole antennas forming the reflecting curtain.

pi network—A network composed of three branches, all connected in series with each other to form a mesh. The three junction points form an input terminal, an output terminal, and a common input and output terminal, respectively.

pinfeed platen—In a computer, a cylindrical platen having integral rings of pins that engage perforated holes in the paper, thus permitting feeding of the paper.

ping—A sonic or ultrasonic pulse of predetermined width.

ping-pong — A programming technique in which two magnetic tape units are used for multiple-reel files. Automatic switching between the two units is carried out until the entire file is processed.

ping-pong–ball effect — The bouncing of sound back and forth between the two sides of a stereophonic reproducing system.

pinhole—A small opening, occurring as an imperfection, extending through the thickness of a substance.

pinhole detector — A photoelectric device that detects extremely small holes and other defects in moving sheets of material, and often actuates sorting equipment that automatically rejects defective sheets.

pinholes—Small holes, occurring as imperfections, which penetrate entirely through an applied material to the substrate (e.g., holes in semiconductor insulating oxides, screened resistors, thin-film elements, etc.).

pinion—Of two gears that mesh, the one with the fewer teeth.

pin jack—A single-conductor jack having a small opening into which a plug tipped with a metal pin can be inserted.

pink noise—Noise whose amplitude is inversely proportional to frequency over a specified range. Equal-energy distribution occurs in any octave bandwidth within that range. Pink noise is very pleasing to the human ear. Many people feel relaxed listening to the patter of rain (a close approximation of pink noise). Other examples include the sound of surf and a shower stream.

PINO—Acronym for positive input, negative output.

pinouts—The external wires or pins on a module (generally having a circuit function).

pin sensing—A process in which a device using a punched card generates digital data by sensing the opening and closing of switches.

pip—*See* Blip, 1.

piped program — A program transmitted

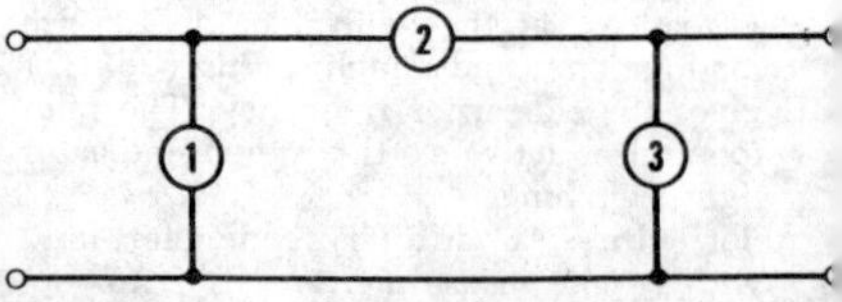

Pi network.

over telephone wires, usually from one studio to another.

pipeline computers—Computers which execute serial programs only.

pip-matching display—A navigational display in which the received signal appears as a pair of blips. The desired quantity is measured by comparing the characteristics.

pi (π) point—The frequency at which the insertion phase shift of an electric structure is 180° or an integral multiple of 180°.

Pirani gage — A bolometric vacuum gage for measuring pressure. Its operation depends on the thermal conduction of the gas present. The pressure being measured is a function of the resistance of a heated filament, ordinarily over a range of 10^{-1} to 10^{-4} mm Hg.

piston — Also called a plunger. In high-frequency communications a conducting plate that can be moved along the inside of an enclosed transmission path to short out high-frequency currents.

piston action—The movement of a speaker cone or diaphragm when driven at the bass audio frequencies.

piston attenuator—An attenuator generally used at microwave frequencies, the amount of attenuation of which can be varied by moving an output coupling device along its longitudinal axis.

pistonphone — A small chamber equipped with a reciprocating piston of measurable displacement. In this way, a known sound pressure can be established in the chamber.

pitch—1. That attribute of auditory sensation by which sounds may be ordered on a scale extending from low to high (e.g., a musical scale). 2. The distance between two adjacent corresponding threads of a screw measured parallel to the axis. 3. The distance between the peaks of two successive grooves of a disc recording. 4. A term applied to a musical tone that is used as a standard for tuning, singing, etc. Standard U.S. and European pitch is based on A = 440Hz. When the pitch is raised one octave, the frequency is twice the original.

pits—Small holes occurring as imperfections which do not penetrate entirely through the printed element.

piv—Abbreviation for peak inverse voltage.

pivot & jewel—A method of suspending the moving coil or moving iron vane of a meter in a magnetic field. Glass jewel and steel pivot.

PLA—Abbreviation for programmable logic array.

place—*See* Column.

planar—1. Lying essentially in a single plane. 2. Constructed in layers or planes.

planar ceramic tube—An electron tube constructed with parallel planar electrodes and a ceramic envelope.

planar devices—*See* Planar Process.

planar diffusion—Technique used to manufacture semiconductors having diffused pn junctions. All the junctions emerge at the top surface of the wafer.

planar diode — A diode containing planar electrodes lying in parallel planes.

planar mask—A shadow or aperture mask which has no curvature; one which is perfectly flat.

planar module — A packaged module wherein the individual components are positioned and terminated flat or parallel with the plane of the substrate.

planar network — A network in which no branches cross when drawn on the same plane.

planar process — The technology used in fabricating semiconductor devices wherein all pn junctions terminate in the same geometric plane. An oxide is formed at the surface for the purpose of stabilizing the parameters (passivating).

planar silicon photoswitch — Abbreviated PSPS. Essentially a complementary SCR. Like the LASCR, it can be triggered by light. In addition, a negative signal (with reference to the anode) at the anode gate terminal can trigger the device.

planar technique—The formation of p-type and/or n-type regions in a semiconductor crystal by diffusing impurity atoms into the crystal through holes in an oxide mask, which is on the surface. The latter is left to protect the junctions so formed against surface contamination.

planar transistor—1. A diffused transistor in which the emitter, base, and collector

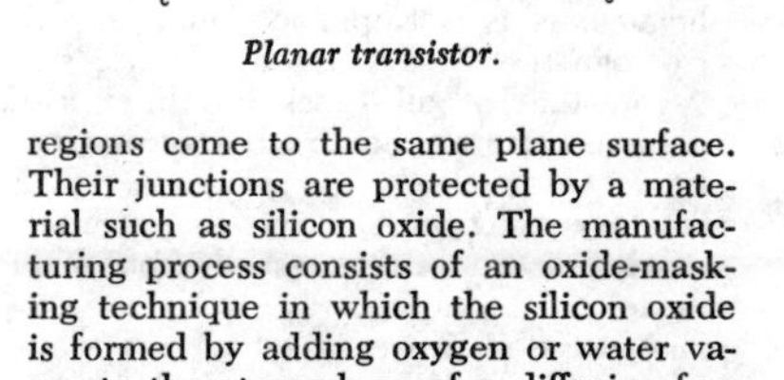

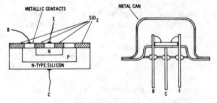

Planar transistor.

regions come to the same plane surface. Their junctions are protected by a material such as silicon oxide. The manufacturing process consists of an oxide-masking technique in which the silicon oxide is formed by adding oxygen or water vapor to the atmosphere of a diffusion furnace. The thickness of the oxide layer is a function of time, temperature, and the amount of oxidizing agent. 2. A junction transistor manufactured by a process in which the surface of a chip is passivated with a thin film of oxide, dopants being introduced by successive etching and diffusion.

planchet—A small metal container or sample holder for radioactive materials undergoing radiation measurements in a proportional counter or scintillation detector.

Planckian locus—A line drawn on a chromaticity diagram to represent light radiation from a reference blackbody at 2000° to 10,000° Kelvin (K).

Planck's constant — Symbolized by h. The constant representing the ratio of the energy of any radiation quantum to its frequency. It has the dimension of action (energy × time) and a numerical value of 6.547×10^{-27} erg-second. Its significance was first recognized by the German physicist Max Planck in 1900.

Planck's distribution—An equation that describes the entire distribution of energy radiated from a blackbody as a function of wavelength, based on quantum mechanics.

Planck's radiation law—An expression representing the spectral radiance of a blackbody as a function of the wavelength and temperature.

plane—A screen of magnetic cores. Planes are combined to form stacks.

plane earth—Earth that is considered to be a plane surface. Used in ground-wave calculations.

plane-earth attenuation—Attenuation of an electromagnetic wave over an imperfectly conducting plane earth in excess of that over a perfectly conducting plane.

plane-earth factor — Ratio of the electric field strength that would result from propagation over an imperfectly conducting plane earth, to that over a perfectly conducting plane.

plane of a loop — An infinite imaginary plane which passes through the center of a loop and is parallel to its wires.

plane of polarization—For a plane-polarized wave, the plane containing the electric field vector and the direction of propagation.

plane-polarized wave—At any point in a homogeneous isotropic medium, an electromagnetic wave with an electric field vector that at all times lies in a fixed plane containing the direction of propagation.

planetary electron—One of the electrons moving in an orbit or shell around the nucleus of an atom.

plane wave—A wave in which the wavefronts are everywhere parallel planes normal to the direction of propagation.

plan-position indicator — Abbreviated ppi. Also called P-display. A type of presentation on a radar indicator. The signal appears as a bright spot, with range indicated by the distance of the spot from the center of the screen, and the bearing by the radial angle of the spot.

plasma — 1. A wholly or partially ionized gas in which the positive ions and negative electrons are roughly equal in number. Hence, the space charge is essentially zero. 2. The region in which gaseous conduction takes place between the cathode and anode of an electric arc. 3. An electrically conductive gas comprised of neutral particles, ionized particles, and free electrons but which, taken as a whole, is electrically neutral.

plasma diode — A thermodynamic engine with electrons as the working fluid, the potential energy of which is converted to a useful output.

plasma engine — A reaction engine using electrically accelerated plasma as the propellant.

plasma frequency—A natural frequency for coherent electron motion in a plasma.

plasma jet—A high-temperature stream of electrons and positive ions produced by the magnetohydrodynamic effect of a strong electrical discharge.

plasma length—*See* Debye Length.

plasma oscillation—Electrostatic or space-charge oscillations in a plasma which are closely related to the plasma frequency. There is usually enough damping due to electron collisions to prevent self-generation of the oscillations. They can be excited, however, by such techniques as shooting a modulated electron beam through the plasma.

plasma physics—The study of highly ionized gases.

plasma sheath—An envelope of ionized gas that surrounds an object moving through an atmosphere at a hypersonic velocity. The plasma sheath affects radio-wave transmission, reception, and diffraction.

plasma thermocouple—An electronic device in which the heat from nuclear fission is converted directly into electric power.

plasmatron—A helium-filled current amplifier that combines the grid-control characteristics and linearity of a vacuum triode with the extremely low internal impedance of a thyratron.

plasticizer—A substance added to a plastic to produce softness and adhesiveness in the finished product.

PLAT—Acronym for pilot landing aid television. A system in which television cameras cover aircraft landings on a carrier from several angles, allowing the landing personnel to "talk" the pilot down with increased precision. Recordings can be made for future reference.

plate—1. Preferably called the anode. The principal electrode to which the electron stream is attracted in an electron tube. 2. One of the conductive electrodes in a capacitor. 3. One of the electrodes in a storage battery. 4. *See* Printed-Circuit Board.

plateau – In the counting-rate-versus-voltage characteristic of a radiation counter tube, that portion in which the counting rate is substantially independent of the applied voltage.

plateau length—The applied-voltage range over which the plateau of a radiation counter tube extends.

plate-bypass capacitor – A capacitor connected between the plate and cathode of a vacuum tube to bypass high-frequency currents and thus keep them out of the load. (*See also* Anode-Bypass Capacitor.)

plate characteristic—A graph showing how changes in plate voltage affect the plate current of a vacuum tube.

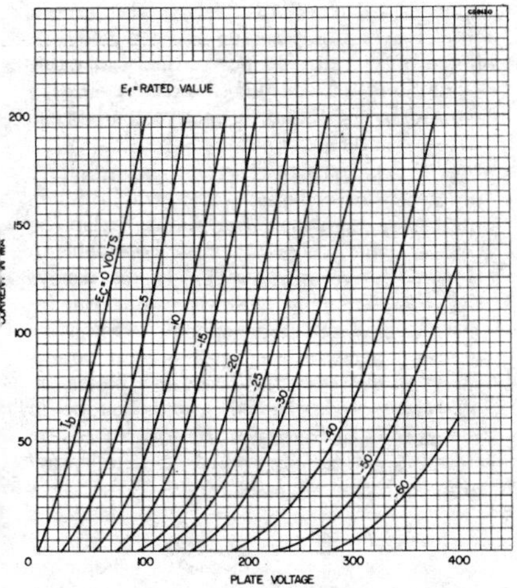

Plate characteristic curves.

plate circuit—The complete external electrical circuit between the plate and cathode of an electron tube.

plate-circuit detector—A detector that functions by virtue of its nonlinear plate-current characteristic.

plate conductance—The in-phase component of the alternating plate current divided by the alternating plate voltage, all other electrode voltages being maintained constant.

plate current—Electron flow from the cathode to the plate inside an electron tube.

plate detection—The operation of a vacuum-tube detector at or near plate-current cutoff so that the input signal is rectified in the plate circuit.

plate dissipation – The amount of power lost as heat in the plate of a vacuum tube.

plated-resist—A material electroplated on conductive areas to make them impervious to etching.

plated thru-hole—*See* Thru-Hole Connection.

plated wire memory—A wire coated with magnetic material on which binary digits can be stored.

plate efficiency—Also called the anode efficiency. Ratio of load-circuit power (alternating current) to plate power input (direct current).

plate impedance – Also called plate-load impedance. The total impedance between the anode and cathode, exclusive of the electron stream.

plate-input power—In the last stage of a transmitter, the direct plate voltage applied to the tubes times the total direct current flowing to their plates, measured without modulation.

plate keying—Keying done by interrupting the plate-supply circuit.

plate-load impedance—Also called the anode-load impedance and plate impedance. The total impedance between the anode and cathode of a vacuum tube, exclusive of the electron stream.

plate modulation—Also called anode modulation. Modulation produced by applying the modulating voltage to the plate of any tube in which the carrier is present.

platen—A backing structure (usually cylindrical) against which a printing mechanism strikes in producing an impression.

plate neutralization—Neutralizing an amplifier by shifting a portion of the plate-to-cathode ac voltage 180° and applying it to the grid-to-cathode circuit through a capacitor. Also called anode neutralization.

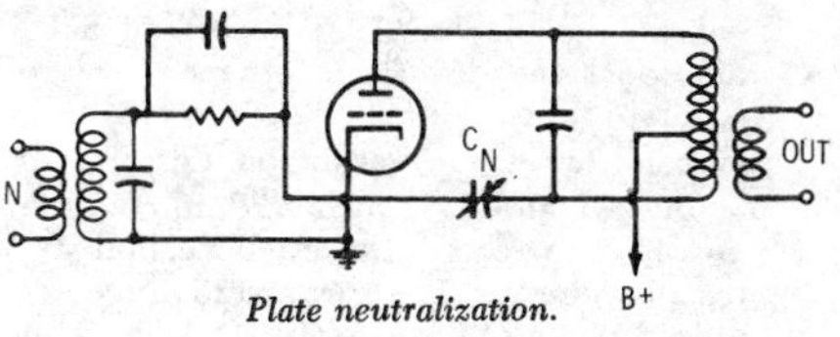

Plate neutralization.

plate power input—Also called the anode power input. The dc power (mean anode voltage times current) delivered to the plate (anode) of a vacuum tube.

plate power supply – *See* Anode Power Supply.

plate pulse modulation—Also called anode pulse modulation. Modulation produced in an amplifier or oscillator by applying externally generated pulses to the plate circuit.

plate resistance—The plate-voltage change divided by the resultant plate-current change in a vacuum tube, all other conditions being fixed.

plate saturation—Also called anode or voltage saturation. The point at which the plate current of a vacuum tube no longer increases as the plate voltage does.

plate supply—*See* Anode Supply.

plate-to-plate impedance—The load impedance between the two plates in a push-pull amplifier stage.

plate voltage—The dc voltage between the plate and cathode of a vacuum tube.

plate winding—A transformer winding connected to the plate circuit of a vacuum tube.

plating resist—Any material which, when deposited on conductive areas, prevents plating of the areas it covers.

platinotron — A cross-field vacuum tube used to generate and amplify microwave energy. It resembles the magnetron, except that it has no resonant circuit and has two external rf connections instead of only one.

platinum — A heavy, almost white metal that resists practically all acids and is capable of withstanding high temperatures.

platinum contacts — Used where currents must be broken frequently (e.g., in induction coils and electric bells). Sparking does not damage platinum as much as it does other metals. Hence a cleaner contact is assured with minimum attention.

platter — A popular term for phonograph records and transcriptions.

playback—The reproduction of a tape recording or disc through an amplifier and loudspeaker or phones.

playback head—The magnetic assembly on a tape recorder that responds to the recorded pattern on the tape, and develops a signal representing that pattern to feed to the preamplifier. In some tape machines, the playback and recording head are the same device; in others they are separate units.

playback loss—*See* Translation Loss.

playing weight—1. Downward force of a pickup on a record. Sometimes called stylus pressure. 2. The downward force required on the pickup stylus to keep it in the groove and to counter the mechanical reactions of replay.

plethysmogram—A graph of the changes in size or volume of an organ or the amount of blood present or passing through a vessel. The transducer for a plethysmogram can be either a pressure transducer or a pair of surface electrodes.

plethysmograph—An instrument for detecting variations of blood volume in the tissues during the cardiac cycle. *See also* Electrical-Impedance Cephalography *and* Finger Plethysmograph.

pliodynatron—A four-element vacuum tube with an additional grid, which is maintained at a higher voltage than the plate to obtain negative-resistance characteristics.

pliotron—An industrial-electronic term for a hot-cathode vacuum tube having one or more grids.

PLL—Abbreviation for phase-locked loop.

plot—*See* Print.

plotter — A device that produces an inscribed visual display of the variation of a dependent variable as a function of one or more other variables.

plotting board—A device which plots one or more variables against one or more other variables.

ploy effect — In surface channel charge-coupled devices (CCDs), the tendency for charges to be captured by surface effects, thus resulting in a loss of signal. By continuously introducing a charge into all CCD channels through a diffusion at the beginning of the channel, the areas that trap charges are filled by the induced charges rather than the signal charges, thus increasing transfer efficiency.

plt—Abbreviation for power-line transients—one kind of conducted noise generally caused by switching inductive loads measured on the power line.

plug—1. The part of the two mating halves of a connector which is movable when not fastened to the other mating half. The plug is usually thought of as the male portion of the connector. This is not always the case; the plug comprises contacts, insert, shell, and any special sealing or holding accessories. (*See also* Connector.) 2. A method of braking a motor by applying partial or full rated voltage in reverse in an attempt to quickly bring the motor to zero speed. Also called plugging and plug reverse.

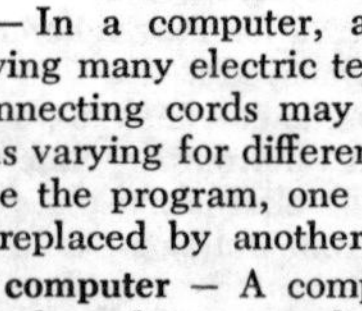

Plug.

plugboard — In a computer, a removable board having many electric terminals into which connecting cords may be plugged in patterns varying for different programs. To change the program, one wired plugboard is replaced by another.

plugboard computer — A computer which has a punch-card input and output, and to which program instructions are delivered by means of interconnecting patch cords on a removable plugboard.

plug connector—An electrical fitting containing male, female, or male and female contacts and constructed so that it can be affixed to the end of a cable, conduit, coaxial line, cord, or wire for convenience in joining with another electrical connector or connectors. It is not designed for mounting on a bulkhead, chassis, or panel.

pluggable unit—A chassis which can be removed from or inserted into the rest of the equipment by merely plugging in or pulling out a plug.

plugging—*See* Plug 2.

plug-in—Any device to which connections can be completed through pins, plugs, jacks, sockets, receptacles, or other ready connectors.

plug-in coil—A coil that can be easily interchanged and used for varying the tuning range of a receiver or transmitter. It is wound around a form often resembling an elongated tube base, with the coil leads connected to pins on the base.

plug-in device—A component or group of components and their circuitry which can be easily installed or removed from the equipment. Electrical connections are made by mating contacts.

plug-in unit — A standard subassembly of components that can be readily plugged into or pulled out of a circuit as a unit.

plug reverse—*See* Plug 2.

Plumbicon — A vidicon with a lead-oxide target; its major advantage is its lack of image retention. It is a tube with the simplicity of a vidicon, and the sensitivity and lag of a glass target image orthicon. The tube is used for live black-and-white and color broadcasting. Trademark of N. V. Philips of Holland.

plumbing—Coaxial lines or waveguides and accessory equipment for transmission of radio-frequency energy.

plume — The hot gaseous material ejected briefly from a highly absorbent material after bombardment by an intense laser pulse. The plume emits broad-band white light and is the most prominent feature in most irradiation experiments.

plunger—*See* Piston.

plunger relay—A relay consisting of a movable core or plunger surrounded by a coil. Solenoid action causes the plunger or core to move and thus energize the relay whenever current flows through the coil.

plunger-type instrument—A moving-iron instrument for measuring current. It consists of a pointer attached to a plunger inside a coil. The current being measured flows through the coil and pulls the plunger down. How far it goes into the coil depends on the magnitude of the current.

plutonium—A heavy element which undergoes fission when bombarded by neutrons. It is a useful fuel in nuclear reactors. Its symbol is *Pu;* its atomic number, 94.

pm—Abbreviation for phase modulation or permanent magnet.

pm erasing head — A head which uses the fields of one or more permanent magnets for erasing.

PMOS—1. P-channel MOS having p-type source and drain regions diffused into an n-type substrate to create a p channel for conduction. 2. MOS devices made on an n-type silicon substrate in which the active carriers are holes (p) flowing between p-type source and drain controls. 3. An MOS (unipolar) transistor in which the working current consists of positive (p) electrical charges.

pm speaker—Abbreviation for permanent-magnet speaker.

pn boundary—The surface where the donor and acceptor concentrations are equal in the transition region between p- and n-type materials.

PNdB — Perceived noise level expressed in decibels. *See* Perceived Noise Level.

pn diode—A diode that has no intrinsic region and a short storage time. It functions as a normal diode rectifier into the high microwave frequencies. If the diode is given a dc bias that is large compared to the rf signal, it ceases to be a rectifier; thus, it can be used as a reflective microwave switch. It also can be employed as a variable reflective attenuator, except in that operating region for which the bias and rf voltages are comparable and rectification occurs.

pneumatic bellows — A gas-filled bellows sometimes used to provide delay time in plunger-type relays.

pneumatic speaker—A speaker in which the acoustic output is produced by controlled variation of an air stream.

pneumogram—A graphic recording of respiratory movements and/or forces.

pneumograph—An instrument that produces a strip-chart respiratory-activity recording from which breathing rate and volume can be determined.

pn hook transistor—Also called hook transistor or hook collector transistor. A junction transistor which secures increased current amplification by means of an extra pn junction.

pnin—A transistor in which its intrinsic region is between two n-regions.

pnip transistor—A pnp transistor in which a layer of high-purity germanium has been placed between the base and collector to extend the frequency range. When the same process is applied to an npn transistor, the resulting device is called an npin transistor.

pn junction—1. The region of transition between p-type and n-type material in a single semiconductor crystal. 2. The boundary surface between p-type and n-type materials.

pnpn (four-layer) diode—A semiconductor device which may be regarded as a two-transistor structure with two separate emitters feeding a common collector. This combination constitutes a feedback loop

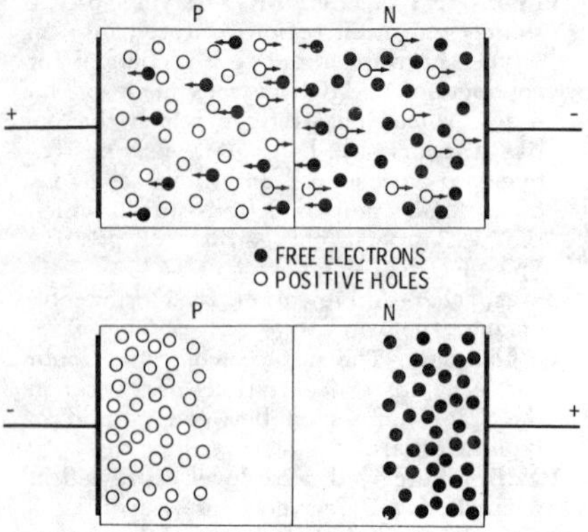

Pn junction.

which is unstable for loop gains greater than unity. The instability results in a current which increases until ohmic circuit resistances limit the maximum value. This gives rise to a negative-resistance region which may be utilized for switching or for waveform generation.

pnpn transistor—A hook transistor with an n-type base, p-type emitter, and a hook collector. The electrodes are connected to the four end layers of the n- and p-type semiconductor materials.

pnpn-type switch – A bistable semiconductor device made up of three or more junctions, at least one of which is able to switch between reverse and forward voltage polarity within a single quadrant of the anode-to-cathode voltage-current characteristic.

pnp transistor—A transistor consisting of two p-type regions separated by an n-type region. When a small forward bias is applied to the first junction and a large reverse bias to the second junction, the system behaves much like a vacuum-tube triode.

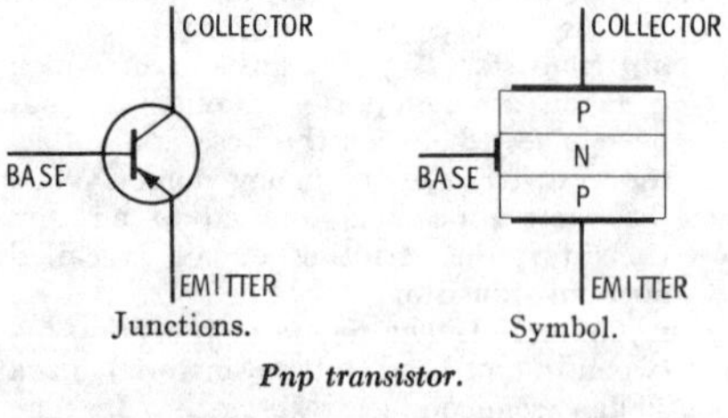

Junctions. Symbol.

Pnp transistor.

Pockel's effect—The alternation in the refractive properties of a transparent piezoelectric crystal by the application of an electric field. (*See also* Modulator Crystal.)

Pockels-effect modulation – A phenomenon that occurs when a transparent dielectric is a piezoelectric crystal. The crystal tends to strain whenever an electric field is applied, rotating the plane of polarization of the incident wave. Some 7500 V/m cause a 90-degree rotation of light.

POGO – Acronym for polar orbiting geophysical observatory.

poid—The curve traced by the center of a sphere when it rolls or slides over a surface having a sinusoidal profile.

point—Called the binary point in binary notation, and the decimal point in decimal notation. In positional notation, the character or location of an implied symbol which separates the integral part of a number from its fractional part.

point availability—The percent of time an equipment is available for use when an operator requires it.

point-based linearity – Nonlinearity expressed as the deviation from a straight line which passes through a given point or points.

point contact—A pressure contact between a semiconductor body and a metallic point.

point-contact diode—A diode that consists of a semiconductor against which the end of a fine wire (catwhisker) is pressed. Such a diode has a very low reactance and can be used as a detector or mixer over most of the microwave range. It has a square-law response at low power levels.

point-contact transistor—A transistor having a base electrode and two or more point-contact electrodes.

point defect – An imperfection caused by the presence of an extra atom or the absence of an atom from its proper place in the crystal.

point effect—The phenomenon whereby a discharge will occur more readily at sharp points than elsewhere on an object or electrode.

pointer—1. Also called a needle. A slender rod that moves over the scale of a meter.

pointer—Registers in a CPU which contain memory addresses. *See also* Program Counter and Data pointer.

pointer address – The address of a core-memory location that contains the actual effective address.

point impedance – Ratio of the maximum E-field to the maximum H-field observed at a given point in a waveguide or transmission line.

point-junction transistor—A transistor having a base electrode and both point-contact and junction electrodes.

point-plane rectifier—*See* Glow-Tube Rectifier.

point source—1. A radiation source whose dimensions are small compared with the distance from which it is observed. 2. Radiation source whose maximum dimension is less than 1/10 the distance between source and receiver.

point-to-point radio communication—Radio communication between two fixed stations.

point-to-point transmission – Direct transmission of data between two points without using an intermediate terminal or computer.

point-to-point wiring—A method of forming circuit paths by connecting the various devices, components, modules, etc., with individual pieces of wire or ribbon. May be soldered, welded, or attached by other means.

point transposition – Transposition, usually in an open-wire line, which is executed within a distance comparable to the wire separation, without material distortion of the normal wire configuration outside this distance.

Poisson's ratio – The ratio of the lateral strain to the longitudinal strain in a specimen subjected to a longitudinal stress.

polar – Pertaining to, measured from, or having a pole (e.g., the poles of the earth or a magnet).

polar circuit – A teletypewriter circuit in which current flows in one direction on a marking impulse and in the opposite direction during a spacing impulse.

polar coordinates—A system of coordinates in which a point is located by its distance and direction (angle) from a fixed point on a reference line (called the polar axis).

polar crystals – Crystals having a lattice composed of alternate positive and negative ions.

polar diagram – A diagram in which the magnitude of a quantity is shown by polar coordinates.

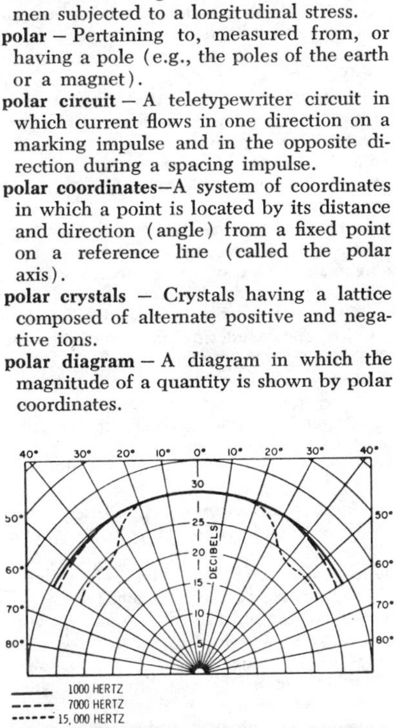

Polar diagram of a speaker enclosure.

polar grid—A type of circular grid on which range and azimuth are represented from a central reference point.

polarity—1. A condition by which the direction of the flow of current can be determined in an electrical circuit (usually batteries and other direct-voltage sources). 2. Having two opposite charges, one positive and the other negative. 3. Having two opposite magnetic poles, one north and the other south.

polarity of picture signal—Stated as black negative or black positive. The particular potential state of a portion of the signal representing a dark area of a scene relative to the potential representing a light area.

polarization – 1. The process of making light or other radiation vibrate perpendicular to the ray. The vibrations are straight lines, circles, or ellipses—giving plane, circular, or elliptical polarization, respectively. 2. The increased resistance of an electrolytic cell as the potential of an electrode changes during electrolysis. In dry cells, this shortens their useful life. 3. The slight displacement of the positive charge in each atom whenever a dielectric is placed into an electric field. 4. The magnetic orientation of molecules in a piece of iron or other magnetizable material placed in a magnetic field, whereby the tiny internal magnets tend to line up with the magnetic lines of force. 5. The direction of the electric vector in a linearly polarized wave radiated from an antenna. 6. A mechanical arrangement of inserts and shell configuration (referred to as clocking in some instances) which prohibits the mating of mismatched plugs and receptacles.

polarization diversity – A term that designates a method of transmission and reception used to minimize the effects of selective fading of the horizontal and vertical components of a radio signal. It is usually accomplished through the use of separate vertically and horizontally polarized receiving antennas.

polarization-diversity antenna—An antenna in which any of a number of types of polarization can be readily selected. The polarization can be horizontal, vertical, right-hand circular, left-hand circular, or any combination of these four.

polarization-diversity reception – Diversity reception which uses separate vertically and horizontally polarized receiving antennas.

polarization error—In navigation, the error arising from the transmission or reception of radiation having other than the intended polarization for the system.

polarization fading—Fading as the result of changes in the direction of polarization in one or more of the propagation paths of waves arriving at a receiving point. (*See also* Faraday Effect, 1.)

polarization in a dielectric—The slight displacement of the positive charge in each atom whenever a dielectric is placed into an electric field.

polarization index—A practical measure of

dielectric absorption expressed numerically as the ratio of the insulation after 10 minutes to the insulation resistance after 1 minute of voltage application.

polarization modulation – A technique in which modulation is produced by changing the direction of polarization of circularly polarized layer energy.

polarization receiving factor–Ratio of the power received by an antenna from a given plane wave of arbitrary polarization, to the power received by the same antenna from a plane wave of the same power density and direction of propagation whose state of polarization has been adjusted for the maximum received power.

polarization unit vector (for a field vector) –(A complex field vector at a point, divided by the magnitude of the vector.

polarize–1. To cause to be polarized. 2. To arrange mating connectors so that they can be joined in only one way.

polarized capacitor–An electrolytic capacitor with the dielectric film formed adjacent to only one metal electrode. The opposition to the passage of current is then greater in one direction than in the other. The polarity is established for minimum current; operation with reversed polarity can result in damage to the parts if excessive current occurs.

polarized double-biased relay–Also called magnetic-latch relay. A relay whose operation depends on the polarity of the energizing current and which is magnetically biased, or latched, in either of two positions. Its coil symbol is usually marked + and DB.

polarized light–Light which has the electric-field vector of all the energy vibrating in the same plane. Looking into the end of a beam of polarized light, one would see the electric-field vectors as parallel or coincident lines.

polarized no-bias relay–A three-position or a center-stable polarized relay. Its coil symbol is usually marked + and NB.

polarized plug–A plug so constructed that it may be inserted in its receptacle only in a predetermined position.

polarized receptacle – A receptacle into which a polarized plug can be inserted only in a predetermined position.

polarized relay–Also called a polar relay. A relay in which the armature movement depends on the direction of the current. Its coil symbol is sometimes marked +.

polarized-vane instrument–*See* Permanent-Magnet, Moving-Iron Instrument.

polarizer–A substance which, when added to an electrolyte, increases the polarization.

polarizing slots–Also called indexing slots. One or more slots placed in the edge of a printed-circuit board to accommodate and align certain types of connectors.

polar keying–A form of telegraph signal in which circuit current in one direction is used for marking, and current in the other direction is used for spacing.

polar modulation – A form of amplitude modulation in which the positive excursions of the carrier are modulated by one signal and the negative ones by another.

polar radiation pattern – A diagram that shows the relative strength of the radiation from a source in all directions in a given plane.

polar relay–1. A relay containing a permanent magnet that centers the armature. The direction of movement of the armature is governed by the direction of the current. (*See also* Polarized Relay.) 2. A permanent-magnet-core relay that is designed to operate only when current flows in a specified direction.

polar response–Polar diagram or circular graph that shows the sensitivity of an antenna or microphone or the output from speakers in an angular mode through 360°.

polar signal–A signal whose information is transmitted by means of directional currents.

pole–1. One end of a magnet. 2. One electrode of a battery. 3. An output terminal on a switch.

pole face–In a relay, the end of the magnetic core nearest the armature.

pole piece – One or more pieces of ferromagnetic material forming one end of a magnet and so shaped that the distribution of the magnetic flux in the adjacent medium is appreciably controlled.

pole shoe–The portion of a field pole facing the armature of the machine. It may be separable from the body of the pole.

poles of a network function–Those real or complex values of p, for which the network function is infinite.

pole-type transformer–A transformer suitable for mounting on a pole or similar structure.

police calls – Broadcasts (usually orders) issued by police radio stations.

poling–1. The adjustment of polarity. Specifically, in wire-line practice, it signifies the use of transpositions between transposition sections of open wire or between lengths of cable, to cause the residual cross-talk couplings in individual sections or lengths to oppose one another. 2. A step in the production of ceramic piezoelectric bodies which orients the axes of the crystallites in the preferred direction. In general, a process similar to magnetizing ferromagnetic materials.

Polish notation–A system for writing logical and arithmetic expressions without the use of parentheses. So called because it was originated by Polish logician, J. Lukasiewicz.

polling – 1. Periodic interrogation of each of the terminals that share a communications line to determine whether it requires servicing. The multiplexer or control station sends a poll that has the effect of asking the selected terminal, "Do you have anything to transmit?" 2. A means of controlling communication lines. The communication control device will send signals to a terminal saying, "Terminal A, have you anything to send?" If not, "Terminal B, have you any thing to send?" and so on. Polling is an alternative to contention. It makes sure that no terminal is kept waiting for a long time.

polycrystalline material—Material, typically an element like silicon or germanium, made up of many single crystals having a random orientation. The term may be applied to a twin crystal as well as to a heterogeneous growth of many crystals.

polycrystalline structure – The granular structure of crystals which have nonuniform shapes and arrangements.

polyergic—A type of emission in which the groups of energies or velocities are produced simultaneously (e.g., simulated micrometeoroids in varying charge states, separated by velocity where accelerated by the same potential).

polyester – Polyethylene glycol terephthalate, the material most often used as a base film for precision magnetic tape. The chief advantages of this material compared to other materials are its stability with respect to humidity and time, its resistance to solvents, and its mechanical strength.

polyester backing – A plastic-film backing added to magnetic tape to make it stronger and more resistant to changes in humidity.

polyesters – A class of thermosetting synthetic resins having great strength and good resistance to moisture and chemicals.

polyethylene—Short for polymerized ethylene, a tough, white plastic insulator with low moisture absorption. It is often used as a dielectric.

polygraph—Also called a lie detector. A recorder of several signals simultaneously, such as blood pressure, respiratory motion, galvanic skin resistance, etc., commonly used for study of emotional reactions involving deception (lie detection).

polyphase – Having or utilizing several phases. Thus, a polyphase motor operates from a power line having several phases of alternating current.

polyphase circuit—A group of alternating-current circuits (usually interconnected) which enter (or leave) a delimited region at more than two points of entry. They are intended to be so energized that, in the steady state, the alternating potential differences between them all have exactly equal periods, but have differences in phase, and may have differences in waveform.

polyphase motor – An induction motor wound for operation on two-or three-phase alternating current.

polyphase synchronous generator – A generator with its ac circuits so arranged that two or more symmetrical alternating electromotive forces with definite phase relationships to each other are produced at its terminals.

polyphase transformer—A transformer designed for use in polyphase circuits.

polyphase voltages—In an ac electrical system, voltages having a definite phase relationship to each other.

polyplexer – Radar equipment that combines the functions of duplexing and lobe switching.

polypropylene—A thermoplastic with good electrical characteristics, high tensile strength, and resistance to heat.

polyrod antenna – An end-fire, dielectric, microwave antenna made of tapered polystyrene rods.

polysilicon—A multicrystalline form of silicon used in silicon-gate MOS technology that is electrically conductive and optically transparent. It is commonly used to form electrodes of solid-state imaging devices.

polystyrene—A clear thermoplastic material having excellent dielectric properties, especially at ultrahigh frequencies.

polystyrene capacitor—A low-loss precision capacitor with a polystyrene dielectric.

polyvinyl chloride – Abbreviated PVC. A general-purpose thermoplastic used for insulations and jackets on components, wire, and cable.

pony circuit—A local, on-base circuit that does not have direct entry into a relay network.

pool cathode—A cathode at which the principal source of electron emission is a cathode spot on a metallic-pool electrode.

pool tube—A gas tube with a pool cathode.

popcorn noise – A type of noise generally associated with operational amplifiers. It is a type of random shot noise and was given its name because of its similarity to the sound made by the popping of corn. When popcorn noise is present, it does not occur in all devices made by a given process or even in all devices on the same wafer. It becomes worse at low temperatures, and it disappears above some threshold temperature.

Pope cell—*See* Sulfonated Polystyrene Sensor.

popi (Post Office Position Indicator) – A British long-distance navigational system for providing bearing information. It is a continuous-wave, low-frequency system in which the phase difference between sequential transmissions on a single frequency is measured.

population—The entire group of items being studied, from which samples are drawn. Sometimes called universe.

population inversion — A nonequilibrium condition that exists when there are more atoms in the excited state than in the ground state. Atoms return to the lower-energy level, and release energy and emit photons.

porcelain—A glazed ceramic insulating material made from clay, quartz, and feldspar.

porcelain capacitor — A fixed electrostatic capacitor, the dielectric of which is a high grade of porcelain, molecularly fused to alternate layers of fine silver electrodes so as to form a monolithic capacitor.

port—1. A place of access to a system or circuit. Through it, energy can be selectively supplied or withdrawn or measurements can be made. Examples are the port in a waveguide or in a base-reflex speaker enclosure. 2. A fluid connection to the servovalve (e.g., supply port, return port, control port). 3. An external opening of an internal passage.

portable data medium—A data medium intended to be transportable easily and independently of the mechanism used in interpreting it.

portable recorder — A sound recorder designed for easy mobility, but which may require connection to a 120-volt ac supply for operation. Also applied to self-powered units which do not require external power for operation.

portable standard meter—A portable meter used principally as a standard for testing other meters.

portable transmitter—A transmitter which can be readily carried on one's person and operated while in motion (e.g., walkie-talkies, Handie-Talkies, and similar personal transmitters). (*See also* Transportable Transmitter.)

portamento — The continuous change of a tone from one pitch to another. (*See also* Glissando.)

pos—Abbreviation for positive.

posistor — A thermally sensitive resistor which has a positive temperature characteristic of resistance.

position—1. The location of an object with respect to a specific reference point or points. 2. In a string, a location that can be occupied by a character or bit and that can be identified by a serial number.

positional cross talk—In a multibeam cathode-ray tube, the deviation of an electron beam from its path under the influence of another electron beam within the tube.

positional notation—One of the schemes for representing numbers. It is characterized by the arrangemennt of digits in sequence, with successive digits forming coefficients of successive powers of an integer called the base of the number system.

position-changing mechanism — The mechanism used to move a removable circuit-breaker unit to and from the connected, disconnected, and test positions.

position control system — A discrete or point-to-point control in which the controlled motion is used as a means of arriving at a given end point without path control during the movement between end points.

position dialing—Dialing over the regular position cord circuits by means of a relay circuit controlled by a dial of the regular cord circuits.

position feedback—A feedback signal which is proportional to the position or deflection of some object.

position of effective short—The distance between a specified reference plane and the apparent location of the short circuit of a fixed switching tube in its mount.

position sensor—A device that measures posion and converts the measurement into a form convenient for transmission as a feedback signal.

position-type telemeter — *See* Ratio-type Telemeter.

positive—Any point to which electrons are attracted—as opposed to negative, from where they come.

positive bias—The condition in which the control grid of a vacuum tube is more positive than the cathode.

positive charge—An electrical charge with fewer electrons than normal.

positive column—The luminous glow, often striated, between the Faraday dark space and the anode in a glow-discharge, cold-cathode tube.

positive electricity — The electricity which predominates in a glass body after it has been electrified by rubbing with silk. *See* Positive Charge.

positive electrode — The conductor that is connected to the positive terminal of a primary cell and serves as the anode when the cell is discharging. Electrons flow to it through the external circuit.

positive electron—*See* Positron.

positive feedback—*See also* Regeneration, 1. 1. The process by which the amplification is increased by having part of the power in the output circuit returned to the input circuit in order to reinforce the input power. 2. Recycling of a signal that is in phase with the input to increase amplification. Used in digital circuits to standardize the waveforms in spite of any anomalies in the input.

positive ghost — A television ghost-signal

display with the same tonal variations as those of the image.

positive-going — Increasing toward a positive direction (e.g., a current or waveform).

positive grid—A grid with a more positive potential than the cathode in a vacuum tube.

positive-grid multivibrator—A multivibrator which has one or more grids connected to the plate-voltage supply, usually through a large resistance.

positive-grid oscillator—*See* Retarding-Field Oscillator.

positive-grid oscillator tube—Also called a Barkhausen tube. An oscillating triode in which the grid has a more positive quiescent voltage than either of the other electrodes.

positive ground — The positive battery terminal of a vehicle is connected to the body and frame.

positive ion—An atom which has lost one or more electrons and thus has an excess of protons, giving it a positive charge.

positive-ion emission—Thermionic emission of positive particles from the cathode of a vacuum tube. They either are made up of ions from the metal in the cathode, or are due to some impurity in it.

positive-ion sheath—A collection of positive ions on the control grid of a gas-filled triode tube. If too high a negative bias is applied to the grid, this positive sheath will block the plate current.

positive light modulation — Modulation in which the transmitted power increases as the light intensity does, and vice versa.

positive logic—A form of logic in which the more positive logic level represents 1 and the more negative level represents 0.

positive magnetostriction—Magnetostriction in which a material expands whenever a magnetic field is applied.

positive modulation — Also called positive picture modulation. In an am television system, modulation in which the brightness increases as the transmitted power does, and vice versa.

positive phase-sequence relay — A relay which is energized by the positive phase-sequence component of the current, voltage, or power of a circuit.

positive picture modulation — *See* Positive Modulation.

positive picture phase — The condition in which the picture-signal voltage goes positive above the zero level whenever a positive scene or picture increases in brilliance.

positive plate—1. A hollow lead grid filled with active material and connected to the positive terminal of a storage battery. When the battery is discharging, electrons flow toward it through the external circuit. 2. In a charged capacitor, the plate that has fewer electrons.

positive ray—*See* Canal Ray.

positive temperature coefficient—The condition whereby the resistance, capacitance, length, or other characteristic of a substance increases as the temperature does.

positive terminal — In a battery or other voltage source, the terminal toward which electrons flow through the external circuit from the negative terminal.

positive transmission—Transmission of television signals in such a way that the transmitted power increases whenever the initial light intensity does.

positron — Fundamental particle equal in mass and energy to an electron, but having a positive charge. It has a very short life, being usually lost in the formation of a photon by combination with an electron. It occurs in the radiations from a few radioactive isotopes.

post—In a computer, to place a unit of information on a record.

postaccelerating electrode — *See* Intensifier Electrode.

postacceleration — Also called post-deflection acceleration. Acceleration of the beam electrons in a tube after they have been deflected.

postconversion bandwidth — In a telemetry receiver, the bandwidth presented to the detector.

postdeflection accelerating electrode — *See* Intensifier Electrode.

postdeflection acceleration — *See* Post Acceleration.

postedit — In a computer, to edit output data resulting from a previous computation.

postemphasis—*See* De-emphasis.

postequalization—*See* De-emphasis.

postmortem — A diagnostic computer routine for locating a malfunction in the computer or an error in coding a problem. Should a problem tape come to a standstill, the computer will print out—either automatically or when called for—any information concerning the contents of all or part of the registers in the computer.

postregulator — A circuit that performs the functions of reference, comparison, and control in power supplies. So called because it follows the transformer, rectifier, and (usually) the ripple filter.

pot—Short for potentiometer.

potassium — An alkali metal having photosensitive characteristics, especially to blue light. It is used on the cathodes of phototubes whenever maximum response to blue light is desired.

pot core—A magnetic structure consisting of a rod and a sleeve arranged so that the rod fits inside a coil and the sleeve fits around the coil. The sleeve and rod are connected at one end by a plate. The open end (opposite the plate) is usually, but not necessarily, ground so that two pot cores or a pot core and a separate plate can be put together around a suitable coil to form a low-reluctance magnetic path and/or shield for the coil.

potential—1. The difference in voltage between two points of a circuit. Frequently one point is assumed to be ground, which has zero potential. 2. In general, the electrical voltage difference between two bodies. When bodies of different potentials are brought into communication, a current is set up between them.

potential barrier—A semiconductor region through which electric charges attempting to pass will encounter opposition and may be turned back.

potential coil—The shunt coil in a measuring instrument or other device having series and shunt coils—i.e., the coil connected across the circuit and affected by changes in voltage.

potential difference—A voltage existing between two points (e.g., the voltage drop across an impedance, from one end to another).

potential divider—*See* Voltage Divider.

potential drop—The difference in potential between the two ends of a resistance with a current flowing through it.

potential energy—Energy due to the position of one body with respect to another or to the relative parts of the same body.

potential galvanometer — A galvanometer with such a high resistance that it takes practically no current. It has been replaced by the vacuum-tube voltmeter.

potential gradient—1. The rate of change of potential with distance. Units such as volts per meter or kilovolts per centimeter may be used. 2. Voltage gradient due to the diffusion of holes and electrons across the space charge region.

potential transformer—Also called a voltage transformer. An instrument transformer, the primary winding of which is connected in parallel with the circuit whose voltage is to be measured or controlled.

potential wells—A voltage placed on MIS capacitor electrodes causes a voltage gradient zone to be formed under the electrode so as to collect minority carriers.

potentiometer—1. A resistor provided with a tap that can be moved along it in such a way as to put the tap effectively at the junction of two resistors whose sum is the total resistance, the ratio of the two effective resistors being a function of the position of the tap. 2. A measuring instrument in which a potentiometer is used as a voltage divider in order to provide a known voltage that can be balanced against an unknown voltage. 3. A variable voltage divider. A resistor which has a variable contact arm so that portion of the potential applied between its ends may be selected. 4. A variable voltage divider used for measuring an unknown electromotive force or potential difference by balancing it, in whole or in part, by a known potential difference. 5. An instrument used to measure or compare voltages.

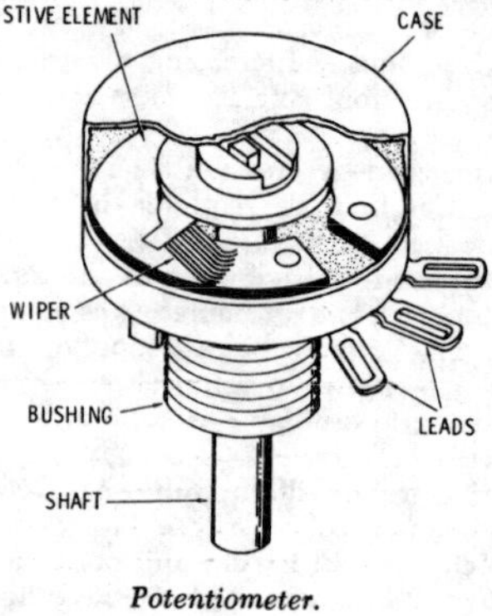

Potentiometer.

potentiometer circuit—A network arranged so that when two or more electromotive forces (or potential differences) are present in as many branches, the response of a suitable detecting device in any branch can be made zero by adjusting the electrical constants of the network.

potentiometer recorder — A null-balance type of recorder using a servo-operated voltage-balancing device; the sliding contact of a precision measuring potentiometer is adjusted automatically by a servomechanism so that the difference in voltage of the circuit becomes zero. Main feature is high sensitivity.

potentiometric transducer—A transducer in which displacement of a force-summing member is transmitted to the slider in a potentiometer, thus changing the ratio of output resistance to total resistance. Transduction is accomplished by changing the ratios of a voltage divider.

potentiometric transduction — The conversion of the measurand into a change in the position of a contact on a resistance element across which excitation is applied, the output usually being given as a voltage ratio.

pothead—An insulator for making a sealed joint between an underground cable and an overhead line.

Potier diagram—A vector diagram showing the voltage and current relationships in an ac generator.

pot life—The period after the addition of a catalyst to a potting compound during which the potting operation must be completed.

potted circuit — A circuit which has been

encapsulated in a nonconductive material.

potted line – A pulse-forming network immersed in oil and enclosed in a metal container.

Potter oscillator–A cathode-coupled multivibrator.

potting – An embedding process for parts that are assembled in a container or can into which the insulating material is poured, with the container remaining an integral part as the outer surface of the finished unit.

potting compound–A sealing material used to fill the case or enclosure in which a component is contained.

powdered-iron core – A core consisting of fine particles of magnetic material mixed with a suitable bonding material and pressed into shape.

power–1. The energy dissipated in an electrical or electronic circuit or component that is conducting either ac or dc. 2. Electrical energy developed to do "work" such as the voltage from an amplifier used to drive a speaker. Also acoustical energy or sound pressure developed in a room by a speaker. 3. Rate of doing work. Some units of power are the foot-poundal per second, or (in the CGS system) 1 erg/second, the watt (joule/second), and the kilowatt.

power amplification–*See* Power Gain, 1.

power amplifier–1. An amplifier intended for driving one or more speakers or other transducers. 2. The final stage in a multistage amplifier circuit, designed to give power to the load, rather than to be used mainly as a voltage amplifier. 3. A fluidic device which causes a change in output power following a change, of sufficient magnitude, in control power.

power-amplifier stage – 1. An audio-frequency amplifier stage capable of handling considerable audio-frequency power without distortion. 2. A radio-frequency amplifier stage used in a transmitter primarily to increase the power of the carrier signal.

power attenuation–*See* Power Loss, 1.

power bandwidth–The range of audio frequencies over which an amplifier can produce half its rated power without exceeding its rated distortion. It is determined by using a measurement procedure standardized by the Institute of High Fidelity (IHF). This specification indicates how much power is available at the critical high and low frequencies. The wider the power bandwidth, the better the amplifier.

power connection – British term used for the constant horsepower connection in a multispeed motor.

power consumption – The maximum wattage used by a device within its operating range during steady-state signal conditions.

power cord–Small, flexible, insulated cable used in such applications as supplying line power to power tools and electronic equipment.

power density–The radiated field strength set up by a radiating source, expressed in microvolts per meter or in dB above 1 μv/m.

power derating–Use of computed curves to determine the correct power rating of a device or component to be used above its reference ambient temperature.

power detection – Detection in which the power output of the detector is used for supplying a substantial amount of power directly to a device such as a speaker or recorder.

power detector – A vacuum tube detector operating with such a high plate voltage that strong input signals can be handled without appreciable distortion.

power dissipation–1. The dispersion of the heat generated within a device or component when a current flows through it. This is accomplished by convection to the air, radiation to the surroundings, or conduction. 2. The supply power consumed by a logic circuit operating with a 50-percent duty cycle (equal times in the logic 0 and logic 1 states).

power dissipation rating – The maximum average power that can be continuously dissipated under stated temperature conditions.

power divider – A device that provides a desired distribution of power at a branch point in a waveguide system.

power dump – The removal of all power either accidentally or intentionally. (*See also* Dump, 1.)

power factor–1. Ratio of the actual power of an alternating or pulsating current, as measured by a wattmeter, to the apparent power, as indicated by an ammeter and voltmeter. 2. Ratio of resistance to impedance–therefore, a measure of the loss in an inductor, capacitor, or insulator. 3. The cosine of the phase angle between the voltage applied to a load and the current passing through it. (Sometimes the cosine is multiplied by 100 and expressed as a percentage.)

power-factor correction–Adding capacitors to an inductive circuit in order to increase the power factor by making the total current more nearly in phase with the applied voltage.

power-factor meter – A direct-reading instrument for measuring power factor. Its scale is graduated directly in power factor.

power-factor regulator–A regulator which maintains the power factor of a line or apparatus at a predetermined value, or varies it according to a predetermined plan.

power-factor relay – A device which operates when the power factor in an ac circuit becomes above or below a predetermined value.

power frequency—The frequency at which electric power is generated and distributed. Throughout most of the United States, this frequency is 60 Hz.

power gain—1. Also called power amplification. 2. The ratio of the signal power developed at the output(s) of a device to the signal power applied at the input(s). 3. Of an antenna in a given direction, 4π times the ratio of the radiation intensity to the total power delivered to the antenna. (The term is also applied to receiving antennas.)

power ground – 1. The ground between units which is part of the circuit for the main source of power to, or from, these units. 2. The potential of the terminal or circuit point to which the output of a power supply and often an amplifier output load is returned (i.e., power-supply "zero").

power-handling capability – 1. A measure of the maximum power input a speaker can absorb without damage or unreasonable distortion. 2. The maximum power rating of a component which determines how much current can be passed safely without adverse effects.

power level—At any point in a transmission system, the difference between the measure of the steady-state power at that point, and the measure of an arbitrarily specified amount of power chosen as a reference.

power-level indicator—An ac voltmeter calibrated to read the audio power level.

power line—Two or more wires conducting electric power from one location to another.

power loss – 1. Also called power attenuation. Ratio of the power absorbed by the input circuit of a transducer, to the power delivered to a specified load under specified operating conditions. 2. Also called watt loss. In the circuit of an instrument for measuring current or voltage, the active power at its terminals for nominal full-scale indication. For other instruments, for example, wattmeters, the power loss is expressed at a stated value of current or voltage.

power modulation factor—Ratio of the maximum positive departure of the envelope of an amplitude-modulation wave from its average value to its average value. This rating is used when the modulating signal wave has unequal positive and negative peaks.

power output—The power in watts delivered by a power amplifier to a load such as a speaker.

power output (continuous watts) – In an amplifier, the power output at a total maximum harmonic distortion of 0.5%, with a pure-tone (sine-wave) input.

power pack – A unit for converting power from an alternating- or direct-current supply into alternating- or direct-current power at voltages suitable for supplying an electronic device.

power programmer – A device for controlling the output power of a radar automatically as a function of the target range.

power rating – The maximum power that can be dissipated in a component or device for a specified period.

power ratio—Ratio of the power output to the power input of a device. Usually expressed as the number of decibels loss or gain.

power-rectifier misfire relay – A device which functions if one or more of the power-rectifier anodes fails to fire.

power relay – A relay that functions at a predetermined value of power. It may be an overpower relay, an underpower relay, or a combination of both.

power response – The frequency-response capabilities of an amplifier running at or near its full rated power.

power-spectral density function – A measure of the power distribution of a signal with respect to frequency.

power supply—1. A unit that supplies electrical power to another unit. It changes ac to dc and maintains a constant voltage output within limits. 2. Energy source that provides power for operating electronic apparatus.

power-supply rejection ratio—The ratio of the change in input offset voltage of an operational amplifier to the change in power-supply voltage that cause it.

power switch – Often called an on-off switch. The switch that connects or disconnects a radio receiver, transmitter, or other equipment from its power line.

power switchboard—Part of a switch gear consisting of a panel or panels on which the switching-control, measuring, protective, and regulatory equipment is mounted. The panel or panel supports also may carry the main switching and interrupting devices and their connections.

power transformer—A transformer used for raising or lowering the supply voltage to the various values required by vacuum-tube plate, heater, and bias circuits.

power transistor—1. A transistor designed to handle large currents and safely dissipate large amounts of power. 2. A transistor that can dissipate more than one watt of power. General-purpose types are used for low-frequency service (below 3 MHz) as amplifiers, switches, or current

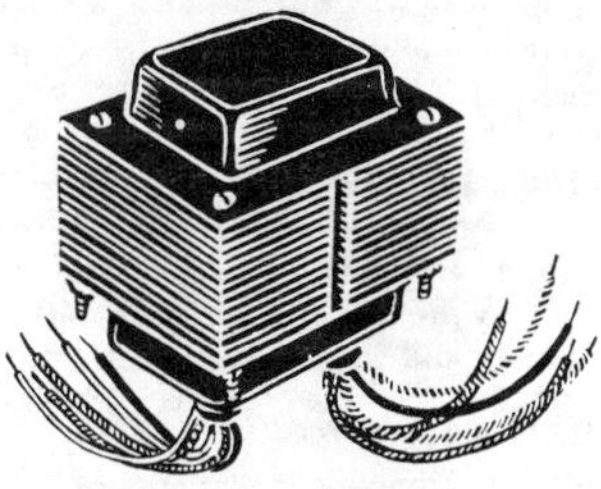

Power transformer.

regulators. Rf types are used to amplify high-frequency signals (above 3 MHz) that reach up to vhf, uhf, and microwave regions. 3. A transistor that handles power levels of about one-quarter of a watt and above. Units handling about one-quarter to ten watts are called medium-power transistors while high-power transistors are those handling above ten watts.

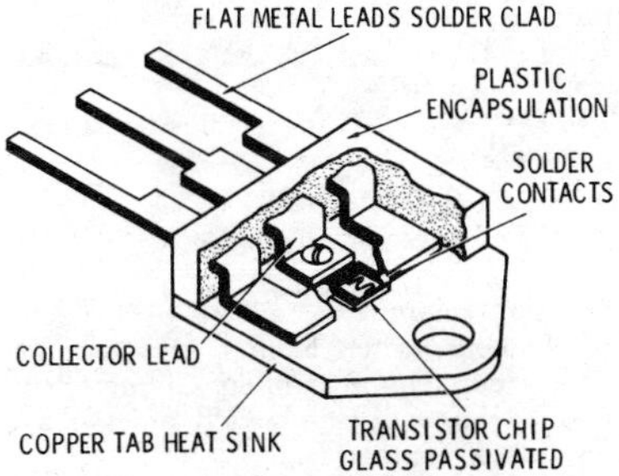

Power transistor.

power tube—An electron tube designed to handle more current and power than a voltage-amplifier tube.

power winding—A saturable-reactor winding to which the power to be controlled is supplied. Commonly, the output and power are furnished by the same winding, then termed the output winding.

Poynting's law—The transfer of energy can be expressed as the product of the values of the magnetic field and of the components of the electric field which are perpendicular to the magnetic field, and the flow of energy at any point is perpendicular to both fields.

Poynting's theorem—The rate of flow of electromagnetic energy into or out of a closed region is at any instant proportional to the surface integral of the vector product of the electric and magnetic intensities.

Poynting's vector—The vector product of the electric and magnetic intensities at one point and at a given instant in a wave.

ppi—Abbreviation for plan-position indicator.

ppi repeater—A unit that repeats a plan-position indicator at a place remote from the radar console. Also called remote plan-position indicator.

ppi scope—A cathode-ray oscilloscope arranged to present a ppi display.

pp junction – A region of transition between two regions having different properties in a p-type semiconducting material.

p+ region—The region created by diffusing into a silicon crystal a group-III element, which creates a deficiency of electrons or an excess of holes.

p+ semiconductor—A p-type semiconductor with an extremely large excess mobile hole concentration.

ppm – 1. Abbreviation for pulse-position modulation. 2. Abbreviation for parts per million.

ppm/am—Amplitude modulation of a carrier by pulses which are position modulated by data.

pps—Abbreviation for pulses per second.

practical system of electrical units—A system in which the units are multiples or submultiples of the units of the centimeter-gram-second electromagnetic system.

praetersonic—The higher region of the sonic spectrum.

praetersonics—The propagation and signal processing of acoustic waves in solids at frequencies that extend into the microwave region.

preamplifier—An amplifier which primarily raises the output of a low-level source so that the signal may be further processed without appreciable degradation in the signal-to-noise ratio. A preamplifier may also include provision for equalizing and/or mixing.

preburning—Stabilizing tubes by operating their heaters continuously for a given number of hours. Cathode current may be drawn and the tubes vibrated at the same time.

precession – The effect resulting when a torque is applied to a rotating body, such

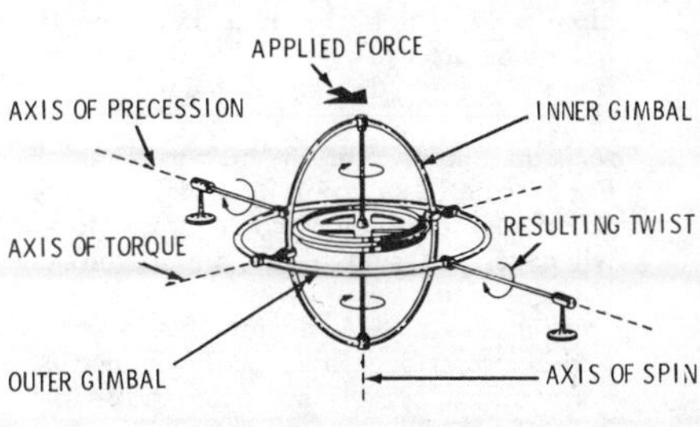

Precession of a gyroscope.

as a gyroscope, causing it to wobble. The wobbling frequency is determined by the gravitational field strength and the mass of the body.

precipitation attenuation–A reduction in radio energy as it passes through a volume of the atmosphere that contains precipitation. Part of the energy loss results from scattering, and part results from absorption.

precipitation noise–Noise generated in an antenna circuit, generally in the form of a relaxation oscillation, caused by the periodic discharge of the antenna or conductors in the vicinity of the antenna into the atmosphere.

precipitation static–A type of interference experienced in a receiver during snow, rain, and dust storms. Often caused by the impact of dust particles against the antenna or the creation of induction fields by nearby corona discharges.

precipitator – Sometimes called a precipitron. An apparatus for removing small particles of smoke, dust, oil, mist, etc., from the air by electrostatic precipitation.

precipitron–*See* Precipitator.

precision–1. The quality of being sharply or exactly defined–i.e., the number of distinguishable alternatives from which a representation was selected. This is sometimes indicated by the number of significant digits the representation contains. (*See also* Accuracy.) 2. The degree with which repeated measurements of a given quantity agree when obtained by the same method and under the same conditions. Also called reproducibility or repeatability.

precision approach radar–1. A rapid-scanning airport radar system so located that aircraft on approach to the runway are presented on the display in terms of linear deviation from a desired glide path and in terms of distance from the point of touchdown on the runway. Abbreviated par. 2. Radar used by traffic controllers in a Ground Controlled Approach to "talk" a pilot on final approach.

precision-balanced hybrid circuit–A circuit for interconnection of a four-wire telephone circuit with a particular two-wire circuit, in which the impedance of the balancing network is adjusted so that a relatively high degree of balance is obtained.

precision comparator–A high-gain amplifier circuit the output of which changes decisively between two definite levels whenever the sum of the input voltages changes sign.

precision device – A device that operates within prescribed limits and will consistently repeat operations within those limits.

precision gate – A circuit that may be switched from closed to open circuit or vice versa without error (time, bias, impedance) in response to a command signal (voltage or current).

precision net–In a four-wire terminating set or similar device employing a hybrid coil, an artificial line designed and adjusted to provide an accurate balance for the loop and subscriber's set or line impedance.

precision potentiometer – A mechanical electrical transducer dependent upon the relative position of a moving contact (wiper) and a resistance element for its operation. It delivers a voltage output that is some specified function of the applied voltage and shaft position, to a high degree of accuracy.

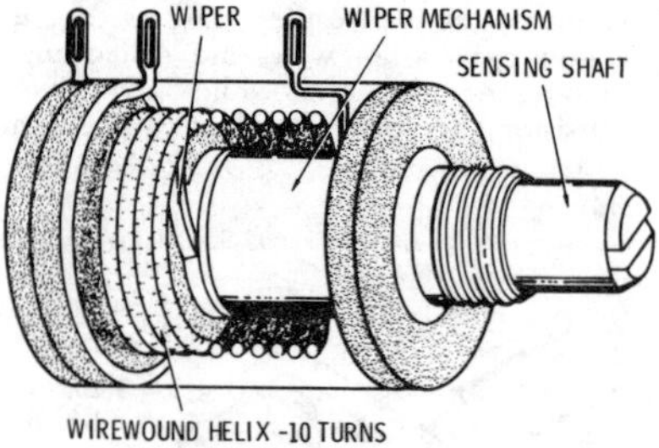

Precision potentiometer.

precision snap-acting switch–An electromechanical switch having predetermined and accurately controlled characteristics and having a spring-loaded quick make-and-break contact action.

precision sweep–A delayed expanded radar sweep for high resolution and range accuracy.

preconduction current–The low value of plate current that flows in a thyratron or other grid-controlled gas tube prior to conduction.

precursor–Also called undershoot. The initial transient response to a unidirectional change in input. It precedes the main transition and is opposite in sense.

predefined process–A named process that consists of one or more operation or program steps that are specified in another part of a routine.

predetection recording – The recording of telemetry receiver intermediate-frequency signals.

predetermined counter–A device that automatically stops an instrument to which it is attached when a preset limit is reached.

predictive control–A type of computer control which allows a digital computer to include a dynamic control loop for repetitive comparison of pertinent factors.

predissociation–The dissociation that occurs in a molecule that has absorbed energy before it has had an opportunity to lose energy by radiation.

predistortion—*See* Preemphasis.

pre-Dolbyed tape—A prerecorded tape with Dolby compression added for low-noise playback in the home via a Dolby B stretcher.

preemphasis—1. In a system, a process designed to emphasize the magnitude of some of the frequency components. Preemphasis is applied at the transmitting end (with de-emphasis at the receiving end) in order to improve the signal-to-noise ratio. 2. Also called pre-equization or predistortion. In recording, an arbitrary change in the frequency response from its basic response (e.g., constant velocity or amplitude) in order to improve the signal-to-noise ratio or to reduce distortion. (*See also* Accentuation.) 3. A scheme sometimes adopted during recording to nullify the effect of tracing distortion on replay. The distortion deliberately introduced is the reciprocal of that produced on replay.

preemphasis network – A network inserted into a system to emphasize one range of frequencies.

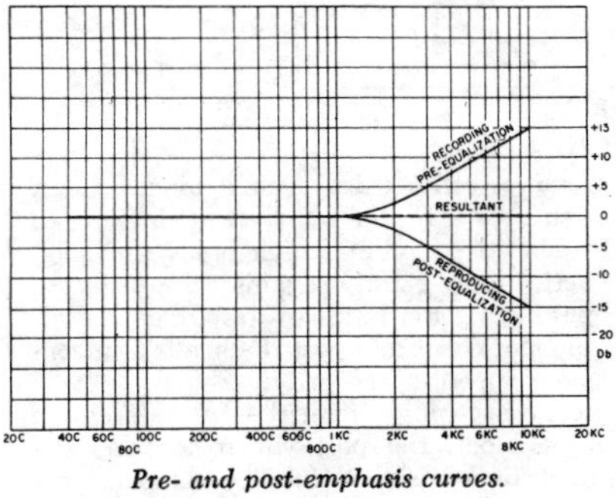

Pre- and post-emphasis curves.

preemption—A feature of some automated networks by which a high-precedence message, call, or transmission pre-empts a line from a use of lower precedence if all other lines are busy.

preequalization—*See* Preemphasis.

preferred tube types—Tube types recommended to designers of electronic equipment, to minimize the number of tube types that must be stocked by the manufacturer or by service agencies.

preferred values—A series of resistor and capacitor values adopted by the EIA and military. In this system, the increase between any two steps is the same percentage as between all other steps. Increases may be in steps of 20%, 10%, or 5% each.

prefix multipliers—Prefixes which designate a greater or smaller unit than the original, by the factor indicated. These prefixes are:

Prefix	*Symbol*	*Factor*
exa	E	10^{18}
peta	P	10^{15}
tera	T	10^{12}
giga	G	10^{9}
mega	M	10^{6}
kilo	k	10^{3}
hecto	h	10^{2}
deka	da	10
deci	d	10^{-1}
centi	c	10^{-2}
milli	m	10^{-3}
micro	μ	10^{-6}
nano	n	10^{-9}
pico	p	10^{-12}
femto	f	10^{-15}
atto	a	10^{-18}

preform—Also called a biscuit. In disc recording, the small slab of record material used in the presses.

preliminary contacts—In a relay, contacts which open or close before other contacts when the relay is actuated.

preohmic alignment—In a semiconductor, the positioning of the oxide opening into which the metallization is placed.

preohmic window – The opening etched through the oxide in a semiconductor for metallization contact to the emitter and base regions.

preprocessor—A device for placing source records into a format that facilitates system processing in a computer.

prerecorded tape—*See* Recorded Tape.

prescaler—A circuit that generates an output signal related to the input signal by a fractional scale factor such as ½, ⅛, ⅒, etc. An example of a digital prescaler is a decade frequency divider, which has an output frequency one tenth of the input frequency.

preselection—1. The use of a preselector. 2. In buffered computers, a time-saving technique in which a block of information is read into the computer memory ahead of time from whichever input tape will next be called on. 3. In digital computers, a technique whereby data from the next input tape is stored while the computer is still processing other data.

preselector—1. A device placed ahead of a frequency converter or other device to pass signals of desired frequencies but reduce all others. 2. In automatic switching a device which makes its selection before seizing an idle trunk.

preselector stage—A radio-frequency amplifier stage in the input of a superheterodyne receiver.

presence – The quality of naturalness in sound reproduction. When the presence of a system is good, the illusion is that the sounds are being produced intimately at the speaker.

presence control—A potentiometer used in

a three-way speaker system for controlling the volume of the middle-range speaker.

presentation—The form which the radar echo signals take on the screen, depending on the nature of the sweep circuit.

presenting – Displaying data in a form which human intelligence can comprehend and use.

preset—1. To establish an initial condition, for example the control values of a loop. 2. An asynchronous input of a flip-flop by which the Q output is set to a logic 1 and the Q output is set to a logic 0.

preset guidance system—A guidance system in which the flight path is determined before the missile is launched and cannot be altered after launch.

preset parameter—In a computer, a parameter that is fixed at a value established by the programmer for each problem.

preshoot—The initial transient response to a unidirectional change in input which precedes the main transmission and may be of the same or opposite polarity.

pressed stem—An obsolete method of vacuum-tube construction in which all support wires are formed into a flattened piece of glass tubing (actually a relic from the lampmaker's era). (*See also* Button Stem.)

pressing—A disc recording produced in a record-molding press from a master or stamper.

press-to-talk switch—Also called a push-to-talk switch. A spring-loaded switch that must be held down as long as the operator talks. Releasing the switch deactivates the microphone. It is used on transmitter and dictating-machine microphones.

pressure—Force per unit area. Measured in pounds per square inch (psi), or by the height (in feet, inches, or centimeters) of a column of water or mercury which the force will support. Absolute pressure is measured with respect to zero pressure. Gage pressure is measured with respect to atmospheric pressure.

pressure amplitude—For a sinusoidal sound wave, the maximum absolute value of the instantaneous sound pressure at a point during any given cycle. The unit is the dyne per square centimeter.

pressure connector—A conductor terminal applied under pressure to make the connection mechanically and electrically more secure. *See also* Solderless Connector.

pressure-gradient hydrophone – A type of hydrophone in which the electric output is essentially determined by a component of the gradient (space derivative) of the sound pressure.

pressure hydrophone – A type of hydrophone in which the electric output is essentially determined by the instantaneous sound pressure of the impressed sound wave.

pressure microphone – A microphone in which the electric output corresponds substantially to the instantaneous sound pressure of the impressed sound waves. It is a gradient microphone of zero order, and is nondirectional when its dimensions are smaller than a wavelength.

pressure pad—In single-motor tape recorders, a device which forces the tape into intimate contact with the head gap, usually by direct pressure at the head assembly. Felt or similar material, occasionally protected with self-lubricating plastic, is used to apply pressure uniformly and with a minimum of drag to the backing side of the tape.

pressure potentiometer—A pressure transducer in which the electrical output is derived by moving a contact arm along a resistance element.

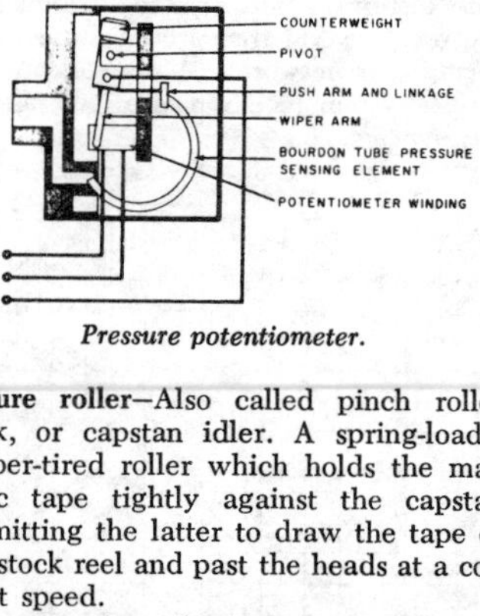

Pressure potentiometer.

pressure roller—Also called pinch roller, puck, or capstan idler. A spring-loaded rubber-tired roller which holds the magnetic tape tightly against the capstan, permitting the latter to draw the tape off the stock reel and past the heads at a constant speed.

pressure-sensing element – In a pressure transducer, the part which converts the measured pressure into mechanical motion.

pressure spectrum level – The effective sound-pressure level for the sound energy contained within a band one hertz wide and centered at a specified frequency. Ordinarily this level is not significant except for sound having a continuous distribution of energy within the frequency range under consideration.

pressure switch—A switch actuated by a change in the pressure of a gas or liquid.

pressure transducer—An instrument which converts a static or dynamic pressure input into the proportionate electrical output.

pressure-type capacitor—A fixed or variable capacitor used chiefly in transmitters. It is mounted inside a metal tank filled with nitrogen at a pressure that may be as great as 300 pounds per square inch. The high pressure permits a voltage rating several times that of air.

pressure unit—A moving-coil speaker drive unit which usually has as its diaphragm a small dome of plastic or metal. It is designed for use in the throat of a horn.

pressurization—The process by which the critical parts of equipment designed for high-altitude operation are surrounded with dry air or an inert gas under pressure (about five pounds per square inch at sea level). Thus, breakdowns from the impaired insulating properties of air at reduced pressure are prevented.

prestore—To store a quantity in an available or convenient location in a computer before it is required in a routine.

pretravel—The distance or angle through which the actuator moves from the actuator free position to the actuator operating position.

pretrigger—1. A timed pulse used to start a sequence of operation prior to the main trigger. 2. In random-sampling oscilloscope technique, a trigger that occurs or arrives prior to a related signal event.

pre-tr—In a radar set, an additional tr box that provides additional attenuation of transmitted pulse to prevent damage to the crystal mixer.

pre-tr tube – A gas-filled radio-frequency switching tube used to protect the tr tube from excessive power and the receiver from frequencies other than the fundamental.

preventive maintenance – Precautionary measures taken on a system to forestall failures rather than to eliminate them after they have occurred.

previous-element coding—A method of signal coding for digital television transmission in which each transmitted picture element depends on the similarity of the preceding element.

prewound core—A motor core (stator laminations) that can be removed and replaced by a factory wound (prewound) stator core.

prf—Abbreviation for pulse-repetition frequency.

pri—Abbreviation for primary.

primaries—*See* Primary Colors.

primary—1. Abbreviated pri and symbolized by P. Also called a primary winding. A transformer winding that carries current and normally sets up a current in one or more secondary windings. 2. Pertaining to the high-voltage conductors of a power-distribution system. 3. Any one of three lights in terms of which a color is specified by giving the amounts required to duplicate it by additive combination.

primary area—*See* Primary Service Area.

primary battery – A battery consisting of primary cells.

primary breakdown—Also called avalanche breakdown. The sustaining mode of a transistor—unlike second breakdown, not a failure mode. The transistor collector-to-emitter voltage is relatively constant for different collector supply voltages.

primary calibration—Calibration in which the transducer output is observed or recorded, while direct stimulus is applied under controlled conditions.

primary-carrier flow—Also called primary flow. The current flow responsible for the major properties of a semiconductor device.

primary cell—A cell that produces electric current through an electrochemical reaction but is not rechargeable. Once discharged, it must be discarded.

primary circuit—The first, in electrical order, of two or more coupled circuits, wherein a change in current will induce a voltage in the other, or secondary, circuit.

primary colors – Also called primaries. A set of colors from which all other colors are derived; hence, any set of stimuli from which all colors may be produced by mixture. A primary color cannot be matched by any combination of other primaries. In color television, the primary colors are red, blue, and green.

primary current – The current flowing through the primary winding of a transformer. Changes in this current cause a voltage to be induced in the secondary winding of the transformer.

primary detector – Also called a sensing, primary, or initial element. The first system element or group of elements that respond quantitatively to the measurand and perform the initial measurement operation. A primary detector performs the initial conversion or control of measurement energy. It does not include those transformers, amplifiers, shunts, resistors, etc., used as auxiliary means.

primary electron—1. After a collision between two electrons, the one with the greater energy. The other is called the secondary electron. 2. The electron produced in a detector or counter tube after ionization.

primary element—*See* Primary Detector.

primary emission – Emission of electrons due to primary causes (e.g., heating of a cathode) rather than secondary effects (e.g., electron bombardment).

primary failure—A failure occurring under normal environmental conditions and having no significant relationship to a previous failure but whose occurrence imposes abnormal stress on some other part or parts which may then undergo a secondary failure.

primary fault—In an electric circuit, initial breakdown of the insulation of a conductor usually followed by a flow of power current.

primary flow—*See* Primary-Carrier Flow.

primary frequency – The frequency assigned for normal use on a particular circuit or communications channel.

primary grid emission – *See* Thermionic Grid Emission.

primary ionizing event–*See* Initial Ionizing Event.

primary power cable–Power service cables connecting the outside power source to the main-office switch and metering equipment.

primary radar–*See* Radar.

primary radiation – Radiation direct from the source without interaction.

primary radiator – The antenna element from which the radiated energy leaves the transmission system.

primary relay – A relay that produces the initial action in a sequence of operations.

primary service area–Also called the primary area. The area within which radio or tv reception is not normally subject to objectionable interference or fading.

primary skip zone – The area beyond the ground-wave range around a transmitter, but within the skip distance. Radio reception is possible in this zone by sporadic and zigzag reflections.

primary standard – A unit directly defined and established by some authority, and against which all secondary standards are calibrated.

primary storage–In a computer, the main internal storage.

primary voltage–1. The voltage applied to the terminals of the primary winding in a transformer. 2. The voltage produced by a primary cell.

primary winding–*See* Primary, 1.

priming illumination – A small, steady illumination applied to a phototube or photoelectric cell to make it more sensitive to variations in the illumination being measured.

primitive period – The smallest increment of time during which a quantity repeats itself.

principal axis – A reference direction for angular coordinates, used in describing the directional characteristics of a transducer employed for sound emission or reception. It is usually an axis of structural symmetry or the direction of maximum response. If these two do not coincide, however, the reference direction must be described explicitly.

principal E-plane – The plane containing the direction of maximum radiation and in which the electric vector lies.

principal focus–For a lens or spherical mirror, the point of convergence of light coming from a source at an infinite distance.

principal H-plane–A plane containing the direction of maximum radiation; the electric vector is everywhere normal to the plane, and the magnetic vector lies in it.

principal mode–*See* Dominant Mode.

print – The possible output formats of a teletypewriter or electric typewriter terminal. Print refers to tabulated output data; plot refers to a graphical arrangement of the output data performed by the typewriter with symbols such as "x" or "*" used to indicate data points–the plot is discontinuous; print-plot refers to the availability of both formats at the same terminal.

printed–Reproduced on a surface by some process (e.g., letterpress, lithography, silk screen, etching).

printed cable–A cable having a thin film of copper laid onto insulation and deriving its strength from the insulation rather than from the conductor.

printed circuit–1. A circuit in which the interconnecting wires have been replaced by conductive strips printed, etched, etc., onto an insulating board. It may also include similarly formed components on the baseboard. 2. A substrate on which a predetermined pattern of printed wiring and printed elements has been formed.

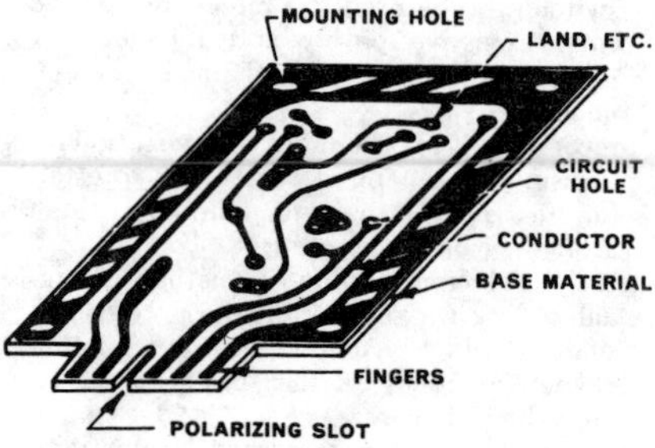

Printed circuit.

printed-circuit assembly–1. A printed-circuit board to which separate components have been attached. 2. An assembly of one or more printed-circuit boards, which may include several components.

printed-circuit board – Also called a card, chassis, or plate. An insulating board onto which a circuit has been printed. (*See also* Printed Circuit.)

printed-circuit switch – A special rotary switch which can be connected directly to a mating printed-circuit board without wires.

printed component–A type of printed circuit intended primarily for electrical and/or magnetic functions other than point-to-point connections or shielding (e.g., printed inductor, resistor, capacitor, transmission line, etc.).

printed contact–The portion of a printed circuit that connects the circuit to a plug-in receptacle and performs the function of a plug pin.

printed element—An element, such as a resistor, capacitor, or transmission line, that is formed on a circuit board by deposition, etching, etc.

printed wiring – A pattern of conductors printed (screened) onto the surface of an insulating base to provide interconnection of active and passive devices to make an electronic circuit.

printed-wiring substrate—A conductive pattern printed on a substrate.

printer – 1. Also called a teleprinter and teletypewriter. A telegraph instrument with a signal-actuated mechanism for automatically typing received messages. It may have a keyboard similar to that of a typewriter for sending messages. The term "receiving-only" is applied to a printer with no keyboard. 2. A device that prints the output from a computer. It ranges from a conventional "singlestick" or "flying typebar" printer (as on a typewriter) to a high-speed unit that prints up to 1000 lines per minute.

printer telegraph code—A five- or seven-unit code used for operation of a teleprinter, teletypewriter, or similar telegraph printer.

printing—The reproduction of a pattern on a surface by any of various processes, such as vapor deposition, photo etching, embossing, or diffusion.

printing demand meter—An integrated demand meter which prints on a paper tape the demand for each interval and indicates the time it occurred.

printing telegraphy – Telegraph operation in which the received signals are automatically recorded as printed characters.

printout—1. The output of a computer program as recorded by a line printer. 2. *See* Display.

print-plot—*See* Print.

print-through—1. The transfer of the magnetic field from layer to layer of tape on the take-up reel. 2. The transfer of magnetically recorded material to adjacent layers of tape on the reel. Print-through causes faint "echoes" preceding and following loudly recorded passages, and is aggravated by recording at excessively high levels or by exposing recorded tapes to alternating magnetic fields, as from nearby power transformers.

print wheel—In a wheel printer, the single element providing the character set at one printing position.

priority indicator—A character group that indicates the relative urgency, and thus the order of transmission, of a message.

priority interrupt—*See* Interrupt.

privacy system – In radio transmission, a system designed to make unauthorized reception difficult.

private-aircraft station—A mobile radio station on board an aircraft not operated as an air carrier.

private automatic branch exchange—A private branch exchange in which remotely controlled switches are employed to make connections.

private automatic exchange—A private telephone exchange in which remotely controlled switches are employed to make connections.

private branch exchange—*See* pbx.

private exchange – A telephone exchange that serves a single organization and has no means of connection to a public telephone system.

private line – A circuit reserved for use solely by one user.

private radio carrier – A radio carrier owned and controlled by the central station organization.

probability distribution – A mathematical model showing a representation of the probabilities for all possible values of a given random variable.

probability of success—The likelihood that an article will function satisfactorily for a stated period of time when subjected to a specified environment.

probable error—The amount of error which, according to the laws of probability, is most likely to occur during a measurement.

probe—1. A resonant conductor which can be placed into a waveguide or cavity resonator to insert or withdraw electromagnetic energy. 2. A test lead which contains an active or passive network and is used with certain types of test equipment. 3. A rod placed into the slotted section of a transmission line to measure the standing-wave ratio or to inject or extract a signal. 4. The method of making a temporary electrical connection to a die so that its electrical properties can be determined.

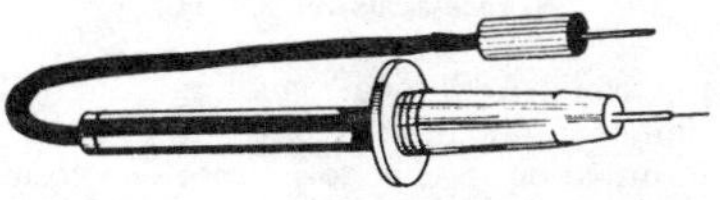

Probe, 2 (high voltage).

probing—1. The determination of radio interference by obtaining the relative interference level in the immediate area of a source by the use of a small insensitive antenna in conjunction with a receiving device. 2. Electrical testing of a semiconductor chip before it is broken out of the wafer. Electrical contact is made to the chip bonding pads so that defective circuits can be marked to eliminate them from further processing. Only low-current dc tests can be carried out by probing.

problem check—A test or tests used to aid in obtaining the correct machine solution to a problem.

problem description—In information processing, a statement of a problem. The statement may include a description of the method of solution.

problem language – The language a computer programmer uses in stating the definition of a problem.

problem-oriented language—In a computer, a source language suited to the description of a specific class of problems.

procedure—In a computer, the course of action taken in solving a problem. Also called an algorithm.

procedure-oriented language – In a computer, a source language suited to describing procedural steps in machine computing.

process—Any operation or sequence of operations involving a change of energy state, composition, dimension, or other property that may be defined with respect to a datum. The term *process* is used in this standard to apply to all variables other than instrument signals.

process control—Automatic control of continuous operations, contrasted with numerical control, which provides automatic control of discrete operations.

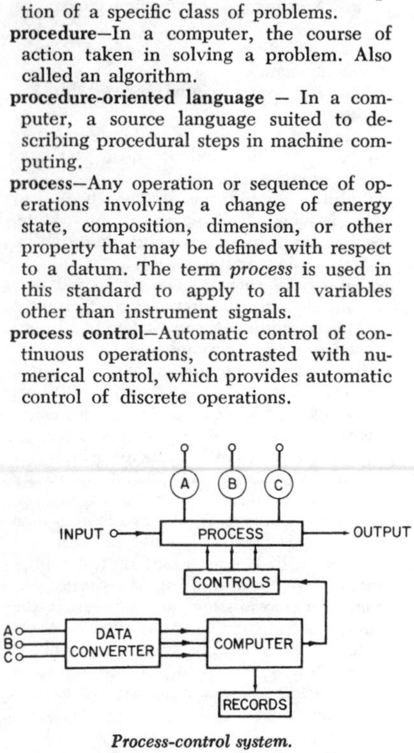

Process-control system.

processing—Additional handling, manipulation, consolidation, compositing, etc., of information to change it from one format to another or to convert it to a manageable and/or intelligible form.

processing section—The portion of a computer that does the actual changing of input into output. This includes the arithmetic and logic sections.

processor—1. In hardware, a data processor. 2. In software, a computer program that includes the compiling, assembling, translating, and related functions for a particular programming language, including logic, memory, arithmetic, and control.

producer's reliability risk—The risk faced by the producer (usually set at 10%) that a product will be rejected by a reliability-acceptance test even though the product is actually equal to or better than a specified value of reliability.

product detector—A demodulator the output of which is the product of the input signal voltage and the signal voltage of a local oscillator operating at the input frequency.

production lot – A group of (electronic) parts manufactured during the same period from the same basic raw materials, processed under the same specifications and procedures, produced with the same equipment, and identified by the documentation defined in the manufacturer's reliability assurance program through all significant manufacturing operations, including final assembly operations. Final assembly operation is considered the last major assembly operation, such as casing, hermetic sealing, or lead attachment, rather than painting or marking.

production sampling tests—Those tests normally made by either the vendor or the purchaser on a portion of a production lot for the purpose of determining the general performance level.

production tests – Those tests normally made on 100% of the items in a production lot by the vendor and normally on a sampling basis by the purchaser.

product modulator—A modulator, the output of which is substantially equal to the carrier times the modulating wave.

profile chart – A vertical cross-sectional drawing of the microwave path between two stations. Terrain, obstructions, antenna-height requirements, etc., are indicated on the drawing.

program—1. A sequence of instructions that tells a computer how to receive, store, process, and deliver information. 2. A plan for solving a problem, including instructions that cause the computer to perform the desired operations and such necessary information as data description and tables. 3. A series of actions proposed in order to achieve a certain result. 4. To prepare such a set of coded instructions. 5. To design, write, and test a program. 6. In a calculator, a sequence of detailed instructions for the operations necessary to solve a problem. Programmable electronic calculators can "learn" the steps of a problem so that, after the first sequence of entries, only the variable numbers need be entered on the keyboard without manual activation of control keys. Some programmable machines store programs on cards or tapes.

program assembly—Also called a translator. A process which translates a symbolic program into a machine-language program before the working program is executed.

If can also integrate several sections or different programs.

programatics—The branch of learning that has to do with the study of programming methods and language.

program break—The length of a program; the first location not used by a program (before relocation); the relocation constant for the following program (after relocation).

program circuit — A telephone circuit that has been equalized to handle a wider range of frequencies than ordinary speech signals require. In this way, musical programs can be transmitted over telephone wires.

program control—A control system which automatically holds or changes its target value on the basis of time, to follow a prescribed program for the process.

program counter — A CPU register which specifies the address of the next instruction to be fetched and executed. Normally it is incremented automatically each time an instruction is fetched.

program-distribution amplifiers — A group of amplifiers fed by a bridging bus from a single source. Each amplifier then feeds a separate line or other service.

program element — The part of a central computer system that performs the sequence of instructions scheduled by the programmer.

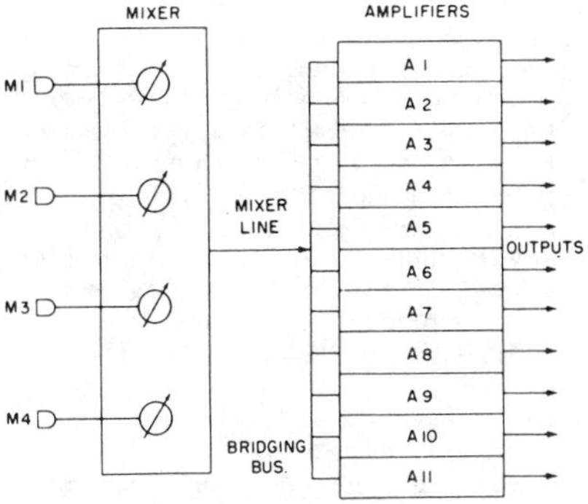

Program-distribution amplifier.

program failure alarm—In broadcasting stations, a relay circuit that gives a visual and aural alarm when a program fails. A delay prevents the relay from giving a false alarm during the silence before and after station-identification or other short breaks.

program flowchart — A flowchart that describes the control flow—the order in which the various program steps are executed—within any computer program or module.

program generator—A program that enables a computer to write other programs automatically.

program level — The measure of the program signal in an audio system. It is expressed in volume units (VU).

program library—A collection of computer program and routines that are available.

program linkage—In a computer, efficient use of all registers and development of subroutines so that there is smooth, economical transition from one program segment to another, and memory capacity is conserved.

program loop—A series of computer instructions that are repeated until a terminal condition is achieved.

programmable calculator — 1. A calculator whose operation is controlled by programs stored in its memory. 2. Electronic calculator capable of performing preset sequences of computations. 3. One that can learn a repetitive series of operations. Can "learn" or be programmed by various means to handle a series of steps so that only variable information need be entered into the calculator.

programmable communications processor—A digital computer that has been specifically programmed to perform one or more control and/or processing functions in a data communications network. As a self-contained system, it may or may not include communications line multiplexors,

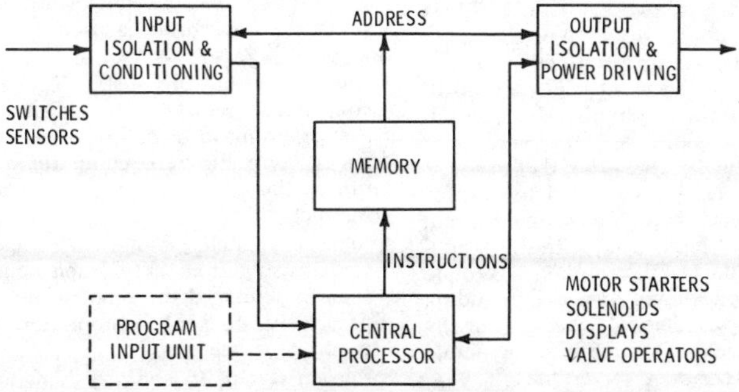

Programmable controller.

line adapters, a computer system interface, and on-line peripherals. It always includes a specific set of user-modifiable software for the communications function.

programmable controller—1. A control machine based on solid-state digital logic and built of computer subsystems, and primarily intended to take the place of electromechanical relay panels in applications in which rewiring is made necessary by periodic changes in sequence. This type of controller is particularly useful in the control of processes, materials handling, and certain machine functions. 2. A controller whose operation is determined by codes or instructions programmed into it by the user.

programmable counter — A device capable of being programmed so that it counts to any number from zero to its maximum possible modulus. Also called modulo-n counter.

programmable logic array — Abbreviated PLA. A general-purpose logic structure consisting of an array of logic circuits. The way in which these circuits are programmed determines how input signals to the PLA are processed. Programming is done on a custom basis at the factory and permanently establishes the functional operation of the PLA.

programmable read-only memory—Abbreviated PROM. An integrated circuit memory array that is manufactured with a pattern of either all logical zeros or ones and has a specific pattern written into it by the user by a special hardware programmer. (Some PROMs, called EAROMs, Electrically Alterable Read-Only Memory, can be erased and reprogrammed.)

programmable unijunction transistor — Abbreviated PUT—A four-layer device similar to an SCR except that the anode gate rather than the cathode gate is brought out. It is used in conventional uni'unction transistor circuits. The characteristics of both devices are similar but the triggering voltage of the PUT is programmable and can be set by an external resistive voltage divider network. The PUT is faster and more sensitive than the UJT. It finds limited application as a phase-control element and is most often used in long duration timer circuits. In general, the PUT is a more versatile and more economical device than the UJT.

programmed check—A means of testing for the correctness of a computer program and machine functioning, either by running a similarly programmed sample problem with a known answer (including mathematical or logical checks) or by building a checking system into the actual program being run.

programmed logic array — An orderly arrangement of logical AND and logical OR functions. Its application is very much like a glorified ROM. It is primarily a combinational logic device. Abbreviated PLA.

programmed marginal check — A computer program that provides voltage variation to check a tube or other computer equipment during a preventive-maintenance check.

programmed operators—Computer instructions that make it possible for subroutines to be accessed with a single programmed instruction.

programmer—1. A person who prepares the sequences of instructions for a computer or other data-handling system. He may or may not convert them into detailed codes. 2. A device for timed switching of several interrelated functions or set of functions.

programming—1. Definition of a computer problem resulting in a flow diagram. 2. Preparing a list of instructions for the computer to use in the solution of a problem. 3. Selecting various circuit patterns by interconnecting or "jumping" the appropriate contacts on one side of a connector plug. 4. The control of a power-supply parameter, such as output voltage, by means of a remotely located or internally located control element (usually resistance) or signal (voltage).

programming device—A device by which a series of mechanical or electrical operations or events may be preset to be performed automatically in a predetermined sequence and at specified time intervals.

programming module—A set of instructions that is discrete and identifiable and usually is handled as a unit by an assembler, compiler, linkage editor, loading routine, or other routine or subroutine.

programming system—Any method of programming problems, other than machine language, consisting of a language and its associated processor(s).

program parameter—In a subroutine of a computing or other data-handling system, an ad'ustable parameter which can be given different values on several occasions when the subroutine is used.

program register — Also called program counter, or control register. The computer control-unit register into which is stored the program instruction being executed, hence controlling the computer operation during the cycles required to execute that instruction.

program-sensitive error—In a computer, an error arising from unforeseen behavior of some circuits, discovered when a comparatively unusual combination of program steps occurs.

program signal—In audio systems and components, the complex electric wave—corresponding to speech, music, and associated sounds—destined for audible reproduction.

program step—An increment, usually one instruction, of a computer program.

program storage—A portion of the internal computer storage reserved for programs, routines, and subroutines as contrasted with temporary storage. In many systems, protective devices are used so that the contents of the program storage cannot be altered inadvertently.

program tape—In a computer, a magnetic or punched paper tape which contains the sequence of instructions for solving a problem.

program time—The phase of computer operation during which an instruction is being interpreted so that the required action can be performed.

program timer – 1. A loosely used term sometimes referring to a complex multicircuit timing device in which the program is readily changed, such as tape-controlled timers, cam timers, time switches, or to any other type of timer. 2. A multiple-circuit repeat-cycle timer which repeats a preset program continuously as long as power is applied.

progressive scanning—A rectilinear process in which adjacent lines are scanned in succession. In television, the scanning process in which the distance from center to center of successively scanned lines is equal to the nominal line width.

project engineer – Engineer in charge of project, may be designer of system, and even in charge of purchasing for project.

projection cathode-ray tube—A cathode-ray tube that produces an intense but relatively small image, which can be projected onto a large viewing screen by an optical system consisting of lenses or a combination of lenses and mirrors.

projection ppi—A unit in which the image of a 4-inch dark-trace cathode-ray tube is projected on a 24-inch horizontal plotting surface. The echoes appear as magenta-colored arcs on a white background. *See also* Dark-Trace Tube.

projection television – A combination of lenses and mirrors for projecting an enlarged television picture onto a screen.

projector—1. A device used in an underwater sound system to radiate sound pulses through the water from the bottom of a ship. 2. A horn designed to direct sound chiefly in one direction from a speaker.

PROM—1. Abbreviation for programmable ROM. A ROM that can be programmed by the user only once. After a PROM is programmed, it effectively becomes a ROM. 2. Similar to the conventional ROM (Read-Only Memory). A write-once memory. When an instruction is written via a memory-write cycle into the programmable ROM, certain kinds of fusing take place and the data are written permanently into the memory.

promethium cell—A low-power cell containing a radioactive isotope called promethium 147, which emits beta particles that strike a phosphor. Two photocells then convert the light output from the phosphor into electrical energy.

prong—*See* Pin.

proof pressure – The maximum pressure which may be applied to the sensing element of a transducer without changing the transducer performance beyond specified tolerances.

propagated error—An error that is carried through succeeding computer operations.

propagation—Also called wave propagation. The travel of electromagnetic waves or sound waves through a medium. Propagation does not refer to the flow of current in the ordinary sense.

propagation anomaly—An irregularity introduced into an electromagnetic or other sensing device by discontinuities in the propagation medium.

propagation constant—1. The transmission characteristic which indicates the effect of a line on a wave being transmitted along the line. It is a complex quantity having a real term called the attenuation constant and an imaginary term called the phase constant. 2. The natural logarithm of the ratio of the current into an electric transducer to the current out of the transducer, with the transducer terminated in its iterative impedance. Also called transfer factor.

propagation delay – 1. A measure of the time required for a logic signal to travel through a device or a series of devices forming a logic string. It occurs as the result of four types of circuit delays—storage, rise, fall, and turn-on delay—and is the time between when the input signal crosses the threshold-voltage point and when the responding voltage at the output crosses the same voltage point. 2. Time delay occurring between the appli-

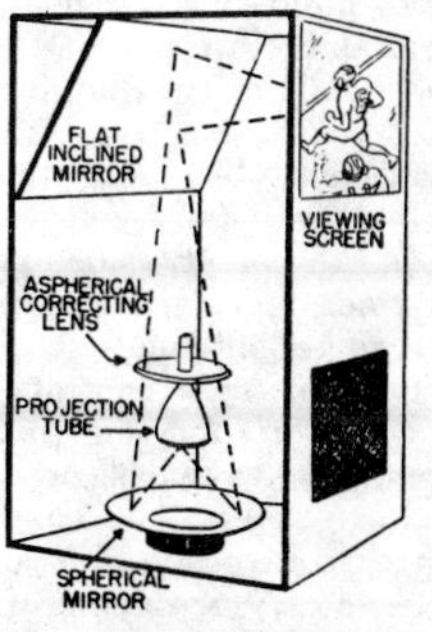

Projection television.

cation of an input to a digital logic circuit and the change of state at the output.

propagation delay time – The time, measured between reference points on the input and output waveforms, between the application of a digital input waveform and the corresponding output change.

propagation error—For ranging systems, the algebraic sum of propagation-velocity error and curved-path error. Except at long ranges and low angles, the curved-path component of propagation error is generally negligible.

propagation factor—*See* Propagation Ratio.

propagation loss—The loss of energy suffered by a signal while passing between two points.

propagation ratio—Also called the propagation factor. For a wave that has been propagated from one point to another, the ratio of the complex electric-field strength at the second point to that at the first point.

propagation time—1. The time it takes for a signal to travel from point to point. In a communications channel, the velocity of signal propagation is less than that of radio. A signal delay of 20 msec per thousand miles is a reasonable maximum. 2. The time required for transmission of a unit of binary information (high or low voltage) from one physical point in a system or subsystem to another, such as from the input to the output of a device.

propagation time delay—The time required for a wave to travel from one point to another along a transmission line. It varies according to the type of line.

propagation velocity—*See* Velocity of Propagation.

property sort—In a computer, a technique for the selection from a file of records meeting a certain criterion.

proportional amplifier—A fluidic unit that maintains a regulated relationship from control input to device output; e.g., an increase in control pressure results in an analogous increase in output pressure, for the case of a pressure-amplifying proportioner.

proportional band—The range of the controlled variable corresponding to the full range of operation of the final control element.

proportional control—1. A method of control in which the intensity of action varies linearly as the condition being regulated deviates from the conditions prescribed. 2. Also called galloping ghost. An advanced type of radio-control system in which the rudder (and sometimes the elevator) can move as much (or as little) as the operator wishes. 3. A control system in which corrective action is always proportionate to any variation of the controlled process from its desired value. For example, instead of snapping directly open-closed in the manner of two-position control, a proportional valve will be always positioned at some point between open and closed, depending on the flow requirement of the system at any given moment.

proportional counter tube – A sealed tube containing an inert gas such as argon, krypton, xenon, methyl bromide, etc. It is used like a Geiger-Mueller counter and operated at about 100 volts in the proportional region.

proportional linearity – A manner of expressing nonlinearity as the deviation from a straight line in terms of a given percentage of the transducer output at the stimulus point under consideration (i.e., as a percentage of the reading).

proportional region—In a radiation counter tube, the applied-voltage range in which the gas amplification is greater than unity and is independent of the charge liberated by the initial ionizing event.

proportional temperature control—A method of stabilizing (an oscillator) by providing heater power that is directly proportional to the difference between the desired operating temperature and the ambient temperature.

protected location—A computer storage location, reserved for special purposes, in which data cannot be stored without being subjected to a screening procedure to establish suitability for storage at that location.

protected wireline distribution system—A communications system to which electromagnetic and physical safeguards have been applied to permit secure electrical transmission of unencrypted, classified information, and which has been approved by the cognizant department or agency. The associated facilities include all equipment and wirelines so safeguarded. Major components are wirelines, subscriber sets, and terminal equipment. Also known as approved circuit.

protective cable—Small-gage quadded cable used in toll cables to serve as fuses, usually at building entrances.

protective device—Any device for keeping an undesirably large current, voltage, or power out of a given part of an electric circuit.

protective gap—A spark gap provided between a conductor and the earth by suitable electrodes. High-voltage surges due to lightning are thus permitted to pass harmlessly to earth through the gap.

protective relay – A relay, the principal function of which is to protect services from interruption or to prevent or limit damage to apparatus.

protective resistance – A resistance placed

in series with a device (e.g., a gas tube) to limit the current to a safe value.

protector—1. A device to protect equipment or personnel from high voltage or current. 2. A protection device used on communication systems to limit the magnitude of extraneous overvoltages. The discharge device within a protector may consist of closely spaced, carbon electrodes discharging in air, or metallic electrodes discharging in a hermetically sealed gaseous atmosphere at reduced pressure. A protector does not contain an element to prevent holdover as in the case of an arrester.

protector block—A rectangular piece of carbon with an insulated metal insert, or porcelain with a carbon insert, which makes one element of a protector. It forms the gap which will break down and provide a path to ground for voltages over 350 volts.

protector tube – A glow-discharge, cold-cathode tube in which a low-voltage breakdown is employed between two or more electrodes to protect the circuit against overvoltage.

protocol—A set of conventions or rules governing the format and timing of message exchanges to control data movements and correct errors. It is important to ensure that the protocol is valid, makes sense, works, and is adhered to by all users of the network in question.

proton—An elementary particle with a positive charge equivalent to the negative charge of the electron, but with approximately 1845 times the mass. The proton is the positive nucleus of the hydrogen atom.

prototype—Original design or first operating model.

prototype model—A working model, usually hand-assembled, and suitable for complete evaluation of mechanical and electrical form, design, and performance. Approved parts are employed throughout, so that it will be completely representative of the final, mass-produced equipment.

proximity detector—A sensing device which gives an indication when approaching or being approached by another object (e.g., a burglar alarm).

proximity effect—The redistribution of current brought about in a conductor by the presence of another current-carrying conductor.

proximity fuse—A fuse designed to detonate a projectile, bomb, mine, or charge when activated by an external influence in the vicinity of a target. The variable time fuse is one type of proximity fuse.

prr—Abbreviation for pulse-repetition rate.

ps—Abbreviation for picosecond (10^{-12} second).

pseudocode – An instruction that is not meant to be followed directly by a computer. Instead, it initiates the linking of a subroutine into the main program.

pseudoinstruction – *See* Instructional Constant.

pseudo-op—An operation that is not part of the computer's operation repertoire as realized by hardware; hence an extension of the set of machine operations.

pseudoprogram—A program that is written in a pseudocode and may include short coded logical routines.

pseudorandom—Having the property of being produced by a definite calculation process while simultaneously satisfying one or more of the standard tests for statistical randomness.

pseudorandom binary sequence – A two-level signal that has a repetitive sequence, but a random pattern within the sequence. Such a signal finds use as a test signal, since it has the basic characteristics of noise, but in terms of parameters that are easily controlled.

pseudostereo—Devices and techniques for obtaining stereo qualities from one channel.

psophometric emf—The electromotive force (or voltage) generated by a source having an internal resistance of 600 ohms and no internal reactance, which, when conected across a standard receiver having 600 ohms of resistance and no reactance, produces the same sinusoidal current as an 800-hertz generator of the same impedance. *See also* Psophometric Voltage.

psophometric voltage – The voltage which would appear across a 600-ohm resistance connected between any two points in a telephone circuit. (This value is one-half the psophometric emf since the latter is essentially the open-circuit potential necessary from a source to produce the psophometric voltage if the source has a 600-ohm internal resistance.)

PSOS—A term used by some companies to denote p-channel silicon gate devices.

PSPS—Abbreviation for planar silicon photoswitch.

psychoacoustics – A relatively new branch of audio that concerns itself with personal and subjective factors in hearing and in evaluating the performance of high-fidelity equipment.

psychogalvanometer—An instrument for recording the electric variations produced by emotional stresses.

ptm—Abbreviation for pulse-time modulation.

ptm/ppm/am—A system in which a number of pulse-position or pulse-time modulated subcarriers are used to amplitude-modulate the carrier.

PTT – Abbreviation for press-to-talk or push-to-talk.

p-type—Pertaining to semiconductor material that has been doped with an excess of acceptor impurity atoms, so that free holes are produced in the material.

p-type conductivity—The conductivity associated with the holes in a semiconductor.

p-type conductor—A positive-type conductor, one with electron holes as the principal carriers. This implies the presence of acceptors.

p-type crystal rectifier — A crystal rectifier in which forward current flows whenever the semiconductor is more positive than the metal.

p-type material—1. A pure crystal of semiconductor material to which an impurity has been added (electron acceptor such as boron or gallium) to give it a deficiency of electrons and alter its electrical characteristics. 2. Semiconductor material having holes as the majority charge carriers; formed by doping with acceptor atoms. 3. Refers to an excess of positive electrical charges in a semiconductor material. Natural silicon is made to be p-type by the addition of an acceptor impurity.

p-type region — Portion of semiconductor material containing a small number of dopant atoms that have an electron deficiency (an empty space) in their outer orbit. The deficiency is called a hole and it behaves like a positively charged particle. The p region is a source of mobile positive charges.

p-type semiconductor — An extrinsic semiconductor in which the hole density exceeds the conduction-electron density. By implication, the net ionized impurity concentration is an acceptor type.

public-address system—Also called a pa system. One or more microphones, an audio-frequency system, and one or more speakers used for picking up and amplifying sounds to a large audience, either indoors or out.

public correspondence — Any telecommunication which the offices and stations, by reason of their being at the disposal of the public, must accept for transmission.

public radiocommunication services—Land mobile or fixed radio services, the stations of which are open to public correspondence.

public-safety radio service—Any radiocommunication service esssential either to the discharge of nonfederal governmental functions relating to public-safety responsibilities or to the alleviation of an emergency endangering life or property. The radio transmitting facilities in this service may be fixed, land, or mobile stations.

puck—*See* Pressure Roller.

puff—British abbreviation for picofarad.

pull—1. To cause an oscillator to depart from its designed frequency of operation. 2. To depart from the designed frequency of operation, as an oscillator.

pull curves — The characteristics relating force to displacement in the actuating system of a relay.

pull-down resistor—1. A resistor connected across the output of a device or circuit to hold the output equal to or less than the 0 input level of the following digital device. Also used to lower the output impedance of a device. 2. A resistor connected to a negative voltage or to ground.

pull-in current (or voltage)—The maximum current (or voltage) required to operate a relay. *See also* Pickup Current.

pull-in torque—Torque that a synchronous motor can exert to bring its driven load into synchronous speed. There is no corresponding term for induction motors.

pulling—1. In an oscillator, the undesired change from the desired frequency. It is caused either by coupling from another source of frequency, or by the influence of the load impedance. 2. In television, partial loss of synchronization.

pulling figure—The difference between the maximum and minimum frequencies of an oscillator whenever the phase angle of the load-impedance reflection coefficient varies through 360°. The absolute value of this coefficient is constant and equal to .20.

pull-out force—The tensile force required to separate a conductor from a contact or terminal, or to separate a contact from a connector.

pull-out torque — Also called breakdown torque, or maximum torque. The maximum torque a motor can deliver without stalling. Usually applied to synchronous motors only.

pull-up—The placing of the output voltage of a logic circuit at the high level by means of an internal current sink or source.

pull-up resistor—A resistor connected to the positive supply voltage of a transistor circuit, as from the collector supply to the output collector.

pull-up torque — The minimum torque developed by an alternating-current motor during the period of acceleration from rest to the speed at which breakdown torque occurs. For motors which do not have a definite breakdown torque, the pull-up torque is the minimum torque developed up to rated speed.

pulsating current — Current that varies in amplitude but does not change polarity.

pulsating direct current — A direct current that changes its value at regular or irregular intervals but flows in the same direction at all times.

pulsating electromotive force — A direct electromotive force and an alternating electromotive force combined.

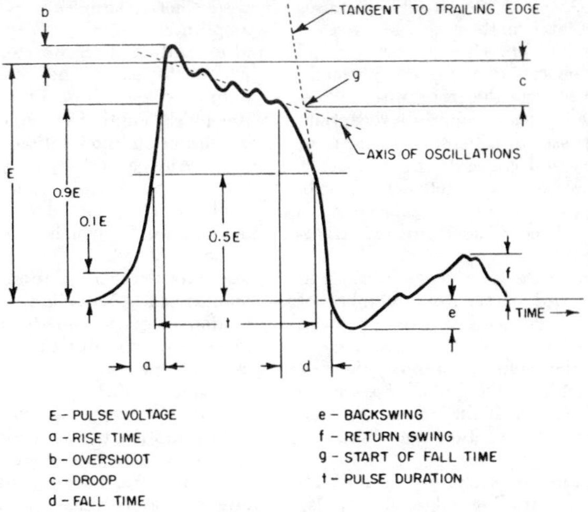

Pulse, 1.

pulsating quantity – A periodic quantity that can be considered the sum of a continuous component and an alternating component in the quantity.

pulsation welding – A form of resistance welding in which the power is alternately applied and removed.

pulse–1. The variation of a quantity having a normally constant value. This variation is characterized by a rise and a decay of a finite duration. 2. An abrupt change in voltage, either positive or negative, which conveys information to a circuit. (*See also* Impulse.)

pulse amplifier–A wideband amplifier used to amplify square waves without appreciably changing their shape.

pulse amplitude – A general term for the magnitude of a pulse. For more specific designation, adjectives such as average, instantaneous, peak, rms (effective), etc., should also be used.

pulse-amplitude modulation–Modulation in which the modulating wave is caused to amplitude-modulate a pulse carrier. Abbreviated pam.

pulse-amplitude modulation / frequency modulation–A system in which a carrier is frequency modulated by pulse-amplitude modulated subcarriers.

pulse analyzer – Equipment for analyzing pulses to determine their time, amplitude, duration, shape, etc.

pulse arc welding – A type of welding in which the material to be welded is positioned together, forming one electrode. The other electrode is positioned to form a gap with one of the workpieces. An arc is struck and the current heats the workpieces to the melting point at their interface. *See also* Arc Percussive Welding.

pulse-average time – The duration of a pulse, measured between two points at 50% of the maximum amplitude on the leading and trailing edges.

pulse bandwidth–The smallest continuous frequency interval outside of which the amplitude of the spectrum does not exceed a prescribed fraction of the amplitude at a specified frequency.

pulse carrier–A carrier consisting of a series of pulses. Usually employed as a subcarrier.

pulse code–1. A pulse or series of pulses which, by means of waveform, pulse width, pulse time, pulse numbers, or pulse sequences, may be used to convey information. 2. Loosely, a code consisting of pulses–e.g., Morse, Baudot, binary.

pulse-code modulation – 1. Abbreviated pcm. Pulsed modulation in which the signal is sampled periodically and each sample is quantized and transmitted as a digital binary code. 2. A digital technique by which information may be carried from one point to another. The signal is carried as a series of separate pulses or digits. No distortion is introduced and no information is lost from a signal unless a complete pulse disappears or unless a spurious noise pulse is formed that is large enough to be accepted by the equipment as a genuine pulse. Thus, many channels of communication can be made available along a single connecting line.

pulse-code modulation/frequency modulation–A system in which pulse-code modulated subcarriers are used to frequency-modulate a second carrier. Binary digits are formed by the absence or presence of a pulse in an assigned position.

pulse coder—A circuit which sets up pulses in an identifiable pattern.

pulse coding and correlation – A general technique concerning a variety of methods used to change the transmitted waveform and then decode upon its reception. Pulse compression is a special form of pulse coding and correlation.

pulse compression – A matched filter technique used to discriminate against signals which do not correspond to the transmitted signal.

pulse counter—A device that gives an indication or record of the total number of pulses that it has received during a given time interval.

pulsed Doppler system—A pulsed radar system which utilizes the Doppler effect to obtain information about the target (not including simple resolution from fixed targets).

pulse decay time—The amount of time required for the trailing edge of a pulse to decay from 90% to 10% of the peak pulse amplitude.

pulse delay time – The time interval between the leading edges of the input and output pulses, measured at 10% of their maximum amplitude.

pulse demoder – Also called a constant-delay discriminator. A circuit which responds only to pulse signals with a specified spacing between them.

pulse digit—A code element comprising the immediately associated train of pulses.

pulse-digit spacing—The time interval between the end of one pulse digit and the start of the next.

pulse discriminator—A device that responds only to pulses having a particular characteristic (e.g., duration, amplitude, period). One that responds to period is also called a time discriminator.

pulsed oscillator—1. An oscillator in which oscillations are sustained by either self-generated or external pulses. 2. An oscillator which generates a carrier-frequency pulse or a train of pulses.

pulsed-oscillator starting time – The interval between the leading-edge times of the pulse at the oscillator control terminals and the related output pulse.

pulse droop—Distortion characterized by a slanting of the top of an otherwise essentially flat-topped rectangular pulse.

pulsed ruby laser—A laser that uses ruby as the active material. The extremely high pumping power required is obtained by discharging a bank of energy storage capacitors through a special high-intensity flash tube.

pulse duration—Also called pulse length or width. The time interval between the points at which the instantaneous value on the leading and trailing edges bears a specified relationship to the peak pulse amplitude.

pulse-duration discriminator – A circuit in which the sense and magnitude of the output is a function of the deviation of the pulse length from a reference.

pulse-duration modulation – Abbreviated pdm. Also called by the less preferred terms "pulse-width modulation" and "pulse-length modulation." Pulse-time modulation in which the duration of a pulse is varied.

pulse-duration modulation / frequency modulation—Also called pulse-width modulation/frequency modulation. A system in which pulse-duration modulated subcarriers are used to frequency-modulate a second carrier.

pulse duty factor – Ratio of the average pulse duration to the average pulse spacing. Equivalent to the average pulse duration times the pulse-repetition rate.

pulse emission—Emission drawn for short periods; it may or may not follow a regular repetition rate.

pulse emitter load—The load seen by the collector of an inverter that drives the pulse input to a flip-flop, pulse amplifier, or delay.

pulse equalizer – A circuit that produces output pulses of uniform size and shape when driven by input pulses that vary in size and shape.

pulse fall time—That time during which the trailing edge of a pulse is decreasing from 90% to 10% of its maximum amplitude.

pulse-forming line—A combination of circuit components used to produce a square pulse of controlled duration.

pulse-forming network – A network which converts either an ac or dc charging source into an approximately rectangular wave output. By means of a high-speed switch, it alternately stores energy from the charging source and releases the energy through the load to which it is connected. This network supplies the accurately shaped pulse required by the magnetron or klystron oscillator of a radar modulator.

pulse-frequency modulation – Abbreviated pfm. More precisely called pulse-repeti-

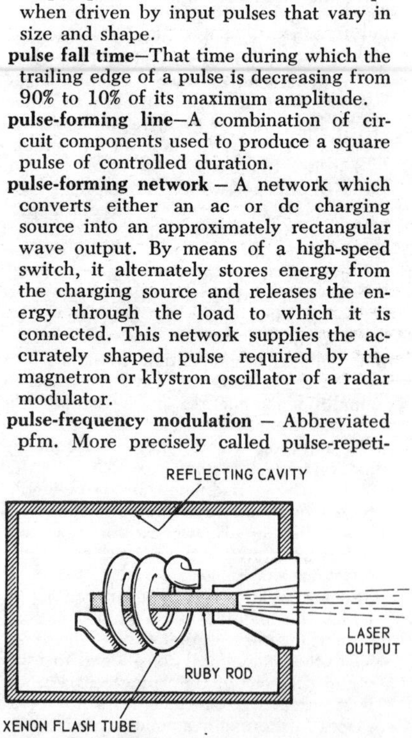

Pulsed ruby laser.

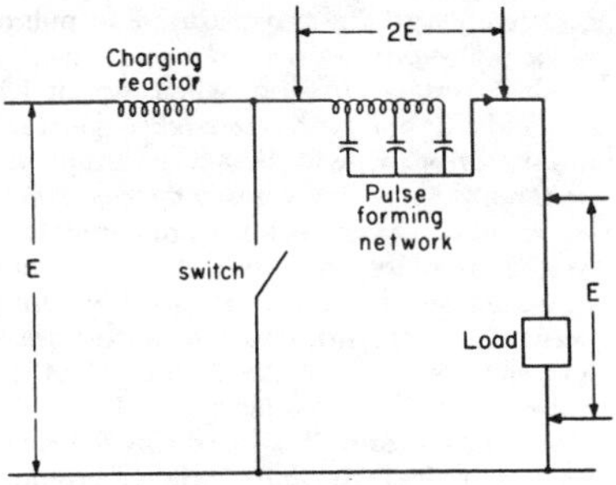

Pulse-forming network.

tion-rate modulation. Pulse-time modulation in which the pulse repetition rate is varied.

pulse-frequency spectrum—*See* Pulse Spectrum.

pulse generator—1. A device for generating a controlled series of electrical pulses. 2. A device that produces a single pulse or a train of repetitive pulses.

pulse group—*See* Pulse Train.

pulse-height analyzer — Also called kick-sorter. An instrument that indicates the number or rate of occurrence of pulses within each of one or more specified amplitude ranges.

pulse-height discriminator — Also called a pulse-height selector. A circuit that selects and passes only those pulses which exceed a certain minimum amplitude.

pulse-height selector—*See* Pulse-Height Discriminator.

pulse-improvement threshold — In a constant-amplitude pulse-modulation system the condition existing when the peak pulse voltage is at least twice the peak noise voltage after selection and before any nonlinear process such as amplitude clipping and limiting. The ratio of peak to rms noise voltage is ordinarily assumed to be 4 to 1.

pulse-interference eliminator — A device that removes pulse signals which are not precisely on the radar operating frequency.

pulse-interference separator and blanker—A circuit that automatically blanks all video signals that are not synchronous with the radar prf.

pulse interleaving — A process in which pulses from two or more time-division multiplexers are systematically combined in time division for transmission over a common path.

pulse interrogation—1. The triggering of a transponder by a pulse or pulse mode. Interrogations by the latter may be employed to trigger one or more transponders. 2. Periodic electrical activation and observation, either synchronous or asynchronous, of the shaft position of an encoder.

pulse interval—*See* Pulse Spacing.

pulse-interval jitter—The time or displacement band within which lie all of the transitions in the same direction of a signal through a specified amplitude. This is a dynamic value to be determined at one or more specified speeds. Pulse-interval jitter is expressed as a percentage of one pulse interval at a specified speed.

pulse-interval modulation—Pulse-time modulation in which the pulse spacing is varied. (*See also* Pulse-Spacing Modulation.)

pulse jitter—A relatively slight variation of the pulse spacing in a pulse train. It may be random or systematic, depending on its origin, and is generally not coherent with any imposed pulse modulation.

pulse length—*See* Pulse Duration.

pulse-length modulation — *See* Pulse-Duration Modulation.

pulse-link repeater—In a telephone signaling system, equipment that receives pulses from one E-and-M signaling circuit and transmits corresponding pulses into another E-and-M signaling circuit.

pulse load—The load presented to a pulse source.

pulse mode—1. A finite sequence of pulses, in a prearranged pattern, used for selecting and isolating a communication channel. 2. The prearranged pattern in (1) above.

pulse-mode multiplex—A process or device for selecting channels by means of pulse

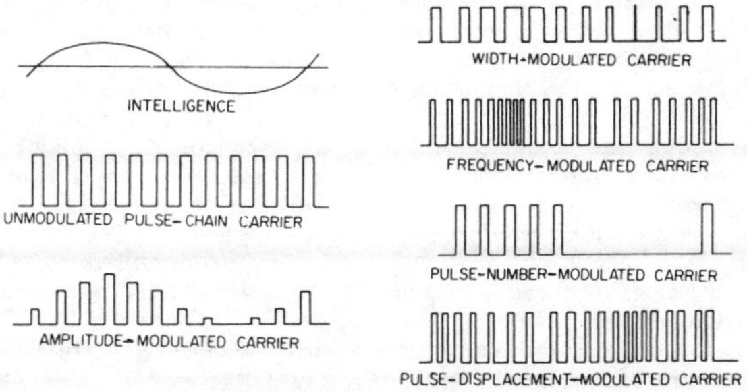

Pulse modulation, 3.

modes. In this way, two or more channels can use the same carrier frequency.

pulse moder – A device for producing a pulse mode. (*See also* Pulse Demoder.)

pulse-modulated jamming–The use of jamming pulses that have various widths and repetition rates.

pulse-modulated radar–Radar in which the radiation consists of a series of discrete pulses.

pulse-modulated waves – Recurrent wave trains used extensively in radar. In general, their duration is shorter than the interval between them.

pulse modulation–1. The use of a series of pulses modulated to convey information. Modulation may involve changes of pulse amplitude (pam), position (ppm), or duration (pdm). 2. Modulation of a carrier by a train of pulses: the generation of carrier-frequency pulses. 3. Modulation of a characteristic or characteristics of a pulse carrier: transmission of information on a pulse carrier.

pulse modulator – A device which applies pulses to the element being modulated.

pulse numbers modulation–A type of modulation in which the pulse density per unit time of a carrier is varied in accordance with a modulating wave; omissions of pulses are made systematically without changing the phase or amplitude of the transmitted pulses. For example, the omission of every other pulse could represent zero modulation; the reinsertion of pulses would then represent positive modulation, and the omission of more pulses would represent negative modulation.

pulse operation–The method whereby the energy is delivered in pulses. Usually described in terms of the shape and the frequency of the pulses.

pulse oscillator–An oscillator in which the oscillations are sustained by self-generated or external pulses.

pulse packet–In radar, the volume of space occupied by the pulse energy.

pulse period – In telephony, the time required for the dial pulse springs to open and close one time. Also called impulse period.

pulse-position modulation – Abbreviated ppm. Pulse-time modulation in which the value of each instantaneous sample of the wave modulates the position in time of a pulse.

pulse-position modulator – A device which converts analog information to variations in pulse position.

pulser – A generator which produces extremely short, high-voltage pulses at definite recurrence rates for use in radar transmitters and similar pulsed systems.

pulse rate–*See* Pulse-Repetition Rate.

pulse ratio – Ratio of the length of any pulse to its total period.

pulse recovery–The time, usually in microseconds, required for electron flow in a diode to start or stop when voltage is suddenly applied or removed.

pulse-recurrence, counting-type frequency meter–A device for measuring frequency. It uses a direct-current ammeter calibrated in pulses per second.

pulse-recurrence time – The time elapsing between the start of one transmitted pulse and the next pulse. The reciprocal of the pulse-recurrence frequency.

pulse regeneration–The restoring of a series of pulses to their original timing, form, and relative magnitude.

pulse repeater – Also called a transponder. A device that receives pulses from one circuit and transmits corresponding pulses at another frequency, waveshape, etc., into another circuit.

pulse-repetition frequency–The rate (usually given in hertz or pulses per second) at which pulses or pulse groups are transmitted from a radar set. Abbreviated prf.

pulse-repetition period – The reciprocal of the pulse-repetition frequency.

pulse-repetition rate–Abbreviated prr. Also called pulse rate. The average number of pulses per unit of time.

pulse-repetition-rate modulation–*See* Pulse-Frequency Modulation.

pulse reply–The transmission of a pulse or pulse mode by a transponder as the result of an interrogation.

pulse resolution–The minimum time separation, usually in microseconds or milliseconds, between input pulses that permits proper circuit or component response.

pulse rise time – The interval of time required for the leading edge of a pulse to rise from 10% to 90% of its peak amplitude, unless some other percentage is stated.

pulse sample-and-hold circuit – A circuit which holds the final amplitude of an integrated pulse until the final amplitude of the succeeding integrated pulse is reached. A less desirable sample-and-hold circuit resets after each hold period to a fixed level before integrating a succeeding pulse.

pulse scaler–A device capable of producing an output signal whenever a prescribed number of input pulses has been received. It frequently includes indicating devices that facilitate interpolation.

pulse selector–A circuit or device that selects the proper pulse from a sequence of (telemetering) pulses.

pulse separation–The interval between the trailing-edge pulse time of a pulse and the leading-edge pulse time of the succeeding pulse.

pulse shaper – Any transducer (including pulse regenerators) used for changing one or more characteristics of a pulse.

pulse shaping – Intentionally changing the shape of a pulse.

pulse spacing–The time interval from one pulse to the next–i.e., between the corresponding times of two consecutive pulses. (The term "pulse interval" is ambiguous –it may be taken to mean the duration of a pulse instead of the space or interval between pulses.)

pulse-spacing modulation–Formerly called pulse-interval modulation. A form of pulse-time modulation in which the pulse spacing is varied.

pulse spectrum – Also called pulse-frequency spectrum. The frequency distribution, in relative amplitude and phase, of the sinusoidal components of a pulse.

pulse spike – A relatively short duration pulse superimposed on the main pulse.

pulse-spike amplitude–The peak amplitude of a pulse spike.

pulse stepper – A stepper motor that responds directly to a pulse of specified length and amplitude. The positioning of the motor shaft is directly proportionate to the number of pulses applied. Rotational direction is controlled by electrical shading.

pulse-storage time–The time interval from a point at 90% of the maximum amplitude on the trailing edge of the input pulse to the same 90% point on the trailing edge of the output pulse.

pulse stretcher–1. A circuit designed to extend the duration of a pulse–primarily so that its pulse modulation will be more readily discernible in an audio presentation. 2. In a computer, a circuit which generates a long pulse when triggered by a short pulse. The width of the output pulse is determined by the value of the coupling capacitor. The maximum width of the output pulse cannot exceed 50% of the clock rate.

pulse tilt–A distortion characterized in an otherwise essentially flat-topped rectangular pulse by either a decline or a rise of the pulse top.

pulse time–The time interval from a point at 90% of the maximum amplitude on the leading edge of a pulse, to the 90% point on the trailing edge.

pulse-time–modulated radiosonde – Also called time-interval radiosonde. A radiosonde which transmits the indications of the meteorological sensing elements in the form of pulses spaced in time. The meteorological data are evaluated from the intervals between the pulses.

pulse-time modulation – Abbreviated ptm. Modulation (e.g., pulse-duration and pulse-position) in which the values of instantaneous samples of the modulating wave are made to modulate the occurrence time of some characteristic of a pulse carrier.

pulse timer – A timer the time cycle of which is started with a continuous input voltage and application or removal of an additional positive input voltage pulse. At the end of the time-delay period, the contacts transfer.

pulse train–Also called pulse group or impulse train. A group or sequence of pulses of similar characteristics.

pulse-train frequency spectrum–*See* Pulse-Train Spectrum.

pulse-train spectrum – Also called pulse-train frequency spectrum. The frequency distribution, in amplitude and in phase angle, of the sinusoidal components of the pulse train.

pulse transformer–A special type of transformer designed to pass pulse waveforms as distinguished from sine waves. The major features of a pulse transformer are high-voltage insulation between windings and to ground, low capacitance between windings, and low reactance in the windings.

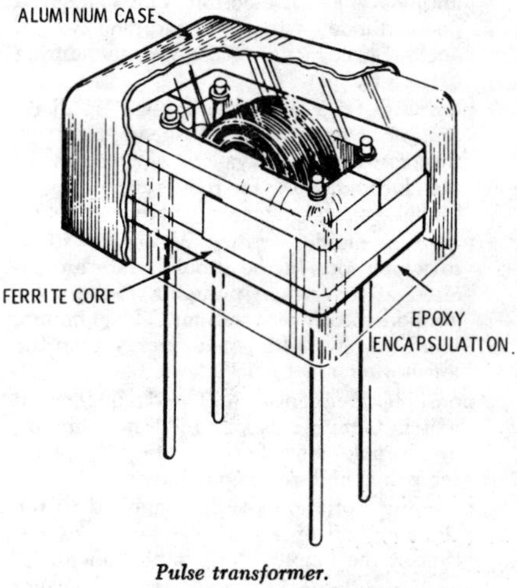

Pulse transformer.

pulse transmitter – 1. A pulse-modulated transmitter in which the peak power-output capabilities are usually larger than the average power-output rating. 2. A transmitter used to generate and transmit pulses over a telemetering or pilot-wire circuit to the remote indicating or receiving device.

pulse-triggered binary–A flip-flop in which a change of state results from application of a pulse or wavefrom of short duration to the input.

pulse valley–In a pulse, the portion between two specified maxima.

pulse width–*See* Pulse Duration.

pulse-width discriminator – A device that measures the pulse length of video signals and passes only those the time duration of which falls into some predetermined design tolerance.

pulse-width modulation – *See* Pulse-Duration Modulation.

pulse-width modulation/frequency modulation–*See* Pulse-Duration Modulation/Frequency Modulation.

pulsing key – 1. A method of transmitting voice-frequency pulses over a line under control of a key at the original office. It is used with E-and-M supervision in intertoll dialing. 2. A system of signaling in which numbered keys are used instead of a dial.

pulsing transformer – A transformer designed to supply pulses of voltage or current.

pump – 1. An external source used to increase the electron population of excited energy states. 2. Of a parametric device, the source of alternating-current power which causes the nonlinear reactor to behave as a time-varying reactance. 3. To supply high-frequency energy to a maser, laser, parametric amplifier, etc.

pumped tube–An electron tube (chiefly a pool-cathode) which is continuously connected to evacuating equipment during operation.

pumping–Of a laser, the application of radiation of appropriate frequency to invert the distribution of systems of electrons of the laser media so that the levels of higher energy states are more populated.

pumping band–A group of energy states to which ions in the ground state are excited at first when pumping radiation is applied to a laser medium. The pumping band usually is higher in energy than the levels that are to be inverted.

pumping frequency – The frequency at which pumping is provided in a maser, quadrupole amplifier, or other amplifier requiring high-frequency excitation.

pumping radiation – Light applied to the sides or end of a laser crystal for excitation of the ions to the pumping band.

pump oscillator – An alternating-current generator that supplies pumping energy for maser and parametric amplifiers. Operates at twice or some higher multiple of the signal frequency.

punch-card machine–*See* Key Punch.

punched card–A heavy, stiff paper of constant size and shape, suitable for punching in a pattern that has meaning, and for being handled mechanically. The punched holes are sensed electrically by wire brushes, mechanically by metal fingers, or photoelectrically by photocells. The standard card measures 3¼″ × 7⅜″ and contains eighty columns and twelve rows in which information may be punched.

punched tape–Also called tape, perforated tape, or punched paper tape. Paper tape punched in a coded pattern of holes, which convey information.

punched-tape recorder–A recorder that records data in the form of holes punched in tape strip.

punch-through voltage–1. That voltage at which two adjacent diffused transistor beds become shorted together, causing a sharp rise in current. 2. The value of the collector-base voltage of a transistor, above which the open-circuit emitter-base voltage increases almost linearly with increasing collector-base voltage. (Reach-through voltage is a term also used in the U.S.A.) 3. A form of transistor failure in which an internal short develops between emitter and collector across the base, usually as a result of excessive voltages.

puncture–A disruptive discharge of current through insulation, which breaks down under electrostatic stress and permits the flow of a sudden, large current through the opening. (*See also* Breakdown.)

puncture voltage – The voltage at which insulation fails by disruptive discharge through the insulation sample. It is assumed that the sample area is large enough to prevent flashover.

Pupin coil – An iron-core loading coil inserted into telephone lines at regular intervals to balance out the effect of capacitance between the lines.

pup jack–*See* Tip Jack.

pure code–Code that is never modified in the process of execution. Hence it is possible to let many users share the same copy of a program.

pure tone–*See* Simple Tone, 2.

purity–Physically complete saturation of a hue–i.e., uncontaminated by white and other colors. (*See also* Excitation Purity.)

purity coil – A coil consisting of two current-carrying windings. In a color television receiver, they produce a magnetic field which directs the three electron beams so that each one will strike only the proper set of phosphor dots.

purity control–A variable resistor that controls the current through the purity coil mounted around the neck of a color picture tube.

purity magnet – Two adjustable magnetic rings used in place of a purity coil.

purple boundary–The straight line drawn between the ends of the spectrum locus on a chromaticity diagram.

purple plague–A compound that forms as a result of intimate contact between gold and aluminum, and appears on silicon planar devices and integrated circuits in which gold leads are bonded to aluminum thin-film contacts and interconnections. It causes serious degradation of the reliability of semiconductor devices.

pushback hookup wire – Tinned copper

wire covered with loosely braided insulation, which can be pushed back with the fingers to expose enough bare wire for making a connection.

push button—Device mounted on a plunger (or actuator) which interfaces the operator's fingertip with the internal mechanism of the switch.

push-button control—Control of equipment by means of push buttons, which in turn operate relays, etc.

push-button dialing pad—A twelve-key device used for originating tone keying signals. Usually, it is attached to a rotary-dial telephone for origination ot data signals.

push-button switch — A switch in which a button must be depressed each time the contacts are to be opened or closed.

push-button tuner—A series of interlocked push-button switches which connect into a circuit, the correct tuning elements for the frequency corresponding to the depressed button.

pushdown dialing—The use of keys or push buttons instead of a rotary dial to generate a signal, usually in the form of tones, representing a sequence of digits and used to set up a circuit connection. Also called tone dialing and touch call.

pushdown list—A list that is made up and maintained in such a way that the next item to be retrieved is the item most recently stored (last in, first out).

pushdown stack—1. A circuit that operates in the reverse of a shift register. Whereas a shift register is a first-in first-out (FIFO) circuit, push-down stacks are last-in first-out (LIFO) memories. When data is requested, the stack will read the last data stored, and all other data will move one step closer to the output. Unless a memory is emptied, the first data in will never be retrieved. 2. A register that receives information from the Program Counter and stores the address locations of the instructions which have been pushed down during an interrupt. This stack can be used for subroutining. Its size determines the level of subroutine nesting (one less than its size or 15 levels of subroutine nesting in a 16-word register). When instructions are returned they are popped back on a last-in first-out (LIFO) basis. Also called P-stack.

pushing figure — The change in oscillator frequency due to a specified change in plate current (excluding thermal effects).

push-pull amplifier — *See* Balanced Amplifier.

push-pull circuit—A circuit containing two like elements which operate in 180° phase relationship to produce additive output components of the desired wave and cancellation of certain unwanted products. Push-pull amplifiers and oscillators use such a circuit.

push-pull configuration—A fundamental oscillator design in which each half of the circuit operates during a portion of the rf cycle. Primary advantages are increased power output over a single or parallel pair of transistors or tubes and reduction of second harmonic content in the output.

push-push configuration—A harmonic-oscillator design in which the signals from each output transistor or tube operating at f_o are combined to produce an output signal at 2 f_o. The main advantage of this configuration is the extension of transistor operating-frequency limits without the use of an extra frequency-doubler circuit.

push-pull currents—Balanced currents.

push-pull doubler — An amplifier used for frequency doubling. It consists of two transistors or vacuum tubes; the latter have their grids (input) connected in push-pull and their plates (output) in push-push or parallel.

push-pull microphone—A microphone comprising two like elements actuated by the same sound waves and operated 180° out of phase.

push-pull oscillator—A balanced oscillator employing two similar tubes or transistors in phase opposition.

push-pull transformer—An audio-frequency transformer that has a center-tapped winding and is used in a push-pull amplifier circuit.

push-pull voltages—Balanced voltages.

push-push circuit—A circuit usually used as a frequency multiplier to emphasize even-order harmonics. Two similar transistors are employed, or two tubes with their grids connected in phase opposition and their plates in parallel to a common load.

push-push currents — Currents which are equal in magnitude and which flow in the same direction at every point in the two conductors of a balanced line.

push-push voltages — Voltages which are equal in magnitude and have the same polarity (relative to ground) at every point on the two conductors of a balanced line.

push rod—A shaft that connects a servo or other actuator with a part of the controlled device.

push-to-talk switch — *See* Press-to-Talk Switch.

pushup list — A list that is made up and maintained in such a way that the next item to be retrieved and removed is the oldest item remaining in the list (first in, first out).

put—To insert a single data record into an output file.

PUT—Abbreviation for Programmable Unijunction Transistor.

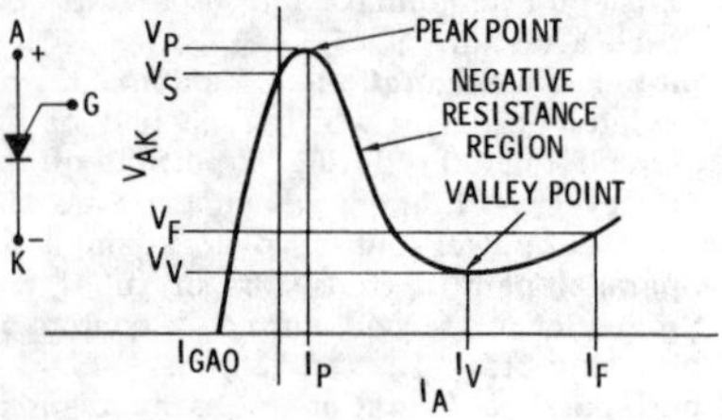

PUT.

PVC—Abbreviation for polyvinyl chloride.

pW—Abbreviation for picowatt.

pwm/fm—A system in which a number of pulse-width—modulated subcarriers are used to frequency-modulate the carrier.

pyrotechnic – Pertaining to explosive-actuated devices especially those that burn rather than producing a shattering effect.

pylon antenna – A vertical antenna constructed of one or more sheet-metal cylinders with a lengthwise slot. The gain depends on the number of sections.

pyramidal horn—An electromagnetic horn, the sides of which form a pyramid. The electromagnetic field in such a horn would be expressed basically in a family of spherical coordinates.

pyramid wave – A triangular wave, the sides of which are approximately equal in length.

pyrheliometer—A device for the measurement of infrared radiation.

pyroelectric effect—Also called pyroelectricity. The redistribution of the charge in a crystal that has been heated. The crystal is left with a net electric dipole moment—i.e., the centers of the positive and negative charges are separated.

Pylon antenna.

pyroelectricity—*See* Pyroelectric Effect.

pyroelectric material – A material which produces an electrical output when subjected to a change in temperature.

pyromagnetic—Pertaining to the effect of heat and magnetism on each other.

pyrometer—An instrument used to measure elevated temperatures (beyond the range of mercury thermometers) by electric means. These include immersion, optical, radiation, resistance, and thermoelectric pyrometers.

pyrone detector – A crystal detector in which rectification occurs between iron pyrites and copper (or other metallic points).

Pythagorean scale – A musical scale in which the frequency intervals are represented by the ratios of integral powers of 2 and 3.

Q

Q—Symbol for quantity of electric charge.

Q—1. A measure of the relationship between stored energy and rate of dissipation in certain electric elements, structures, or materials. In an inductor or capacitor, the ratio of its reactance to its effective series resistance at a given frequency. Also called quality factor or *Q* factor. 2. A measure of the sharpness of resonance or frequency selectivity of a mechanical or electrical system.

Q-8—Formerly known as Quad-8, term applied to tape cartridges when they contain four-channel programming.

Q antenna—A dipole matched to its transmission line by stub matching.

Q band—A band of frequencies extending from 36 to 46 GHz, corresponding to wavelengths of 0.834 to 0.652 cm.

Q (Q-bar) output – The second output of a flip-flop; its logic level is always opposite to that of the Q output.

Q channel—The 0.5-MHz-wide band used in the American NTSC color-television system to transmit green-magenta color information.

QCW—A 3.58-MHz continuous-wave signal having Q phase. The term is generally limited to reference to the color-television receiver local oscillator and associated circuits.

QCW signal – *See* Quadrature-Phase Subcarrier Signal.

Q demodulator—A demodulator circuit the inputs of which are the chrominance signal and the signal from the local 3.58-MHz oscillator after it has been shifted 90°. This phase shift is necessary so that the local signal will be an accurate representation of the Q subcarrier that was suppressed at the transmitter. The output of the Q demodulator is a color-video signal representing colors in the televised scene.

***Q* factor**—Of a tuned circuit the ratio of the inductive reactance of the circuit at the resonant frequency to its radio-frequency resistance. It is a measure of the increase in voltage which is developed across the turned circuit at resonant frequency, and so the term "magnification factor" is sometimes used for "Q." If the "Q" factor of a tuned circuit is high, the voltages developed across it are high and its selectivity is good. 2. *See* *Q*, 1.

***Q*-meter**—Also called a quality-factor meter. An instrument for measuring the *Q*, or quality factor, of a circuit or circuit element.

***Q*-multiplier**—A special filter which has a sharply peaked response curve or a deep rejection notch at a particular frequency.

Q output—The reference output of a flip-flop. That is, the flip-flop is said to be in the 1 state when this output is 1, and it is said to be in the 0 state when this output is 0.

Q-phase—Also called quadrature carrier. A color-television signal carrier having a phase difference of 147° from the color subcarrier.

QRM—An obsolete term for any type of man-made interference.

QRS complex — That portion of the waveform in an electrocardiogram extending from point Q to point S; it includes the maximum amplitude shown in the trace.

QS—A matrix system developed by Sansui Electronics.

Q-signal—1. In color television, the signal formed by the combination of R—Y and B—Y color-difference signals having positive polarities of 0.48 and 0.41, respectively. It is one of the two signals used to modulate the chrominance subcarrier, the other being the I signal. (*See also* Coarse-Chrominance Primary.) 2. One of a special group of abbreviations used in radiocommunications.

QSL card—A card exchanged by radio amateurs to confirm radiocommunications with each other.

Q-switch—A device used to rapidly change the Q of an optical resonator. Used in the optical resonator of a laser to prevent lasing action until a high level of inversion (optical gain and energy storage) is achieved in the laser rod. A giant pulse is generated when the switch rapidly increases the Q of the cavity.

Q-switched pulse—The output of a laser when the Q of the cavity resonator initially is kept very low so that the population inversion achieved is much larger than that which normally characterizes laser operation. Upon restoration of the Q to its normal high value, a high-power, short-duration pulse of coherent radiation (called a giant pulse) is emitted. Used most often in conjunction with pulsed pump radiation.

quad—1. A structural unit employed in cables. A quad consists of four separately insulated conductors twisted together. These conductors may take the form of two twisted pairs. 2. A combination of four elements, either electronic components or complete circuits, in a series-parallel or parallel-series arrangement. 3. A (series-parallel) combination of four transistors.

Quad-8—*See* Q8.

quadded cable—A cable in which some or all of the conductors are in the form of quads.

quadding—Connecting transistors in a series-parallel configuration, to achieve greater reliability.

quad latch—A group of four flip-flops, each of which has the capability of storing a true or false logic level, and all of which normally are enabled by a single-control line. When the flip-flops are all enabled, new information may be stored in each of them.

quadradisc—Another name for CD-4 disc.

quadrant—1. A sector, arc, or angle of 90°. 2. An instrument for measuring or setting vertical angles.

quadrantal error—The error caused in magnetic-compass readings by the magnetic field of the steel hull of a ship, or by metal structures near the loop antenna of radio direction finders aboard a vessel or aircraft.

quadrant electrometer — An electrometer for measuring voltages and charges by means of electrostatic forces. A metal plate or needle is suspended horizontally inside a vertical metal cylinder that is divided into four insulated parts, each connected electrically to the one opposite it. The two parts of quadrants are connected to the two terminals between which the potential difference is to be measured. The resultant electrostatic forces displace the suspended indicator a certain amount, depending on the voltage.

quadraphonic—*See* Quadriphonic.

quadraphony — A scheme of extended "stereo" whereby ambient and dimensional information is fed directly or via a matrix to a set of four speaker systems suitably orientated in the listening room. Various modulation or matrix systems are sometimes used so that four channels can be obtained by using some two-channel (stereo) equipment. The signals are then "decoded" so that four channels

Quadraphony.

of sound can be reproduced through four speakers.

quadrasonic—*See* Quadriphonic.

quadrature—The state or condition of two related periodic functions or two related points separated by a quarter of a cycle, or 90 electrical degrees.

quadrature amplifier—A stage used to supply two signals of the same frequency but with phase angles that differ by 90 electrical degrees.

quadrature carrier—*See* Q-phase.

quadrature component—1. The reactive current or voltage component due to inductive or capacitive reactance in a circuit. 2. A vector representing an alternating quantity which is in quadrature (at 90°) with some reference vector.

quadrature modulation—The modulation of two carrier components 90° apart in phase by separate modulating functions.

quadrature-phase subcarrier signal – That portion of the chrominance signal which leads or lags the in-phase portion by 90°. Abbreviated qcw signal.

quadrature portion – In the chrominance signal, the portion with the same or opposite phase from that of the subcarrier modulated by the Q-signal. This portion of the chrominance signal may lead or lag the in-phase portion by 90 electrical degrees.

quadrature sensitivity – Also called side sensitivity, lateral sensitivity, or cross-talk sensitivity. The sensitivity of a transducer to motion normal to the principal axis. Commonly expressed in percent of the sensitivity in the principal axis.

quadriphonic—A term used to describe 4-channel sound systems and equipment. Also spelled quadriphonic, quadrasonic, etc., sometimes contracted to quad. Sounds recorded and reproduced from four different directions, to produce a field of sounds coming from an apparent 360 degrees around the listener. Generally, any system of sound reproduction using more than the two usual stereo signals to recreate an impression of sounds coming from the rear of the listener as well as from the front.

quadripole network – *See* Two-Terminal-Pair Network.

quadruple-diversity system – A receiving system in which space-diversity and frequency-diversity techniques are employed simultaneously.

quadruple play—Magnetic recording tape which is thinner than "standard play" tape and consequently makes possible recordings four times longer than with standard play tape.

quadruplex circuit—A telegraph circuit designed for carrying two messages in each direction simultaneously.

quadrupole—A combination of two dipoles that produces a force varying in inverse proportion to the fourth power of the distance from the generating charge.

quadrupole network – *See* Two-Terminal-Pair Network.

qualification – The entire procedure by which electronic parts are examined and tested to obtain and maintain approval at specified failure rate levels, and then identified on the qualified products lists.

qualifying activity – The military activity or its agent delegated to administer the qualification program.

quality—The extent of conformance to specifications of a device or the proportion of satisfactory devices in a lot.

quality assurance – A planned, systematic pattern of actions necessary to provide suitable confidence that an item will perform satisfactorily in actual operation.

quality control—The control of variation of workmanship, processes, and materials in order to produce a consistent, uniform product.

quality engineering – An engineering program the purposes of which are to establish suitable quality tests and quality acceptance criteria and to interpret quality data.

quality factor—*See* Q, 1.

quality-factor meter—*See* Q-meter.

quantity—1. Any positive or negative number. It may be a whole number, a fraction, or a whole number and a fraction. 2. A constant, variable, function name, or expression. 3. In computers, a positive or negative real number; the term "quantity" is preferred to the term "number."

quantity of electricity – *See* Electrical Charge.

quantization – The process whereby the range of values of a wave is divided into a finite number of subranges, each represented by an assigned (quantized) value.

quantization distortion—Also called quantization noise. Inherent distortion introduced during quantization.

quantization error—The difference between the actual values of data and corresponding discrete values resulting from quantization.

quantization level — 1. A particular subrange in quantization. 2. The symbol designating the subrange of (1) above.

quantization noise—*See* Quantization Distortion.

quantize—To convert a continuous variable, such as a waveform, into a series of levels or steps. There are no "in-between" values in such a quantized waveform. All values of signal are represented by the nearest standard value or code position.

quantized pulse modulation—Pulse modulation which involves quantization (e.g., pulse numbers or pulse code modulation).

quantitizer — A device which partitions a continuum of analog values into discrete ranges to be represented by a digital code. An analog-to-digital converter.

quantizing—Expressing an analog value as the nearest one of a discrete set of prechosen values.

quantitizing error — The basic uncertainty associated with digitizing an analog signal, due to the finite resolution of an a/d converter. An ideal converter has a maximum quantizing error of $\pm\frac{1}{2}$ lsb.

quantum—1. A discrete portion of energy of a definite amount. It was first associated with intra-atomic or intermolecular processes involving changes among the electrons, and the corresponding radiation. 2. If the magnitude of a quantity is always an integral multiple of a definite unit, then that unit is called the quantum of the quantity. 3. The angular increment of input-shaft rotation of an encoder, subtended by one code position. 4. The unit carrier of energy: the photon for light, and the electron for electricity.

quantum efficiency—1. In a phototube, the average number of electrons photoelectrically emitted from the photocathode per incident photon of a given wavelength. 2. The fraction of those ions or atoms excited to a higher energy level by pumping radiation that decays with light emission in a particular desired range of frequencies. The remaining ions decay by various radiationless mechanisms (such as phonon emission) or by undesired radiative transitions. 3. The ratio of the number of carriers generated to the number of photons incident upon the active region.

quantum mechanics — The study of atomic structure and other related problems in terms of quantities that can actually be measured.

quantum noise — A random variation or noise signal due to fluctuations in the average rate of incidence of quanta on a detector. The basic electromagnetic quantum of noise power is just 1 photon per electromagnetic mode.

quantum number—Any of a set of numbers assigned to the particular values of a quantized quantity in its discrete range. The state of a particle or system may be described by a set of compatible quantum numbers.

quantum theory—The theory that an atom or molecule does not emit or absorb energy continuously. Rather, it does so in a series of steps, each step being the emission or absorption of an amount of energy called the quantum. The energy in each quantum is directly proportionate to the frequency.

quantum transition — A transition between two quantum states.

quarter phase—*See* Two-Phase.

quarter-squares multiplier—An analog multiplier unit that makes use of the relationship $xy = \frac{1}{4}[(x+y)^2 - (x-y)^2]$.

quarter track—*See* Four-Track Tape.

quarter-track tape—*See* Four-Track Tape.

quarter wave—One-quarter cycle of a wave.

quarter-wave antenna — An antenna, the electrical length of which is one-quarter the wavelength of the transmitted or received signal.

quarter-wave attenuator — Two energy-absorbing grids or other structures placed in a transmission line and separated by an odd number of quarter wavelengths. As a result, the wave reflected from the first grid annuls the wave reflected from the second grid.

quarter-wavelength — That distance which corresponds to an electrical length of a quarter of a wavelength of the frequency under consideration.

quarter-wave line—*See* Quarter-Wave Stub.

quarter-wave plate—A mica or other double-refracting crystal plate of such thickness that a phase difference of one-quarter cycle is introduced between the ordinary and extraordinary components of light passing through.

quarter-wave resonance—In a quarter-wave antenna, the condition in which its resonant frequency is equal to the frequency at which it is to be used.

quarter-wave stub — Also called a quarter-wave line or quarter-wave transmission line. A section of transmission line equal to one-quarter of a wavelength at the fundamental frequency. It is commonly used to suppress even harmonics. This is done by shorting the far end so that the open end presents a high impedance to the fundamental frequency and all odd harmonics, but not to the even-order harmonics.

quarter-wave support—A quarter-wave metallic stub, used in place of dielectric insulators between the inner and outer conductors of a coaxial transmission line.

quarter-wave termination — A waveguide termination consisting of a metal plate and a wire grating (or semiconducting film) spaced one-quarter wavelength

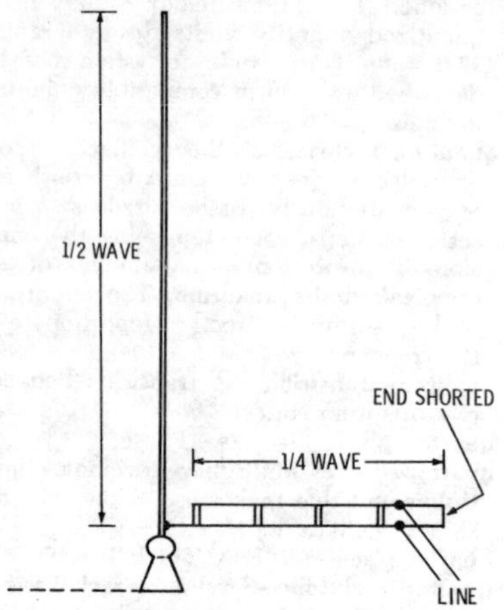

Quarter-wave stub.

apart. The plate is the terminating element. The wave reflected by the grating (or film) is cancelled by the wave reflected by the plate.

quarter-wave transformer – A one-quarter-wavelength section of transmission line used for impedance matching.

quarter-wave transmission line–*See* Quarter-Wave Stub.

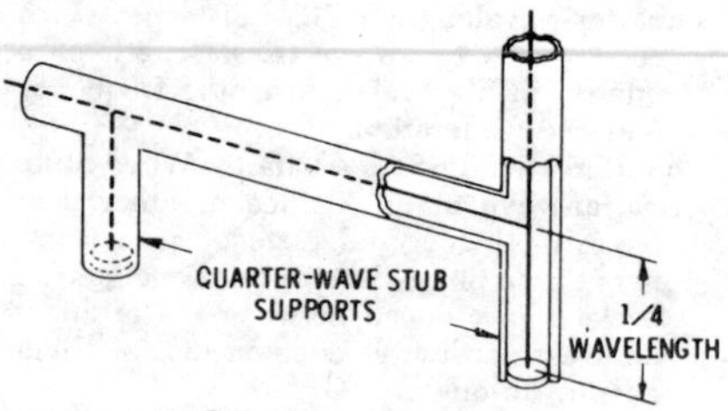

Quarter-wave supports.

quartz–A mineral (silicon dioxide) occurring in hexagonal crystals in nature and having piezoelectric properties that are highly useful in radio and carrier communication. The crystals from which slabs are cut for oscillators are transparent and almost colorless. When excited electrically, they vibrate and maintain extremely accurate and stable frequencies.

quartz crystal–Also called a crystal. A thin slab cut from quartz and ground to the thickness at which it will vibrate at the desired frequency when supplied with energy. It is used to accurately control the frequency of an oscillator.

quartz delay line – A delay line in which fused quartz is the medium for delaying sound transmission or a train of waves.

quartz lamp–A mercury-vapor lamp having a transparent envelope made from quartz instead of glass. Quartz resists heat (permitting a higher current) and passes ultraviolet rays, which glass will absorb.

quartz plate – A crystalline-quartz section completely finished to specifications, with its two major faces essentially parallel.

quartz resonator–A piezoelectric resonator with a quartz plate.

quasi-linear feedback-control system – A feedback-control system in which the relationships between the pertinent measures of the system input and output signals are substantially linear despite the existence of nonlinear elements.

quasi-optical–Having properties similar to those of light waves. The propagation of waves in the television spectrum is said to be quasi-optical (i.e., cut off by the horizon).

quasi-random code generator–A high-speed pcm information source that provides a means of closed-loop testing for use in designing and evaluating wideband communications links.

quasi-rectangular wave – A wave nearly, but not, rectangular in shape.

quasi-single–sideband – Simulated single-sideband transmission done by transmitting parts of both sidebands.

quasi-steady–state vibration–A nearly periodic vibration in which the amplitude and phase relationships of the component sinusoids vary slowly with time.

quaternary signaling–The communications mode in which information is passed by the presence and absence, or plus and minus variations, of four discrete levels of one parameter of the signaling medium.

quenched spark–A spark consisting of only a few sharply defined oscillations, because the gap is deionized almost immediately after the initial spark has passed.

quenched spark gap–A spark gap with provision for producing a quenched spark. One form consists of many small gaps between electrodes that have a relatively large mass and thus are good radiators of heat. As a result, they cool the gaps rapidly and thereby stop conduction.

quenched spark-gap converter–A spark-gap generator or other power source in which the oscillatory discharge of a capacitor through an inductor and a spark gap provides the radio-frequency power. The spark gap comprises one or more closely spaced gaps in series.

quench frequency – 1. An ac voltage applied to an electrode of a tube used as a superregenerative detector to alternately vary its sensitivity and thereby prevent sustained oscillations. The quench frequency is usually lower than the signal frequency to be received. 2. The number of times per second a circuit goes in and out of oscillation.

quenching – 1. The terminating of a discharge in a radiation-counter tube by inhibiting the re-ignition. 2. A process of rapid cooling from an elevated temperature, in contact with liquids, gases, or solids.

quenching circuit—A circuit which inhibits multiple discharges from an ionizing event by suppressing or reversing the voltage applied to a counter tube.

quenching frequency – That frequency at which oscillations in a superregenerative receiver are suppressed (quenched).

quench oscillator—A superregenerative receiver circuit which produces the quench-frequency signal.

queue—1. A line of items waiting for service in a system, such as messages to be transmitted in a message-switching system. 2. To arrange in or form a queue.

queue control block – A control block for regulating the sequential use of a programmer-defined facility for a number of tasks.

queued access method – An access method that provides automatic synchronization of data transfer between the program using the access method and input/output devices. Delays for input/output operations are thereby eliminated.

queuing theory—A research technique having to do with the correct order of moving units, such as sequence assignments for bits of information, whole messages, assembly-line products, or automobiles in traffic.

quick-break—A characteristic of a switch or circuit breaker, whereby it has a fast contact-opening speed that is independent of the operator.

quick-break fuse—A fuse that draws out the arc and rapidly breaks the circuit when its wire melts. Usually a spring or weight is used to quickly separate the broken ends.

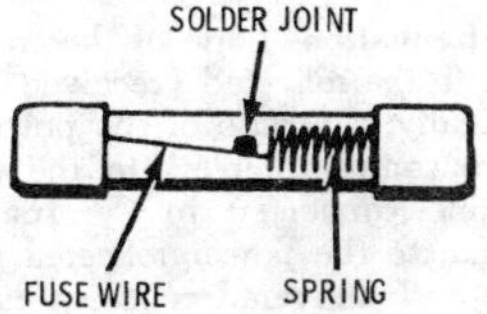

Quick-break fuse.

quick-break switch – A switch that minimizes arcing by breaking a circuit rapidly, independent of the rate at which the switch handle is moved.

quick-connect terminal—A plug-in type of terminal designed to make possible rapid wiring.

quick disconnect—A type of connector designed to facilitate rapid locking and unlocking of two contacts or connector halves.

quick-make—A characteristic of a switch or circuit breaker, whereby it has a fast contact-closing speed that is independent of the operator.

quick-make switch – A switch or circuit breaker which has a high contact-closing speed independent of the operator.

quick-stop control—On some tape recorders, and on all recorders used for dictation, a control with which the operator can stop the tape without taking the machine out of the play or record position.

quiescence—1. The state of a transistor amplifier with no signal applied. 2. The operating condition that exists in a circuit when no input signal is applied to the circuit.

quiescent – At rest—specifically, the condition of a circuit when no input signal is being applied to it.

quiescent-carrier modulation – A modulation system in which the carrier is suppressed during intervals when there is no modulation.

quiescent-carrier telephony – Telephony in which the carrier is suppressed whenever no modulating signals are to be transmitted.

quiescent current—*See* Idling Current.

quiescent dissipation—The power dissipated by a component or circuit in the absence of dynamic activating signals applied to the input or inputs.

quiescent input voltage—The dc voltage at the input of an amplifier that has one input terminal when that terminal is not connected to any source.

quiescent operating point – *See* Quiescent Point.

quiescent output voltage – The dc voltage at the output terminals of an amplifier when the input is grounded for ac through a resistance equal to the resistance of the signal source.

quiescent period—The resting period—e.g., the period between pulses in pulse transmissions.

quiescent point – On the characteristic curve of an amplifier, the point representing the conditions existing when there is no input signal. (*See also* Operating Point.)

quiescent push-pull—In a radio receiver, a push-pull output stage in which practically no current flows when no signal is being received. Thus, there is no noise while the radio is being tuned between stations.

quiescent state—The time during which a tube or other circuit element is not performing its active function in the circuit.

quiescent value – The voltage or current value of a vacuum-tube electrode when no signals are present.

quiescing—The process of stopping a multiprogrammed system by rejection of new jobs.

quiet automatic volume control – *See* Delayed Automatic Volume Control.

quiet avc—*See* Delayed Automatic Volume Control.

quiet battery – Also called talking battery. A source of energy of special design, or with added filters, which is sufficiently quiet and free from interference that it may be used for speech transmission.

quieting—The decrease in noise voltage at the output of an fm receiver in the presence of an unmodulated carrier.

quieting sensitivity—In an fm receiver, the minimum input signal that will give a specified output signal-to-noise ratio.

quiet tuning—In a radio receiver, a form of tuning in which the output is silenced except when the receiver is tuned to the precise frequency of the incoming carrier wave.

R

R—Symbol for resistor, resistance, or reluctance.

r—Abbreviation for roentgen.

race—1. The condition that exists when a signal is propagated through two or more memory elements during the same clock period. 2. The condition that occurs when changing the state of a system requires a change in two or more state variables. If the final state is affected by which variable changes first the condition is a critical race. 3. An improper condition in which data which is supposed to move in steps, as in a shift register, goes through a whole string of stages at one step. Usually caused by incorrect timing pulses.

raceway—Any channel designed and used solely for holding wires, cables, or bus bars.

rack—1. A bar with teeth, which engage a pinion, worm, etc., to provide straight-line motion. 2. A vertical frame on which equipment, relays, etc., are mounted.

rack and panel connector—A connector that is attached to a panel or side of equipment so that when these members are brought together, the connector is engaged.

rack and pinion – A toothed bar (rack) which engages a gear (pinion) to convert the back-and-forth motion of the rack into rotary motion, or the rotary motion of the pinion into back-and-forth motion.

racon—*See* Radar Beacon.

rad—Radiation absorbed dose. The amount of radiation that delivers 100 ergs of energy to 1 gram of a substance. It is approximately equivalent to 1 roentgen in the case of body tissue.

radar – Acronym for radio detecting and ranging. A system that measures distance (and usually the direction) to an object by determining the amount of time required by electromagnetic energy to travel to and return from an object. Called primary radar when the signals are returned by reflection. Called secondary radar when the incident signal triggers a responder beacon and causes it to transmit a second signal.

radar altitude – Also called radio altitude. Absolute altitude measured by a radar altimeter.

radar and television aid to navigation – A device which converts a circular scan radar presentation to a horizontally scanned television presentation. It provides a continuous bright display with target trails for course and speed indications of moving targets.

radar antenna—Any of the many types of antennas used in radar.

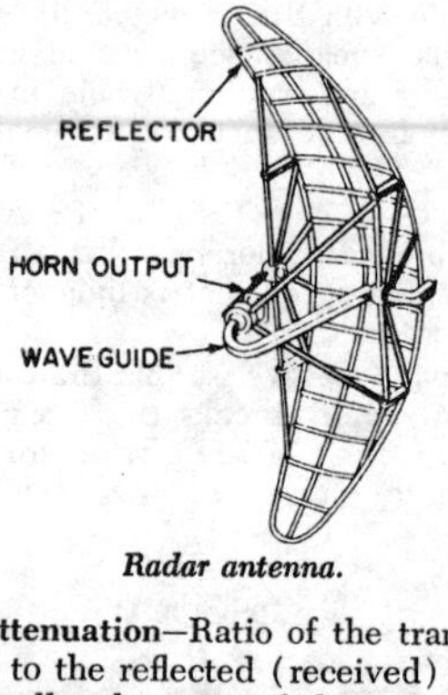

Radar antenna.

radar attenuation—Ratio of the transmitted power to the reflected (received) power—specifically, the ratio of the power which the transmitter delivers to the transmission line connected to the transmitting antenna, to the power reflected from the target and delivered to the transmission line connected to the receiving antenna.

radar beacon—Also called a racon. An automatic transmitter-receiver which receives signals from a radar transmitter, and retransmits coded signals that enable the radar operator to determine his position.

radar beam—The space where a target can be effectively detected and/or tracked in front of a radar. Its boundary is defined as the locus of points measured radially from the beam center at which the power has decreased to one-half.

radar calibration—Taking measurements on various parts of electronic equipment (e.g., radar, iff, communications) to determine its performance level.

radar camouflage—The use of coverings or surfaces on an object to considerably reduce the reflected radio energy and thus "conceal" the object from the radar beam.

radar cell—The volume enclosed by dimensions of one radar-pulse length by one radar beamwidth.

radar clutter—The image produced on a radar indicator screen by sea or ground return. If not of particular interest, it tends to obscure the target indication.

radar-confusion reflectors—Metallic devices (e.g., chaff, corner reflectors) employed to return false signals in order to confuse enemy radar receivers. The use of radar-confusion reflectors is termed reflective jamming.

radar control area — The designated space within which aircraft approach, holding, stacking, and similar operations are performed under guidance of a surveillance-radar system.

radar countermeasures — Interception, jamming, deception, and evasion of enemy radar signals to obtain information about the enemy from his radar and to prevent him from obtaining accurate, usable information from his radar.

radar-coverage indicator—A device showing how far a radar station should track a given aircraft. It also provides a reference (detection) range for quality control. Aircraft size and altitude, screening angle, site elevation, type of radar, antenna radiation pattern, and antenna tilt are all taken into account.

radar cross section — That portion of the back-scattering cross section of a target associated with a specified polarization component of the scattered wave.

radar data filtering—A quality-analysis process in which the computer rejects certain radar data and alerts personnel at the mapping and surveillance consoles to the rejection.

radar deception — Radiation or reradiation of radar signals in order to confuse or mislead an enemy operator when he interprets the data shown on his scope.

radar decoy—An object that has the same reflective characteristics as a target and is used in radar deception.

radar distribution switchboard — A switching panel for connecting video, trigger, and bearing from any 1 of 5 systems, to any or all of 20 repeaters. Also contains order lights, bearing cutouts, alarms, test equipment, etc.

radar dome—A weatherproof cover for protection of a primary radiating element of a radar or radio device; the cover is transparent to radio-frequency energy and permits operation of the radiating element, including rotation or other physical movement. *See also* Radome.

radar echo—1. The radio-frequency energy received after it has been reflected from an object. 2. The deflection or change of intensity which a radar echo produces in the display of a cathode-ray tube.

radar equation—A mathematical expression relating the transmitted and received powers and antenna gains of a primary-radar system to the echo area and distance of the target.

radar fence—A network of radar warning stations which maintain constant watch against surprise attack (e.g., the DEW line).

radar field gradient—The rate at which the strength of the field from a primary radar decreases as the distance from the transmitting antenna increases. Abbreviated RFG.

radar homing—Missile guidance, the intelligence for which is provided by a radar aboard the missile.

radar horizon — The most distant point (from the radar antenna along a given azimuth) on the earth's surface illuminated by the radar on purely geometric conditions. The conditions are that the illumination occurs along a straight-line path, where the path is taken over an effective earth's radius of 4/3 its true radius and where the illuminating power of the radar is considered unlimited.

radar illumination—The subjection of an object (target) to electromagnetic radiation from a radar.

radar indicator — A cathode-ray tube with its associated equipment that provides a visual indication of the echo signals picked up by the radar set.

radar marker—A fixed facility which continuously emits a radar signal so that a bearing indication appears on a radar display.

radar nautical mile—The time interval of approximately 12.367 microseconds required for radio-frequency energy to travel one nautical mile and return, a total of two nautical miles.

radar paint—A radar-energy–absorbent material that can be applied to an object to reduce the possibility of detection.

radar-performance figure — Ratio of the pulse power of the radar transmitter to the power of the minimum signal detectable by the receiver.

radar picket—Early-warning aircraft which flies at a distance from a ship or other force beting protected, to increase the radar detection range.

radar pulse modulator — A modulator that turns an rf energy source off and on in a precise, known manner. In essence, the modulator supplies the energy which causes an rf source to oscillate or amplify, thus creating a burst or "pulse" of rf energy.

radar range—The maximum range at which a radar can ordinarily detect objects.

radar receiver—The receiver which amplifies the returned radar signal and demodulates the rf carrier before further amplifying the desired signal and delivering it—in a form suitable for presentation—to the indicator. Unlike a radio receiver, it is more sensitive, has a lower noise level, and is designed to pass a pulse signal.

radar-reflection interval — The length of time required for a radar pulse to reach a target and return.

radar reflectivity—A measure of the ability of a radar target to intercept and return a radar signal.

radar relay—Equipment for relaying radar video and appropriate synchronizing signals to a remote location.

radar repeaters—Remote indicators used to reproduce radar data from a primary source.

radar resolution—The ability of a radar to distinguish between the desired target and its surroundings.

radarscope—A cathode-ray tube serving as an oscilloscope, the face of which is the radar viewing screen.

radar selector switch—A manual or motor-driven switch which transfers a plan-position-indicator repeater from one system to another, switching video, trigger, and bearing data.

radar shadow—An area shielded from radar illumination by an intervening reflecting or absorbing medium. This region appears as an area that is void of targets on a radar display.

radar silence—1. A period of time during which radar operations are stopped. 2. An imposed discipline under which the transmission by radar of electromagnetic signals on some or all frequencies is prohibited.

radar target—Any reflecting object of particular interest in the path of the radar beam (usually, but not necessarily, the object being tracked).

radar trace—The pattern produced on the screen of the cathode-ray tube in a radar unit.

radar transmitter—The transmitter portion of a radar system. The unit of the radar system in which the rf power is generated and keyed.

radiac—Acronym for radioactive detection, identification, and computation, a descriptive term referring to the detection, identification and measurement of nuclear radiation.

radiac instrument—*See* Radiac Set.

radiac meter—*See* Radiac Set.

radiac set — Also called radiac meter or radiac instrument. Equipment for detecting, identifying, and measuring the intensity of nuclear radiations.

radiac test equipment—Equipment for testing radiac sets.

radial—Pertaining to or placed like a radius (i.e., extending or moving outward from a central point, like the spokes of a wagon wheel).

radial-beam tube—A vacuum tube producing a flat, radial electron beam which can be rotated about the axis of the tube by an external magnetic field.

radial component—A component that acts along (parallel to) a radius—as contrasted to a tangential component, which acts at right angles (perpendicular) to a radius.

radial component of the electric field—The component of an electric field in the direction of the slant-range vector or the radius vector at which an antenna pattern is measured. The radial component of the electric field is relatively small in the far field and is normally neglected.

radial field—A field of force that is directed toward or away from a point in space.

radial field cathode-ray tube — A crt in which a fine-grained, curved, high-transmission metallic mesh is placed on the exit side of the deflection area. The mesh establishes a ground plane for the post-accelerating field so that the resulting equipotential surfaces are truly spherical, creating a radial electrostatic field.

radial grating—A conformal grating consisting of wires arranged radially in a circular frame, like the spokes of a wagon wheel. The radial grating is placed inside a circular waveguide to obstruct E waves of zero order, but not the corresponding H waves.

radial lead—A lead extending out the side of a component, rather than from the end. (The latter is called an axial lead.) Some resistors have radial leads.

radial transmission line—A pair of parallel conducting planes used for propagating uniform cylindrical waves whose axes are normal to the planes.

radian — In a circle, the angle included within an arc equal to the radius of the circle. Numerically it is equal to 57°, 17′, 44.8″. A complete circle contains 2π radians. One radian equals 57.3 degrees, and 1 degree equals 0.01745 radian.

radiance — 1. The apparent radiation of a surface. It is the same as luminance except radiance applies to all kinds of radiation instead of only light flux. 2. The radiant intensity per unit solid angle per unit of projected area of an extended source.

radian frequency—*See* Angular Velocity.

radian length—The distance, in a sinusoidal wave, between phases differing by an angle of one radian. It is equal to the wavelength divided by 2π.

radian per second—A unit of angular velocity.

radiant energy—Energy transmitted in the form of electromagnetic radiation (e.g., radio, heat, or light waves). It is measured in units of energy such as kilowatt-hours, ergs, joules, or calories.

radiant-energy detecting device—A device employing radiant energy to detect flaws in the surface and/or volume of solids.

radiant flux—Time rate of flow of radiant energy, expressed in watts or in ergs per second.

radiant heat—Infrared radiation from a body not hot enough to emit visible radiation.

radiant heater—An electric heating appliance with an exposed incandescent heating element.

radiant intensity — The energy emitted within a certain length of time, per unit solid angle about the direction considered.

radiant power—The rate of transfer of radiant energy.

radiant reflectance—The ratio of reflected radiant power to incident radiant power.

radiant sensitivity—The output current of a phototube or camera tube divided by the incident radiant flux of a given wavelength at constant electrode voltages. The term "output current" as here used does not include the dark current.

radiant transmittance — The ratio of transmitted radiant power to incident radiant power.

radiate—To emit rays from a center source —e.g., electromagnetic waves emanating from an antenna.

radiated—Energy transfer by propagation of electromagnetic fields.

radiated interference—1. Interference which is transmitted through the atmosphere according to the laws of electromagnetic wave propagation. (The term "radiated interference" is generally considered to include the transfer of interfering energy by inductive or capacitive coupling.) 2. Any unwanted electrical signal which is radiated from the equipment under test, or from any lines connected to that equipment under test.

radiated power—The total energy, in the form of Hertzian waves, radiated from an antenna.

radiating curtain—An array of dipoles in a vertical plane, positioned to reinforce each other. Usually placed one-quarter wavelength ahead of a reflecting curtain of corresponding half-wave reflecting antennas.

radiating element—Also called radiator. A basic subdivision of an antenna. It, by itself, is capable of radiating or receiving radio-frequency energy.

radiating guide—A waveguide designed to radiate energy into free space. The waves may emerge through slots or gaps in the guide, or through horns inserted into its wall.

radiation—1. The propagation of energy through space or through a material. It may be in the form of electromagnetic waves or corpuscular emissions. The former is usually classified according to frequency—e.g., Hertzian, infrared (visible), light, ultraviolet, X-rays, gamma rays, etc. Corpuscular emissions are classified as alpha, beta, or cosmic. 2. A general term for energy emitted from a substance and traveling across space in straight lines. (Originally used only for electromagnetic waves, it now includes streams of particles, such as alpha particles, beta particles, etc.) 3. The transfer of heat from a hot body to a cooler body, through space, without heating any medium that may be in the space.

radiation angle — The vertical angle at which maximum energy is radiated by an antenna with respect to the earth. Low-angle radiation delivers more energy in the local area. High angle radiation augments skip transmission and is wasteful of power.

radiation belt — *See* Van Allen Radiation Belts.

radiation characteristic — An identifying feature, such as frequency or pulse width, of a radiated signal.

radiation counter — A device for counting radiation particles (alpha, beta, gamma, neutrons, etc.) or photons of energy (X-rays, etc.) usually using either scintillation or ionization resulting from the presence of the particle or photon to be measured.

radiation counter tube—*See* Counter Tube.

radiation damage—A loss in certain physical properties of organic substances, such as elastomers, caused principally by ionization of the long chain molecule. This ionization process (i.e., loss of electrons) is believed to result in redundant cross-linking and possible scission of the molecule. The effect is cumulative.

radiation-detector tube—A tube in which current passes between its electrodes whenever the tube is exposed to penetrating radiation. The amount of this current corresponds to the intensity of radiation.

radiation dosage—The total radiation energy absorbed by a substance, usually expressed in ergs per gram. *See also* Rad, REP, Roentgen, Roentgen Equivalent Man.

radiation efficiency—In an antenna, the ratio of the radiated power to the total power supplied to the antenna at a given frequency.

radiation field—The electromagnetic field that breaks away from a transmitting antenna and radiates outward into space as electromagnetic waves.

radiation filter—A transparent body which transmits only selected wavelengths.

radiation hardening—Manufacturing techniques applied to a device so that its performance is not degraded significantly by exposure to high gamma and neutron radiation environments. Examples are the use of dielectric isolation techniques and nichrome thin-film resistors.

radiation hazard – The health hazard caused by exposure to ionizing radiation.

radiation intensity—In a given direction, the power radiated from an antenna per unit solid angle in that direction.

radiation lobe—*See* Lobe.

radiation loss – In a transmission system, the portion of the transmission loss due to radiation of the radio-frequency power.

radiation monitor—A device for determining amount of exposure to radioactivity. May be periodic or continuous, may monitor an area or an individual's breath, clothing, etc.

radiation pattern—*See* Directional Pattern.

radiation potential—The voltage required to excite an atom or molecule and cause the emission of one of its characteristic radiation frequencies.

radiation pyrometer – Also called a radiation thermometer. A pyrometer which uses the radiant power from the object or source whose temperature is being measured. Within wide- or narrow-wavelength bands filling a definite solid angle, the radiant power impinges on a suitable detector—usually a thermocouple, thermopile, or a bolometer responsive to the heating effect of the radiant power, or a photosensitive device connected to a sensitive electric instrument.

radiation report—A formal report of radiation measurements made by an engineer skilled in interference control techniques. Usually required by the FCC prior to certification of industrial heating equipment.

radiation resistance—1. The power radiated by an antenna, divided by the square of the effective antenna current referred to a specified point. 2. The resistance which, if inserted in place of the antenna, would consume the same amount of power radiated by the antenna. 3. The characteristic of a material that enables it to retain useful properties during or after exposure to nuclear radiation.

radiation sickness – An illness resulting from exposure to radiation.

radiation survey meter—An instrument that measures instantaneous radiation.

radiation temperature—The temperature to which an ideal blackbody must be heated so it will have the same emissive power as a given source of thermal radiation.

radiation thermometer—*See* Radiation Pyrometer.

radiative equilibrium – The constant-temperature condition that exists in a material when the radiant energy absorbed and emitted are equal.

radiator—Any device which emits radiation. (*See also* Radiating Element.)

radio—1. Communication by electromagnetic waves transmitted through space. 2. A general term, principally an adjective, applied to the use of electromagnetic waves between 10 kHz and 3000 GHz. 3. Electronic equipment for the wireless transmission or reception, or both, of electromagnetic waves, especially when used to transmit and receive sounds, activate a remote-control mechanism, etc.; a radio set.

radioacoustic position finding—A method of determining distance through water. This is done by closing a circuit at the same instant a charge is exploded under water. The distance to the observing station can then be calculated from the difference in arrival times between the radio signal and the sound of the explosion.

radioacoustics—A study of the production, transmission, and reproduction of sounds

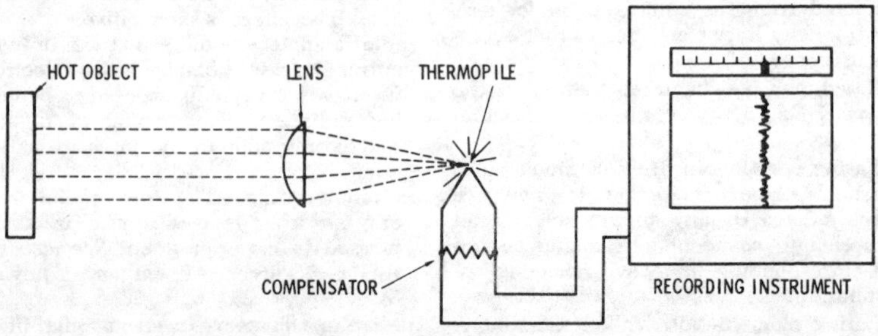

Radiation pyrometer.

carried from one place to another by radio-telephony.

radioactive – Pertaining to or exhibiting radioactivity.

radioactive isotope–*See* Radioisotope.

radioactive series–A succession of radioactive elements, each derived from the disintegration of the preceding element in the series. The final element, known as the end product, is not radioactive.

radioactivity–A property exhibited by certain elements, the atomic nuclei of which spontaneously disintegrate and gradually transmute the original element into stable isotopes of that element or into another element with different chemical properties. The process is accompanied by the emission of alpha particles, beta particles, gamma rays, positrons, or similar radiations.

radio altitude–*See* Radar Altitude.

radio approach aids – Equipment making use of radio to determine the position of an aircraft with considerable accuracy from the time it is in the vicinity of an airfield or carrier until it reaches a position from which a landing can be carried out.

radioastronomy–The branch of astronomy where the radio waves emitted by certain celestial bodies are used for obtaining data about them.

radio attenuation – For one-way propagation, the ratio of the power delivered by the transmitter to the transmission line connecting it with the transmitting antenna, to the power delivered to the receiver by the transmission line connecting it with the receiving antenna.

radio beacon–Also called a radiophone or, in air operations, an aerophare. A radio transmitter, usually nondirectional, which emits identifiable signals for direction finding.

radio-beacon station – In the radionavigation service, a station the emissions of which are intended to enable a mobile station to determine its bearing or direction in relation to the radio-beacon station.

radio beam – 1. A radio wave in which most of the energy is confined within a relatively small angle. 2. A low-frequency radio transmitter used in direction finding, for determining fixed and homing—a process of navigation whereby the pilot directs the aircraft toward the station to which he is tuned.

radio bearing–The angle between the apparent direction of a source of electromagnetic waves, and a reference direction determined at a radio direction-finding station. In a true radio bearing, this reference direction is true north. Likewise, in a magnetic radio bearing, it is magnetic north.

radiobiology–The study of the effects on living matter (or substances derived therefrom) of high energy radiation extending from x-rays to gamma rays, including high energy beams of neutrons and charged particles, e.g., alpha particles, electrons, protons, deuterons.

radio breakthrough – The breakthrough of modulated radio signals into the channels of an audio amplifier due to the presence of high-level radio signal fields. The effect is that the base/emitter junction of the low-level input transistor rectifies the signals picked up by the wiring or circuit components and the resulting audio is then handled by the amplifier in the ordinary way so that the radio program appears as a disconcerting background on the wanted source signal.

radio broadcast – A program of music, voice, and/or other sounds broadcast from a radio transmitter for reception by the general public.

radio broadcasting–*See* Radio Broadcast.

radio channel–A band of frequencies wide enough to be used for radiocommunication. The width of a channel depends on the type of transmission and on the tolerance for the frequency of emission.

radio circuit–A means for carrying out one radiocommunication at a time in either direction between two points.

radiocommunication – An overall term for transmission by radio of writing, signs, signals, pictures, and sounds of all kinds.

radiocommunication circuit – A radio system for carrying out one communication at a time in either direction between two points.

radiocommunication guard–A communication station designated to listen for and record transmission and to handle traffic on a designated frequency for a certain unit or units.

radio compass–*See* Direction Finder.

radio control–Remote control of apparatus by radio waves (e.g., model airplanes, boats).

radio deception – Sending false dispatches, using deceptive headings or enemy call signs, etc., by radio to deceive the enemy.

radio detection–Also called radio warning. Determining the presence of an object by radiolocation, but not its precise position.

radio detection and location – Use of an electronic system to detect, locate, and predict future positions of an earth satellite.

radio detection and ranging–1. Any of certain methods or systems of using beamed and reflected electromagnetic energy for detecting and locating objects; for measuring distance, velocity, or altitude; or for other purposes such as navigating, homing, bombing, missile tracking, mapping, etc. 2. In Federal Communications

Commission regulations, a radiodetermination system based on the comparison of reference signals with radio signals reflected or retransmitted from the position to be determined.

radio direction finder – A radio receiver which pinpoints the line of travel of the received waves.

radio direction finding – Radiolocation in which only the direction, not the precise location, of a source of radio emission is determined by means of a directive receiving antenna.

radio direction-finding station – A radiolocation station that determines only the direction of other stations, not their location, by monitoring their transmission.

radio doppler – A device for determining the radial component of the relative velocity of an object by observing the frequency change due to such velocity.

radioelectrocardiograph–A transmitter for broadcasting electrocardiograph signals from the subject to a remote receiver. It makes an ecg practical while the subject is exercising or during natural work or home activities.

radioelectroencephalograph–An electroencephalograph in which a radio link is used so that the patient may move about while the electroencephalogram is being recorded.

radio engineering–The branch of engineering concerned with the generation, transmission, and reception of radio (and now, television) waves, and with the design, manufacture, and testing of associated equipment.

radio fadeout–Also called the Dellinger effect. The partial or complete absorption of substantially all radio waves normally reflected by the ionospheric layers in or above the E-region.

radio field intensity–Also called radio field strength. The maximum (unless otherwise stated) electric or magnetic field intensity at a given location associated with the passage of radio waves. It is commonly expressed as the electric field intensity in microvolts, millivolts, or volts per meter. For a sinusoidal wave, its root-mean-square value is commonly stated instead.

radio field strength–*See* Radio Field Intensity.

radio field-to-noise ratio–The ratio of the radio field intensity of the desired wave, to the noise field intensity at a given location.

radio fix–1. A method by which the position source of radio signals can be determined. Two or more radio direction finders monitor the transmissions and obtain cross bearings. The position can then be pinpointed by triangulation. 2. The method by which a ship, aircraft, etc., equipped with direction finding equipment can determine its own position. This it does by obtaining radio bearings from two or more transmitting stations of known location. The position can then be pinpointed by triangulation as in (1) above.

radio-fixing aids–Equipment making use of radio to assist a user in determining his geographical position.

radio frequency – 1. Abbreviated rf. Any frequency at which coherent electromagnetic radiation of energy is possible. 2. A term describing incoming radio signals to a receiver or outgoing signals from a radio transmitter. There are no finite limits in the rf range but it is usually considered to denote frequency above 150 kHz and extending up to infrared range.

rado-frequency alternator–A rotating generator that produces radio-frequency power.

radio-frequency amplification – Amplification of a signal by a receiver before detection, or by a transmitter before radiation.

radio-frequency choke – An inductor used to impede the flow of radio-frequency currents. Its core is generally air or pulverized iron.

radio-frequency component–In a signal or wave, the portion consisting of the rf alternations only–not its audio rate of change in amplitude or frequency.

radio-frequency converter–A power source for producing electrical power at frequencies of 10 kHz and above.

radio-frequency generator – In industrial and dielectric heaters, a power source comprising an electron-tube oscillator, an amplifier (if used), a power supply, and associated control equipment.

radio-frequency heating – The process of heating a substance by subjecting it to a high-frequency energy field. *See also* Dielectric Heating.

radio-frequency interference – Any electrical signal capable of being propagated into and interfering with the proper operation of electrical or electronic equipment. The frequency range of such interference may be taken to include the entire electromagnetic spectrum.

radio-frequency oscillator – Abbreviated rf oscillator. An oscillator that generates alternating current at radio frequencies.

radio-frequency pacemaker – A pacemaker that consists of an inplanted circuit designed to receive pacing signals from an extracorporeal transmitter.

radio-frequency preheating – A method of preheating used in the molding of materials so that the molding operation may be facilitated or the molding cycle reduced. The frequencies used most often are between 10 and 100 MHz.

radio-frequency pulse – A radio-frequency

carrier which is amplitude-modulated by a pulse. Between pulses, the modulated carrier has zero amplitude. (The coherence of the carrier with itself is not implied.)

radio-frequency resistance—*See* Skin Effect.

radio-frequency signal generator – *See* RF Signal Generator.

radio-frequency suppressor—A device that absorbs radiated energy which might interfere with radio reception.

radio-frequency transformer – Abbreviated rf transformer. A transformer used with radio-frequency currents.

radio-frequency welding—Also called high-frequency welding. A method of welding thermoplastics using a radio-frequency field to apply the necessary heat.

radiogoniometer – In a radio direction finder, the part that determines the phase difference between the two received signals. The Bellini-Tose system has two loop antennas, both at right angles to each other and connected to two field coils in the radiogoniometer. Bearings are obtained by rotating a search coil inductively coupled to the field coils.

radiogram—A message sent via radio telegraphy.

radiograph – 1. An X-ray film image that shows internal structural features of the body. 2. *See* Radiophoto.

radiography – Any nondestructive method of internal examination in which metal objects are exposed to a beam of X-ray or gamma radiation. Differences in thickness, density, or absorption caused by internal defects or inclusions are apparent in the shadow image, either on a fluorescent screen or on photographic film placed behind the object.

radio guard—A ship, aircraft, or radio station designated to listen for and record transmissions, and to handle traffic on a designated frequency for a certain unit or units.

radio guidance system – A system that makes use of radio signals in guiding a missile or vehicle in flight. The system includes both the flightborne equipment and the guidance-station equipment on the ground.

radio homing aids—Radio equipment used to assist in the location of an area with sufficient accuracy that an approach may be effected.

radio horizon—The boundary line beyond which direct rays of the radio waves cannot be propagated over the earth's surface. This distance is not a constant; rather, it is affected by atmospheric refraction of the waves.

radio inertial-guidance system – A command type of guidance system consisting essentially of: a. A radar tracking unit, comprising radar equipment on the ground, one or more transponders in the missile, and necessary communications links to the guidance station. b. A computer that accepts missile position and velocity information from the tracking system and furnishes to the command link appropriate signals to steer the missile. c. The command link, which consists of a transmitter on the ground and an antenna and receiver on the missile; actually the command link is built into the tracking unit. d. An inertial system for partial guidance in case of radio guidance failure.

radio influence – Radio-frequency interference that originates on and from power lines.

radio intelligence—Interception and interpretation of enemy radio transmissions.

radio intercept—1. An act or instance of interception of a radio message. 2. An intercepted radio mesage. 3. A service or agency that intercepts radio messages.

radio interference—Undesired conducted or radiated electrical disturbances, including transients, which can interfere with the operation of electrical or electronic equipment. These disturbances fall between 14 kHz and 10 GHz.

radioisotope—1. Also called a radioactive isotope. It is the isotope produced when an element is placed into a nuclear reactor and bombarded with neutrons. Radioisotopes are used as tracers in many areas of science and industry. Like all isotopes, they decay spontaneously with the emission of their radiation, at a definite rate measured by their half-lives. 2. Tracer form of element having similar chemical behavior but "tagged" with radioactive substance (chromium, iodine, phosphor, etc.) so that it emits gamma rays that can be counted with a scintillation counter.

radio jamming – Blocking communications by sending overpowering interference signals.

radio knife—A form of surgical knife that uses a high-frequency electric arc at its tip to cut tissue and, at the same time, also sterilizes the edges of the wound.

radio landing aids—Radio equipment used in assisting an aircraft in making its actual landing.

radio landing beam—A distribution of vertical, directional radio waves used for guiding aircraft into a landing.

radiolocation—1. Use of the constant-velocity or rectilinear-propagation characteristics of radio waves to detect an object or to determine its direction, position, or motion. 2. With respect to Federal Communications Commission regulations, radiodetermination used for purposes other than those of radionavigation.

radiolocation service – A radio service in which radiolocation is used.

radiolocation station–A radio station in the radiolocation service.

radio log–A record of all messages sent and received, transmitter tests made, and other important information pertaining to the operation of a particular station.

radiologist–A specialist in the field of radiology.

radiology–The branch of medical science that deals with the use of radiant energy in the diagnosis and treatment of disease. While radiology originally involved only the use of radiant energy in visualizing tissues, this science now covers a wide spectrum of diagnostic and therapeutic applications, such as diagnosis using roentgen rays, diagnosis using radioisotope scanning techniques, genetics radioisotopes in research, roentgen cinematography, angiography, fluoroscopy, pneumoencephalography, spectroscopy, and treatment of tumors using roentgen rays, implanted radioactive pellets, cyclotrons, or lasers.

radioluminescence – Luminescence produced by radiant energy (e.g., by X-rays, radioactive emissions, alpha particles, or electrons).

radiomagnetic indicator – Abbreviated RMI. A navigational instrument used by land vehicles. It presents a display combining the heading and the relative and magnetic bearings of the vehicle with the relative bearing of a radio station whose location is known.

radio marker beacon–In the aeronautical radionavigation service, a land station which provides a signal to designate a small area above the station.

radio marker station – Station marking a definite location on the grounds as an aid to air navigation.

radiometallography–X-ray examination of the crystalline structure and other characteristics of metals and alloys.

radiometeorograph–A meteorograph which, when carried into the stratosphere by an unmanned gas-filled rubber balloon, automatically reports atmospheric conditions by radio as it ascends into the stratosphere. The ultrahigh-frequency signals are transmitted so that they can be recorded and interpreted in terms of pressure, temperature, and humidity. (*See also* Radiosonde.)

radiometer–1. A device for measuring radiant flux density. Generally, a blackened thermocouple or bolometer is employed, but the simplest type employs a rotating vane. 2. An instrument of measuring intensity of radiant energy, including X-ray.

radiometric – Pertaining to the measurement of radiation.

radiometry – 1. The measurement of the spectral emission characteristics of sources of electromagnetic radiation. 2. The science of radiation measurement concerned with the detection and measurement of radiant energy either at separate wavelengths or integrated over a broad wavelength band, and the interaction of radiation with matter such as absorption, reflectance and emission. 3. The measurement of radiation in the infrared, visible, and ultraviolet portion of the spectrum.

radiomicrometer–*See* Microradiometer.

radionavigation–Navigational use of radiolocation for determining position or direction, or for providing a warning of obstructions. Radionavigation includes use of radio direction finding, radio ranges, radio compasses, radio homing beacons, etc.

radionavigation land station – A fixed station in the radionavigation service–i.e., one not intended for mobile operation.

radionavigation mobile station – A radionavigation station operated from a vehicle.

radionavigation service – A radiolocation service used for radionavigation.

radio net–A system of radio stations that communicate with each other. A military net usually is made up of a radio station of a superior unit and stations of all subordinate or supporting units.

radio-noise field intensity – A measure of the field intensity of interfering electromagnetic waves at some point (e.g., a radio receiving station). In practice, the

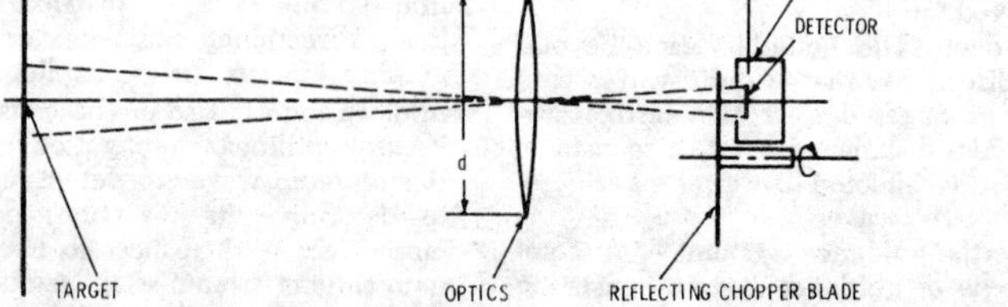

Simple radiometer.

field intensity itself is not measured, but some proportionate quantity.

radiopaque – Not penetrable by X-rays or other radiation. (The opposite of radioparent.)

radioparent—Penetrable by X-rays or other radiation. (The opposite of radiopaque.)

radiophone—*See* Radiotelephone *and* Radio Beacon.

radiophoto—Also called radiophotography, radiophotograph, radiophotogram, facsimile, or radiograph. The transmission of photographs and other illustrations by radio.

radiophotogram—*See* Radiophoto.

radiophotoluminescence – The property whereby the previous exposure of certain materials to nuclear radiation enables them to give off visible light when irradiated with ultraviolet light.

radio position finding—Determining the location of a radio station by using two or more direction finders and a process of triangulation.

radiopositioning land station – A station, other than a radionavigation station, in the radiolocation service not intended to be operated while in motion.

radiopositioning mobile station—A station, other than a radionavigation station, in the radiolocation service intended to be operated while in motion or while stopped at unspecified points.

radio prospecting—The use of radio equipment to locate mineral or oil deposits.

radio proximity fuse—A radio device which detonates a missile by electromagnetic interaction within a predetermined distance from the target.

radio pulse—An intense, split-second burst of electromagnetic energy.

radio range – A radionavigational facility, the emissions of which provide radial lines of position by having special characteristics that are recognizable as bearing information and useful in lateral guidance of aircraft.

radio range beacon – A radionavigation land station in the aeronautical radionavigation service providing radio equisignal zones.

radio range finding – Determination of range by means of radio waves.

radio range leg—The space within which an aircraft will receive an on-course signal from a radio-range station.

radio-range monitor – An instrument that automatically monitors the signal from a radio-range beacon and warns attending personnel when the transmitter deviates from its specified current bearings. It also transmits a distinctive warning to apoaching planes whenever trouble exists at the beacon.

radio-range station—A land station that operates in the aeronautical radionavigation service and provides radial equisignal zones.

radio receiver—A device for converting radio waves into signals perceptible to humans.

radio reception—Reception of radioed messages, programs, or other intelligence.

radio recognition—The use of radio means to determine the friendly or enemy character or the individuality of another.

radio relay—*See* Radio-Relay System.

radio-relay system—Also called radio relay. A radio transmission system in which the signals are received and retransmitted from point to point by intermediate radio stations.

radio set – A radio transmitter, radio receiver, or a combination of the two.

radio shielding—A metallic covering over all electric wiring and ignition apparatus which is grounded at frequent intervals for the purpose of eliminating electric interference with radiocommunications.

radio signal—A signal that is transmitted by radio.

radio silence—A period during which all or certain radio equipment capable of radiation is kept inoperative.

radiosonde—A balloonborne instrument for the simultaneous measurement and transmission of meteorological data. The instrument consists of transducers for the measurement of pressure, temperature, and humidity; a modulator for the conversion of the output of the transducers to a quantity which controls a property of the radio-frequency signal; a selector switch which determines the sequence in which the parameters are to be transmitted; and a transmitter which generates the radio-frequency carrier.

radiosonde recorder—An instrument which is located at the surface observing station

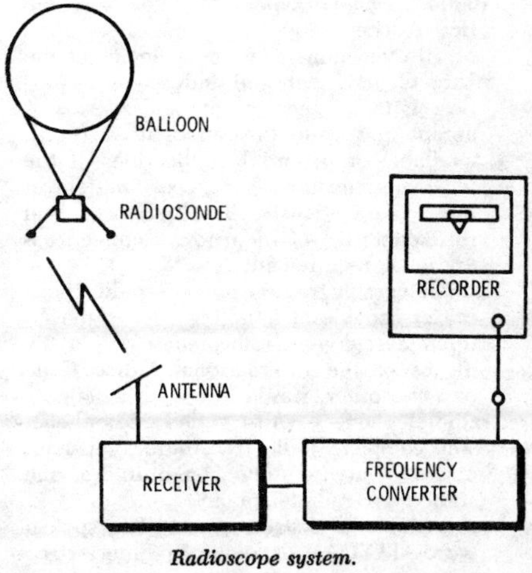

Radioscope system.

and is used to record the data presented by a radiosonde aloft. The mechanism of the recorder depends upon the type of radiosonde system used.

radiosonde transmitter—The component of the radiosonde which includes the modulating blocking oscillator and the radio-frequency carrier oscillator.

radiosonobuoy—*See* Sonobuoy.

radio spectrum—The range of frequencies of electromagnetic radiation usable for radiocommunication. The radio spectrum may range from about 3 kHz to over 300 GHz. Corresponding wavelengths are 100 km to 1 mm.

Radio Spectrum

Band Number	Frequency Range*	Metric Subdivision	Band Designation
4	3-30 kHz	Myriametric waves	VLF
5	30-300 kHz	Kilometric waves	LF
6	300-3000 kHz	Hectometric waves	MF
7	3-30 MHz	Decametric waves	HF
8	30-300 MHz	Metric waves	VHF
9	300-3000 MHz	Decimetric waves	UHF
10	3-30 GHz	Centimetric waves	SHF
11	30-300 GHz	Millimetric waves	EHF
12	300-3000 GHz or 3 THz	Decimillimetric waves	—

* Lower limit is exclusive, upper limit inclusive.

radio station—An assemblage of equipment for radio transmission, reception, or both.

radio-station interference — Interference caused by reception of radio waves from other than the desired station.

Radio Technical Commission for Aeronautics—A cooperative, nonprofit association of all telecommunications agencies of the U.S. Government and industry. Its purpose is to advance the art and science of aeronautics through investigation of all available or potential applications of the telecommunications art, coordination of these applications with allied arts, and the adaptation of them to recognized operational requirements.

radiotelegraph transmitter — A radio transmitter capable of handling code signals.

radiotelegraphy — Radiocommunication by means of the International Morse Code or some other, similarly coded signal.

radiotelephone — Also called radiophone. The complete radio transmitter, receiver, and associated equipment required at one station for radiotelephony.

radiotelephone distress signal—The spoken word MAYDAY (phonetic spelling of the French expression *m' aidez,* or "help me"). It corresponds to S O S in radiotelegraphy and is used by aircraft, ships, etc., needing help.

radiotelephone transmitter—A radio transmitter capable of handling audio-frequency modulation (e.g., voice and music).

radiotelephony—Two-way transmission and reception of sounds by radio.

radio telescope—A very sensitive radio receiver that is used in conjunction with a large and highly directional antenna to receive signals from radio stars.

radiotherapy — The treatment of disease with radiations, especially ultraviolet, infrared, X-rays and gamma rays.

radiothermics—The application of heat generated by radio waves (e.g., in diathermy or electronic heating).

radio transmission—Transmission of signals by electromagnetic waves other than light or heat waves.

radio transmitter—A device capable of producing radio-frequency power and used for radio transmission.

radiotransparent—Permitting the passage of X-rays or similar radiation.

radio tube—A general term for any type of electron tube used in electronic equipment.

radiovision—An early name for television.

radio warning—*See* Radio Detection.

radio watch—Also called watch. The vigil maintained by an operator when on duty in the radio room of a vessel and listening for signals—especially on the international distress frequencies.

radio wavefront distortion—A change in the direction of advance of a radio wave.

radio-wave propagation — The transfer of energy by electromagnetic radiation at frequencies below approximately 3×10^{12} hertz.

radio waves — Electromagnetic waves at a frequency lower than 3000 GHz and propagated through space without an artificial guide.

radist—A radionavigation system in which the position of a vehicle is determined by comparing the arrival times of transmitted pulses at three or more ground stations.

radix—1. Also called the base. The total number of distinct marks or symbols used in a numbering system. For example, since the decimal numbering system uses ten symbols (0, 1, 2, 3, 4, 5, 6, 7, 8, 9), the radix is 10. In the binary numbering system the radix is 2, because there are only two marks or symbols (0, 1). 2. In positional representation, that integer, if it exists, by which the significance of the digit place must be multiplied to give the significance of the next higher digit place. For example, in decimal notation, the radix of each place is ten; synonymous with base.

radix complement—A number obtained by subtracting each digit of the given number from one less than the radix, then adding to the least significant digit, executing all required carries. Examples are the tens complement in decimal notation and the twos complement in binary notation. Also called true complement.

radix-minus-one complement — A number obtained by subtracting each digit of the given number from one less than the radix. Examples are the nines complement in decimal notation and the ones complement in binary notation. Also called diminished-radix complement.

radix notation—A positional representation in which any two ad'acent digit positions have significances in an integral ratio called the radix of the least significant of the two positions; the digit in any position may have a value from zero to one less than the radix of that position.

radix point—Also called base point, and binary point, decimal point, etc., depending on the numbering system. The index which separates the integral and fractional digits of the numbering system in which the quantity is represented.

radome — Also called a radar dome. The housing which protects a radar antenna from the elements, but does not block radio frequencies.

radux—A long-distance, low-frequency navigational system which provides hyperbolic lines of position. It is of the continuous-wave, phase-comparison type.

railing—Radar pulse jamming at high recurrence rates (50 to 150 kHz). On a radar indicator it results in an image resembling fence railing.

railings—The pattern produced on an A-scope by cw modulated with a high-frequency signal. It appears as a series of vertical lines resembling target echoes along the baseline.

rain-barrel effect—The characteristic sound noted on an equalized line that is overcompensated.

rainbow — The technique which applies pulse-to-pulse frequency changing to identifying and discriminating against decoys and chaff.

rainbow generator—A signal generator with which the entire color spectrum can be produced on the screen of a color television receiver. The colors merge together as in a rainbow. (*See also* Keyed Rainbow Generator.)

rake angle—The angle that a stylus makes with a phonograph record when viewed from the side.

RAM — Abbreviation for random-access memory.

ram — The moving part in the head of a crimping tool.

ramark — Also called a radar marker. A fixed radar transmitter which emits a continuous signal that is used as a bearing indication on a radar display.

ramp input—A change, in an input signal, which varies at a constant rate.

Ramsauer effect — The absorption of slow-moving electrons by intervening matter.

random—Irregular; having no set pattern.

random access — 1. Access to a computer storage under conditions whereby there is no rule for predetermining the position from where the next item of information is to be obtained. 2. The process of obtaining information from computer storage with the access time independent of the location of the information.

random-access device — A data-storage device in which the access time is effectively independent of the location of the stored data. Also called direct-access device.

random-access discrete address—A communications technique in which radio users share a single wide band instead of having a narrow band for each user.

random-access memory—Abbreviated RAM. 1. A storage arrangement from which information can be retrieved with a speed that is independent of the location of the information in the storage. For example, a core memory is a random-access memory, but a magnetic-tape memory is not. 2. A memory that can be written into or read by locating any data address. 3. A device that permits individual interrogation of any memory cell in a completely random sequence. Any point in the total memory system can be accessed without looking at any other bit. 4. A memory that has the stored information immediately available when addressed, regardless of the previous memory address location. As the memory words can be selected in any order, there is equal access time to all. 5. A memory which may be written to, or read from, any address location in any order. May refer specifically to the integrated circuit method of implementation. 6. A read/write memory that stores information in such a way that each bit of information may be retrieved within the same amount of time as any other bit. As opposed to serial memory.

randam-access programming—Programming a problem for a computer without regard to the access time to the information in the registers called for in the program.

random-access storage — A form of storage where information can be recovered immediately, regardless of when it was stored. For example, magnetic-core memory devices will usually yield any bit of information with almost no time penalty regardless of where it is located, while magnetic tape must be run until the required information can be found.

random bundle—A fiber-optics bundle that scrambles an image but can convey light from a source to a relatively inaccessible point, and then again from that location to a photocell.

random error—The inherent imprecision of a given process of measurement; the unpredictable component of repeated independent measurements on the same object under sensible uniform conditions, usually of an approximately normal (Gaussian) frequency distribution, other than systematic or erratic errors and mistakes, sometimes called short-period errors.

random experiment – An experiment that can be repeated a large number of times but may yield different results each time, even when performed under similar circumstances.

random failure—Also called chance failure. Any chance failure the occurrence of which at a given time is unpredictable.

random-function generator—A device which generates nonrepetitive signals which are distributed over a broad frequency range.

random-impulse generator—A generator of electrical impulses which occur at random rather than at specific intervals.

random interlace – A condition in which there is no fixed relationship between adjacent scanning lines and successive fields.

randomized jitter – Jitter produced by means of noise modulation.

randomness—A condition of equal chance for the occurrence of any of the possible outcomes.

random noise – 1. Also called fluctuation noise. A signal the instantaneous amplitude of which is determined at random and therefore is unpredictable. It contains no periodic frequency components and its spectrum is continuous. (*See also* Broad-Band Electrical Noise.) 2. Noise generated in a circuit by random movement of electrons caused by thermal agitation. In tape recording it can be caused by uneven distribution of magnetized particles and is reproduced as a background "hiss."

random-noise generator—A generator of a succession of random signals which are distributed over a wide frequency spectrum.

random number—1. A set of digits such that each successive digit is equally likely to be any of n digits to the base n of the number. 2. A number composed of digits selected from an orderless sequence of digits. 3. Array of independent digits having no logical interrelationship, so that the occurrence of any particular one is totally unpredictable.

random-number generator – A special machine routine or hardware that produces a random number or series of random numbers in accordance with specified limitations.

random pard—Pertains to that portion of the total pard in an electronic power supply which is not periodic. This phenomenon is frequently referred to as noise.

random processing—The treatment of information without respect to where it is located in external storage and in an arbitrary sequence determined by the input against which it is to be processed.

random pulsing—Varying the repetition rate of pulses by noise modulation or continuous frequency change.

random sample—A sample in which every item in the lot is equally likely to be selected in the sample.

random sampling – A sampling process in which there is significant time uncertainty between the signal being sampled and the taking of samples.

random-sampling oscilloscope – An oscilloscope that functions by constructing a coherent display from samples taken at random.

random sequential memory—A memory in which one reference can be found immediately; the other reference is found in a fixed sequence.

random signals—Waveforms having at least one parameter (usually amplitude) that is a random function of time (e.g., thermal noise or shot noise).

random variable – 1. Also called variate. The result of a random experiment. 2. A discrete or continuous variable which may assume any one of a number of values, each having the same probability of occurernce. 3. Also called stochastic variable. Any signal the amplitude or phase of which cannot be predicted by a study of previous values of the signal.

random velocity—The instantaneous velocity of a particle without regard to direction. It may be characterized by its distribution function or by its average, root-mean-square, or most probable value.

random vibration – A vibration generally composed of a broad, continuous spectrum of frequencies, the instantaneous magnitude of which cannot be specified to any given moment of time. (If random vibration has instantaneous magnitudes distributed according to the Gaussian distribution, it is called Gaussian random vibration.) (*See also* White Noise.)

random winding—A coil winding in which the turns and layers are not regularly positioned or spaced but are positioned haphazardly.

range—1. The maximum useful distance of a radar or radio transmitter. 2. The difference between the maximum and the minimum value of a variable. 3. The set of values that may be assumed by a quantity or function. 4. *See* Receiving Margin.

range-amplitude display – A radar display in which a time base provides the range scale from which echoes appear as deflections normal to the base.

range calibration – Adjustment of radar-range indications by use of known range targets or delayed signals so that, when on target, the radar set will indicate the correct range.

range coding–A method of coding a beacon response so that the response appears as a series of pulses on a radarscope. The coding provides identification.

range finder–A movable, calibrated unit of the receiving mechanism of a teletypewriter that can be used to move the selecting interval relative to the start signal.

range gate–A gate voltage used to select radar echoes from a very short-range interval.

range-height indicator–A radar display on which an echo appears as a bright spot on a rectangular field. The slant range is indicated along the X-axis, and the height above the horizontal plane (on a magnified scale) along the Y-axis. A cursor shows the height above the earth.

range mark–*See* Distance Mark.

range marker – A variable or movable discontinuity in the range time base of a radar display (in the case of a ppi, a ring). It is used for measuring the range of an echo or calibrating the range scale.

range of an instrument–*See* Total Range of an Instrument.

range resolution–The minimum difference in range between two radar targets along the same line of bearing for which an operator can distinguish between targets.

range ring–An accurate, adjustable ranging mark on a plan-position indicator corresponding to a range step on a type-M indicator.

range step – The vertical displacement on an M-indicator sweep to measure range.

range unit–A radar-system component used for control and indication (usually counters) of range measurements.

range zero – Alignment of the start of a sweep trace with zero range.

ranging oscillator – An oscillator circuit containing an LC resonant combination in the cathode circuit, usually used in radar equipment to provide range marks.

rank–To arrange in a series in ascending or descending order of importance.

rapid memory–*See* Rapid Storage.

rapid storage–Computer storage in which the access time is very short; rapid access usually is gained by limiting the storage capacity. Also called rapid memory, fast-access storage, and high-speed storage.

rare gas–*See* Noble Gas.

raser–Acronym for Radio Amplification by Stimulated Emission of Radiation, a chemical "pumping" process that is accomplished without external radiation.

raster–1. On the screen of a cathode-ray tube, a predetermined pattern of scanning lines which provide substantially uniform coverage of an area. 2. The illuminated area produced by the scanning lines on a television picture tube when no signal is being received. 3. Rectangular line pattern of light produced on the screen of cathode ray tube with no signal present. It is formed by deflecting the electron beam rapidly from left to right and relatively slowly from top to bottom.

raster burn–In camera tubes, a change in the characteristics of the area that has been scanned. As a result, a spurious signal corresponding to that area will be produced when a larger or tilted raster is scanned.

raster scanning – Radar antenna scanning similar to electron-beam scanning in a television picture tube; a horizontal sector scan that is change in elevation.

ratchet relay–A stepping relay actuated by an armature-driven ratchet.

rate action – Also called derivative action. Corrective action, the rate of which is determined by how fast the error being corrected is increasing.

rated contact current – The current which contacts are designed to handle for their rated life.

rated operational voltage – The voltage on which a specific application rating of a device is based.

rated output – The output power, voltage, current, etc., at which a machine, device, or apparatus is designed to operate under normal conditions.

rated power output – 1. The normal radio-frequency, power-output capability (peak or average) of a transmitter under optimum adjustment and operation conditions. 2. The maximum power that an amplifier will deliver continuously without exceeding its specified distortion rating. Also called continuous power output or rms power output.

rated range – The nominal range within which a device can be operated and still maintain the level of performance specified for it by the manufacturer.

rated thermal current–In a contactor, that current at which the permissible temperature rise is reached.

rated voltage–The voltage at which a device or component is designed to operate under normal conditions.

rate effect – The anode-voltage transients that cause pnpn devices to switch into high conduction.

rate generator – A proportional element which converts angular speed into a constant-frequency output voltage. (*See also* Angular Velocity.)

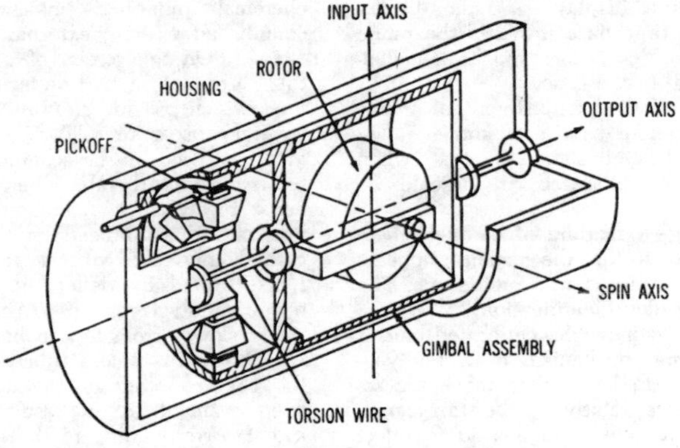

Rate gyro.

rate-grown junction—A grown junction, in a semiconductor, produced by varying the rate of crystal growth periodically so that n-type impurities alternately predominate.

rate-grown transistor – Also called graded-junction transistor. A variation of the double-doped transistor, in which n- and p-type impurities are added to the melt.

rate gyro—A particular kind of gyroscope used for measuring angular rates. A system of three rate gyros, each oriented to one of three mutually perpendicular axes—roll, pitch, and yaw—can control a missile or aircraft by detecting angular rates and then generating proportional corrective signals.

rate limit—*See* Slew Rate, 2.

rate limiting—Nonlinear behavior in an amplifier due to its limited ability to produce large, rapid changes in output voltage (slewing), restricting it to rates of change of voltage lower than might be predicted by observing the small signal frequency response.

ratemeter—1. An instrument for measuring the rate at which counts are received, usually in counts per minute. 2. A type of radiation detector whose output is proportional to instantaneous radiation intensity (rate of radioactivity emission).

rate of decay—The rate at which the sound-pressure level (velocity level, or sound-energy density level) is decreasing at a given point and at a given time. The practical unit is the decibel per second.

rate-of-rise relay—*See* Instantaneous Overcurrent Relay.

rate of transmission—*See* Speed of Transmission.

rate receiver—A device for receiving a signal giving the rate of speed of a launched missile.

rate signal – A signal proportional to the time derivative of a specified variable.

rate test – A test to verify that the time constants of the integrators in an analog computer are correct.

rate transmitter—A guidance antenna used to signal the desired speed for a missile in flight.

rating—A value which establishes either a limiting capability or a limiting condition for an electron device. It is determined for specified values of environment and operation, and may be stated in any suitable terms. (Limiting conditions may be either maxima or minima.)

rating system—The set of principles upon which ratings are established and by which the interpretation of the ratings is determined.

ratio—The value obtained by dividing one number by another. This value indicates their relationship to each other.

ratio arms—Two adjacent arms of a Wheatstone bridge, both having an adjustable resistance and so arranged that they can be set to have any of several fixed ratios to each other.

ratio calibration – 1. The calibration of a dimensionless quantity that represents the ratio of one of its values to another. 2. A method by which potentiometric transducers may be calibrated, in which the value of the measurand is expressed in terms of decimal fractions representing the ratio of output resistance to total resistance.

ratio detector – An fm detector which inherently discriminates against amplitude modulation. A pair of diodes are connected in such a manner that the audio output is proportionate to the ratio of the fm voltages applied to them.

ratio meter—1. An instrument which measures electrically the quotient of two quantities. It generally has no mechanical control means such as springs. Instead, it operates by the balancing of electromagnetic forces, which are a function of the position of the moving element.

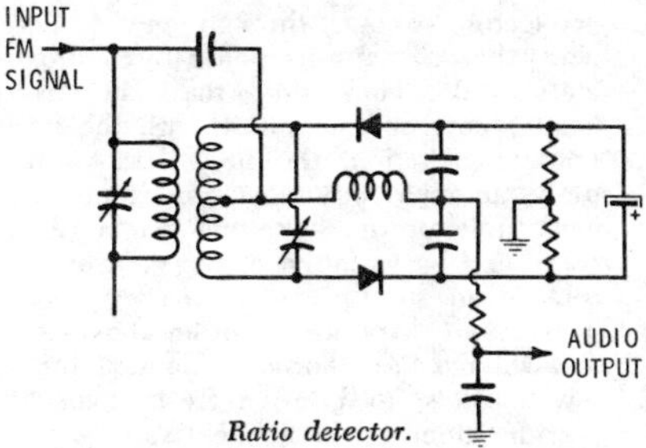

Ratio detector.

2. Instrument for measuring the ratio of transformation of a transformer by means of a resistance bridge arrangement. 3. Moving-coil type of instrument in which the deflection is proportional to the ratio of the current sent through two coils.

ratio of transformation—The ratio of the secondary voltage of a transformer to the primary voltage under no-load conditions, or the corresponding ratio of currents in a current transformer.

ratio squelch—*See* Squelch Circuit.

ratio-type telemeter—Also called a position-type telemeter. A telemeter in which the relative phase position between, or magnitude relation of, two or more electrical quantities is the translating means.

rat race — A magic-T modification for the acceptance of higher power. A circular loop of coaxial line closed upon itself and having four branching connections. *See also* Hybrid Ring.

raw data — Information that has not been processed or reduced.

raw tape—Also called virgin tape or blank tape. A term sometimes used to describe tape that has not been recorded. *See also* Blank Tape.

ray—A line of propagation of any form of radiant energy.

ray-control electrode — An electrode that controls the position of the electron beam on the screen of a cathode-ray tuning-indicator tube.

Raydist — A system using cw transmission to provide hyperbolic lines of position through rf phase comparison techniques. The system is used for surveying or ship positioning in a two-dimensional array. The frequency bands is 1.7 to 2.5 MHz. Similar to LORAC in principle.

Rayleigh disc—A special acoustic radiometer used for the fundamental measurement of particle velocity.

Rayleigh distribution—Frequency distribution for an infinite number of quantities of the same magnitude, but of random phase relationships. Sky-wave field intensities follow the Rayleigh distribution for intervals of one minute or less.

Rayleigh distribution (fading) — A statistical distribution describing the magnitude of a phasor composed of the sum of many component phasors randomly distributed in amplitude and phase. Fading of signals caused by cancellation and reinforcement of contributions received over separate paths often exhibits a Rayleigh distribution.

Rayleigh line — In scattered radiation, a spectrum line which has the same frequency as the corresponding incident radiation.

Rayleigh reciprocity theorem — The reciprocal relationship for an antenna when transmitting or receiving. The effective heights and the radiation resistance and pattern are alike, whether the antenna is transmitting or receiving.

Rayleigh scattering—Scattering of radiation by minute particles suspended in air (e.g., by dust).

Rayleigh wave—A surface wave associated with the free boundary of a solid. The wave is of maximum intensity at the surface, but diminishes quite rapidly as it proceeds into the solid.

ray path—An imaginary line, perpendicular to the wavefront, which describes the path along which the energy associated with a point on a wavefront moves.

Raysistor—Raytheon trade name for a device that contains a photosensitive semiconductor element and a light source that can be used to control the conductivity of the semiconductor.

RC—Symbol for resistance-capacitance, resistance-coupled, or ray-control electrode.

RC amplifier — Abbreviation for resistance-capacitance coupled amplifier.

RC circuit—A time-determining network of resistors and capacitors in which the time constant is defined as resistance times capacitance.

RC constant—The time constant of a resistor-capacitor circuit, equal in seconds to the resistance value in ohms multiplied by the capacitance value in farads.

RC coupling — *See* Resistance-Capacitance Coupling.

RC filter—*See* Resistance-Capacitance filter.

RC network — A circuit containing resistances and capacitances arranged in a particular manner to perform a specific function.

RC oscillator — An oscillator in which the frequency is determined by resistance-capacitance elements. *See also* Resistance-Capacitance Oscillator.

RCTL—Abbreviation for resistor-capacitor-transistor logic.

rd—Abbreviation for rutherford.

rdf—Abbreviation for radio direction finder (or finding).

R-display — Essentially an expanded radar A-display in which the trace of an echo can be expanded for more detailed examination.

reach-through voltage – That value of reverse voltage for which the depletion layer in a reverse-biased pn junction spreads sufficiently to make electrical contact with another junction.

reacquisition time—The time required for a tracking radar to relock on the target after the radar's automatic tracking mechanism has been disengaged. *See also* Punch-Through Voltage.

reactance—1. Symbolized by X. Opposition to the flow of alternating current. Capacitive reactance (X_C) is the opposition offered by capacitors, and inductive reactance (X_L) is the opposition offered by a coil or other inductance. Both reactances are measured in ohms. 2. The opposition offered an alternating flow by a capacitance or inductance related to the frequency of the alternating current. The reactance of a capacitor decreases with increasing frequency, but the reactance of an inductance increases with frequency.

reactance drop—The voltage drop in quadrature with the current.

reactance factor—In a conductor, the ratio of its ac resistance to its ohmic resistance.

reactance frequency multiplier – A frequency multiplier the essential element of which is a nonlinear reactor, where the nonlinearity of the reactor is used to generate harmonics of a sinusoidal source.

reactance grounded – Grounded through a reactance.

reactance modulator – A modulator the reactance of which can be varied in accordance with the instantaneous amplitude of the modulating electromotive force applied. It is normally an electron-tube circuit that usually modulates the phase or frequency.

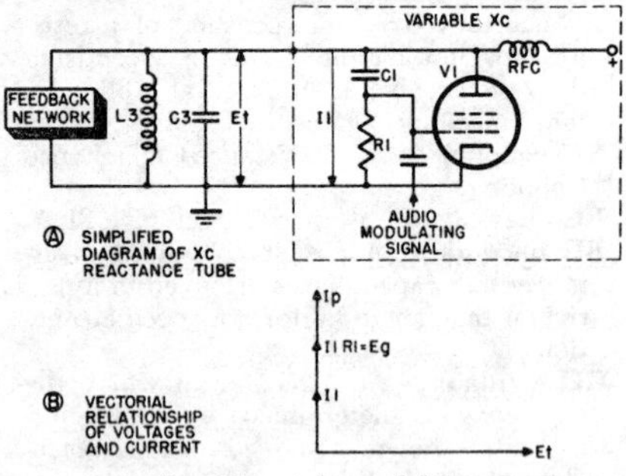

Reactance-tube modulator.

reactance relay—A form of impedance relay the operation of which is a function of the reactance of the circuit.

reactance tube (transistor) – A vacuum tube or transistor connected so as to appear as a reactance to the rest of the circuit. This is accomplished by deriving the grid (base) excitation from the plate (collector) voltage through an RC network that changes its phase by approximately 90°. Thus, since the plate (collector) current is in phase with the grid (base) excitation, the tube (transistor) draws an apparently reactive current. The magnitude of the apparent reactance is controlled by variation of the gain of the tube (transistor).

reactivation – Application of an above-normal voltage to a thoriated filament for a few seconds, to bring a fresh layer of thorium atoms to the filament surface and thereby improve electron emission.

reactive—Pertaining to either inductive or capacitive reactance. A reactive circuit has a higher reactance than resistance.

reactive attenuator—An attenuator that dissipates almost no energy.

reactive balance—1. The capacitive or inductive balance which is often required to null the output of certain transducers or systems when the excitation and/or the output is given in terms of alternating current. 2. The condition of an ac circuit where the phase angle between the voltage and current is zero. 3. The amount of corrective capacitance or inductance required to null the output of certain transducers or systems having ac excitation.

reactive factor—The ratio of reactive power to total power in a circuit.

reactive-factor meter – An instrument for measuring reactive factor.

reactive load – A load having reactance (i.e., a capacitive or inductive load), as opposed to a resistive load.

reactive near-field region—That region of the field of an antenna immediately surrounding the antenna wherein the reactive field predominates.

reactive power—Also called wattless power. The reactive voltage times the current, or the voltage times the reactive current, in an ac circuit. Unit of measurement is the var.

reactive sputtering—A sputtering technique that involves the introduction of reactive gases such as oxygen, nitrogen, and hydrogen into the glow discharge so that oxides, nitrides, and hydrides of the evaporant are deposited on the substrate.

reactive volt-ampere—*See* Volt-Ampere Reactive.

reactive volt-ampere-hour meter—*See* Varhour Meter.

reactive volt-ampere meter—*See* Varmeter.

reactor—1. A physical device used primarily to introduce reactance or susceptance into a branch. 2. *See* Nuclear Reactor.

reactor-start motor – A form of split-phase motor designed for starting with a reactor in series with the main winding. The reactor is short-circuited (or otherwise made ineffective) and the auxiliary circuit is opened as soon as the motor attains a predetermined speed.

read—In a computer: 1. To copy, usually from one form of storage to another. 2. To sense the meaning of an arrangement of hardware representing information. 3. To exact information. 4. To transmit data from an input device to a computer.

readability—The ability to be understood—specifically, the understandability of signals sent by any means of telecommunications.

read-around—*See* Read-Around Ratio.

read-around ratio—Also called read-around. In electrostatic storage tubes, the number of successive times information can be recorded as an electrostatic charge on a single spot in the array without producing the necessity for restoring the charge on surrounding spots.

reader—In a computer a device which converts information in one form of storage to information in another form of storage.

read head—In a computer, a device that converts digital information stored on a magnetic tape or on a drum into electrical signals usable by computer arithmetic.

read-in—To sense information in a source and transmit it to an internal storage.

reading access time—In a computer, the time before a word may be used during the reading cycle.

reading rate—In a computer, the number of characters, cards, etc., that can be sensed by an input unit in a given time.

read-in program—A program that itself can be put into a computer in simple binary form but that makes it possible for other programs to be read into the computer in more complex forms.

read-mainly memory—*See* RMM.

read-mostly memory — An integrated array of amorphous and crystalline semiconductor devices that is capable of being programmed, read, and reprogrammed repeatedly. Once programmed, this type of memory retains data unless it is altered intentionally.

read-only memory—1. A storage arrangement primarily for information-retrieval applications. The information may be wired in when the storage device is made, or it may be written in at a speed much less than the retrieval speed. Abbreviated ROM. 2. A memory that cannot be altered in normal use of a computer. Usually a small memory that contains often-used instructions such as microprograms or system software as firmware. Peripheral equipment uses ROM for character generation, code translation and for designing peripheral processors. 3. A memory in which information is stored permanently, e.g., a math function or a microprogram. A ROM is programmed according to the user's requirements during memory fabrication and cannot be reprogrammed. A ROM is analogous to the dictionary where a certain address results in predetermined information output.

readout—The manner in which a computer displays the processed information—e.g., digital visual display, punched tape, punched cards, automatic typewriter, etc. *See also* Display.

readout device—In a computer, a device, consisting usually of physical equipment, that records the computer output either as a curve or as a set of printed numbers or letters.

readout equipment—The electronic apparatus that provides indications and/or recordings of transducer output.

readout station — A recording or receiving radio station at which information is received as it is read out by the transmitter in a missile, probe, satellite, or other spacecraft. (The same station may serve also as a tracking station.)

read pulse—A pulse applied to one or more binary cells to determine whether a bit of information is stored there.

readthrough — The continuous recovery in an audio channel of the target modulation, making possible rapid evaluation of the effectiveness of a jamming effort.

read time—With respect to a memory, the interval between the time the read control and the address or location are present and the time the data output changes state. More commonly called access time.

read-write check indicator—A device incorporated in some computers to indicate, upon interrogation, whether there was an error in reading or writing. The machine can be made to stop, attempt the operation again, or follow a special subroutine, depending on the result of the interrogation.

read-write cycle time—*See* Cycle Time.

read-write head—The device that reads and writes information on tape, drum, or disc storage devices.

read-write memory—A memory whose contents can be continuously changed quickly and easily during system operation. It differs from a read-only memory (ROM), whose contents are fixed and not subject to change, and a reprogrammable ROM, whose contents can be changed but only periodically.

ready-to-receive signal—In a facsimile system, a signal returned to the transmitter to indicate that the receiver is ready to accept a transmission.

real estate—Slang for the area on a printed-circuit board or the surface of a wafer on which circuits can be built.

real power — The component of apparent power that represents true work in an ac circuit. It is expressed in watts and is equal to the apparent power times the power factor.

real time—1. Having to do with the actual time during which physical events take place. 2. The performance of a computation during the actual time that the related physical process transpires in order that results of the computations are useful in guiding the physical process.

real-time clock—A clock that indicates the passage of actual time, such as elapsed time in the flight of a missile, as opposed to some fictitious time established by a computer program.

real-time input—Input information inserted into a system at the time it is generated by another system.

real-time operation — 1. Operations performed on a computer in time with a physical process so that the answers obtained are useful in controlling that process. 2. The use of a computer to control a process as it is actually occurring, necessitating, in general, relatively rapid operation on the part of the computer.

real-time output — Output information removed from a system at the time it is needed by another system.

real-time spectrum analyzer — A device in which analysis of the spectrum of the incoming signal is performed continuously with the time sequence of events preserved between input and output.

real-time system — A computer system in which data processing is performed so that the results are available in time to influence the controlled or monitored process.

rear projection—A projection television system in which the picture is projected on a ground-glass screen to be viewed from the opposite side of that screen.

rear suspension—In moving-coil loudspeakers, a pliable support situated near the apex of the cone. Assists in keeping the coil in a concentric position in the air gap between the magnet poles.

rebatron — A relativistic electron-bunching accelerator which produces a very tightly bunched beam with little velocity modulation, but high harmonic content. The beam can be used to excite structures which are large compared to a wavelength.

rebecca—An airborne interrogator-responsor of the British rebecca-eureka navigation system. It can also be used with a special ground beacon known as babs to provide low-approach facilities.

rebecca-eureka system — A British radar navigational system employing an airborne interrogator (rebecca) and a ground transponder beacon (eureka). It provides homing to an airfield from distances of up to 90 miles.

rebroadcast—The reception and the simultaneous or subsequent retransmission of a radio or television program by a broadcast station.

recalescent point — The temperature at which heat is suddenly liberated as the temperature of a heated metal drops.

recall—In a calculator, to retrieve from a register a previously entered number, for checking or use in further calculations.

receive current — The amount of current drawn by a transceiver when receiving radio signals.

received power—The power of a returned target signal received at the radar antenna.

received signal level — The strength of an intercepted radio signal at the antenna terminals of the receiver, expressed in microvolts or dBm.

receive only — Abbreviated ro. A teletypewriter-type terminal having no keyboard or tape reader.

receive-only typing reperforator — A teletypewriter receiver the output of which is a perforated tape that has characters along the edge of the tape. Also called rotor.

receiver—1. The portion of a communications system that converts electric waves into a visible or audible form. 2. An electromechanical device for converting electrical energy into sound waves. (*See also* Earphone.) 3. A device for the reception and, if necessary, demodulation of electronic signals.

receiver bandwidth — The spread in frequency between the half-power points on the response curve of a receiver.

receiver gating — Application of operating voltages to one or more stages of a receiver, only during the part of a cycle when reception is desired.

receiver incremental tuning—A control feature to permit receiver tuning (of a transceiver) up to 3 kHz to either side of the transmitter frequency.

receiver lockout system — In mobile communications, an arrangement of control circuits whereby only one receiver can feed the system at one time, to avoid distortion.

receiver noise figure — The ratio of noise voltage in a given receiver to that of a theoretically perfect receiver.

receiver noise threshold — The level that must be exceeded by the minimum discernible signal. External noise reaching the front end of a receiver and the noise added by the receiver itself determine the noise threshold.

receiver primaries — Constant-chromaticity, variable-luminance colors that are produced by a television receiver and that when mixed in proper proportions produce other colors. Usually three primaries, red, green, and blue, are used.

receiver radiation—Radiation of interfering electromagnetic signals by any oscillator of a receiver.

receiver sensitivity—The lower limit of useful signal input to the receiver. It is set by the signal-to-noise ratio at the output.

receiving amplifier—The amplifier used at the receiving end of a system to raise the level of the signal.

receiving antenna—A device for converting received space-propagated electromagnetic energy into electrical energy.

receiving circuit — An apparatus and connections used exclusively for the reception of messages at a radiotelephone or radiotelegraph station.

receiving equipment—The equipment (amplifiers, filters, oscillator, demodulator, etc.) associated with incoming signals.

receiving-loop loss—That part of the repetition equivalent assignable to the station set, subscriber line, and battery-supply circuit on the receiving end of a telephone line.

receiving margin—In telegraphy, the usable range of adjustment of the range finder; for a machine that is adjusted properly, approximately 75 points on a 120-point scale. Also called range or operating range.

receiving perforator—In printing telegraph systems, an apparatus that punches a paper strip automatically, in accordance with the arriving signals. When the paper strip is later passed through a printing telegraph machine, the signals will be reproduced as printed messages, ready for delivery to the customer.

receptacle—Usually the fixed or stationary half of a two-piece multiple-contact connector. Also, the connector half usually mounted on a panel and containing socket contacts.

reception—Listening to, copying, recording, or viewing any form of emission.

rechargeable—Capable of being recharged. Usually used in reference to secondary cells or batteries.

rechargeable primary cell—A cell which is ordinarily used in "one-shot" service, but which is capable of a limited number of charge/discharge cycles.

reciprocal—The number 1 (unity) divided by a quantity—e.g., the reciprocal of 2 is ½; of 4, ¼, etc.

reciprocal-energy theorem—If an electromotive force E_1 in one branch of a circuit produces a current I_2 in any other branch, and if an electromotive force E_2 inserted into this other branch produces a current I_1 in the first branch, then $I_1E_1 = I_2E_2$.

reciprocal ferrite switch — A ferrite switch that can be placed in a waveguide to route an input signal to either of two output waveguides. Switching is accomplished by a Faraday rotator under the influence of an external magnetic field.

reciprocal impedance—Two impedances Z_1 and Z_2 are said to be reciprocal impedances with respect to an impedance Z (invariably a resistance) if they are so related as to satisfy the equation $Z_1Z_2 = Z^2$.

reciprocal transducer — A transducer that satisfies the principles of reciprocity—i.e., if the roles of excitation and response are interchanged, the ratio of excitation to response will remain the same.

reciprocation—The process of deriving a reciprocal impedance from a given impedance or finding a reciprocal network for a given network.

reciprocity theorem — In any system composed of linear bilateral impedances, if an electromotive force E is applied between any two terminals and the current I is measured in any branch, their ratio (called the "transfer impedance") will be equal to the ratio obtained if the positions E and I are interchanged.

reclosing relay — Any voltage, current, power, etc., relay which recloses a circuit automatically.

recognition device — A device which can identify any number of a set of distinguishable entities.

recognition differential—For a specified listening system, the amount by which the signal level exceeds the noise level that is presented to the ear when a 50% probability of detection of the signal exists.

recombination—1. The simultaneous elimination of both an electron and a hole in a semiconductor. 2. A process in which current carriers of opposite signs combine and form stable, neutral entities.

recombination coefficient — The quantity that results from dividing the time rate of recombination of ions in an ionized gas by the product of the positive-ion density and the negative-ion density.

recombination radiation — The radiation produced in a semiconductor when electrons in the conduction band recombine with holes in the valence band. If an actual population inversion between portions of the valence and conduction bands (or between adjacent localized states of acceptors or donors near these bands) is achieved, stimulated emission and laser amplification or oscillation can take place. This is the radiation process of importance in injection lasers.

recombination velocity—On a semiconductor surface, the normal component of the electron (or hole) current density at the surface divided by the excess electron (or hole) charge density at the surface.

reconditioned-carrier reception—Also called exalted-carrier reception. Reception in which the carrier is separated from the sidebands in order to eliminate amplitude variations and noise, and then is increased and added to the sidebands in order to provide a relatively undistorted output. This method is frequently employed with a reduced-carrier single-sideband transmitter.

record—1. A character or characters that are grouped together in the flow of data in a system; for example, one line of type of the contents of a punched card. A record may be of fixed length, as with punched cards, or of variable length, as with a line of type. 2. A group of related facts or fields of information handled as a unit; thus a listing of information, usually printed or in printable form. 3. The process of putting data into a computer storage device. 4. To preserve for later reproduction.

record changer—A device which will automatically play a number of phonograph records in succession.

record code—A special control code used to mark the separation between adjacent records.

record compensator – Also called a record equalizer. An electrical network that compensates for different frequency-response curves in various recording techniques.

recorded tape – Also called a prerecorded tape. A tape that contains music, dialogue, etc., and is sold to audiophiles and others for their listening pleasure.

recorded value—The value recorded by the marking device on a chart with reference to the division lines marked on the chart.

recorded wavelength—In a phonograph record, the length of groove required for a signal of given frequency to complete one cycle. At any particular distance from the record center, i.e., at a particular groove velocity, the recorded wavelength decreases with increasing frequency. Similarly for a given frequency, the recorded wavelength decreases with progress toward the record center (i.e., as groove velocity decreases).

record equalizer—*See* Record Compensator.

recorder – Also called a recording instrument. An instrument that makes a permanent record of varying electrical impulses —e.g., a code recorder, which punches code messages into a paper tape; a sound recorder, which preserves music and voices on disc, film, tape, or wire; a facsimile recorder, which reproduces pictures and text on paper; and a video recorder, which records television pictures on film or tape.

record gap—In a computer, a space between records on a tape. It is usually produced by acceleration or deceleration of the tape during the write operation.

recording ammeter—An ammeter that provides a permanent recording of the value of either an alternating or a direct current.

recording blank—*See* Recording Disc.

recording channel – One of several independent recorders in a recording system, or independent recording tracks on a recording medium.

recording-completing trunk – A trunk for the purpose of extending a connection from a local line to a toll operator; it is used for recording the call and completing the toll connection.

recording curve—See Equalization.

recording demand meter – Also called demand recorder. An instrument that records the average value of the load in a circuit during successive short periods.

recording density—The number of bits recorded per unit of length in a single linear track in a recording medium.

recording disc – Also called a recording blank. A blank (unrecorded) disc made for recording purposes.

recording head – A magnetic head that transforms electrical variations into magnetic variations for storage on magnetic media. (*See also* Cutter.)

Recording head.

recording instrument – Also called a recorder or graphic instrument. An instrument which makes a graphic record of the value of one or more quantities as a function of another variable (usually time).

recording lamp—A light source used in the variable-density system of sound recording on movie film. Its intensity varies in step with the variations of the audio-frequency signal sent through it.

recording level – The amplifier output required to provide a satisfactory recording.

recording loss – In mechanical recording, the loss that occurs in the recorded level because the amplitude executed by the recording stylus differs from the amplitude of the wave in the recording medium.

recording noise—Noise induced by the amplifier and other components of a recorder.

recording preamplifier—*See* Preamplifier.

recording-reproducing head – A dual-purpose head used in magnetic recording.

recording spot – An instantaneous area acted on by the registering system of a facsimile recorder.

recording storage tube—A type of cathode-ray tube in which the equivalent of an image can be stored in the form of a pattern of electrostatic charge on a storage surface. There is no visual display, but the stored information can be read out at any later time in the form of an electric output signal.

recording stylus—The tool which inscribes a groove into the recording medium.

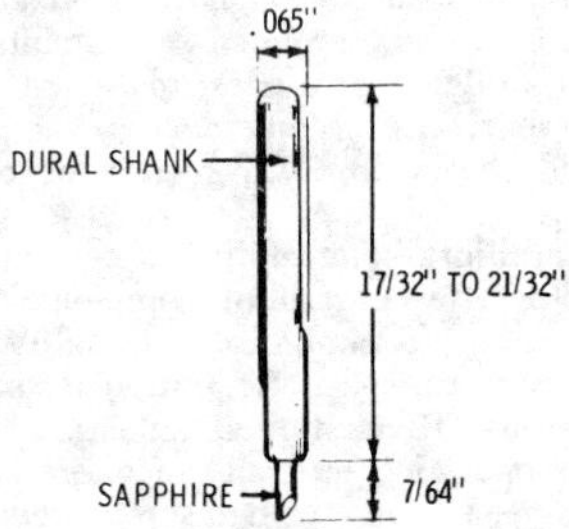

Recording stylus.

recording trunk—A trunk that extends between a local central office or private branch exchange and a toll office and that is used only for communication with toll operators and not for the completion of toll connections.

recording voltmeter—A voltmeter that provides a permanent record of the value of either alternating or direct voltage.

record layout—The arrangement and structure of information in a record, including the sequence and size of the components. By extension, the description of such an arrangement.

record length—A measure of the size of a record, usually expressed in terms of such units as words or characters.

record mark—A means of marking the separation between adjacent records: on magnetic tape, a record gap; on paper tape, a record code; in data transmission, a record pause. Often, the record code is used along with a gap or a pause to allow for different devices throughout a data-transmission system, each of which can recognize only one of the three types of record marks.

record medium – In a facsimile recorder, the physical medium onto which the image of the subject copy is formed.

record player – A motor-driven turntable, pickup arm, and stylus, for converting the signals impressed onto a phonograph record into a corresponding af voltage. This voltage is then applied to an amplifier (usually contained within the record player cabinet) for amplification and conversion to sound waves.

record separator—A character intended as an identifier of a logical boundary between records.

record sheet – In a facsimile recorder, a sheet or medium upon which the image of the subject copy is recorded.

recovered audio – The value of the audio voltage measured at the detector output under the specified circuit conditions.

recovery—1. In an electronic device, the time required to enable the device to react to new signals. 2. In fluidic devices, a generally percentile representation of output capture as related to supply, such as output pressure versus input pressure.

recovery time—1. The time required for a fired atr tube in its mount to deionize to the level where its normalized conductance and susceptance are within the specified ranges. 2. The time required for the control electrode in a gas tube to regain control after the anode current has been interrupted. 3. In Geiger-Mueller counters, the minimum time from the start of a counted pulse to the instant a succeeding pulse can attain a specified percentage of the maximum amplitude of the counted pulse. 4. The time required for a fired tr or pre-tr tube to deionize to the level where the attenuation of a low-level radio-frequency signal transmitted through the tube drops to the specified value. 5. In a radar or its component, the time required—after the end of the transmitted pulse—for recovery to a specified relation between receiving sensitivity or received signal and the normal value. 6. The interval required, after a sudden decrease in input-signal amplitude to a system or component, for a specified percentage (usually 63%) of the ultimate change in amplification or attenuation to be attained. 7. In a thermal time-delay relay, the cooling time required from heater de-energization to re-energization such that the new time delay is 85% of that exhibited from a cold start. 8. In a power supply, the time required for recovery of the load voltage from a step change in load current or line voltage. Also called response time.

rectangular array—*See* Matrix.

rectangular scanning – A two-dimensional sector scan in which a slow sector scan in one direction is superimposed perpendicularly onto a rapid sector scan.

rectangular wave—A periodic wave which alternately assumes one of two fixed values, the time of transition being negligible in comparison with the duration of each fixed value.

rectangular waveguide—A hollow enclosure of rectangular cross section, normally

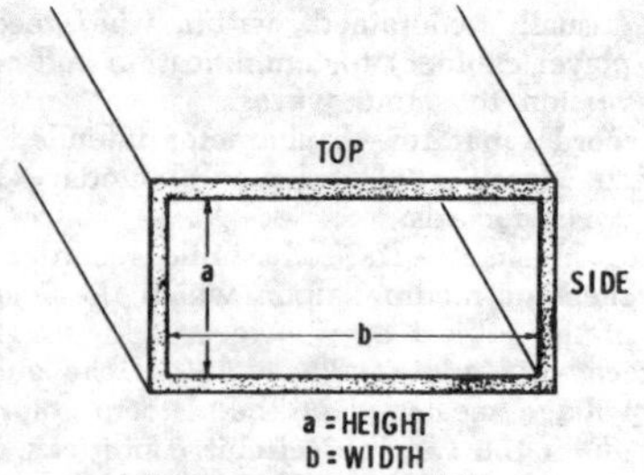

Rectangular waveguide.

with the dimensions of the sides in a ratio of 2:1. With dimensions so proportioned, the dominant mode has a free-space wavelength range between one and two times the longer side dimension. Rectangular waveguide normally can be used only over less than octave ranges.

Rectenna – A device that converts microwave power into dc power. It consists of a number of small dipoles, each of which has an associated diode rectifier network connected to a dc bus.

rectification—Conversion of alternating current into unidirectional or direct current by means of a rectifier.

rectification efficiency—Ratio of the direct-current power output to the alternating-current power input of a rectifier.

rectification factor—The change in average current of an electrode divided by the change in amplitude of the alternating sinusoidal voltage applied to the same electrode, the direct voltages of all electrodes being maintained constant.

rectified value—The average of all the positive or negative values of an alternating quantity over a whole number of periods.

rectifier—1. Device having an asymmetrical conduction characteristic employed in a way to convert alternating current into unidirectional current. In amplitude modulation detection, recovery of original signals is frequently accomplished by a rectifier. 2. Device that converts alternating current into unidirectional current by permitting appreciable current flow in one direction only. 3. A two-element tube or a solid-state device which is used to convert alternating current to direct current. Usually rated above one-half ampere.

rectifier instrument—The combination of an instrument sensitive to direct current, and a rectifying device whereby alternating currents or voltages can be measured.

rectifier meter—*See* Rectifier Instrument.

rectifier stack—A dry-disc rectifier made up of layers of individual rectifier discs (e.g., a selenium or copper-oxide rectifier).

rectifier transformer—A transformer the primary of which operates at the fundamental frequency of the ac system and the secondary of which has one or more windings conductively connected to the main electrodes of the rectifier.

rectifying element—A circuit element which conducts current in one direction only.

rectigon – A hot-cathode gas-filled diode that operates at a high pressure. Used most frequently in battery-charging circuits.

rectilineal compliance – A mechanical element that opposes a change in the applied force (e.g., the springiness that opposes a force on the diaphragm of a speaker or microphone).

rectilinear—In a straight line—specifically, moving, forming, or bounded by a straight line.

rectilinear scanning – The scanning of an area in a predetermined sequence of narrow, straight, parallel strips.

rectilinear writing recorder – An oscillograph that records in rectilinear coordinates.

recuperability—The ability to continue operating after partial or complete loss of the primary communications facility.

recurrence rate—*See* Repetition Rate, 1.

recursion—The continued repeating of the same operation or group of operations.

recursive—Capable of being repeated.

recyclability – The capability of a battery system to be recharged after it has been discharged.

recycling modulo-n counter—A counter that has n distinct states and that counts to a maximum number and returns, when the next input pulse is applied, to its minimum number.

red gun—In a three-gun color-television picture tube, the electron gun whose beam strikes only the phosphor dots that emit the red primary.

redistribution—In a charge-storage tube or television camera tube, the alteration of charges on one area of a storage surface by secondary electrons from any other area of the same surface.

redox cell—A cell designed to convert the energy of the reactants to electrical energy. An intermediate reductant, in the form of a liquid electrolyte, reacts at the anode in a conventional manner; it is then regenerated by reaction with a primary fuel.

red-tape operation—In a computer, operations which do not directly contribute to the results, i.e., those internal operations which are necessary to process data, but do not in themselves contribute to any final answer.

reduced coefficient of performance—The ratio of a given coefficient of performance to the corresponding coefficient of performance of a Carnot cycle.

reduced generator efficiency – The ratio of a given thermoelectric-generator efficiency to the corresponding Carnot efficiency.

reduced telemetry—Telemetry data transformed from raw form into a usable form.

reduction technique—A technique for simplifying or restructuring a Boolean expression for easier, lower-cost implementation in circuitry.

redundancy—1. The employment of several devices, each performing the same function, in order to improve the reliability of a particular function. 2. Added or repeated information employed to reduce ambiguity or error in a transmission of information. As the signal-to-noise ratio decreases, redundancy may be employed to prevent an increase in transmission error. 3. In information transmission, the fraction of the total information content of a message that can be eliminated without losing essential information.

redundancy check—In a computer, an automatic or programmed check that makes use of components or characters inserted especially for checking purposes.

redundant data—A data sample so similar to the preceding sample from the same source that it is of no interest in connection with subsequent analysis of the experiment or test, except for the fact of the similarity.

red video voltage—The signal voltage that controls the grid of the red gun in a three-gun picture tube. This signal is a reproduction of the output from the red camera at the transmitter.

reed—A thin bar located in a narrow gap and made to vibrate electrically, magnetically, or mechanically by forcing air through the gap.

reed frequency meter—*See* Vibrating-Reed Meter.

reed relay—1. A relay in which two flat magnetic strips mounted inside a coil are attracted to each other when the coil is energized. The relay contacts are mounted on the strips. 2. A device that uses two (sometimes three) strips of magnetizable metal, enclosed in glass, as the contacts. The control member is a coil surrounding the glass capsule. 3. One or more reed switches operated by a single coil.

reel-to-reel—*See* Open Reel.

re-entrancy — 1. A type of feedback employed in microwave oscillators. In most magnetrons a circuit is used that can be described as a slow-wave structure that feeds back into itself. In beam re-entrancy the beam may be circulated repeatedly through the interaction space. 2. That characteristic of a computer subroutine that permits a second task to enter the subroutine before completion of its execution by the first task.

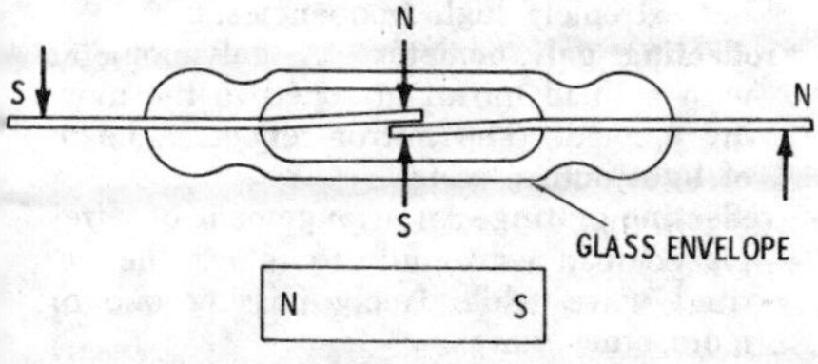

Reed switch.

re-entrant cavity — A resonant cavity in which one or more sections are directed inward with the result that the electric field is confined to a small area or volume.

re-entrant subroutines — Computer subroutines that can be executed from any of several application programs operating on different levels of priority.

re-entrant winding — An armature winding that returns to its starting point, thus forming a closed circuit.

reference acoustic pressure — That magnitude of a complex sound that produces a sound-level meter reading equal to the reading that results from a sound pressure of 0.0002 dyne per square centimeter at 1000 hertz. Also called reference sound level.

reference address—An address used in digital-computer programming as a reference for a group of relative addresses.

reference angle—The angle formed between the center line of a radar beam as it strikes a reflecting surface and the perpendicular drawn to that reflecting surface.

reference black level — The picture-signal level corresponding to a specified maximum limit for black peaks.

reference boresight—A direction defined by an optical, mechanical, or electrical axis of an antenna, established as a reference for the purpose of beam direction or tracking axis alignment.

reference burst—*See* Color Burst.

reference dipole—A half-wave straight dipole tuned and matched for a given frequency and used as a unit of comparison in antenna measurement work.

reference electrode—In pH measurements, an electrode, usually hydrogen-filled, used to provide a reference potential. (*See also* Glass Electrode.)

reference frequency—A frequency coinciding with, or having a fixed and specified relation to, the assigned frequency. It does not necessarily correspond to any frequency in an emission.

reference level—The starting point for designating the value of an alternating quantity or a change in it by means of decibel units. For sound loudness, the reference level is usually the threshold of hearing. For communications receivers, 60 microwatts is normally used. A common reference in electronics is one milliwatt, and power is stated as so many decibels above or below this figure.

reference line—A line from which angular measurements are made.

reference monitor — A receiver (or other similar device of known performance capabilities) used for judging the transmission quality.

reference noise—The magnitude of circuit noise that will produce a noise-meter reading equal to that produced by 10^{-12} watt of electric power at 1000 hertz.

reference oscillator — The high stability, usually crystal and temperature controlled, rf signal source used as a phase reference in phase-locked oscillators.

reference phase — The phase of the color burst transmitted with color-television carriers. It is used in synchronizing the receiver reference oscillator with the transmitted color signals.

reference point—A terminal that is common to both the input and the output circuits.

reference record—In digital-computer programming, a compiler output that lists the operations and their positions in the final specific routine, plus information describing the segmentation and storage allocation of the routine.

reference recording—A recording of a radio program for future reference or checking.

reference sound level — *See* Reference Acoustic Pressure.

reference time—In a computer, an instant chosen near the beginning of switching as an origin for time measurements. It is taken as the first instant at which either the instantaneous value of the drive pulse, the voltage response of the magnetic cell, or the integrated voltage response reaches a specified fraction of its peak pulse amplitude.

reference tone—A stable tone of known frequency continuously recorded on one track of multitrack signal recordings and intermittently recorded on signal-track recordings by the collection-equipment operators for subsequent use by the data analysts as a frequency reference.

reference voltage—Alternating-current voltage in a synchro servosystem used to determine the in-phase or 180° out-of-phase condition to provide directional sense.

reference volume — The magnitude of a complex electric wave, such as that corresponding to speech or music, which gives a reading of zero vu on a standard volume indicator. The sensitivity of the volume indicator is adjusted so that the reference volume or zero vu is read when the instrument is connected across a 600-ohm resistance to which there is delivered a power of 1 milliwatt at 1000 hertz.

reference white — The light from a nonselective diffuse reflector as a result of the normal illumination of the scene to be televised.

reference white level — In television, the picture-signal level corresponding to a specified maximum limit for white peaks.

reflectance—*See* Reflection Factor.

reflected impedance—1. The apparent impedance across the primary of a transformer when current flows in the secondary. 2. The impedance at the input terminals of a transducer as a result of the impedance characteristics at the output terminals. 3. The impedance seen at the input of a network when its output is terminated in an impedance of a specified value. (Most often, the 'network' referred to in this connection is a transformer.)

reflected power or signal—The power flowing back to the generator from the load.

reflected resistance — The apparent resistance across the primary of a transformer when a resistive load is across the secondary.

reflected wave—The wave which has been reflected from a surface, a junction of two different media, or a discontinuity in the medium it is traveling in (e.g., the echo from a target in radar, the sky wave in radio, the wave traveling toward the source from the termination of a transmission line).

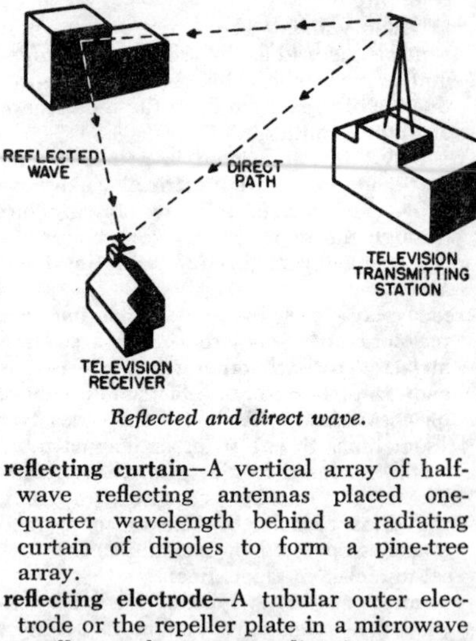

Reflected and direct wave.

reflecting curtain—A vertical array of half-wave reflecting antennas placed one-quarter wavelength behind a radiating curtain of dipoles to form a pine-tree array.

reflecting electrode—A tubular outer electrode or the repeller plate in a microwave oscillator tube corresponding in construction but not function to the plate of an ordinary triode. It is capable of generating extremely high frequencies.

reflecting galvanometer — A galvanometer with a small mirror attached to the moving element. The mirror reflects a beam of light onto a scale.

reflecting grating—An arrangement of wires placed in a waveguide to reflect the desired wave while freely passing one or more other waves.

reflection—The phenomenon in which a

wave that strikes a medium of different characteristics is returned to the original medium with the angles of incidence and reflection equal and lying in the same plane.

reflection altimeter—An aircraft altimeter that determines altitude by the reflection of sound, supersonic, or radio waves from the earth.

reflection coefficient—1. At the junction of a uniform transmission line and a mismatched terminating impedance, the vector ratios between the electric fields associated with the reflected and the incident waves. 2. At any specified plane in a uniform transmission medium, the vector ratios between the electric fields associated with the reflected and the incident waves. 3. At any specified plane in a uniform transmission line between a source and an absorber of power, the vector ratio between the electric fields associated with the reflected and the incident waves. It is given by the formula $(Z_2 - Z_1)/(Z_2 + Z_1)$, where Z_1 and Z_2 are the impedances of the source and load, respectively.

reflection color tube—A color picture tube that produces an image electron reflection in the screen region.

reflection Doppler—A system utilizing the Doppler frequency shift to measure the position and/or velocity of an object not carrying a transponder.

reflection error—In navigation, the error due to the wave energy that reaches the receiver as a result of undesired reflections.

reflection factor—1. Also called mismatch factor, reflectance, reflectivity, or transition factor. The ratio of the current delivered to a load whose impedance is not matched to the source, to the current that would be delivered to a load of matched impedance. Expressed as a formula:

$$\frac{\sqrt{4Z_1Z_2}}{Z_1 + Z_2}$$

where Z_1 and Z_2 are the unmatched and the matched impedances, respectively. 2. A measure of the effectiveness of a surface in reflecting light; the ratio of reflected lumens to the incident lumens. Also called reflectance and reflectivity.

reflection grating—A wire grating that is placed inside a waveguide so as to reflect a desired wave while at the same time allowing one or more other waves to pass freely.

reflection law — For any reflected object, the angle of incidence is equal to the angle of reflection.

reflection loss—1. That part of transmission loss due to the reflection of power at the discontinuity. 2. The ratio (in decibels) of the power incident upon the discontinuity, to the difference between the powers incident upon and reflected from the discontinuity.

reflection sounding—Echo depth sounding in which the depth is measured by reflecting sound or supersonic waves off the bottom of the ocean.

reflective code—A code that appears to be the mirror image of a normal counting code. The most useful property of reflective codes is that only one digit changes at a time in increasing or decreasing by 1. The reflective binary code is called a Gray code.

reflective jamming — *See* Radar-Confusion Reflectors.

reflective optics—A system of mirrors and lenses used in projection television.

reflectivity—*See* Reflection Factor, 2.

reflectometer — A microwave system arranged to measure the incidental and reflected voltages and indicate their ratio.

reflector—Also called a reflector element. 1. One or more conductors or conducting surfaces for reflecting radiant energy—specifically, a parasitic antenna element located in other than the general direction of the major lobe of radiation. 2. *See* Repeller. 3. An element in a vhf antenna which is situated to the rear of the main dipole. Reflects radio waves on to the dipole. 4. The "dish" often employed to reflect quiet or distant sounds in an open air on to a microphone.

reflector element—*See* Reflector.

reflector satellite—A satellite so designed that radio or other waves bounce off its surface.

reflector voltage—The voltage between the reflector electrode and the cathode in a reflex klystron.

reflex baffle—A speaker baffle in which a portion of the radiation from the rear of the diaphragm is propagated forward after a controlled phase shift or other modification. This is done to increase the overall radiation in some portion of the frequency spectrum.

reflex bunching—In a microwave tube, a type of bunching that is brought about when the velocity-modulated electron stream is made to reverse its direction by means of an opposing dc field.

reflex cabinet—A type of speaker enclosure fitted with a vent or port through which out-of-phase signals from the rear of the cone are "reflexed" by allowing the enclosed air in the cabinet to be tuned for a coupled resonance effect with the cone of the drive unit. The signals are then brought into phase with the front radiation from the cone of the speaker so as to reduce the "boomy" effect of resonance.

reflex circuit—1. A circuit through which

the signal passes for amplification both before and after detection. 2. A single stage of amplification that operates on two signals in widely separated frequency ranges.

reflex klystron—A klystron with a reflector (repeller) electrode in place of a second resonant cavity, to redirect the velocity-modulated electrons through the resonant cavity which produced the modulation. Such klystrons are well suited for use as oscillators, because the frequency is easily controlled by repositioning the reflector.

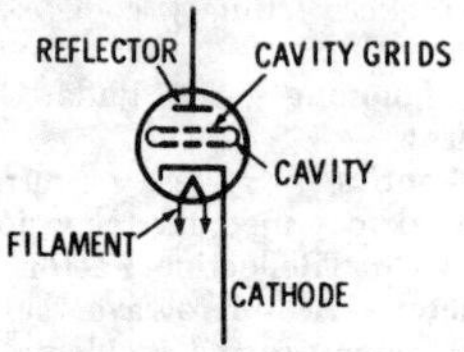

Reflex-klystron schematic.

reflowing—The melting of an electrodeposit followed by solidification. The surface has the appearance and physical characteristics of being hot-dipped (especially tin or tin alloy plates).

refracted wave—Also called the transmitted wave. In an incident wave, the portion that travels from one medium into a second medium.

refraction—The change in direction of propagation of a wavefront due to its passing obliquely from one medium into another in which its speed is different. Refraction may also occur in a single medium of varying characteristics.

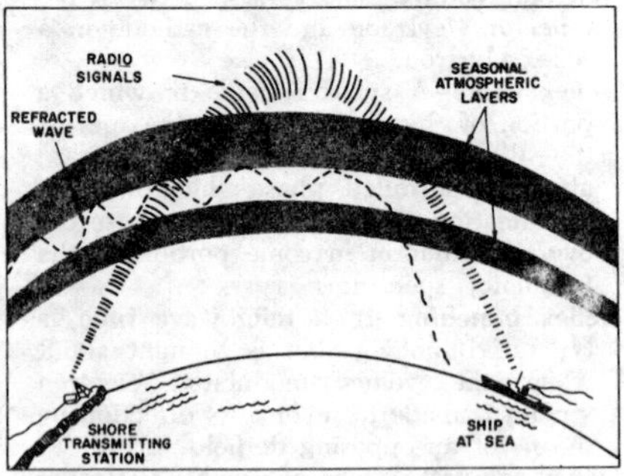

Refraction (of radio waves).

refraction error—In navigation, the error due to the bending of one or more wave paths by undesired refraction.

refraction loss—That part of the transmission loss due to refraction resulting from nonuniformity of the medium.

refractive index—Of a wave-transmission medium, the ratio between the phase velocity in free space and in the medium.

refractive modulus—Also called the excess modified index of refraction. The excess over unity of the modified index of refraction. It is expressed in millionths and is given by the equation:

$$M = (n + h/a - 1)10^6$$

where M is the refractive modulus, n is the index of refraction at a height h above sea level, and a is the radius of the earth.

refractivity—Ratio of phase velocity in free space to that in the medium, minus 1.

refractometer—An instrument for measuring the refractive index of a liquid or solid, usually from the critical angle at which total reflection occurs.

refractor metal—A metal that has an extremely high melting point; in the broad sense, a metal that has a melting point above those of iron, nickel, and cobalt.

refractory-metal-oxide semiconductor—*See* RMOS.

refrangible—Capable of being refracted.

refresh—The periodic renewing or restoring of data, or data-carrying electrical charge, in a semiconductor memory. Without refresh the data in certain MOS memories would quickly be lost.

refresh display—A crt device that requires the refresh of its screen presentation at a high rate in order that the image will not fade or flicker. The refresh rate is proportional to the decay rate of the phosphor. Refresh displays require continuous interaction with the host computer and display controller.

regeneration—1. The gain in power obtained by coupling from a high-level point back to a lower-level point in an amplifier or in a system which encloses devices having a power-level gain. Also called regenerative feedback and positive feedback. 2. In a computer storage device whose information storing state may deteriorate, the process of restoring the device to its latest undeteriorated state. (*See also* Rewrite.) 3. The replacement of a charge in a charge-storage tube to overcome decay effects, including a loss of charge by reading.

regeneration control—A variable capacitor, variable inductor, potentiometer, or rheostat used in a regenerative receiver to control the amount of feedback and thereby keep regeneration within useful limits.

regenerative amplification—Amplification where increased gain and selectivity are given by a feedback arrangement similar to that in a regenerative detector. However, the operation is always kept just below the point of oscillation.

regenerative braking—Dynamic braking in which the momentum, as the equipment is being braked, causes the traction motors to act as generators. A retarding

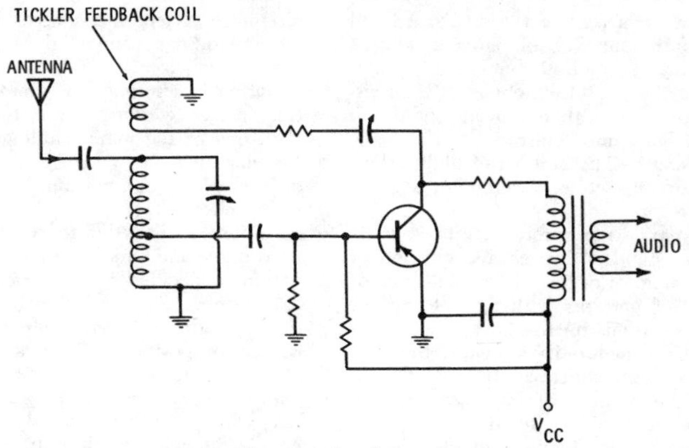

Regenerative detector.

force is then exerted by the return energy to the power-supply system.

regenerative detector—A detector circuit in which regeneration is produced by positive feedback from the output to the input circuit. In this way, the amplification and sensitivity of the circuit are greatly increased.

regenerative divider—Also called a regenerative modulator. A frequency divider in which the output wave is produced by modulation, amplification, and selective feedback.

regenerative feedback—*See* Regeneration, 1.

regenerative modulator—*See* Regenerative Divider.

regenerative receiver—A receiver in which controlled regeneration is used to increase the amplification provided by the detector stage.

regenerative repeater – 1. A repeater that regenerates pulses to restore the original shape. Used in teletypewriter and other code circuits, each code element is replaced by a new code element with specified timing, waveform, and magnitude. 2. Normally, a repeater utilized in telegraph applications. Its function is to retime and retransmit the received signal impulses restored to their original strength. These repeaters are speed- and code-sensitive, and are intended for use with standard telegraph speeds and codes.

regional channel – A standard broadcast channel within which several stations may operate at five kilowatts or less. However, interference may limit the primary service area of such stations to a given field-intensity contour.

regional interconnections – *See* Interconnection, 1.

region of limited proportionality – The range of applied voltage, below the Geiger-Mueller threshold, where the gas amplification depends on the charge liberated by the initial ionizing event.

register—1. A short-term storage circuit the capacity of which usually is one computer word. Variations may include provisions for shifting, calculating, etc. *See also* Static Shift Register *and* Dynamic Shift Register. 2. The relative position of all or part of the conductive pattern with respect to a mechanical feature of the board or to another pattern on the obverse side of the printed-circuit board (e.g., pattern-to-hole register or pattern-front-to-pattern-back register). 3. Also called registration. The accurate matching of two or more patterns such as the three images in color television. 4. A range of notes used for playing a particular piece or part of it (e.g., melody or harmony), particularly the range covered by a clavier or manual. 5. In an automatic-switching telephone system, the part of the system that receives and stores the dialing pulses that control the additional operations necessary to establish a telephone connection. 6. A computer memory on a smaller scale. The words stored therein may involve arithmetical, logical, or transferral operations. Storage in registers may be temporary, but even more important is their accessibility by the CPU. The number of registers in a microprocessor is considered one of the most important features of its architecture.

register constant—Symbolized by Kr. The factor by which the register reading must be multiplied in order to provide proper consideration of the register (gear) ratio and the instrument transformer ratios, to obtain the registration in the desired units.

register control—Any device that provides automatic register. In photoelectric register control, a light source and phototube form a scanning head. Whenever a special mark or a part of the design printed on a continuous web of paper arrives at the scanning head, the amount of light reaching the phototube changes. If necessary, the web is then moved slightly to bring it back into register.

register length—The number of digits, characters, or bits which a computer register can store.

register mark—In printed circuits, a mark used to establish the relative position of one or more printed-wiring patterns or portions of patterns with respect to their desired locations on the base.

register of a meter—In a meter, the part which registers the revolutions of the rotor, or the number of impulses received from or transmitted to the meter and gives the answer in units of electric energy or other quantity measured.

registration—The accuracy of relative position or concentricity of all functional patterns on any mask with the corresponding patterns of any other mask of a given device series of masks when the masks are properly superimposed.

registration, conductive pattern-to-board outline—The location of the printed pattern relative to the overall outline dimensions of the printed-circuit board.

registration, front-to-back—On a printed-circuit board, the location of the printed pattern on one side relative to the printed pattern on the opposite side.

registration of a meter — The apparent amount of electric energy (or other quantity being measured) that has passed through the meter, as shown by the register reading. It is equal to the register reading times the register constant. During a given period, it is equal to the register constant times the difference between the register readings at the beginning and end of the period.

registry—The superposition of one image onto another (e.g., in the formation of an interlaced scanning raster).

regular—Pertaining to reflection, refraction, or transmission in a definite direction rather than in a diffused or scattered manner.

regulated power supply — A unit which maintains a constant output voltage or current for changes in line voltage, output load, ambient temperature, or time.

regulating device—A device that functions to regulate a quantity or quantities such as voltage, current, power, speed, frequency, temperature, and load, at a certain value or between certain limits for machines, tie lines, or other apparatus.

regulating transformer—A transformer for adjusting the voltage or the phase relation (or both) in steps, usually without interrupting the load. It comprises one or more windings excited from the system circuit or a separate source, and one or more windings connected in series with the system circuit.

regulating winding — A supplementary transformer winding connected in series with one of the main windings and used for changing the ratio of transformation or the phase relationship, or both, between circuits.

regulation—1. The difference between the maximum and minimum voltage drops within a specified anode-current range in a gas tube. 2. The holding constant of some condition (e.g., voltage, current, power, or position). 3. In a power supply, the ability to maintain a constant load voltage or current despite changes in line voltage or load impedance. 4. The change in value of dc output voltage of a power supply resulting from a change in ac input voltage over the specified range from low (105 Vac) to high (125 Vac) or from high line to low line. Normally specified as the + or − change around the nominal ac input voltage.

regulation of a constant-current transformer—The maximum departure of the secondary current from its rated value, expressed in percent of the rated secondary current, with the rated primary voltage and frequency applied and at the rated secondary power factor and with the current variation taken between the limits of a short circuit and rated load.

regulation of a constant-potential transformer—The change in secondary voltage, expressed in percent of rated secondary voltage, which occurs when the rated kVA output at a specified power factor is reduced to zero, with the primary impressed terminal voltage maintained constant. In the case of a multiwinding transformer, the loads on all windings at specified power factors are to be reduced from the rated kVA to zero simultaneously.

regulator — 1. A device, the function of which is to maintain a designated characteristic at a predetermined value or to vary it according to a predetermined plan. 2. A device used to maintain a desired output voltage or current constant regardless of normal changes to the input or to the output load.

regulator tube—A two-electrode glow-discharge gas tube that has an essentially constant voltage drop. When series-connected with a resistance across a dc source, the tube will maintain a constant dc voltage across its terminal, with wide variations in the dc source voltage.

reignition — The generation of multiple counts, within a radiation-counter tube, by the atoms or molecules excited or ionized in the discharge accompanying a tube count.

reignition voltage – Also called restriking voltage. That voltage which is just sufficient to reestablish conduction of a gas tube if applied during the deionization period. It varies inversely with time during the deionization period.

Reike diagram – A polar-coordinate load diagram for microwave oscillators, particularly for klystrons and magnetrons.

Reinartz crystal oscillator–A crystal-controlled vacuum-tube oscillator in which the crystal current is kept low by placing in the cathode lead a resonant circuit turned to half the crystal frequency. The resultant regeneration at the crystal frequency improves the efficiency, but without the problem of uncontrollable oscillations at other frequencies.

reinforced insulation–An insulation providing protection against electrical shock hazard; equivalent to double insulation.

reinsertion of a carrier–Combining a locally generated carrier signal in a receiver with an incoming suppressed-carrier signal.

rejection band–The frequency range below the cutoff frequency of a uniconductor waveguide.

rejector–Filter or part of a circuit which rejects a particular frequency or band of frequencies.

rejector circuit–A circuit that suppresses or eliminates signals of the frequency to which it is tuned.

rejuvenator–A device or instrument for restoring the emissivity of a thermionic cathode by running it at an elevated temperature for a short period of time.

rel–A unit of reluctance, equal to one ampere-turn per magnetic line of force.

relative accuracy–1. The possible deviation among the standards in a group. 2. The input to output error as a fraction of full scale with gain and offset errors adjusted to zero. Relative accuracy is a function of linearity.

relative address–1. A designation used to identify the position of a memory location in a computer routine or subroutine. 2. A label used to identify a word in a routine or subroutine with respect to its position in that routine or subroutine.

relative bearing–The bearing in which the direction of the reference line is the heading of the vehicle.

relative binary–The primary form in which information is generated by a link editor and as such has all internal links resolved and is capable of holding address references such that they are relative, usually to one single base address.

relative coding–In a computer, coding in which all addresses refer to an arbitrarily selected position, or in which all addresses are represented symbolically.

relative damping of an instrument–Also called specific damping. Ratio of the actual damping torque at a given angular velocity of a moving element, to the damping torque which would produce critical damping at this same angular velocity.

relative detector response–A plot showing how the response (ability to detect a signal) varies with wavelength.

relative dielectric constant – Ratio of the dielectric constant of a material to that of a vacuum. The latter is arbitrarily given a value of 1.

relative gain of an antenna–The gain of an antenna in a given direction when the reference antenna is a half-wave loss-free dipole isolated in space and the equatorial plane of which contains the given direction.

relative humidity–1. Ratio of the quantity of water vapor in the atmosphere, to the quantity which would saturate at the existing temperature. 2. Ratio of the pressure of water vapor to that of saturated water vapor at the same temperature.

relative interference effect–With respect to a single-frequency electric wave in an electroacoustic system, the ratio (usually expressed in decibels) of the amplitude of a wave of specified reference frequency to that of the given wave when the two waves produce equal interference effects.

relative luminosity – Ratio of the actual luminosity at a particular wavelength, to the maximum luminosity at the same wavelength.

relative Peltier coefficient–The Peltier coefficient of a couple made up of the given material as the first-named conductor and a specified standard conductor, commonly platinum, lead, or copper.

relative permeability–Ratio of the magnetic permeability of one material to that of another, or of the same material under different conditions.

relative plateau slope – The average percentage change in the counting rate of a radiation-counter tube, near the midpoint of the plateau, per increment of applied voltage. It is usually expressed as the percentage change in counting rate per 100-volt change in applied voltage.

relative power–A power level referred to another power level.

relative power gain (of one transmitting or receiving antenna over another)–The measured ratio of the signal power one antenna produces at the receiver input terminals, to the signal power produced by the other, the transmitting power level remaining fixed.

relative refractive index–Ratio of the refractive indices of two media.

relative response – The ratio, usually expressed in decibels, of the response under some particular conditions to the response under reference conditions (which should be stated explicitly).

relative Seebeck coefficient—The Seebeck coefficient of a couple made up of the given material as the first-named conductor and a specified standard conductor such as platinum, lead, or copper.

relative spectral response — The output or response of a device as a function of wavelength normalized to the maximum value.

relative velocity (of a point with respect to a reference frame)—The rate at which a position vector of that point changes with respect to the reference frame.

relaxation—An action requiring an observable length of time for initiation in response to a sudden change in conditions.

relaxation circuit—A circuit arrangement, usually of vacuum tubes, reactances, and resistances, which has two states or conditions, one, both, or neither of which may be stable. The transient voltage produced by passing from one to the other, or the voltage in the state of rest, can be used in other circuits.

relaxation inverter—An inveter that uses a relaxation-oscillator circuit to convert dc power to ac power.

relaxation oscillator—1. An oscillator which generates a nonsinusoidal wave by gradually charging and quickly discharging a capacitor or an inductor through a resistor. The frequency of a relaxation oscillator may be self-determined, or determined by a synchronizing voltage derived from an external source. 2. An oscillator characterized by two semistable states such that when changed to either state the system will, after a time, recover the other state without external excitation.

relaxation time—1. The time an exponentially decaying quantity takes to decrease in amplitude by a factor of 0.3679. 2. The average time between collisions of an electron with the lattice.

relay—1. An electromechanical device in which contacts are opened and/or closed by variations in the conditions of one electric circuit and thereby affect the operation of other devices in the same or other electric circuits. 2. A transmission forwarded by way of an intermediate action.

relay bias—Bias produced by a spring on an electromagnet. By acting on the relay armature, the spring tends to hold it in a given position.

relay broadcast station—A station licensed to retransmit, from points where wire facilities are not available, the programs from one or more broadcast stations.

relay center — A central point at which switching of messages takes place.

relay contacts—Contacts that are closed or opened by the movement of a relay armature.

relay driver—A circuit with the high-voltage and high-current switching capability necessary for actuation of electromechanical relays.

relay drop—A relay activated by an incoming ringing current, to call an operator's attention to a telephone subscriber's line.

relay flutter—Erratic rather than positive operation and release of a relay.

relay function—Control of power in one circuit by means of a low power, isolated signal in another circuit.

relay magnet—A coil and iron core forming an electromagnet which, when energized, attracts the armature of a relay and thereby opens or closes the relay contacts.

relay receiver—A specific assembly of apparatus which accepts a sound or television relay signal at its input terminals, and delivers the amplified signal at its output terminals.

relay selector — A relay circuit associated with a selector, consisting of a magnetic impulse counter, for registering digits and holding a circuit.

relay station—*See* Relay Transmitter.

relay-station satellite — An artificial earth satellite intended to receive radio signals from the earth and retransmit them on command to other receiving stations.

relay system — Dial switching equipment made up principally of relays instead of mechanical switches.

relay transmitter—Also called a repeater or relay station (but only if the signal is reduced to a composite picture signal at a standard impedance level and polarity between the receiver and transmitter). The specific assembly of apparatus which accepts a sound or television-relay input signal from the relay receiver, and rebroadcasts it to another station outside the range of the operating station.

release—1. An electromagnetic device that

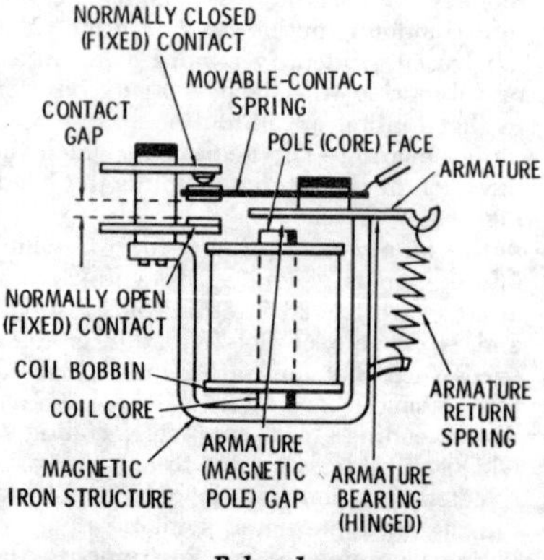

Relay, 1.

opens a circuit breaker automatically or allows a motor starter to return to its off position when tripped by hand, by an interruption of power-supply operation, or by an excessive current. 2. The condition attained by a relay when it has been de-energized, all contacts have functioned, and the armature (if applicable) has attained a fully opened position. 3. *See* Disconnect, 3.

release current – The maximum current needed to fully release a relay, after it has been fully closed.

release force—The value to which the force on the actuator of a momentary-contact switch must be reduced to allow the contacts to snap from the operated contact position to the normal contact position.

release time – 1. The time interval from relay coil de-energization to the functioning of the last contact to function. Where not otherwise stated, the functioning time of the contact in question is taken as its initial actuation time. 2. The time interval during which data must not change after a flip-flop has been clocked. Also called hold time.

releasing position—That position of the actuator of a momentary-contact switch at which the contacts snap from the operated contact position to the normal contact position.

reliability—The probability that a device will perform adequately for the length of time intended and under the operating environment encountered.

reliability assurance – The management and technical integration of the reliability activities essential in maintaining reliability achievements, including design, production and product assurance.

reliability control—The scientific coordination and direction of technical reliability activities from a system viewpoint.

reliability data—1. Data related to the frequency of failure of an item, equipment, or system. These data may be expressed in terms of "failure rate," "mean time between failures" (MTBF), or "probability of success." 2. Data contained in comprehensive documents that provide a detailed history of the reliability evaluation of component parts, component assemblies, etc., or the entire program during the design, development, production, and major product improvement phases of an equipment, weapon, or weapon system in which engineering studies have been performed to select the most reliable product for the intended application.

reliability engineering—The establishment, during design, of an inherently high reliability in a product.

reliability test—Tests and analyses carried out in addition to other type tests and designed to evaluate the level of reliability in a product, etc., as well as the dependability, or stability, of this level relative to time and use under various environmental conditions.

relieving anode – In a pool-cathode tube, an auxiliary anode which provides an alternative conducting path to reduce the current to another electrode.

relocatability – That characteristic of a compiled or assembled computer program segment that makes possible loading (locating) it into any region of memory.

relocatable binary—The form in which information is generated by a compiler or assembler and which is the primary input to a link editor.

relocate – In computer programming, to move a routine from one part of storage to another and adjust the necessary address references to permit execution of the routine in its new location.

relocation dictionary – In a computer, the part of an object or load module that provides identification of all relocatable address constants in the module.

reluctance – The resistance of a magnetic path to the flow of magnetic lines of force through it. It is the reciprocal of permeance and is equal to the magnetomotive force divided by the magnetic flux.

reluctance-type synchronous motor—An ac motor that runs at synchronous speed without excitation because of the salient pole rotor punchings of the laminations. The rotor has a squirrel cage type winding. Made in single- and 3-phase types.

reluctive transduction—The conversion of the measurand into a change in ac voltage by changing the reluctance path between two or more coils when ac excitation is applied.

reluctivity—The ability of a magnetic material to conduct magnetic flux. It is the reciprocal of permeability.

rem—*See* Roentgen Equivalent Man.

remanence—1. The extent to which a body remains magnetized after removal of a magnetizing field that has brought the body to its saturation (maximum) magnetization. A substance with remanence is known as ferromagnetic. 2. The magnetic induction which remains in a magnetic circuit after the removal of an applied magnetomotive force. (If there is an air gap in the magnetic circuit, the remanence will be less than the residual induction.)

remanent magnetization—The magnetization retained by a substance after the magnetizing force has been removed.

remodulator—A device for converting amplitude modulation to audio-frequency-shift modulation for transmission over a voice radio-frequency channel.

remote—*See* Field Pickup.

remote access—Having to do with communication with a data-processing facility by stations at a distance from the facility.

remote control—Any system of control performed from a distance. The control signal may be conveyed by intervening wires, sound (ultrasonics), light, or radio.

remote-cutoff tube—Also called a variable-mu or extended-cutoff tube. An electron tube used mainly in rf amplifiers. The control-grid wires are farther apart at the center than at the ends. Therefore, the amplification of the tube does not vary in direct proportion to the bias. Also, some plate current will flow, regardless of the negative bias on the grid.

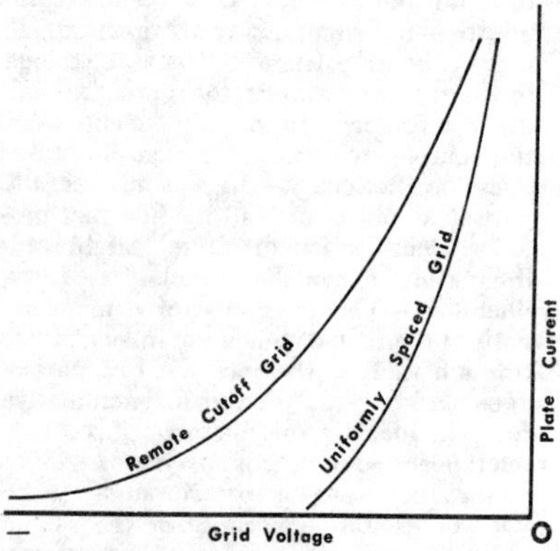

Remote-cutoff tube characteristic.

remote error sensing—A means by which the regulator circuit of a related power supply senses the voltage directly at the load. This connection is used to compensate for a voltage drop in the connecting wires.

remote indicator — 1. A radar indicator which is connected in parallel with a primary indicator. 2. An indicator located some distance away from the data-gathering element.

remote line—A program transmission line between a remote-pickup point and the studio or transmitter site.

remotely adjustable timer — A time delay having external leads to which a variable or fixed resistor can be attached.

remote metering—*See* Telemetering.

remote pickup—A program which originates away from the studio and is transmitted to the studio or transmitter over telephone lines or a radio link.

remote plan-position indicator — *See* PPI Repeater.

remote programming—A power supply feature whereby the controlled output parameter, voltage or current may be controlled through the application of external resistance, voltage signals, or current signals to designated external terminals.

remote sensing—In a power supply, terminations which allow the regulator to sense and regulate the output voltage at a remote location, usually the load. This connection is used to compensate for voltage drops in the power leads. The maximum voltage drop in both power leads must be specified.

remote station — Data terminal equipment for communication with a data-processing system that is distant electrically or in terms of time or space.

remote subscriber — A network subscriber without direct access to the switching center, but with access to the circuit through a facility such as a base message center.

remote terminal—A terminal that is a substantial distance from the central computer. Usually it accesses the computer through a telephone line or other type of communication link.

renewable fuse — A fuse which may be readily restored to operation by replacing the fused link.

reoperate time—The release time of a thermal relay.

rep—Abbreviation for roentgen equivalent physical.

repeatability—1. The ability of a device or an instrument to come back to the same reading after a certain length of time. 2. The ability of an instrument to repeat its readings taken when deflecting the pointer upscale, compared to the readings taken when deflecting the pointer downscale, expressed as a percentage of the full-scale deflection. 3. *See* Precision, 2.

repeatability error—The inability to reproduce the same output readings when a given level of measurand is applied repeatedly in the same direction.

repeat-cycle timer—A timer which has any number of load contacts and which continues to repeat its time program as long as power is applied.

repeater—1. An fm, tv, facsimile, or similar station that receives a signal on some input frequency and automatically transmits the received signal on some output frequency. The purpose of a repeater is to extend the communication range between a group of stations. The repeater generally consists of a receiver with its antenna, a control unit, and a transmitter, with its antenna. 2. Switch by which originating central office, calling telephone, dialed pulses are repeated to switches at a distant office. 3. Relay circuit in dial signaling, which amplifies and repeats dial pulses received from one circuit into another. 4. A device used to amplify and/or reshape signals.

repeater facility—Radio equipment needed to relay radio signals between central station, satellite station, and/or protected premises.

repeater jammer—Equipment intended to confuse or deceive the enemy by causing his equipment to present false information regarding azimuth, range, number of targets, etc. This result is achieved by a system that intercepts and reradiates a signal on the frequency of the enemy equipment.

repeater station—*See* Relay Tansmitter.

repeating coil—An audio-frequency transformer, usually with a 1:1 ratio, for connecting two sections of telephone line inductively, to permit the formation of simplex and phantom circuits.

repeating-coil bridge cord—A way of connecting the common office battery to the the midpoints of a repeating coil, which is bridged across the cord circuit.

repeating flash tube—A flash tube which, by producing rapid, brilliant flashes, permits night aerial photographs to be taken from as high as two miles.

repeating timer – A timer which repeats each operating cycle automatically until excitation is removed.

repeller—Sometimes called a reflector. An electrode, the primary function of which is to reverse the direction of an electron stream.

reperforator—1. A device that converts teletypewriter signals into perforations on tape instead of the usual typed copy on a roll of paper or ticker tape. 2. A machine that reads one punched paper tape or card and punches the same information into another paper tape or card.

reperforator/transmitter – An integrated unit for temporarily storing traffic for retransmission. It consists of a paper-tape punch and a paper-tape reader.

repertory dialer—A device that automatically places a telephone call to any one of the phone numbers stored therein.

repertory instruction—1. A set of instructions that a computing or data-processing system can perform. 2. A set of instructions assembled by an automatic coding system.

repetition equivalent – A measure of the quality of transmission experienced by the subscribers usng a complete telephone connection. It represents a combination of the effects of volume, distortion, noise, and all other subscriber reactions and usage.

repetition frequency—The number of repetitions of an event per unit time.

repetition instruction—A computer instruction that calls for one or more instructions to be executed an indicated number of times.

repetition rate – 1. The number of repetitions of an event per unit time. For example, in radar, the rate (usually expressed in pulses per second) at which pulses are transmitted (also called pulse repetition frequency, recurrence rate, or repetition frequency). 2. The number of repetitions per unit time requested by users of a telephone connection.

repetitive peak-inverse voltage—The maximum allowable instantaneous value of reverse (negative) voltage that may be repeatedly applied to the anode of an SCR with the gate open. This value of peak-inverse voltage does not represent a breakdown voltage, but it should never be exceeded (except by the transient rating if the device has such a rating).

repetitive unit—A type of circuit that appears more than once in a computer.

replacement theory – The mathematics of deterioration and failure, used in estimating replacement costs and selecting optimum replacement policies.

reply—In transponder operation, the radio-frequency signal or signals transmitted as a result of an interrogation.

repolarization—The process by which the normal resting potential of a biological cell is restored after the cell has "fired" or depolarized.

report—The output document produced by a data-processing system.

report generation—In a computer, production of complete output reports from only a specification of the desired content and arrangement and from specifications regarding the input file.

report generator—A special computer routine designed to prepare an object routine that, when later run on the computer, produces the desired report.

representative calculating time—A measure of the performance speed of a computer; it is the time required for performance of a specified operation or series of operations.

reproduce—In a computer, to prepare a duplication of stored information.

reproduce head—An electromagnetic transducer which converts the remanent flux pattern in a magnetic tape into electric signals during the reproduce (playback) process.

reproducer – A device used to translate electrical signals into sound waves.

reproducibility – 1. The exactness with which measurement of a given value can be duplicated. 2. *See* Precision.

reproducing stylus—A mechanical element that follows the modulations of a record groove and transmits the mechanical motion thus derived to the pickup mechanism.

reproduction speed—In facsimile, the area of copy recorded per time.

reprogrammable ROM—A ROM that can be programmed any number of times. Generally, however, the information stored in a reprogrammable ROM is changed very seldom.

repulsion–A mechanical force tending to separate bodies having a like electric charge or magnetic polarity–or in the case of adjacent conductors, currents flowing in opposite directions.

repulsion-induction motor–A constant or variable-speed repulsion motor with a squirrel-cage winding in the rotor, in addition to the regular winding.

repulsion motor–A single-phase motor in which the stator winding is connected to the source of power and the rotor winding to the commutator. Brushes on the commutator are short-circuited and are placed so that the magnetic axis of the rotor winding is inclined to that of the stator winding. This type of motor has a varying speed characteristic.

repulsion-start, induction-run motor – A single-phase motor which has the same windings as a repulsion motor but operates at a constant speed. The rotor winding is short-circuited (or otherwise connected) to give the equivalent of a squirrel-cage winding. It starts as a repulsion motor, but operates as an induction motor with constant-speed characteristics.

request repeat system–A system that uses an error-detecting code and is arranged so that when a signal is detected as being in error, a request for retransmission of the signal is initiated automatically.

reradiation–1. Scattering of incident radiation. 2. Radiation from a radio receiver resulting from insufficient isolation between the antenna circuit and the local oscillator and causing undesirable interference in other receivers.

rerecording–The process of making a recording by reproducing a recorded sound source and recording this reproduction. *See also* Dubbing.

rerecording system–An association of reproducers, mixers, amplifiers, and recorders capable of being used for combining or modifying various sound recordings to provide a final sound record. Recording of speech, music, and sound effects may be so combined.

rering locked in–A universal cord circuit feature by means of which, on magnetic lines, the called party or the calling party may ring the operator again, with the result that the supervisory lamps of the cord circuit remain lighted until the operator answers.

rerun–1. Also called rollback. To run a computer program (or a portion of it) over again. 2. To repeat an entire transmission. 3. In a computer a system that will restart the running program after a system failure. Snapshots of data and programs are stored at periodic intervals and the system rolls back to restart at the last recorded snapshot.

rerun point–In a computer program, one of a set of preselected points located in a computer program such that if an error is detected between two such points, the problem may be rerun by returning to the last such point instead of returning to the start of the problem.

rerun routine – A computer routine designed to be used, in the event of a malfunction or mistake, to reconstitute a routine from the previous rerun point.

reset–1. To restore a storage device to a prescribed state. 2. To place a binary cell in the initial, or zero, state. (*See also* Clear.) 3. An input to a binary, counter, or register that causes all binary elements to assume the zero logic state or the minimum binary state.

reset action–A type of control in which correction is proportional to both the length of time and the amount that a controlled process has deviated from the desired value, and provision is made to ensure that the process is returned to its setpoint.

reset pulse–A drive pulse which tends to reset a magnetic cell.

reset rate–The number of corrections made per unit of time by a control system; it usually is expressed in terms of the number of repeats per minute.

reset terminal – In a flip-flop, the input terminal used to trigger the circuit from its second state back to its original state. Also called clear terminal or zero-input terminal.

reset time–The period required from the time of the reset command until a timer is fully returned to the before-start conditions ready for the next cycle.

reset time–1. A timer which can be reset by electrical means. May be either an "on delay" or "off delay" type. 2. A timer with one or more circuits which spring-reset to zero when the clutch is disengaged.

residual–1. The difference between any value and some estimate of the mean of such values; e.g., residuals from a curve of regression. (Residual unbalance.) 2. The complement of a set is called a residual set. 3. (adjective) Pertaining to a measure of the output of a transducer under static conditions and with no stimulus applied.

residual charge–The charge remaining on the plates after an initial discharge of a capacitor.

residual current–1. The vector sum of the currents in the several wires of an electric supply circuit. 2. Current through a thermionic diode in the absence of anode voltage; it results from the velocity of the electrons emitted by the heated cathode.

residual deviation – Apparent modulation

due to noise and/or distortion in the transmitter.

residual discharge—A discharge of the residual charge of a capacitor remaining after the initial discharge.

residual error—The direction-finding errors remaining after errors due to site and antenna effects have been minimized.

residual field – The magnetic field left in an iron field structure after excitation has been removed.

residual flux—1. The value of magnetic induction that remains in a magnetic circuit when the magnetomotive force is reduced to zero. 2. In a uniformly magnetized sample of magnetic material the product of the residual flux density and the cross-sectional area. Residual flux is indicative of the output that can be expected from a tape at long wavelengths.

residual flux density – The magnetic flux density that exists at zero magnetizing field strength when a sample of magnetic material has undergone symmetrically cyclical magnetization.

residual fm—*See* Incidental FM.

residual frequency modulation—In a klystron, frequency modulation of the fundamental frequency due to shot and ion noises, ac heater voltage, etc.

residual gases—The small amounts of gases remaining in a vacuum tube despite the best possible exhaustion by vacuum pumps.

residual induction—The magnetic induction which remains in a magnetized material when the effective magnetizing force has been reduced to zero. When the material is in a symmetrically cyclic magnetic condition, the residual induction is termed the "normal residual induction." Abbreviated B_r.

residual ionization—Ionization of air or other gas not accounted for in a closed chamber by recognizable neighboring agencies.

residual losses—In a magnetic core, the difference between the total losses and the sum of the eddy-current and hysteresis losses.

residual magnetic induction—Magnetic induction remaining in a ferromagnetic object after the magnetizing force has been removed. The amount depends on the material, shape, and previous magnetic history.

residual magnetism—The magnetism which remains in the core of an electromagnet after the operating circuit has been opened.

residual modulation – *See* Carrier Noise Level.

residual screw—A brass screw in the center of a relay armature. It is used to adjust the residual air gap between the armature and the coil core, to prevent residual magnetism from holding the armature operated after the relay operating circuit has opened.

residual voltage—The vector sum of the voltages to ground of the several phase wires in an electric supply circuit.

resin—One of a class of solid or semisolid, natural or synthetic, organic products, generally with a high molecular weight and no definite melting point. Most resins are polymers.

resist—1. A material placed on the surface of a copper-clad base material to prevent the removal by etching of the conductive layer from the area covered. 2. A material deposited on conductive areas to prevent plating of the areas covered.

resistance – 1. A property of conductors which—depending on their dimensions, material, and temperature – determines the current produced by a given difference of potential; that property of a substance which impedes current and results in the dissipation of power in the form of heat. The practical unit of resistance is the ohm. It is defined as the resistance through which a difference of potential of one volt will produce a current of one ampere. 2. A circuit element designed to offer a predetermined resistance to current. 3. Ratio of the applied electromotive force to the resulting current in a circuit. It is a measure of the resistance of the circuit to the passage of an electric current and is measured in ohms. Its value is determined from:

$$R = \frac{E}{I}$$

where, E is voltage in volts, and I is current in amperes.

resistance balance—The amount of resistance which is required to null the output of certain transducers or input systems.

resistance box—An assembly of resistors and the necessary switching or other means for changing the resistance connected across its output terminals by known, fixed amounts. (*See also* Decade Box.)

resistance brazing – Brazing by resistance heating, the joint being part of the electrical circuit.

resistance bridge – A common form of Wheatstone bridge employing resistances in three arms. (*See also* Wheatstone Bridge.)

resistance-bridge pressure pickup—A pressure transducer in which the electrical output is derived from the unbalance of a resistance bridge, which is varied according to the applied pressure.

resistance-capacitance–coupled amplifier—An amplifier, the stages of which are connected by a suitable arrangement of resistors and capacitors.

resistance-capacitance coupling—Also called RC coupling. Coupling between two or more circuits, usually amplifier stages, by a combination of resistive and capacitive elements.

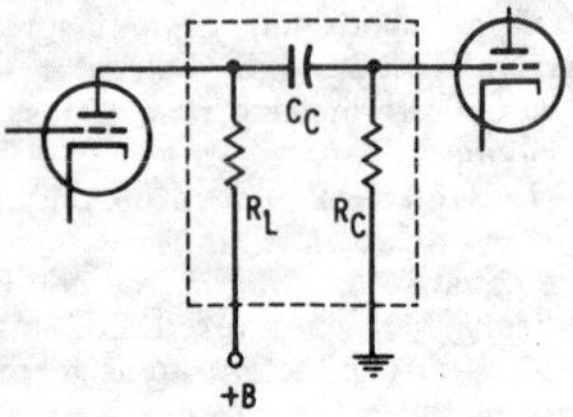

Resistance-capacitance coupling.

resistance-capacitance filter—A filter made up only of resistive and capacitive elements. Abbreviated RC filter.

resistance-capacitance oscillator—An oscillator the output frequency of which is determined by resistance and capacitance elements. Abbreviated RC oscillator.

resistance coupling — Also called resistive coupling. The association of circuits with one another by means of the mutual resistance between circuits.

resistance drop—The voltage drop occurring across two points on a conductor when current flows through the resistance between those points. Multiplying the resistance in ohms by the current in amperes gives the voltage drop in volts.

resistance furnace—An electric furnace in which the heat is developed by the passage of current through a suitable resistor, which may be the charge itself or a resistor imbedded in or surrounding the charge.

resistance grounded—Grounded through a resistance.

resistance lamp—An electric lamp used as a resistance to limit the amount of current in a circuit.

resistance loss—The power lost when current flows through a resistance. Its value in watts is equal to the resistance in ohms multiplied by the square of the current in amperes ($W = R \times I^2$).

resistance magnetometer—A magnetometer which depends for its operation on the variation in the electrical resistance of a material immersed in the field to be measured.

resistance material—A material having sufficiently high resistance per unit length or volume to permit its use in the construction of resistors.

resistance noise—*See* Thermal Noise, 1.

resistance pad—A network employing only resistances. It is used to provide a fixed amount of attenuation without altering the frequency response.

resistance ratio—In a thermistor, the ratio of the resistances measured at two specified reference temperatures with zero power in the thermistor.

resistance standard—*See* Standard Resistor.

resistance-start motor — A form of split-phase motor having a resistance connected in series with the auxiliary winding. The auxiliary circuit opens whenever the motor attains a predetermined speed.

resistance strain gage—A strain gage consisting of a small strip of resistance material cemented to the part under test. Its resistance changes when the strip is compressed or stretched.

resistance temperature coefficient—The ratio of the resistance change of an element between two temperatures to the product of the temperature change and the original resistance. A positive value of the coefficient indicates an increase of resistance with an increase in temperature; a negative value indicates a decrease in resistance with an increase in temperature, a zero value indicates no resistance change with temperature.

resistance temperature detector — Also called resistance-thermometer detector or resistance-thermometer resistor. A resistor made of some material for which the electrical resistivity is a known function of the temperature. It is intended for use with a resistance thermometer and is usually in such a form that it can be placed in the region where the temperature is to be determined.

resistance temperature meter—*See* Resistance Thermometer.

resistance thermometer—1. Also called resistance temperature meter. An electric thermometer which has a temperature-responsive element called a resistance temperature detector. Since the resistance is a known function of the temperature, the latter can be readily determined by measuring the electrical resistance of the resistor. 2. Thermometer, using variation of resistance with temperature of some material, usually platinum, considered the standard for temperature measurement over its range.

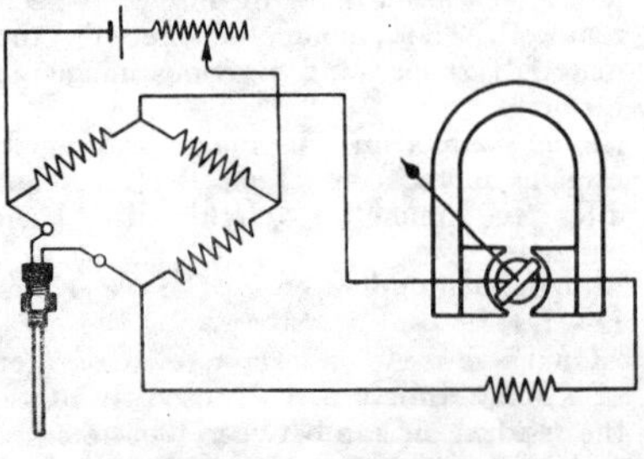

Resistance thermometer.

resistance-thermometer detector—*See* Resistance Temperature Detector.

resistance-thermometer resistor – *See* Resistance Temperature Detector.

resistance thermometry – A temperature-measuring technique that utilizes the temperature coefficient of a wirewound resistor. Known as a resistance thermometer, this resistor consists of a spiral of nickel or platinum wire. Since the ohmic value of the wire varies with temperature, the resistance of the spiral is thus an indication of temperature.

resistance weld—The junction produced by heat obtained from the resistance of the work to the flow of electric current in a circuit of which the work is a part, and by the applications of pressure before and during the flow of current. The term includes all types of bonds produced by the process, which may or may not be classified metallurgically as welds.

resistance welding—Welding in which the metals to be joined are heated to melting temperature at their points of contact by a localized electric current while pressure is applied.

resistance wire—A wire made from a metal or alloy having a high resistance per unit length (e.g., Nichrome). It is used in wirewound resistors, heating elements, and other high-resistance circuits.

resist-etchant—Any material deposited onto a copper-clad base material to prevent the conductive area underneath from being etched away.

resistive conductor—A conductor used primarily because of its high electrical resistance.

resistive coupling – *See* Resistance Coupling.

resistive load—A load in which the voltage is in phase with the current.

resistive transduction—The conversion of the measurand into a change in resistance.

resistive unbalance—Unequal resistance in the two wires of a transmission line.

resistivity—1. A measure of the resistance of a material to electric current either through its volume or on a surface. The unit of volume resistivity is the ohm-centimeter; the unit of surface resistivity is the ohm. 2. The ability to resist current; the reciprocal of conductivity.

resistor—A component made of a material (like carbon) that has a specified resistance, or opposition to the flow of electrical current. Resistors are used to control (or limit) the amount of current flowing in a circuit or to provide a voltage drop.

resistor-capacitor-transistor logic – Abbreviated RCTL. A logic circuit design that employs a resistor and a speedup capacitor in parallel for each input of the gate. A transistor's base is connected to one end of the RC network. A positive voltage on the RC input will energize the transistor and turn it on, so that the output voltage is nearly zero volts. This circuit is a positive NOR or negative NAND when npn transistors are used in the circuit.

resistor color code—A code adopted by the Electronic Industries Association to mark the values of resistance on resistors in a readily recognizable manner. The first color represents the first significant figure of the resistor value, the second color the second significant figure, and the third color represents the number of zeros following the first two figures. A fourth color is sometimes added to indicate the tolerance of the resistor.

resistor core—An insulating support around which a resistor element is wound or otherwise placed.

resistor element—1. That portion of a resistor which possesses the property of electric resistance. 2. That portion of a potentiometer which provides the change in resistance as the shaft is rotated.

resistor housing—The enclosure around the resistance element and the core of a resistor.

resistor starting (or starter) – A motor starter using resistance to limit inrush current. The resistors are shorted by a paralleling contactor on the final step. A nontransition type of starting.

resistor-transistor logic—Abbreviated RTL. A form of logic that has a resistor as the input component that is coupled to the base of an npn transistor. As in RCTL, the transistor is an inverting element that produces the positive NOR gate or the negative NAND gate function.

resist plating—Any material which, when deposited on a conductive area, prevents the areas underneath from being plated.

resnatron – A high-power cavity-resonator tetrode for high-efficiency operation in the very-high-frequency and ultrahigh-frequency bands. It is water cooled, and the cavities form an integral part of the tube.

resolution—1. The deriving of a series of discrete elements from a sound, scene, or other form of intelligence so that the original may subsequently be synthesized. 2. The degree to which nearly equal values of a quantity can be discriminated. 3. The degree to which a system or a device distinguishes fineness of detail in a spatial pattern. 4. In facsimile, a measure of the narrowest line width that may be transmitted and reproduced. 5. A measure of the smallest possible increment of change in the variable output of a device. 6. In a potentiometer, the smallest possible incremental resistance change. 7. The reciprocal number of steps per revolution of a motor shaft, expressed in degrees per step.) 8. The degree to which the dis-

RESISTOR COLOR CODES			
COLOR	DIGIT	MULTIPLIER	TOLERANCE
BLACK	0	1	±20%
BROWN	1	10	±1%
RED	2	100	±2%
ORANGE	3	1000	±3%
YELLOW	4	10000	GMV
GREEN	5	100000	±5% (EIA ALTERNATE)
BLUE	6	1000000	±6%
VIOLET	7	10000000	±12-1/2%
GRAY	8	.01 (EIA ALTERNATE)	±30%
WHITE	9	.1 (EIA ALTERNATE)	±10% EIA ALTERNATE
GOLD		.1 (JAN AND EIA PREFERRED)	±5% (JAN AND EIA PREFERRED)
SILVER		.01 (JAN AND EIA PREFERRED)	±10% (JAN AND EIA PREFERRED)
NO COLOR			±20%

GMV=GUARANTEED MINIMUM VALUE, OR - 0 + 100% TOLERANCE
±3, 6, 12-1/2, AND 30% ARE ASA 40, 20, 10 and AND 5 STEP TOLERANCES.

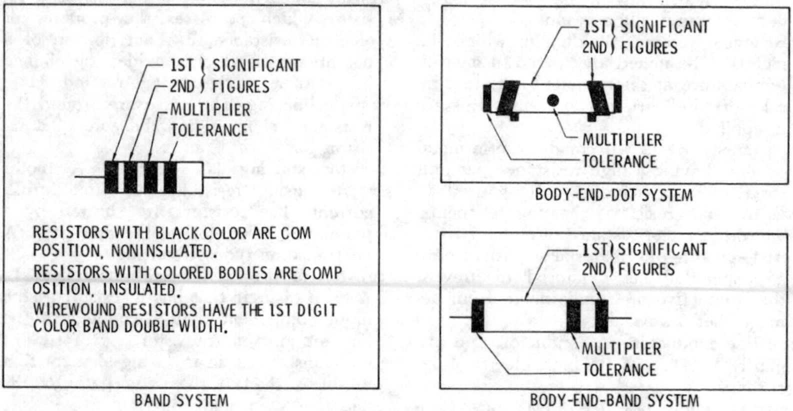

Resistor color code.

tance separating different states of magnetization recorded along a tape can be reduced and still permit these states to be distinguished usefully on reproduction. 9. In radar, the minimum angular or distance separation between two targets that permits them to be distinguished on the radar screen. 10. In television, the maximum number of lines discernible on the screen in a distance equal to the tube height. 11. The degree to which significant signals can be extracted from comparatively random signals, as with a radio telescope.

resolution chart—1. A pattern of black and white lines used to determine the resolution capabilities of equipment. 2. A chart used to examine the definition, linearity and contrast of television systems.

resolution noise – The noise due to the stepped character of the resistance element in wirewound potentiometric transducers.

resolution wedge—A narrow-angled, wedge-shaped pattern calibrated for the measurement of resolution. It is composed of alternate contrasting strips which gradually converge and taper individually to preserve equal widths along a line drawn perpendicular to the axis of the wedge.

resolver—1. A means for resolving a vector into two mutually perpendicular components. 2. A transformer, the coupling between primary and secondary of which can be varied. 3. A small section with a faster access than the remainder of the magnetic-drum memory in a computer. 4. A device which separates or breaks up a quantity into constituent parts or elements. 5. An electromechanical transducing device which develops an output voltage proportional to the product of an input voltage and the sine of the shaft angle.

resolving cell—In radar, a volume in space the diameter of which is the product of slant range and beamwidth and the length of which is the pulse length.

resolving power—1. The reciprocal of the beam width in a unidirectional antenna, measured in degrees. It may differ from the resolution of a directional-radio system, since the latter is affected by other factors as well. 2. The ability of an optical instrument to distinguish closely spaced points in an optical image, small angles between light beams, or components of light beams with small wavelength differences.

resolving time—1. The minimum time interval by which two events must be separated to be distinguishable. 2. In computers, the shortest time interval between trigger pulses for which reliable operations of a binary cell can be obtained.

resonance—1. A circuit condition whereby the inductive- and capacitive-reactance (or -impedance) components of a circuit have been balanced. In usual circuits, resonance can be obtained for only a comparatively narrow frequency band or range. 2. In a mechanical system, the frequency at which the maximum displacement occurs.

resonance bridge—A four-arm, alternating-current bridge normally used for measuring inductance, capacitance, or frequency. An inductor and a capacitor are both present in one arm, the other three arms being (usually) nonreactive resistors. The adjustment for balance includes the establishment of resonance for the applied frequency. Two general types—series or parallel—can be distinguished, depending on how the inductor and capacitor are connected.

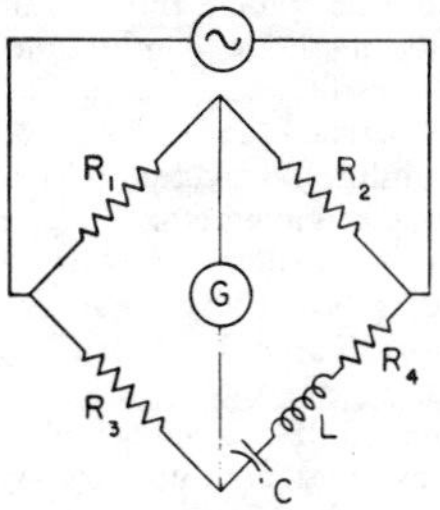

$$R_1R_4 = R_2R_3$$

$$\omega^2LC = 1$$

Resonance bridge (series).

resonance characteristics – *See* Resonance Curve.

resonance curve – Also called a resonant curve or resonance characteristic. A graphical representation of how a tuned circuit responds to the various frequencies at and near resonance.

resonance indicator—A meter, neon lamp, headphone, etc., that indicates when a circuit is at resonance.

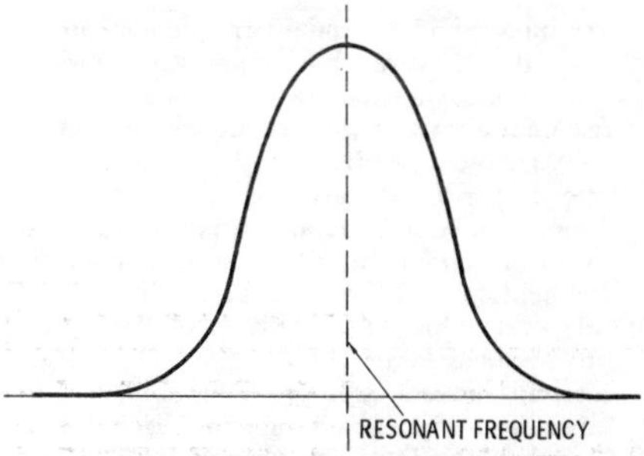

Resonance curve.

resonance radiation—Radiation from a gas or vapor due to excitation and having the same frequency as the exciting source (e.g., sodium vapor irradiated with sodium light).

resonant capacitor – A tubular capacitor that is purposely wound so as to have inductance in series with its capacitance to resonate at a predetermined (if) frequency. Used as a bypass capacitor in if amplifiers for more effective bypassing.

resonant cavity—A form of resonant circuit in which the current is distributed on the inner surface of an enclosed chamber. By making the chamber of the proper dimensions, it is possible to give the circuit a high *Q* at microwave frequencies. The resonant frequency can be changed by adjusting screws which protrude into the cavity, or by changing the shape of the cavity.

resonant-chamber switch – A waveguide switch in which a tuned cavity is placed in each waveguide branch; detuning of a cavity prevents the flow of energy in the associated branch.

resonant charging choke—A modulator inductor which sets up an oscillation of a given charging frequency with the effective capacitance of a pulse-forming network in order to charge a line to a high voltage.

resonant circuit—A circuit which contains both inductance and capacitance and is therefore tuned to resonance at a certain

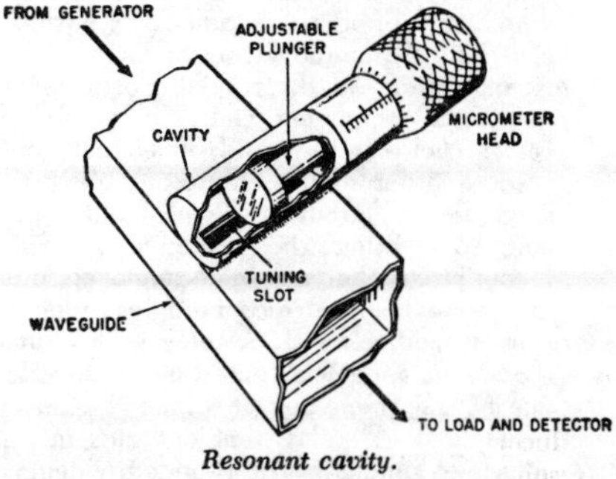

Resonant cavity.

frequency. The resonant frequency can be raised or lowered by changing the inductance and/or capacitance values.

resonant-circuit–type frequency indicator —A frequency-indicating device which depends for its operation on the frequency-versus-reactance characteristics of two series-resonant circuits. The circuit is arranged so that the deflecting torque is independent of the amplitude of the signal to be measured.

resonant-current step-up—The ability of a parallel-resonant circuit to circulate a much higher current through its inductor and capacitor than the current fed into the circuit.

resonant curve—*See* Resonance Curve.

resonant diaphragm – In waveguide technique, a diaphragm so proportioned that it does not introduce reactive impedance at the design frequency.

resonant frequency – 1. The frequency at which a given system or object will respond with maximum amplitude when driven by an external sinusoidal force of constant amplitude. For an LC circuit, the resonant frequency is determined by the formula:

$$f = \frac{1}{2\pi\sqrt{LC}}$$

where, f is in hertz, L is in henrys, and C is in farads. 2. The frequency of a crystal unit for a particular mode of vibration to which, discounting dissipation, the effective impedance of the crystal unit is zero.

resonant gap—The small region where the electric field is concentrated in the resonant structure inside a tr tube.

resonant gate transistor—A surface field-effect transistor incorporating a cantilevered beam which resonates at a specific frequency to provide high-Q frequency discrimination.

resonant line—A transmission line in which the distributed inductance and capacitance are such that the line is resonant at the frequency it is handling.

resonant-line oscillator – An oscillator in which one or more sections of transmission line are employed as tanks.

resonant mode—In the response of a linear device, a component characterized by a certain field pattern and, when not coupled to other modes, representable as a single-tuned circuit. When modes are coupled together, the combined behavior is similar to that of single-tuned circuits that have been correspondingly coupled.

resonant-reed relay—1. A relay with multiple contacts, each actuated by an ac voltage of the frequency at which the reeds resonate. 2. A relay that operates in response to signals of the proper frequency, power level and time duration. Usually used for multicircuit control.

resonant resistance – The resistance value to which a resonant circuit is equivalent.

resonant voltage step-up—The ability of an inductor and a capacitor in a series-resonant circuit to deliver a voltage several times greater than the input voltage.

resonant window—A parallel combination of inductive and capacitive diaphragms used in a waveguide structure so that transmission occurs at the resonant frequency and reflection occurs at other frequencies.

resonate – To bring to resonance—i.e., to maximize or minimize the amplitude or other characteristic of a steady-state quantity.

resonating cavity—A waveguide that is adjustable in length and terminates in a metal piston, diaphragm, or other wave-reflecting device at either or both ends. It is used as a filter, a means of coupling between guides of different sizes, or an impedance network.

resonator – An apparatus or system in which some physical quantity can be made to oscillate by oscillations in another system.

resonator cavity—A section of coaxial line or waveguide completely enclosed by conductive walls.

resonator grid—A grid attached to a cavity resonator in a velocity-modulated tube to couple the resonator and the electron beam.

resonator mode—A condition of operation corresponding to a particular field configuration for which the electron stream introduces negative conductance into the coupled circuit.

resonator wavemeter – A resonant circuit for determining wavelength (e.g., a cavity-resonator wavemeter).

resonistor—An experimental resonating device that is essentially a cantilevered chip of silicon mounted, like a tiny diving board, on a substrate. It measures 0.0350 inch long, 0.090 inch wide, and 0.008 inch thick, and contains two special diffused resistors. When an input signal is applied, it causes the silicon chip to vibrate, and the resonistor then delivers an output voltage with a specific stable frequency.

resource—Any facility of a computing system or operating system that is necessary for a job or task; for example, main storage, input-output devices, central processing unit, data sets, and control processing programs.

resource-sharing—The sharing of one central processor by both several users and several peripheral devices. Principally used in connection with the sharing of time and memory.

responder—The part of a transponder that automatically transmits a reply to the interrogator-responser. By contrast, the responser is the receiver which accepts and interprets the signals from the transponder.

response—A quantitative expression of the output of a device or system as a function of the input, under conditions which must be explicitly stated. The response characteristic, often presented graphically, gives the response as a function of some independent variable such as frequency or direction.

response curve—1. A plot of output versus frequency for a specific device. 2. A plot of stimulus versus output. 3. A graphical representation of frequency response. Usually measured in decibels, with reference a given level on a vertical scale. When the response curve of an amplifier, pickup, microphone, etc., is accurately plotted it represents the relative levels of amplitude at all frequencies within a specified bandwidth.

responser—*See* Responsor.

response speed—The time for a control action to start after a temperature change has occurred at the sensor.

response time—1. The time (usually expressed in cycles of the power frequency) required for the output voltage of a magnetic amplifier to reach 63% of its final average value in response to a step-function change of signal voltage. 2. The time required for the pointer of an instrument to come to apparent rest in its new position after the measured quantity abruptly changes to a new, constant value. 3. *See* Recovery Time, 6. 4. The time required to reach a specified percentage (e.g., 90, 98, 99%) of the final output value. 5. *See* Transient Recovery Time, 6. The elapsed time between generation of an inquiry at a computer terminal and receipt of a response at the same terminal. This includes the time for transmission to the computer, processing at the computer, and transmission back to the terminal.

responsitivity—In a photosensor, the ratio of the change in photocurrent to the change in incident radiant flux density.

responsor—1. Also spelled responser. The receiver used to receive and interpret the signals from a transponder. 2. An electronic device used to receive an electronic challenge and to display a reply thereto.

restart—To re-establish performance of a computer routine, using the information recorded at a checkpoint.

resting frequency—*See* Center Frequency, 1.

resting potential — The voltage (typically about 80 millivolts) between the inside and outside of a nerve cell or muscle cell.

restore—1. To do periodic charge regeneration of a volatile computer storage system. 2. In computers, to regenerate. 3. In a computer, to return a cycle index or variable address to its initial value. 4. To store again.

restorer—*See* DC Restorer.

restorer pulses — In a computer, pairs of complement pulses applied to restore the charge of the coupling capacitor in an ac flip-flop.

restoring spring—Also called return spring. A spring that moves the armature to the normal position and holds it there when the relay is de-energized.

rest potential—The residual potential difference remaining between an electrode and an electrolyte after the electrode has become polarized.

restriking voltage—*See* Reignition Voltage.

resultant—The effect produced by two or more forces or vectors.

retained image—Also called image burn. A change that is produced on the target of a television camera tube by a stationary light image and that results in the production of a spurious electrical signal corresponding to the light image for a large number of frames after the image is removed.

retardation coil — A high-inductance coil used in telephone circuits to permit passage of dc or low-frequency current while blocking audio-frequency currents. *See also* Inductor.

retarding-field oscillator—Also called a positive-grid oscillator. An oscillator tube in which the electrons move back and forth through a grid which is more positive than the cathode and plate. The frequency depends on the electron-transit time and sometimes on the associated circuit parameters. The field around the grid retards the electrons and draws them back as they pass through it in either direction. Barkhausen-Kurz and Gill-Morell oscillators are examples of a retarding-field oscillator.

retarding magnet — Also called a braking magnet or drag magnet. A magnet used for limiting the speed of the rotor in a motor-type meter.

retard transmitter—A transmitter in which a delay is introduced between the time it is actuated and the time transmission begins.

retention time—The maximum time after writing into a storage tube that an acceptable output can be obtained by reading.

retentivity — That property of a material measured by the normal residual induction remaining after the removal of an applied magnetizing force corresponding to the saturation induction for the material.

retentivity of vision – The image retained momentarily by the mind after the view has left the field of vision. (*See also* Persistence of Vision.)

RETMA–Abbreviation for Radio-Electronics-Television Manufacturers Association, now the Electronic Industries Association (EIA).

retrace–*See* Flyback, 2.

retrace blanking–The blanking of a television picture tube during vertical-retrace intervals to prevent the retrace lines from being visible on the screen.

retrace interval–*See* Return Interval.

retrace line–Also called the return line. The line traced by the electron beam in a cathode-ray tube as it travels from the end of one line or field to the start of the next line or field.

retrace time–*See* Return Interval.

retractile spring–The spring which tends to open the armature of an electromagnetic device and which holds the armature open when its force is not overcome by magnetic attraction.

retransmission unit–A control unit used at an intermediate station for automatically feeding one radio receiver-transmitter unit from another for two-way communications.

retrieve–In a computer, to select specific information.

retrodirective reflector – A reflector which redirects incident flux back toward the point of origin of the flux.

retrofit–To fit an earlier system so it can be compatible with later technology.

return–1. A received radar signal. 2. To go back to a planned point in a computer program and run part of the program again (usually because an error has been detected).

return code–In a computer, a code used to influence the carrying out of following programs.

return-code register–In a computer, a register used for storage of a return code.

return interval–Also called retrace interval, retrace time, or return time. The interval corresponding to the direction of sweep not used for delineation.

return line–*See* Retrace Line.

return lines – Conductors connecting the loads to the lowest-potential power-supply terminal. Lowest potential means most nearly zero with respect to the ground point, regardless of polarity. The term "ground" is reserved for a single point; return lines include all means of connecting the low-potential terminal of the power supply to the load such as ground busses and chassis.

return loss – 1. At a discontinuity in a transmission system, the difference between the power incident upon, and the power reflected from, the discontinuity. 2. The ratio in decibels of (1) above.

return spring–*See* Restoring Spring.

return time–*See* Return Interval.

return trace–The path of the scanning spot in a cathode-ray tube during the return interval.

return transfer function – In a feedback control device, the transfer function that relates a loop return signal to the corresponding loop input signal.

return wire–The ground, common, or negative wire of a direct-current circuit.

reusable routine–A computer routine that can be used by more than one task.

reverberation–1. The persistence of sound due to the repeated reflections from walls, ceiling, floor, furniture, and occupants in a room or auditorium. 2. A slight, tapering prolongation of sounds due to multiple reflections in a large auditorium. As distinguished from echo, which is (acoustically) a sudden return of sound rather than a smooth decay. 3. The act of sound or pressure waves being reflected by the surfaces of an enclosure.

reverberation chamber – An enclosure in which all surfaces have been made as sound-reflective as possible. It is used for certain acoustic measurements.

reverberation period – The time required for the sound in an enclosure to die down to one millionth (60 dB) of its original intensity.

reverberation strength–The difference between (a.) the level of a plane wave that produces in a nondirectional transducer a response equal to that produced by the reverberation corresponding to a 1-yard range from the effective center of the transducer and (b.) the index level of the pulse transmitted by the same transducer on any bearing.

reverberation time–The time required for sound energy in an enclosure to decay by 60 dB, i.e., fall to one-millionth of its initial intensity. This time is directly proportional to the volume of the enclosure and inversely proportional to the total absorption in the enclosure. It should be less than 1 second for speech, and 2 to 3.5 seconds for a large concert orchestra.

reverberation-time meter – An instrument for measuring the reverberation time of an enclosure.

reverberation unit–A circuit or device that adds an artificial echo to a sound being reproduced or transmitted.

reverberator – Any of various electromechanical devices that process an audio signal in such a way as to simulate the effects of reverberation.

reversal – A change in the direction of transmission or polarity.

reverse–As distinguished from high speed

rewind, a distinguished function that allows a tape recorder to change head configuration and tape direction at the end of a tape so as to continue playing (or recording) on the reverse tracks without switching reels or turning over the cassette.

reverse bias—Also called back bias. An external voltage applied to a semiconductor pn junction to reduce the flow the of current across the junction and thereby widen the depletion region. It is the opposite of forward bias.

reverse-blocking pnpn-type switch — A pnpn-type switch which exhibits a reverse-blocking state when its anode-to-cathode voltage is negative and does not switch in the normal manner of a pnpn-type switch.

reverse-blocking diode thyristor—A two-terminal thyristor which for negative anode-to-cathode voltage does not switch, but exhibits a reverse-blocking state.

reverse-blocking triode thyristor—A three-terminal thyristor which for negative anode-to-cathode voltage does not switch, but exhibits a reverse blocking state. This is the device that is often known in the power field as an SCR (semiconductor controlled rectifier).

reverse-breakdown voltage — The voltage that produces a sharp increase in reverse current in a semiconductor, without a significant increase in voltage.

reverse-conducting diode thyristor—A two-terminal thyristor which for negative anode-to-cathode voltage does not switch, but conducts large currents at voltages comparable in magnitude to the on-state voltage.

reverse-conducting triode thyristor — A three-terminal thyristor which for negative anode-to-cathode voltage does not switch, but conducts large currents at voltages comparable in magnitude to the on-state voltage.

reverse coupler—A directional coupler used for sampling reflected power.

reverse current—*See* Back Current.

reverse-current relay—A relay that operates whenever current flows in the reverse direction.

reversed feedback—*See* Negative Feedback.

reversed feedback amplifier—An amplifier in which inverse feedback is employed to reduce harmonic distortion and otherwise improve fidelity.

reverse direction—Also called inverse direction. The direction of greater resistance to current through a diode or rectifier.

reverse-direction flow—In flowcharting, a flow in a direction other than from left to right or top to bottom.

reverse emission—*See* Back Emission.

reverse gate-to-source breakdown voltage—The breakdown voltage between the gate and source terminals of an insulated gate field effect transistor with a reverse gate-to-source voltage applied, and all other terminals short-circuited to the source terminal.

reverse key—A key used in a circuit to reverse the polarity of that circuit.

reverse leakage current — The current through a device when a voltage of polarity opposite to that normally specified is impressed across the device. The term is commonly used with electrolytic capacitors.

reverse open-circuit voltage amplification factor—In a transistor, the ratio of incremental values of input voltage to output voltage measured with the input ac open-circuited.

reverse-phase current relay — Also called phase-balance current relay. A device which functions when the polyphase currents are of reverse-phase sequence, or when the polyphase currents are unbalanced or contain negative phase-sequence components above a given amount.

reverse recovery time—In a semiconductor diode, the time required for the current or voltage to reach a specified state after being switched instantaneously from a specified reversed bias condition.

reverse resistance — The resistance measured at a specified reverse voltage or current in a diode or rectifier.

reverse saturation current—The reverse current that flows in a semiconductor because of a specified reverse voltage.

reverse voltage—1. The voltage applied in the reverse direction to a diode or rectifier. 2. In the case of two opposing voltages, the voltage with the polarity that results in the smaller current.

reversible booster — A booster capable of adding to or subtracting from the voltage of a circuit.

reversible capacitance—For a capacitor, the limit, as the amplitude of the applied sinusoidal voltage approaches zero, of the ratio of the amplitude of the in-phase, fundamental-frequency component of transferred charge to the amplitude of the applied voltage, with a specified constant bias voltage superimposed on the sinusoidal voltage.

reversible capacitance characteristic—The function giving the relation of reversible capacitance to bias voltage.

reversible counter—*See* Up-Down Counter.

reversible motor—A motor in which the rotation can be reversed by a switch that changes the motor connections.

reversible permeability — The limit approached by the incremental permeability as the alternating field strength approaches zero.

reversible transducer—*See* Bilateral Transducer.

reversing switch—A switch used for chang-

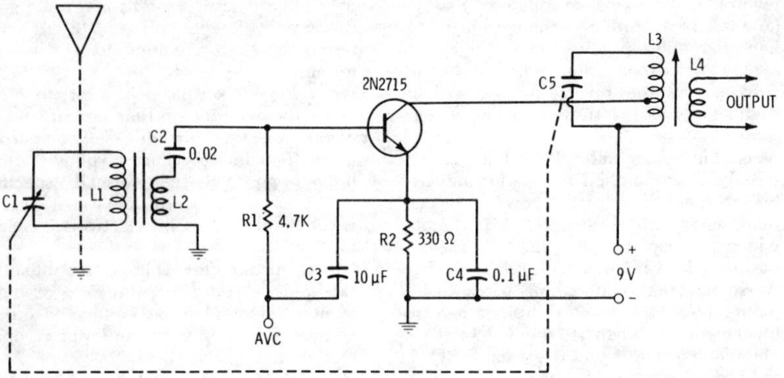

Rf amplifier.

ing the direction of any form of motion—specifically, the direction of motor rotation or the polarity of circuit connections.

reverting call—In telephony, a call made by one party on a line to another party on the same line.

rewind—To return the tape to its starting point in a magnetic recorder.

rewind control—A button or lever for rapidly rewinding magnetic recording tape from the take-up reel to the feed reel.

rewrite—Also called regeneration. In a storage device where the information is destroyed by being read, the restoring of information into the storage.

rf—Abbreviation for radio frequency.

rf amplifier—An amplifier capable of operation in the radio-frequency portion of the spectrum.

rf bandwidth — The band of frequencies comprising 99% of the total radiated power extended to include any discrete frequency on which the power is at least 0.25% of the total radiated power.

rfc — Abbreviation for radio-frequency choke.

rf cavity preselector — An ultrahigh-frequency circuit component which is similar in function to a tuned resonant circuit. A tunable cavity.

rf choke—A coil designed to have a high inductive reactance at radio frequencies and used to prevent currents at these frequencies from passing from one circuit to another.

rf component—The portion of a signal or wave which consists only of the radio-frequency alternations, and not including its audio rate of change in amplitude or frequency.

rf current—Alternating current having a frequency higher than 10,000 hertz.

rf energy—Alternating-current energy generated at radio frequencies.

RFG—Abbreviation for radar field gradient.

rf generator — A generator that produces sufficient rf energy at its assigned frequency for induction or dielectric heating.

rf head—A unit consisting of a radar transmitter and part of a radar receiver, the two contained in a package for ready removal and installation.

rfi (radio-frequency interference)—Radio-frequency energy of sufficient magnitude to have a possible influence on the operation of other electronic equipment.

rf indicator—A device that shows the presence of rf energy. It may consist of a tuned circuit or parallel line connected to an incandescent lamp or other indicator.

rf interference-shield ground—The grounding technique for all shields that are used to suppress the radiation of interference from leads.

rf intermodulation distortion—Intermodulation distortion that has its origin in the rf stages of a receiver.

rf line — 1. A system of metallic tubes (waveguides and/or coaxial lines) which conduct radio-frequency energy from one point to another. 2. A metallic conductor used to transmit radio-frequency energy from one point to another.

rf oscillator—*See* Radio-Frequency Oscillator.

rf plumbing—Radio-frequency transmission lines and associated equipment in the form of waveguides.

rf power supply—A high-voltage power supply consisting of an rf oscillator the output voltage of which is stepped up and then rectified. Used in television receivers or other equipment to supply the high dc voltage required by the second anode of cathode-ray tubes.

rf preheating—*See* Radio Frequency Preheating.

rf preselectors—Bandpass filters which improve the selectivity by rejecting unwanted frequencies at the radio-frequency input stage.

rf probe—1. A resonant conductor which is

placed in a waveguide or cavity resonator for the purpose of inserting or withdrawing electromagnetic energy. 2. A detecting device used with a vacuum-tube voltmeter to measure rf voltages.

rf pulse—A radio-frequency carrier amplitude-modulated by a pulse. The carrier amplitude is zero before and after the pulse. Coherence of the carrier with itself is not implied.

rf resistance—*See* High-Frequency Resistance.

rf shift—*See* Frequency Shift.

rf signal generator—Also called service oscillator. A test instrument that generates several bands of radio frequencies necessary for the alignment and servicing of radios, television, and other electronic equipment.

rf tolerance—The amount of rf energy the human body can receive without injury.

rf transformer — *See* Radio-Frequency Transformer.

r/h—Abbreviation for roentgens per hour.

rheo—Abbreviation for rheostat.

rheoencephalography — *See* Electrical-Impedance Cephalography.

rheostat—A variable resistor which has one fixed terminal and a movable contact (often erroneously referred to as a "two-terminal potentiometer"). Potentiometers may be used as rheostats, but a rheostat cannot be used as a potentiometer because connections cannot be made to both ends of the resistance element.

rhi — Abbreviation for range-height indicator. A radar display in which the abscissa represents the range to the target, and the ordinate indicates height.

rhombic antenna—An antenna composed of long-wire radiators comprising the sides of a rhombus. The antenna usually is terminated in an impedance. The sides of the rhombus and the angle between them, the elevation, and the termination are proportioned to give the desired directivity.

rho-theta system—1. Any electronic navigation system in which position is defined in terms of distance (ρ) and bearing (θ) relative to a transmitting station. 2. A polar-coordinate navigational system providing sufficiently accurate data so that a computer can be used to provide arbitrary course lines anywhere within the coverage area of the system.

rhumbatron — A resonant cavity consisting of lumped inductance and capacitance. It is used, instead of circuits, to act as an oscillator capable of giving an output of several kilowatts at frequencies of several thousand megahertz.

rhythm bar—In some organs, a bar used to permit rhythmic playing of a chord without interrupting the pressure on the keys or chord button.

RIAA—Abbreviation for Recording Industry Association of America. The official association of the disc recording field.

RIAA curve—1. A standard recording characteristic curve approved for long-playing records by the Recording Industry Association of America. 2. The equalization curve for playback of records recorded as in (1).

ribbon — A common intraconnecting material, usually nickel, of rectangular cross section, which is used to connect electronic component parts or modules together into the form of a functioning circuit.

ribbon cable—Cable made of more than one conductor, laid parallel.

ribbon microphone — 1. A microphone in which the moving conductor is in the form of a ribbon driven directly by the sound waves. 2. A microphone that uses a narrow corrugated aluminum alloy strip suspended in a magnetic field. Sound makes the strip vibrate in a direction perpendicular to the magnetic field, resulting in an ac current being induced in a coil. The natural response of a ribbon unit is bidirectional and a ribbon microphone has

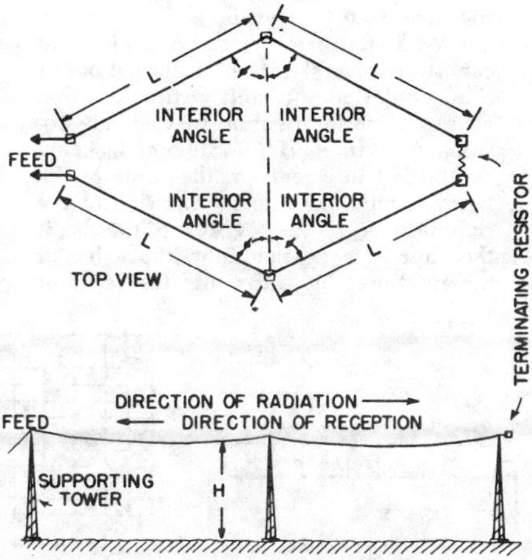

Rhombic antenna.

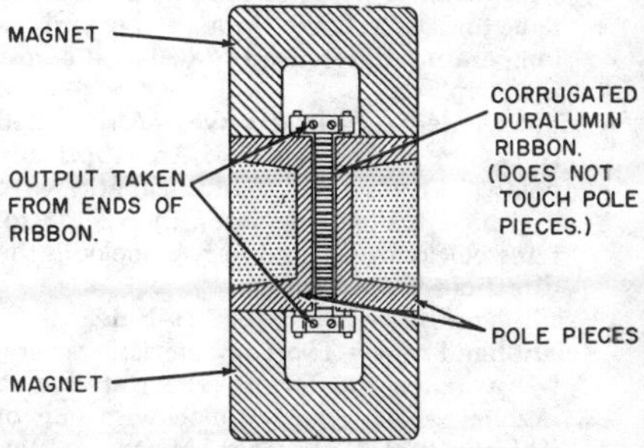

Ribbon microphone.

certain areas of minimum sensitivity. Only a very small unit of response is produced by quite loud noises from the sides. The polar diagram for this type is known as "figure of eight." Some ribbon microphones have cardioid and hypercardioid responses.

ribbon tweeter—A high-frequency speaker, usually horn loaded, in which a stretched, straight flat ribbon is used instead of a conventional voice coil. The magnetic gap is a straight slit that can be made quite narrow so that a maximum amount of flux is concentrated in it. The ribbon serves both as an extremely light driven element and as a diaphragm.

Rice neutralizing circuit — A radio-frequency amplifier circuit that neutralizes the grid-to-plate capacitance of the amplifier tube.

Richardson effect—*See* Edison Effect.

Richardson equation—An expression for the density of the thermionic emission at saturation current, in terms of the absolute temperature of the filament.

ride gain—To continually adjust the volume level of a program while observing a volume indicator so that the resulting audio-frequency signal will have the necessary magnitude for proper operation of the transmission equipment.

ridge waveguide—A circular or rectangular waveguide with one or more longitudinal ridges projecting inwardly from one or both sides. The ridges increase the transmission bandwidth by lowering the cutoff frequency.

Rieke diagram—A special polar-coordinate chart where load conditions can be determined for oscillators such as klystrons and magnetrons. Information from the chart indicates where optimum operation is located, as well as the limitations oscillators have with various loads.

rig—1. A system of components. 2. An amateur station consisting of receiver, transmitter, and all the necessary accessory equipment.

Righi-Leduc effect — The phenomenon whereby when a metal strip is placed with its plane perpendicular to a magnetic field and heat flows through the strip, a temperature difference is developed across the strip.

right-handed polarized wave — Also called clockwise polarized wave. An elliptically polarized transverse electromagnetic wave in which the electric intensity vector rotates clockwise as an observer looks in the direction of propagation.

right-hand rule—*See* Fleming's Rule.

right-hand taper—The characteristic whereby a potentiometer or rheostat has a higher resistance in the clockwise half of its rotational range than in its counterclockwise half (looking at the shaft end).

rigid metal conduit—A raceway specially constructed for the purpose of the pulling in or the withdrawing of wires or cables after the conduit is in place. It is made of metal pipes of standard weight and thickness, permitting the cutting of standard threads.

rim drive—The method of driving a phonograph or sound-recorder turntable by means of a small, rubber-covered wheel that contacts the shaft of an electric motor and the rim of the turntable.

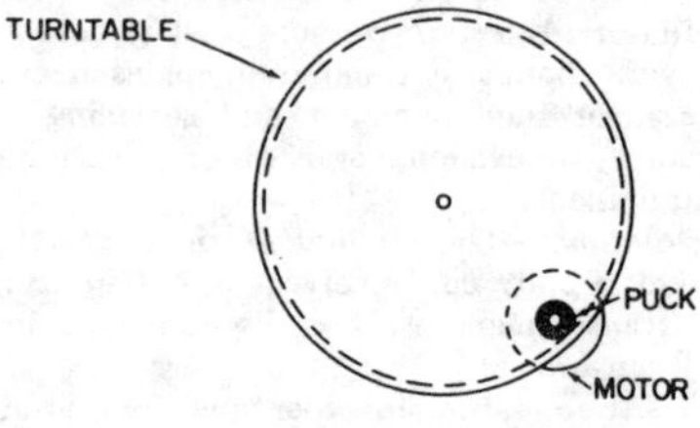

Rim drive.

rim magnet—*See* Field-Neutralizing Magnet.

ring—1. A ring-shaped contacting part of a plug usually placed in back of but insulated from the tip. 2. An audible alerting signal on a telephone line.

ring-around—In a secondary radar: 1. The undesired triggering of a transponder by its own transmitter. 2. The triggering of a transponder at all bearings, causing a ring presentation on a ppi.

ring circuit—In waveguide practice, a hybrid-T having the physical configuration of a ring with radial branches.

ring counter—1. A loop of interconnected bistable elements arranged so that only one is in a specified state at any given time. As input signals are counted, the specified state moves in an ordered sequence around the loop. 2. A device that can store several bits of information. A ring counter accepts shift instructions that cause all the information to shift one position at a time. If information is being shifted left in a register, the value of the leftmost bit shifts into the position of the rightmost bit of the register. Similarly, if the shift is to be the right, the value of the rightmost bit shifts into the leftmost

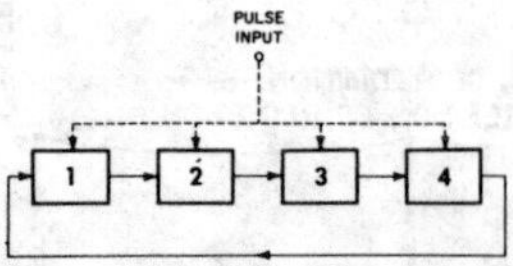

Block diagram.

PULSES	STAGE 1	2	3	4
0	ON			
1		ON		
2			ON	
3				ON
4	ON			
5		ON		
6			ON	
7				ON
8	ON			

Sequence.

Ring counter, 1.

bit position in the register. In a ring counter, the information recycles every n shift pulses, where n is the number of bits in the ring counter.

ringdown – A method of signaling subscribers and operators using either a 20-hertz ac signal, a 135-hertz ac signal, or a 1000-hertz signal interrupted 20 times per second.

ring head–A magnetic head in which the magnetic material forms an enclosure with one or more air gaps. The magnetic-recording medium bridges one of these gaps and contacts or is close to the pole pieces on one side only.

ringing–1. The production of an audible or visible signal at a station or switchboard by means of an alternating or pulsating current. 2. A damped oscillation in the output signal of a system, as a result of a sudden change in the input signal. 3. High-frequency damped oscillations caused by shock excitation of high-frequency resonances. 4. Transient decaying oscillation about high or low limit induced by unmatched impedance reflections.

ringing current – An alternating current which may or may not be superimposed onto a direct current for telephone ringing.

ringing key–A key which, when operated, sends a ringing current over its circuit.

ringing signal–Any ac or dc signal transmitted over a line or trunk for the purpose of alerting a party at the distant end of an incoming call. The signal may operate a visual or aural device.

ring magnet–A ceramic permanent magnet in which the axial length is no greater than the wall thickness and the wall thickness is no less than 15% of the outside diameter.

ring modulator – A modulator used as a balanced modulator, demodulator, or phase detector that has four diodes connected in series to form a ring around which current can easily flow in one direction. Input and output connections are made at the four nodal points of the ring.

ring oscillator–A circuit configuration in which two or more pairs of tubes are operated as push-pull oscillators in a ringlike arrangement. Usually alternate successive pairs of plates and grids are connected to tank circuits, and the load is coupled to the plate circuits.

ring retard – Also called slug retard. A heavy conductor surrounding the iron flux path in a relay to retard the establishment or the decay of flux in the path.

ring time–In radar, the time during which the output of an echo box remains above a specified level. It is used in measuring the performance of radar equipment.

RIOMETER – Acronym for relative ionospheric opacity meter. An instrument for recording the level of extra-terrestrial cosmic noise at selected frequencies in the hf and vhf regions.

ripple–1. That portion of the output voltage of a power supply harmonically related in frequency to the input power, and to any internally generated switching frequency. Normally ripple is expressed as an rms percentage, but it can also be expressed as peak to peak. Ripple is replaced by a new term, pard, which includes hum and noise as well as spikes in the output. 2. The wavelike variations in the amplitude response of a filter. Ideally, Tchebychev and elliptic-function filters have characteristics such that the differences in peaks and valleys of the amplitude response in the passband are always the same. Butterworth, Gaussian, and Bessel functions do not have ripple. Ripple usually is measured in dB. 3. The serial transmission of data; a serial reaction that may be compared to a bucket brigade or a row of falling dominoes.

ripple adder–A binary adding system in which the column of lowest order is added, the resulting carry is added to the column of the next highest order, and so on for all columns. It is necessary to wait for propagation of the signal even though all columns are present at the same time (parallel).

ripple counter – An asynchronously controlled counter; the clock is derived from a previous-stage output.

ripple current–The alternating component of a substantially steady current.

ripple-current rating–The rms value of the ac component of the current through a capacitor.

ripple filter–A low-pass filter designed to reduce the ripple current while freely passing the direct current from a rectifier or generator. Also called smoothing circuit and smoothing filter.

ripple frequency–The frequency of the ripple current. In a full-wave rectifier it is twice the supply frequency. In a generator it is a function of the speed and the number of poles.

ripple quantity–The alternating component of a pulsating quantity when this component is small relative to the continuous component.

ripple-through counter–*See* Serial Counter.

ripple voltage – The alternating component of a unidirectional voltage (this component is small relative to the continuous component).

rise cable–In communication practice: 1. The vertical portion of a house cable extending from one floor to another. 2. Sometimes, any other vertical sections of cable.

rise time–1. The time required for the

leading edge of a pulse to rise from 10% to 90% of its final value. It is proportionate to the time constant and is a measure of the steepness of the wavefront. 2. The measured length of time required for an output voltage of a digital circuit to change from a low voltage level (0) to a high voltage level (1), after the change has started. 3. The time required for the pointer of an electrical indicating instrument to attain 90% (within a specified tolerance) of end-scale deflection following sudden application of constant electric power from a source with sufficiently high impedance so as not to influence damping (100 times the impedance of the instrument). 3. For a switching transistor, the time interval between the instants at which the magnitude of the pulse at the output terminals reaches specified lower and upper limits respectively when the transistor is being switched from its nonconducting to its conducting state. (The lower and upper limits are usually 10% and 90%, respectively, of the amplitude of the output pulse.)

rising-sun magnetron—A multicavity vane-type magnetron in which resonators of two different resonant frequencies are arranged alternately for the purpose of mode separation.

rising-sun resonator—A magnetron anode structure in which large and small cavities alternate around the perimeter of the structure.

RMA – Abbreviation for Radio Manufacturers Association, now Electronic Industries Association (EIA).

RMA color codes—A term formerly used to designate the EIA color codes.

RMI—Abbreviation for radio-magnetic indicator.

r/min – Abbreviation for roentgens per minute.

R − Y signal—In color television, the red-minus-luminance color-difference signal. When combined with the luminance (Y) signal, it produces the red primary signal.

RMM—Abbreviation for read-mainly memory. A nonvolatile memory used much as a ROM or PROM except that the data contained therein may be altered through the use of special techniques (often involving external action) which are much too slow for read/write use.

RMOS—Abbreviation for refractory-metal-oxide semiconductor. An MOS that uses refractory metals like molybdenum instead of aluminum or silicon as the gate metal.

rms—Abbreviation for root-mean-square.

rms amplitude – Root-mean-square amplitude, also called effective amplitude. The value assigned to an alternating current or voltage that results in the same power dissipation in a given resistance as dc current or voltage of the same numerical value. The rms value of a periodic quantity is equal to the square root of the average of the squares of the instantaneous values of the quantity taken throughout one period. If the quantity is a sine wave, its rms amplitude is 0.707 of its peak amplitude.

rms pard—The value of the output waveform of a regulated power supply, omitting the dc component value.

rms power—*See* Continuous Power.

rms pulse amplitude—The square root of the average of the squares of the instantaneous amplitudes taken over the duration of the pulse. Also called effective pulse amplitude.

rms value – The "root-mean-square" value of ac voltage, current, or power. Calculated as 0.707 of peak amplitude of a sine wave at a given frequency.

rms voltage—The effective value of a varying or alternating voltage. That value which would produce the same power loss as if a continuous voltage were applied to a pure resistance. In sine-wave voltages, the rms voltage is equal to 0.707 times the peak voltage.

ro—*See* Receive Only.

Roberts rumble—The nickname given to a phenomenon whereby certain radio disturbances appear to be connected with the passage of satellites through the earth's ionosphere. Investigations have been made to determine the feasibility of exploiting this effect for the tracking of satellites.

Robinson antenna—A microwave scanning antenna consisting of an astigmatic reflector and a feed system in which a parallel-plate region is fed by a waveguide. The parallel-plate region is made so that the feed waveguide end is circular, to permit rotation of the feed guide, and is in approximately the same plane as the output end, or larger aperture of the parallel-plate region.

robot pilot—*See* Autopilot.

Rochelle-salt crystal – A crystal made of sodium potassium tartrate. Because of its pronounced piezoelectric effect, it is used extensively in crystal microphones and phonograph pickups. Perfect Rochelle-salt crystals up to four inches and even more in length can be grown artificially.

rock—To move a control back and forth (or make other adjustments as necessary), in order to obtain the best alignment or on-station tuning.

rocking—Rotating the tuning control in a superheterodyne receiver back and forth while adjusting the oscillator padder near the low-frequency end of the tuning dial, to obtain more accurate alignment.

rocky-point effect – Transient but violent

discharges between electrodes in high-voltage transmitting tubes.

rod gap—A spark gap in which the electrodes are two coaxial rods, with ends between which the discharge takes place, cut perpendicularly to the axis.

roentgen—1. The unit of radioactive dose of exposure. It is the amount of gamma radiation that will produce one electrostatic unit of charge in one cubic centimeter of air which is surrounded by an infinite mass of air at standard temperature and pressure conditions. 2. A quality of gamma or X-ray radiation equal to approximately 83 ergs of absorbed energy per gram of air.

roentgen densitometer — A device for recording changes in concentration of a radiopaque indicator injected into circulation for evaluating circulatory function.

roentgen equivalent man—A radiation-exposure dose which produces the same effects on human tissue as one roentgen of X-ray radiation. Abbreviated rem.

roentgen equivalent physical — An amount of ionizing radiation that results in an absorption of energy of approximately 83 to 93 ergs per gram of tissue. Abbreviated rep.

roentgen meter—Also called a roentgenometer. An instrument for measuring the quantity or intensity of roentgen rays (X-rays or gamma rays).

roentgenogram—Also called an X-ray photograph or an X-ray. A photograph taken by showering an object or the human body with X-rays (roentgen rays). Depending on the transparency of the object or body, the interior can thus be seen and recorded.

roentgenology—That branch of science related to the application of roentgen rays (X-rays) for diagnostic or therapeutic purposes.

roentgenometer—*See* Roentgen Meter.

roentgen rays—*See* X-rays.

roger—A code word used in communications to mean: 1. Your message has been received and is understood. 2. OK—an expression of agreement.

Roget spiral—A helix of wire that contracts in length, when a current is sent through, as a result of the mutual attraction between adjacent turns.

roll—1. Also called flip-flop, especially when intermittent. The upward or downward movement of a television picture due to lack of vertical synchronization. 2. The process in which alphanumeric text moves across a crt screen to the left or right. As a character disappears on one end of the screen, a new character appears at the other end. Roll can also be extended to include nontextual graphical constructions, although this is more properly called translation.

rollback—*See* Rerun, 1.

roll bonding—*See* Yield-Strength-Controlled Bonding.

roll in—In a computer, to restore in main storage information previously transferred from main to auxiliary storage.

rolling transposition—The method by which two or more conductors of an open-wire circuit are spiral-wound. With two wires, a complete transposition can be executed utilizing two consecutive suspension points.

rolloff—1. A gradual increase in attenuation over a range of frequencies; sometimes called slope. 2. An attenuation that varies with frequency, generally increasing at a constant rate beyond the corners of the amplitude-frequency characteristic of a system. 3. The rate of attenuation of a filter at the bass end (high-pass) or treble end (low-pass). The crossover frequency is that frequency where the response is −3 dB of the midfrequency response. The rolloff of an amplifier is similarly defined but in terms of decrease in amplification.

roll out—To read out of a computer storage by simultaneously increasing by 1 the value of the digit in each column, repeating this r times (where r is the radix) and, at the instant the representation changes from (r−1) to zero, either generating a particular signal, terminating a sequence of signals, or originating a sequence of signals.

roll over indexing—In a calculator, allows depression of a second key before releasing first key.

ROM—Abbreviation for read-only memory. Read-only memory (fixed memory) is any type of memory which cannot be readily rewritten. RO requires a masking operation during production to permanently record program or data patterns in it. The information is stored on a permanent basis and used repetitively. Such storage is useful for programs or tables of data that remain fixed and is usually randomly accessible.

Romex cable — A moisture-resistant, flame-resistant, flexible cable that contains one or more wires.

roof filter — A low-pass filter used in a carrier telephone system to limit the frequency response to that needed for normal transmission, thereby blocking unwanted high-frequency interference induced in the circuit by external sources. Use of such a filter gives improved run-around crosstalk suppression and minimized high-frequency singing.

room acoustics—The quality of a room that affects how sounds will be heard in it. Room acoustics are a function of the room's size, geometry, structural materials, and furnishings. A "live" room is one in which sounds are fairly reverberant; a "dead" room is one in which sounds are fairly absorbed.

room noise—*See* Ambient Noise.

root mean square—The square root of the average of the squares of the values of a periodic quantity taken throughout one complete period. It is the effective value of a periodic quantity. Abbreviated rms.

root-mean-square amplitude—*See* RMS Amplitude.

root mean square power – *See* Continuous Power.

root-mean-square value—Of alternating currents and voltages, the effective current or voltage applied. It is that value of alternating current or voltage that produces the same heating effect as would be produced by an equal value of direct current or voltage. For a sine wave, it is equal to 0.707 times the peak value.

root segment—1. In a computer, the segment of an overlay program that remains in main storage throughout the execution of the overlay program. 2. The first segment in an overlay program.

root-sum square—The square root of the sum of the squares. A common expression of the total harmonic distortion.

rope—Similar to chaff, but longer. Electromagnetic-wave reflectors used to confuse enemy radar. They consist of long strips of metal foil, to which small parachutes may be attached to reduce their rate of fall.

rope-lay conductor or cable—A cable consisting of one or more layers of helically laid groups of wires surrounding a central core.

rope lay strand—A conductor made of multiple groups of filaments. A 7 × 19 rope lay strand has 19 wires laid into a group and then 7 such groups laid cabled into a conductor.

rosin connection—Also called a rosin joint. A defective connection of a conductor to a piece of equipment or to another conductor. Supposedly the joint is tightly soldered, but actually it is held together only by unburnt rosin flux.

rosin-core solder – Self-fluxing solder consisting of a hollow center filled with rosin.

rosin joint—*See* Rosin Connection.

rotary-beam antenna—A highly directional short-wave antenna system. It is mounted on a mast and can be rotated manually or by an electric-motor drive to any desired position.

rotary converter—*See* Dynamotor.

rotary coupler—*See* Rotating Coupler.

rotary generator (induction-heating usage) – An alternating-current generator adapted to be rotated by a motor or other prime mover.

rotary joint—*See* Rotating Coupler.

rotary phase converter—A machine which converts power from an alternating-current system of one or more phases to an alternating-current system of a different number of phases, but of the same frequency.

rotary plunger relay—A relay in which the linear motion of the plunger is converted mechanically into rotary motion.

rotary relay—1. A relay in which the armature rotates to close the gap between two or more pole faces (usually with a balanced armature). 2. A term sometimes used for stepping relay.

rotary-solenoid relay – A relay in which the linear motion of the plunger is converted into rotary motion by mechanical means.

rotary spark gap—A device used to produce periodic spark discharges. It consists of several electrodes which are mounted on a wheel and rotate past a fixed electrode.

rotary stepping relay—*See* Stepping Relay.

rotary stepping switch—*See* Stepping Relay.

rotary switch—An electromechanical device which is capable of selecting, making, or breaking an electrical circuit. It is actuated by a rotational torque applied to its shaft.

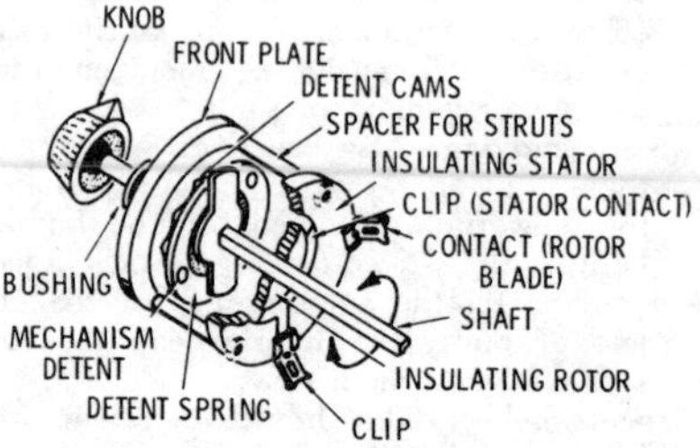

Rotary switch.

rotary transformer—A term sometimes applied to a rotating machine used to transform direct-current power from one voltage to another.

rotary-vane attenuator—A device designed to introduce attenuation into a waveguide circuit through variation of the angular position of a resistive material placed in the guide.

rotary voltmeter – *See* Generating Voltmeter.

rotatable phase-adjusting transformer – A transformer in which the secondary voltage may be adjusted to have any desired phase relation with the primary voltage by mechanically orienting the secondary winding with respect to the primary. The latter winding consists usually of a distributed symmetrical polyphase winding and is energized from a polyphase circuit. (*See also* Phase-Shifting Transformer.)

rotating-anode tube – An X-ray tube in

which the anode rotates continually to bring a fresh area of its surface into the beam of electrons. This procedure allows a greater output without melting the target.

rotating coupler—Also called rotary coupler and rotary joint. A joint that permits one section of a waveguide to rotate while passing rf energy.

rotating disc—*See* Drum Memory.

rotating element of a meter—*See* Rotor of a Meter.

rotating field — The magnetic field in the stator of induction motors. Because of excitation from a polyphase source, the field appears to rotate around the stator from pole to pole.

rotating joint—A device that permits one section of a transmission line to rotate continuously with respect to the other while still maintaining radio-frequency continuity.

rotating radio beacon—A radio transmitter that rotates a concentrated beam horizontally at a constant speed. Different signals are transmitted in each direction so that ships and aircraft without directional receiving equipment can determine their bearings.

rotational life—The ability of a potentiometer to be operational after a defined number of wiper sweeps across the element.

rotational wave—*See* Sheer Wave, 1.

rotation spectrum — An X-ray spectrum of the diffraction pattern obtained when X-rays are sent through a rotating crystal.

rotator—1. A motor-driven assembly which turns an antenna so that it can be aimed in the direction of best reception. 2. In waveguides, a means of rotating the plane of polarization. In rectangular waveguides it is done by simply twisting the guide itself.

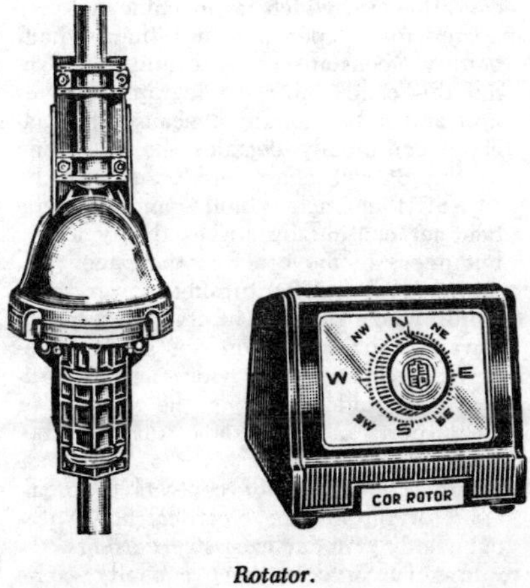

Rotator.

rotoflector—In radar, an elliptically shaped rotating reflector used to divert a vertically directed radar beam at right angles so that it radiates horizontally.

rotor—1. The rotating member of an electric machine. In a motor, it is connected to and turns the drive shaft. In a generator, the rotor is turned to produce electricity by cutting magnetic lines of force. 2. The movable plates of a variable capacitor. 3. *See* Receive-Only Typing Reperforator.

rotor of a meter—Also called the rotating element of a meter. The portion driven directly by electromagnetic action.

rotor plates—The movable plates of a variable capacitor.

round conductor—A solid or stranded conductor with a substantially circular cross section.

rounding—A lack of a sharp corner of a waveform, or a smooth transition from the leading or trailing edge to the limiting final value.

rounding error—The error that results when the less significant digits of a number are dropped and the most significant digits are then adjusted.

round off—To delete less significant digits from a number and possibly apply some rule of correction to the part retained. For example, if the discarded part is 5, 6, 7, 8, or 9 (or 50 . . . , etc.), the new final digit is raised by 1 (e.g., 30.7 would be rounded off to 31, and 519.2 to 519). This is done for ease of calculation, where an estimate will suffice. Also known as truncation.

round-off error—*See* Rounding Error.

round-trip echoes—Multiple-reflection echoes produced when the radar pulse is reflected from a target strongly enough that the echo is reflected back to the target, where it produces a second echo.

round up—In a calculator, the last digit displayed in an answer is increased by one if the following digit would have been a one or greater.

routine—A set of computer instructions arranged in a correct sequence and used to direct a computer in performing one or more desired operations.

routine library—An ordered set of standard and proven computer routines that may be used to solve problems or parts of problems.

routing—The assignment of the communications path by which information is carried to its destination.

routing indicator—An address, or group of characters, used in the header of a message to specify the final circuit or terminal to which the message is to be delivered.

row — 1. A horizontal arrangement of a number of characters or other expressions. 2. In computers, the characters or corresponding bits of binary-coded characters that make up a word. 3. A path, perpendicular to the edge of a tape, along which storage of information may be accomplished by means of the presence or absence of holes or magnetized areas. 3. A predetermined number of consecutive functional patterns lying along a line parallel to the X axis of a photomask.

row binary—Having to do with the binary representation of data on cards by a method in which adjacent positions in a row correspond to adjacent bits of data; for example, the representation of 80 consecutive bits of two 40-bit words may be contained in each row of an 80-column card.

row pitch — The distance between corresponding points in adjacent rows.

rpm—Abbreviation for revolutions per minute.

rps—Abbreviation for revolutions per second.

R-S flip-flop—A flip-flop having two inputs, designated R and S. At the application of a clock pulse, a 1 on the S input will set the flip-flop to the 1 or "on" state, and 1 on the R input will reset it to the 0 or "off" state. It is assumed that 1's will never appear simultaneously at both inputs.

R-S-T flip-flop—A flip-flop having three inputs, R, S, and T. The R and S inputs produce states as described for the R-S flip-flop; the T causes the flip-flop to change states.

RTL — Abbreviation for resistor-transistor logic.

RTMA—Abbreviation for Radio-Television Manufacturers Association, now Electronic Industries Association (EIA).

rubber-covered wire—A wire with rubber insulation.

ruby — A type of aluminum-oxide crystal used to produce one form of solid-state laser.

ruby laser—An optically pumped solid-state laser in which a ruby crystal produces an extremely narrow and intense beam of coherent red light. Used for localized heating and for light-beam communication.

ruby maser—A maser that has a ruby crystal in the cavity resonator.

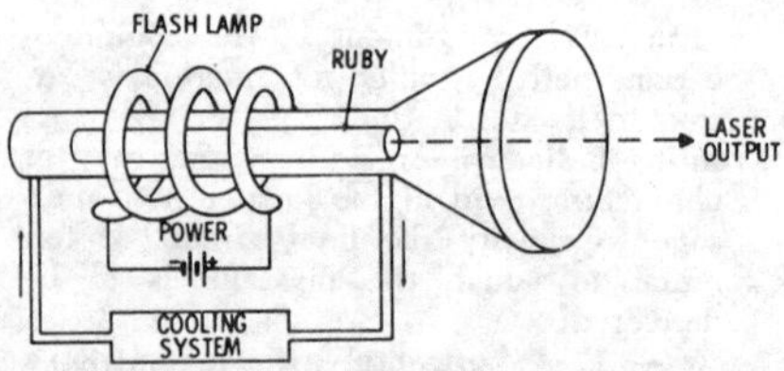

Ruby laser.

ruggedization—The redesign of a piece of equipment or its components, to make them able to withstand prolonged vibration and mechanical shock.

Ruhmkorff coil—An induction coil having a magnetic interrupter. It is used to produce a spark discharge across an air gap.

rumble—1. Also called turntable rumble. A descriptive term for a low-frequency vibration, which is mechanically transmitted to the recording or reproducing turntable and superimposed onto the reproduction. 2. Low-frequency noise caused by a tape transport.

run — A single, continuous execution of a program by a computer.

runaround cross talk—Cross talk resulting from the coupling of the high-level end of one repeater to the low-level end of another repeater. Often a third repeater or line is the means of coupling; therefore, runaround cross talk may be a form of interaction cross talk.

runaway — 1. Any additive condition to which continued exposure will eventually destroy a device. 2. A condition in which one of the dynamic variables of a system makes an unintended increase to a level beyond the design limits, often with destructive consequences.

run book—All the material needed to document a run of a program on a computer.

run-in—The start of a groove at the beginning of a side of a gramophone record which "runs in" to the recorded section.

run motor—In facsimile equipment, a motor which supplies the power to drive the scanning or recording mechanisms. A synchronous motor is used to limit the speed.

running circuit breaker—A device the principal function of which is to connect a machine to its source of running voltage after having been brought up to the desired speed on the starting connection.

running open—In telegraph applications, a condition in which a machine is connected to an open line or a line without battery (constant space condition). Under this condition, the telegraph receiver appears to be running, because the machine continually decodes the open line as the Baudot character "blank" or the USASCII character "null," and the type hammer continually strikes the type box but does not move across the page.

running torque—The turning power of a motor when running at its rated speed.

runway-localizing beacon — A small radio-range beacon that provides accurate directional guidance along the runway of an airport and for some distance beyond it.

rupture—The ability of contacts to break apart or rupture the electrical flow without welding under excessive currents.

rupture (or interrupting) capacity — The

maximum current that a protective device will interrupt. Specified as the number of interruptions in amperes (adjusted circuit) without a change in calibration or a failure of dielectric strength.

rush-box—A superregenerative receiver.

rutherford—A quantity of radioactive material that produces one million disintegrations per second. Abbreviated rd.

ryotron—A thin-film inductive superconductive device. An inductive switch capable of inductance variation of better than three orders of magnitude.

S

S—Symbol for secondary and source electrode.

s—Abbreviation for second.

sabin (square-foot unit of absorption)—A measure of the sound absorption of a surface. It is equivalent to 1 square foot of a perfectly absorptive surface. (*See also* Equivalent Absorption.)

saddle—Insulation placed under a splice in a coil lead.

safety factor – The amount by which the normal operating rating of a device can be exceeded without causing failure of the device.

safety service—A radiocommunication service used permanently or temporarily for the safeguarding of human life and property.

SAG MOS—Abbreviation for self aligning-gate MOS.

SAGE system—*See* Semiautomatic Ground Environment.

St. Elmo's fire—A visible electric discharge sometimes seen at the tips of aircraft propellers or wings, the mast of a ship, or any other metal point where there is considerable atmospheric difference of potential due to concentration of the electric field at the points of the conductor.

sal-ammoniac cell – A cell in which the electrolyte consists primarily of a solution of ammonium chloride.

salient pole—1. A pole consisting of a separate radial projection having its own iron pole piece and its own field coil, used in the field system of a generator or motor. 2. In an electric motor or generator, a magnetic field pole projecting toward the armature.

Salisbury darkbox – An isolating chamber used for test work in connection with radar equipment. The walls of the chamber are specially constructed to absorb all impinging microwave energy at a certain frequency.

SAM—Acronym for sequential access memory.

sample – One or more units of product drawn from a lot, the units being selected at random without regard to their quality.

sample and hold—1. A circuit used in an analog-to-digital converter whenever it is desirable to make a measurement of a signal and to know precisely when the input signal corresponds to the results of the measurement. It is also used to increase the duration of a signal. 2. A system in which a "sample" of an analog input signal is frozen in time (is stored in a capacitor) and held while it is converted to a digital representation or otherwise processed. 3. A circuit that holds or "freezes" a changing analog input signal voltage. Usually, the voltage thus frozen is then converted into another form, either by a voltage-controlled oscillator, an analog-to-digital (a/d) converter or some other device.

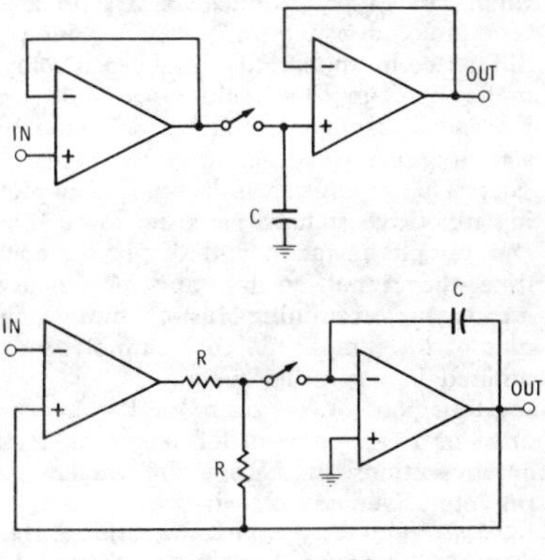

Sample and hold.

sampled data—Data in which the information content is determined only at discrete time intervals. Sampled data may be either analog or digital in form.

sample pulse—*See* Strobe Pulse, 1.

sampler—1. A directional coupler that has a detector attached to the auxiliary arm so that a video output sample proportional to the input power level is obtained. When the sampler is used to monitor power or drive a closed-loop source leveling system, the directional coupler must have a flat coupling coefficient. 2. *See* Sampling Circuit.

sample rate—The rate at which the analog sample is measured and/or displayed per second.

sampling—Obtain values of a function that correspond to discrete, regularly or irregularly spaced values of the independent variable.

sampling circuit—Also called a sampler. A circuit the output of which is a series of discrete values representative of the values of the input at a series of points in time.

sampling distribution — In random-sampling-oscilloscope technique, a function that describes the manner in which the density of a large number of randomly placed samples varies across the signal period.

sampling gate—A device which must be activated by a selector pulse before it will extract information from the input waveform.

sampling oscilloscope—An oscilloscope technique very similar in principle to the use of stroboscopic light to study fast mechanical motion or other very-high frequency occurrences. Progressive samples of adjacent portions of successive waveforms are taken; then they are "stretched" in time, amplified by relatively low-bandwidth amplifiers, and finally shown, one sample at a time, on the screen of a cathode-ray tube. The graph produced is a replica of the sampled waveforms. The principal difference in appearance between displays made by sampling techniques and conventional displays is that those made by sampling comprise separate segments or dots. This technique is limited to depicting repetitive signals, since no more than one sample is taken and displayed each time the signal occurs, and provides a means for examining fast-changing signals of low amplitude that cannot be examined in any other way.

sampling plan—A program for the acceptance or rejection of a lot based on tests or inspections indicating the quality of predetermined sample sizes.

sampling rate—The number of times that a particular data channel is sampled by a commutator in one second.

sampling theorem—A theorem (developed by Nyquist in 1928) which states that two samples per cycle will completely characterize a band-limited signal; that is, the sampling rate must be twice the highest-frequency component. (In practice, the sampling rate is ordinarily from five to ten times the highest frequency.)

sand load—An attenuator used as a terminating section on a transmission line to dissipate power. The space between inner and outer conductors is filled with a sand and carbon mixture which acts as the dissipative element.

sandwich—A packaging method in which components are placed between boards or layers.

sanitary motor—A type of motor used in the food industry. It usually has a frame that is so shaped that deposits of material cannot collect to contaminate nearby food, and can be easily kept clean.

sapphire—A gem used on the tip of quality phonograph needles, and also for bearings in precision instruments.

SARAH — Acronym for search and rescue and homing. A radio homing device, originally designed for personnel rescue, used in operations for the recovery of spacecraft at sea.

satellite—A tv station licensed to rebroadcast the programming of a parent station. It differs from a translator in that satellite power limits are much higher, and satellites may also originate some programming.

SATO—Abbreviation for self-aligned thick oxide.

saturable-core magnetometer — A magnetometer in which the change in permeability of a ferromagnetic core provides a measure of the field.

saturable-core oscillator—A relaxation oscillator in which the occurrence of saturation in a magnetic core initiates a change in the conductive state of amplifying or switching elements.

saturable-core reactor—*See* Saturable Reactor.

saturable reactor—Also called a saturable-core reactor. A magnetic-core reactor, the reactance of which is controlled by changing the saturation of the core by varying a superimposed unidirectional flux.

saturable transformer—A saturable reactor with an additional winding to provide voltage transformation or isolation from the ac supply.

saturated—That operating state of a transistor in which there is no further increase in collector current when the base current increases; in this state, the collector-emitter voltage is low, typically less than 1 volt.

saturated color—A pure color—i.e., one not contaminated by white.

saturated logic—A type of logic in which one output stage is the saturation voltage of a transistor. Examples are resistor-transistor logic (RTL), diode-transistor logic (DTL), and transistor-transistor logic (TTL). *See also* Unsaturated Logic.

saturated recovery time—The recovery time of a thermal relay measured when the relay is de-energized after temperature saturation (equilibrium) has been reached.

saturated reoperate time — Reoperate time of a thermal relay when temperature saturation (equilibrium) is reached before the relay is de-energized.

saturating reactor—A magnetic-core reactor capable of operating in the region of saturation without independent control means.

saturating signal — In radar, a signal of

greater amplitude than the dynamic range of the receiving system.

saturation — 1. The degree of purity of color. The less white light there is in a given color, the greater is the saturation of that color. 2. The operating condition of a transistor when an increase in base current produces no further increase in collector current. 3. The state of magnetism beyond which a metal or alloy is incapable of further magnetization—i.e., the point beyond which the BH curve is a straight line. 4. A circuit condition whereby an increase in the driving or input signal no longer produces a change in the output. 5. The condition when a transistor is driven so hard that it becomes biased in the forward direction. In a switching application, the charge stored in the base region prevents the transistor from turning off quickly under saturation conditions. 6. The condition in an electron tube when maximum current is passing through the cathode circuit.

saturation absorption — The condition obtainable between pairs of electron energy levels in which, under vary high radiation intensity, the absorption coefficient gradually decreases to zero because of the absence of empty levels to which transition can occur.

saturation control—In a color television receiver, a control which regulates the amplitude of the chrominance signal. The latter, in turn, determines the color saturation.

saturation current—1. The current in the plate circuit of a vacuum tube when all electrons emitted by the cathode pass on to the plate. 2. The current that flows between the base and collector of a transistor when an increase in the emitter-to-base voltage causes no further increase in the collector current. 3. The maximum current obtainable as the applied voltage is increased.

saturation curve — A magnetization curve for a ferromagnetic material.

saturation flux density—*See* Saturation Induction.

saturation induction—Sometimes loosely referred to as saturation flux density. The maximum intrinsic induction possible in a material.

saturation limiting—Limiting the minimum output voltage of a vacuum-tube circuit by operating the tube in the region of plate-current saturation.

saturation magnetization — The magnetic condition of a body when an increase in the magnetizing force produces practically no change in the intensity of magnetization.

saturation moment—The greatest magnetic moment possible in a sample of magnetic material.

saturation noise—The noise arising when a uniformly saturated tape is reproduced. This is often some 15 dB higher than the bulk-erased noise and is associated with imperfect particle dispersion.

saturation point—The point beyond which an increase in one of two quantities produces no increase in the other.

saturation resistance — Ratio of voltage to current in a saturated semiconductor.

saturation value—1. The highest value that can be obtained under given conditions. 2. The value of magnetic-flux density beyond which increases in the magnetizing force have no appreciable effect on the flux density in a particular sample of magnetic material.

saturation voltage—1. The voltage drop appearing across a switching transistor that is fully turned on. 2. Generally, the voltage excursion at which a circuit self-limits (i.e., is unable to respond to excitation in a proportional manner). In operational amplifiers, the output-voltage saturation limits may be imposed by any stage, from the input to the output, depending in part on the external loading and feedback parameters.

SAW—Abbreviation for surface acoustic-wave device. A technology for broadbandwidth signal delay, custom-designed filters, and complex signal generation and correlation at if frequencies. SAW devices make use of a single crystal, planar substrate with aluminum or gold electrode patterns fabricated by photolithography of a substrate, of piezoelectric material. The electrode patterns are used to excite and detect minute acoustic waves that travel over the surface of the substrate much like earthquake waves travel over the crust of the earth.

sawtooth—A waveform increasing approximately linearly as a function of time for a fixed interval, returning to its original state sharply, and repeating the process periodically.

sawtooth current — A current that has a sawtooth waveform.

sawtooth generator — An oscillator providing an alternating voltage with a sawtooth waveform.

sawtooth-modulated jamming—An electronic-countermeasures technique in which a high-level jamming signal is transmitted so that large agc voltages are developed at the radar receiver and the target pip and receiver noise are caused to disappear completely.

sawtooth voltage—A voltage that varies between two values in such a manner that the waveshape resembles the teeth of a saw.

sawtooth wave—A periodic wave, the amplitude of which varies linearly between two values. A longer interval is required

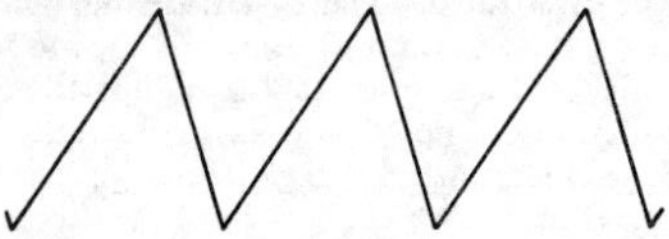

Sawtooth wave.

for one direction of progress than for the other.

saxophone—A linear-array antenna with a cosecant-squared radiation pattern.

SBA—Abbreviation for standard beam approach.

S-band—A radio-frequency band of 1550 to 5200 MHz, with wavelengths of 19.35 to 5.77 cm.

SBT—Abbreviation for surface-barrier transistor.

SCA—Abbreviation for Secondary (or Subsidiary) Communications Authorization. Permission granted by the FCC for an fm broadcaster to send out, on the same carrier frequency, a program in addition to the one heard with ordinary receivers, simultaneously with the regularly heard program. The purpose is to provide special programming to a limited audience without the cost of an entire new transmitter. SCA is primarily for the transmission of programs which are of a broadcast nature, but which are of interest primarily to limited segments of the public wishing to subscribe. This includes background music, storecasting, detailed weather forecasting, special time signals, and other material of a broadcast nature expressly designed and intended for business, professional, educational, religious, trade, labor, agriculture, or other groups engaged in any lawful activity. SCA facilities may also be used in transmission of signals directly related to the operation of fm broadcast stations.

scalar—1. A quantity that has magnitude but no direction (e.g., real numbers). 2. A circuit with two stable states, which can be triggered to the opposite state by appropriate means (a bistable circuit).

scalar function—A function which has magnitude only. Thus, the scalar product of two vectors is a scalar function, as is a real function of a real variable.

scalar quantity — Any quantity which has magnitude only—e.g., time, temperature, quantity of electricity.

scale—1. A series of musical notes, symbols, sensations, or stimuli arranged from low to high by a specified scheme of intervals suitable for musical purposes. 2. The theoretical basis of a numerical system. 3. A series of markings used for measurement or computation. 4. A defined set of values, in terms of which different quantities of the same nature can be measured. 5. In a computer, to change the units of a variable so that the problem is within its capacity.

scale division—The space between two adjacent markings on a scale.

scale factor — 1. In analog computing, a proportionality factor which relates the magnitude of a variable to its representation within a computer. 2. In digital computing, the arbitrary factor which may be associated with numbers in a computer to adjust the position of the radix point so that the significant digits occupy specified columns. 3. The factor by which the number of scale divisions indicated or recorded by an instrument must be multiplied to compute the value of the measurand. 4. A value used to convert a quantity from one notation to another. 5. The amount by which a measured quantity must change in order to produce unit deflection of a recording pen.

scale length of an indicating instrument—The length of the path described by the tip of the pointer (or other indicating means) in moving from one end of the scale to the other. Pointers that extend beyond the scale division marks are considered to end at the outer end of the shortest division marks; in multiscale instruments, the longest scale is used in determining the scale length.

scale-of-ten circuit—*See* Decade Scaler.

scale-of-two counter—A flip-flop circuit in which successive similar pulses are applied at a common point, causing the circuit to alternate between its two conditions of permanent stability.

scaler—Also called a scaling circuit. A circuit which produces an output after a predetermined number of input pulses have been received.

scaler frequency meter—A frequency meter in which electronic circuits are used for counting and gating electrical signals to indicate their number and/or rate.

scale span — The algebraic difference between the values of the actuating electrical quantity corresponding to the two ends of the scale of an instrument.

scaling—1. An electronic method of counting electrical pulses occurring too fast to be handled by mechanical recorders. 2. The changing of a quantity from one notation to another. 3. Adjusting the coefficient of a circuit to each of its one or more input-signal terminals. The relative scaling of one input to another is called weighting. 4. In computing, relating problem variables to machine variables.

scaling circuit—*See* Scaler.

scaling factor—Also called the scaling ratio. The number of input pulses per output pulse required by a scaler.

scaling ratio—*See* Scaling Factor.

scalloping distortion—In video tape recording, a series of small, vertical curves in the recorded image. Caused by unequal stretching across the width of the tape.

scan—1. In facsimile, to analyze the density of successive elemental areas of the subject copy in a predetermined pattern at the transmitter, or to record these areas at the recorder. 2. To examine point by point—e.g., in converting a televised scene or image into a methodical sequence of elemental areas. 3. One sweep of the mosaic in a camera tube or of the screen in a picture tube. 4. To examine point-by-point in logical sequence. 5. To sample each of a number of inputs intermittently. A scanning device may provide additional functions such as record or alarm.

scan-coded tracking system — *See* Monopulse Tracking.

scan converter — Equipment that samples radar images at a 3-kHz to 10-kHz rate that can be sent over telephone lines or narrow bandwidth radio circuits and converted into a slow-scan image by a similar converter. *See also* Slowed-Down Video.

scan-converter tube—A device consisting of a cathode-ray tube and a vidicon imaging tube assembled face-to-face in the same envelope.

scanistor—An integrated semiconductor optical-scanning device that converts images into electrical signals. The output analog signal represents both the amount and the position of the light shining on its surface.

scanner—1. An instrument which automatically samples or interrogates the state of various processes, conditions, or physical states and initiates action in accordance with the information obtained. 2. In a facsimile transmitter, the part which systematically translates the densities of the subject copy into the signal waveform. 3. The moving parts of an antenna that cause the beam to scan.

scanner amplifier—A vacuum-tube amplifier in a facsimile transmitter used to amplify the output-signal voltage of the scanner.

scanning—1. In television, facsimile, or picture transmission, the successive analyzing or synthesizing, according to a predetermined method, of the light values or equivalent characteristics of elements constituting a picture area. 2. In radar, the directing of a beam of radio-frequency energy successively over the elements of a given region, or the corresponding process in reception. 3. The comparison of input variables with some reference to determine a particular action. 3. The successive exposure of small portions of an object to a sensing device of some type. Television, radioactive scanning, facsimile transmission, and photoelectric scanning are all examples of this technique.

scanning-antenna mount—An antenna support which provides a mechanical means for scanning or tracking with the antenna, and a means for taking off information and using it for indication and control.

scanning beam—A beam of light, a radar beam, or an electron beam that is used in scanning.

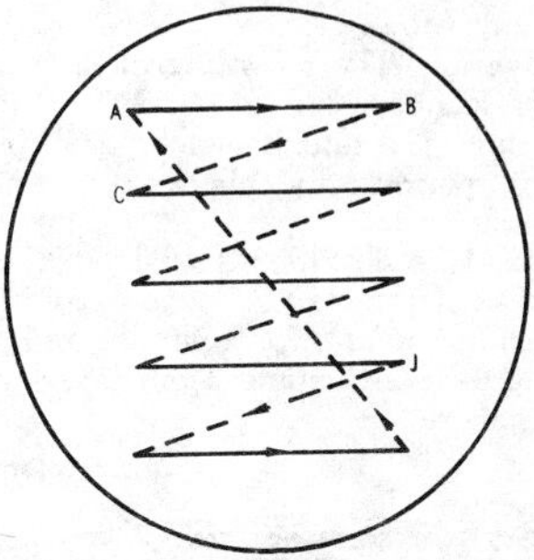

Scanning beam.

scanning circuit—A circuit which produces a linear, circular, or other movement of the beam in a cathode-ray tube at regular intervals.

scanning disc—1. A rotating tricolor wheel used between the camera lens and subject, or picture tube and veiwer, in field-sequential color television. 2. A Nipkow disc.

scanning frequency—*See* Stroke Speed.

scanning head—A light source and a phototube combined as a single unit for scanning a moving strip of paper, cloth, or metal in photoelectric side-register control systems.

scanning line—A single, narrow, continuous strip containing highlights, shadows, and halftones of the picture area, as determined by the scanning process.

scanning linearity—In television, the uniformity of the scanning speed during the trace interval.

scanning-line frequency — The number of scanning lines per second. *Also see* Stroke Speed.

scanning loss—In a radar system employing a scanning antenna, the reduced sensitivity that occurs in scanning across a target compared with the sensitivity when a constant beam is directed at the target. This loss is expressed in decibels.

scanning sonar—An echo-ranging system in which the sound pulse is transmitted simultaneously through the entire angle to be searched and a rapidly rotating transducer having a narrow beam angle scans for the returning echoes.

scanning speed—The number of inches per second explored by the spot of light or other source of energy in television, facsimile, radar, etc.

scanning spot—The immediate area being explored at any instant by a spot of light or other energy source in television, facsimile, radar, etc. (*See also* Picture Element.)

scanning yoke – A yoke-shaped iron core that supports the electromagnetic deflecting coils around the neck of some cathode-ray tubes.

scannogram—The recording made on paper by a scanner.

scan rate—The rate at which a control computer periodically checks a controlled quantity.

scatter—1. A disordered change in the direction of propagation when radio waves encounter matter. 2. Spurious radar echoes due to reflections from layers of the ionosphere.

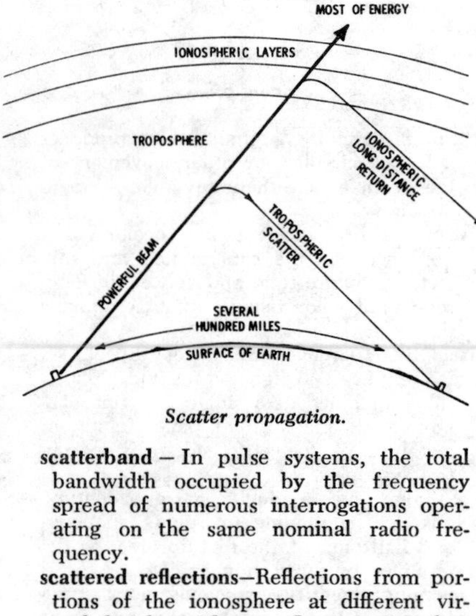

Scatter propagation.

scatterband – In pulse systems, the total bandwidth occupied by the frequency spread of numerous interrogations operating on the same nominal radio frequency.

scattered reflections—Reflections from portions of the ionosphere at different virtual heights. These reflections interfere with each other and cause rapid fading of the signal.

scattering—1. The change in direction, frequency, or polarization of radio waves when they encounter matter. 2. In a narrower sense, a disordered change in the incident energy of (1) above. 3. The change of direction of particles or photons colliding with other particles or systems. 4. The diffusion of a sound or light beam due to discontinuities in the transmitting medium.

scattering loss—That part of the transmission lost because of scattering within the medium or the roughness of the reflecting surface.

scatter loading—In a computer, the form of fetch that may result in placement of the control sections of a load module in nonadjoining main-storage positions.

scatterometer—A wide-sweep radar for terrain mapping.

scatter propagation – Also called beyond-the-horizon propagation, beyond-the-horizon transmission, or over-the-horizon transmission. Transmission of high-power radio waves beyond line-of-sight distances by reflecting them from the troposphere or ionosphere.

scatter read—The ability of a computer to distribute data into several memory areas as it is being entered into the system from magnetic tape.

scc wire – Abbreviation for single-cotton-covered wire.

SCEPTRON – Acronym for spectral comparative pattern recognizer. A device which automatically classifies complex signals derived from any type of information that can be changed into an electrical signal.

sce wire—Wire with a single cotton covering over enamel insulation.

scheduled maintenance—Maintenance performed according to an established plan.

scheduling – Determining the order in which job programs will use the available computer facilities.

schematic circuit diagram—*See* Schematic Diagram.

schematic diagram—1. Also called a schematic circuit diagram, diagram, or schematic. A diagram of the electrical scheme of a circuit, with components represented by graphical symbols. 2. A stylized drawing of a circuit in which the various elements are represented by conventional symbols.

Schering bridge – A four-arm alternating-current bridge used for measuring capacitance and dissipation factor. The unknown capacitor and a standard loss-free capacitor form two adjacent arms, the arm adjacent to the standard capacitor

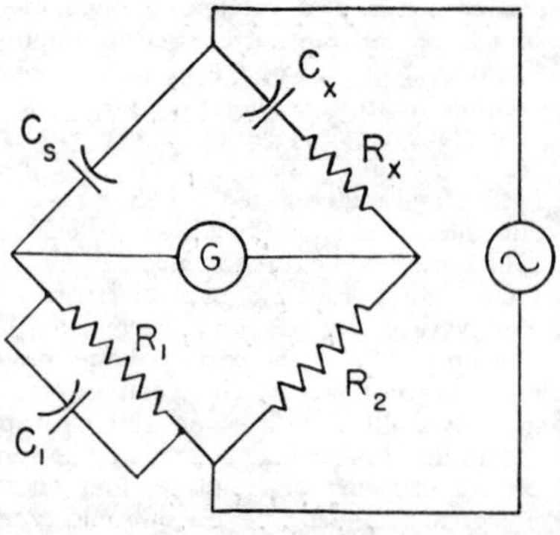

$$C_X R_2 = C_S R_1$$
$$C_X R_X = C_1 R_1$$

Schering bridge.

consists of a resistor and capacitor in parallel, and the fourth arm is a nonreactive resistor.

schlieren—An optical system that produces images in which the illumination or hue at a given point is related to the angular deflection a light ray undergoes in passing through the corresponding point in the object. The object is back-illuminated, and a straightedge, circular aperature, or graded density or multicolored filter is employed in the system to discriminate between deflected and undeflected rays.

Schmidt antenna – A microwave scanning antenna similar in principle to the optical Schmidt camera. The spherical reflector has a spheric microwave lens at the center of curvature and a scanner is located approximately halfway between these elements.

Schmidt optical system—An optical system for magnifying and projecting a small, brilliant image from a projection-type cathode-ray tube onto a screen.

Schmitt limiter—*See* Schmitt Trigger.

Schmitt trigger – Also called Schmitt limiter. A bistable pulse generator in which an output pulse of constant amplitude exists only as long as the input voltage exceeds a certain dc value. The circuit can convert a slowly changing input waveform to an output waveform with sharp transitions. Normally, there is hysteresis between an upper and a lower triggering level. 2. A regenerative circuit which changes state abruptly when the input signal crosses specified dc triggering levels.

Schottky barrier – A simple metal-to-semiconductor interface that exhibits a nonlinear impedance.

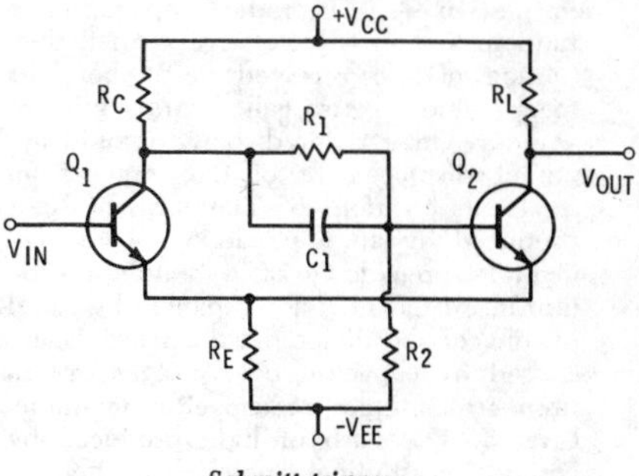

Schmitt trigger.

Schottky barrier diode – A junction diode with the junction formed between the semiconductor and a metal contact rather than between dissimilar semiconductor materials, as in the case of an ordinary pn diode.

Schottky transistor logic – (Abbreviated STL.) An improved version of integrated injection logic that has a power-delay product that is three times lower than that previously reported for I^2L.

Schottky TTL—A TTL circuit that incorporates Schottky diodes to greatly speed up TTL circuit operation.

scientific notation – In a calculator, the number is entered or a result is displayed in terms of a power of ten. For example, the number 1234 is entered as 1.234×10^3 and the number 0.001234 would appear as 1.234×10^{-3}.

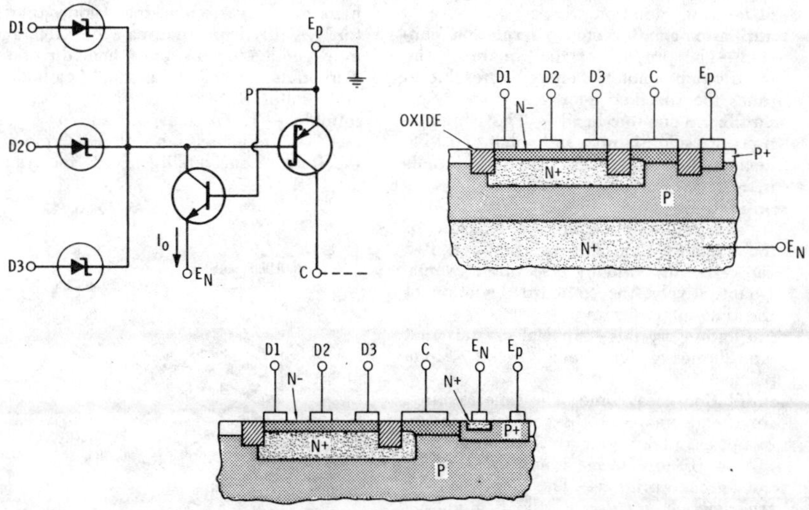

Schottky transistor logic.

scintillate—To emit flashes of light.

scintillation – 1. In radio propagation, a random and usually relatively small fluctuation of the received field about its mean value. 2. Also called target glint or wander. On a radar display, a rapid apparent displacement of the target from its mean position. 3. The flash of light produced by an ionic action. 4. A momentary breakdown of a tantalum-oxide film in a capacitor, accompanied by rapid healing of the dielectric. Such events are caused by capacitor overvoltages or improper techniques of capacitor manufacture. 5. The flash of light produced by certain crystalline materials when a charged particle is passed through them.

scintillation conversion efficiency – In a scintillator, the ratio of the optical photon energy emitted to the energy of the incident particle or photon of ionizing radiation.

scintillation counter – A device that indirectly detects charged particles and gamma rays and neutrons by using a photomultiplier tube to convert the short flashes of light, produced as the particle passes through a transparent scintillating material, into electric signals which can be recorded. One advantage of scintillation counters is that they are very fast; i.e., they have a very small resolving time.

scintillation-counter cesium resolution – The scintillation-counter energy resolution for the gamma ray or conversion electron emitted from cesium-137.

scintillation-counter energy resolution – In a scintillation counter, a measure of the smallest discernible difference in energy between two particles or photons of ionizing radiation.

scintillation-counter energy-resolution constant—The product of the square of the scintillation-counter energy resolution times the specified energy.

scintillation-counter head – The combination of scintillators and photosensitive devices that produces electrical signals in response to ionizing radiation.

scintillation-counter time discrimination—In a scintillation counter, a measure of the smallest time interval between two successive individually discernible events. Quantitatively, the standard deviation of the time-interval curve.

scintillation crystals – Special crystals that emit flashes of light when struck by alpha particles.

scintillation decay time—The time required for the decrease of the rate of emission of optical photons in a scintillation from 90% to 10% of the maximum value.

scintillation duration – The interval from the time of emission of the first optical photon of a scintillation to the time when 90% of the optical photons of the scintillation have been emitted.

scintillation rise time – The time interval occupied by the increase of the rate of emission of optical photons of a scintillation from 10% to 90% of the maximum value.

scintillator—The combination of the body of scintillator material and its container.

scintillator material – A material that exhibits the property of emitting optical photons in response to ionizing radiation.

scintillator-material total-conversion efficiency—In a scintillator material, the ratio of the produced optical photon energy to the energy of a particle or photon of ionizing radiation that is entirely absorbed in the scintillator material.

scissoring—The ability of the vector generator to blank the beam whenever it is moved outside of the screen (where the image becomes distorted).

scope—Slang for a cathode-ray oscilloscope.

scophony television system—A mechanical television projection system developed in England. The light-storage phenomenon of a supersonic light value is utilized, and ingenious optical and mechanical methods provide large, bright images suitable for theater installation as well as home television receivers. The apparent screen brightness is multiplied several hundred times because several hundred picture elements are projected simultaneously.

scoring system—In motion-picture production, a system for recording music in time with the action on the film.

Scott connection—A method of connecting transformers to convert two-phase power to three-phase or vice versa.

SCR – Silicon controlled rectifier. Formal name is "reverse-blocking triode thyristor." A thyristor that can be triggered into conduction in only one direction. Terminals are called "anode," "cathode," and "gate."

scramble – 1. To transpose and/or invert bands of frequencies, or otherwise modify the form of the intelligence at the trans-

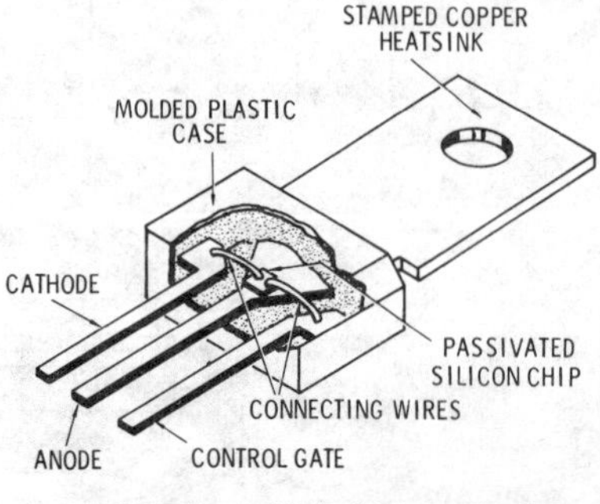

SCR (in molded plastic case).

mitting end, according to a prearranged scheme, to obtain secrecy. 2. To mix, in cryptography, in a random or quasi-random fashion.

scrambled speech — Also called inverted speech. Speech that has been made unintelligible (e.g., for secret transmission) by inverting its frequency. At the receiving end, it can then be converted back into intelligible speech by reinverting the frequency.

scrambler circuit — Also called a speech scrambler. A circuit where essential speech frequencies are divided into several ranges by filters and then inverted to produce scrambled speech. (*See also* Speech Inverter.)

scratch filter—A low-pass filter, often an integral part of an amplifier circuit, which attenuates the higher frequency noise derived from disc recordings. The scratch filter is also suitable for the suppression of background noise produced by tape background hiss.

scratch pad—Information which the processing unit of a computer stores or holds temporarily. It is a memory containing subtotals for various unknowns which are needed for final results.

scratch-pad memory—1. A high-speed, limited-capacity computer information store that interfaces directly with the central processor. It is used to supply the central processor with the data for the immediate computation, thus avoiding the delays that would be encountered by interfacing with the main memory. (The function of the scratch-pad memory is analogous to that of a pad of paper used for jotting down notes.) 2. A high-speed memory used to temporarily store a small amount of data so that the data can be retrieved quickly when needed. Interim calculations are stored in a scratch pad memory. 3. Small high-speed RAM used to hold data and instructions having immediate use.

screen—1. The surface upon which the visible pattern is produced in a cathode-ray tube. 2. A metal partition which isolates a device from external electric or magnetic fields. 3. *See* Screen Grid.

screen angle—A vertical angle bounded by a straight line from the radar antenna to the horizon and the horizontal at the antenna, assuming a 4/3 earth's radius.

screen dissipation — The power which the screen grid dissipates as heat after bombardment by the electron stream.

screen grid—Also called a screen. A grid placed between a control grid and an anode and usually maintained at a fixed positive potential. By reducing the electrostatic influence of the anode, it prevents the electrons from bunching in the space between the screen grid and the cathode.

screen-grid modulation — Modulation produced by introducing the signal into the screen-grid circuit of any multigrid tube where the carrier is present.

screen-grid tube—A vacuum tube in which a grid is placed between the control grid and the anode to prevent the latter from reacting with the control grid.

screen-grid voltage — The direct-voltage value applied between the screen grid and the cathode of a vacuum tube.

scribe projection—A method of automatic information presentation in which information is placed on a small metallic-coated glass slide by using a movable, servocontrolled, fine-pointed scribe to remove the coating. Light passed through the scribed area is projected onto a screen.

scribing—A process, similar to glass cutting, in which a slice of semiconductor devices is scored in rows and columns so that it may be separated easily into individual devices. The process is performed in a machine called a scriber by repeated movement of a weighted diamond stylus across the slice to form the scored pattern.

SCS — Abbreviation for silicon controlled switch.

S-curve — An S-shaped frequency-response curve showing how the output of a frequency-modulation detector or circuit varies with frequency.

sea clutter—*See* Sea Return.

seal—Any device used to prevent gases or liquids from passing through.

sealed contacts — A contact assembly enclosed in a sealed compartment separate from the other parts of the relay.

sealed-gage pressure transducer — A pressure transducer that has the sensing element sealed in its case at room ambient pressure. The sealing method holds the original internal pressure for long periods of time.

sealed meter—A meter constructed so that moisture or vapor cannot enter the meter under specified test conditions.

sealed tube—Used chiefly for pool-cathode tubes. A hermetically sealed electron tube.

sealing compound—A type of wax or pitch compound used in dry batteries, capacitor blocks, transformers, or circuit units to keep out air and moisture.

sealing off—The final closing of the bulb of a vacuum tube or lamp after evacuation.

seam welding—A resistance welding process in which overlapping spot welds are made progressively along a joint by means of circular electrodes. The circular seam-welding wheels roll along the overlapping edges to be welded, and the control circuit is arranged to pass current at sufficiently close intervals to produce the desired degree of overlapping of the spot welds. The primary purpose of a seam-welding joint is to produce liquid-tight or airtight containers from comparatively thin sheet metal.

search—1. In radar operation, the directing of the lobe (beam of radiated energy) in order to cover a large area. A broad-beam antenna may be used, or a rotating or scanning antenna. 2. A systematic examination of the available information in a specific field of interest. 3. To scan available stored information. 4. The process of applying a sweeping tuning signal to the free running oscillator portion of a phase-locked oscillator, causing the oscillator output frequency to pass within the capture bandwidth of the feedback network and ensuring a locked condition in response to a new reference frequency.

search coil—*See* Magnetic Test Coil.

search gate—A gate pulse which is made to search back and forth over a certain range.

searchlighting — In radar, the opposite of scanning. Instead, the beam is projected continuously at an object.

searchlight-type sonar — An echo-ranging system employing the same narrow beam pattern for both transmission and reception.

search radar—A radar intended primarily for displaying targets as soon as possible after their entrance into the coverage area.

search receiver—*See* Intercept Receiver.

search time—The time required for location of a particular data field in a computer storage device. The process involves comparison of each field with a predetermined standard until an identity is obtained. Contrasted with access time.

sea return—Also called sea clutter. In radar, the aggregate received echoes reflected from the sea.

seasonal factors—Factors which are used to adjust sky-wave absorption data for seasonal variations. Those variations are due primarily to seasonal fluctuations in the heights of the ionospheric layers.

seasoning — Overcoming a temporary unsteadiness of a component which may appear when the component is first installed.

seating time — The elapsed time after the coil of a relay has been energized until the armature of the relay is seated.

sec—1. Abbreviation for second or for secondary winding of a transformer. 2. Abbreviation for secondary electron conduction.

secondary — 1. The transformer output winding where the current flow is due to inductive coupling with another coil called the primary. 2. Low-voltage conductors of a power distributing system.

secondary area — *See* Secondary Service Area.

secondary calibration — Also called sense step. Calibration of accessory equipment in which a transducer is deliberately unbalanced electrically to change the output voltage, current, or impedance. Generally performed by means of a calibration resistor which is placed across one leg of the bridge.

secondary cell—A voltaic cell which, after being discharged, may be restored to a charged condition by an electric current sent through the cell in a direction opposite that of the discharge current. *See also* Storage Cell, 1.

secondary color—A color produced by combining any two primary colors in equal proportions. In the light-additive process, the three secondary colors are cyan, magenta, and yellow.

Secondary Communications Authorization —*See* SCA.

secondary electron — An electron emitted from a material as a result of bombardment by electrons or of the collision of a charged particle with a surface.

secondary electron conduction tube—A sensitive tv tube that uses a two-step process to convert the invisible image to a charge image. In the image intensifier stage light ejects electrons from a photoemitter. After imaging and amplification in the middle or imaging section the primary photoelectrons fall on the thin film face of the secondary target. This charge image modulates the scanning beam current from the reading section. The electric field applied across the film causes a majority of the secondary electrons to be transported through a potassium-chloride low-density film layer to produce a secondary conduction current.

secondary-electron multiplier—An amplifier tube in which the electron stream is focused onto a succession of targets, each

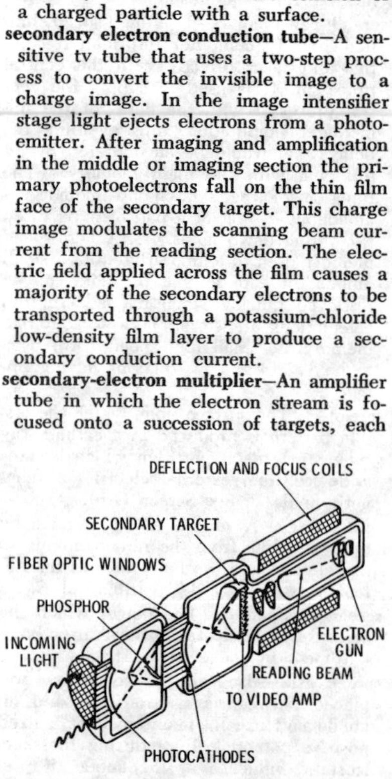

Secondary electron conduction.

of which adds its secondary electrons to the stream. In this way, considerable amplification is provided.

secondary emission—The liberation of electrons from an element, other than the cathode, as a result of being struck by other high-velocity electrons. In a vacuum tube there are usually more secondary than primary electrons—a desirable phenomenon in electron-multiplier or dynatron-oscillator tubes. However, pentodes have a suppressor grid to nullify the undesirable effect of secondary emission.

secondary-emission ratio — The average number of secondary electrons emitted from a surface per incident primary electron.

secondary-emission tube — A tube which makes use of secondary emission to achieve a useful end. The photomultiplier is an example.

secondary failure — 1. A failure occurring as a direct result of the abnormal stress on a component brought about by the failure of another part or parts. 2. Any failure that is the direct or indirect result of a primary failure.

secondary grid emission — Emission from the grid of a tube as a result of high-velocity electrons being driven against it and knocking off additional electrons. The effect is the same as for primary grid emission.

secondary line—The conductors connected between the secondaries of distribution transformers and the consumer service entrances.

secondary radar—*See* Radar.

secondary radiation — Random reradiation of electromagnetic waves.

secondary service area—Also called the secondary area. The service area of a radio or television broadcast station within which satisfactory reception can be obtained only under favorable conditions.

secondary standard—A unit (e.g., length, capacitance, weight) used as a standard of comparison in individual countries or localities, but checked against the one primary standard in existence somewhere in the world.

secondary storage—Storage which is not an integral part of a computer, but which is directly linked to and controlled by it.

secondary voltage—The voltage across the secondary winding of a transformer.

secondary winding — The winding on the output side of a transformer.

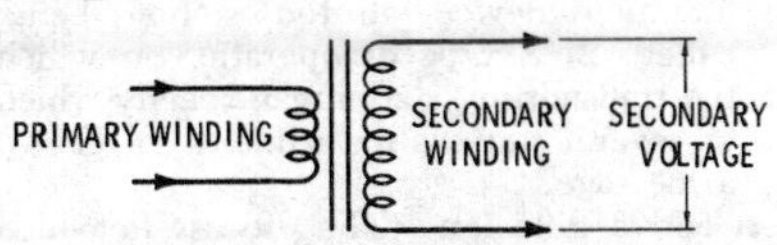

Secondary voltage.

secondary X-rays—X-rays given off by an object irradiated with X-rays. Their frequency depends on the material in the object.

second breakdown—1. A condition in which the output impedance of a transistor changes almost instantaneously from a large value to a small limiting value. It may be distinguished from normal transistor operation by the fact that once it occurs, the base no longer controls normal collector characteristics. Second breakdown is associated with imperfections in the device structure, usually being more severe in multiple-diffused, high-speed devices. 2. Lateral current instability through a transistor when operating at relatively high voltages and current. It has its greatest effect under dc conditions, but falls off with increasing temperature and frequencies; the breakdown caused is usually permanent.

second-channel attenuation — Alternate-channel attenuation. *See also* Selectance, 2.

second-channel interference — Also called alternate-channel interference. Interference in which the extraneous power originates from an assigned (authorized) signal two channels away from the desired channel.

second detector—Also called a demodulator. In a superheterodyne receiver, the portion that separates the audio component from the modulated intermediate frequency.

second-generation computer — A computer in which solid-state components are used.

second-time-around echo—An echo received after a time interval greater than the pulse interval.

section—1. A four-terminal network which cannot be divided into a cascade of two simpler four-terminal networks. 2. One individual span of a radio relay system; the number of sections in a system is one more than the number of repeaters.

sectional center—A toll switching point to which are connected a number of primary outlets.

sectionalized vertical antenna — A vertical antenna separated into parts by insulators at one or more points along its length. When suitable reactances or driving voltages are placed across the insulated points, the current distribution is modified to give a more desirable radiation pattern in the vertical plane.

sectoral horn—A horn with two parallel and two diverging sides. (*See* illustration, page 644.)

sector cable — A multiconductor cable in which the cross section of each conductor is essentially a sector of a circle, an ellipse, or some figure intermediate between them. Sector cables are used in

Sectoral horn.

order to make possible the use of larger conductors in a cable of given diameter.

sector display—A range-amplitude display used with a radar set. The antenna system rotates continuously, and the screen (of the long-persistence type) is excited only while the beam is within a narrow sector centered on the object.

sector scan—A scan in which the antenna oscillates through a selected angle.

sector scanning—Modified circular scanning in which only a portion of the plane or flat cone is generated.

secular variation—A slow variation in the strength of the earth's magnetic field.

SED—Abbreviation for spectral energy distribution.

Seebeck coefficient of a couple—For homogeneous conductors, the limit, as the difference in temperature approaches zero, of the quotient of the Seebeck emf divided by the temperature difference between the junctions. By convention, the Seebeck coefficient of a couple is considered positive if, at the cold junction, the first named conductor has a positive potential with respect to the second. It is the algebraic difference between either the relative or absolute Seebeck coefficients of the two conductors.

Seebeck effect—The production of an emf in a circuit composed of two dissimilar metals when their two junctions are at different temperatures. The emf is considered to be the resultant of the Peltier and Thomson emfs around the circuit.

Seebeck emf—Also called thermal emf. The emf produced by the Seeback effect.

seed—A special single crystal from which large single crystals are grown by the Czochralski technique.

seek — With reference to a computer, to look for data according to information given with respect to those data.

segment—1. In a routine, the part short enough to be stored entirely in the internal storage of a computer, yet containing all the coding necessary to call in and jump automatically to other segments. 2. To divide a program into an integral number of parts, each of which performs a part of the total program and is short enough to be completely stored in internal memory.

segmented thermoelectric arm—A thermoelectric arm made up of two or more materials that have different compositions.

segmenting—*See* Partitioning.

seismic mass—The force-summing member for applying acceleration and/or gravitational force in an accelerometer.

seismograph—An instrument for recording the time, direction, and intensity of earthquakes or of earth shocks produced by explosions.

Seitz breakdown theory—Breakdown is due to the attainment of a critical avalanche size which leads to a conducting path.

selectance—1. A measure of the drop in response as a resonant device loses its resonance. It is the ratio of the amplitude of response at the resonant frequency, to the response at some other, specified frequency. 2. Often expressed as adjacent-channel attenuation (aca) or second-channel attenuation (sca). The reciprocal of the ratio of the sensitivity of a receiver tuned to a specified channel, to its sensitivity at another channel a specified number of channels away.

selected mode—A mode of operation for an encoder selector circuit in which one set of brushes is selected to be read and another inhibited from being read; also a mode of operation for a system controlling several encoder outputs in which the encoder is selected to be read and all others inhibited from being read.

selection check—A verification of a computer instruction, usually automatic, to ensure that the correct register or device has been chosen.

selection ratio—The ratio of the least magnetomotive force used to select a cell or core to the maximum magnetomotive force used which is not intended to select a cell or core.

selective—The characteristic of responding to a desired frequency to a greater degree than to other frequencies.

selective absorption—Absorption of rays of a certain group of frequencies only.

selective calling—1. A means of calling in which code signals are transmitted for the purpose of activating the automatic attention device at the station being called. 2. A type of operation in which the transmitting station can specify which of several stations on a line is to receive a message.

selective diffusion — The process in which specified isolated regions in a semiconductor material are doped. The components in a silicon integrated circuit are formed in this way.

selective dump—A dump of a selected area of internal storage.

selective fading—Fading in which the received signal does not have the same variation in strength for all frequencies in the band. Selective fading usually occurs during multipath transmission.

selective interference—Interference the energy of which is concentrated within narrow frequency bands.

selective ringing—An arrangement used on telephone party lines so that only the bell of the called subscriber rings.

selective squelch—*See* Squelch Circuit.

selectivity—1. The characteristic which determines the extent to which the desired signal can be differentiated from disturbances of other frequencies. 2. A tuner's ability to discriminate between a wanted signal and an interfering signal on adjacent frequency settings of the tuning dial.

selectivity control—The control for making a receiver more selective.

select lines—In a core memory array, the wires which pass through magnetic cores and carry the selecting coincident currents.

selector—1. On a punch-card machine, a mechanism which reports a condition and accordingly causes a card or an operation to be selected. 2. In a telephone system, the switch or relay-group switching systems that select the path the call is to take through the system. It operates under the control of the dial at the calling station. 3. A sequential switch, usually multicontact or motor driven.

selector pulse—A pulse used to identify one event of a series.

selector relay—A relay capable of automatically selecting one or more circuits.

selector switch—A multiposition switch that permits one or more conductors to be connected to any of several other conductors.

selectron—A computer-memory tube capable of storing 256 binary digits and permitting very rapid selection and access.

selenium—A chemical element with marked photosensitive properties and a resistance that varies inversely with illumination. It is used as a rectifier layer in metallic rectifiers.

selenium cell—A photoconductive cell consisting of a layer of selenium on a substrate whose electrical resistance varies with the illumination falling on the cell. (Selenium cells have been largely replaced by photocells of one kind or another.)

selenium rectifier – A metallic rectifier in which a thin layer of selenium is deposited one one side of an aluminum plate and a highly conductive metal is coated over it. Electrons flows more freely from the coating to the selenium than in the opposite direction, thereby providing rectification.

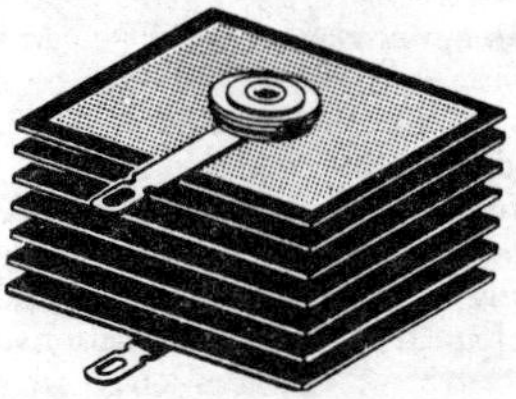

Selenium rectifier.

self-adapting—Pertaining to the ability of a system to change its performance characteristics in response to its environment.

self-adaptive system—A system which can exhibit the qualities of reorganization and/or learning.

self-adjusting communication – *See* Adaptive Communication.

self-aligned-thick oxide – A term used to describe a proprietary low-voltage, self-aligned gate process.

self-aligning-gate MOS—Abbreviated SAG MOS. 1. MOS device where a polycrystalline silicon layer is substituted for the usual aluminum metal gate. The key feature is a different processing technology in which the gate is automatically aligned. 2. A process in which materials like polycrystalline silicon or refractory metals are used in place of aluminum at the gate. These materials act as a mask and result in the gate being automatically aligned between source and drain regions.

self and systems testing and checkout – Logical and numerical processing for the purpose of exercising and monitoring responses of the system and the functioning of the computer itself.

self-balancing recorder—A recording device operating on the servomechanism principle.

self-bias – The voltage developed by the flow of vacuum-tube current through a resistor in a grid or cathode lead. Also called automatic bias.

self-capacitance – *See* Distributed Capacitance.

self-checking code—In computers, a code in which errors produce forbidden combinations. A single error-detecting code produces a forbidden combination if a digit gains or loses a single bit. A double error-detecting code produces a forbidden combination if a digit gains or loses either one or two bits, and so forth.

self-cleaning contact—*See* Wiping Contact.

self-complementing code—A machine language in which the code of the complement of a digit is the complement of the code of the digit.

self-contained instrument – An instrument that has all the necessary equipment built into the case or made a corporate part thereof.

self-demagnetization – The process by which a magnetized sample of magnetic material tends to demagnetize itself by virtue of the opposing fields created within it by its own magnetization. Self-demagnetization inhibits the successful recording of signal components having short wavelengths or sharp transitions.

self-erasure – The tendency for strongly magnetized areas of the tape coating to erase adjacent areas of opposite-polarity magnetization. This is a major cause of loss of high frequencies at reduced tape speeds.

self-excitation–The supplying of required exciting voltages by a device itself rather than from an external source.

self-excited oscillator – An oscillator that operates without external excitation and solely by the direct voltages applied to the electrodes. It depends on its resonant circuits for frequency determination (i.e., not crystal controlled).

self-extinguishing – Material which ignites and burns when exposed to flame or elevated temperature, but which stops burning when the flame or high temperature is removed.

self-focused picture tube–A television picture tube with an automatic electrostatic focus designed into the electron gun.

self-generating transducer – A transducer which requires no external electrical excitation to provide an output.

self-healing – The characteristic of metallized capacitors by which faults or shorts occurring during operation are removed, or healed, by an internal clearing action, and the part continues to function. *See also* Clearing.

self-healing capacitor–A capacitor that restores itself to operation after a breakdown caused by excessive voltage.

self-heated thermistor – A thermistor the body temperature of which is significantly higher than the temperature of its ambient medium as a result of the power being dissipated in it.

self-heating coefficient of resistivity – The maximum change in resistance due to temperature change caused by power dissipation, at constant ambient temperature. Usually expressed in percent or per-unit (ppm) change in nominal resistance per watt of dissipation. This parameter is actually the product of the power coefficient and the resistor temperature coefficient.

self-impedance–At any pair of terminals of a network, the ratio of an applied potential difference to the resultant current at these terminals (all other terminals open).

self-inductance–1. The property which determines the amount of electromotive force induced in a circuit whenever the current changes in the circuit. 2. At any pair of terminals of a network, the ratio of an applied potential difference to the resultant current at these terminals, all other terminals being open.

self-induction–The property that causes a counterelectromotive force to be produced in a conductor when the magnetic field produced by the conductor collapses or expands with a change in current.

self-inductor–An inductor used for changing the self-inductance of a circuit.

self-instructed carry–A system of executing the carry process in a computer by allowing information to propagate to succeeding places as soon as it is generated, without receipt of a specific signal.

self-latching relay–A relay in which the armature remains mechanically locked in the energized position until deliberately reset.

self-locking nut–A nut with an inherent locking action, so it cannot readily be loosened by vibration.

self-optimizing communication–*See* Adaptive Communication.

self-organizing – Having to do with the ability of a system to arrange its own internal structure.

self-organizing machines – Machines that can recognize, or learn to recognize, such stimuli as patterns, characters, and sound, and which can then adapt to a changing environment.

self-powered – Equipment containing its own power supply. It may be either a combination of wet and dry cells, or dry cells in conjunction with a spring-driven motor.

self-pulse modulation–Modulation accomplished by using an internally generated pulse. (*See also* Blocking Oscillator, 1.)

self-pulsing–A special type of grid-pulsing circuit which automatically stops and starts the oscillations at the pulsing rate.

self-quenched counter tube – A radiation-counter tube in which reignition of the discharge is inhibited.

self-quenched detector – A superregenerative detector in which the grid-leak–grid-capacitor time constant is sufficiently large to cause intermittent oscillation above audio frequencies. As a result, normal regeneration is stopped just before it spills over into a squealing condition.

self-quenching oscillator – An intermittent self-oscillator producing a series of short trains of rf oscillations separated by intervals of quiescence. The quiescence is caused by rectified oscillatory currents, which build up to the point where they cut off the oscillations.

self-rectifying X-ray tube—An X-ray tube operating with an alternating anode potential.

self-regulation—The tendency of a component or system to resist change in its condition or state of operation.

self-repeating timer—A form of time-delay circuit in which relay contacts are used to restart the time delay.

self-reset — Automatically returning to the original position when normal conditions are resumed (applied chiefly to relays and circuit breakers).

self-saturating rectifier—A half-wave rectifying-circuit element connected in series with the output windings of a saturable reactor in a self-saturating magnetic-amplifier circuit.

self-saturation—The saturation obtained in a magnetic amplifier by rectifying the output current of a saturable reactor.

self-screening range—The range at which a target can be detected by a radar in the midst of its jamming mask, with a certain specified probability.

self-selecting V scan—The V-scan method of reading a polystrophic code (primarily the binary code) in which diode logic circuits are used internally in the encoder to perform the necessary bit-to-bit selection to prevent ambiguity in the encoder output data.

self-starting synchronous motor—A synchronous motor provided with the equivalent of a squirrel-cage winding so it can be started like an induction motor.

self-stopping modulo-n counter—A counter that has n distinct states that stops when it reaches a predetermined maximum number; it then does not accept count pulses until it is reset to a number less than the maximum number.

self-sustained oscillations — Oscillations maintained by the energy fed back from the output to the input circuit.

self-testing — The ability of a piece of equipment to automatically verify the proper operation of its components or subsystems.

self-threading reel—A device for spooling and storing tape; in particular, one which does not require external aid to affix or start the first turn of tape on its winding surface or hub. Flanges on such reels can be continuous and free of windows; winding surfaces are continuous and free of distortion-producing threading slots.

self-wiping contact—*See* Wiping Contact.

selsyn—Electrical remote-indicating instrument operating on direct current, in which the angular position of the transmitter shaft carrying a contact arm moving on a resistance strip, controls the pointer on the indicator dial.

semantics—The relationships between symbols and their meanings.

semiactive homing guidance—A system of homing guidance in which radiations used by the receiver in the missile are reflected from a target being illuminated by an outside source.

semiactive repeater — A communications satellite which uses a minimum of onboard electronics to take a modulated signal beamed at it from a ground station and transfer its information (modulation) into an unmodulated beam (on a different frequency) set up by the receiving station. For this transfer it uses Van Atta or other directive arrays and nonlinear elements.

Semiautomatic Ground Environment — An air-defense system in which data from air surveillance are processed for transmission to computers at direction centers. Abbreviated SAGE.

semiautomatic keying circuits—Mechanization that provides torn-tape switching systems in teleprinter links. Incoming and outgoing messages are placed on tapes that are inserted manually into a distributor that provides automatic mechanical keying of the circuit.

semiautomatic message switching center—A center at which messages are routed by an operator on the basis of information contained in them.

semiautomatic starter—A starter in which some of the operations are not automatic, but selected portions are automatic.

semiautomatic tape relay — A method of communication whereby messages are received and retransmitted in teletypewriter tape form involving manual intervention in the transfer of the tape from the receiving reperforator to the automatic transmitter.

semiautomatic telephone system — A telephone system in which operators receive orders from the calling parties verbally, but they use automatic apparatus in making connections.

semiconducting material—A solid or liquid having a resistivity midway between that of an insulator and a metal.

semiconductor—1. A solid or liquid electronic conductor, with resistivity between that of metals and that of insulators, in which the electrical charge carrier concentration increases with increasing temperature over some temperature range. Over most of the practical temperature range, the resistance has a negative temperature coefficient. Certain semiconductors possess two types of carriers, negative electrons and positive holes. The charge carriers are usually electrons, but there may be also some ionic conductivity. 2. An electronic device the main functioning parts of which are made from semiconductor materials. Examples include germanium, lead sulfide, lead telluride, selenium, silicon, and silicon carbide. Used in diodes, photocells, thermistors, and transistors.

semiconductor chip—A single piece of semiconductor material of any dimension.

semiconductor device — 1. A device in which the characteristic that distinguishes electronic conduction takes place within a semiconducting material. 2. A device where n-type and p-type materials used in combination to obtain specific characteristics for controlling the flow of current. 3. A device whose essential characteristics are due to the flow of charge carriers within a semiconductor.

semiconductor diode—1. A device consisting of n-type and p-type semiconductor material joined together to form a pn junction, which passes current in the forward direction (from anode to cathode) and blocks current in the reverse direction. *See also* Crystal Diode. 2. A light-emitting diode that emits coherent light by suitably arranged geometry. Gallium arsenide is used for lasers of this type.

semiconductor-diode parametric amplifier—A parametric amplifier using one or more varactors.

semiconductor integrated circuit—Complex circuits fabricated by suitable and selectively modifying areas on and within a wafer of semiconductor material to yield patterns of interconnected passive as well as active elements. The circuit may be assembled from several chips and use thin film elements or even discrete components to achieve a specified performance when the necessary device parameters cannot be achieved by materials modification.

semiconductor intrinsic properties—Properties of a semiconductor that are characteristic to the ideal crystal.

semiconductor junction — The region of transition between semiconducting regions of different electrical properties, usually between p-type and n-type materials.

semiconductor laser—A light-emitting diode that uses stimulated emission to produce a coherent-light output. *See also* Diode Laser.

semiconductor material—1. A material in which the conductivity ranges between that of a conductor and an insulator. The electrical characteristics of semiconductor materials such as silicon are dependent upon the small amounts of added impurities or dopants. 2. A chemical element, like silicon or germanium, which has a crystal lattice whose atomic bonds are such that the crystal can be made to conduct an electric current by means of free electrons or holes.

semiconductor memory — 1. A memory whose storage medium is a semiconductor circuit. Often used for high-speed buffer memories and for read-only memories. 2. A memory in which semiconductors are used as the storage elements, and characterized by low-to-moderate cost storage and a wide range of memory operating speed, from very fast to relatively slow. Almost all semiconductor memories are volatile.

semiconductor rectifier diode—A semiconductor diode designed for rectification and including its associated mounting and cooling attachments if integral with it.

semidirectional microphone—A microphone the field response of which is determined by the angle of incidence in part of the frequency range but is substantially independent of the angle of incidence in the remaining part.

semiduplex—In a communications circuit, a method of operation in which one end is duplex and one end simplex. This type of operation is sometmies used in mobile systems with the base station duplex and the mobile station or stations simplex. A semiduplex system requires two operating frequencies.

semimagnetic controller—An electrical controller in which not all its basic functions are performed by electromagnets.

semimetals—Materials such as bismuth, antimony, and arsenic having characteristics that class them between semiconductors and metals.

semiremote control—Radio-transmitter control performed near the transmitter by devices connected to but not an integral part of the transmitter.

semiselective ringing—An arrangement in which the bells of two stations on a telephone party line are rung simultaneously; differentiation is made by the number of rings.

semitone—Also called half step. The interval between two sounds. Its basic frequency ratio is equal to approximately the twelfth root of 2.

semitransparent photocathode — A photocathode in which radiant flux incident on one side produces photoelectric emission from the opposite side.

sender—That part of an automatic-switching telephone system that receives pulses originated by a dial or other source and, in accordance with the pulses received, controls the further operations necessary to establish the connection.

sending-end impedance — Also called the driving-point impedance. The ratio of an applied potential difference of a transmission line, to the resultant current at the point where the potential difference is applied.

sending filter—A filter used at the transmitting terminal to restrict the transmitted frequency band.

sensation level—*See* Level Above Threshold.

sense—1. In navigation, the relationship between the change in indication of a radio-navigational facility and the change in the navigational parameter being indicated. 2. In some navigational equipment, the property of permitting the resolution of 180° ambiguities. 3. To examine or determine the status of some system components. 4. To read holes in punched tape or cards.

sense amplifier—1. A circuit used to sense low-level voltages such as those produced by magnetic or plated-wire memories and to amplify these signals to the logic voltage levels of the system. 2. A circuit used in communications-electronics equipment to determine a change of phase or voltage and to provide an automatic control function.

sense finder — In a direction finder, that portion which permits determination of direction without 180° ambiguity.

sense-reversing reflectivity—The characteristic of a reflector that reverses the sense of an incident ray. (For example, a perfect corner reflector is invisible to a circularly, polarized radar because it reverses the sense.)

sense step—*See* Secondary Calibration.

sense switch—One of a series of switches on the console of the digital computer that permits the operator to control some parts of a program externally.

sensing — The process of determining the sense of an indication.

sensing element—*See* Primary Detector.

sensing field—The zone in which an object can be sensed by a proximity switch.

sensistor—A silicon resistor the resistance of which varies with temperature, power, and time.

sensitive relay—A relay requiring only a small current. It is used extensively in photoelectric circuits. A small amount of current is usually defined as 100 milliwatts or less.

sensitive volume — In a radiation-counter tube, the portion responding to a specific radiation.

sensitivity — 1. The minimum input signal required in a radio receiver or similar device, to produce a specified output signal having a specified signal-to-noise ratio. This signal input may be expressed as power or voltage at a stipulated input network impedance. 2. Ratio of the response of a measuring device to the magnitude of the measured quantity. It may be expressed directly in divisions per volt, milliradians per microampere, etc., or indirectly by stating a property from which sensitivity can be computed (e.g., ohms per volt for a stated deflection). 3. The signal current developed in a camera tube per unit incident radiation density (i.e., per watt per unit area). 4. The degree of response of an instrument or control unit to a change in the incoming signal. 5. In tape recording the relative intensity of the magnetic signal recorded within the tape coating's linear region by a magnetizing field of a given intensity.

sensitivity adjustment — Also called span adjustment. The control of the ratio of output signal to excitation voltage per unit measurand. Generally accomplished in a system by changing the gain of one or more amplifiers. The practice of placing excitation control components (such as potentiometers or rheostats) in series with the excitation to a transducer is a sensitivity adjustment for the system. However, in the latter case no significant change is introduced in the output-to-input ratio of the transducer.

sensitivity control — The control that adjusts the amplification of the radio-frequency amplifier stages and thereby makes the receiver more sensitive.

sensitivity-time control — Also called gain-time control or time gain. The portion of a system which varies the amplification of a radio receiver in a predetermined manner.

sensitizing (electrostatography) — The establishing of an electrostatic surface charge of uniform density on an insulating medium.

sensitometer—An instrument used to measure the sensitivity of light-sensitive materials.

sensitometery — Measurement of the light-response characteristics of photographic film.

sensor — 1. In a navigational system, the portion which perceives deviations from a reference and converts them into signals. 2. A component that converts mechanical energy into an electrical signal, either by generating the signal or by controlling an external electrical source. 3. *See* Primary Detector. 4. An information-pickup device. 5. A transducer designed to produce an electrical output proportional to some time-varying quantity, as temperature, illumination, pressure, etc. 6. The component of an instrument that converts an input signal into a quantity which is measured by another part of the instrument.

sentinel—1. A symbol marking the beginning or end of some piece of information in digital-computer programming. 2. *See* Tag.

separate excitation — Excitation in which generator field current is provided by an independent source, or motor field current is provided from a source other than the one connected across the armature.

separately instructed carry—Executing the carry process in a computer by allowing carry information to propagate to succeeding places only when a specific signal is received.

separate parts of a network – The unconnected parts.

separation—The degree to which two stereo signals are kept apart. Stereo realism is dependent on the successsful prevention of their mixture before reaching the output terminals of the power amplifier. Tape systems have a separation capability inherently far superior to that of the disc systems.

separation circuit—A circuit that separates signals according to their amplitude, frequency, or some other selected characteristic.

separation filter—A combination of filters used to separate one band of frequencies from another—often, to separate carrier and voice frequencies for transmission over individual paths.

separation loss—The loss that occurs in output when the surface coating of a tape fails to make perfect contact with the surfaces of either the record or reproduce head.

separator—1. An insulating sheet or other device employed in a storage battery to prevent metallic contact between plates of opposite polarity within a cell. 2. An insulator used in the construction of convolutely wound capacitors. 3. *See* Delimiter, 2.

septate coaxial cavity – A coaxial cavity with a vane or septum added between the inner and outer conductors. The result is a cavity that acts as if it had a rectangular cross section bent transversely.

septate waveguide—A waveguide with one or more septa placed across it to control microwave power transmission.

septum—A thin metal vane which has been perforated with an appropriate wave pattern. It is inserted into a waveguide to reflect the wave.

sequence—1. The order in which objects or items are arranged. 2. To place in order. 3. A succession of terms so related that each may be derived from one or more of the preceding terms in accordance with some fixed law.

sequence checking routine – A checking routine which examines every instruction executed and prints certain data concerning this check.

sequence control – Automatic control of a series of operations in a predetermined order.

sequencer—A mechanical or electronic device that may be set to initiate a series of events and to make the events follow in sequence.

sequence relay—A relay which controls two or more sets of contacts in a predetermined sequence.

sequencer register—In a computer, a counter which is pulsed or reset following the execution of an instruction to form the new memory address which locates the next instruction.

sequence timer—A succession of time-delay circuits arranged so that completion of the delay in one circuit initiates the delay in the following circuit.

sequencing equipment—A special selecting device by means of which messages received from several teletypewriter circuits may be subsequently selected and retransmitted over a smaller number of trunks or circuit.

sequency of operation—A detailed written description of the order in which electrical devices and other parts of the equipment should function.

sequential-access memory – A serial-type memory in which words are selected in a fixed order. The addressing circuit steps from word to word in a predetermined order, with the result that the access time for the stored information (words) is variable. Abbreviated SAM.

sequential color television – A color television system in which the three primary colors are transmitted in succession and reproduced on the receiver screen in the same manner.

sequential color transmission – The transmission of television signals that originate from variously colored parts of an image in a particular sequential order.

sequential computer—A computer in which events occur in time sequence with little or no simultaneous occurrence or overlap of events.

sequential control—Digital-computer operation in which the instructions are set up in sequence and fed to the computer consecutively during the solution of a problem.

sequential element – A device having at least one output channel and one or more input channels, all characterized by discrete states, such that the state of each output channel is determined by the previous states of the input channels.

sequential interlace—A method of interlacing in which the lines of one field are placed directly under the corresponding lines of the preceding field.

sequential lobing—A direction-determining technique utilizing the signals of overlapping lobes exsiting at the same time.

sequential logic—A circuit arrangement in a computer in which the output state is determined by the previous state of the input. *See also* Combination Logic.

sequential logic element—A device that has one or more output channels and one or more input channels, all of which have discrete states, such that the state of each output channel depends on the previous states of the input channel.

sequential operation—The carrying out of operations one after the other.

sequential relay—A relay that controls two or more sets of contacts in a predetermined sequence.

sequential sampling — Sampling inspection in which the decision to accept, reject, or inspect another unit is made following the inspection of each unit.

sequential scanning—In television, rectilinear scanning in which the distance from center to center of successively scanned lines is equal to the nominal line width.

sequential switcher—A device which automatically permits the viewing of pictures from a number of cctv cameras on one cctv monitor in a selected sequence.

sequential timer—A timer in which each interval is initiated by the completion of the preceding interval. All intervals may be independently adjusted.

serial — 1. Pertaining to time-sequential transmission of, storage of, or logical operations on the parts of a word in a computer—the same facilities being used for successive parts. 2. The technique for handling a binary data word which has more than one bit. The bits are acted upon one at a time, analogous to a parade passing a review point.

serial access—1. Pertaining to transmission of data to or from storage in a sequential or consecutive manner. 2. Pertaining to the process in which information is obtained from or placed into storage with the time required for such operations dependent on the location of the information most recently obtained or placed in storage. *See also* Random Access.

serial adder—A device in which additions are performed in a series of steps: the least significant addition is performed first, and progressively more significant additions are performed in order until the sum of the two numbers is obtained.

serial arithmetic unit—One in which the digits are operated on sequentially. (*See also* Parallel Arithmetic Unit.)

serial bit—Pertaining to computer storage in which the individual bits making up a word appear in time sequence.

serial computer—A computer having a single arithmetic and logic unit.

serial counter—Also called ripple-through counter. A counter in which each flip-flop cannot change state until after the preceding flip-flop has changed state; relatively long delays after an input pulse is applied to the counter can occur before all flip-flops reach their final states.

serial digital computer—One in which the digits are handled serially. Mixed serial and parallel machines are frequently called serial or parallel, according to the way the arithmetic processes are performed. An example of a serial digital computer is one which handles decimal digits serially, although the bits which comprise a digit might be handled either serially or in parallel. (*See also* Parallel Digital Computer.)

serialize—To convert from parallel-by-bit to serial-by-bit.

serially reusable routine—A computer routine in main storage that can be used by another task following conclusion of the current use.

serial memory—A memory in which information is stored in series and reading or writing of information is done in time sequence, as with a shift register. Compared to a RAM, a serial memory has slow to medium speed and lower cost. *See also* Sequential Access Memory.

serial mode—A type of computer operation that is performed bit by bit, generally with the least significant bit handled first. Readin and readout are accomplished bit after bit by shifting the binary data through the register.

serial operation—In a digital computer, information transfer such that the bits are handled sequentially, rather than simultaneously as they are in parallel operation. Serial operation is slower than parallel operation, but it is accomplished with less complex circuitry.

serial-parallel—Having the property of being partially serial and partially parallel.

serial printer—A device that can print characters one at a time across a page.

serial processing—The sequential or consecutive execution of more than one process into a single device such as a channel or processing unit. Opposed to parallel processing.

serial programming — Programming of a digital computer in such a manner that only one arithmetical or logical operation can be executed at one time.

serial storage—In a computer, storage in which time is one of the coordinates used in the location of any given bit, character, or word.

serial transfer—Data transfer in which the characters of an element of information are transferred in sequence over a single path.

serial transmission—Information transmission in which the characters of a word are transmitted in sequence over a single line.

series—1. The connecting of components end to end in a circuit, to provide a single path for the current. 2. An indicated sum of a set of terms in a mathematical expression (e.g., in an alternating or arithmetic series).

series circuit—A circuit in which resistances

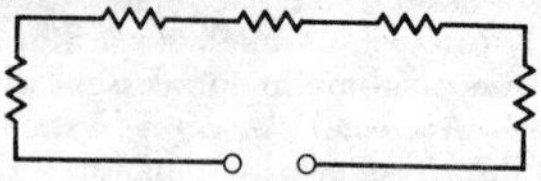

Series circuit.

or other components are connected end to end so that the same current flows throughout the circuit.

series connection — A way of making connections so as to form a series circuit.

series excitation—The field excitation obtained in a motor or generator by allowing the armature current to flow through the field winding.

series-fed vertical antenna—A vertical antenna which is insulated from ground and energized at the base.

series feed—The method by which the dc voltage to the plate or grid of a vacuum tube is applied through the same impedance in which the alternating current flows.

series field—In a machine, the part of the total magnetic flux due to the series winding.

series loading — Loading in which reactances are inserted in series with the conductors of a transmission circuit.

series modulation — Modulation in which the plate circuits of a modulating tube and a modulated amplifier tube are in series with the same plate-voltage supply.

series motor—A motor in which the field and armature circuits are connected in series. In small motors with laminated field frames, the performance will be similar when operated on direct current or alternating current. For this reason, the series motor is frequently called a "universal" motor.

series operation—The connection of two or more power supplies together to obtain an output voltage of the combination equal to the sum of the individual supplies. A common current passes through all the supplies.

series-parallel network—Any network which contains only resistors, inductors, and capacitors and in which successive branches are connected in series and/or in parallel.

series-parallel switch — A switch which changes the connections of lamps or other devices from series to parallel or vice versa.

series peaking—*See* Peaking Network.

series regulator—A device that is placed in series with a source of power and is able to automatically vary its series resistance, thereby controlling the voltage or current output.

series resistor — A resistor generally used for adapting an instrument so that it will operate on some designated voltage or voltages. It forms an essential part of the voltage circuit and may be either internal or external to the instrument.

series resonance—The condition existing in a circuit when the source of electromotive force is in series with an inductance and capacitance the reactances of which cancel each other at the applied frequency, thereby reducing the impedance to minimum.

series-resonant circuit—A circuit in which an inductor and capacitor are connected in series and have values such that the inductive reactance of the inductor will be equal to the capacitive reactance of the capacitor at the desired resonant frequency. At resonance, the current through a series-resonant circuit is at maximum.

series-shunt network—*See* Ladder Network.

series T-junction—*See* E-plane T-junction.

series winding—In a motor or generator, a field winding that carries the same current as the armature; i.e., this winding is in series with the armature rather than in parallel with it. Series-wound motors are used in fractional horsepower, ac-dc applications, such as fans and electric mixers. Their other chief use is in heavy-duty dc traction equipment such as electric locomotives, because of their extremely high starting torque.

series-wound motor—A commutator motor in which the field and armature circuits are in series.

serrated pulse — A vertical synchronizing pulse divided into a number of small pulses, each acting for the duration of half a line in a television system.

serrated rotor plate—Also called a slotted or split rotor plate. A rotor plate with radial slots which permit different sections of the plate to be bent inward or outward so that the total capacitance of a variable-capacitor section can be adjusted during alignment.

serration—1. The sawtooth appearance of vertical and near-vertical lines in a television picture. This is caused by their starting at different points during the horizontal scan. 2. A designed irregular surface used as a reservoir to retain excess infiltrating material and/or multiple

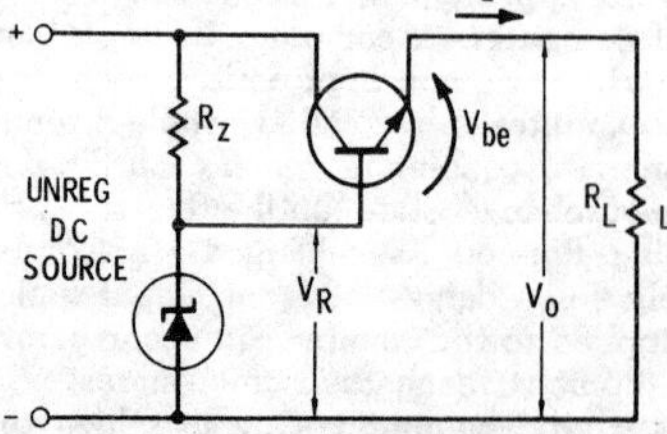

Series regulator.

points to obtain high-current-density resistance welding or resistance brazing.

serrodyne—A frequency translator or frequency converter, based on linear sawtooth modulation of phase shift or time delay. One convenient modulable device for serrodyne use is a traveling-wave tube, which provides gain as well as frequency translation.

serve – With reference to cable construction, a type of separator applied directly over the conductor or conductors. The serve may consist of one or more materials such as paper, cotton, silk, nylon, or rayon. These materials may be applied spirally or laterally.

serviceability – Those properties of an equipment design that facilitate service and repair in operation.

service area—1. The area within which a navigational aid is of use. 2. The area, surrounding a broadcasting station, where the signal is strong enough for satisfactory reception at all times (i.e., not subpect to objectionable interference or fading).

service band—The band of frequencies allocated to a class of radio service.

service channel—A band of frequencies, usually including a voice channel, utilized for maintenance and fault indication on a communication system.

service life—1. The period of time during which a device is expected to perform in a satisfactory manner. 2. The length of time a primary cell or battery needs to reach a specified final electrical condition on a service test that duplicates normal usage.

service oscillator—*See* RF Signal Generator.

service routine – In digital computer programming, a routine designed to assist in the actual operation of the computer.

service switch—A switch, usually in a box, for disconnecting the line voltage from the circuits it services.

service unit—In a microwave system, the equipment or facilities used for maintenance communications and transmission of fault indications.

serving (of a cable)—A wrapping applied around the core before a cable is leaded, or over the lead if the cable is armored. Some common materials are jute, cotton, or duck tape.

servo—Short for servomotor. A device that contains and delivers power to move a control or controls.

servoamplifier—A servo unit in which information from a synchro is amplified to control the speed and direction of the servomotor output.

servo loop—In a servoamplifier, the entire closed loop formed by feedback from output to input. In a position servo, the output position is compared to a command signal at the input.

servomechanism – 1. An automatic feedback-control system in which one or more of its signals represent mechanical motion. 2. A system in which output is compared to input to control error according to desired relationship, or feedback. 3. A self-contained system (except for inputs) in which the feedback signal is subtracted from a desired value so that the difference is reduced to zero.

servomotor—A motor used in a servo system. Its rotation or speed (or both) are controlled by a corrective electric signal that has been amplified and fed into the motor circuit.

servo noise—The hunting of the tracking servomechanism of a radar as a result of backlash and compliance in the gears, shafts, and structures of the mount.

servo oscillation—An unstable condition in which the load tends to hunt back and forth about the ordered position.

servo system—An automatic control system for maintaining a condition at or near a predetermined value by activation of an element such as a control rod. It compares the required condition (desired value) with the actual condition and adjusts the control element in accordance with the difference (and sometimes the rate of change of the difference).

servo techniques—Methods of studying the performance of servomechanisms or other control systems.

servovalve – (Electrohydraulic flow control.) An electrical-input, fluid-control valve capable of continuous control.

sesquisideband transmission – A system in which the carrier, one full sideband, and half of the other sideband are transmitted.

set—1. To place a storage device in a prescribed state. 2. To place a binary cell in the *1* state. 3. A permanent change, attributable to any cause, in a given parameter. 4. *See* Equipment. 5. Pertaining to a flip-flop input used to affect the Q output. Through this input, signals can be entered to change the Q output from 0 to 1. It cannot be used to change Q to 0. Opposite of clear. 6. An input to a binary, counter, or register that forces all binaries to assume the maximum binary state.

set analyzer—A test instrument designed to permit convenient measurement of voltages and currents.

set composite—A signaling circuit in which two signaling or telegraph legs may be superimposed on a two-wire interoffice trunk by means of one of the balanced pairs of high-impedance coils connected to each side of the line with an associated capacitor network.

set input—An asynchronous input to a flip-flop used to force the Q output to its high state.

set noise—Inherent random noise caused in a receiver by thermal currents in resistors and by variations in the emission currents of vacuum tubes.

set point—In a feedback control loop, the point which determines the desired value of the quantity being controlled.

set pulse—A drive pulse which tends to set a magnetic cell.

set-reset flip-flop—A standard flip-flop except that if both the set and reset inputs are a 1 at the same time, the flip-flop will assume a prescribed state.

set terminal—The flip-flop input terminal that triggers the circuit from its first state to its second state. *See* Input-Output Terminal.

settling time—1. The time interval, following the initiation of a specified stimulus to a system, required for a specified variable to enter and remain within a specified narrow band centered on the final value of the variable. 2. In an operational amplifier, the interval between the time of application of an ideal step input and the time at which the closed-loop amplifier output enters and remains within a specified band of error, usually symmetrical about the final value. Settling time includes a propagation delay and the time needed for the output to slew to the vicinity of the final value, recover from the overload condition associated with slewing, and settle within the specified error range. 3. In a feedback control system, the time required for an error to be reduced to a specified fraction, usually 2 percent or 5 percent, of its original magnitude. 4. The time required for the output frequency of a voltage or current-tuned oscillator to change from the initial value to within a specified window around the final value in response to a voltage or current step on the tuning input port.

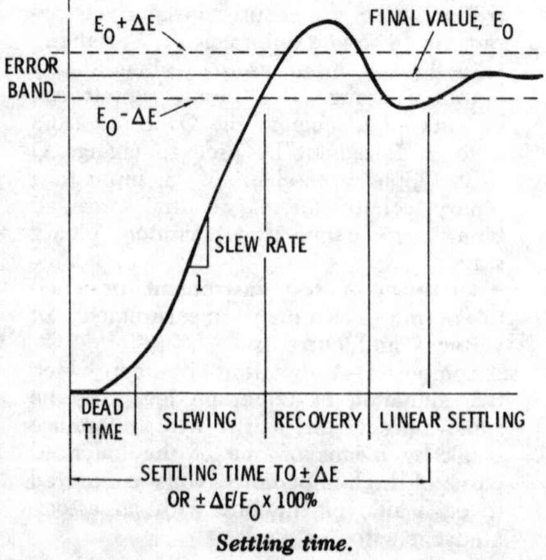

Settling time.

setup—1. In television, the ratio between the reference black and reference white levels, both measured from the blanking level. It is usually expressed in percent. 2. An arrangement of data or devices for the solution of a particular problem.

setup diagram—A diagram that specifies a given computer setup.

setup time—1. The time, measured from the point of 10% input change, required for a capacitor-diode gate to open or close after the occurrence of a change of input level. 2. The length of time that data must be present and unchanging before a flip-flop is clocked.

seven-segment display — A display format consisting of seven bars so arranged that each digit from 0 to 9 can be displayed by energizing two or more bars. LED, LCD, and gas discharge displays all use seven-segment display formats.

73—Abbreviation for "best regards" in radio communications.

seven-unit teleprinter code — Frequently called Teletype code. A code that represents the letters of the alphabet, the numerals, the punctuation marks, and the various control functions necessary on a teleprinter (such as line feed, carriage return, upshift, and downshift) by five-unit combinations of *mark* and *space* conditions. In addition to the five units which indicate the letter or other data, the code contains a "start" unit, which is always a "space," and a "stop" unit, which is always a "mark," to indicate the beginning and ending of each character.

sexadecimal — Pertaining to the number system that has a radix of sixteen.

sexadecimal notation—Also called dekahexadecimal notation. A scale of notation for numbers in which the base is 16.

sferics—Contraction of the term "atmospherics," meaning interference.

sferics receiver—Also called lightning recorder. A type of radio direction finder which measures electronically the direction of arrival, intensity, and rate of occurrence of atmospherics. In its simplest form the instrument consists of two orthogonally crossed antennas. Their output signals are connected to an oscillograph so that one loop measures the east-west components. These are combined vertically to give the azimuth.

sg—Abbreviation for screen-grid electrode (of a vacuum tube).

shaded-pole motor—A single-phase induction motor provided with one or more auxiliary short-circuited stator windings that are displaced magnetically from the main winding.

shading—1. A brightness gradient in the reproduced picture not present in the original scene, but caused by the camera tube. 2. Compensating for the spurious signal generated in a camera tube during the trace intervals. 3. Controlling the directivity pattern of a transducer through the distribution of phase and amplitude of the transducer action over the active face.

shading coil—*See* Shading Ring.

shading ring—1. A heavy copper ring sometimes placed around the central pole piece of an electrodynamic loudspeaker to act as a shorted turn for cancellation of the hum voltage of the field coil. 2. A copper ring set into part of the pole piece of a small alternating-current motor to produce the lagging component of a rotating magnetic field for starting purposes.

shading signal—A signal that increases the gain of the amplifier in a television camera while the electron beam is scanning a dark portion.

shadow attenuation—Attenuation of radio waves over a sphere in excess of that over a plane when the distance over the surface and other factors are the same.

shadow factor—The ratio of the electric field strength which would result from propagation over a sphere, to that which would result from propagation over a plane (other factors being the same).

shadow mask—*See* Aperture Mask.

shadow region—A region in which, under normal propagation conditions, an obstruction reduces the field strength from a transmitter to the point where radio reception or radar detection is ineffective or virtually so.

shadow tuning indicator—A vacuum tube in which a moving shadow shows how accurately a radio receiver is tuned.

shaft—The axial member to which torque is applied to cause rotation of an adjustable component.

shaft angle encoder—An electromechanical device that has a means for counting equally spaced radii that represent angular increments around the periphery of a disk. Usually, the measurement is in degrees, minutes, and seconds, since 2π radians is not a prime number. Usually the disc is divided into an even number of equal increments.

shaft position encoder—Also called converter or coder. An analog-to-digital converter which transduces a mechanical analog shaft rotation to an electrical digital representation.

shakedown test—An equipment test carried out during the installation work.

shaker—An electromagnetic device capable of imparting known and/or controlled vibratory acceleration to a given object.

shake table—A laboratory tester in which an instrument or component is subjected to vibration that simulates operating conditions.

shake-table test—A laboratory test in which a device or component is placed in a vibrator to determine the reliability of the device or component when subjected to vibration.

shank—1. The part of a phonograph needle which is clamped into position in the pickup or cutting head. 2. The cylindrical or rodlike portion of a connector or contact. 3. That part of a fastener lying between the head and the extreme opposite end.

Shannon limit—The maximum signal-to-noise ratio improvement which can be achieved by the best modulation technique, as implied by Shannon's theorem relating channel capacity to signal-to-noise ratio.

shaped-beam antenna—An antenna whose directional pattern over a certain angular range is designed to a special shape for some particular use.

shaped-beam display tube—A cathode-ray tube in which the beam is first deflected through a matrix, then repositioned along the axis of the tube, and finally deflected into the desired position on the faceplate. A typical tube is the Charactron.

shape factor—1. For a filter, the ratio (usually maximum) comparing a high-attenuation–level bandwidth and a low-attenuation–level bandwidth. 2. The ratio of the 60-dB bandwidth to the 6-dB bandwidth. Defines the selectivity of an amplifier stage. 3. A factor used to take the shape of a coil into account when its inductance is computed.

shaping—Adjustment of a plan-position-indicator pattern set up by a rotating magnetic field.

shaping network — An electrical network designed to be inserted into a circuit to improve its transmission or impedance properties, or both. (*See also* Corrective Network.)

shared file—A direct-access device that two systems may use at the same time; a shared file may link two systems.

sharp-cutoff tube—The opposite of a remote-cutoff tube. A tube in which the control-grid spirals are uniformly and closely spaced. As the grid voltage is made more and more negative, the plate current decreases steadily to cutoff.

sharp tuning—Response to a limited range of frequencies.

shaving—In mechanical recording, the removal of material from the surface of a recording medium for the purpose of obtaining a new surface.

shear wave—1. Also called a rotational wave. A wave, usually in an elastic solid, which causes an element of the solid to change its shape but not its volume. 2. A wave in which particle displacement is at right angles to the direction of propagation.

sheath—1. The external conducting surface of a shielded transmission line. 2. A metal wall of a waveguide. 3. Part of a discharge in a rarefied gas, in which there is a space charge due to an accumulation of electrons or ions.

sheath-reshaping converter—A converter in which the pattern of the wave is changed by gradual reshaping of the waveguide sheath and the metal sheets mounted longitudinally in the guide.

sheet grating—A three-dimensional grating consisting of thin metal sheets extending along the inside of a waveguide for about one wavelength. It is used to stop all but the predetermined wave, which passes unimpeded.

sheet resistivity — Also called ohms per square. The electrical resistance measured across the opposite sides of a square pattern of deposited film material.

Sheffer-stroke function—The Boolean operator that gives a truth-table value of true only when both of the variables that the operator connects are not true.

shelf aging—The change with time of the properties of a stored component or material.

shelf corrosion—Consumption of the negative electrode of a dry cell as a result of local action.

shelf life—The length of time under specified conditions that a material or component retains its usability.

shell—1. A group of electrons having a common energy level that forms part of the outer structure of an atom. 2. The outer section of a plug or receptacle that mechanically supports the assembly and in some cases provides coupling and locking. 3. That part of a phonograph pickup which carries the cartridge. The head shell can often be detached from the pickup arm.

shell-type transformer — A transformer in which the magnetic circuit completely surrounds the windings.

shf—Abbreviation for superhigh frequency.

shield—A device designed to protect a circuit, transmission line, etc., from stray voltages or currents induced by electric or magnetic fields, consisting, in the case of an electric field, of a grounded conductor surrounding the protected object. At high frequencies this will provide magnetic shielding as well. At low frequencies (through the audio range) magnetic shielding is accomplished by surrounding the object with a material of high magnetic permeability. *See* Braid, 2.

shield coverage—*See* Shield Percentage.

shielded building—In modern practice, a building in which shielding was incorporated in the basic architectural design. Shielded buildings often employ structural steel members as an integral part of rfi shielding.

shielded cable—A single- or multiple-conductor cable surrounded by a separate conductor (the shield) intended to minimize the effects of adjacent electrical circuits.

shielded-conductor cable—A cable in which the insulated conductor or conductors are enclosed in a conducting envelope or envelopes, almost every point on the surface of which is at ground potential or at some predetermined potential with respect to ground.

shielded enclosure — A room, hangar, or box, shielded or screened so as to provide a controlled electromagnetic environment.

shielded joint—A cable joint having its insulation so enveloped by a conducting shield that substantially every point on the surface of the insulation is at ground potential, or at some predetermined potential with respect to ground.

shielded line—A transmission line the elements of which confine propagated radio waves to an essentially finite space inside a tubular conducting surface called the sheath, thus preventing the line from radiating radio waves.

shielded pair — A two-wire transmission line surrounded by a metallic sheath.

shielded room—An enclosed area made free from electrical interference that would affect the sensitivity of electrical equipment.

shielded transmission line—A transmission line the elements of which confine the propagated electrical energy inside a conducting sheath.

shielded wire—An insulated wire covered with a metal shield—usually of tinned, braided copper wire.

shielded X-ray tube—An X-ray tube enclosed in a grounded metal container, except for a small opening through which the X-rays emerge.

shield effectiveness—The relative ability of a shield to screen out undesirable radiation.

shield factor—Ratio between the noise (or induced current or voltage) in a telephone circuit when a source of shielding is present and when it is not.

shield grid—In a glass tube, a structure which shields the control electrode from the anode or cathode, or both. It prevents the radiation of heat from and the depositing of thermionic activating material on them. It also reduces the electrostatic influence of the anode, and may be used as a control electrode in some applications.

shield-grid thyratron – A thyratron that contains a shield grid, usually operated at the same potential as the cathode.

shielding–1. The practice of confining the dielectric field of an electric cable to the inside of the cable insulation or insulated conductor assembly by surrounding the insulation or assembly with a grounded conducting medium called a shield. 2. Metal covering used on a cable; also a metal can, case partition or plates enclosing an electronic circuit or component. Shielding is used to prevent undesirable radiation, pickup of signals, magnetic induction, stray current, ac hum, or radiation of an electrical signal.

shielding effectiveness–The relative reduction of radiated electromagnetic energy levels occasioned by the use of an enclosure either to contain or exclude the energy.

shield percentage–Also called shield coverage. The physical area of a circuit or cable actually covered by shielding material, expressed in percent.

shield wire–A wire employed for reducing the effects of extraneous electromagnetic fields on electric supply or communication circuits.

shift–1. Displacement of an ordered set of computer characters one or more places to the left or right. If the characters are the digits of a numerical expression, a shift is equivalent to multiplying a power of the base. 2. The process of moving information from one place to another in a computer; generally, a number of bits are moved at once. A word can be shifted sequentially (generally referred to as shifting left or right), or all bits of a word can be shifted at the same time (called parallel load or parallel shift). 3. In a computer, an operation whereby a number is moved one or more places to the left or right. The number 110, for instance, becomes 1100 if shifted one place left or 11 if shifted one place right. The operation is of considerable use in digital computer operations.

shift counter–1. A shift register in which the first stage, through logic feedback, produces a pattern of ones or zeros as a function of the state of other stages in the register. The pattern of ones and zeros so produced is termed a ring code.

shift counter 2–*See* Ring Counter.

shift-frequency modulation–A form of frequency modulation in which the modulating wave shifts the output frequency between predetermined values and the output wave is coherent, with no phase discontinuity.

shift-in character–In a computer, a code extension character that can be used by itself to bring about a return to the character set in effect before substitution of another set was caused by a shift-out character.

shift out–In a computer, to move information within a register toward one end so that, as the information leaves this one end, 0's are entered into the other end.

shift-out character–In a computer, a code extension character that can be used by itself to cause substitution of some other character set for the standard set, usually to access additional graphic characters.

shift pulse–A drive pulse which initiates the shifting of characters in a register.

shift register–A digital storage circuit in which information is shifted from one flip-flop of a chain to the adjacent flip-flop upon application of each clock pulse. Data may be shifted several places to the right or left, depending on additional gating and the number of clock pulses applied to the register. Depending on the number of positions shifted, the rightmost characters are lost in a right shift, and the leftmost characters are lost in a left shift. *See also* Dynamic Shift Register *and* Static Shift Register.

ship error – A radio direction-finder error that occurs when radio waves are reradiated by the metal structure of a ship.

ship-heading marker–On a ppi scope, an electronic radial sweep line indicating the heading of the ship on which the equipment is installed.

ship's emergency transmitter – A ship's transmitter to be used exclusively on a distress frequency for distress, urgency, or safety purposes.

ship station–A radio station operated in the maritime-mobile service and located on board a vessel which is not permanently moored.

ship-to-shore communication–Communication by radio between a ship at sea and a shore station.

shock–1. An abrupt impact applied to a stationary object. It is usually expressed in gravities (g). 2. An acceleration transient of short duration and nonrepetitive occurrence.

shock absorber – A device for dissipating vibratory energy, to modify the response of a mechanical system to an applied shock.

shock excitation–*See* Impulse Excitation.

shock-excited oscillations–*See* Free Oscillations.

shock hazard–A hazardous condition that exists at a part of a ground-fault circuit interrupter if: (a) there would be current of 5 milliamperes or more in a resistance of 500 ohms connected between the part in question, and the grounded supply conductor, and (b) the device would not operate to open the circuit to the 500-ohm resistor within a specified time.

shock isolator—Also called a shock mount. A resilient support which tends to isolate a system from applied mechanical shock.

Shockley diode — A four-layer controlled semiconductor rectifier diode without a base connection, used as a trigger or switching diode.

shock motion — In a mechanical system, transient motion characterized by suddenness and by significant relative displacements.

shock mount—*See* Shock Isolator.

shockover capacitor—A capacitor connected between the grid and cathode of a thyratron to prevent premature firing.

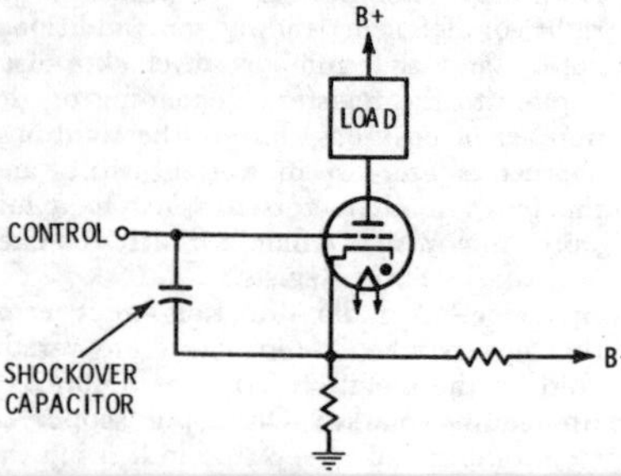

Shockover capacitor.

shock pulse—Usually a single disturbance that is characterized by an increase and a decrease of acceleration in a relatively short period of time.

shodop — Acronym for SHOrt-range DOPpler.

shoran — Acronym for SHOrt Range Air Navigation. A precision distance-measuring system employing the pulse timing principle to measure the distance from an aircraft to one or more fixed responder ground stations. Fundamentally, the system consists of a mobile transmitter-receiver-indicator unit and a fixed receiver-transmitter unit (transponder). Pulses are sent from the mobile transmitter and returned to the originating point by the transmitter. The indicator measures the time interval required for the travel of a pulse between stations and converts this information into distance to the nearest thousandth of a mile.

shore effect—The bending of radio waves toward the shore line when traveling over water, due presumably to the slightly greater velocity of radio waves over water than over land. This effect causes errors in radio direction-finder indications.

shore-to-ship communication—Communication by radio between a shore station and a ship at sea.

short—*See* Short Circuit.

short base-line system — A system which uses continuous waves and which has a base-line length that is short in comparison to the target distance.

short circuit—Also called a short. An abnormal connection of relatively low resistance between two points of a circuit. The result is a flow of excess (often damaging) current between these points.

short-circuit current—The current a power supply delivers when its output terminals are short-circuited.

short-circuit driving-point admittance—The driving-point admittance between the j terminal of an n-terminal network and the reference terminal when all other terminals have zero alternating components of voltage with respect to the reference point.

short-circuit feedback admittance (of an electron-device transducer) — The short-circuit transfer admittance from the output terminals to the input terminals of a specified socket, the associated filters and the electron device.

short-circuit forward admittance (of an electron-device transducer) — The short-circuit transfer admittance from the input terminals to the output terminals of a specified socket, the associated filters and the electron device.

short-circuit impedance—The driving-point impedance of a line or four-terminal network when its far end is short-circuited.

short-circuiting or grounding device — A power or stored-energy operated device which functions to short-circuit or ground a circuit in response to automatic or manual means.

short-circuit input admittance (of an electron-device transducer)—The short-circuit driving-point admittance at the input terminals of a specified socket, the associated filters, and the electron device.

short-circuit output admittance (of an electron-device transducer)—The short-circuit driving-point admittance at the output terminals of a specified socket, the associated filters, and the electron device.

short-circuit output capacitance (of an n-terminal electron device)—The effective capacitance determined from the short-circuit output admittance.

short-circuit parameters—In an equivalent circuit of a transistor, the resultant parameters when independent variables are selected for the input and output voltages.

short-circuit protection (automatic) — A current-limiting system which enables a power supply to continue operating without damage into any output overload, including short circuits. The output voltage must be restored to normal when the overload is removed, as distinguished from a fuse or circuit-breaker system, which opens at overload and must be reclosed to restore power.

short-circuit transfer admittance – The transfer admittance from terminal j to terminal l of an n-terminal network when all terminals except j have zero complex alternating components of voltage with respect to the reference point.

short-circuit transfer capacitance (of an electron device) – The effective capacitance determined from the short-circuit transfer admittance.

short code—A system of instructions that causes an automaton to behave as if it were another, specified automaton.

short-contact switch—A selector switch in which the movable contact is wider than the distance between its clips, so that the new circuit is made before the old one is broken.

short-distance navigational aid—An equipment or system which provides navigational assistance to a range not exceeding 200 miles.

shorted out—Made inactive by connecting a heavy wire or other low-resistance path around a device or portion of a circuit.

short-gate gain—Video gain on a short-range gate.

shorting noise—A noise which occurs in wirewound potentiometric transducers, even when no current is drawn from the device. It is due to the shorting out of adjacent turns of the wire as the slider traverses the winding. The portion of the interturn current which flows through the slider appears as noise.

shorting switch – A switch type in which contact is made for a new position before breaking contact with the previous position. Classified as a "make-before-break" switch.

short plug—A plug designed to connect the springs of a jack together or to short them.

short-range navigation – A precision position-fixing system using a pulse transmitter and receiver in connection with two transponder beacons at fixed points.

short-range navigation aid – A navigation aid that is usable only at distances within radio line of sight.

short-slot coupler—*See* Three-dB Coupler.

short-time duty—A service requirement that demands operation at a substantially constant load for a short, specified time.

short time limits—Values of minimum and maximum trip time measured at various percentages of overload.

short-time rating—The rating that defines the load which a machine, apparatus, or device can carry at approximately the room temperature for a short, specified time.

short wave—Radio frequencies from 1.6 to 30 MHz, which fall above the commercial broadcasting band and are used for sky-wave communication over long distances.

short-wave converter – An electronic unit designed to be connected between a receiver and its antenna system to permit reception of frequencies higher than those the receiver ordinarily handles.

short-wave transmitter—A radio transmitter that radiates short waves, which ordinarily are shorter than 200 meters.

shot effect – Noise voltages developed by the random travel of electrons within a tube. The effect is characterized by a steady hiss from a radio, and by snow or grass in a television picture.

shotgun—An extremely unidirectional microphone, used for "spotting" a speaker or soloist from a considerable distance. Also called a hyperdirectional or long-reach mike.

shot noise—1. Noise generated due to the random passage of discrete current carriers across a barrier or discontinuity, e.g., a semiconductor junction. Shot noise is characteristic of all transistors and diodes and is directly proportional to the square root of the applied current. 2. Noise voltages developed in a thermionic tube as a result of random variations in the number and velocity of electrons emitted by the cathode. The effect is characterized by the presence of a steady hiss in audio reproduction and of "snow" or "grass" in video reproduction.

shunt – 1. A precision low-value resistor placed across the terminals of an ammeter to increase its range. The shunt may be either internal or external to the instrument. 2. Any part connected, or the act of connecting any part, in parallel with some other part. 3. In an electric circuit, a branch the winding of which is in parallel with the external or line circuit.

shunt calibration—A procedure in which a parallel resistance is placed across a (similar) element to obtain a known and deliberate electrical change.

shunt-fed vertical antenna—A vertical an-

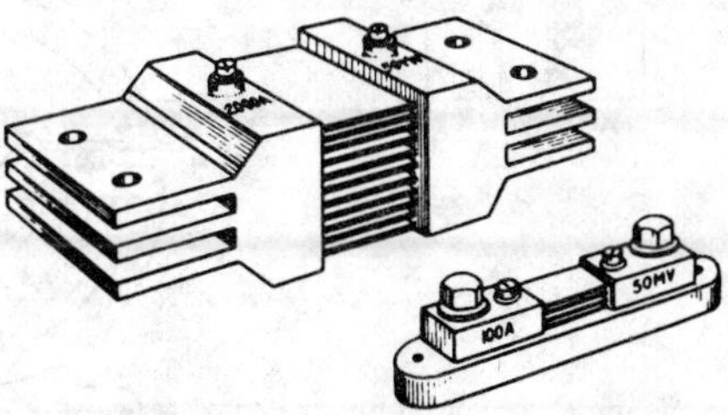

Shunts, 1.

tenna connected to the ground at the base, and energized at a suitable point above the grounding point.

shunt feed—*See* Parallel Feed.

shunt field—Part of the magnetic flux produced in a machine by the shunt winding connected across the voltage source.

shunt-field relay—A special polarized relay with two coils on opposite sides of a closed magnetic circuit. The relay operates only when the currents in its two windings flow in the same direction.

shunting effect—A reduction in signal amplitude caused by the load which an amplifier or measuring instrument imposes on the signal source. For dc signals the shunting effect is directly proportional to the output impedance of the signal source and inversely proportional to the input impedance of the amplifier.

shunting or discharge switch — A switch which serves to open or to close a shunting circuit around any piece of apparatus (except a resistor) such as a machine field, a machine armature, a capacitor, or a reactor.

shunt leads—Those leads which connect the circuit of an instrument to an external shunt. The resistance of these leads must be taken into account when the instrument is adjusted.

shunt loading — Loading in which reactances are applied in parallel across the conductors of a transmission circuit.

shunt neutralization — *See* Inductive Neutralization.

shunt peaking—*See* Peaking Network.

shunt regulator—A device placed across the output of a regulated power supply to control the current through a series-dropping resistance in order to maintain a constant output voltage or current.

shunt T-junction—*See* H-plane T-junction.

shunt-wound generator — A direct-current generator in which the field coils and armature are connected in parallel.

shunt-wound motor—A direct current motor which has its field (stationary member) and armature (rotating member) circuit connected in parallel. Its speed can be regulated by varying either the applied armature or field voltage.

shutoff — A provision whereby a recorder will automatically go into the stop mode at the end of a tape. In some recorders, the automatic shutoff can be made to turn off the entire unit as well as any other components powered by it.

shutter — A movable cover that prevents light from reaching the film or other light-sensitive surface in a still, movie, or television camera except during the exposure time.

shuttle — A high-speed tape-running mode that permits fast cuing or rewinding of the tape.

sibilance—The strong emphasis in pronunciation of the letter "S" and "SH" in speech. It can be exaggerated by microphones having peaks in their high frequency response.

side armature—An armature which rotates about an axis parallel to that of the core, with the pole face on a side surface of the core of a relay.

sideband attenuation—Attenuation in which the relative transmitted amplitude of one or more components of a modulated signal (excluding the carrier) is smaller than the amplitude produced by modulation.

sideband power—The power contained in the sidebands. This is the power to which a receiver responds when receiving a modulated wave, not the carrier power.

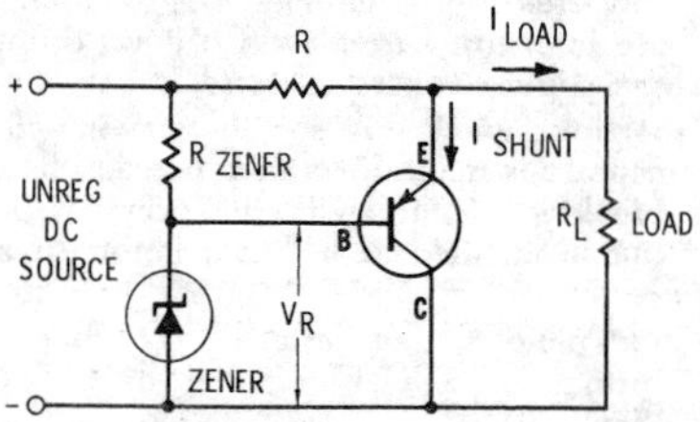

Shunt regulator.

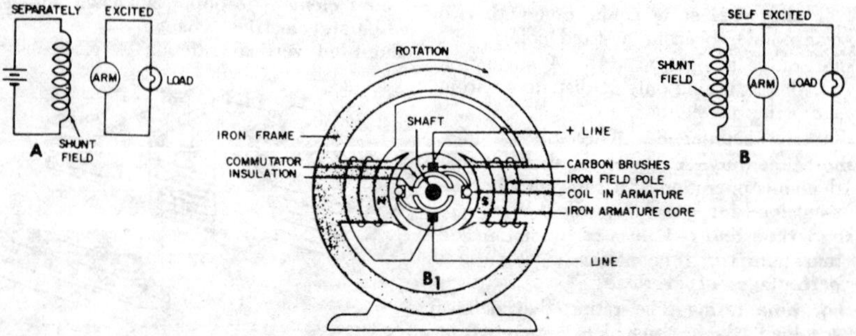

Shunt-wound dc generator.

sidebands—1. The frequency bands on both sides of the carrier frequency. The frequencies of the wave produced by modulation fall within these bands. 2. The wave components lying within such bands. During amplitude modulation with a sine-wave carrier, the upper sideband includes the sum (carrier plus modulating) frequencies, and the lower sideband includes the difference (carrier minus modulating) frequencies.

sideband splatter—1. Those portions of the modulation sidebands that lie beyond the limits of the assigned channel. 2. In radio communications, interference on other channels caused by spurious sidebands resulting from overmodulation.

side circuit—A circuit arrangement for deriving a phantom circuit. In four-wire circuits, the two wires associated with the "go" channel form one "side circuit" and those associated with the return channel form another. (*See also* Phantom Circuit.)

side-circuit loading coil—A loading coil for introducing a desired amount of inductance into a side circuit while introducing a minimum amount of inductance into the associated phantom circuit.

side-circuit repeating coil — A device that functions as a transformer at a terminal of a side circuit, and acts simultaneously as a device for superposing one side of a phantom circuit on the side circuit. Also called side-circuit repeat coil.

side echo—An echo due to a side lobe of an antenna.

side frequency—One of the frequencies of a sideband.

side lobe—A portion of the beam from an antenna, other than the main lobe. It is usually much smaller than the main lobe.

side-lobe blanking — A technique which compares relative signal strengths between an omnidirectional antenna and the radar antenna.

side-lobe cancellation — A technique designed to exclude or greatly attenuate jamming signals introduced through the side or back lobes of a receiving antenna.

side-looking airborne radar—A high-resolution airborne radar system in which the beam from the antenna is directed at right angles to the direction of flight.

sideswiper—A telegraph key which operates from side to side rather than up and down.

side thrust—1. In disc recording, the radial component of force on a pickup arm caused by the stylus drag. 2. The tendency of a stylus to "skate" toward the center of a record causing increased wear on the inner groove wall. With low tracking weight, side thrust can cause the stylus to "jump" the record's groove.

sidetone—The reproduction, in a telephone receiver, of sounds received by the transmitter of the same telephone set (e.g., hearing one's own voice in the receiver of a telephone set when speaking into the mouthpiece).

sidetone telephone set — A telephone set with no balancing network for reducing sidetone.

siemens—The new and preferred term for mho.

sight check—To verify the sorting or punching of punched cards by looking through the pattern of punched holes.

sign—1. A symbol which distinguishes negative from positive quantities. 2. A symbol which indicates whether a quantity is greater or less than zero. 3. A binary indicator of the position of the magnitude of a number relative to zero.

signal—1. A visible, audible, or other conveyor of information. 2. The intelligence, message, or effect to be conveyed over a communication system. 3. A signal wave. 4. The physical embodiment of a message.

signal attenuation — The reduction in the strength of electrical signals.

signal-averaging computer — An electronic averager that filters out signals of interest from background "noise."

signal bias—A form of teletypewriter signal distortion brought about by the lengthening or shortening of pulses during transmission. When marking pulses are all

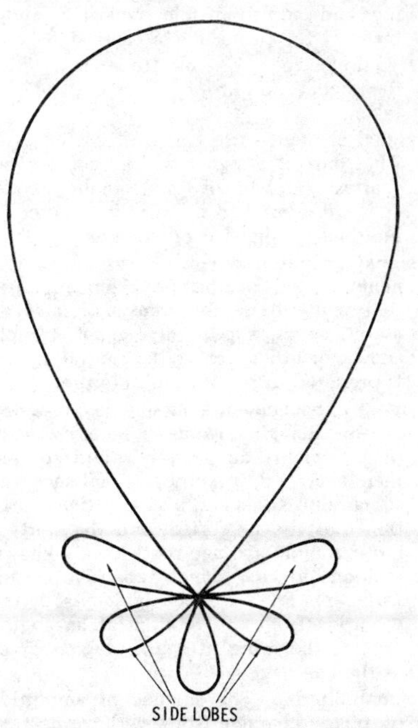

Side lobe.

lengthened, a marking-signal bias results; when marking pulses are all shortened, a spacing-signal bias results.

signal-carrier fm recording – A method of recording in which the input signal is frequency modulated onto a carrier, and the carrier is recorded on a single track at saturation and without bias.

signal conditioner – A device placed between a signal source and a readout device for the purpose of conditioning the signal. Some examples are damping networks, attenuator networks, preamplifiers, excitation and demodulation circuits, converters for changing one electrical quantity into another (such as voltage to current), instrument transformers, equalizing or matching networks, and filters.

signal conditioning–To process the form or mode of a signal so as to make it intelligible to, or compatible with, a given device, including such manipulation as pulse shaping, pulse clipping, digitizing, and linearizing.

signal converter – A circuit that reduces, filters, and (if necessary) rectifies incoming signals to logic system levels.

signal delay – The transmission time of a signal through a network. The time is always finite, and it may be undesired or purposely introduced.

signal-distortion generator–An instrument furnished and designed to apply distortion on a signal for the purpose of ranging and adjusting teletypewriter equipment or for furnishing a clear signal.

signal electrode–The electrode from which the signal output of a camera tube is taken.

signal element–Also called a unit interval. That part of a signal which occupies the shortest interval of the signaling code. It is considered to be of unit duration in building up signal combinations.

signal encoding device–A system component located at the protected premises that will initiate the transmission of an alarm signal, supervisory signal, trouble signal, or other signals the central station is prepared to receive and interpret.

signal enhancement – Ensemble averaging of time-domain signals, where a set of time domain samples are digitized and then averaged. In order to enhance the signal due to averaging, the time function must be repetitive, and the start of the ensemble average must have a known relationship to some repetitive event (trigger). Such a repetitive signal is the vibration from one rotation of an engine (where the firing of spark plug 1 serves as the trigger.)

signal filtering–The shaping of amplitude or phase characteristics with respect to frequency, for the purpose of meeting an operational requirement. This usually is accomplished by analog methods.

signal-frequency shift–In a frequency-shift facsimile system, the numerical difference between the frequencies corresponding to the white and black signals at any point in the system.

signal generator – Also called a standard voltage generator. A device which supplies a standard voltage of known amplitude, frequency, and waveform for measuring purposes.

signal ground–The ground return for low-level signals such as inputs to audio amplifiers or other circuits that are susceptible to coupling through ground-loop currents.

signaling – The process by which a caller on the transmitting end of a line informs a particular party at the receiving end that a message is to be communicated. Signaling is also that supervisory information that lets the caller know that the called party is ready to talk, that his line is busy, or that he has hung up. Signaling also holds the voice path together while a conversation goes on.

signaling channel–A tone channel used for signaling purposes.

signaling key–A key used in wire or radiotelegraphy to control the sequence of current impulses that form the code signals.

signal injector–A test instrument, usually small, which contains an audio-frequency pulse oscillator. A signal injected at points in the circuitry aids in troubleshooting.

signal intelligence–A generic term which includes both communications intelligence and electronics intelligence.

signal interpolation–*See* Interpolation.

signal lamp–A lamp that indicates, when lit or out, the existence of certain conditions in a circuit (e.g., signal lamps on switchboards, or pilot lamps in radio sets).

signal leakage – Interference in a given playback channel which has its origin in the recording system. Such interference occurs during simultaneous record/reproduce, and has a leading time displacement with reference to the signal on the tape.

signal level – The difference between the measure of the signal at any point in a transmission system, and the measure of an arbitrary reference signal. (Audio signals are often stated in decibels–thus their difference can be conveniently expressed as a ratio.)

signal-muting switch–A switch used on a record changer to ground (mute) the signal from the pickup during a change cycle.

signal-noise ratio—*See* Signal-to-Noise Ratio.

signal plate—A metal plate that backs up the mica sheet containing the mosaic in one type of cathode-ray television camera tube. The electron beam acts on the capacitance between this plate and each globule of the mosaic to produce the television signal.

signal plus noise and distortion—A radio-receiver sensitivity measurement based on the signal input required to produce 50% of the rated output at a 12-decibel ratio (4:1 voltage ratio) of signal plus noise and distortion to noise and distortion alone.

signal-separation filter — A bandpass filter which selects the desired signal or channel from a composite signal.

signal-shaping network — An electric network inserted into a telegraph circuit, usually at the receiving end, to improve the waveshape of the signals.

signal-shield ground — A ground technique for all shields used for the protection from stray pickup of leads carrying low-level, low-frequency signals.

signal shifter—A variable-frequency oscillator for shifting amateur transmitters to a less crowded frequency within a given band.

signal strength—The strength of the signal produced by a transmitter at a particular location. Usually it is expressed as so many millivolts per meter of the effective receiving-antenna length.

signal-strength meter — Also called an S meter. A meter connected in the avc circuit of a receiver and calibrated in dB or arbitrary "S" units to read the strength of a received signal.

signal-to-noise ratio — Also called signal-noise ratio. 1. Ratio of the magnitude of the signal to that of the noise (often expressed in decibels). 2. In television transmission, the ratio in decibels of the maximum peak-to-peak voltage of the video television signal (including the synchronizing pulse), to the rms voltage of the noise at any point. 3. The ratio of the amplitude of a signal after detection to the amplitude of the noise accompanying the signal. It may also be considered as the ratio, at any specific point of a circuit, of signal power to total circuit-noise power. Abbreviated s/n ratio or snr. 4. Ratio of the root-mean-square facsimile signal level to the root-mean-square noise level. 5. The difference measured in decibels, between a specified signal reference level and the level of unwanted noise. The higher the ratio, the better the equipment. 6. The span, measured in decibels, of signal intensity between a device's overload point at the upper limit and its background noise at the lower limit. (In tape recording, the s/n ratio usually lies between the permissible limit of saturation distortion and the tape's background hiss.)

signal tracer—A test instrument used for tracing a signal through the circuit in order to find faulty wiring or components.

signal tracing—The process of locating a fault in a circuit by injecting a test signal at the input and checking each stage, usually from the output backwards.

signal voltage — The effective (root-mean-square) voltage value of a signal.

signal wave — A wave with characteristics that permit it to carry intelligence.

signal-wave envelope — The contour of a signal wave which is composed of a series of wave cycles.

signal winding—Also called an input winding. In a saturable reactor, the control winding to which the independent variable (signal wave) is applied.

signature (target)—The characteristic pattern of the target displayed by detection and classification equipment.

sign bit—In complementary arithmetic, the leftmost bit of a number. If the sign bit is 1, the number is negative; if it is 0, the number is positive.

sign-control flip-flop—In computers, a flip-flop in the arithmetic unit used for storing the sign of the result of an operation.

sign digit—A character (+ or −) used to designate the algebraic sign of a number.

significance — Weight. In positional representation, the factor by which a digit must be multiplied to obtain its additive contribution to the value of a number; the factor is determined by the digit position.

significant digits (of a number)—1. A set of digits from consecutive columns, beginning with the most significant digit other than zero, and ending with the least significant digit the value of which is known or assumed to be relevant. The digits of a number can be ordered according to their significance, which is greater when occupying a column corresponding to a higher power of the radix. 2. A digit that contributes to the precision of a numeral. Significant digits are counted from the first digit on the left that is not zero, and continue to the last accurate digit on the right. (A righthand zero may be counted if it is an accurate part of the numeral.) For example, 2500.0 has five significant digits, 2500 probably has only two (it is not known that the last two digits are accurate) but 2501 has four, and 0.0025 has two.

sign position—A position, normally at one end of a number, that contains an indicator of the algebraic sign of the number.

silent discharge—The gradual and nondisruptive discharge of electricity from a conductor into the atmosphere. It is sometimes accompanied by the production of ozone.

silent period – An hourly period during which ship and shore radio stations must remain silent and listen for distress calls.

silica gel – A moisture-absorbent chemical used for dehydrating waveguides, coaxial lines, pressurized components, shipping containers, etc.

silicon – A metallic element often mixed with iron or steel during smelting to provide desirable magnetic properties for transformer-core materials. In its pure state, it is used as a semiconductor.

silicon bilateral switch–A device that has characteristics similar to those of the silicon unilateral switch, but exhibits the same characteristics in both directions.

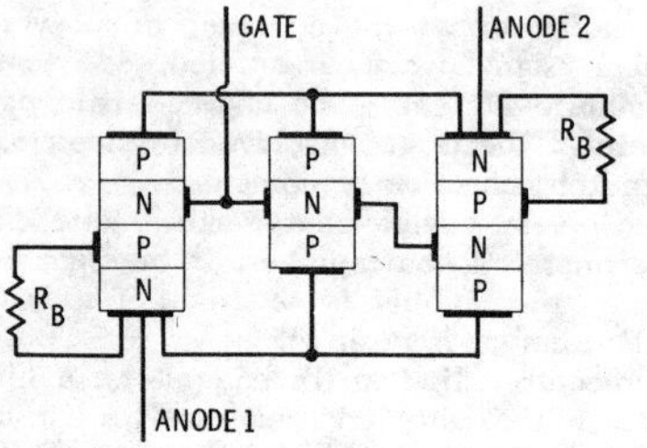

Junctions.

Symbol.

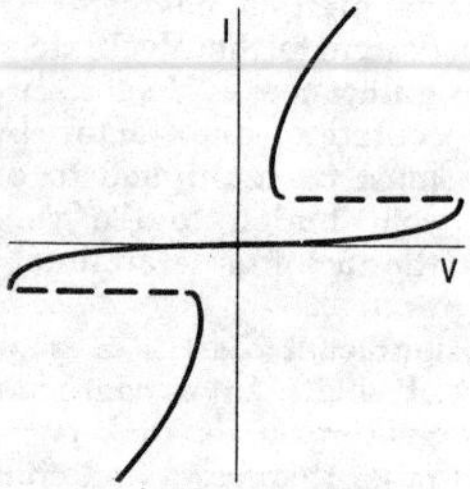

Anode 2 characteristic.

Silicon bilateral switch.

silicon capacitor–*See* Varactor.

silicon controlled rectifier–1. Abbreviated SCR. A four-layer pnpn semiconductor device that, when in its normal state, blocks a voltage applied in either direction. The device is enabled to conduct in the forward direction when an appropriate signal is applied to the gate electrode. When such conduction is established, it continues even with the control signal removed until the anode supply is removed, reduced, or reversed. The SCR is the solid-state equivalent of the thyratron tube. 2. A semiconductor device that functions as an electrically controlled switch for dc loads. The SCR is one type of thyristor. 3. A reverse-blocking triode thyristor, that can be triggered into conduction in only one direction. Terminals

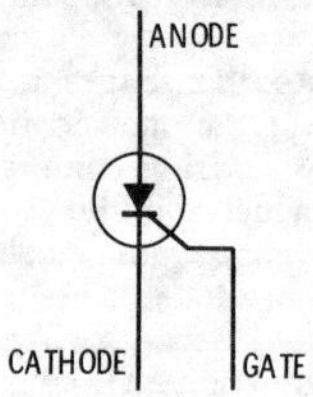

Symbol.

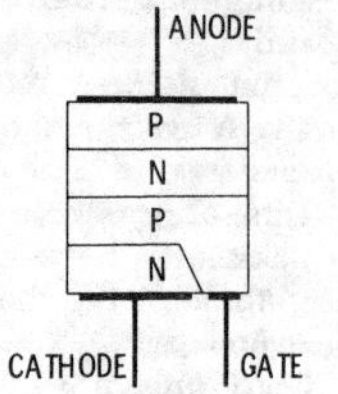

Junction.

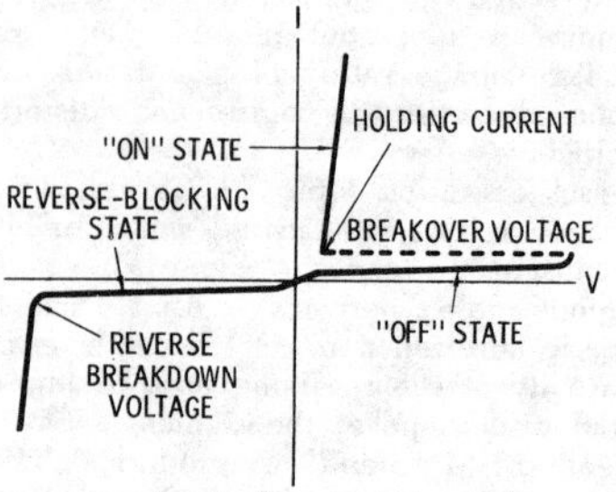

Anode characteristic.

Silicon controlled rectifier.

are anode cathode and gate. 4. A silicon rectifier having a pnpn structure that blocks current in both directions unless it is triggered into forward conduction by a pulse applied to its gate electrode. The rectifier stops conducting only if the forward current falls below the holding current.

silicon controlled switch–A four-terminal pnpn semiconductor switching device; it can be triggered into conduction by the application of either a positive or negative pulse. (Abbreviated SCS.)

silicon detector–*See* Silicon Diode.

silicon diffused epitaxial mesa transistor–A silicon transistor that has high voltage

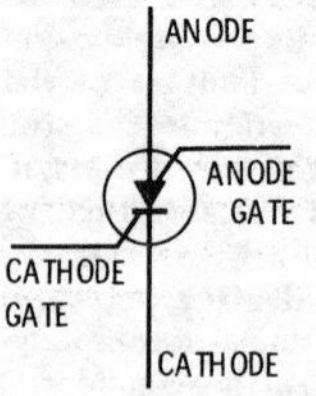

Symbol.

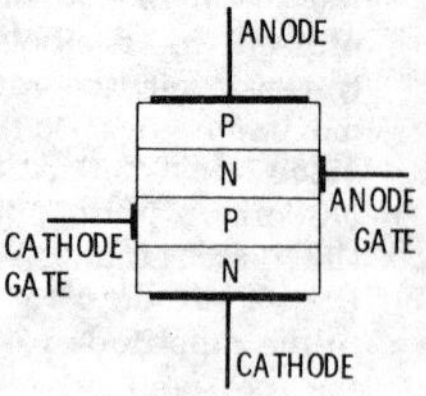

Junctions.

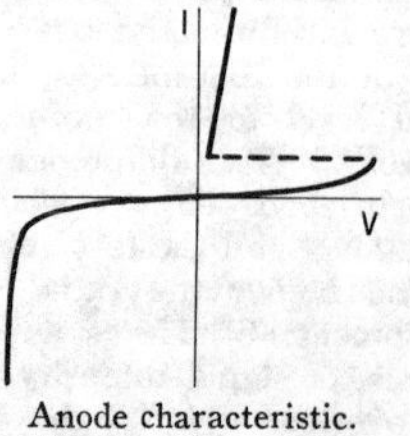

Anode characteristic.

Silicon controlled switch.

and power ratings and low storage time and saturation voltage.

silicon diode—Also called a silicon detector. A crystal detector used for rectifying or detecting uhf and shf signals. It consists of a metal contact held against a piece of silicon in a particular crystalline state.

silicon dioxide—A compound that results from oxidizing silicon quartz. Selective etching of silicon dioxide makes possible selective doping for the generation of components in monolithic integrated circuits.

silicon diode array tube—A highly sensitive vidicon-type tube used in cctv cameras designed for low light level applications.

silicon double-base diode—*See* Unijunction Transistor.

silicone—A member of the family of polymeric materials characterized by a recurring chemical group that contains silicon and oxygen atoms as links in the main chain. These compounds are presently derived from silica (sand) and methyl chloride. One of their important properties is resistance to heat.

silicon gate—A MOS process that uses silicon rather than metal as one of the transistor elements. This permits the use of lower operating voltages and increases the dynamic response of the device.

silicon-gate MOS — A process using polycrystalline silicon to replace the metal layer as the gate electrode. It offers high speed and low threshold.

silicon monoxide—A dielectric material often used in the fabrication of a microelectronic device to form an insulator, substrate, or a thin-film capacitor dielectric.

silicon nitride—A compound that is deposited on the surface of a silicon monolithic integrated circuit to improve the stability of the integrated circuit. Silicon nitride is relatively impervious to some ions that penetrate silicon dioxide; best stability is obtained through the use of a combination of silicon nitride and silicon dioxide. Charge storage at the interface between layers of silicon nitride and silicon dioxide has resulted in memory devices in which the retention times are extremely long. 2. A semiconductor manufacturing process which uses an insulated material (sapphire) instead of silicon as a substrate on which the epitaxial layer is grown. With the process, MOS or bipolar performance can be significantly improved over that of conventional devices. Also called spinel. 3. A fabrication technique in which thin crystalline films of silicon are deposited on a single crystal alumina (sapphire) substrate. The thickness of SOS films is comparable with diffusion depths commonly used in MOS/LI fabrication. Consequently doping impurities penetrate completely through the silicon, so that the only component of the pn junction is that normal to the surface. Since the principal area contributions to a pn junction come from the underside and side walls of a diffusion well, the SOS vertical junction area—and hence capacitance—is reduced considerably.

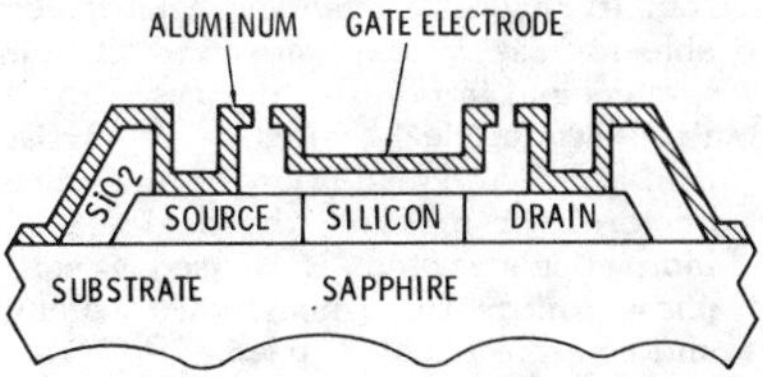

Silicon on sapphire transistor.

silicon oxide — A dielectric material commonly used in the surface passivation of microelectronic circuits. Silicon oxide contains various combinations of silicon monoxide and silicon dioxide.

silicon planar transistor—A silicon transistor produced by the planar process and consisting of a series of etchings and diffusions to produce a transistor with a thin oxide layer within the planes of a silicon substrate.

silicon rectifier—1. One or more silicon rectifying cells or cell assemblies. 2. Semiconductor diode that converts alternating current to direct current and which can be designed to withstand large currents and high voltages.

silicon rectifying cell—An elementary two-terminal silicon device which consists of a positive and a negative electrode and conducts current effectively in only one direction.

silicon solar cell—A photovoltaic cell designed to convert light energy into power for electronic and communication equipment. It consists essentially of a thin wafer of specially processed silicon.

silicon steel—Steel containing 3% to 5% sili-

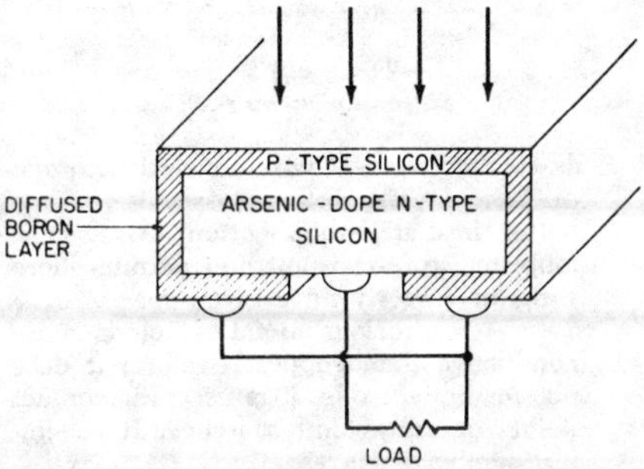

Silicon solar cell.

con. Its magnetic qualities make it desirable for use in the iron cores of transformers and in other ac devices.

silicon symmetrical switch – A thyristor modified by the addition of a semiconductor layer to make the device into a bidirectional switch. It is used as an ac phase control for synchronous switching and control of motor speed.

silicon transistor – A transistor in which silicon is used as the semiconducting material.

silicon unijunction transistor–*See* Unijunction Transistor.

silicon unilateral switch–Abbreviated sus. A device similar to the silicon controlled switch, except that a zener junction is added to the anode gate so that the silicon unilateral switch is triggered into conduction at approximately 8 volts. The sus also can be triggered by application of a negative pulse to the gate.

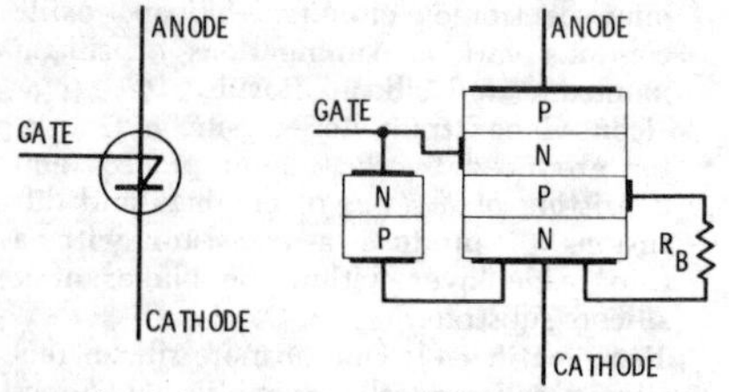

Symbol. Junctions.

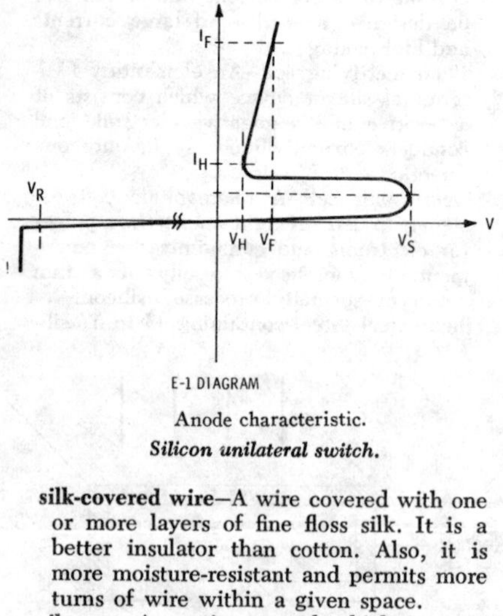

Anode characteristic.

Silicon unilateral switch.

silk-covered wire–A wire covered with one or more layers of fine floss silk. It is a better insulator than cotton. Also, it is more moisture-resistant and permits more turns of wire within a given space.

silver – A precious metal which is more conductive than copper. Because it does not readily corrode, it is used for contact points of relays and switches. Its chemical symbol is Ag.

silvered-mica capacitor–A mica capacitor that has a coating of silver deposited directly on the mica sheets instead of using conducting metal foil.

silvering–*See* Silver Spraying.

silver migration–A process by which silver, in contact with an insulating surface, under conditions of high humidity, and with an electrical potential applied, is removed ionically from one location and redeposited as a metal in another location. This transfer results in reduced insulation resistance and dielectric failure.

silver solder–A solder that is composed of copper, silver, and zinc. It has a melting point lower than that of silver, but higher than that of lead-tin solder.

silver spraying–Also called silvering. Metalizing the surface of an original master disc recording by using a dual spray nozzle in which ammoniated silver nitrate and a reducer are combined in an atomized spray to precipitate the metallic silver.

silverstat – An arrangement of closely spaced contactors. Sometimes used as a step-by-step device to unbalance the arms of a resistance bridge.

silverstat regulator–A multitapped resistor, the taps of which are connected to single-leaf silver contacts. Variation of voltage causes a solenoid to open or close these contacts, shorting out more or less of the resistance in the exciter circuit as a means of regulating the output voltage to the desired value.

simple buffering – A technique for buffer control such that the buffers are assigned to a single data control block and remain assigned to it until it is closed.

simple-gate IC–An integrated circuit that consists of one or more gate circuits formed on a single chip. The input and output of each gate are brought out to separate pins on the integrated-circuit package.

simple harmonic current–Also called sinusoidal current. A symmetrical alternating current, the instantaneous value of which is equal to the product of a constant and the sine or cosine of an angle having a value varying linearly with time.

simple harmonic electromotive force – Instantaneous values of a symmetrical alternating electromotive force which are equal to the product of a constant and the cosine or sine of an angle, the values of which vary linearly with time.

simple quad–*See* S-quad.

simple scanning – Scanning of only one scanning spot at a time.

simple sound source–A source which radiates sound uniformly in all directions under free-field conditions.

simple steady-state vibration – A periodic motion made up of a single sinusoid.

simple target–In radar, a target the reflecting surface of which does not cause the amplitude of the reflected signal to vary with the aspect of the target (e.g., a metal sphere).

simple tone—1. A sound wave, the instantaneous sound pressure of which is a simple sinusoidal function of time. 2. Also called a pure tone. A sound sensation characterized by its singleness of pitch.

simplex—A form of communication satellite operation which involves communication in only one direction at a time (mainly for facsimile, television, and some data).

simplex channel — A path for electrical transmission of information in one direction between two or more terminals.

simplex coil—A repeating coil used on a pair of wires to derive a commercial simplex circuit.

simplexed circuit—A two-wire metallic circuit from which a simplex circuit is derived, the metallic and simplex circuits being capable of simultaneous use.

simplex mode—Operation of a communication channel in one direction only with no capability for reversing.

simplex operation — Communication that takes place in only one direction at a time between two stations. Included in this classification are ordinary transmit-receive or press-to-talk operation, voice-operated carrier, and other forms of manual or automatic switching from transmit to receive.

simulate—1. To use the behavior of another system to represent certain behavioral features of a physical or abstract system. 2. To represent the functioning of a device, system, or computer program by another; e.g., to represent one computer by another, to represent the behavior of a physical system by the execution of a computer program, to represent a biological system by a mathematical model.

simulation—1. A type of problem in which a physical model and the conditions to which the model may be subjected are all represented by mathematical formulas. 2. The substitution of instrumentation (often a computer) for actual operational conditions, so that valid data can be obtained.

simulator—1. A device which represents a system or phenomenon and which reflects the effects of changes in the original so that it may be studied, analyzed, and understood from the behavior of that device. 2. Software simulators are sometimes used in the debug process to simulate the execution of machine language programs using another computer (often a timesharing system). These simulators are especially useful if the actual computer is not available. They may facilitate the debugging by providing access to internal registers of the CPU which are not brought out to external pins in the hardware.

simulcast — 1. To broadcast a program simultaneously over more than one type of broadcast station, e.g., to broadcast a stereophonic program over an am and fm station. 2. A program so broadcast.

simulcasting — Broadcasting a stereo program over an am and fm station. An am and fm tuner are required for stereo reception.

simultaneous—Pertaining to the occurrence of events at the same instant of time.

simultaneous access—*See* Parallel Access.

simultaneous computer — A computer in which there is a separate unit to perform each portion of the complete computation concurrently, the units being interconnected in a manner that depends on the computation. At different times during a run, a given interconnection carries signals that represent different values of the same variable. For example, the simultaneous computer is a differential analyzer.

simultaneous lobing—In radar, a direction-determining technique utilizing the received energy of two concurrent and partially overlapped signal lobes. The relative phase of power of the two signals received from a target is a measure of the angular displacement of the target from the equiphase or equisignal direction.

simultaneous transmission — Transmission of control characters or data in one direction at the same time that information is being received in the opposite direction.

sine—The sine of an angle of a right triangle is equal to the side opposite that angle, divided by the hypotenuse (the long side opposite the right angle).

sine galvanometer—An instrument resembling a tangent galvanometer except that its coil is in the plane of the deflecting needle. The sine of the angle of deflection will then be proportionate to the current.

sine law—The law which states that the intensity of radiation in any direction from a linear source varies in proportion to the sine of the angle between a given direction and the axis of the source.

sine potentiometer — A dc voltage divider (potentiometer), the output of which is proportionate to the sine of the shaft-angle position.

sine wave — 1. A wave which can be ex-

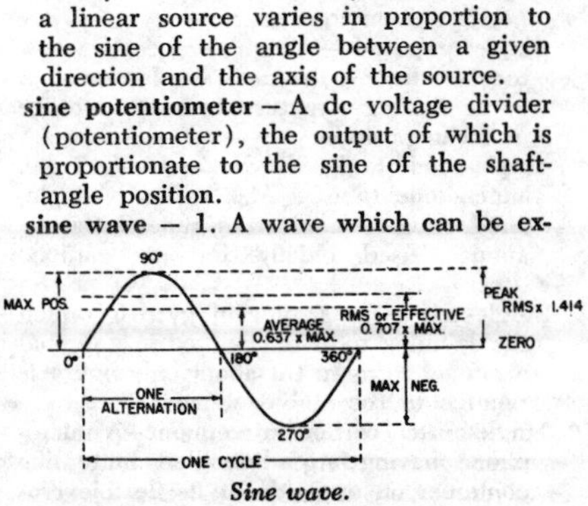

Sine wave.

pressed as the sine of a linear function of time, space, or both. 2. A waveform (often viewed on an oscilloscope) of a pure alternating current or voltage. It is drawn on a graph of amplitude versus time or radial degrees and follows the rules of sine and cosine values in relation to angular rotation of an alternator. It can be simulated by means of an electronic oscillator.

sine-wave modulated jamming—A jamming signal consisting of a cw signal modulated by one or more sine waves.

singing—An undesired self-sustained oscillation at a frequency in or above the passband of a system or component.

singing margin—Also called gain margin. The excess of loss over gain around a possible singing path at any frequency, or the minimum value of such excess over a range of frequencies.

singing point—1. The condition of a circuit or transmission path where the sum of the gains exceeds the sum of the losses. When expressed in decibels, it is the gain that can be added to the circuit equivalent before singing will begin. 2. The amount of total gain in a transmission system (most commonly used in connection with two-wire repeaters) which causes the system to begin to lose efficiency of performance because the self-oscillating point is too closely approached. 3. The singing point of a circuit which is coupled back to itself is the point at which the gain is just sufficient to make the circuit break into oscillation.

singing-stovepipe effect—Reception and reproduction of radio-signal modulation by ordinary pieces of metal, such as sections of stovepipe, in contact with each other. It is caused by mechanically poor connections, such as rusty bolts or faulty welds, that act as nonlinear diodes and produce intermodulation distortion when subjected to strong radiated fields near transmitters.

single-address code—An instruction which contains the location of the data and the operation or sequence of operations to be performed on these data.

single amplitude—With reference to vibratory conditions, the peak displacement of an oscillating structure from its average or mean position.

single-anode tank—*See* Single-Anode Tube.

single-anode tube – Also called a single-anode tank. An electron tube with one anode (used chiefly for pool-cathode tubes).

single-axis gyro—A type of gyro in which the spinning rotor is mounted in a gimbal arranged so as to tilt about only one axis relative to the stable element.

single-button carbon microphone—A microphone having a carbon-filled buttonlike container on one side of its flexible diaphragm. As the sound waves move the diaphragm, the resistance of the carbon changes, and the microphone current constitutes the desired audio-frequency signal.

single-carrier fm recording – The method of recording in which the input signal is frequency-modulated onto a carrier and the carrier is recorded on a single track at saturation and without bias.

single-channel – A carrier-only or single-tone modulated radio control transmitter and matching receiver installation.

single-channel monopulse tracking system—*See* Monopulse Tracking.

single-channel simplex – Nonsimultaneous communication between stations over the same frequency channel.

single circuit—A telegraph circuit capable of nonsimultaneous two-way communication.

single-conversion receiver—A receiver employing a superheterodyne circuit in which the input signal is downconverted once.

single crystal—A piece of material in which the crystallographic orientation of all the basic groups of atoms is the same.

single-degree-of-freedom system—A system for which only one co-ordinate is required to define the configuration of the system.

single-dial control—Control of a number of different devices or circuits by means of a single adjustment (e.g., in tuning all variable-capacitor sections of a radio receiver).

single-ended—Unbalanced such as grounding one side of a circuit or transmission line.

single-ended amplifier – An amplifier in which only one tube or transistor normally is employed in each stage—or if more than one is used, they are connected in parallel so that operation is asymmetric with respect to ground.

single-ended input impedance—The impedance between one amplifier input terminal and ground (with the other input terminal, if any, grounded for ac) when the amplifier is balanced.

single-ended input voltage – The signal voltage applied to one amplifier input terminal with the other input terminal at signal ground.

single-ended output voltage – The signal voltage between one amplifier output terminal and ground.

single-ended push-pull amplifier circuit – An amplifier circuit having two transmission paths designed to operate in a complementary manner and connected to provide a single unbalanced output. (No transformer is used.)

single-ended tube—A metal tube in which all electrodes—including the control grid—

are connected to base pins and there is no top connection. The letter *S* after the first numerals in a receiving-tube designation (e.g., 6SN7) indicates a single-ended tube.

single-ended voltage gain—Within the linear range of an amplifier, the ratio of a change in output voltage to the corresponding change in single-ended input voltage.

single-frequency duplex — A method that provides communications in opposite directions over a single-frequency carrier channel, but not at the same time. The change between transmitting and receiving conditions is controlled automatically by the voices of the communicating parties.

single-frequency simplex—A system of single-frequency carrier communications in which the change from transmission to reception is accomplished by manual rather than automatic means.

single-grip terminal—A solderless terminal designed to permit a crimp to the wire only.

single-groove stereo — *See* Monogroove Stereo.

single-gun color tube—A color picture tube with a single electron gun that produces only one beam, which is sequentially deflected across the phosphor dots.

single harmonic distortion — The ratio of the power at the fundamental frequency measured at the output of the transmission system considered, to the power of any single harmonic observed at the output of the system because of its nonlinearity, when a single-frequency signal of specified power is applied to the input of the system. It is expressed in dB.

single-hop propagation — Transmission in which the radio waves are reflected only once in the ionosphere.

single inline package—Abbreviated sip. A package for electronic components that is suited for automated assembly into printed-circuit boards. The sip is characterized by a single row of external connecting terminals, or pins, which are inserted into the holes of the printed-circuit board.

single-junction photosensitive semiconductor — Two layers of semiconductor materials with an electrode connection to each material. Light energy controls the amount of current flow.

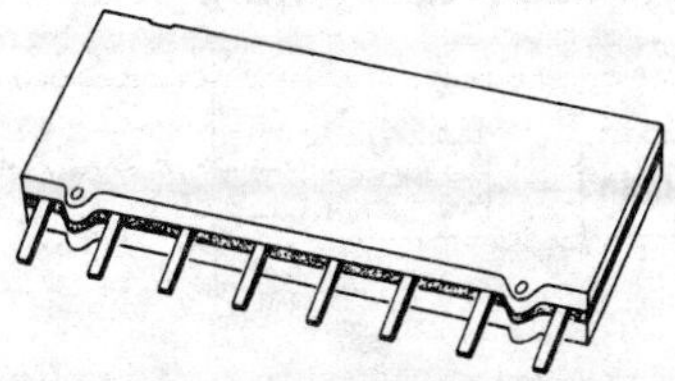

Single inline package.

single-line diagram—A form of schematic diagram in which single lines are used to show component interconnections even though two or more conductors are required in the actual circuit.

single-loop feedback — A loop in which feedback may occur only through one electrical path.

single-phase circuit—Either an alternating-current circuit with only two points of entry, or one with more than two points of entry but energized in such a way that the potential differences between all pairs of points of entry are either in phase or 180° out of phase. A single-phase circuit with only two points of entry is called a single-phase, two-wire circuit.

single-phase synchronous generator — A generator which produces a single alternating electromotive force at its terminals.

single phasing—The tendency of the rotor (of a motor tach generator) to continue to rotate when one winding is opened and the other winding remains excited.

single-point ground—*See* uniground.

single-point grounding — A grounding system that attempts to confine all return currents to a network which serves as the circuit reference. It does not imply that the grounding system is limited to one earth connection. To be effective, no appreciable current is allowed to flow in the circuit reference; i.e., th sum of the above return currents is zero.

single-polarity pulse — A pulse which departs from normal in one direction only.

single pole — A contact arrangement in which all contacts in the arrangement connect, in one position or another, to a common contact.

single-pole, double-throw — Abbreviated spdt. A three-terminal switch or relay contact for connecting one terminal to either of two other terminals.

single-pole-piece magnetic head — A magnetic head with only one pole piece on one side of the recording medium.

single-pole, single-throw—Abbreviated spst. 1. A two-terminal switch or relay contact which either opens or closes one circuit. 2. A switch with only one moving and one stationary contact. Available either normally open (no) or normally closed (nc).

single rail—The method of data transfer in a computer on only one line or wire. The device at the destination must be able to handle the data in either the high-level or low-level value. The return path is by way of common or ground.

single-rank binary—A flip-flop that requires no more than one full clock pulse from a single clock system to transfer the logic from a synchronous input to the output of the binary. It contains only one memory stage.

single sampling plan—The plan which consists of a single sample size with associated acceptance and rejection criteria.

single-shield solid enclosure—An all-metal enclosure providing higher attenuation than cell-type units. It is usually a rigid, free-standing enclosure.

single shot—*See* monostable.

single-shot blocking oscillator—A blocking oscillator modified to operate as a single-shot trigger circuit.

single-shot multivibrator—1. Also called a single-trip multivibrator. A multivibrator modified to operate as a single-shot trigger circuit. (*See also* Monostable Multivibrator.) 2. A monostable multivibrator which, after being triggered to the quasi-stable state, will "flop" back by itself to the stable state after a certain period of time.

single-shot trigger circuit — Also called a single-trip trigger circuit. A trigger circuit in which the pulse initiates one complete cycle of conditions ending with a stable condition.

single sideband—Abbreviated ssb. An am radio transmitter technique in which only one sideband is transmitted. The other sideband, and the carrier are suppressed.

single-sideband filter—A bandpass filter in which the slope on one side of the response curve is greater than on the other side. So-called because it is used in systems to suppress a carrier frequency and transmit one or both sidebands.

single-sideband modulation — Modulation whereby the spectrum of the modulating wave is translated in frequency by a specified amount, either with or without inversion.

single-sideband suppressed carrier—Modulation resulting from the partial or complete elimination of the carrier and all components of one sideband from an amplitude-modulated wave.

single-sideband system — A type of radiotelephone service in which one set of sidebands (either the upper or lower) is completely suppressed and the transmitted carrier is partially suppressed.

single-sideband transmission—Transmission of only one sideband, the other sideband being suppressed. The carrier wave may be transmitted or suppressed.

single-sideband transmitter—A transmitter in which only one sideband is transmitted.

single-signal receiver — A superheterodyne receiver equipped for single-signal reception. A highly selective filter is placed in the intermediate-frequency amplifier, and provision is included for varying the selectivity of the receiver to suit the requirements of the band condition.

single-signal reception—Use of a piezoelectric quartz crystal and associated coupling circuits as a crystal filter, to provide the high degree of selectivity required for reception in a crowded band.

single step—Pertaining to a method of computer operation in which each step is carried out in response to a single manual operation.

single-stub transformer — A shorted section of coaxial line connected to a main coaxial line near a discontinuity so that impedance matching at the discontinuity is achieved.

single-stub tuner—A section of transmission line that is terminated by a movable short-circuitry plunger or bar, and that is attached to a main transmission line to provide impedance matching.

single sweep — The operating mode for a triggering-sweep oscilloscope in which the sweep must be reset for each operation, thus preventing unwanted multiple display; it is particularly useful for trace photography. In the interval after the sweep is reset and before it is triggered, the oscilloscope is said to be armed.

single throw — A contact arrangement in which each contact form included is a single contact pair.

single-throw circuit breaker — A circuit breaker in which only one set of contacts need be moved to open or close the circuit.

single-throw switch — A switch in which only one set of contacts need be moved to open or close the circuit.

single-tone keying — Keying in which the carrier is modulated with a single tone for one condition, either marking or spacing, but is unmodulated for the other condition.

single-track magnetic system—A magnetic-

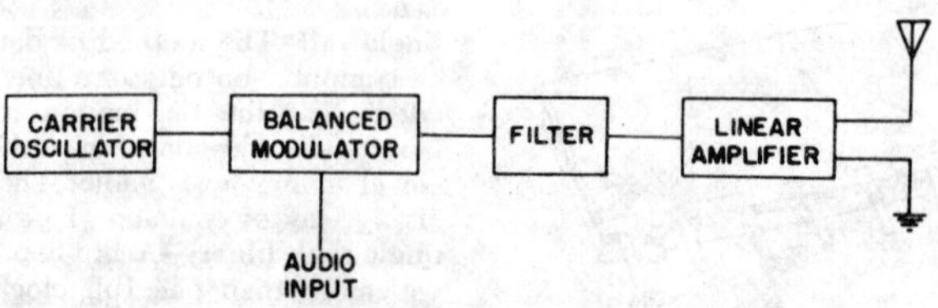

Single-sideband transmitter.

recording system, the medium of which has only one track.

single-track recorder – A tape recorder which records or plays only one track at a time on or from the tape. *See also* Monaural Recorder.

single-trip multivibrator – *See* Single-Shot Multivibrator.

single-trip trigger circuit – *See* Single-Shot Trigger Circuit.

single-tuned amplifier – An amplifier characterized by resonance at a single frequency.

single-tuned circuit – A circuit which may be represented by a single inductance and capacitance, together with associated resistances.

single-turn potentiometer—A potentiometer in which the slider travels the complete length of the resistive element with only one revolution of the shaft.

single-unit semiconductor device – A semiconductor device having one set of electrodes associated with a single carrier stream.

single-wire line – A transmission line that uses the ground as one side of the circuit.

single-wound resistor—A resistor in which only one layer of resistance wire or ribbon is wound around the base or core.

sink—In communication practice: 1. A device which drains off energy from a system. 2. A place where energy from several sources is collected or drained away. 3. Anything into which power of some kind is dissipated. 4. The component or network into which energy (usually current) flows.

sinker – An n+ region that extends down from the collector contact area on an integrated transistor to the n+ island under the collector for the purpose of reducing the collector resistance.

sink load—A load with a current in the direction out of its input. A sink load must be driven by a current sink.

sins—Acronym for a Ship's Inertial marine Navigational System especially applicable to submarine use.

sinter—A ceramic material or mixture fired so that it is not completely fused but is a coherent mass.

sintered plate—A powder which holds the active plate material used for both the anode and cathode in secondary cells. This provides a large surface area for the active material, allowing better cycle life, higher discharge rates, and better efficiency than the pocket-type plate design.

sintering – The process in which metal or other powders are bonded by cold-pressing them into the desired shape and then heating them so that a strong, cohesive body is formed.

sinusoid—A curve having ordinates proportional to the sine of the abscissa.

sinusoidal – Varying in proportion to the sine of an angle or time function (e.g., ordinary alternating current).

sinusoidal current – *See* Simple Harmonic Current.

sinusoidal electromagnetic wave—In a homogeneous medium, a wave with an electric field strength proportionate to the sine (or cosine) of an angle that is a linear function of time, distance, or both.

sinusoidal field—A field in which the magnitude of the quantity at any point varies as the sine or cosine or an independent variable such as time, displacement, or temperature.

sinusoidal quantity—A quantity that varies in the manner of a sinusoid.

sinusoidal vibration—A cyclical motion in which the object moves linearly. The instantaneous position is a sinusoidal function of time.

sinusoidal wave—A wave the displacement of which varies as the sine (or cosine) of an angle that is proportional to time, distance, or both.

sip—Abbreviation for single inline package.

site error—In navigation, the error that occurs when the radiated field is distorted by objects near navigational equipment.

situation-display tube—A large cathode-ray tube used for displaying tubular and vector information having to do with the various functions of an air-defense mission.

six-phase circuit—A combination of circuits energized by alternating electromotive forces which differ in phase by one-sixth of a cycle (60°).

size control – On a television receiver, a control which varies the size of the picture either horizontally or vertically.

skein winding—A method of winding single phase motors where each pole is a long skein of wire, formed by winding around two headless nails or bolts (smooth) set some distance apart on a piece of wood. A pin coil. The entire pole is wound in place by twisting the coil to form a concentric chain. No internal connection is made between coils of the same pole. The winding is measured with a single turn of wire, then using this endless wire, the location of the winding pins can be found.

skeletal coding – Sets of computer instructions in which some addresses and other parts are undetermined. These items usually are determined by routines designed to modify them according to given parameters.

skew – 1. In facsimile, the nonrectangular received frame due to asynchronism between the scanner and recorder. Numerically it is the tangent of the angle of this deviation. 2. The motion characterized on a magnetic tape by an angular velocity between the gap center line and a line perpendicular to the tape center line. 3. In magnetic thin film, the deviation of the easy axis during fabrication. 4. The angular displacement of a printed character, character group, or other data from the intended or ideal placement.

skewed distribution—A frequency distribution of any natural phenomenon in which zero or infinity is one of its limits.

skewing—The time delay or offset between two signals with respect to each other.

skewness – A statistical measure of the asymmetry existing in a distribution.

skiatron—1. A dark-trace oscilloscope tube. (*See also* Dark-Trace Tube.) 2. A display employing an optical system with a dark-trace tube.

skin antenna—A flush-mounted aircraft antenna made by isolating a portion of the metal skin of the aircraft with insulating materials.

skin depth—In a current-carrying conductor, the depth below the surface at which the current density has decreased one neper below the current density at the surface; that is, the field has decreased to $1/\epsilon$ (36.8%) of its surface value. Also called depth of penetration.

skin effect—Also called radio-frequency resistance. The tendency of rf currents to flow near the surface of a conductor. Thus they are restricted to a small part of the total sectional area, which has the effect of increasing the resistance.

skinner—A wire brought out at the end of a cable prepared for soldering to a terminal.

skinning—Peeling the insulation from a wire.

skin tracking—Radar tracking of an object without the aid of a beacon or other signal device on board the object.

skip—1. A digital-computer instruction to proceed to the next instruction. 2. In a computer, a "blank" instruction. 3. To ignore one or more of the instructions in a sequence. 4. Term referring to propagation of radio signals over considerable distances due to reflection back to earth from the ionosphere.

skip distance—The distance separating two points on the earth between which radio waves are transmitted by reflection from the ionized layers of the ionosphere.

skip fading—Fading due to fluctuations of ionization density at the place in the ionosphere where the wave is reflected, which causes the skip distance to increase or decrease.

skip-if-set instructions – In computers, a class of instructions in which provision is made for examining particular logic conditions. Usually they are used in conjunction with a jump (branch) instruction. For example, a skip-if-word-register-ready instruction would allow the program to check for a ready condition of the word register and then permit the program to continue along one of two different paths, depending on the condition of the word register.

skip keying—The reduction of the radar pulse-repetition frequency to a submultiple of that normally used, to reduce the mutual interference between radars or to increase the length of the radar time base.

skip zone—Also called zone of silence. A ring-shaped space or region within the transmission range wherein signals from a transmitter are not received. It is the distance between the farthest point reached by the ground wave and nearest point at which the refracted sky waves come back to earth.

sky error—*See* Ionospheric Error.

sky hook—Amateur term for antenna.

sky noise—Noise produced by radio energy from stars.

sky wave—*See* Ionospheric Wave *and* Indirect Wave.

sky-wave correction—In navigation, a correction for sky-wave propagation errors applied to measured positional data. The amount of the correction is established on the basis of an assumed position and on the height of the ionosphere.

sky-wave station error—In sky-wave–synchronized loran, the station-synchronization error due to the effect of the ionosphere on the synchronizing signal transmitted from one station to the other.

sky-wave–synchronization loran – A loran system in which the range is extended by using ionosphere-reflected signals for synchronizing the two ground stations.

sky-wave transmission delay – The longer time taken by a transmitted pulse when carried by sky waves reflected once from the E-layer, compared with the same pulse carried by ground waves.

slab—A relatively thick crystal from which blanks are cut.

slab line—A double-slotted coaxial line the outer shield of which has been unwrapped and extended to infinity in both directions so that the resulting configuration is a cylindrical conductor between two parallel conductors.

slab wafer—A slice of semiconductor material that has straight edges, as opposed to a conventional rounded wafer that has 21 percent less area than a square with comparable dimensions.

slant range—In radar, the line-of-sight distance from the measuring point to the target, particularly an aerial target.

slave—A component in a system that does not act independently, but only under the control of another similar component.

slave antenna—A directional antenna that is positioned in azimuth and elevation by a servo system. The information controlling the servo system is supplied by a tracking or positioning system.

slave drive—*See* Follower Drive.

slaved tracking—A method of interconnecting two or more regulated power supplies so that the master supply operates to control other power supplies called slaves.

slave operation—A method of interconnecting two or more stabilized power supplies so that coordinated control of the assembly by controlling the master supply alone is achieved, and essentially proportional outputs are obtained from all units.

slave relay—*See* Auxiliary Relay, 2.

slave station—A radionavigational station, the emissions of which are controlled by a master station.

slave sweep—A time base which is synchronized or triggered by a waveform from an external source. It is used in navigational systems for displaying or utilizing the same information at different locations, or in displaying or utilizing different information with a common or related time base.

slaving—The use of a torque to maintain the orientation of the spin axis of a gyro relative to an external reference such as a pendulum or magnetic compass.

sleeping sickness—In transistors, the gradual appearance of leakage.

sleeve—1. A cylindrical contacting part usually placed in back of the tip or ring of a plug and insulated from it. 2. An iron core (usually a thin-walled cylinder) used as an electromagnetic shield around an inductor. 3. A lead tube placed over cable conductors that have been spliced. 4. A tube of woven cotton pushed over a twisted wire joint in a cable. 5. A brass or copper tube or paired tubes for fastening line or drop wires together by twisting, crimping, or rolling. 6. A tube of copper or iron placed over a relay winding to make the relay slow acting.

sleeve antenna—A vertical half-wave antenna the lower half of which is a metallic sleeve through which the concentric feed line runs. The upper radiating portion, which is one-quarter wavelength, is connected to the center of the line.

sleeve-dipole antenna — A dipole antenna with a coaxial sleeve around the center.

sleeve-stub antenna—An antenna consisting of half of a sleeve-dipole antenna projecting from an extended conducting surface.

sleeve wire—1. A third conductor when associated with a pair. 2. Wire which connects to the sleeve of a plug or jack. By extension, it is common practice to designate by this term the conductors having similar functions or arrangements in circuits where plugs or jacks may not be involved.

slewing—1. Rapid change of a mechanism associated with either end of a data-transmission system when it stops following one target and takes up another. 2. In random-sampling-oscilloscope technique, the process of incrementally delaying successive samples or a set of samples with respect to the signal under examination.

slew range—The high-speed range in which a motor can run continuously but cannot stop, start, or reverse without losing step count.

slew rate—1. The maximum rate of change of the output voltage of an amplifier operated within its linear region. 2. The maximum rate of change of the output voltage of a closed-loop amplifier under large-signal conditions (the conditions that exist when an ac input voltage causes saturation of an amplifier stage, resulting in current limiting of that stage). Also called rate limit or voltage velocity limit.

slice—1. A single wafer cut from a silicon ingot and forming a thin substrate on which have been fabricated all the active and passive elements for multiple integrated circuits. A completed slice usually contains hundreds of individual circuits. (*See also* Chip, 3.) 2. Those parts of a waveform between two given amplitude limits on the same side of the zero axis. 3. A type of chip architecture which permits the cascading or stacking of devices to increase word bit size.

slicer—Also called an amplitude gate or a clipper-limiter. A transducer which transmits only portions of an input wave lying between two amplitude boundaries.

slicked switch — An alacritized mercury switch in which the rolling surface has been treated with an oily material.

slideback—The technique of applying a dc voltage to one input of a differential amplifier in order to change the vertical position on the crt screen of the signal applied to the other input.

slideback voltmeter—A vacuum-tube voltmeter that measures effective voltage values indirectly by measuring the change in grid-bias voltage required to restore the plate current of the vacuum tube to the value it had before the unknown voltage was applied to the grid circuit.

slider—A sliding contact.

slide-rule dial—A tuning dial in which a pointer moves in a straight line over a straight scale. So called because it resembles a slide rule.

slide switch—A switch that is actuated by

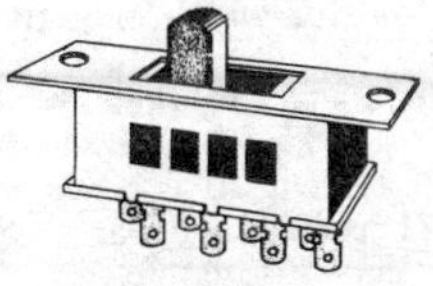

Slide switch.

sliding a control lever from one position to another.

slide wire—A bare resistance wire and a slider that can be set anywhere along the wire to provide a continuously variable resistance.

slide-wire bridge—A simplified Wheatstone bridge in which the resistance ratio is determined by the position of a slider on a resistance wire.

slide-wire rheostat—A long single-layer coil of a resistance wire with a sliding contact. The resistance is varied by moving the slider.

sliding contact—*See* Wiping Contact.

sliding load—A length of transmission line containing a matched electrical load at a distance from the connector end that can be varied.

sliding short—A length of transmission line containing an electrical short at a distance from the connector end that can be varied.

slip—1. The difference between the synchronous speed of a motor and the speed at which it operates. Slip may be expressed as a percent or decimal fraction of synchronous speed or directly in revolutions per minute. 2. Distortion produced in a recorded facsimile image as a result of slippage in the mechanical drive system. 3. A method of interconnecting multiple wiring between switching units so that trunk 1 becomes the first choice for the first switch, trunk 2 becomes the first choice for the second switch, and so on.

slip clutch — A protective device used in gear trains to disengage the load when it exceeds a specified value.

slip process—*See* Wet Process.

slip ring—A device for making electrical connections between stationary and rotating contacts. *See also* Collector Rings.

slip ring motor—Term usually applied to an induction motor with a wound secondary. (The correct term is a wound rotor motor.)

slip speed—The speed difference between speed at any load and the synchronous speed.

slope—1. The essentially linear portion of the grid-voltage, plate-current characteristic curve of a vacuum tube. This is where the operating point is chosen when linear amplification is desired. 2. *See* Rolloff.

slope-based linearity—A manner of expressing nonlinearity as the deviation from a straight line for which only the slope is specified.

slope detection—A discriminator operation on one of the slopes of the response curve for a tuned circuit. It is rarely used in fm receivers because the linear portion of the response curve is too narrow for large-signal operation.

slope detector—A detector in which slope detection is employed.

slot—One of the grooves formed in the iron core of a motor or generator armature for the conductors forming the armature winding.

slot antenna—A radiating element formed by a slot in a conducting surface.

slot armor—An insulator in the slot of a magnetic core of a machine; it may be on the coil or separate from it.

slot cell—A formed sheet of insulation that is separate from the coil and placed in the slot of a magnetic core.

slot coupling—A method of transferring energy between a coaxial cable and a waveguide by means of two coincident narrow slots, one in the sheath of the coaxial cable. E- or H-waves are launched into the guide, depending on whether the cable and guide are parallel or perpendicular to each other.

slot-discharge resistance—*See* Corona Resistance.

slot effect—The minimum voltage of rated frequency applied to the control-voltage winding of a motor tach generator necessary to start the rotor turning at no-load conditions with rated voltage and frequency on the fixed-voltage winding.

slot insulation — Flexible sheet-type insulation inserted into the slots of armatures and stators to insulate the windings from the core.

slot radiator—A primary radiating element in the form of a slot cut in the walls of a metal waveguide or cavity resonator or in a metal plate.

slotted line—*See* Slotted Section.

slotted rotor plate — *See* Serrated Rotor Plate.

slotted section—Also called a slotted line or slotted waveguide. A section of a waveguide or shielded transmission line, the shield of which is slotted to permit examination of the standing waves with a traveling probe.

slotted swr measuring equipment—A device in which standing and/or reflected waves are measured with a slotted line and a detecting probe.

slotted waveguide—*See* Slotted Section.

slow-acting relay—*See* Slow-Operating Relay.

slow-action relay—*See* Time-Delay Relay.

slow death—The gradual change of transistor characteristics with time. This change is attributed to ions which collect on the surface of the transistor.

slowed-down video—A technique of transmitting radar data over narrow-bandwidth circuits. The radar video is stored over the time required for the antenna to move through one beamwidth, and is subsequently sampled at such a rate that all range intervals of interest are sampled at least once each beamwidth or once per azimuth quantum. The radar-return information is quantized at the gap-filler radar site.

slow memory—*See* Slow Storage.

slow-operate, fast-release relay — A relay designed specifically for a long make and short release time.

slow-operate, slow-release relay—A slow-speed relay designed specifically for both a long make and a long release time.

slow-operating relay—Also called a slow-acting relay. One which is slow to attract its armature after its winding is energized. A copper slug, or collar, at the armature end of the core delays the operation momentarily after the operating circuit is completed. Such a relay is often marked SO on circuit diagrams.

slow-release relay—*See* Slow-Releasing Relay.

slow-releasing relay—Also called a slow-release relay. A slow-acting relay in which a copper slug, or collar, at the heelpiece end of the core delays the restoration momentarily after the operating circuit is opened. Such a relay is often marked SR on circuit diagrams.

slow-scan television — A television system that employs a slow rate of horizontal scanning suitable for the transmission of printed matter, photographs, and illustrations.

slow-speed relay—A relay designed specifically for long operate or release time, or both.

slow storage—Computer storage in which the access time is relatively long. Also called slow memory. *See also* Secondary Storage.

slow-wave circuit—A microwave circuit in which the phase velocity of the waves is considerably below the speed of light. Such waves are used in traveling-wave tubes.

slow-wave structure—A circuit composed of selected inductance and capacitance that causes a wave to be propagated at a speed slower than the speed of light.

slug—1. A heavy metal ring or short-circuited winding used on a relay core to delay operation of the relay. 2. A metallic core which can be moved along the axis of a coil for tuning purposes.

slug tuner—A waveguide tuner containing one or more longitudinally adjustable pieces of metal or dielectric.

slug tuning—Varying the frequency of a resonant circuit by introducing a slug of material into the electric or magnetic fields, or both.

slumber switch — A circuit arrangement whereby a radio or a recorder automatic shutoff provision can be made to turn off the apparatus itself as well as any other equipment plugged into its ac outlet.

small-scale integration—*See* SSI.

small signal—That value of an ac voltage or current which, when halved or doubled, will not affect the characteristic being measured beyond the normal accuracy of the measurement of that characteristic.

small-signal analysis—Consideration of only small excursions from the no-signal bias, so that a vacuum tube or transistor can be represented by a linear equivalent circuit.

small-signal characteristics—The characteristics of an amplifier operating in the linear amplification region.

small-signal current gain (current-transfer ratio)—The output current of a transistor with the output circuit shorted, divided by the input current. The current components are understood to be small enough that linear relationships hold between them.

small-signal drain-to-source on-state resistance — The small-signal resistance between the drain and source terminals of an FET with a specified gate-to-source voltage applied to bias the device to the on state. For a depletion-type device, this gate-to-source voltage may be zero.

small-signal, open-circuit forward-transfer impedance — In a transistor, the ratio of the ac output voltage to the ac input current when the ac output current is zero.

small-signal, open-circuit input impedance —In a transistor, the ratio of the ac input voltage to the ac input current when the ac output current is zero.

small-signal, open-circuit output admittance—In a transistor, the ratio of the ac output current to the ac voltage applied to the output terminals when the ac input current is zero.

small-signal, open-circuit output impedance—In a transistor, the ratio of the ac voltage applied to the output terminals to the ac output current when the ac input current is zero.

small-signal, open-circuit reverse-transfer impedance—In a transistor, the ratio of the ac input voltage to the ac output current when the ac input current is zero.

small-signal, open-circuit, reverse voltage transfer ratio—In a transistor, the ratio of the ac input voltage to the ac output voltage when the ac input current is zero.

small-signal power gain—In a transistor, the ratio of the ac output power to the ac input power under specified small-signal conditions. Usually expressed in dB.

small-signal, short-circuit, forward current transfer ratio—In a transistor, the ratio of the ac output current to the ac input current when the ac output voltage is zero.

small-signal, short-circuit forward-transfer admittance—In a transistor, the ratio of the ac output current to the ac input voltage when the ac output voltage is zero.

small-signal, short-circuit input admittance—In a transistor, the ratio of the ac input current to the ac input voltage when the ac output voltage is zero.

small-signal, short-circuit input impedance—In a transistor, the ratio of the ac input voltage to the ac input current when the ac output voltage is zero.

small-signal, short-circuit output admittance—In a transistor, the ratio of the ac output current to the ac output voltage when the ac input voltage is zero.

small-signal, short-circuit reverse-transfer admittance—In a transistor, the ratio of the ac input current to the ac output voltage when the ac input voltage is zero.

small-signal transconductance—In a transistor, the ratio of the ac output current to the ac input voltage when the ac output voltage is zero.

smart terminal—*See* Intelligent Terminal.

smear – 1. Television-picture distortion in which objects appear stretched out horizontally and are blurred. 2. Small frequency and time distortion introduced into a radio signal by a dispersive reflector such as the moon.

smectic phase—A parallel arrangement of liquid crystal molecules arranged in layers. The physical appearance of the smectic state is that of a highly viscous, turbic fluid.

S-meter – A meter provided in some communications receivers to give an indication of the relative strength of the received signal in terms of arbitrary units. *See also* Signal-Strength Meter.

Smith chart—A special polar diagram used in the solution of transmission-line and waveguide problems. It consists of constant-resistance circles, constant-reactance circles, circles of constant standing-wave ratio, and radius lines that represent constant line-angle loci.

smooth – To apply procedures that bring about a decrease in or the elimination of rapid fluctuations in data.

smoothing choke—An iron-core choke coil that filters out fluctuations in the output current of a vacuum-tube rectifier or direct-current generator.

smoothing circuit—A combination of induc-

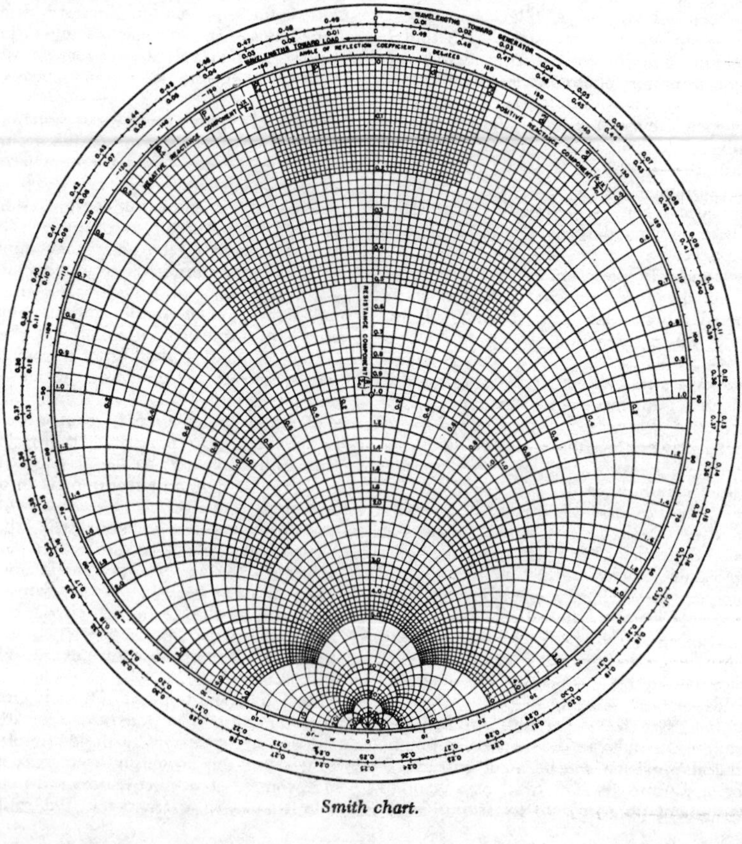

Smith chart.

tance and capacitance employed as a filter circuit to remove fluctuations in the output current of a vacuum-tube or semiconductor rectifier or direct-current generator. Also called ripple filter.

smoothing factor—The factor expressing the effectiveness of a filter in smoothing out ripple voltages.

smoothing filter—1. A filter used to remove fluctuations in the output current of a vacuum-tube or semiconductor rectifier or direct-current generator. Also called ripple filter. 2. A low-pass filter in the vertical-deflection amplifier of a spectrum analyzer. It is used to smooth amplitude fluctuations, in order to display spectral density and the average level of random signals as single lines. (The time constant of a smoothing filter is generally variable.)

SMPTE — Abbreviation for the Society of Motion Picture and Television Engineers.

snake — A tempered steel wire, usually of rectangular cross-section. The snake is pushed through a run of conduit or through an inaccessible space such as a partition and used for drawing in wires.

snap-action — 1. In a mercury switch, the rapid motion of the mercury pool from one position to another. 2. A rapid motion of the contacts from one position to another position, or their return. This action is relatively independent of the rate of travel of the actuator.

snap-action contacts — A contact assembly such that the contacts remain in one of two positions of equilibrium with substantially constant contact pressure during the initial motion of the actuating member until a point is reached at which stored energy causes the contacts to move abruptly to a new position of equilibrium.

snap magnet — A permanent magnet used in thermostatic, pressure, and other control instruments to provide quick make-and-break action at the contact and thereby minimize sparking. The magnet pulls the armature in suddenly against the spring to close the contacts and hold them closed until the spring is compressed enough to make them fly apart.

snap-off diode — A planar epitaxial passivated silicon diode that is processed in such a way that a charge is stored close to the junction when the diode conducts. Upon application of a reverse voltage, the stored charge forces the diode to switch quickly to its blocking state, or "snap off."

snapshot—In a computer, a dynamic printout of selected data in storage that occurs at breakpoints and checkpoints during the computing operations as opposed to a static printout.

snapshot dump—A selective dynamic dump carried out at various points in a machine run.

snapshot routines—Special types of debugging routines that include provisions for dynamic printout of selected data at various checkpoints in a computing operation.

snap switch—A switch (e.g., a light switch) in which the contacts are separated or brought together suddenly as the operating knob or lever compresses or releases a spring.

snap varactor—*See* Step-Recovery Diode.

sneak circuit—That part of a complete electrical circuit which carries an unintentional (sneak) current. Sneak currents may prevent proper operation of interconnected equipment.

sneak current—A leakage current that enters telephone circuits from other circuits; it is not strong enough to cause immediate damage, but it can produce harmful heating effects if permitted to continue.

sneak path—In computers, an unwanted circuit through a series-parallel configuration.

Snell's law—The sine of the angle of incidence, divided by the sine of the angle of refraction, equals a constant called the index of refraction when one of the mediums is air.

snivet—A straight, jagged, or broken vertical black line that appears near the right edge of the screen of a television receiver caused by a discontinuity in the plate-current characteristic of the horizontal amplifier tube under conditions of zero bias.

snooperscope—A night viewing device to permit a user to see objects in total darkness. It consists of an infrared source, an infrared image converter, and a battery-operated high-voltage dc-to-dc converter. Infrared radiation sent out is reflected back to the snooperscope, where it is then converted into a visible image on the fluorescent screen of the image converter tube.

snow—A speckled background caused by random noise on an intensity-modulated display, such as white specks in a television picture (usually indicative of a weak signal).

snowflake transistor—A high-speed medium-power switching transistor for use as a thin-film or core driver. It employs a six-pointed emitter geometry that permits the optimum ratio of emitter periphery to emitter area.

snow static — Precipitation static caused by falling snow.

snr—Abbreviation for signal-to-noise ratio.

snubber capacitor — A capacitor incorporated in a rapidly switched LC circuit to reduce emi by lowering the circuit resonant frequency and characteristic impedance.

snubber circuit—A form of suppression network which consists basically of a series connected resistor and capacitor connected in shunt with an SCR. The snubber circuit combined with the effective circuit series inductance controls the maximum rate of change of voltage and the peak voltage across the device when a stepped forward voltage is applied to it.

soak—In an electromagnetic relay, the condition that exists when the core is approximately saturated.

soakage—The disability of a capacitor to come up to voltage instantaneously without voltage lag or creep during or after charging. The lower the soakage, the lower the lag and creep.

soak time—The period of time required following activation for the electrolyte in a cell or battery to be sufficiently absorbed into the active materials.

soak timer—A reset timer, usually dial-adjustable, as applied in a temperature-control system for controlling the length of time the temperature is held at a predetermined level.

soak value—The voltage, current, or power applied to the coil of the relay coil to ensure that a condition approximating magnetic saturation exists.

socket—An opening that supports and electrically connects to vacuum tubes, bulbs, or other devices or components when they are inserted into it.

socket adapter—A device placed between a tube and its socket so that the tube can be used in a socket designed for some other base, or so that current or voltage can be measured at the electrodes while the tube is in use.

socket connector — A connector that contains socket contacts and that receives a plug connector containing male contacts.

socket contact — A hollow female contact designed to mate with a male contact. It is normally connected to the "live" side of a circuit.

sodar — Acronym for sound detecting and ranging. A device which detects large changes in temperature overhead by the amount of sound returned as echoes (the colder the atmosphere, the louder the echoes). The sound, which is within the range of human hearing, is launched upward, and the echoes are changed into oscilloscope patterns.

sodium amalgam-oxygen cell — A fuel-cell system in which materials functioning in the dual capacity of fuel and anode are consumed continuously. Low operating temperatures and high power-to-weight ratios are significant characteristics of the system.

sodium-vapor lamp—A gas-discharge lamp containing sodium vapor. It is used chiefly for highway illumination.

sofar—Acronym for sound fixing and ranging. An underwater sound system with which air and ship survivors can be located within a square mile and as far as 2000 miles away. Survivors drop a tnt charge into the water. The charge, which is timed to explode at 3000 to 4000 feet, sets up underwater sound waves that can be picked up by hydrophones at shore stations.

soft magnetic material—Also called a low-energy material. Ferromagnetic material which, once having been magnetized, is very easily demagnetized (i.e., requires only a slight coercive force to remove the resultant magnetism).

soft phototube—A gas phototube.

soft tube—1. A high-vacuum tube which has become defective because of the entry of a small amount of gas. 2. An electronic tube into which a small amount of gas has purposely been put to obtain the desired characteristics.

software—1. Programs, routines, codes, and other written information for use with digital computers, as distinguished from the equipment itself, which is referred to as "hardware." 2. A set of computer programs, procedures, rules, and associated documentation concerned with the operation of a data processing system, e.g., compilers, monitors, editors, utility programs. 3. Coded instructions which direct the operation of a computer. A set of such instructions for accomplishing a particular task is called a program.

softwire or computer numerical control—A numerical control system wherein a dedicated stored program computer is used to perform some or all of the basic numerical control functions. The control program can be read in and stored from data on tape, cards, manual switches, etc. Changes in the response, sequence, and/or functions can be made by reading in a different control program.

soft X-rays—X-rays with comparatively long wavelengths and hence poor penetrating power.

solar absorber—A suface that has the property of converting solar radiation into thermal energy.

solar absorption index—A quantity that relates the angle of the sum at different latitudes and local times wtih ionospheric absorption.

solar cell—1. A device capable of converting light or other radiant energy directly into electrical energy. 2. Silicon photovoltaic cell that can be used to generate electricity from direct sunlight. They are especially useful in space vehicles where no other source of electricity is available.

solar concentrator—A device that increases the intensity of solar energy by optical means.

solar-energy conversion – The process of changing solar radiation into electrical or mechanical power, either directly or by using a heat engine.

solar noise–Electromagnetic radiation from the sun at radio frequencies.

solar radiation – Radiation from the sun that is made up of a very wide range of wavelengths, from the long infrared to the short ultraviolet, with its greatest intensity in the visible green at about 5000 angstroms. The solar radiation received on the earth's surface is restricted to the visible and near infrared, since the air strongly absorbs the wavelengths located at either end of the spectrum.

solder–A readily meltable metal or alloy that produces a bond at a junction of two metal surfaces. True solder must have a lower melting point than the metals being joined and must also be capable of uniting with the metals to be joined.

solderability–The property of a metal surface which allows it to be wetted by solder.

solder cup–The end of a terminal or similar device into which a contact is inserted before being soldered.

solder eye–A solder-type terminal provided with a hole at its end through which a wire can be inserted prior to being soldered. A ring-shaped contact termination of a printed-circuit connector for the same purpose.

solder ground – A conducting path to ground due to dripping or overhanging solder.

soldering–The joining of metallic surfaces (e.g., electrical contacts) by melting a metal or an alloy (usually tin and lead) over them.

soldering iron–A soldering tool consisting of a heating element to heat the tip and melt the solder, plus a heat-insulated handle.

solderless connection – The joining of two metallic parts by pressure only, without soldering, brazing, or using any method that requires heat.

solderless connector–A device for clamping two wires firmly together to provide a good connection without solder. A common form is a cap with tapered internal threads, which are twisted over the exposed ends of the wires.

solderless contact–*See* Crimp Contact.

solderless terminals–Small metal parts used for joining a wire to another wire or to a stud by the method of crimping.

solderless wrap–Also called wire wrap. A method of connection in which a solid wire is tightly wrapped around a rectangular, square, or V-shaped terminal by means of a special tool.

solder short–A defect which occurs when solder forms a short-circuit path between two or more conductors.

sole–In a magnetron or a backward-wave oscillator, an electrode used to carry a current that produces a magnetic field in the desired direction.

solenoid – 1. An electric conductor wound as a spiral with a small pitch, or as two or more coaxial spirals. 2. An electromagnet having an energized coil approximately cylindrical in form and an armature the motion of which is reciprocating within and along the axis of the coil. 3. A coil of wire surrounding a movable iron bar that is located in such a way that when the coil is energized the core is drawn into it.

Solenoid.

solenoid valve–A combination of an electromagnet plunger and an orifice to which a disc or plug can be positioned to either restrict or completely shut off a flow. (Orifice closure or restriction occurs when the electromagnet actuates a magnet plunger.)

solid–A state of matter in which the motion of the molecules is restricted. They tend to remain in one position, giving rise to a crystal structure. Unlike a liquid or gas, a solid has a definite shape and volume.

solid circuit–A semiconductor network fabricated in one piece of material by alloying, diffusing, doping, etching, cutting, and the use of necessary jumper wires.

solid electrolyte–A solid semiconductor in direct contact with a thin nonconductive oxide coating.

solid-electrolyte fuel cell–A self-contained fuel cell in which oxygen is the oxidant and hydrogen is the fuel. The oxidant and fuel are kept separated by a solid electrolyte which has a crystalline structure and a low conductivity.

solid-electrolyte tantalum capacitor – Also called solid tantalum capacitor. A tantalum capacitor with a solid semiconductor electrolyte instead of a liquid. A wire anode is used for low capacitance values and a sintered pellet for higher values.

solid logic technology–The use in computers of miniaturized modules that make possible faster circuitry because of the reduced distances current must travel.

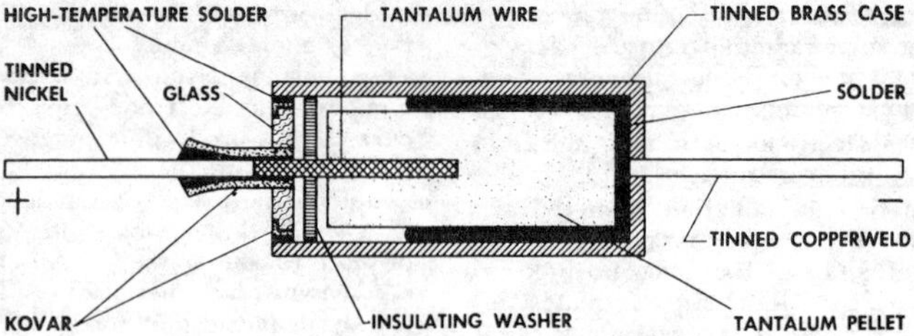

Solid-electrolyte tantalum capacitor.

solidly grounded – Also called directly grounded. Grounded through an adequate grounded connection in which no impedance has been inserted intentionally.

solid silicon circuit—Semiconductor circuit that employs a single piece of silicon material in which the various circuit elements (transistors, diodes, resistors and capacitors) are formed by diffusion in the planar configuration. By combining oxide masking, diffusion, metal deposition and alloying, a complex network with active and passive components is made completely within a die that is part of a single semiconductor wafer. Often thin film devices are applied to the surface of the silicon wafer to provide passive circuit elements beyond the range of solid silicon technology. External connections are made through small wires soldered, welded, or thermocompression-bonded to selected points of the surface.

solid-state – 1. Pertaining to circuits and components using semiconductors. 2. The physics of materials in their solid form. Examples of solid-state materials are transistors, diodes, solid-state lasers, metals and alloys, etc.

solid-state atomic battery – A device in which a radioactive material and a solar cell are combined. The radioactive material emits particles that enter the solar cell, which in turn produces electrical energy.

solid-state bonding—The process of forming a metallurgical joint between similar or dissimilar metals by causing adjoining atoms at the joint interface to combine by interatomic attraction in the solid state. (This process is different from diffusion bonding in that no atomic diffusion is required.) The adjoining surfaces to be bonded must be atomically clean and must be brought within atomic distances before such a bond can become established.

solid-state circuit – A complete circuit formed from a single block of semiconductor material. *See also* Monolithic Integrated Circuit.

solid-state component – A component the operation of which depends on the control of electric or magnetic phenomena in solids (e.g., a transistor, crystal diode, or ferrite).

solid-state computer—A computer built primarily from solid-state electronic circuit elements.

solid-state device—1. Any element that can control current without moving parts, heated filaments, or vacuum gaps. All semiconductors are solid-state devices, although not all solid-state devices (e.g., transformers) are semiconductors. 2. An electronic device which operates by virtue of the movement of electrons within a solid piece of semiconductor material.

solid-state integrated circuits—The class of integrated components in which only solid-state materials are used.

solid-state lamp—A pn junction which emits light when forward biased. Made from a complex compound of gallium, arsenic and phosphorus called gallium arsenide phosphide. Its light output is typically at 670 nanometers. It characteristically looks like a forward-biased diode with a breakdown voltage in the region of 1.6 V.

solid-state laser—A laser using a transparent substance (crystalline or glass) as the active medium, doped to provide the energy states necessary for lasing. The pumping mechanism is the radiation from a powerful light source, such as a flashtube. Ruby lasers are solid-state lasers.

solid-state physics—The branch of physics that deals with the structure and properties of solids, including semiconductors (i.e., a material the electrical resistivity of which is between that of insulators and conductors). Generally used semiconductors are silicon and germanium.

solid-state relay – A relay that employs solid-state semiconductor devices as components.

solid-state watch—A timepiece that uses a quartz-crystal or other precise frequency resonator in conjunction with low-power MOS integrated circuits. Employs liquid crystals or light-emitting diodes to indicate hours, minutes, seconds, data, months, days of the week, etc., in a digital and/or alphanumeric format.

solid tantalum capacitor—*See* Solid-Electrolyte Tantalum Capacitor.

solion—Contraction of solution ion. An electrochemical sensing and control device in which ions in solution carry electric charges to give amplification corresponding to that of vacuum tubes and transistors.

solion integrator—A precision electrochemical cell housed in glass and containing four small platinum electrodes in a solution of potassium iodide and iodine. The integrator anode and cathode make up the covers of a small cylindrical volume (less than 0.00025 cubic inch) for storing electrical information in the form of ions. The integrator cathode contains a fixed amount of hydraulic porosity for completing the internal-solution path to the other two electrodes.

solo manual—*See* Swell Manual.

Sommerfeld formula – An approximate wave-propagation relationship that may be used when distances are short enough that the curvature of the earth may be neglected in the computations.

Sonalert—A solid-state tone-emitting device. (P. R. Mallory & Co., Inc.)

sonar—Acronym for sound navigation and ranging. Also called active sonar if it radiates underwater acoustic energy, or passive (listening) sonar if it merely receives the energy generated from a distant source. Apparatus or technique of obtaining information regarding objects or events underwater through the transmission and reception of acoustic energy. Two well-known uses are to detect submarines and fish.

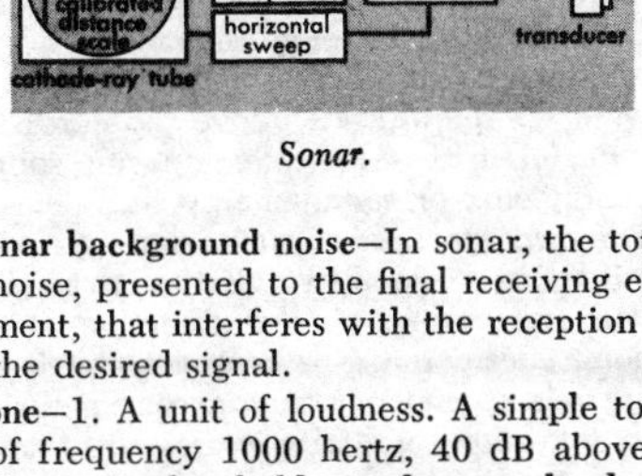

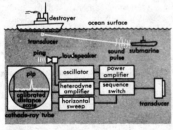

Sonar.

sonar background noise—In sonar, the total noise, presented to the final receiving element, that interferes with the reception of the desired signal.

sone—1. A unit of loudness. A simple tone of frequency 1000 hertz, 40 dB above a listener's threshold, produces a loudness of 1 sone. The loudness of any sound that is judged by the listener to be n times that of the 1-sone tone is n sones. 2. A value for loudness. May be used for overall evaluation of a sound or of a frequency band. The sone scale is linear (in contrast to decibels which are logarithmic).

sonic—1. Pertaining to the speed of sound. 2. Utilizing sound waves.

sonic altimeter – An altimeter that determines the height of an aircraft above the earth by measuring the time the sound waves take to travel from the aircraft to the ground and back, based on the fact that the velocity of sound at sea level is 1080 feet per second through dry air at 0°C (32°F).

sonic applicator – A self-contained electromechanical transducer for local application of sound for therapeutic purposes.

sonic boom—An explosionlike sound heard and felt when the shock wave that is generated by aircraft flying at supersonic speed reaches the ground.

sonic cleaning – The cleaning of contaminated materials by the action of intense sound waves produced in the liquid into which the material is immersed.

sonic delay line—A device in which electroacoustic transducers and the propagation of an elastic wave through a medium are used to produce the delay of an electrical signal.

sonic depth sounder—*See* Fathometer.

sonic drilling – The cutting or shaping of materials with an abrasive slurry driven by a reciprocating tool attached to an electromechanical transducer.

sonic frequencies—Vibrations which can be heard by the human ear (from about 15 hertz to approximately 20,000 hertz).

sonic soldering – The method of joining metals by the use of mechanical vibration to break up the surface oxides.

sonic speed—*See* Speed of Sound.

sonic thermocouple—A thermocouple so designed that gas moves past the junction with a velocity of mach 1 or greater, resulting in maximum heat transfer to the junction.

sonne – Also called consol. A radionavigational aid that provides a number of rotating characteristic signal zones. A bearing may be determined by observation (and interpolation) of the instant when transition occurs from one zone to the following zone.

sonobuoy—Also called a radiosonobuoy. A device used to locate a submerged target (e.g., a submarine). By means of a hydrophone system in the water, a sonobuoy detects the noises and converts them into radio signals, which are transmitted to a receiver in an airplane. Each sonobuoy transmits on one of several possible frequencies, and the receiver in the airplane has a channel selector so the operator can switch from one to another.

sonoluminescence—The creation of light in liquids by sonically induced cavitation.

sonometer – A frequency meter that depends for its operation on mechanical resonance with the vibrations of a variable length of stretched wire.

sonoptography—The use of sound waves to obtain a 3-D image of an object. No lenses are required. In essence, a two stage process in which the diffraction pattern of an object irradiated by sound waves is biased by a coherent sound wave and recorded. The resultant pattern, like the hologram, then is interrogated with a suitable coherent light source to obtain a three-dimensional image.

sophisticated—A piece of equipment, system, etc., which is complex and intricate, or requires special skills to operate.

sophisticated vocabulary—An advanced and elaborate set of computer instructions, enabling the computer to perform such intricate operations as linearizing, extracting square roots, selecting the highest number, etc.

sorption — The combination of absorptive and absorptive processes in the same material.

sort—To arrange items of information according to rules which depend on a key or field contained by the items.

sorter — A machine which sorts cards according to the position of coded holes.

S O S — 1. A distress signal used in radiotelegraphy. 2. Acronym for Silicon-On-Sapphire.

sound—1. Also called a sound wave. An alteration in pressure, stress, particle displacement or velocity, etc., propagated in an elastic material, or the superposition of such propagated alterations. 2. Also called a sound sensation. The auditory sensation usually evoked by the alterations described in (1) above.

sound absorption—The conversion of sound energy into some other form (usually heat) in passing through a medium or on striking a surface.

sound-absorption coefficient — The incident sound energy absorbed by a surface or medium, expressed in the form of a fraction.

sound analyzer—A device for measuring the amplitude and frequency of the components of a complex sound. It usually consists of a microphone, an amplifier, and a wave analyzer.

sound articulation—The percent of articulation obtained when the speech units are fundamental sounds (usually combined into meaningless syllables).

sound bars—Alternate dark and light horizontal bars caused in a television picture by af voltage reaching the video-input circuit of the picture tube.

sound carrier — The frequency-modulated carrier which transmits the sound portion of television programs.

sound concentrator — A parabolic reflector used with a microphone at its focus to obtain a highly directive pickup response.

sound-effect filter—A filter, usually adjustable, designed to reduce the passband of a system at low and/or high frequencies in order to produce special effects.

sound energy—The total energy in a given part of a medium, minus the energy which would exist there if no sound waves were present.

sound-energy density — At a point in a sound field, the sound energy contained in a given infinitesimal part of the medium, divided by the volume there. The commonly used unit is the erg per cubic centimeter.

sound-energy flux — The average rate at which sound energy flows through any specified area for a given period. The commonly used unit is the erg per second.

sound-energy flux density — *See* Sound Intensity.

sounder—*See* Telegraph Sounder.

sound field—A region in any medium containing sound waves.

sound film — Motion-picture film having a sound track along one side of the picture frames, for simultaneous reproduction of the sounds that accompany the film. A beam of light is projected through the sound track and is modulated at an audio rate by the variations in the width or density of the track. A phototube and amplifier then convert these modulations into sound.

sound gate—A mechanical device through which film is passed in a projector, to convert the sound track into audio signals that can be amplified and reproduced. In a television camera used for pickups, a sound gate provides the sound accompaniment for the motion picture being televised. Associated with the sound gate are an exciter lamp, a lens assembly, and a phototube.

sound head—The part of a sound motion-picture projector which converts the photographic or magnetic sound track on the film into audible sound signals.

sounding—Determination of the depth of water or the altitude above the earth.

sound intensity—Also called specific sound-energy flux or sound-energy flux density. The average rate of sound energy transmitted in a specified direction through a unit area normal to this direction at the point considered. The common unit is the erg per second per square centimeter although sound intensity expressed in watts per square centimeter may occasionally be used.

sound intensity level—The amount of sound power passing through a unit area.

sound level—A measure of the overall loudness of sounds on the basis of approximations of equal loudness of pure tones. It is expressed in dB with respect to 0.0002 microbar.

sound-level meter – An instrument–including a microphone, amplifier, output meter, and frequency-weighting networks – for the measurement of noise and sound levels. The measurements approximate the loudness level obtained for pure tones by the more elaborate ear-balance method.

sound-on-sound recording–1. A method by which material previously recorded on one track of a tape may be recorded on another track while simultaneously adding new material to it. 2. A method of tape recording in which an original sound track may be impressed with an added sound track for special effects, such as one performer appearing to play two instruments, etc.

sound-powered telephone set–A telephone set in which the transmitter and receiver are passive transducers; operating power is obtained from the speech input only.

sound-power level–1. The ratio, expressed in dB, of the sound power emitted by a source to a standard reference power of 10^{-13} watt. 2. The number of watts of acoustic power radiated by a noise source.

sound power of a source–The total sound energy radiated by the source per unit of time. The common unit is the erg per second, but the power may also be expressed in watts.

sound pressure–The instantaneous pressure minus the static pressure at some point in a medium when a sound wave is present.

sound-pressure level–Abbreviated spl. 1. In decibels, 20 times the logarithm of the ratio of the pressure of a sound to the refence pressure, which must be explicitly stated (usually, either 2×10^{-4} or 1 dyne per square centimeter). 2. The pressure of an acoustic wave stated in terms of newtons/square meter, dynes/square centimeter, or microbars. (One microbar is approximately equal to one millionth of the standard atmospheric pressure.)

sound probe–A small microphone (or tube added to a conventional microphone) for exploring a sound field without significantly disturbing it.

sound recordings–Records, tapes, or other sonic components upon which audio intelligence is inscribed or recorded or can be reproduced.

sound-recording system–A combination of transducing devices and associated equipment for storing sound in a reproducible form.

sound-reflection coefficient – Also called acoustical reflectivity. Ratio at which the sound energy reflected from a surface flows on the side of incidence, to the incident rate of flow.

sound-reproducing system–A combination of transducers and associated equipment for reproducing prerecorded sound.

sound sensation–*See* Sound, 2.

sound spectrum – The frequency components included within the range of audible sound.

sound stage–The area between the two or more loudspeakers of a stereo or quad setup where subjective sound images or imaginary loudspeakers seem to be, providing a wide area of apparent sound source.

sound takeoff–The connection or coupling at which the 4.5-MHz frequency-modulated sound signal in a television receiver is obtained.

sound track–The narrow band which carries the sound in a movie film. It is usually along the margin of the film, and more than one band may be used (e.g., for stereophonic sound).

sound-transmission coefficient (of an interface or septum) – Also called acoustical transmittivity. The ratio of the transmitted to the incident sound energy. Its value is a function of the angle of incidence of the sound.

sound wave–*See* Sound, 1.

sound-with-sound – A special provision in some recorders which allows the record head for one channel to be used for listening to that track while adding new material in exact synchronism on the adjacent track. Playing and mixing both simultaneously produces a composite sound without the degradation of one sound otherwise caused by a dubbing step in sound-on-sound mixing.

source–1. The device which supplies signal power to a transducer. 2. In a field-effect transistor, the electrode that corresponds to the cathode of a vacuum tube. 3. Supply of energy, or device upstream from a sink. (*See also* Sink). 4. Terminal which usually sources carriers. In MOS devices which are usually symmetrical, it can be interchanged with the drain terminal in a circuit. 5. The working-current terminal (at one end of the channel in an FET) that is the source of holes (p-channel) or free electrons (n-channel) flowing in the channel. Corresponds to emitter in bipolar transistor.

source-cutoff current–The current into the source terminal of a depletion-type transistor with a specified gate-to-drain voltage applied to bias the device to the off state.

source data automation – The methods of recording information in coded forms on paper tapes, punched cards, or tags that can be used repeatedly to produce many other records without rewriting.

source electrode–One of the electrodes in a field-effect transistor. It is analogous to the emitter in a transistor or the cathode in a vacuum tube. Represented by the symbol S.

source impedance – 1. The impedance which a source of energy presents to the input terminals of a device. 2. The impedance which a meter or other instrument "sees," i.e., the impedance of the driving circuit when measured from the input terminals of the meter.

source language–1. The language used to prepare a problem as the input for a computer operation. 2. In a computer, the language from which a statement is translated.

source load–A load with a current in the direction into its input. A source load must be driven by a current source.

source machine – The computer used to translate the source program into the object program.

source module–In a computer, a series of statements expressed in the symbolic language of an assembler or compiler and constituting the entire input to a single execution of the assembler or compiler.

source program–A computer program written in a language designed for ease of expression of a class of problems or procedures by humans. A generator, assembler, translator, or compiler routine translates the source program into an object program in machine language. 2. The original program, as written by the programmer, from which a working program system is derived.

source recording–The recording of information in machine-readable form, such as punched cards or tape, magnetic tape, etc. Once in this form, the information may be transmitted, processed, or reused without a need for manual processing.

source statement – A computer program written in other than machine language, usually in three-letter mnemonic symbols that suggest the definition of the instruction. There are two kinds of source statements, "executive instruction" which translate into operating machine code (opcode), and "assembly directives" which are useful in documenting the source program, but generate no code.

source terminal–The terminal electrically connected to the region from which majority carriers flow into the channel of a FET.

sourcing–Redesign or modification of existing equipment to eliminate a source of emi. When sourcing is not feasible, engineers are forced to resort to suppression, filtering, or shielding.

south pole–In a magnet, the pole into which magnetic lines of force are assumed to enter after emerging from the north pole.

space–1. An impulse which, in a neutral circuit, causes the loop to open or causes the absence of a signal, while in a polar circuit it causes the loop current to flow in a direction opposite to that for a mark impulse. A space impulse is equivalent to a binary 0. 2. In some codes, a character which causes a printer to leave a character width with no printed symbol.

space attenuation–The loss of energy, expressed in decibels, of a signal in free air caused by such factors as absorption, reflection, scattering, and dispersion.

space charge – 1. The negative charge caused by the cloud of electrons that forms in the space between the cathode and plate of a vacuum tube because the cathode emits more electrons than are attracted immediately to the plate. 2. An electrical charge distributed throughout a volume or space.

space-charge debunching–In a microwave tube, a process in which the bunched electrons are dispersed due to the mutual interactions between electrons in the stream.

space-charge effect–Repulsion of electrons emitted from the cathode of a thermionic vacuum tube by electrons accumulated in the space charge near the cathode.

space-charge field – The electric field that occurs inside a plasma due to the net space charge in the volume of the plasma.

space-charge grid–A grid, usually positive, that controls the position, area, and magnitude of a potential minimum, or of a virtual cathode adjacent to the grid.

space-charge region–The region around a pn junction in which holes and electrons recombine leaving no mobile charge carriers and a net charge density different from zero. (*See also* depletion layer.)

space-charge tube – A tube in which the space charge is used to greatly increase the transconductance. A positively charged grid is placed next to the cathode, in front of the control grid. This enlarges the space charge, moving it out to where the control grid can have a greater effect on it and hence on the plate current.

space co-ordinates – A three-dimensional system of rectangular co-ordinates. The x and y co-ordinates lie in a reference plane tangent to the earth, and the z co-ordinate is perpendicular.

space current – The total current between the cathode and all other electrodes in a vacuum tube.

spaced antenna – An antenna system used for minimizing local effects of fading at short-wave receiving stations. So called because it consists of several antennas spaced a considerable distance apart.

spaced-antenna direction finder – A direction finder comprising two or more similar but separate antennas coupled to a common receiver.

space detection and tracking system – A system that can detect and track space vehicles from the earth, and report the orbital characteristics of such vehicles to a central control facility.

space diversity—*See* Space-Diversity Reception.

space-diversity reception—Also called space diversity. Diversity reception from receiving antennas placed in different locations.

spaced-loop direction finder—A spaced-antenna direction finder in which the individual antennas are loops.

space factor – Ratio of the effective area utilized, to the total area in a winding section.

space harmonics—Harmonics in the distribution of flux in the air gap of a resolver. They may be determined as a percent of fundamental by a Fourier analysis of the flux distribution antenna. Space harmonics cause angular inaccuracy.

space-hold—In a computer, the normal no-traffic line condition by which a steady space is transmitted.

space pattern—On a test chart, a pattern designed for the measurement of geometric distortion. The EIA ball chart is an example.

space permeability – The factor that expresses the ratio of magnetic induction to magnetizing force in a vacuum. In the cgs electromagnetic system of units, the permeability of a vacuum is arbitrarily taken as unity.

space phase—Reaching corresponding peak values at the same point in space.

space quadrature – The difference in the position of corresponding points of a wave in space, the points being separated by one-quarter of the wavelength in question.

spacer cable—A means of primary power distribution that consists of three partially insulated or covered phase wires and a high-strength messenger-ground wire, all mounted in plastic or ceramic insulating spacers.

space-to-mark transition—The transition, or switching, from a spacing impulse to a marking impulse. (Teletypewriter term.)

space wave—The radiated energy consisting of the direct and ground waves.

spacing—The distance between stereo microphones or speakers.

spacing end distortion – End distortion which lengthens the spacing impulse by advancing the mark-to-space transition. (Teletypewriter term.)

spacing interval—The interval between successive telegraph signal pulses. During this interval, either no current flows or the current has the opposite polarity from that of the signal pulses.

spacing pulse—In teletypewriter operation, the signal interval during which the selector unit does not operate.

spacing wave – Also called back wave. In telegraphic communication, the emission which takes place between the active portion of the code characters or while no code characters are being transmitted.

spacistor—A semiconductor device consisting of one pn junction and four electrode connections. It is characterized by a low transient time for carriers to flow from the input to the output.

spade bolt—A bolt with a threaded section and one spade-shaped flat end through which there is a hole for a screw or rivet. It is used for fastening shielded coils, capacitors, and other components to the chassis.

spade contact—A contact with fork-shaped female members designed to dovetail with spade-shaped male members. Alignment in this type of connection is very critical if good conductivity is to be achieved.

spade tips—Notched, flat metal strips connected to the end of a cord or wire so that it can be fastened under a binding screw.

spade-tongue terminal – A slotted-tongue terminal designed to be slipped around a screw or stud without removal of the nut.

spaghetti—1. Heavily varnished cloth tubing sometimes used to provide insulation for circuit wiring. 2. A form of tubular insulation that can be slipped over wires before they are connected to terminals.

span—1. The part or space between two consecutive points of support in a conductor, cable, suspension strand, or pole line. 2. The reach or spread between two established limits such as the difference between high and low values in a given range of physical measurements.

spark—1. The abrupt, brilliant phenomenon which characterizes a disruptive discharge. 2. A single, short electrical discharge between two electrodes.

spark capacitor – A capacitor connected across a pair of contact points, or across the inductance which causes the spark, for the purpose of diminishing sparking at these points.

spark coil—An induction coil used to produce spark discharges.

spark duration—The time between the moment when the electrons first jump a gap and the moment when the current ceases to flow across it.

spark electrodes—The conductive element on each side of a spark gap through which current flows to and from the gap.

spark energy—The amount of energy dissipated between the electrodes of a spark gap. This is normally expressed as a steady-state wattage as though dissipated for a full second. Normally expressed in milliwatt seconds or millijoules.

spark frequency – The total number of sparks occurring per second in a spark transmitter (not the frequency of the individual waves).

spark gap—The arrangement of two electrodes between which a disruptive discharge of electricity may occur, and such that the insulation is self-restoring after the passage of a discharge.

spark-gap modulation – Modulation in which a controlled spark-gap breakdown produces one or more pulses of energy for application to the element in which the modulation is to take place.

spark-gap modulator – A modulator employed in certain radar transmitters. A pulse-forming line is discharged across either a stationary or a rotary spark gap.

spark-gap oscillator – A type of oscillator consisting essentially of an interrupted high-voltage discharge and a resonant circuit.

sparking—Intentional or accidental spark discharge, as between the brushes and commutator of a rotating machine, between the contacts of a relay or switch, in a solid tantalum capacitor, or at any other point in which a circuit is broken.

sparking voltage—The minimum voltage at which a spark discharge occurs between electrodes of a given shape, at a given distance apart, under given conditions.

spark killer—An electric network, usually a capacitor and resistor in series, connected across a pair of contact points (or across the inductance which causes the spark) to diminish sparking at these points.

spark lag—The interval between attainment of the sparking voltage and passage of the spark.

sparkover—Breakdown of the air between two eletrical conductors, permitting the passage of a spark.

spark plate – In an automobile radio, a metal plate insulated from the chassis by a thin sheet of mica. It bypasses the noise signals picked up by the wiring under the hood.

spark-quenching device – *See* Spark Suppressor.

spark recorder—A recorder in which the recording paper passes through a spark gap formed by a metal plate underneath and a moving metal pointer above the paper. Sparks from an induction coil pass through the paper, periodically burning small holes that form the record trace.

spark source—A device used to produce a short-circuit pulse of luminous energy by an electrical discharge between two closely spaced electrodes either in air or in a controlled atmosphere at a pressure usually greater than half an atmosphere.

spark spectrum—The spectrum produced in a substance when the light from a spark passes between terminals made of that substance or through an atmosphere of that substance.

spark suppressor – Also called a spark-quenching device or an arc suppressor. An electric network, such as a capacitance and resistance in series, or a diode connected across a pair of contacts to diminish sparking (arcing) at these contacts.

spark test—A test performed on wire and cable to determine the amount of detrimental porosity or defects in the insulation.

spark transmitter – A radio transmitter in which the source of radio-frequency power is the oscillatory discharge of a capacitor through an inductor and a spark gap.

spatial coherence – The phase relationship of two wave trains in space.

spatial distribution—The directional properties of a speaker, transmitting antenna, or other radiator.

spdt—Abbreviation for single-pole, double-throw.

speaker – Abbreviated spkr. Also called a loudspeaker. An electroacoustic transducer that radiates acoustic power into the air with essentially the same waveform as that of the electrical input.

speaker efficiency—Ratio of the total useful sound radiated from a speaker at any frequency, to the electrical power applied to the voice coil.

speaker impedance – The rated impedance of the voice coil of a speaker.

speaker-reversal switch—A switch for connecting the left channel to the right speaker and vice versa on a stereo amplifier. It is a means of correcting for improper left-right orientation in the program source.

speaker system—A combination of one or

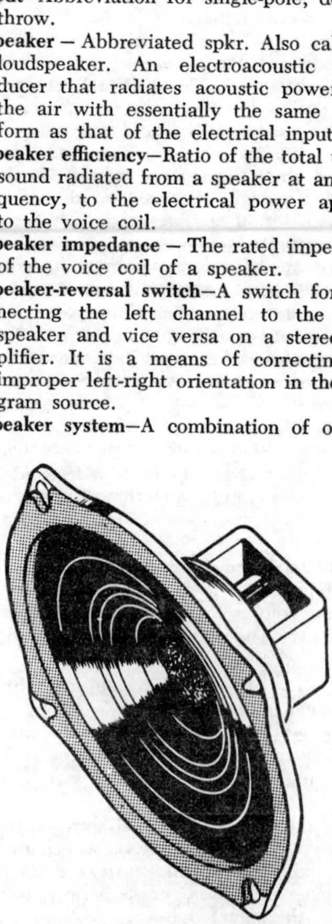

Speaker.

more speakers and all associated baffles, horns, and dividing networks used to couple the driving electric circuit and the acoustic medium together.

speaker voice coil – In a moving-coil speaker, the part which is moved back and forth by electric impulses and is fastened to the cone in order to produce sound waves.

speaking arc – A dc arc on which audio-frequency currents have been superimposed. As a result, the arc reproduces sounds in a manner similar to a speaker, and its light output will vary at the audio rate required for sound-film recording.

special effects generator – An apparatus used in the production of videotapes, this unit makes possible smooth switching of camera inputs, and provides a wide variety of screen techniques such as split and wipes.

special-purpose computer—A computer designed to solve a restricted class of problems, as contrasted with a general-purpose computer.

special-purpose motor—A motor possessing special operating characteristics and/or special mechanical construction, designed for a particular application, and not included in the definition of a general-purpose motor.

special-purpose relay—A relay the application of which requires special features not characteristic of general-purpose or definite-purpose relays.

specific acoustic impedance – Also called unit-area acoustic impedance. The complex ratio of sound pressure to particle velocity at a point in a medium.

specific acoustic reactance—The imaginary component of the specific acoustic impedance.

specific acoustic resistance—The real component of the specific acoustic impedance.

specific coding—Digital-computer coding in which all addresses refer to specific registers and locations.

specific conductance—*See* Electrolytic Conductivity.

specific conductivity—The conducting ability of a material in mhos per cubic centimeter. It is the reciprocal of resistivity.

specific damping of an instrument – *See* Relative Damping of an Instrument.

specific dielectric strength – The dielectric strength per millimeter of thickness of an insulating material.

specific gravity—The weight of a substance compared with the weight of the same volume of water at the same temperature.

specific heat—1. The capacity of a material to be heated at a given temperature (expressed as calories per degree C per gram), compared to water, which has a specific heat of 1. 2. The amount of heat required to raise a specified mass by one unit of a specified temperature.

specific inductive capacity – *See* Dielectric Constant.

specific magnetic moment – The saturation moment of a magnetic material per unit weight. It is expressed in terms of emu/-gram.

specific program – Digital-computer programming for solving a specific problem.

specific repetition rate—In loran, one of a set of closely spaced repetition rates derived from the basic rate and associated with a specific set of synchronized stations.

specific resistance—The resistance of a conductor. It is expressed in ohms per unit length per unit area, usually circular mil feet. (*See also* Resistivity, 1.)

specific routine—A digital-computer routine expressed in specific computer coding and used to solve a specific mathematical, logical, or data-handling problem.

specific sound-energy flux – *See* Sound Intensity.

spectral characteristic—The relationship between the radiant sensitivity of a phototube and the wavelength of the incident radiant flux. It is usually shown by a graph.

spectral coherence – A measure of the extent to which the output of a photodetector is restricted to a single wavelength or band of wavelengths; color response.

spectral contour plotter – A spectrum analyzer that presents a three-dimensional contour plot of analog signals, heart sounds, brain waves, etc.

spectral density—A value of a function the integral of which over a frequency interval represents the contributions of the signal components within that frequency interval.

spectral energy distribution – Abbreviated SED. A plot of energy as a function of wavelength for a given light.

spectral intensity—A function that precisely defines the spectrum and has the units of voltage squared per unit frequency.

spectral output (of a light-emitting diode) —A description of the radiant-energy or light-emission characteristic versus wavelength. This information is usually given by stating the wavelength at peak emission and the bandwidth between half-power points or by means of a curve.

spectral radiant reflectance—That fraction of the power in a light beam that is reflected from a surface. In less precise usage, called reflectivity.

spectral response – 1. Also called spectral sensitivity characteristic. The relative amount of visual sensation produced by one unit of radiant flux of any one wavelength. The human eye or a photocell exhibits greatest spectral response to the wavelengths producing yellow-green light. 2. The variation of responsivity of the detector with the wavelength of the impinging radiation.

spectral sensitivity—The color response of a photosensitive device.

spectral sensitivity characteristic—*See* Spectral Response.

spectral voltage density – The rms voltage corresponding to the energy contained in a frequency band having a width of one hertz. For the spectral voltage density at a given frequency, the band is centered on the given frequency.

spectrograph – An instrument with an entrance slit and dispersing device that uses photography to obtain a record of the spectral range. The radiant power passing through the optical system is integrated over time, and the quantity recorded is a function of the radiant energy.

spectrometer—A test instrument that determines the frequency distribution of the energy generated by any source and displays all components simultaneously.

spectrophotoelectric—Pertaining to the dependence of photoelectric phenomena on the wavelength of the incident radiation.

spectroradiometer—An instrument for measuring the radiant energy from a source at each wavelength through the spectrum. Spectral regions are separated either by calibrated filters or a calibrated monochromator. The detector is usually an energy receiver such as a thermocouple.

spectroscope – An instrument used to disperse radiation into its component wavelengths and to observe or measure the resultant spectrum.

spectroscopy – The branch of optics that deals with radiations in the infrared, visible and ultraviolet regions of the spectrum.

spectrum—1. A continuous range of electromagnetic radiations, from the longest known radio waves to the shortest known cosmic rays. Light, which is the visible portion of the spectrum, lies about midway between these two extremes. 2. The frequency components that make up a complex waveform. The band of frequencies necessary for transmission of a given type of intelligence. 3. The range of frequencies considered in a system.

spectrum analysis—1. The study of energy distribution across the frequency spectrum for a given electrical signal. 2. The process in determining the magnitude of frequency components of a signal, i.e., magnitude of the Fourier transform.

spectrum analyzer—1. A scanning receiver that automatically tunes through a selected frequency spectrum and displays on a crt or a chart a plot of amplitude versus frequency of the signals present at its input. A spectrum analyzer is, in effect, an automatic Fourier analysis plotter. 2. A test instrument that shows the frequency distribution of the energy emitted by a pulse magnetron. It also is used in measuring the Q of resonant cavities and lines, and in measuring the cold impedance of a magnetron.

spectrum intervals—Frequency bands represented as intervals on a frequency scale.

spectrum level—For a specified signal at a particular frequency, the level of that part contained within a band one hertz wide, centered at the particular frequency.

spectrum locus—The locus of a point representing the colors of the visible spectrum in a chromaticity diagram.

spectrum-selectivity characteristic—A measure of the increase in the minimum input-signal power over the minimum detectable signal required to produce an indication on a radar indicator, if the received signal has a spectrum different from that of the normally received signal.

spectrum signature analysis—The evaluation of emi from transmitting and receiving equipment in order to determine operational and environmental compatibility.

spectrum utilization characteristics – The compiled data of either transmitters or receivers which describe operational parameters such as bandwidth, sensitivity, stability, antenna pattern, power output, etc. Also describes the capabilities of the equipment to either reject or suppress unwanted electromagnetic energy.

specular reflection – Reflection of light, sound, or radio waves from a surface so smooth that its inequalities are small in comparison with the wavelength of the incident rays. As a result, each incident ray produces a reflected ray in the same plane.

SPEDAC—Acronym for solid-state, parallel, expandable, differential-analyzer computer. A high-speed digital differential analyzer using parallel logic and arithmetic, solid-state circuitry, and modular construction and capable of being expanded in computing capacity, precision, and operating speed.

speech amplifier—A voltage amplifier made specifically for a microphone.

speech audiometer – An audiometer for measuring either live or recorded speech signals.

speech clipper—1. A speech-amplitude-limiting circuit which permits the average modulation percentage of an amplitude-modulated transmitter to be increased. 2. Circuit using one or more biased diodes to limit the wave crests of speech frequency signals. Used in speech amplifiers of transmitters to maintain a high average modulation percentage.

speech compression – A modulation technique that makes use of certain properties of the speech signal in transmitting adequate information regarding quality, characteristics, and the sequential pattern of a speaker's voice over a narrower frequency band than otherwise would be necessary.

speech frequency—*See* Voice Frequency.

speech-interference level—A value for rating the effect of background noise on the intelligibility of speech.

speech interpolation — The method of obtaining more than one voice channel per voice circuit by giving each subscriber a speech path in the proper direction only at the times when his speech requires it.

speech inverter—An apparatus that interchanges high and low speech frequencies by removing the carrier wave and transmission of only one sideband in a radiotelephone. This renders the speech unintelligible unless picked up by apparatus capable of replacing the carrier wave in the correct manner. (*See also* Scrambler Circuit.)

speech level — The energy of speech (or music), measured in volume units on a volume indicator.

speech scrambler—*See* Scrambler Circuit.

speech synthesizer—A system used in research to generate speech from electrical signals in order to study human vocal patterns.

speed limit—A control function that prevents the controlled speed from exceeding prescribed limits.

speed of light—The speed at which light travels, or 186,284 miles per second.

speed of sound — Also called sonic speed. The speed at which sound waves travel through a medium (in air and at standard sea-level conditions, about 750 miles per hour or 1080 feet per second).

speed of transmission—Also called rate of transmission. The instantaneous rate of processing information by a transmission facility. Usually measured in characters or bits per unit time.

speed-ratio control — A control function which maintains a preset ratio of the speeds of two drives.

speed regulation—A figure of merit indicating the change in motor speed from no load to full load expressed as a percentage of full load speed. Speed regulation is generally established at rated speed.

speed regulator—A regulator which maintains or varies the speed of a motor at a predetermined rate.

speed-up capacitor — A capacitor used in RCTL to permit faster turn-on of the transistor in response to a change in input; it also helps overcome the storage delay of the transistor itself.

sphere gap—A spark gap with spherical electrodes. It is used as an excess-voltage protective device.

sphere-gap voltmeter — An instrument for measuring high voltages. It consists of a sphere gap, and the electrodes are moved together until the spark will just barely pass. The voltage can be calculated from gap spacing and the electrode diameter, or read directly form a calibrated scale.

spherical aberration—Image defects (e.g., blurring) due to the spherical form of a lens or mirror. These defects cause a blurred image because the lens or mirror brings the central and marginal rays to different focuses. Common types of spherical aberration are astigmatism and curvature of the field.

spherical candlepower—In a lamp, the average candlepower in all directions in space. It is equal to the total luminous flux of the lamp, measured in lumens, divided by 4π.

spherical coordinates—A system of polar coordinates which originate in the center of a sphere. All points lie on the surface of the sphere, and the polar axis cuts the sphere at its two poles.

spherical-earth attenuation — Attenuation of radio waves over an imperfectly conducting spherical earth in excess of the attenuation that would occur over a perfectly conducting plane.

spherical-earth factor—Ratio between the electrical field strengths that would result from propagation over an imperfectly conducting spherical earth and a perfectly conducting plane.

spherical wave—A wave in which the wavefronts are concentric spheres.

sphygmocardiograph—A device for simultaneous recording of heart beat and pulse.

sphygmogram—A graphic recording of the movements, forms, and forces of an arterial pulse.

sphygmomanometer — An instrument for measuring blood pressure, especially that in the arteries.

spider — highly flexible ring, washer, or punched flat member used in a dynamic

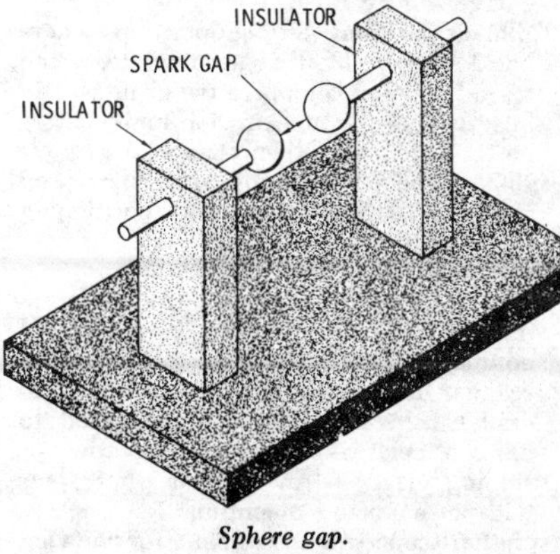

Sphere gap.

speaker to center the voice coil on the pole piece without appreciably hindering the in-and-out motion of the voice coil and its attached diaphragm.

spider bonding—A method used for connecting an integrated-circuit chip to its package leads or to a film-type substrate for hybrid constructions. Instead of running individual wires from each bonding pad on the chip to the corresponding package lead, a preformed lead frame is placed over the chip, and all connections are made by a single operation of a bonding machine.

spider-web antenna—An all-wave receiving antenna having several lengths of doublets connected somewhat like the web of a spider, to give favorable pickup characteristics over a wide range of frequencies.

spider-web coil—A flat coil having an open weave somewhat like the bottom of a woven basket. It was used in older radio receivers.

spike—An abrupt transient which comprises part of a pulse but exceeds its average amplitude considerably.

spike discriminator—A circuit used in a transponder to discriminate against pulses of extremely short duration, such as might be caused by ignition noise.

spike-leakage energy—The radio-frequency energy per pulse transmitted through tr and pre-tr tubes before and during the establishment of the steady-state radio-frequency discharge.

spike noise—In a chopper, the static field noise caused by insulating material; can be observed if the chopper is followed by a wideband amplifier.

spiking—Short, multiple, irregular bursts of laser-output radiation. Spiking is characteristic of pulse lasers, especially flash-pumped solid-dielectric types (e.g., ruby, neodymium in glass). The spike duration is typically 0.2 to 2 μs.

spill—The redistribution and hence loss of information from a storage element of a charge-storage tube.

spillover positions—Storage positions where backlogged traffic that accumulates when a send channel is inoperative or unusually busy is held for transmission immediately on the availability of a channel.

spindle—The upward-projecting shaft used on a phonograph turntable for positioning and centering the record.

spinel—*See* Silicon-on-Sapphire.

spinner—1. An automatically rotatable radar antenna, together with its associated equipment. 2. The part of a mechanical scanner that is rotated about an axis; generally, use of the term is restricted to cases of relatively high-speed rotation.

spinning electron—An electron that spins with an angular momentum.

spinthariscope—An instrument for viewing the scintillations of alpha particles on a luminescent screen.

spin wave—A moving magnetic disturbance in a ferrite. The presence of an alternating magnetic field with its frequency near that of the natural precessional frequency of a spin tends to increase the precession angle. This increase in precession angle is passed from atom to atom by dipole and exchange interactions.

spin-wave amplitude — The difference between the precession angles of two spins.

spiral distortion—In camera tubes or image tubes using magnetic focusing, a form of distortion in which image rotation varies with distance from the axis of symmetry of the electron optical system.

spiral four—A quad in which the four conductors are twisted about a common axis, the two sets of opposite conductors being used as pairs.

spiral scanning — Scanning in which the maximum radiation describes a portion of a spiral, with the rotation always in one direction.

spkr—Abbreviation for speaker.

spl—Abbreviation for sound-pressure level.

splashproof—A device or machine so constructed and protected that external splashing will not interfere with its operation.

splashproof motor—An open motor in which the ventilating openings are so constructed that drops of liquid or solid particles falling on it, or coming toward it in a straight line at any angle not greater than 100° from the vertical, cannot enter either directly or by striking and running along a surface of the motor.

splatter—Adjacent-channel interference due to overmodulation of a transmitter by abrupt peak audio signals. It is particularly noticeable for sounds containing high-frequency harmonics.

splice—A device used for joining two or more conductors.

splice insulation—Insulation used over a splice.

splicer—Any device which holds magnetic recording tape ends in place for a properly aligned splice. Many splicers are automatic to some extent, emitting the tape and trimming the edges of the splice by means of built-in cutting edges.

splicing block — A nonautomatic recording tape splicer consisting of an elongated block of metal or plastic with a shallow groove to hold the tape, and a narrow slot, usually diagonal, across the middle of the block to guide the cutting blade. Splices are made with a special splicing tape which generally needs no edge trimming.

splicing tape—A special adhesive tape designed for joining magnetic tape, and characterized by high flexibility and an adhesive which will not flow out from under the splice and thus cause adhesion between adjacent layers on the reel.

split—1. Of radar tracks, the separation of radar data from a single track to such a degree that one or more additional tracks can be initiated manually or automatically; a similar condition exists when a raid is separated to such an extent that it can be represented on a situation display as two or more raids. 2. Initiation of split tracks, raids, or groups by a direction center track monitor or by programming means.

split-anode magnetron—A magnetron with an anode divided into two segments, usually by parallel slots.

split-conductor cable – A cable in which each conductor is composed of two or more insulated conductors normally connected in parallel.

split fitting—A conduit fitting, bend, elbow, or tee split longitudinally so that it can be positioned after the wires have been drawn into the conduit. The two parts are held together usually by screws.

split gear—A type of gear designed to minimize backlash. The method consists of splitting one gear of a meshing pair and so connecting a spring between the two halves that pressure is exerted on both sides of the teeth of the other gear.

split hydrophone—A direction hydrophone in which the electroacoustic transducers are divided and arranged so that each division can induce a separate electromotive force between its own terminals.

split image—Two or more scenes appearing on a television screen as a result of trick "photography" at the studio.

split-phase motor—A single-phase induction motor having an auxiliary winding connected in parallel with the main winding, but displaced in magnetic position from the main winding so as to produce the required rotating magnetic field for starting. The auxiliary circuit is generally opened when the motor has reached a predetermined speed.

split projector—A directional projector in which electroacoustic transducing elements are divided and arranged so that each division can be energized separately through its own terminals.

split rotor plate—*See* Serrated Rotor Plate.

split-sound system—An early television receiver i-f system in which the audio and video i-f signals are separated right after the mixer stage and are amplified in separate i-f stages. Replaced by the more current intercarrier sound system.

split-stator variable capacitor—A variable capacitor with a rotor section common to two separate stator sections. Used for balancing in the grid and plate tank circuits of transmitters.

splitter—A passive device similar to an antenna coupler but designed to match a 75-ohm impedance.

split transducer – A directional transducer in which electroacoustic transducing elements are divided and arranged so that each division is electrically separate.

split winding—An equal division of a winding which will allow series or parallel external connection of the divided winding (four external leads) of a servomotor.

spoiler—A rod grating mounted on a parabolic reflector to cause the radiation pattern to change from a pencil beam to a cosecant-squared pattern. When the reflector and grating are rotated through 90° with respect to the feed antenna, one pattern changes to the other.

spontaneous emission – 1. Emission occurring without stimulation or quenching after excitation. 2. Radiation produced when a quantum mechanical system falls spontaneously from an excited state to a lower state. Emission of this radiation occurs in accordance with the laws of probability and without regard for the presence of similar radiation at the same time.

spool—A flanged form serving as the foundation on which a coil is wound.

sporadic-E ionization – Ionization that appears in the atmosphere at E-layer heights, is more noticeable at higher latitudes, and occurs at all times of the day. It may be caused by particle radiation from the sun.

sporadic E-layer—A portion that sometimes breaks away from the normal E-layer in the ionosphere and exhibits unusual erratic characteristics.

sporadic reflections – Also called abnormal reflections. Sharply defined, intense reflections from the sporadic E-layer. Their frequencies are higher than the critical frequency of the layer, and they occur anytime, anywhere, and at any frequency.

spot—1. The area instantaneously affected by the impact of an electron beam of a cathode-ray tube. 2. *See* Land, 2.

spot bonding—*See* Yield-Strength-Controlled Bonding.

spot jamming—Jamming of a specific frequency or channel.

spot noise factor—*See* Spot Noise Figure.

spot noise figure – Also called spot noise factor. Ratio of the output noise of a transducer, to the portion attributable to the thermal noise in the input termination when the termination has a standard noise temperature (290°K). The spot noise figure is a point function of input frequency.

spot projection—In facsimile: 1. An optical method in which the scanning or recording spot is delineated by an aperture between the light source and the subject copy or record sheet. 2. The optical system in which the scanning or recording spot is the size of the area being scanned or reproduced.

spot speed—1. In facsimile, the length of the scanning line times the number of lines per second. 2. In television, the product of the length (in units of elemental area, i.e., in spots) of the scanning line and the number of scanning lines per second.

spottiness — Bright spots scattered irregularly over the reproduced image in a television receiver, due to man-made or static interference entering the television system at some point.

spot welding—A resistance welding process whereby welds are made between two or more overlapping sheets of metal by pressing them together between two electrodes arranged to conduct current to the outer surfaces of the overlapped sheets. The tips of one or both of the electrodes are restricted in area to approximately the diameter of the spot weld desired.

spot wobble—An externally produced oscillating movement of an electron beam and its resultant spot. Spot wobble is used to eliminate the horizontal lines across the screen and thus make the picture more pleasing.

spreader—1. An insulating crossarm used to hold the wires of a transmission line apart. 2. The crossarm separating the parallel wire elements of an antenna.

spread groove—A groove cut between recordings. The groove, which has an abnormally high pitch, separates the recorded material but still enables the stylus to travel from one to the next.

spreading anomaly—That part of the propagation anomaly that is identifiable with the geometry of the ray pattern.

spreading loss — The transmission loss suffered by radiant energy. The effect of spreading, or divergence, is measured by this loss.

spread spectrum transmission—A communications technique in which many different signal waveforms are transmitted in a wide band. Power is spread thinly over the band so narrow-band radios can operate within the wide band without interference.

spring—A resilient, flat piece of metal forming or supporting a contact member in a jack or a key.

spring-actuated stepping relay—A stepping relay in which cocking is done electrically and operation is produced by spring action.

spring contact—A relay or switch contact, usually of phosphor bronze and mounted on a flat spring.

spring curve—A plot of the spring force on the armature of a relay versus armature travel.

spring-finger action—The design of a contact, as used in a printed-circuit connector or a socket contact, permitting easy stress-free spring action to provide contact pressure and/or retention.

spring pile-up—An assembly of all contact springs operated by one armature lever.

spring-return switch — A switch which returns to its normal position when the operating pressure is released.

spring stop—In a relay, the member used to control the position of a pretensioned spring.

spring stud—In a relay, an insulating member that transmits the armature motion from one movable contact to another in the same pileup.

sprocket holes—Holes punched on each line of a perforated tape used as a timing reference and for driving certain transports.

sprocket pulse — 1. A pulse generated by one of the magnetized spots that accompany every character recorded on magnetic tape. During read operations, sprocket pulses permit regulation of the timing of the read circuits, and they also provide a count of the number of characters read from the tape. 2. A pulse generated by a sprocket or driving hole in a paper tape; this pulse serves as the timing pulse for reading or punching the tape.

spst — Abbreviation for single-pole, single-throw.

spurious counts—*See* Spurious Tube Counts.

spurious emanations—Unintentional and undesired emissions from a transmitting circuit.

spurious emission—*See* Spurious Radiation.

spurious modulation — Undesired modulation of an oscillator; for example, frequency modulation resulting from mechanical vibration.

spurious pulse—In a scintillation counter, a pulse not purposely generated or directly due to ionizing radiation.

spurious pulse mode — An unwanted pulse mode which is formed by the chance combination of two or more pulse modes and is indistinguishable from a pulse interrogation or reply.

spurious radiation — Also called spurious emission. Emissions from a radio transmitter at frequencies outside its assigned or intended emission frequency. Spurious emissions include harmonic emissions, parasitic emissions, and intermodulation products, but exclude emissions in the immediate vicinity of the necessary band, which are a result of the modulation process for the transmission of information.

spurious response — 1. Any undesired response from an electric transducer or similar device. 2. The sensitivity of a circuit to signals of frequencies other than the frequency to which the circuit is tuned. 3. In electronic warfare, undesirable signal images in the intercept receiver as a result of mixing of the intercepted signal with harmonics of the receiver local oscillators.

spurious-response attenuation – The ability of a receiver to discriminate between a desired signal to which it is resonant and an undesired signal at any other frequency to which it is simultaneously responsive.

spurious-response ratio – Ratio of the field strength at the frequency which produces a spurious response, to the field strength at the desired frequency, each field being applied in turn to produce equal outputs. Image ratio and intermediate-frequency response ratio are special forms of spurious-response ratio.

spurious response rejection–The ability of an fm tuner to reject spurious signals falling outside the tuned frequency, or the immunity of the tuner itself to the production of spurious signals as the result of intermodulation, etc.

spurious signal – 1. An unwanted signal generated either in the equipment itself or externally and heard (or seen) as noise. 2. Undesired signals appearing external to an equipment or circuit. They may be harmonics of existing desired signals, high frequency components of complex wave shapes, or signals produced by incidental oscillatory circuits.

spurious transmitter output – Any component of the radio-frequency output that is not implied by the type of modulation and the specified bandwidth.

spurious transmitter output, conducted–A spurious output of a radio transmitter that is conducted over a tangible transmission path such as a power line, control circuit, radio-frequency transmission line, waveguide, etc.

spurious transmitter output, extraband–A spurious transmitter output that lies outside the specified band of transmission.

spurious transmitter output, inband – A spurious transmitter output that lies within the specified band of transmission.

spurious transmitter output, radiated – A spurious output radiated from a radio transmitter. (The associated antenna and transmission lines are not considered part of the transmitter.)

spurious tube counts–Also called spurious counts. The counts in radiation-counter tubes, other than background counts and those caused directly by the radiation to be measured. They are caused by electrical leakage, failure of the quenching process, etc.

spurt tone – A short-duration audio-frequency tone used for signaling or dialing selection.

sputtering–1. Also called cathode sputtering. A process sometimes used in the production of the metal master disc. In this process the original is coated with an electric conducting layer by means of an electric discharge in a vacuum. 2. A thin-film technique in which material for the film is ejected from the surface of the bulk source when the source is subjected to ion bombardment. 3. Dislocation of surface atoms of a material bombarded by high energy atomic particles.

SQ–A matrix system developed by CBS, Inc.

s-quad–Also called simple quad. An arrangement of two parallel paths, each of which contains two elements in series.

square-law demodulator – *See* Square-Law Detector.

square-law detection – Detection in which the output voltage is substantially proportional to the square of the input voltage over the useful range of the detector.

square-law detector – A detector in which the output signal current is proportional to the square of the radio-frequency input voltage. Operation of this circuit depends on nonlinearity of the detector characteristic, rather than on rectification. Also called square-law demodulator.

square-law scale meter–A meter in which the deflection is proportional to the square of the applied energies.

square-loop ferrite–A ferrite with a rectangular hysteresis loop.

squareness ratio – 1. For a magnetic material in a symmetrically cyclically magnetized condition, the ratio of the flux density at zero magnetizing force, to the maximum flux density. 2. The ratio of the flux density to the maximum flux density when the magnetizing force has changed halfway from zero toward its negative limiting value.

square wave–1. A square- or rectangular-shaped periodic wave which alternately assumes two fixed values for equal lengths of time, the transition time being negligible in comparison with the duration of each fixed value. 2. An ac periodic waveform in which voltage alternates rapidly from a positive peak value to the negative peak value, and vice versa, after a delay.

square-wave amplifier–A resistance-coupled amplifier (in effect, a wideband video amplifier) which amplifies a square wave with a minimum of distortion.

square-wave generator–A signal generator

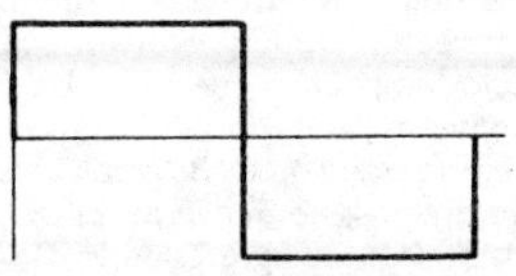

Square wave (*ideal*).

for producing square or rectangular waves.

square-wave response—In camera tubes, the ratio of the peak-to-peak signal amplitude given by a test pattern consisting of alternate black and white bars of equal widths, to the difference in signal between large-area blacks and whites having the same illuminations as the bars. Horizontal square-wave response is measured if the bars are perpendicular to the horizontal scan, and vertical square-wave response is measured if they are parallel.

squaring circuit—1. A circuit which changes a sine wave or other wave into a square wave. 2. A circuit which contains nonlinear elements and which produces an output voltage proportional to the square of the input voltage.

squawker – The midrange speaker of a three-way system.

squeal—1. Audible tape vibrations, primarily in the longitudinal node, caused by a frictional excitation at the heads and guides. 2. In a radio receiver, a high pitched tone heard together with the wanted signal.

squealing – The high-pitched noise heard along with the desired intelligence in a radio receiver. It is due to interference between stations or to oscillation in one of the receiver circuits.

squeezable waveguide—In radar, a variable-width waveguide for shifting the phase of the radio-frequency wave traveling through it.

squeeze section – A length of waveguide the critical dimension of which can be altered to correspond to changes in the electrical length.

squeeze track – A variable-density sound track in which variable width with greater signal-to-noise ratio is obtained by means or adjusting masking of the recording light beam and simultaneous increase of the electric signal applied to the light modulator.

squegger – A self-quenching oscillator in which the suppression occurs in the grid circuit.

squegging—A self-blocking condition in an oscillator circuit.

squegging oscillator—*See* Blocking Oscillator, 1.

squelch—To automatically quiet a receiver by reducing its gain in response to a specified characteristic of the input.

squelch circuit—A circuit for preventing a radio receiver from producing an audio-frequency output in the absence of a signal having predetermined characteristics. A squelch circuit may be operated by signal energy in the receiver passband, by noise quieting, or by a combination of the two (ratio squelch). It may also be operated by a signal having special modulation characteristics (selective squelch).

squint—In radar, an ambiguous term, meaning either the angle between the two major-lobe axes in a lobe-switching antenna, or the angular difference between the axis of antenna radiation and a selected geometric axis such as the axis of the reflector.

squint angle—The angle between the physical axis of the antenna center and the axis of the radiated beam.

squirrel-cage induction motor – An induction motor in which the secondary circuit, usually the rotor, consists of a squirrel-cage winding (two discs connected along their circumference with copper bars) arranged in slots in the iron core.

squirrel-cage winding – A permanently short-circuited winding which is usually uninsulated, has its conductors uniformly distributed around the periphery of the machine, and is joined by continuous end rings.

squitter – In radar, random firing (intentional or otherwise) of the transponder transmitter in the absence of interrogation.

S/rf meter—An indicator on some CB transceivers to indicate relative strength of an intercepted signal when receiving, and the relative rf power output when transmitting.

ssb—*See* Single-Sideband.

ssfm—A system of multiplex in which single-sideband subcarriers are used to frequency-modulate a second carrier.

SSI—Abbreviation for small-scale integration. Having less complexity than MSI, i.e., less than the equivalent of 12 gates.

SS loran—Sky-wave synchronized loran. Loran in which the slave station is controlled by the sky wave from the master station rather than the ground wave. This method is used with unusually long base lines.

sspm—A system of multiplex in which single-sideband subcarriers are used to phase-modulate a second carrier.

SSR—Abbreviation for solid-state relay.

stability—1. The ability of a component or device to maintain its nominal operating characteristics after being subjected to changes in temperature, environment, current, and time. It is usually expressed in either percent or parts per million for a given period of time. 2. The ability of a power supply to maintain a constant output voltage (or current) over a period of time under fixed conditions of input, load, and temperature. Usually expressed in terms of a voltage (or current) change over a fixed length of time. 3. The ability to maintain effectiveness within reasonable bounds in spite of large changes in environment. 4. For a feedback-control system or element, the property such that its output will ultimately attain a steady-state dc level within the linear range and without continuing external stimuli.

stability factor – The measure of the bias stability of a transistor amplifier. It is defined as the change in collector current, I_c, per change in cutoff current, I_{co}.

stabilivolt – A gas-filled tube containing a number of concentric, coated iron electrodes. It is used as a source of practically constant voltage for apparatus drawing only small currents.

stabilization–1. The introducing of stability into a circuit. 2. A process, by means of which the output of electromechanical transducers is optimized, by adjusting magnetic and mechanical parameters for its maximum magnetic permanency, to maintain its output stable under changing environmental and external conditions. 3. The reduction of variations in voltage or current not due to prescribed conditions.

stabilization network – A network used to prevent oscillation in an amplifier with negative feedback.

stabilized feedback – *See* Negative Feedback.

stabilized flight–A type of flight in which control information is obtained from inertia-stabilized references such as gyroscopes.

stabilized local oscillator – An extremely stable radio-frequency oscillator used as a local oscillator in the superheterodyne radar receiver in a moving-target indicator system.

stabilized master oscillator–The master oscillator in complex microwave systems that acts as the frequency reference for all rf signals within the system. Usually crystal and temperature controlled for maximum frequency stability.

stabilized shunt-wound motor – A shunt-wound motor to which a light series winding has been added to prevent a rise in speed, or to reduce the speed when the load increases.

stabilized winding – Also called tertiary winding. An auxiliary winding used particularly in star-connected transformers: a. To stabilize the neutral point of the fundamental frequency voltage. b. To protect the transformer and the system from excessive third-harmonic voltages. c. To prevent telephone interference caused by third-harmonic currents and voltages in the lines and earth.

stabistor–A voltage-limiting semiconductor. A diode designed to break over and conduct at a certain voltage. This is the normal forward conduction of a diode and is also characteristic of zener diodes, which avalanche into conduction when breakdown (backward) voltage is exceeded.

stable element – In navigation, an instru-

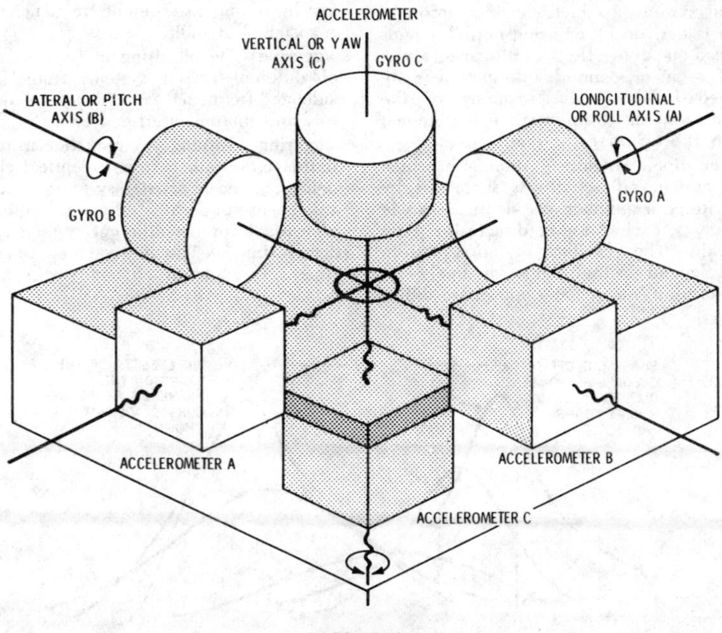

Stable platform.

ment or device which maintains a desired orientation independently of the vehicle motion.

stable oscillation – A response which does not increase indefinitely with time; the opposite of an unstable oscillation.

stable platform—Also called a gyrostabilized platform. A gyro instrument which provides accurate azimuth, pitch, and roll attitude information. In addition to serving as reference elements, they are used for stabilizing accelerometers, star trackers, and similar devices in space.

stable strobe – A series of strobes that behaves as if caused by a single jammer.

stack—1. That portion of a computer memory and/or registers used to temporarily hold information. 2. *See* Pileup, 1. 3. A block of successive memory locations which is accessible from one end on a last-in first-out basis (LIFO). The stack is coordinated with the stack pointer which keeps track of storage and retrieval of each byte of information in the stack. A stack may be any block of successive information locations in the read/write memory.

stack pointer—The counter or register used to address a stack in the memory.

stacked array—An antenna system consisting of two or more antennas connected together, and placed with respect to each other, to increase the gain in a specific direction or directions.

stacked-beam radar – A three-dimensional radar system in which elevation information is derived by emitting narrow beams placed one above the other to cover a vertical segment, azimuth information is obtained by horizontal scanning of the beam, and range information is obtained from the echo-return time.

stacked dipole antenna – Antenna in which the antenna directivity is increased by providing a number of identical dipole elements, excited either directly or parasitically. The resultant radiation pattern will depend on the number of dipole elements used, the spacing and phase difference between the elements, and the relative magnitudes of the currents.

stacked heads – Also called in-line heads. An arrangement of magnetic recording heads used for stereophonic sound. The two heads are directly in line, one above the other.

stage—1. A term usually applied to an amplifier to mean one step, especially if part of a multistep process; or the apparatus employed in such a step. 2. A hydraulic amplifier used in a servovalve. Servovalves may be single-stage, two-stage, three-stage, etc.

stage-by-stage elimination – A method of locating trouble in electronic equipment by using a signal generator to introduce a test signal into each stage, one at a time, until the defective stage is found.

stage efficiency—Ratio of useful power (alternating current) delivered to the load, to the power at the input (direct current).

stagger—Periodic positional error of the recorded spot along a recorded facsimile line.

staggered heads—An infrequently used arrangement of magnetic recording heads for stereophonoic sound. The heads are $1\frac{7}{32}$ inch apart. Stereo tapes recorded with staggered heads cannot be played on recorders using stacked heads, and vice versa.

staggered tuning—A means of producing a wide bandwidth in a multistage if amplifier by tuning to different frequencies by a specified amount.

staggering—The offsetting of two channels of different carrier systems from exact sideband-frequency coincidence, in order to avoid mutual interference.

staggering advantage—A reduction in intelligible cross talk between identical channels of adjacent carrier systems as a result of using slightly different frequency allocations for the different systems.

stagger time – The interval between the times of actuation of any two contact sets.

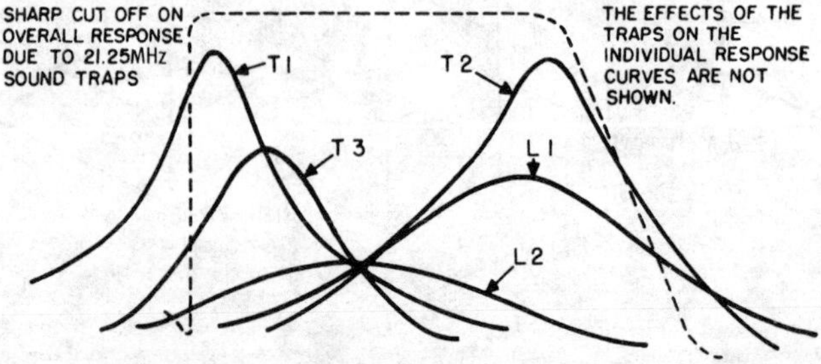

Staggered tuning.

stagger-tuned amplifier—An amplifier consisting of two or more stages, each tuned to a different frequency.

stagnation thermocouple—A type of thermocouple in which a high recovery factor is achieved by stagnating the flow in a space surrounding the junction. This results in a high response time as compared with an exposed junction.

staircase – A video test signal containing several steps at increasing luminance levels. The staircase signal is usually amplitude modulated by the subcarrier frequency and is useful to checking amplitude and phase linearities in video systems.

staircase generator—A special-purpose signal generator that produces an output which increases in steps; thus its output waveform has the appearance of a staircase.

staircase signal—A waveform consisting of a series of discrete steps resembling a staircase.

stalled-torque control – A control function used to control the drive torque at zero speed.

stall torque—1. The torque which the rotor of an energized motor produces when restrained from motion. 2. The torque developed by a servomotor at speed in excess of 1 rpm but less than ½% of the synchronous speed with a rated voltage and frequency of the proper phase relationship applied to both windings.

stalo—1. Acronym for stabilized local oscillater used as part of a moving-target indication device in conjunction with a radar. 2. A highly stable oscillator, usually stabilized by feedback from a very high-*Q* LC circuit such as a high-*Q* cavity.

stamper – A negative (generally made of metal by electroforming) from which finished records are molded.

stand-alone system—A microcomputer software development system which runs on a microcomputer without connection to another computer or a timesharing system. This system includes an assembler, editor and debugging aids. It may include some of the features of a prototyping kit.

standard—An exact value, or a concept established by authority, custom, or agreement, to serve as a model or rule in the measurement of a quantity or in the establishment of a procedure.

standard antenna—An open single-wire antenna, including the lead-in wire, having an effective height of four meters.

standard beam approach—Abbreviated sba. A vhf, 40-MHz, continuous-wave, low-approach system using a localizer and markers. The two main-signal lobes are tone-modulated with the Morse-code letters *E* and *T* (· and –). These modulations form a continuous tone when the aircraft is on its course. The airborne equipment is usually instrumented for visual reference, but may be used aurally in some applications.

standard broadcast band—The band of frequencies extending from 535 to 1605 kilohertz.

standard broadcast channel – The band of frequencies occupied by the carrier and two sidebands of a broadcast signal. The carrier frequency is at the center, with the sidebands extending 5 kHz on either side.

standard broadcast station—A radio station operated on a frequency between 535 and 1605 kHz for the purpose of transmitting programs intended for reception by the general public.

standard candle – A unit of candlepower equal to a specified fraction of the visible light radiated by a group of 45 carbon-filament lamps preserved at the National Bureau of Standards, the lamps being operated at a specified voltage. The standard candle was originally the amount of light radiated by a tallow candle of specified composition and shape.

standard capacitor – A capacitor in which its capacitance is not likely to vary. It is used chiefly in capacitance bridges.

standard cell—A primary cell which serves as a standard of voltage.

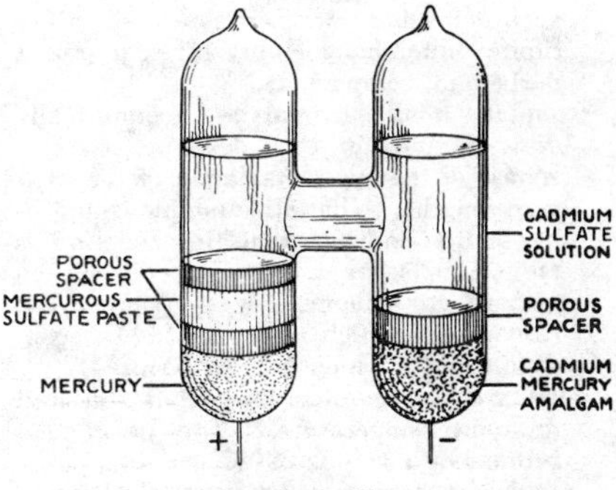

Standard cell.

standard component – A component which is regularly produced by some manufacturer and is carried in stock by one or more distributors.

standard deflection – 1. In a galvanometer having an attached scale, one scale division. 2. In a galvanometer without an attached scale, 1 millimeter when the scale distance is 1 meter.

standard deviation—A measure of the variation of data from the average. It is equal to the root mean square of the individual deviations from the average.

standard eye—An observer that has red and infrared luminosity functions.

standard facility—In programming actions,

a basic communications electronics functional entity which is engineered to satisfy a specific communications electronics operational requirement. An associated standard facility equipment list describes the facility functionally and indicates the material required for the standard facility.

standard-frequency service—A radiocommunication service that transmits for general reception specified standard frequencies of known high accuracy.

standard-frequency signal – One of the highly accurate signals broadcast by the National Bureau of Standards radio station WWV on 2.5, 5, 10, 15, 20, and 25 MHz.

standard-gain horn – A waveguide device that has essentially flared out its waveguide dimensions to specific lengths that match (with certain gain) the incoming energy to the atmosphere. The applications include using these horns for reflectors and lenses, pickup horns for sampling power, and receiving and transmitting antennas.

standardization—The process of establishing by common agreement engineering criteria, terms, principles, practices, materials, items, processes, equipment, parts, subassemblies, and assemblies to achieve the greatest practicable uniformity of items of supply and engineering practices, to ensure the minimum feasible variety of such items and practices, and to effect optimum interchangeability of equipment parts and components.

standard luminosity curve—An empirically derived function that describes the response of the eye to radiation of different wavelengths. (The terms luminous and illumination indicate that this function is taken into account.)

standard microphone – A microphone, the response of which is known for the condition under which it is to be used.

standard noise temperature – A standard reference temperature (T) for noise measurements, taken as 290K.

standard observer—A hypothetical observer who requires standard amounts of primaries in a color mixture to match every color.

standard play – An arbitrary description given to identify a spool of "thick" recording tape with a specified playing time according to reel size. For example, a 7-inch spool will hold 1200 ft. of "standard play" tape; a 5-inch spool, 600 ft. The actual playing time is then dependent on tape speed.

standard pitch—The tone *A* at 440 hertz.

standard propagation—Propagation of radio waves over a smooth, spherical earth of uniform dielectric constant and conductivity, under standard atmospheric refraction.

standard reference temperature—In a thermistor, the body temperature for which the nominal zero-power resistance is specified.

standard refraction – The refraction which would occur in an idealized atmosphere; the index of refraction decreases uniformly with height at the rate of 39×10^{-6} per kilometer.

standard register of a motor meter – Also called a dial register. A four- or five-dial register, each dial being divided into ten equal parts numbered from zero to nine. The dial pointers are geared so that adjacent ones move in opposite directions at a 10-to-1 ratio.

standard resistor—Also called a resistance standard. A resistor which is adjusted to a specified value is only slightly affected by variations in temperature, and is substantially constant over long periods of time.

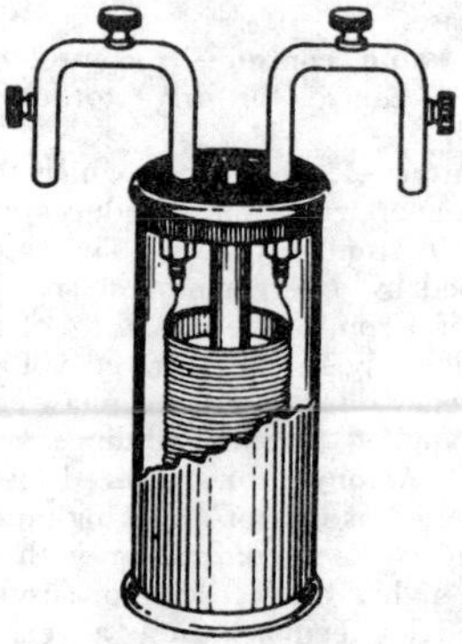

Standard resistor.

standard rod gap—A gap between the ends of two half-inch square rods. Each rod is cut off squarely and mounted on supports so that it overhangs the inner edge of each support by a length equal to or greater than half the gap spacing. It is used for approximate measurements of crest voltages.

standard sea-water conditions – Sea water with a static pressure of 1 atmosphere, a temperature of 15°C, and a salinity such that the velocity of sound propagation is exactly 1500 meters per second.

standard sphere gap—A gap between two metal spheres of standard dimensions. It is used for measuring the crest value of a voltage by observing the maximum gap spacing at which sparkover occurs when a voltage is applied under known atmospheric conditions.

standard subroutine—In a computer, a subroutine which is applicable to a class of problems.

standard television signal—A signal which conforms to accepted specifications.

standard test conditions—The environmental conditions under which measurements should be made when disagreement of data obtained by various observers at different times and places may result from making measurements under other conditions.

standard test-tone power—One milliwatt (0 dBm) at 1000 hertz.

standard voltage generator — *See* Signal Generator.

standard volume indicator—A volume indicator with the characteristics prescribed by the American Standards Association.

standby — 1. The condition of equipment which will permit complete resumption of stable operation within a short period of time. 2. A duplicate set of equipment to be used if the primary unit becomes unusable because of malfunction.

standby battery—A storage battery held in reserve to serve as an emergency power source in event the regular power facilities at a radio station, hospital, etc., fail.

standby register — A register in which accepted or verified information can be stored to be available for a rerun in the event of a mistake in the program or a malfunction in the computer.

standby transmitter—A transmitter installed and maintained ready for use whenever the main transmitter is out of service.

standing current — The current present in a circuit in the absence of signal.

standing-on-nines carry — A system of executing the carry process in a computer. If a carry into a given place produces a carry from there, the incoming carry information is routed around that place.

standing wave—The distribution of current and voltage on a transmission line formed by two sets of waves traveling in opposite directions, and characterized by the presence of a number of points of successive maxima and minima in the distribution curves. Also called stationary wave.

standing waves—The behavior of air pressure waves in an enclosed room or box, giving rise to resonances which occur. They are created by the effects of multiple sound reflections between opposite walls, and cycle at frequencies determined by the distance between them. In effect, the room acts as a resonator.

standing-wave detector — *See* Standing-Wave Meter.

standing-wave indicator — *See* Standing-Wave Meter.

standing-wave loss factor — Ratio of the transmission loss in a waveguide when it is unmatched, compared with the loss when it is matched.

standing-wave meter—Also called a standing-wave indicator or detector. An instrument for measuring the standing-wave ratio in a transmission line. It may also include means for finding the location of maximum and minimum amplitudes.

standing-wave ratio—Abbreviated swr. The ratio of current (or voltage) at a loop (maximum) in the transmission line to the value at a (minimum) node. It is equal to the ratio of the characteristic impedance of the line to the impedance of the load connected to the output end of the line.

standing-wave-ratio bridge — Abbreviated swr bridge. A bridge for measuring the standing-wave ratio on a transmission line to check the impedance match.

standoff insulator — An insulator used to hold a wire or other radio component away from the structure on which it is mounted.

star chain—A group of navigational radio transmitting stations comprising a master station about which three or more slave stations are symmetrically located.

star-connected circuit—A polyphase circuit in which all current paths within the region that limits the circuit extend from each of the points of entry of the phase conductors to a common conductor (which may be the neutral conductor).

star connection—*See* Wye Connection.

stark effect — The splitting or shifting of spectral lines or energy levels due to an applied electric field.

starlight—A scene illumination of 1/10,000 of a footcandle.

star network — A set of three or more branches with one terminal of each connected at a common node.

star-quad cable — Four wires laid together and twisted as a group.

start dialing signal — A signal transmitted from the incoming end of a circuit, after receipt of a seizing signal, to indicate that the circuit conditions necessary for receiving the numerical routine information have been established.

start element — In certain serial transmissions, the initial element of a character, used for the purposes of synchronization. In Baudot teletypewriter operation, the start element is one space bit.

starter—1. An auxiliary electrode used to initiate conduction in a glow-discharge, cold-cathode tube. 2. Sometimes referred to as a trigger electrode. A control electrode, the principal function of which is to establish sufficient ionization to reduce the anode breakdown voltage in a gas tube. 3. An electric controller for accelerating a motor from rest to normal speed.

starter breakdown voltage—The voltage required to initiate conduction across the starter gap of a glow-discharge, cold-cathode tube, all other tube elements being held at cathode potential before breakdown.

starter gap—The conduction path between a starter and the other electrode to which the starting voltage is applied in a glow-discharge, cold-cathode tube.

starter voltage drop – The voltage drop across the starter gap after conduction is established there in a glow-discharge, cold-cathode tube.

starting anode—The anode that establishes the initial arc in a mercury-arc rectifier tube.

starting circuit breaker—A device the principal function of which is to connect a machine to its source of starting voltage.

starting current of an oscillator—The value of oscillator current at which self-sustaining oscillations will start under specified loading.

starting electrode—The electrode that establishes the cathode spot in a pool-cathode tube.

starting reactor – A reactor for decreasing the starting current of a machine or device.

starting torque – 1. Also called pull-in torque. The maximum load torque with which motors can start and come to synchronous speed. 2. The torque necessary to initiate motion of a system.

starting-to-running transition contactor—A device which operates to initiate or cause the automatic transfer of a machine from the starting to the running power connection.

starting voltage—The voltage necessary for a gaseous voltage regulator to become ionized or to start conducting. As soon as this happens, the voltage drops to the operating value.

start lead—Also called inside lead. The inner termination of a winding.

startover—A program function that causes an inactive computer to become active.

start-record signal – In facsimile transmission, the signal that starts the converting of the electrical signal to an image on the record sheet.

start signal—The signal that converts facsimile-transmission equipment from standby to active.

start-stop multivibrator – *See* Monostable Multivibrator *and* Flip-Flop Multivibrator.

start-stop-printing telegraphy—Printing telegraphy in which the signal-receiving mechanisms are started and stopped at the beginning and end of each transmitted character.

start-stop system—A system in which each group of code elements that represents an alphabetical signal is preceded by a start signal and followed by a stop signal. The start signal prepares the receiving mechanism to receive and register the character. The stop signal causes the receiving mechanism to come to rest in preparation for reception of the next character.

start-stop transmission – The method of transmission used in a start-stop system.

starved amplifier—An amplifier employing pentode tubes in which the screen voltage is set 10% below the plate voltage and the plate-load resistance is increased to 10 times the normal value. Thus, the amplification factor is greatly increased—often a stage gain of 2000 is achieved.

stat—A prefix used to identify electrostatic units in the cgs system. (*See also* Statampere, Statcoulomb, Statfarad, Stathenry, Statmho, Statohm, *and* Statvolt.)

statampere – The cgs electrostatic unit of current, equal to 3.3356×10^{-10} ampere (absolute).

statcoulomb—The cgs electrostatic unit of charge equal to 3.3356×10^{-10} coulomb (absolute).

state—1. The condition of a circuit, system, etc. 2. The condition at the output of a circuit that represents logic 0 or logic 1. 3. A condition or set of conditions considered together, especially one of the two normal sets of operating conditions of a gate or flip-flop.

state code – A coded indication of what state the CPU is in—responding to an interrupt, servicing a DMA request, executing an i/o instruction, etc.

statement – In computer programming, a meaningful expression or generalized instruction written in a source language.

state of charge—The condition of a storage cell or battery in terms of the remaining capacity.

statfarad—The cgs electrostatic unit of capacitance equal to 1.11263×10^{-12} farad (absolute).

stathenry—The cgs electrostatic unit of inductance equal to 8.98766×10^{11} henrys (absolute).

static—1. *See* Atmospherics. 2. A form of information storage in shift registers and memories whereby information will be retained as long as power is applied. 3. Capable of maintaining the same state indefinitely (with power applied) without any change of condition. Not requiring a continuous refreshing.

static behavior—The behavior of a control system or an individual unit under fixed conditions (as contrasted to dynamic behavior, under changing conditions).

static characteristic – The relationship between a pair of variables such as electrode voltage and electrode current, all other voltages being maintained constant. This relationship is usually represented by a graph.

static charge – The accumulated electric charge on an object.

static check—Of a computer, consists of one or more tests of computing elements, their interconnections, or both, performed under static conditions.

static control – A control system in which control functions are performed by solid-state devices.

static convergence – Convergence of the three electron beams at the center of the aperture mask in a color picture tube. The term "static" applies to the theoretical paths the beams would follow if no scanning forces were present.

static decay—In a storage tube, decay that is a function only of storage surface properties such as lateral or transverse leakage.

static detector – A device used to detect presence of static charges of electricity, which could cause explosions in hazardous atmospheres.

static device—As associated with electronic and other control or information-handling circuits, the term "static" refers to devices with switching functions that have no moving parts.

static dump—In a computer, a dump performed at a particular time with respect to a machine run, often at the end of the run.

static electricity – Stationary electricity—i.e., in the form of a charge in equilibrium, or considered independently of the effects of its motion.

static eliminator—A device for reducing atmospheric static interference in a radio receiver.

static error—An error that does not depend on the time-varying nature of a variable.

static field—A field that is present between the poles of either a permanent magnet or an electromagnet that has a direct current passing through its coils.

static focus—The focus attained when the electron beam is theoretically at rest or is at the position it would occupy if scanning energy were not applied.

static forward-current transfer ratio – In a transistor, the ratio, under specified test conditions, of the dc output current to the dc input current.

static input resistance—In a transistor, the ratio of the dc input voltage to the dc input current.

staticize – 1. In a computer, to perform a conversion of serial or time-dependent parallel data into a static form. 2. Occasionally, to retrieve an instruction and its operands from storage prior to executing the instruction.

staticizer—A storage device which is able to take information sequentially in time and put it out in parallel.

static machine – A machine for generating an electric charge, usually by induction.

static measurement—A measurement taken under conditions where neither the stimulus nor the environmental conditions fluctuate.

static memory – A type of semiconductor memory where the basic storage element can be set to either of two states, in which it will remain so long as the power stays on. *See also* dynamic memory.

static MOS array – A circuit made up of MOS devices which does not require a clock signal.

static power conversion equipment – Any equipment which converts electrical power from one form to another without the use of moving parts such as rotors or vibrators. "Static" implies the use of semiconductors.

static pressure – Also called hydrostatic pressure. The pressure that would exist at a certain point in a medium with no sound waves present. In acoustics, the commonly used unit is the microbar.

static printout—In a computer, a printout of data that is not one of the sequential operations and occurs after conclusion of the machine run.

static register—A computer register which retains its information in static form.

static regulator – A transmission regulator in which the adjusting mechanism is in self-equilibrium at any setting and control power must be applied to change the setting.

static sensitivity—In phototubes, the direct anode current divided by the incident radiant flux of constant value.

static shift register – A shift register in which logic flip-flops are used for storage. This technique, in integrated form, results in greater storage-cell size and consequently in shorter shift-register lengths. Its primary advantage is that information is retained as long as power is supplied to the device. A minimum clock rate is not required, and, in fact, the device can be unclocked.

static skew—A measure of the distance that the output from one track is ahead or behind (i.e., leading or lagging) the output of another track as a tape is transported over the read head.

static storage – In computers, storage in which the information does not change position (e.g., electrostatic storage, flip-flop storage, binary magnetic-core storage, etc.). The opposite of dynamic storage.

static subroutine—A digital-computer subroutine involving no parameters, other than the address of the operands.

static switch – A semiconductor switching device in which there are no moving parts.

static torque—*See* Locked-Rotor Torque.

static transconductance—In a transistor, the ratio of the dc output current to the dc input voltage.

station—1. One or more transmitters, receivers, and accessory equipment required to carry on a definite radio communication service. The station assumes the classification of the service in which it operates. 2. An input or output point in a communications system, such as the telephone set in a telephone system or the point at which a business machine interfaces the channel on a leased private line.

stationarity—The absence of variations with time in the spectral intensity and amplitude distribution of random noise.

stationary battery — A storage battery designed for service in a permanent location.

stationary contacts—Those members of contact pairs that are not moved directly by the actuating system.

stationary field—A constant field—i.e., one where the scalar (or vector) at any point does not change during the time interval under consideration.

stationary wave—*See* Standing Wave.

station battery—The electrical power source for signaling in telegraphy.

station break—1. A cue given by the station originating a program, to notify network stations that they may identify themselves to their audiences, broadcast local items, etc. 2. The actual time taken in (1) above.

statmho—The cgs electrostatic unit of conductance equal to 1.1126×10^{-12} mho (absolute).

statohm—The cgs electrostatic unit of resistance equal to 8.98766×10^{11} ohms (absolute).

stator—1. The nonrotating part of the magnetic structure in an induction motor. It usually contains the primary winding. 2. The stationary plates of a variable capacitor. 3. The conducting surfaces of a switch. Similar to the commutator in electrical rotating mechanisms.

stator of an induction watthour meter—A voltage circuit, one or more current circuits, and a magnetic circuit combined so that the reaction with currents induced in an individual, or a common, conducting disc exerts a driving torque on the rotor.

stator plates—The fixed plates of a variable capacitor.

status word register — A group of binary numbers which informs the user of the present condition of the microprocessor.

statvolt—The cgs electrostatic unit of voltage equal to 299.796 volts (absolute).

stave — One of the number of individual longitudinal elements which comprise a sonar transducer.

stay cord—A component of a cable, usually a high-tensile textile used to anchor the cable ends at their points of termination, to keep any pull on the cable from being transferred to the electrical connection.

steady state—A condition in which circuit values remain essentially constant, occurring after all initial transients or fluctuating conditions have settled down.

steady-state deviation — The difference between the final value assumed by a specified variable after the expiration of transients and its ideal value.

steady-state oscillation—Also called steady-state vibration. Oscillation in which the motion at each point is a periodic quantity.

steady-state regulation — Slow changes in the output voltage of a power supply following an input-voltage and/or load-impedance variation; it is usually expressed in volts (ΔV) or as a percentage of the nominal output.

steady-state vibration—*See* Steady-State Oscillation.

steatite—A ceramic consisting chiefly of a silicate of magnesium. Because of its excellent insulating properties—even at high frequencies—it is used extensively in insulators.

steer — To adjust by electrical means the polar response pattern of an antenna or the direction of current flow in a circuit.

steerable antenna — An antenna the major lobe of which can be readily shifted in direction.

Stefan-Boltzmann constant — The constant of proportionality between power radiated by a blackbody and temperature: 5.672×10^{-5} ergs per cm^2 per T^4 per sec.

Stefan-Boltzmann law — The total emitted radiant energy per unit of a blackbody is proportionate to the fourth power of its absolute temperature.

Steinmetz coefficent — A factor by which the 1.6th power of the magnetic flux density must be multiplied to give the approximate hysteresis loss of an iron or steel sample in ergs per cubic centimeter per cycle when that sample is undergoing successive magnetization cycles having the same maximum flux density.

stenode circuit—A superheterodyne receiving circuit in which a piezoelectric unit is used in the intermediate-frequency amplifier to balance out all frequencies except signals at the crystal frequency, thereby giving very high selectivity.

step and repeat—A method of dimensionally positioning multiples of the same or intermixed functional patterns on a given area of a photoplate or a film by repetitions or contact or projection printing of a single original pattern of each type.

step-and-repeat technique — A mechanical technique that provides for linear indexing of a movable platform carrying a wafer or photographic plate. Applications of this technique are in the testing of a device wafer, and in the masking operation that is part of the process of fabricating microelectronic devices.

step-and-repeat printers – Projection printers capable of reproducing a multiplicity of images from a master transparency on a single support coated with a photosensitive layer by indexing the receiving material from position to position. Used in microcircuit production.

step-by-step automatic telephone system– A switching system that employs successive step-by-step selector switches that are actuated by current impulses from a telephone dial.

step-by-step switch – A bank-and-wiper switch in which the wipers are moved by individual electromagnetic ratchets.

step calibration–Also called interval calibration. Often confused with sense step, that is, the application of a calibration resistor to produce a deliberate electrical unbalance.

step counter – In a computer, a counter used in the arithmetical unit to count the steps in multiplication, division, and shift operation.

step-down transformer–A transformer in which the voltage is reduced as the energy is transferred from its primary to its secondary winding.

step function–A signal characterized by instantaneous changes between amplitude levels. The term usually refers to a rectangular-front waveform used for making tests of transient response.

step-function response–*See* Transient Response, 2.

step generator–A device for testing the linearity of an amplifier. A step wave is applied to the amplifier input and the step waveform observed, on an oscilloscope, at the output.

step input–A suddent but sustained change in an input signal.

step load change–An instantaneous change in the magnitude of the load current.

stepped oxide–A technique of forming the SiO_2 layer of each electrode in two thicknesses so that a two-level potential well can be formed with one voltage. (MIS technology term.)

stepper motor–1. A motor the normal operation of which consists of discrete angular motions of essentially uniform magnitude, rather than continuous rotation. 2. A digital device which converts electrical pulses into proportionate mechanical movement. Each revolution of the motor shaft is made in a series of discrete identical steps. The design of the motor usually provides for clockwise and/or counterclockwise rotation. (Thus, the stepper is ideally suited for many positional and control applications.) 3. Electromechanical prime mover which rotates through fixed angles in response to applied pulses. The motor accordingly permits use of digital signals to control mechanical motion or position. In addition, the high holding torque associated with each step permits a stepping motor to replace devices such as brakes and clutches, with a gain in system reliability.

stepping–*See* Zoning, 1.

stepping relay–Also called rotary stepping switch (or relay) or stepping switch. A multiposition relay in which moving wiper contacts mate with successive sets of fixed contacts in a series of steps, moving from one step to the next in successive operations of the relay.

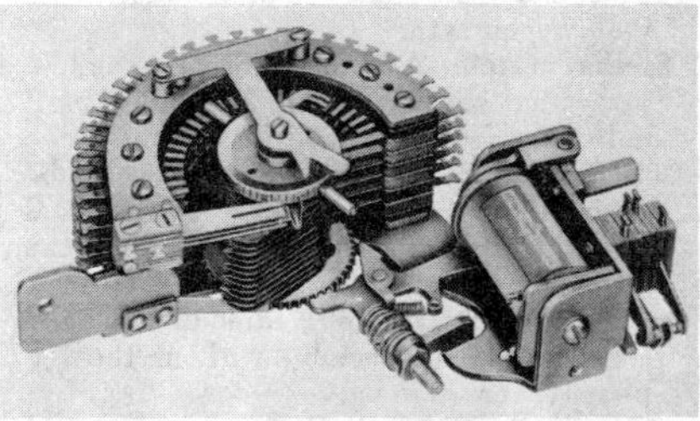

Stepping relay.

stepping switch–*See* Stepping Relay.

step-recovery diode – 1. A varactor in which forward voltage injects carriers across the junction, but before the carriers can combine, the voltage reverses and carriers return to their origin in a group. The result is the abrupt cessation of reverse current and a harmonic-rich waveform. 2. A special form of pn junction in which the charge storage and switching characteristics are optimized for use in microwave frequency multipliers and comb generators. The device stores charge while in forward conduction and a large reverse bias current can be obtained until all the charge is removed. The device impedance then goes from a low value to a very high one in transition times as low as 50 picoseconds. Also called snap varactor.

step-servo motor – A device that, when properly energized by dc voltage, indexes in definite angular increments.

step-strobe marker – A form of strobe marker in which the discontinuity is in the form of a step in the time base.

step-up transformer – A transformer in which the voltage is increased as the energy is transferred from the primary to the secondary winding.

step voltage–The potential difference between two points on the earth's surface separated by a distance of one pace, or about 3 feet, in the direction of maximum potential gradient.

step-voltage regulator–A device consisting of a regulating transformer and a means for adjusting the voltage or the phase relation of the system circuit in steps, usually without interrupting the load.

steradian—1. A solid spherical angle which encloses a surface equal, on a sphere, to the square of the radius of the sphere. 2. The unit solid angle subtended at the center of a sphere by an area on its surface equivalent to the square of the radius. The unit of solid angular measurement often used in problems of illumination.

Sterba antenna—A series-fed array of adjacent, broadside-firing, transposed square loops that have half-wave sides and are spaced a distance of approximately one-half wavelength.

Sterba curtain—A stacked dipole antenna array that consists of one or more phased half-wave sections and a quarter-wave section at each end. The array can be oriented for either vertical or horizontal radiation and can be fed at either the center or the end.

stereo—A prefix meaning three-dimensional—specifically (especially without the hyphen), stereophonic.

stereo adapter—Also called a stereo control unit. A device used with two sets of monophonic equipment to make them act as a single stereo system.

stereo amplifier—An audio-frequency amplifier, with two or more channels, for a stereo sound system.

stereo broadcasting—*See* Stereocasting.

stereo cartridge—A phonograph pickup for reproduction of stereophonic recordings. Its high-compliance needle is coupled to two independent voltage-producing elements.

stereocasting—Also called stereo broadcasting. Broadcasting over two sound channels to provide stereo reproduction. This may be done by simulcasting, multicasting, or multiplexing.

stereocephaloid microphone—Two or more microphones arranged to simulate the acoustical patterns of human hearing.

stereo control unit—*See* Stereo Adapter.

stereo microphones—Two or more microphones spaced as required for stereo recording.

stereophonic—1. Designating a sound reproduction system in which sound is delivered to the listener through at least two channels, creating the illusion of depth and of locality of source. 2. A two-channel recording and reproduction system more popularly referred to as "stereo." At the recording studio separate microphones are used for each recorded channel. The correct reproduction of stereo signals in the home gives to the listener a sense of direction of sound and thus the realism. 3. A multiple-channel sound system or recording in which each channel carries a unique version of the total original performance. When the channels are blended, acoustically, they recreate the breadth and depth of the original, adding a new dimension to reproduced sound. At least two channels are required for playback, although more than two may be used in recording.

stereophonic reception—Reception involving the use of two receivers having a phase difference in their reproduced sounds. The sense of depth given to the received program is analogous to the listener's being in the same room as the orchestra or other medium.

stereophonic separation—The ratio of the electrical signal caused in the right (or left) stereophonic channel to the electrical signal caused in the left (or right) stereophonic channel by the transmission of only a right (or left) signal.

stereophonic sound system—A sound system with two or more microphones, transmission channels, and speakers arranged to give depth to the reproduced sound.

stereophonic subcarrier—A subcarrier having a frequency which is the second harmonic of the pilot subcarrier frequency and which is employed in fm stereophonic broadcasting.

stereophonic subchannel—The band of frequencies from 23 to 53 kilohertz containing the stereophonic subcarrier and its associated sidebands.

stereophonic system—A sound-reproducing system in which a plurality of microphones, transmission channels, and speakers (or earphones) are arranged to afford a listener a sense of the spatial distribution of the sound sources.

stereo pickup—A phonograph pickup used with single-groove, two-channel stereo records.

stereo recording—The impressing of signals from two channels onto a tape or disc in such a way that the channels are heard separately on playback. The result is a directional, three-dimensional effect.

stereoscopic television—A system of television broadcasting in which the images appear to be three-dimensional.

stereosonic system—A recording technique using two closely spaced directional microphones with their maximum directions of reception 45° from each other. In this way, one picks up sound largely from the right and the other from the left, similar to mid-side recording.

stick circuit—A circuit used to maintain energization of a relay or similar unit through its own contacts.

stickiness—The condition of physical interference with the operation of the moving part of an electrical indicating instrument.

sticking—In computers, the tendency of a flip-flop to remain in, or to spontaneously switch to, one of its two stable states.

stiff – A voltage source whose value is largely independent of the current drawn having a relatively low impedance.

stiffness factor—The angular lag between the input and output of a servosystem.

stilb (sb)—The unit of luminance (photometric brightness) equal to one candela per square centimeter.

still—Photographic or other stationary illustrative material used in a television broadcast.

stimularity—An arbitrary measure of sensitivity to stimulation. It is proportional to the quantum efficiency relative to incident radiation.

stimulated emission—The emission of radiation by a system going from an excited electron energy level to a lower energy level under the influence of a radiation field. The emitted radiation is in phase with the stimulating radiation and produces a negative-absorption condition.

stimulus—*See* Excitation, 1 *and* Measurand.

stirring effect—The circulation in a molten conductive charge due to the combined motor and pinch effects.

stitch bonding—A bonding technique where wire is fed through a capillary tube. A bent section of the wire is bonded to the contact area by the capillary. The capillary is removed and a cutter severs the wire, forming a new bend for the next bonding operation.

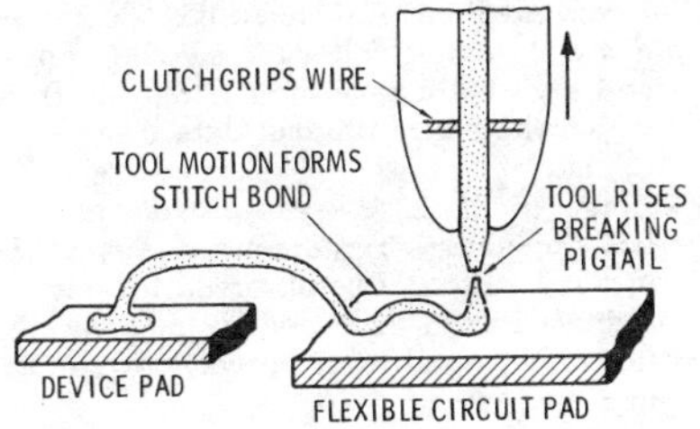

Stitch bonding.

stitching—The process of welding thermoplastic materials by successive applications of two small electrodes that are connected to the output of a radio-frequency generator; a mechanism similar to that of a conventional sewing machine is used.

stitch wire—A semiautomatic system of point-to-point interconnections in which gold-plated steel pins are pressed into holes in conventional printed-circuit boards. Teflon-insulated 30 AWG nickel wire is bonded to the pins. Electronic components are then soldered to terminal projections on the opposite side of the board.

STL—Abbreviation for Schottky transistor logic.

stochastic – The characteristic of events changing the probabilities of vairous responses.

stochastic process—A random process.

stock reel—Also called supply reel or storage reel. On a tape recorder, the reel from which unrecorded or unplayed tape is taken as the machine records on or plays it.

stoichiometric impurity—A crystalline imperfection caused in a semiconductor by a deviation from the stoichiometric composition.

stopband—That part of the frequency spectrum that is subjected to specified attenuation of signal strength by a filter. (The part of the spectrum between the passband and the stopband is called the transition region.)

stop-cycle timer – A timer which runs through a single cycle and then stops until the starting signal is reinitiated.

stop element—In certain types of serial transmission, the last element of a character used to ensure that the next start element will be recognized. Also called stop signal.

stop instruction—A machine operation or routine that requires some manual action other than operation of the start key to continue processing.

stop opening—In a camera, the size of the aperture that controls the amount of light passing through the lens.

stopping potential—The voltage required to stop the outward movement of electrons emitted by photoelectric or thermionic action.

stop-record signal—A facsimile signal used for stopping the conversion of the electric signal into an image on the record sheet.

stop signal—The signal that transfers facsimile equipment from active to standby.

storage—1. The act of storing information (*See also* Store, 1 and 2.) 2. A computer section used primarily for storing information in electrostatic, ferroelectric, magnetic, acoustic, optical, chemical, electronic, electrical, mechanical, etc., form. Such a section is sometimes called a memory, or a store, in British terminology. 3. In an oscilloscope, the ability to retain the image of an electrical event on the cathode-ray tube (crt) for further analysis after that event ceases to exist. This image retention may be for only a few seconds with variable persistence storage, or it may be for hours with bistable storage.

storage access time—In a computer, the time required to transfer information from a storage location to the local storage register or other location, where the information then becomes available for processing.

storage allocation—The assignment of specific sections of computer memory to blocks of data or instructions.

storage battery—Two or more storage cells connected in series and used as a unit.

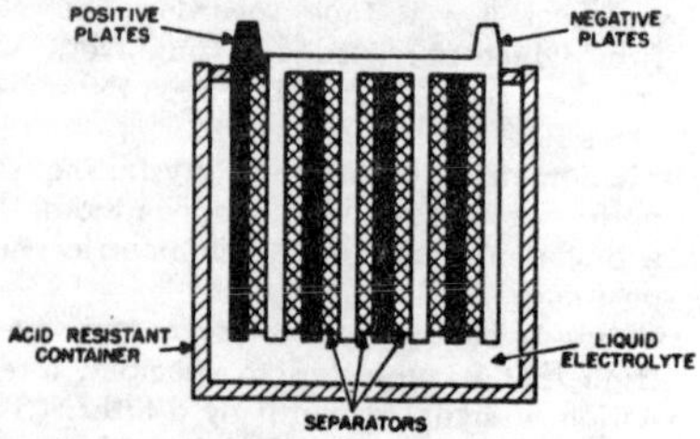

Storage battery.

storage capacity—The amount of information that a storage (memory) device can retain. It is often expressed as the number of words the device can retain (given the number of digits and the base of the standard word). In the case of comparisons among devices that use different bases and word lengths, the capacity is customarily expressed in bits.

storage cell—1. Also called a secondary cell. A cell which, after being discharged, can be recharged by sending an electric current through it in the opposite direction from the discharging current. 2. An elementary unit of storage (e.g., binary cell, decimal cell).

storage counter—A counter in which a series of current pulses charge a capacitor with each pulse raising the voltage to a higher level. A comparator circuit determines when the capacitor voltage reaches a predetermined level. Special techniques are frequently used to linearize the charging curve of the capacitor.

storage cycle—1. The periodic sequence of events that occur when information is transferred to or from a computer storage device. 2. Storing, sensing, and regeneration from parts of the storage sequence.

storage device—A device into which data can be inserted, in which it can be retained, and from which it can be retrieved.

storage element—1. An area which retains information distinguishable from that of adjacent areas on the storage area of charge-storage tubes. 2. The smallest part of a digital-computer storage facility used for storage of a single bit.

storage integrator—In an analog computer, an integrator used to store a voltage in the hold condition for future use while the rest of the computer assumes another computer control state.

storage key—In a computer, an indicator that is associated with a storage block or blocks and that requires tasks to have a matching protection key in order to use the blocks.

storage laser—Any laser which stores unusually high energy prior to discharge. For example, a storage diode laser is a laser in which some carriers are electrically excited for a time longer than the lasing period.

storage life—The minimum length of time over which a device, system, or transducer can be exposed to specified environmental storage conditions without changing the performance beyond a specified tolerance.

storage location—A computer storage position that holds one machine word and usually has a specific address.

storage medium—Any recording device or medium into which data can be copied and held until some later date, and from which the entire original data can be obtained.

storage oscilloscope — An instrument that has the ability to store a crt display in order that it may be observed for any required time. This stored display may be instantly erased to make way for storage of a later event.

storage print—In computers, a utility program that causes the requested core im-

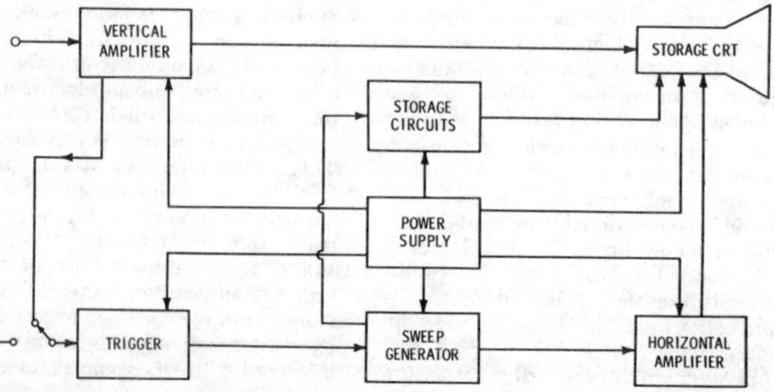

Storage oscilloscope.

age, core memory, or drum locations to be recorded in either absolute or symbolic form on either the line printer or the delayed printer tape.

storage protection—In a computer, an arrangement by which access to storage is prevented for reading, writing, or both. Also called memory protection.

storage surface—The part of an electrostatic storage tube on which storage of information takes place.

storage temperature—The temperature of the medium, immediately adjacent to the device, at which the device, without power applied, may be stored indefinitely without deterioration.

storage time—1. The time during which the output current or voltage of a pulse is failing from maximum to zero after the input current or voltage has been removed. 2. An increase in the time needed to turn off a transistor that has been driven into saturation. It results from the fact that a transistor in heavy conduction has many excess charge carriers moving in the collector region. When the base signal is changed to the cutoff level, collector current continues until all the excess charge carriers have been removed from the collector region.

storage tube—1. An electron tube into which information can be stored and later read out, usually a cathode ray tube with a storage screen which will retain charges impressed on it and which can control an electron beam in some way, allowing changes to be read out. 2. A crt that stores images on a separate storage screen behind the viewing screen in the tube. Images then remain on the viewing screen until the storage screen is erased. Since a storage tube does not have to be refreshed, it can display an extremely large amount of data without flicker.

store—1. To retain information in a device from which the information can later be withdrawn. 2. To introduce information into the device in (1) above. 3. A British synonym for storage. (*See* Storage, 2).

store-and-forward — Process of message handling used in a message-switching system.

stored-energy welding—A method of welding in which electric energy is accumulated (stored) electrostatically, electromagnetically, or electrochemically at a relatively slow rate and is then released at the required rate for welding.

stored program—A set of instructions in the computer memory specifying the operations to be performed and the location of the data on which these operations are to be performed.

stored-program computer — 1. Also called general-purpose computer. A computer in which the instructions specifying the program to be performed are stored in the memory section along with the data to be operated on. 2. A digital computer which, under control of its own instructions, can synthesize, alter, and store instructions as though they were data and can subsequently execute these new instructions.

stored program logic—A program stored in a memory unit containing logical commands to the remainder of the memory so that the same processes are performed on all problems.

stored routine—In computers, a series of stored instructions for directing the step-by-step operation of the machine.

store transmission bridge—A transmission bridge which consists of four identical impedance coils (the two windings of the back-bridge relay and the live relay of a connector, respectively) separated by two capacitors. It couples the calling and called telephones together electrostatically for the transmission of voice-frequency (alternating) currents, but separates the two lines for the transmission of direct current for talking purposes (talking current).

storm loading—The mechanical loading imposed on the components of a pole line by wind, ice, etc., and by the weight of the components themselves.

straight dipole—A half-wave antenna consisting of one conductor, usually center-fed.

straightforward circuit—A circuit in which signaling is performed automatically and in one direction.

straightforward trunking — In a manual telephone switchboard system, that method of operation in which one operator gives the order to another operator over the trunk that later carries the conversation.

straight-line capacitance — The variable-capacitor characteristic obtained when the rotor plates are shaped so that the capacitance varies directly with the angle of rotation.

straight-line code—The repetition of a sequence of instructions, with or without address modification, by explicitly writing the instructions for each repetition. Generally straight-line coding will require less execution time and more space than equivalent loop coding. If the number of repetitions is large, this type of coding is tedious unless a generator is used. The feasibility of straight-line coding is limited by the space required as well as the difficulty of coding a variable number of repetitions.

straight-line frequency — The variable-capacitor characteristics obtained when the rotor plates are shaped so that the resonant frequency of the tuned circuit containing the capacitor varies directly with the angle of rotation.

straight-line wavelength – The variable-capacitor characteristic obtained when the rotor plates are shaped so that the wavelength of resonance in the tuned circuit containing the capacitor varies directly with the angle of rotation.

strain–The physical deformation, deflection, or change in length resulting from stress (force per unit area). The magnitude of strain is normally expressed in microinches per inch.

strain anisotropy–A force that directs the magnetization of a particle along a preferred direction relative to the strain.

strain gage–1. A resistive transducer the electrical output of which is proportional to the amount it is deformed under strain. 2. A measuring element for converting force, pressure, tension, etc., into an electrical signal. 3. A device for measuring the expansion or contraction of an object under stress, comprising wires that change resistance with expansion or contraction. *See also* Load Cell.

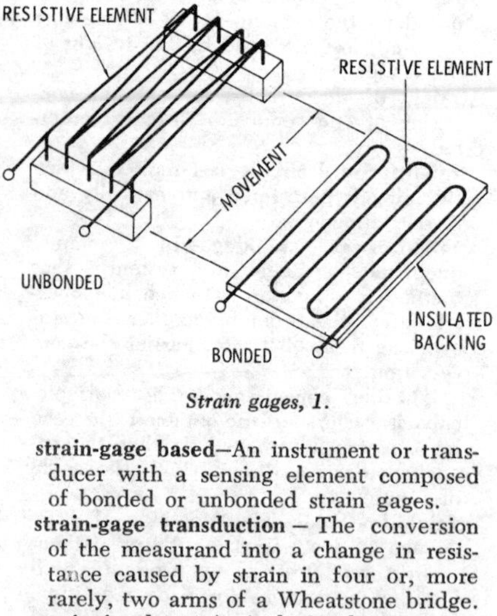

Strain gages, 1.

strain-gage based–An instrument or transducer with a sensing element composed of bonded or unbonded strain gages.

strain-gage transduction – The conversion of the measurand into a change in resistance caused by strain in four or, more rarely, two arms of a Wheatstone bridge.

strain insulator–A single insulator, an insulator string, or two or more strings in parallel designed to transmit the entire pull of the conductor to, and insulate the conductor from, the tower or other support.

strain pickup–A phonograph pickup cartridge using the principle of the strain gage.

strain wire–Wire having a composition such that it exhibits favorable strain-gage performance. Also the strain-sensitive filaments which constitute the transfer electrical elements in certain transducers.

strand–One of the wires or groups of wires of any stranded conductor.

stranded conductor–*See* Stranded Wire.

stranded wire–Also called stranded conductor. A conductor composed of a group of wires or any combination of groups of wires. The wires in a stranded conductor are usually twisted or braided together.

strap–A wire or strip connected between the ends of the segments in the anode of a cavity magnetron to promote operation in the desired mode.

strapping–In a multicavity magnetron, the connecting together of resonator segments that have the same polarity, so that undesired modes of oscillation are suppressed.

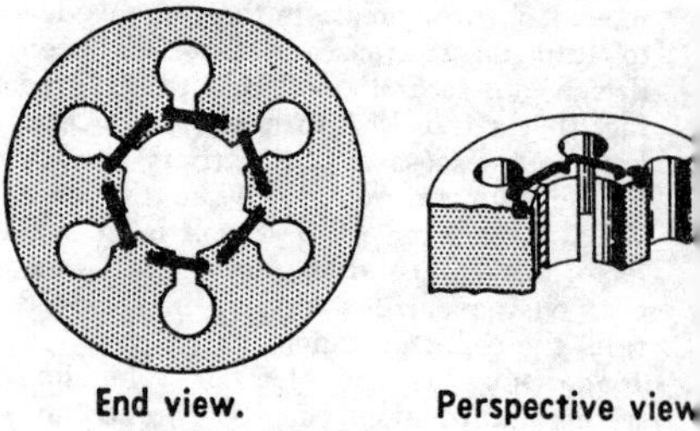

Straps on cavity-magnetron anode.

stratosphere–A calm region of the upper atmosphere characterized by little or no temperature change throughout. It is separated from the lower atmosphere (troposphere) by a region called the tropopause.

stray capacitance–The capacitance introduced into a circuit by the leads and wires connecting the circuit components.

stray current–A portion of the total current that flows over paths other than the intended circuit.

stray-current corrosion–Corrosion that results when a direct current from a battery or other external source causes a metal in contact with an electrolyte to become anodic with respect to another metal in contact with the same electrolyte. Accelerated corrosion will occur at the electrode where the current direction is from the metal to the electrolyte and will generally be in proportion to the total current.

stray field–The leakage magnetic flux that spreads outward from an inductor and does no useful work.

strays–*See* Atmospherics.

streaking – Distortion in which televised objects appear stretched horizontally beyond their normal boundaries. It is most apparent at the vertical edges, where there is a large transition from black to white or white to black, and is usually expressed as short, medium, or long streaking. Long streaking may extend as far as the right edge of the picture, and in extreme low-frequency distortion, even over a whole line interval.

stream deflection amplifier—A fluidic device which utilizes one or more control streams to deflect a power stream, altering the output.

streamer breakdown — Breakdown caused by an increase in the field due to the accumulation of positive ions produced during electron avalanches.

streaming—The production of a unidirectional flow of currents in a medium where sound waves are present.

strength of a simple sound source—The rms magnitude of the total air flow at the surface of a simple source in cubic meters per second (or cubic centimeters per second), where a simple source is taken to be a spherical source the radius of which is small compared with one-sixth wavelength.

strength of a sound source—The maximum instantaneous rate of volume displacement produced by the source when emitting a sinusoidal wave.

stress — The force producing strain in a solid.

stretched display — A ppi display having the polar plot expanded in one rectangular dimension. The equal-range circles of the normal ppi display become ellipses.

striation technique — Rendering sound waves visible by using their individual ability to refract light waves.

striking an arc—Starting an electric arc by touching two electrodes together momentarily.

striking distance—The effective separation of two conductors having an insulating fluid between them.

striking potential—1. The voltage required to start an electric arc. 2. The lowest grid-to-cathode potential at which plate current begins flowing in a gas-filled triode.

string—1. In a list of items, a group of items that are already in sequence according to a rule. 2. A connected sequence of entities, such as characters, in a command string.

string electrometer—An electrostatic voltage-measuring instrument consisting of a conducting fiber stretched midway between and parallel to two conducting plates. The electrostatic field between the plates displaces the fiber laterally in proportion to the voltage between the plates.

string-shadow instrument — An instrument in which the indicating means is the shadow (projected or viewed through an optical system) of a filamentary conductor the position of which in a magnetic or an electric field depends on the measured quantity.

strip—To remove insulation from a wire or cable.

strip chart recorder — An instrument for recording variations in the measurement of a quantity by time, using a moving pen on a long strip of paper.

strip contacts—Formed contacts in a continuous length, or strip, for use in an automatic installation machine.

striping—In flowcharting, the use of a line across the upper portion of a symbol to indicate the presence of a detailed representation elsewhere in the same set of flow-charts.

stripline — Layout and interconnection method used for circuits that operate at very high frequencies, usually above 1 gigahertz. Also called microstrip.

stripper—A hand-operated or motor-driven tool for removing insulation from wires.

strip transmission line—A microwave transmission line in the form of a thin, narrow rectangular strip adjacent to a wide ground-plane conductor or between two wide ground-plane conductors. Separation of the conductors usually is achieved by using a low-loss dielectric material on which the conductors are formed by etching.

strobe—1. An intensified spot in the sweep of a deflection-type indicator, used as a reference mark for ranging or expanding the presentation. 2. An intensified sweep on a plan-position indicator or B-scope. Such a strobe may result from certain types of interference or it may be purposely applied as a bearing or heading marker. 3. On a console oscilloscope, a line representing the azimuth data generated by a jammed radar site. 4. *See* Electronic Flash. 5. An input to a counter or register that permits the asynchronous entry of parallel data.

strobe hold time—The time necessary to strobe parallel data into a counter or register completely.

strobe marker—A small bright spot, a short gap, or other discontinuity produced on the line trace of a radar display to indicate the part of the time base which is receiving attention.

strobe pulse—1. Also called sample pulse. A pulse used to gate the output of a core-memory sense amplifier into a trigger in a register. 2. A pulse of a duration less than the time period of a recurrent phenomenon used for making a close investigation of that phenomenon. The frequency of the strobe pulse bears a simple relation to that of the phenomenon, and the relative timing is usually adjustable. 3. A pulse which enables a system for a fixed period only.

strobe release time—The time that must elapse after the strobe input is disabled before a clocking transition will be recognized by a counter.

Strobolume—A trade name used to describe a high-intensity electronic stroboscope manufactured by General Radio Company.

stroboscope—1. A light that flashes at a frequency that can be adjusted to coincide with any repeating motion. When the flashing rate is synchronized with the motion, the moving device appears to be stopped. Since the flashing rate is known, the speed of the device can be determined without physically contacting the device. A flashing rate slightly different from the synchronized rate makes the device appear to operate in slow motion. With the stroboscope, such characteristics of the motion as whip, vibration, or chatter can be observed readily. 2. A device for producing a flickering light of controlled frequency, used to observe movements, vibrations, rotation.

stroboscopic disc—A printed disc having several rings, each with a different number of dark segments. The pattern is placed on a rotating phonograph turntable and illuminated at a known frequency by a flashing discharge tube. The speed can then readily be determined by noting which pattern appears to stand still or rotate the slowest.

Stroboscopic disc.

stroboscopic tachometer — A stroboscope with a scale calibrated in flashes or in revolutions per minute. The stroboscopic lamp is directed onto the rotating device being measured, and the flashing rate is adjusted until the device appears to be standing still. The speed can then be read directly from the scale.

strobotron—A type of glow lamp that produces intense flashes of light when fed with accurately timed voltage pulses. It is used in electronic stroboscopes for visual inspection of high-speed moving parts.

stroke—In character recognition, a straight line segment or arc that forms a part of a graphic character.

stroke centerline—In character recognition, a line equidistant from the two stroke edges.

stroke edge—In character recognition, the region of discontinuity between a side of a stroke and the background, determined by averaging, over the length of the stroke, the irregularities produced by the printing and detection processes.

stroke speed—Also called scanning-line frequency or scanning frequency. The number of times per minute that a fixed line, perpendicular to the direction of scanning, is crossed in one direction by a scanning or recording spot in a facsimile system.

stroke width—In character recognition, the distance, measured in the direction perpendicular to the stroke centerline, between the two edges of a stroke.

structureborne noise—Undesired vibration in, or of, a solid body.

stub—1. A short length of transmission line or cable joined as a branch to another transmission line or cable. 2. A short section of transmission line, open or shorted at the far end, connected in parallel with a transmission line to match the impedance of the line to that of an antenna or transmitter.

stub angle—A right-angle elbow for a coaxial rf transmission line, the inner conductor being supported by a quarter-wave stub.

stub cable—A short branch from a principal cable. The end is often sealed until used. Pairs in the stub are referred to as stubbed-out pairs.

stub matching—Using a stub to match a transmission line to an antenna or load.

stub-supported coaxial—A coaxial cable the inner conductor of which is supported by short-circuited coaxial stubs.

stub tuner—A stub terminated by movable short-circuiting means and used for matching impedance in the line to which it is joined as a branch.

stunt box—A device for controlling the nonprinting functions of a teletype terminal.

stutter—In facsimile, a series of undesired black and white lines sometimes produced when the signal amplitude changes sharply.

stylus—1. Also called a needle. The needle-like object used in a sound recorder to cut or emboss the record grooves. Generally it is made of sapphire, stellite, or steel. The plural is styli. 2. The pointed element that contacts the record sheet in a facsimile recorder. 3. A small piece of industrial grade diamond or artificial sapphire, conically shaped, which tracks the groove in a phonograph record.

stylus alignment—The position of the stylus with respect to the record. The correct position is perpendicular.

stylus drag—Also called needle drag. The

friction between the reproducing stylus and the surface of the recording medium.

stylus force – Also called vertical stylus force, tracking force, and formerly called needle pressure or stylus pressure. The downward force, in grams or ounces, exerted on the disc by the reproducing stylus.

stylus oscillograph – An instrument in which a pen or stylus records, on paper or another suitable medium, the value of an electrical quantity as a function of time.

stylus pressure–*See* Stylus Force.

stylus radius–The radius in mils of the spherical tip that contacts the groove wall of a phonograph record. In an elliptical stylus there are two radii. The smaller radius applies to the sides of the stylus looking down on it, and the larger radius applies to the curvature that contacts the V groove, looking along the groove.

subassembly–Parts and components combined into a unit for convenience in assembling or servicing. A subassembly is only part of an operating unit; it is not complete in itself.

subatomic–Smaller than atoms–i.e., electrons or protons.

subatomic particles – The particles that make up the atom–i.e., protons, electrons, and neutrons.

subcarrier–1. A carrier used to generate a modulated wave which is applied, in turn, as a modulating wave to modulate another carrier. 2. The carrier used in stereo broadcasting to accommodate the subchannel stereo components. Frequency is 38 kHz and is suppressed at the transmitter leaving only the sidebands, but is re-formed at the receiver for detection of the stereo components by doubling the synchronized 19-kHz pilot tone.

subcarrier band–A band associated with a given subcarrier and specified in terms of maximum subcarrier deviation.

subcarrier channel–The channel required to convey the telemetric information of a subcarrier band.

subcarrier discriminator tuning unit – A device which tunes the discriminator to a particular subcarrier.

subcarrier frequency shift–The use of an audio-frequency shift signal to modulate a radio transmitter.

subcarrier oscillator–1. In a telemetry system, the oscillator which is directly modulated by the measurand or its equivalent in terms of changes in the transfer elements of a transducer. 2. In a color television receiver, the crystal oscillator operating at the chrominance subcarrier frequency of 3.58 MHz.

subcarrier transmission – A subdivision of carrier transmission designed to increase the capacity of the telephone system; used in other circumstances in a similar fashion to indicate a transmission modulated upon a carrier or subcarrier of a communication channel.

subchannel – In a telemetry system, the route for conveying the magnitude of one subcommutated measurand.

subchassis–The chassis on which closely associated components such as those of an amplifier or power supply are mounted. A subchassis is a building block, easily changed and usable in a variety of systems.

subclutter visibility – The characteristic that relates to how well a radar equipped with a moving-target indicator can see through clutter.

subcommutation–In a computer, the act of connecting one data source to a sampled data system less frequently than other data sources.

subcommutation frame–In pcm systems, a recurring integral number of subcommutator words which include a single subcommutation frame synchronization word. The number of subcommutator words in a subcommutation frame is equal to an integral number of primary commutator frames. The length of a subcommutation frame is equal to the total number of words or bits generated as a direct output of the subcommutator.

subcycle generator–A frequency-reducing device used in telephone equipment to furnish ringing power at a submultiple of the power-supply frequency.

subdivided capacitor–A capacitor in which several capacitors known as sections are mounted so that they may be used individually or in combination.

subelement–A distinguishable portion of a circuit element (e.g., the emitter, collector, and base are subelements of an integrated bipolar transistor).

subframe–A complete sequence of frames during which all subchannels of a specific channel are sampled once.

subharmonic – A sinusoidal quantity the frequency of which is an integral submultiple of the fundamental frequency of its related periodic quantity. A wave with half the frequency of the fundamental of another wave is called the second subharmonic of that wave; one with a third of the fundamental frequency is called a third subharmonic, etc.

subjamming visibility–The characteristic that relates to the ability of a particular radar antijam technique to see through jamming signals.

subject copy–Also called copy. In facsimile, the material in graphic form to be transmitted for reproduction by the recorder.

submarine cable – A cable designed for

service underwater; usually a lead-covered cable with a steel armor applied between layers of jute.

submerged-resistor induction furnace – A furnace for melting metal. It comprises a melting hearth, a descending melting channel closed through the hearth, a primary induction winding, and a magnetic core which links the melting channel and primary winding.

submersible transformer – A transformer so constructed that it will operate when submerged in water under predetermined conditions of pressure and time—e.g., a subway transformer.

subminiature tube—A small electron tube used generally in miniaturized equipment.

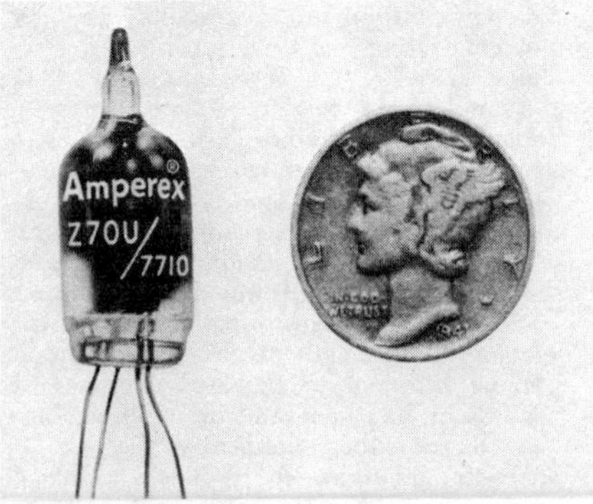

Subminiature tube.

subminiaturization—The technique of packaging discrete miniaturized parts, using assembly techniques which result in increased volumetric efficiency (e.g., a hybrid circuit).

submultiple resonance—Resonance at a frequency that is a submultiple of the frequency of the exciting impulses.

subnanosecond—Less than a nanosecond.

subpanel—An assembly of electrical devices connected together which forms a simple functional unit in itself.

subprogram—An independently compilable part of a larger computer program.

subrefraction—Atmospheric refraction that is less than standard refraction.

subroutine—1. In computer technology, the portion of a routine that causes a computer to carry out a well-defined mathematical or logical operation. 2. Usually called a closed subroutine. One to which control may be transferred from a master routine, and returned to the master routine at the conclusion of the subroutine. 3. One step of a computer program, or routine. A side function in which the computer carries out a mathematical or logical portion of a complete routine. 4. A set of instructions and constants necessary to direct a computer to carry out a well-defined mathematical or logical operation and so constructed that it can be incorporated with another routine to provide a useful result with little effort. 5. A subprogram (group of instructions) reached from more than one place in a main program. The process of passing control from the main program to a subroutine is a subroutine call, and the mechanism is a subroutine linkage. Often data or data addresses are made available by the main program to the subroutine. The process of returning control from subroutine to main program is subroutine return. The linkage automatically returns control to the original position in the main program or to another subroutine. *See also* Nesting.

subroutine call—*See* Subroutine.

Subroutine linkage—*See* Subroutine.

subroutine re-entry—In a computer, initiation of a subroutine by a program before the subroutine has completed its response to another program that called for it. This can occur when a control program is subjected to a priority interrupt.

subroutine return—*See* Subroutine.

subscriber line—A line that connects a central office and a telephone station, private branch exchange, or other end equipment.

subscriber multiple – A bank of manual-switchboard jacks that provides outgoing access to subscriber lines and usually has more than one appearance across the switchboard.

subscriber's drop – The line from a telephone cable to a subscriber's building.

subscriber set—Also called a customer set. An assembly of apparatus for originating or receiving calls on the premises of a subscriber to a communication or signaling service.

subscript—A notation for use in a computer to specify one member of an array where each member is referenced only in terms of the array name.

subscripted variable—A variable with one or more following subscripts enclosed in parentheses.

subscription television—*See* Pay Television.

subscription television broadcast program —A television broadcast program intended to be received in intelligible form by members of the public only for a fee or charge.

subset—1. In a telephone system, the handset or deskset at the station location. 2. Also known as a modem, data set, or subscriber set. A modulation/demodulation device designed to make business-machine signals compatible with communications facilities.

Subsidiary Communications Authorization —*See* SCA.

subsidiary conduit—The terminating branch

of an underground conduit run extending from a manhole or handhole to a nearby building, handhole, or pole.

subsonic frequency – *See* Infrasonic Frequency.

subsonic speed – A speed less than the speed of sound.

substation—Any building or outdoor location at which electric energy in a power system is transformed, converted, or controlled.

substep—A part of a computer step.

substitute—In a computer, to replace one element of information by another.

substitute character – An accuracy-control character, intended to be used in place of a character determined to be invalid, in error, or not representable on a particular device.

substitution interference measurement – A measurement in which the noise level of the source being measured is compared to a known level from a calibrated source; impulse generator for broad-band noise or sine-wave generator for cw.

substitution method—A three-step method of measuring an unknown quantity in a circuit. First, some circuit effect dependent on the unknown quantity is meassured or observed. Then a similar but measurable quantity is substituted in the circuit. Finally, the latter quantity is adjusted to produce a like effect. The unknown value is then assumed to be equal to the adjusted known value.

substrate—1. The supporting material on or in which the parts of an integrated circuit are attached or made. The substrate may be passive (thin film, hybrid) or active (monolithic compatible). 2. A material on the surface of which an adhesive substance is spread for bonding or coating; any material which provides a supporting surface for other materials, especially materials used to support printed-circuit patterns.

subsurface wave—An electromagnetic wave propagated through water or land. Operating frequencies for communications may be limited to approximately 35 kHz due to attenuation of high frequencies.

subsynchronous—Having a frequency that is a submultiple of the driving frequency.

subsynchronous reluctance motor—A form of reluctance motor with more salient poles in the primary winding. As a result, the motor operates at a constant average speed which is a submultiple of its apparent synchronous speed.

subsystem—A major, essential, functional part of a system. The subsystem usually consists of several components.

subtractive process—A printed-circuit manufacturing process in which a conductive pattern is formed by the removal of portions of the surface of a metal clad insulator by chemical means (etching).

subtractive filter—An optical filter which is of a certain color and eliminates that color when placed in the path of white light.

subtractor – An operational amplifier circuit in which the output is proportional to the difference between its two input voltages or between the net sums of its positive and negative inputs.

subvoice-grade channel – A channel the bandwidth of which is less than that of a voice-grade channel. Such a channel usually is a subchannel of a voice-grade line. (According to common usage, a telegraph channel is excluded from this definition.)

subway transformer—A transformer of submersible construction.

success ratio—The ratio of the number of successful attempts to the total number of trials. It is frequently used as a reliability index.

suckout—A hole in the response pattern of a tuned circuit due to the self-resonance of components at certain frequencies.

sudden commencement—Magnetic storms which start suddenly (within a few seconds) and simultaneously all over the earth.

sudden ionospheric disturbances—The sudden increase in ionization density in lower parts of the ionosphere, caused by a bright solar chromospheric eruption. It gives rise to a sudden increase of absorption in radio waves propagated through the low parts of the ionosphere, and sometimes to simultaneous disturbances of terrestrial magnetism and earth current. The change takes place within one or a few minutes, and conditions usually return to normal within one or a few hours.

Suhl effect—When a strong transverse magnetic field is applied to an n-type semiconducting filament, the holes injected into the filament are deflected to the surface. Here they may recombine rapidly with electrons and thus have a much shorter life, or they may be withdrawn by a probe as though the conductance had increased.

suicide control—A control function which uses negative feedback to reduce and automatically maintain the generator voltage at approximately zero.

sulfating—The accumulation of lead sulfate on the plates of a lead-acid storage battery. This reduces the energy-storing ability of the battery and causes it to fail prematurely.

sulfonated polystyrene sensor—An ion-exchange device with good response, accuracy, and long-term stability whose resistance changes exponentially with humidity and temperature. Also called Pope cell.

sum—The combination of two electrical signals of the same electrical polarity. The total electrical energy produced by combining the two different signals of a stereo program.

sum channel—A combination of left and right stereo channels identical to the program, which may be recorded or transmitted monophonically.

summary punch—A punch-card machine which may be attached to another machine in such a way that it will punch information produced, calculated, or summarized by the other machine.

summary recorder—In computers, output equipment which records a summary of the information handled.

summation check—A redundant computer check in which groups of digits are summed, usually without regard to overflow. The sum is then checked against a previously computed sum to verify the accuracy of the computation.

summation frequency—A frequency which is the sum of two other frequencies that are produced simultaneously

summation tone—A combination tone, heard under certain circumstances, the pitch of which corresponds to a frequency equal to the sum of the frequencies of the two components.

summing point—1. A mixing point the output of which is obtained by adding its inputs (with the prescribed signs). 2. The input terminal to which the feedback loop from the output of an amplifier is returned.

sum of products—A general form of Boolean-algebraic expression that can be implemented readily through the use of electronic gate circuits.

sum operand—In a calculator, automatically adds first factors of any sequence of multiplication or division problems. Used when obtaining average unit price and standard deviation.

supercardioid microphone — A microphone having a cardioid "polar pattern" with an unusually high discrimination between sounds from the front and rear.

supercommutation — 1. Commutation at a higher rate by connection of a single data input source to equally spaced contacts of the commutator (cross patching). Corresponding cross patching is required at the decommutator. 2. The connection of one data source of a computer to a sampled-data system more frequently than other data sources are connected.

superconducting — Exhibiting superconductivity.

superconductivity—The decrease in resistance of certain materials (lead, tin, thallium, etc.) as their temperature is reduced to nearly absolute zero. When the critical (transition) temperature is reached, the resistance will be almost zero.

superconductor — A material that exhibits superconductivity.

superemitron—*See* Image Iconoscope.

supergroup — In carrier telephony, five groups (60 voice channels) multiplexed together and handled as a single unit. A basic supergroup occupies the 312-552 kHz band.

superhet—Slang for a superheterodyne receiver.

superheterodyne receiver — A receiver in which the incoming modulated rf signals are usually amplified in a preamplifier and then fed into the mixer for conversion into a fixed, lower carrier frequency (called the intermediate frequency). The modulated if signals undergo very high amplification in the if-amplifier stages and are then fed into the detector for demodulation. The resultant audio or video signals are usually further amplified before being sent to the output.

superheterodyne reception — A method of receiving radio waves in which heterodyne reception converts the voltage of the received wave into a voltage having an intermediate, but usually superaudible, frequency which is then detected.

superhigh frequency—Abbreviated shf. The frequency band extending from 3 to 30 GHz (100 to 10 mm).

superimpose — In a tape recorder, to record one or more signals over another without erasure, so that when a tape is played back, all recordings can be heard simultaneously. (Particularly useful if one wishes to have a spoken commentary with a musical background.)

Supermalloy—Trade name of Arnold Engineering Company for a magnetic alloy with a maximum permeability greater than 1,000,000.

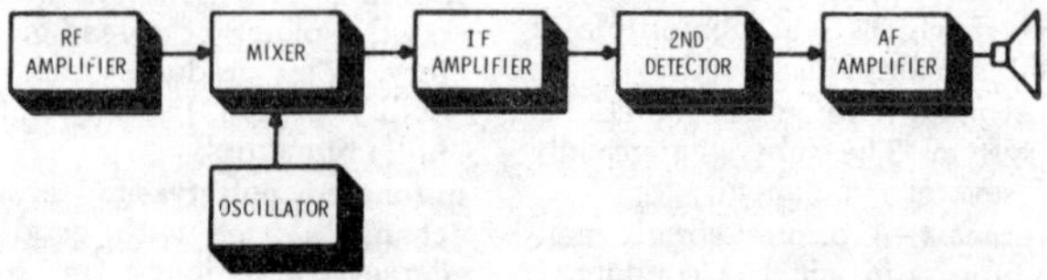

Superheterodyne receiver.

supermode laser – A frequency-modulated laser, the output of which is passed through a second phase modulator driven 180° out of phase and with the same modulation index as the first modulator. All of the energy of the previously existing laser modes is compressed into a single frequency, with nearly the full power of the laser concentrated in that signal.

superposed circuit—An additional channel obtained in such a manner from one or more circuits normally provided for other channels that all channels can be used simultaneously, without mutual interference.

superposed ringing – Also called superimposed ringing. Party-line telephone ringing accomplished by using a combination of alternating and direct currents; direct current of both polarities is used to provide selective ringing.

superposition theorem—When a number of voltages (distributed in any manner throughout a linear network) are applied to the network simultaneously, the current that flows is the sum of the component currents that would flow if the same voltages had acted individually. Likewise, the potential difference that exists between any two points is the component potential difference that would exist there under the same conditions.

superpower—A comparatively large power (sometimes over 1,000,000 watts) used by a broadcasting station in its antenna.

superradiance – A rapid increase in intensity of fluorescent-line emission with increasing excitation (pump) power. This intensity increase and associated line narrowing are attributed to coherent reinforcement of spontaneously emitted photons during a single pass through the active region. (Laser term.)

superrefraction – Abnormally large refraction of radio waves in the lower layers of the atmosphere, leading to abnormal ranges of operation.

superregeneration—1. A form of regenerative amplification frequently used in radio-receiver detecting circuits. Oscillations are alternately allowed to build up and are quenched at a superaudible rate. 2. Method used to produce greater regeneration than otherwise possible without the harmful effects of oscillation. *See also* Quench Frequency.

superregenerative detector – A detector that functions on superregeneration to achieve extremely high sensitivity with a minimum number of amplifier stages.

superregenerative receiver – A receiver in which the regeneration is varied in such a manner that the circuit is periodically rendered oscillatory and nonoscillatory.

supersensitive relay—A relay that operates on extremely small currents (usually less than 250 microamperes).

supersonic—Faster than the speed of sound (approximately 750 mph). These speeds are usually referred to by the term "mach" or "mach number." Mach 1 equals the speed of sound; mach 2, twice the speed of sound, etc.

supersonic communication—Communication through water by manually keying the sound output of echo-ranging equipment used on ships.

supersonic frequency—*See* Ultrasonic Frequency.

supersonics—1. The general term covering phenomena associated with speeds higher than that of sound (e.g., aircraft and projectiles which travel faster than sound). 2. The general term covering the use of frequencies above the range of normal hearing.

supersonic sounding—A system of determining ocean depths by measuring the time interval between the production of a supersonic wave just below the surface of the water, and the arrival of the echo reflected from the bottom. The sounds are transmitted and received by either magnetostriction or piezoelectric units, and electronic equipment is employed to provide a continuous indication of depth (sometimes with a permanent recording).

supersync signal—A combination horizontal- and vertical-sync signal transmitted at the end of each scanning line in commercial television.

superturnstile antenna—A stacked antenna array in which each element is a batwing antenna.

supervisor – A routine or routines carried out in response to a requirement for changing or interrupting the flow of operation through a central processing unit, or for performance of input-output operations; therefore, the medium for coordinating the use of resources and maintaining the flow of operations through the central processing unit. Hence, a control routine executed in supervisor state.

supervisory control—A system by which selective control and automatic indication of remote units is provided by electrical means over a relatively small number of common transmission lines. (Carrier-current channels on power lines can be used for this purpose.)

supervisory control signalling—Characters or signals that actuate equipment or indicators at a remote terminal automatically.

supervisory signal—A signal for attracting the attention of an attendant in connection with switching apparatus, etc.

supervoltage—A voltage applied to X-ray tubes operating between 500 and 2,000 kilovolts.

supplementary insulation—An independent

insulation provided in addition to the functional insulation to ensure protection against electrical shock hazard in the event that functional insulation should fail.

supplementary group—In wire communications, a group of trunks that provides direct connection of local or trunk switching centers by way of other than a fundamental route.

supply port—In a fluidic device, the port at which power is provided to an active device.

supply reel—In a tape recorder, the "feed" reel from which the tape unwinds while playing or recording.

supply voltage—The voltage obtained from a power supply to operate a circuit.

suppressed carrier — That type of system which results in the suppression of the carrier frequency from the transmission medium. (The intelligence of a carrier wave after modulation is contained in either sideband, and normally only one sideband is transmitted; the other sideband and carrier frequency are suppressed. The intelligence is recovered at the receiving end by inserting a carrier frequency from a local source which, when combined with the incoming signal, produces the original frequencies with which the transmitting carrier was modulated.)

suppressed-carrier operation — *See* Suppressed-Carrier Transmission.

suppressed-carrier transmission — Transmission in which the carrier frequency is either partially or totally suppressed. One or both sidebands may be transmitted.

suppressed time delay—Deliberate displacement of the zero of the time scale with respect to the time of emission of a pulse, in order to simulate electrically a geographical displacement of the true position of a transponder.

suppressed-zero instrument—An indicating or recording instrument in which the zero position is below the end of the scale markings.

suppression—1. Elimination of any component of an emission—e.g., a particular frequency or group of frequencies in an audio- or radio-frequency signal. 2. Reduction or elimination of noise pulses generated by a motor or motor generator. 3. Elimination of unwanted signals or interference by means of shielding, filtering, grounding, component relocation, or sometimes redesign. 4. An optional function in on-line or off-line printing devices by which they can ignore certain characters or character groups transmitted through them.

suppression pulse—The pulse generated in an airborne transponder by coincidence of the first interrogation pulse and the control pulse. This pulse, also known as a killer pulse, is used to suppress unwanted interrogations from the side lobes.

suppressor—1. A resistor used in an electron-tube circuit to reduce or prevent oscillation or the generation of unwanted rf signals. 2. A resistor in the high-tension lead of the ignition system in a gasoline engine.

suppressor grid—A grid interposed between two positive electrodes (usually the screen grid and the plate) primarily to reduce the flow of secondary electrons from one to the other.

suppressor pulse—The pulse used to disable an ionized flow field or beacon transponder during intervals when interference would be encountered.

surface acoustic wave—*See* SAW.

surface analyzer—An instrument that measures or records irregularities in a surface. As a crystal-pickup stylus or similar device moves over the surface, the resulting voltage is amplified and fed to an indicator or recorder that magnifies the surface irregularities as high as 50,000 times.

surface asperities—Small projecting imperfections on the surface coating of a tape that limit and cause variations in head-to-tape contact. A term useful in discussions of friction and modulation noise.

surface barrier—A barrier formed automatically at a surface by the electrons trapped there.

surface-barrier transistor — Abbreviated SBT. A wafer of semiconductor material into which depressions have been etched electrochemically on opposite sides. The emitter- and collector-base junction or metal-to-semiconductor contacts are then formed by electroplating a suitable metal onto the semiconductor in the etched depressions. The original wafer constitutes the base region.

surface-controlled avalanche transistor — A transistor in which the avalanche breakdown voltage is controlled by an external field applied through surface insulating layers, and which permits operation at frequencies up to the 10 GHz range.

surface duct — An atmospheric duct for which the lower boundary is the surface of the earth.

surface insulation—Also called oxalizing or insulazing. A coating applied to magnetic-core laminations to retard the passage of current from one lamination to another.

surface leakage—The passage of current over the surface of an insulator rather than through it. Surface leakage in new components is very low, but when a component is installed in equipment and exposed to dust, dirt, moisture and other degrading environments, leakage current can increase and cause problems.

surface noise—1. Also called needle scratch. In mechanical recording, the noise caused in the electrical output of a pickup by irregular contact surfaces in the groove. 2. Noise generated by contact of a phonograph stylus with minute particles of dust or other irregularities in a record groove. Can also be caused by excessive wear of a disc or by poor quality coating on recording tape.

surface of position – Any surface defined by a constant value of some navigational coordinate.

surface recombination rate—The rate at which free electrons and holes recombine at the surface of a semiconductor.

surface recording—Storage of information on a coating of magnetic material such as that which is used on magnetic tape, magnetic drums, etc.

surface reflection—Also called Fresnel loss. The part of the incident radiation that is reflected from the surface of a refractive material. It is directly proportional to the refractive index of the material and is reduced for a given wavelength by application of an appropriate surface coating.

surface resistance—The ratio of the direct voltage applied to an insulation system to the current that passes across the surface of the system. In this case, the surface consists of the geometric surface and the material immediately in contact with it. The thin layer of moisture at the interface between a gas and a solid usually has the greatest effect on surface resistance.

surface resistivity—The resistance between opposite edges of a surface film 1 cm square. It is measured by determining the resistance between two straight conductors 1 cm apart, pressed upon the surface of a slab of the material. Water-absorbent materials usually show a lower resistivity than nonabsorbent ones.

surface states – Discontinuities and contaminants at the surface of a semiconductor device which tend to change the surface characteristics and promote device parameter instability.

surface-temperature-resistor – A platinum-resistance thermometer designed for installation directly on the surface whose temperature is being measured.

surface wave – A subclassification of the ground wave. So called because it travels along the surface of the earth.

surface-wave filter—A filter whose operation is based on the use of the interlaced, deposited electrode pairs and acoustic surface waves on a single crystal substrate. Device operates on traveling-wave principles, not lumped-element concept.

surface-wave transmission line – Ideally, a nonradiating broad-band transmission line that functions by guiding electromagnetic energy in the surrounding air. A surface line allows coupling to the field anywhere along the guide (in contrast to an ordinary waveguide, in which the internal fields are completely screened from external space by metal walls). A noncontacting device can be used to couple some or all of the energy to or from the line. Abbreviated SWTL.

surge—Sudden current or voltage changes in a circuit.

surge admittance—Reciprocal of surge impedance.

surge-crest ammeter—A special magnetometer used with magnetizable links to measure the crest value of transient electric currents.

surge generator—*See* Impulse Generator, 1.

surge impedance – *See* Characteristic Impedance.

surge suppressor – A two-terminal device (pnp) which will conduct in either direction above a specified voltage and polarity, but otherwise acts as a blocking device to current. It essentially is a back-to-back diode with avalanche characteristics for protecting circuitry from high alternating voltage peaks or transients. Also called voltage clipper or thyrector.

surge voltage (or current)—A large, sudden change of voltage (or current), usually caused by the collapse of a magnetic field or by a shorted or open circuit element.

surge-voltage recorder – *See* Lichtenberg Figure Camera.

surround—The part of a speaker cone by which its outside edge is anchored, usually corrugated.

surveillance—Systematic observation of air, surface, or subsurface areas by visual, electronic, photographic, or other means.

surveillance radar – In air-traffic control systems, a radar set or system used in a ground-controlled approach system to detect aircraft within a certain radius of an airport and to present continuously to the

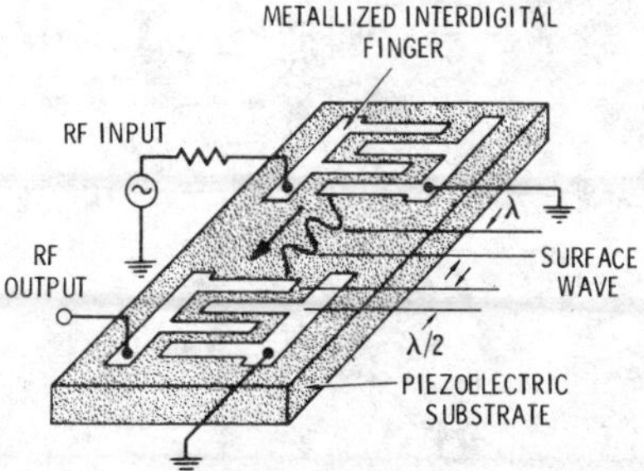

Surface wave device.

radar operator information as to the position, in distance and azimuth, of these aircraft.

surveillance radar station—In the aeronautical radionavigation service, a land station employing radar to detect the presence of aircraft.

sus—*See* Silicon Unilateral Switch.

susceptance—The reciprocal of reactance, and the imaginary part of admittance. It is measured in mhos.

susceptance standard — A standard with which small, calibrated values of shunt capacitance are introduced into 50-ohm coaxial transmission arrays.

susceptibility — 1. Ratio of the induced magnetization to the inducing magnetic force. 2. The undesired response of an equipment to emissions, interference, or transients, or to signals other than those to which the equipment is intended to be responsive.

susceptibility meter—A device for measuring low values of magnetic susceptibility.

susceptiveness — The tendency of a telephone system to pick up noise and low-frequency induction from a power system. It is determined by telephone-circuit balance, transpositions, wire spacing, and isolation from ground.

suspension—A wire that supports the moving coil of a galvanometer or similar instrument.

suspension galvanometer—An early type of moving-coil instrument in which a coil of wire was suspended in a magnetic field and would rotate when it carried an electric current. A mirror attached to the coil deflected a beam of light, causing a spot of light to travel on a scale some distance from the instrument. The effect was a pointer of greater length but no mass.

suspension lines—The main lines leading from the canopy to the suspended load. Often mistakenly called "shroud" lines.

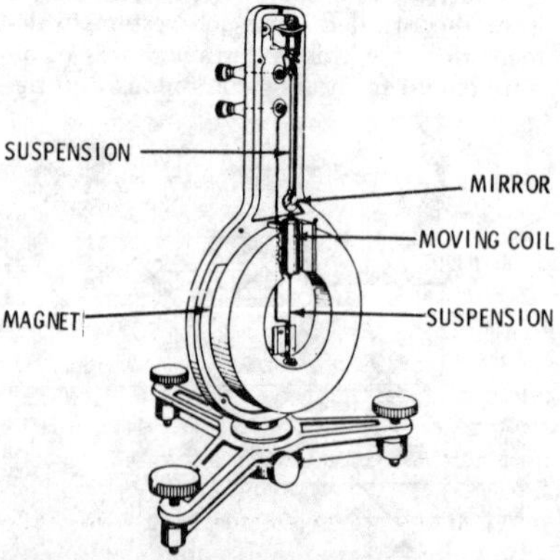

Suspension galvanometer.

sustain—In an organ, the effect produced by circuitry which causes a note to diminish gradually after the key controlling the note has been released.

sustained oscillation — 1. Oscillation in which forces outside the system but controlled by it maintain a periodic oscillation at a period or frequency that is nearly the natural period of the system. 2. Continued oscillation due to insufficient attenuation in the feedback path.

sustained start — An electrical signal for starting a timer which is of any duration longer than the timer setting.

sustaining current — The current required to maintain ionization across a spark gap.

sw—Abbreviation for short wave.

swamping resistor—In transistor circuits, a resistor placed in the emitter lead to mask (minimize the effects of) variations caused in the emitter-base junction resistance by temperature variations.

swamp resistance—A small amount of resistance, provided by a resistor with a negative or small positive resistance-temperature coefficient, placed in series with the coil of an electrical indicating instrument to reduce the overall temperature coefficients of the instrument. The resistor also may be used for adjustment of the terminal resistance of the meter.

sweep—The crossing of a range of values of a quantity for the purpose of delineating, sampling, or controlling another quantity. Examples of swept quantities are the displacement of a scanning spot on the screen of a cathode-ray tube, and the frequency of a wave.

sweep accuracy—The accuracy of the trace horizontal displacement in an oscilloscope compared with the reference independent variable, usually expressed in terms of average rate error as a percent of full scale.

sweep amplifier — An amplifier stage designed to increase the amplitude of the sweep voltage.

sweep circuit — A circuit which produces, at regular intervals, an approximately linear, circular, or other movement of the beam in a cathode-ray tube.

sweep delay—The time between the application of a pulse to the sweep-trigger input of an oscilloscope and the start of the sweep.

sweep-delay accuracy—The accuracy of an indicated sweep delay in an oscilloscope, usually specified in error terms.

sweep expander—*See* Sweep Magnifier.

sweep-frequency generator—A signal source capable of changing frequency automatically and in synchronism with a display device. The frequency sweep can be obtained by either mechanical or electronic means.

sweep-frequency record—A test record on

which a series of constant-amplitude frequencies have been recorded. Each frequency is typically repeated 20 times per second, starting at 50 Hz and continuing up to 10 kHz or higher.

sweep generator – Also called timing-axis oscillator. A circuit which applies voltages or currents to the deflection elements in a cathode-ray tube in such a way that the deflection of the electron beam is a known function of time, against which other periodic electrical phenomena may be examined, compared, and measured.

sweeping receivers – Automatically and continuously tuned receivers designed to stop and lock on when a signal is found or to continually plot band occupancy.

sweep jammer–An electric jammer which sweeps a narrow band of electronic enrgy over a broad bandwidth.

sweep linearity – The maximum displacement error of the independent variable between specified points on the display area in an oscilloscope.

sweep lockout – A means for preventing multiple sweeps when operating an oscilloscope in a single-sweep mode.

sweep magnifier – Also called sweep expander. A circuit or control for expanding part of the sweep display of an oscilloscope.

sweep oscillator–An oscillator used to develop a sawtooth voltage which can be amplified to deflect the electron beam of a cathode-ray tube. (*See also* Sweep Generator.)

sweep switching–The alternate display of two or more time bases or other sweeps using a single-beam crt. Comparable to dual- or multiple-trace operation of a deflection amplifier.

sweep test–Pertaining to cable, checking the frequency response by generating an rf voltage, the frequency of which is varied back and forth through a given frequency range at a rapid constant rate while observing the results on an oscilloscope.

sweep-through–A jamming transmitter that sweeps through a radio-frequency band and jams each frequency briefly, producing a sound like that of an aircraft engine.

sweep voltage–The voltage used for deflecting an electron beam. It may be applied to either the magnetic deflecting coils or the electrostatic plates.

swell manual–Also called solo manual. In an organ, the upper manual normally used to play the melody. (*See also* Manual, 2.)

swept resistance–The portion of the total resistance of a potentiometric transducer over which the slider travels when the device is operated through its total range.

swim–The phenomenon in which the constructs on a crt screen appear to move about their normal positions. It can be observed when the refresh rate is slow and is not some multiple or submultiple of line frequency. In some cases, swim is a result of instability in the digital-to-analog converters in the display controller.

swing–The variation in frequency or amplitude of an electrical quantity.

swinging – Momentary variations in frequency of a received wave.

swinging arm – A type of mounting and feed used to move the cutting head at a uniform rate across the recording disc in some recorders. All phonograph pickups are of the swinging-arm type.

swinging choke–A filter inductor designed with an air gap in its magnetic circuit so its inductance decreases as the current through it increases. When used in a power-supply filter, a swinging choke can maintain approximately critical inductance over wide variation in load current.

"swiss-cheese" packaging–Also called imitation 2-D. A high-density packaging technique in which passive and active components are inserted into holes punched in printed-circuit-board substrates and attached by soldering or thermocompression bonding or by means of conductive epoxy adhesive.

switch–1. A mechanical or electrical device that completes or breaks the path of the current or sends it over a different path. 2. In a computer, a device or programming technique by means of which selections are made. 3. A device that connects, disconnects, or transfers one or more circuits and is not designated as a controller, relay or control valve. The term is also applied to the functions performed by switches.

switcher–A device that allows the pictures from a number of cameras to be viewed on one monitor.

switchboard–1. A manually operated apparatus at a telephone exchange. The various circuits from subscribers and other exchanges terminate here, so that operators can establish communications between two subscribers on the same exchange, or on different exchanges. 2. A single large panel or an assembly of panels on which are mounted the switches, circuit breakers, meters, fuses, and terminals essential to the operation of electrical equipment.

switching center–*See* Switching Office.

switch detector–A detector which extracts information from the input waveform only at instants determined by a selector pulse.

switched network–The network by which switched telephone service is provided to the public. Also called public switched network and switched message network.

switch gear – A general term covering switching, interrupting, control, metering, protective, and regulating devices; also assemblies of these devices and associated interconnections, accessories, and supporting structures, used primarily in connection with the generation, transmission, and distribution of electric power.

switch hook—A switch associated with the structure on a telephone set that supports the receiver or handset. The switch is operated when the receiver or handset is removed from or replaced on the support.

switching—Making, breaking, or changing the connections in an electrical circuit.

switching center—1. A location at which data from an incoming circuit are routed to the proper outgoing circuit. 2. A group of equipment within a relay station for automatically or semiautomatically relaying communications traffic.

switching characteristics—An indication of how a device responds to an input pulse under specified driving conditions.

switching circuit—A circuit which performs a switching function. In computers, this is performed automatically by the presence of a certain signal (usually a pulse signal). When combined, switching circuits can perform a logical operation.

switching coefficient—The derivative of applied magnetizing force with respect to the reciprocal of the resultant switching time. It is usually determined as the reciprocal of the slope of a curve of reciprocals of switching times versus the values of applied magnetizing forces, which are applied as step functions.

switching control—An installation in a wire system where telephone or teletypewriter switchboards are installed to interconnect circuits.

switching current—The current through a device at the switching voltage point.

switching device – Any device or mechanism, either electrical or mechanical, which can place another device or circuit in an operating or nonoperating state.

switching diode—A diode that has a high resistance (corresponding to an open switch) below a specified applied voltage but changes suddenly to a low resistance (closed switch) above that voltage.

switching mode—A way of utilizing a vacuum tube or transistor so that (except for negligibly small transition times) it is either in cutoff or saturation. A transistor operated in this mode can switch large currents with little power dissipation.

switching office – A location where either toll or local telephone traffic is switched or connected from one line or circuit to another. Also called switching center.

switching pad—A transmission-loss pad automatically inserted into or removed from a toll circuit for different desired operating conditions.

switching power supply—A power supply (usually dc output) which achieves its output regulation by means of one or more active power handling devices which are alternately placed in the "off" and "on" states. Distinguished from "linear" or "dissipative" power supplies, in which regulation is achieved by power handling devices whose conduction is varied continuously over a wide range which seldom (if ever) includes the full "off" or full "on" condition.

switching time—1. The interval between the reference time and the last instant at which the instantaneous-voltage response of a magnetic cell reaches a stated fraction of its peak value. 2. The interval between the reference time and the first instant at which the instantaneous integrated-voltage response reaches a stated fraction of its peak value.

switching trunk—A trunk that runs between a long-distance office and a local exchange office and is used for completing a long-distance call.

switching voltage—The maximum forward voltage a device can sustain without breaking over into full conduction.

switchplate – A small plate attached to a wall to cover a push-button or other type of switch.

switch room—That part of a telephone cen-

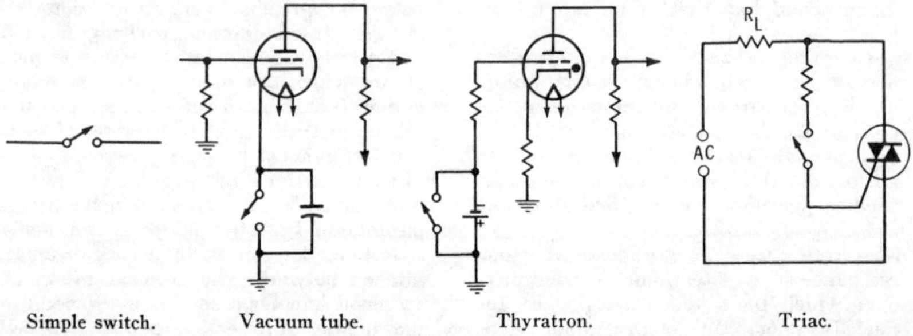

Simple switch. Vacuum tube. Thyratron. Triac.

Switching circuits.

tral office building that houses switching mechanism and associated apparatus.

switchtail ring counter – A type of ring counter in which the output of one stage is inverted before being applied as an input to the next stage. An even number of states equal to 2n (where n is the number of flip-flops) normally is produced. For example, a modulo-10 counter can be made from five flip-flops. Each flip-flop changes states on every fifth count. Decoding of all ten states is accomplished conveniently with ten two-input gates. A switchtail ring counter will contain the complement of the information it contained initially after n clock pulses, and will contain the initial information again after 2n clock pulses.

switch train – A sequence of switches through which connection must be made when a circuit between a calling telephone and a called telephone is established.

swr—Abbreviation for standing-wave ratio.

swr bridge – *See* Standing-Wave-Ratio Bridge.

swr meter—An external or built-in circuit which measures the standing-wave ratio at the transceiver end of the antenna transmission line.

SWTL – Abbreviation for surface-wave transmission line.

syllabic companding – Companding in which the effective gain variations are made at speeds allowing response to the syllables of speech but not to individual cycles of the signal wave.

syllable articulation – Also called percent of syllabic articulation. The percent of articulation obtained when the speech units considered are syllables (usually meaningless and usually of the consonant-vowel-consonant type).

symbol—1. A simplified design representing a part in a schematic circuit diagram. 2. A letter representing a particular quantity in formulas.

symbolic—Having to do with the representation of something by a conventional sign.

symbolic address—Also called a floating address. In digital-computer programming, a label chosen in a routine to identify a particular word, function, or other information independent of the location of the information within the routine.

symbolic code—A code by which programs are expressed in source language; that is, storage locations and machine operations are referred to by symbolic names and addresses that do not depend on their hardware-determined names and addresses. Also called pseudocode. Contrasted with computer code.

symbolic coding—In digital computer programming, any coding system using symbolic rather than actual computer addresses.

symbolic deck—A deck of cards punched in programmer coding language rather than binary language.

symbolic-language programming – The writing of program instructions in a language that facilitates the translation of programs into binary code through the use of mnemonic convention. Also called assembly-language programming.

symbolic logic—A special computer or control system language composed of symbols that the instrumentation can accept and handle. Combinations of these symbols can be fed in to represent many complex operations.

symbolic programming – A program using symbols instead of numbers for the operations and locations in a computer. Although the writing of a program is easier and faster, an assembly program must be used to decode the symbol into machine language and assign instruction locations.

symmetrical—Balanced – i.e., having equal characteristics on each side of a central line, position, or value.

symmetrical alternating quantity—An alternating quantity for which all values separated by a half period have the same magnitude but opposite sign.

symmetrical avalanche rectifier – An avalanche rectifier which can be triggered in either direction. After triggering, it presents a low impedance in the triggered direction.

symmetrically, cyclically magnetized condition—The condition of a cyclically magnetized material when the limits of the applied magnetizing forces are equal and of opposite sign.

symmetrical transducer (with respect to specified terminations)—A transducer in which all possible pairs of specified terminations can be interchanged without affecting the transmission.

symmetrical transistor – A transistor in which the collector and emitter are made identical, so either can be used interchangeably.

sync—Short for synchronous, synchronization, synchronizing, etc.

sync compression – The reduction in gain applied to the sync signal over any part of its amplitude range with respect to the gain at a specified reference level.

sync generator—An electronic device that supplies pulses to synchronize a television system.

synchro—1. A small motorlike device containing a stator and a rotor and capable of transforming an angular-position input into an electrical output or an electrical input into an angular output. When several synchros are correctly connected together, all rotors will line up at the same

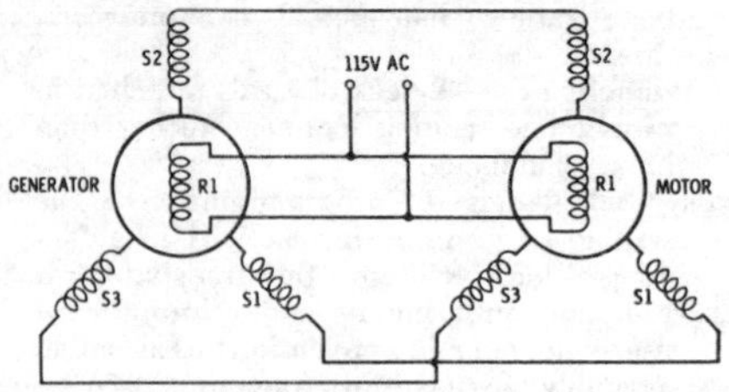

Synchro.

angle of rotation. 2. A range of ac electromechanical devices which are used in data transmission and computing systems. A synchro provides mechanical indication of its shaft position as the result of an electrical input or an electrical output which represents some function of the angular displacement of its shaft. Such components are basically variable transformers. As the rotor of a synchro rotates it causes a change in synchro voltage outputs. Major types or classes of synchros include torque synchros, control synchros, resolvers, and induction potentiometers (linear synchro transmitters).

synchro-control differential generator – A rotary component for modifying the synchro-control generator output signal to correspond to the addition or subtraction from the generator shaft angle. Usually used with a synchro-control generator and synchro-control transducer.

synchro-control generator–A rotary component for transforming the shaft angle to a corresponding set of electrical signals for ultimate retransformation to the shaft position in a remote location.

synchro-control transformer–A rotary component which accepts signals from a generator or differential generator for reconversion to the shaft angle with the aid of a servomechanism. Often used by itself as an angle-to-signal transducer, but usually with a synchro-control generator and synchro-control differential generator.

synchro differential generator–A synchro unit which receives an order from a synchro generator at its primary terminals, modifies this order mechanically by any desired amount according to the angular position of the rotor, and transmits the modified order from its secondary terminals to other synchro units.

synchro differential motor–A motor which is electrically similar to the synchro differential generator except that a damping device is added to prevent oscillation. Its rotor and stator are both connected to synchro generators, and its function is to indicate the sum of/or difference between the two signals transmitted by the generators.

synchro generator–A synchro which has an electrical output proportional to the angular position of its rotor.

Synchroguide–A type of control circuit for horizontal scanning in which the sync signal, oscillator voltage pulse, and scanning voltage are compared and kept in synchronism.

synchro motor – A synchro in which the rotor-shaft position is dependent on the electrical input.

synchronism–1. The phase relationship between two or more quantities of the same period when their phase difference is zero. 2. Applied to the synchronous motor, the condition under which the motor runs at a speed which is directly related to the frequency of the power applied to the motor and is not dependent upon other variables.

synchronization – 1. The precise matching of two waves or functions. 2. The process of keeping the electron beam on the television screen in the same position as the scanning beam at the transmitter. 3. In a carrier, that degree of matching, in frequency, between the carrier used for modulation and the carrier used for demodulation which is sufficiently accurate to permit efficient functioning of the system.

synchronization error – In navigation, the error due to imperfect timing of two operations (may or may not include the signal transmission time).

synchronization pulses – Pulses originated by the transmitting equipment and introduced into the receiving equipment to keep the equipment at both locations operating in step.

synchronize–1. To adjust the periodicity of an electrical system so that it bears an integral relationship to the frequency of the periodic phenomenon under investigation. 2. To lock one element of a system into step with another. The term usually refers to locking a receiver to a transmitter, but it can refer to locking the data-terminal equipment bit rate to the data set frequency.

synchronized sweep–A sweep which would free-run in the absence of an applied signal, but in the presence of the signal is synchronized by it.

synchronizer–1. The component of a radar set which generates the timing voltage for the complete set. (*See also* Timer, *3.*) 2. A computer storage device used to compensate for a difference in a rate of flow of information or time of occurrence of events when information is being transmitted from one device to another.

synchronizing (in television)–Maintaining two or more scanning processes in phase.

synchronizing-pulse selector–A circuit used to separate synchronizing pulses from commutated pulse trains.

synchronizing reactor – A current-limiting reactor that is connected momentarily across the open contacts of a circuit-interrupting device for synchronizing purposes.

synchronizing relay–A relay which functions when two alternating-current sources are in agreement within predetermined limits of phase angle and frequency.

synchronizing separator – *See* Amplitude Separator.

synchronizing signal–*See* Sync Signal.

synchronous–1. In step or in phase, as applied to two devices or machines. 2. A term applied to a computer, in which the performance of a sequence of operations is controlled by equally spaced clock signals or pulses. 3. Having a constant time interval between successive bits, characters, or events. The term implies that all equipment in the system is in step. Operation of a switching network by a clock pulse generator. More critical than asynchronous timing but requires fewer and simpler circuits.

synchronous booster converter–A synchronous converter connected in series with an ac generator and mounted on the same shaft. It is used for adjusting the voltage at the commutator of the converter.

synchronous capacitor–A rotating machine running without mechanical load and designed so that its field excitation can be varied in order to draw a leading current (like a capacitor) and thereby modify the power factor of the ac system, or influence the load voltage through such change in power factor.

synchronous clock–An electric clock driven by a synchronous motor, for operation on an ac power system in which the frequency is accurately controlled.

synchronous communications satellite – A communications satellite the orbital speed of which is adjusted so that the satellite remains above a particular point on the surface of the earth.

synchronous computer–A digital computer in which all ordinary operations are controlled by clock pulses from a master clock.

synchronous converter–A synchronous machine which converts alternating current to direct current or vice versa. The armature winding is connected to the collector rings and commutator.

synchronous demodulator – Also called a synchronous detector. A demodulator in which the reference signal has the same frequency as the carrier or subcarrier to be demodulated. It is used in color television receivers to recover either the I or the Q signals from the chrominance sidebands.

synchronous detector – *See* Synchronous Demodulator.

synchronous gate – A time gate in which the output intervals are synchronized with the incoming signal.

synchronous generator–A circuit designed to synchronize an externally generated signal with a train of clock pulses. The generator produces precisely one output pulse for each cycle of the input signal. The output pulse thus has a width equal to that of the period of the clock-pulse train.

synchronous idle character–A communication control character used to provide a signal for synchronization of the equipment at the data terminals.

synchronous induction motor – A motor with the rotor laminations cut away exposing the rotor and definite poles. Also, a motor that has definite poles but the poles are permanent magnets. This term is also applied to a wound rotor (slip ring) motor, that is started as an induction motor, and when near synchronism, dc is applied to two of the rings to operate as a true synchronous motor.

synchronous inputs–Those inputs of a flip-flop that do not control the output directly, as do those of a gate, but only when the clock permits and commands. Called J and K inputs or ac set and reset inputs.

synchronous inverter–*See* Dynamotor.

synchronous logic–The type of digital logic used in a system in which logical operations take place in synchronism with clock pulses.

synchronous machine – A machine which has an average speed exactly proportionate to the frequency of the system to which it is connected.

synchronous motor–1. An induction motor which runs at synchronous speed. Its stator windings have the same arrangement as in nonsynchronous induction motors, but the rotor does not slip behind the rotating magnetic stator field. 2. Type of ac electric motor in which rotor speed is related directly to frequency of power supply.

synchronous multiplexer – A multiplexer which can time interleave two data streams into one higher-speed stream. In a system using these types of multiplexers, all peripheral equipment in the system must be under the control of a master synchronizing device or "clock."

synchronous operation–Operation of a system under the control of clock pulses.

synchronous rectifier–A rectifier in which contacts are opened and closed at the correct instant by either a synchronous vibrator or a commutator driven by a synchronous motor.

synchronous speed–A speed value related to the frequency of an ac power line and the number of poles in the rotating equipment. Synchronous speed in revolutions per minute is equal to the frequency in hertz divided by the number of poles, with the result multiplied by 120.

synchronous system—A system in which the sending and receiving instruments are operating continuously at substantially the same frequency, and are maintained in a desired phase relationship.

synchronous torque — The maximum load torque with which a motor can be loaded after it comes to synchronous speed. These torques are usually higher than starting torques.

synchronous transmission—Transmission in which the sending and receiving instruments operate continuously at the same frequency and are held in a desired phase relationship by correction devices.

synchronous vibrator—An electromagnetic vibrator that simultaneously converts a low dc voltage to a low alternating voltage and applies it to a power transformer, from which a high alternating voltage is obtained and rectified. In power packs, it eliminates the need for a rectifier tube.

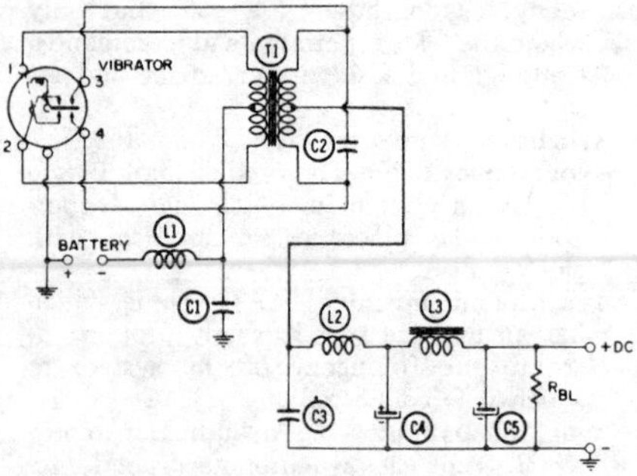

Synchronous-vibrator power supply.

synchroscope—1. An instrument used to determine the phase difference or degree of synchronism of two alternating-current generators or quantities. 2. An oscilloscope on which recurrent pulses or waveforms may be observed, and which incorporates a sweep generator that produces one sweep for each pulse.

synchro system—A system for obtaining remote indication or control by means of self-synchronizing motors such as selsyns and equivalent types.

synchro-torque receiver—A relatively low-impedance positioning device that generates its own torque when driven by a suitable synchro-torque transmitter.

synchro-torque transmitter — A positioning device that generates electrical information of sufficient power to drive a suitable torque receiver.

synchrotron — A device for accelerating charged particles (e.g., electrons) in a vacuum. The particles are guided by a changing magnetic field while being accelerated many times in a closed path by a radio-frequency electric field.

synchrotron radiation — Also called magnetic *Bremsstrahlung*. The radiation produced by relativistic electrons as they travel in a region of space containing magnetic fields.

sync level—The level of the sync peaks.

sync limiter — A circuit used in television circuits to prevent sync pulses from exceeding a predetermined amplitude.

sync pulse—Part of the sync signal in a television system.

sync section—A color tv circuit comprising a keyer, burst amplifier, phase detector, reactance tube, subcarrier oscillator, and quadrature amplifier.

sync separator — The circuit which separates the picture signals from the control pulses in a television system.

sync signal — Also called a synchronizing signal. The signal employed for synchronizing the scanning. In television it is composed of pulses at rates related to the line and field frequencies.

sync-signal generator—A synchronizing signal generator for a television receiver or transmitter.

syntax—1. The make-up of expressions in a language. 2. The rules that govern the structure of expressions in a language.

synthesis — The combination of parts to form a whole.

synthesizer—A device that can generate a number of crystal-controlled frequencies for multichannel communications equipment.

synthesizer frequency meter—A device for measuring frequency by utilizing a synthesized crystal-based signal for the internally generated signal.

synthetic display generation—Logical and numerical processing to display collected or calculated data in symbolic form.

syntony—The condition in which two oscillating circuits have the same resonant frequency.

system—1. An assembly of component parts linked together by some form of regulated interaction into an organized whole. 2. A collection of consecutive operations and procedures required to accomplish a specific objective.

systematic distortion—Distortion of a periodic or constant nature, such as bias or characteristic distortion; the opposite of fortuitous distortion.

systematic error—1. The magnitude and direction of the tendency of a measuring process to measure some quantity other than the one intended. 2. An error of the type that have an orderly character that can be corrected by calibration.

systematic inaccuracies—Those inaccuracies due to inherent limitations in the equipment.

system deviation—The instantaneous difference between the value of a specified system variable and the ideal value of the same system variable.

system element — One or more basic elements, together with other components necessary to form all or a significant part of one of the general functional groups into which a measurement system can be classified.

system engineering—A method of engineering analysis whereby all the elements in a system, including the process itself, are considered.

system failure rate—The number of occasions during a given time period on which a given number of identical systems do not function properly.

system input unit—A device defined as a source of an input job stream.

system layout—In a microwave system, a chart or diagram showing the number, type, and terminations of circuits used in the system.

system library—The assemblage of all cataloged data sets at an installation.

system macroinstruction — A predefined macroinstruction that makes available access to operating system facilities.

system master tapes—Magnetic tapes that contain programmed instructions necessary for preparation of a computer before programs are run.

system noise—The output of a system when it is operating with zero input signal.

system of beams—The three electron beams emitted by the triple electron-gun assembly in a color tube. They occupy positions equidistant from a common axis and are spaced 120° apart around the axis.

system of units—An assemblage of units for expressing the magnitudes of physical quantities.

system output unit—An output device that is shared by all jobs and onto which specified output information is transcribed.

system overshoot — The largest value of system deviation following the dynamic crossing of the ideal value as a result of a specified stimulus.

system reliability — The probability that a system will perform its specified task properly under stated conditions of environment.

system residence volume — The volume in which are located the nucleus of an operating system and the highest-level index of the catalog.

system resonance—The fundamental resonance of the woofer/enclosure combination. Related to the low frequency performance of the systems, but not without ambiguity, especially when comparing different types of enclosures. Not to be confused with free-air resonance of the woofer.

systems analysis — The examination of an activity, technique, or business to determine what and how necessary operations may best be accomplished.

T

T—1. Symbol for transformer or absolute temperature. 2. Abbreviation for the prefix "tera" (10^{12}).

tab—1. *See* Land, 2. 2. A nonprinting spacing action on a typewriter or tape preparation device, the code of which is necessary to the tab sequential format method of programming.

table—A collection of data each item of which is uniquely identified by a label, its position with respect to the other items, or some other means.

table lockup—In a computer, a method of controlling the location to which a jump or transfer is made. It is used especially when there is a large number of alternatives, as in function evaluation in scientific computations.

tab sequential format—A means for identification of a computer word by the number of tab characters in the block preceding the word. The initial character in each word is a tab character. Words must be presented in a certain order, but all characters in a word except the tab character may be omitted when the command that word represents is not desired.

tabulate—1. To arrange data into a table. 2. To print totals.

tabulator—A machine (e.g., a punch-card machine) which reads information from one medium and produces lists, totals, or tabulations on separate forms or continuous paper strips.

tachometer—1. An instrument used to measure the frequency of mechanical systems by the determination of angular velocity. 2. A transducer that gives an electric output signal proportional to the rotational speed of a shaft. 3. A device for measuring rate by counting the number of pulses that occur in a given time period.

tachometer generator — A small generator attached to a rotating shaft for the purpose of generating a voltage proportional to the shaft speed. In speed-control circuits, a tachometer may be coupled to the shaft of the motor whose speed is to

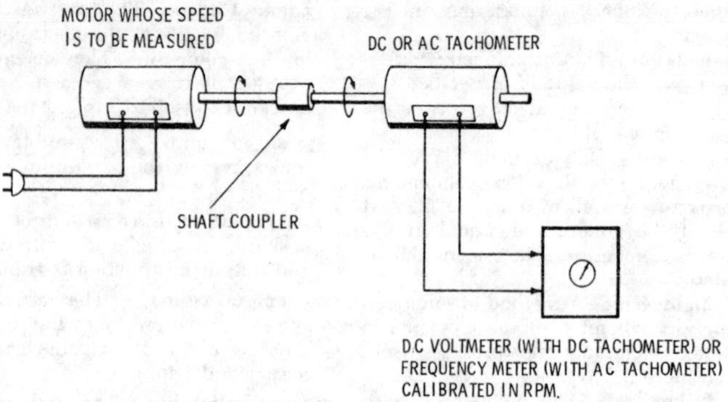

Tachometer.

be controlled. A change of motor speed will produce a change of tachometer output voltage. This change of voltage can be used as an error signal to restore the speed to the desired value.

tactical air navigation—A short-range uhf air-navigation system that presents accurate information to a pilot in two dimensions, distance and bearing from a selected ground station.

tag—Also called a sentinel. In digital-computer programming, a unit of information the composition of which differs from that of other members of the set so that it can be used as a marker or label.

tail — A small pulse following the main pulse and in the same direction, or the slow decay following the main body of the pulse.

tail clipping—A method of sharpening the trailing edge of a pulse.

tailing—*See* Hangover, 1.

tail pulse—A pulse in which the decay time is much longer than the rise time.

tails-out—1. Storage of a nonreversing tape on the takeup reel rather than on the supply reel, to avoid the tape-distorting interval winding stresses and uneven wind of high-speed rewinding. 2. In the case of reversing tapes (i.e., cassettes or four-track open-reel tapes), storing the tape wound on whichever reel (or hub) it ends up on after having been played or recorded.

tail-warning radar set—A radar set placed in the tail of an aircraft to warn of aircraft approaching from the rear.

take-up reel—The reel which accumulates the tape as it is recorded or played on a tape recorder.

talbot—A unit of luminous energy in the mksa system equal to 1 lumen-second.

talker echo—An echo which reaches the ear of the person who originated the sound.

talking battery — The dc voltage supplied by the central office to the subscriber's loop to operate the carbon transmitter in the handset.

talking path — In a telephone circuit, the transmission path consisting of the tip and ring conductors.

talk-listen switch — A switch on an intercommunication unit to switch the speaker as required to function either as a reproducer or as a microphone.

tandem—*See* Cascade.

tandem office—In a telephone system, an office that interconnects the local end offices over tandem trunks in a densely settled exchange area where it is not economical to provide direct interconnection between all end offices. The tandem office completes all calls between the end offices but is not connected directly to subscriber's stations.

tandem transistor—Two transistors in one package and internally connected together.

tangent—A straight line which touches the circumference of a circle at one point.

tangent galvanometer—A galvanometer consisting of a small compass mounted horizontally in the center of a large vertical coil of wire. The current through the coil is proportional to the tangent of the angle at which the compass needle is deflected.

tangential component—A component acting at right angles to a radius.

tangential pickup arm—A pickup arm that maintains the longitudinal axis of the stylus tangent to the record grooves throughout the entire movement of the arm across the record.

tangential sensitivity—A term generally applied as an indication of quality in a receiving system. This term can be used to define the minimum signal level that can be detected above the background noise. However, it is usually expressed as that signal power level which causes a 3-dB rise above the noise-level reading.

tangential sensitivity on look-through—The strength of the target signal, measured at the receiver terminals, required to produce a signal pulse having twice the apparent height of the noise.

tangential wave path—In radio-wave propagation of a direct wave over the earth, a path which is tangential to the surface of the earth. The tangential wave path is curved by atmospheric refraction.

tangent sensitivity—The slope of the line tangent to the response curve at the point being measured.

tank – 1. A unit of acoustically operating delay-line storage containing a set of channels. Each channel forms a separate recirculation path. 2. *See* Tank Circuit, 2.

tank circuit—1. A circuit capable of storing electrical energy over a band of frequencies continuously distributed about a single frequency at which the circuit is said to be resonant or tuned. The selectivity of the circuit is proportionate to the ratio between the energy stored in the circuit and the energy dissipated. This ratio is often called the *Q* of the circuit. 2. Also called a tank. A parallel-resonant circuit connected in the plate circuit of an electron-tube generator.

tantalum (electrolytic) capacitor—An electrolytic capacitor with a tantalum foil or sintered-slug anode.

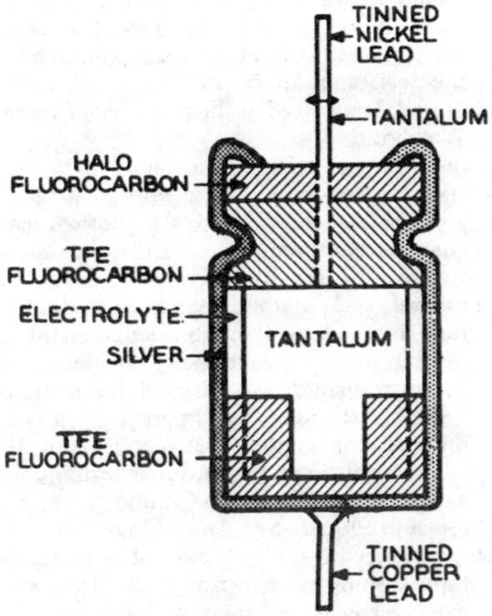

Tantalum capacitor.

tantalum-foil electrolytic capacitor—A capacitor which consists of two tantalum foil electrodes with an oxide on the anode, and separated by layers of absorbent paper saturated with an operating electrolyte.

tantalum oxide—A dielectric material used in capacitors; it is formed electrochemically in a thin film on surfaces of tantalum metal.

T-antenna—Any antenna consisting of one or more horizontal wires, with the lead-in connected approximately in the center.

tap—1. A fixed electrical connection to a specified position on the element of a potentiometer, transformer, etc. 2. A branch. Applies to conductors, such as a battery tap, and to miscellaneous general use.

tap crystal – A compound semiconductor that stores current when stimulated by light and then gives up energy in the form of flashes of light when subjected to mechanical tapping.

tape – 1. Plastic ribbons, with one side coated metallically to receive impressions from a recording head, or, if already recorded, to induce signals in a playback head. Tape is regularly wound on reels or packaged in magazines or cartridge form. 2. *See* Punched Tape. 3. A ribbon of flexible material—e.g., friction, magnetic, punched, etc.

tape cable—Also called flat, flexible cable. A form of flexible multiple conductor in which parallel strips of metal are imbedded in an insulating material.

tape cartridge—A magazine or holder for a length of magnetic tape which by its design avoids the necessity for manual threading or handling. Usually compatible only with one specific type of machine. *See also* Tape Magazine.

tape character—Information consisting of bits stored across the several longitudinal channels of a tape.

tape-controlled carriage – A paper-feeding device automatically controlled by a punched paper tape.

tape copy – A message received in tape form as the result of a transmission.

tape deck—The basic assembly of a tape recorder, consisting of the tape-moving mechanism (the tape transport) and a

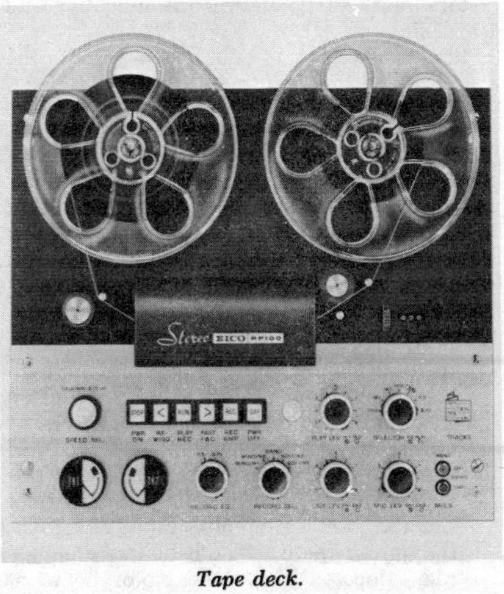

Tape deck.

head assembly. Some decks also include recording and playback preamplifiers; these properly are called tape recorders. Some have playback-only preamplifiers; these have no standard name but are often called "tape players."

tape drive—A mechanism for moving tape past a head; tape transport.

tape feed—A mechanism which feeds the tape to be read or sensed by a computer or other data-handling system.

tape guides—Grooved pins of nonmagnetic material mounted at either side of the recording-head assembly. Their function is to position the magnetic tape on the head as it is being recorded or played.

tape head—The transducer on a tape recorder past which the tape runs during record or replay. It applies a magnetic field to the tape during the recording process and provides electrical output during replay.

tape lifters — A system of movable guides which automatically divert tape from contact with the recorder heads during the fast forward or rewinding mode of operation.

tape limited—Pertaining to a computer operation in which the time required for the reading and writing of tapes is greater than the time required for computation.

tape loop—A length of magnetic tape with the ends joined together to form an endless loop. Used either on a standard recorder, a special message-repeater unit, or in conjunction with a cartridge device, it enables a recorded message to be played back repetitively; there is no need to rewind the tape.

tape magazine — Also called a tape cartridge. A container holding a reel of magnetic recording tape, which can be played without being threaded manually.

tape mark—A special record indicating end of file.

tape monitor—A circuit which permits the checking of recordings by taking the signal directly from the tape a moment after the recording is made. This is only possible on three-head recorders.

tape-on surface-temperature resistor—A surface-temperature resistor installed by adhering the sensing element to the surface with a piece of pressure-sensitive tape.

tape parity—Parity error that occurs when information is transferred to or from magnetic tape.

tape-path center line — The locus or path traced by an imaginary point, located on the recording tape midway between its edges, as it travels from reel to reel through guides, past heads, and between capstan and pressure rollers. For correct tracking and minimum tape distortion, the entire path should lie in a single plane located above the motor board at a height compatible with head-gap locations.

tape phonograph—*See* Tape Player.

tape player—Sometimes called a tape phonograph or a tape reproducer. A unit for playing recorded tapes. It has no facilities for recording.

taper—In communication practice, a continuous or gradual change in electrical properties with length—e.g., as obtained by a continuous change of cross section of a waveguide or by the distribution and change of resistance of a potentiometer or rheostat.

tape recorder—A mechanical-electronic device for recording voice, music, and other audio-frequency material. Sound is converted to electrical energy, which in turn sets up a corresponding magnetic pattern on iron-oxide particles suspended on paper or plastic tape. During playback, this magnetic pattern is reconverted into electrical energy and then changed back to sound through the medium of headphones or a speaker. The recorded material may be converted to a visual display by the use of an oscilloscope, other visual indicator, or a graphic recorder. (*See also* Video Tape Recording.)

tapered potentiometer—A continuously adjustable potentiometer, the resistance of which varies nonuniformly along the element—being greater or less for equal slider movement at various points along the resistance element.

tapered transmission line — *See* Tapered Waveguide.

tapered waveguide—Also called a tapered transmission line. A waveguide in which a physical or electrical characteristic changes continuously with distance along the axis of the guide.

tape relay—A method in which perforated tape is used as the intermediate storage in the process of relaying messages between transmitting and receiving stations.

tape-relay station—A component of a communications center that carries out the function of receiving and forwarding messages by means of tape relay.

tape reproducer—*See* Tape Player.

tape reservoir — That part of a magnetic tape system used to isolate the tape storage inertia (i.e., tape reels, etc.) from the drive system.

tape skew—The deviation of a tape from following a linear path when transported across the heads, causing a time displacement between signals recorded on different tracks and amplitude differences between the outputs from individual tracks owing to variations in azimuth alignment. The adjectives "static" and "dynamic" are used to distinguish between the steady and fluctuating components of the tape skew.

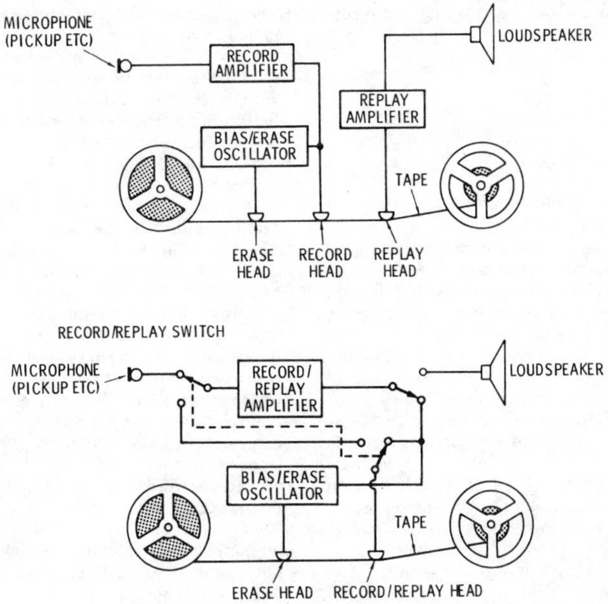

Tape recorder.

tape speed—The speed at which tape moves past the head in the recording or playback mode. The standard tape speed for home use is 7½ ips or half this speed (3¾ ips). One-fourth or even one-eighth this speed is also used, but usually only for music when special high-quality tape is utilized. The professional recording speed for music mastering is usually 15 ips.

tape-speed errors – Any variation in tape speed from the normal speed over the record or reproduce head, regardless of cause.

tape-speed variations—*See* Flutter, 1.

tape splicer – A device for splicing magnetic tape automatically or semiautomatically.

tape station—*See* Tape Unit.

tape threader—A device that makes easier the threading of magnetic recording tape onto the reel.

tape-to-card—Pertaining to equipment or methods used to transfer data from magnetic or punched tape to punched cards.

tape-to-head speed—The relative speed of the tape and head during normal recording or replay. (The tape-to-head speed coincides with the tape speed in conventional longitudinal recording, but it is considerably greater than the tape speed in systems where the heads are scanned across or along the tape.)

tape-to-tape converter—A device for changing from one form of input/output medium or code to another, i.e., magnetic tape to paper tape (or vice versa) or eight-channel code to five-channel code.

tape transmitter—1. A machine actuated by previously punched paper tape and used for high-speed code transmission. 2. A facsimile transmitter designed for transmission of subject copy printed on narrow tape.

tape transport—*See* Transport.

tape-transport mechanism – *See* Motor Board.

tape unit—A device that contains a tape drive and the associated heads and controls. Also called tape station.

tape-wound core—A magnetic core consisting of a plastic or ceramic toroid around which is wound a strip of thin magnetic tape possessing a square-hysteresis-loop characteristic. Also known as bimag, a tape-wound core is used principally as a shift-register element.

tap lead—The lead connected to a tap on a coil winding.

tapped control—A rheostat or potentiometer having a fixed tap at some point along the resistance element, usually to provide fixed grid bias or automatic tone compensation.

tapped line—A delay line in which more than two terminal pairs are associated with a single sonic-delay channel.

tapped resistor—A wirewound fixed resistor having one or more additional terminals along its length, generally for voltage-divider applications.

tapped winding—A coil winding with connections brought out from turns at various points.

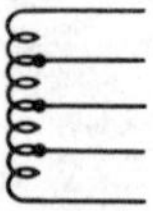

Tapped winding.

tap switch — A multicontact switch used chiefly for connecting a load to any one of a number of taps on a resistor or coil.

target—1. In a camera tube, a structure employing a storage surface which is scanned by an electron beam to generate an output-signal current corresponding to the charge-density pattern stored thereon. 2. Also called an anticathode. In an X-ray tube, an electrode or part of an electrode on which a beam of electrons is focused and from which X-rays are emitted. 3. In radar, a specific object of radar search or surveillance. 4. Any object which reflects energy back to the radar receiver.

target acquisition—In radar operation, the first appearance of a recognizable and useful signal returned from a new target.

target capacitance—In camera tubes, the capacitance between the scanned area of the target and backplate.

target cutoff voltage—In camera tubes, the lowest target voltage at which any detectable electrical signal, corresponding to a light image on the sensitive surface of the tube, can be obtained.

target discrimination—The characteristic of a guidance system that permits it to distinguish between two or more targets in close proximity.

target fade—The loss or decrease of signal from the target due to interference or other phenomena.

target glint—*See* Scintillation, 2.

target identification—A visual procedure by which a radar target is positively identified as either hostile or friendly.

target language—The language into which some other language is to be properly translated.

target noise—Reflections of a transmitted radar signal from a target that has a number of reflecting elements randomly oriented in space.

target reflectivity—The degree to which a target reflects electromagnetic energy.

target scintillation—The apparent random movement of the center of reflectivity of a target observed during the course of an operation.

target seeker—In a missile homing system, the element that senses some feature of the target so that the resulting information can be used to direct appropriate maneuvers to maintain a collision course.

target signature—The characteristic pattern of a given target when displayed by detection and classification equipment.

target voltage—In a camera tube with low-velocity scanning, the potential difference between the thermionic cathode and the backplate.

task—A unit of work for the central processing unit as determined by the control program; therefore, the basic multiprogramming unit under the control program.

task control block—The consolidation of the control information that has to do with a task.

task dispatcher—The control-program function that selects a task from the task queue and gives control of the central processing unit to that task.

task management—Those functions of the control program by which the use of the central processing unit and other resources by tasks is regulated.

task queue—A queue of all the task control blocks present in a system at a given time.

taut-band galvanometer — A galvanometer whose moving coil is suspended between two taut ribbons.

taut-band suspension—In an indicating instrument, a mechanical arrangement in which the moving element is suspended by means of a thin, flat conducting ribbon at each end. The ribbons normally are in tension sufficient to maintain the lateral motion of the moving element within limits that permit the freedom of useful motion for any mounting position of the instrument. A restoring torque is produced within the ribbons when the moving element rotates.

Tchebychev filters—Filter networks that are designed to exhibit a predetermined ripple in the passband (ripple amplitudes from 0.01 dB to 3 dB are common) in exchange for which they provide a more rapid attenuation above cutoff—which, unlike their passband response, is monotonic.

Tchebychev function — A mathematical function the curve of which ripples within certain bounds (*see* Ripple, 2.). This produces an amplitude response more square than that of the Butterworth function, but with less desirable phase and time-delay characteristics. There is an entire family of Tchebychev functions (0.1 ripple, 0.5 ripple, etc.)

T-circulator — A circulator consisting of three identical rectangular waveguides joined asymmetrically to form a T-shaped structure, with a ferrite post or wedge at the center of the structure. Power that enters any waveguide emerges from only one adjacent waveguide.

tdm—*See* Time-Division Multiplex.

tearing—Distortion observed on the television screen when the horizontal synchronization is unstable.

teaser transformer—A transformer of two T-connected, single-phase units for three-phase to two-phase or two-phase to three-phase operation; it is connected between the midpoint of the main transformer and the third wire of the three-phase system.

teasing—In the life-testing of switches, the slow movement of rotor contacts making and breaking with stator contacts.

technical control board—In a switch center or relay station, a testing position at which there are provisions for making tests on switches and associated access lines and trunks.

technical load—The portion of the operational power load of a facility that is required for communications-electronics, tactical-operations, and ancillary equipment. It includes power for lighting, air conditioning, or ventilation necessary for full continuity of communications-electronics operation.

technician—A person who works directly with scientists, engineers, and other professionals in every field of science and technology. Technicians' duties vary greatly, depending on their field of specialization. But in general, the scientist or engineer does the theoretical work, and the technician translates theory into action.

technician license—A class of amateur radio license issued in the United States by the FCC for the primary purpose of operation and experimentation on frequencies above 50 MHz.

tecnetron—A high-power multichannel field-effect transistor similar to a triode tube in that it has anode and cathode connections (and a grid connection between) on opposite ends of a small germanium rod.

tee junction — Also spelled T-junction. A junction of waveguides in which the longitudinal guide axes form a T. The guide which continues through the junction is called the main guide; the one which terminates at a junction, the branch guide.

telautograph—Also called a telewriter. A writing telegraph instrument in which the movement of a pen in the transmitting apparatus varies the current and thereby causes the corresponding movement of a pen at the remote receiving instrument.

telecamera—Acronym for a television camera.

telecardiophone — An amplifying stethoscope which permits heart sounds to be heard at a distance.

telecast—Acronym for television broadcasting—specifically, a television program, or the act of broadcasting a television program.

telecasting — The broadcasting of a television program.

telecommunication—1. All types of systems in which electric or electromagnetic signals are used to transmit information between or among points. Transmission media may be radio, light, or waves in other portions of the electromagnetic spectrum; wire; cable; or any other medium. 2. Data transmission between a computing system and remotely located devices via a unit that performs the necessary format conversion and controls the rate of transmission.

teleconference—A conference between persons who are remote from one another but linked together by a telecommunications system.

telegenic—The suitability of a subject or model for televising.

telegraph—A system that employs interruptions or polarity changes of direct current for the transmission of signals.

telegraph channel—The transmission media and intervening apparatus involved in the transmission of telegraph signals in a given direction between two intermediate telegraph installations. A means of one-way transmission of telegraph signals.

telegraph circuit—A complete circuit over which signal currents flow between transmitting and receiving apparatus in a telegraph system. It sometimes consists of an overhead wire or cable and a return path through the ground.

telegraph concentrator — A switching arrangement by means of which a number of branch or subscriber lines or station sets may be connected to a lesser number of trunk lines, operating positions, or instruments through the medium of manual or automatic switching devices to obtain more efficient use of facilities.

telegraph distributor—A device which effectively associates one direct current or carrier telegraph channel in rapid succession with the elements of one or more signal-sending or signal-receiving devices.

telegraph grade circuit—A circuit suitable for transmission by teletypewriter equipment. Normally the circuit is considered to employ dc signalling at a maximum speed of 75 bauds.

telegraph key—A hand-operated device for opening and closing contacts to modulate the current with telegraph signals.

telegraph-modulated waves — Continuous waves, the amplitude or frequency of which is varied by means of telegraphic keying.

telegraph repeater—Apparatus that receives telegraph signals from one line and retransmits corresponding signals on another line.

telegraph selector—A device which performs a switching operation in response to a definite signal or group of successive signals received over a controlling circuit.

Telegraph key.

telegraph signal distortion—The time displacement of transitions between conditions such as marking and spacing, with respect to their proper relative positions in perfectly timed signals. The total distortion is the algebraic sum of the bias and the characteristic and fortuitous distortions.

telegraph sounder — A telegraph receiving instrument in which an electromagnet attracts an armature each time a pulse arrives. This armature makes an audible sound as it hits against its stops at the beginning and end of each current impulse, and the intervals between these sounds are translated from code into the received message by the operator.

telegraph transmission speed—The rate at which signals are transmitted. This may be measured by the equivalent number of dot-cycles per second or by the average number of letters or words transmitted and received per minute.

telegraph transmitter—A device for controlling a source of electric power in order to form telegraph signals.

telegraph wave—A completely random two-level signal.

telegraphy—1. A system of telecommunication for the transmission of graphic symbols, usually letters or numerals, by the use of a signal code. It is used primarily for record communication. 2. Any system of telecommunication for the transmission of graphic symbols or images for reception in record form, usually without gradation of shade values.

telelectrocardiograph—A device for transmission and remote reception of electrocardiograph signals.

telemeter—1. To transmit analog or digital reports of measurements and observations over a distance (e.g., by radio transmission from a guided missile to a control or recording station on the ground). 2. A complete measuring, transmitting, and receiving apparatus for indicating, recording, or integrating the value of a quantity at a distance by electric translating means.

telemetering—A measurement accomplished with the aid of intermediate means which allow perception, recording, or interpretation of data at a distance from a primary sensor. The most widely employed interpretation of telemetering restricts its significance to data transmitted by means of electromagnetic propagation.

telemeter service—Metered telegraph transmission between paired telegraph instruments over an intervening circuit adapted to serve a number of such pairs on a shared-time basis.

telemetry—1. The science of sensing and measuring information at some remote location and transmitting the data to a convenient location to be read and recorded. 2. The transmission of measurements, obtained by automatic sensors and the like, over communications channels. 3. The practice of transmitting and receiving the measurement of a variable for readout or other uses. The term is most commonly applied to electric signal systems.

telemetry beacon — A system whereby two or more reply pulses are transmitted by the beacon for the transmission of data from the test vehicle to the ground station.

telemetry cable—Cable used for the transmission of information from instruments to the peripheral recording equipment.

telemetry frame—In pcm systems, one complete sampling of words or channels of information at a given rate; in time-division multiplexing, one complete commutator revolution.

telemetry-frame rate — The frequency derived from the period of one frame.

telephone—Combination of apparatus for converting speech energy to electrical waves, transmitting the electrical energy to a distant point, and there reconverting the electrical energy to audible sounds.

telephone capacitor—A fixed capacitor connected in parallel with a telephone receiver to bypass rf and higher audio frequencies and thereby reduce noise.

telephone carrier current—A carrier current used for telephone communication so that more than one channel can be obtained on a single pair of wires.

telephone central office—A switching unit, installed in a telephone system that provides service to the general public, that has the necessary facilities for terminating and interconnecting lines and trunks.

telephone channel—A channel suitable for the transmission of telephone signals.

telephone circuit—A complete circuit over which audio and signaling currents travel between two telephone subscribers communicating with each other in a telephone system.

telephone current—An electric current produced or controlled by the operation of a telephone transmitter.

telephone jack—*See* Phone Jack.

telephone pickup—Any of several devices used to monitor telephone conversations, usually without direct connection to the telephone line and operating on the principle of mutual magnetic coupling.

telephone plug—*See* Phone Plug.

telephone receiver—The earphone used in a telephone system.

telephone repeater—An assemblage of amplifiers and other equipment employed at points along the line to rebuild the signal strength in a telephone circuit.

telephone ringer—An electric bell that operates on low-frequency alternating or pulsating current and is used for indicating a telephone call to a station being alerted.

telephone system—A group of telephones plus the lines, trunks, switching mechanisms, and all other accessories required to interconnect the telephones.

telephone transmitter—A microphone used in a telephone system. (*See also* Microphone.)

telephony—1. The transmission of speech current over wires, enabling two persons to converse over almost any distance. 2. A telecommunications system for transmitting speech or other sounds.

telephoto—Also called telephotography. A photoelectrical transmission system for point-to-point or air-to-ground transmission of high-definition pictorial information.

telephotography—*See* Telephoto.

telephoto lens—A lens system that is physically shorter than its rated focal length. It is used in still, movie, and television cameras to enlarge images of objects photographed at comparatively great distances.

teleprinter — 1. *See* Printer, 1. 2. Trade name used by Western Union for its telegraph terminal equipment. 3. *See* Teletypewriter.

teleprocessing—A form of information handling in which the data-processing system operates in conjunction with communication facilities (originally a trademark of International Business Machines Corp.).

teleran — A navigational system in which radar and television transmitting equipment are employed on the ground, with television receiving equipment in the aircraft, to televise the image of the ground radar ppi scope to the aircraft along with map and weather data.

telering—A frequency-selector device for the production of ringing power.

telesynd — Telemeter or remote-control equipment which is synchronous in both speed and position.

Teletype—A trademark of Teletype Corporation for a series of teleprinter equipment such as tape punches, reperforators, page printers, etc., used in communications systems.

teletypewriter—1. See Printer, 1. 2. A generic term referring to the basic equipment made by Teleytpe Corporation and to teleprinter equipment. The teletypewriter uses electromechanical functions to generate codes (Baudot) in response to a human input to a manual keyboard. 3. Also known as a teleprinter. A keyboard machine that can transmit and receive alphabetical, numerical, and certain control (nonprinting) characters as a train of pulses on two wires. Attachments can be fitted for punching paper tape and printing on a roll of paper at the same time, also for reading tape and printing the message that is read.

teletypewriter code — A special code in which each code group is made up of five units, or elements, of equal length which are known as marking or spacing impulses. The five-unit start-stop code consists of five signal impulses preceded by a start impulse and followed by a stop impulse. Each impulse except the stop impulse is 22 milliseconds in length; the stop impulse is 32 milliseconds (based on 60-word-per-minute operation).

teletypewriter exchange service—1. A commercial service that provides teletypewriter communication on the same basis as telephone service, through central switchboards to stations in the same city or other cities. 2. Abbreviated twx. An AT&T public switched teletypewriter service in which teletypewriter stations are provided with lines to a central office for access to other such stations throughout the U.S.A. and Canada.

teletypewriter signal distortion—With respect to a stop-start teletypewriter signal, a shift of the transition points of the signal pulses from their proper positions in relation to the beginning of the start pulse. The magnitude of the distortion is expressed as a percentage of a perfect unit pulse length.

teletypewriter switching system — A total message switching system the terminals of which are teletypewriter equipment.

teletypewriter test tape—A tape perforated so that it contains the identification of the transmitting station followed by repetitions of the letters RY and a test that consists of letters and figures.

televise—The act of converting a scene or image field into a television signal.

television—Abbreviated tv. A telecommunication system for transmission of transient images of fixed or moving objects.

television and radar navigation—A navigational system which; a. employs ground-based search radar equipment along an airway to locate aircraft flying near that airway, b. transmits, by television means, information pertaining to these aircraft and other information to the pilots of properly equipped aircraft, and c. provides information to the pilots appropriate for use in the landing approach.

television broadcast band—The frequencies assignable to television broadcast stations in the band extending from 54 to 806 MHz. These frequencies are grouped into channels, as follows: Channel 2 through 4, 54 to 72 MHz; Channels 5 and 6, 76 to 88 MHz; Channels 7 through 13, 174 to 216 MHz; and Channels 14 through 69, 470 to 806 MHz.

television broadcast station — A radio station for transmitting visual signals, and usually simultaneous aural signals, for general reception.

television camera—A camera that contains an electronic image tube in place of a photographic film. The image formed on the tube face by a lens is scanned rapidly by a moving electron beam. The beam current varies with the local brightness of the image, which is transmitted to the viewer's set where it controls the brightness of the scanning spot in a cathode ray tube. The scanning spots at the camera and the viewing tube must be accurately synchronized.

television channel—A channel suitable for the transmission of television signals. The channel for associated sound signals may or may not be considered part of the television channel.

television engineering — *See* Radio Engineering.

television interference—Interference in the reception of the sound and/or video portion of a television program by a transmitter or another device.

television pickup station — A land mobile station used for the transmission of television program material and related communications from the scenes of events occurring at points removed from television broadcast station studios, to the television broadcast stations.

television picture monitor — A special-purpose television set for displaying picture signals in broadcast or closed-circuit television systems. Applications are in studio master control, for tape monitoring, for control of picture quality in studios and intercity network relays, and for the display of pictures for audiences.

television radar air navigation—A system in which aircraft positions are determined by ground radar and the resulting ppi display, superimposed on a map, is transmitted to the aircraft by television. By this means, each pilot can observe the position of his aircraft in relation to others.

television receiver — A radio receiver for converting incoming electric signals into television pictures and the associated sound.

television reconnaissance — Air reconnaissance by optical or electronic means to supplement photographic and visual reconnaissance.

television relay system—A system of two or more stations for transmitting television relay signals from point to point, using radio waves in free space as a medium. Such transmission is not intended for direct reception by the public.

television repeater—A repeater used in a television circuit.

television screen—In a television receiver, the fluorescent screen of the picture tube.

television signal — The audio signal and video signal that are broadcast simultaneously to produce the sound and picture portions of a televised scene.

television transmitter—The aggregate radio-frequency and modulating equipment necessary to supply, to an antenna system, the modulated radio-frequency power by which all component parts of a complete television signal (including audio, video, and synchronizing signals) are concurrently transmitted.

televoltmeter — A telemeter that measures voltage.

telewattmeter—A telemeter that measures power.

telewriter—*See* Telautograph.

telex — 1. An audio-frequency teleprinter system used in Great Britain to provide teletypewriter service over telephone lines. 2. An automatic teleprinter exchange service provided by Western Union; it is similar to twx, but is worldwide. Only Baudot equipment is provided, but business machines may be used also. 3. A dial-up telegraph service enabling its subscribers to communicate directly and temporarily among themselves by means of start-stop apparatus and circuits of the public telegraph network. The service operates world-wide. Computers can be connected to the Telex network.

telluric current—*See* Earth Current.

telpak — A communications-carrier service for the leasing of wide-band channels between points.

Telstar—A low-altitude active communications satellite used for microwave communication and satellite tracking.

temperature coefficient—1. A factor used to calculate the change in the characteristics of a substance, device, or circuit element with changes in its temperature. 2. The percentage change in the output voltage (or current) of a regulated power supply due to a variation of ambient temperature. The values are usually expressed as a percentage per degree Celsius and restricted to the specified ambient range of the unit.

temperature coefficient of frequency—The rate at which the frequency changes with temperature, generally expressed in hertz per megahertz per degree Celsius at a given temperature.

temperature coefficient of permeability—A coefficient expressing the change in permeability as the temperature rises or falls. It is expressed as the rate of change in permeability per degree.

temperature coefficient of resistance – 1. The ratio of the change in resistance (or resistivity) to the original value for a unit change in temperature. The temperature coefficient over the temperature range from t to t_1, referred to the resistance R_t at temperature t, is determined by the following ratio:

$$\frac{R_{t1} - R_t}{R_t(t_1 - t)}$$

where t is the temperature, preferably in degrees C. The value will be positive unless otherwise indicated by a negative sign. 2. The maximum change in resistance per unit change in temperature, usually referred to in parts per million (ppm) per degree Celsius and specified over a temperature range. The temperature is that of the resistor itself, not ambient temperature.

temperature coefficient of voltage drop – The change in the voltage drop of a glow-discharge tube, divided by the change in ambient temperature or in the temperature of the envelope.

temperature coefficient value – The expected percentage change per degree of temperature difference from a specified temperature.

temperature-compensated zener diode – A positive-temperature-coefficient reversed-bias zener diode (pn junction) connected in series with one or more negative-temperature forward-biased diodes within a single package.

temperature-compensating capacitor—A capacitor the capacitance of which varies with temperature in a known and predictable manner. Normally this characteristic is specified with a P or N (to indicate the direction of change) followed by a number that indicates the change in parts per million per degree Celsius (centigrade). Such capacitors are used extensively in oscillator circuits to compensate for changes due to temperature variations in the values of other components.

temperature compensation – The process whereby the effects of an increase or decrease in ambient temperature are canceled (e.g., as in the case of an oscillator that is required to maintain a stable output frequency regardless of ambient temperature changes).

temperature control—1. A switch actuated by a thermostat responsive to changes in temperature, and used to maintain temperature within certain limits. 2. A control device responsive to temperature.

temperature cycling—A type of accelerated test in which systems or devices are subjected alternately to high and low temperatures to simulate diurnal temperature fluctuation.

temperature derating—Lowering the voltage, current, or power rating of a device or component when it is used at elevated temperatures.

temperature detector—An instrument used to measure the temperature of a body. Any physical property that is dependent on temperature may be employed, such as the differential expansion of two bodies, thermoelectromotive force at the junction of two metals, change of resistance of a metal, or the radiation from a hot body.

temperature-limited – The condition of a cathode when all the electrons emitted from it are drawn away by a strong positive field. The only way to increase the flow of electrons is to raise the cathode temperature.

temperature relay—A relay which functions at a predetermined temperature.

temperature rise—The difference between the initial and final temperature of a component or device. Temperature rise is expressed in degrees C or F, usually referred to an ambient temperature, and equals the hot-spot temperature minus the ambient temperature.

temperature saturation—*See* Filament Saturation.

temperature sensor – *See* Thermistor and Thermocouple.

temperature shock – A rapid change from one temperature extreme to another.

temperature-wattage characteristic – In a thermistor, the relationship, for a specific ambient temperature, between the temperature of the thermistor and the applied steady-state power.

temporary magnet—A magnetized material having a high permeability and low retentivity.

temporary storage – Internal storage locations in a computer reserved for intermediate and partial results.

tem wave—Abbreviation for transverse electromagnetic wave.

10-Code – Abbreviations used by CB'ers and other radio communications users to minimize use of air time.

tens complement—1. An arithmetic process employed in a computer to perform decimal subtractions through the use of addition techniques. The tens complement negative of a number is obtained by individually subtracting each digit in the number from 9 and adding 1 to the result. 2. The radix complement in decimal notation.

tensilized tape – A variety of polyester backing that has been "prestretched" to prevent further severe elongation when subjected to excessive tension. A tape that does not stretch before breaking can be spliced back together without loss of program material.

tensiometer–A device for determining the tautness of a supporting wire or cable.

tension – 1. Mechanical–the condition of strain which tends to stretch. 2. Electrical –the potential or electrostatic voltage.

tenth-power width–In a plane containing the direction of the maximum of a lobe, the full angle between the two directions in that plane, about the maximum, in which the radiation intensity is one-tenth the maximum value of the lobe.

tera–Prefix for the numerical quantity of 10^{12}. Abbreviated T.

terahertz – One million megahertz, or 10^{12} hertz. Abbreviated THz.

teraohm – One million megohms, or 10^{12} ohms.

teraohmmeter – An instrument used to measure extremely high resistance.

terminal–1. A point of connection for two or more conductors in an electrical circuit. 2. A device attached to a conductor to facilitate connection with another conductor. 3. A point in a system or communication network at which data can be either inserted or removed. 4. A device that permits access to a central computer; for example, a teletypewriter, electric typewriter, or graphic terminal. 5. Any device capable of sending and/or receiving information over a communications channel; the means by which data are entered into a computer system and by which the decisions of the system are communicated to the environment it affects. A wide variety of terminal devices have been built, including teleprinters, special keyboards, light displays, cathode-ray tubes, thermocouples, pressure gauges and other instrumentation, radar units, and telephones.

terminal area–1. A portion of a microelectronic circuit used for making electrical connections to the conductor pattern–e.g., an enlarged pad (area) on a semiconductor die. 2. A portion of a printed circuit used for making electrical connections to the conductive pattern, such as the enlarged portion of conductor material surrounding a component mounting hole.

terminal block–An insulating base or slab equipped with one or more terminal connectors for the purpose of making electrical connections thereto.

terminal board – Also called a terminal strip. An insulating base or slab equipped with terminals for connecting wiring.

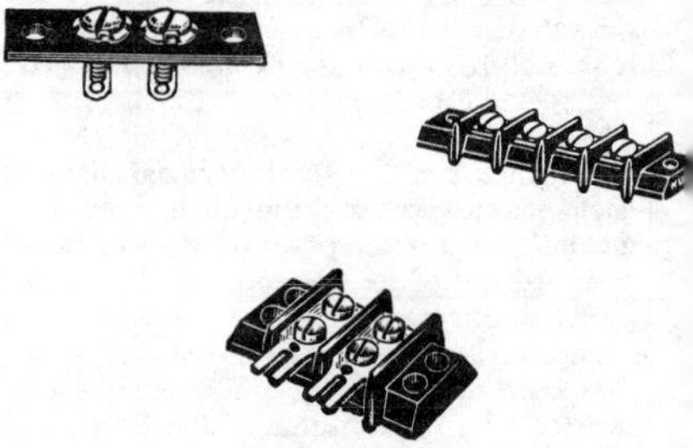

Terminal boards.

terminal box–A housing where cable pairs are brought out to terminations for connections.

terminal brush–A brush with long bristles for cleaning fuses and terminals in a terminal box.

terminal cutout pairs – Numbered, designated pairs brought out of a cable at a terminal.

terminal equipment – 1. At the end of a communications channel, the equipment essential for controlling the transmission and/or reception of messages. 2. Telephone and teletypewriter switchboards and other centrally located equipment to which wire circuits are terminated. 3. Assemblage of communications-type equipment required to transmit and/or receive a signal on a channel or circuit, whether it be for delivery or relay. 4. In radio relay systems, usually refers to equipment used at points where intelligence is inserted or derived, as distinct from equipment used to relay a reconstituted signal.

terminal guidance – 1. Guidance applied to a guided missile between midcourse guidance and arrival at the target. 2. Electronic, mechanical, visual, or other assistance given an aircraft pilot to facilitate arrival at, landing upon, or departure from an air landing or air-drop facility.

terminal impedance–1. The complex impedance seen at the unloaded output or input terminals of transmission equipment or a line in otherwise normal operating condition. 2. *See* Terminal Resistance.

terminal leg–*See* Terminal Stub.

terminal lug–1. A threaded lug to which a wire may be fastened in a terminal box. 2. A cylindrical piece of metal, either solid or hollow and of two or more diameters, which can be stacked, flared, swaged, or pressed into a hole for the purpose of connecting leads or external wires to the conductive pattern.

terminal pad–An alternate term for "terminal area" or "pad."

terminal pair—An associated pair of accessible terminals (e.g., the input or output terminals of a device or network).

terminal repeater — 1. An assemblage of equipment designed specifically for use at the end of a communication circuit—as contrasted with the repeater, which is designed for an intermediate point. 2. Two microwave terminals arranged to provide for the interconnection of separate systems, or separate sections of a system.

terminal resistance — The total resistance measured between the input terminals of a meter. For an ac meter, it is the effective dc resistance measured by the voltage-doubling or substitution technique with rated end-scale input of the appropriate frequency applied. Also called terminal impedance.

terminal room — In telephone practice, a room associated with a central office, private branch exchange, or private exchange, which contains distributing frames, relays, and similar apparatus.

terminal station — The microwave equipment and associated multiplex equipment employed at the ends of a microwave system.

terminal strip—*See* Terminal Board.

terminal stub—Also called terminal leg. A piece of cable which comes with a cable terminal for splicing into the main cable.

terminal unit—Equipment usable on a communication channel for either input or output.

terminal vhf omnirange — Very-high-frequency omnirange, normally low powered, complete with a local monitoring device which will automatically shut down the facility if it is not operating properly.

terminated line — A transmission line terminated in a resistance equal to the characteristic impedance of the line, so that there is no reflection or standing waves.

terminating—The closing of the circuit at either end of a line or transducer by connection of some device. Terminating does not imply any special condition, such as the elimination of reflection.

termination — 1. A load connected to a transmission line or other device. To avoid wave reflections, it must match the characteristic impedance of the line or device. 2. A waveguide technique; the point at which energy flowing along a waveguide continues in a nonwaveguide mode of propagation. 3. The terminals at an antenna to which the transmission line is connected (screw terminals, solder connections, coaxial connector, etc.).

termination block—A nonconductive material on which are provided several termination points.

ternary—1. A numerical system of notation using the base 3 and employing the characters 0, 1, and 2. 2. Able to assume three distinct states.

ternary code—A code in which each element may be any one of three distinct kinds or values.

ternary pulse-code modulation—A form of pulse-code modulation in which each element of information is represented by one of three distinct values, e.g., positive pulses, negative pulses, and spaces.

terrain-avoidance radar — Airborne radar which provides a display of terrain ahead of a low-flying airplane to permit horizontal avoidance of obstacles.

terrain-clearance indicator — A device for measuring the distance from an aircraft to the surface of the sea or earth.

terrain error—In navigation, the error resulting from distortion of the radiated field by the nonhomogeneous characteristics of the terrain over which the radiation in question has been propagated.

terrain-following radar — Airborne radar which provides a display of terrain ahead of a low-flying aircraft to permit manual control, or signals for automatic control, to maintain constant altitude above the ground.

terrestrial-reference flight—Stabilized flight in which control information is obtained from terrestrial phenomena (e.g., flight in which basic information derived from the magnetic field of the earth, atmospheric pressure, and the like is fed into a conventional automatic pilot).

tertiary coil—A third coil used in the output transformer of an audio amplifier to supply a feedback voltage.

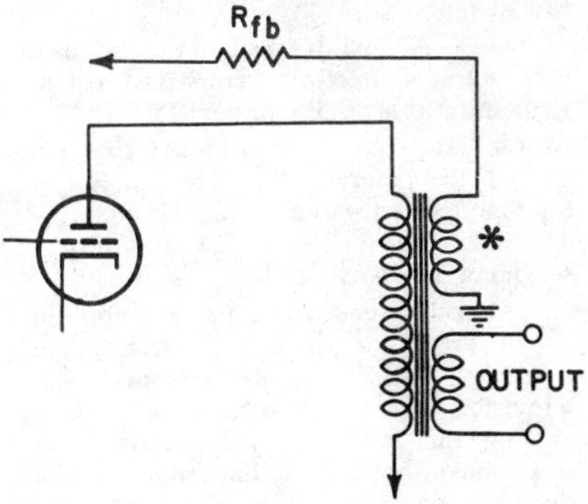

Tertiary coil.

tertiary winding—1. A winding added to a transformer, in addition to the conventional primary and secondary windings, to suppress third harmonics or to make connections to a power-factor–correcting device. 2. *See* Stabilized Winding.

tesla—A unit of magnetic induction equal

to 1 weber per square meter in the mksa system.

Tesla coil—An air-core transformer used for developing high-voltage discharge at a very high frequency. It has a few turns of heavy wire as the primary and many turns of fine wire as the secondary.

test—A procedure or sequence of operations for determining the manner in which equipment is functioning or the existence, type, and location of any trouble.

test bench — Equipment designed specifically for making overall bench tests on equipment in a particular test setup under controlled conditions.

test board—A switchboard equipped with testing apparatus, arranged so that connections can be made from it to telephone lines or central office equipment for testing purposes.

test clip—A spring clip fastened to the end of an insulated wire to enable quick temporary connections when circuits or devices are being tested.

test lead — A flexible, insulated lead wire that usually has a test prod on one end. It is ordinarily used for making temporary electrical connections. The insulation normally is rubber; the standard colors are red and black.

testing level—The value of power used for reference represented by 0.001 watt working into 600 ohms.

test loop—A cycle of tests that can be repeated over and over, e.g., to locate intermittent faults.

test jack—1. A jack that makes a circuit or circuit element available for testing purposes. 2. In recent practice, a jack that is multiplied with the operating jack on the switchboard.

test lead — A flexible insulated lead used chiefly for connecting meters and test instruments to a circuit under test.

test oscillator—A test instrument that can be set to generate an unmodulated or tone-modulated radio-frequency signal at any frequency needed for aligning or servicing receivers and/or amplifiers.

test pattern—A geometric pattern containing a group ot lines and circles, and used for testing the performance of a television receiver or transmitter by revealing the following video-signal characteristics: horizontal linearity, vertical linearity, contrast, aspect ratio, interlace, streaking, ringing, vertical resolution, and horizontal resolution.

test point—A connection to which no instrument is permanently connected, but which is intended for temporary, intermittent, or future connection of an instrument.

test prod — A sharp metal point used for making a touch connection to a circuit terminal. It has an insulated handle and a means for electrically connecting the point to a test lead.

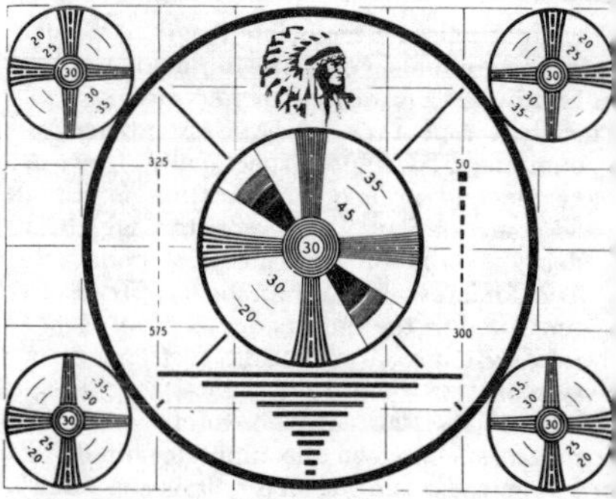

Test pattern.

test record—A phonograph disc designed to test the quality and characteristics of turntables, pickups, amplifiers, etc.

test routine (in a computer)—1. A synonym for check routine. 2. Generally both the check and the diagnostic routines.

test set—One or more instruments required for servicing of a particular type of equipment.

test tone—A tone used in circuit identification for purposes of locating trouble or making adjustments.

tetrad—A group of four, especially a group of four pulses used to express a digit in the scale of 10 or 16.

tetrode—A four-electrode electron tube containing an anode, a cathode, a control electrode (grid), and one additional electrode that is ordinarily a screen grid.

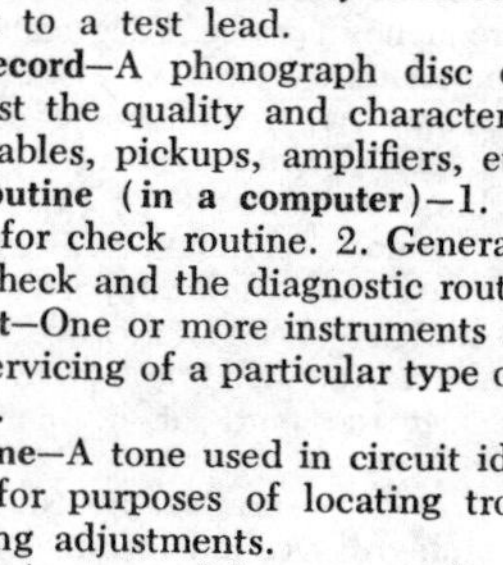

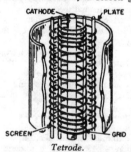

Tetrode.

tetrode junction transistor — *See* Double-Base Junction Transistor.

tetrode transistor — A junction transistor with two electrode connections to the base to reduce interelement capacitance (in addition to the normal emitter and collector elements, each having one connection).

Te value—The temperature at which the resistance of a centimeter cube is 1 megohm.

te wave—Abbreviation for transverse electric wave.

text—1. In USASCII and communications, a sequence of characters treated as an entity if it is preceded by one STX communication control character and terminated by one EXT communication control character. 2. The control sections of an object or load module, considered together.

T flip-flop—1. Also called binary. A type of flip-flop the outputs of which change state each time the input-signal voltage falls from 1 to 0 and remain unchanged when the input-signal voltage rises from 0 to 1. Thus, there is one change in output state for every two changes in input signal. Thus the frequency of the output is half the frequency of the input. 2. A flip-flop with only one input. When a pulse appears on the input, the flip-flop changes states. (Used in ripple counters.)

thallofide cell – A photoconducting cell which has thallium oxysulfide as the light-sensitive agent.

thd—Abbreviation for total harmonic distortion.

theoretical acceleration at stall – A figure of merit derived from the stall-torque to rotor-inertia ratio, which indicates how rapidly a motor will accelerate from stall.

theoretical cutoff – *See* Theoretical Cutoff Frequency.

theoretical cutoff frequency – Also called theoretical cutoff. The frequency at which, disregarding the effects of dissipation, the attenuation constant of an electric structure changes from zero to a positive value or vice versa.

theoretical electrical travel – The shaft travel over which the theoretical function characteristic of a precision potentiometer extends, as determined from the index point.

theremin—An electronic musical instrument consisting of two radio-frequency oscillators which beat against each other to produce an audio-frequency tone, in a manner similar to a beat-frequency audio oscillator. The pitch and volume are varied by hand capacitance.

thermal—A general term for all forms of thermoelectric thermometers, including a series of couples, thermopiles, and single thermocouples.

thermal agitation—1. Movement of the free electrons in a material. In a conductor they produce minute pulses of current. When these pulses occur at the input of a high-gain amplifier in the conductors of a resonant circuit, the fluctuations are amplified together with the signal currents and heard as noise. 2. Also called thermal effect. Minute voltages arising from random electron motion, which is a function of absolute temperature expressed in degrees Kelvin. 3. In a semiconductor, the random movement of holes and electrons within a crystal due to the thermal (heat) energy.

thermal-agitation voltage – The potential difference produced in circuits by thermal agitation of the electrons in the conductor.

thermal ammeter—*See* Hot-Wire Ammeter.

thermal breakdown – 1. A form of breakdown in which decomposition or melting occurs due to the temperature rise resulting from the applied electric stress. 2. A runaway condition in a dielectric, the loss factor of which increases with temperature. Dielectric loss heats the material, producing an increase in temperature. Therefore, the dielectric loss increases still more, producing a further increase in temperature, and so on.

thermal circuit breaker—A circuit breaker the operation of which depends on temperature expansion due to electrical heating.

thermal coefficient of resistance – The change in the resistivity of a substance due to the effects of temperature only. Usually expressed in ohms per ohm per degree change in temperature.

thermal compensation – A method employed to reduce or eliminate the thermal effects on one or more of the performance parameters of a transducer.

thermal compression bonding – Diffusion bonding where two carefully prepared surfaces are brought into intimate contact under carefuly controlled conditions of temperature, time, and clamping pressure. Plastic deformation is induced by the combined effects of pressure and temperature, which in turn results in atom movement causing the development of a crystal lattice bridging the gap between the facing surfaces and results in bonding. (Time is a critical factor in controlling the ambient temperature at the area to be bonded and the size of the bond that is formed.) Generally, the process is performed under a protective atmosphere of inert gas to keep the surfaces to be bonded clean while they are being heated.

thermal conduction – 1. The transfer of thermal energy by processes having no net movement of mass and having rates proportional to the temperature gradient. 2. The rate of flow of heat through a material by thermal conduction.

thermal conductivity – A measure of the ability of a substance to conduct heat. Expressed in terms of calories of heat conducted per second per sq cm per cm of thickness per degree C difference in temperature from one surface to the other.

thermal conductor—A material which readily transmits heat by conduction.

thermal contraction—The shrinkage exhibited by most metals when cooled.

thermal converter—Also called thermocouple converter, thermoelectric converter, thermoelectric generator, or thermoelement. On or more thermojunctions in thermal contact with an electric heater or integral, so that the electromotive force developed by thermoelectric action at the output terminals gives a measure of the input current in the heater.

thermal cutout—1. An overcurrent protective device which contains a heater element that affects a fusible member and thereby opens the circuit. 2. A heat-sensitive switch that automatically opens the circuit of an electrical device when the operating temperature of the device exceeds a predetermined value.

thermal detector—*See* Bolometer.

thermal drift—A change in the output of a regulated power supply over a period of time, due to changes in internal ambient temperatures not normally related to environmental changes. Thermal drift is usually associated with changes in line voltage and/or load changes.

thermal effect—*See* Thermal Agitation, 2.

thermal emf—The electromotive force generated when the junction of two dissimilar metals is heated. (*See also* Seebeck Emf.)

thermal endurance — An indication of the relative life expectancy of a product when exposed to operating temperatures much higher than normal room temperature.

thermal equilibrium — The condition that exists when a system and its surroundings are at the same temperature.

thermal expansion—Physical expansion resulting from an increase in temperature; it may be linear and volumetric.

thermal flasher — An electric device that automatically opens and closes a circuit at regular intervals, owing to alternate heating and cooling of a bimetallic strip heated by a resistance element in series with the circuit being controlled.

thermal generation—The creation of a hole and a free electron by freeing a bound electron through the addition of heat energy.

thermal instrument — An instrument that depends on the heating effect of an electric current for its operation (e.g., thermocouple and hot-wire instruments).

thermal ionization—Ionization due to high temperature (e.g., in the electrically conducting gases of a flame).

thermal junction—*See* Thermocouple.

thermal lag—The time expended in raising the entire mass of a cathode structure to the temperature of the heater.

thermal life—The operating life of a device under varying ambient temperatures.

thermal microphone — A microphone depending for its action on the variation in the resistance of an electrically heated conductor that is being alternately increased and decreased in temperature by sound waves.

thermal noise — 1. Also called resistance noise. Random circuit noise associated with the thermodynamic interchange of energy necessary to maintain thermal equilibrium between the circuit and its surroundings. (*See also* Johnson Noise). 2. Noise generated by the random thermal motion of charged particles.

thermal noise level — The equivalent rms voltage value, over a stated bandwidth, of all energy components generated by a resistor at a stated resistor temperature with no externally supplied current flowing through the resistor.

thermal protector—A current- and temperature-responsive device used to protcet another device against overheating due to overload.

thermal radiation — Commonly known as heat. Radiation produced by the action of heat on molecules or atoms. Its frequency extends between the extremes of infrared and ultraviolet.

thermal rating—A statement of the permissible temperature rating beyond which unsatisfactory performance occurs.

thermal regenerative cell—A fuel-cell system in which there is continuous regeneration of the reactants from the products formed during the cell reaction.

thermal relay—A relay that responds to the heating effect of an energizing current, rather than to the electromagnetic effect. Delay between the start of the energizing current and the switching response is generally predictable and sometimes adjustable.

thermal resistance—1. Ratio of the temperature rise to the rate at which heat is generated within a device under steady-

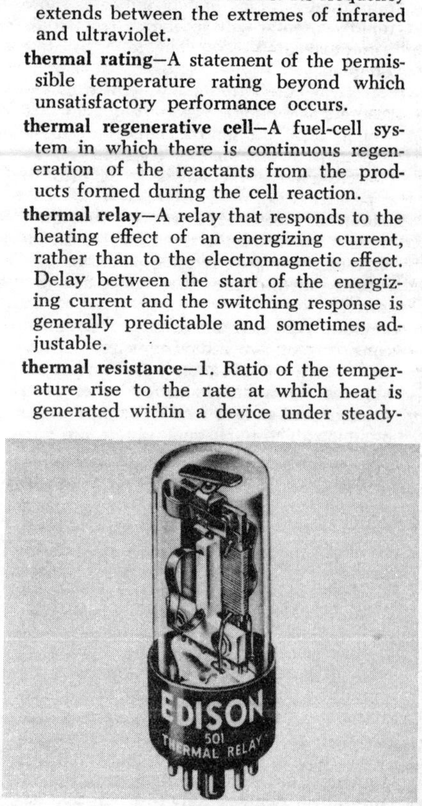

Thermal relay.

state conditions. 2. The resistance of a substance to the conductivity of heat.

thermal resistor — An electronic device which makes use of the change in resistivity of a semiconductor with changes in temperature.

thermal response time—The time from the occurrence of a step change in power dissipation until the junction temperature reaches 90% of the final value of junction-temperature change, when the device-case or ambient temperature is held constant.

thermal runaway—A regenerative condition in a transistor, where heating at the collector junction causes collector current to increase, which in turn causes more heating, etc. The temperature can rapidly approach levels that are destructive to the transistor.

thermal sensitivity set — A permanent change in sensitivity due to temperature effects only. Usually expressed as the difference in sensitivity at room temperature before and after a temperature cycle over the operating temperature range of the transducer.

thermal shock — 1. A sudden, marked change in the temperature of the medium in which a component or device operates. 2. The effect of heat or cold applied at such a rate that nonuniform thermal expansion or contraction occurs within a given material or combination of materials.

thermal telephone receiver — Also called thermophone. An electroacoustic transducer, such as a telephone receiver, in which the temperature of a conductor is caused to vary in response to the current input, thereby producing sound waves as a result of the expansion and contraction of the adjacent air.

thermal time constant—1. The time from the occurrence of a step change in power dissipation until the junction temperature reaches 63.2% of the final value of junction-temperature change, when the device-case or ambient temperature remains constant. 2. In a thermistor, the time required for 63.2% of the change from initial to final body temperature after the application of a step change in temperature under zero-power conditions.

thermal time-delay relay—A type of relay in which the time interval between energization and actuation is determined by the thermal storage capacity of the actuator critical operating temperature, power input, and thermal insulation.

thermal time-delay switch—An overcurrent-protective device containing a heater element and thermal delay.

thermal tuning—Adjusting the frequency of a cavity resonator by using thermal expansion to vary its shape.

thermal tuning time (cooling)—In microwave tubes, the time required to tune through a specified frequency range when the tuner power is instantaneously changed from the specified maximum to zero. The initial condition must be one of equilibrium.

thermal tuning time (heating)—In microwave tubes the time required to tune through a specified frequency range when the tuner power is instantaneously changed from zero to the specified maximum. The initial condition must be one of equilibrium.

thermic—Pertaining to heat.

thermion—An ion, either positive or negative, which has been emitted from a heated body. Negative thermions are electrons (thermoelectrons).

thermionic—Pertaining to the emission of electrons by heat.

thermionic cathode—*See* Hot Cathode.

thermionic converter — Also called thermionic generator or thermoelectron engine. A device which produces electrical power directly from heat. One type contains a heated cathode to emit electrons and a cold anode to collect them, thereby causing a flow of current. Both electrodes are enclosed in a vacuum or gas-filled envelope.

thermionic current — Current due to directed movements of thermions (e.g., the flow of electrons from the cathode to the plate in a thermionic vacuum tube).

thermionic detector—A detector circuit in which a thermionic vacuum tube delivers an audio-frequency signal when fed with a modulated radio-frequency signal.

thermionic diode—A diode electron tube which has a heated cathode.

thermionic emission—Emission of electrons from a solid body as a result of elevated temperature. *See also* Edison Effect.

thermionic energy conversion—The direct production of electricity by means of the electron emission from a heated substance.

thermionic generator — *See* Thermionic Converter.

thermionic grid emission—Also called primary grid emission. The current produced by the electrons thermionically emitted from a grid. Generally it is due to excessive grid temperatures or to contamination of the grid wires by cathode-coating material.

thermionic rectifier—A rectifier utilizing a thermionic vacuum tube to convert alternating current into unidirectional current.

thermionic tube—*See* Hot-Cathode Tube.

thermionic work function—The energy required to transfer electrons from a given metal to a vacuum or some other adjacent medium during thermionic emission.

thermistor — A thermally sensitive solid-

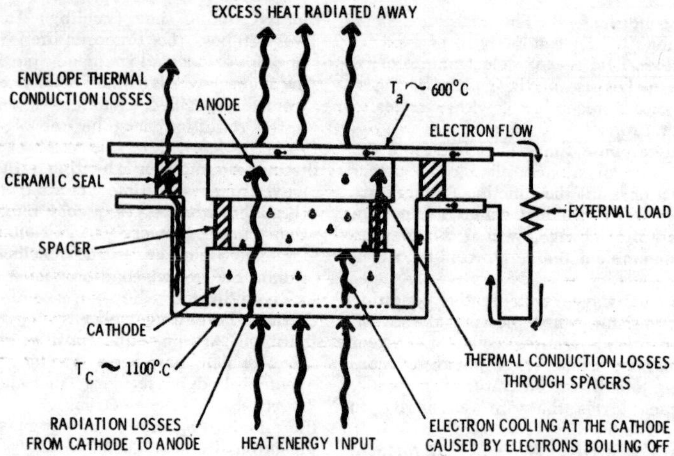

Thermionic converter.

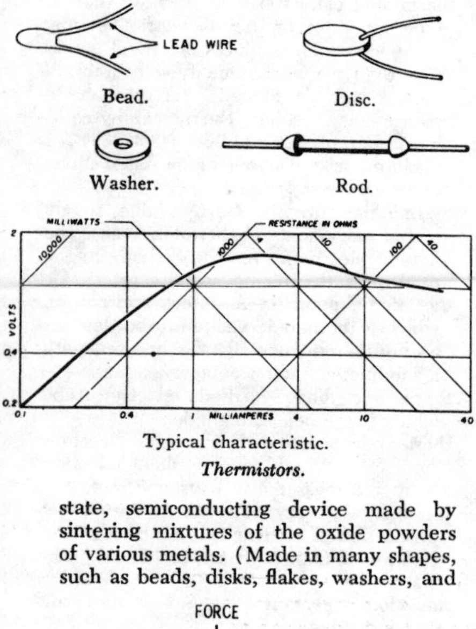

Typical characteristic.

Thermistors.

state, semiconducting device made by sintering mixtures of the oxide powders of various metals. (Made in many shapes, such as beads, disks, flakes, washers, and rods, to which contact wires are attached.) As its temperature is increased, the electrical resistance (typically) decreases. The associated temperature coefficient of resistance is extremely high, nonlinear, and most frequently negative. The large temperature coefficients and the nonlinear resistance temperature characteristics of thermistors enable them to perform many unique regulatory functions.

thermoammeter—Also called a thermocouple ammeter. An ammeter that is actuated by the voltage generated in a thermocouple through which the current to be measured is sent. It is used chiefly for measuring radio-frequency currents.

thermocompensator—In pH meters, a temperature-sensitive device sometimes used to make electronic adjustments in the circuit that are required due to changes in the temperature of the solution.

thermocompression bond – *See* Thermal Compression Bond.

thermocompression bonding—1. A method

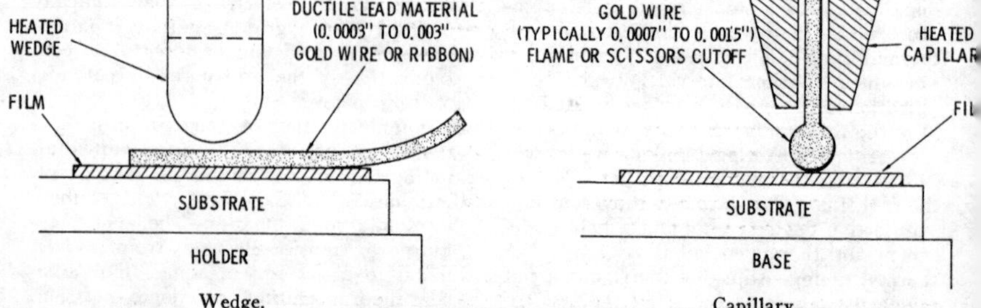

Wedge. Capillary.

Thermocompression bonding.

of interconnecting ICs in a circuit by bonding thin gold wires between conducting patterns of a circuit and to the IC chip's metal preforms by means of heat and pressure. 2. A bond formed by two elements through the simultaneous application of heat and pressure. No additional materials are used to assist the fusing. Common types of thermocompression bonds are wedge, ball (nailhead), and stitch.

thermocouple – Also called thermal junction. Temperature transducer comprising a closed circuit made of two different metals. If the two junctions are at different temperatures, an electromotive force is developed that is proportional to the temperature difference between the junctions. This is called the Seebeck effect.

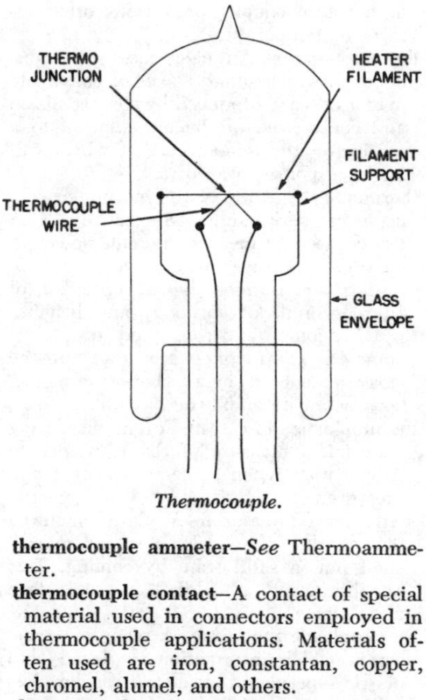

Thermocouple.

thermocouple ammeter–*See* Thermoammeter.

thermocouple contact–A contact of special material used in connectors employed in thermocouple applications. Materials often used are iron, constantan, copper, chromel, alumel, and others.

thermocouple converter–*See* Thermal Converter.

thermocouple instrument – An electrothermic instrument in which one or more thermojunctions are heated by an electric current, causing a direct current to flow through the coil of a suitable direct-current mechanism such as one of the permanent-magnet, moving-coil type.

thermocouple lead wire–An insulated pair of wires used from the thermocouple to a junction box or to the recording instrument.

thermocouple thermometer – *See* Thermoelectric Thermometer.

thermocouple vacuum gage – A vacuum gage which depends for its operation on the thermal conduction of the gas present. The pressure being measured is a function of the electromotive force of a thermocouple, the measuring junction of which is in thermal contact wtih a heater carrying a constant current. Thermocouple vacuum gages ordinarily are used over a pressure range of 10^{-1} to 10^{-3} mm Hg.

thermocouple wire – A wire drawn from special metals or alloys and calibrated to established specifications for use as a thermocouple pair (e.g., iron, constantan, alumel, etc.).

thermodynamics–The study of the relationship between heat and other forms of energy.

thermoelectric arm – Also called thermoelectric leg. The portion of a thermoelectric device having the electric current density and the temperature gradient approximately parallel or antiparallel and having electrical connections made at its extremities to a part in which the opposite relation between the direction of the temperature gradient and the electric current density exists.

thermoelectric converter–A device capable of converting heat energy directly into electrical energy (bismuth telluride). *See also* Thermal Converter.

thermoelectric cooler–1. A device utilizing the Peltier phenomenon to provide a silent, nonmoving cooler having a controllable cooling rate. 2. A solid-state device that cools as current passes through it.

thermoelectric couple – A thermoelectric device in which there are two arms of unlike composition.

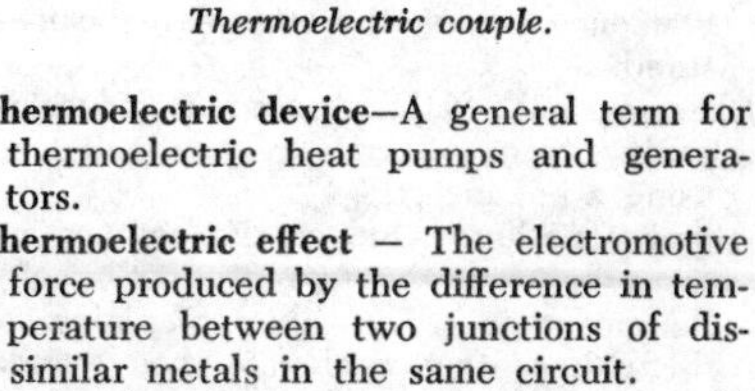

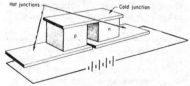

Thermoelectric couple.

thermoelectric device–A general term for thermoelectric heat pumps and generators.

thermoelectric effect – The electromotive force produced by the difference in temperature between two junctions of dissimilar metals in the same circuit.

thermoelectric generator – *See* Thermal Converter.

thermoelectric heating device–A thermoelectric heat pump used for adding thermal energy to a body.

thermoelectric heat pump – A device in which the direct interaction of an electrical current and heat flow is used to transfer thermal energy between bodies.

thermoelectricity—1. The direct conversion of heat into electricity. 2. The reciprocal use of electricity to create heat or cold. (*See also* Seebeck Effect, Peltier Effect, *and* Thomson Effect.) 3. Electricity produced by the agency of heat alone. *See also* Thermocouple.

thermoelectric junction—A thermojunction, as in a thermocouple.

thermoelectric leg – *See* Thermoelectric Arm.

thermoelectric manometer – A manometer (pressure-measuring instrument) that depends on the variation of thermoelectromotive force (voltage due to heat) with pressure.

thermoelectric series – A series of metals arranged in the order of their thermoelectric powers.

thermoelectric thermometer—Also called a thermocouple thermometer. A thermometer employing one or more thermocouples, of which one set of measuring junctions is in thermal contact with the body whose temperature is to be measured, while the temperature of the reference junctions is either known or otherwise taken into account.

thermoelectromotive force—1. The voltage developed due to the differences in temperature between parts of a circuit containing two or more different metals. 2. The algebraic sum of the Peltier emf at a thermocouple junction and the Thomson emf's in the thermocouple metals.

thermoelectron—The electron emitted from a heated body.

thermoelectron engine – *See* Thermionic Converter.

thermoelement – A device consisting of a thermocouple and a heating element arranged for measuring small currents. (*See also* Thermal Converter.)

thermogalvanometer – An instrument for measuring small high-frequency currents from their heating effect. Generally it consists of a dc galvanometer connected to a thermocouple that is heated by a filament carrying the current to be measured.

thermogram – High-resolution images resulting from a series of thermal scans using a bolometer.

thermojunction – One of the contact surfaces between the two conductors of a thermocouple. The thermojunction in thermal contact with the body under measurement is called the measuring junction, and the other thermojunction is called the reference junction.

thermojunction battery – A nuclear-type battery which converts heat into electrical energy directly by the thermoelectric or the Seebeck effect.

thermograph – Radiation chart that plots heat emitted by the body.

thermography – The process of recording the distribution of temperature over the surface of an object by detecting the heat radiation from it. The detection may be by effect of radiation on an infrared-sensitive phosphor, or by the use of cholestric liquid crystals that exhibit brilliant colors with temperature change.

thermoluminescence – The production of light in a material by moderate heat.

thermomagnetic—1. Pertaining to the effect of temperature on the magnetic properties of a substance. 2. Pertaining to the effect of a magnetic field on the temperature distribution in a conductor.

thermometer—An instrument for measuring temperature. Electrical versions depend on the change in resistance of a material with temperature, the voltage produced in a thermocouple, or various other effects of temperature.

thermophone – An electroacoustic transducer in which sound waves of calculable magnitude are produced by the expansion and contraction of the air adjacent to a conductor, the temperature of which varies in response to a current input.

thermopile—A group of thermocouples connected in series-aiding. Specifically: 1. A device used to measure radiant power or energy. 2. A source of electric energy. 3. Battery of thermocouples, consisting of alternate rods of antimony and bismuth suitably joined and connected to a galvanometer. Thermopiles are used for the measurement of heat where thermometers cannot be employed.

thermoplastic flow test—For an insulating material, a measure of the resistance to deformation when subjected to heat and pressure.

thermoplastic material—A plastic material that can be softened by heat and rehardened into a solid state by cooling. This remelting and remolding can be done many times.

thermoplastic recording—A recording process in which information is placed onto plastic tape electronically. A special electron gun, fed by a digital or scanner input, writes a charge in narrow bands on a moving film coated with a plastic that has a low melting point. The film is heated by an rf heater that melts the plastic coating, permitting it to be deformed by electrostatic and surface-tension forces in proportion to the charge laid down by the beam of the electron gun. The ridges cool quickly and form a diffraction grating that can then be viewed, projected by suitable optics, or read out by a flying-spot scanner. The system operates in a high vacuum.

thermosetting material—Plastic which hardens when heat and pressure are applied. Unlike a thermoplastic, it cannot be remelted or remolded.

thermostat—A mechanism that can be set to operate at definite temperatures and can convert the expansion of heated metal or fluid into sufficient movement and power to operate small devices, control electric circuits or small valves, etc.

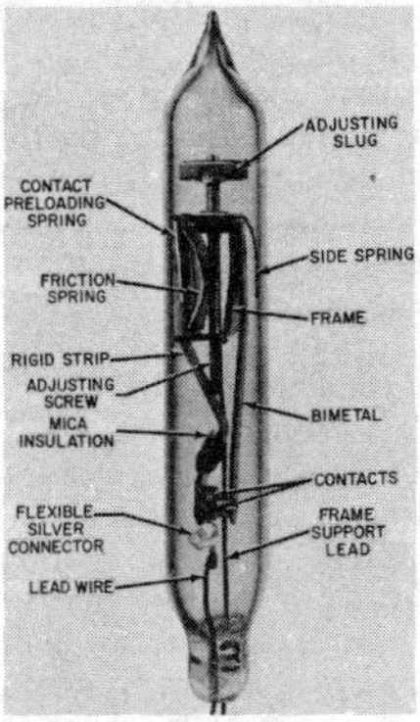

Thermostat.

thermostatic switch – A temperature-operated switch that receives its operating energy by thermal conduction or convection from the device being controlled or operated.

theta—Brain wave signals whose frequency is approximately 3.5 to 7.5 Hz. The associated mental state is fuzzy, unreal, uncertainty, daydreaming, ambiguity.

theta polarization—The state of the wave whereby the E vector is tangential to the meridian lines of some given spherical frame of reference.

Thevenin's theorem—The current that will flow through an impedance Z_1 when connected to any two terminals of a linear network between which an open-circuit voltage E and impedance Z previously existed, is equal to the voltage E divided by the sum of Z and Z_1.

thick-film—Pertaining to a film pattern usually made by applying conductive and insulating materials to a ceramic substrate by a silk-screen process. Thick films can be used to form conductors, resistors, and capacitors.

thick-film circuit—A microelectronic assembly in which the passive circuit elements and their interconnections are defined on a ceramic substrate by using the silk-screen process. The active elements are added as discrete chips.

thick-film hybrid integrated circuits—The physical realization of a hybrid integrated circuit fabricated on a thick-film network.

thick-film process—A method in which electronic circuit elements—resistors, conductors, capacitors, etc.—are produced by applying specially formulated pastes to a ceramic substrate in a defined pattern and sequence. The substrate containing the "green pattern" is fired at a relatively high temperature to mature the circuit elements and bond them integrally to the substrate. The term "thick film" distinguishes this technology from thin-film practice, in which circuit elements are made by evaporation or sputtering in a high-vacuum environment. Circuits produced by this technology are identified by several descriptive phrases: thick-film passive circuits or thick-film hybrid circuits when active devices are included.

thick-film resistor—A fixed resistor the resistance element of which is a film considerably more than one-thousandth of an inch thick.

thick films—Layers of resistive, dielectric, and conductive inks that are deposited on a substrate. The deposition process, similar to graphic silk screening, employs a fine-mesh screen to hold the pattern for the components that are to be deposited. The pattern is produced by photographic means, and wherever the inks are not to be deposited, the holes in the mesh are blocked by an emulsion.

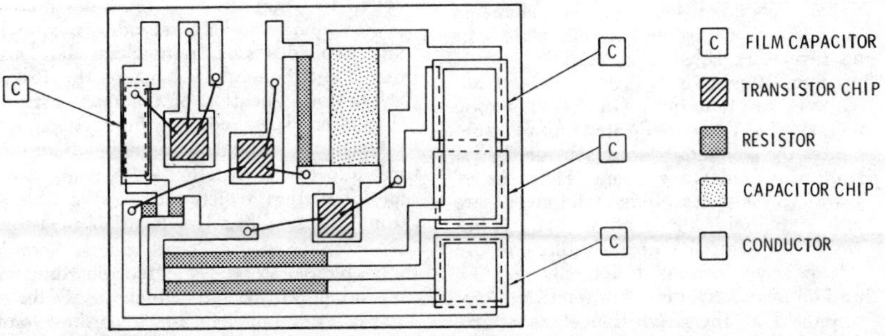

Thick-film hybrid integrated circuit.

thickness vibration – Vibration of a piezoelectric crystal in the direction of its thickness.

thin-film—A film of conductive or insulating material, usually deposited by sputtering or evaporation, that may be made in a pattern to form electronic components and conductors on a substrate or used as insulation between successive layers of components.

thin-film capacitor—A capacitor utilizing a metal oxide as the dielectric or insulating material. Both the electrodes and the dielectric are deposited in layers on a substrate. This device is usually associated with microelectronics and integrated and thin-film circuits.

thin-film formation – A process that is either additive by pattern formation through masks, or subtractive by selective etching of predeposited films from a substrate.

thin-film hybrid integrated circuits – The physical realization of a hybrid integrated circuit fabricated on a thin-film network.

thin-film integrated circuit – 1. An integrated circuit consisting of a passive substrate on which the various passive elements (resistors and capacitors) are deposited in the form of thin-patterned films of conductive or nonconductive material. Active components (transistors and diodes) are attached separately as individually packaged devices or in unpackaged (chip) form, or may be formed integrally by thin-film techniques. (*See also* Hybrid Thin-Film Circuit.) 2. A device consisting of a number of electrical elements entirely in the form of thin films deposited in a pattern on a supporting material.

thin-film memory—In a computer, a storage device made of thin discs of magnetic material deposited on a nonmagnetic base. Its operation is similar to the core memory. (*See also* Storage, 2 *and* Core Memory, 1.)

thin-film microelectronics – Circuits made up of two-dimensional passive and essentially two-dimensional active elements mounted or deposited on thin wafers of an insulating substrate material.

thin films—Resistive materials made of nichrome or tantalum (or other exotic metal). They are deposited onto a substrate by vacuum deposition or by rf sputtering under a vacuum. Thin-film resistors provide excellent tolerances, are very stable with temperature changes, but are costly to apply, necessitating the use of expensive vacuum equipment.

thin-film semiconductor – A semiconductor produced by the deposition of an appropriate single-crystal layer on a suitable insulator.

thin-walled conduit—Metallic tubing used to enclose insulated wires in an electrical circuit.

thin-wall ring magnet – A type of ceramic permanent magnet in which the axial length is greater than the wall thickness or the wall thickness is less than 15% of the outside diameter.

third harmonic – A sine-wave component having three times the fundamental frequency of a complex wave.

third-harmonic distortion – The rms third-harmonic voltage divided by the rms fundamental voltage. This value often is used as a measure of distortion in an essentially symmetrical system, such as ac-biased recording.

Thomson bridge—*See* Kelvin Bridge.

Thomson coefficient – The ratio of the voltage between two points on a metallic conductor to the difference in temperature between the same points.

Thomson effect—The production or absorption of heat (in addition to the I^2R loss) by a current between two points within a temperature gradient in a homogeneous conductor. Whether the heat is given off or absorbed depends on the direction of the current.

Thomson electromotive force—The voltage that exists between two points that are of different temperatures in a conductor.

Thomson heat – The thermal energy absorbed or produced due to the Thomson effect.

thoriated filament – A tungsten vacuum-tube filament to which a small amount of thorium has been added to improve emission. The thorium comes to the surface and is primarily responsible for the electron emission.

thread—*See* Chip, 1.

threading slot—A slot, in the cover plate of a recording-head assembly, into which the tape is slipped in threading the reels of a recorder.

three-address—Pertaining to an instruction format that contains three address parts.

three-address code—Also called instruction code. In computers, a multiple-address code which includes three addresses, usually two addresses from which data are taken and one address where the result is entered. Location of the next instruction is not specified, and instructions are taken from storage in preassigned order.

three-address instruction – In computers, an instruction which includes an operation and specifies the location of three registers. (*See also* Three-Address Code.)

three-channel stereo—A stereo recording or reproduction system which uses three spaced microphones for recording and three sound reproducers for playback.

three conductor jack – Receptacle having

three through circuits: tip, ring, and sleeve.

three-layer diode – Also called Diac. A two-terminal voltage-controlled device exhibiting a bilateral negative resistance characteristic. The device has symmetrical switching voltages ranging from 20 to 40 volts and is specifically designed for use as a trigger in ac power-control circuits such as those using triacs.

three-level maser (or laser)–A maser or laser system that involves the ground state and two other energy levels. Laser action usually takes place between the intermediate and ground states. The pump populates the intermediate state by way of the highest state. When applied to a maser, this pump brings the population of the highest state and the ground state into equilibrium. Maser action may take place either between the upper and intermediate levels or between the intermediate and ground levels, depending on the relaxation times of the different transitions.

three-phase circuit–A combination of circuits energized by alternating electromotive forces which differ in phase by one-third of a cycle, or 120 electrical degrees. In practice, the phases may vary several degrees from the specified angle.

three-phase current–A current delivered through three wires–each wire serving as the return for the other two, and the three current components differing in phase successively by one-third of a cycle, or 120 electrical degrees.

three-phase, four-wire system–An ac supply system comprising four conductors–three connected as in a three-phase, three-wire system and the fourth to the neutral point of the supply, which may be grounded.

three-phase motor–An ac motor operated from a three-phase circuit.

three-phase, seven-wire system–A system of ac supply from groups of three single-phase transformers connected in a Y. Thus, a three-phase, four-wire, grounded-neutral system of a higher voltage for power is obtained, the neutral wire being common to both systems.

three-phase, three-wire system–An ac supply system comprising three conductors, between successive pairs of which are maintained alternating differences of potential successively displaced in phase by one-third of a cycle.

three-plus-one instruction–In digital computer programming, a four-address instruction in which one of the addresses always specifies the location of the next instruction to be performed.

three-pole switch–An arrangement of three single-pole switches coupled together to operate three contacts simultaneously.

three-position relay – Sometimes called a center-stable polar relay.

three-pulse cascade canceler–A moving-target indicator technique in which two "two-pulse cancelers" are cascaded together. This improves the velocity response.

three-quarter bridge–A bridge connection in which one of the diode rectifiers has been replaced by a resistor.

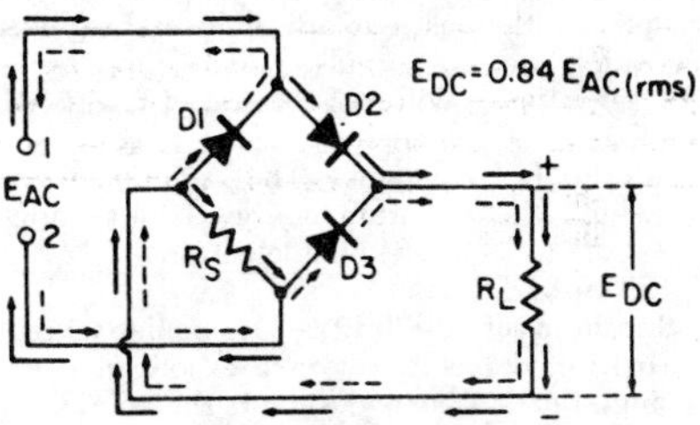

Three-quarter bridge.

three-way speaker system–A sound-reproducing system using three separate speakers, each designed for a specific portion of the audio spectrum (high, low, and middle frequencies). The high- and low-frequency speakers are known as the tweeter and woofer, respectively.

three-way switch–A switch which can connect one conductor to any one of two other conductors.

three-wire system–A system of electric supply comprising three conductors, one of which (known as the neutral wire) is maintained at a potential midway between the potential of the other two (referred to as the outer conductors). Part of the load may be connected directly between the outer conductors, the remainder being divided as evenly as possible into two parts, each of which is connected between the neutral and one outer conductor. There are thus two distinct supply voltages, one being twice the voltage of the other.

threshold–1. The point at which an effect is first produced, observed, or otherwise indicated. 2. In a modulation system, the smallest value of carrier-to-noise ratio at the input of the demodulator for all values above which a small percentage change in the input carrier-to-noise ratio produces a substantially equal or smaller percentage change in the output signal-to-noise ratio. 3. That point at which an indication exceeds the background or ambient.

threshold current–The minimum current at which a gas discharge becomes self-sustaining.

threshold decoding–A decoding procedure so arranged that the decision on the symbol that was transmitted is based on a majority count of the parity-check equations involving that symbol.

threshold element—A device that performs the logic threshold operation, but in which the contribution to the output determination by the truth of each input statement has a weight associated with that statement.

threshold field – The least magnetizing force in a direction which tends to decrease the remanence, which, when applied either as a steady field of long duration or as a pulsed field appearing many times, will cause a stated fractional change of remanence.

threshold frequency – The frequency at which the quantum energy is just sufficient to release photoelectrons from a given surface.

threshold of audibility—Also called threshold of detectability or threshold of hearing. For a specified signal, the minimum effective sound pressure of a signal capable of evoking an auditory sensation in a specified fraction of the trials. The characteristics of the signal, the manner in which it is presented to the listener, and the point at which the sound pressure is measured must all be specified. This threshold is usually expressed in decibels relative to 0.0002 microbar.

threshold of detectability—*See* Threshold of Audibility.

threshold of discomfort—Also called threshold of feeling. For a specified signal, the minimum effective sound-pressure level which, in a specified fraction of the trials by a battery of listeners, will stimulate the ear to the point where the sensation of feeling becomes uncomfortable. This threshold is customarily expressed in decibels relative to 0.0002 microbar.

threshold of feeling – *See* Threshold of Discomfort.

threshold of hearing – *See* Threshold of Audibility.

threshold of luminescence—*See* Luminescence Threshold.

threshold of sensitivity—The smallest stimulus or signal that will result in a detectable output. This phrase is frequently used to describe the voltage point at which an operations monitor or event marker will trigger.

threshold signal—In navigation, the smallest signal capable of producing a recognizable change in the positional information.

threshold-triggered flip-flop—A flip-flop the state of which changes when the actuating signal passes through a certain voltage level, regardless of the rate at which the voltage changes.

threshold value—1. The minimum input that produces a corrective action in an automatic control system. 2. The minimum level for which there is a measurable output.

threshold voltage—1. The level of input voltage at which a binary logic circuit changes from one logic state to the other. 2. The voltage at which a pn junction begins to conduct current. 3. In a solid-state lamp, the voltage at which emission of light begins. 4. The minimum gate voltage needed to turn on an MOS enhancement-mode device.

throat – 1. Part of the flare or tapered parallel-plate guide immediately adjacent to and connected to the main run of a waveguide. 2. The smaller cross-sectional area of a horn.

throat microphone – A microphone worn around the throat and actuated by vibrations of the larynx as the user talks. It is used in jet airplanes, tanks, and other places where background noise would drown out the conversation.

through path—The transmission path from the loop input signal to the loop output signal in a feedback control loop.

throughput—1. A measure of the efficiency of a system; the rate at which the system can handle work. 2. The speed with which problems or segments of problems are performed in a computer. Throughput varies from application to application and is meaningful only in terms of a specific application.

throughput rate—The highest rate at which a multiplexer can switch from channel to channel at its specified accuracy. This rate is determined by the settling time.

through repeater – A microwave repeater that is not equipped to provide for connections to any local facilities other than the service channel.

through transfer function – The transfer function of the through path in a feedback control loop.

throw—In an electric motor or generator, the number of core slots spanned between the bottom leg of a coil and the top leg of the same coil.

throwing power—Referring to the ability of an anode used in an impressed-current cathodic-protection system to distribute its current over a large surface. It is principally dependent upon the anode voltage, current, surface area, position of the anode with respect to the cathode, and the salinity and flow velocity of the water.

throw-out spiral—*See* Lead-Out Groove.

thru-hole connection—Also called feed-thru connection and plated thru hole. A conductive material used to make electrical and mechanical connection between the conductive patterns on opposite sides of a printed-circuit board.

thru repeater—In a microwave system, a

repeater station that is not equipped to be connected to any local facilities other than the service channel.

thump—1. A low-frequency transient disturbance in a system or component. 2. The noise caused in a receiver by telegraph currents when the receiver is connected to a telephone circuit on which a direct-current telegraph channel is superimposed.

thyratron – 1. A hot-cathode gas tube in which one or more control electrodes initiate the anode current, but do not limit it except under certain operating conditions. 2. A gas-filled triode in which a sufficiently large positive pulse applied to the control grid ionizes the gas and initiated conduction. Thereafter, the grid has no further effect, but conduction can be halted by reducing the plate voltage to zero (or less), or by reducing plate current to a value too small to maintain ionization.

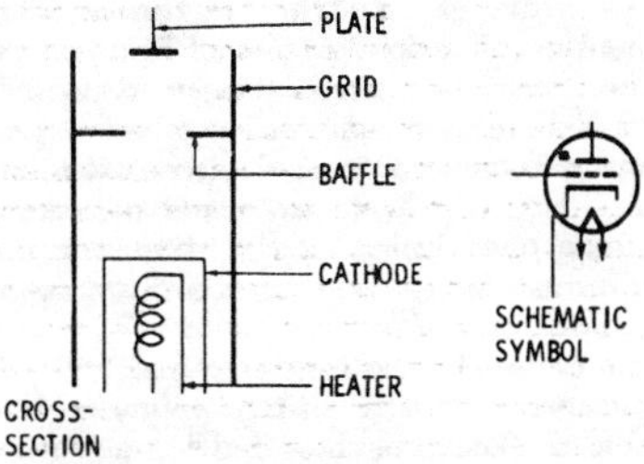

Thyratron.

thyratron gate—In computers, an AND gate consisting of a multielement gas-filled tube in which conduction is initiated by the coincident application of two or more signals. Conduction may continue after one or more of the initiating signals have been removed.

thyratron inverter—An inverter circuit in which thyratron tubes convert the dc power to ac power.

thyrector—A silicon diode that acts as an insulator until its rated voltage is reached, and as a conductor above that voltage. It is used for ac surge-voltage protection.

thyristor – A bistable device comprising three or more junctions. At least one of the junctions can switch between reverse- and forward-voltage polarity within a single quadrant of the anode-to-cathode voltage-current characteristics. Used in a generic sense to include silicon controlled rectifiers and gate-control switches, as well as multilayer two-terminal devices. (Many countries use the term "thyristor" in place of the term "pnpn-type switch.")

thyrite—A silicon-carbide ceramic material with nonlinear resistance characteristics.

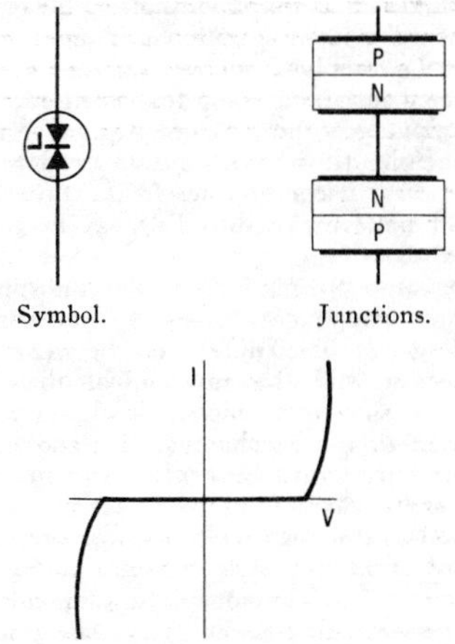

Symbol. Junctions.

Characteristic.

Thyrector.

Above a critical voltage, the resistance falls considerably.

THz – Abbreviation for terahertz (10^{12} hertz).

tickler—A small coil connected in series with the anode circuit of an electron tube and inductively coupled to a grid-circuit coil. The tickler coil is used chiefly in regenerative detector circuits, to establish feedback or regeneration.

tie cable—1. A cable between two distributing frames or distributing points. 2. A cable between two private branch exchanges. 3. A cable between a private branch exchange switchboard and the main office. 4. A cable connecting two other cables.

tie line—1. *See* Interconnection, 1. 2. A private line communication channel of the type communications common carriers provide for joining two or more points.

tie point—An insulated distributing point (other than an active terminal connection) where junctions of component leads are made in circuit wiring.

tier array—An array of antenna elements, one above the other.

ties—1. Electrical connections or straps. 2. Tie wires.

tie trunk—A telephone line or channel directly connecting two private branch exchanges.

tie wires—Short pieces of wire used to tie open-line wires to insulators.

tight coupling—*See* Close Coupling.

tilt—1. In radar, the angle between the axis of radiation in the vertical plane, and a reference axis which is normally the horizontal. 2. *See* Wave Tilt. 3. The angle an antenna makes with the horizontal.

tilt angle—In radar, the angle between the vertical axis of radiation and a reference axis (normally the horizontal).

tilt controls—In a color television receiver employing the magnetic-convergence principle, the three controls used to tilt the vertical center rows in the three colored patterns produced by a dot-generator signal.

tilt error—1. In navigation, the ionospheric error component due to nonuniform height. 2. The difference between the true tilt and the mechanical tilt. The sign is such that when it is algebraically added to the mechanical tilt, the result is the true tilt. (*See also* Antenna Tilt Error.)

tilting—1. Forward inclination of the wave front of radio waves traveling along the ground. The amount of tilt depends on the electrical constants of the ground. 2. Changing the angle of a television camera to follow a moving object being televised. 3. Changing the vertical angle of a directional antenna.

timbre—Also called tone color or musical quality. The character of a musical tone that distinguishes one musical instrument from another playing the same note. The difference between two steady tones having the same pitch and degree of volume is called the difference in timbre. Timbre depends mostly on the relative intensity of the different harmonics and the frequencies of the most prominent harmonics.

time—The measure of the duration of an event. The fundamental unit of time is the second.

time assignment speech interpolation—Telephone switching equipment whereby a person is connected to idle circuits when he starts talking, and is disconnected when he stops talking.

time base — A voltage generated by the sweep circuit of a cathode-ray–tube indicator. Its waveshape is such that the trace is either linear with respect to time or, if nonlinear, is still at a known timing.

time clock—A loosely used term sometimes referring to time switches, sometimes to interval timers, or to any type of timer.

time-code generator—A timer with an absolute time reference having digital outputs suitable for machine interpretation. Dial or numeric display may be included.

time constant—The time required for an exponential quantity to change by an amount equal to 0.632 times the total change that will occur. Specifically: 1. In a capacitor-resistor circuit, the number of seconds required for the capacitor to reach 63.2% of its full charge after a voltage is applied. The time constant of a capacitor having a capacitance C in farads in series with a resistance R in ohms is equal to $R \times C$. 2. In an inductor-resistor circuit, the number of seconds required for the current to reach 63.2% of its final value. The time constant of an inductor having an inductance L in henrys and resistance R in ohms is equal to L/R. 3. The time required for a motor to accelerate from 0 to 63.2% of its final no-load speed when the rated voltage, with the proper phase relationship, is applied.

time constant of fall—The time required for a pulse to fall from 70.7% to 26.0% of its maximum amplitude, excluding spikes.

time-current characteristics of a fuse—The relation between the root-mean-square alternating current or direct current and the time for the fuse to perform the whole or some specified part of its interrupting function. Usually shown as a curve.

timed acceleration—A control function that automatically controls the speed increase of a drive as a function of time.

timed deceleration—A control function that automatically controls the speed decrease of a drive as a function of time.

time date generator—A device used in surveillance which electronically produces a single row display of day, time, and date information on any standard tv screen (or raster).

time delay—1. The time required for a signal to travel between two points in a circuit. 2. The time required for a wave to travel between two points in space. 3. The total elapsed time or lag required for a given command to be effected after the command is given. 4. The slope of the phase versus frequency curve at a specified frequency. In a loose sense, this is the time it takes a designated point in a wave to pass through a filter. Also called envelope delay.

time-delay circuit—A circuit that delays the transmission of an impulse signal, or the performance of a transducer, for a definite desired length of time.

time-delay closing relay—*See* Time-Delay Starting Relay.

time-delay generator—A device which accepts an input signal and provides a delay in time before the initiation of an output signal.

time-delay relay—1. Also called slow-action relay. A relay in which there is an appreciable interval of time between the energizing or de-energizing of the coil and the movement of the armature (e.g., slow-operating relays and slow-release relays). 2. A device in which a defined delay occurs between the application or removal of the "control" current and the switching of the "load" current. 3. An electronic device with either relay or solid-state output that performs a retarding function upon receival of instruction.

time-delay starting relay—Also called time-delay closing relay. A device which functions to give a desired amount of time delay before or after any point or operation in a switching sequence or protective relay system.

time-delay stopping or opening relay—A time-delay device which serves in conjunction with the device that initiates the shut-down, stopping, or opening operation in an automatic sequence.

time-derived channel—Any of the channels obtained by time-division multiplexing of a channel.

time discriminator—A circuit in which the sense and magnitude of the output is a function of the time difference of, and relative time sequence between, two pulses.

time-distribution analyzer — Also called time sorter. An instrument that indicates the number or rate of occurrence of time intervals falling within one or more specified ranges. The time interval is defined by the separation between members of a pulse pair.

time-division data link—Radiocommunications which use time-division techniques for channel separation.

time-division multiplexer—A device which samples all data input from different low-speed devices, and retransmits all the samples in an equal amount of time.

time division multiplex channel — A path established for a segment of time during which the central station equipment obtains a signal indicating the status of a signal-encoding device.

time-division multiplexing — 1. A process by which two or more channels of information are transmitted over the same link by allocating a different time interval for the transmission of each channel. 2. A signaling method characterized by the sequential and noninterfering transmission of more than one signal in a communication channel. Signals from all terminal locations are distinguished from one another by each signal occupying a different position in time with reference to synchronizing signals. 3. A system of multiplexing in which channels are established by connecting terminals one at a time at regular intervals by means of an automatic distribution.

time-edit—To prevent the time code in a computer from being recorded or transmitted when no data samples are being passed, or, conversely, to record or transmit the time code with each data sample passed.

time flutter—A variation in the synchronization of components of a radar system, leading to variations in the position of the observed pulse along the time base, and reducing the accuracy with which the time of arrival of a pulse may be determined.

time frame—In telemetry, the time period containing all elements between corresponding points of two successive reference markers.

time gain—*See* Sensitivity-Time Control.

time gate—A transducer which has an output during chosen time intervals only.

time harmonics — High-frequency components generated by nonlinearity in a resolver magnetic circuit. In ac resolvers with no dc components present, time harmonics are odd multiples of the fundamental frequency. Time harmonics appear as a component of null voltage.

time-interval selector—A circuit that functions to produce a specified output pulse when and only when the time interval between two input pulses is between set limits.

time lag—1. The interval between application of any force and full attainment of the resultant effect. 2. The interval between two phenomena.

time lapse vtr—A videotape recorder which provides a continuous record of events over a period of from 12 to 48 hours on one reel of videotape.

time-mark generator—A circuit that produces accurately spaced pulses for display on the screen of an oscilloscope.

time modulation—Modulation in which the time of occurrence of a definite part of a waveform is made to vary in accordance with a modulating signal.

time-out—The interval of time allotted for certain operations to occur (for example, response to polling or addressing) before operation of the system is interrupted and must be started again.

time pattern — A picture-tube presentation of horizontal and vertical lines or rows of dots generated by two stable frequency sources operating at multiples of the line and field frequencies.

time phase—Reaching corresponding peak values at the same instants of time, though not necessarily at the same points in space.

time pulse distributor—A device or circuit for allocating timing pulses or clock pulses to one or more conducting paths or control lines in a specified sequence.

time quadrature—Differing by a time interval corresponding to one-fourth the time of one cycle of the frequency in question.

timer — 1. A special clock mechanism or motor-operated device used to perform switching operations at predetermined time intervals. 2. An assembly of electric circuits and associated equipment which provides the following: trigger pulses, sweep circuits, intensifier pulses, gate voltages, blanking voltages, and power supplies. 3. The part of a radar set that initiates pulse transmission and synchronizes this with the beginning of indicator sweeps, timing of gates, range markers, etc. *See also* Synchronizer, 1.

time resolution – The smallest interval of time that can be measured with a given system.

time response–An output, expressed as a function of time, that results from a specified input applied under specified operating conditions.

timer motor – A synchronous clock motor having a definite output speed determined by the number of poles in the stator and the reduction of the associated gear train.

time-sequencing–In a computer, switching signals generated by a program purely as a function of accurately measured elapsed time.

time-share – To use a device for two or more purposes on a time-sharing basis.

time sharing–1. A method of operation in which a computer facility is shared by several users for different purposes at (apparently) the same time. Although the computer actually services each user in sequence, the high speed of the computer makes it appear that the users are all handled simultaneously. 2. A means of making more efficient use of a facility by allowing more than one using activity access to the facility on a sequential basis. 3. The simultaneous use of one large-scale computer by several operators engaged in unrelated tasks and located at terminals remote from the central processing unit.

time signals–Time-controlled radio signals broadcast by government-operated radio station WWV at regular intervals each day on several frequencies.

time sorter – *See* Time-Distribution Analyzer.

time stability–The degree to which a component value is maintained to a stated degree of certainty (probability) under stated conditions of use, over a stated period of time. It is usually expressed in ±percent or± per unit (ppm) change per 1000 hours of continuous use.

time switch–A clock-controlled switch used to open or close a circuit at one or more predetermined times. Usually refers to repeat cycle timers with a dial graduated with time of day (in 24-hour type) or with time and day in 7-day type. Has means of adjusting the time when circuits are turned on or off, usually with pins or clips. Used for turning lights on at night, ringing bells in offices, schools, etc.

time-to-digital conversion–The process of converting an interval of time into a digital number.

time window–In random-sampling-oscilloscope technique, the part of the signal period that is displayed.

timing axis oscillator–*See* Sweep Generator.

timing-pulse distributor–Also called waveform generator. A computer circuit driven by pulses from the master clock. It operates in conjunction with the operation decoder to generate timed pulses needed by other machine circuits to perform the various operations.

timing relay–A form of auxiliary relay used to introduce a definite time delay in the performance of a function.

timing signal–Any signal recorded simultaneously with data to provide a time index.

timing tape–A variety of leader tape having printed indications at intervals of 7½ inches along its length, to facilitate measuring off a desired number of seconds' worth of leader.

tinkertoy – An attempt at modularization where wafers, with one or more component parts printed or mounted on them, are stacked vertically, with interconnecting wiring stiff enough to provide support running through holes around the periphery of the wafer.

tinned–Covered with metallic tin to permit easy soldering.

tinned wire–A copper wire that has been coated with a layer of tin or solder to prevent corrosion and to simplify soldering.

tinsel conductor–A type of electrical conductor comprising a number of tiny threads, each thread having a fine, flat ribbon of copper or cadmium bronze closely spiraled about it. Used for small cables requiring limpness and extra-long flex life (e.g., used with headsets or handsets).

tip–1. The contacting part at the end of a plug or probe. 2. Also called tip side. The end of the plug used to make circuit connections in a manual switchboard. The tip is the connector attached to the positive side of the common battery which powers the station equipment. By extension, it is the positive battery side of a communications line.

tip jack–Also called a pup jack. A small single-hole jack for a single-pin contact plug.

tip mass–The effective mass at the tip of the stylus of a pickup cartridge. (Modern techniques have reduced this mass toward 1 milligram or below.)

tipoff–The last portion of a vacuum-tube bulb to be melted and sealed after evacuation of the bulb.

tip side–Also called a tip wire. The conductor of a circuit associated with the tip of a plug or the tip spring of a jack. *See also* Tip, 2.

tip wire–*See* Tip Side.

T-junction—*See* Tee Junction.

$tm_{m,n}$ wave—In a rectangular waveguide, the transverse magnetic wave for which *m* and *n* are the number of half-period variations of the magnetic field along the longer and shorter transverse dimensions, respectively.

tm wave—Abbreviation for transverse magnetic wave.

T-network—A network composed of three branches. One end of each branch is connected to a common junction point. The three remaining ends are connected to an input terminal, an output terminal, and a common input and output terminal, respectively.

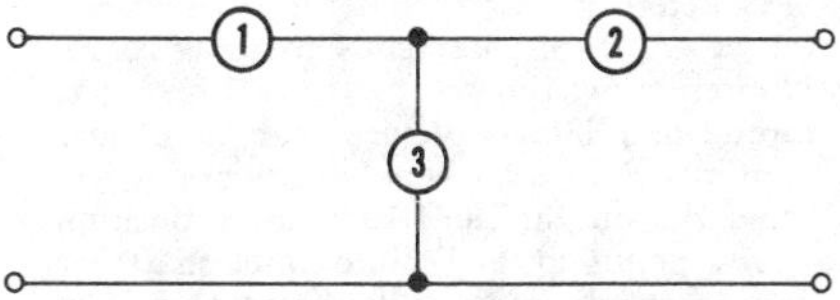

T-network.

TO can — Abbreviation for transistor-outline metal-can package.

to-from indicator — An instrument that forms part of the omnirange facilities and is used for resolving the 180° ambiguity.

toggle—1. A flip-flop. The term implies that the flip-flop will change state upon receipt of a clock pulse, and is used mainly with reference to flip-flops connected as a counter, in which the process of changing state is called toggling. 2. To use switches to enter data into the memory of a computer. 3. To change states abruptly in such a way that the process goes to completion once a critical point is reached, even if the actuating cause is removed; in logic circuits, a change of state of a flip-flop or Schmitt trigger.

toggle frequency—In a digital circuit, the number of times per second that the circuit changes state.

toggle rate—Twice the frequency at which a flip-flop completes a full cycle encompassing both states. Usually used to denote the maximum input frequency that a flip-flop can follow.

toggle switch—A switch with a projecting lever whose movement through a small arc opens or closes one or more electric circuits.

Toggle switch.

tolerance—1. A permissible deviation from a specified value. A frequency tolerance is expressed in hertz or as a percentage of the nominal frequency; an orientation tolerance, in minutes of arc; a temperature tolerance, in degrees Celsius; and a dimensional tolerance, in decimals or fractions. 2. Maximum error or variation from the standard permissible in a measuring instrument. 3. Maximum electrical or mechanical variation from specifications which can be tolerated without impairing the operation of a device.

toll call—Telephone call to points beyond the area within which telephone calls are covered by a flat monthly rate or are charged for on a message unit basis.

toll center—Also called toll office and toll point. The basic toll switching entity; a central office where channels and toll-message circuits terminate. While this is usually one particular central office in a city, larger cities may have several central offices where toll-message circuits terminate.

toll-terminal loss — On a toll connection, that part of the overall transmission loss attributable to the facilities from the toll center, through the tributary office to and including the subscriber's equipment.

tone—1. A sound wave capable of exciting an auditory sensation having pitch. 2. A sound sensation having pitch.

tone arm—The pivoted arm of a record player that extends over the record and holds the pickup cartridge. Wires from the cartridge run through the arm to the preamplifier, usually located on the underside of the turntable mounting board.

tone burst—A single sine-wave frequency, 50 to 500 microseconds long having a rectangular envelope, used for testing the transient response of speakers.

tone channel—An intelligence or signalling circuit in which on-off or frequency-shift modulation of a frequency (usually an audio frequency) is used as a means of transmission.

tone control—A control, usually part of a resistance-capacitance network, used to alter the frequency response of an amplifier so that the listener can obtain the most pleasing sound. In effect, a tone control accentuates or attenuates the bass or treble portion of the audio-frequency spectrum.

tone dialing—*See* Pushdown Dialing.

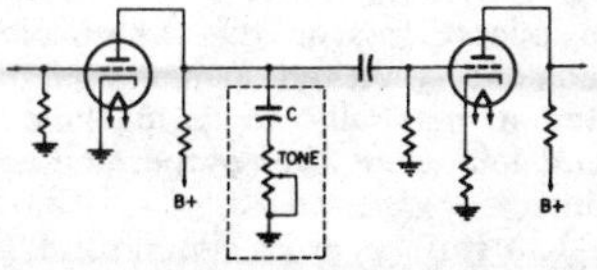

Tone control.

tone generator—A device for providing an audio-frequency current suitable for testing audio-frequency equipment or for signaling.

tone keyer – An instrument device which converts direct-current impulses to audio tones, for line transmission or for keying a transmitter.

tone localizer—*See* Equisignal Localizer.

tone-modulated waves – Waves obtained from continuous waves by amplitude-modulating them at an audio frequency in a substantially periodic manner.

tone modulation – A type of code-signal transmission obtained by causing the radio-frequency carrier amplitude to vary at a fixed audio frequency.

toner—Charged carbon particles, dry or suspended in a liquid solvent, used to produce a dark image on a light medium.

tone reversal – Distortion of the recorder copy in facsimile. It causes the various shades of black and white not to be in the proper order.

tone signaling—The transmission of supervisory, address and alerting signals over a telephone circuit by means of voice frequency tones. Also used in "touch tone" dialing.

tongue—The portion of a solderless terminal that projects from the barrel.

Tonotron—A multimode, selective-erasure storage tube.

tool function—In automatic control of machine tools, a command that identifies a specific tool and calls for the use of that tool.

top cap—A terminal in the form of a metal cap at the top of some vacuum tubes and connected to one of the electrodes.

top-loaded vertical antenna—A vertical antenna that is larger at the top, resulting in a modified current distribution, which gives a more desirable radiation pattern vertically. A series reactor may be connected between the enlarged portion of the antenna and the remaining structure.

topology—The surface layout of the elements comprising an IC.

tornadotron – A millimeter wave device which generates radio-frequency power from an enclosed, orbiting electron cloud excited by a radio-frequency field when subjected to a strong, pulsed magnetic field.

torn-tape relay—A method of receiving messages in tape form, breaking the tape, and retransmitting the message in tape form.

torn-tape switching center – A location at which operators tear off the incoming printed and punched paper tape and transfer it manually to a machine for transmission over the proper outgoing circuit.

toroid—1. A surface, or its closed solid, generated by any closed plane rotating about a straight line in its own plane—the resulting configuration being doughnut shaped. 2. A highly efficient type of coil wound upon a ring or "doughnut" type of core. The toroid provides for high concentrated magnetic field within itself, and has a minimum magnetic flux leakage (external field).

toroidal coil—A coil wound in the form of a toroidal helix.

toroidal core—A ring-shaped core.

toroidal permeability—Under stated conditions, the relative permeability of a toroidal body of the given material. The permeability is determined from measurements of a coil wound on the toroid such that stray fields are minimized or can be neglected.

torque—A force that tends to produce rotation or twisting.

torque amplifier—A device with input and output shafts and supplying work to rotate the output shaft so that its position corresponds to that of the input shaft but does not impose any significant torque on the latter.

torque-coil magnetometer—A magnetometer which depends for its operation on the torque developed by a known current in a coil capable of turning in the field to be measured.

torque gradient – The torque required in inch-ounces to pull a specific energized synchro 1° away from its normal position.

torque motors—A motor that is designed to provide its maximum torque under the condition of "stall" or "locked-rotor." A second criteria of torque motors is that they must be capable of remaining in a stalled condition for prolonged periods.

torque of an instrument—Also called deflecting torque. The turning moment produced on the moving element by the quantity to be measured or by some quantity dependent thereon acting through the mechanism.

torquer—A device which produces torque about an axis of freedom in response to a signal input.

torque-to-inertia ratio—*See* Acceleration At Stall.

torr—The unit of pressure used in the measurement of a vacuum. It is equal to 1/760 of a standard atmosphere, and for practical purposes may be considered equivalent to one millimeter of mercury (mm Hg).

torsiometer – An instrument for measuring the amount of power which a rotating shaft is transmitting.

torsion galvanometer—A galvanometer in which the force between the fixed and moving systems is measured by the angle through which the supporting head of the moving system must be rotated to return the moving system to zero.

torsion-string galvanometer – A sensitive galvanometer in which the moving system is suspended by two parallel fibers that tend to twist around each other.

total capacitance – The capacitance between a given conductor and all other conductors in a system when all other conductors are connected together.

total combined regulation—The change in output of a regulated power supply arising from simultaneous changes in all of the specified operating conditions, when the direction of such changes is such as to make their effects additive. It may be stated as a percentage of the specified output and/or an absolute value.

total distortion—The sum total of all forms of signal distortions.

total emission—The magnitude of the current produced when electrons are emitted from a cathode under the influence of a voltage such that all the electrons emitted are drawn away from the cathode.

total emissivity – The ratio of radiation emitted by a surface to the radiation emitted by the surface of a blackbody under identical conditions. Important conditions which affect emissivity of a material are surface finish, color, temperature, and wavelength of radiation. Emissivity may be expressed for radiation of a single wavelength (monochromatic emissivity), for total radiation of a specified range of wavelengths (total spectral emissivity), or for total radiation of all wavelengths (total emissivity).

total excursion—The application of a stimulus, in a controlled manner, over the span of an instrument.

total harmonic distortion—Abbreviated thd. 1. The ratio of the power at the fundamental frequency measured at the output of the transmission system considered, to the power of all harmonics observed at the output of the system because of its nonlinearity when a single frequency signal of specified power is applied to the input of the system. It is expressed in decibels. 2. The square root of the sum of the squares of the rms harmonic voltages divided by the rms fundamental voltage.

total internal reflection—When light passes from one medium to another which is optically less dense (e.g., from glass to air), the ray is bent away from the normal. If the incident ray meets the surface at such an angle that the refracted ray must be bent away at an angle of more than 90°, the light cannot emerge at all, and is totally internally reflected.

totalizing—To register a precise total count from mechanical, photoelectric, electromagnetic, or electronic inputs or detectors.

total losses of a ferromagnetic part—Under stated conditions, the power absorbed and then dissipated as heat when a body of ferromagnetic material is placed in a time-varying magnetic field.

total losses of a transformer—The losses represented by the sum of the no-load and load loses.

total luminous flux—The total light emitted in all directions by a light source.

totally enclosed motor – A motor so enclosed as to prevent the free exchange of air between the inside and the outside of the case, but not sufficiently enclosed to be termed airtight.

totally unbalanced currents—*See* Push-Push Currents.

total range of an instrument—Also called the range of an instrument. The region between the limits within which the quantity measured is to be indicated or recorded.

total regulation – The arithmetic sum of changes in output of a regulated power supply arising from changes in each of the specified operating conditions (current, voltage, or power) when such changes are applied individually, and in a manner to make their effects additive. It may be stated as a percentage of the specified output and/or absolute value.

total resistance – The dc resistance of a (precision) potentiometer between the input terminals with the shaft positioned so as to give a maximum resistance value.

total spectral emissivity—*See* Total Emissivity.

total telegraph distortion—Telegraph transmission impairment expressed in terms of time displacement of mark-space and space-mark transitions from their proper positions and given in percent of the shortest perfect pulse called the unit pulse.

total transition time—In a circuit, the time interval between the point of 10% input change and the point of 90% output change. It is equal to the sum of the delay time and rise (or fall) time.

totem pole amplifier—A push-pull amplifier circuit that provides a single-ended output signal without the use of a transformer. Used mainly in the output stages of transistor audio amplifiers.

touch call—*See* Pushdown Dialing.

touch control—A control circuit that actuates a circuit when two metal areas or a preselected area are bridged by one's finger or hand.

touch-tone—A service mark of the American Telephone and Telegraph Company which identifies its pushbutton dialing service. The signal form is multiple tones in the audio-frequency range.

touch-tone dialing—Push-button telephone dialing in which the direct-current pulsing technique of the rotating dial is replaced by a combination of tones to provide that proper automatic switching.

touch voltage—The potential difference between a grounded metallic structure and a point on the earth's surface equal to the normal maximum horizontal reach (approximately 3 feet).

tourmaline—A strongly piezoelectric natural or synthetic crystal.

tower—A structure usually used when an antenna must be mounted higher than 50 feet.

tower loading—The load placed on a tower by its own weight, the weight of the wires and insulators with or without ice covering, the wind pressure acting on both the tower and the wires, and the tension in the wires.

tower radiator — A metal tower structure used as a transmitting antenna.

Townsend criterion — The relationship expressing the minimum requirement for breakdown in terms of the ionization coefficients.

Townsend discharge — An electrical discharge in a gas at moderate pressure (above about 0.1 millimeter of mercury). It corresponds to corona, and is free from space charges.

Townsend ionization coefficient—The average number of ionizing collisions made by an electron as it drifts a unit distance in the direction of an applied electric force.

T-pad—A pad made up of resistance elements arranged in a T-network (two resistors inserted in one line, with a third between their junction and the other line).

tp tape—*See* Triple-Play Tape.

tr—1. Abbreviation for transient response. 2. Abbreviation for transmit-receive.

trace—1. The pattern on the screen of a cathode-ray tube. 2. Software used for extremely detailed testing of the validity of an application program or other software.

trace interval—The interval corresponding to the direction of sweep used for delineation.

tracer—1. A radioisotope which is mixed with a stable material to trace the material as it undergoes chemical and physical charges. 2. A thread of contrasting color woven into the insulation of a wire for identification purposes. 3. In automatic machine control, a sensing element that is made to follow the outline of a contoured template and produce the desired control signals. 4. *See* Signal Tracer.

trace routines—Special types of debugging routines that develop a sequential record of the execution of programs to be checked or traced.

trace width — The distance between two points on opposite sides of a trace on an oscilloscope at which luminance is 50% of maximum. If the trace departs from a well-behaved (approximately Gaussian) form, it should be smoothed for the purpose of measurement.

tracing distortion—The nonlinear distortion introduced in the reproduction of a mechanical recording when the curve traced by the reproducing stylus is not an exact replica of the modulated groove. For example, in the case of sine-wave modulation in vertical recording, the curve traced will be a poid.

tracing routine — A routine that makes available a historical record of specified events in the carrying out of a program.

track—1. A path which contains reproducible information left on a medium by recording means energized from a single channel. 2. In electronic computers, that portion of a moving-type storage medium accessible to a given reading station (e.g., film, drum, tapes, discs). (*See also* Band, 3 *and* Channel, 3.) 3. To follow a point of radiation to obtain guidance.

trackability—How well the pickup system will track high amplitude and velocity modulation on a phonograph record at a given tracking weight.

trackball—A data-entry device consisting of a sphere mounted in a small open-topped box. This sphere is hand operated by rotating it about either its X or Y axis. The rotation is translated into digital data and stored in special input registers. The registers are then read by the host computer. The most common use of a trackball is in the control of cursor and tracking-symbol movements, although it can be used wherever it is desired to continuously vary display or process parameters.

track-command guidance—A missile guidance method in which both target and missile are tracked by separate radars and commands are sent to the missile to correct its course.

track configuration—The number of separate recorded tracks that a machine can play or record. Thus, a two-track machine makes or plays two parallel tracks, each occupying slightly less than half the width of the tape. A four-track machine makes or plays four parallel tracks, two in the same playing direction on a two-channel machine, or four in the same playing direction on a quadraphonic machine. Monophonic machines can be full-track, half-track or quarter-track.

track hold memory—A circuit that, in its "track" mode, develops an output that follows (ideally) the input exactly, or is proportional to it; and then, in its "hold" mode, maintains the output constant (ideally) at the value it had at the instant the circuit was commanded to change from "track" to "hold."

track homing—The process of following a

line of position known to pass through an object.

tracking—1. The process of keeping a radio beam, or the cross hairs of an optical system, set on a target—usually while determining the range of the target. 2. The maintenance of proper frequency relationships in circuits designed to be simultaneously varied by ganged operations. 3. The accuracy with which the stylus of a phonograph pickup follows a prescribed path over the surface of the record. 4. The ability of a multigun tube to superimpose simultaneously information from each gun. Tracking error is the maximum allowable distance between the displays of any two guns. 5. The ability of an instrument to indicate at the division line being checked when energized by corresponding proportional values of actual end-scale excitation. 6. The interconnecting of power supplies in such a manner that one unit serves to control all units in series operation. In this manner, the output voltage of the slave units follows the variations of the master. 7. The difference at any shaft position between the output ratios of any two commonly actuated similar electrical elements expressed as a percentage of the single total voltage applied to them. 8. The process of following the movement of a satellite or rocket by radar, radio, and photographic observations. 9. The moving of a tracking symbol on the face of a crt with a lightpen. The meaning can also be extended to include the movement of any graphical image on the screen, using any of the data entry devices.

tracking accuracy—The ability to indicate at a division line being checked when a meter is energized by corresponding proportional values of actual end-scale excitation, expressed as a percentage of actual end-scale value. Tracking accuracy is usually within one and one-half times the rated full-scale accuracy of the meter.

tracking antenna – A directional antenna system which automatically changes in position or characteristics to follow the motion of a moving signal source.

tracking error—1. In lateral mechanical recording, the angle between the vibration axis of the mechanical system of the pickup and a plane which contains the tangent to the unmodulated record groove and is perpendicular to the recording surface at the point of needle contact. 2. In horizontal recording, the angle between the body of a phonograph cartridge and the tangent of a groove at the point of a stylus contact. A conventional fixed-pivot tone arm can be adjusted to produce zero horizontal tracking angle error at only one or two points on the span of grooves.

tracking filter—An electronic device that

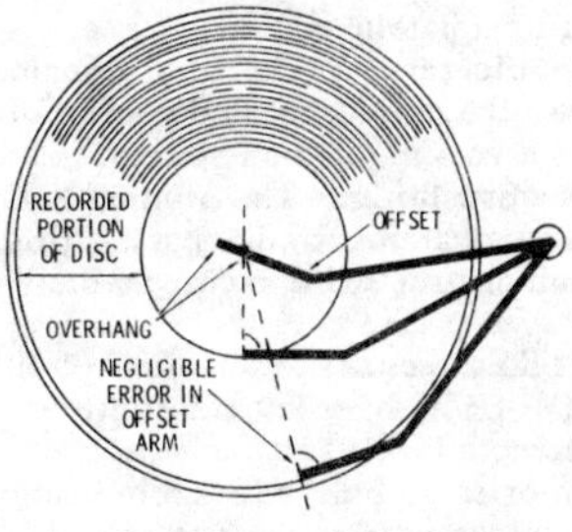

Tracking error.

attenuates unwanted signals and passes the desired signal, by the use of phase-lock techniques that reduce the effective bandwidth of the circuit and eliminate amplitude variations.

tracking force—The downward force applied to a stylus by means of an adjustment on the tone arm. Sometimes specified as a recommended range or a single force, in grams. It is usually advisable to track near the upper rather than the lower suggested value, since distortional record damage from mistracking at too low a force is more objectionable than the slightly increased wear from the higher force. A large stylus has a large groove contact area, and thus can tolerate a higher tracking force. *See also* Stylus Force.

tracking jitter—*See* Jitter, 1.

tracking resistance—*See* Arc Resistance.

tracking spacing – The distance between the center line of adjacent tracks on a recording tape. In the typical case of longitudinal tracks 50 mils wide, the tracking spacing is 70 mils.

tracking spot—A moving spot used for target indication on a radar.

tracking weight—The downward force of a pickup stylus which ensures optimum reproduction of recorded groove modulation with minimum wear of groove wall and stylus.

track-while scan—A radar system utilizing electronic-computer techniques whereby raw data are used to track an assigned target, compute target velocity, and predict its future position without interfering with the scanning rate.

track width—The width of the track on a recording tape, corresponding to a given record gap. The most common track widths encountered in longitudinal recording are 48 and 50 mils, several such tracks being accommodated on a half-inch-wide tape.

tractive force—The force which a permanent magnet exerts on a ferromagnetic object.

traffic—Messages handled by communication or amateur stations.

traffic diagram—A chart or drawing that shows the movement and control of traffic over a communication system.

traffic distribution—The routing by which communications traffic passes through a terminal to a switchboard or dialing center.

traffic-flow security—Protection of the contents of messages by transmitting an uninterrupted flow of random text over a wire or radio link, with no indication that an interceptor may use to determine which portions of this steady stream constitute encrypted messages and which portions are merely random filler.

trailer—Tough, nonmagnetic tape spliced at the end of the recorded material on a tape that is expected to receive rough or frequent handling. Usually has one matte-finished surface for writing, and is often available in a variety of colors for coding purposes.

trailer record—A record that follows one or more records and contains information related to those records.

trailers—Bright streaks at the right of large dark areas, or dark areas or streaks at the right of bright areas, in the televised picture. The usual cause is insufficient gain at low video frequencies.

trailing blacks—*See* Following Blacks.

trailing edge—The transition of a pulse that occurs last, such as the high-to-low transition of a high clock pulse.

trailing-edge pulse time – The time at which the instantaneous amplitude last reaches a stated fraction of the peak pulse amplitude.

trailing reversal – *See* Following Whites *and* Following Blacks.

trailing whites—*See* Following Whites.

train – A sequence of units of apparatus linked together to forward or complete a telephone call.

trainer – 1. The representation of an operating system by computers and its associated equipment and personnel. 2. Electronic equipment, used for training operators of radar or sonar apparatus by simulating signals received under operating conditions in the field.

tramlines – A pattern on an *A* scope appearing as a number of horizontal lines above the baseline. The effect is produced by cw modulated with a low-frequency signal.

trans—Abbreviation for transmitter. Also abbreviated xmtr or xmitter.

transaction file—In a computer, a file containing information relating to current activities or transactions.

transadmittance – From one electrode to another, the alternating components of the current of the second electrode divided by the alternating component of the voltage of the first electrode, all other electrode voltages being maintained constant. As most precisely used, the term refers to infinitesimal amplitudes.

transadmittance compression ratio—Ratio of the magnitude of the small-signal forward transadmittance of the tube to the magnitude of the forward transadmittance at a given input-signal level.

transceiver – 1. The combination of radio transmitting and receiving equipment in a common housing, usually for portable or mobile use and employing some common circuit components for both transmitting and receiving. 2. A device which transmits and receives data from punched card to punched card. It is essentially a conversion device which, at the sending end, reads the card and transmits the data over the wire. At the receiving end, it punches the data into a card.

transceiver data link—Integrated data processing by means of punched cards, using transceivers as terminal equipment. The transmission path can be wire or radio.

transconductance – Symbolized g_m. Also called mutual conductance. An electron-tube rating equal to the change in anode current divided by the change in grid voltage causing the anode-current change. The unit of transconductance is the mho; a more commonly used unit is the micromho (10^{-6}mho).

transconductance amplifier – An amplifier that supplies an output current proportional to its input voltage. From its output terminals the amplifier appears to be a current source with a high output impedance Z_o. Its input impedance $Z_{in} \gg Z_g$ and the output impedance $Z_o \gg Z_L$, where Z_g and Z_L are the source and load impedances.

transconductance meter—Also called a mutual-conductance meter. An instrument for indicating the transconductance of a grid-controlled electron tube.

transconductance tube tester – Also called a mutual-conductance or dynamic mutual-conductance tube tester. A tube tester with circuits set up in such a manner that the test is made by applying an ac signal of known voltage to the control grid and the tube amplification factor or mutual conductance is measured under dynamic operating conditions.

transconductor – An active or passive network whose short-circuit output current is a specific, accurately known, linear or nonlinear function of the input voltage, thereby establishing a predetermined relationship between input voltage and output current.

transcribe – 1. To copy, with or without translating, from one external storage medium of a computer to another. 2. To copy from a computer into a storage medium or vice versa.

transcriber—Equipment associated with a computing machine for the purpose of transferring input or output data from a record of information in a given language to the medium and language used by a digital computing machine, or from a computing machine to a record of information.

transcription—An electrical recording (e.g., a high-fidelity, 33⅓-rpm record) containing part or all of a (radio) program. It may be either an instantaneous recording disc or a pressing.

transdiode—A transistor so connected that the base and collector are actively maintained at equal potentials, although not connected together. The logarithmic transfer relationship between collector current and base-emitter voltage very closely approximates that of an ideal diode.

transducer—1. A device that when actuated by signals from one or more systems or media can supply related signals to one or more other systems or media. 2. A device that converts energy from one form to another.

transducer-coupling system efficiency—The power output at the point of application, divided by the electrical power input into the transducer.

transducer efficiency—Ratio of the power output to the electrical power input at the rated power.

transducer equivalent noise pressure (of an electroacoustic transducer or system used for sound reception) — Also called equivalent noise pressure. For a sinusoidal plane-progressive wave, the rms sound pressure which, if propagated parallel to the principal axis of the transducer, would produce an open-circuit signal voltage equal to the rms of the inherent open-circuit noise voltage of the transducer in a transmission band having a bandwidth of 1 hertz and centered on the frequency of the plane sound wave.

transducer gain—Ratio of the power that the transducer delivers to the load under specified operating conditions, to the available power of the source.

transducer insertion loss – *See* Insertion Loss.

transducer loss – Ratio of the available power of the source, to the power that the transducer delivers to the load under specified operating conditions.

transducer pulse delay – The interval of time between a specified point on the input pulse and on its related output pulse.

transfer—1. To transmit, or copy, information from one device to another. 2. To jump. 3. The act of transferring. 4. In electrostatography, the act of moving a developed image, or a portion thereof, from one surface to another without altering its geometrical configuration (e.g., by electrostatic forces or by contact with an adhesive-coated surface). 5. In a computer, to terminate one sequence of instructions and start another.

transfer accuracy—The input-to-output error as a percentage of the input. Transfer accuracy depends on the source impedance, switch resistance, load impedance if the multiplexer is not buffered, and the signal frequency.

transfer admittance—The complex ratio of the current at the second pair of terminals of an electrical transducer, to the electromotive force applied between the first pair, all pairs of terminals being terminated in any specified manner.

transfer characteristic—1. The relationship, usually shown by a graph, between the voltage of one electrode and the current to another electrode, all other electrode voltages being maintained constant. 2. The characteristics pertaining to the degree of output activity for an input stimulus, or vice versa. 3. Function which multiplied by an input magnitude will give a resulting output magnitude. 4. Relation between the illumination on a camera tube and the corresponding output-signal current, under specified conditions of illumination.

transfer check—In a computer, verification of transmitted data by temporarily storing, retransmitting, and comparing. Also a check to see if the transfer or jump instruction was properly performed.

transfer circuit—A circuit which connects communication centers of two or more separate networks in order to transfer the traffic between the networks.

transfer constant—Also called the image-transfer constant. One-half the natural logarithm of the complex ratio of the voltage times current entering a transducer, to that leaving the transducer when it is terminated in its image impedances.

transfer contact—*See* Break-Make Contact.

transfer control—*See* Jump, 1.

transfer current – The starter-gap current required to cause conduction across the main gap of a glow-discharge, cold-cathode tube.

transfer efficiency—The percentage of total charge that is transferred in a CCD from one position to another during readout.

transfer factor—Number relating the input and output currents of a transducer; *see* Propagation Constant, 2.

transfer function—1. A characterization of a system (for example, mechanical structure of filter) by determining how the system responds to input energy at different frequencies. It may be measured by one of three methods: (a.) applying a single sine wave and measuring the response to that sine wave, then varying the frequency of the sine wave and noting the change in the system's response. (b.) applying spectrally shaped random noise (many frequencies simultaneously) and observing the system response, or (c.) applying an impulse (also containing a broad range of frequencies). Transfer function can be derived by dividing the cross-spectral density by the power spectral density of the input or driven signal. 2. A mathematical expression that describes the relationship between the values of a set of conditions at two different times, typically at the beginning and end of a process. 3. The mathematical expression that relates the output and input of any characteristics of a filter as a function of frequency. The function usually is complex and therefore usually contains components corresponding to both attenuation and phase.

transfer impedance—Ratio of the potential difference applied at one pair of terminals of a network to the resultant current at another pair of terminals (all terminals being terminated in any specified manner).

transfer instruction—An instruction, signal, or operation which conditionally or unconditionally specifies the location of the next instruction and directs a computer to that instruction.

transfer operation — An operation which moves data from one storage location or one storage medium to another. Transfer is sometimes taken to refer specifically to movement between different media; storage to movement within the same medium.

transfer rate—The rate at which a transfer of data between the computer registers and storage, input, or output devices may be performed. It is usually expressed as a number of characters per second.

transfer ratio—From one point to another in a transducer at a specified frequency, the complex ratio of the generalized force or velocity at the second point to that applied at the first point.

transferred charge—The net electric charge moved from one terminal of a capacitor to another through an external circuit.

transferred charge characteristic — The mathematical function that relates transferred charge to capacitor voltage.

transfer switch—A form of air switch arranged so that a conductor connection can be transferred from one circuit to another without interrupting the current.

transfer time—1. The time required for a transfer to be made in a digital computer. 2. In a relay, the total time—after contact bounce has ceased—between the breaking of one set of contacts and the making of another.

Transfluxor—A trade name (RCA) for a binary magnetic core having two or more openings. Control of the transfer of magnetic flux between the three or more legs of the magnetic circuits provides ways to store information, gate electrical signals, etc.

transform—1. To convert a current or voltage from one magnitude to another, or from one type to another. 2. In digital-computer programming, to change information in structure or composition without significantly altering the meaning or value.

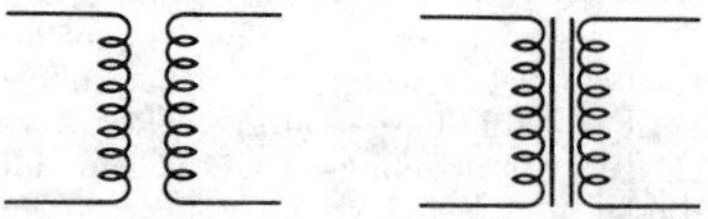

Air core. Iron core.

Transformer.

transformation point—The temperature at which an alloy or metal changes from one crystal state to another as it heats or cools.

transformation ratio—The ratio between electrical output and input under certain specified conditions.

transformer—An electrical device which, by electromagnetic induction, transforms electric energy from one or more circuits to one or more other circuits at the same frequency, but usually at a different voltage and current value.

transformer build—The amount of window area used in constructing a transformer.

transformer-coupled amplifier—An amplifier, the stages of which are coupled together by transformers.

transformer coupling—Use of a transformer between stages of an amplifier, i.e., to connect the anode circuit of one stage to the grid circuit of the following stage.

transformerless receiver — A receiver in which the power-line voltage is applied directly to series-connected tube heaters or filaments and to a rectifier circuit instead of first being stepped up or down in voltage by a power transformer.

transformer load loss—Losses in a transformer which are incident to the carrying of the load. Load losses include I^2R loss in the windings due to load current, stray loss due to stray fluxes in the winding core, clamps, etc., and to circulating current, if any, in parallel windings.

transformer loss—Expressed in decibels, the ratio of the signal power that an ideal transformer of the same impedance ratio would deliver to the load impedance, to the signal power delivered by the actual transformer.

transformer oil—A high-quality insulating oil in which windings of large power transformers are sometimes immersed to provide high dielectric strength, insulation resistance, and flash point, plus freedom from moisture and oxidation.

transformer read-only store—In computers, a type of read-only store in which the presence or absence of mutual inductance between two circuits is the condition that determines whether a binary 1 or 0 is stored.

transformer vault—An isolated enclosure, either above or below ground, with fire-resistant walls, ceiling, and floor for unattended transformers and their auxiliaries.

transformer voltage ratio—The ratio of the root-mean-square primary terminal voltage to the root-mean-square secondary terminal voltage under specified conditions of load.

transforming section—A length of waveguide or transmission line of modified cross section, or with a metallic or dielectric insert, used for impedance transformation.

transhybrid loss—The transmission loss at a given frequency measured across a hybrid circuit when connected to a given two-wire termination and balancing network.

transient—1. A phenomenon caused in a system by a sudden change in conditions, and which persists for a relatively short time after the change. 2. A distinct line or series of lines perpendicular to the direction of scanning produced in the recorded copy immediately following a sudden change in density. 3. A momentary surge on a signal or power line. It may produce false signals or triggering impulses and cause insulation or component breakdowns and failures. 4. Signal component of fast rising leading side and usually of short duration. 5. A pulse, damped oscillation, or other temporary phenomenon occurring in a circuit or system.

transient analyzer – An electronic device for repeatedly producing a succession of equal electric surges of small amplitude and of adjustable waveform in a test circuit and presenting this waveform on the screen of an oscilloscope.

transient behavior—In a power supply, the general attitude of response in terms of amplitude and time.

transient distortion—1. Distortion due to the inability of a system to reproduce or amplify transients linearly. 2. Distortion of a transient signal component such as produced by resonance and lack of damping, the effect being overshoot or ringing following the fast rising leading side of the signal component.

transient magnistor—A high-speed saturable reactor in which an alternating electric current in the form of a sine-wave carrier or pulses is passed through a signal winding and modulated by variations of current passing through a control coil.

transient motion – Any motion which has not reached, or has ceased to be, a steady state.

transient oscillation—A momentary oscillation which occurs in a circuit during switching.

transient overshoot—The largest transient deviation of the measured quantity following the dynamic crossing of the final value as a result of the application of a specified stimulus.

transient peak-inverse voltage—Under specified conditions, the maximum allowable instantaneous value of nonrecurrent reverse (negative) voltage that may be applied to the anode of an SCR with gate open.

transient phenomena – Rapidly changing actions occurring in a circuit during the interval between closing of a switch and settling to steady-state conditions, or any other temporary actions occurring after some change in a circuit or its constants.

transient recovery time—Also called recovery time, transient response time, or response time. 1. The interval between the time a transient deviated from a specified amplitude range and the time it returns and remains within the specified amplitude range. The amplitude range is centered about the average of the steady-state values that exist immediately before and after the transient. 2. The time required for the output voltage of a power supply to come back to within a level approximating the normal dc output following a sudden change in load current.

transient response – 1. The transient of a dependent variable resulting from an abrupt change of an independent variable with all other variables constant. 2. Also called step-function response. Commonly, the characteristic response of a system to a unit sweep or unit impulse. The elements of transient response most commonly specified are: rise time, fall time, overshoot, undershoot, preshoot, and ringing. 3. (Input line changes) The transient of an output variable (voltage, current, or power) of an electronic power supply resulting from an abrupt specified change of input line voltage, with all other conditions held constant. 4. (Load changes) The transient of an output variable (voltage, current, or power) of an electronic power supply resulting from an abrupt specified change of load, with all other conditions held constant. 5. Ability to respond to percussive signals cleanly and instantly. 6. The ability of an amplifier or transducer to handle and faithfully reproduce rapid changes in signal amplitudes. A very short "rise-time" is a measure of this characteristic. 7. The closed-loop step function response of an amplifier under small signal conditions.

transient response time—*See* Transient Recovery Time.

transient state—The condition in which a variable temporarily behaves in an erratic manner instead of changing smoothly and predictably.

transinformation (of an output symbol about an input symbol)—Also called mutual information. The difference between the information content of the input symbol and the conditional information content of the input symbol which is given the output symbol.

transistance – 1. The characteristic of an electrical element which makes possible the control of voltages, currents, or flux so as to produce gain or switching action in a circuit. Examples of the physical realization of transistance occur in transistors, diodes, saturable reactors, etc. 2. An electronic characteristic exhibited in the form of voltage or current gain or in the ability of control voltages or currents in a precise and nonlinear manner. (Examples of parts showing transistance: Transistors, diodes, vacuum tubes.) 3. A function of transistors. An electrical property that causes applied voltage to create amplification or accomplish switching.

transistor—An active semiconductor device having three or more electrodes, and capable of performing almost all the functions of tubes, including rectification and amplifications. Germanium and silicon are the main materials used, with impurities introduced to determine the conductivity type (n-type as an excess of free electrons; p-type, a deficiency). Conduction is by means of electrons (elementary particles having the smallest negative electrical charge that can exist) and holes mobile electron vacancies equivalent to a positive charge).

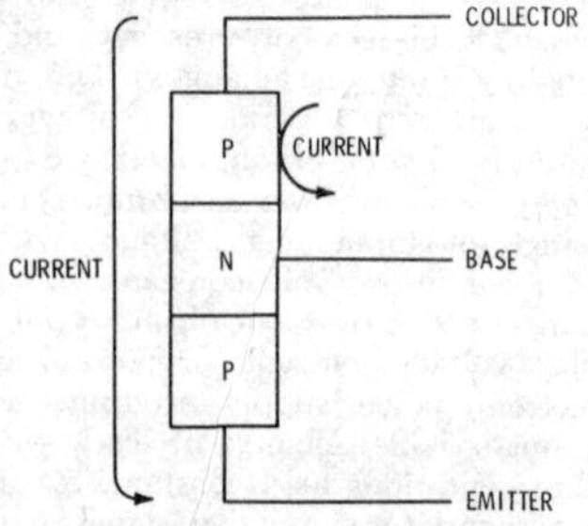

Typical transistor.

transistor action—The physical mechanism of amplification in a junction transistor.

transistor amplifier—An amplifier in which the required amplification is produced by one or more transistors.

transistor base – The region between an emitter and a collector of a transistor, into which minority carriers are injected.

transistor chip—An unencapsulated transistor element of very small size used in microcircuits.

transistor dissipation – The power dissipated in the form of heat by the collector. The difference between the power supplied to the collector and the power delivered by the transistor to the load.

transistorized—Pertaining to equipment or a design in which transistors instead of vacuum tubes are used.

transistor oscillator – An oscillator which uses a transistor in place of an electron tube.

transistor-outline metal-can package – Abbreviated TO-can. A type of package that resembles a transistor can, but generally is larger and has more leads. The pins are arranged in a circular pattern in the base. This type of package is often used for MSI, LSI, and MOS ICs.

transistor parameters – The performance characteristics of a transistor or class of transistors.

transistor pentode – A transistor designed for mixing, modulating, or switching, and containing the equivalent of three emitters, a base, and a collector.

transistor radio – A radio receiver which uses transistors in place of electron tubes.

transistor region—The region around a pn junction in which the majority carriers from each side of the junction diffuse across it to recombine with their respective counterparts.

transistor-resistor logic—Abbreviated TRL. An early form of logic-circuit design that was employed prior to the advent of monolithic circuits. The input elements are resistors; however, unlike RTL there is only one active output transistor. In discrete circuit designs, this logic form was the least expensive since it required a minimum number (one) of active devices.

transistor seconds – Also called fallouts. Those transistors that remain after the firsts (units meeting rigid specifications for a specific application) have been removed from the production line.

transistor symbols—Symbols used to represent transistors on schematic diagrams.

transistor-transistor logic – Abbreviated TTL or T^{2}L. Also called multiemitter transistor logic. A logic-circuit design similar to DTL, with the diode inputs replaced by a multiple-emitter transistor. In a four-input DTL gate, there are four

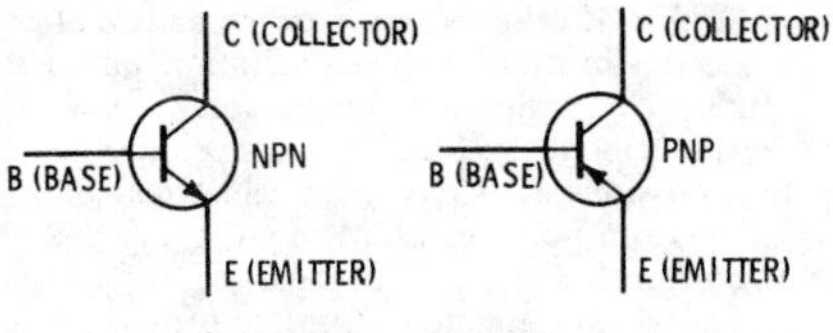

Transistor symbols.

diodes at the input. A four-input TTL gate will have four emitters of a single transistor as the input element. TTL gates using npn transistors are positive-level NAND gates or negative-level NOR gates.

transit angle—The product of angular frequency and the time taken for an electron to cross a given path.

transition—1. The process in which a quantum-mechanical system makes a change between energy levels. During this process, energy is either emitted or absorbed, usually in the form of photons, phonons, or kinetic energy of particles. Transitions that involve photons alone are called direct radiative transitions, whereas those that require a combination of a photon and a phonon are called indirect transitions. 2. The instance of changing from one state (such as a positive voltage) to a second state (such as a negative voltage) in a serial transmission.

transitional function—A transfer function that represents a compromise made between the classical types of functions (Tchebychev, Butterworth, Gaussian, etc.) as a result of the fact that each classical function has certain advantages and disadvantages. In general, however, better functions than the transitional functions are available if the transfer function is optimized with respect to its own particular requirement.

transition card—A card which signals the computer that the reading-in of a program has ended and that the carrying out of the program has started.

transition element — An element used for coupling different types of transmission systems, e.g., for coupling a coaxial line to a waveguide.

transition error—In an encoder, the difference between the shaft angle at which a code-position change should occur and the angle at which it actually does occur.

transition factor—*See* Reflection Factor, 1.

transition frequency—1. Also called crossover frequency. In a disc-recording system, the frequency corresponding to the point of intersection of the asymptotes to the constant-amplitude and the constant-velocity portions of its frequency-response curve. This curve its plotted with output-voltage ratio in decibels as the ordinate, and the logarithm of the frequency as the abscissa. 2. In a transistor, the product of the magnitude of the small-signal, common-emitter, forward-current transfer ratio times the frequency of measurement when this frequency is high enough that the magnitude is decreasing with a slope of approximately 6 dB per octave.

transition layer—*See* Transition Region, 1.

transition-layer capacitance — *See* Depletion-Layer Capacitance.

transition loss—At any point in a transmission system, the ratio of the available power from that part of the system ahead of the point under consideration, to the power delivered to that part of the system beyond the point under consideration.

transition point—A point at which the circuit constants change in such a way that a wave being propagated along the circuit is reflected.

transition region—1. Also called transition layer. The region, between two homogeneous semiconductor regions, in which the inpurity concentration changes. 2. The area, between the passband and the stopband, in which the attenuation of a filter is neither great nor small. The narrower the transition region, the more difficult is the design of the filter. 3. The region around a pn junction in which the majority carriers of each side diffuse across the junction to recombine with their respective counterparts.

transition temperature — The temperature below which the electrical resistance of a material becomes too small to be measured.

transition time—The time required for a voltage change from one logic level to the opposite level, as measured between specified points on the transition waveform. The transition times from high to low level and from low to high level generally are different.

transitron—A thermionic tube circuit the action of which depends on the negative transconductance of the suppressor grid of a pentode with respect to the screen grid.

transitron oscillator—A negative-transconductance oscillator employing a pentode tube with a capacitor connected between the screen and suppressor grids. The sup-

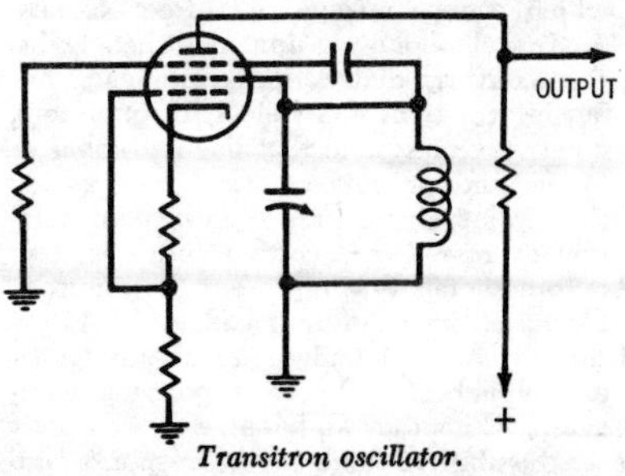

Transitron oscillator.

pressor grid periodically divides the current between the screen grid and anode, thereby producing oscillations.

transit time—1. The time taken for a charge carrier to cross a given path. 2. The average time a minority carrier takes to diffuse from emitter to collector in a junction transistor. 3. The time an electron takes to cross the distance between the cathode and anode.

transit-time mode—A condition of oscillator operation, corresponding to a limited range of drift-space transit angle, for which the electron stream introduces a negative conductance into the coupled circuit.

translate—1. To change computer information from one language to another without significantly affecting the meaning. 2. To change one binary word to another, or to change to a different pair of binary signal levels, for the purpose of achieving compatibility between parts of a system.

translation—The conversion of a message to a different set of symbols. For example, in instrumentation, the changing of conventional algebraic expressions into machine language for computer programming.

translational morphology — The structural characterization of an electronic component in which it is possible to identify the areas of patterns of resistive, conductive, dielectric, and active materials in or on the surface of the structure with specific corresponding devices assembled to perform an equivalent function.

translation loss—Also called playback loss. The loss in the reproduction of a mechanical recording, whereby the amplitude of motion of the reproducing stylus differs from the recorded amplitude in the medium.

translator—1. A device which transforms signals from the form in which they were generated, into a useful form for the purpose at hand. (*See also* Program Assembly.) 2. A television receiver and low-power transmitter which receives television signals on one channel and retransmits them on another channel (usually a uhf channel) to valleys and like areas which cannot receive the direct signals. 3. In telephone equipment, the device that converts dialed digits into call-routing information. 4. In electric computers, a network or system that has a number of inputs and outputs and is connected so that input signals representing information expressed in a certain code result in output signals that represent the input information in a different code.

transliterate—To change the characters of one alphabet to the corresponding characters of another alphabet.

transmissibility—Ratio of the response amplitude of the system in steady-state forced vibration, to the excitation amplitude. The ratio may be between forces, displacements, velocities, or accelerations.

transmission—1. Conveying electrical energy from point to point along a path. 2. The transfer of a signal, message, or other form of intelligence from one place to another by electrical means.

transmission anomaly—The difference, in decibels, between the total loss in intensity and the reduction in intensity that would be due to an inverse-square divergence.

transmission band—The range of frequencies above the cutoff frequency of a waveguide, or the comparable useful frequency range for any other type of transmission device.

transmission coefficient—For a transition or discontinuity between two transmission media at a given frequency, the ratio of some quantity associated with the transmitted wave at a specified point in the second medium, to the same quantity associated with the incident wave at a specified point in the first medium.

transmission facility—The transmission medium and all the associated equipment required to transmit a message.

transmission function—The ratio of the output voltage to the input voltage of a filter expressed in terms of magnitude and phase (or delay).

transmission gain—*See* Gain.

transmission gate—The solid-state equivalent of a relay, having terminals that are connected to each other or not depending on the application of a separate control voltage. (In CMOS logic it is bidirectional.)

transmission level — The level of signal power at any point in a transmission system. It is equal to the ratio of the power at that point to the power at some point in the system chosen as a reference point. This ratio is usually expressed in decibels.

transmission line—1. A material structure forming a continuous path from one place to another, and used for directing the transmission of electromagnetic energy along this path. 2. A conductor or series of conductors used for conveying electrical energy from a source to a load. 3. One or more insulated conductors arranged to transmit electrical energy signals from one locality to another.

transmission-line coupler — A coupler that allows the passage of electric energy in either direction between balanced and unbalanced transmission lines.

transmission-line loss—1. The difference in the amount of energy delivered at the output of a transmitter and the amount of energy absorbed by the antenna. For example, if a transmitter delivers 3 watts and the transmission line (coaxial cable) loss is 1 dB, the antenna will absorb approximately 2.4 watts since 20% of the available power is lost in the transmission line. 2. The power lost when a radio signal is fed from the antenna to the receiver through coaxial cable; expressed in dB.

transmission-line-tuned frequency meter—A frequency meter using a tuned length of wire or a coaxial cavity as the frequency-determining element.

transmission measuring set—A measuring instrument comprising a signal source and a signal receiver having known impedances, for measuring the insertion loss or gain of a network or transmission path connected between those impedances.

transmission mode—A form of waveguide propagation along a transmission line characterized by the presence of transverse magnetic or transverse electromagnetic waves. Waveguide transmission modes are designated by integers (modal numbers) associated with the orthogonal functions used to describe the waveform. These integers are known as waveguide-mode subscripts. They may be assigned from observations of the transverse field components of the wave and without reference to mathematics.

transmission modulation—Amplitude modulation of the current in the reading beam of a charge-storage tube as the beam passes through apertures in the storage surface. The degree of modulation depends on the stored charge pattern.

transmission primaries — The set of three primaries, either physical or nonphysical, each chosen to correspond in amount to one of the three independent signals contained in the color-picture signal. The I, Q, and Y signals.

transmission regulator — A device which maintains the transmission substantially constant over a system.

transmission response—The conversion of electrical energy traveling at the speed of light to sonic energy traveling at the speed of sound.

transmission security—That part of communications security resulting from all measures intended for protection of transmissions from unauthorized interception, traffic analysis, and imitative deception.

transmission selective—The transmission of electromagnetic energy at wavelengths other than those that are reflected or absorbed.

transmission speed—The number of information elements sent per unit time, usually expressed as bits, characters, word groups, or records per second or per minute.

transmission system—1. An assembly of elements capable of functioning together to transmit signal waves. 2. One of two broad groups of communication means, distinguished by their information carrying bandwidth and referred to as either wideband or narrowband transmission systems. The wideband system includes tropospheric scatter radio relay, and submarine cables, and has an information carrying bandwidth of more than four voice-frequency (3 to 4 kHz bandwidth) channels. The narrowband system includes high frequency radio (single sideband, amplitude modulated, ionospheric scatter) and landlines, and has an information carrying bandwidth of four voice channels or less.

transmission time—The absolute time interval between transmission and reception of a signal.

transmission-type frequency meter—A frequency meter in which a tuned electrical circuit or a cavity is used to transmit the energy from the signal source under test to a detecting load.

transmission wavemeter — A device that makes use of a cavity to transmit maximum power at resonance so that maximum deflection is obtained on a readout meter at the frequency of resonance.

transmit—To send a program, message, or other information from one location to another.

transmit current—The current drawn by a transceiver when in the transmit mode.

transmit-on-alarm—A security device which activates only when triggered. Most security devices are of this nature; however, in multiplexing systems, it is usually impossible to determine whether or not transmit-on-alarm devices are actuated or whether the transmitter has failed or the communication channel is inoperative.

transmit-on-interrogation—The response of a security device which stores an alarm until it is interrogated in an appropriate manner. Used in multiplexed systems to avoid the problem of two simultaneous transmissions of alarm by two or more different stations.

transmit-receive switch—Also called a tr switch, tr box, tr tube, or duplexing assembly. An automatic device employed in a radar to prevent the transmitted energy from reaching the receiver, but allowing the received energy to do so without appreciable loss.

transmit-receive tube—*See* tr Tube.

transmittal mode — The method by which the contents of an input buffer are made available to the program and the method by which records are made available by a program for output.

transmittance—1. Of a material, the ratio of the radiant power transmitted through the material, to the incident radiant

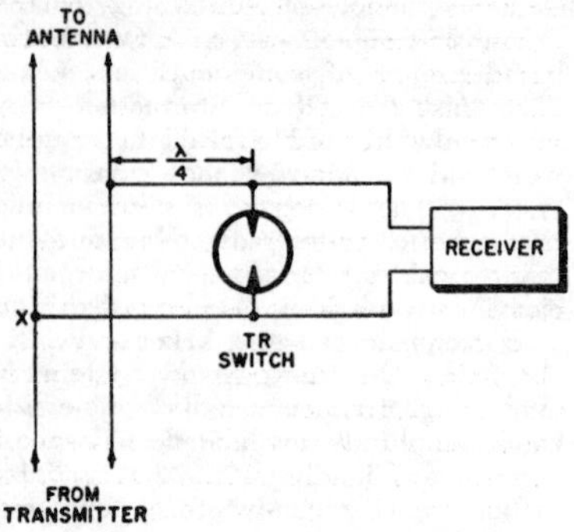

Transmit-receive switch.

power. 2. As a transfer function, a response function for which the variables are measured at different ports (terminal pairs).

transmitted-carrier operation—Amplitude-modulated carrier transmission in which the carrier wave is transmitted.

transmitted wave—*See* Refracted Wave.

transmitter—1. Equipment used to generate and amplify an rf carrier signal, modulate this carrier with intelligence, and radiate the modulated rf carrier into space. 2. In telephony, the microphone that converts sound waves into electrical signals at an audio-frequency rate.

transmitter distributor — In teletypewriter operations, a motor-driven device which translates teletypewriter code combinations from perforated tape into electrical impulses, and transmits these impulses to one or more receiving stations.

transmitter frequency tolerance—The extent to which the carrier frequency of a transmitter may legally depart from the assigned frequency.

transmitter start code—Usually a two-letter call that is sent to an outlying machine to automatically turn on its tape transmitter.

transmitting antenna—A device for converting electrical energy into electromagnetic radiation capable of being propagated through space.

transmitting current response—Of an electroacoustic transducer used for sound emission, the ratio of the sound pressure apparent at a distance of 1 meter in a specified direction from the effective acoustic center of the transducer, to the current flowing at the electric input terminals. Usually expressed in decibels above a reference-current response of 1 microbar per ampere.

transmitting efficiency (projector efficiency)—Ratio of the total acoustic power output of an electroacoustic transducer, to the electric power input.

transmitting loop loss—That part of the repetition equivalent assignable to the station set, subscriber line, and battery supply circuit which are on the transmitting end.

transmitting power response (projector power response) — Of an electroacoustic transducer used for sound emission, the ratio of the effective sound pressure apparent at a distance of 1 meter in a specified direction from the effective acoustic center of the transducer, to the electric power input. Usually expressed in decibels above a reference response of 1 microbar square per watt of electric input.

transmitting station—A location at which the transmitter of a radio system and its antenna and associated equipment are grouped.

transmitting voltage response—Of an electroacoustic transducer used for sound emission, the ratio of the sound pressure apparent at a distance of 1 meter in a specified direction from the effective acoustic center of the transducer, to the signal voltage applied at the electric input terminals. Usually expressed in decibels above a reference-voltage response of 1 microbar per volt.

transonic speed—A speed of 600 to 900 miles per hour, corresponding to about Mach 0.8 to Mach 1.2.

transponder — A radio transmitter-receiver which transmits identifiable signals automatically when the proper interrogation is received. (*See also* Pulse Repeater.)

transponder dead time — The time interval between the start of a pulse and the earliest instant at which a new pulse can be received or produced by a transponder.

transponder efficiency—A ratio expressed as a percentage of the number of replies to the number of interrogations from a transponder.

transponder suppressed time delay — The overall fixed delay between reception of an interrogation and the transmission of a reply.

transport—Platform or deck of a tape recorder on which the motor (or motors), reels, heads and controls are mounted. It includes those parts of the recorder other than the amplifier, preamplifier, loudspeaker and case. Also called tape transport.

transportable transmitter—Sometimes called a portable transmitter. A transmitter designed to be readily carried from place to place, but normally not operated while in motion.

transport time—In an automatic system, the time required to move an object, element, or information between two predetermined positions.

transposition—The interchanging of the relative positions of the conductors in an open-wire line to reduce noise, interference, and cross talk.

transposition blocks – Spreaders used to space and reverse the relative positions of two conductors at fixed intervals.

transposition section – A length of open-wire line to which a fundamental transposition design or pattern is applied as a unit.

transradar – A bandwidth-compression system for use in long-range, narrow-band transmission of radio signals from a radar receiver to a remote location.

transrectification–The rectification that occurs in one circuit when an alternating voltage is applied to another circuit.

transrectification characteristic–The graph obtained by plotting the direct voltage values for one electrode of a vacuum tube as abscissas against the average current values in the circuit of that electrode as ordinates, for various values of alternating voltage applied to another electrode as a parameter. The alternating voltage is held constant for each curve, and the voltages on other electrodes are maintained constant.

transrectification factor – The change in average current of an electrode, divided by the change in the amplitude of the alternating sinusoidal voltage applied to another electrode (the direct voltages of this and other electrodes being maintained constant).

transrectifier–A device, ordinarily a vacuum tube, in which rectification occurs in one electrode circuit when an alternating voltage is applied to another electrode.

transresistance amplifier–An amplifier that supplies an output voltage proportional to an input current. The transfer function of the amplifier is $e_o/i_{in} = R_m$, where R_m is the transresistance.

transverse-beam traveling-wave tube – A traveling-wave tube in which the electron beam intersects the signal wave rather than moving in the same direction.

transverse cross-talk coupling–Between a disturbing and a disturbed circuit in any given section, the vector summation of the direct couplings between adjacent short lengths of the two circuits, without dependence on intermediate flow in nearby circuits.

transverse electric wave–Abbreviated te wave. In a homogeneous isotropic medium, an electromagnetic wave in which the electric field vector is everywhere perpendicular to the direction of propagation. The dominant mode in a rectangular waveguide is te_{10}.

transverse electromagnetic wave – Abbreviated tem wave. In a homogeneous isotropic medium, an electromagnetic wave in which the electric and magnetic field vectors both are everywhere perpendicular to the direction of propagation. This is the normal mode of propagation in coaxial line, open-wire line, and stripline.

transverse-field traveling-wave tube – A traveling-wave tube in which the traveling electric fields which interact with the electrons are essentially transverse to the average motion of the electrons.

transverse interference–Interference occurring across terminals or between signal leads.

transverse magnetic e-mode – A type of mode in which the longitudinal component of the magnetic field is zero and the longitudinal component of the electric field is not zero.

transverse magnetic wave–Abbreviated tm wave. In a homogeneous isotropic medium, an electromagnetic wave in which the magnetic field vector is everywhere perpendicular to the direction of propagation.

transverse magnetization – Magnetization of a recording medium in a direction perpendicular to the line of travel and parallel to the greater cross-sectional dimension.

transverse recording–The technique of recording with rotating heads which are oriented perpendicular to the edge and surface of the tape.

transverse wave–A wave in which the direction of displacement is perpendicular to the direction of propagation at each point of the medium. When the direction of displacement forms an acute angle with the direction of propagation, the wave is considered to have both a longitudinal and a transverse component.

trap–1. A selective circuit that attenuates undesired signals but does not affect the desired ones. (*See also* Wave Trap.) 2. A crystal imperfection which can trap carriers. 3. A unprogrammed conditional jump to a known location, the jump being activated automatically by hardware; the location from which the jump occurs is recorded.

TRAPATT diode–Abbreviation for trapped plasma avalanche transit time. A microwave avalanche diode that has either an n+pp+ or a p+nn+ structure. It may be manufactured from either silicon or germanium. When the diode is biased into breakdown, an electron-hole plasma fills the entire p or n region and the voltage across the diode drops. A large current, induced by the low residual electric field, extracts the plasma from the diode. After the plasma is removed, the current drops and voltage rises. Energy stored in the resonant circuits of an oscillator raises the voltage above breakdown and the cycle repeats.

trapezoidal distortion–Distortion in which a televised picture has the shape of a trapezoid (wide at top or bottom) instead of a rectangle. It is due to the interaction between the vertical- and horizontal-deflection coils (or plates) of the cathode-ray tube.

trapezoidal generator—An electronic circuit that produces a trapezoidal voltage wave.

trapezoidal pattern – An oscilloscope pattern which indicates the percentage of modulation in an amplitude-modulated system.

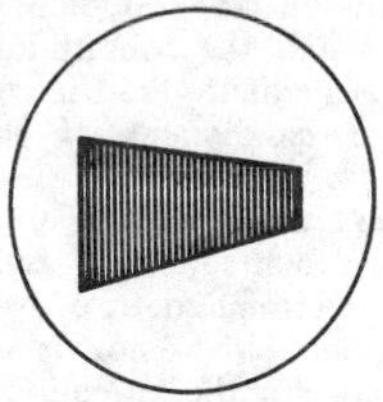

Trapezoidal pattern.

trapezoidal wave—1. A trapezoidal-shaped waveform. 2. A square wave onto which a sawtooth has been superimposed. It is the voltage wave necessary to give a linear deflection current through the coils of a magnetically deflected cathode-ray tube.

trapped flux – In a material in the superconducting state, magnetic flux linked with a closed superconducting loop.

trapping – 1. The holding of electrons or holes by any of several mechanisms in a crystal, thereby preventing them from moving. 2. In a computer, instructions that cause initiation by the central processing unit of an internal interrupt that transfers control to a subroutine which activates the desired operation of the instruction. Also, the subroutine can be changed, thereby causing the operation on the instruction to be changed.

trap wire – A low-voltage wire used at hinge points, where severe flexing occurs, usually in burglar alarm systems. It is made with tinsel conductor.

traveling detector—A probe mounted on a slider and free to move along a longitudinal slot cut into a waveguide or coaxial transmission line. The traveling detector is connected to auxiliary measuring apparatus and used for examining the relative magnitude of any standing-wave system.

traveling plane wave – A plane wave in which each frequency component has an exponential variation of amplitude and a linear variation of phase in the direction of propagation.

traveling wave—The resulting wave when the electric variation in a circuit takes the form of translation of energy along a conductor, such energy being always equally divided between current and potential forms.

traveling-wave amplifiers – A two-port amplifier where the input signal excites a space-charge wave at one end of a slab of negative-differential-resistivity bulk material. The wave travels through the material between ohmic contacts which serve as a transmission line. The charge builds as it moves, so the output is a larger version of the input.

traveling-wave magnetron – A traveling-wave tube in which the electrons move in crossed static electric and magnetic fields that are substantially normal to the direction of wave propagation.

traveling-wave magnetron oscillations—Oscillations sustained by the interaction between the space-charge cloud of a magnetron and a traveling electromagnetic field with approximately the same phase velocity as the mean velocity of the cloud.

traveling-wave parametric amplifier – A parametric amplifier which has a continuous iterated structure incorporating nonlinear reactors and in which the signal, pump, and difference-frequency waves are propagated along the structure.

traveling-wave phototube – A traveling-wave tube that has a photocathode and a window that admits a modulated laser beam, which causes the emission of a current-modulated photoelectron beam. This beam is then accelerated by an electron gun and directed into the helical slow-wave structure of the tube.

traveling-wave tube – Abbreviated twt. A tube in which a stream of electrons interacts continuously or repeatedly with a guided electromagnetic wave moving sub-

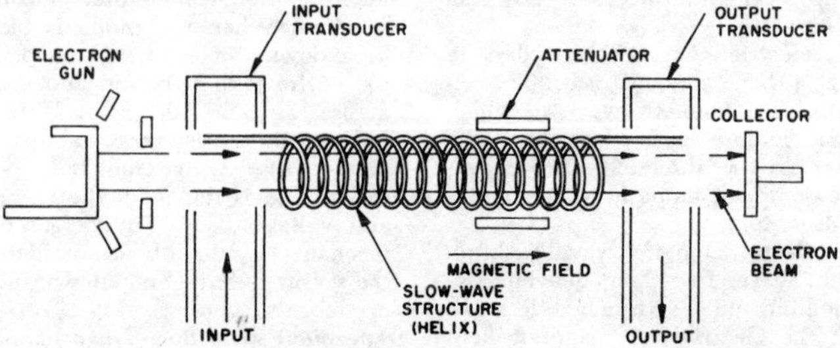

Traveling-wave tube.

stantially in synchronism with it, and in such a way that there is a net transfer of energy from the stream to the wave.

traveling-wave-tube amplifier – A power-amplifying piece of equipment yielding about 30 dB of gain over broad bandwidths. The units are built around certain twt's in the most commonly used frequency bands, and they have external modulation circuitry incorporated.

traveling-wave-tube interaction circuit—An extended electrode arrangement used in a traveling-wave tube to propagate an electromagnetic wave in such a manner that the traveling electromagnetic fields are retarded to the point where they extend into the space occupied by the electron stream.

tr box—*See* Transmit-Receive Switch.

tr cavity – The resonant portion of a tr switch.

treating—Any of the processes of applying varnishes or other insulating compounds to insulation, coils, or windings. This includes the process of impregnation, soaking, and surface coating by any of the various methods.

treble—The higher part in harmonic music or voice; of high or acute pitch. In music, the frequencies from middle *C* (261.63 hertz) upward.

treble boost—Deliberate adjustment of the amplitude-frequency response of a system or component to accentuate the higher audio frequencies.

tree—A set of connected branches without meshes.

treeing – A progressive type of insulation failure in which branching hollow channels slowly penetrate the insulation at rather low applied voltages.

tremolo—1. The amplitude modulation of an audio tone. 2. A warbling or fluctuating effect, approximately seven times per second, in the tone of an instrument characterized by a variation in intensity rather than pitch. (*See also* Vibrato.)

trf – Abbreviation for tuned radio frequency.

tri—Abbreviation for triode.

triac – 1. A bidirectional rectifier (essentially two SCRs in parallel) that function as an electrically controlled switch for ac loads and having an npnpn structure that can be triggered into either forward or reverse conduction by a pulse applied to its gate electrode. A triac will pass an alternating current. 2. (Formal name is "bidirectional triode thyristor.") A thyristor that can be triggered into conduction in either direction. Terminals are called "main terminal 1," and "gate."

triad—1. Three radio stations operated as a group for determining the position of aircraft or ships. 2. A group of three dots, one of each color-emitting phosphor, on

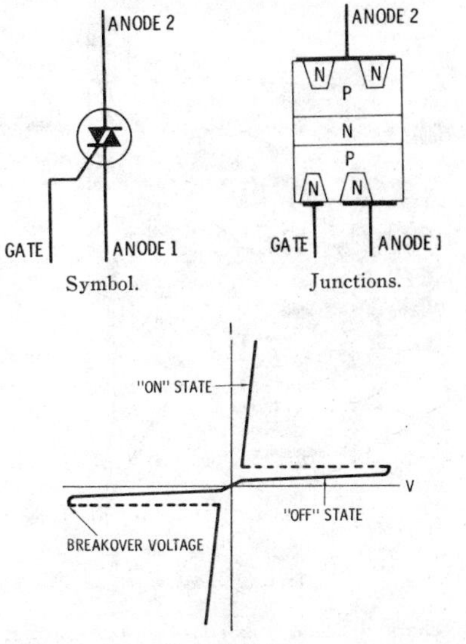

Symbol. Junctions.

Characteristic.

Triac.

the screen of a color picture tube. 3. A group of three bits or three pulses, usually in sequence on one wire or simultaneously on three wires. 3. A group of three insulated conductors twisted together without (or with) a sheath overall. Usually color-coded for identification. Also called a triplex.

triangulation—A method of finding the location of a third point by taking bearings from two fixed points a known distance apart. The third point will be at the intersection of the two bearing lines.

triax—A type of shielded conductor that employs a shield and jacket over the primary insulation, plus a second shield and jacket over all. Aside from applications requiring maximum attenuation of radiated signals or minimum pickup of external interference, this cable can also be used to carry two separate signals.

triaxial cable—A special form of coaxial cable containing three conductors.

triaxial speaker – A dynamic speaker unit consisting of three independently driven units combined into a single speaker.

tribo—A prefix meaning due to or pertaining to friction.

triboelectric—Pertaining to electricity generated by friction.

triboelectricity—Electrostatic charges generated due to friction between different materials.

triboelectric series—A list of substances arranged so that any of them can become positively electrified when rubbed with one farther down the list, or negatively

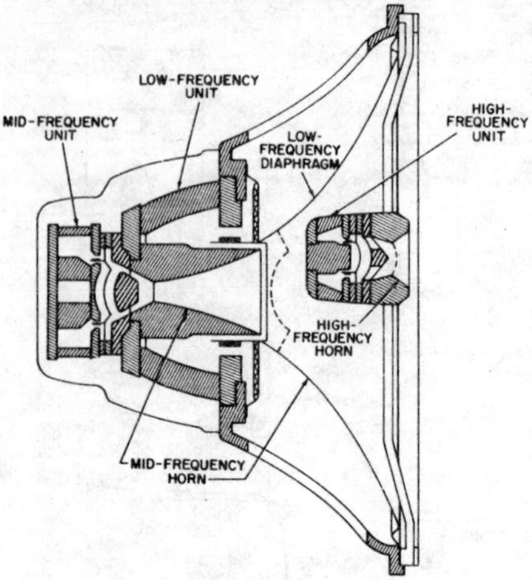

Triaxial speaker.

charged when rubbed with one farther up the list.

Triboelectric Series

Positive
Asbestos
Rabbit fur
Glass
Mica
Nylon
Wool
Cat fur
Silk
Paper
Cotton
Wood
Lucite
Sealing wax
Amber
Polystyrene
Polyethylene
Rubber balloon
Sulfur
Celluloid
Hard rubber
Vinylite
Saran Wrap
Negative

triboluminescence — Luminescence that arises from friction. Usually occurs in crystalline materials.

tributary circuit—A circuit which connects an individual drop, or drops, to a switching center.

tributary station—A communications terminal consisting of equipment compatible for the introduction of messages into, or reception from, its associated relay station.

trickle charge—1. A continuous charge of a storage battery at a slow rate approximately equal to the internal losses and suitable to maintain the battery in a fully charged condition. 2. Very slow rates of charge suitable not only in compensating for internal losses, but in restoring small, intermittent discharges to the load circuit delivered from time to time.

trickle charger — A device for charging a storage battery at a low rate continuously, or for several hours at one time.

tricolor camera — A television camera designed to separate reflected light into three frequency groups, each corresponding to the light energies of the three primary colors. The camera transforms the intensity variations of each primary into amplitude variations of an electrical signal.

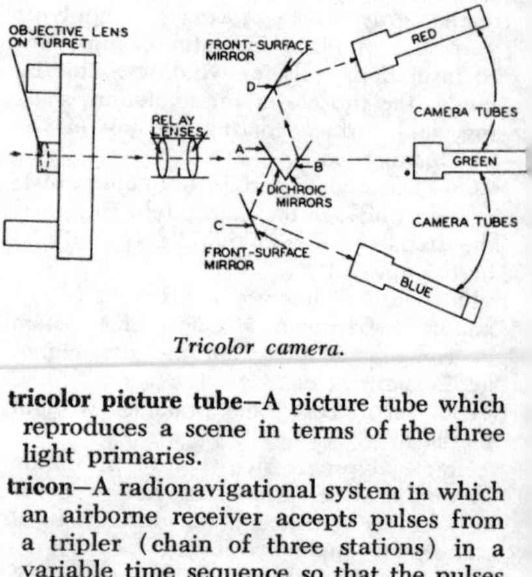

Tricolor camera.

tricolor picture tube—A picture tube which reproduces a scene in terms of the three light primaries.

tricon—A radionavigational system in which an airborne receiver accepts pulses from a tripler (chain of three stations) in a variable time sequence so that the pulses arrive at the same time even though traveling over paths of various lengths.

trigatron — An electronic switch in which breakdown of an auxiliary gap initiates conduction.

trigger—1. To cause, by means of one circuit, action to start in another circuit, which then functions for a certain length of time under its own control. 2. A pulse that starts an action.

trigger action—The instantaneous initiation of main current flow by a weak controlling impulse in a device.

trigger circuit—1. A circuit with two conditions of stability, with means for passing spontaneously or through application of an external stimulus from one to the other when certain conditions are satisfied. *See also* Flip-Flop, 1. 2. A circuit in which an electron tube performs the function of a relay. Impulses applied to the input of the tube produce corresponding impulses in the output circuit, starting a chain of events.

trigger countdown—A process that reduces the repetition rate of a triggering signal.

trigger diode — A symmetrical three-layer avalanche diode used to control SCRs and triacs. It has a symmetrical switching mode, and therefore fires whenever the breakover voltage is exceeded in either polarity of applied voltage.

triggered blocking oscillator — A blocking oscillator that can be reset to its starting condition by the application of a trigger voltage.

triggered spark gap—A fixed spark gap in which the discharge passes between two electrodes and is struck (started) by an auxiliary electrode called the trigger, to which low-power pulses are applied.

triggered sweep—In a cathode-ray oscilloscope, a sweep initiated by a signal pulse.

trigger electrode—*See* Starter, 2.

trigger gap — A gas discharge device with three electrodes which can be triggered upon command. Two electrodes provide the main conduction path and the third electrode serves as the trigger.

triggering—The starting of circuit action, which then continues for a predetermined time under its own control.

triggering signal—The signal from which a trigger is derived.

trigger level—1. In a transponder, the minimum receiver input capable of causing the transmitter to emit a reply. 2. The instantaneous level of a triggering signal at which a trigger is to be generated. Also the name of the control which selects the level.

trigger point — The amplitude point on the input pulse at which triggering of the sweep of a cathode-ray oscilloscope occurs.

trigger pulse—A pulse used for triggering.

trigger-pulse steering — In transistors, the routing or directing of trigger signals (usually pulses) through diodes or transistors (called steering diodes or transistors) so that the signals affect only one of several associated circuits.

trigger recognition — In random-sampling-oscilloscope technique, the process of making a response to a suitably applied trigger such as the time reference for the time window.

trigistor — A bistable pnpn semiconductor component with characteristics comparable to those of a flip-flop or bistable multivibrator.

tri-gun color picture tube—*See* Three-Gun Color Picture Tube.

trim—To make a fine adjustment in a circuit or a circuit element.

trim control — On some regulated power supplies, a control used to make minor adjustments of output voltage.

trimmer—1. *See* Trimmer Capacitor. 2. A small adjustable circuit element connected in series or parallel with a circuit element of the same kind that its adjustment sets the combination of the two to a desired value.

trimmer capacitor—Also called a trimmer. A small variable capacitor associated with another capacitor and used for fine adjustment of the total capacitance of the combination.

Trimmer capacitor.

trimmer potentiometer—A lead-screw-actuated potentiometer.

trimmer resistor—A small rheostat used in place of a fixed resistor to permit adjustment of resistance values in a circuit during initial calibration of an equipment or when recalibration is required.

trimming—1. The fine adjustment of capacitance, resistance, or inductance in a circuit. 2. The process of using abrasive or laser technique to accurately remove film material from a substrate to change (increase) the resistance value.

trimming potentiometer—An electrical mechanical device with three terminals. Two terminals are connected to the ends of a resistive element and one terminal is connected to a movable conductive contact which slides over the element, thus allowing the input voltage to be divided as a function of the mechanical input. It can function as either a voltage divider or rheostat.

trinistor—A three-terminal silicon semiconductor device with characteristics similar to those of a thyratron and used for controlling large amounts of power.

trinoscope—1. Any assembly of three kinescopes producing the red, green, and blue images required for tricolor television optical projection (e.g., for theater tv).

triode—A three-electrode electron tube con-

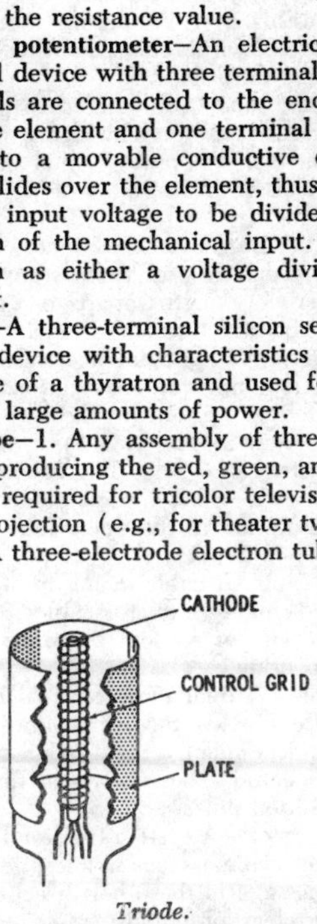

Triode.

taining an anode, a cathode, and a control electrode (grid).

triode amplifier—An amplifier in which only triode tubes are used.

triode-heptode converter – A superheterodyne converter circuit which uses a triode local oscillator and a heptode converter, both contained in one envelope.

triode-hexode converter – A superheterodyne converter circuit which uses a triode local oscillator and a hexode converter, both contained in one envelope.

triode laser—A gas laser the light output of which may be modulated by applying signal voltages to an integral grid.

triode-pentode – A dual-purpose vacuum tube containing a triode and a pentode in the same envelope.

triode pnpn-type switch – A pnpn-type switch having an anode, cathode, and gate terminal.

trip coil – An electromagnet in which a moving armature trips a circuit breaker or other protective device and thereby opens a circuit under abnormal conditions.

triple-conversion receiver – A communications receiver in which three different intermediate frequencies are employed to give better adjacent-channel selectivity and greater image-frequency suppression.

triple detection – *See* Double Superheterodyne Reception.

triple diffused – Pertaining to transistors fabricated within a monolithic substrate by three diffusion steps.

triple-play tape—Abbreviated tp tape. Very thin recording tape of which it is possible to wind 3600 ft. onto a 7-inch spool. This would give 180 minutes of playing time at 3¾ ips.

triple-stub transformer—A microwave transformer consisting of three stubs placed a quarter-wavelength apart on a coaxial line. The stubs are adjusted in length to compensate for impedance mismatch.

triplet—1. Three radionavigational stations operated as a group for the determination of position. 2. The waveform of the output voltage of a delay line when the input pulse has a width approximately equal to the resolution of the delay line.

triplex—*See* Triad.

triplex cable—A cable made up of three insulated single-conductor cables twisted together with or without a common insulating covering.

triplexer – A dual duplexer which permits the use of two receivers simultaneously and independently in a radar system by disconnecting the receivers during the transmitted pulse.

triplex system—A system for simultaneously sending two messages in one direction and one message in the other direction over a single telegraph circuit.

tripod—A three-legged camera support.

tripping device – A mechanical or electromagnetic device used for opening (turning off) a circuit breaker or starter, either when certain abnormal electrical conditions occur or when a catch is actuated manually.

tripping relay – Also called trip-free relay. A device which functions to trip a circuit breaker, contactor, or equipment or to permit immediate tripping by other devices; or to prevent immediate reclosure of a circuit interrupter, in case it should open automatically even though its closing circuit is maintained closed.

trip protection circuit—A protective circuit that electrically interrupts the output when an overload occurs.

trip voltage—The voltage at which ionization occurs under any circumstances (also referred to as firing voltage).

trisistor – A fast-switching semiconductor consisting of an alloyed junction pnp device in which the collector is capable of electron injection into the base. Its characteristics resemble those of a thyratron electron tube, and the switching time is in the nanosecond range.

tri-state—Logic systems utilizing three conditions on one line: a definitely applied high voltage (logic 1); a definite low voltage (logic 0); and an open circuit or underfined state, permitting another part of the circuit to determine whether the line will be high or low. Usually refers to device outputs or systems using outputs of the tri-state type. Useful in a buss-organized system.

tristimulus values—The amounts of each of the three primary colors that must be combined to match a sample.

tri-tet oscillator—A crystal-controlled, electron-coupled vacuum-tube oscillator which is isolated from the output circuit through use of the screen-grid electrode as the oscillator anode. Used for multiband operation because it generates strong harmonics of the crystal frequency.

tritium—A radioactive isotope of hydrogen with an atomic number of 3.

TRL – Abbreviation for transistor-resistor logic.

trombone—An adjustable U-shaped coaxial-line matching assembly.

tropicalization – A chemical treatment developed to combat the fungi that ruin electronic equipment in hot, humid jungle regions.

tropo—*See* Tropospheric Scatter Communication.

troposphere—The lower layer of the earth's atmosphere, extending to about 60,000 feet at the equator and 30,000 feet at the poles. In this area, the temperature generally decreases with altitude, clouds form, and convection is active.

tropospheric scatter – The propagation of radio waves by scattering as a result of irregularities or discontinuities in the physical properties of the troposphere.

tropospheric scatter communication – Also called tropo. A method or system of transmitting, within the troposphere, microwaves in the uhf or shf bands to effect radiocommunication between two points on the earth's surface separated by moderate distances of from 70 to 600 miles. Such a span or hop may be augmented by other spans in tandem to permit end-to-end or through circuits up to many thousands of miles. More specifically, this method of communication is now generally understood to embrace a radio system that permits communication over the distances indicated, with excellent reliability and good information capacity, using relatively high transmitted power, frequency modulation, and highly sensitive receiving apparatus.

tropospheric superrefraction—The phenomenon occurring in the troposphere whereby radio waves are bent sufficiently to be returned to the earth.

tropospheric wave – A radio wave that is propagated by reflection from a place of abrupt change in the dielectric constant of its gradient in the troposphere. In some cases, the ground wave may be so altered that new components appear to arise from reflections in regions of rapidly changing dielectric constants; when these components are distinguishable from the other components, they are called tropospheric waves.

trouble—Failure of a circuit or element to perform in a standard manner.

trouble-location problem—A computer test problem, the incorrect solution of which supplies information on the location of faulty equipment. It is used after a check problem has shown that a fault exists.

troubleshooting – Locating and diagnosing malfunctions or breakdowns in equipment by means of systematic checking or analysis.

trouble unit—A weighting figure applied to indicate the expected performance of a telephone circuit or circuits in a given period of time.

tr switch—*See* Transmit-Receive Switch.

tr tube – (Transmit-receive tube.) A gas-filled rf switching tube that permits use of the same antenna for both transmitting and receiving by preventing the transmitted power from damaging the receiver. (Usually, a tr unit consists of a cavity containing a discharge gap which connects the transmitter to the antenna, and a coupling circuit which connects the antenna to the receiver when the discharge gap is not fired, indicating that the transmitting tube is quiescent.

true bearing—A bearing given in relation to geographic north, as opposed to a magnetic bearing.

true complement—*See* Radix Complement.

true course—A course in which the direction of the reference line is true rather than magnetic north.

true credit balance—In a calculator, when the answer is negative, the minus sign automatically appears in the display.

true homing—The following of a course in such a way that the true bearing of an aircraft or other vehicle is held constant.

true north—Geographic north.

true ohm—The actual value of the practical unit of resistance. It is equal to 10^9 absolute electromagnetic units of resistance.

true power—The average power consumed by a circuit during one complete cycle of alternating current.

true radio bearing—*See* Radio Bearing.

true random noise – A noise characterized by a normal or Gaussian distribution of amplitudes.

true value—The value of a physical quantity that would be attributed to an object or physical system if that value could be determined with no error.

truncate—1. To drop digits of a number of terms in a series, thereby lessening precision. For example, the value 3.14159265 (π) when truncated to five figures is 3.1415, whereas it could be rounded off to 3.1416. 2. To conclude a computational process according to some rule; for example, to stop the evaluation of a power series at a specified term.

truncated paraboloid—A paraboloid reflector in which a portion of the top and bottom have been cut away to broaden the main radiated lobe in the vertical plane.

truncation—The process of dropping one or more digits at the left or right of a number without changing any of the remaining digits. For example, in most operations the number 3847.39 would become 3847.3 when truncated one place at the right, whereas the same number would become 3847.4 when rounded correspondingly. *See also* Roundoff.

truncation error—The error resulting from the use of only a finite number of terms of an infinite series, or from the approximation of operations in the infinitesimal calculus by operations in the calculus of finite differences.

trunk—1. A single message circuit between two points, both of which are switching centers and/or individual message distribution points. 2. A communications channel between two different offices, or between groups of equipment within the same office. 3. A telephone line or channel between two central offices or switching devices, which is used in providing telephone connections between subscribers.

trunk circuit—A circuit which connects two switching centers.

trunk facility—That part of a communication channel connecting two or more leg facilities to a central or satellite station.

trunk group—Those trunks connecting two points both of which are switching centers and/or individual message distribution points and which make use of the same multiplex terminal equipments.

trunk hunting—A method by which an incoming call is switched to the next consecutive number if the first called number is busy.

trunk loss — That part of the repetition equivalent assignable to the trunk used in the telephone connection.

truth table—A tabulation that shows the relation of all output logic levels of a digital circuit to all possible combinations of input logic levels in such a way as to characterize the circuit functions completely.

T_s—Abbreviation for color temperature.

TSC — Abbreviation for transmitter start code.

T^2L or TTL — Abbreviation for transistor-transistor logic.

TTL or T^2L compatible—The ability of a device or circuit to be connected directly to the input or output of TTL logic devices. Such compatibility eliminates the need for interfacing circuitry.

tuba — A powerful land-based radar jamming transmitter operated between 480 and 500 MHz. It was developed during World War II for use against night fighter planes.

tube—A hermetically sealed glass or metal envelope in which conduction of electrons takes place through a vacuum or gas.

tube bridge — An instrument used in the precise measurement of vacuum-tube characteristics. It contains one or more bridge-type measuring circuits, plus power supplies and signal sources for all possible electrode combinations.

tube coefficients — Constants that describe the characteristics of a thermionic vacuum tube (e.g., amplification factor, mutual conductance, ac plate resistance, etc.).

tube complement — The number and types of electron tubes required in an electronic equipment.

tube count — A terminated discharge produced by an ionizing event in a radiation-counter tube.

tube drop—The voltage measured across a tube, from plate to cathode, when the tube is conducting at its normal current rating.

tube electrometer — A thermionic vacuum tube adapted for use as an electrometer, to measure potential difference.

tube heating time — The time required for the coolest portion of a mercury-vapor tube to attain its operating temperature.

tube noise—Noise originating in a vacuum tube (e.g., from shot effect, thermal agitation, etc.).

tube shield — A metallic enclosure placed over a vacuum tube to prevent external fields from interfering with the function of the tube.

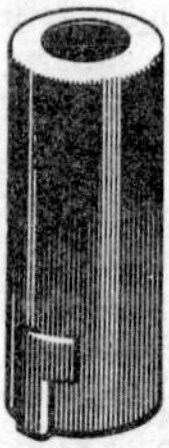

Tube shield.

tube socket — A receptacle which provides mechanical support and electrical connection for a vacuum tube.

tube tester—A test instrument for indicating the condition of vacuum tubes used in electronic equipment.

tube voltage drop—The anode voltage in an electron tube during conduction.

tubing—Extruded nonsupported plastic materials designed as protection and electrical insulation for exposed components of electrical and electronic assemblies, as opposed to a coated, braided, or woven tube termed sleeving.

tubular capacitor — A paper, ceramic, or electrolytic capacitor shaped like a cylinder; leads or lugs project from one or both ends.

tunable-cavity filter—A microwave filter in which tuning can be accomplished by adjustment of one or more screws that project into the cavity or by adjustment of the positions of one or more rectangular or circular irises in the cavity or waveguide.

tunable echo box—An echo box consisting of an adjustable cavity operating in a single mode. When the echo box is calibrated, the setting of the plunger at resonance will indicate the wavelength.

tunable magnetron — A magnetron that can be tuned mechanically or electronically over a limited band of frequencies.

tuned—Adjusted to resonate or operate at a specified frequency.

tuned amplifier—An amplifier in which the load is a tuned circuit. Thus, the load impedance and amplifier gain vary with the frequency.

tuned-anode oscillator — *See* Tuned-Plate Oscillator.

tuned antenna – An antenna designed, by means of its own inductance and capacitance, to provide resonance at the desired operating frequency.

tuned-base oscillator–A transistor oscillator comparable to a tuned-grid electron-tube oscillator. The frequency-determining device (resonant circuit) is located in the base circuit.

tuned circuit–A circuit consisting of inductance and capacitance which can be adjusted for resonance at the desired frequency.

tuned-collector oscillator–A transistor oscillator comparable to the tuned-plate electron-tube oscillator. The frequency-determining device is located in the collector circuit.

tuned dipole–A dipole antenna that resonates at its operating frequency.

tuned filter – A resonant arrangement of electronic components which either attenuates signals at a particular frequency and passes signals at other frequencies, or vice versa.

tuned-filter oscillator – An oscillator in which a tuned filter is used.

tuned-grid oscillator–An oscillator, the frequency of which is determined by a parallel-tuned tank in the grid circuit. The tank is coupled to the plate to provide the required feedback.

tuned-grid–tuned-anode oscillator – *See* Tuned-Grid–Tuned-Plate Oscillator.

tuned-grid–tuned-plate oscillator – Also called tuned-grid–tuned-anode oscillator. An oscillator having parallel-resonant circuits in both the plate and the grid circuits. The necessary feedback is provided by the plate-to-grid interelectrode capacitance.

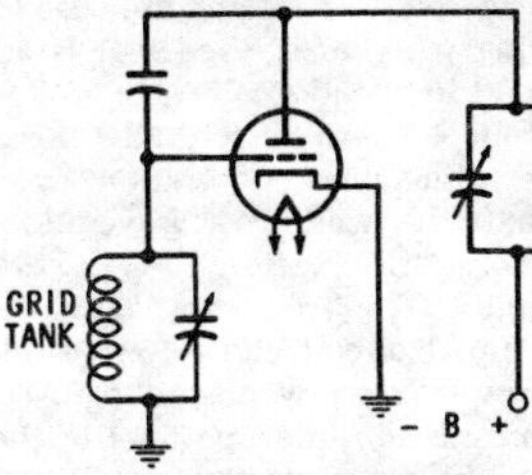

Tuned-grid–tuned-plate oscillator.

tuned-plate oscillator – Also called tuned-anode oscillator. An oscillator, the frequency of which is determined by a parallel-tuned tank in the plate circuit. The tank is coupled to the grid to provide the required feedback.

tuned radio-frequency amplifier – A tuned amplifier using resonant-circuit coupling and designed to operate at radio frequencies.

tuned radio-frequency receiver – A radio receiver consisting of several amplifier stages, which are tuned to resonance at the carrier frequency of the desired signal by a ganged variable-tuning capacitor. The amplified signals at the original carrier frequency are fed directly into the detector for demodulation. The resultant audio-frequency signals are again amplified, and are then reproduced by a speaker.

tuned radio-frequency transformer – A transformer used for selective coupling in radio-frequency stages.

tuned-reed frequency meter – A vibrating-reed instrument for measuring the frequency of an alternating current.

tuned relay–A relay having mechanical or other resonating arrangements that limit the response to currents at one particular frequency.

tuned resonating cavity–A resonating cavity half a wavelength long or some multiple of a half wavelength, used in connection with a waveguide to produce a resultant wave with the amplitude in the cavity greatly exceeding that of the wave in the guide. For reception of waves, a detecting grating can be placed at the point of maximum amplitude in the cavity, to convert the energy to a form suit-

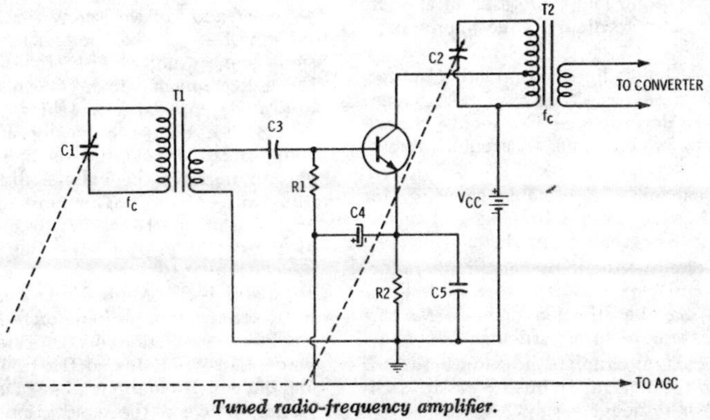

Tuned radio-frequency amplifier.

able for amplification in a telephone or television circuit. A tuned cavity is a nonreflecting termination for a guide.

tuned rope—Long lengths of chaff cut to the various lengths necessary for tuning to different wavelengths.

tuned transformer—A transformer, the associated circuit elements of which are adjusted as a whole to be resonant at the frequency of the alternating current supplied to the primary, thereby causing the secondary voltage to build up to higher values than would otherwise be obtained.

tuner—In the broad sense, a device for tuning. Specifically, in radio-receiver practice: 1. A packaged unit capable of producing only the first portion of the functions of a receiver and delivering either rf, if, or demodulated information to some other equipment. 2. That portion of a receiver which contains the circuits that are tuned to resonance at the received-signal frequency and those which are tuned to the local-oscillator frequency. 3. A radio, or tv, receiving circuit; a high fidelity component containing such circuits.

tungar rectifier—A gaseous rectifier containing argon gas. It is employed in battery chargers and low-voltage power supplies.

tungar tube — A phanotron (hot-cathode, gas-filled rectifier tube) having a heated filament serving as the cathode and a graphite disc as the anode in a bulb filled with low-pressure argon. Used chiefly in battery chargers.

tungsten—A metal used in the manufacture of filaments for vacuum tubes and in making contact points for switches and other parts where sparking may occur. After the tungsten is made ductile by rolling, swaging, and hammering, it is very tough.

tungsten filament—A filament used in incandescent lamps, and in thermionic vacuum tubes and other tubes requiring an incandescent cathode. Smaller tungsten filaments are operated in a vacuum, while those for larger lamps are used in an inert gas at about ordinary atmospheric pressure.

tuning — The adjustment relating to frequency of a circuit or system to secure optimum performance. Commonly, the adjustment of a circuit or circuits to resonance.

tuning capacitor—A variable capacitor for adjusting the natural frequency of an oscillatory or resonant circuit.

tuning circuit—A circuit containing inductance and capacitance, either or both of which may be adjusted to make the circuit responsive to a particular frequency.

tuning coil—A variable inductance for adjusting the natural frequency of an oscillatory or resonant circuit.

tuning control—A control knob that adjusts all tuned circuits simultaneously.

tuning core—Normally a molded iron core for permeability tuning, into which an adjusting screw has been cemented or molded.

tuning eye—Slang for a cathode-ray tuning indicator.

tuning fork—A two-pronged hard-steel device that vibrates at a definite natural frequency when struck or when set in motion by electromagnetic means. Used in some electronic equipment as an accurately controllable source of signals, because its vibrations can be formed readily into audio-frequency signals by means of pickup coils.

tuning-fork drive—Control of an oscillator by continuous vibrations of a tuning fork. A high harmonic of the oscillating signal obtained from the fork is selected by filter circuits and is strongly amplified to determine the main-oscillator frequency in a transmitter or other equipment.

tuning in—Adjusting the tuning controls of a receiver to obtain maximum response to the signals of the station it is desired to receive.

tuning indicator — A device that indicates whether or not a receiver is tuned accurately. It is connected to some circuit in which current or voltage is maximum or minimum when the receiver is accurately tuned to give the strongest output signal.

tuning meter—A direct-current meter connected to a receiver circuit and used for determining whether the receiver is accurately tuned to a station.

tuning probe — An essentially lossless probe of adjustable penetration extending through the wall of a waveguide or cavity resonator.

tuning range — The frequency range over which a tuned circuit can be adjusted.

tuning screw — A screw or probe inserted into a transmission line (parallel to the E field) to produce susceptance of magnitude and sign that depend on the depth of penetration of the screw.

tuning stub—1. A short length of transmission line, usually with the free end shorted, connected to a transmission line to provide impedance matching. 2. A type of inductor element, usually adjustable, connected to a transmission line at intervals to improve the voltage distribution.

tuning susceptance — The normalized susceptance of an atr tube in its mount due to its deviation from the desired resonance frequency.

tuning wand—*See* Neutralizing Tool.

tunnel action (in a pn junction)—A process whereby conduction occurs through the potential barrier due to the tunnel effect and in which electrons pass in either direction between the conduction band in the n region and the valence band in the p region. (Tunnel action, unlike the diffusion of charge carriers, involves electrons only and for all practical purposes the transit time is negligible.)

tunnel cathode – A metal-insulator-metal sandwich. Electrons tunnel from the metal substrate and appear in the metal film as hot electrons. Some of the hot electrons have sufficient energy to pass over the cathode surface barrier into the vacuum.

tunnel diode–1. A pn diode to which has been added a large amount of impurity. The tunnel diode has high-speed charge movement and a negative-resistance region above a minimum level of applied voltage. With the addition of suitable external circuits, it can be used as an oscillator or amplifier. Also called Esaki diode. 2. A two-layer device similar to the rectifier diode. As a small voltage is applied, current starts to flow. Increase the voltage a little more and current drops to zero. Add still a little more voltage and the current increases again. At still higher voltages it responds like an ordinary diode. That first surge of current is called tunneling.

tunnel effect–The piercing of a potential hill by a carrier, which would be impossible according to classical mechanics, but the probability of which is not zero according to wave mechanics, if the width

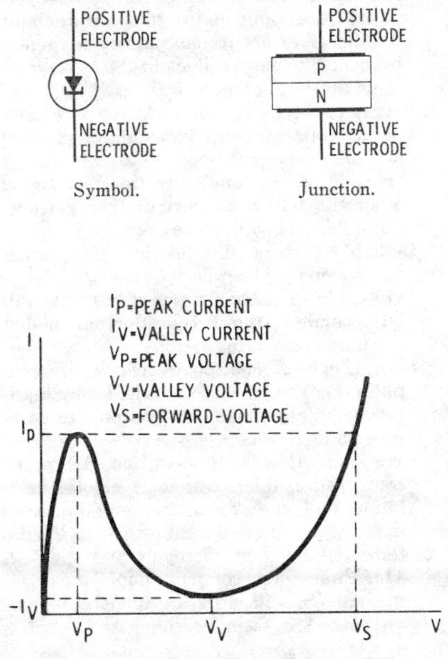

Characteristic.

Tunnel diode.

of the hill is small enough. The wave associated with the carrier is almost totally reflected on the first slope, but a small fraction crosses the hill.

tunneling–*See* Tunnel Diode.

tunnel rectifier–A tunnel diode that has a relatively low peak-current rating in comparison with other tunnel diodes employed in memory-circuit applications.

tunnel resistor–A resistor containing a thin layer of metal plated across a tunneling junction, so that the characteristics of a tunnel diode and an ordinary resistor are combined.

tunnel triode – A transistorlike device in which the emitter-base junction is a tunnel diode and the collector-base junction is a conventional diode.

tunneluminescence–The emission of light from a phosphor film deposited on the surface of a thin-film metal-oxide-metal sandwich.

turbidimeter–*See* Opacimeter.

turbulence amplifier–1. Fluidic digital element using laminar-to-turbulent flow transition to create the control effect. 2. A fluidic device in which the power jet is at a pressure such that it is in the transition region of laminar stability and can be caused to become turbulent by a secondary jet or by sound.

turn–One complete loop of wire.

turnaround time–1. The interval between the time at which a job is submitted to a computing center and the time at which the results are returned. 2. The actual time required to reverse the direction of transmission in a half-duplex circuit. For most communications facilities, time is required by line machine reaction. (A typical time is 200 milliseconds on a half-duplex telephone connection.)

turn factor – Under stated conditions, the number of turns that a coil of given shape and dimensions placed on a core in a given position must have for a coefficient of self-inductance of 1 henry to be obtained for the core.

turn-off delay time–The time interval between occurrence of the trailing edge of a fast input pulse and the occurrence of the 90% point of the negative-going output waveform.

turn-off reversal–A polarity reversal of the output of the electronic power supply occurring from the turning off of the power of a regulated power supply.

turn-off thyristor–A thyristor which can be turned from the on state to the off state, and vice versa, with appreciable gain by applying control signals of appropriate polarities to the gate terminal.

turn-off time–The time that a switching circuit (gate) takes to stop the flow of current in the circuit it is controlling.

turn-on delay time–The time interval from

the occurrence of the leading edge of a fast input pulse to the occurrence of the 10% point of the positive-going output waveform, assuming that the rise time of the incoming pulses is 1/10 of the rise time of the element to be measured under loaded conditions.

turn-on overshoot – Overshoots occurring from the turning on of power of a regulated power supply.

turn-on reversal–A polarity reversal of the output of the power supply occurring from the turning on of the power supply of a regulated power supply.

turnover–*See* Equalization.

turnover cartridge–A phonograph cartridge adapted, by the use of two styli, to play both large- and fine-groove records.

Turnover cartridge.

turnover frequency–In disc recording, the frequency below which constant-amplitude recording is used and above which constant-velocity recording is employed.

turnover pickup–Also called dual pickup. A pickup designed for playing both standard and microgroove records, using a single magnetic structure. (*See also* Turnover Cartridge.)

turns ratio – The ratio of the number of turns in the primary winding to the number in the secondary winding of a transformer.

turnstile antenna–An antenna composed of two dipole antennas normal to each other and with their axes intersecting at their midpoints. Usually the currents are equal and in phase quadrature.

turntable–1. The round platter on which a phonograph record rests during cutting or playback. Also refers to the platter and its driving motor and associated parts, as a high fidelity component. 2. In tape recording, the rotating flat disc on which the reel (or, in some professional machines, the tape "pie") lies slightly raised above the recorder's front or top panel. The turntable is usually fitted with a center spindle and gripping keys of some kind to keep the reel centered and prevent it from slipping against the rotation of the turntable.

turntable rumble–*See* Rumble, 1.

turret–A revolving plate mounted at the front of some television cameras and carrying two or more lenses of different types, to permit rapid interchange of lenses.

turret tuner – A television-receiver tuner containing a separate set of resonating circuit elements for each channel. Each set is mounted on an insulating strip or strips placed on a drum rotated from the channel-position for the desired channel.

tv–Abbreviation for television.

tv camera–An optical device consisting of lens, electron beam tube, and preamplifier, which converts an optical image into an electrical signal.

tvi – Abbreviation for television interference.

tv recording–A permanent record of video signals recorded photographically, electronically, or by other means, and which may be displayed through a television system or projected as a motion-picture film.

tweeter–Also called a high-frequency unit. A speaker intended to reproduce the very high frequencies, usually those above 3000 Hz, in a high-fidelity audio system. Units may be ionic, ribbon, electrostatic, or dynamic types.

twelve punch–A punch in the top row of a card.

twenty-one type repeater–A two-wire telephone repeater in which one amplifier serves to amplify the telephone currents in both directions. The circuit is arranged so that the input and output terminals of the amplifier are in one pair of conjugate branches, while the lines in the two directions are in another pair.

twenty-two type repeater–A two-wire telephone repeater with two amplifiers. One amplifies the telephone currents being transmitted in one direction, and the other the telephone currents being transmitted in the other direction.

twilight–A scene illumination of approximately one footcandle.

twin cable–A cable composed of two parallel insulated stranded conductors having a common covering.

twin check–A continuous check of computer operations accomplished by duplication of equipment and automatic comparison of results.

twin lead–Also called twin line. A type of transmission line covered by a solid insulation and comprising two parallel conductors, the impedance of which is determined by their diameter and spacing. The three most common impedance values are 75, 150, and 300 ohms.

twin line–*See* Twin Lead.

Twin lead.

twinning—The intergrowth of two crystal regions having opposite oriented axes. Two types of twinning may occur, electrical and optical. In electrical twinning the electrical senses of the crystal axes are reversed and the twinned regions will interfere with one another piezoelectrically. It is this effect which limits the high-temperature utility of quartz as a piezoelectric material.

twin-T network—*See* Parallel-T Network.

twin triode—Two triode vacuum tubes in a single envelope.

twin wire—A cable composed of two small, parallel insulated conductors having a common covering.

twist – 1. The progressive rotation of the cross section of a waveguide about the longitudinal axis. 2. The deviation from a plane surface measured from one corner to the corner diagonally opposite.

twisted joint—A union of two conductors wound tightly around each other. A sleeve may be used, and it and the conductors twisted.

twisted pair— A cable composed of two small insulated conductors twisted together without a common covering. The two conductors of a twisted pair are usually substantially insulated, so that the combination is a special case of a cord.

twister—A piezoelectric crystal that generates a voltage when twisted.

twistor—A computer memory element containing inclined helical windings of magnet wire on a nonmagnetic wire, with another winding over the helix. Information is stored in the form of polarized helical magnetization.

two-address – In a computer, having the property that each complete instruction includes an operation and specifies the location of two registers, usually one containing an operand and the other the result of the operation.

two-address code—A computer code that uses two address instructions.

two-address instruction – A computer instruction that includes an operation and specifies the location of two registers.

two-conductor jack—Receptacle having two through circuits, tip and sleeve.

two-dimensional circuitry – *See* Thin-Film Integrated Circuit.

two-fluid cell—A cell having unlike electrolytes at the positive and negative electrodes.

two-hole directional coupler—A directional coupler that consists of two parallel coaxial lines in contact, with holes or slots through their contacting walls at two points one-quarter wavelength apart. With this device, a portion of the rf energy traveling in one direction through the main line may be extracted while energy traveling in the opposite direction is rejected. It is necessary that one end of the secondary line be terminated in its characteristics impedance.

two-level system—A laser which uses only two electron energy levels. Electrons in the ground state (level 1) are pumped to the excited state (level 2). The electrons then surrender their energy by stimulated emission and return to the ground state.

two-out-of-five code—A type of positional notation in which each decimal digit is represented by five binary digits; two of the five are of one kind (for example, ones) and three are of the other kind (for example, zeros).

two-part code—A randomized code with an encoding section and a decoding section. In the encoding section, the plain-text groups are arranged in alphabetical or other significant order accompanied by their code groups in nonalphabetical or random order. In the decoding section, the code groups are arranged in alphabetical or numerical order and accompanied by their meanings given in the encoding section.

two-phase—Also called quarter phase. Having a phase difference of 90 electrical degrees, or one quarter-cycle.

two-phase current—Two currents delivered through two pairs of wires at a phase difference of one quarter-cycle (90°) between them.

two-phase dynamic—Pertaining to a dynamic logic circuit that uses two clock signals to control the processing of information through the circuit or logic system.

two-phase, five-wire system—An alternating-current supply in which four of its conductors are connected as in a four-wire, two-phase system and the fifth is connected to the neutral points of each phase and usually grounded. Despite its name, it is strictly a four-phase, five-wire system.

two-phase, four-wire system—A system of alternating-current supply comprising two pairs of conductors, between one pair of which is maintained an alternating difference of potential displaced in phase by one-quarter of a period from an alternating difference of potential of the same frequency maintained between the other pair.

two-phase, three-wire system—An alternating-current supply consisting of three conductors. Between one conductor (known as the common return) and each of the other two, alternating differences of potential which are 90° out of phase with each other are maintained.

two-pilot regulation—The use of two pilot frequencies within a transmitted band so that the change in attenuation due to twist can be detected and compensated for by a regulator.

two-plus-one address—Pertaining to an instruction that contains two operand addresses and one control address.

two-port network – A network with two ports.

two-pulse canceler—A moving-target indicator canceler which compares the phase variation of two successive pulses received from a target. It discriminates against signals with radial velocities which produce a Doppler frequency equal to a multiple of the pulse-repetition frequency.

two-quadrant multiplier – Of an analog computer, a multiplier in which operation is restricted to a single sign of one input variable only.

twos complement—Pertaining to a form of binary arithmetic used in a computer to perform binary subtractions with addition techniques. The twos-complement negative of a binary number is formed by complementing each bit in the number and adding 1 to the result.

two-source frequency keying – Keying in which the modulating wave shifts the output frequency between predetermined values derived from independent sources.

two-state device – A mechanical or electronic device that, except during the time it is changing between states, is intended to be operated in either of two states or conditions.

two-terminal network – A network that is connected by only two terminals to an external system.

two-terminal–pair network—Also called a four-pole, quadripole, or quadrupole network. A network with four accessible terminals grouped in pairs. One terminal of each pair may coincide with a network node.

two-tone keying—Keying in which the modulating wave causes the carrier to be modulated with one frequency for the marking condition, and a different frequency for the spacing condition.

two-tone modulation—A method of modulation in which two different carrier frequencies are used for the two signaling conditions.

two-track recorder – *See* Dual-Track Recorder.

two-track recording—On quarter-inch-wide tape, the arrangement by which only two channels of sound may be recorded, either as a stereo pair in one direction or as separate monophonic tracks (usually in opposite directions).

two-value capacitor motor – A capacitor motor that uses different values of effective capacitance for starting and running.

two-wattmeter method—A method of measuring total power in a balanced or unbalanced three-phase system by adding the readings of two wattmeters, each with its current coil in one phase and its voltage coil connected between it and the third phase.

two-way amplifier—An amplifier in which the right and left channels of a stereo system are both amplified simultaneously by the same tubes, using push-pull circuitry but feeding one signal to the input grids in parallel instead of push-pull. The parallel and push-pull signals are then separated by two output transformers in a matrixing circuit.

two-way communication – Communication between radio station, each having both transmitting and receiving equipment.

two-way repeater—*See* Repeater.

two-way switch—A switch used for controlling electrical or electronic equipment, components, or circuits from either of two positions.

two-way system – A speaker in which the low and the high frequencies are reproduced separately by two electrically independent speaker elements, each of which is provided with a suitable sound-radiating system.

two-wire channel—A two-way circuit for transmission in either direction.

two-wire circuit—A metallic circuit formed by two conductors insulated from each other. It is possible to use the two conductors as either a one-way transmission path, a half-duplex path, or a duplex path. Also used in contrast with a four-wire circuit to indicate a circuit using one line or channel for transmission of electric waves in both directions.

two-wire repeater—A repeater that can be used for transmission in both directions over a two-wire circuit. In carrier operation, it usually makes use of the principle of frequency separation for the two directions of transmission.

two-wire system—1. A system of electric supply comprising two conductors, with the load connected between them. 2. A system in which all communication takes place over a two-wire circuit or the equivalent.

twt—Abbreviation for traveling-wave tube.

twx—*See* Teletypewriter Exchange Service, 2.

Twystron – A very high power, hybrid microwave tube combining the input section of a high-power klystron with the output section of a traveling-wave tube. It is characterized by high operating efficiency and wide bandwidths.

type-A facsimile—Facsimile communication in which the images are built up of lines of constant-intensity dots.

type-A waves—Continuous waves.

type A_1 waves—Unmodulated, keyed, continuous waves.

type A_2 waves—Modulated, keyed, continuous waves.

type A_3 waves—Continuous waves modulated by music, speech, or other sounds.

type A_4 waves—Superaudio-frequency modulated continuous waves, as used in a facsimile system.

type A_5 waves—Superaudio-frequency modulated continuous waves as used in television.

type A_9 waves – Composite transmissions and cases not covered by type-A through type-A_5 waves.

typebar—A linear type element that contains all printable symbols.

type-B facsimile—Facsimile communication in which the images are built up of lines of dots having a varying intensity (e.g., in telephotography and photoradio).

type-B waves—Keyed, damped waves.

type-printed telegraphy – Telegraphy in which the message is automatically printed at the receiving station.

typing reperforator – A reperforator that types on chadless tape about one-half inch beyond the place at which the corresponding characters are punched. In some units, typing is done on the edge of special-width tape.

U

UART – Abbreviation for universal asynchronous receiver transmitter.

ubitron—An amplifier or oscillator in which an undulating electron beam interacts with an rf wave. The kinetic energy of the beam is converted into rf energy (0-type interaction). The undulation of the beam is produced by a periodic magnetic field. This field gives the beam a transverse-velocity component which interacts with the rf wave.

U-bolt—A U-shaped bolt threaded on both ends, for fastening antennas to masts.

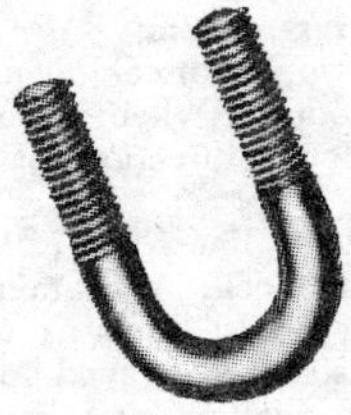

U-bolt.

uhf—Abbreviation for ultrahigh frequency.

UJT—Abbreviation for unijunction transistor.

UL – Abbreviation for Underwriters' Laboratories, Inc., a corporation supported by some underwriters for the purpose of establishing safety standards on types of equipment or components.

ultimately controlled variable – The quantity whose control is the end purpose of an automatic control system.

ultimate sensitivity or threshold—One-half the deadband in a graphic recorder. When the instrument is balanced at the center of the deadband, it denotes the minimum change in measured quantity required to initiate pen response.

ultimate trip current—The smallest value of current that will cause tripping of a circuit breaker under a given set of ambient conditions.

ultimate trip limits – The values of overload of a circuit breaker at which the minimum and maximum limits of the time-current curve become asymptotic; i.e., the limits of current that will trip or not trip the breaker "ultimately." Minimum and maximum limits of ultimate trip are often called calibration.

ultor—An adjective used to identify the picture-tube anode or element farthest from the cathode, to anode to which the highest voltage is applied, or the voltage itself (e.g., the ultor anode is the second anode of the picture tube, and the ultor voltage is the voltage applied to it).

ultor element—The element which receives the highest dc voltage in a cathode-ray tube.

ultra-audible frequency – *See* Ultrasonic Frequency.

ultra-audion – Any of several special vacuum-tube circuits employing regeneration.

ultra-audion circuit—A regenerative detector circuit in which a parallel-resonant circuit is connected between the grid and the plate of a vacuum tube, and a variable capacitor is connected between the plate and cathode to control the amount of regeneration.

ultra-audion oscillator—A variation of the Colpitts oscillator in which the resonant circuit employs a transmission-line section.

Ultrafax—A trade name of RCA for a system in which printed information is transmitted by radio, facsimile, and television at high speeds.

ultrahigh frequency—Abbreviated uhf. Frequency band: 300 to 3000 MHz. Wavelength: 100 to 10 centimeters.

ultrahigh-frequency converter – A circuit used to convert uhf television signals to vhf, to permit uhf television reception on a vhf receiver.

ultrahigh-frequency generator—Any device for generating ultrahigh-frequency alternating currents (e.g., a conventional negative-grid generator; a positive-grid, or Barkhausen, generator; a magnetron; and a velocity-modulation, or electron-beam, generator such as the klystron).

ultrahigh-frequency loop—Generally a single-loop antenna used in ultrahigh-frequency work to secure a nondirectional radiation pattern in the plane of the loop. The doughnut-shaped pattern is perpendicular to the loop.

ultrahigh-frequency translator – A television-broadcast translator station that transmits on a uhf television-broadcast channel.

ultralinear amplifier—A class-AB or -B audio amplifier using pentodes or high-power output beam-power tubes whose screen voltages are taken from taps on a specially wound output transformer rather than from a fixed dc source. This form of operation results in a considerable decrease in distortion in high fidelity systems.

ultramicrometer—An instrument for measuring very small displacements by electrical means (e.g., by the variation in capacitance produced by the movement being measured).

ultramicroscope – A dark-field microscope with which very small particles are illuminated by refracted light so that diffraction rings may be observed.

ultramicrowave – Having wavelengths of about 10^{-1} to 10^{-4} cm.

ultrashort waves—Radio waves shorter than 10 meters in wavelength (about 30 MHz in frequency). Waves shorter than a meter are called microwaves.

ultrasonic—1. Having a frequency above that of audible sound—i.e., between sonic and hypersonic. 2. Sound waves that vibrate at frequencies, beyond the hearing power of human beings (above 16,000 hertz). Commercial and military applications include ultrasonic cleaning, gauging, cutting, detection instruments, and welding.

ultrasonic bonding – A process for joining metal parts by the scrubbing action and energy transfer of a tool vibrating at an ultrasonic rate. This method is used to attach leads to pads on silicon devices.

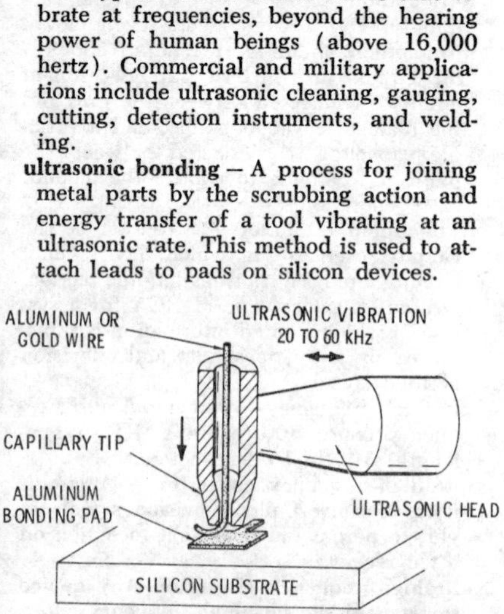

Ultrasonic bonding.

ultrasonic brazing—*See* Ultrasonic Soldering.

ultrasonic cleaner – A device using ultrasonic pressure waves to clean objects.

ultrasonic cleaning tank – A heavy-gage, polished stainless-steel tank with transducers mounted on the bottom or sides.

ultrasonic coagulation – The bonding together of small particles by the action of ultrasonic waves.

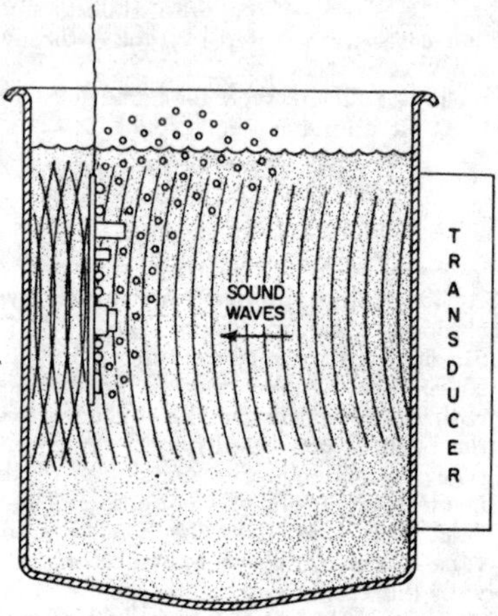

Ultrasonic cleaning tank.

ultrasonic cross grating – Also called grating. The two- or three-dimensional space grating produced when ultrasonic beams having different directions of propagation intersect.

ultrasonic delay line—Also called an ultrasonic storage cell. A contained medium (usually a liquid such as mercury) in which the signal is delayed because of the longer propagation time of the sound waves in the medium.

ultrasonic densitometer—An instrument device for determining the thickness or density of an object or material, based on the time required for an ultrasonic signal to penetrate to a receiver and/or "echo" back to a receiver adjacent to another transmitter.

ultrasonic detector—A device – either mechanical, electrical, thermal, or optical—for detecting and measuring ultrasonic waves.

ultrasonic diagnosis—A method of obtaining information from within the body in a visual presentation without employing ionizing radiation. It differs from X-ray techniques in that the form of energy used is high-frequency sound. Also, in contrast to X-ray techniques, in which the film is placed behind the tissue being examined, ultrasonic information is picked up at the original point of transmission in the form of echoes from internal structures.

ultrasonic disintegrator—An apparatus for using the pressure wave produced by an ultrasonic generator to tear cells apart.

ultrasonic drill—A special type of drill that has a magnetostrictive transducer attached to a tapered cone that serves as a velocity transformer. With an appropriate tool, practically any shape of hole can be drilled in brittle and hard material.

ultrasonic flaw detector — Equipment comprising an ultrasonic generator, transducer, detector, and display and used to detect flaws or cracks in solids from the reflection pattern of ultrasonic signals observed on a cathode-ray tube.

ultrasonic frequency—Also called an ultra-audible frequency. Any frequency above the audio range, but commonly applied to elastic waves propagated in gases, liquids, or solids.

ultrasonic generator—A device for producing mechanical vibrations at frequencies above the range of human hearing. Typically, such a device consists of an rf oscillator the output of which is applied to a piezoelectric crystal.

ultrasonic grating constant — The distance between diffracting centers of the sound wave producing particular light-diffraction spectra.

ultrasonic inspection — A nondestructive testing method of locating internal defects in a part by sending ultrasonic impulses (inaudible high-frequency sound waves of 0.5 to 11 megahertz) into the part and measuring the time required for these impulses to penetrate the material, be reflected from the opposite side or from the defect, and return to the sending point.

ultrasonic level detector—A level detector consisting of an ultrasonic receiver and transmitter located in one wall of a container or vessel. With nothing to obstruct the beam, it is reflected from the opposite wall. When the level of the liquid or other material in the container reaches the beam, the liquid or material acts as a reflector, thus reducing the reflection time and indicating that a given level has been reached.

ultrasonic light diffraction—The formation of optical diffraction spectra when a beam of light is passed through a longitudinal sound-wave field. The diffraction results from the periodic variation of the light refraction in the sound field.

ultrasonic light modulator—A device containing a fluid which, by action of ultrasonic waves passing through the fluid, modulates a beam of light passed transversely through the fluid.

ultrasonic material dispersion—The production of suspensions or emulsions of one material in another by the action of high-intensity ultrasonic waves.

ultrasonic plating—The chemical or electrochemical deposition and bonding of one or more solid materials to the surface of another material by the use of vibrational wave energy.

ultrasonic probe—A rod for directing ultrasonic force, used in a disintegration or foreign-body location application.

ultrasonics—The general subject of sound in the frequency range above 15 kilohertz.

ultrasonic sealing — A film sealing method based on the application of vibratory mechanical pressure at ultrasonic frequencies (20 to 40 kHz). Electrical energy is converted into ultrasonic vibrations by a magnetostrictive or piezoelectric transducer. The vibratory pressures at the film interface in the sealing area produce localized heat losses that cause the plastic surfaces to melt, thereby forming the seal.

ultrasonic soldering—A method of forming a nonporous, continuously metallic connection between metal or alloy parts without necessarily employing chemicals or mechanical abrasives. Instead, vibrational wave energy, heat, and a separate alloy or metal having a melting point below 800°F and also below that of the metals or alloys being joined is used.

ultrasonic space grating—Also called grating. A periodic spatial variation in the index of refraction caused by the presence of acoustic waves within the medium.

ultrasonic storage cell—*See* Ultrasonic Delay Line.

ultrasonic stroboscope—A light interrupter in which the light beam is modulated by an ultrasonic field.

ultrasonic therapy — The use of ultrasonic vibrations for therapeutic purposes.

ultrasonic thickness gage—A thickness gage in which the propagation time of an ultrasonic beam through a sheet of material is translated into a measure of the thickness of the material.

ultrasonic transducer — A device which takes the electrical oscillations produced by the ultrasonic generators and transforms them into mechanical oscillations. Typical transducer materials are piezoelectric (e.g., quartz or barium titanate) or magnetostrictive (e.g., nickel).

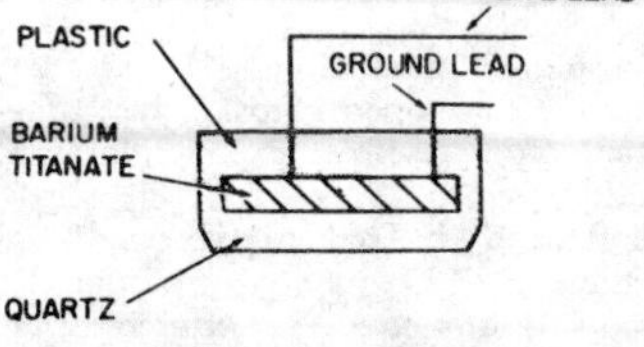

Ultrasonic transducer.

ultrasonic waves – Waves having a frequency in the ultrasonic range.

ultrasonic welding–A process which joins two pieces of metal by a form of diffusion bonding. The metals to be welded are clamped between a rigid anvil and a probe which is vibrated at ultrasonic frequency. The vibration removes any surface oxide film by a simple mechanical scrubbing action, thus exposing the metal surfaces. Then plastic deformation caused by the imposed mechanical clamping load causes the atom movement necessary to joint the two crystal lattices creating a strong, bonded, monolithic structure at the joint line. Some heat is generated by friction caused by the rubbing of the one surface upon the other. This heat undoubtedly aids the diffusion mechanism which occurs but is obviously insufficient to cause welding by itself.

ultrasonography – A medical diagnostic technique in which pulses of ultrasonic energy are directed into the body, and returning echoes are detected.

ultraviolet – Pertaining to electromagnetic radiations at wavelengths beyond the violet end of the spectrum of visible radiation. Because of the shorter wavelengths (200 to 4000 angstron units), the photons of ultraviolet light have enough energy to initiate some chemical reactions and to degrade most plastics.

ultraviolet lamp–A lamp providing a high proportion of ultraviolet radiation (e.g., arc lamps, mercurcy-vapor lamps, or incandescent lamps in bulbs of a special glass that is transparent to ultraviolet rays).

ultraviolet rays–Radiation in the ultraviolet region.

umbilical cable–A lifeline cable used for the main power supply to a missile in order to launch it. It is attached by means of a connector, which detaches as the missile becomes airborne. (Usually seen as a cable waving like a snake alongside the missile as it moves off the launching pad.)

umbilical connector–A device for connecting cables to a rocket or missile prior to launch. It is removed (unmated) from the missile at the time of launching.

umbrella antenna – An antenna in which the wires are guyed downward in all directions from a central pole or tower to the ground, somewhat like the ribs of an open umbrella.

unamplified back bias – A degenerative voltage developed across a fast time-constant circuit within an amplifier stage itself.

unbalanced–1. Lacking the conditions for balance. 2. Frequently, a circuit having one side grounded. 3. Differential mutual impedance or mutual admittance between two circuits which ideally would have no coupling.

unbalanced circuit–A circuit, the two sides of which are electrically unlike.

unbalanced line – A transmission line in which the voltages on the two conductors are not equal with respect to ground (e.g., a coaxial line).

unbalanced output–An output where one of the two output terminals is substantially at ground potential.

unbalanced wire circuit–A circuit, the two sides of which are electrically unlike.

unblanking – The turning on of the crt beam.

unblanking generator–A circuit for producing pulses that turn on the beam of a cathode-ray tube.

unblanking pulse–A pulse that turns on the beam of a cathode-ray tube.

unblocked record–A record contained in a file in which each block contains only one record or record segment.

unbonded strain gage – A pressure-sensing element made up of resistance strain-gage wire elements arranged in a Wheatstone bridge. It can have two or four active arms, which respond to a pressure applied to the transducer. The unbonded strain-gage wires are suspended in air and are activated by a mechanism attached to a diaphragm or other pressure-responding element.

unbundling–Pricing certain types of software and services separately from the hardware.

uncertainty – A number or numbers assigned to a measurement as an assessment of all the errors associated with the process producing the measurement. *See also* Accuracy.

uncharged–Having a normal number of electrons and hence no electrical charge.

unconditional–In a computer, not subject to conditions external to the specific instruction.

unconditional jump–A computer instruction which interrupts the normal process of obtaining the instructions in an ordered sequence and specifies the address from which the next instruction must be taken.

unconditional transfer of control – In a digital computer which obtains its instructions serially from an ordered sequence of addresses, an instruction which causes the following instruction to be taken from an address that becomes the first of a new sequence.

undamped natural frequency – The frequency at which a system with a single degree of freedom will oscillate, in the absence of damping, upon momentary displacement from the rest position by a transient force.

undamped oscillations – Oscillations that

have a constant amplitude for their duration.

undamped wave—A wave the amplitude of which does not change.

undefined record—A record contained in a file in which the records have not been defined as being fixed-length records or variable-length records.

underbunching – The condition whereby the buncher voltage of a velocity-modulation tube is lower than the value required for optimum bunching of the electrons.

undercompounded—A generator in which the output voltage drops as the load is increased.

undercurrent relay—A relay that functions when its coil currrent falls below a predetermined value.

undercut—In a printed-circuit board, the reduction of the cross section of a metal-foil conductor due to the removal of metal from beneath the edge of the resist by the etchant.

undercutting—A cutting with too shallow a groove or with insufficient lateral movement of the stylus during sound disc recordings.

underdamped—A degree of damping that is not sufficient to prevent oscillation in the output of a system following application of an abrupt stimulus.

underdamping—1. In a system, the condition whereby the amount of damping is so small that the system executes one or more oscillations when subjected to a single disturbance (either constant or instantaneous). 2. Oscillation of the transducer output about a final steady value in response to a step change in the measurand. After an initial overshoot, the oscillation amplitude decreases. 3. *See* Periodic Damping.

underflow—1. In a computer, the generation of a quantity smaller than the accepted minimum (e.g., floating-point underflow). 2. Pertaining to the condition that arises when a machine computation yields a nonzero result that is smaller than the smallest nonzero quantity that the intended unit of storage is capable of storing. (Contrast with overflow.) 3. When the calculator's capacity is exceeded, some of the least significant digits are discarded and the resulting display is sometimes zero.

underground cable—A cable installed below the surface of the earth.

under insulation – The insulation under wire that is brought from the center of a coil over the top or bottom wall.

underlap—Recorded elemental areas that are smaller than normal—specifically, the space between the recorded elemental area in one recording line of a facsimile system and the adjacent elemental area in the next recording line, or the elemental areas in the direction of the recording line.

underload relay – A relay that operates when the load in a circuit drops below a certain value.

undermodulation – Insufficient modulation of a transmitter, due to misadjustment or to insufficient modulation signal.

underpass – A semiconductor component that permits two conductors to cross each other without a short circuit between them. Generally, it is in the form of a low-value resistor covered by a silicon-dioxide layer that isolates the top conductor; the resistor is part of the bottom conductor.

underpower relay—A relay which functions when the power decreases below a predetermined value.

undershoot – 1. The initial transient response to a unidirectional change in input which precedes the main transition and is opposite in sense. (*See also* Precursor.) 2. The crossing of the base line in the direction opposite to that of the principal pulse, but with insufficient amplitude to be considered a bipolar pulse.

underthrow distortion—Distortion resulting when the maximum amplitude of the signal wavefront is less than the steady-state amplitude which would be attained by a prolonged signal wave.

undervoltage protection—Also called low-voltage protection. The effect of a device to cause and maintain the interruption of power to the main circuit upon the reduction or failure of voltage.

undervoltage relay—A relay which operates when its coil voltage falls below a predetermined value.

underwater sound projector – An electroacoustic transducer designed to convert electric waves into sound waves, which are radiated in water for reception at a distance.

Underwriters' Laboratories, Inc.—Abbreviated UL. An independent laboratory that tests equipment to determine whether it meets certain safety standards when properly used.

undistorted wave – A periodic wave in which both the attenuation and the velocity of propagation are the same for all sinusoidal components, and in which the same sinusoidal component is present at all points.

undisturbed-one output—A 1 output of a magnetic cell to which no partial-read pulses have been applied since that cell was last selected for writing.

undisturbed-zero output—A 0 output of a magnetic cell to which no partial-write pulses have been applied since that cell was last selected for reading.

unfired tube—The condition of tr, atr, and pre-tr tubes when there is no radio-frequency glow discharge at either the resonant gap or the resonant window.

unfurlable antenna—A device which can be unfolded to form a larger antenna.

ungrounded—Not intentionally connected to ground except through high-impedance devices.

ungrounded system—A system in which no point is directly connected to earth except through potential or ground-detecting transformers or other very-high-impedance devices.

uniaxial magnetic anistropy—A property of magnetic thin film in which the direction of magnetization is always parallel to the easy axis unless an external force acts upon it.

uniconductor waveguide – A waveguide consisting of a rectangular or cylindrical metallic surface surrounding a uniform dielectric medium.

unidirectional—Flowing in only one direction (e.g., direct current).

unidirectional antenna—An antenna with a single, well-defined direction of maximum gain.

unidirectional coupler—A directional coupler which samples only one direction of transmission.

unidirectional current—A direct current—i.e., one that is always positive or always negative—never alternating.

unidirectional log-periodic antenna – A broad-band antenna in which the cut-out portions of a log-periodic antenna are placed at an angle to each other, to produce a unidirectional radiation pattern the major lobe of which is in the backward direction, off the apex of the antenna. The impedance and the radiation pattern are essentially constant for all frequencies.

unidirectional microphone—A microphone which is most sensitive to sounds arriving at it from one direction.

unidirectional pulses—Single-polarity pulses which all rise in the same direction.

unidirectional pulse train—A pulse train in which all pulses rise in the same direction.

unidirectional transducer – A transducer that responds to stimuli in only one direction from a reference zero or rest position.

uniform field—A field in which the scalar (or vector) has the same value at every point in the region under consideration at that instant.

uniformity – In terms of magnetic tape properties, a figure of merit relating to the ability of the tape to deliver a steady and consistent output level upon being recorded with a constant input. Usually expressed in dB variation from average at a midrange frequency.

uniform line – A line with substantially identical electrical properties throughout its length.

uniform plane wave – A plane wave with constant-amplitude electric and magnetic field vectors over the equiphase surfaces. Such a wave can only be found in free space, at an infinite distance from the source.

uniform precession – The condition in which the magnetic moments of all atoms in a sample are parallel and precess in phase about the magnetic field. Uniform precession occurs in regions where the magnetic field is uniform. A spin wave is a phase distortion of this condition.

uniform waveguide—A waveguide in which the physical and electrical characteristics do not change with distance along its axis.

uniground—A single point in an electrical system connected to ground to eliminate noise currents. Also called single-point ground.

unijunction transistor – Abbreviated UJT. Formerly called a double base diode. A three-terminal semiconductor device which exhibits a stable negative-resistance characteristic between two of its terminals. It is this negative resistance feature that makes the UJT suitable for the applications with which it is associated—thyristor trigger circuits, oscillator circuit, timing circuits, bistable circuits, etc.

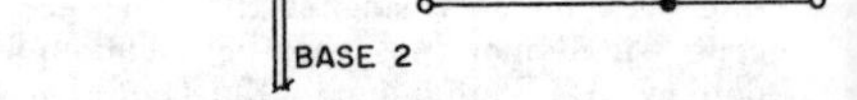

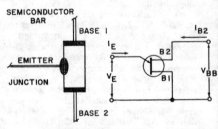

Unijunction transistor.

unilateral area track—A sound track in which only one edge of the opaque area is modulated in accordance with the recorded signal. However, there may be a second edge modulated by a noise-reduction device.

unilateral bearing – A bearing obtained with a radio direction finder having a unilateral response, eliminating the chance of a 180° error.

unilateral conductivity – Conductivity in only one direction (e.g., in a perfect rectifier).

unilateral element—A two-terminal element with a zero voltage-to-current characteristic (or the equivalent) on one side of the origin.

unilateralization—A special case of neutralization in which the feedback parameters are completely balanced out. In transistors, these feedback parameters include a resistive in addition to a capacitive component. Unilateralization changes a network from bilateral to unilateral.

unilateral network—A network in which any driving force applied at one pair of terminals produces a response at a second pair, but yields no response when the driving force is applied in the other direction.

unilateral switch—A semiconductor device similar to a miniature SCR. It switches at a fixed voltage that depends on its internal construction.

unilateral transducer – *See* Unidirectional Transducer.

uninterruptible power systems – A solid-state power conversion system to provide regulated ac power to critical loads. System provides uninterupted power even during brownouts and blackouts.

unipolar—1. Having but one pole, polarity, or direction. With respect to amplifier or power supplies, having an output that varies in only one polarity from zero and, therefore, always contains a dc component. 2. *See* Neutral Transmission. 3. Refers to transistors in which the working current flows through only one type of semiconductor material, either n type or p type. In unipolar transistors, the working current consists of either positive or negative electrical charges, but never both. All MOS IC transistors are unipolar. Unipolar (MOS) IC transistors operate slower than bipolar IC transistors, but take up much less space on a chip and are much more economical to manufacture.

unipolar field-effect transistor—A field-controlled majority-carrier device wherein the conductance of a semiconductor channel is modulated by a transverse electric field. The field is controlled by the combination of gate bias voltage V_{gs} and the net voltage V_{ds} between channel drain and source.

unipolar pulse—A pulse that has appreciable amplitude in only one direction.

unipolar transistor—A transistor in which charge carriers are of only one polarity. (*See also* Field-Effect Transistor.)

unipole—1. An all-pass filter section with one pole and one zero. 2. A hypothetical antenna which radiates and receives equally in all directions. (*See also* Isotropic Antenna.)

unipotential cathode – *See* Indirectly Heated Cathode.

unit—1. A computer portion or subassembly which constitutes the means of accomplishing some inclusive operation or function (e.g., an arithmetic unit). 2. The specific magnitude of a quantity set apart by appropriate definition and serving as a basis for the comparison or measurement of like quantities. 3. The lowest standard quantity in any system of measurement. The unit of electrical energy, for example, is the kilowatt-hour. 4. One of the transceivers covered by a CB station license when more than one transceiver is used.

unit-area acoustic impedance—*See* Specific Acoustic Impedance.

unitary code—A code having only one digit, the number of times it is repeated determining the quantity it represents.

unit charge—The electrical charge which will repel a force of one dyne on an equal and like charge one centimeter away in a vacuum, assuming each charge is concentrated at a point.

United States of America Standards Institute – *See* American National Standards Institute.

unit interval—*See* Signal Element.

unit length—The basic element of time for determining code speeds in message transmission.

unit magnetic pole—A pole with a strength such that when it is placed 1 cm away from a like pole, the force between the two is 1 dyne.

unitor—In computers, a device or circuit which performs a function corresponding to the Boolean operation of union. (*See also* OR Gate.)

unit pulse—*See* Baud.

unit record equipment – Equipment using punched cards as input data, such as collators, tabulating machines, etc.

unit sequence starting relay – A device which functions to start the next available unit in a multiple-unit equipment on the failure or on the nonavailability of the normally preceding unit.

unit sequence switch—A switch used to change the sequence in which units may be placed in and out of service in multiple unit equipment.

unit step current (or voltage)—A current (or voltage) which undergoes an instantaneous change in magnitude from one constant level to another.

unit substation transformer—A transformer which is mechanically and electrically connected to and coordinated in design with one or more switch-gear or motor-controlled assemblies or combinations thereof.

unit torque gradient—The torque gradient of a synchro, measured when the synchro is electrically connected to another synchro of the same size.

unitunnel diode—A diode similar to a tunnel diode, but specially treated to give peak reverse currents in the micoampere region while providing high forward conductance at low voltage levels.

unity coupling—Perfect magnetic coupling

between two coils, so that all the magnetic flux produced by the primary winding passes through the entire secondary winding.

unity-gain bandwidth – The frequency at which the open-loop gain reaches unity, based on a 6-dB-per-octave crossing. It is a measure of the gain-frequency product of an amplifier.

unity-gain crossover frequency – The frequency at which the curve of open-loop voltage gain of an amplifier crosses through unity gain, or zero dB.

unity power factor–A power factor of 1.0. It is obtained only when current and voltage are in phase (e.g., in a circuit containing only resistance, or in a reactive circuit at resonance).

universal asynchronous receiver transmitter – Abbreviated UART. A device that will interface a word-parallel controller or data terminal to a bit serial communication network.

universal motor–A series-wound motor designed to operate at approximately the same speed and output on direct current or on a single-phase alternating current of not more than 60 hertz and approximately the same rms voltage.

universal output transformer – An output transformer having a number of taps on its winding. By proper choice of connections, it can be used between the audio-frequency output stage and the speaker of practically any radio receiver or audio amplifier.

universal receiver – Also called an ac/dc receiver. A receiver with no power transformer and thus capable of operating from either ac or dc power lines, without changes in its internal connections.

universal shunt–*See* Ayrton Shunt.

Universal Time–A standard based on the rotation of the earth on its axis, with reference to the position of the sun. Also called Greenwich Mean Time and Greenwich Civil Time.

universe–*See* Population.

unload–In a computer: 1. To remove the tape from the columns of a recorder by raising or lowering the recording head. 2. To remove a portion of the address part of an instruction. 3. (*See also* Dump.)

unloaded antenna – An antenna with no added inductance or capacitance.

unloaded applicator impedance (dielectric heaters)–The complex impedance measured at the point of application and at a specified frequency without the load material in position.

unloaded *Q* (switching tubes)–Also called the intrinsic *Q*. The *Q* of a tube unloaded by either the generator or termination.

unloading amplifier–An amplifier capable of reproducing or amplifying a given voltage signal while drawing negligible current from the voltage source.

unloading circuit–In an analog computer, a computing element or combination of computing elements capable of reproducing or amplifying a given voltage signal while drawing negligible current from the voltage source, thus decreasing the loading errors.

unmodulated–Having no modulation–e.g., a carrier that is transmitted during moments of silence in radio programs, or a silent groove in a disc recording.

unmodulated groove–Also called a blank groove. In mechanical recording, the groove made in the medium with no signal applied to the emitter.

unoriented–A structure in which the crystallographic axes of the grains of a metal are not aligned to give directional magnetic properties.

unpack–In a computer, to separate combined items of information, each into a separate machine word.

unsaturated logic–A form of logic containing transistors operated outside the region of saturation; for example, current-mode logic (CML) and emitter-coupled logic (ECL).

untuned–Not resonant at any of the frequencies being handled.

unusable samples–In random-sampling-oscilloscope technique, those samples not falling within the time window.

unwind–In a computer, to code all the operations of a cycle, at length and in full, for the express purpose of eliminating all red-tape operations.

up-converter–A type of parametric amplifier which is characterized by the frequency of the output signal being greater than the frequency of the input signal.

update–1. To search the file (such as a particular record in a computer tape) and select one entry, then perform some operation to bring the entry up to date. 2. In a computer, to modify an instruction so that the address numbers in it are increased by a specified amount each time the instruction is executed.

updating–The act of bringing information up to the current value.

up-down counter–A counter with the capability of counting in an ascending or descending order, depending on the logic present at the up-down inputs. Also called reversible counter.

upper operating temperature–The maximum temperature to which a material can be subjected and still maintain specified operating characteristics within limits.

upper sideband–The higher frequency or group of frequencies produced by an amplitude-modulation process.

upset-duplex system–A direct-current telegraph system in which a station between any two pieces of duplex equipment may transmit signals by opening and closing the line circuit and thereby upsetting the duplex balance.

upset welding—A resistance-welding process wherein the weld is made simultaneously over the entire area of abutting surfaces or progressively along the joint with the aid of rolls or clamps which force the abutting surfaces together. The pressure is applied before heating starts and is maintained throughout the heating period.

up time—The time during which an equipment is either operating or available for operation as opposed to down time when no productive work can be accomplished.

urea plastic material—A thermosetting plastic material, with good dielectric qualities, used for radio-receiver cabinets, instrument housing, etc.

usable samples—In random-sampling-oscilloscope technique, those samples falling within the time window.

USASCII—Abbreviation for USA Standard Code for Information Interchange. The standard code, using a coded character set consisting of 7-bit coded characters (bits including parity check), used for information interchange among data processing communication systems, and associated equipment. The USASCII set consists of control characters and graphic characters. Synonymous with ASCII.

USASCSOCR—The United States of America Standard Character Set for Optical Characters.

USASI—Abbreviation for United States of America Standards Institute, the successor to ASA (American Standard Association).

"U" scan—A parallel reading method to prevent ambiguity in the readout of polystrophic codes at the code-position transitions by reading one of two sets of brushes, depending on the state of a control or selector bit.

useful life—The total time a device operates between debugging and wearout.

user-to-user service — A switching method that permits direct user-to-user connection that does not include provision for message store-and-forward service.

utility program—A program providing basic conveniences, such as capability for loading and saving programs, for observing and changing values in a computer, and for initiating program execution. The utility program eliminates the need for "re-inventing the wheel" every time a designer wants to perform a common function.

utilization factor—In electrical power distribution, the ratio of the maximum demand of a system (or part of a system) to the rated capacity of the system (or part) under consideration.

Utilogic—A line of digital IC's built around a basic AND and a basic NOR circuit. The AND has multiple-emitter inputs; the NOR has emitter-follower inputs. The output for the AND is an emitter follower, and for the NOR, a totem-pole arrangement. JK binary element is also included in the line.

V

V – 1. Symbol for volt or voltmeter. 2. Schematic symbol for vacuum tube.

VA—Abbreviation for volt-ampere.

vac—Abbreviation for vacuum.

vacuum – Abbreviated vac. Theoretically, an enclosed space from which all air and gases have been removed. However, since such a perfect vacuum is never attained, the term is taken to mean a condition whereby sufficient air has been removed so that any remaining gas will not affect the characteristics beyond an allowable amount.

vacuum capacitor—A capacitor consisting usually of two concentric cylinders enclosed in a vacuum to raise the breakdown voltage.

vacuum deposition—A process in which a substance is heated in a vacuum enclosure until the substance vaporizes and condenses (deposits) on the surface of another material in the enclosure. This process is used in the manufacture of resistors, capacitors, microcircuits, and semiconductor devices. The deposited material is called a thin film.

vacuum envelope – The airtight envelope which contains the electrodes of an electron tube.

vacuum evaporation—A process in which a material is vaporized and the vapor deposits itself, through openings in a mask, onto a substrate to form a thin film.

vacuum gage—A device that indicates the absolute gas pressure in a vacuum system (e.g., in the evacuated parts of a mercury-arc rectifier).

vacuum impregnation—Filling the spaces between electric parts or turns of a coil with an insulating compound while the coil or parts are in a vacuum.

vacuum level—The degree of a vacuum, as determined by the pressure: rough vacuum (760 torr to 1 torr), medium vacuum (1 torr to 10^{-3} torr), high vacuum (10^{-3} torr to 10^{-6} torr), very high (hard) vacuum (10^{-6} torr to 10^{-9} torr), ultrahigh (ultrahard) vacuum (below 10^{-9} torr).

vacuum metalizing – A process in which surfaces are given a thin coating of metal by exposing them to metallic vapor produced by evaporation under vacuum (one (millionth of normal atmospheric pressure).

vacuum phototube—A phototube which is evacuated to such a degree that its electrical characteristics are essentially unaffected by gaseous ionization.

vacuum range—For a communications system, the maximum range computed for an atmospheric attenuation of zero.

vacuum seal—An airtight junction between component parts of an evacuated system.

vacuum switch—A switch in which the contacts are enclosed in an evacuated bulb, usually to minimize sparking.

vacuum tank – An airtight metal chamber which contains the electrodes and in which the rectifying action takes place in a mercury-arc rectifier.

vacuum tight—*See* Hermetic.

vacuum tube—An electron tube evacuated to such a degree that its electrical characteristics are essentially unaffected by the presence of residual gas or vapor.

vacuum-tube amplifier – An amplifier in which electron tubes are used to control the power from the local source.

vacuum-tube characteristics – Data that show how a vacuum tube will operate under various electrical conditions.

vacuum-tube keying – A code-transmitter keying system in which a vacuum tube is connected in series with the plate-supply lead going to the winding in the plate circuit of the final stage. The grid of the tube is connected to its filament through the transmitting key so that when the key is open, the tube is blocked, interrupting the plate supply to the output stage. Closing the key allows plate current once more to flow through the keying tube and the output tubes.

vacuum-tube modulator – A modulator in which a vacuum tube is the modulating element.

vacuum-tube oscillator—A circuit in which a vacuum tube is used to convert dc power into ac power at the desired frequency.

vacuum-tube rectifier – A tube which changes an alternating current to a unidirectional pulsating direct current.

vacuum-tube transmitter – A radio transmitter in which electron tubes are utilized to convert the applied electric power into radio-frequency power.

vacuum-tube voltmeter—Abbreviated vtvm. *See* Electronic Voltmeter.

valence – A number representing the proportion in which the atom is able to combine with other atoms. It generally depends on the number and arrangement of electrons in the outermost shell of each type of atom.

valence band—In the spectrum of a solid crystal, the range of energy states containing the energies of the valence electrons which bind the crystal together. In a semiconductive material, it is just below the conduction band separated from it by the forbidden gap.

valence bond—Also called a bond. The bond formed between the electrons of two or more atoms.

valence electrons – The electrons of an atom in the outer shell that determine the chemical valency of the atom.

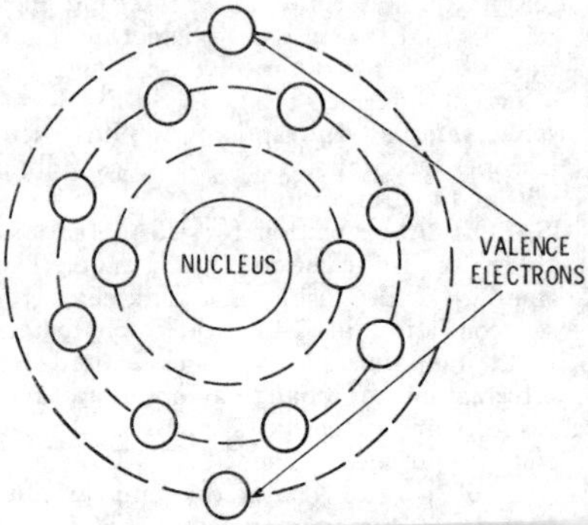

Valence electrons.

valence shell—The electrons which form the outermost shell of an atom.

validate – To ensure correctness of data which has been (previously) or is being entered by any of a number of means, including check digit, batch total, numeric only field, and verification.

validity – Correctness – specifically, how closely repeated approximations approach the desired (i.e., correct) result.

validity check – 1. A check to determine that a code group actually represents a character in the particular code being used. 2. A computer input-data check based on known limits for variables in given fields.

valley—A dip between two peaks in a curve.

valley current—In a tunnel diode, the current measured at the positive voltage for which the current has a minimum value from which it will increase if the voltage is further increased.

valley point emitter current—The current flowing in the emitter of a UJT when the device is biased to the valley point.

valley voltage—In a tunnel diode, the voltage corresponding to the valley current.

value—The magnitude of a physical quantity.

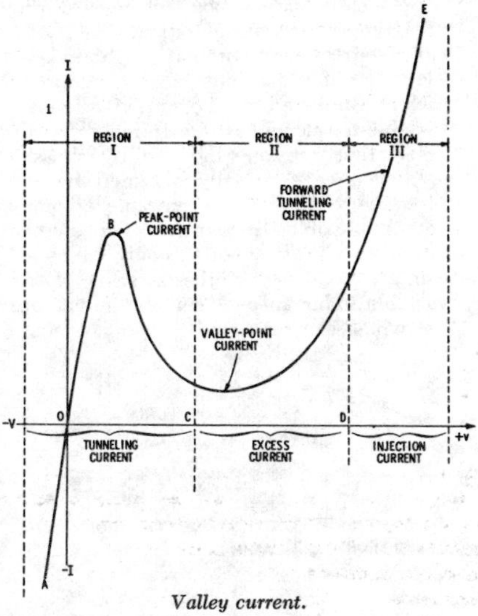

Valley current.

value theory—The assignment of numerical significance to the worths of alternative choices.

valve—1. A British term for a vacuum tube. 2. A device permitting current flow in one direction only (e.g., a rectifier). 3. A device or system which is capable of flow diversion, cutoff, or modulation.

valve tube—*See* Kenotron, 1.

Van Allen radiation belts—Two doughnut-shaped belts of high-energy particles which surround the earth and are trapped in its magnetic field. They were first discovered by Dr. James A. Van Allen of Iowa State University.

Van Atta array—An antenna array designed so that the received signal is reflected back toward its source in a narrow beam to provide signal enhancement without amplification. It consists of pairs of corner reflectors or other elements equidistant from the center of the array and connected together by means of low-loss transmission line.

Van de Graaff accelerator—An electrostatic-generator type of particle accelerator from which the voltage is obtained by picking up static electricity at one end of the machine (on a rubber belt) and carrying it to the other end, where it is stored.

vane-anode magnetron—A cavity magnetron in which the walls between adjacent cavities have parallel plane surfaces.

vane attenuator—A waveguide device designed to present attenuation in a circuit by sliding a resistive element from the side wall of the waveguide to the center for maximum attenuation. This method of attenuation is used in precision calibrated attenuation readings, and resetting must be made. Some of its countless applications are calibrations of other attenuators, directional couplers, filters, and other lossy components, in atenna-pattern measurements, noise-level measurements, and for setting power levels to desired values.

vane-type instrument—A measuring instrument in which the pointer is moved by the force of repulsion between fixed and movable magnetized iron vanes, or by the force between a coil and a pivoted vane-shaped piece of soft iron.

vane-type magnetron—A cavity magnetron in which the walls between adjacent cavities have plane surfaces.

V-antenna — A V-shaped arrangement of conductors, the two branches being fed equally in opposite phase at the apex.

vapor pressure—The pressure of the vapor accumulated above a confined liquid (e.g., in a mercury-vapor rectifier tube).

var—1. Abbreviation for volt ampere reactive. The unit of reactive power, as opposed to real power in watts. One var is equal to one reactive volt-ampere. 2. Abbreviation for visual-aural range.

varactor—Also called varactor diode, silicon capacitor, voltage-controlled capacitor, and voltage-variable capacitor. A two-terminal solid-state device that utilizes the voltage-variable capacitance of a pn junction. In the normal semiconductor diode, efforts are made to minimize inherent capacitance, while in the varactor, this capacitance is emphasized. Since the capacitance varies with the applied voltage, it is possible to amplify, multiply, and switch with this device.

varactor-tuned oscillator—An oscillator in which a varactor diode is used in the frequency-determining networks that encompass the circuit's active device(s).

varhour meter—Also called a reactive volt-ampere-hour meter. An electricity meter which measures and registers the integral (usually in kilovarhours) of the reactive power of the circuit into which the meter is connected.

variable—1. Any factor or condition which can be measured, altered, or controlled (e.g., temperature, pressure, flow, liquid level, humidity, weight, chemical composition, color, etc.). 2. A quantity that can take on any of a given set of values. 3. In a computer, a symbol the numeric value of which changes from one iteration of a program to the next or within each iteration of a program.

variable-area track—A sound track divided laterally into opaque and transparent areas. A sharp line of demarcation between these areas forms an oscillographic trace of the waveshape of the recorded signal.

variable-capacitance diode — Abbreviated

vcd. A semiconductor diode in which the junction capacitance present in all semiconductor diodes has been accentuated. An appreciable change in the thickness of the junction-depletion layer and a corresponding change in the capacitance occur when the dc voltage applied to the diode is changed.

variable-capacitance transducer—A transducer which measures a parameter or a change in a parameter by means of a change in capacitance.

variable capacitor—A capacitor which can be changed in capacitance by varying the useful area of its plates, as in a rotary capacitor, or by altering the distance between them, as in some trimmer capacitors.

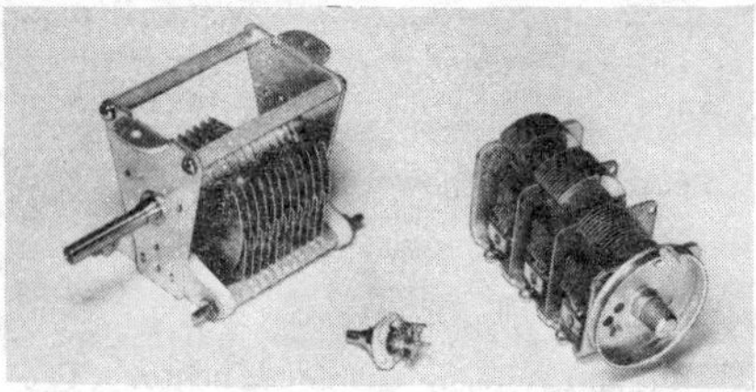

Variable capacitors.

variable-carrier modulation – *See* Controlled-Carrier Modulation.

variable connector—1. A flowchart symbol representing a sequence connection which is not fixed, but which can be varied by the flowcharted procedure itself. 2. The device which inserts instructions in a program corresponding to the selection of paths appearing in a flowchart. 3. The computer instructions which cause a logical chain to take one of several alternative paths.

variable coupling—Inductive coupling that can be varied by moving the windings.

variable-cycle operation—Computer operation in which any cycle is started at the completion of the previous cycle, instead of at specified clock times.

variable-density track—A sound track of constant width and usually, but not necessarily, of uniform light transmission on any instantaneous transverse axis. The average light transmission varies along the longitudinal axis in proportion to some characteristic of the applied signal.

variable-depth sonar—A sensor that can be lowered by cable from a ship or helicopter, through thermal layers, to detect submarines operating at deeper levels.

variable-erase recording—The method of recording on magnetic tape by selective erasure of a prerecorded signal.

variable field—A field in which the scalar (or vector) at any point changes during the time under consideration.

variable-frequency oscillator—Abbreviated vfo. A stable oscillator, the frequency of which can be adjusted over a given range.

variable-frequency synthesizer—An instrument that translates the stability of a single frequency, usually obtained from a frequency standard, to any one of many other possible frequencies. In common usage, such instruments usually are called simply frequency synthesizers.

variable inductance—A coil the inductance of which can be varied.

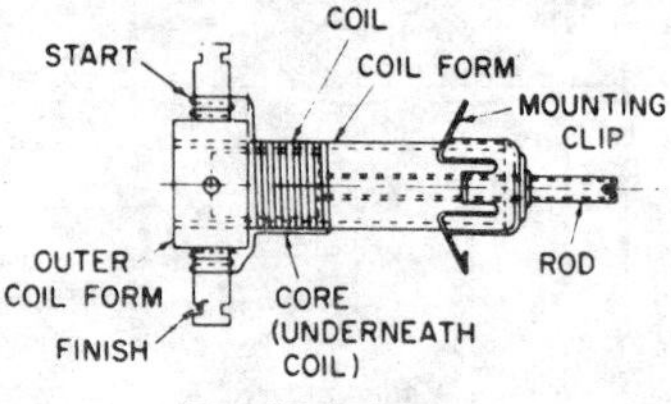

Variable inductance.

variable-inductance pickup—A phonograph pickup in which the movement of a stylus causes the inductance to vary accordingly.

variable-inductance transducer – A transducer in which the output voltage is a function of the change in a variable-inductance element.

variable-length record—In a computer, pertaining to a file in which there is no constraint on the record length. (Opposite of fixed-length record.)

variable monoergic—A type of emission in which the magnitude of the homogeneous particle or radiation energy is continuously variable over broad limits; e.g., the proton energy from some types of accelerators can be controlled by varying the high voltage.

variable-mu pressure transducer—A device which converts mechanical input pressure to a proportional electrical output based on the change of the mu of its magnetic circuit due to the applied pressure (Villari effect).

variable-mu tube—*See* Remote-Cutoff Tube.

variable point—Pertaining to a system of numeration in which the position of the radix point is indicated by a special character at that position.

variable radio-frequency radiosonde – A radiosonde the carrier frequency of which is modulated by the magnitude of the meteorological variables being sensed.

variable reluctance—Principle employed in certain phonograph pickups. Deflections of the stylus when playing a record make an armature vibrate between the poles of an electromagnet. The reluctance is the ratio of magnetic force to magnetic flux in a magnetic field. Variations due to stylus movement create variations in the current through the electromagnet.

variable-reluctance microphone—Also called a magnetic microphone. A microphone which depends for its operation on the variations in reluctance of a magnetic circuit.

variable reluctance pickup—A phonograph cartridge that derives its electrical output signal from change effected in a magnetic circuit by means of some mechanical device such as a moving coil or magnet.

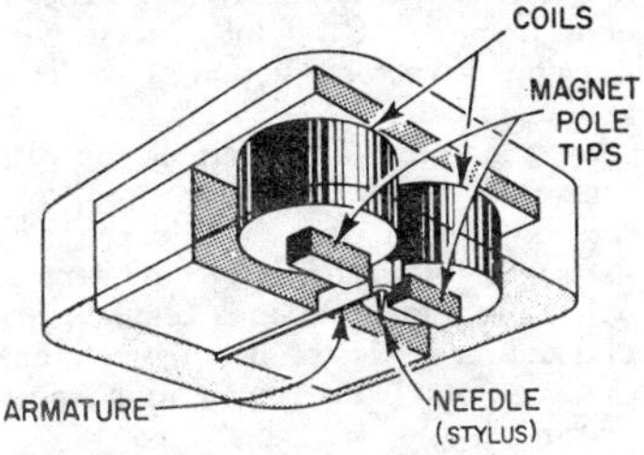

Variable-reluctance pickup.

variable-reluctance stepping motor—A motor with a soft-iron rotor that is made to step by sequential excitation of stator coils. It requires complementary equipment to furnish the sequencing pulses.

variable-reluctance transducer—Also called a magnetic transducer. A transducer that depends for its operation on the variations in reluctance of a magnetic circuit.

variable-resistance pickup—A phonograph pickup which depends for its operation on the variation of a resistance.

variable-resistance transducer – A transducer in which the signal output depends on the change in a resistance element.

variable resistor—A wirewound or composition resistor, the resistance of which may be changed. (*See also* Rheostat *and* Potentiometer, 1.)

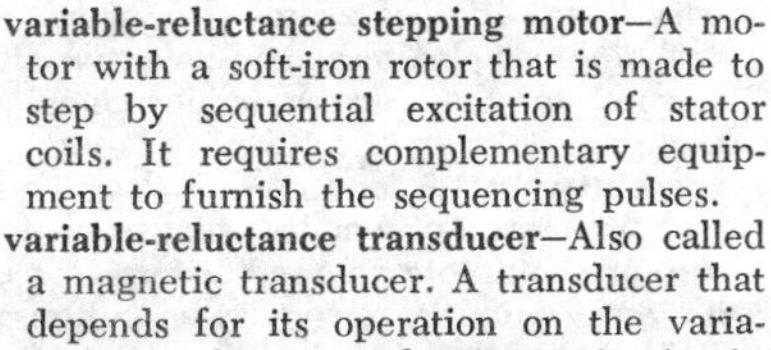

Variable resistors.

variable-speed motor—A motor the speed of which can be adjusted within certain limitations, regardless of load.

variable-speed scanning – A scanning method whereby the optical density of the film being scanned determines the speed at which the scanning beam in the cathode-ray tube of a television camera is deflected.

variable transformer – An iron-core transformer with provision for varying its output voltage over a limited range, or continuously from zero to maximum—generally by the movement of a contact arm along exposed turns of the secondary winding.

Variac—An autotransformer that contains a toroidal winding and a rotating carbon brush so that the output voltage is continuously adjustable from zero to line voltage plus 17%. Trade name of General Radio Company.

variate—*See* Random Variable, 1.

variations—The angular difference between a true and a magnetic bearing or heading.

varicap—*See* Varactor.

varindor—An inductor, the inductance of which varies markedly with the current in the winding.

variocoupler – A radio transformer with windings that have an essentially constant self-impedance, but the mutual impedance between them is adjustable.

variolosser—A device whose loss can be controlled by a voltage or current.

variometer—A variable inductor consisting of a pair of series or paralled-connected coils whose axes can be varied one with respect to the other. The change in mutual inductance causing a change in the total inductance.

varioplex – A telegraph switching system that establishes connections on a circuit-sharing basis between a number of transmitters in one locality and corresponding receivers in another locality over one or more intervening channels. Maximum use of channel capacity is obtained by employing momentary storage of signals and allocating circuit time in rotation among the transmitters that have information in storage.

varistor—1. A two-electrode semiconductor device with a voltage-dependent nonlinear resistance that drops markedly as the applied voltage is increased. 2. A passive resistorlike circuit element whose resistance is a function of the current through it. The current through it is a nonlinear function of the voltage across its terminals, hence, a self-varying resistance. *Also see* Voltage-Dependent Resistor.

Varley loop—A type of Wheatstone-bridge circuit which gives, in one measurement, the difference in resistance between two wires of a loop.

varmeter—Also called a reactive volt-ampere meter. An instrument for measuring reactive power in either vars, kilovars, or megavars. If the scale is graduated in kilovars or megavars, the instrument is sometimes designated a kilovarmeter or megavarmeter.

varnished cambric—A linen or cotton fabric that has been impregnated with varnish or insulating oil and baked. It is used as insulation in coils and other radio parts.

varying duty—A requirement of service that demands operation at loads and for intervals of time, both of which may be subject to wide variation.

varying-speed motor—A motor which slows down as the load increases (e.g., a series motor, or an induction motor with a large amount of slip).

varying-voltage control—A form of armature-voltage control obtained by impressing, on the armature, a voltage which varies considerably with a change in load and consequently changes the speed of the motor (e.g., by using a differentially compound-wound generator or a resistance in the armature circuit).

V-beam system—A radar system for measuring elevation. The antenna emits two fan-shaped beams, one vertical and the other inclined, which intersect at ground level. Each beam rotates continuously about a vertical axis, and the time elapsing between the two echoes from the target provides a measure of its elevation.

V_{CC}—Symbol for the supply voltage to an integrated circuit with respect to ground.

vcd—Abbreviation for variable-capacitance diode.

V-cut—A type of oscillator-crystal cut in which the major plane surfaces are not parallel to the X, Y, or Z planes.

V_{DD}, V_{SS}, V_{CC}, V_{EE}—In an MOS circuit, the designation of the power-supply terminal serving the drain, source, collector, or emitter. The double subscript refers to the power-supply terminal, while a single subscript references the parameter at the element of a device. For example, V_C is the voltage measured on the collector itself, while V_{CC} is the (constant) voltage supplied to the collector circuit.
Note: In CMOS, the term V_{DD} has been adopted as a convention referring to the positive power-supply terminal, although it is actually applied to the source of a p-channel transistor.

vdr—*See* Voltage-Dependent Resistor.

vector—A quantity that has both magnitude and direction. Vectors commonly are represented by a line segment with a length that represents the magnitude and an orientation in space that represents the direction.

vector admittance—The ratio for a single sinusoidal current and potential difference in a portion of a circuit of the corresponding complex harmonic current to the corresponding complex potential difference.

vector-ampere — The unit of measurement of vector power.

vector cardiograph — An instrument that measures both the magnitude and the direction of heart signals by displaying cardiograph signals on any desired set of axes, usually X, Y, and Z.

vector diagram—An arrangement of vectors showing the relationships between alternating quantities having the same frequency.

vector field—In a given region of space, the total value of some vector quantity which has a definite value at each point of the region (e.g., the distribution of magnetic intensity in a region surrounding a current-carrying conductor).

vector function—A function which has both magnitude and direction (e.g., the magnetic intensity at a point near an electric circuit is a vector function of the current in that circuit).

vector generator—That part of the display controller which draws vectors on the screen. Control codes within the display list specify whether the vector generators will move the beam in a blanked or unblanked mode. If the beam is unblanked, the vector will be drawn in a specified texture.

vector impedance—The ratio for a simple sinusoidal current and potential difference in a portion of a circuit of the corresponding complex harmonic potential difference to the corresponding complex current.

vector interrupt—A term used to describe a microprocessor system in which each interrupt, both internal and external, has its own uniquely recognizable address. This enables the microprocessor to perform a set of specified operations which are preprogrammed by the user to handle each interrupt in a distinctively different manner.

vector power—A vector quantity equal to the square root of the sum of the squares of the active and reactive powers. The unit is the vector-ampere.

vector power factor — Ratio of the active power to the vector power. In sinusoidal quantities, it is the same as power factor.

vector quantity—A quantity that has both magnitude and direction. Examples of quantities that are vectors are: displacement, velocity force, and magnetic intensity.

vectorscope—An oscilloscope with a circular time base of extreme stability (determined by the frequency of the color subcarrier). The instrument can be used to check the time delay between two signals because the phase difference at a particular frequency can be related to time difference.

velocimeter—A cw reflection Doppler system used to measure the radial velocity of an object.

velocity — A vector quantity that includes both magnitude (speed) and direction in relation to a given frame of reference.

velocity error—The amount of angular displacement existing between the input and output shafts of a servomechanism when both are turning at the same speed.

velocity filter—A storage-tube device which blanks all targets that do not move more than one resolution cell in less than a predetermined number of antenna scans.

velocity hydrophone—A type of hydrophone in which the electric output is substantially proportional to the instantaneous particle velocity in the incident sound wave.

velocity-lag error—A lag, between the input and output of a device, that is proportional to the rate of variation of the input.

velocity level—In decibels of a sound, 20 times the logarithm to the base 10 of the ratio of the particle velocity of the sound to the reference particle velocity. The latter must be stated explicitly.

velocity microphone – A microphone in which the electric output corresponds substantially to the instantaneous particle velocity in the impressed sound wave. It is a gradient microphone of order one, and is inherently bidirectional.

velocity-modulated amplifier—Also called a velocity-variation amplifier. An amplifier in which velocity modulation is employed for amplifying radio frequencies.

velocity-modulated oscillator—Also called a velocity-variation oscillator. An electron-tube structure in which the velocity of an electron stream is varied (velocity-modulated) in passing through a resonant cavity called a buncher. Energy is extracted from the bunched electron stream at a higher level in passing through a second cavity resonator called the catcher. Oscillations are sustained by coupling energy from the catcher cavity back to the buncher cavity.

velocity modulation – Also called velocity variation. Modification of the velocity of an electron stream by the alternate acceleration and deceleration of the electrons with a period comparable to that of the transit time in the space concerned.

velocity of light—A physical constant equal to 2.99796×10^{10} centimeters per second. (More conveniently expressed as 186,280 statute miles per second, 161,750 nautical miles per second, or 328 yards per microsecond.)

velocity of propagation—1. The speed at which a disturbance (sound, radio, light, etc., waves) is radiated through a medium. 2. The ratio of the speed of the flow of an electric current in an insulated cable to the speed of light, expressed in percentage. In the case of coaxial cables, this ratio is 65-66%, where the insulation is polyethylene.

velocity pickup—A magnetic pickup whose output increase is a function of recorded velocity. (This differs from the ceramic pickup whose output is a function of amplitude or deflection of the stylus.)

velocity resonance—*See* Phase Resonance.

velocity sorting—The selecting of electrons according to their velocity.

velocity spectrograph – An apparatus for separating an emission of electrically charged particles into distinct streams, in accordance with their speed, by means of magnetic or electric deflection.

velocity transducer – A transducer which generates an output proportionate to the imparted velocities.

velocity variation – *See* Velocity Modulation.

velocity-variation amplifier—*See* Velocity-Modulated Amplifier.

velocity-variation oscillator—*See* Velocity-Modulated Oscillator.

Venn diagrams—Diagrams in which circles or ellipses are used to give a graphic representation of basic logic relations. Logic relations between classes, operations on classes, and the terms of the propositions are illustrated and defined by the inclusion, exclusion, or intersection of these figures. Shading indicates empty areas, crosses indicate areas that are not empty, and blank spaces indicate areas that may be either. Named for English logician John Venn, who devised them.

vent—A controlled weakness somewhere in the enclosure of an aluminum electrolytic capacitor to permit pressure relief in case of failure due to shorting or improper installation of the capacitor.

vented baffle—An enclosure designed to properly couple a speaker to the air.

ventilated transformer—A dry-type transformer which is so constructed that the ambient air may circulate through its enclosure to cool the transformer core and windings.

venturi tube—A short tube with flaring ends and a constricted throat. It is used for measuring flow velocity by measurement of the throat pressure, which decreases as the velocity increases.

verification – The process of checking the results of one data transcription against those of another, both transcriptions usually involving manual operations. (*See also* Check.)

verifier—Of computers, a device on which a record can be compared or tested for identity character by character with a retranscription or copy as it is being prepared.

verify—1. To check, usually with an automatic machine, one recording of data against another in order to minimize the number of human errors in the data transcription. 2. To make certain that the information being prepared for a computer is correct.

vernier – 1. An auxiliary scale comprising subdivisions of the main measuring scale and thus permitting more accurate measurements than are possible from the main scale alone. 2. An auxiliary device used for obtaining fine adjustments.

vernier capacitor – A variable capacitor placed in parallel with a larger tuning capacitor and used to provide a finer adjustment after the larger one has been set to the approximate desired position.

vernier dial—A type of tuning dial used chiefly for radio equipment. Each complete rotation of its control knob moves the main shaft only a fraction of a revolution and thereby permits fine adjustment.

vernitel – A precision device which makes possible the transmission of data with high accuracy over standard frequency-modulated–frequency-modulated telemetering systems.

Versatile Automatic Test Equipment – A computer-controlled tester for missile electronic systems. It troubleshoots faults by deductive logic and isolates them to the plug-in module or component level.

vertex—*See* Node, 1.

vertex plate—A matching plate placed at the vertex of a reflector.

vertical—In 45/45 recording, the signal produced by a sound arriving at the two mirophones simultaneously and 180° out of phase, causing the cutting stylus to move vertically.

vertical amplification—Signal gain in the circuits of an oscilloscope that produce vertical deflection on the screen.

vertical-amplitude controls—*See* Parabola Controls.

vertical antenna—A vertical metal tower, rod, or suspended wire that is used as a receiving and/or transmitting antenna.

vertical-blanking pulse – In television, a pulse transmitted at the end of each field to cut off the cathode-ray beam while it returns to start the next field.

vertical-centering control—A control provided in a television receiver or cathode-ray oscilloscope to shift the entire image up or down on the screen.

vertical compliance—The ability of a reproducing stylus to move vertically while in the reproducing position on a record.

vertical-deflection electrodes—The pair of electrodes that move the electron beam up and down on the screen of a cathode-ray tube employing electrostatic deflection.

vertical dynamic convergence – Convergence of the three electron beams at the aperture mask of a color picture tube during the scanning of each point along a vertical line at the center of the tube.

vertical field-strength diagram—A representation of the field strength at a constant distance from, and in a vertical plane passing through, an antenna.

vertical frame transfer—A ccd configuration where all charges accumulated during an integration period are rapidly moved out of the optically active area and to an identical optically shielded ccd area where it is read out at a slower pace during the next integration period.

vertical-frequency response—In an oscilloscope, the band of frequencies passed, with amplification between specified limits, by the amplifiers that produce vertical deflection on the screen.

vertical-hold control—*See* Hold Control.

vertical-incidence transmission—The transmission of a radio wave vertically to the ionosphere and back. The transmission remains practically the same for a slight departure from the vertical (e.g., when the transmitter and receiver are a few kilometers apart).

vertical-lateral recording—A technique of making stereo phonograph discs by recording one signal laterally, as in monophonic records, and the other vertically, as in hill-and-dale-transcriptions.

vertical-linearity control – A control that permits adjustment of the spacing of the horizontal lines on the upper portion of the picture to effect linear vertical reproduction of a television scene.

vertically polarized wave—1. An electromagnetic wave with a vertical electric vector. 2. A linearly polarized wave with a horizontal magnetic field vector.

vertical polarization – 1. Transmission in which the transmitting and receiving antennas are placed in a vertical plane, so that the electrostatic field also varies in a vertical plane. 2. Transmission of radio waves whose undulations vary vertically with respect to the earth. (Nearly all CB base stations and mobile units employ vertically polarized antennas.)

vertical quarter-wave stub – An antenna with a vertical portion that is electrically one quarter-wavelength long. It is used generally with a ground plane at the base of the stub.

vertical radiator—A transmitting antenna perpendicular to the earth's surface.

vertical recording – Also called hill-and-dale recording. Mechanical recording in which the groove modulation is perpendicular to the surface of the recording medium.

vertical redundance – In a computer, an error condition which exists when a character fails a parity check, i.e., has an even number of bits in an odd-parity system or vice versa.

vertical resolution – On a television test pattern, the number of horizontal wedge lines that can be clearly discerned by the eye before they merge together.

vertical retrace—The return of the electron beam from the bottom of the image to the top after each vertical sweep.

vertical speed· transducer—An instrument which furnishes an electrical output that is proportionate to the vertical speed of the aircraft or missile in which it is installed.

vertical stylus force—*See* Stylus Force.

vertical sweep—The downward movement of the scanning beam from top to bottom of the televised picture.

vertical-synchronizing pulse – Also called picture-synchronizing pulse. One of the six pulses transmitted at the end of each field in a television system. It maintains the receiver in field-by-field synchronism with the transmitter.

very high frequency—Abbreviated vhf. Frequency band: 30 to 300 MHz. Wavelength: 10 to 1 meters. Metric waves.

very long range—A classification of ground radar sets by slant range, applied to those with a maximum range exceeding 250 miles.

very low frequency—Abbreviated vlf. Frequency band: below 30 kHz. Wavelength: above 10,000 meters.

very short range—Classification of ground radar sets by slant range, applied to those with a maximum range of less than 25 miles.

vestigial—Pertaining to a remnant or remaining part.

vestigial sideband – Amplitude-modulated transmission in which a portion of one sideband has been largely suppressed by a transducer having a gradual cutoff in the neighborhood of the carrier frequency.

vestigial-sideband filter—A filter that is inserted between an am transmitter and its transmitting antenna to suppress part of one of the sidebands.

vestigial-sideband transmission—Also called asymmetric-sideband transmission. Signal transmission in which one normal sideband and the corresponding vestigial sideband are utilized.

vestigial-sideband transmitter – A transmitter in which one sideband and only a portion of the other are transmitted.

vfo – Abbreviation for variable-frequency oscillator.

vhf—Abbreviation for very high frequency.

vhf omnirange—Abbreviated VOR. A specific type of range operating at vhf and providing radial lines of position in a direction determined by the bearing selection within the receiving equipment. A nondirectional reference modulation is emitted, along with a rotation pattern which develops a variable modulation of the same frequency as the reference modulation. Lines of position are determined by comparing the phase of the variable with that of the reference.

vibrating bell—A bell having a mechanism designed to strike repeatedly and as long as it is actuated.

vibrating-reed meter—A frequency meter consisting of a row of steel reeds, each having a different natural frequency. All are excited by an electromagnet fed with the alternating current whose frequency is to be measured. The reed whose frequency corresponds most nearly with that of the current vibrates, and the frequency is read on a scale beside the row of reeds. Also called reed frequency meter.

vibrating-reed relay—1. A type of relay in which an alternating or a self-interrupted voltage is applied to the driving coil so as to produce an alternating or pulsating magnetic field that causes a reed to vibrate. 2. A type of relay that is actuated by sound frequency. Can be triggered by an electrical resonant circuit, or simply by a mechanically induced sound vibration.

vibrating-wire transducer – A transducer which utilizes a thin wire suspended in a magnetic field; the change in tension of the wire reflects a frequency-modulating output.

vibration – 1. A continuously reversing change in the magnitude of a given force. 2. A mechanical oscillation or motion

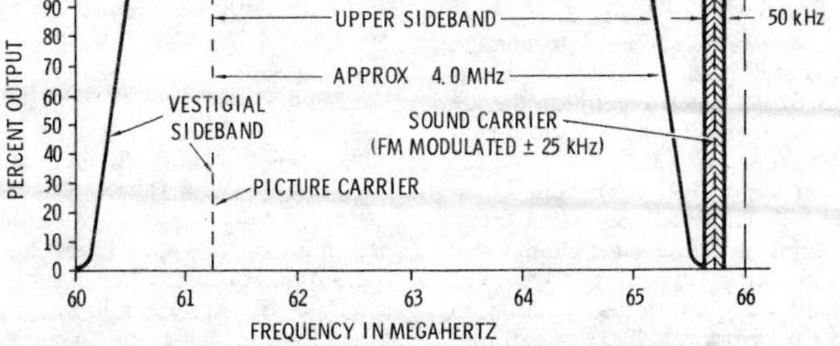

Vestigial-sideband television channel.

about a reference point of equilibrium.

vibration analyzer—A device used to analyze mechanical vibrations.

vibration galvanometer—An ac galvanometer in which a reading is obtained by making the natural oscillation frequency of the moving element equal to the frequency of the current being measured.

vibration isolator—A resilient support that tends to isolate a system from steady-state excitation or vibration.

vibration meter—Also called a vibrometer. An apparatus comprising a vibration pickup, calibrated amplifier, and output meter, for the measurement of displacement, velocity, and acceleration of a vibrating body.

vibration pickup — A microphone that responds to mechanical vibrations rather than to sound waves. In one type a piezoelectric unit is employed; the twisting or bending of a Rochelle-salt crystal generates a voltage that varies with the vibration being analyzed.

vibration sensitivity — The peak instantaneous change in output at a given sinusoidal vibration level for any one stimulus value within the range of an instrument or equipment. It is usually expressed in percentage of full-scale output per vibratory "g" over a given frequency range. It may also be specified as a total error in percentage of full-scale output for a given vibratory acceleration level.

vibration survey—A method of determining the natural frequency of a transducer by observation of the output waveform upon the application of a shock or tapping of sufficient magnitude to initiate oscillation of the instrument.

vibrato — A musical embellishment which depends primarily on periodic variations of frequency, often accompanied by variations in amplitude and waveform. The quantitative description of vibrato is usually in terms of the corresponding modulation of frequency (5-7 Hz), amplitude, or waveform, or all three.

vibrator — 1. A vibrating reed which is driven like a buzzer and has contacts arranged to interrupt direct current to the winding(s) of a transformer resulting in an alternating current being supplied from another winding to the load. 2. Electromagnetic device which is used to change a continuous steady current into a pulsating current. 3. An electromagnetic device for converting a direct voltage into an alternating voltage.

vibrator power supply—A power supply incorporating a vibrator, step-up transformer, rectifier, and filters for changing a low dc voltage to a high dc voltage.

vibratron—A triode with an anode that can be moved or vibrated by an external force. Thus, the anode current will vary in proportion to the amplitude and frequency of the applied force.

vibrocardiography — The recording of the acceleration of the chest wall during the cardiac cycle.

vibrograph — An apparatus for recording mechanical vibrations.

vibrometer—*See* Vibration Meter.

vibrophonocardiograph—An instrument for recording heart vibrations and sounds.

video—1. Pertaining to the bandwidth and spectrum position of the signal resulting from radar or television scanning. In current usage, video means a bandwidth on the order of several megahertz. 2. A prefix to the name of television parts or circuits which carry picture signals. 3. Radar or television signals which actuate the cathode-ray tube.

video amplifier — An amplifier which provides wideband operation in the frequency range of approximately 15 hertz to 5 megahertz.

video carrier—The television signal whose modulation sidebands contain the picture, sync, and blanking signals.

videocast—1. To broadcast a program by means of television. 2. A program so broadcast.

video correlator — A radar circuit that enhances the capability for automatic target detection; supplies data for digital target plotting; and provides improved immunity to noise, interference, and jamming.

video data digital processing—Digital processing of video signals for pictures transmitted by way of a television link. A computer is used to compare each scanned line with adjacent lines so that extreme changes resulting from electromagnetic interference can be eliminated.

video detector — The demodulator circuit which extracts the picture information from the amplitude-modulated intermediate frequency in a television receiver.

video discrimination — A radar circuit that reduces the frequency band of the video-amplifier stage in which it is used.

video frequency—1. The frequency of the signal voltage containing the picture information which arises from the television scanning process. In the present United States television system, these frequencies are limited from approximately 30 Hz to 4 MHz. 2. A band of frequencies extending from approximately 100 hertz to several megahertz.

video-frequency amplifier—A device capable of amplifying those signals that comprise the periodic visual presentation.

video-gain control—A control for adjusting the amplitude of a video signal. Two such controls are provided in the matrix section of some color television receivers so that the proper ratios between the amplitudes of the three color signals can be obtained.

video integration—A method of improving the output signal-to-noise ratio by utilizing the redundancy of repetitive signals to sum the succesive video signals.

video integrator—A device which uses the redundancy of repetitive signals to improve the output signal-to-noise ratio by summing the successive video signals.

video mapping—The procedure whereby a chart of an area is electronically superimposed on a radar display.

video masking—A method for the removal of chaff echoes and other extended clutter from radar displays.

video mixer — A circuit or device used to combine the signals from two or more television cameras.

video recording (magnetic tape) — The methods of recording data having a bandwidth in excess of 100 kHz on a single track.

video signal — 1. The picture signal in a television system—generally applied to the signal itself and the required synchronizing and equalizing pulses. 2. In television, the signal that conveys all of the intelligence present in the image together with the necessary synchronizing and equalizing pulses.

video stretching — In navigation, a procedure whereby the duration of a video pulse is increased.

video synthesizer — A video analog computer that accepts standard video signals from a camera, film chain, video-tape or graphics generator, which it then processes and applies a combination of effects to change positions, size and aspect ratios to reshape and add motion (animation) to fixed graphics or live scenes.

video tape—A wide magnetic tape designed for recording and playing back a composite black and white or color television signal.

video tape recorder—Abbreviated VTR. A device which permits audio and video signals to be recorded on magnetic tape and then played back without any processing, as with films, on a cctv monitor.

video tape recording — Abbreviated vtr. A method of recording television picture and sound signals on tape for reproduction at some later time.

vidicon—A camera tube in which a charge-density pattern is formed by photoconduction and stored on that surface of the photoconductor which is scanned by an electron beam, usually of low-velocity electrons.

viewfinder — An auxiliary optical or electronic device attached to a television camera so the operator can see the scene as the camera sees it.

viewing area — The area of the crt face which can be directly seen by the user. Its size is determined by the size of the crt and the area covered by the face mask.

viewing mirror—A mirror used in some television receivers to reflect the image formed on the screen of the picture tube at an angle convenient to the viewer.

viewing screen—The face of a cathode-ray tube on which the image is produced.

viewing time — The time during which a storage tube presents a visible output that corresponds to the stored information.

Villari effect — A phenomenon in which a change in magnetic induction occurs when a mechanical stress is applied along a specified direction to a magnetic material having magnetostrictive properties.

vinyl resin—A soft plastic used for making phonograph records.

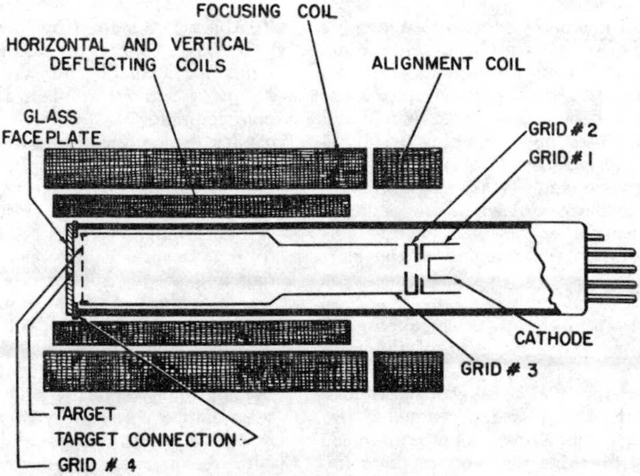

Vidicon.

virgin tape—*See* Raw Tape.

virtual address—In a computer, an immediate, or real-time, address.

virtual cathode — An electron cloud that forms around the outer grid in a thermionic vacuum tube when the inner grid is maintained slightly more positive than the cathode.

virtual height—The height of the equivalent reflection point that will cause a wave to travel to the ionosphere and back in the same time required for an actual reflection. In determining the virtual height, the wave is assumed to travel at uniform speed and the height is determined by the time required to go to the ionosphere and back at the assumed velocity of light.

virtual image—The optical counterpart of an object, formed at imaginary focuses by prolongations of light rays (e.g., the image that appears to be behind an ordinary mirror).

virtual memory—1. The use of techniques by which the computer programmer may use the memory as though the main memory and mass memory were available simultaneously. 2. A technique that permits the user to treat secondary (disk) storage as an extension of core memory, thus giving the "virtual" appearance of a larger core memory to the programmer.

virtual ppi reflectoscope — A device for superimposing a virtual image of a chart onto the ppi pattern. The chart is usually prepared with white lines on a black background to the scale of the ppi range scale.

viscometer—Also called a viscosimeter. A device for measuring the degree to which a liquid resists a change in shape.

viscosimeter—*See* Viscometer.

viscosity — The frictional resistance offered by one part or layer of a liquid as it moves past an adjacent part or layer of the same liquid.

viscous and magnetic damping — Damping by virtue of the viscosity of a fluid around the sensing element or of a magnetic field.

viscous-damped arm—A phonograph pickup arm mounted on a liquid cushion of oil, which provides high damping to eliminate arm resonances. It also protects the record groove and stylus; the arm does not fall on the record when dropped, but floats down gently.

visible spectrum—That region of the electromagnetc spectrum to which the retina is sensitive and by which the eye sees. It extends from about 400 to about 750 millimicrons in wavelengths of the radiation.

visibility factor — Also called display loss. Ratio of the minimum input-signal power detectable by ideal instruments connected to the output of a receiver, to the minimum signal power detectable by a human operator through a display connected to the same receiver. The visibility factor may include the scanning loss.

visible radiation — Radiation with wavelengths ranging from about 4000 to 8000 angstrom units, corresponding to the visible spectrum of light.

visual-aural range—Abbreviated var. A special type of vhf range providing a pair of radial lines of position which are reciprocal in bearing and are displayed to the pilot on a zero-center, left-right indicator. This facility also provides a pair of reciprocal radial lines of position located 90° from the above visually indicated lines. These are presented to the pilot as aural A-N radio-range signals, which provide a means for differentiating between the two visually indicated lines (and vice versa).

visual carrier frequency—The frequency of the television carrier which is modulated by the picture information.

visual communication—Communication by optical signs such as flags and lights.

visual radio range—Abbreviated vrr. A radio range, the course of which is followed by means of visual instruments.

visual scanner—1. A device that generates an analog or digital signal by optically scanning printed or written data. 2. *See* Scanner, 2.

visual telephony—The transmission of picture information (television) by means of telephone lines.

visual transmitter — Also called a picture transmitter. In television, the radio equipment for transmission of the picture signals only.

visual transmitter power—The peak power output during transmission of a standard television signal.

vitreous—Having the nature of glass.

vitrification—The progressive reduction in porosity of a ceramic material as a result of heat treatment or some other process.

vlf—Abbreviation for very low frequency.

vocabulary—A list of operating codes or instructions available for writing the program for a given problem and for a specific computer.

vocoder (voice operated coder)—A device used to compress the frequency-bandwidth requirement of voice communications. It consists of an electronic speech analyzer which converts the speech waveform to several simultaneous analog signals and an electronic speech synthesizer which produces artificial sounds in accordance with analog control voltages.

vodas—Acronym for voice-operated device, anti-sing. A system for preventing the overall voice-frequency singing of a two-way telephone circuit by disabling one direction of transmission at all times.

voder—Acronym for voice-operation demonstrator. An electronic device capable of artificially producing voice sounds. It uses active devices in connection with electrical filters controlled through a keyboard.

vogad – Acronym for voice-operated gain-adjusting device. A voice-operated device used to give a substantially constant volume output for a wide range of inputs.

voice analyzer – An electronic instrument for printing out waveforms corresponding to vocal characteristics; an aid in identifying speech problems as well as speaker.

voice coder – A device that converts a speech signal into digital form prior to encypherment for secure transmission, and converts the digital signals back into speech at the receiving point.

voice coil—Also called a speaker voice coil. A coil attached to the diaphragm of a dynamic speaker and moved through the air gap between the pole pieces.

voice/data system—An integrated communications system for transmission of both voice and digital-data signals.

voice filter—A parallel-resonant circuit connected in series with a line feeding several speakers. Its purpose is to remove the tubbiness of the male voice. The frequency of resonance is adjusted somewhere between 125 and 300 Hz.

voice frequency – Also called speech frequency. The audible range of frequencies (32-16,000 Hz). In telephony, the voice range of speech is about 100-3500 Hz.

voice-frequency carrier telegraphy—Carrier telegraphy in which the carrier currents have frequencies such that the modulated currents may be transmitted over a voice-frequency telephone channel.

voice-frequency dialing—A method of dialing by which the direct-current pulses from the dial are transformed into voice-frequency alternating-current pulses.

voice-frequency telegraph system – A telegraph system by which many channels can be carried on a single circuit. A different audio frequency is used for each channel and is keyed in the conventional manner. Each frequency is generated by a vacuum-tube oscillator controlled by a tuning fork. At the receiving end, the various audio frequencies are separated by filter circuits and are fed to their respective receiving circuits.

voice-frequency telephony – Telephony in which the frequencies of the components of the transmitted electric waves are substantially the same as the frequencies of corresponding components of the actuating acoustical waves.

voice grade – A telephone circuit suitable for transmitting a bandpass from 300 to 2700 Hz, or greater, with certain standards of noise and interference such that intelligible speech can be transmitted.

voice-grade channel—A channel suitable for the transmission of speech, digital or analog data, or facsimile, generally with a frequency range of about 300 to 3000 hertz.

voice-grade circuit – A circuit suitable for the transmission of speech, digital or analog data, or facsimile, generally with a frequency range of about 300 to 3000 hertz.

voice-operated coder—*See* Vocoder.

voice-operated device—A device which permits the presence of voice or sound signals to effect a desired control.

voice-operated device antising – A system for preventing the overall voice-frequency singing of a two-way telephone circuit by disabling one direction of transmission at all times.

voice-operated gain-adjusting device – A voice-operated device used to give a substantially constant-volume output for a wide range of inputs.

voice-operated loss control and suppressor —A voice-operated device which switches the loss out of the transmitting branch and inserts the loss into the receiving branch under control of the subscriber's speech.

voiceprint – A speech spectrograph sufficiently sensitive and detailed to identify individual human voices.

volatile—1. A computer storage medium in which information cannot be retained without continuous power dissipation. 2. Capable of evaporating.

volatile memory – 1. In computers, any memory which can return information only as long as energizing power is applied. The opposite of nonvolatile memory. 2. A memory whose contents are irretrievably lost when operating power is removed. Practically all semiconductor memories are volatile.

volatile storage—A computer storage device in which the stored information is lost when the power is removed (e.g., acoustic delay lines, electrostatics, capacitors, etc.).

volatile store—A storage device in which stored data are lost when the applied power is removed (e.g., an acoustic delay line).

Voldicon—The trade name of Adage, Inc., for a family of high-speed, all-semiconductor, current-balancing devices that use digital logic and readout for high-speed precision measurement of analog signals.

volt—Abbreviated v. The unit of measurement of electromotive force. It is equivalent to the force required to produce a current of 1 ampere through a resistance of 1 ohm.

Volta effect—*See* Contact Potential.

voltage – 1. Electrical pressure—i.e., the force which causes current to flow through an electrical conductor. 2. Symbolized by E. The greatest effective difference of potential between any two conductors of a circuit. 3. The term most often used in place of electromotive force, potential, potential difference, or voltage drop, to designate electric pressure that exists between two points and is capable of producing a flow of current when a closed circuit is connected between the two points.

voltage amplification – Also called voltage gain. Ratio of the voltage across a specified load impedance connected to a transducer, to the voltage across the input of the transducer.

voltage amplifier—An amplifier used specifically to increase a voltage. It is usually capable of delivering only a small current.

voltage-amplifier tube—A tube that is designed primarily as a voltage amplifier. It has high gain, but delivers very little output power.

voltage and power directional relay—A device which permits or causes the connection of two circuits when the voltage difference between them exceeds a given value in a predetermined direction, and causes these two circuits to be disconnected from each other when the power flowing between them exceeds a given value in the opposite direction.

voltage attenuation – Ratio of the voltage across the input of a transducer, to the voltage delivered to a specified load impedance connected to the transducer.

voltage balance relay—A device which operates on a given difference in voltage between two circuits.

voltage breakdown—1. The voltage necessary to cause insulation failure. 2. A rapid increase of current flow, from a relatively low value to a relatively high value, upon the application of a voltage to a pn junction or dielectric.

voltage-breakdown test—A test whereby a specified voltage is applied between given points in a device, to ascertain that no breakdown will occur at that specified voltage.

voltage calibrator – Test equipment which supplies accurate ac voltages for comparison on a scope screen with other waveforms to determine their voltage level.

voltage circuit of a meter—The combination of conductors and windings of the meter itself, excluding multipliers, shunts, or other external circuitry, to which is applied the voltage to be measured, a definite fraction of that voltage, or a voltage dependent on it.

voltage clipper—*See* Surge Suppressor.

voltage coefficient of capacitance – Also called voltage sensitivity. The quotient of the derivative with respect to voltage of a capacitance characteristic at a point divided by the capacitance at that point.

voltage coefficient of resistivity—The maximum change in nominal resistance value due to the application of a voltage across a resistor, after correcting for self-heating effects usually expressed in percent or per-unit (ppm) change in nominal resistance per volt applied.

voltage comparator—1. An amplifying device with a differential input that will provide an output polarity reversal when one input signal exceeds the other. When operating with open loop and without phase compensation, operational amplifiers make fast and accurate voltage comparators. 2. A circuit that compares two analog voltages and develops a logic output when the voltages being compared are equal or one is greater or less than the reference level.

voltage control—A method of varying the magnitude of voltage in a circuit by means of amplitude control, phase control, or both.

voltage-controlled capacitor—*See* Varactor.

voltage-controlled crystal oscillator – A crystal oscillator the operating frequency of which can be changed by applying a controlling voltage to introduce a phase shift in the oscillator circuit.

voltage-controlled oscillator—Any oscillator for which a change in tuning voltage results in predetermined change in output frequency. Frequency tuning is accomplished by either changing the bias voltage on a varactor diode in the frequency determining resonant network or the bias voltage to the active device. The former approach, although more complex than the latter method of tuning, is capable of multi-octave bandwidth.

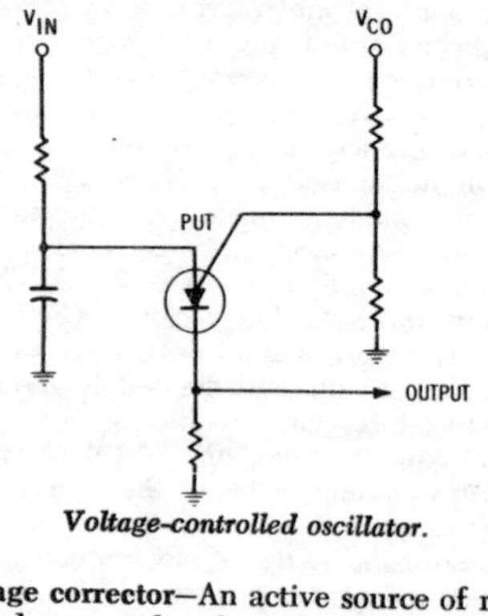

Voltage-controlled oscillator.

voltage corrector—An active source of regulated power placed in series with the output of an unregulated supply. The voltage corrector senses changes in the output voltage (or current) and corrects for these changes automatically by varying its own output in the opposite direction so as to maintain the total output voltage constant.

voltage/current crossover—The characteristic of a power supply that automatically converts the mode of operation of a power supply from voltage regulation to current regulation (or vice versa) as required by preset limits. The constant-current and constant-voltage settings are independently adjustable over specified limits. The region near the intersection of the constant-voltage and constant-current curves is described by the term "crossover characteristics."

voltage-dependent resistor — Abbreviated vdr. Also called varistor or metal oxide varistor (MOV). A special type of resistor whose resistance changes appreciably, nonlinearly, and consistently in response to voltage across its terminals.

voltage-directional relay—1. A relay which functions in conformance with the direction of an applied voltage. 2. A device which operates when the voltage across an open circuit breaker or contactor exceeds a given value in a given direction.

voltage divider—Also called a potential divider. A resistor or reactor connected across a voltage and tapped to make a fixed or variable fraction of the applied voltage available. (*See also* Potentiometer, 1 *and* Rheostat.)

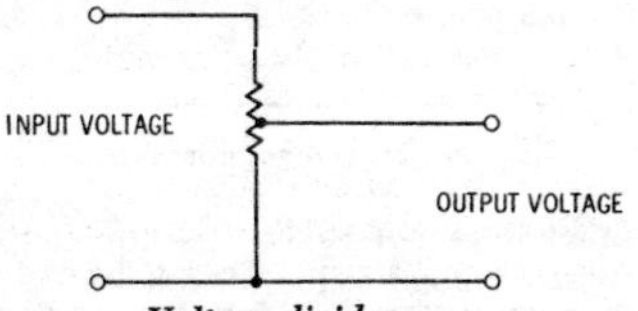

Voltage divider.

voltage doubler—A voltage multiplier which rectifies each half cycle of the applied alternating voltage separately, and then adds the two rectified voltages to produce a direct voltage having approximately twice the peak amplitude of the applied alternating voltage.

voltage drop—The difference in voltage between two points, due to the loss of electrical pressure as a current flows through an impedance.

voltage endurance—*See* Corona Resistance.

voltage feed—Excitation of a transmitting antenna by applying voltage at a point of maximum potential (at a voltage loop or antinode).

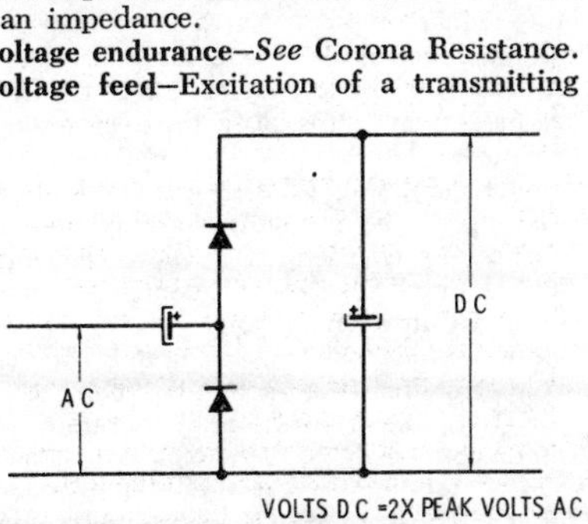

Voltage doubler.

voltage feedback—A form of amplifier feedback in which the voltage drop across part of the load impedance is put in series with the input-signal voltage.

voltage frequency converter—A circuit that produces an output frequency which varies with the voltage applied to its input.

voltage gain—*See* Voltage Amplification.

voltage generator — A two-terminal circuit element with a terminal voltage independent of the current through the element.

voltage gradient — The voltage per unit length along a resistor or other conductive path.

voltage inverter — A circuit having a response (output) proportional to a constant (the gain) times the input signal, but opposite in sign to it. In a unit-gain inverter, the output is (−1) times the input.

voltage jump—An abrupt change or discontinuity in the tube voltage drop during operation of glow-discharge tubes.

voltage level—Ratio of the voltage at any point in a transmission system, to an arbitrary value of voltage used as a reference. In television and other systems where waveshapes are not sinusoidal or symmetrical about a zero axis and where the sum of the maximum positive and negative excursions of the wave is important in system performance, the two voltages are given as peak-to-peak values. This ratio is usually expressed in dBV, signifying decibels referred to 1 volt peak-to-peak.

voltage limit—A control function that maintains a voltage between predetermined values.

voltage loop—A point of maximum voltage in a stationary wave system. A voltage loop exists at the ends of a half-wave antenna.

voltage loss—The voltage between the terminals of a current-measuring instrument when the applied current has a magnitude corresponding to nominal end-scale deflection. In other instruments, the voltage loss is the voltage between the terminals at rated current.

voltage-measuring equipment — Equipment for measuring the magnitude of an alternating or direct voltage.

voltage multiplier—1. A rectifying circuit which produces a direct voltage approximately equal to an integral multiple of the peak amplitude of the applied alternating voltage. 2. A series arrangement of capacitors charged by rapidly rotating brushes in sequence, giving a high direct voltage equal to the source voltage multiplied by the number of capcitors in series.

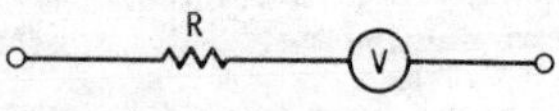

Voltage multiplier.

3. A precision resistor used in series with a voltmeter to extend its measuring range.

voltage node—A point having zero voltage in a stationary-wave system (e.g., at the center of a half-wave antenna).

voltage quadrupler – A rectifier circuit in which four diodes are employed to produce a dc voltage of four times the peak value of the ac input voltage.

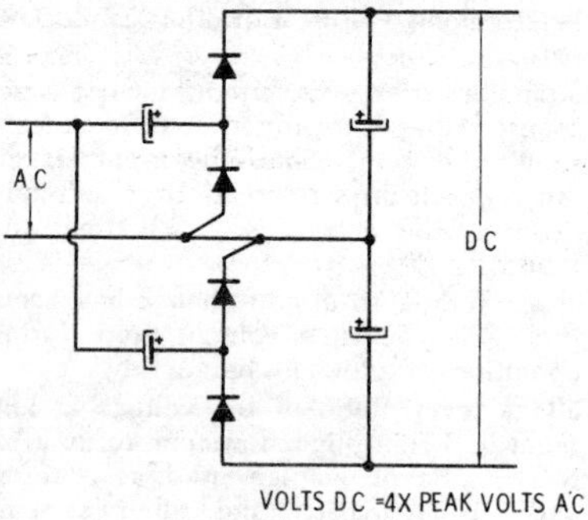

Voltage quadrupler.

voltage-range multiplier—Also called an instrument multiplier. A series resistor installed external to the measurement device to extend its voltage range.

voltage rating – Also called the working voltage. The maximum voltage which an electrical device or component can sustain without breaking down.

voltage ratio (of a transformer)—Ratio of the rms primary terminal voltage to the rms secondary terminal voltage under specified load conditions.

voltage-ratio box – *See* Measurement Voltage Divider.

voltage reference—A highly regulated voltage source used as a standard to which the output voltage of a power supply is continuously compared for purposes of regulation.

voltage reference diode—A diode which develops across its terminals a reference voltage of specified accuracy, when biased to operate within a specified current range.

voltage-reference tube—A gas tube in which the voltage drop is essentially constant over the operating range of current, and is relatively stable at fixed values of current and temperature.

voltage reflection coefficient—The ratio of the complex electric field strength or voltage of a reflected wave to that of the incident wave.

voltage-regulating transformer – A saturated-core type of transformer which holds the output voltage to within a few percent, with input variations up to ±20%. Considerable harmonic distortion results unless extensive filters are employed.

voltage regulation – A measure of the degree to which a power source maintains its output voltage stability under varying load conditions.

voltage regulator—1. A circuit that holds an output voltage at a predetermined value or causes it to vary according to a predetermined plan, regardless of normal input-voltage changes or changes in the load impedance. 2. A gas-filled electronic tube which has the property of maintaining a nearly constant voltage across its terminals over a considerable range of current through the tube. It is used in an electronic voltage regulator.

voltage-regulator tube—Also called a VR tube. A glow-discharge, cold-cathode tube in which the voltage drop is essentially constant over the operating range of current, and which is designed to provide a regulated direct-voltage output.

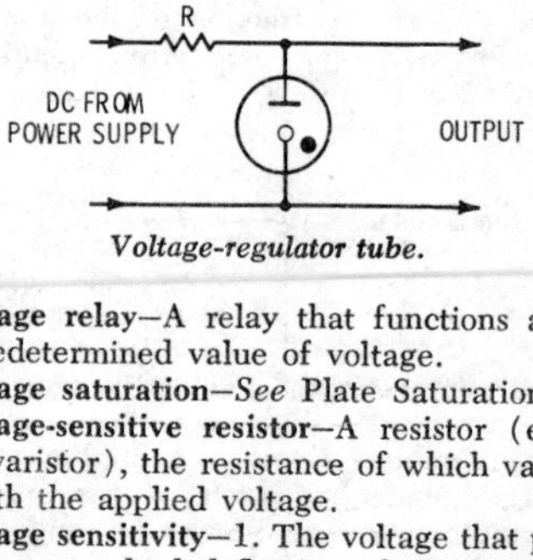

Voltage-regulator tube.

voltage relay—A relay that functions at a predetermined value of voltage.

voltage saturation—*See* Plate Saturation.

voltage-sensitive resistor—A resistor (e.g., a varistor), the resistance of which varies with the applied voltage.

voltage sensitivity—1. The voltage that produces standard deflection of a galvanometer when impressed on a circuit made up of the galvanometer coil and the external critical-damping resistance. The voltage sensitivity is equal to the product of the current sensitivity and the total circuit resistance. 2. *See* Voltage Coefficient of Capacitance.

voltage spectrum – A function that is the square root of the spectral intensity; it is expressed in terms of voltage in a unit frequency band.

voltage-stabilizing tube—A gas-filled glow-discharge tube normally working in that part of its characteristic where the voltage is practically independent of current drop within a given range.

voltage standard – An accurately known voltage source (e.g., a standard cell) used for comparison with or calibration of other voltages.

voltage standing-wave ratio – Abbreviated vswr. In a stationary wave system (such as in a waveguide or coaxial cable), the ratio of the amplitude of the electric field or voltage at a voltage maximum to that at an adjacent voltage minimum.

voltage to ground – The voltage between any live conductor of a circuit and earth (or common reference plane).

voltage transformer – *See* Potential Transformer.

voltage tripler—A rectifier circuit in which three diodes are employed to produce a dc voltage equal to approximately three times the peak ac input voltage.

voltage-tunable magnetron – A high-frequency, continuous-wave oscillator operating in the microwave region. Power outputs begin in the milliwatts range and extend through hundreds of watts.

voltage-tunable tube—An oscillator tube the operating frequency of which can be changed by varying one or more of its electrode voltages (e.g., a backward-wave magnetron).

voltage-tuned cavity oscillator—Primarily a cavity oscillator with the addition of a varactor diode to slightly modify the cavity's resonant frequency. These oscillators are capable of fm and/or phase-locking.

voltage-tuned crystal oscillator—A crystal-controlled oscillator with a varactor diode in the frequency-determining network that is used to slightly vary the crystal frequency.

voltage-type telemeter – A telemeter in which the translating means is the magnitude of a single voltage.

voltage-variable capacitor—*See* Varactor.

voltage velocity limit—*See* Slew Rate, 2.

voltaic cell – An electric cell having two electrodes of unlike metals immersed in a solution that chemically affects one or both of them, thus producing an electromotive force. The name is derived from Volta, a physicist who discovered this effect.

voltaic couple – Two dissimilar metals in contact, resulting in a contact potential difference.

voltaic pile—A voltage source consisting of alternate pairs of dissimilar metal discs separated by moistened pads, forming a number of elementary primary cells in series.

volt-ammeter—An instrument, calibrated to read both voltage and current.

volt-ampere—Abbreviated VA. A unit of apparent power in an ac circuit containing reactance. It is equal to the potential in volts multiplied by the current in amperes, without taking phase into consideration.

volt-ampere-hour meter—An electricity meter which measures the integral, usually in kilovolt-ampere-hours, of the apparent power in the circuit where the meter is connected.

volt-ampere loss—*See* Apparent Power Loss.

volt-ampere meter—An instrument for measuring the apparent power in an alternating-current circuit. Its scale is graduated in volt-amperes or kilovolt-amperes.

volt-ampere reactive—Component of the apparent power in an alternating current circuit which is delivered to the circuit during part of a cycle, but is returned to the source during another part of the cycle. The practical unit of reactive power is the var, equal to one reactive volt ampere. Also called wattless power.

Volta's law—When two dissimilar conductors are placed in contact, the same contact potential is developed between them, whether the contact is direct or through one or more intermediate conductors.

volt box – *See* Measurement Voltage Divider.

volt-electron – An obsolete expression for electronvolt.

voltmeter—1. An instrument for measuring potential difference. Its scale is usually graduated in volts. If graduated in millivolts or kilovolts, the instrument is usually designated as a millivoltmeter or a kilovoltmeter. 2. An instrument used for the measurement of electric voltage. The instrument may be of the electrostatic or tube type, but usually consists of a moving coil ammeter connected in series with a high resistance. The resistance of the meter being fixed, the current passing through it will be directly proportional to the voltage at the points where it is connected and so the instrument can be calibrated in volts.

voltmeter-ammeter – A voltmeter and an ammeter combined into a single case, but with separate circuits.

voltmeter sensitivity—The ratio, expressed in ohms per volt, of the total resistance of a voltmeter to its full-scale reading.

volt-ohm-milliammeter – A test instrument with several ranges, for measuring voltage, current, and resistance.

volume—1. Also called power level. 2. The magnitude (measured on a standard volume indicator) of a complex audio-frequency wave, expressed in volume units. In addition, the term "volume" is used loosely to signify either the intensity of a sound or the magnitude of an audio-frequency wave. 3. The amount or a measure of energy in an electrical or acoustical train of waves.

volume compression—Also called automatic volume compression. The limiting of the volume range to about 30 to 40 decibels at the transmitter, to permit a higher average percentage modulation without overmodulation. Also used in recording to raise the signal-to-noise ratio.

volume compressor—Audio frequency control circuit that limits the volume range of a (radio) program at the transmitter to permit using a higher average percent modulation without risk of overmodulation.

volume conductivity—*See* Conductivity, 1.

volume control—A variable resistor for adjusting the loudness of a radio receiver or amplifying device.

volume equivalent—A measure of the loudness of speech reproduced over a complete telephone connection. It is expressed numerically in terms of the trunk loss of a working reference system which has been adjusted to give equal loudness.

volume expander—A circuit which provides volume expansion.

volume expansion—*See* Automatic Volume Expansion.

volume indicator — An instrument for indicating the volume of a complex electric wave such as that corresponding to speech.

volume lifetime—The average time interval between the generation and recombination of minority carriers in a homogeneous semiconductor.

volume limiter—An amplifier the gain of which is automatically reduced when the average input volume to the amplifier exceeds a predetermined value, so that the output for all inputs in excess of this value is substantially constant. A volume limiter differs from a peak limiter in that it is controlled by the average volume instead of by the instantaneous peaks.

volume-limiting amplifier — An amplifier which reduces the gain whenever the input volume exceeds a predetermined level, so that the output volume is maintained substantially constant. The normal gain is restored whenever the input volume drops below the predetermined limit.

volume magnetostriction—The relative volume change of a body of ferromagnetic material when the magnetization of the body is increased from zero to a specified value (usually saturation) under specified conditions.

volume range—1. Of a transmission system, the difference, expressed in dB, between the maximum and minimum volumes which the system can satisfactorily handle. 2. Of a complex audio-frequency signal, the difference, expressed in dB, between the maximum and minimum volumes occurring over a specified period.

volume recombination rate — The rate at which free electrons and holes recombine within the volume of a semiconductor.

volume resistance—Ratio of the dc voltage applied to two electrodes in contact with or embedded in a specimen, to that portion of the current between them distributed through the specimen.

volume resistivity—1. Ratio of the potential gradient parallel to the current in a material, to the current density. (*See also* Resistivity, 1.) 2. Also called specific insulation resistance. The electrical resistance between opposite faces of a 1-cm cube of insulating material, commonly expressed in ohm-centimeters.

volumetric displacement — The change in volume required to displace the diaphragm of a pressure transducer from its rest position to a position corresponding to the application of a stimulus equal to the rated range of the transducer.

volumetric efficiency—Also called packing factor. The ratio of parts volume to total equipment volume, expressed in percent. In modules, it is usually taken as the volume of component bodies only (not including leads, other interconnecting media, insulators, heat sinks, and so on) and based on nominal sizes of components and nominal outside module dimensions.

volumetric radar—A radar capable of producing three-dimensional position data on several targets.

volume unit—Abbreviated vu. 1. The unit of transmission measurement for measuring the level of nonsteady-state currents. Zero level is the steady-state reference power of 1 milliwatt in a circuit of 600 ohms characteristic impedance. 2. A measure of the power level of the voice wave. Zero volume unit is equivalent to +4 dBm for simple electrical waves (single frequencies).

volume-unit indicator—Also called a VU indicator or volume-unit meter. An instrument calibrated to read audio-frequency power levels directly in volume units.

volume-unit meter—Abbreviated VU meter. *See* Volume-Unit Indicator.

volume velocity—The rate at which a medium flows through a specified area due to a sound wave.

Von Hippel breakdown theory—Also called low-energy criterion. Breakdown occurs at fields for which the rate of recombination of electrons and positive holes is less than the rate of collisional ionization. Assumes no distribution in electron energies.

VOR—Abbreviation for vhf omnirange.

vortex amplifier — 1. A fluidic device in which the angular rate of a vortex is controlled to alter the output. 2. A fluidic amplifier using momentum interaction to produce stream rotation and consequent output modification.

vowel articulation — The percent of articulation obtained when the speech units considered are vowels, usually combined with consonants into meaningless syllables.

V-ring—In commutator construction, a specially shaped insulating structure having one or more V-shaped sections.

vrr—Abbreviation for visual radio range.

vr tube—Abbreviation for voltage-regulator tube.

vswr – Abbreviation for voltage standing-wave ratio.

VTL–Abbreviation for variable threshold logic.

vtm–*See* voltage-tunable magnetron.

vtr–Abbreviation for video tape recording.

vtvm–Abbreviation for vacuum-tube voltmeter.

vu–Abbreviation for volume unit.

vu indicator–Abbreviation for volume-unit indicator.

vulcanized fiber – A dense, homogeneous, cellulosic material that has been partially gelatinized by swelling with a zinc chloride solution. It may be made in the form of sheets, rods, coils, or tubes, and it may be used for electrical insulation as well as in mechanical applications.

vu meter–1. Abbreviation for volume-unit meter. A volume indicator with a decibel scale and specified dynamic and other characteristics. It is used to obtain correlated readings of speech power necessitated by the rapid fluctuations in level of voice currents. 2. Strictly, a recording-level meter whose indicator needle's motion is damped according to a specified standard, to allow it to respond at a certain speed to sudden impulses, without overshooting the mark by more than a certain amount. The term is loosely applied to practically any record-level indicator that uses an indicator needle.

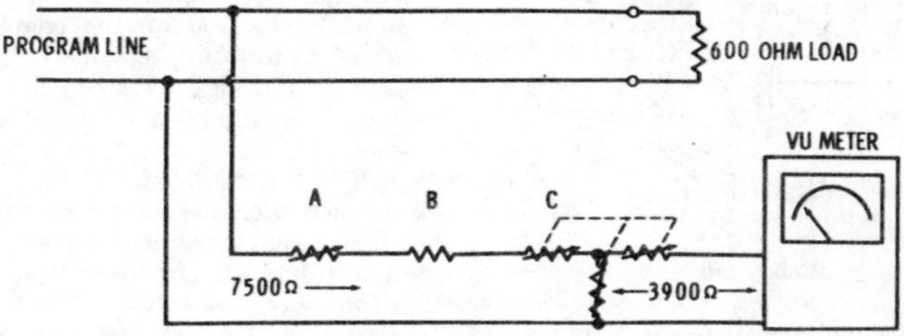

VU meter.

W

W–Symbol for energy, watt, or work.

wafer–1. A thin semiconductor slice of silicon or germanium with parallel faces on which matrices of microcircuits or individual semiconductors can be formed. After processing, the wafer is separated into dice or chips containing individual circuits. 2. A single section of a wafer switch. 3. *See also* Slice.

wafer fabrication–The process of epitaxial growth, impurity diffusion, oxide deposition and metalization that forms the semiconductor devices on the wafer.

wafer socket–A vacuum-tube socket that consists of two punched sheets or wafers of an insulating material, separated by spring-metal clips that grip the terminal pins of the inserted tube.

wafer switch–A rotary multiposition switch with fixed terminals on ceramic or Bakelite wafers. The rotor arm, in the center, is positioned by a shaft. Several decks may be stacked onto one switch and rotated by a common shaft.

Wagner ground–A bridge with an additional pair of ratio arms, onto which the ground connection to the bridge is moved in order to effect a perfect balance, free from error.

waiting time–1. In certain tubes (e.g., thyratrons), the time that must elapse between the turning on of their heaters and the application of plate voltage. 2. *See* Access Time, 1.

walkie-lookie–A compact, portable televi-

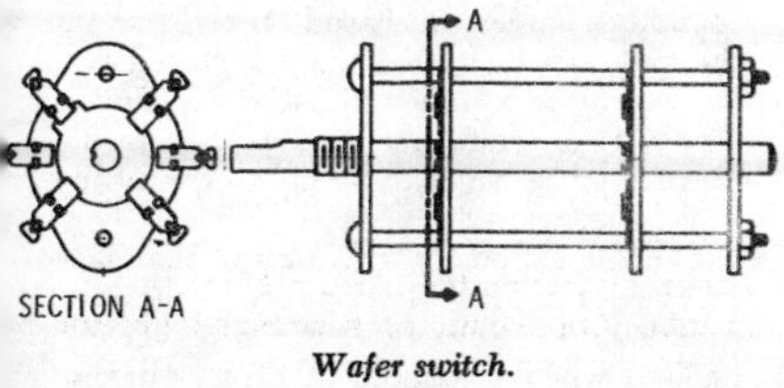

Wafer switch.

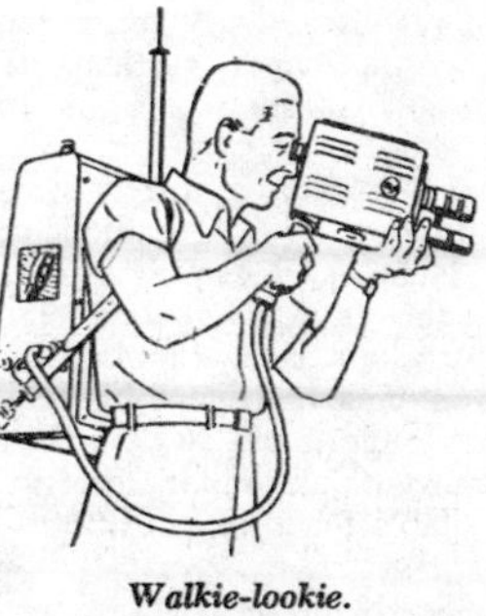

Walkie-lookie.

sion camera used for remote broadcasts. The resultant electrical pulses are transmitted by microwave radio to a local control point for retransmission over a standard television station.

walkie-talkie—1. A two-way radio communication set designed to be carried by one person, usually strapped to his back, and capable of being operated while in motion. 2. A hand-held transceiver.

Walkie-talkie.

wall-attachment amplifier—1. A family of fluidic elements which make use of flow-created low-pressure regions, causing fluid to adhere to an amplifier wall in a controlled manner. 2. A fluidic device in which the control of the attachment of a stream to a wall(s) (Coanda effect) alters the output.

wall box—A metal box placed in the wall and containing switches, fuses, etc.

wall outlet — A spring-contact device to which a portable lamp or appliance is connected by means of a plug attached to a flexible cord. The wall outlet is installed in a box, and connected permanently to the powerline wiring of a home or building.

walls — The sides of the groove in a disc record.

Walmsley antenna — An array of vertical rectangular loops with a height of one wavelength and a spacing of one-half wavelength. The loops are arranged in parallel planes and are series fed at the centers of the longer, vertical sides, with appropriate transpositions of the feed lines.

wamoscope — Acronym for a wave-modulated oscilloscope. A cathode-ray tube which includes detection, amplification, and display of a microwave signal in a single envelope, thus eliminating the local oscillator, mixer, if amplifier, detector, video amplifier, and associated circuitry in a conventional radar receiver. Tubes are available for a range of 2000 to 4000 MHz.

wander—*See* Scintillation, 2.

warble-tone generator — An oscillator, the frequency of which is varied cyclically at a subaudio rate over a fixed range. It is usually used with an integrating detector to obtain an averaged transmission or cross-talk measurement.

warm-up time—1. In an indirectly heated tube, the time which elapses, after the heater is turned on, before the cathode reaches its optimum operating temperature. 2. The time, following power application to a device, required for the output to stabilize within specifications.

washout emitter process—A process used in the manufacture of semiconductor devices. It involves use of the same mask for both the emitter diffusion and the deposition of the ohmic contact. The aluminum metalization for the contact is permitted to form in the same hole that was used for the diffusion.

waste instruction — *See* No-Operation Instruction.

watch—*See* Radio Watch.

water-activated battery—A primary battery which contains the electrolyte but requires the addition of (or immersion in) water before it becomes usable.

water-cooled tube—A vacuum tube having an anode structure projecting through the glass envelope and constructed to permit circulation of water around the anode for cooling purposes during operation.

water load—A matched waveguide termination in which the electromagnetic energy is absorbed in water. The output power is calculated from the difference in temperature between the water at the input and output.

waterproof motor—A totally enclosed motor so constructed that it will exclude water applied in the form of a stream from a hose. Leakage may occur around the shaft, provided it is prevented from entering the oil reservoir and provision is made for automatically draining the motor. The means for automatic draining may be a check valve, or a tapped hole at the lowest part of the frame which will serve for application of a drain pipe.

WATS—Abbreviation for wide area telephone service. A service provided by telephone companies which permits a customer, by use of an access line, to make calls to telephones in a specific zone on a dial basis for a flat monthly charge. Monthly charges are based on the size of the area in which the calls are placed not on the number or length of calls. Under the WATS arrangement, the U.S. is divided into six zones to be called on a full-time or measured-time basis.

watt—Abbreviated W. 1. A unit of the electric power required to do work at the rate of 1 joule per second. It is the power expended when 1 ampere of direct current flows through a resistance of 1 ohm. In an alternating-current circuit, the true power in watts is effective volt-amperes multiplied by the circuit power factor. (There are 746 watts in 1 horsepower.) 2. A measure of electrical or acoustical power. The electrical wattage rating of an amplifier describes the power it can develop to drive a loudspeaker. Acoustical wattage describes the actual sound produced by a loudspeaker in the given environment. (The two figures, in any given amplifier-speaker system are necessarily very widely divergent inasmuch as the low efficiency of speakers necessitates their receiving relatively large amounts of amplifier power in order to produce satisfactory sound levels).

wattage rating—The maximum power that a device can safely handle.

watt-hour—A unit of electrical work indicating the expenditure of 1 watt of electrical power for 1 hour. Equal to 3600 joules.

watt-hour capacity—The number of watt-hours delivered by a storage battery at a specified temperature, rate of discharge, and final voltage.

watt-hour constant of a meter—The registration, expressed in watt-hours, corresponding to one revolution of the rotor.

watt-hour–demand meter — A combined watt-hour meter and demand meter.

watt-hour meter — An electricity meter which measures and registers the integral, usually in kilowatt-hours, of the active power of the circuit into which the meter is connected. This power integral is the energy delivered to the circuit during the integration interval.

wattless component—A reactive component.

wattless power—*See* Reactive Power.

watt loss—*See* Power Loss, 2.

wattmeter—An instrument for measuring the magnitude of the active power in an electric circuit. Its scale is usually graduated in watts. If graduated in kilowatts or megawatts, the instrument is usually designated as a kilowattmeter or megawattmeter.

watt-second—The amount of energy corresponding to 1 watt acting for 1 second. It is equal to 1 joule.

watt-second constant of a meter—The registration, in watt-seconds, corresponding to one revolution of the rotor.

wave—1. A physical activity that rises and falls, or advances and retreats, periodically as it travels through a medium. 2. Propagated disturbance, usually periodic, such as a radio wave or sound wave. If the periodic motion is regular and recurring, it is said to be a periodic or damped.

wave amplitude — The maximum change from zero of the characteristic of a wave.

wave analyzer—An electric instrument for measuring the amplitude and frequency of the various components of a complex current or voltage wave.

wave angle—The angle at which a wave is propagated from one point to another.

wave antenna—Also called a Beverage antenna. A directional antenna composed of parallel horizontal conductors one-half to several wavelengths long, and terminated to ground in its characteristic impedance at the far end.

wave band—A band of frequencies, such as that assigned to a particular type of communication service.

wave-band switch—A multiposition switch for changing the frequency band tuned by a receiver or transmitter.

wave clutter — Clutter caused on a radar screen by echoes from sea waves.

wave converter—A device for changing a wave from one pattern to another (e.g., baffle-plate, grating, and sheath-reshaping converters for waveguides).

wave duct — 1. A tubular waveguide capable of concentrating the propagation of waves within it. 2. A natural duct formed in air by atmospheric conditions. Waves of certain frequencies travel through it with more than average efficiency.

wave equation—An equation that gives a mathematical specification of a wave process, or describes the performance of a medium through which a wave is passing.

wave filter — A transducer for separating waves on the basis of their frequency. It introduces a relatively small insertion loss to waves in one or more frequency bands, and a relatively large insertion loss to

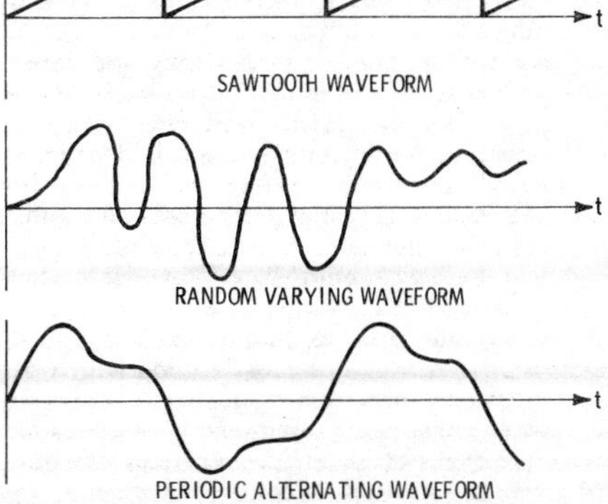

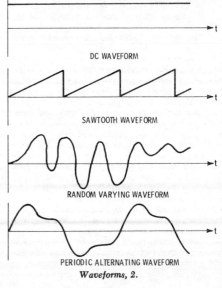

Waveforms, 2.

waves of other frequencies. (*See also* Filter, 1.)

waveform—1. The shape of an electromagnetic wave. 2. A graphical representation of the relationship between voltage, current or power against time. It also provides a picture of the behavior of signals at given frequencies.

waveform-amplitude distortion—Sometimes called amplitude distortion. Nonlinear waveform distortion caused by unequal attenuation or amplification between the input and output of a device.

waveform analyzer – An instrument that measures the amplitude and frequency of the components in a complex waveform.

waveform generator – *See* Timing-Pulse Distributor.

waveform influence—The change in meter indication, caused solely by a change in waveform from a specified waveform, of the applied current and/or voltage.

waveform monitor – An oscilloscope designed for viewing signal waveforms.

waveform synthesizer—Equipment for generating a signal of a desired waveform.

wavefront—1. With respect to a wave in space, a continuous surface at every point of which the displacement from zero in the positive or negative direction is the same at any instant. In a periodic wave the displacements of the points on a wavefront are in phase. For a surface wave, the wavefront is a continuous line, the points of which have the same properties as do the points in the wavefront of a wave in space. 2. That part of a signal-wave envelope between the initial point of the envelope and the point at which the envelope reaches its crest.

wave function—In a wave equation, a point function that specifies the amplitude of a wave.

waveguide—1. A system of material boundaries capable of guiding electromagnetic waves. 2. A transmission line comprising a hollow conducting tube within which electromagnetic waves are propagated on a solid dielectric or dielectric-filled conductor.

waveguide attenuator—A waveguide device for producing attenuation by some means (e.g., by absorption and reflection).

waveguide connector—Also called a waveguide coupling. A mechanical device for electrically joining parts of a waveguide system together.

waveguide coupling—*See* Waveguide Connector.

waveguide critical dimension—The dimension of waveguide cross section which determines the cutoff frequency.

waveguide cutoff frequency – Also called the critical frequency. The frequency limit of propagation, along a waveguide, for waves of a given field configuration.

waveguide dummy load—Sections of waveguide for dissipating all the power entering the input flange.

waveguide elbow—A bend in a waveguide.

Waveguide elbows.

waveguide flange—*See* Flange.

waveguide lens – A microwave device in which the required phase changes are produced by refraction through suitable waveguide elements acting as lenses.

waveguide mode suppressor—A filter used to suppress undesired modes of propagation in a waveguide.

waveguide phase shifter—A device for adjusting the phase of the output current or voltage relative to the phase at the input of a device.

waveguide plunger—1. In a waveguide, a plunger used for reflecting the incident energy. 2. A movable shorting plate used to adjust the length of a resonant waveguide section.

waveguide post—In a waveguide, a rod placed across the waveguide and behaving substantially like a shunt susceptance.

waveguide propagation – Long-range communications in the 10-kHz to 35-kHz frequency range by the waveguide characteristics of the atmospheric duct formed by the ionospheric D-layer and the surface of the earth. (*See also* Atmospheric Duct.)

waveguide resonator—A waveguide device intended primarily for storing oscillating electromagnetic energy.

waveguide shim—A thin metal sheet inserted between waveguide components to ensure electrical contact.

waveguide shutter—A vane within a waveguide, used to protect the receiver system from adjacent radar power by establishing an electrical short across the waveguide when a companion transmitter is operating.

waveguide stub—An auxiliary section of waveguide that has an essentially nondissipative termination and is joined to the main section of the waveguide.

waveguide switch – A transmission-line switch for connecting a transmitter or receiver from one antenna to another or to a dummy load.

waveguide taper – A section of tapered waveguide.

waveguide tee—A junction for connecting a branch section of waveguide in series or parallel with the main transmission line.

waveguide transformer—A device, usually fixed, added to a waveguide for the purpose of impedance transformation.

waveguide tuner — An adjustable device added to a waveguide for the purpose of impedance transformation.

waveguide twist—A waveguide section in which the cross section rotates about the longitudinal axis.

waveguide wavelength—Also called a guide wavelength. For a traveling plane wave at a given frequency, the distance along the waveguide, between the points at which a field component (or the voltage or current) differs in phase by 2π radians.

wave heating—The heating of a material by energy absorption from a traveling electromagnetic wave.

wave impedance (of a transmission line)—At every point in a specified plane, the complex ratio between the transverse components of the electric and magnetic fields. (Incident and reflected waves may both be present.)

wave interference—The phenomenon which results when waves of the same or nearly same type and frequency are superimposed. It is characterized by variations in the wave amplitude which differ from that of the individual superimposed waves.

wavelength—1. In a periodic wave, the distance between points of corresponding phase of two consecutive cycles. The wavelength (λ) is related to the phase velocity (v) and frequency (f) by the formula $\lambda = v/f$. 2. The physical distance between cycles. Wavelengths also may be stated as being equal to the distance traveled by a wave in the time required for one cycle. Wavelength equals the speed divided by the frequency. (For most purposes, we assume that all electromagnetic waves travel at the speed of light in a vacuum, approximately 186,000 miles per second.) 3. The distance between the beginning and the end of a complete cycle of any spatial periodic phenomenon. In acoustics, it is the distance occupied by one cycle of a repetitive sound traveling through the air at a velocity of about 1100 feet per second. (An 1100-Hz tone has a wavelength of one foot.) In magnetic recording, it refers to the length of tape occupied by a full cycle of recorded signal (at 7½ ips tape speed, a recorded frequency of 1000 Hz has a wavelength, on the tape of 0.0075 inches).

wavelength constant—The imaginary part of the propagation constant—i.e., the part that refers to the retardation in phase of an alternating current passing through a length of transmission line.

wavelength shifter — A photofluorescent compound employed with a scintillator material. Its purpose is to absorb photons and emit related photons of a longer wavelength, thus permitting more efficient use of the photons by the phototube or photocell.

wave mechanics—A general physical theory whereby wave characteristics are assigned to the components of atomic structure, and all physical phenomena are interpreted in terms of hypothetical waveforms.

wavemeter—An instrument for measuring the wavelength of a radio-frequency wave. Resonant-cavity, resonant-circuit, and standing-wave meters are representative types.

wave normal—A unit vector normal to an equiphase surface, with its positive direction taken on the same side of the surface as the direction of propagation. In isotropic media, the wave normal is in the direction of propagation.

wave number—The number of waves per unit length. The usual unit of wave number is the reciprocal centimeter, cm^{-1}. In terms of this unit, the wave number is the reciprocal of the wavelength when the latter is in centimeters in a vacuum.

wave packet—A short pulse of waves (e.g., spin waves).

wave propagation—*See* Propagation.

waveshape—A graph of a wave as a function of time or distance.

wave soldering—A manufacturing process for connecting components to a printed-circuit board. After the components are inserted and the copper side of the board is fluxed, the board moves on a conveyor across a tank which is so adjusted that the peak of a standing wave of molten solder just makes contact with the copper side of the board and thus solders all components to the pc board in the pan.

wave tail—That part of a signal-wave envelope between the steady-state value (or crest) of an envelope and the end.

wave tilt—The forward inclination of a radio wave due to its proximity to ground.

wave train—A limited series of wave cycles caused by periodic short-duration disturbances.

wave trap—Also called a trap. A device used to exclude unwanted signals or interference from a receiver. Wave traps are usually tunable to enable the interfering signal to be rejected or the true frequency of a received signal to be determined.

wave velocity—A quantity which specifies the speed and direction at which a wave travels through a medium.

wax—In mechanical recording, a blend of waves with metallic soaps.

wax master—*See* Wax Original.

wax original—Also called a wax master. An original recording on a wax surface, from which the master is made.

way-operated circuit – A circuit shared by three or more stations on a "party-line" basis. One of the stations may be a switching center. May be a single or duplex circuit.

way point–A selected point having some particular significance on a radionavigational course line.

way station–In telegraphy, one of the stations on a multipoint network.

weak coupling–Loose coupling in a radiofrequency transformer. (*See also* Loose Coupling.)

wearout–The point at which the continued operation and repair of an item becomes uneconomical because of the increased frequency of failure. The end of the useful life of the item.

wearout failure–A failure that is predictable on the basis of known wearout characteristics. This type of failure is due to deterioration processes or mechanical wear, the probability of occurrence of which increases with time.

weather-protected motor–An open motor the ventilating passages of which are so designed as to minimize the entrance of rain, snow, and airborne particles to the electric parts.

weber–The practical unit of magnetic flux equal to the amount which, when linked at a uniform rate with a single-turn electric circuit during an interval of 1 second, will induce an electromotive force of 1 volt in the circuit. One weber equals 10^8 maxwells.

wedge–1. The fan-shaped pattern of equidistant black and white converging lines in a television test pattern. 2. A waveguide termination consisting of a tapered length of dissipative material, such as carbon, introduced into the guide.

wedge bonding–1. A type of thermocompression bonding used in integrated-circuit manufacturing where a wedge-shaped tool is used to press a small section of the lead wire onto the bonding pad. 2. A bond formed when a heated wedge is brought down on a wire prepositioned on a heated contact. The wedge's heat and pressure in combinations with heat applied to the mounting contact form the bond.

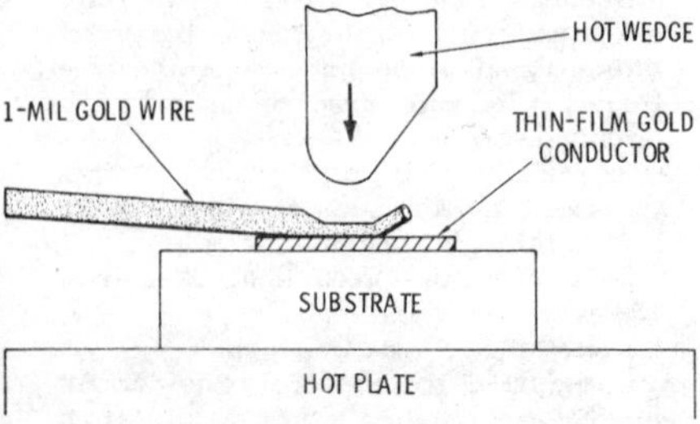

Wedge bonding.

Wehnelt cathode–A hot cathode that consists of a metallic core coated with alkaline-earth oxides. It is widely used in vacuum tubes. *See also* Oxide-Coated Cathode.

weight–1. The force with which a body is attracted toward the earth. 2. *See* Significance.

weight coefficient of a thermoelectric generator – The quotient of the electrical power output of the thermoelectric generator divided by the weight of the generator.

weight coefficient of a thermoelectric generator couple–The quotient of the electrical power output of the thermoelectric couple divided by the weight of the couple.

weighted distortion factor–The weighting of harmonics in proportion to their harmonic relationship.

weighted noise level – The noise level weighted in accordance with the 70-decibel equal-loudness contour of the human ear and expressed in dBm.

weighting–1. The artificial adjustment of measurements in order to account for factors which, during normal use of a device, would otherwise differ from the conditions during measurement. For example, background-noise measurements may be weighted by applying factors or introducing networks to reduce the measured values in inverse ratio to their interference. 2. Any correction factor added to a measurement to make it correlate more accurately with subjective perceptions. A noise measurement may be weighted at various parts of the audio spectrum to reflect the ear's acute sensitivity around 3000 Hz and relative lack of sensitivity at 60 Hz.

weightlessness–1. A condition in which no acceleration, whether of gravity or other force, can be detected by an observer within the system in question. 2. A condition in which gravitational and other external forces acting on a body produce no stress, either internal or external, in the body. Weightlessness occurs when gravity forces are exactly balanced by other forms of acceleration (zero g).

weld–The consolidation of two metals, usually by application of heat to the proposed joint.

weldgate pulse–A waveform used in controlling the flow of welding current.

welding transformer–A power transformer with a secondary winding consisting of only a few turns of very heavy wire, It is used to produce high-value alternating currents at low voltages for welding purposes.

weld junction–A junction formed by heat or metallurgical fusion of conductors. It provides a strong electrical connection with good conductivity. It is widely used in microelectronic packaging. Wires, ribbons, or films as small as 0.0005-inch thick can be joined by resistance and electron-beam welding methods.

weld-on surface-temperature resistor – A surface-temperature resistor installed by welding the sensing element to the surface being measured.

weld polarity—Certain material combinations have a different resistance to a weld current, depending on the direction of the current. In dc welding, a suitable weld may be possible in only one direction of current. A weld schedule must define the proper polarity for such cases.

weld time—The interval during which current is allowed to flow through the work during the performance of one weld. In pulsation welding, the weld period includes the "cool" time intervals.

Wertheim effect—When a wire placed in a longitudinal magnetic field is twisted, there will be a transient voltage difference between the ends of the wire.

Western Union joint—A strong, highly conductive splice made by crossing the cleaned ends of two wires, twisting them together, and soldering.

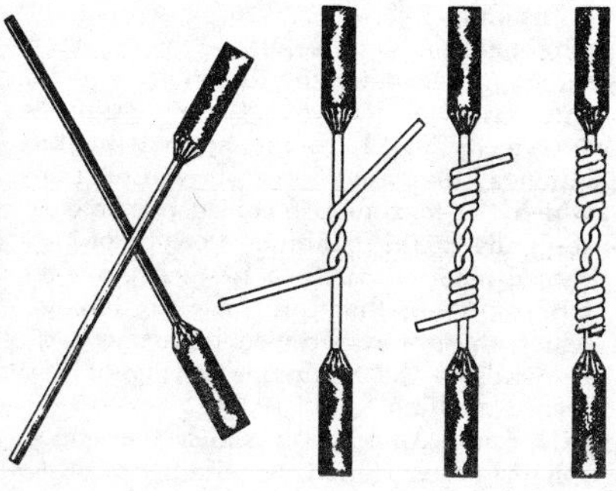

Western Union joint.

Weston normal cell—A standard cell of the saturated cadmium type in which the positive electrode is cadmium and the electrolyte is a cadmium-sulfate solution.

Westrex system – *See* Forty-Five/Forty-Five.

wet – Term describing the condition in which the liquid electrolyte in a cell is free-flowing.

wet cell—A cell the electrolyte of which is in liquid form and free to flow and move.

wet-charged stand—The period of time that a charged, wet secondary cell can stand before losing a specified, small percentage of its capacity.

wet circuit—1. A circuit which carries direct current. 2. Circuit having current flow to melt (microscopically) contact material at point of contact, thereby dissolving and evaporating away contaminants.

wet contact—A contact through which direct current flows.

wet electrolytic capacitor—An electrolytic capacitor that has a liquid electrolyte.

wet flashover voltage – The voltage at which the air surrounding a clean, wet insulator shell breaks down completely between electrodes. This voltage will depend on the conditions under which the test is made.

wet process—A method of preparing a ceramic body in which the constituents are blended in a form that is sufficiently liquid to produce a suspension for use as is or in subsequent processing. Also called slip process.

wet-reed relay—A reed-type relay containing mercury at the relay contacts to reduce arcing and contact bounce.

wet shelf life—The period of time that a wet secondary cell can remain discharged without deteriorating to a point where it cannot be recharged.

wet tantalum capacitor—A polar capacitor the cathode of which is a liquid electrolyte (a highly ionized acid or salt solution). Characteristics: highest capacitance per unit volume, low impedance, lowest dc leakage, excellent shelf life.

wetted surface—A surface on which solder flows uniformly to make a smooth, continuous, adherent layer.

wetting—1. The formation of a uniform, smooth, unbroken, and adherent film of solder to a base metal. 2. A phenomenon involving a solid and a liquid in such intimate contact that the adhesive force between the two phases is greater than the cohesive force within the liquid. Thus, a solid that is wet, on being removed from the liquid bath, will have a thin continuous layer of liquid adhering to it. Foreign substances such as grease may prevent wetting. (Other agents, such as detergents, may induce wetting by lowering the surface tension of the liquid.)

Wheatstone bridge—Also called resistance bridge. A null-type resistance-measuring circuit in which resistance is measured by direct comparison with a standard resistance.

wheel static—Auto-radio interference due to a static charge building up between the brake drum and the wheel spindle.

whiffletree switch—In computers, a multiposition electronic switch composed of gate tubes and flip-flops. It is so named because its circuit diagram resembles a whiffletree.

whip antenna—A simple vertical antenna consisting of a slender whiplike conductor supported on a base insulator and used mainly on motor vehicles.

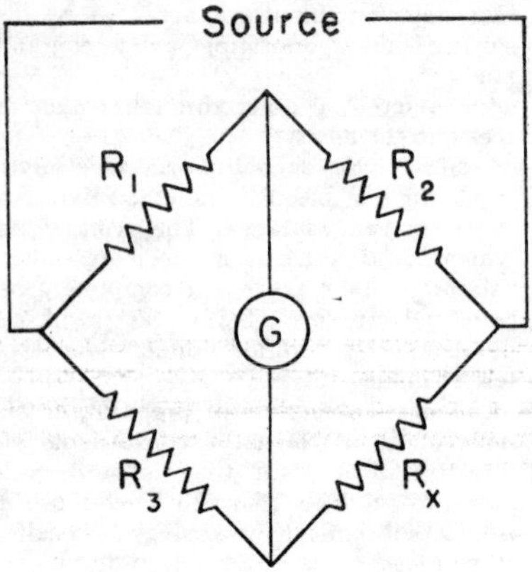

$R_X = R_3R_2/R_1$

Wheatstone bridge.

whisker – 1. *See* Catwhisker. 2. A very small, hairlike metallic growth (a micron-size single crystal with a tensile strength of the order of one million psi) on a metallic circuit component. 3. Ultrapure elongated metal and ceramic filaments of extremely high tensile strengths.

whistler mode propagation – 1. The transmission of radio waves between conjugate points with respect to the geomagnetic equator (i.e., points of opposite geomagnetic latitude and equal geomagnetic longitude) by the apparent ducting of waves along the flux lines of the geomagnetic field. 2. The transmission of radio signals along the flux lines of the earth's magnetic field from the northern hemisphere to the southern hemisphere.

whistlers–1. High-frequency atmospherics that decrease in pitch and then tend to rise again. 2. Audio-frequency waves from lightning-stroke radiation that have penetrated the ionosphere.

white–1. For color tv, white is a mixture of red, green, and blue. In the picture this is produced by exciting all three dots in each phosphor trio. Since the eye cannot distinguish the individual dots, the mixture appears white. 2. The facsimile signal produced when an area of subject copy having minimum density is scanned.

white circuit–A cathode follower with another tube replacing the cathode resistor. By driving this additional tube with a signal that is out of phase with the original signal, low-impedance, broad-band characteristics are obtained.

white compression–In facsimile or television, a reduction in gain (relative to the gain at the level for a midrange light value) at signal levels corresponding to light areas of the picture. The overall effect of white compression is reduced contrast in the highlights of the picture. Also called white saturation.

white-dot pattern–*See* Dot Pattern.

white level–The carrier-signal level which corresponds to maximum picture brightness in television and facsimile.

white light–Radiation that was a spectral energy distribution that produces the same color stimulus to the unaided eye as that of noon sunlight.

white noise–1. Random noise (e.g., shot and thermal noise) whose constant energy per unit bandwidth is independent of the central frequency at the band. The name is taken from the analogous definition of white light. 2. The random motion of electrons in a conductor which, when reproduced through a loudspeaker or phones, sounds like noise and covers a wide frequency range. It is used to test loudspeakers and phones for resonance and sensitivity. 3. Noise whose amplitude (strength) is a random (Gaussian) variable but which has equal energy distribution over all frequencies of interest, regardless of the center frequency of the frequency range being considered.

white object–An object which reflects all wavelengths of light with substantially equal high efficiencies and considerable diffusion.

white peak–A peak excursion of the picture signal in the white direction.

white raster–*See* Chroma-Clear Raster.

white recording–1. In an amplitude-modulation system, that form of recording in which the maximum received power corresponds to the minimum density of the record medium. 2. In a frequency-modulation system, that form of recording in which the lowest received frequency corresponds to the minimum density of the record medium.

white room–An area in which the atmosphere is controlled to eliminate dust, moisture, and bacteria. It is used in the production and assembly of components and systems, the reliability or functions of which might be adversely affected by the presence of foreign matter.

white saturation–*See* White Compression.

white signal – The facsimile signal produced when a minimum-density area of the subject copy is scanned.

white-to-black amplitude range–1. In a facsimile system employing positive amplitude modulation, the ratio of signal voltage (or current) for picture white to that for picture black at any point in the system. 2. In a facsimile system employing negative amplitude modulation, the ratio of the signal voltage (or current) for picture black to that for picture white.

white-to-black frequency swing–In a facsimile system employing frequency modulation, the numerical difference between the signal frequencies corresponding to picture white and picture black at any point in the system.

white transmission – 1. In an amplitude-modulation system, that form of transmission in which the maximum transmitted power corresponds to the minimum density of the subject copy. 2. In a frequency-modulation system, that form of transmission in which the lowest transmitted frequency corresponds to the minimum density of the subject copy.

whole step—*See* Whole Tone.

whole tone—Also called a whole step. The interval between two sounds with a basic frequency ratio approximately equal to the sixth root of 2.

wicking—1. The flow of solder up under the insulation on covered wire. 2. The act of drawing moisture through a fabric or thread, like the action of a wick in an oil lamp when it draws oil up to the flame.

wide-angle lens—A lens that picks up a wide area of a television stage setting at a short distance.

wide area data service – Automatic wide teletypwriter data exchange service by way of leased commercial lines.

wideband—1. Capable of passing a broad range of frequencies (said of a tuner or amplifier). Especially vital to good multiplex reception, and for faithful audio reproduction. 2. Having a bandwidth greater than a voice band.

wideband amplifier—An amplifier capable of passing a wide range of frequencies with equal gain.

wideband axis – In phasor representation of the chrominance signal, the direction of the phasor representing the fine-chrominance primary.

wideband communications system—A communications system which provides numerous channels of communications on a highly reliable and secure basis; the channels are relatively invulnerable to interruption by natural phenomena or countermeasures. Included are multichannel telephone cable, tropospheric scatter, and multichannel line-of-sight radio systems such as microwave.

wideband improvement—Ratio of the signal-to-noise ratio of the system in question, to the signal-to-noise ratio of a reference system.

wideband ratio—Ratio of the occupied-frequency bandwidth to the intelligence bandwidth.

wideband repeater – An airborne system that receives an rf signal, and conditions it, translates it in frequency, and amplifies it for transmission. Such a repeater is used in reconnaissance missions when low-altitude aircraft require an airborne relay platform for transmission of data to a readout station beyond the line of sight.

wide open—Refers to the untuned characteristic or lack of frequency selectivity.

wide-open receiver—A receiver that has essentially no tuned circuits and is designed to receive all frequencies simultaneously in the band of coverage.

width—1. The distance between two specified points of a pulse. 2. The horizontal dimension of a television or facsimile display.

width coding – Modifying the duration of the pulses emitted from the transponder according to a prearranged code for recognition in the display.

width control—A television-receiver or an oscilloscope control which varies the amplitude of the horizontal sweep and hence the width of the picture.

Wiedemann effect—The direct Wiedemann effect is the twist produced in a wire placed in a longitudinal magnetic field when a current flows through the wire; the twist is due to the helical resultant of the impressed longitudinal field and the circular field of the wire. The magnetic material expands (or contracts) parallel to the helical lines of force and hence the twist. The inverse Wiedemann effect is the axial magnetization of a current-carrying wire when twisted.

Wiedemann-Franz law – A theoretical result which states that the ratio of thermal conductivity to electrical conductivity is the same for all metals at the same temperature.

Wien bridge—An alternating-current bridge used to measure inductance or capacitance in terms of resistance and frequency. (*See also* Wien Capacitance Bridge *and* Wien Inductance Bridge.

Wien-bridge oscillator—An oscillator, the frequency of which is controlled by a Wien bridge.

Wien capacitance bridge—A four-arm, alternating-current capacitance bridge used for measuring capacitance in terms of resistance and frequency. Two adjacent arms contain capacitors—one in series and the other in parallel with a resistor—while the other two are normally nonreactive resistors. The balance depends on the frequency, but the capacitance of either or both capacitors can be computed from the resistances of all four arms and from the frequency.

Wien displacement law – The relationship between the temperature of a blackbody and the wavelength for its emission maximum. The wavelength of maximum emission may be found from the expression:

$$\lambda_{max} = \frac{2898 \text{ micron degrees}}{T}$$

where T is in degrees Kelvin.

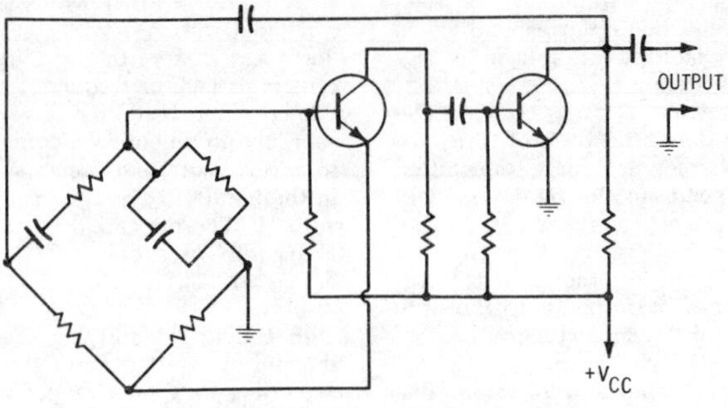

Wien-bridge oscillator.

Wien inductance bridge—A four-arm alternating-current inductance bridge used for measuring inductance in terms of resistance and frequency. Two adjacent arms contain inductors—one in series and the other in parallel with a resistor—while the other two are normally nonreactive resistors. The balance depends on the frequency, but the inductances of either or both inductors can be computed from the resistances of the four arms and from the frequency.

Wien's law—The wavelength of maximum radiation intensity is inversely proportional to the absolute temperature of a blackbody, and the intensity of radiation at this maximum wavelength varies as the fifth power of the absolute temperature. *See also* Wien Displacement Law.

Wien radiation law—An expression representing approximately the spectral radiance of a blackbody as a function of its wavelength and temperature.

Williamson amplifier—A high-fidelity, push-pull, audio-frequency amplifier using tri-

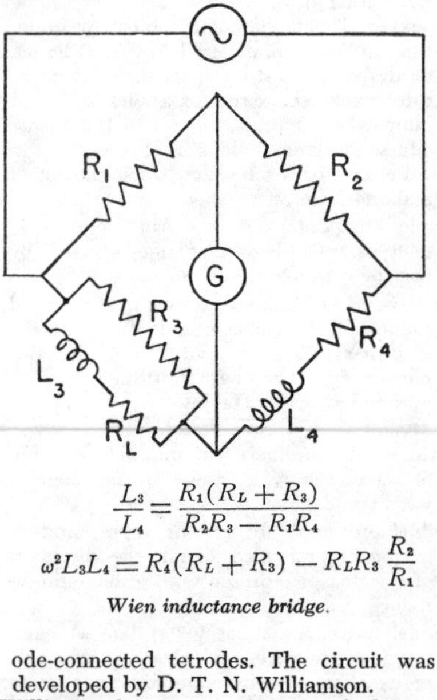

$$\frac{L_3}{L_4} = \frac{R_1(R_L + R_3)}{R_2R_3 - R_1R_4}$$

$$\omega^2 L_3 L_4 = R_4(R_L + R_3) - R_L R_3 \frac{R_2}{R_1}$$

Wien inductance bridge.

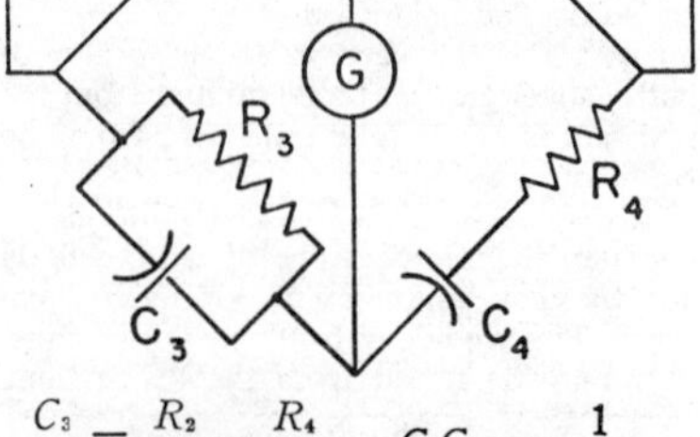

$$\frac{C_3}{C_4} = \frac{R_2}{R_1} - \frac{R_4}{R_3} \qquad C_3C_4 = \frac{1}{\omega^2 R_3 R_4}$$

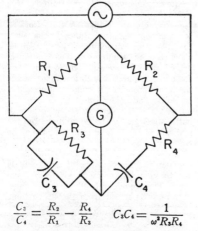

Wien capacitance bridge.

ode-connected tetrodes. The circuit was developed by D. T. N. Williamson.

Williams-tube storage—A type of electrostatic storage using a cathode-ray tube.

Wimshurst machine—A common static machine or electrostatic generator consisting of two coaxial insulating discs rotating in opposite directions. Sectors of tinfoil are arranged, with respect to a connecting rod and collecting combs, so that static electricity is produced for charging Leyden jars or discharging across a gap.

wind—The way in which recording tape is wound onto a reel. An A wind is one in which the tape is wound so that the coated surface faces toward the hub; a B wind is one in which the coated surface faces away from the hub. A uniform, as opposed to an uneven, wind is one giving a flat side tape pack, free from laterally displaced, protruding layers.

wind charger—A wind-driven dc generator

Wimshurst inductance static machine.

for changing batteries (e.g., 32-volt batteries formerly used on many farms).

winding—A conductive path, usually wire, inductively coupled to a magnetic core or cell. Windings may be designated according to function—e.g., sense, bias, drive, etc.

winding arc—In an electrical machine, the length of a winding stated in terms of degrees.

winding factor—The ratio of the total area of wire in the center hole of a toroid to the window area of a toroid or transformer core.

wind loading—The maximum wind an antenna is rated to withstand without being damaged. Expressed in mph.

Windom antenna—A horizontal half-wave dipole located above ground and fed by a vertical or nearly vertical single wire connected at a point approximately one-twelfth wavelength from the center of the dipole.

window—1. Strips of metal foil, wire, or bars dropped from aircraft or fired from shells or rockets as a radar countermeasure. 2. The small area through which beta rays enter a Geiger-Mueller tube. 3. Aperture in a photoresist coating produced by exposure and development.

window area—The opening in the laminations of a transformer.

window corridor—Also called the infected area or lane. An area where window has been sown.

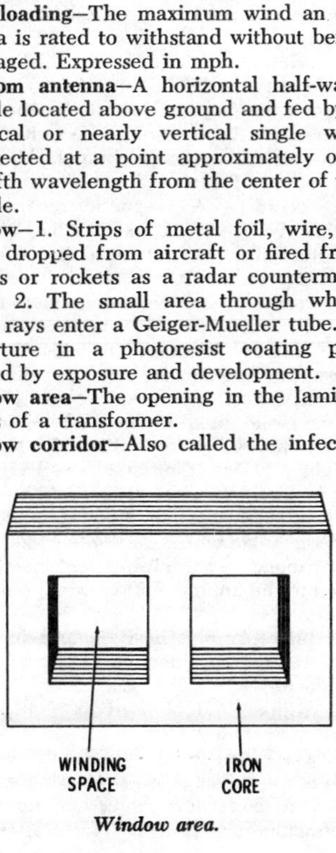

Window area.

window jamming—Reradiation of electromagnetic energy by reflecting it from a window to jam enemy electronic devices.

windshield—In radar, a streamlined cover placed in front of airborne paraboloid antennas to minimize wind resistance. The cover material is such as to present no appreciable attenuation to the radiation of the radar energy.

wing spot generator—An electronic circuit that grows wings on the video target signal of a type-G indicator. These wings are inversely proportional in size to the range.

wiper—1. The moving contact which makes contact with a terminal in a stepping relay or switch. 2. In a potentiometer, the contact that moves along the element, dividing the resistance according to its mechanical position.

wiper arm—In a pressure potentiometer, the movable electrical contact that is driven by the sensing element and moves along the coil.

wiping action—The action which occurs when contacts are mated with a sliding action. Wiping has the effect of removing small amounts of contamination from the contact surfaces, thus establishing better conductivity.

wiping contact — Also called self-cleaning contact, sliding contact, and self-wiping contact. A switch or relay contact designed to move laterally with a wiping motion when engaging with or disengaging from a mating contact.

wire—1. A solid or stranded group of solid cylindrical conductors having a low resistance to current flow, together with any associated insulation. 2. A conductor of round, square, or rectangular section, either bare or insulated. 3. A slender rod or filament of drawn metal. The term is a generally used one, which may refer to any single conductor. If larger than 9 AWG or multiple conductors, it is usually referred to as cable.

wire bond—The method by which very fine wires are attached to semiconductor components for interconnection of those components with each other or with package leads. *See also* Beam Lead.

wire bonding—A lead-covered tie used for connecting two cable sheaths until a splice is closed and covered permanently.

wire communication — Transmission of signs, signals, pictures, and sounds of all kinds over wire, cable, or other similar connections.

wired AND—The external connection of separate circuits or functions in such a way that the combination of their outputs results in an AND function. The logic level at the point at which the separate circuits are wired together is 1 if all circuits feed 1's into this point. Also called dot AND or implied AND.

wired OR—The external connection of separate circuits or functions in such a way that the combination of their outputs results in an OR function. The logic level at the point at which the separate circuits are wired together is 1 if any of the circuits feeds a 1 into this point. Also called dot OR or implied OR.

wired OR and AND—The connection of two or more (open-collector or tri-state) logic outputs to a common bus so that any 1 can pull the bus down to 0 level. Depending on the logic convention used (positive or negative), it will be the logical OR or the logical AND function.

wired-program computer—A computer in which nearly all instructions are determined by the placement of interconnecting wires held in a removable plugboard. This arrangement allows for changes of operations by simply changing plugboards. If the wires are held in permanently soldered connections, the computer is called a fixed-program type.

wired radio—Communication whereby the radio waves travel over conductors.

wire drawing—The pulling of wire through dies made of tungsten carbide or diamond with a resultant reduction in the diameter of the wire.

wire dress—Arranging of wires or conductors in preparation for a mechanical hookup.

wire gage — Also called American Wire Gage (AWG), and formerly Brown and Sharpe (B&S) Gage. A system of numerical designations of wire sizes, starting with 0000 as the largest size and going to 000, 00, 0, 1, 2, and beyond for the smaller sizes.

wire grating—An arrangement of wires set into a waveguide to pass one or more waves while obstructing all others.

wire-guided — In missile terminology, guided by electrical impulses sent over a closed wire circuit between the guidance point and the missile.

wire-lead termination — The method by which wire leads are fastened at a circuit termination; for example, soldering, wire wrapping, or crimping.

wireless—1. A British term for radio. 2. Used in the United States, in the sense of (1) above, when the word "radio" might be misinterpreted (e.g., wireless record player).

wireless device—Any apparatus (e.g., a wireless record player) that generates a radio-frequency electromagnetic field for operating associated apparatus not physically connected and at a distance in feet not greater than 157,000 divided by the frequency in kilohertz. Legally, the total electromagnetic field produced at the maximum operating distance cannot exceed 15 microvolts per meter.

wireless microphone—A microphone connected to a small (frequently hidden) radio transmitter that sends a signal to a suitable receiver located a short distance away.

wireless record player—*See* Wireless Device.

wire-link telemetry—Also called hard-wire telemetry. Telemetry in which a hard-wire link is used as the transmission path, no radio link being used.

wire mile—The unit in which the length of two-conductor wire between two points is expressed. The number of wire miles is obtained by multiplying the length of the route by the number of circuits. This figure does not include the slack for ties, overheads, etc.; for computer purposes, the slack accounts for an additional 50% per wire mile.

wirephoto—Transmission of a photograph or other single image over a telegraph system. The image is scanned into elemental areas in orderly sequence, and each area is converted into proportional electric signals which are transmitted in sequence and reassembled in correct order at the receiver.

wirephoto facsimile — A facsimile photograph.

wire printer — A high-speed printer that prints characterlike configurations of dots through the proper selection of wire ends from a matrix of wire ends, rather than conventional characters through the selection of type faces.

wire recorder — A magnetic recorder in which the recording medium is a round stainless-steel wire about 0.004 inch in diameter rather than magnetic tape.

wire recording—A recording method in which the medium is a thin stainless-steel wire (instead of a tape or disc).

wiresonde—An atmospheric sounding instrument which is supported by a captive balloon and used to obtain temperature and humidity data from ground level to a height of a few thousand feet. Height is determined by means of a sensitive altimeter, or from the amount of cable released and the angle which the cable makes with the ground. The information is telemetered to the ground station through a wire cable.

wire splice—An electrically sound and mechanically strong junction of two or more conductors.

wire stripper—A tool used to remove the insulation from a wire.

wireways—Sheet-metal troughs with hinged covers for housing and protecting electrical conductors and cable, and in which conductors are laid in place after the wireway has been installed as a complete system.

wirewound resistor—A resistor in which the resistance element is a length of high-resistance wire or ribbon wound onto an insulating form.

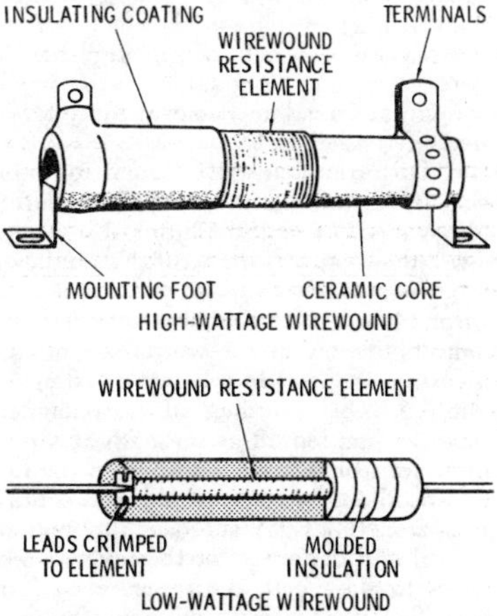

Wirewound resistor.

wirewound trimming potentiometer – A trimming potentiometer characterized by a resistance element made up of turns of wire on which the wiper contacts only a small portion of each turn.

wire wrapping—A technique for terminating conductors.

wiring connector—A device for joining one or more wires together.

wiring diagram—A drawing that shows electrical equipment and/or components, together with all interconnecting wiring.

wobble—*See* Flicker, 3.

wobble bond—A thermocompression, multicontact bond produced by rocking (wobbling) a bonding head on the beam leads. *See also* Beam Lead.

wobble stick—A rod extending from a pendant station to operate the "stop" contacts; it will function when pushed in any direction.

wobbulator – More commonly called a sweep generator. A signal generator, the frequency of which is varied automatically and periodically over a definite range. It is used, together with a cathode-ray tube, for testing frequency response. One form consists of a motor-driven variable capacitor, which is used to vary the output frequency of a signal generator periodically between two limits.

woofer—1. A speaker designed primarily for reproduction of the lower audio frequencies. Woofers may operate up to several thousand hertz, but their output becomes quite directional at these frequencies. Woofers are characterized by large, heavy diaphrams and large voice coils that overhang the magnetic gap. 2. A low-frequency or bass loudspeaker for reproducing musical notes in the approximate range of 25-2500 Hz. Employed with a "tweeter" and crossover network to reproduce a range of frequencies for audio reproduction.

word—1. A group of characters occupying one storage location in a computer. It is treated by the computer circuits as an entity, by the control unit as an instruction, and by the arithmetic unit as a quantity. 2. In telegraphy, six operations or characters (five characters and one space). Also called group. 3. The number of bits needed to represent a computer instruction, or the number of bits needed to represent the largest data element normally processed by a computer. 4. A number of consecutive characters.

word-address format—In a computer, the addressing of each word in a block of information by a character or characters that identify the meaning of the word.

word code – A word that, by prearrangement, conveys a meaning other than its conventional one.

word format—The way in which characters are arranged in a word, with each position or group of positions in the word containing certain specified information.

word generator—An instrument that generates a data stream of ones and zeros with bit position, bit frequency, etc., completely under the control of the operator. It may be considered to be a read-only memory, a substitute for a paper-tape reader, a computer simulator, a tester for a data-transmission line, a programming device, or a programmable pulse generator.

word length—1. The number of bits in a sequence that is handled as a unit and that normally can be stored in one location in a memory. A greater word length implies high precision and more intricate instructions. 2. The size of a field.

word pattern – The smallest meaningful language unit recognized by a machine. It is usually composed of a group of syllables and/or words.

word processing—A system for converting ideas and concepts into hard copy.

word rate—The frequency derived from the elapsed period between the beginning of the transmission of one word and the beginning of transmission of the next word.

word size – In computer terminology, the number of decimal or binary bits comprising a word.

word time—1. The time required to move one word from one storage device to a second storage device. 2. In a storage device providing serial access to storage positions, the interval of time between the appearance of corresponding parts of successive words.

work—1. The magnitude of a force times the distance through which that force is applied. 2. *See* Load, 6.

work area—A portion of computer storage in which an item of data may be processed or temporarily stored. Often, the term work area is used to refer to a place in storage used to retain intermediate results of a calculation, particularly those results that will not appear directly as output from the program.

work coil—*See* Load Coil.

work function – 1. The minimum energy (commonly expressed in electrovolts) required to remove an electron from the Fermi-level of a material and send it into field-free space. 2. A general term applied to the energy required to transfer electrons or other particles from the interior of one medium, across a boundary, into an adjacent medium.

working memory—*See* Working Storage.

working *Q*—*See* Loaded *Q*, 1.

working storage—Also called the working memory. In a computer storage (internal), a portion reserved by the program for the data upon which the operations are being performed.

working voltage—1. *See* Voltage Rating. 2. The recommended maximum voltage of operation for an insulated conductor. It is usually set at approximately one-third of the breakdown voltage. 3. The voltage rating of a fixed capacitor. It is the recommended maximum voltage at which the unit should be operated.

worst-case circuit analysis—A type of circuit analysis used to determine the worst possible effect on the output parameters due to changes in the values of circuit elements. The circuit elements are set to the values within their anticipated ranges that produce the maximum detrimental changes in the output.

worst-case design—An extremely conservative design approach in which the circuit is designed to function normally, even though all component values have simultaneously assumed the worst possible condition that can be caused by initial tolerance, aging, etc. Worst-case techniques are also applied to obtain conservative derating of transient and speed specifications.

worst-case noise pattern—Maximum noise appearing when half of the half-selected cores are in a 1 state and the other half are in a 0 state. Sometimes called checkerboard or double-checkerboard pattern.

wound capacitor – A capacitor made by winding foils and dielectric material on a mandrel.

wound-rotor induction motor—An induction motor in which the secondary circuit consists of a polyphase winding or coils with either short-circuited terminals or ones closed through suitable circuits.

wound rotor motor—*See* Slip Ring Motor.

woven-screen storage – A digital storage plane woven from wires coated with thin films of magnetic material. When currents are passed through a selected pair of wires that lie at right angles in the screen, storage and readout occur at the intersection of those wires.

wow—Distortion caused in sound reproduction by variations in speed of the turntable or tape. (*See also* Flutter, 1.)

wow meter—An instrument that indicates the instantaneous speed variation of a turntable or similar equipment.

wpm—Abbreviation for words per minute, measure of speed in telegraph systems.

wrap – 1. One winding of ferromagnetic tape. 2. The length of the path of a magnetic recording tape along which the tape and head are in intimate physical contact. It is sometimes measured as the angle of arrival and departure of the tape with respect to the head. 3. A measure of the length of recording tape which is in intimate contact with the surface of the record or play head. The better the tape-to-head contact over the head gap, the better the high-frequency response; the better the contact over the rest of the area of the head, the better the response at middle and low frequencies.

wrap and fill—A method of capacitor encasement in which the capacitor element is wrapped with plastic tape and sealed on the ends with an epoxy resin.

wrap-around—The amount of curvature exhibited by the magnetic tape or film in passing over the pole pieces of the magnetic heads.

wrapper—An insulating barrier applied to a coil by wrapping a sheet of insulating material around the coil periphery so as to form an integral part of the coil.

wrapping—A method of applying insulation to wire by serving insulating tapes around the conductor.

wratten filter—An optical filter used for filtering a given band of light. It is used in film-recorder optical systems when recording directly on color film (direct positive). It is also extensively used in photography.

wrinkle finish—An exterior paint that dries to a wrinkled surface when applied to cabinets or panels.

write—1. In a computer, to copy, usually from internal to external storage. 2. In a computer, to transfer elements of information to an output medium. 3. In a computer, to record information in a register, location, or other storage device or medium. 4. In a charge-storage tube, to establish a charge pattern corresponding to the input.

write head—A device that stores digital information by placing coded pulses on a magnetic drum or tape.

write pulse—In a computer, a pulse that is used to enter information into one or more magnetic cells for storage purposes.

write time—*See* Access Time, 2.

writing rate—The maximum speed at which the spot on a cathode-ray tube can move and still produce a satisfactory image.

writing speed—1. The rate of writing on successive storage elements in a charge-storage tube. 2. In a cathode-ray tube, the maximum linear speed at which the electron beam can produce a visible trace.

Wullenweber antenna – An antenna array that consists of two concentric circles of masts so connected as to permit electronic steering.

wvdc—Abbreviation for working voltage, direct current. This is the maximum safe dc operating voltage that can be applied across the terminals of a capacitor at its maximum operating temperature.

WWV—Call letters of the radio station of the National Bureau of Standards at Ft. Collins, Colo. WWV provides radio-broadcast technical services, including time signals, standard radio and audio frequencies, and radio-propagation disturbance warnings at carrier frequencies of 2.5, 5, 10, 15, 20, and 25 megahertz.

WWVH—Call letters of the National Bureau of Standards radio station at Maui, Hawaii. It broadcasts on 2.5, 5, 10, and 15 MHz for many locations not served by WWV.

wye—A network consisting of three branches meeting at a common node; an alternate form of TEE network.

wye connection—Also called a star connection. A Y-shaped winding connection.

wye junction – A Y-shaped junction of waveguides.

X

X—Symbol for reactance.

X and Z demodulation—A system of color-tv demodulation in which the two reinserted 3.58-MHz subcarrier signals differ by approximately 60° rather than the usual 90°. The R – Y, B – Y, and G – Y voltages are derived from the demodulated signals, and these voltages control the three guns of the picture tube. An important advantage of this system is that receiver circuitry is simpler than that required with I and Q demodulation.

X-axis—1. The reference axis in a quartz crystal. 2. The horizontal axis in a system of rectangular coordinates.

X-band—A radio-frequency band of 5200 to 11,000 MHz, with wavelengths of 5.77 to 2.75 cm.

X-bar—A rectangular crystal bar, usually cut from a Z-section, elongated parallel to X and with its edges parallel to X, Y, and Z.

X_C—Symbol for capacitive reactance.

X-cut crystal—A crystal cut so that its major surfaces are perpendicular to an electrical (X) axis of the original quartz crystal.

xenon—A rare gas used in some thyratron and other gas tubes.

xenon flashtube—A high-intensity source of incoherent white light; it operates by discharging a capacitor through a tube of xenon gas. Such a device is used frequently as a source of pumping radiation for various optically excited lasers.

xerographic printer—A device for printing an optical image on paper; light and dark areas are represented by electrostatically charged and uncharged areas on the paper. Powdered ink, dusted on the paper, adheres to the charged areas and is subsequently melted into the paper by the application of heat.

xerographic recording – A recording produced by xerography.

xerography—1. That branch of electrostatic electrophotography in which images are formed onto a photoconductive insulating medium by infrared, visible, or ultraviolet radiation. The medium is then dusted with a powder, which adheres only to the electrostatically charged image. Heat is then applied in order to fuse the powder into a permanent image. 2. A printing process of electrostatic electrophotography that uses a photoconductive, insulating medium, in conjunction with infrared, visible or ultraviolet radiation, to produce latent electrostatic charge patterns for achieving an observable record.

xeroprinting—That branch of electrostatic electrophotography in which a pattern of insulating material on a conductive medium is employed to form electrostatic charge patterns for use in duplicating.

xeroadiography—That branch of electrostatic electrophotography in which a photoconductive insulating medium is employed, with the aid of X-rays or gamma rays, to form latent electrostatic charge patterns for use in producing a viewable record.

xeroradiography equipment – Equipment employing principles of electrostatics and photoconductivity to record X-ray images

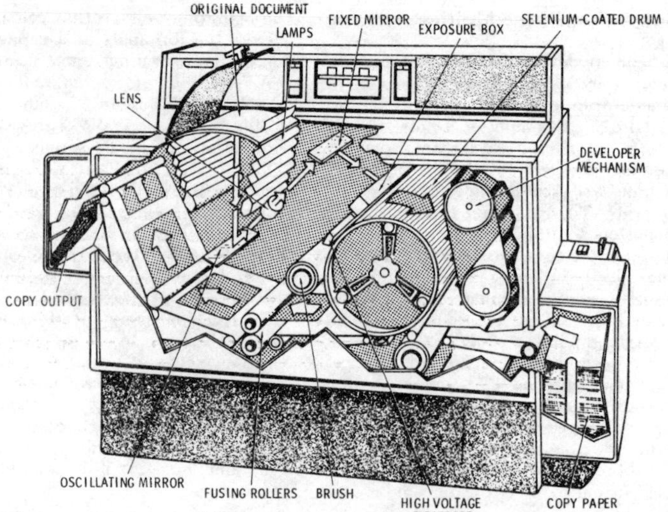

Xerographic printer.

on a sensitized plate in a short time after exposure.

xistor—Abbreviation for transistor.

X_L—Symbol for inductive reactance.

XLR connector—A shielded three-conductor microphone plug or socket with a finger-release lock to prevent accidental removal. The standard connector for professional microphone users.

xmitter—Abbreviation for transmitter. Also abbreviated trans or xmtr.

xmsn—Abbreviation for transmission.

xmtr—Abbreviation for transmitter. Also abbreviated trans or xmitter.

X-off—Transmitter off.

X-on—Transmitter on.

X-particle—A particle having the same negative charge as an electron, but a mass between that of an electron and a proton. It is produced by cosmic radiation impinging on gas molecules or actually forming a part of cosmic rays.

X-ray apparatus — An X-ray tube and its accessories, including the X-ray machine.

X-ray crystallography — Use of X-rays in studying the arrangement of the atoms in a crystal.

X-ray detecting device—A device which detects surface and volume discontinuities in solids by means of X-rays.

X-ray diffraction camera — A camera that directs a beam of X-rays into a sample of unknown material and allows the resultant diffracted rays to act on a strip of film.

X-ray diffraction pattern—The pattern produced on film exposed in an X-ray diffraction camera. It is made up of portions of circles having various spacings, depending on the material being examined.

X-ray goniometer—An instrument that determines the position of the electrical axes of a quartz crystal by reflecting X-rays from the atomic planes of the crystal.

X-rays—Also called roentgen rays. Penetrating radiation similar to light, but having much shorter wavelengths (10^{-7} to 10^{-10} cm). They are usually generated by bombarding a metal target with a stream of high-speed electrons.

X-ray spectrometer—An instrument for producing an X-ray spectrum and measuring the wavelengths of its components.

X-ray spectrum—An arrangement of a beam of X-rays in order of wavelength.

X-ray thickness gage—A contactless thickness gage used to measure and indicate the thickness of moving cold-rolled sheet steel during the rolling process. An X-ray beam directed through the sheet is absorbed in proportion to the thickness of the material and its atomic number, and measurement of the amount of absorption

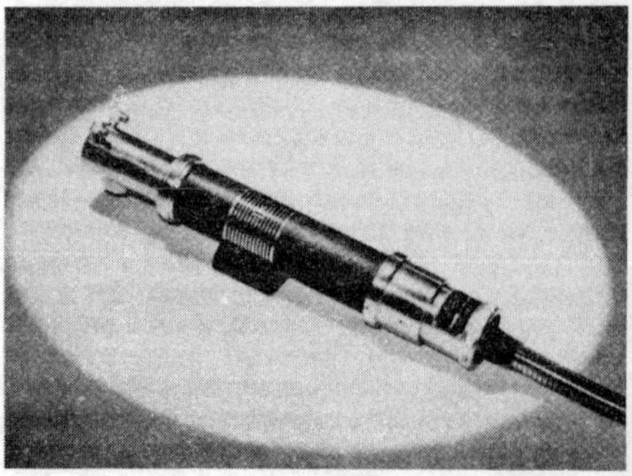

X-ray tube.

gives a continuous indication of sheet thickness.

X-ray tube—A vacuum tube in which X-rays are produced by bombarding a target with high-velocity electrons accelerated by an electrostatic field.

xtal—Abbreviation for crystal.

X-wave—One of the two components into which the magnetic field of the earth divides a radio wave in the ionosphere. The other component is the ordinary, or O-, wave.

XY-cut crystal—A crystal cut so that its characteristics fall between those of an X- and a Y-cut crystal.

XY plotter – A device used in conjunction with a computer to plot coordinate prints in the form of a graph.

XY recorder—1. A recorder that traces, on a chart, the relationship between two variables, neither of which is time. Sometimes the chart moves and one of the variables is controlled so that the relationship does increase in proportion to time. 2. A recorder in which two signals are recorded simultaneously by one pen, which is driven in one direction (X-axis) by one signal, and in the other direction (Y-axis) by the second signal.

XY switch—A remote-controlled bank-and-wiper switch arranged so that the wipers move back and forth horizontally.

Y

Y—Symbol for admittance.

Yagi antenna – An end-fire antenna that consists of a driven dipole (usually a folded dipole), a parasitic dipole reflector, and one or more parasitic dipole directors. All the elements usually lie in the same plane; however, the parasitic elements need not be coplanar, but can be distributed on both sides of the plane of symmetry. Also called Yagi-Uda antenna.

Yagi antenna.

Y antenna—*See* Delta Matched Antenna.

Y-axis – 1. A line perpendicular to two parallel faces of a quartz crystal. 2. The vertical axis in a system of rectangular coordinates.

Y-bar—A crystal bar cut in Z-sections, with its long direction parallel to Y.

Y circulator – A circulator consisting of three identical rectangular waveguides joined in a symmetrical Y-shaped configuration with a ferrite post or wedge at the center. Power that enters any waveguide emerges from only one adjacent waveguide.

Y-connected circuit – A star-connected, three-phase circuit.

Y-connection—*See* Y-network.

Y-cut crystal—A crystal cut in such a way that its major flat surfaces are perpendicular to the Y axis of the original quartz crystal.

yield – 1. In a production process, the quantity or percentage of finished parts which conform to specifications, relative to either the quantity started into production, or to time. 2. The ratio of useable chips to the total number available on a single wafer of semiconductor material. The greater the yield, the more efficient the manufacturing process and the greater its profitability.

yield map—A microcircuit or semiconductor wafer on which dots indicate those devices that failed the test criteria.

yield strength—*See* Yield Value.

yield-strength-controlled bonding – The method of diffusion bonding based on the use of pressures that exceed the yield stress of the metal at the bonding temperature. The process is characterized by the use of high unit loads for brief time cycles ranging from fractions of a second to a few minutes. Typical examples are spot bonding and roll bonding.

yield value—Also called yield strength. The lowest stress at which a material undergoes plastic deformation. Below this stress, the material is elastic; above it, viscous.

yig—Abbreviation for yttrium iron garnet. Crystalline material which resonates at microwave frequencies when immersed in a magnetic field. Small spheres of yig material are mounted in resonant structures for tuning applications.

yig devices—Small solid-state filters, discriminators, and multiplexers that contain yttrium-iron garnet crystals used in combination with a variable magnetic field to accomplish wideband tuning in microwave circuits.

yig filter—A filter that consists of an yttrium-iron garnet crystal positioned in the field of a permanent magnet and a solenoid. Tuning is accomplished by controlling the direct current through the solenoid. The bias magnet tunes the filter to the center of the band and thus minimizes the solenoid power required to tune over wide bandwidths.

yig-tuned parametric amplifier – A parametric amplifier in which tuning is accomplished by controlling the direct current through the solenoid of a yig filter.

yig-tuned tunnel-diode oscillator–A microwave oscillator in which precisely controlled wideband tuning is accomplished through control of the current through a tuning solenoid that acts on a yig filter in the circuit of a tunnel-diode oscillator.

Y-junction – A junction of waveguides in which their longitudinal axes form a Y.

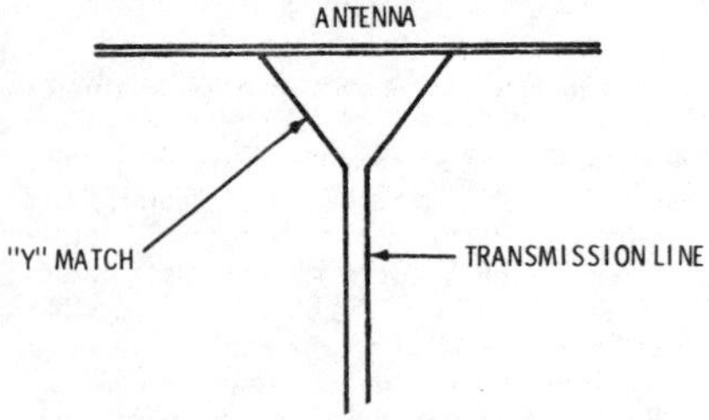

Y-match.

Y-match–Also called a delta match. A method of connecting to an unbroken dipole. The transmission line is fanned out and connected to the dipole at the points where the impedance is the same as that of the line.

Y-network–Also called a Y-connection. A star network of three branches.

yoke–1. A set of coils placed over the neck of a magnetically deflected cathode-ray tube to deflect the electron beam horizontally and vertically when suitable currents are passed through them. 2. A piece of ferromagnetic material that does not have windings and that is used to connect two or more magnet cores permanently.

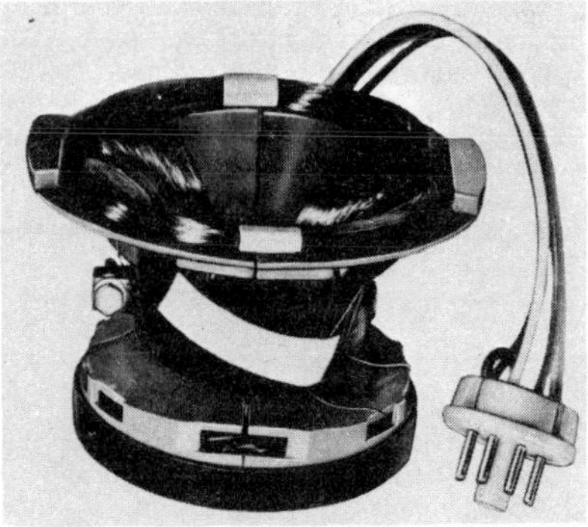

Yoke.

Young's modulus – A constant that expresses the ratio of unit stress to unit deformation for all values within the proportional limit of the material.

Y-punch–On a Hollerith punched card, a punch in the top row, two rows above the zero row.

Y-signal – A luminance transmission primary which is 1.5 to 4.2 MHz wide and equivalent to a monochrome signal. For color pictures, it contributes the finest details and brightness information. *See also* Luminance Signal.

yttrium iron garnet–See Yig.

Z

Z–Symbol for impedance.

Zamboni pile–A primary electrochemical system capable of supplying very high electrical potentials in comparatively little space. The anode material in the pile is aluminum, and the cathode is manganese dioxide and carbon black. The electrolyte in the chemical system is aluminum chloride.

Z-angle meter – An electronic instrument for measuring impedance in ohms and phase angle in electrical degrees.

Z-axis modulation–Also called beam modulation or intensity modulation. Varying the intensity of the electron stream of a cathode-ray tube by applying a pulse or square wave to the control grid or cathode.

Z-bar–A rectangular crystal bar usually cut from X sections and elongated parallel to Z.

Zebra time–An alphabetic expression denoting Greenwich mean time.

Zeeman effect – If an electric discharge tube or other light source emitting a bright-line spectrum is placed between the poles of a powerful electromagnet, a very powerful spectroscope will show that the action of the magnetic field has split each spectrum line into three or more closely spaced but separate lines, the amount of splitting or the separation of the lines being directly proportional at the strength of the magnetic field.

zener breakdown–A breakdown caused in a semiconductor device by the field emission of charge carrier in the depletion layer.

zener diode–A two-layer device that, above a certain reverse voltage (the zener value), has a sudden rise in current. If forward-biased, the diode is an ordinary rectifier. But, when reversed-biased, the diode exhibits a typical knee, or sharp break, in its current-voltage graph. The voltage across the device remains essentially constant for any further increase of reverse current, up to the allowable dissipation rating. The zener diode is a good voltage regulator, overvoltage protector, voltage reference, level shifter, etc. True zener breakdown occurs at less than 6 volts.

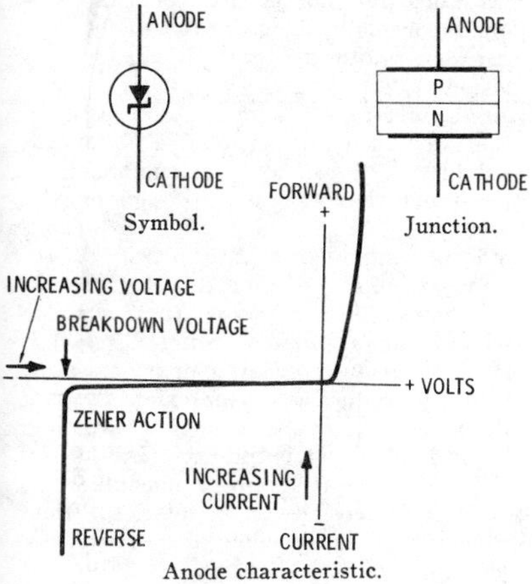

Symbol. Junction.

Anode characteristic.

Zener diode.

zener effect—A reverse-current breakdown due to the presence of a high electric field at the junction of a semiconductor or insulator.

zener impedance—*See* Breakdown Impedance.

zener knee current — The reverse current which flows through a zener diode at the breakdown point or zener knee. Typically knee currents range from 0.25 mA to 5 mA.

zener voltage—*See* Breakdown Voltage, 2.

zeppelin antenna — A horizontal antenna that is a multiple of a half-wavelength long. One end is fed by one lead of a two-wire transmission line that is also a multiple of a half-wavelength long.

zero (in a computer)—Positive binary zero is indicated by the absence of a digit or pulse in a word. In a coded-decimal computer, decimal zero and binary zero may not have the same configuration. In most computers, there are distinct representations for plus and minus zero conditions.

zero-access storage—Computer storage for which the waiting time is negligible (e.g., flip-flop, trigger, or indicator storage).

zero-address instruction — A digital-computer instruction specifying an operation in which the locations of the operands are defined by the computer code; no explicit address is required.

zero adjuster — A device for adjusting a meter so that the pointer will rest exactly on zero when the electrical quantity is zero.

zero adjustment—1. The act of nulling out the output from a system or device. 2. The circuit or other means by which a no-output condition is obtained from an instrument when properly energized.

zero-axis symmetry—A type of symmetry in which a waveform is symmetrical about an axis and does not exhibit a net dc component.

zero beat—The condition whereby two frequencies being mixed are exactly the same and therefore produce no beat note.

zero-beat reception—*See* Homodyne Reception.

zero bias—1. The absence of a potential difference between the control grid and the cathode. 2. When the received teletypewriter signal is equal to the transmitted signal (neither longer nor shorter), the circuit is said to have zero bias.

zero-bias tube—A vacuum tube designed to be operated as a class-B amplifier with no negative bias applied to its control grid.

zero compensation—A method by which, in certain transducers, the effects of temperature on the output at zero measurand may be minimized and maintained within known limits.

zero compression—In computers, any of several techniques used to eliminate the storage of nonsignificant leading zeros.

zero-cut crystal — A quartz crystal cut in such a direction that its temperature coefficient with respect to the frequency is essentially zero.

zero drift—*See* Zero Shift.

zero elimination—In a computer, the editing or deleting of nonsignificant zeros appearing to the left of the integral part of a quantity.

zero error—The delay time occuring within the transmitter and receiver circuits of a radar system. For accurate range data, this delay time must be compensated for when the range unit is calibrated.

zero-field emission — Thermionic emission from a hot conductor which is surrounded by a region of uniform electric potential.

zero-gravity switch—Also called weightlessness switch. A switch that closes when weightlessness or zero gravity is approached.

zero-input terminal—Also called a reset terminal. The terminal which, when triggered, will put a flip-flop in the zero (starting) condition—unless the flip-flop is already in a zero condition, in which event it will not change.

zero level—A reference level for comparing sound or signal intensities. In audio-frequency work, it is usually a power of

.006 watt; and in sound, the threshold of hearing.

zero-line stability—An absence of drift in an indicating instrument when it is registering zero.

zero method—*See* Null Method.

zero-modulation noise — The noise arising when reproducing an erased tape with the erase and record heads energized as they would be in normal operation, but with zero input signal (usually 3 to 4 dB higher than the bulk-erased noise).

zero output—The voltage response obtained by a reading or resetting process from a magnetic cell that is in a zero state.

zero-output terminal—The terminal which produces an output (of the correct polarity to trigger a following circuit) when a flip-flop is in the zero condition.

zero phase-sequence relay—A relay which functions in conformance with the zero phase-sequence component of the current, voltage, or power of the circuit.

zero pole—1. A reference point for an open-wire pole line. 2. The dead-end pole at the origin of the line. 3. The lowest-numbered pole.

zero potential—The potential of the earth, taken as a convenient reference for comparison.

zero-power resistance—In a thermistor, the resistance at a specified temperature when the electrical power dissipation is zero.

zero-power resistance-temperature characteristic—In a thermistor, the function relating the zero-power resistance and body temperature.

zero-power temperature coeffiicent of resistance—In a thermistor, the ratio, at a specified temperature, of the rate of change with temperature of zero-power resistance to the zero-power resistance.

zero set—1. A permanent change in the output of a device at zero measurand due to any cause. 2. A control on a vtvm to set the pointer to zero. 3. A control for adjusting a range counter to give the correct range.

zero shift — Also called zero drift. The amount by which the zero or minimum reading of an instrument deviates from the calibrated point as a result of aging or the application of an external condition to the instrument.

zero-shift error—In an electrical indicating instrument error manifested as a difference in deflection between an initial position of the pointer, such as zero, and the deflection after the instrument has remained deflected up-scale for an extended length of time. The error is expressed as a percentage of the end-scale deflection.

zeros of a network function—Those values of p (real or complex) for which the network function is zero.

zero stability—The ability of an instrument to withstand effects which might cause zero shift. Usually expressed as a percentage of full scale.

zero state—In a magnetic cell, the state wherein the magnetic flux through a specified cross-sectional area has a negative value, from an arbitrarily specified direction. (*See also* One State.)

zero-subcarrier chromaticity—The chromaticity normally displayed when the subcarrier amplitude is zero in a color television system.

zero suppression—1. In a recording system, the injection of a controllable voltage to balance out the steady-state component of the input signal. 2. In a computer, the elimination of zeros to the left of the significant integral part of a quantity. 3. Internal circuits in a calculator that prevent the nonsignificant zeros that precede whole numbers from being displayed; thus, an uncluttered number is shown. 4. In a calibrated zero-suppression system, the magnitude of the component which is being bucked out is indicated by the setting of the zero suppression control.

zero time reference—In a radar, the time reference of the schedule of events during one cycle of operation.

zero transmission-level reference point—An arbitrarily chosen point in a circuit, the level which is used as a reference for all relative transmission levels.

zero-voltage switch—A circuit designed to switch on at the instant the ac supply voltage passes through zero, thereby minimizing the radio-frequency interference generated at switch closure.

Z factor—In thermoelectricity, an accepted figure of merit which denotes the quality of the material.

zigzag reflections — From a layer of the ionosphere, high-order multiple reflections which may be of abnormal intensity. They occur in waves which travel by multihop ionospheric reflections and finally turn back toward their starting point by repeated reflections from a slightly curved or sloping portion of an ionized layer.

zinc—A bluish-white metal which, in its pure form, is used in dry cells.

Z-marker—Also called a zone marker. A marker beacon that radiates vertically and is used for defining a zone above a radio range station.

Z_0—Symbol for characteristic impedance—i.e., the ratio of the voltage to the current at every point along a transmission line on which there are no standing waves.

zone—1. Any of the three top positions (12, 11, or 0) on a punch card. 2. A part of internal computer storage allocated for a particular purpose.

zone bits—1. The two leftmost binary digits in a digital computer in which six binary digits are used for characters and the four rightmost are used for decimal digits. 2. The bits in a group of bit positions that are used to indicate a specific class of items (e.g., numbers, letters, special signs, and commands).

zone blanking—A method of turning off the cathode-ray tube during part of the sweep of the antenna.

zone leveling—In semiconductor processing, the passage of one or more molten zones along a semiconductor body, for the purpose of uniformly distributing impurities throughout the material.

zone marker—*See* Z-marker.

zone of silence – An area – between the points at which the ground wave becomes too weak to be detected and the sky wave first returns to earth—where normal radio signals cannot be heard. (*See also* Skip Zone.)

zone-position indicator—An auxiliary radar set for indicating the general position of an object to another radar set with a narrower field.

zone punch—*See* Overpunch.

zone purification—In semiconductor processing, the passage of one or more molten zones along a semiconductor to reduce the impurity concentration of part of the ingot.

zone refining – A purification process in which an rf heating coil is used to melt a zone, or portion, of a silicon billet. The molten section is moved the length of the billet, and the impurities are deposited at the end.

zoning—1. Also called stepping. Displacement of the various portions of the lens or surface of a microwave reflector, so that the resulting phase front in the near field remains unchanged. 2. Purifying a metal by passing it through an induction coil; the impurities are swept ahead of the heating effect. Specifically used in purifying semiconductor crystals.

zoom—To enlarge or reduce on a continuously variable basis the size of a televised image. It may be done electronically or optically.

zoom lens—An optical lens with some elements made movable so that the focal length or angle of view can be adjusted continuously without losing the focus.

International System of Units (SI)

The SI is constructed from seven base units for independent quantities plus two supplementary units for plane angle and solid angle as listed in Table 1.

Table 1. SI Base and Supplementary Units

Quantity	Name	Symbol
	Base Units	
length	meter	m
mass	kilogram	kg
time	second	s
electric current	ampere	A
thermodynamic temperature	kelvin	K
amount of substance	mole	mol
luminous intensity	candela	cd
	Supplementary Units	
plane angle	radian	rad
solid angle	steradian	sr

Units for all other quantities are derived from these nine units. Table 2 lists 17 SI derived units with special names that were derived from the base and supplementary units in a coherent manner. In brief, they are expressed as products and ratios of the nine base and supplementary units without numerical factors.

Table 2. SI Derived Units With Special Names

	SI Unit		
Quantity	Name	Symbol	Expression in Terms of Other Units
frequency	hertz	Hz	s^{-1}
force	newton	N	$kg \cdot m/s^2$
pressure, stress	pascal	Pa	N/m^2
energy, work, quantity of heat	joule	J	$N \cdot m$
power, radiant flux	watt	W	J/s
quantity of electricity, electric charge	coulomb	C	$A \cdot s$
electric potential, potential difference, electromotive force	volt	V	W/A
capacitance	farad	F	C/V
electric resistance	ohm	Ω	V/A
conductance	siemens	S	A/V
magnetic flux	weber	Wb	$V \cdot s$
magnetic flux density	tesla	T	Wb/m^2
inductance	henry	H	Wb/A
luminous flux	lumen	lm	$cd \cdot sr$
illuminance	lux	lx	lm/m^2
activity of ionizing radiation source	becquerel	Bq	s^{-1}
absorbed dose	gray	Gy	J/kg

All other SI derived units, such as those in Tables 3 and 4, are similarly derived in a coherent manner from the 26 base, supplementary, and special-name SI units.

Table 3. Examples of SI Derived Units Expressed in Terms of Base Units

Quantity	SI Unit	Unit Symbol
area	square meter	m^2
volume	cubic meter	m^3
speed, velocity	meter per second	m/s
acceleration	meter per second squared	m/s^2
wave number	1 per meter	m^{-1}
density, mass density	kilogram per cubic meter	kg/m^3
current density	ampere per square meter	A/m^2
magnetic field strength	ampere per meter	A/m
concentration (of amount of substance)	mole per cubic meter	mol/m^3
specific volume	cubic meter per kilogram	m^3/kg
luminance	candela per square meter	cd/m^2

Table 4. Examples of SI Derived Units Expressed by Means of Special Names

Quantity	Name	Unit Symbol
dynamic viscosity	pascal second	Pa·s
moment of force	newton meter	N·m
surface tension	newton per meter	N/m
heat flux density, irradiance	watt per square meter	W/m^2
heat capacity, entropy	joule per kelvin	J/K
special heat capacity, specific entropy	joule per kilogram kelvìn	J/(kg·K)
specific energy	joule per kilogram	J/kg
thermal conductivity	watt per meter kelvin	W/(m·K)
energy density	joule per cubic meter	J/m^3
electric field strength	volt per meter	V/m
electric charge density	coulomb per cubic meter	C/m^3
electric flux density	coulomb per square meter	C/m^2
permittivity	farad per meter	F/m
permeability	henry per meter	H/m
molar energy	joule per mole	J/mol
molar entrophy, molar heat capacity	joule per mole kelvin	J/(mol·K)

For use with the SI units there is a set of 16 prefixes (see Table 5) to form multiples and submultiples of these units. It is important to note that the kilogram is the only SI unit with a prefix. Because double prefixes are not to be used, the prefixes of Table 5, in the case of mass, are to be used with gram (symbol g) and not with kilogram (symbol kg).

Table 5. SI Prefixes

Factor	Prefix	Symbol	Factor	Prefix	Symbol
10^{18}	exa	E	10^{-1}	deci	d
10^{15}	peta	P	10^{-2}	centi	c
10^{12}	tera	T	10^{-3}	milli	m
10^{9}	giga	G	10^{-6}	micro	μ
10^{6}	mega	M	10^{-9}	nano	n
10^{3}	kilo	k	10^{-12}	pico	p
10^{2}	hecto	h	10^{-15}	femto	f
10^{1}	deka	da	10^{-19}	atto	a

Schematic Symbols

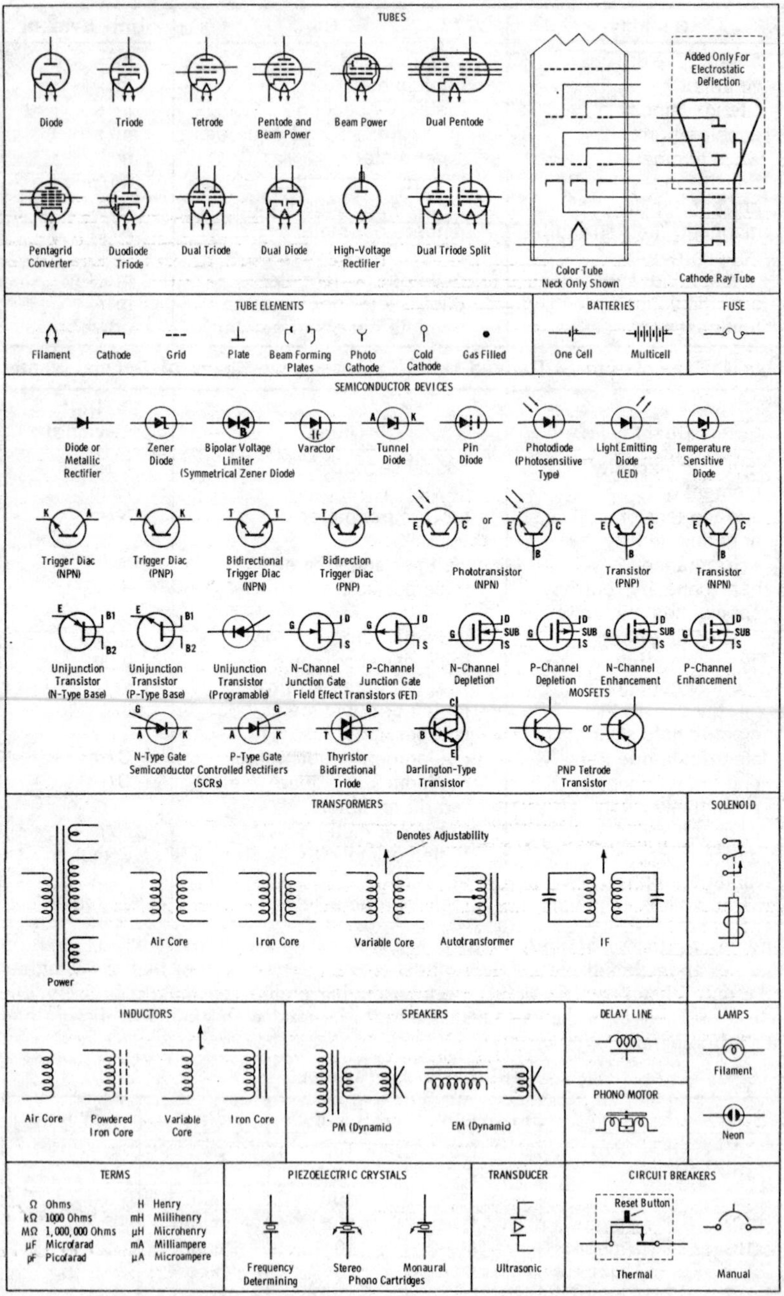
TUBES
Diode
Triode
Tetrode
Pentode and Beam Power
Beam Power
Dual Pentode
Added Only For Electrostatic Deflection
Pentagrid Converter
Duodiode Triode
Dual Triode
Dual Diode
High-Voltage Rectifier
Dual Triode Split
Color Tube Neck Only Shown
Cathode Ray Tube
TUBE ELEMENTS
Filament
Cathode
Grid
Plate
Beam Forming Plates
Photo Cathode
Cold Cathode
Gas Filled
BATTERIES
One Cell
Multicell
FUSE
SEMICONDUCTOR DEVICES
Diode or Metallic Rectifier
Zener Diode
Bipolar Voltage Limiter (Symmetrical Zener Diode)
Varactor
Tunnel Diode
Pin Diode
Photodiode (Photosensitive Type)
Light Emitting Diode (LED)
Temperature Sensitive Diode
Trigger Diac (NPN)
Trigger Diac (PNP)
Bidirectional Trigger Diac (NPN)
Bidirection Trigger Diac (PNP)
or
Phototransistor (NPN)
Transistor (PNP)
Transistor (NPN)
Unijunction Transistor (N-Type Base)
Unijunction Transistor (P-Type Base)
Unijunction Transistor (Programable)
N-Channel Junction Gate
P-Channel Junction Gate
Field Effect Transistors (FET)
N-Channel Depletion
P-Channel Depletion
N-Channel Enhancement
P-Channel Enhancement
MOSFETS
N-Type Gate
P-Type Gate
Semiconductor Controlled Rectifiers (SCRs)
Thyristor Bidirectional Triode
Darlington-Type Transistor
or
PNP Tetrode Transistor
TRANSFORMERS
Denotes Adjustability
Power
Air Core
Iron Core
Variable Core
Autotransformer
IF
SOLENOID
INDUCTORS
Air Core
Powdered Iron Core
Variable Core
Iron Core
SPEAKERS
PM (Dynamic)
EM (Dynamic)
DELAY LINE
PHONO MOTOR
LAMPS
Filament
Neon
TERMS
Ω Ohms
kΩ 1000 Ohms
MΩ 1,000,000 Ohms
µF Microfarad
pF Picofarad
H Henry
mH Millihenry
µH Microhenry
mA Milliampere
µA Microampere
PIEZOELECTRIC CRYSTALS
Frequency Determining
Stereo
Monaural
Phono Cartridges
TRANSDUCER
Ultrasonic
CIRCUIT BREAKERS
Reset Button
Thermal
Manual

Schematic Symbols

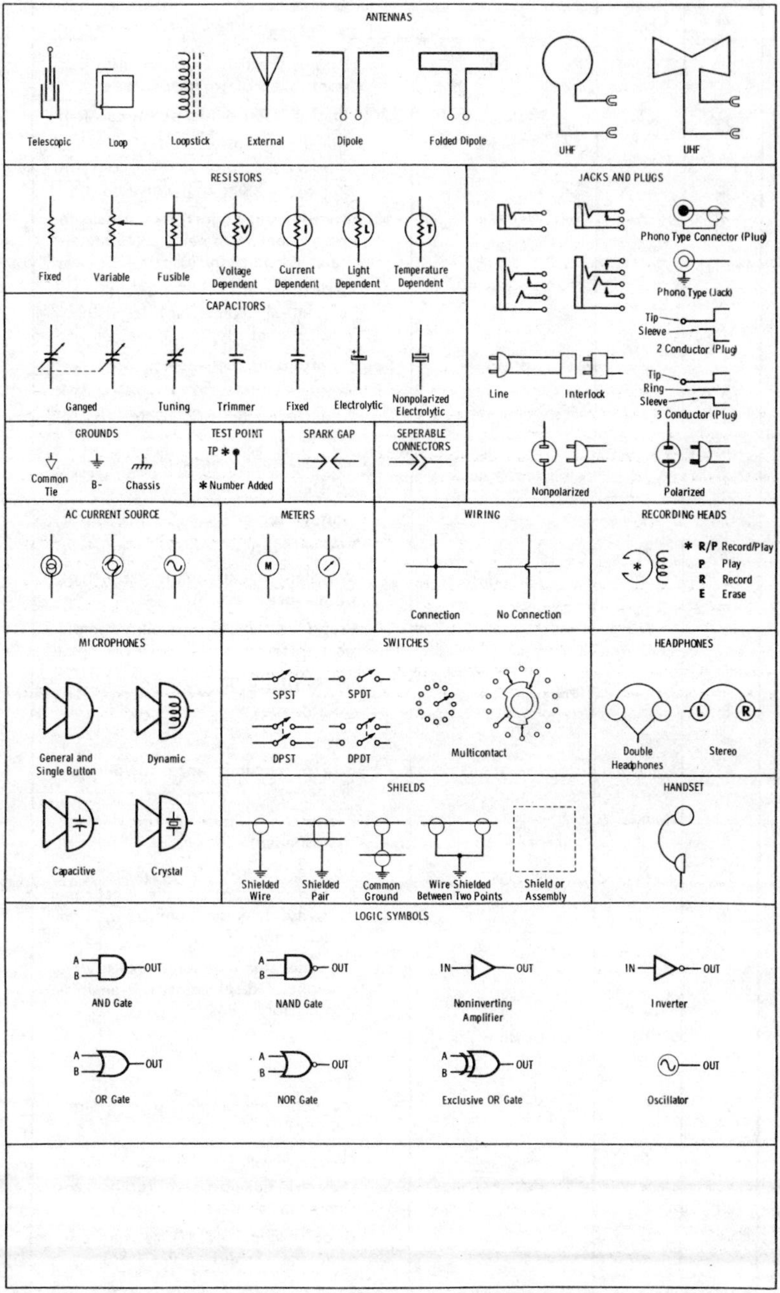

Greek Alphabet

Letter		Name	Designates
Small	Capital		
α	A	Alpha	Angles, coefficients, attenuation constant, absorption factor, area.
β	B	Beta	Angles, coefficients, phase constant.
γ	Γ	Gamma	Specific quantity, angles, electrical conductivity, propagation constant, complex propagation constant (cap).
δ	Δ	Delta	Density, angles, increment or decrement (cap or small), determinant (cap), permittivity (cap).
ϵ	E	Epsilon	Dielectric constant, permittivity, base of natural (Napierian) logarithms, electric intensity.
ζ	Z	Zeta	Co-ordinate, coefficients.
η	H	Eta	Intrinsic impedance, efficiency, surface charge density, hysteresis, coordinates.
θ	Θ	Theta	Angular phase displacement, time constant, reluctance, angles.
ι	I	Iota	Unit vector.
κ	K	Kappa	Susceptibility, coupling coefficient.
λ	Λ	Lambda	Wavelength, attenuation constant, permeance (cap).
μ	M	Mu	Prefix *micro-*, permeability, amplification factor.
ν	N	Nu	Reluctivity, frequency.
ξ	Ξ	Xi	Co-ordinates.
ο	O	Omicron	——
π	Π	Pi	3.1416 (circumference divided by diameter).
ρ	P	Rho	Resistivity, volume charge density, co-ordinates.
σ	Σ	Sigma	Surface charge density, complex propagation constant, electrical conductivity, leakage coefficient, sign of summation (cap).
τ	T	Tau	Time constant, volume resistivity, time-phase displacement, transmission factor, density.
υ	ϒ	Upsilon	——
φ	Φ	Phi	Magnetic flux, angles, scalar potential (cap).
χ	X	Chi	Electric susceptibility, angles.
ψ	Ψ	Psi	Dielectric flux, phase difference, coordinates, angles.
ω	Ω	Omega	Angular velocity ($2\pi f$), resistance in ohms (cap), solid angles (cap).